Formulas

Distance formula

The distance between two points (x_1, y_1) and (x_2, y_2) is

$$d = \sqrt{(x_2 - x_1)^2 + (y_2 - y_1)^2}$$

Midpoint formula

The midpoint of the line segment connecting two points (x_1, y_1) and (x_2, y_2) is the point $(\bar{x}, \bar{y})$ where

$$\bar{x} = \frac{x_1 + x_2}{2} \quad \text{and} \quad \bar{y} = \frac{y_1 + y_2}{2}$$

Quadratic formula

If $a \neq 0$, then the solution(s) of $ax^2 + bx + c = 0$ are $x = \dfrac{-b \pm \sqrt{b^2 - 4ac}}{2a}$

Simple interest

If a principal of P dollars is deposited in an account paying simple interest at a rate r, then the balance B after t years is given by $B = P(1 + rt)$.

Compound interest

If a principal of P dollars is deposited in an account paying interest at a rate r compounded n times per year, then the balance B after t years is given by

$$B = P\left(1 + \frac{r}{n}\right)^{nt}$$

Arithmetic series

If $a_1, a_2, a_3, \ldots, a_n$ is an arithmetic sequence, then $\sum_{k=1}^{n} a_k = \dfrac{n}{2}(a_1 + a_n)$.

Geometric series

If $a_1, a_2, a_3, \ldots, a_n$ is a geometric sequence with common ratio $r \neq 1$, then

$$\sum_{k=1}^{n} a_k = \frac{a_1(1 - r^n)}{1 - r}$$

Binomial theorem

$$(a + b)^n = \binom{n}{n}a^n + \binom{n}{n-1}a^{n-1}b + \binom{n}{n-2}a^{n-2}b^2 + \cdots + \binom{n}{1}ab^{n-1} + \binom{n}{0}b^n$$

Height of a falling object

If an object is thrown into the air from an initial height s_0 feet with an initial velocity v_0 feet per second, its height in feet after t seconds is given by $h(t) = -16t^2 + v_0 t + s_0$.

Universal law of gravitation

The force of attraction between two objects with masses m_1 and m_2 kilograms that are a distance r meters apart is given by $F = G\dfrac{m_1 m_2}{r^2}$, where $G = 6.672 \times 10^{-11}$ Nm^2/kg^2.

Time dilation formula

If a person travels at a velocity equal to v times the speed of light for an elapsed time of T_0 seconds (according to her watch), then the elapsed time on Earth is given by

$$T = \frac{T_0}{\sqrt{1 - v^2}}$$

Algebraic Formulas

Sum of cubes
$$a^3 + b^3 = (a + b)(a^2 - ab + b^2)$$

Difference of cubes
$$a^3 - b^3 = (a - b)(a^2 + ab + b^2)$$

Difference of squares
$$a^2 - b^2 = (a + b)(a - b)$$

Precalculus

David Dwyer
Mark Gruenwald
University of Evansville

THOMSON
BROOKS/COLE

Australia • Canada • Mexico • Singapore • Spain
United Kingdom • United States

THOMSON

™

BROOKS/COLE

Executive Editor: *Bob Pirtle*
Math Editor: *John-Paul Ramin*
Development Editor: *Leslie Lahr*
Assistant Editor: *Lisa Chow*
Editorial Assistant: *Darlene Amidon-Brent*
Technology Project Manager: *Christopher Delgado*
Marketing Manager: *Karin Sandberg*
Marketing Assistant: *Jennifer Gee*
Advertising Project Manager: *Bryan Vann*
Project Manager, Editorial Production: *Janet Hill*
Print/Media Buyer: *Kristin Waller*

Production Service: *Nancy Shammas, New Leaf Publishing Services*
Text Designer: *Patrick Devine Design*
Art Editor: *Lisa Torri*
Photo Researcher: *Kathleen Olson*
Copy Editor: *Carol Dondrea*
Illustrator: *Lori Heckelman*
Cover Design/Image: *Larry Didona*
Cover Printer: *CTPS*
Compositor: *New England Typographic Service*
Printer: *CTPS*

Printed in China by CTPS

2 3 4 5 6 7 06 05

For more information about our products, contact us at:
Thomson Learning Academic Resource Center
1-800-423-0563
For permission to use material from this text, contact us by:
Phone: 1-800-730-2214 **Fax:** 1-800-730-2215
Web: http://www.thomsonrights.com

Library of Congress Control Number: 2003111890

Student Edition: ISBN 0-534-35287-1
Annotated Instructor's Edition: ISBN 0-534-40311-5

Brooks/Cole–Thomson Learning
10 Davis Drive
Belmont, CA 94002-3098
USA

Asia
Thomson Learning
60 Albert Street, #15-01
Albert Complex
Singapore 189969

Australia/New Zealand
Thomson Learning
102 Dodds Street
Southbank, Victoria 3006
Australia

Canada
Nelson
1120 Birchmount Road
Toronto, Ontario M1K 5G4
Canada

Europe/Middle East/Africa
Thomson Learning
Berkshire House
168-173 High Holborn
London WC1V 7AA
United Kingdom

Latin America
Thomson Learning
Seneca, 53
Colonia Polanco
11560 Mexico D.F.
Mexico

Spain/Portugal
Paraninfo Thomson Learning
Calle/Magallanes, 25
28015 Madrid, Spain

To our families, Margaret, Carrie,
David, Benjamin, Matthew, and ??

About the Authors

David Dwyer

Born in Fort Wayne, Indiana, in the 1960's, Dave spent much of his early childhood playing basketball and reading compulsively, especially reference works such as *The World Almanac* and *The Guinness Book of World Records*. As a high school student, he began to believe that the secrets of the universe lie in mathematics and Kung Fu, a belief still held to this day, except for the part about Kung Fu. Dave studied mathematics at Purdue University, where he received a B.S. in 1983 and a Ph.D. in 1990. As a tutor and teaching assistant, Dave developed a passion for the compassionate and creative teaching of mathematics, believing that nearly all students could learn to appreciate and master the fundamentals of algebra, trigonometry, and calculus if properly motivated and encouraged.

Dave was hired by the Mathematics Department of the University of Evansville in 1990, where he met his future co-author and best man, Mark Gruenwald. Upon meeting Gruenwald, Dave is said to have had two thoughts: "I think I could work with this guy," and "I'm pretty sure I could take him." Inspired by their common vision of bringing precalculus mathematics to life, Dwyer and Gruenwald have been writing together ever since.

Dwyer's hobbies include crossword puzzles, reading, computer programming, writing of all kinds, and watching way too much television. Among his many influences, he cites his late father, Danny Dwyer, the mathematician Karl Gauss, Bob Dylan, Ernest Hemingway, and peer pressure. Dwyer is a professor of mathematics at the University of Evansville. He lives in Evansville with his wife, Carrie.

Mark Gruenwald

Mark was born in Swift Current, Saskatchewan, Canada in the (very late) 1950's, making him only slightly older than his co-author, though blessed with significantly more hair. His family moved to Villard, Minnesota, in 1968, where Mark soon developed a life-long love of both music and mathematics. He earned a B.S. in Mathematics Education from Concordia University in River Forest, Illinois, in 1981, and a Ph.D. in Mathematics from Northern Illinois University in 1989, narrowly escaping a career as a professional trumpeter. While working on his doctoral dissertation, Mark discovered that the process of clearly and elegantly writing the results of his research was as rewarding as the research itself, a realization that led to a desire to write mathematics textbooks.

Gruenwald accepted a position as assistant professor of mathematics at the University of Evansville in 1989. One year later, he helped hire his future co-author, Dave Dwyer. Upon meeting Dwyer, Mark is said to have had two thoughts: "I think I could work with this guy," and "Maybe if he gets this job, he'll be able to afford a real suit." With their shared passion for teaching and their unique combination of skills and interests, the two have completed several joint projects and, in the process, have become great friends.

Mark's hobbies include running, genealogy research, and the partial completion of home improvement projects. He is currently chairman of the Mathematics Department at the University of Evansville. He lives in Evansville with his wife, Margaret, and sons, Matthew, Benjamin, and David.

Contents

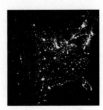

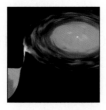

Chapter 9

Systems of Equations and Inequalities 675

Chapter 10

Integer Functions and Probability 789

Preface

Mathematics, the science of pattern and the universal language of quantification, is a living, breathing subject with profound implications for nearly every aspect of life. It is woven into the fabric of art, astronomy, chemistry, cinema, law, literature, music, religion, philosophy, physics, politics, sports, and technology. Moreover, mathematics is a work in progress, with its frontiers ever expanding and new areas of application arising daily. Ozone depletion, rain forest destruction, global warming, the spread of contagious disease, nuclear proliferation, poverty, and overpopulation are just a few of the world's problems illuminated by mathematical techniques. Arguably, the destiny of Earth itself hinges upon our ability to make intelligent decisions based on mathematical models of these real-world phenomena.

But far too many students view mathematics in an entirely different light. To them math is a soulless game of symbolic manipulation, a lifeless subject passed down through the generations without alteration, its arcane formulas and rules to be memorized and quickly forgotten. They see mathematics as inaccessible, insignificant, and utterly irrelevant to their majors, their careers, and their lives. Not surprisingly, mathematics courses are often loathed like few others in the college curriculum, for who among us enjoys struggling to learn mind-numbingly boring and yet oppressively difficult material?

In writing this text, we attempt to overcome such attitudes by sharing our vision of mathematics as a vital discipline that is not only widely applicable, but also interesting, entertaining, and even—at times—breathtakingly beautiful. In addition to developing fundamental concepts and skills, we want our readers to experience the thrill of conducting penetrating mathematical explorations, with paper, pencil, a graphing calculator and sheer force of reason as their only sources of illumination. And though precalculus can be challenging, we believe that most of us come equipped with brains hardwired to think mathematically, though such circuitry may lie dormant from disuse and neglect. It is our goal to tap into the student's natural affinity for mathematics by providing motivation, inspiration, and intellectual stimulation through the explanations, examples, and exercises that constitute this text.

What Is Precalculus?

Many quantities in mathematics and physics involve limiting processes of some sort. For example, the area under a curve can be viewed as the limiting value of the total area of a collection of approximating rectangles, and the instantaneous velocity of a projectile can be viewed as the limiting value of its average velocity over progressively shorter time intervals. In fact, any rate of change, from the rate of decay of Carbon-14 to the rate of growth of the population of India, from the velocity of a baseball to the rate at which a new skill is mastered, can be defined using the language of limits. Thus **calculus**, the branch of mathematics that deals with limiting processes, can be applied in countless fields, including all of the physical and social sciences. Because of its broad applicability and the sheer brilliance of its arguments, calculus has been described as the crowning glory of the human intellect. It is no wonder that the study of calculus is required or recommended in many academic disciplines, especially the sciences, engineering, technology, and business.

Calculus courses, though greatly rewarding, are notoriously challenging. While it is true that modern calculus students are asked to master material developed over centuries by some of history's greatest minds, the most imposing obstacle to student comprehension is not the intrinsic difficulty of the subject matter, but rather a shaky understanding of the algebraic and trigonometric concepts that are foundational to calculus. **Precalculus**, as the name suggests, consists of topics from algebra and trigonometry that form the foundation of calculus. Moreover, precalculus emphasizes functions, not only because of their usefulness in modeling the real world, but also because of the central role they play in the study of calculus.

In addition to providing a thorough treatment of topics prerequisite to calculus, this text employs graphing technology to help foreshadow important calculus concepts. With the graphing calculator, a student can estimate maximum and minimum values of functions, determine where a function is increasing and where it is decreasing, and even steal a glance at a tangent line. In addition, the *Preview of Calculus* feature alerts the reader to connections between the material under discussion and important concepts from calculus.

Features of the Text

Chapter Opening Photographs Each chapter begins with a photograph and an extended caption that connects the image to the subject matter of the chapter. These striking images are designed to generate interest by underscoring the relevance and applicability of the material in that chapter.

Motivating Questions Each section begins with stimulating questions designed to pique the reader's curiosity about the material in the section. For example, Section 3 of Chapter 4 begins with the questions: *How powerful would the "bombquake" be that would result from simultaneously detonating all of the world's nuclear warheads? How can multiplication and division be performed simply by sliding pieces of paper? How many boom boxes would it take to cause a fatal musical overdose? If the entire population of China simultaneously jumped a foot off the ground, how strong would the resulting earthquake be?* Some answers can be found simply by reading the text, whereas others require the reader to complete one of the exercises at the end of the section.

Examples Examples are used to illustrate the techniques and theory discussed in the text, or in some cases to introduce a topic or theoretical concept. Every attempt is made to include the reader in the problem-solving process, to explain not just *how* each step is executed but also *why* a particular problem-solving strategy is chosen. Multi-step examples often include explanatory notes beside each step, or detailed explanations between important steps. In addition, examples using a graphing calculator are indicated by an icon in the Annotated Instructor's Edition. Ⓖ

Graphing Calculators The graphing calculator is an invaluable tool for solving equations, exploring mathematical concepts graphically, and tackling complex applications. To make optimal use of the power of the graphing calculator, one must know two things: *how* to use the calculator and, of equal importance, *when* to use the calculator. Throughout the text, we discuss when to use the graphing calculator, when to use symbolic methods, and when to use a combination of symbolic and graphical techniques. All of the important graphing calculator functions are discussed in boxes titled "Calculator Keys."

| Art and Graphics | The text includes hundreds of graphs, all of which are computer generated for accuracy. In addition, graphing calculator screen images have been incorporated wherever students will be performing a task with the graphing calculator. Many applied examples and exercises are accompanied by color illustrations or bar graphs. The text also includes dozens of color photographs, selected for their relevance, novelty, and beauty. |

Previews of Calculus Wherever possible, the text establishes links between the algebraic or trigonometric topic at hand and its natural extension in calculus. For example, Chapter 2 includes an overview of graphs of functions. In an accompanying Preview of Calculus box, we discuss the slope of a curve, a central concept in the study of calculus.

Mathematical Notes In order to underscore the vitality of the subject, the text includes brief accounts of recent developments in mathematics, such as Andrew Wiles' proof of Fermat's Last Theorem and recent investigations of dynamical systems. In addition, we explore several intriguing applications of mathematics, such as the use of mathematical morphing in movies and television, radiocarbon dating to test the authenticity of the Shroud of Turin, and mathematical models assessing the risk of a cometary collision with Earth.

Warnings and Rules of Thumb Warning boxes are provided to point out common mathematical pitfalls and how they can be avoided. For example, when solving equations, students are cautioned to avoid dividing by expressions containing a variable. Rule of Thumb boxes describe helpful hints and tricks of the trade.

Understanding and Mastery Checklists Each section concludes with a checklist of newly introduced topics and terminology, along with a separate list of the skills students are expected to have mastered upon completion of the section. The checklists are designed to help students assess whether they are ready to start solving exercises.

Exercise Sets There are some tasks for which pencil and paper techniques are superior, others where the graphing calculator is much more efficient, and still others that are best solved using some combination of algebraic and graphical tools. To develop the ability to select the mathematical tool most appropriate for a given task, many exercises have been included in which the student must first decide whether or not to use a graphing calculator. Thus, exercises that require the use of a graphing calculator are integrated within many of the following categories of exercises and are indicated in the Annotated Instructor's Edition.

- **Standard exercises**, graded in difficulty, are included to reinforce the development of important skills and concepts. Many of the standard exercises are similar to examples and are intended primarily to test the student's manipulative ability. Some are unlike any example in the text in order to ensure that the student has synthesized the underlying concepts and isn't just blindly using examples as templates.
- **Applications**, or applied problems, vary in scope from explorations of topical issues with important social, environmental, and economic ramifications, to problems with whimsical, thought-provoking, and (we hope) entertaining premises. Many involve modeling with real-world data from such diverse sources as governmental agencies, the *World Almanac*, the *Environmental Almanac*, and the *Guinness Book of World Records*. Some involve important notions from the physical and social sciences, such as gravity, projectile motion, earthquakes, the speed of light, time dilation, body mass index, the TNT equivalent of a nuclear warhead, population dynamics, and the spread of disease. On the lighter side are problems involving whimsical premises

such as two cows tormenting a train engineer, a cyclist powering a small city, and a hot-air balloon raising the Statue of Liberty. A fictional detective, Inspector Magill, appears throughout the text solving mysteries such as a kidnapping, cracking a secret code, a carjacking, murders, and a medical mystery.

- **Questions for Discussion or Essay** are intended for use as thought-provoking discussion questions, or as subjects for short essays. Some ask students to analyze the premises of application problems. Others require that students connect the material with their experiences in other math classes, other disciplines, and everyday life.
- **Concepts and Critical Thinking** exercises, including true/false and "give an example of" questions, are designed to help students assimilate the most important ideas and terminology of a section. These conceptually oriented exercises complement more computationally oriented standard exercises.
- **Projects for Enrichment** are nontrivial multistep problems designed for either group work or as especially challenging problems, possibly for extra credit. Some explore related mathematical topics that go beyond the classical boundaries of precalculus. Examples of such topics include the cubic formula, fractal geometry, and iterates of a function. Other projects are in-depth application problems. Examples here include an investigation of the relationship between alcohol consumption and blood alcohol level, a development of Kepler's third law for planetary motion, an application of systems of equations to balancing chemical equations, and a technique for computing the day of the week for any date in history.

Chapter Reviews and Tests   Each chapter concludes with a chapter review and a chapter test. The chapter reviews consist of comprehensive selections of exercises representative of those previously encountered in the chapter. The chapter tests include a sampling of representative problems but are not intended to be comprehensive. Each includes a section of true and false and "give an example of" problems that are designed to test the student's mastery of important concepts and key definitions.

Ancillaries for the Text   The following ancillaries are available to accompany this text:

Instructor's Solutions Manual, by Ignacio Alarcon of Santa Barbara City College, includes worked-out solutions to all of the even-numbered text exercises. (ISBN 0-534-35288-X)

Student Solutions Manual, by Ignacio Alarcon of Santa Barbara City College, includes worked out solutions to all of the odd-numbered text exercises. (ISBN 0-534-35289-8)

Test Bank contains eight printed test forms per chapter with answers for the instructor. (ISBN 0-534-35293-6)

BCA Instructor Version contains BCA Testing and BCA Tutorial. (ISBN 0-534-35291-X)

BCA Tutorial Student Version is text-specific interactive tutorial software. (ISBN 0-534-40316-6)

Text-Specific Videos features worked-out examples from each chapter. (ISBN 0-534-38289-4)

WebTutor ToolBox features lecture notes, discussion threads, and quizzes on WebCT and Blackboard. (bundle ISBN WebCT 0-534-11644-2, Blackboard 0-534-11662-0)

Interactive Video Skillbuilder CD-ROM contains hours of video instruction, section quizzes, chapter tests, and MathCue Tutorial. (ISBN 0-534-38377-7)

Acknowledgments

The candid comments and insightful suggestions of our colleagues were instrumental in the development of this text. We would like to thank the following reviewers: Mary E. Bradley, University of Louisville; Marie Donigan, Utah State University; Joseph Ediger, Portland State University; Tom Foy, Florida Gulf Coast University; Daniel Harned, Michigan State University; James Hart, Middle Tennessee State University; John Jaroma, Lake Superior State University; Laura Keane, University of Tampa; David Legg, Indiana University-Purdue University at Fort Wayne; Peter Loth, Sacred Heart University; Tim Loughlin, New York Institute of Technology; Thomas Post, Grand Rapids Community College; Mike Prophet, University of Northern Iowa; Donald Ransford, Edison Community College; Michael Trappuzzano, Arizona State University; Lynn Weathers, Mercer University; Rebecca Wooten, University of South Florida; and Fred Worth, Henderson State University. We are particularly grateful to Sudhir Goel, Ian Crewe, Mike Prophet, and Fred Worth for their invaluable corrections and helpful suggestions.

We are grateful for the assistance provided by the editorial, marketing, and production professionals on the Brooks/Cole team who have contributed to our texts. Thanks to Karin Sandberg, marketing manager; Lisa Chow, ancillary editor; Janet Hill, production project manager; Christopher Delgado, technology project manager; Darlene Amidon-Brent, editorial assistant; and Vernon Boes, design director. Thank you Larry Didona for the stunning artwork that graces our cover. Our appreciation also goes to Carol Dondrea, copyeditor; Patrick Devine, designer; and Kathleen Olson, photo researcher, for their dedication to the project. Special thanks to Nancy Shammas for her tireless efforts in coordinating the production of this text. We are indebted to our talented developmental editor, Leslie Lahr, and our insightful acquisitions editor, John-Paul Ramin, for their invaluable guidance and support. They helped us transform a good manuscript into what we believe is a much better book.

David Dwyer
Mark Gruenwald

Chapter 1 Foundations and Fundamentals

Shown above is an artist's rendering of the X-43C, an air-breathing craft capable of speeds of up to 5,000 mph. Currently under development by NASA, with testing scheduled for as early as 2008, the X-43C could cut the flight time from Los Angeles to New York to under 40 minutes. This cutting-edge technology is built on a foundation of sophisticated mathematical models encompassing aerodynamics, fluid flow, gravitation, and more. But it doesn't take a rocket scientist to appreciate the power of mathematical modeling. In Exercise 100 of Section 1.5 we investigate an elementary mathematical model of projectile motion.

Section 1.1

The Real Number System

- ✤ Which is larger: 0.999. . . or 1?
- ✤ How many families could Bill Gates feed for a year if he were willing to liquidate his entire fortune?
- ✤ What property do addition, multiplication, and marriage have that subtraction, division, and love do not?
- ✤ Which discount would you rather have: 30% off or 10% off of a price already reduced 20%?

Number Systems

Number systems have developed as a language for expressing how many, how much, and how far. A summary of the various number systems we will encounter in this text is given in the following box. The relationships among the systems are diagrammed in Figure 1.

Number Systems

Natural numbers are the numbers 1, 2, 3, . . .

Whole numbers are the numbers 0, 1, 2, 3, . . .

Integers are the numbers . . . , $-2, -1, 0, 1, 2, 3, \ldots$

Rational numbers are fractions of the form p/q, where p and q are integers and $q \neq 0$. Rational numbers have terminating or repeating decimal expansions.

Real numbers are the numbers that may be written as decimals. The real numbers that are not rational are called **irrational**, and they have nonterminating, nonrepeating decimal expansions. Examples include π and $\sqrt{2}$.

Complex numbers are numbers of the form $a + bi$, where a and b are real numbers and $i = \sqrt{-1}$. The term **imaginary** is used to describe complex numbers of the form bi, where $b \neq 0$.

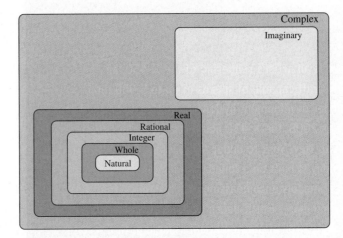

Figure 1

·····EXAMPLE 1

Identifying Numbers

List all the terms that describe the given number. Choose from natural, whole, integer, rational, irrational, real, and complex.

a. 22.47

b. -2

c. $0.24343434\ldots$

d. $0.101001000100001000001\ldots$

e. The number of atoms that constitute the planet Jupiter

f. A number that, when multiplied by itself, gives -25

Solution

a. 22.47 is a terminating decimal, so it is a rational number. Written as a fraction, $22.47 = \frac{2247}{100}$. The number 22.47 is also real and complex.

b. -2 is an integer, as well as rational, real, and complex.

c. Since $0.24343434\ldots$ is a repeating decimal, it is a rational number and so is also real and complex.

d. Since $0.101001000100001000001\ldots$ neither repeats nor terminates, it is not a rational number. Hence, it is irrational, real, and complex.

e. Although the number of atoms constituting the planet Jupiter is unknown (and no doubt enormous), this number is a natural number, and hence a whole number, integer, rational number, real number, and complex number as well.

f. There is no real number whose square is negative; only imaginary numbers have negative squares. Thus, any number that, when multiplied by itself, gives -25 is imaginary and so also complex.

The Real Number Line

The real number system can be represented geometrically by the real number line, shown in Figure 2. The point labeled 0 is called the **origin**. The points to the left of the origin correspond to the negative real numbers, and the points to the right of the origin correspond to the positive real numbers. The arrow is used to indicate the direction of increasing real numbers.

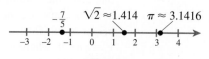

Figure 2

The number line establishes a **one-to-one correspondence** with the real numbers in the sense that every point on the number line corresponds to a real number, and every real number corresponds to a point on the number line. As a consequence, we can order the real numbers according to their position on the number line.

Ordering the Real Numbers

For two real numbers a and b, we say that a **is less than** b (denoted $a < b$) if a lies to the left of b on the number line, and a **is greater than** b (denoted $a > b$) if a lies to the right of b.

The symbols < and > are called **inequality symbols**. Two other inequality symbols are ≤, which denotes **less than or equal to**, and ≥, which denotes **greater than or equal to**.

-------EXAMPLE 2

Ordering Real Numbers

Place the appropriate inequality symbol in the box between the numbers.

a. $\sqrt{2}$ ☐ 5   **b.** $-\frac{2}{3}$ ☐ $-\frac{5}{6}$ **c.** -2.34 ☐ 1.03

Solution

a. Since $\sqrt{2} \approx 1.4$ lies to the left of 5 on the number line, we write $\sqrt{2} < 5$.

b. Since $-\frac{2}{3} = -\frac{4}{6}$ and $-\frac{4}{6}$ lies to the right of $-\frac{5}{6}$ on the number line, we write $-\frac{2}{3} > -\frac{5}{6}$.

c. Every negative number is less than every positive number; thus, we have $-2.34 < 1.03$.

An important component of numeracy (facility in using and understanding numbers) is intuition regarding the relative sizes of the objects that make up our world. Most of us do respectably well when estimating the sizes of human-sized objects such as cars, refrigerators, desks, and so forth, but tend to ridiculously underestimate the dimensions of large, naturally occurring features such as canyons, volcanos, and mountains. The answer to the following estimation problem may surprise you.

-------EXAMPLE 3

Leveling Everest–An Estimation Problem

If it were possible to grind up Mt. Everest, fill semitrailers with the contents, and then form the semis into a convoy, how long would the convoy be?

Solution   First we estimate the volume of Mt. Everest. We look in a reference book (such as an almanac or *The Guinness Book of World Records*) and find that Mt. Everest is approximately 29,000 feet tall. Now, unless we are lucky enough to have a reference that gives the volume of Mt. Everest, we are going to have to make an educated guess as to its shape. Let us assume that the mountain is roughly the shape of a cone, with a height of 29,000 feet. We look up the formula for the volume of a cone and find that $V = \frac{1}{3}\pi r^2 h$, where r is the radius of the circle that forms the base of the cone, and h is the height. We are taking h to be 29,000, but we don't know r. Here again, we make an educated guess that the diameter of the base is roughly twice the height, so the radius equals the height. Thus, we take r to be 29,000 feet. Upon substituting the values of r and h into the formula for V, we find that

$$V = \frac{1}{3}\pi(29{,}000\ \text{ft})^2\,(29{,}000\ \text{ft}) \approx 25{,}540{,}000{,}000{,}000 \text{ cubic feet}$$

A team of scientists headed by B. Chamoux and A. de Polenza measures Mt. Everest.

Next we estimate the capacity of a semitrailer. They appear to be about 40 feet long, 8 feet wide, and 9 feet tall. This would make the volume of a semitrailer equal to 40 feet × 8 feet × 9 feet = 2880 cubic feet. Hence, it would take approximately $25{,}540{,}000{,}000{,}000 \div 2880 \approx 8{,}868{,}000{,}000$ semitrailers to haul off Mt. Everest. Now we must figure out how long a convoy this would make. The cab of a semi is about 10 feet in length, so the entire truck would be about 10 feet + 40 feet = 50 feet long.

Assuming two truck lengths (or 100 feet) between trucks, the convoy would be about $8,868,000,000 \times 150$ feet $= 1,330,200,000,000$ feet long. Using the fact that a mile is 5280 feet, we determine that the convoy is $1,330,200,000,000 \div 5280 \approx 251,900,000$ miles long. This, by the way, is long enough to reach to the sun and back! Although our estimates were crude, even if we are off by a factor of 10 or even 100, it is clear that such a convoy would be impossible to form.

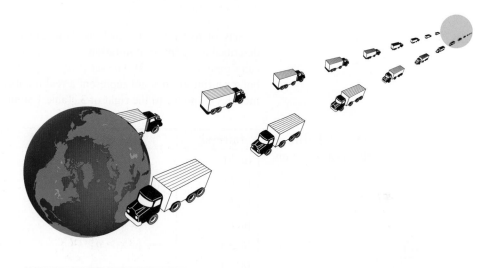

Sets of Real Numbers

The most direct way to write a set is simply to list all of its elements within braces. For example, the set consisting of the positive even numbers less than 10 could be written as {2, 4, 6, 8}. The set of all prime numbers less than 30 is {2, 3, 5, 7, 11, 13, 17, 19, 23, 29}. The **empty set**, the set with no elements, may be written as {} or ∅.

Although listing is convenient for small sets, it is impractical for large sets and impossible for sets with infinitely many elements. **Set-builder notation** is useful for describing large or infinite sets that are defined by a property. For example, consider the set of real numbers less than 2. In set-builder notation, we write $\{x \mid x < 2\}$. The set of real numbers between and including -3 and 3 can be written $\{x \mid -3 \le x \le 3\}$.

It is often helpful to graph sets of real numbers on a number line. To do this, we simply draw a number line and then draw a heavy line through all values that are included in the set. For example, the set $\{x \mid -2 < x \le 4\}$ is graphed as shown in Figure 3. Note that a parenthesis, (, is used at -2 to indicate that -2 is *not* included, whereas a bracket,], is used at 4 to indicate that 4 *is* included.

Figure 3

EXAMPLE 4

Graphing Sets of Real Numbers

Graph the following sets:

a. $\{x \mid x \ge 3\}$ **b.** $\left\{x \mid -\dfrac{3}{2} \le x < 3\right\}$

Solution

a.

b.

Sets of real numbers, such as those graphed in Example 4, can be conveniently described using **interval notation**. For example, instead of writing $\{x \mid -\frac{3}{2} \le x < 3\}$, we can simply write $[-\frac{3}{2}, 3)$. The set $\{x \mid x \ge 3\}$ can be written as $[3, \infty)$. Note that the symbol ∞ (infinity) does not represent a real number but rather indicates that the interval includes all points to the right of 3. Table 1 summarizes the nine types of intervals.

Table 1
Interval notation

Put in notes

Interval	Set-builder notation	Graph
(a, b)	$\{x \mid a < x < b\}$	
$[a, b]$	$\{x \mid a \le x \le b\}$	
$(a, b]$	$\{x \mid a < x \le b\}$	
$[a, b)$	$\{x \mid a \le x < b\}$	
(a, ∞)	$\{x \mid x > a\}$	
$[a, \infty)$	$\{x \mid x \ge a\}$	
$(-\infty, b)$	$\{x \mid x < b\}$	
$(-\infty, b]$	$\{x \mid x \le b\}$	
$(-\infty, \infty)$	$\{x \mid -\infty < x < \infty\}$	

Intervals that include their endpoints, such as $[a, b]$, $[a, \infty)$, and $(-\infty, b]$, are said to be **closed**. Intervals that do not include their endpoints, such as (a, b), (a, ∞), and $(-\infty, b)$, are **open**.

EXAMPLE 5

Using Interval Notation

Write the given set using interval notation and sketch its graph.

a. $\{x \mid -7 \le x < 2\}$ **b.** $\{s \mid s < \sqrt{5}\}$

Solution

a. $[-7, 2)$

b. $\left(-\infty, \sqrt{5}\right)$

EXAMPLE 6

Converting from Interval Notation to Set-Builder Notation

Express each of the following sets using set-builder notation:

a. $(2, 5]$ **b.** $(-\infty, 3)$ **c.** $[-4, \infty)$

Solution

a. $\{x|2 < x \le 5\}$　　**b.** $\{x|x < 3\}$　　**c.** $\{x|x \ge -4\}$

Arithmetic Operations　The operations of addition and multiplication have properties that can be used to simplify many mathematical problems. For example, addition is *commutative*; that is, the order in which numbers are added is immaterial—we get the same answer either way. We can state the commutative property of addition algebraically as $a + b = b + a$. The most significant of the addition and multiplication properties for real numbers are summarized in Table 2.

Table 2　Properties of the real number system

Name of property	Verbal description	Algebraic description
Commutative property of addition	The order doesn't matter when adding.	$a + b = b + a$
Commutative property of multiplication	The order doesn't matter when multiplying.	$ab = ba$
Associative property of addition	The grouping doesn't matter when adding.	$a + (b + c) = (a + b) + c$
Associative property of multiplication	The grouping doesn't matter when multiplying.	$a(bc) = (ab)c$
Distributive property	The product of a number and the sum of two other numbers is the same as the sum of the products of the first number with each of the other two numbers.	$a(b + c) = ab + ac$
Existence of the additive identity	There is a number called 0 (zero) whose sum with any number is that number.	$a + 0 = a$
Existence of the multiplicative identity	There is a number called 1 (one) whose product with any number is that number.	$a \cdot 1 = a$
Existence of the additive inverse	Every real number has an additive inverse; that is, for each number there is a second number that, when added to the first, gives 0.	$a + (-a) = 0$
Existence of the multiplicative inverse	Every real number has a multiplicative inverse; that is, for each number there is a second number that, when multiplied by the first, gives 1.	$a\left(\dfrac{1}{a}\right) = 1$

Note that all of the properties in Table 2 can be extended to the complex number system, with $0 + 0i$ and $1 + 0i$ playing the roles of the additive and multiplicative identities, respectively. We will consider arithmetic operations on complex numbers in Section 1.3 and again in Chapter 7. Many of the properties in Table 2 can also be considered for operations other than addition and multiplication, and even outside of the context of number systems, as we see in the following example and in Exercises 79–80.

EXAMPLE 7

The Property of Commutativity in Other Contexts

Indicate whether the given operation is commutative.

a. Taking the minimum of two real numbers
b. Forming a hyphenated surname from a couple's surnames prior to their marriage

Solution

a. Taking the minimum of two numbers is commutative: The minimum of x and y is the same as the minimum of y and x.
b. The formation of a hyphenated surname is not commutative; Jackie Kersey-Joiner is not the same as Jackie Joiner-Kersey.

Subtraction and division can be defined in terms of addition and multiplication. Subtraction is just addition of the additive inverse, and division is nothing more than multiplication by the multiplicative inverse. More formally, we define

$$x - y = x + (-y)$$
$$x \div y = x \cdot \frac{1}{y} \qquad \text{for } y \neq 0$$

Note that division by 0 is not defined. Why is this? Let us consider the expression $\frac{1}{0}$. According to the definition, $\frac{1}{0}$ would be the multiplicative inverse of zero. In other words, if we let $c = \frac{1}{0}$, then $0 \cdot c = 1$. But we know that $0 \cdot c = 0$. This is a contradiction, so there is no such number as $\frac{1}{0}$.

Order of Operations

The associative properties of addition and multiplication tell us that, when either adding or multiplying a string of numbers together, we may group them in any way that we wish. For example, we could evaluate $1 + 2 + 3$ in either of the following ways:

$$(1 + 2) + 3 = 3 + 3 = 6$$
$$1 + (2 + 3) = 1 + 5 = 6$$

But suppose that we need to evaluate $1 - 2 - 3$. Depending on how the subtractions are grouped, we could obtain either of the following two results:

$$(1 - 2) - 3 = -1 - 3 = -4$$
$$1 - (2 - 3) = 1 - (-1) = 2$$

Since the grouping affects the result, we conclude that there is no associative property for subtraction. Similarly, there is no associative property for division. In fact, the grouping also matters if we have both addition and multiplication in the same expression. For example, if we wish to evaluate $2 \times 3 + 5$, we might obtain either of the following:

$$(2 \times 3) + 5 = 6 + 5 = 11$$
$$2 \times (3 + 5) = 2 \times 8 = 16$$

The question is, Which result is correct? The answer is simple. We agree that unless grouping symbols (parentheses, brackets, and so forth) indicate otherwise, we will evaluate expressions involving the fundamental operations of negation, addition, subtraction, multiplication, and division according to the following rule.

Order of Arithmetic Operations

If no grouping symbols are used, perform operations in the following order:

1. Negations (taking the negative)
2. Multiplications and divisions from left to right
3. Additions and subtractions from left to right

If grouping symbols (such as parentheses) are present, the operations within the grouping symbols are performed first, again following the order just given.

EXAMPLE 8

Order of Operations

Evaluate the following expressions:

a. $10 - 6 \div 2 \times 5$ **b.** $(10 - 6) \div (2 \times 5)$

Solution

a. We first perform divisions and multiplications in left-to-right order.

$$10 - 6 \div 2 \times 5 = 10 - (6 \div 2) \times 5$$
$$= 10 - 3 \times 5$$
$$= 10 - (3 \times 5)$$
$$= 10 - 15$$
$$= -5$$

b. We perform the operations in parentheses first.

$$(10 - 6) \div (2 \times 5) = 4 \div 10$$
$$= 0.4$$

Calculator Keys

Arithmetic Operations

With most scientific and graphing calculators, arithmetic expressions—even those involving complex numbers—are entered intuitively: They are typed the way they are written. The only rule for entering arithmetic expressions that might at first seem unnatural concerns the distinction between the negation and subtraction symbols. When we subtract one expression from another, as in $5 - 3$, for example, we are using the **minus sign**. On the other hand, when we express the negative of a number, such as -2, for example, we are using the **negation symbol**. Distinguishing between the two may seem fussy when writing, but it is a crucial distinction to the calculator. Thus, for example, to enter -2, we must key in $\boxed{(-)}\,\boxed{2}$ (or $\boxed{2}\,\boxed{+/-}$ on some scientific calculators), *not* $\boxed{-}\,\boxed{2}$.

Absolute Value   The notion of absolute value is used whenever we are interested in the size or magnitude of a real number, without regard to whether it is positive or negative. Thus, 3 and -3 have different values but have the same *absolute* value—namely, 3. In general, we define the absolute value of a number x to be the distance from zero to x. More precisely, we make the following definition.

Definition of Absolute Value

The absolute value of a real number x, denoted $|x|$, is its distance from zero. In symbols,

$$|x| = \begin{cases} x & \text{if } x \geq 0 \\ -x & \text{if } x < 0 \end{cases}$$

Note that since the absolute value of a number represents a distance, it cannot be negative. Thus, the absolute value of a negative number can be obtained by changing the sign of the number, whereas the absolute value of a positive number is the number itself. In other words, we have the following rule.

Rule of Thumb

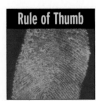

When finding the absolute value of a negative number, remove the negative sign. When finding the absolute value of a positive number, leave it alone.

EXAMPLE 9

Finding Absolute Value

Evaluate each of the following:

a. $|14.5|$ **b.** $|-\pi|$ **c.** $|0|$

d. $-|-4|$ **e.** $|x|$, where x is 27 units from the origin

Solution

a. Since $14.5 \geq 0$, $|14.5| = 14.5$.

b. Since $-\pi < 0$, $|-\pi| = -(-\pi) = \pi$.

c. Since $0 \geq 0$, $|0| = 0$.

d. $-|-4|$ means $-(|-4|)$ or "the opposite of $|-4|$." Since $-4 < 0$, $|-4| = 4$. Thus, $-|-4| = -4$. Note that this does not contradict the rule of thumb because it is the negative sign outside of the absolute value that makes the result negative.

e. Since the absolute value of a number is its distance from the origin, $|x| = 27$. Of course, this means that either $x = 27$ or $x = -27$.

EXAMPLE 10

Rewriting an Absolute Value Expression

Write the expression $|x - 2|$ without using absolute value, given that $x < 2$.

Solution   Since $x < 2$, $x - 2 < 0$. Thus, $|x - 2| = -(x - 2) = -x + 2$.

The distance between two real numbers can be found by subtracting the smaller from the larger. Using absolute value, the distance between two numbers can be defined without regard to which of the two is larger. For example, the distance between 2 and 6 can be found by computing $|2 - 6| = 4$. Generally, we have the following definition.

Distance

The **distance** between two real numbers a and b is given by $|a - b|$.

EXAMPLE 11

Finding Distance

Find the distance between $\frac{4}{5}$ and $\frac{9}{11}$.

Solution   The distance is $\left|\frac{4}{5} - \frac{9}{11}\right| = \left|\frac{44}{55} - \frac{45}{55}\right| = \left|-\frac{1}{55}\right| = \frac{1}{55}$.

EXAMPLE 12

Expressing Distance With Absolute Value

Find an expression for the distance between a real number x and -3.

Solution   The distance between x and -3 is given by $|x - (-3)| = |x + 3|$.

Understanding and Mastery Checklists

Concepts to Understand

Natural number system

❖

Whole numbers

❖

Integers

❖

Rational number system

❖

Irrational numbers

❖

Real number system

❖

Imaginary numbers

❖

Complex number system

❖

The number line

❖

Inequality symbols

❖

Set-builder notation

❖

Interval notation

❖

Properties of real numbers

❖

Order of operations

❖

Absolute value

Skills to Master

Determine to which systems a
given number belongs.

❖

Determine which of two real numbers is greater.

❖

Use the symbols $<$, $>$, $\leq$, and $\geq$ appropriately.

❖

Express sets of real numbers in
set-builder notation.

❖

Express sets of real numbers using
interval notation.

❖

Estimate quantities by making
reasonable assumptions.

❖

Evaluate expressions by performing
arithmetic operations in the correct order.

❖

Compute the absolute value of a real number.

Exercises 1.1

Exercises 1-14 *List all the terms that describe the given number. Choose from natural, whole, integer, rational, irrational, real, and complex.*

1. $\dfrac{6}{3}$

2. -47.134

3. $\sqrt{7}$

4. $\dfrac{12}{5}$

5. $3 - i$

6. $-\sqrt{3}$

7. $0.8888888\ldots$

8. 0.8888888

9. $\sqrt{-4}$

10. The ratio of the circumference of a circle to its diameter

11. $-\sqrt{25}$

12. The number of men named Bob living as of 9:43:26.24 A.M. EST on June 17, 2003

13. The height in miles of a 1-foot-tall rooster

14. The hypotenuse of an isosceles right triangle with legs of length 1

Exercises 15-18 *Place the most appropriate order symbol between each pair of quantities.*

15. $3 \;\square\; 0$

16. $-4 \;\square\; -5$

17. $-\dfrac{6}{5} \;\square\; -\dfrac{4}{5}$

18. $\dfrac{5}{6} \;\square\; \dfrac{2}{3}$

Exercises 19-22 *Place the given list of numbers in ascending order (the smallest first, the largest last). Estimate as needed.*

19. $23, -22.9, -23, 22.9$

20. $-\dfrac{1}{4}, -0.26, -\dfrac{3}{16}, -0.2$

21. $\dfrac{22}{7}, \pi, 3.15, 3 + \dfrac{1}{10} + \dfrac{4}{100}$

22. $\dfrac{1}{3}, 0.33333, 0.3334, \dfrac{1001}{3000}$

Exercises 23-30 *Write each set using interval notation and sketch the graph.*

23. $\{x \mid -2 < x \le 4\}$

24. $\{x \mid 3 \le x < 8\}$

25. $\left\{x \mid -7 < x < -\dfrac{1}{2}\right\}$

26. $\{x \mid -2 \le x \le \sqrt{3}\}$

27. $\{x \mid x < -3\}$

28. $\{x \mid x \ge -1\}$

29. $\{x \mid x \ge \sqrt{2}\}$

30. $\left\{x \mid x \le \dfrac{3}{2}\right\}$

Exercises 31-36 *Write each set using set-builder notation.*

31. $(3, 5)$
32. $(-\infty, 2)$

33. $(0, \infty)$
34. $[1, 2]$

35. $(100, \infty)$
36. $(1, 2)$

Exercises 37-44 *Evaluate the given expression.*

37. $5 - 3 \times 4$
38. $6 + 3 \div 2$

39. $1.2 - 2.5 - 3.4 - 1.8$
40. $3 \div 1 + 2 - 4 + 2 \times 3$

41. $4 \times [3 - (4 - 5)]$
42. $4 \times 3 - 4 - 5$

43. $3 \times 4 - 2 + 5 \times 2 - 1$
44. $3 \times (4 - 2) + 5 \times (2 - 1)$

Exercises 45-50 *Evaluate the given expression.*

45. $|-6|$
46. $-|4|$

47. $|3 - 1|$

48. $|-1 - 4|$

49. $\dfrac{|-3.7|}{-3.7}$

50. $\dfrac{|2 - 5|}{2 - 5}$

Exercises 51-54 *Write the expression without using absolute values.*

51. $|x - 1|$ if $x > 1$

52. $|t + 4|$ if $t < -4$

53. $|y + 2|$ if $y < -2$

54. $|x - 7|$ if $x > 7$

Exercises 55-60 *Find an expression for the distance between the given quantities. Simplify when possible.*

55. -3 and 7

56. $\dfrac{3}{4}$ and $\dfrac{5}{6}$

57. $3x$ and 2

58. $-x$ and 1

59. $\left(x + \dfrac{2}{3}\right)$ and $-\dfrac{2}{3}$

60. $(2x - 3)$ and $(x + 2)$

Applications

61. Cigarette Expenses Estimate the amount of money that a 3-pack-a-day smoker will spend on cigarettes in 30 years.

62. Lifting a Van Estimate the number of kindergartners required to lift a minivan. What practical difficulties would arise in performing such a stunt?

63. Sharing Wealth As of July 1, 2003, Bill Gates's personal wealth was estimated to be $58.3 billion. (*Data source:* Bill Gates's Personal Wealth Clock, http://philip.greenspun.com/WealthClock.) Estimate the number of families that Bill Gates could feed for a year if he were willing to spend his entire fortune. If Bill Gates had not existed, would the rest of the world be $58.3 billion richer?

64. Papering the Nation Total U.S. circulation of daily newspapers in 2001 was 55 million and for Sunday newspapers was 59 million. (*Data source:* Newspaper Association of America.) At this rate, how many years would it take to paper over the United States, with an area of 3,618,770 square miles?

65. Billion Dollar Cube Estimate the side length of a cube that could contain $1 billion in $100 bills.

66. Shrinking Earth The Great Wall of China, with a length of approximately 4000 miles, is the only man-made structure visible from space with the naked eye. If you were constructing an extremely detailed globe with a diameter of 10 feet, how long would the Great Wall of China be? How tall would a human be? How much area would Canada cover?

The Great Wall of China appears as an orange line in this satellite image.

67. Prolific Author Dame Barbara Cartland, once listed in *The Guinness Book of World Records* as the world's best-selling living author, wrote roughly 723 books from 1925 until her death in 2000. Estimate the average number of words written per day by Dame Barbara during this period.

68. The Book of Pi Suppose that all the known digits of pi (approximately 1.24 trillion as of early 2003) were written on pages the size of this page and these pages were then bound to form a book. How thick would such a book be?

69. Percentage Discount One of the authors (really!) went to a fast-food Mexican restaurant with a friend and ordered two meals on the same ticket. The clerk interrupted, informing them that there was a special unadvertised 10% discount on everything in the restaurant and suggesting that, in light of this special, the author and his friend should order their meals separately. Seeing the puzzled look on the author's face, the clerk explained that by ordering separately, there would be two 10% discounts—one for each order—for a total of 20%! Suppose that the bill for the author's meal is $3.50 and that of his friend is $2.80.

a. Compute the bill (not counting tax) if both meals are placed on the same ticket and then 10% is discounted from the total.

b. Compute the total bill if the meals are on separate tickets, each of which is discounted by 10%.

c. Use the distributive law to explain the puzzled look on the author's face.

70. Employee Discount An applicant for a position at a clothing store is told by the interviewer that employees receive a 10% discount on all purchases, including sales items. The interviewer gives as an example the following scenario: "Suppose that we're having a 20% clearance sale. Then you would get an additional 10% off, for a total discount of 30%." Explain the flaw in the interviewer's reasoning and compute the actual total discount. Does the answer depend on which discount—the 10% employee discount or the 20% clearance discount—is taken first? What properties of real number arithmetic are you using to answer these questions?

Concepts and Critical Thinking

Exercises 71-74 *Answer true or false.*

71. All rational numbers have a finite number of decimal places.

72. For all real numbers a, b, and c, $\dfrac{a}{b+c} = \dfrac{a}{b} + \dfrac{a}{c}$.

73. 0 is the multiplicative identity.

74. For all real numbers a and b satisfying $a < b$, $|a| < |b|$.

Exercises 75-78 *Give an example of each.*

75. An integer that is not a natural number

76. A real number that is not a rational number

77. A property that addition and multiplication have but that subtraction and division do not

78. Two numbers a and b such that $a - b = b - a$

79. An operation is said to be commutative if "the order doesn't matter." For example, addition is commutative because $a + b = b + a$. Subtraction, however, can be shown to be noncommutative by considering, for example, $a = 0$ and $b = 1$. Here, $a - b = 0 - 1 = -1$, whereas $b - a = 1 - 0 = 1$. The pair of numbers $a = 0$, $b = 1$ are a *counterexample* to the statement that subtraction is commutative. State whether each of the following operations is commutative; if it is not, give a counterexample.

  a. Division

  b. Multiplication

  c. Marriage (Is the outcome of Bob marrying Sue the same as that of Sue marrying Bob?)

  d. Love (Is the outcome of Dirk loving Camille the same as Camille loving Dirk?)

  e. Taking the maximum of two real numbers

  f. Addressing (Is "Jamal speaks to Ann" the same as "Ann speaks to Jamal"?)

  g. Implication (Is "If A, then B" the same as "If B, then A"?)

  h. Finding the greatest common divisor of two numbers

80. Indicate whether the given operation is associative. If not, give a counterexample. [*Hint:* An operation is associative if the grouping doesn't matter. For example, addition is associative because $a + (b + c) = (a + b) + c$.]

  a. Multiplication

  b. Division

  c. Subtraction

  d. Linking chains

  e. Connecting Lego blocks together

81. Place a one dollar bill in front of you right side up. Rotate the front of the bill 90° clockwise so that George Washington's nose points toward you. Now flip the bill over (top to bottom) so that you now see the reverse side of the dollar bill with the seal of the United States nearest you. Now, return the dollar bill to its original position, and this time, flip the bill first and then rotate it 90° clockwise. Do rotations and flips commute with one another?

82. Explain how you could compute $999,999 \times 75$ in your head. (*Hint:* Think of 999,999 as $1,000,000 - 1$ and use the distributive property.)

83. Explain how you could compute $1,000,003 \times 90$ in your head.

84. Use the properties of real numbers to show that $a \cdot 0 = 0$. [*Hint:* Use $0 = 1 + (-1)$ and the distributive property.]

85. Suppose that we were unaware of the rules regarding order of operations. List all possible results of the computation $3 \cdot 5 + 6 \div 2$. (*Hint:* There are four in all.)

Questions for Discussion or Essay

86. It is a fact (see Exercise 90) that $0.999\ldots = 1$. Although mathematicians are in universal agreement, many other people refuse to believe it. What makes this fact so difficult to believe? Does the statement that $0.999\ldots = 1$ seem to defy common sense? Why?

87. The definition of absolute value as stated in the text is

$$|x| = \begin{cases} x & \text{if } x \geq 0 \\ -x & \text{if } x < 0 \end{cases}$$

Thus, under some circumstances, $|x| = -x$. On the other hand, absolute values are never negative. Explain this apparent contradiction.

88. Which of the properties discussed in this section do you think are so transparently obvious that virtually everyone who does arithmetic with some regularity—clerks, schoolchildren, bookkeepers, waiters, and so forth—employs them correctly? Which properties are most likely to be misused?

89. To prove that subtraction is *not* commutative, it is sufficient to produce a single pair of numbers (such as 0 and 1) for which we obtain a different answer depending on the order in which we perform the subtraction. Why then *can't* we establish the commutative property of addition by simply producing a single pair of numbers such that the order in which we add these numbers doesn't matter? For example, what is wrong with this "proof" of the commutativity of addition? "Consider the numbers 2 and 3. Since $2 + 3 = 5$, and also $3 + 2 = 5$, addition is commutative."

Projects for Enrichment

90. Decimal Expansions   The text states that rational numbers have terminating or repeating decimal expansions. The converse of this statement is also true; that is, if a decimal expansion terminates or repeats, then it represents a rational number. In this project, we learn how to express terminating and repeating decimals as fractions.

Terminating decimals are converted to fractions as follows. The numerator of the fraction is the integer obtained by removing the decimal point; the denominator is an appropriate power of 10 (10, 100, 1000, and so on). For example, 0.123 may be written as $\frac{123}{1000}$, 1.03 as $\frac{103}{100}$, and so on.

a. Express each of the terminating decimals as a fraction.

   i. 87.096   **ii.** 232.1

   iii. 0.0034   **iv.** 0.000048

The following examples illustrate a method for converting repeating decimals to fractions.
Example: Convert $13.454545\ldots$ to a fraction.
Solution: Let $x = 13.454545\ldots$. Then $100x = 1345.454545\ldots$. We subtract x from $100x$ as follows:

$$\begin{aligned} 100x &= 1345.454545\ldots \\ -x &= -13.454545\ldots \\ \hline 99x &= 1332.0 \end{aligned}$$

Dividing both sides by 99, we obtain $x = \dfrac{1332}{99}$.

Example: Convert $2.512012012\ldots$ to a fraction.

Solution: Let $x = 2.512012012\ldots$. Then $10,000x = 25,120.12012012\ldots$ and $10x = 25.12012012\ldots$. Subtracting, we obtain

$$\begin{aligned} 10,000x &= 25,120.120120120\ldots \\ -10x &= -25.120120120\ldots \\ \hline 9,990x &= 25,095.0 \end{aligned}$$

Thus, $x = \dfrac{25,095}{9990} = \dfrac{1673}{666}$.

b. Express each of the following repeating decimals as a fraction.

   i. $0.252525\ldots$ **ii.** $0.999\ldots$

   iii. $12.0010101\ldots$ **iv.** $344.123123123\ldots$

91. The Irrationality of $\sqrt{2}$ In this project, we show that $\sqrt{2}$ is irrational. We use the method called *reductio ad absurdum*, or reduction to an absurdity. That is, we assume the opposite of what we think is true and show that this leads to a contradiction.

a. Suppose that $\sqrt{2}$ is rational—that is, that $\sqrt{2} = p/q$ for a pair of integers p and q with no common factors. (If p and q have a common factor, then p/q is not in lowest terms.) Show that $p^2 = 2q^2$.

b. Use part a to show that 2 must be a factor of p^2. If 2 is a factor of p^2, does it follow that 2 is a factor of p?

c. Suppose that 2 divides evenly into p. Then $p = 2m$ for some integer m. Show that $4m^2 = 2q^2$.

d. Show that $q^2 = 2m^2$ and hence that 2 divides into q.

e. The result to part d is an absurdity. Why?

Since the assumption that $\sqrt{2}$ is rational leads to an absurdity, it follows that $\sqrt{2}$ is irrational.

92. Stupid Human Tricks   Although most of us find it difficult to multiply numbers larger than 12 in our heads, there are individuals—some with IQs in the mentally retarded range—capable of extraordinary feats of mental computation. There are documented cases of so-called idiot savants who can mentally multiply 10-digit numbers but are unable to solve a logic problem that would be child's play for a normal 3-year-old. In such cases, the method by which the computation is performed is concealed by the verbal limitations of the performer, although recent research suggests that these individuals rely on a combination of prodigious memories—often *eidetic*, or visual memory—and a collection of specialized rules. Fortunately, not all human calculators are mentally challenged. Scott Flansburg ("The Human Calculator") has written several books [including *Math Magic* (New York: Harper, Perennial, 1994)] in which he describes many of his techniques for rapid mental computation. In this project, we investigate how and why two of Scott's tricks work.

a. To multiply two numbers that are almost 100—say, 92 and 97—do the following. First write the two factors in a column, and then form a second column by subtracting each number from 100, as shown in Figure 4. In this case, we obtain 8 and 3. Next, compute the "cross-difference," $92 - 3$ or $97 - 8$, and write the result, 89, at the bottom of the first column. Finally, multiply the entries in the second column, 8 and 3, and write the product, 24, at the bottom of the second column. The product of the two numbers is then given by the digits at the bottom of the two columns—in this case, 8924.

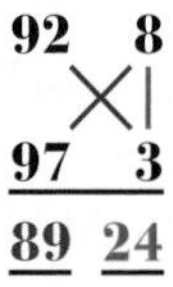

$$
\begin{array}{cc}
92 & 8 \\
\times \\
97 & 3 \\
\hline
89 & 24
\end{array}
$$

Figure 4

Figure 5 shows how this technique can be used to multiply 94 and 91.

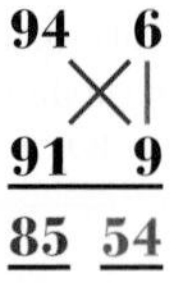

$$
\begin{array}{cc}
94 & 6 \\
\times \\
91 & 9 \\
\hline
85 & 54
\end{array}
$$

Figure 5

i. Perform the following multiplications using the technique illustrated in Figures 4 and 5: $93 \cdot 96$, $89 \cdot 98$, $93 \cdot 95$.

ii. Attempt to do the following multiplications using the same technique, but this time do the computations in your head: $94 \cdot 97$, $92 \cdot 99$, $96 \cdot 92$.

iii. Suppose we wish to multiply x and y (both of which are close to but less than 100) using Scott's technique. Complete the computations shown in Figure 6.

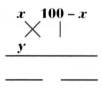

Figure 6

iv. Referring to our first example, the number 8924 is equivalent to $8900 + 24$. Thus, $92 \cdot 97 = 89 \cdot 100 + 24$. Similarly, $94 \cdot 91 = 85 \cdot 100 + 54$. Use this idea and your work from part iii to find an expression for xy.

v. Now use basic properties to confirm the formula found in part iv.

vi. Use algebra to confirm that it does not matter which cross-difference is used to compute the first two digits of a product using this technique.

vii. Explain why this technique is most useful for multiplying numbers that are close to 100.

b. A similar technique allows us to multiply two numbers that are just larger than 100, as shown in Figure 7.

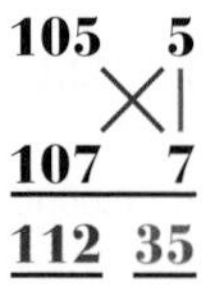

Figure 7

i. Using Figure 7 as a guide, write out instructions for multiplying two numbers just larger than 100.

ii. Use algebra to explain why this technique works.

Section 1.2 | Exponents and Radicals

- If 2^3 is 2 multiplied by itself 3 times, then what is 2^{-3}?
- An individual doubles in weight between the ages of birth and 6 months. If he continued to double in weight every 6 months, how much would he weigh on his forty-first birthday?
- What does it mean to multiply a number by itself one-third times?
- The skid marks at the scene of an accident measure 200 feet. How fast was the car traveling?
- A woman on her way to a local restaurant to celebrate her thirtieth birthday picks up her great-great-great-great-grandson at his retirement home and then stops off at the bank to withdraw $1,000,000 from the account that she had established by depositing $1000 on the previous day. How can this be?

Natural Number Exponents

In much the same way that multiplication expresses repeated additions, exponential notation expresses repeated multiplications. We define natural number exponents as follows.

Definition of Natural Number Exponents

If n is a natural number, then $a^n = a \times a \times \cdots \times a$, where there are n factors of a. The number a is called the **base** and the number n is called the **exponent**.

⋯⋗EXAMPLE 1

Evaluating Expressions Involving Natural Number Exponents

Evaluate each of the following:

a. 2^5 **b.** $(-3)^3$ **c.** $\left(\dfrac{3}{4}\right)^2$

Solution

a. $2^5 = 2 \cdot 2 \cdot 2 \cdot 2 \cdot 2 = 32$

b. $(-3)^3 = (-3) \cdot (-3) \cdot (-3) = -27$

c. $\left(\dfrac{3}{4}\right)^2 = \dfrac{3}{4} \cdot \dfrac{3}{4} = \dfrac{9}{16}$

Note that the parentheses in parts b and c of Example 1 indicate that the base is the entire quantity inside the parentheses. What about expressions such as -3^2, where no parentheses are given? Should we assume that -3^2 means $(-3)^2 = 9$ or $-(3)^2 = -9$? In Section 1.1, we discussed the fact that whenever an algebraic expression can have more than one meaning, conventions determine the order in which operations are performed. For example, if there are no grouping symbols, then multiplication and division are performed before addition and subtraction. In the case of exponents, the rule is that we perform all exponentiations before we perform negations, multiplications, divisions, additions, and subtractions. Thus, $-3^2 = -(3^2) = -9$.

Order of Operations

If no grouping symbols are used, perform operations in the following order:

1. Exponentiations
2. Negations
3. Multiplications and divisions from left to right
4. Additions or subtractions from left to right

If grouping symbols are present, the operations within the grouping symbols are performed first, again using the order just given.

EXAMPLE 2

A Doubling Problem

Suppose that a child who weighs 7 pounds at birth doubles in weight in the first 6 months of his life and continues to double in weight every 6 months. How much will he weigh on his forty-first birthday? His fiftieth birthday?

Solution Because the child weighs 7 pounds at birth, after one 6-month period he will weigh $2 \cdot 7$ pounds, after two 6-month periods he will weigh $2 \cdot 2 \cdot 7 = 2^2 \cdot 7$ pounds, and so forth. Thus, if n represents the number of 6-month periods that have elapsed, he will weigh $2^n \cdot 7$ pounds after n 6-month periods. On his forty-first birthday, $2 \cdot 41 = 82$ 6-month periods will have elapsed. Thus, he will weigh

$$2^{82} \cdot 7 = 33{,}849{,}922{,}949{,}209{,}616{,}891{,}772{,}928 \text{ pounds}$$

more than double the weight of Earth. On his fiftieth birthday, he will weigh

$$2^{100} \cdot 7 = 8{,}873{,}554{,}201{,}597{,}605{,}810{,}476{,}922{,}437{,}632 \text{ pounds}$$

more than double the weight of the sun!

The following properties are used for simplifying expressions involving exponents. Each can be proven using the definition of natural number exponents, as suggested by the accompanying illustrations.

Properties of Exponents

Assume that m and n are natural numbers.

Property	Example	Illustration
1. $a^m \cdot a^n = a^{m+n}$	$2^3 2^4 = 2^{3+4} = 2^7$	$(2 \cdot 2 \cdot 2) \cdot (2 \cdot 2 \cdot 2 \cdot 2)$ $= 2 \cdot 2 \cdot 2 \cdot 2 \cdot 2 \cdot 2 \cdot 2 = 2^7$
2. $\dfrac{a^m}{a^n} = a^{m-n}$	$\dfrac{3^5}{3^3} = 3^{5-3} = 3^2$	$\dfrac{3 \cdot 3 \cdot 3 \cdot 3 \cdot 3}{3 \cdot 3 \cdot 3} = \dfrac{\cancel{3} \cdot \cancel{3} \cdot \cancel{3} \cdot 3 \cdot 3}{\cancel{3} \cdot \cancel{3} \cdot \cancel{3}}$ $= 3^2$
3. $(a^m)^n = a^{mn}$	$(5^2)^3 = 5^{2 \cdot 3} = 5^6$	$(5 \cdot 5) \cdot (5 \cdot 5) \cdot (5 \cdot 5) = 5^6$
4. $(ab)^n = a^n b^n$	$(3 \cdot 5)^2 = 3^2 \cdot 5^2$	$(3 \cdot 5) \cdot (3 \cdot 5) = (3 \cdot 3) \cdot (5 \cdot 5)$ $= 3^2 \cdot 5^2$
5. $\left(\dfrac{a}{b}\right)^n = \dfrac{a^n}{b^n}$	$\left(\dfrac{2}{3}\right)^3 = \dfrac{2^3}{5^3}$	$\dfrac{2}{5} \cdot \dfrac{2}{5} \cdot \dfrac{2}{5} = \dfrac{2 \cdot 2 \cdot 2}{5 \cdot 5 \cdot 5} = \dfrac{2^3}{5^3}$

EXAMPLE 3

Simplifying Expressions Involving Natural Number Exponents

Simplify the given expression using the rules of exponents.

a. $3^4 \cdot 3^{17}$ **b.** $\left(x^3\right)^5$ **c.** $\dfrac{3^4 x^6 y^2}{3^2 x^3 y} \cdot \dfrac{3x^3 y^3}{\left(3x^2 y\right)^2}$

Solution

a. $3^4 \cdot 3^{17} = 3^{4+17}$ Using Property 1

      $= 3^{21}$

b. $\left(x^3\right)^5 = x^{3 \cdot 5}$ Using Property 3

      $= x^{15}$

c. $\dfrac{3^4 x^6 y^2}{3^2 x^3 y} \cdot \dfrac{3x^3 y^3}{\left(3x^2 y\right)^2} = \dfrac{3^4 x^6 y^2 \cdot 3x^3 y^3}{3^2 x^3 y \cdot 3^2 x^4 y^2}$ Multiplying fractions and using Property 4 to remove parentheses

      $= \dfrac{3^5 x^9 y^5}{3^4 x^7 y^3}$ Using Property 1 in the numerator and denominator

      $= 3x^2 y^2$ Using Property 2

The properties of exponents also hold for complex numbers—that is, for numbers of the form $a + bi$, where a and b are real and $i = \sqrt{-1}$. In the following example, we consider natural number powers of imaginary numbers (those of the form bi). In Section 7.4, we consider powers of complex numbers more generally.

EXAMPLE 4

Natural Number Powers of Imaginary Numbers

Simplify $(2i)^7$, given that $i = \sqrt{-1}$.

Solution   In the following steps, note how an odd power of i is rewritten as i times an even power of i. This allows us to take advantage of the fact that $i^2 = -1$.

$(2i)^7 = 2^7 i^7$ Using Property 4

      $= 128 i^6 i$ Using Property 1

      $= 128\left(i^2\right)^3 i$ Using Property 3

      $= 128(-1)^3 i$ Using the fact that $i^2 = -1$

      $= -128 i$

Integer Exponents

If exponentiation is repeated multiplication, then what is 2^0? That is, how do you multiply 2 by itself 0 times? The answer is that, although "multiplying 2 by itself 0 times" *doesn't* have any meaning, we can still define 2^0 in a sensible way using the properties of exponents just given. For example, according to Property 2 for natural number exponents, 2^0 should have the same value as $2^1/2^1 = 1$; therefore, we *define* 2^0 to be 1. Similarly, we can define 2^{-3} to be $2^0/2^3 = 1/2^3$. More generally, we give the following definitions.

Definition of Negative and Zero Exponents

For $a \neq 0$, and n a positive integer,

$$a^{-n} = \frac{1}{a^n}$$

For $a \neq 0$,

$$a^0 = 1$$

EXAMPLE 5

Evaluating Expressions With Nonpositive Exponents

Evaluate each of the following expressions:

a. 3^{-2} **b.** 10^{-1} **c.** $(57892.09593)^0$

Solution

a. $3^{-2} = \dfrac{1}{3^2} = \dfrac{1}{9}$ **b.** $10^{-1} = \dfrac{1}{10^1} = \dfrac{1}{10}$

c. $(57892.09593)^0 = 1$, since any number (except 0) raised to the zero power is 1.

Exponentiation is now defined for integer exponents in such a way that all of the properties of natural number exponents are valid. We use these properties in the following examples.

EXAMPLE 6

Simplifying Expressions With Negative Exponents

Simplify the given expression. Express your answer using only positive exponents.

a. $x^3 \cdot x^{-5}$ **b.** $\dfrac{2^3 a^{-4} b^2}{2^{-2} a^{-2} b^5}$ **c.** $\left(\dfrac{x^3 y^{-2} z^{-3}}{x^{-5} y z^{-1}} \right)^{-1}$

Solution

a. $x^3 \cdot x^{-5} = x^{3+(-5)}$ Using Property 1

              $= x^{-2}$ Simplifying

              $= \dfrac{1}{x^2}$ Using the definition of negative exponents

b. $\dfrac{2^3 a^{-4} b^2}{2^{-2} a^{-2} b^5} = \dfrac{2^3}{2^{-2}} \cdot \dfrac{a^{-4}}{a^{-2}} \cdot \dfrac{b^2}{b^5}$ Grouping like factors

            $= 2^{3-(-2)} \cdot a^{-4-(-2)} \cdot b^{2-5}$ Using Property 2

            $= 2^5 a^{-2} b^{-3}$ Simplifying

            $= 2^5 \cdot \dfrac{1}{a^2} \cdot \dfrac{1}{b^3}$ Using the definition of negative exponents

            $= \dfrac{32}{a^2 b^3}$ Simplifying

c. $\left(\dfrac{x^3 y^{-2} z^{-3}}{x^{-5} y z^{-1}}\right)^{-1} = \left(\dfrac{x^3}{x^{-5}} \cdot \dfrac{y^{-2}}{y} \cdot \dfrac{z^{-3}}{z^{-1}}\right)^{-1}$ Grouping like factors

$\qquad\qquad = \left(x^8 y^{-3} z^{-2}\right)^{-1}$ Using Property 2

$\qquad\qquad = x^{-8} y^3 z^2$ Using Property 4

$\qquad\qquad = \dfrac{y^3 z^2}{x^8}$ Using the definition of negative exponents

---

Scientific Notation

Collectively, the sciences provide us with a window on the universe through which we steal fleeting glimpses of wonders unthinkably small and unimaginably large. From ephemeral subatomic particles dancing into and out of existence to violently colliding clusters of galaxies; from the minuscule bacteria that teem on this very page to ecosystems spanning entire continents—science deals with the microscopic, the macroscopic, and everything in between. The decimal system of notation, so useful when dealing with human-scale entities, is stretched past its functional limit when used to express quantities at such vastly different scales, as Table 3 illustrates.

Table 3

Quantity	Approximation
Distance light travels in a year in miles	5,878,000,000,000
Weight of a bloodsucking banded louse in pounds	0.00000001101
Mass of an electron in grams	0.000000000000000000000000000910953
Number of atoms in the observable universe	10,000,000,000,000,000,000,000,000,000, 000,000,000,000,000,000,000,000,000,000, 000,000,000,000,000,000,000,000,000

Scientific notation was developed as a more compact way of expressing such very large and very small numbers. For example, using scientific notation, we would say that there are 5.878×10^{12} miles in a light-year, that the louse weighs 1.1101×10^{-8} pound, an electron weighs 9.10953×10^{-28} gram, and there are about 1×10^{85} atoms in the observable universe. Note that in each case, we have expressed a real number as the product of two numbers: a number between 1 and 10 and an integer power of 10. More precisely, we have the following definition.

Scientific Notation

A number is said to be expressed in scientific notation if it is written in the form $a \times 10^n$, where $1 \le a < 10$, and n is an integer.

Rule of Thumb

When converting from decimal notation to scientific notation, the number between 1 and 10 is obtained by shifting the decimal point to the position immediately after the first nonzero digit. The power of 10 is obtained by counting the number of places that the decimal point has been shifted. If the original number is larger than 1, the exponent is nonnegative. If the original number is smaller than 1, then the exponent is negative.

EXAMPLE 7

Converting to Scientific Notation

Express the following numbers in scientific notation:

a. 234.0023   **b.** 0.00012

Solution

a. $234.0023 = 2.340023 \times 10^2$

b. $0.00012 = 1.2 \times 10^{-4}$

EXAMPLE 8

Converting from Scientific Notation to Decimal Notation

Express the following numbers in standard decimal notation:

a. 7.03×10^{-3} **b.** 9.05×10^5

Solution

a. $7.03 \times 10^{-3} = 0.00703$ The decimal point has been moved 3 units to the left.

b. $9.05 \times 10^5 = 905,000$ The decimal point has been moved 5 units to the right.

Calculator Keys

Scientific Notation

Numbers expressed in scientific notation can be entered into a calculator using a key that is usually labeled either EE or EXP. For example, to enter the number 2.04×10^{32}, we first enter the number 2.04, then press EE (or EXP), and finally enter the number 32. The resulting display would show 2.04E32 (or perhaps 2.04 ^{32} or 2.04 32).

Radicals

As we have seen, *exponentiation* is a shorthand notation for repeated multiplication of a given base, so that, for example, $2^5 = 2 \cdot 2 \cdot 2 \cdot 2 \cdot 2 = 32$. *Roots* result from reversing this process: An *n*th root of x is a number whose *n*th power is x. Thus, for example, since 32 is the fifth power of 2, it follows that 2 is a fifth root of 32.

Definition of *n*th Root

A number y is an ***n*th root** of x if $y^n = x$. If $n = 2$, the root is called a **square root**. If $n = 3$, the root is called a **cube root**.

Since $2^2 = 4$, and also $(-2)^2 = 4$, both 2 and -2 are square roots of 4. The positive root, 2, is called the **principal square root** of 4. Whenever n is even, there will be two *n*th roots, one positive and one negative.

Definition of Principal Root

For n odd and x any real number, the **principal *n*th root** of x is the real number y such that $y^n = x$. For n even and $x > 0$, the **principal *n*th root** of x is the *positive* real number y such that $y^n = x$. The principal *n*th root of x is denoted by $\sqrt[n]{x}$.

EXAMPLE 9 | Finding *n*th Roots

Find the given roots of 16.

a. 4th   **b.** principal 4th

Solution

a. Since $2^4 = 16$ and $(-2)^4 = 16$, both 2 and -2 are 4th roots of 16.

b. The principal 4th root is the *positive* 4th root—namely, 2.

Radical Notation and Terminology

In the expression $\sqrt[n]{x}$, the symbol $\sqrt{}$ is called a **radical**, the number n is called the **index** of the radical, and x is called the **radicand**. If no index is given, it is understood to be 2. Thus, $\sqrt{x} = \sqrt[2]{x}$. Note that $\sqrt[n]{}$ always refers to the principal root.

EXAMPLE 10 | Evaluating Radicals

Evaluate each of the following radical expressions:

a. $\sqrt{121}$ **b.** $\sqrt[5]{-32}$ **c.** $\sqrt[4]{\dfrac{1}{81}}$

Solution

a. $\sqrt{121} = 11$ since $11^2 = 121$ and $11 > 0$.

b. $\sqrt[5]{-32} = -2$ since $(-2)^5 = -32$.

c. $\sqrt[4]{\dfrac{1}{81}} = \dfrac{1}{3}$ since $\left(\dfrac{1}{3}\right)^4 = \dfrac{1}{81}$, and $\dfrac{1}{3} > 0$.

WARNING!

The *principal root* always refers to a single number. For example, $\sqrt{25}$ is **not** equal to ± 5, but rather $\sqrt{25}$ is equal to 5.

The definition of the principal square root can be extended to the case where the radicand is negative by making use of the complex number $i = \sqrt{-1}$.

EXAMPLE 11 | Simplification of Square Roots of Negative Numbers

Simplify each of the following complex numbers:

a. $\sqrt{-16}$

b. $\sqrt{-8}$

Solution

a. $\sqrt{-16} = \sqrt{16}\,\sqrt{-1} = 4i$

b. $\sqrt{-8} = \sqrt{4}\,\sqrt{2}\,\sqrt{-1} = 2\sqrt{2}i$

The rules for simplifying radical expressions can be obtained from the definition of principal root and the properties of exponents. The most commonly used properties of radicals are listed here.

Properties of Radicals

Assume that m and n are positive integers and that x and y are such that all radicals are real.

Property	Examples	Comment
1. $\left(\sqrt[n]{x}\right)^n = x$	$\left(\sqrt[3]{2}\right)^3 = 2$	This is just the definition of nth root.
2. For n odd: $\sqrt[n]{x^n} = x$	$\sqrt[3]{7^3} = 7$ $\sqrt[5]{x^{15}} = x^3$	In general, $\sqrt[n]{x^{mn}} = \sqrt[n]{(x^m)^n} = x^m$.
For n even: $\sqrt[n]{x^n} = \lvert x \rvert$	$\sqrt{3^2} = 3$ $\sqrt[4]{(-3)^4} = 3$ $\sqrt{x^2} = \lvert x \rvert$	The absolute value arises because of the fact that the *principal root* of a positive number is positive.
3. $\sqrt[n]{x}\,\sqrt[n]{y} = \sqrt[n]{xy}$	$\sqrt{3}\,\sqrt{5} = \sqrt{15}$	The indices of the radicals must be the same.
4. $\dfrac{\sqrt[n]{x}}{\sqrt[n]{y}} = \sqrt[n]{\dfrac{x}{y}}$	$\dfrac{\sqrt[5]{30}}{\sqrt[5]{5}} = \sqrt[5]{6}$	The indices of the radicals must be the same.
5. $\left(\sqrt[n]{x}\right)^m = \sqrt[n]{x^m}$	$\left(\sqrt[3]{5}\right)^2 = \sqrt[3]{5^2}$	This can be derived from Property 3.
6. $\sqrt[m]{\sqrt[n]{x}} = \sqrt[mn]{x}$	$\sqrt[4]{\sqrt[3]{x}} = \sqrt[12]{x}$	This property reminds us of $\left(x^m\right)^n = x^{mn}$.

Generally, a radical expression is considered to be completely simplified whenever

- The radicand is as "small" as possible; that is, the radicand contains as few factors as possible. For example, $2\sqrt[3]{2}$ is simpler than $\sqrt[3]{16}$. In general, this means that the radicand should contain no factors that are perfect nth powers, where n is the index.
- There are as few **nested radicals** (radicals inside of other radicals) as possible. For example, $\sqrt[15]{x}$ is simpler than $\sqrt[3]{\sqrt[5]{x}}$.
- The index of the radical is as small as possible. For example, $\sqrt{5}$ is simpler than $\sqrt[4]{25}$.
- There are no radicals in denominators. For example, $\sqrt{2}/2$ is simpler than $1/\sqrt{2}$. (See Exercise 116 for further discussion of this point.)

····**EXAMPLE 12**

Simplifying Radical Expressions

Simplify the following expressions. Assume all variables represent positive quantities.

a. $\sqrt{8}$ **b.** $\sqrt[3]{16x^7}$ **c.** $\sqrt{75a^3b^5c^6}$

Solution

a. $\sqrt{8} = \sqrt{4 \cdot 2}$ Factoring out the largest perfect square

$\quad\quad = \sqrt{4}\,\sqrt{2}$ Using Property 3

$\quad\quad = 2\sqrt{2}$ Simplifying

b. $\sqrt[3]{16x^7} = \sqrt[3]{8x^6 \cdot 2x}$ Factoring out the largest perfect cubes

$\quad\quad = \sqrt[3]{8}\,\sqrt[3]{x^6}\,\sqrt[3]{2x}$ Using Property 3

$\quad\quad = \sqrt[3]{8}\,\sqrt[3]{(x^2)^3}\,\sqrt[3]{2x}$ Rewriting x^6 as $(x^2)^3$

$\quad\quad = 2x^2\,\sqrt[3]{2x}$ Using Property 2

c. $\sqrt{75a^3b^5c^6} = \sqrt{25a^2b^4c^6 \cdot 3ab}$ Factoring out the perfect squares

$\quad\quad = \sqrt{25}\sqrt{a^2}\sqrt{b^4}\sqrt{c^6}\sqrt{3ab}$ Using Property 3

$\quad\quad = 5|a||b^2||c^3|\sqrt{3ab}$ Using Property 2

$\quad\quad = 5ab^2c^3\sqrt{3ab}$ Assuming all variables represent positive quantities

---

Rational expressions with radicals in the denominator are simplified by **rationalizing the denominator**. Rationalizing the denominator consists of multiplying both the numerator and denominator of an expression by a factor *chosen to eliminate any radicals in the denominator.*

EXAMPLE 13

Rationalizing the Denominator

Rationalize the denominator.

a. $\dfrac{1}{\sqrt{2}}$ **b.** $\dfrac{6}{\sqrt[3]{9}}$

Solution

a. $\dfrac{1}{\sqrt{2}} = \dfrac{1}{\sqrt{2}} \cdot \dfrac{\sqrt{2}}{\sqrt{2}}$ $\sqrt{2}$ is chosen as the factor since multiplication by $\sqrt{2}$ eliminates the radical in the denominator.

$\quad\quad = \dfrac{\sqrt{2}}{2}$

b. $\dfrac{6}{\sqrt[3]{9}} = \dfrac{6}{\sqrt[3]{9}} \cdot \dfrac{\sqrt[3]{3}}{\sqrt[3]{3}}$ $\sqrt[3]{3}$ is chosen to obtain a perfect cube (27) in the radicand of the denominator.

$\quad\quad = \dfrac{6\sqrt[3]{3}}{\sqrt[3]{27}}$ Using Property 3

$\quad\quad = \dfrac{6\sqrt[3]{3}}{3}$ Simplifying

$\quad\quad = 2\sqrt[3]{3}$ Simplifying

When radical expressions are to be added or subtracted, we simplify the individual expressions first and then combine like terms.

Expressions involving the roots of sums or differences don't generally simplify. For example:

$\sqrt{x + y}$ does not equal $\sqrt{x} + \sqrt{y}$ ($\sqrt{16 + 9} = 5$, but $\sqrt{16} + \sqrt{9} = 7$)

$\sqrt{x^2 + y^2}$ does not equal $x + y$ ($\sqrt{4^2 + 3^2} = 5$, but $4 + 3 = 7$)

$\sqrt[3]{x^3 - y^3}$ does not equal $x - y$ ($\sqrt[3]{2^3 - 1^3} = \sqrt[3]{7}$, but $2 - 1 = 1$)

Time Dilation

Of all the otherworldly phenomena predicted by Einstein's Theory of Relativity, perhaps none is so mind-bending as **time dilation**: the tendency for physical processes—including clocks—to run more slowly in systems accelerated to high speeds. This experimentally confirmed fact contradicts our intuitive notion of time as an absolute quantity, independent of observer, independent of frame of reference. Of course, our intuition is forged from countless everyday experiences with time, experiences taking place at—to borrow from *Star Trek* terminology—subwarp speeds. It is only at speeds approaching that of light that time dilation emerges as a measurable effect with bizarre consequences.

A full development of time dilation is well beyond the scope of this text, but we can express the relationship between Earth time and traveler time with a simple algebraic formula. In fact, it can be shown that

$$T = \frac{T_0}{\sqrt{1 - v^2}}$$

where T is time elapsed on Earth, T_0 is the time as measured by the traveler, and v is the velocity of the traveler relative to Earth (expressed as a fraction of the speed of light, 186,000 miles per second). For example, if we were to travel at half the speed of light for 2 years (according to our own clocks), then the time elapsed on Earth would be given by

$$T = \frac{2}{\sqrt{1 - \left(\frac{1}{2}\right)^2}} \approx 2.309 \text{ years}$$

The faster we travel, the greater the discrepancy between Earth time and traveler time. For example, consider the famous **twin paradox**. A 30-year-old astronaut travels to a nearby star and back at great speeds, while her identical twin sister remains on Earth. Because of time dilation, the astronaut's clocks, including the aging processes of her own body, run more slowly than those of her terrestrial twin. Specifically, if our astronaut travels at 99.5% of the speed of light, then after 5 years (astronaut time), approximately $5/\sqrt{1 - 0.995^2} \approx 50$ years will have elapsed on Earth. Thus, when our astronaut triumphantly returns from her celestial sojourn, she will emerge from her starship as a 35-year-old woman with an 80-year-old identical twin sister!

Fractional Exponents

Many of the properties of radicals bear a striking resemblance to those of exponents. For example, the property $\sqrt[m]{a}\,\sqrt[m]{b} = \sqrt[m]{ab}$ is very similar to the exponential property $a^m b^m = (ab)^m$. There is a good reason for this similarity—radicals can, in fact, be represented by expressions involving *fractional exponents*. To see this, consider the expression $2^{1/3}$. If the properties of integer exponents are to hold for fractional exponents, then we have

$$\left(2^{1/3}\right)^3 = 2^{(1/3)\cdot 3} = 2^1 = 2$$

Thus, since the cube of $2^{1/3}$ is 2, $2^{1/3}$ is nothing more than the cube root of 2, or $\sqrt[3]{2}$. More generally, we give the following definitions.

Definition of Fractional Exponents

Let $a > 0$. For a natural number n,

$$a^{1/n} = \sqrt[n]{a}$$

For natural numbers m and n,

$$a^{m/n} = \left(\sqrt[n]{a}\right)^m = \sqrt[n]{a^m}$$

and

$$a^{-(m/n)} = \frac{1}{a^{m/n}}$$

In the examples and exercises that follow, assume that all variables are positive in expressions involving fractional exponents.

⋯⋯▷EXAMPLE 14

Converting from Fractional Exponents to Radicals

Write each of the following as a radical:

a. $3^{1/10}$ **b.** $x^{1/2}$ **c.** $a^{3/7}$

Solution

a. $3^{1/10} = \sqrt[10]{3}$ **b.** $x^{1/2} = \sqrt{x}$ **c.** $a^{3/7} = \sqrt[7]{a^3}$

⋯⋯▷EXAMPLE 15

Expressing Radicals in Exponential Form

Write the following radical expressions in exponential form:

a. $\sqrt{5}$ **b.** $\sqrt[7]{a^2}$ **c.** $\sqrt[4]{x^6}$

Solution

a. $\sqrt{5} = 5^{1/2}$ **b.** $\sqrt[7]{a^2} = a^{2/7}$ **c.** $\sqrt[4]{x^6} = x^{6/4} = x^{3/2}$

⋯⋯▷EXAMPLE 16

Simplifying an Expression Involving a Fractional Exponent

Evaluate $4^{-3/2}$.

Solution

$$4^{-3/2} = \frac{1}{4^{3/2}} = \frac{1}{\sqrt{4^3}} = \frac{1}{\sqrt{64}} = \frac{1}{8}$$

We now have defined the expression a^q (with $a > 0$) for all rational numbers q. All of the properties for natural number exponents given earlier hold for rational number exponents as well. We use these properties in the following example.

⋯⋯▷EXAMPLE 17

Simplifying Expressions Involving Fractional Exponents

Simplify the following.

a. $x^{1/2} \cdot x^{3/4}$ **b.** $\left(a^{2/3}\right)^{3/4}$ **c.** $\dfrac{p^{2/5}q^{-1/3}}{pq^{-2/3}}$

Solution

a. $x^{1/2} \cdot x^{3/4} = x^{(1/2)+(3/4)}$

$\qquad = x^{5/4}$

b. $\left(a^{2/3}\right)^{3/4} = a^{(2/3)\cdot(3/4)}$

$\qquad = a^{1/2}$

c. $\dfrac{p^{2/5}q^{-1/3}}{pq^{-2/3}} = p^{2/5-1}q^{(-1/3)-(-2/3)}$

$\qquad = p^{-3/5}q^{1/3}$

$\qquad = \dfrac{q^{1/3}}{p^{3/5}}$

Calculator Keys

Computing Roots

If your calculator does not have an *n*th root key, roots can be computed by first converting the radical to a fractional exponent and then using the exponent key $\boxed{y^x}$ or $\boxed{\wedge}$. For example, $\sqrt[5]{2}$ can be computed by entering $2\boxed{\wedge}(1/5)$.

Understanding and Mastery Checklists

Concepts to Understand

Exponential notation

◇

Negative and zero exponents

◇

Scientific notation

◇

Principal *n*th root

◇

Radical, radicand, and index

◇

Properties of radicals

◇

Completely simplified radical expressions

◇

Fractional exponents

Skills to Master

Evaluate expressions involving integer exponents.

◇

Simplify expressions involving negative exponents.

◇

Convert from decimal to scientific notation.

◇

Convert from scientific notation to decimal notation.

◇

Evaluate the principal *n*th root of perfect *n*th powers.

◇

Simplify radical expressions.

◇

Rationalize the denominator.

◇

Add, subtract, multiply, and divide radical expressions.

◇

Convert from radical to exponential notation.

◇

Convert from exponential to radical notation.

Exercises 1.2

Exercises 1–12 *Evaluate the given expression.*

1. $\left(\frac{2}{5}\right)^3$

2. $(-2)^4$

3. $\left(-\frac{1}{10}\right)^3$

4. 3^{-2}

5. 4^{-3}

6. $(-2)^{-1}$

7. $(-1)^{-4}$

8. -5^2

9. -3^{-2}

10. $(2^3)^2$

11. $(4^{-2})^3$

12. $2^{(3^2)}$

Exercises 13-24 *Simplify the given expression. Write your final answer without using negative exponents.*

13. $a^3 a^6$

14. $\dfrac{2x^8}{x^3}$

15. $\dfrac{r^5}{r^7}$

16. $\dfrac{y^{-1}}{y^{-2}}$

17. $\left(a^3 b^{-2}\right)\left(a^{-2} b^{-4}\right)$

18. $\left(x^2 y^3\right)^2$

19. $\dfrac{p^{10} q^6}{p^4 q^2}$

20. $\left(a^3 b^{-1} c^2\right) \cdot \left(a^{-2} b^{-3} c^3\right)$

21. $\dfrac{3^{-1} x^4 y^{-2}}{3^{-2} x^{-2} y}$

22. $\left(\dfrac{p^2}{q^3}\right)^{-1}$

23. $\left(y^{-1} z^2\right) \cdot \left(yz^{-1}\right)^{-1}$

24. $\left(\dfrac{a^2 b^3 c}{a^2 b^{-1} c^2}\right)^2$

Exercises 25-28 *Simplify. Assume $i^2 = -1$.*

25. i^{11}

26. $-i^{14}$

27. $i^{701} \cdot i^3$

28. $i^{2001} \cdot i$

Exercises 29-32 *Convert the given number to standard decimal notation.*

29. 3.04×10^{-3}

30. 8.38×10^3

31. 2.7×10^{-1}

32. 7.5×10^0

Exercises 33-40 *Convert the given number to scientific notation.*

33. $4{,}000{,}000$

34. $10{,}000.001$

35. 0.0943

36. 0.0008456

37. $\left(2.5 \times 10^{-45}\right) \cdot \left(4 \times 10^{52}\right)$

38. $\left(8.1 \times 10^{12}\right) \div \left(2.7 \times 10^{-5}\right)$

39. $13{,}176{,}000{,}000{,}000{,}000{,}000{,}000{,}000$ (the weight of Earth in pounds)

40. 0.000007 (the probability of being struck by lightning in a given year)

Exercises 41-44 *Evaluate the given radical expression.*

41. $\sqrt[3]{27}$

42. $\sqrt{\dfrac{9}{4}}$

43. $\sqrt[4]{\dfrac{1}{16}}$

44. $\sqrt[3]{-8}$

Exercises 45-60 *Simplify the given expression. Assume that all variables represent positive real numbers.*

45. $\sqrt{18}$

46. $\sqrt[3]{16}$

47. $\sqrt{5^8}$

48. $\sqrt[3]{3^{17}}$

49. $\sqrt{x^3 y^7}$

50. $\sqrt{28x^4 y^3 z^2}$

51. $\sqrt{50a^{30} b^{20} c^{11}}$

52. $\sqrt{(a+b)^6}$

53. $\sqrt{x^7 (y+z)^4}$

54. $\sqrt[3]{a^7}$

55. $\sqrt[3]{54s^6 t^7}$

56. $\sqrt{\dfrac{a^4}{b^2}}$

57. $\dfrac{3}{\sqrt{7}}$

58. $\dfrac{6}{\sqrt[3]{4}}$

59. $\sqrt[3]{\sqrt{x}}$

60. $\sqrt[5]{\sqrt[3]{x^5}}$

Exercises 61-62 *Simplify. Assume $i = \sqrt{-1}$.*

61. $\sqrt{-25}$

62. $\sqrt{-\dfrac{1}{27}}$

Exercises 63-70 *Perform the indicated operation and simplify.*

63. $\sqrt{3} \cdot \sqrt{15}$

64. $\sqrt{5a^3 b} \cdot \sqrt{10ab^3}$

65. $\sqrt[3]{2xyz} \cdot \sqrt[3]{20x^2 y^4 z}$

66. $\left(\sqrt{8}\right)^3$

67. $\dfrac{\sqrt{10}}{\sqrt{2}}$

68. $\left(\sqrt[3]{a}\right)^5$

69. $\sqrt{6} + \sqrt{24}$

70. $\sqrt{8} + \dfrac{1}{\sqrt{2}}$

Exercises 71-74 *Convert from radical notation to exponential notation. Assume all variables are positive.*

71. $\sqrt[3]{x}$

72. $\sqrt[5]{y^2}$

73. $\dfrac{1}{\sqrt{x}}$

74. $\dfrac{1}{\sqrt[3]{5}}$

Exercises 75-78 *Convert the given exponential expression into radical form.*

75. $2^{3/2}$

76. $3^{-1/5}$

77. $b^{2/3}$

78. $x^{4/7}$

Exercises 79-82 *Simplify.*

79. $16^{-1/2}$

80. $27^{1/3}$

81. $32^{2/5}$

82. $8^{-2/3}$

Exercises 83-88 *Simplify the given exponential expression. Express the answer in exponential form without negative exponents.*

83. $x^{1/3} x^{2/3}$

84. $a^{1/4} a^{1/3}$

85. $\left(81a^4 b^{12}\right)^{1/4}$

86. $\left(\dfrac{1}{8} u^6 v^9\right)^{2/3}$

87. $\dfrac{z^2}{z^{3/2}}$

88. $\dfrac{x^2 y^{1/3}}{x^{1/2} y}$

Applications

89. Salary Options You are hired as a temporary worker to begin work on the first of May; the job is scheduled to be completed on May 31. Your employer gives you the choice of either receiving $300,000 a day for a month, or of receiving 1¢ on the first day of the month, 2¢ on the second day, 4¢ on the third, doubling the pay on each successive day of the month. Which deal should you take? Justify your answer.

90. Faded Jeans Suppose that a pair of jeans fades in such a way that only $\frac{3}{4}$ of the dye remains after every washing. Assuming that the jeans are washed once a week for a year, how much of the original dye remains after one year?

91. The Bouncing Ball Whenever a certain ball is dropped, it bounces to two-thirds of its original height. Suppose the ball is dropped off the Sears Tower in Chicago (height 1454 feet).

  a. How high would the ball rise after one bounce?

  b. How high would the ball rise after two bounces?

  c. How high would the ball rise after n bounces?

  d. Suppose that a film of the bouncing ball is run backward. The first visible bounce appears to be to a height of about 1 inch. How many additional bounces will occur before the ball is seen leaping over the edge of the Sears Tower?

92. Cross-Country Trip Suppose that you resolve to drive from Los Angeles to New York (driving distance 2774 miles) according to the following scheme. On the first day you will drive half the distance, and on each successive day you will drive half of the remaining distance. How many miles will remain after 3 weeks of travel?

Exercises 93-94 *If inflation were to remain at a steady rate of i%, then, on the average, the price of an item will increase according to the formula $N = P\left(1 + \frac{i}{100}\right)^n$, where P is the original price of the item and N is the price after n years.*

93. CD Inflation If a CD costs $18.00 in 2003, how much will it cost in 2033 assuming 4% inflation?

94. Television Inflation Given that inflation for the period 1955–2002 averaged about 4.1%, estimate the price in 1955 dollars for a color television that sold for $300 in 2002. In fact, a new color television would have cost over $500 in 1955. Does this fact suggest that our inflationary model is not legitimate?

95. Terminal Velocity According to *The Guinness Book of World Records*, Vesna Vulovic of the former Yugoslavia survived a fall of 33,330 feet after the DC-9 on which she was a flight attendant blew up on January 26, 1972. If gravity were the only force acting on her, Vulovic's velocity at impact would be given by $v = 8\sqrt{s}$, where s is the distance fallen in feet and v is the velocity in feet per second. Find Vulovic's velocity at impact using this formula. The terminal velocity of a human in free-fall is roughly 180 feet per second. How

do you explain the discrepancy between this value and your answer? In fact, Vulovic fell in a portion of the fuselage. How might this have affected her terminal velocity?

96. Braking Distance The velocity at which an automobile is traveling immediately before braking can be estimated from the braking distance using the formula $v = \sqrt{20d}$, where v is the velocity of the car in miles per hour and d is the distance in feet required to brake. If skid marks measure 200 feet, then how fast was the car traveling?

97. Hang Time The maximum height h of a jump, the hang time t, and the initial speed v_0 with which a jumper leaves the ground are related by the formulas $v_0 = \sqrt{64h}$ and $h = 4t^2$, where h is in feet, t is in seconds, and v_0 is in feet per second.

  a. Use these equations to find a formula for v_0 in terms of t.

  b. It is estimated that Michael Wilson, former University of Memphis basketball star and Harlem Globetrotter, had a hang time of 1.036 seconds. Find his initial speed and maximum height.

  c. Wilson is reputed to have dunked on a 12-foot goal. Is this consistent with your findings in part b?

98. Unknown Interest Rate If a deposit of P dollars grows to a balance of B dollars after n years, then the annually compounded interest rate r is given by

$$r = \sqrt[n]{\frac{B}{P}} - 1$$

At what interest rate will $1 grow to $1000 in 15 years?

Exercises 99-100 *Time dilation is the tendency of physical processes to run more slowly in systems accelerated to speeds approaching that of light. Consequently, a person traveling at speeds approaching that of light would age more slowly than a counterpart on Earth. The precise relationship between Earth time and the time of the traveler is given by*

$$T = \frac{T_0}{\sqrt{1 - v^2}}$$

where T is time elapsed on Earth, T_0 is the time as measured by the traveler, and v is the ratio of the traveler's speed to the speed of light. (For further discussion, refer to the math note on page 26.)

99. Fountain of Youth? The official airspeed record of 2200 miles per hour (3.28554×10^{-6} the speed of light) is held by the Lockheed SR-71 Blackbird. If a pilot were to stay aloft at this speed for 50 years (according to his watch), how much younger would he be when he lands than if he had stayed on the ground? (Give your answer in seconds.)

100. Banking on Time Suppose that a woman deposits $1000 into a bank account earning 4% interest compounded annually and she then travels at 99.99999999% of the speed of light for 1 day (according to her watch).

a. Use the time-dilation formula to compute the amount of time that would have elapsed on Earth.

b. The balance B after t years in an account with principal P and annual compound interest rate r is given by the formula $B = P(1 + r)^t$. Compute the balance that the woman would find in her account after returning to Earth.

Concepts and Critical Thinking

Exercises 101–106 *Answer true or false.*

101. For all real numbers x and integers a and b, $x^a x^b = x^{ab}$.

102. $-3^2 = 9$

103. 11.72×10^{13} is not in scientific notation.

104. $\sqrt{x^2 + 25}$ simplifies to $x + 5$.

105. $(\pm 6)^2 = 36$

106. $\sqrt{36} = \pm 6$

Exercises 107–110 *Give an example of each.*

107. A number greater than 100 expressed in scientific notation

108. A number less than 1 expressed in scientific notation

109. A number x such that $\sqrt{x^2} \neq x$

110. A number x such that the principal square root of x is equal to the principal cube root of x

111. Explain why $\sqrt{25} \neq \pm 5$.

Questions for Discussion or Essay

112. In this section, we have dealt with many very large numbers. But do such numbers have any significance? Can you think of any context (other than those mentioned in the text) in which numbers larger than a million arise? In particular, what topics discussed on the nightly news might involve very large numbers?

113. In the text, we define $a^{-n} = 1/a^n$ for positive integers n and $a \neq 0$. Why do we exclude $a = 0$?

114. Example 2 involved the premise that a child continues to double in weight every 6 months for the rest of his life. This, of course, led to ridiculous results. What factors prevent a child from continuing to double in weight every 6 months? The hypothesis that a child's weight doubles every 6 months is an example of a *mathematical model*. This particular mathematical model was a complete failure. What characteristics should a good mathematical model have? What information would be helpful in constructing a good mathematical model of weight gain?

115. It is not uncommon for algebra students to view algebra as a sort of game with arbitrary rules that the "player" must memorize. From their point of view, the statements $\sqrt{ab} = \sqrt{a}\sqrt{b}$ and $\sqrt{a + b} = \sqrt{a} + \sqrt{b}$ seem equally valid. In fact, when told that the second equality is not, in general, true, they might respond, "Why not? It's just like the first one." How would you answer that question? Why isn't the second equality true? What would it take to prove that it isn't true? Who decides which algebraic statements are true and which aren't?

116. Historically, students were encouraged to eliminate radicals from denominators (rationalize denominators) for computational reasons. In the precalculator era, it was more difficult to perform $3 \div \sqrt{2}$ than the equivalent $\left(3 \times \sqrt{2}\right) \div 2$. Why? Now that calculators are cheap and widely available, is there any justification for continuing to insist that denominators be rationalized? How could you simplify an expression like $\dfrac{\sqrt{2}}{2} - \dfrac{1}{\sqrt{2}}$ without first rationalizing the denominator?

Projects for Enrichment

117. Repeated Exponentiation As was mentioned in the text, multiplication represents repeated addition, and exponentiation represents repeated multiplication. Suppose that we take this one step further and define "hyper-exponentiation" to be repeated exponentiation. For example, just as $2 \cdot 3$ represents 2 added to itself three times, and 2^3 represents 2 multiplied by itself 3 times, we could define $2 \boxed{h} 3$ to be 2 "exponentiated 3 times," or $2^{(2^2)} = 2^4 = 16$. More generally, we could define $m \boxed{h} n$ to be

$$m^{\left(m^{\left(m^{(\cdots m)}\right)}\right)}$$

where m appears n times in all.

Compute the following quantities, and express the answer in scientific notation.

a. $5 \boxed{h} 2$ **b.** $2 \boxed{h} 4$

c. $1 \boxed{h} 10$ **d.** $10 \boxed{h} 3$

118. Heron's Formula Heron's Formula gives the area of a triangle in terms of its side lengths. If the triangle has sides of length a, b, and c, and A represents the area of the triangle, then

$$A = \sqrt{s(s-a)(s-b)(s-c)}, \text{ where } s = \frac{a+b+c}{2}$$

a. Find the area of a triangle with sides of length 3, 4, and 5.

b. Find the area of a triangle with sides of length 1, 1, and 1.

c. Use Heron's Formula to show that the area of an equilateral triangle of side length x is given by $\left(\sqrt{3}/4\right)x^2$.

d. What is the area of a triangle with sides of length 1, 2, and 5?

e. Explain your answer to part d.

Section 1.3 Algebraic Expressions

- If you are traveling at a rate of 65 mph, and you suddenly notice an obstruction in the road 200 feet ahead of you, will you be able to stop in time?
- Is there any truth to the saying, It's not the heat; it's the humidity?
- How can polynomial factorizations be used to find a factor of 1,000,001 without a calculator in just seconds?
- What possible effect could the planet Mars have on the birth of a child?
- How is hydrostatic weighing used to calculate body fat percentage?

Definition of Polynomial

Expressions involving variables raised to natural number powers arise frequently in the study of real-world phenomena. For example, the laws of physics tell us that if an object is dropped near the surface of Earth, it will fall $16t^2$ feet in t seconds. The expression $16t^2$ is an example of a *monomial*.

Definition of Monomial

A **monomial** is the product of a number and one or more variables raised to non-negative integer powers. The **coefficient** of a monomial is the number preceding the variable(s). The **degree** of a monomial is the sum of the powers.

Table 4 lists some examples of expressions that are monomials and some that are not. For those that are monomials, we identify the coefficient and the degree.

Table 4

Expression	Monomial?	Coefficient	Degree	Comment
$2x^3$	Yes	2	3	
$\frac{1}{3}p^2q^5$	Yes	$\frac{1}{3}$	7	The degree is $2 + 5 = 7$, the sum of the powers.
$5y^{-2}$	No			The exponent is negative.
-4	Yes	-4	0	Numbers can be considered to be monomials since, for example, $-4 = -4x^0$.
$1.5t^{1/3}$	No			The exponent is not an integer.

Definition of Polynomial

A **polynomial** is a sum of monomials. The monomials that make up a polynomial are called the **terms** of the polynomial. The **degree** of a polynomial is the highest degree of its terms.

A monomial can be viewed as a polynomial with only one term. Two other categories of polynomials are *binomials* and *trinomials*. As the prefixes *bi* and *tri* suggest, a **binomial** is a polynomial with two terms, and a **trinomial** is a polynomial with three terms.

EXAMPLE 1

Identifying Polynomials

Identify each of the following as a monomial, binomial, trinomial, polynomial, or none of these, and determine the degree of any that are polynomials:

a. $-2a^5 + 4a^3 - 3a + 4$ **b.** $3x - 5$

c. $3x^2y + 4y^{1/3} + 2xy^3$ **d.** $\frac{1}{2}p^2q^3 + \frac{2}{3}p^3q^5 - pq^6$

Solution

a. Polynomial of degree 5
b. Binomial (polynomial also) of degree 1
c. This is not a polynomial because the exponent in $y^{1/3}$ is not an integer.
d. Trinomial (polynomial also) of degree 8

Arithmetic Operations on Polynomials

The operations of addition and subtraction of polynomials are performed by *combining like terms*—that is, combining terms with identical variables raised to identical powers. The mechanics of this process involve the associative, commutative, and distributive properties.

EXAMPLE 2 **Simplifying Polynomials**

Simplify.

a. $(2x^3 - 5x^2 + x - 4) - (8x^2 - 3x + 4)$

b. $(7a^2 - 4ab + b^2 + 5b) + (3a^2 + 2ab - 2b^2)$

Solution

a. $(2x^3 - 5x^2 + x - 4) - (8x^2 - 3x + 4)$

$= 2x^3 - 5x^2 + x - 4 - 8x^2 + 3x - 4$ Distributing

$= 2x^3 + (-5x^2 - 8x^2) + (x + 3x) + (-4 - 4)$ Combining like terms

$= 2x^3 - 13x^2 + 4x - 8$

b. $(7a^2 - 4ab + b^2 + 5b) + (3a^2 + 2ab - 2b^2)$

$= (7a^2 + 3a^2) + (-4ab + 2ab) + (b^2 - 2b^2) + 5b$ Combining like terms

$= 10a^2 - 2ab - b^2 + 5b$

Polynomials are multiplied by repeated application of the distributive property, as the following example demonstrates.

EXAMPLE 3 **Multiplying Polynomials**

Find the given product.

a. $3mn(8m^3 - 12m^2n - n^3)$

b. $(2y - 5)(3y^2 - y + 2)$

Solution

a. $3mn(8m^3 - 12m^2n - n^3)$

$= 3mn(8m^3) - 3mn(12m^2n) - 3mn(n^3)$ Distributing

$= 24m^4n - 36m^3n^2 - 3mn^4$

b. $(2y - 5)(3y^2 - y + 2)$

$= 2y(3y^2 - y + 2) + (-5)(3y^2 - y + 2)$ Distributing

$= (6y^3 - 2y^2 + 4y) + (-15y^2 + 5y - 10)$ Distributing

$= 6y^3 + (-2y^2 - 15y^2) + (4y + 5y) - 10$ Combining like terms

$= 6y^3 - 17y^2 + 9y - 10$

When multiplying binomials, it is convenient to use a shortcut that is often referred to as the FOIL method. Consider, for example, the product $(x - 1)(2x + 3)$ and the diagram in Figure 8.

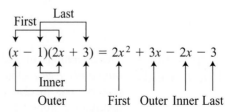

Figure 8

Here, $2x^2$ is the product of the **F**irst terms, $3x$ is the product of the **O**uter terms, $-2x$ is the product of the **I**nner terms, and -3 is the product of the **L**ast terms. Putting these terms together, we obtain **F**irst, **O**uter, **I**nner, and **L**ast, or FOIL for short.

EXAMPLE 4

Multiplying a Binomial Using the FOIL Method

Perform the multiplication $(3t - 5)(t + 3)$ using the FOIL method.

Solution

$$(3t - 5)(t + 3) = 3t^2 + 9t - 5t - 15$$
$$= 3t^2 + 4t - 15$$

Note that if the FOIL method is applied to $(a + b)(a - b)$, we obtain $a^2 - ab + ab - b^2$, which simplifies to $a^2 - b^2$. In the next example, we see how the special product $(a + b)(a - b) = a^2 - b^2$ can be used to rationalize a denominator. In Example 6, we use a similar method to divide complex numbers.

EXAMPLE 5

Rationalizing the Denominator

Rationalize the denominator of $\dfrac{7}{4 - \sqrt{2}}$.

Solution If we multiply $4 - \sqrt{2}$ by its conjugate $4 + \sqrt{2}$, we obtain $\left(4 - \sqrt{2}\right) \cdot \left(4 + \sqrt{2}\right) = 4^2 - \left(\sqrt{2}\right)^2 = 16 - 2 = 14$. Thus, we multiply both numerator and denominator by $4 + \sqrt{2}$ as follows:

$$\frac{7}{4 - \sqrt{2}} = \frac{7}{4 - \sqrt{2}} \cdot \frac{4 + \sqrt{2}}{4 + \sqrt{2}} \qquad \text{Multiplying numerator and denominator by } 4 + \sqrt{2}$$

$$= \frac{7 \cdot \left(4 + \sqrt{2}\right)}{4^2 - \left(\sqrt{2}\right)^2} \qquad \text{Using the special product } (a + b)(a - b) = a^2 - b^2$$

$$= \frac{7 \cdot \left(4 + \sqrt{2}\right)}{16 - 2} \qquad \text{Simplifying}$$

$$= \frac{7 \cdot \left(4 + \sqrt{2}\right)}{14} \qquad \text{Simplifying}$$

$$= \frac{4 + \sqrt{2}}{2} \qquad \text{Simplifying}$$

Arithmetic operations on complex numbers (that is, numbers of the form $a + bi$, where $i = \sqrt{-1}$) can be handled very much like the corresponding operations on polynomials. We simply treat the complex numbers as though they were binomials involving the variable i and use the fact that $i^2 = -1$. In the case of division, we use the complex conjugate of the denominator to write the quotient in standard form. The **complex conjugate** of the complex number $a + bi$ is defined to be $a - bi$.

⸺⸻**EXAMPLE 6**

Arithmetic Operations on Complex Numbers

Perform the indicated operation. Recall $i = \sqrt{-1}$. Express the final answer in the form $a + bi$.

a. $(2 + i) + (13 + 4i)$ **b.** $(2 - i)(3 + 2i)$ **c.** $\dfrac{5}{4 - 2i}$

Solution

a. By combining like terms—namely, the real and imaginary parts—we obtain

$$(2 + i) + (13 + 4i) = (2 + 13) + (i + 4i) = 15 + 5i$$

b. Applying the FOIL method, we have

$$(2 - i)(3 + 2i) = 6 + 4i - 3i - 2i^2 = 6 + i - 2i^2$$

Since $i^2 = -1$, we can simplify further to obtain

$$6 + i - 2(-1) = 8 + i$$

c. We multiply the numerator and denominator by $4 + 2i$, the complex conjugate of the denominator $4 - 2i$. Note how the i in the denominator is eliminated.

$$\frac{5}{4 - 2i} = \frac{5}{4 - 2i} \cdot \frac{4 + 2i}{4 + 2i} \qquad \text{Multiplying by } 1 = \frac{4 + 2i}{4 + 2i}$$

$$= \frac{20 + 10i}{16 + 8i - 8i - 4i^2} \qquad \text{Multiplying numerators and denominators}$$

$$= \frac{20 + 10i}{16 - 4(-1)} \qquad \text{Substituting } i^2 = -1$$

$$= \frac{20 + 10i}{20}$$

$$= \frac{20}{20} + \frac{10}{20}i$$

$$= 1 + \frac{1}{2}i$$

Factoring Polynomials

We consider a polynomial to be **completely factored over the integers** when it cannot be factored any further using polynomials with integer coefficients. For example, $4x^2y^3 - 6xy^2$ is not completely factored over the integers because each term has $2xy^2$ as a common monomial factor. Factoring, we obtain $2xy^2(2xy - 3)$, which is completely factored over the integers since no polynomial with integer coefficients can be factored out of $2xy - 3$.

 ⸺⸻**EXAMPLE 7**

Factoring Out the Greatest Common Monomial

Factor out the greatest common monomial in $12a^6 - 3a^5 + 6a^3 + 9a^2$.

Solution The greatest common factor of each term is $3a^2$. Factoring out $3a^2$ from each term, we obtain

$$12a^6 - 3a^5 + 6a^3 + 9a^2 = 3a^2(4a^4 - a^3 + 2a + 3)$$

Sometimes it is difficult to tell whether a polynomial is completely factored, especially if the polynomial has many terms. However, for polynomials with three or fewer terms, certain clues make the task much easier. We begin with three special binomial forms.

Special Binomial Forms

Difference of squares:	$a^2 - b^2 = (a + b)(a - b)$
Difference of cubes:	$a^3 - b^3 = (a - b)(a^2 + ab + b^2)$
Sum of cubes:	$a^3 + b^3 = (a + b)(a^2 - ab + b^2)$

EXAMPLE 8

Factoring Binomials

Factor the given binomial.

a. $8z^3 + 125$ **b.** $100x^2 - 4y^2$

Solution

a.
$$\begin{aligned}
8z^3 + 125 &= (2z)^3 + 5^3 &&\text{Rewriting } 8z^3 \text{ as } (2z)^3 \\
&= (2z + 5)[(2z)^2 - (2z)5 + 5^2] &&\text{Applying the sum of cubes form} \\
&= (2z + 5)(4z^2 - 10z + 25) &&\text{Simplifying}
\end{aligned}$$

b.
$$\begin{aligned}
100x^2 - 4y^2 &= 4(25x^2 - y^2) &&\text{Factoring out the common monomial} \\
&= 4[(5x)^2 - y^2] &&\text{Rewriting } 25x^2 \text{ as } (5x)^2 \\
&= 4(5x + y)(5x - y) &&\text{Applying the difference of squares form}
\end{aligned}$$

Some trinomials can be factored by recognizing them as *perfect square trinomials*. A perfect square trinomial is one that can be written as the square of a binomial; there are two forms you should recognize. They are summarized as follows.

Perfect Square Trinomials

$$a^2 + 2ab + b^2 = (a + b)^2$$
$$a^2 - 2ab + b^2 = (a - b)^2$$

EXAMPLE 9

Factoring Perfect Square Trinomials

Factor the given trinomial.

a. $x^2 - 8x + 16$ **b.** $9y^2 + 30yz + 25z^2$

Solution

a. Notice that $x^2 - 8x + 16$ is of the form $a^2 - 2ab + b^2$ with $a = x$ and $b = 4$. Thus, factoring as a perfect square trinomial, we obtain

$$\begin{aligned}
x^2 - 8x + 16 &= x^2 - 2(4)(x) + 4^2 \\
&= (x - 4)^2
\end{aligned}$$

b. Note that $9y^2 = (3y)^2$, $25z^2 = (5z)^2$, and the middle term, $30yz$, is twice the product of $3y$ and $5z$. Thus, we have

$$\begin{aligned}
9y^2 + 30yz + 25z^2 &= (3y)^2 + 2(3y)(5z) + (5z)^2 \\
&= (3y + 5z)^2
\end{aligned}$$

Trinomials that are not perfect squares can often be factored by "educated guessing." Here we rely on experience (and sometimes a little trial and error) to guide us to the proper choice of factors.

EXAMPLE 10

Factoring Trinomials

Factor the given trinomial.

a. $x^2 - 5x - 6$ **b.** $2t^4 + 5t^3 - 3t^2$

Solution

a. Since the coefficient of x^2 is 1, we are looking for a factorization of the form $(x + \square)(x + \triangle)$, where $\square$ and $\triangle$ are factors of the constant term, -6, that add up to the coefficient of the middle term, -5. The correct values are -6 and 1. Thus, $x^2 - 5x - 6 = (x - 6)(x + 1)$.

b. First we factor out the common factor t^2.

$$2t^4 + 5t^3 - 3t^2 = t^2(2t^2 + 5t - 3)$$

Next we factor $2t^2 + 5t - 3$. We are looking for a factorization of the form $(2t + \square)(t + \triangle)$, where $\square$ and $\triangle$ are factors of the constant term, -3. By trial and error, we discover that $2t^2 + 5t - 3 = (2t - 1)(t + 3)$. Thus, $2t^4 + 5t^3 - 3t^2 = t^2(2t - 1)(t + 3)$.

Rational Expressions

A quotient of two polynomials, such as

$$\frac{x^2 - 4}{x^2 - 3x + 2}$$

is called a **rational expression**. We say that a rational expression is in **lowest terms** if the numerator and denominator have no common factors. We simplify rational expressions by dividing out common factors in the numerator and denominator, thus writing them in lowest terms.

EXAMPLE 11

Simplifying a Rational Expression by Factoring

Simplify $\dfrac{x^2 - 4}{x^2 - 3x + 2}$.

Solution   We factor both numerator and denominator and divide out common factors, as follows:

$$\frac{x^2 - 4}{x^2 - 3x + 2} = \frac{(x + 2)(x - 2)}{(x - 1)(x - 2)} \qquad \text{Factoring both numerator and denominator}$$

$$= \frac{x + 2}{x - 1} \qquad \text{Dividing out the common factor } x - 2$$

Note that the original expression is not defined for $x = 2$, whereas the simplified expression is. We will adopt the convention that the simplified expression is understood to be equal to the original expression only for values for which the original expression is defined.

Although we can divide out common factors of the numerator and denominator of a rational expression, we cannot "cancel" out common terms. Thus, for example, we cannot begin the solution of Example 11 by "canceling" x^2 in both numerator and denominator.

Do divide out common factors	Don't "cancel" common terms
$\dfrac{a \cdot b}{a \cdot c} = \dfrac{\cancel{a} \cdot b}{\cancel{a} \cdot c} = \dfrac{b}{c}$	$\dfrac{a + b}{a + c} \neq \dfrac{\cancel{a} + b}{\cancel{a} + c}$, and so $\dfrac{a + b}{a + c} \neq \dfrac{b}{c}$
$\dfrac{x^2 \cdot y^2}{x \cdot y} = \dfrac{\cancel{x^2} \cdot \cancel{y^2}}{\cancel{x} \cdot \cancel{y}} = xy$	$\dfrac{x^2 + y^2}{x + y} \neq \dfrac{\cancel{x^2} + \cancel{y^2}}{\cancel{x} + \cancel{y}}$, and so $\dfrac{x^2 + y^2}{x + y} \neq x + y$
$\dfrac{xy}{xyz} = \dfrac{\cancel{xy}}{\cancel{xy}z} = \dfrac{1}{z}$	$\dfrac{x + y}{x + y + z} \neq \dfrac{\cancel{x} + \cancel{y}}{\cancel{x} + \cancel{y} + z}$, and so $\dfrac{x + y}{x + y + z} \neq \dfrac{1}{z}$

Dividing out common factors is just one of several properties we use when performing arithmetic operations on rational expressions. These properties, which bear a striking resemblance to the arithmetic properties of fractions, are summarized as follows.

Properties of Rational Expressions

Assume that a, b, c, and d represent polynomial expressions.

Property	Comment
1. $\dfrac{ac}{bc} = \dfrac{a}{b}$	Divide out the common factor.
2. $\dfrac{a}{b} \cdot \dfrac{c}{d} = \dfrac{ac}{bd}$	Multiply numerators and denominators.
3. $\dfrac{a}{b} \div \dfrac{c}{d} = \dfrac{a}{b} \cdot \dfrac{d}{c}$	Multiply by the reciprocal of the divisor. (Invert and multiply.)
4. $\dfrac{a}{b} + \dfrac{c}{b} = \dfrac{a + c}{b}$	Add numerators over a common denominator.
5. $\dfrac{a}{b} - \dfrac{c}{b} = \dfrac{a - c}{b}$	Subtract numerators over a common denominator.

EXAMPLE 12

Multiplying Rational Expressions

Express $\dfrac{y - 1}{y + 1} \cdot \dfrac{y^2 - y - 2}{3y^2 - 3y}$ in lowest terms.

Solution

$$\frac{y - 1}{y + 1} \cdot \frac{y^2 - y - 2}{3y^2 - 3y} = \frac{(y - 1)(y^2 - y - 2)}{(y + 1)(3y^2 - 3y)}$$

Multiplying numerators and denominators

$$= \frac{(y - 1)(y + 1)(y - 2)}{(y + 1)3y(y - 1)}$$

Factoring to check for common factors

$$= \frac{y - 2}{3y}$$

Dividing out the common factors $(y + 1)$ and $(y - 1)$

EXAMPLE 13

Dividing Rational Expressions

Express the given quotient in lowest terms.

a. $\dfrac{r^2 - 36}{r^2 + r} \div \dfrac{6 - r}{r}$ **b.** $\dfrac{\dfrac{1}{x} - \dfrac{1}{y}}{\dfrac{1}{x} + \dfrac{1}{y}}$

Solution

a. $\dfrac{r^2 - 36}{r^2 + r} \div \dfrac{6 - r}{r} = \dfrac{r^2 - 36}{r^2 + r} \cdot \dfrac{r}{6 - r}$ Multiplying by the reciprocal of the divisor

$= \dfrac{(r + 6)(r - 6)}{r(r + 1)} \cdot \dfrac{r}{-(r - 6)}$ Factoring and writing $(6 - r)$ as $-(r - 6)$

$= \dfrac{r + 6}{-(r + 1)}$ Dividing out common factors

$= -\dfrac{r + 6}{r + 1}$

b. Instead of a two-step process of simplifying the main numerator and denominator and then inverting and multiplying, we can clear all fractions in one step by multiplying the main numerator and denominator by the least common multiple of all denominators in the expression. In this case, the least common multiple is xy, and so we proceed as follows:

$\dfrac{\dfrac{1}{x} - \dfrac{1}{y}}{\dfrac{1}{x} + \dfrac{1}{y}} = \dfrac{\left(\dfrac{1}{x} - \dfrac{1}{y}\right)xy}{\left(\dfrac{1}{x} + \dfrac{1}{y}\right)xy}$ Multiplying numerator and denominator by xy

$= \dfrac{\dfrac{1}{x}xy - \dfrac{1}{y}xy}{\dfrac{1}{x}xy + \dfrac{1}{y}xy}$ Distributing

$= \dfrac{y - x}{y + x}$ Simplifying

When adding or subtracting rational expressions, we must first find a common denominator. As is the case with adding fractions, we are looking for the least common multiple of the denominators. This can be found by factoring the denominators, choosing the largest power of each factor that appears, and forming the product of all the factors with the appropriate powers.

EXAMPLE 14

Adding Rational Expressions

Express $\dfrac{1}{t^2 + t} + \dfrac{2t}{t^2 - 1}$ in lowest terms.

Solution   Since $t^2 + t = t(t + 1)$ and $t^2 - 1 = (t + 1)(t - 1)$, the least common denominator is $t(t + 1)(t - 1)$. Thus, we proceed as follows:

$$\frac{1}{t^2 + t} + \frac{2t}{t^2 - 1} = \frac{1}{t(t + 1)} + \frac{2t}{(t + 1)(t - 1)} \qquad \text{Factoring the denominators}$$

$$= \frac{1}{t(t + 1)} \cdot \frac{t - 1}{t - 1} + \frac{2t}{(t + 1)(t - 1)} \cdot \frac{t}{t} \qquad \text{Obtaining a common denominator}$$

$$= \frac{t - 1}{t(t + 1)(t - 1)} + \frac{2t^2}{t(t + 1)(t - 1)}$$

$$= \frac{2t^2 + t - 1}{t(t + 1)(t - 1)} \qquad \text{Adding numerators}$$

To write this expression in lowest terms, we must factor the numerator and divide out any common factors.

$$\frac{2t^2 + t - 1}{t(t + 1)(t - 1)} = \frac{(2t - 1)(t + 1)}{t(t + 1)(t - 1)}$$

$$= \frac{2t - 1}{t(t - 1)}$$

Understanding and Mastery Checklists

Concepts to Understand

Polynomials: coefficients, terms, and degree

Monomials, binomials, and trinomials

Operations with polynomials

Complete factorization

Greatest common factors

Special binomial factoring forms

Factoring trinomials

Rational expression

Lowest terms

Operations with rational expressions

Common denominator

Skills to Master

Recognize a polynomial and determine its degree.

Add, subtract, and multiply polynomials.

Recognize when an algebraic expression is completely factored.

Factor out the greatest common factor.

Use special binomial forms to factor polynomials.

Factor trinomials by informed trial and error.

Simplify a rational expression by factoring.

Add, subtract, multiply, and divide rational expressions.

Exercises 1.3

Exercises 1–8 *Determine whether or not each expression is a polynomial. For each polynomial, indicate if it is a monomial, binomial, or trinomial, and give its degree.*

1. $2a^2 - 5a + 6$

2. $-z + 2z^3 - 4z^5 + 3$

3. $\dfrac{7}{2}k^4$

4. $3x^2 - 2x + x^{1/2} - 1$

5. $2u^5 - u^{1/3} + 4u$

6. -10

7. $2x^3y^2 + x^4 - y^4$

8. $x^{-2}y + y^{-2}x$

Exercises 9–26 *Perform the indicated operation. Simplify whenever possible.*

9. $(3x^2 + 5x - 9) + (x^2 - 2x + 1)$

10. $(-6t^4 - t^2) - (3t^3 + 7t^2 - 1)$

11. $x\left(\dfrac{1}{2}x - 1\right) + 3x\left(x + \dfrac{5}{3}\right)$

12. $k^2\left(k - \dfrac{3}{2}\right) - k(k^2 + 2k)$

13. $(2a + 1)(3a - 2)$

14. $(4x - 2)(x + 6)$

15. $\left(k - \dfrac{2}{3}\right)\left(k + \dfrac{2}{3}\right)$

16. $(x + \sqrt{3})(x - \sqrt{3})$

17. $(2x^2 + 1)(2x^2 - 1)$

18. $(t + 4)^2$

19. $(3a + 2b)^2$

20. $(z^2 - 4)^2$

21. $(2x + 1)(x^2 - 6x + 2)$

22. $(t^2 + 2)(2t^3 + t^2 - 5t)$

23. $(x - y + 2)(x + y - 2)$

24. $(x - 2)(x^3 + 2x^2 + 4x + 8)$

25. $(a - b)^3$

26. $(a + b)^3$

Exercises 27–44 *Factor completely.*

27. $w(w + 2) + 5(w + 2)$

28. $(x - 3)^2 + 4(x - 3)$

29. $2t^2 - 8$

30. $r^2 - rs - 2s^2$

31. $6n^2 - 19n + 10$

32. $8x^3 - 27$

33. $z^4 - 2z^2 + 1$

34. $t^6 - 64$

35. $x^2 + x - 42$

36. $p^6 - 125q^3$

37. $16a^2 + 20a + 6$

38. $81x^3 - 3$

39. $t^2 - 7t + 12$

40. $4r^3 + 4s^3$

41. $4x^2 - 12xy + 9y^2$

42. $2ab^2 - 32ac^2$

43. $6x^3y^3 + 48$

44. $x^6 - x^3 - 20$

Exercises 45–52 *Simplify whenever possible.*

45. $\dfrac{4t^6 - 10t^4}{2t^2}$

46. $\dfrac{21n^4 + 15n^3}{6n^3}$

47. $\dfrac{x^2 - 4}{x + 2}$

48. $\dfrac{4z^2 - 9}{2z + 3}$

49. $\dfrac{r^2 + 10r + 25}{2r^2 + 4r - 30}$

50. $\dfrac{3a^2 - 6a - 24}{a^2 - 6a + 8}$

51. $\dfrac{y^2 - 3y}{y - 3}$

52. $\dfrac{s^2t^3 - s^5}{st^2 - s^3}$

Exercises 53–72 *Perform the indicated operation(s). Simplify whenever possible.*

53. $\dfrac{15x^2}{16y^3} \cdot \dfrac{12y^3}{20x^3}$

54. $\dfrac{8st^2}{25r^2t} \cdot \dfrac{10r^2t}{32s^3t^2}$

55. $\dfrac{1}{a + 1} + \dfrac{1}{a - 1}$

56. $\dfrac{u + v}{u - v} - \dfrac{u - v}{u + v}$

57. $\dfrac{2q + 1}{q^2 + q} - \dfrac{q + 2}{q + 1}$

58. $\dfrac{x - 1}{x + 2} \cdot \dfrac{x^2 - 4}{x^2 - 1}$

59. $\dfrac{x + 3}{x + 4} \div \dfrac{x^3 + 27}{x^2 - 16}$

60. $\dfrac{1}{x^2 + 2x} - \dfrac{1}{x^2 + 4x + 4}$

61. $(2a^2 - 8b^2) \cdot \dfrac{b - a}{a^2 - 4ab + 4b^2}$

62. $y + \dfrac{1}{y + 1}$

63. $\dfrac{4x + 20}{x^2 + 8x + 16} \cdot \dfrac{x^2 + 4x}{x^2 + 7x + 10}$

64. $\dfrac{\dfrac{r - s}{r + s}}{s^2 - r^2}$

65. $\dfrac{t + 2}{t^3 + t^2} - \dfrac{1}{t + 1}$

66. $\dfrac{a + 1}{a - 2} \cdot \dfrac{a^2 - 3a + 2}{a - 3} \cdot \dfrac{3 - a}{a^2 - 1}$

67. $\dfrac{x + 5}{x^3 + 125} - \dfrac{1}{x + 5} + \dfrac{x - 5}{x^2 - 25}$

68. $\dfrac{\dfrac{1}{a} - b}{\dfrac{1}{b} - a}$

69. $\dfrac{\dfrac{1}{x} + \dfrac{1}{y}}{\dfrac{1}{xy}}$

70. $\dfrac{\dfrac{1}{3 + h} - \dfrac{1}{3}}{h}$

71. $\dfrac{\dfrac{1}{x} - \dfrac{1}{c}}{x - c}$

72. $\dfrac{\dfrac{1}{x + h} - \dfrac{1}{x}}{h}$

Exercises 73–74 *Rationalize the denominator.*

73. $\dfrac{2}{3 - \sqrt{5}}$

74. $\dfrac{6}{2 + \sqrt{3}}$

Exercises 75–84 *Perform the indicated operation. Write your answer in the form a + bi.*

75. $(6 + 2i) - (3 + 4i)$

76. $(9 + 7i) + (5 - 2i)$

77. $(2 + 3i)(3 + 4i)$

78. $(7 - 2i)(5 + 4i)$

79. $(\sqrt{2} - 4i)(\sqrt{2} + 4i)$

80. $\left(3 - i\sqrt{5}\right)^2$

81. $\dfrac{5}{2 + i}$

82. $\dfrac{-13i}{2 - 3i}$

83. $\dfrac{-18 + 13i}{5 - 2i}$

84. $\dfrac{3 - 2i}{5i}$

Applications

85. Engine Displacement The displacement of a gas-powered engine is computed using the formula $\dfrac{\pi}{4} CB^2 S$, where C is the number of cylinders, B is the bore (the diameter of each cylinder in inches), and S is the stroke (the distance in inches that the piston travels one way within the cylinder). Find the displacement of an 8-cylinder Oldsmobile "Big Block" engine with a bore of 4.126 inches and a stroke of 4.254 inches.

86. Gas Tank Volume An underground gas tank is formed by attaching a hemisphere to each end of a cylinder with radius r and height h. The volume of each hemisphere is half that of a sphere of radius r, or $\dfrac{1}{2}\left(\dfrac{4}{3}\pi r^3\right)$. Write out a polynomial that gives the total volume of the tank. What is the volume of a tank formed from a cylinder with height 15 feet and hemispherical ends of radius 5 feet?

87. Board Feet A board foot is a unit of lumber measurement equal to 1 foot square by 1 inch thick. The number of board feet of lumber in a ponderosa pine can be approximated by $0.0015c^3$, where c is the circumference of the tree in inches at waist height. How many board feet of lumber are in a ponderosa pine with a circumference of 60 inches? What about a ponderosa pine with a *diameter* of 60 inches?

88. Falling Marble A marble is dropped from the top of the Sears Tower in Chicago. After t seconds, the distance it has traveled is $16t^2$ feet, and its velocity is $32t$ feet per second. If the marble hits the ground after 9.533 seconds, how high is the Sears Tower and what is the velocity of the marble upon impact?

89. Height of a Ball A ball thrown into the air with a velocity of 80 feet per second has height $-16t^2 + 80t$ after t seconds. What is the height of the ball after 1 second? After 2 seconds? Continue computing the height for integer values of t until you find the time at which the ball hits the ground.

90. Braking Distance Suppose you are driving on a freeway when the brake lights of the car ahead of you suddenly go on. This begins an important chain of events: Your brain receives the signal to stop the car; it sends a message to your foot, which slams on the brake pedal; and the brakes dissipate energy to bring your car to a stop. From the moment the signal reaches your brain to the moment your car comes to a complete stop, your car will have traveled a certain distance. Studies have shown that this distance (measured in feet) can be approximated with the polynomial $1.1v + 0.05v^2$, where v represents the speed of the car (measured in miles per hour) before the brakes are applied. How far will your car have traveled if your speed was 65 miles per hour?

91. Triathlon Calorie Expenditure The total number of calories, C, expended by a triathlete weighing W pounds who swims S hours, cycles B hours, and runs for R hours is approximated by

$$C = (5.0S + 5.4B + 6.4R)W$$

The Ironman triathlon in Kailua-Kona, Hawaii, consists of a 2.4-mile swim, a 112-mile bike ride, and a 26.2-mile run (marathon). A competitive time for the Ironman would combine about 1 hour for the swim, 5 hours for the bike ride, and 3 hours for the marathon. About how many calories would be expended by a 150-pound Ironman competitor?

[*Data sources*: www.outsidemag.com/events/ironman96/ overall.html, Web page active as of April 2003; Henry J. Montoye, ed., et al. *Measuring Physical Activity and Energy Expenditures* (Champaign, IL: Human Kinetics, 1996).]

92. Basketball Lineup Suppose that a basketball coach wishes to try every possible starting lineup of 5 players, without regard to the positions (center, forward, guard) of the players. It can be shown that for a team with n players, the number of possible 5-player starting lineups is given by

$$\frac{n(n-1)(n-2)(n-3)(n-4)}{120}$$

a. Find the number of possible starting lineups if the coach has 6 players.

b. If a team with 12 players played one game per day, how long would it take for the coach to try every possible starting lineup?

93. Heat Index At high temperatures, our level of discomfort is determined not only by the temperature but also by the amount of moisture in the air (humidity). The *heat index* is a measure of perceived temperature that accounts for humidity, in much the same way that the wind chill factor takes into account the cooling effects of wind. One formula for apparent temperature (in degrees Fahrenheit) is given by

$$T_A = -42.38 + 2.049t + 10.14r - 0.2248tr$$
$$- 0.00684t^2 - 0.0548r^2 + 0.00123t^2r$$
$$+ 0.000853tr^2 - 0.000002t^2r^2$$

where t is the actual temperature (in degrees Fahrenheit), and r% is the relative humidity. Compute the apparent temperature corresponding to the given temperature-humidity combinations.

a. 90°, 60%   **b.** 100°, 80%

Sometimes it's not the humidity; it really is the heat. Death Valley has temperatures of up to 130°F and very little moisture.

94. Gravitational Pull Astrologists claim that the alignment of the planets at the time a person is born has an influence on the birth. Let's see what Newton's law of gravitation has to say about this. We will compute the gravitational pull between the planet Mars and a child born on February 12, 1995, which is the date in 1995 on which Mars was closest to Earth. According to Newton's law, the force of attraction in newtons between two bodies of mass m_1 and m_2 kilograms that are a distance r meters apart is

$$\frac{Gm_1m_2}{r^2}$$

where $G = 6.672 \times 10^{-11}$ N · m^2/kg^2 is the universal gravitational constant.

a. Given that the mass of Mars is 6.42×10^{23} kilograms and its distance from Earth on February 12, 1995, was 10^{11} meters, compute the gravitational pull between Mars and a 3.4-kilogram baby born on February 12, 1995.

b. Compute the gravitational pull between an obstetrician with mass 70 kilograms and a 3.4-kilogram baby if the obstetrician is standing 1 meter away.

c. Which of the two bodies (Mars or the obstetrician) exerts more gravitational pull on the baby?

95. Waiting Time The manager of a fast-food restaurant is concerned about the amount of time a customer must wait in the drive-through line. One model suggests that if s denotes the average number of minutes for the drive-through worker to process an order and if c denotes the average time between customer arrivals, then the average waiting time is given by the rational expression

$$\frac{1}{\dfrac{1}{s} - \dfrac{1}{c}}$$

[*Data source*: Michael Mesterton-Gibbons, *A Concrete Approach to Mathematical Modeling* (Reading, MA: Addison-Wesley, 1989).]

a. Determine the average waiting time for the following values of c and s:

    i. $c = 10, s = 5$ **ii.** $c = 10, s = 6$

    iii. $c = 10, s = 7$ **iv.** $c = 10, s = 8$

    v. $c = 10, s = 9$

    What happens to the waiting time as s and c get closer together? Does this make sense?

b. Try some examples to see what happens to the length of time if c is less than s. Does this agree with your intuition about waiting time? Explain. What can you conclude about the waiting-time model?

96. Percentage Body Fat Hydrostatic weighing is among the most accurate methods for measuring a person's percentage of body fat. First, the mass of the subject is measured using an ordinary scale. Next, the subject is asked to expel as much as air as possible from her lungs and is submerged in water. The submerged weight is then recorded. The subject's body density is then approximated by

$$d = \frac{m}{m - w - 1}$$

where m is the body mass in kilograms and w is the underwater "weight" in kilograms (technically, a kilogram is a unit of mass). According to one widely used model of body composition, the percentage of body fat is approximated by

$$\text{Percentage body fat} = \frac{495}{d} - 450$$

a. Find a simplified expression for the percentage of body fat in terms of m and w.

b. Estimate the percentage of body fat for a 100-kilogram (220-pound) man with a submerged weight of 7 kilograms.

[*Data source*: William D. McArdle, Frank I. Katch, Victor L. Katch, *Exercise Physiology*, 4th ed. (Baltimore: Williams and Wilkins, 1996), p. 554.]

Concepts and Critical Thinking

Exercises 97-100 *Answer true or false.*

97. The degree of the product of two polynomials is the sum of the degrees of the polynomials.

98. The degree of the sum of two polynomials is the sum of the degrees of the polynomials.

99. $(a + b)^2 = a^2 + b^2$

100. $a^3 - b^3 = (a + b)(a^2 - ab + b^2)$

Exercises 101-104 *Give an example of each.*

101. A fourth-degree binomial

102. A third-degree trinomial

103. A polynomial of degree 2 for which $x - 3$ is a factor

104. A perfect square trinomial for which $2x + 1$ is a factor

105. How does the degree of the sum or difference of two polynomials compare to the degrees of the two polynomials? Is the product of two polynomials a polynomial? If so, is there a connection between the degrees of the original polynomials and their product?

106. Show that one less than the square of a number is divisible by one less than that number.

107. Show that the sum of two perfect cubes is never a prime number.

Questions for Discussion or Essay

108. Explain how the sum of cubes factorization formula can be used to find a factor of 1,000,001 in just seconds without a calculator.

109. Compare and contrast factoring polynomials to "decomposing" chemical substances into elements. Can you think of any processes in other disciplines that involve decomposing complex structures into their basic building blocks?

110. In what instances will the quotient of polynomials be a polynomial?

Projects for Enrichment

111. Measuring Inflation with the CPI   The Consumer Price Index (CPI) is computed each month by the Bureau of Labor Statistics. It is a measure of the retail cost of a representative sample of goods and services purchased by American consumers. Table 5 shows the CPI for each month of 2001. We see, for example, that in January the CPI was 175.1. Roughly speaking, this means that goods costing $100 in 1982, which is the base year—the year against which other years are compared—would cost $175.1 in January 2001.

Table 5

Month (2001)	CPI
January	175.1
February	175.8
March	176.2
April	176.9
May	177.7
June	178.0
July	177.5
August	178.3
September	177.7
October	177.4
November	176.7
December	177.1

Changes in the CPI from one month to the next provide a measure of changes in retail prices paid by consumers. As can be seen in Table 5, the CPI for most months of 2001 increased from the previous month, suggesting that retail prices were rising throughout most of the year. The average of the monthly CPI values for a given year provides an annual measure of retail prices, which can then be used to compare retail prices from one year to the next.

a. Find the average CPI for 2001, and use it to help you complete Table 6. Note that the percent change is computed as follows:

$$\frac{|(\text{Current year CPI}) - (\text{Previous year CPI})|}{(\text{Previous year CPI})} \times 100\%$$

Table 6

Year	Average CPI	Percent change
1995	152.4	
1996	156.9	2.95%
1997	160.5	
1998	163.0	
1999	166.6	
2000	172.2	
2001		

Monthly and annual CPI trends are used by government agencies to aid in the formulation of fiscal and monetary policies. They are also used by various agencies, organizations, and companies to keep such things as wages, salaries, pensions, rents, alimony payments, and child support payments in line with changing prices.

Suppose that a company gives salary raises that are based on the CPI. More specifically, suppose that in January of a given year, the percent increase in the average CPI from the previous two calendar years is used to compute new salaries for the coming year. For example, in January 1997 the company would have determined that the average CPI for 1996 had increased 2.95% from that of 1995, and so salaries for 1997 would be increased by 2.95% over the 1996 level.

b. Find the salary in 1997 for an employee who made $18,000 in 1996.

c. Show that if S_0 denotes the salary for a given year and if r denotes the percent increase that is to be applied to that salary, then the salary for the following year, S_1, can be found using the formula $S_1 = S_0(1 + r)$.

d. Compute the salary for each of the years 1998 through 2002 for the employee from part b.

In part d, we were able to find the salary in 2002 by computing the salaries for all of the years between 1996 and 2002. Suppose instead that we wished to find the salary in 2002 without computing any of the intermediate salaries. This can be done with the help of polynomial expressions.

Let $r_1, r_2, \ldots, r_n$ denote n consecutive percent increases in the annual CPI, let S_i denote the annual salary after the percent increase r_i has been applied, and let S_0 denote the salary for the year prior to that in which the percent increase r_1 has been applied. Then, using the formula from part c, we have

$$S_1 = S_0(1 + r_1)$$
$$S_2 = S_1(1 + r_2) = S_0(1 + r_1)(1 + r_2)$$
$$S_3 = S_2(1 + r_3) = S_0(1 + r_1)(1 + r_2)(1 + r_3)$$
$$\vdots$$
$$S_n = S_0(1 + r_1)(1 + r_2) \cdots (1 + r_n)$$

So the final salary, S_n, can be found by multiplying the initial salary, S_0, by the polynomial $(1 + r_1)(1 + r_2) \cdots (1 + r_n)$.

e. Use a polynomial expression of the form just given to find the salary in 2002 for employees who had the following salaries in 1996:

i. $20,000   **ii.** $34,000

Note that you need to compute the product $(1 + r_1)(1 + r_2) \cdots (1 + r_6)$ only once.

f. Use the CPI values in 1995 and 2001 to find the overall percent increase from 1995 to 2001. How does this value compare to the value of the polynomial $(1 + r_1)(1 + r_2) \cdots (1 + r_6)$ from part e? What does this suggest about finding someone's salary in 2002 if the 1996 salary is known?

g. Do you think the CPI is a reasonable tool for determining raises? Why or why not?

112. Averages The media and advertisers use the term *average* in many different contexts. It's not uncommon to hear or read phrases such as "the average annual income is $40,000" or "the average teenager spends 2 hours each day on the phone." What exactly is meant by the term *average*, and how is it computed in each case? It may surprise you to learn that this term is subject to multiple interpretations. The measure that is most commonly referred to as an average is more properly called the *arithmetic mean*. The arithmetic mean of n numbers $a_1, a_2, \ldots, a_n$ is given by

$$A = \frac{a_1 + a_2 + \cdots + a_n}{n}$$

a. During the first three basketball games of the season, Stephanie scored 12, 16, and 13 points. During the next four games, she scored 16, 22, 17, and 19 points. Compute her average for the first three games and then for the next four. How does the average of the two averages compare to the average over the seven games?

b. Assuming that A is the arithmetic mean of m numbers $a_1, a_2, \ldots, a_m$ and B is the arithmetic mean of n numbers $b_1, b_2, \ldots, b_n$, find an expression for the arithmetic mean of A and B. Use the operations on rational expressions to show that the arithmetic mean of A and B is not necessarily the same as the arithmetic mean of the $m + n$ numbers $a_1, a_2, \ldots, a_m, b_1, b_2, \ldots, b_n$.

Another type of average is called the *weighted mean*. The weighted mean of n numbers $a_1, a_2, \ldots, a_n$ that are weighted by the factors $w_1, w_2, \ldots, w_n$ is given by

$$W = \frac{a_1 w_1 + a_2 w_2 + \cdots + a_n w_n}{w_1 + w_2 + \cdots + w_n}$$

c. David's scores on three exams in college algebra were 76, 84, and 80, respectively. On the final, he scored 66. If each exam counts 20% and the final counts 40%, David's weighted average is

$$\frac{76 \cdot 20 + 84 \cdot 20 + 80 \cdot 20 + 66 \cdot 40}{20 + 20 + 20 + 40} = \frac{7440}{100} = 74.4$$

Find his weighted average if each exam counts 30% and the final counts 10%.

d. Assuming that each of the weighting factors is the same, w say, find a simplified expression for the weighted mean of n numbers $a_1, a_2, \ldots, a_n$. Does it look familiar?

A third type of average is called the *harmonic mean*. The harmonic mean of n numbers $a_1, a_2, \ldots, a_n$ is given by

$$H = \frac{n}{\dfrac{1}{a_1} + \dfrac{1}{a_2} + \cdots + \dfrac{1}{a_n}}$$

e. Suppose that you and a friend are given the task of purchasing the beverages for a rather large party. You spend $48 on beverages costing $3 per 12-pack and your friend spends $48 on beverages costing $4 per 12-pack. The average price is *not* ($3 + $4)/2 = $3.50 per 12-pack. Instead, since a total of $96 is spent on $\frac{48}{3} + \frac{48}{4} = 28$ 12-packs, the actual cost is $\frac{96}{28} = \$3.43$ per 12-pack. Show that the harmonic mean of $3 and $4 is equal to $3.43.

f. Show that the harmonic mean of three numbers a, b, and c is

$$H = \frac{3abc}{bc + ac + ab}$$

113. Installment Loans Have you ever wondered how much the monthly payments would be for a certain car loan? Or have you ever wondered how much you could borrow if you can afford to pay back only a certain amount each month? It may not surprise you to hear that there are standard formulas for determining such things. And it may not surprise you to hear that these formulas are rather complex and would likely intimidate the "algebraically challenged" among us. In a book entitled *The Complete How to Figure It* (New York: W. W. Norton, 1996), author Darrell Huff avoids the use of complex algebraic formulas and instead describes verbally how certain loan calculations can be made. In this project, we compare these verbal descriptions with their algebraic counterparts.

An installment loan is a loan that is paid back in regular—usually monthly—equal payments. Each payment includes interest on the remaining unpaid balance. Two examples of installment loans are mortgage loans (for the purchase of a house) and auto loans. The following steps paraphrase those given by Huff to determine the amount of an installment payment if the annual interest rate, the number of payments per year, and the number of years for the loan are known (Huff, p. 123). (Note that we will assume the payments are made at the end of each payment period and that interest is compounded when payments are due.)

1. Divide the annual interest rate by the number of payments per year. Save the result for the next step and a future step.

2. Add 1 to the result of step 1.

3. Raise the result of step 2 to the power equal to the total number of payments to be made (the product of the number of years and the number of payments per year). Save the result for the next step and for a future step.

4. Subtract 1 from the result of step 3.

5. Divide the result of step 3 by the result of step 4.

6. Multiply the result of step 5 by the amount of the original loan.

7. Multiply the result of step 6 by your result from step 1. The answer is the amount of the installment payment.

a. Use these steps to determine the amount of a *monthly* installment payment if you borrow $20,000 at an annual interest rate of 8% for 10 years (a total of 120 payments). Write down the result of each step.

b. Let P denote the original amount of the loan, let r denote the annual interest rate, let n denote the number of payments per year, and let t denote the number of years. Use the seven steps just given to find an algebraic formula for computing the payment amount A.

c. Use the formula you found in part b to find the amount of an installment payment for the following loans:

  i. Monthly payments on $10,000 at 9% for 5 years

  ii. Bimonthly (twice per month) payments on $70,000 at 8.25% for 15 years

The following formula gives the amount of a loan P in terms of the payment amount A, the annual interest rate r, the number of payments per year n, and the number of years t.

$$P = \frac{An}{r}\left[1 - \frac{1}{\left(1 + \frac{r}{n}\right)^{nt}}\right]$$

d. Find the amount of a loan with the following payments:

   i. A \$300 monthly payment for 5 years at 7.75%

   ii. A \$250 bimonthly payment for 15 years at 8%

e. Write out a step-by-step verbal procedure for computing the amount of a loan if the payment amount, the annual interest rate, the number of payments per year, and the number of years are known. Avoid the use of any algebraic symbols.

f. Compare and contrast verbal and algebraic methods for describing loan calculations.

Section 1.4   Graphs of Equations

- In what way does a city's system of streets and avenues suggest a link between algebra and geometry?
- How can the ideas of a Greek mathematician from several hundred years B.C. and a French philosopher from the 17th century A.D. be combined to determine distances without a tape measure?
- If the life expectancy for an American is 76 years, then how much longer can someone 75 years and 364 days old expect to live?
- What doesn't your graphing calculator show that you may need to know?
- How can a tiny island have a coastline more than a billion miles long?

The Cartesian Plane

In Section 1.1, we saw that real numbers can be represented by points on the number line. René Descartes, a brilliant 17th-century mathematician and philosopher, extended this connection between the real number system and geometry by associating with every point in the plane a pair of real numbers. This simple yet profound step enables us to apply powerful algebraic techniques to solve difficult geometric problems, extending greatly the range of practical problems that can be solved with algebra.

We form the **Cartesian plane** by placing a horizontal number line, called the **x-axis**, and a vertical number line, called the **y-axis**, so that they intersect at the zero point of each line. The point of intersection of the two axes is called the **origin**, and the axes divide the plane into four regions, called **quadrants**, as shown in Figure 9.

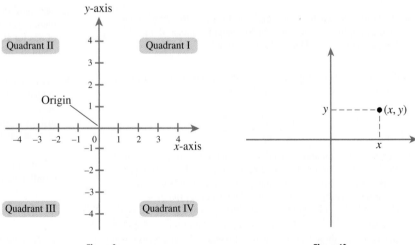

Figure 9   Figure 10

Each point in the Cartesian plane corresponds to an ordered pair of numbers (x, y), where x is the number on the x-axis directly above or below the point, and y is the number on the y-axis directly to the right or left of the point, as shown in Figure 10. In other words, the number x indicates how many units to move horizontally from the origin (to the right if x is positive and to the left if it is negative); the number y indicates how many units to move vertically (up if y is positive and down if it is negative). The numbers x and y are called the **coordinates** of the point.

EXAMPLE 1

Plotting Points

Plot the points $(2, 3)$, $(-1, -4)$, $(-3, 0)$, and $(\pi, -\sqrt{2})$.

Solution To plot the point $(2, 3)$, we move 2 units to the *right* of the origin and 3 units *up*. For the point $(-1, -4)$, we move 1 unit to the *left* of the origin and four units *down*. The point $(-3, 0)$ is on the x-axis, 3 units to the *left* of the origin. Finally, the point $(\pi, -\sqrt{2})$ is π units to the *right* of the origin and $\sqrt{2}$ units *down*. See Figure 11.

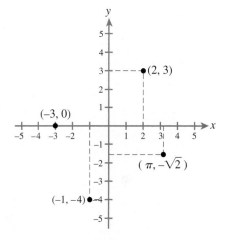

Figure 11

The Distance and Midpoint Formulas

The distance between two points in the Cartesian plane can be found using the Pythagorean Theorem, which states that if a right triangle has side lengths a and b and hypotenuse length c, then $c^2 = a^2 + b^2$ (see Figure 12). Suppose $P(x_1, y_1)$ and $Q(x_2, y_2)$ are two points in the plane, as shown in Figure 13. Then the right triangle with hypotenuse $\overline{PQ}$ has side lengths $|x_2 - x_1|$ and $|y_2 - y_1|$. By the Pythagorean Theorem, the length d of the hypotenuse $\overline{PQ}$ is given by $d = \sqrt{(x_2 - x_1)^2 + (y_2 - y_1)^2}$.

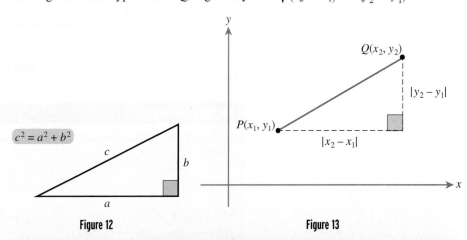

Figure 12   **Figure 13**

The Distance Formula

The distance d between the points (x_1, y_1) and (x_2, y_2) is given by

$$d = \sqrt{(x_2 - x_1)^2 + (y_2 - y_1)^2}$$

⋯⋯▷**EXAMPLE 2**

Finding the Distance Between Points

Find the distance between $(3, 5)$ and $(-1, -2)$.

Solution   According to the distance formula, we compute the square of the differences between the x- and y-coordinates, add them, and take the square root of the sum.

$$\begin{aligned}
d &= \sqrt{(-1 - 3)^2 + (-2 - 5)^2} \\
&= \sqrt{(-4)^2 + (-7)^2} \\
&= \sqrt{16 + 49} \\
&= \sqrt{65} \approx 8.06
\end{aligned}$$

Now suppose we are interested in finding the coordinates $(\bar{x}, \bar{y})$ of the midpoint of the line segment $\overline{PQ}$ joining the two points $P(x_1, y_1)$ and $Q(x_2, y_2)$, shown in Figure 14. Intuitively, we might expect that $\bar{x}$ and $\bar{y}$ are the averages of the x- and y-coordinates, respectively, of the two points. Thus, we suspect that

$$\bar{x} = \frac{x_1 + x_2}{2} \qquad \text{and} \qquad \bar{y} = \frac{y_1 + y_2}{2}$$

By applying the distance formula, we can confirm that these formulas are valid (see Exercise 77).

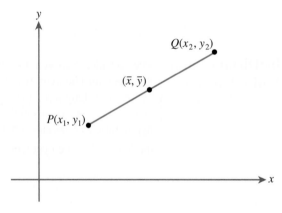

Figure 14

The Midpoint Formula

The midpoint of the line segment joining the points (x_1, y_1) and (x_2, y_2) is $(\bar{x}, \bar{y})$, where

$$\bar{x} = \frac{x_1 + x_2}{2} \qquad \text{and} \qquad \bar{y} = \frac{y_1 + y_2}{2}$$

Note that $\bar{x}$ is the average of the x-coordinates, and $\bar{y}$ is the average of the y-coordinates.

⋯⋯▷**EXAMPLE 3**

Using the Midpoint Formula

Find the coordinates of the midpoint of the line segment connecting $(3, 5)$ and $(-1, -2)$.

Solution According to the midpoint formula, we have

$$\bar{x} = \frac{3 + (-1)}{2} = 1 \quad \text{and} \quad \bar{y} = \frac{5 + (-2)}{2} = \frac{3}{2}$$

Thus, the midpoint has coordinates $\left(1, \frac{3}{2}\right)$.

Graphs of Equations

Many of the most profound mathematical and scientific results are relationships among quantities that can be expressed as simple algebraic equations. Examples include Pythagoras's $c^2 = a^2 + b^2$, Newton's $F = ma$, and Einstein's $E = mc^2$. Because of the central role played by equations in the application of mathematics to the real world, we will explore them in great depth in much of the remainder of this text.

It is often quite useful to represent an equation graphically. To this end, we begin by restricting ourselves to equations involving only two variables, which we call x and y. We define the **graph of an equation in x and y** to be the set of points (x, y) in the coordinate plane such that x and y satisfy the equation. For example, the graph of $y = 2x - 4$ is the set of points (x, y) in the coordinate plane such that x and y satisfy $y = 2x - 4$.

The simplest technique for graphing an equation is **plotting points**. To do this, we choose a convenient value for one of the variables and then use the equation to determine the value of the other variable. We then graph, or "plot," the corresponding point. The process is repeated as necessary, and then the points are connected by drawing a smooth curve through the plotted points.

EXAMPLE 4

Graphing an Equation by Plotting Points

Sketch a graph of the equation $y = 2x - 4$.

Solution We choose a few convenient values of x, and compute the corresponding y-values.

x	$y = 2x - 4$
0	-4
1	-2
2	0
3	2
4	4

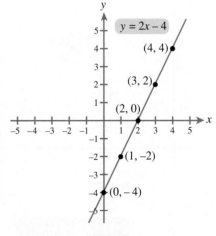

Figure 15

Thus, the points $(0, -4)$, $(1, -2)$, $(2, 0)$, $(3, 2)$, and $(4, 4)$ are on the graph of $y = 2x - 4$. After plotting these points and "connecting the dots," we obtain the graph shown in Figure 15.

The points at which the graph of an equation crosses an axis are called the **intercepts**. More specifically, we make the following definitions.

Intercepts

The points where the graph of an equation in x and y crosses the x-axis are called the **x-intercepts**, and the points where the graph crosses the y-axis are called the **y-intercepts**. Note that an x-intercept is of the form $(a, 0)$. In practice, we often refer to the number a as the x-intercept. Similarly, a y-intercept is of the form $(0, b)$, and we often refer to the number b as the y-intercept. See Figure 16.

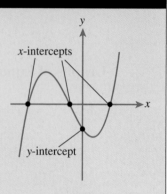

Figure 16

In Example 4, $(2, 0)$ is the only x-intercept and $(0, -4)$ is the only y-intercept. If intercepts are not obvious from the graph, they can be found by using the fact that a point on the x-axis has y-coordinate 0, and a point on the y-axis has x-coordinate 0. Thus, to find any x-intercepts, we set $y = 0$ and find the corresponding x-values. Similarly, to find y-intercepts, we set $x = 0$ and find the corresponding y-values.

Circles

Certain equations are more easily graphed, not by plotting many points, but by determining key characteristics of the graph from a special form of the equation. Examples of such equations include those for lines and circles. We consider circles here and will provide a thorough treatment of lines in Section 1.7.

A circle is defined to be the set of points that are an equal distance (the length of the radius) from a fixed point (the center). If we denote the center by (h, k) and the distance from the center to an arbitrary point on the circle (x, y) by r, as in Figure 17, we can apply the distance formula to obtain

$$r = \text{The distance from } (h, k) \text{ to } (x, y)$$
$$= \sqrt{(x - h)^2 + (y - k)^2}$$

Squaring both sides, we arrive at the standard equation of a circle.

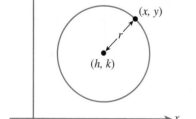

Figure 17

Standard Equation of a Circle

The standard form for the equation of a circle with radius r and center (h, k) is

$$(x - h)^2 + (y - k)^2 = r^2$$

EXAMPLE 5

Finding the Equation of a Circle

Find an equation of the circle with radius 2 and center $(3, -5)$.

Solution   We use the standard equation of a circle with $h = 3$, $k = -5$, and $r = 2$.

$$(x - 3)^2 + (y - (-5))^2 = 2^2$$
$$(x - 3)^2 + (y + 5)^2 = 4$$

·····⟩EXAMPLE 6

Graphing a Circle

Graph the circle with equation $(x + 4)^2 + (y - 3)^2 = 16$.

Solution   We begin by rewriting our equation in standard form.

$$(x - (-4))^2 + (y - 3)^2 = 4^2$$

We then simply "read off" $h = -4$, $k = 3$, and $r = 4$. Thus, our circle is centered at $(-4, 3)$ and has radius 4. The graph is shown in Figure 18.

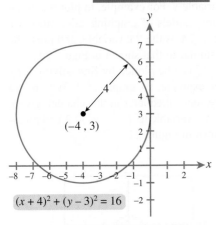

$(x + 4)^2 + (y - 3)^2 = 16$

Figure 18

Consider the equation $(x + 4)^2 + (y - 3)^2 = 16$ from Example 6. If we square the terms on the left-hand side, we obtain

$$(x + 4)^2 + (y - 3)^2 = 16$$
$$[x^2 + 2(4)x + 4^2] + [y^2 + 2(-3)y + (-3)^2] = 16$$
$$x^2 + 8x + 16 + y^2 - 6y + 9 = 16$$
$$x^2 + y^2 + 8x - 6y = -9$$

Now in this form, the center and radius of the circle are difficult to identify. Thus, if we were given the task of graphing the equation $x^2 + y^2 + 8x - 6y = -9$, our first step would be to write it in standard form by somehow reversing the squaring-out process just shown. This can be accomplished by **completing the square**.

Completing the Square

To complete the square on an expression of the form $x^2 + Dx$, first divide the coefficient D of x by 2, then square this result, and add to the original expression to obtain

$$x^2 + Dx + \left(\frac{D}{2}\right)^2$$

Naturally, if the expression is part of an equation, we must ensure that whatever is added to one side of the equation is added to the other as well.

·····⟩EXAMPLE 7

Completing the Square to Graph a Circle

Graph the circle with equation $x^2 + y^2 - 2x + 4y - 20 = 0$.

Solution   To find the center and radius, we must write the equation in standard form. To do this, we complete the square in both x and y as follows:

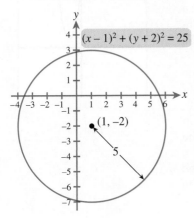

$(x - 1)^2 + (y + 2)^2 = 25$

Figure 19

$$x^2 - 2x + y^2 + 4y = 20 \qquad \text{Grouping the } x \text{ and } y \text{ terms}$$
$$(x^2 - 2x + \boxed{}) + (y^2 + 4y + \boxed{}) = 20 \qquad \text{Preparing to complete the square in } x \text{ and } y$$
$$(x^2 - 2x + \boxed{1}) + (y^2 + 4y + \boxed{}) = 20 + 1 \qquad \text{Adding } \left(\frac{-2}{2}\right)^2 = 1 \text{ to both sides}$$
$$(x^2 - 2x + \boxed{1}) + (y^2 + 4y + \boxed{4}) = 20 + 1 + 4 \qquad \text{Adding } \left(\frac{4}{2}\right)^2 = 4 \text{ to both sides}$$
$$(x - 1)^2 + (y + 2)^2 = 25 \qquad \text{Expressing in standard form}$$

Thus, the circle is centered at $(1, -2)$, with radius 5. Its graph is shown in Figure 19.

Graphing Calculators

Graphing calculators are capable of plotting almost any equation of the form $y = \square$, where $\square$ is an expression involving only the variable x. For example, to plot the graph of the equation $y = 3x^2 - 12x + 14$ with many models of graphing calculators, we would simply enter the expression $3\text{x}^2 - 12\text{x} + 14$ as the Y1 variable, and press the graph button. The resulting graph should look similar to that shown in Figure 20.

Many equations that are not of the form $y = \square$ can be plotted by first solving for y and then graphing the resulting equation. For example, to graph $x^2 + 3y = 6$, we would first solve for y using algebraic techniques described later in this chapter, giving us $y = (6 - x^2)/3$, and then plot by setting the Y1 variable to $(6 - \text{x}^2)/3$ and pressing the graph button. The resulting graph is shown in Figure 21.

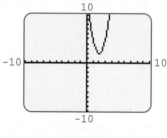

Figure 20

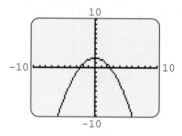

Figure 21

Viewing Windows

The **viewing window** is the portion of the coordinate system that is displayed by the graphing calculator. The viewing window can be selected by indicating the range of x- and y-values that are to be shown. This is done by adjusting the values of the **window variables** Xmin, Xmax, Ymin, and Ymax, which indicate the minimum and maximum values of x and y, respectively, as suggested by Figure 22.

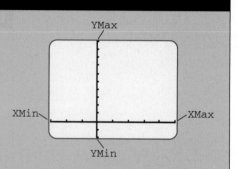

Figure 22

The first step to producing a graph is to choose appropriate initial settings for the window variables. A good place to start is with the calculator's default settings, which produce what we will refer to as the **standard viewing window**. Consider, for example, the equation $y = 3x^2 - 12x + 14$. When the equation is plotted with the standard viewing window on one popular brand of graphing calculator ($\text{Xmin} = -10$, $\text{Xmax} = 10$, $\text{Ymin} = -10$, and $\text{Ymax} = 10$), we obtain the graph shown in Figure 23. After noting the general shape of the graph, we refine the viewing window to $\text{Xmin} = -5$, $\text{Xmax} = 5$, $\text{Ymin} = 0$, and $\text{Ymax} = 20$, which yields the graph shown in Figure 24.

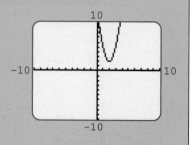

Figure 23

Although the standard viewing window often provides an adequate initial view of a graph, there may be instances when the standard view is distorted. Consider the equations $y = 2x + 1$ and $y = -\frac{1}{2}x - 2$. Using information about lines and slope (which we will review in Section 1.7), we can show that these are the equations of perpendicular lines. However, when the graphs are produced in the standard viewing window (see Figure 25), they do not appear to be perpendicular. This is because the calculator screen is wider than it is high, and so in the standard viewing window the units on the x-axis are spaced further apart than those on the y-axis. To an extent, we can compensate for this effect by changing to the **square** viewing window, where units are equally spaced along the x- and y-axes. This produces the graph shown in Figure 26.

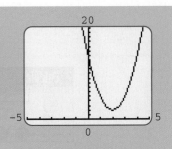

Figure 24

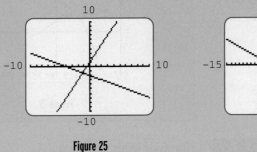

Figure 25   **Figure 26**

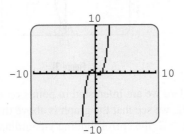

EXAMPLE 8

Narrowing the Range of the Viewing Window

Use a graphing calculator to graph $y = x^3 - \frac{1}{2}x^2 - x + \frac{1}{2}$ with a view that shows the approximate locations of "humps" and intercepts.

Solution   First, we enter the equation $y = x^3 - \frac{1}{2}x^2 - x + \frac{1}{2}$ and set the window variables to the values $\texttt{Xmin} = -10$, $\texttt{Xmax} = 10$, $\texttt{Ymin} = -10$, and $\texttt{Ymax} = 10$. The plot is shown in Figure 27. Although this plot does give some idea of what the graph looks like, the intercepts and "humps" do not stand out. We can remedy this by selecting a smaller viewing window. We adjust the window variables so that $\texttt{Xmin} = -2$, $\texttt{Xmax} = 2$, $\texttt{Ymin} = -2$, and $\texttt{Ymax} = 2$. This yields the plot in Figure 28.

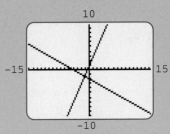

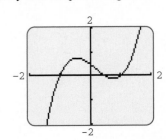

Figure 27   **Figure 28**

The coordinates of interesting points on the graph—such as the "humps" and intercepts from the previous example—can be approximated using the trace feature of a graphing calculator. We describe the general procedure as follows.

Tracing a Graph

When we select the trace feature on a graphing calculator, the cursor traces points on the graph within the viewing window, and the coordinates of the points on the graph are displayed. Figures 29 and 30 show graphs of $y = x^3 - \frac{1}{2}x^2 - x + \frac{1}{2}$ in which the trace feature has been enabled. In Figure 29, the cursor is located roughly at the point where the graph turns, and we see from the coordinates at the bottom of the viewing window that this point is approximately $(-0.43, 0.76)$. In Figure 30, we see that one of the x-intercepts has an x-coordinate of approximately 0.51. Note that the accuracy of these approximations depends on the viewing window chosen. We will discuss accuracy in greater detail in Section 1.5.

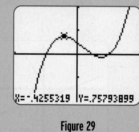

Figure 29

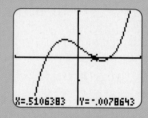

Figure 30

EXAMPLE 9

Identifying Points on a Graph

For the equation $y = x^3 - 8$, use a graphing calculator to estimate

a. the x-intercept to the nearest integer
b. the x-values of points on the graph having positive y-coordinates
c. the x-values of points on the graph having negative y-coordinates

Solution

a. The graph of $y = x^3 - 8$ is shown in Figure 31. Using the trace feature, we estimate the x-intercept to be 2, as shown in Figure 32.

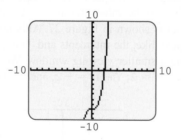

Figure 31

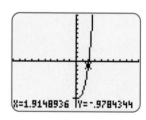

Figure 32

b. Here we are interested in points on the graph that lie *above* the x-axis. From Figure 32, we see that the graph is above the x-axis when x is greater than 2. Thus, for each $x > 2$, the corresponding y-coordinate is positive.

c. In this case, we are interested in points on the graph that lie *below* the x-axis. Since the graph is below the x-axis for x-values less than 2, the y-coordinates are negative for $x < 2$.

Understanding and Mastery Checklists

Concepts to Understand	Skills to Master
Cartesian plane	Plot points in the coordinate plane.
Coordinates	Find the distance between two points.
Distance formula	Find the midpoint between two points.
Midpoint formula	Sketch the graph of an equation by plotting points.
Graph of an equation	Determine the intercepts of a graph.
Intercepts	Find the equation of a circle given its radius and center.
Standard equation of a circle	Complete the square to place the equation of a circle in standard form and then graph it.
Completing the square	Choose an appropriate scale when plotting points.
Scale	Select an appropriate viewing window by adjusting window variables.
Viewing window	Identify key features of a graph by using the trace feature.
Window variables	
Trace feature	

Exercises 1.4

Exercises 1-4 *Plot the pair of points, find the distance between them, and find the coordinates of the midpoint of the line segment connecting them.*

1. $(1, 2)$ and $(5, 4)$

2. $(-2, 3)$ and $(4, -1)$

3. $\left(-\frac{4}{3}, 2\right)$ and $\left(\frac{8}{3}, 1\right)$

4. $\left(-2, -\frac{7}{2}\right)$ and $\left(5, \frac{1}{2}\right)$

Exercises 5-8 *Show that the given points form the vertices of the indicated polygon.*

5. Parallelogram: $(2, 3), (4, 5), (6, 4), (8, 6)$ (*Hint*: Show that opposite sides are of equal length.)

6. Rhombus: $(-1, -3), (1, 5), (8, -1), (-8, 3)$ (*Hint*: Show that all four sides are of equal length.)

7. Right triangle: $(1, 0), (2, 3), (4, -1)$ (*Hint*: Use the Pythagorean Theorem.)

8. Square: $(1, 10), (-5, 2), (3, -4), (9, 4)$ (*Hint*: The diagonals must be the same length, as must the adjacent sides.)

Exercises 9-12 *Show that the given points are on the graph of the equation.*

9. $x^2 + y^2 = 100$; $(8, 6), (0, -10), \left(5\sqrt{2}, 5\sqrt{2}\right)$

10. $x^2 + xy + y^2 = 7$; $(1, 2), (-2, 3), \left(\sqrt{7}, 0\right)$

11. $y = \dfrac{x}{x + 1}$; $(-2, 2), \left(-\frac{4}{3}, 4\right), \left(\sqrt{2}, 2 - \sqrt{2}\right)$

12. $x = \sqrt{y^2 - 3y}$; $(0, 3)$, $(2, -1)$, $\left(\sqrt{10}, 5\right)$

Exercises 13-14 *Use the coordinates of the given points to construct a graph of the relevant portion of the equation. Be sure to choose an appropriate scale.*

13. $y = 1000(x^3 + 1)$

x	y
-0.1	999
0.0	1000
0.1	1001
0.2	1008
0.3	1027

14. $y = x^3 - 3x^2$

x	y
-1	-4
0	0
1	-2
2	-4
3	0
4	16

Exercises 15-20 *Sketch the graph of each of the following equations by plotting points.*

15. $y = 2x - 4$

16. $y = x^2 - 4$

17. $y = -x^2 + 9$

18. $y = -3x + 6$

19. $y = \sqrt{x - 2}$

20. $y = \dfrac{10}{x^2 + 1}$

Exercises 21-30 *Find the center and radius of the circle and sketch its graph.*

21. $x^2 + y^2 = 25$

22. $x^2 + y^2 = 10$

23. $2x^2 + 2y^2 = 16$

24. $9 - x^2 = y^2$

25. $(x + 4)^2 + (y - 1)^2 = 4$

26. $(x - 3)^2 + (y + 1)^2 = 16$

27. $x^2 - 4x + y^2 + 6y + 13 = 4$

28. $x^2 + y^2 + 6x + 6y + 8 = 0$

29. $4x^2 + 12x + 4y^2 + 8y = 3$ (*Hint:* First divide through by the coefficient of x^2.)

30. $9x^2 + 9y^2 + 36x - 12y + 31 = 0$ (*Hint:* First divide through by the coefficient of x^2.)

Exercises 31-36 *Find the equation of the circle with the given properties.*

31. Center $(-1, 3)$ and radius 4

32. Center $(4, 0)$ and radius 2

33. Center $(4, -3)$ and passes through the origin

34. Center $(-3, -6)$ and passes through the point $(-3, 0)$

35. A diameter with endpoints $(-4, 2)$ and $(2, 8)$

36. A diameter with endpoints $(-3, 6)$ and $(5, -6)$

Exercises 37-42 *Find the graph that matches the given equation. Choose from i–vi.*

i.

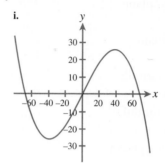

ii.

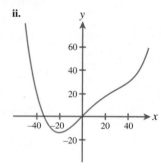

iii.

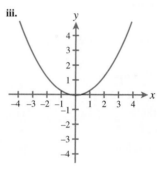

iv.

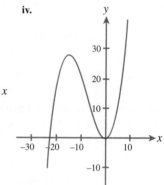

v.

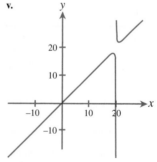

vi.

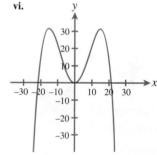

37. $y = \dfrac{x^4}{108000} - \dfrac{x^3}{1800} + x$

38. $y = x - \dfrac{x^3}{4500}$

39. $y = \dfrac{x^2}{3}$

40. $y = \dfrac{x^3}{60} + \dfrac{3x^2}{8}$

41. $y = \dfrac{9x^2}{32} - \dfrac{x^4}{1600}$ **42.** $y = x + \dfrac{1}{x - 20}$ **44.** $y = x - 2$

Exercises 43-46   *Match each graph in parts a–d with a viewing window chosen from i–iv.*

   i. Xmin $= -2$, Xmax $= 2$, Ymin $= -2$, and Ymax $= 2$

   ii. Xmin $= -8$, Xmax $= 8$, Ymin $= -8$, and Ymax $= 8$

   iii. Xmin $= -2$, Xmax $= 2$, Ymin $= -4$, and Ymax $= 4$

   iv. Xmin $= -4$, Xmax $= 4$, Ymin $= -2$, and Ymax $= 2$

43. $x^2 + y^2 = 4$

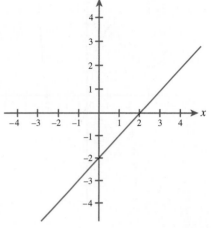

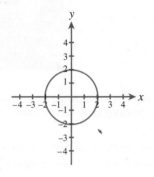

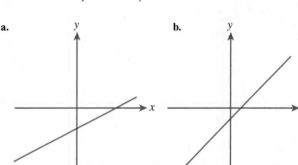

a.   **b.**

c.

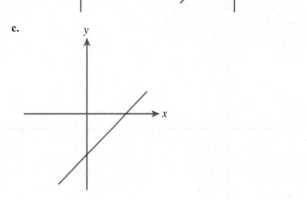

d.

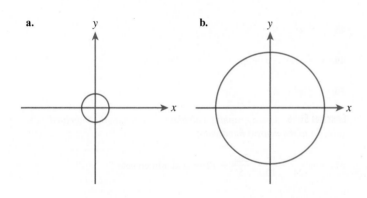

a.   **b.**

c.   **d.**

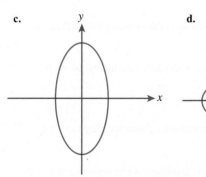

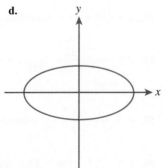

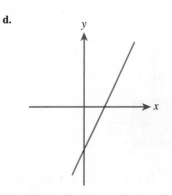

45. $y = x^3 - 4x$

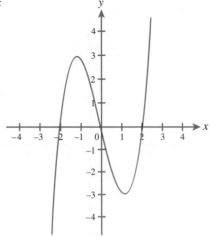

a.

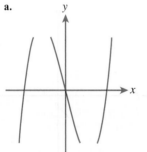

b.

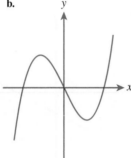

c.

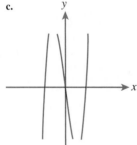

d.

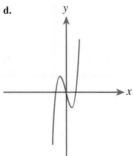

46.

a.

b.

c.

d.

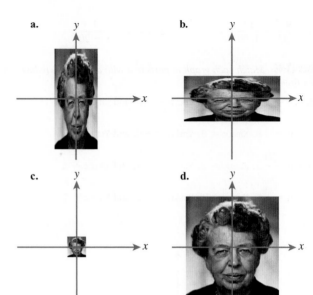

Exercises 47–50 *For the given equation, use a graphing calculator to estimate the indicated value(s) to the nearest hundredth.*

a. *the x-intercept(s)*

b. *the x-values of points on the graph having positive y-coordinates*

c. *the x-values of points on the graph having negative y-coordinates*

47. $y = -\dfrac{1}{2}x^3 - 4$

48. $y = x^2 - 9$

49. $y = -x^2 - x + 6$

50. $y = x^3 - 4x$

Exercises 51–56 *Use a graphing calculator to estimate the indicated value(s) to the nearest hundredth.*

51. $y = \dfrac{x^3}{24} - \dfrac{x^2}{10} - \dfrac{22x}{5} + 12$; x- and y-intercepts

52. $y = \dfrac{x^4}{324} - \dfrac{17x^2}{36} + 15$; x- and y-intercepts

53. $y = \dfrac{1}{2}x^2 + 4x - 4$ coordinates of the lowest point on the graph

54. $y = \dfrac{12}{\sqrt{x^2 - 30x + 226}}$; coordinates of the highest point on the graph

55. $y = \dfrac{10}{x^2 - 16x + 64}$; x-coordinate(s) corresponding to $y = 2$

56. $y = -2x^2 - 16x - 20$; x-coordinate(s) corresponding to $y = 5$

Applications

57. Minimum Wage The U.S. Department of Labor has recorded the following data on the minimum hourly wage from 1955 to 2000.

Year	Wage
1955	0.75
1960	1.00
1965	1.25
1970	1.60
1975	2.10
1980	3.10
1985	3.35
1990	3.80
1995	4.25
2000	5.15

Use the year as the *x*-coordinate and the wage as the *y*-coordinate to plot the data on a rectangular coordinate system. On the basis of the data, what would you predict for the minimum wage in 1991? The actual value was $4.25. How far off is your estimate?

58. Life Expectancy In the graph in Figure 33, we see that the expected age at death increases with age. For example, a 20-year-old can expect to live to the age of 77.1, whereas a 60-year-old can expect to live to the age of 81.2. Use the data in Figure 33 to find data points (x, y), with *x* representing current age and *y* representing life expectancy—the expected number of years until death. Plot these points. Use your plot to estimate the life expectancy and the expected age at death of a 25-year-old and a 90-year-old. (*Data source:* U.S. Centers for Disease Control.)

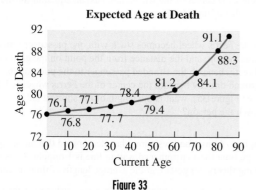

Figure 33

59. City Street Map A coordinate system is placed on a city street map, with the origin located at the intersection of Main Street and Center Avenue, as shown in Figure 34. All streets and avenues, except Lloyd Avenue, run north/south or east/west. What is the shortest route from point $A(-6, -2)$ to point $B(5, 4)$? Compute the total distance for this shortest route (using a standard city block as 1 unit), and compare it to the distance one would have to travel if Lloyd Avenue were closed for repairs.

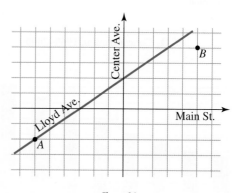

Figure 34

60. City Street Map Refer to the city street map in Figure 34. A bicycle messenger, currently at point $A(-6, -2)$, receives a message to pick up a package from an office building at the intersection of Main and Center and deliver it to a building at point $B(5, 4)$. Find the shortest route for her to take, and compute the total distance (using a standard city block as 1 unit).

61. U.S. Population The U.S. population for the years 1960–2000 can be approximated with the equation

$$y = 0.0008x^3 - 0.04x^2 + 2.8x + 180$$

where *y* represents the population (in millions) and *x* denotes the year (with $x = 0$ corresponding to 1960). Use an appropriate scale to plot the graph of the equation, and approximate the year when the population reached 260 million.

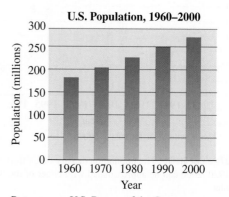

Data source: U.S. Bureau of the Census.

62. **CO_2 Concentration** The concentration of CO_2 in the atmosphere during the years 1960–2000 can be approximated with the equation

$$y = -0.00035x^3 + 0.033x^2 + 0.56x + 317$$

where y represents the concentration of CO_2 (measured in parts per million) and x denotes the year (with $x = 0$ corresponding to 1960). Use an appropriate scale to plot the graph of the equation, and approximate the year when the concentration first exceeded 380 parts per million.

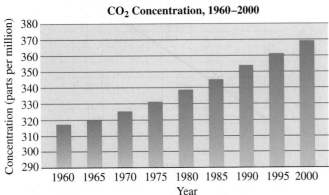

CO_2 Concentration, 1960–2000

Data source: The Carbon Dioxide Information Analysis Center, U.S. Department of Energy.

63. **AIDS Cases** The number of new AIDS cases reported during the years 1985–1999 can be approximated with the equation

$$y = 13x^4 - 440x^3 + 3700x^2 - 1000x + 15,000$$

where y represents the number of new cases and x denotes the year (with $x = 0$ corresponding to 1985). Use an appropriate scale to plot the graph of the equation, and approximate the year when the number of new cases was largest.

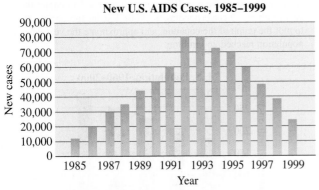

New U.S. AIDS Cases, 1985–1999

Data source: U.S. Centers for Disease Control.

64. **California Population Density** The population density of California for the years 1920–2000 can be approximated with either of the two equations

$$y = 0.0005x^3 - 0.04x^2 + 2.8x + 22$$

or

$$y = 0.00054x^3 + 0.072x^2 - 0.11x + 22$$

In both, y represents the density (people per square mile) and x the year (with $x = 0$ corresponding to 1920). For which years do these equations predict the same density?

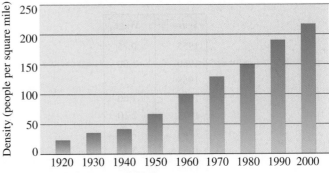

California Population Density, 1920–2000

Data source: U.S. Bureau of the Census.

65. **Ferris Wheel** An amusement park has a Ferris wheel with a radius of 30 feet. The center of the wheel is 35 feet above the ground (see Figure 35).

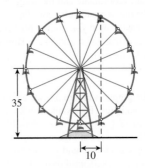

Figure 35

a. Set up a coordinate system with the origin on the ground directly below the center of the wheel, and find an equation for the circular edge of the wheel.

b. If the Ferris wheel becomes stuck with one passenger remaining near the top, and the distance from the point on the ground directly beneath the stranded passenger to the point on the ground directly beneath the center of the Ferris wheel is 10 feet (see Figure 35), what is the length of the shortest ladder that will reach the person?

66. **Waiting Time** A fast-food restaurant has determined that an acceptable average time for a customer to wait in line is 1 minute. A result from queuing theory suggests that the average length of time in minutes that a customer waits in line is given by $\dfrac{1}{x - y}$, where x is the number of customers served per hour and y is the number of customers who arrive per hour. Thus, the equation

$$1 = \frac{1}{x - y}$$

(or equivalently, $y = x - 1$) describes the relationship between x and y when the average waiting time is 1 minute.

a. Plot the graph of this equation and describe its shape.

b. During the busiest time of the day, customers arrive at a rate of 30 per hour. How many customers must be served per hour so that the time in line is still 1 minute? Illustrate this on your graph from part a.

Concepts and Critical Thinking

Exercises 67–70   *Answer true or false.*

67. The distance between two points is the sum of the differences of the x- and y-coordinates.

68. A point with coordinates $(x, -x)$ (with $x \neq 0$) must lie in either the second or fourth quadrant.

69. The x-intercepts of the graph of an equation can be found by setting $x = 0$.

70. The y-intercepts of the graph of an equation can be found by setting $y = 0$.

Exercises 71–74   *Give an example of each.*

71. A point lying in the third quadrant

72. The coordinates of a point in the fourth quadrant that is 2 units from the y-axis

73. Four points lying on the graph of $x^2 + y^2 = 16$

74. An equation in x and y whose graph has no x-intercept

75. A point (x, y) is 4 units from the point $(2, -3)$. Write an equation that expresses this fact.

76. How many coordinates are required to specify a position on Earth? How many coordinates must be specified in order to schedule a meeting?

77. Suppose $P(x_1, y_1)$ and $Q(x_2, y_2)$ are two points and let $(\bar{x}, \bar{y})$ be the point with coordinates

$$\bar{x} = \frac{x_1 + x_2}{2} \quad \text{and} \quad \bar{y} = \frac{y_1 + y_2}{2}$$

Show that the distance between P and $(\bar{x}, \bar{y})$ is the same as the distance between Q and $(\bar{x}, \bar{y})$ and that the distance is half the distance between P and Q, thus verifying that $(\bar{x}, \bar{y})$ is the midpoint of $\overline{PQ}$.

Questions for Discussion or Essay

78. Figures 36 and 37 illustrate the growth in newspaper circulation for a certain city over a 3-year period. Discuss how the choice of scale and the use of a broken vertical axis could lead to differences in interpretation.

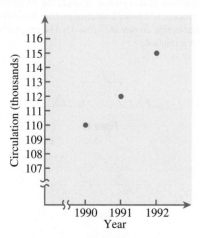

Figure 36

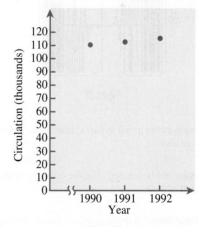

Figure 37

79. Why is it usually difficult for you to graph by hand an equation that *cannot* be solved for either x or y? Why do you suppose a graphing calculator cannot graph such equations either?

80. A television image of a person is, in a sense, a graph. Of course, as with all graphs, the appearance depends on the choice of scale. It is often said that television "adds 10 pounds" to a subject; explain this effect.

Projects for Enrichment

81. Hidden Intercepts

    a. Graph the equation $y = 10x^4 \sin\left(\dfrac{0.1}{x}\right)$ with a graphing calculator.

    (The expression "sin" denotes the trigonometric function sine, but you needn't have any knowledge of trigonometry to solve this problem. Just use the $\boxed{\text{SIN}}$ key on your calculator when entering the equation.) If you use range settings of Xmin $= -10$, Xmax $= 10$, Ymin $= -10$, and Ymax $= -10$, your graph should look like the one shown in Figure 38.

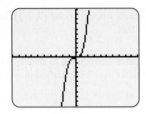

Figure 38

    b. Use the graph you found in part a to estimate the number of x-intercepts of the equation.

    c. Adjust the viewing window to obtain a plot similar to the one shown in Figure 39. List the values of the window variables.

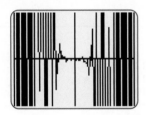

Figure 39

    d. On the basis of the graph in part c, how many x-intercepts do you think there are?

    e. What lesson can be learned from the disparity of your answers to parts b and d?

82. Relations and Fractal Curves In this section, we considered equations in the variables x and y. Such equations define relations between x and y. More formally, a **relation** is any set of ordered pairs (x, y). For example, the sets

$A = \{(9, -3), (4, -2), (1, -1), (0, 0), (1, 1), (4, 2), (9, 3)\}$

$B = \{(x, y) \mid y = x^2\}$

$C = \{(x, x^3) \mid x = 0, \pm 1, \pm 2, \ldots\}$

$D = \{(x, y) \mid x = 0, 1, 2, 3 \text{ and } y \geq x\}$

$E = \{(x, y) \mid x^2 + y^2 \leq 1\}$

are all relations. Thus, a relation can be a finite set of points, an infinite set of points defined by an equation, or a set of points arising from some other description. In any case, the graph of a relation R is the plot of the points in the set R.

    a. Plot the graphs of the relations A–E just given.

It may not always be possible to describe a relation with concise set notation, as we did with the relations A–E. Consider the set of points obtained by the following process. Start with a line segment such as that shown in Figure 40. Divide the line segment into thirds, as shown in Figure 41. Take out the middle third of the segment and replace it with two line segments that form two sides of an equilateral triangle, as in Figure 42.

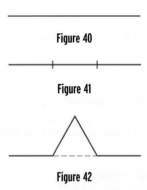

Figure 40

Figure 41

Figure 42

Notice that the total length of the new figure is $\frac{4}{3}$ times the length of original line segment. Now repeat this process on each of the four line segments. That is, take out the middle third of each line segment and replace it with two line segments, as shown in Figure 43. Another repetition of this process leads to the curve shown in Figure 44. If this process is repeated indefinitely, the result is a *fractal curve* similar to that shown in Figure 45. The set of points that form this curve is a relation.

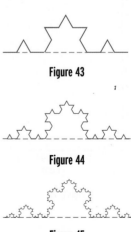

Figure 43

Figure 44

Figure 45

An interesting fact about the resulting fractal curve is that it has infinite length. This is because the total length of the curve at each stage is $\frac{4}{3}$ times the length at the preceding stage, and the process is repeated infinitely many times.

b. Assuming the length of the line segment in Figure 40 is 1 unit, show that the area of the triangle in Figure 42 is $\sqrt{3}/36$.

c. Compute the total area of the regions in Figures 43 and 44 bounded above by the solid lines and below by the dashed line.

d. Even though the total area at each successive stage increases, the area of the region bounded by the fractal curve and the dashed line is finite. In fact, the area is less than $\sqrt{3}/6$. Explain how we know this.

e. On the basis of what you've seen, explain how the area of an island could be finite even though the coastline has infinite length.

There is a sense in which a coastline, like that of this island in Lake Victoria, has infinite length.

Section 1.5 | Techniques for Solving Equations

- How is searching with a graphing calculator like exploring the skies with a view finder and a powerful telescope?
- If a fly begins flying back and forth between two trains that are traveling toward each other on parallel tracks, what distance will the fly have traveled when the trains meet?
- How large would a snowball be if it consisted of all the water from the Great Lakes?
- If there is a quadratic formula for solving second-degree polynomial equations, is there a cubic formula for solving third-degree equations?
- If a company of 3000 employees, 30% of whom are women, mandates that 60% of new hires be female, then how many employees must be added to achieve a 50-50 gender balance?

Solutions of Equations

A **solution** to an equation involving one variable is a value of the variable that makes the equation true. For example, the number 2 is a solution to $3x + 5 = 11$ since, if we replace x with 2, we obtain the true statement $3(2) + 5 = 11$. The collection of all solutions of an equation is called the **solution set**. Three of the most common algebraic techniques for solving—that is, finding the solution sets of—equations are as follows.

Algebraic Techniques for Solving Equations

1. Produce an **equivalent** but simpler equation by adding, subtracting, multiplying, or dividing the same number (the number must be nonzero when multiplying or dividing) on both sides of the equation. Note that two equations are said to be equivalent if they have the same solution set.
2. Apply the **zero-product property**: For real numbers a and b, $ab = 0$ if and only if $a = 0$ or $b = 0$.

(continued)

3. Apply the **quadratic formula**: If $ax^2 + bx + c = 0$ and $a \neq 0$, then

$$x = \frac{-b \pm \sqrt{b^2 - 4ac}}{2a}$$

The quantity $b^2 - 4ac$ is called the **discriminant**. If $b^2 - 4ac > 0$, there are two distinct real solutions; if $b^2 - 4ac = 0$, there is one real solution; and if $b^2 - 4ac < 0$, there are two nonreal, complex solutions.

A **linear equation** in the variable x is an equation that can be put in the form $ax + b = 0$. Such equations are among the most common and are easily solved by producing equivalent equations.

> **EXAMPLE 1**

Solving a Linear Equation

Solve $5x - 7 = \frac{1}{3}(4x + 1)$.

Solution

$$5x - 7 = \frac{1}{3}(4x + 1) \qquad \text{Original equation}$$

$$15x - 21 = 4x + 1 \qquad \text{Multiplying both sides by 3}$$

$$15x = 4x + 22 \qquad \text{Adding 21 to both sides}$$

$$11x = 22 \qquad \text{Subtracting } 4x \text{ on both sides}$$

$$x = 2 \qquad \text{Dividing both sides by 11}$$

The zero-product property is useful for solving equations of the form $\square = 0$, where $\square$ is a polynomial that factors easily.

> **EXAMPLE 2**

Applying the Zero-Product Property

Solve $4x^3 = 16x$.

Solution

$$4x^3 = 16x \qquad \text{Original equation}$$

$$4x^3 - 16x = 0 \qquad \text{Writing in standard form } \square = 0$$

$$4x(x^2 - 4) = 0 \qquad \text{Factoring out } 4x$$

$$4x(x + 2)(x - 2) = 0 \qquad \text{Factoring the difference of squares}$$

$$4x = 0 \quad \text{or} \quad x + 2 = 0 \quad \text{or} \quad x - 2 = 0 \qquad \text{Applying the zero-product property}$$

$$x = 0 \quad \text{or} \quad x = -2 \quad \text{or} \quad x = 2$$

WARNING!

Incorrect solutions to an equation can result from multiplying or dividing both sides of the equation by an expression involving the unknown variable. For example, if in Example 2 we were to begin by dividing both sides by $4x$, we would lose the solution $x = 0$.

All quadratic equations—that is, polynomial equations of degree 2—can be solved using the quadratic formula. This is especially useful for quadratics that don't factor easily.

EXAMPLE 3

Using the Quadratic Formula

Solve the given equation using the quadratic formula.

a. $2x^2 - 3x = 10$

b. $x^2 - 6x + 13 = 0$

Solution

a.
$$2x^2 - 3x = 10 \qquad \text{Original equation}$$
$$2x^2 - 3x - 10 = 0 \qquad \text{Writing in standard quadratic form}$$
$$x = \frac{-(-3) \pm \sqrt{(-3)^2 - 4(2)(-10)}}{2(2)} \qquad \text{Applying the quadratic formula with } a = 2, b = -3, \text{ and } c = -10$$
$$= \frac{3 \pm \sqrt{89}}{4} \qquad \text{Simplifying}$$
$$x \approx 3.108 \quad \text{or} \quad x \approx -1.608 \qquad \text{Approximating}$$

b. $x^2 - 6x + 13 = 0 \qquad \text{Original equation}$
$$x = \frac{-(-6) \pm \sqrt{(-6)^2 - 4(1)(13)}}{2(1)} \qquad \text{Applying the quadratic formula with } a = 1, b = -6, \text{ and } c = 13$$
$$= \frac{6 \pm \sqrt{-16}}{2} \qquad \text{Simplifying}$$
$$= \frac{6 \pm 4i}{2} \qquad \text{Using } \sqrt{-1} = i$$
$$= 3 \pm 2i$$

WARNING!

Note that in part a of Example 3 it would not be correct to simply factor $2x^2 - 3x = 10$ as $x(2x - 3) = 10$ and then claim that $x = 10$ or $2x - 3 = 10$. The factoring approach of the zero-product property can be applied only to equations of the form $\square = 0$.

Certain other types of equations can be solved using specialized techniques in combination with those already discussed. Note, however, that some of these specialized techniques produce extraneous solutions, and so it is essential to check the solutions that are obtained.

EXAMPLE 4

Clearing Denominators

Solve $\dfrac{2}{x} + \dfrac{1}{2} = \dfrac{7}{6}$.

Solution

$$\frac{2}{x} + \frac{1}{2} = \frac{7}{6} \qquad \text{Original equation}$$

$$6x \cdot \left(\frac{2}{x} + \frac{1}{2}\right) = 6x \cdot \frac{7}{6} \qquad \text{Multiplying through by the least common denominator, } 6x$$

$$6x \cdot \frac{2}{x} + 6x \cdot \frac{1}{2} = 7x \qquad \text{Distributing}$$

$$12 + 3x = 7x \qquad \text{Simplifying}$$
$$12 = 4x \qquad \text{Subtracting } 3x \text{ on both sides}$$
$$x = 3 \qquad \text{Dividing both sides by 4 and rearranging}$$

Checking in the original equation, we see that $x = 3$ is indeed a solution.

 EXAMPLE 5

Eliminating Radicals

Solve $\sqrt{4x + 17} - 2x = 1$.

Solution

$$\sqrt{4x + 17} - 2x = 1 \qquad \text{Original equation}$$
$$\sqrt{4x + 17} = 2x + 1 \qquad \text{Isolating the radical}$$
$$\left(\sqrt{4x + 17}\right)^2 = (2x + 1)^2 \qquad \text{Squaring both sides to eliminate the radical}$$
$$4x + 17 = 4x^2 + 4x + 1 \qquad \text{Expanding}$$
$$-4x^2 + 16 = 0 \qquad \text{Collecting all expressions on the left side}$$
$$x^2 - 4 = 0 \qquad \text{Dividing both sides by } -4$$
$$(x - 2)(x + 2) = 0 \qquad \text{Factoring}$$
$$x = 2, x = -2 \qquad \text{Applying the zero-product property}$$

Checking in the original equation, we see that $x = 2$ is a solution, but $x = -2$ is not. Thus, the only solution is $x = 2$.

Fermat's Last Theorem

In this text, we are primarily interested in real number solutions to equations. **Diophantine equations**, named after Diophantus of Alexandria, are equations for which only integer solutions are desired. For example, the equation $a^2 + b^2 = c^2$ has integer solutions consisting of Pythagorean triples, such as $a = 3$, $b = 4$, and $c = 5$. In fact, it is easily shown that $a^2 + b^2 = c^2$ has infinitely many integer solutions.

Sometime in the early 17th century, the amateur mathematician Pierre Fermat was reading a copy of *Arithmetic*, written by Diophantus himself, when he made what is undoubtedly the most significant marginal note of all time: "It is impossible to separate a cube into two cubes or a fourth power into two fourth powers or, in general, any power greater than the second into powers of like degree. I have discovered a truly marvelous demonstration, which this margin is too narrow to contain." Fermat was asserting that no equation of the form $a^3 + b^3 = c^3$, or $a^4 + b^4 = c^4$, or indeed $a^n + b^n = c^n$ for any $n > 2$ has integer solutions. In subsequent years, Fermat, in correspondence with the leading mathematicians of his time, proved that the equations $a^3 + b^3 = c^3$ and $a^4 + b^4 = c^4$ have no integer solutions, but his proof of the more general case referred

Andrew Wiles presenting his results.

to in his marginal note was never found. His claim was dubbed "Fermat's last theorem."

Over the course of the next three centuries, entire branches of mathematics arose from ill-fated attempts at proving Fermat's last theorem. In spite of the best efforts of some of the greatest intellects the world has seen, the theorem stubbornly refused to yield a proof. The best that could be done was to establish the theorem for particular cases. By 1990, it had been shown that Fermat's last theorem holds for all equations $a^n + b^n = c^n$, where $n < 4{,}000{,}000$.

In June 1993, Andrew Wiles of Princeton University shocked the mathematical community when he announced that he had proven Fermat's last theorem. A brilliant mathematician with a distinguished record of research, Wiles had worked in relative seclusion for some 7 years on proving a conjecture, the truth of which would imply Fermat's last theorem. Upon close examination of the 200-page paper in which this conjecture is established, referees found several gaps in his proof, all but one of which was easily corrected. Wiles and his colleague Richard Taylor successfully bridged the last gap in 1994, and the complete proof has been verified by many prominent mathematicians.

Applications   Mathematical techniques provide insight into problems that arise in many different areas: art, astronomy, chemistry, cinema, economics, finance, law, literature, music, religion, philosophy, physics, politics, sports, technology, and many others. In the following examples, we illustrate the use of equation-solving techniques.

···⋮EXAMPLE 6

Highway 70 Rendezvous

A man attending the Columbus campus of Ohio State University is arranging a date for a Friday afternoon in mid-April with his girlfriend, a student at Indiana State University in Terre Haute. To maximize their time together, he suggests that they meet somewhere on I-70, the highway connecting Columbus and Terre Haute, with the exact location to be arranged via cell phone en route. Since his last class gets out at 2:00 P.M. and hers at 3:00 P.M., he points out that if they each leave immediately afterward, he will be driving for an extra hour. His girlfriend, aware of the 1-hour time difference between Columbus and Terre Haute at that time of the year, realizes that he will actually be driving 2 hours longer, and so quickly agrees to the plan. If the campuses are 255 miles apart, and the average speeds of the woman and her boyfriend are 70 and 60 miles per hour, respectively, where and when will they meet?

Solution   The information given in the problem is depicted in Figure 46, where d_1 and d_2 represent the distances traveled by the woman and the man, respectively, at the point where they meet.

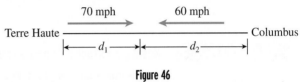

Figure 46

We assign the variable t for the travel time (in hours) of the woman. Since the man leaves 2 hours earlier, his travel time is $t + 2$ hours. Sometimes a table is a helpful way to organize all the information.

	Distance	Rate	Time
Woman (ISU student)	d_1	70	t
Man (OSU student)	d_2	60	$t + 2$

Since distance equals rate times time, we obtain the following equations:

$$d_1 = 70t$$
$$d_2 = 60(t + 2)$$

But we also know that Terre Haute and Columbus are 255 miles apart. This leads to the following equation:

Distance traveled by the woman + Distance traveled by the man = 255

$$d_1 + d_2 = 255$$

By substituting for d_1 and d_2, we obtain an equation that can be solved for t.

$$70t + 60(t + 2) = 255$$
$$130t + 120 = 255$$
$$130t = 135$$
$$t = \frac{135}{130} \approx 1.038 \text{ hours}$$

In 1.038 hours (just after 3:00 P.M. Terre Haute time), the woman will have traveled $70(1.038) \approx 72.7$ miles from Terre Haute. A quick look at a road atlas shows that a point 72.7 miles east of Terre Haute is somewhere on the west side of Indianapolis, near Indianapolis International Airport. As a check, we determine that the man travels for 3.038 hours, which yields a distance of $60(3.038) \approx 182.3$ miles from Columbus. The two distances total 255 miles, as required.

The steps we followed to arrive at the solution to Example 6 are summarized as follows.

Problem-Solving Strategy

1. Read the problem carefully. Make note of any information. Draw a sketch if possible.
2. Define a variable to represent the unknown quantity. Represent other quantities in terms of the variable. Label the sketch.
3. Verbalize an equation and then write it using the variable.
4. Solve the equation.
5. Check the answer.
6. Answer the stated problem completely.

> **EXAMPLE 7**

Meeting a Personnel Goal—A Mixture Problem

An international conglomerate currently has 3,000 employees, only 30% of whom are women. The company personnel director has dictated that, beginning immediately, 60% of the new hires are to be women until the company's target of 50% women is met. How many new employees must be hired for the company to meet its hiring goal? (Assume that no employees leave the company during the hiring period.)

Solution Let x be the total number of new employees to be hired. Since 30% of the current employees are women and 60% of the new hires are to be women, we have

Total number of women = Current number of women + Number of newly hired women

$$= 30\% \cdot 3000 + 60\% \cdot x$$

$$= 900 + 0.6x$$

The total number of employees is just the sum of the original number of employees and the number of new employees. Thus, we have

$$\text{Total number of employees} = 3000 + x$$

Now the percentage of women in the company is given by the formula

$$\text{Percentage of women} = \frac{\text{Total number of women}}{\text{Total number of employees}} \cdot 100\%$$

or in this case

$$50\% = \frac{900 + 0.6x}{3000 + x} \cdot 100\%$$

$$0.5 = \frac{900 + 0.6x}{3000 + x}$$

Solving for x, we have

$$0.5(3000 + x) = 900 + 0.6x$$
$$1500 + 0.5x = 900 + 0.6x$$
$$-0.1x = -600$$
$$x = \frac{-600}{-0.1}$$
$$x = 6000$$

Thus, 6000 employees must be hired for the company to achieve its desired gender balance.

Approximating Solutions Graphically

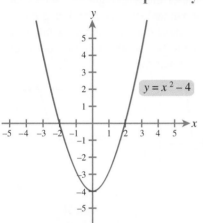

$y = x^2 - 4$

Figure 47

Many equations are extremely difficult—or even impossible—to solve by purely algebraic techniques. In such cases, we may still be able to *approximate* a solution with the aid of a graphing utility such as a graphing calculator or a computer. To do so, we must first understand the connection between graphs of equations and solutions.

Recall from Section 1.4 that the x-intercepts of the graph of an equation are the points where the graph intersects the x-axis. In other words, an x-intercept of a graph is a point on the graph with y-coordinate 0. Thus, the graph of $y = x^2 - 4$, which is shown in Figure 47, has exactly two x-intercepts, $(-2, 0)$ and $(2, 0)$.

We've seen how to find intercepts using the trace feature of the calculator. A second approach is to plot the graph and instruct the calculator to compute the x-intercepts. Regardless of which method we use, it is usually necessary to adjust the view of the graph shown by the calculator so that the intercepts are clearly visible. In the following Calculator Keys box, we explain some of the features of graphing calculators that are invaluable for adjusting the viewing window and for finding intercepts.

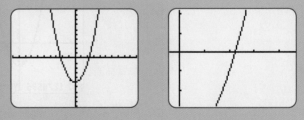

Calculator Keys

Adjusting the View

Zoom In This feature provides us with a close-up view of a region of interest. By zooming in on an intercept repeatedly, we can approximate it with great accuracy. Figures 48 and 49 show the graph of $y = x^2 - 5$ before and after zooming in on the positive x-intercept.

Figure 48	**Figure 49**

Zoom Out Zooming out is the opposite of zooming in. When we zoom out, we obtain a broader view of the graph; it is as if we are viewing the graph from a greater

(continued)

distance. For example, Figure 50 shows a view of the graph of $y = 0.05x^2 - 8$ in which no intercept is visible. After zooming out (Figure 51), we see two x-intercepts.

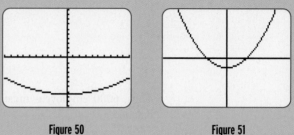

Figure 50   Figure 51

Zoom Box With the zoom-box feature, we can draw a box around a region of interest that then becomes our viewing window. Typically, the box is drawn by indicating the locations of diagonally opposite corners of the box. Figure 52 shows the graph of an equation that appears to have an x-intercept near $x = 4$. Figure 53 shows the zoom box that we have drawn around the region in which the x-intercept appears to lie. Figure 54 shows the viewing window that results from the zoom box of Figure 53. Note that there are actually three x-intercepts!

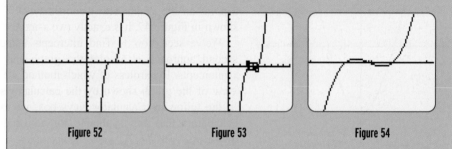

Figure 52   Figure 53   Figure 54

Trace The trace feature is especially useful for estimating intercepts. Figure 55 shows a graph of $y = x^2 - 5$ in which the trace feature has been enabled. The cursor has been moved just to the left of the x-intercept, and the x-coordinate of the cursor position is displayed as 2.1276596. In Figure 56, the cursor has been moved just to the right of the x-intercept, and the x-coordinate of the cursor is displayed as 2.3404255. Thus, the x-intercept lies between 2.1276596 and 2.3404255. For greater accuracy, we could zoom in on the intercept and repeat this process.

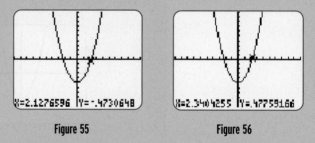

Figure 55   Figure 56

Calculate-Zero Many graphing calculators are capable of "automatically" calculating intercepts (also called zeros or roots). With most models, the intercepts must be found one at a time by plotting the graph with a view that clearly shows an intercept, issuing the calculate-zero command, and then specifying a **left bound** (a number to the left of the intercept), a **right bound** (a number to the right of the intercept), and an **initial guess** (a rough approximation for the intercept, one that is located between the two bounds). The bounds and initial guess can usually be given either by positioning the trace cursor or by entering a number. Figures 57–60 illustrate the process for estimating the negative intercept of $y = x^2 - 5$. In general, the accuracy of the approximation depends on both the equation and the calculator being used. The calculator used for this illustration "guarantees" an approximation that is within 10^{-5} of the actual value.

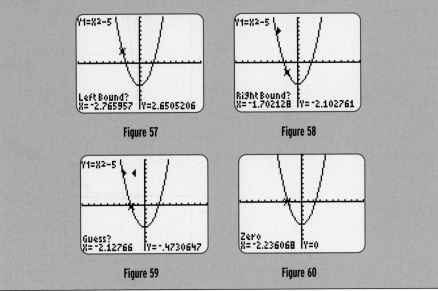

Figure 57   Figure 58

Figure 59   Figure 60

Since even the "automatic" calculation of intercepts with a graphing calculator requires an appropriate view of the graph, we place most of our emphasis in the following examples on adjusting the viewing window.

EXAMPLE 8

Finding Intercepts Using Zoom Boxes

Approximate the x-intercepts of the equation $y = 1000x^3 - 15x^2 + 0.0002$.

Solution   The plot of this equation, with a viewing window defined by Xmin $= -10$, Xmax $= 10$, Ymin $= -10$, and Ymax $= 10$, is given in Figure 61. Although it appears that the graph crosses the x-axis somewhere near the origin, it is impossible to tell *exactly* where the graph crosses or even how many times it crosses.

The sequence of plots shown in Figures 62–67 illustrates how zoom boxes can give us a better view of the graph near the x-intercepts. Notice that at each step we construct a zoom box around the region where the graph appears to cross the x-axis.

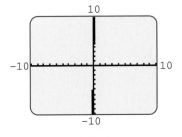

Figure 61

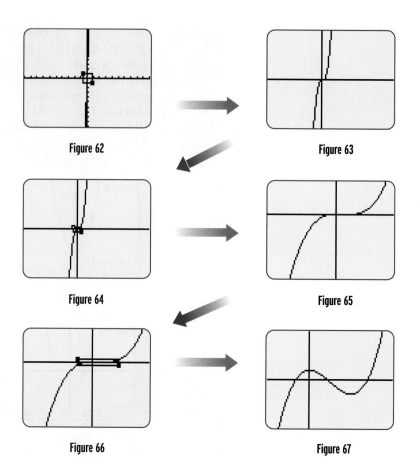

Figure 62

Figure 63

Figure 64

Figure 65

Figure 66

Figure 67

Now, using the trace feature, we can obtain not only approximations for the *x*-coordinates of the intercepts, but also information concerning the accuracy of these approximations. For example, in Figure 68, the cursor has been moved just to the left of the first positive *x*-intercept, and the *x*-coordinate of the cursor position is displayed as 0.00404255. In Figure 69, the cursor has been moved just to the right of that same *x*-intercept (note that the *y*-coordinate has changed from positive to negative), and the *x*-coordinate of the cursor is displayed as 0.0043617. Thus, the *x*-intercept lies between 0.00404255 and 0.0043617, and so we are certain that the value 0.004 is correct to the nearest thousandth.

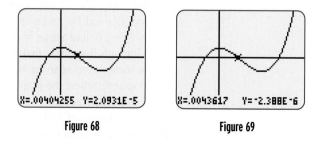

Figure 68

Figure 69

Repeating this process for the other two intercepts, we obtain -0.003 and 0.014, both of which are correct to the nearest thousandth. Thus, the three intercepts are -0.003, 0.004, and 0.014.

Note that if we had attempted to use the calculate-zero feature as soon as an intercept was clearly visible—after Figure 63, for example—then we would have missed two of the three x-intercepts. Certainly, once all three intercepts had become visible—at Figure 67, for example—we could have employed calculate-zero to quickly and accurately approximate their values.

In the previous example, we approximated the values of x such that the y-coordinate, $y = 1000x^3 - 15x^2 + 0.0002$, was equal to 0. Thus, we found approximate solutions to the equation

$$1000x^3 - 15x^2 + 0.0002 = 0$$

This suggests the following strategy for solving equations graphically.

Steps for Solving Equations Graphically

1. Algebraically rearrange the equation so that it is of the form $\square = 0$, where $\square$ is an expression involving the variable.
2. Graph the equation $y = \square$.
3. Find the x-intercepts of the graph. These are the solutions of the original equation. If there are no x-intercepts, then the equation has no real-valued solutions.

EXAMPLE 9

Solving an Equation Graphically

Solve $x^3 - 7x^2 = 14 - 17x$ graphically. Approximate to the nearest hundredth.

Solution   We begin by rewriting the equation so that it is of the form $\square = 0$.

$$x^3 - 7x^2 = 14 - 17x$$
$$x^3 - 7x^2 + 17x - 14 = 0 \qquad \text{Subtracting 14 and adding } 17x \text{ to both sides}$$

Next we use a graphing calculator to produce a graph of the equation $y = x^3 - 7x^2 + 17x - 14$. The graph in Figure 70 suggests that there is an x-intercept somewhere between 1 and 3. We zoom in on the apparent intercept by placing a zoom box around the region where the graph appears to cross the x-axis, as in Figure 71. The result is the viewing window shown in Figure 72.

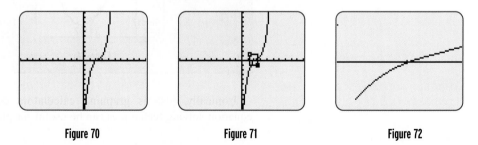

Figure 70   Figure 71   Figure 72

At this point, we are confident that there is only one intercept in the viewing window, and we use the automatic calculate-zero feature to obtain an approximation of 2. By zooming out repeatedly, we conclude that there are no other places where the graph crosses the x-axis. Thus, we have $x \approx 2$ as the only solution of the equation.

Figure 73

An alternative graphical technique can be used to solve equations like that of Example 9. If we graph $y = x^3 - 7x^2$ and $y = 14 - 17x$ on the same set of coordinate axes, then the point of intersection of the two graphs is a point (x, y) such that $y = x^3 - 7x^2$ *and* $y = 14 - 17x$. Thus, $x^3 - 7x^2 = 14 - 17x$. In other words, the x-coordinate of the point of intersection of the two graphs is a solution to the original equation. The coordinates of the point of intersection can be found either by using the zoom and trace features, as suggested by Figure 73, or by applying the calculate-intersect feature described in the following Calculator Keys box. Although it is generally easier to rearrange an equation so that it is of the form $\square = 0$ and then find its x-intercepts, as demonstrated in Example 9, there are occasions when it is more convenient to graph the left and right sides of the equation and then find any point(s) of intersection.

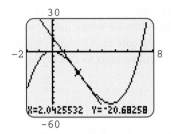

Calculator Keys

Finding Intersection Points

Many graphing calculators are capable of automatically calculating the coordinates of the intersection points of two curves. The process typically involves plotting the two curves with a view that shows an intersection point, selecting the **calculate-intersect** command from a menu, identifying the two curves on which the point lies, and finally specifying an initial guess for the point. Figures 74–77 illustrate this process for estimating the coordinates of the point of intersection of the curves $y = x^3 - 7x^2$ and $y = 14 - 17x$.

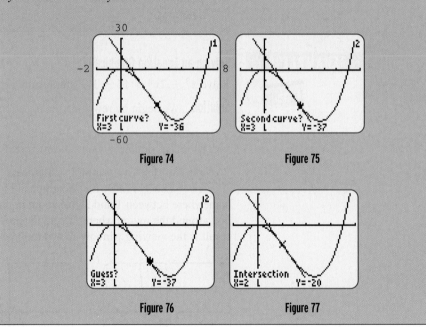

Figure 74   **Figure 75**

Figure 76   **Figure 77**

Ironically, just as graphing calculators can be useful in solving equations, equation-solving techniques can be useful for plotting graphs of certain equations.

EXAMPLE 10   **Using the Quadratic Equation to Plot a Parabola**

Solve the equation $x = y^2 + 3y + 1$ for y. Use a graphing calculator to plot the graph(s) of the resulting equation(s) and estimate the intercept(s) to the nearest hundredth.

Solution   We first rewrite the equation in standard quadratic form (in y) with 0 on the right-hand side.

$$x = y^2 + 3y + 1$$
$$y^2 + 3y + 1 - x = 0$$

Here $a = 1$, $b = 3$, and $c = 1 - x$. Substituting into the quadratic formula and simplifying, we obtain

$$y = \frac{-b \pm \sqrt{b^2 - 4ac}}{2a}$$

$$= \frac{-3 \pm \sqrt{3^2 - 4 \cdot 1 \cdot (1 - x)}}{2 \cdot 1}$$

$$= \frac{-3 \pm \sqrt{9 - 4 + 4x}}{2}$$

$$= \frac{-3 \pm \sqrt{5 + 4x}}{2}$$

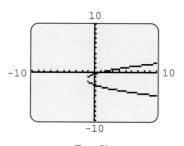

Figure 78

Now we enter and plot the two equations $y = \left(-3 + \sqrt{5 + 4x}\right)/2$ and $y = \left(-3 - \sqrt{5 + 4x}\right)/2$. The result is shown in Figure 78. The intercepts could be estimated from these graphs using the trace feature, but all are easily found algebraically. To find the x-intercept, we set $y = 0$ in the original equation. This yields $x = 1$. To find the y-intercepts, we set $x = 0$ as follows:

$$y = \frac{-3 \pm \sqrt{5 + 4x}}{2}$$

$$= \frac{-3 \pm \sqrt{5 + 4 \cdot 0}}{2}$$

$$= \frac{-3 \pm \sqrt{5}}{2}$$

Approximating, we have $y \approx -0.38$ and $y \approx -2.62$.

Understanding and Mastery Checklists

Concepts to Understand	Skills to Master
Solution set of an equation	Solve linear equations.
❖	❖
Equivalent equations	Solve polynomial equations by factoring.
❖	❖
Zero-product property	Solve quadratic equations with the quadratic formula.
❖	
Quadratic equations	❖
❖	

(continued)

The quadratic formula

◆

Strategies for applied problem solving

◆

Connection between x-intercepts and solutions

◆

Graphical approximations of solutions

◆

Zoom features of a graphing calculator

Define appropriate variables when setting up applied problems.

◆

Approximate solutions to equations graphically.

Exercises 1.5

Exercises 1–32 *Find the real solutions of the given equation. Give exact values.*

1. $3x - 4 = 2$

2. $-2x + 7 = 9$

3. $-2(m + 3) = 9$

4. $-3(n - 4) = 8$

5. $5(-2w + 9) = 3(w - 18) + 4$

6. $2z + 3(z - 6) = 5(z + 8)$

7. $(a + 3)(a - 4)(a + 6) = 0$

8. $(2y + 1)(y + 3)(5y - 2) = 0$

9. $(x - 2)(x^2 + 5x + 6) = 0$

10. $x(x - 2) = 3$

11. $s^2 = 3s$

12. $x^2 = 24$

13. $(y + 1)^2 = 3$

14. $x^3 - 27 = 0$

15. $u^4 - 16 = 0$

16. $6t^2 - t - 2 = 0$

17. $\dfrac{7}{y + 4} = \dfrac{3}{y - 4}$

18. $\dfrac{2}{2x + 1} - \dfrac{3}{4x + 1} = 0$

19. $\dfrac{s}{s^2 - 1} + \dfrac{2}{s + 1} = \dfrac{4}{s - 1}$

20. $\dfrac{x}{2x - 1} + \dfrac{3}{x} = \dfrac{2(x^2 + 1)}{2x^2 - x}$

21. $\dfrac{1}{x + 2} + \dfrac{3}{x} = \dfrac{2x + 2}{x(x + 2)}$

22. $\dfrac{x + 2}{x - 1} + \dfrac{3x + 3}{x + 1} = \dfrac{2x^2 + 4x + 2}{x^2 - 1}$

23. $\dfrac{7}{x + 5} - \dfrac{3}{x + 1} = -1$

24. $\dfrac{x}{x + 1} + \dfrac{3}{x - 1} = \dfrac{6}{x^2 - 1}$

25. $\sqrt{x} = 3$

26. $\sqrt{x} = -2$

27. $\sqrt{8x - 7} = 0$

28. $\sqrt{u + 1} = \sqrt{2u - 2}$

29. $\sqrt{4t + 1} = -\sqrt{4t + 1}$

30. $\sqrt{\dfrac{3}{r}} - \sqrt{\dfrac{r}{3}} = 0$

31. $\sqrt{x + 5} - \sqrt{x} = 1$

32. $\sqrt{x + 13} - x = 1$

Exercises 33-38   *Solve the given quadratic equation.*

33. $x^2 + 6x = -9$

34. $u^2 - 6u + 3 = 0$

35. $2x^2 + 4x + 1 = 0$

36. $x^2 + x = 1$

37. $w^2 - 4w + 5 = 0$

38. $3y^2 - 2y + 1 = 0$

Exercises 39-48   *Solve for the indicated variable.*

39. $E = mc^2$ for m

40. $P = 2w + 2l$ for l

41. $A = P(1 + r)$ for r

42. $F = G\dfrac{m_1 m_2}{r^2}$ for m_1

43. $d_1(t + 1) = d_2(t - 1)$ for t

44. $s = -16t^2 + v_0 t + s_0$ for v_0

45. $s = -16t^2 + v_0 t + s_0$ for t

46. $w^2 + pw + q = 0$ for w

47. $\dfrac{1}{R} = \dfrac{1}{R_1} + \dfrac{1}{R_2}$ for R_1

48. $\dfrac{l}{h} = \dfrac{s + d}{s}$ for s

Exercises 49-56   *An equation and the number of solutions are given. Use a graphing calculator to approximate the solution(s) to the nearest hundredth.*

49. $x^3 - 6x^2 + 14x - 11 = 0$ (1 solution)

50. $x^3 = 15x^2 + 88x + 100$ (3 solutions)

51. $\sqrt{3x + 7} = \sqrt{x + 2} + 1$ (2 solutions)

52. $(x - 10)^3 = \dfrac{x - 10}{25}$ (3 solutions)

53. $x^5 + x = 1$ (1 solution)

54. $x^3 - \dfrac{13}{3}x^2 - x + 15 = 0$ (2 solutions)

55. $x^4 - 143x^2 = 100$ (2 solutions)

56. $\sqrt{x} + \sqrt{x + 1} = x$ (1 solution)

Exercises 57-62   *Solve the given equation for y. Use a graphing calculator to plot the graph of the resulting equation(s) and estimate the x-intercept(s) to the nearest hundredth.*

57. $y^2 + 3x - 4 = 0$

58. $x^4 + 16y^4 = 9$

59. $4x^2 - 15 = 8x + 4y$

60. $x^2 y - x^2 - 3x + 4y = 0$

61. $x = 2y^2 + 4y + 5$

62. $4x^2 - y^2 + 8x + 6y = 6$

Applications

63. Exam Scores   Juan scored 85 and 88 on the first two exams of a course. He must have an average of 90 to receive an A. What must he score on the third exam to get an A for the course?

64. Bowling Scores   A world-record season bowling average of 251.0 was set by Ross Hung during the 1995–1996 season. (*Data source: The Guinness Book of World Records.*) If Ross were to bowl 230 and 240 in the first two games of a 3-game series, what must he bowl in the third game to average 251 for the 3-game series?

65. Highway Rendezvous   Two groups of students from the same school are traveling in different cars to Florida during spring break. The first travels at an average speed of 55 miles per hour. The second leaves 30 minutes after the first and travels at an average speed of 65 miles per hour. How long will it take the second car to catch up to the first?

66. Traveling Fly   Two trains are on parallel tracks headed toward each other, one at a rate of 6 miles per hour and the other at a rate of 4 miles per hour. At the instant the trains are 2 miles apart, a fly begins flying back and forth between the two trains at a rate of 10 feet per second. What total distance has the fly traveled by the time the trains meet?

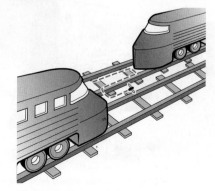

67. Falling Ball   A ball dropped from the top of the Sears Tower in Chicago has height $s = -16t^2 + 1454$ feet after t seconds. How long before the ball hits the ground?

68. Circumference of a Tree   A board foot is a unit of lumber measurement equal to 1 foot square by 1 inch thick. The number of board feet in a ponderosa pine can be approximated with the polynomial $B = 0.0015c^3$, where c is the circumference of the tree in inches at waist height. What is the circumference of a ponderosa pine with 2500 board feet?

69. Great Lakes Snowball   Approximate the radius of a snowball that contains enough moisture to fill the Great Lakes. The total volume of water in the Great Lakes is approximately 5440 cubic miles, and the amount of water in each cubic inch of loosely packed snow is approximately $\frac{1}{6}$ cubic inch.

70. Global Temperature   The average global temperature during the years 1950–2000 can be modeled by the equation

$$y = -0.00002x^3 + 0.0014x^2 - 0.0004x + 61.3$$

where y represents the average temperature (measured in degrees Fahrenheit) and x denotes the year (with $x = 0$ corresponding to 1950). Approximate the year in this time range in which the temperature reached 62.1°F.

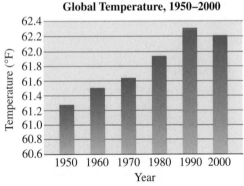

Global Temperature, 1950–2000

Data source: U.K. Meteorological Office.

71. Intel Stock Performance   The value of Intel stock from January 1996 through July 1998 can be approximated by

$$y = 21.7x^4 - 111x^3 + 162x^2 - 29.2x + 28.6$$

where x denotes the year ($x = 0$ corresponds to January 1996) and y is the value in dollars per share for that year. If the equation continues to hold, estimate the month in which Intel stock will be worth $100.

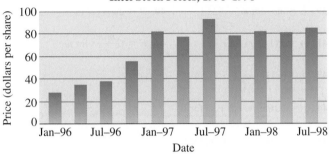

Intel Stock Prices, 1996–1998

Data source: CSI.

72. Dimensions of a Can   The height of a cylindrical soft drink can is 1 inch more than three times its radius. Its lateral surface area (the area of the "side" of the can, as opposed to the top or the bottom) is 30π square inches. Find the height and the radius of the can. (The lateral surface area L of a cylinder of radius r and height h is given by $L = 2\pi rh$.)

73. Vertical Jump   If a basketball player has a vertical jump of J inches, then it can be shown that her elevation in inches t seconds after jumping is given by $h = -192t^2 + 16\sqrt{3J}\,t$.

a. If a basketball player has a vertical jump of 24 inches, how long does it take her to reach an elevation of 1 foot?

b. Find the time required for a basketball player with a vertical jump of 48 inches to attain an elevation of 1 foot.

One consequence of global warming would be widespread floods, such as the one shown here in Bangladesh.

USC and 1996 Olympic basketball star Lisa Leslie dunking.

74. Profit   The price p (in dollars) of a product is related to the number of units sold, x, by the relationship $p = 11 - 0.01x$. Each unit costs $4.00 to produce. If x units are sold, the *cost* of production is the total amount of money it costs to produce x units. The *revenue* generated by the product is the total amount of money obtained from sales of x units. The *profit* is the revenue minus the cost.

   a. Explain why the cost to produce x units is $4x$, whereas the revenue from the sale of x units is $11x - 0.01x^2$.

   b. How many units must be sold to obtain a revenue of $1000?

   c. What is the profit when the revenue is $1000?

   d. How many units must be sold to generate a profit of $1000?

75. Shadow Length   Suppose that a woman h feet tall is standing d feet from an l-foot-tall lamppost (Figure 79). It can be shown that if the length of the woman's shadow is s, then $\dfrac{l}{h} = \dfrac{s + d}{s}$.

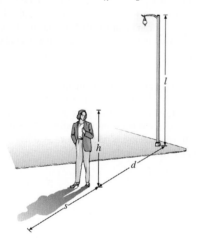

Figure 79

   a. Solve this equation for s.

   b. How long would the shadow of a woman be if the woman is 5 feet 6 inches and is standing 30 feet from a 40-foot-tall lamppost?

76. River Current   Suppose that the record-holding racing boat *Miss Freei*, capable of traveling 205 miles per hour in still water, travels a total of 40 miles upstream and back.

   a. If x denotes the speed of the current in miles per hour, find an expression for the time spent traveling upstream.

   b. Find an expression for the time traveling downstream.

   c. Find the speed of the current given that the total time was 24 minutes.

77. Roofing Estimate   A roofing contractor observes that his new employee, Brandon, works at half the rate of his foreman, Craig. Based on previous experience, the contractor knows that Craig can shingle a certain roof in 8 hours working alone. He concludes that if Brandon were allowed to work alone, he would take 16 hours to shingle the roof.

   a. To determine the time required to shingle a roof when Craig and Brandon work together, the contractor simply assumes an average of 12 hours per roof per man. How long would it take the two to complete one roof using this assumption?

   b. Find the actual time required to complete one roof, using the assumption that their rate working together is the sum of their rates working alone. (*Hint*: If x denotes the number of hours required when working together, then $1/x$ is their combined rate in roofs per hour.)

   c. By how much will the contractor overestimate the time required for Craig and Brandon to shingle 20 roofs working together?

78. Running Cable   Fiber-optic cable is to be run between two office buildings that are separated by a 20-yard roadway, as shown in Figure 80. The cable must start at point A, pass under the roadway, and then continue under the ground to point B. It costs $500 per yard to pass the cable under the roadway, and it costs $100 per yard to bury it under the ground the rest of the way to point B. If the total cost of laying the cable is $30,000 and if the distance from point B to the point directly opposite A is 100 yards, find the point where the cable reaches the other side of the road.

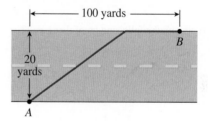

Figure 80

79. Average Speed   Compute the average speed of a trip consisting of two legs: driving 40 miles per hour for 50 miles and then cycling 20 miles per hour for another 50 miles. Why isn't the average speed 30 miles per hour?

80. Growth Rate   The rate of growth for a boy between the ages of 1 and 19 is approximated by

$$y = \frac{4}{x^2} - \frac{x}{2} + 9 + \frac{32}{3(x - 14)^2 + 4}$$

where y is measured in centimeters per year and x in years.

   a. Estimate the age at which the average rate of growth is 0 centimeters per year. What interpretation can be given to your answer?

   b. Estimate the two ages at which the average rate of growth is 10 centimeters per year.

   [*Data source:* Robert M. Malina and Claude Bouchard, *Growth, Maturation, and Physical Activity* (Champaign, IL: Human Kinetics, 1991), p. 52.]

Concepts and Critical Thinking

Exercises 81-84   *Answer true or false.*

81. The x-intercepts of an equation of the form $y = \square$ correspond to solutions of $\square = 0$.

82. If $x(x + 2) = 4$, then we can conclude that either $x = 4$ or $x + 2 = 4$.

83. If $x(x + 2) = 0$, then we can conclude that either $x = 0$ or $x + 2 = 0$.

84. Some quadratic equations cannot be solved by using the quadratic formula.

Exercises 85-88   *Give an example of each.*

85. A reason why "Let d = Dimes" would not be an appropriate variable definition

86. A 1000th-degree polynomial equation having exactly two real solutions

87. A quadratic equation having 3 as its only solution

88. A number that couldn't possibly be a solution of an equation of the following form, where $\square$ and $\triangle$ are expressions involving x:

$$\frac{\square}{x - 1} = \triangle$$

89. What's wrong with the following proof that $1 = 2$?

$$x = 0$$
$$x + x = 0 + x$$
$$2x = x$$
$$\frac{2x}{x} = \frac{x}{x}$$
$$2 = 1$$

90. A view of the graph of an equation is shown in Figure 81. Discuss the number of possible x-intercepts. In particular, is it possible for the graph to have two x-intercepts within the given viewing window? Could it have three x-intercepts? How about four or five?

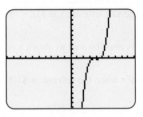

Figure 81

91. The polynomial equation $x^3 - 7x - 6 = 0$ factors as $(x - 3)(x + 1)(x + 2) = 0$ and hence has solutions $x = 3$, $x = -1$, and $x = -2$. Do you think this works in reverse? In other words, if you are given that $x = -1$, $x = 2$, and $x = 3$ are the solutions of $x^3 - 4x^2 + x + 6 = 0$, what can you conclude about how $x^3 - 4x^2 + x + 6$ can be factored? How do you think $2x^2 - 5x + 2$ should factor given that $2x^2 - 5x + 2 = 0$ has solutions $x = \frac{1}{2}$ and $x = 2$? Explain.

92. Use a graphing calculator to plot the graphs of $y = x^2$ and $y = x^4$. Then sketch $y = x^3$ and $y = x^5$. What do these graphs suggest about the shape of $y = x^n$? Use your results to explain why, if n is even, then $x^n = a$ has two solutions when $a > 0$ and no solutions when $a < 0$, whereas if n is odd, then $x^n = a$ has exactly one solution no matter what the value of a.

Questions for Discussion or Essay

93. Compare the process by which an intercept of the graph of an equation is located using a graphing calculator with the method that one would use to find a particular crater on the moon using a viewfinder (a small wide-angled telescope) and a powerful telescope. Explain why the powerful telescope alone would be useless.

94. In Example 6, we made the assumption that the rate is constant. What does your experience driving (or riding in) a car tell you about such an assumption? If the rate is not constant, what meaning does "rate" have in the equation Distance = Rate × Time?

95. Some hand-held calculators can instantly solve any quadratic equation. One could argue that this eliminates the need for students to learn the quadratic formula. What do you think? If inexpensive calculators can instantly solve quadratic equations, should students be forced to learn the quadratic formula? For that matter, why should any of us learn to add, subtract, multiply, or divide when $5 calculators can easily accomplish the same task?

96. Describe the most efficient method for solving the equation $ax^2 + bx + c = 0$ in each of the following cases:

   a. $a = 0$ **b.** $b = 0$ **c.** $c = 0$

97. In Example 10, we sketched the graph of $x = y^2 + 3y + 1$ by first solving the equation for y, obtaining $y = -\frac{3}{2} + \frac{1}{2}\sqrt{5 + 4x}$ and $y = -\frac{3}{2} - \frac{1}{2}\sqrt{5 + 4x}$, and then plotting both of these branches in the same viewing window. Suppose that we had instead begun by graphing $y = x^2 + 3x + 1$, the equation obtained by exchanging the roles of x and y. How is this graph related to that of $x = y^2 + 3y + 1$? In particular, how could we use the graph of $y = x^2 + 3x + 1$ as an aid in graphing $x = y^2 + 3y + 1$? What connection is there between the x- and y-intercepts of the two graphs?

Projects for Enrichment

98. The Effects of Shape on Surface Area

a. Human beings have a density approximately that of water, which weighs about 62.4 pounds per cubic foot. Estimate the volume of a 200-pound man.

b. Estimate the radius of a ball that has the volume of a 200-pound man, and compute the surface area of such a ball.

c. For each of the following heights, estimate the radius and surface area of the cylinder such that the resulting volume is that of a 200-pound man.

   i. $h = 3$ feet   **ii.** $h = 4$ feet

   iii. $h = 5$ feet   **iv.** $h = 6$ feet

   v. $h = 7$ feet

d. Based on your results from parts b and c, what type of 200-pound man would most likely have the greatest surface area: a very short obese man, a muscular 6-footer, or a slender 7-footer?

e. Given that heat is dissipated through the skin, comment on the ideal bodily proportions for cold versus warm climates. What other factors besides surface area might affect the ability of the body to dissipate or retain heat?

The short, stout build of the Inuit helps conserve body heat in the Arctic. Members of the Dinka tribe tend to be long-limbed and slender, an adaptation to the searing Sudanese heat.

99. Cubic Formulas   A general **cubic equation** is of the form $ax^3 + bx^2 + cx + d = 0$ and has three complex solutions, either one or three of which are real. Just as the quadratic formula gives the solution to a quadratic equation in terms of its coefficients, the **cubic formula** gives the solution to a cubic equation in terms of its coefficients. However, the cubic formula is considerably more difficult to use than the quadratic formula. Before the equation can be solved, it first must be placed in *normal form* and then in *reduced form*. Even after the equation is in *reduced form*, certain complica-

tions can arise (taking square roots of imaginary numbers, for example). In this project, we deal only with "nice" examples in which there are no complications. Furthermore, we find only one of the three solutions.

I. Finding the Normal and Reduced Forms

An equation of the form $x^3 + rx^2 + sx + t = 0$ is said to be in **normal form**. Note that the coefficient of the cube term is 1. A cubic equation can be written in normal form by dividing through by the coefficient of the cube term.

a. Express $5x^3 + 4x^2 - 10x + 9 = 0$ in normal form.

b. Express $-x^3 - 3x^2 - 4x - 7 = 0$ in normal form.

A cubic equation of the form $y^3 + py + q = 0$ is said to be in **reduced form**. Note that the coefficient of the cube term is 1 and there is no square term. To transform the normal equation $x^3 + rx^2 + sx + t = 0$ into reduced form, we make the substitution $x = y - \frac{r}{3}$.

c. Express $x^3 - 3x^2 + x - 6 = 0$ in reduced form.

d. Express $2x^3 - 6x^2 + 4x + 10 = 0$ in reduced form.

II. The Cubic Formula

A real solution to the equation $y^3 + py + q = 0$ (provided that $\frac{q^2}{4} + \frac{p^3}{27} > 0$), is given by

$$y = \sqrt[3]{-\frac{q}{2} + \sqrt{\frac{q^2}{4} + \frac{p^3}{27}}} + \sqrt[3]{-\frac{q}{2} - \sqrt{\frac{q^2}{4} + \frac{p^3}{27}}}$$

e. Find a solution to the equation $y^3 + 2y - 1 = 0$ using the cubic formula. Evaluate your answer using a calculator.

f. Find a solution of the equation $3x^3 - 6x^2 + 3x - 6 = 0$ by first writing it in normal form and then in reduced form. Solve the resulting equation for y and then find x. Evaluate your answer using a calculator and check it in the original equation.

100. Investigating Equations of Motion

a. Drop a coin from each of the following heights and measure the time required for the coin to strike the ground. For each height, perform four trials, and then compute the average time.

   i. 4 feet   **ii.** 6 feet

   iii. 8 feet   **iv.** 10 feet

If we assume that the only force acting on a projectile is a constant gravitational force, then, when the projectile is released from an initial height of h with an initial upward velocity of v_0, the height after t seconds is given by $y = -\frac{1}{2}gt^2 + v_0t + h$, where the constant g represents the acceleration due to gravity. If we use feet for our unit of distance, seconds for our unit of time, and feet per second as our unit of velocity, then $g \approx 32$ ft/sec^2.

b. For each of the initial heights given in part a, use the equation of motion to compute the time required for the coin to drop.

c. Explain any discrepancies between your answers from part a and your answers from part b.

d. Since we are assuming that the gravitational force is constant, an object tossed into the air will slow down by 32 ft/sec each second. Thus, if v represents the velocity of the object in feet per second, then $v = v_0 - 32t$. Compute the velocity upon impact of each of the coins from part a.

e. Toss a coin into the air and observe how its velocity changes. At what point does the velocity appear to be zero?

f. Find a formula for the maximum height of an object in terms of its initial height and initial velocity. (*Hint*: Begin by using the result of part e and the formula from part d to find an expression for the time at which the maximum height occurs.)

g. Now use the result of part f to find a formula for the initial velocity in terms of the initial height and maximum height.

h. What is the initial upward velocity of an outfielder who leaps 3 feet into the air to prevent a home run?

i. How fast must a ball be thrown if it is to rise 1 mile in the air?

j. Throughout this project, we have assumed that gravity was the only force acting on the projectile and that the gravitational force was constant. What other forces might be relevant, and under what circumstances would the gravitational force vary?

Section 1.6   Inequalities

- How many years would it take a half-ton man to reach a weight of 200 pounds on a standard weight-loss program?
- How much variation in measured distance will improperly inflated car tires cause?
- How can distance be estimated with a bicycle tire?
- If a missile is shot straight up at 4000 feet per second, for how long will it be out of the stratosphere?

Linear Inequalities

In Section 1.5, we solved linear equations—those of the form $ax + b = 0$—by using arithmetic operations to produce successively simpler but equivalent equations. The procedure for solving linear inequalities is virtually identical. The only difference, although a significant one, occurs when both sides of the inequality are multiplied or divided by a *negative* number. In this case, the direction of the inequality must be reversed. This is perhaps easiest to see through an example. If we begin with the true statement $1 < 2$ and multiply both sides by -1 without reversing the inequality, we would obtain $-1 < -2$, which is false. However, if we reverse the inequality, we obtain the true statement $-1 > -2$. We summarize the technique as follows.

Solving Linear Inequalities

A linear inequality can be transformed into an equivalent inequality by performing any combination of the following steps:

1. Adding or subtracting the same expression on both sides of the inequality
2. Multiplying or dividing by the same **positive** number on both sides of the inequality
3. Multiplying or dividing by the same **negative** number on both sides and **reversing the inequality**

Because it is generally not clear whether a variable represents a positive or negative value, it is best to avoid multiplying or dividing by an expression containing the variable.

EXAMPLE 1

Solving a Linear Inequality

Solve $4(t - 1) \geq t$ and graph its solution set.

Solution

$$4(t - 1) \geq t$$
$$4t - 4 \geq t \qquad \text{Distributing}$$
$$4t \geq t + 4 \qquad \text{Adding 4 to both sides}$$
$$3t \geq 4 \qquad \text{Subtracting } t \text{ from both sides}$$
$$t \geq \frac{4}{3} \qquad \text{Dividing both sides by 3}$$

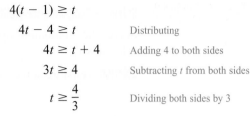

Figure 82

So the solution set is $\left[\frac{4}{3}, \infty\right)$, and the graph is shown in Figure 82.

EXAMPLE 2

Solving a Linear Inequality

Solve $3t - 7 < 6t + 8$ and graph its solution set.

Solution

$$3t - 7 < 6t + 8$$
$$3t < 6t + 15 \qquad \text{Adding 7 to both sides}$$
$$-3t < 15 \qquad \text{Subtracting } 6t \text{ from both sides}$$
$$t > -5 \qquad \text{Dividing both sides by } -3 \text{ and reversing the inequality}$$

Figure 83

So the solution set is $(-5, \infty)$, and the graph is shown in Figure 83.

Compound Inequalities

A compound inequality is formed by joining two linear inequalities with the words *and* or *or*. A simple example of a compound inequality is the statement

$$x > -3 \quad \text{and} \quad x < 2$$

Figure 84

The solution set for this statement is the **intersection** (the portion common to both) of the intervals $(-3, \infty)$ and $(-\infty, 2)$—namely, the interval $(-3, 2)$. The graph of this solution set is shown in red in Figure 84. To solve more complicated compound inequalities using the word *and*, we simply solve each inequality separately and take the intersection of the solution sets.

EXAMPLE 3

Solving a Compound Inequality With *and*

Solve the compound inequality $6u - 8 \leq 16$ *and* $1 - 4u \leq -3$, and graph the solution set.

Solution   We first solve each inequality separately.

$$6u - 8 \leq 16 \quad \text{and} \quad 1 - 4u \leq -3$$
$$6u \leq 24 \quad \text{and} \quad -4u \leq -4$$
$$u \leq 4 \quad \text{and} \quad u \geq 1$$

Figure 85

By intersecting the two sets $(-\infty, 4]$ and $[1, \infty)$, we obtain the solution set $[1, 4]$. The graph of this set is shown in red in Figure 85.

Compound inequalities formed with the word *or* are solved in a similar way, except instead of intersecting the two sets, we combine them; that is, we form their **union**. The union of two sets A and B is denoted $A \cup B$.

⋯⋯⟩EXAMPLE 4

Solving a Compound Inequality With *or*

Solve the compound inequality $\frac{1}{5}y + 4 < 3$　*or*　$-5y - 3 \leq y$, and graph the solution set.

Solution　First we solve each part separately.

$$\frac{1}{5}y + 4 < 3 \quad \text{or} \quad -5y - 3 \leq y$$

$$\frac{1}{5}y < -1 \quad \text{or} \quad -5y \leq y + 3$$

$$y < -5 \quad \text{or} \quad -6y \leq 3$$

$$y < -5 \quad \text{or} \quad y \geq -\frac{1}{2}$$

Figure 86

The solution is formed by taking the union of these two solutions sets to obtain $(-\infty, -5) \cup \left[-\frac{1}{2}, \infty\right)$, as shown in the graph in Figure 86.

Rule of Thumb

The solution of a compound inequality formed with the word *and* is found by solving each part separately and then taking the intersection of the two sets. The solution of a compound inequality formed with the word *or* is found by solving each part separately and then taking the union of the two sets.

Some compound inequalities are disguised as **double inequalities**, such as $-4 \leq 3x - 2 < 7$. This statement is equivalent to $-4 \leq 3x - 2$ *and* $3x - 2 < 7$, although it is more convenient to leave it as a double inequality so both inequalities can be solved together, as shown in Example 5.

⋯⋯⟩EXAMPLE 5

Solving a Double Inequality

Solve the double inequality $-4 \leq 3x - 2 < 7$.

Solution　We must isolate x as the middle term.

$$-4 \leq 3x - 2 < 7$$
$$-2 \leq 3x < 9 \qquad \text{Adding 2 to all "sides"}$$
$$-\frac{2}{3} \leq x < 3 \qquad \text{Dividing all "sides" by 3}$$

Figure 87

The solution set is $\left[-\frac{2}{3}, 3\right)$ as shown in the graph in Figure 87.

Compound inequalities can often be used to solve application problems involving a range of solution values. You can spot such problems by the presence of such phrases as "at least," "at most," "no more than," "no less than," and so forth.

----->EXAMPLE 6

Weight Loss

A company is marketing a diet and exercise program that it claims will result in a weight loss of at least 1.5 pounds per week. For health reasons, they also warn that one should not lose more than 5 pounds per week. If Robert Earl Hughes (1926–1958) had started this program after he reached his world record weight of 1069 pounds, over what range of weeks would his weight have dropped to 203 pounds (his weight at age 6)?

Solution   For each week, Robert would lose at least $\frac{3}{2}$ pounds and at most 5 pounds, and so we have

Weight loss $\geq \frac{3}{2} \cdot$ (Numbers of weeks) *and* Weight loss $\leq 5 \cdot$ (Numbers of weeks)

Let x represent the number of weeks that Robert remained on the program. In dropping from 1069 to 203, he would have a weight loss of 866 pounds in x weeks. Thus, we have

$$866 \geq \frac{3}{2}x \quad \text{and} \quad 866 \leq 5x$$

Solving for x, we obtain

$$\frac{2}{3} \cdot 866 \geq x \quad \text{and} \quad \frac{866}{5} \leq x$$

$$577\frac{1}{3} \geq x \quad \text{and} \quad 173\frac{1}{5} \leq x$$

Robert Hughes, shown just after he passed the 1000-pound mark.

If the company's claims are correct, it will take anywhere from $173\frac{1}{5}$ to $577\frac{1}{3}$ weeks for Robert to reach 203 pounds.

Absolute Value Equations and Inequalities

Equations involving absolute value, such as $|x| = 3$, can be solved by considering distances. According to the definition of absolute value, the equation $|x| = 3$ is equivalent to the statement "the distance between x and 0 is 3." Since there are two values for x that are a distance 3 from 0 (namely, 3 and -3), the solution set for the equation $|x| = 3$ is $\{3, -3\}$. These observations generalize to the following rule for solving absolute value equations.

Solving Absolute Value Equations

If $c > 0$, then $|u| = c$ if and only if $u = -c$ or $u = c$.
If $c < 0$, then $|u| = c$ has no solution since the absolute value of a number can never be negative.

----->EXAMPLE 7

Solving an Absolute Value Equation

Solve $|2t| = 10$.

Solution   Applying the rules just given, we obtain two equations to solve separately.

$$|2t| = 10$$
$$2t = -10 \quad \text{or} \quad 2t = 10$$
$$t = -5 \quad \text{or} \quad t = 5$$

Check

	$t = -5$			$t = 5$	
$\|2t\|$	10		$\|2t\|$	10	
$\|2(-5)\|$			$\|2(5)\|$		
$\|-10\|$			$\|10\|$		
10		✓	10		✓

Inequalities involving absolute value can also be solved by considering distance, and the solutions can be expressed using compound inequalities, as we demonstrate in Table 7.

Table 7

Original inequality	Distance interpretation	Compound inequality	Interval and graph
$\|x\| < 3$	Numbers whose distance from 0 is less than 3	$-3 < x < 3$	$(-3, 3)$ -5-4-3-2-1 0 1 2 3 4 5
$\|x\| > 3$	Numbers whose distance from 0 is greater than 3	$x < -3$ or $x > 3$	$(-\infty, -3) \cup (3, \infty)$ -5-4-3-2-1 0 1 2 3 4 5

The inequalities in Table 7 lead us to the following general rules for dealing with absolute value inequalities.

Solving Absolute Value Inequalities

If $c > 0$, then

Property 1. $\|u\| < c$ if and only if $-c < u < c$
Property 2. $\|u\| > c$ if and only if $u < -c$ or $u > c$

These properties are also valid if $<$ is replaced by $\leq$ and $>$ is replaced by $\geq$.

EXAMPLE 8

Solving an Absolute Value Inequality

Solve $|2x + 1| < 3$.

Solution We first apply Property 1 and write $|2x + 1| < 3$ as the compound inequality $-3 < 2x + 1 < 3$. Next we solve for x.

$$-3 < 2x + 1 < 3$$
$$-4 < 2x < 2$$
$$-2 < x < 1$$

The solution is the interval $(-2, 1)$.

EXAMPLE 9

Solving an Absolute Value Inequality

Solve $|4 - y| \leq -\frac{1}{2}$.

Solution The rules for solving absolute value inequalities do not apply to $|4 - y| \leq -\frac{1}{2}$ since the number on the right-hand side is negative. This inequality is a contradiction since absolute values are never negative, so the solution is the empty set.

> **EXAMPLE 10**

Solving an Absolute Value Inequality

Solve $|1 - x| - 5 \geq -2$.

Solution The inequality must be in the form $|u| \geq c$ before we apply Property 2.

$$|1 - x| - 5 \geq -2$$
$$|1 - x| \geq 3$$

According to Property 2, this is equivalent to the compound inequality $1 - x \leq -3$ *or* $1 - x \geq 3$. Solving each inequality separately, we obtain

$$1 - x \leq -3 \quad \text{or} \quad 1 - x \geq 3$$
$$-x \leq -4 \quad \text{or} \quad -x \geq 2$$
$$x \geq 4 \quad \text{or} \quad x \leq -2$$

We form the union of these two intervals to obtain the solution $(-\infty, -2] \cup [4, \infty)$.

Absolute value inequalities are useful for expressing measurement errors. For example, if the length l of a metal pin is to be 4 centimeters, with an error of no more than 0.01 centimeter, then we can write $|l - 4| \leq 0.01$ to indicate that l and 4 may not be more than 0.01 centimeter apart. By solving the inequality, we obtain the allowable range of l values.

$$|l - 4| \leq 0.01$$
$$-0.01 \leq l - 4 \leq 0.01 \qquad \text{Property 1}$$
$$3.99 \leq l \leq 4.01 \qquad \text{Adding 4}$$

The answer could also have been expressed as 4 ± 0.01 centimeter and could have been obtained without even considering absolute values. In the following example, however, the solution is not as obvious and would be difficult to obtain without using absolute values.

> **EXAMPLE 11**

Bicycle Tires and Measurement Error

The circumference of a bicycle tire is to be 85 inches, with an error of no more than 0.3 inch. Find the acceptable range of values for the radius of the tire.

Solution The circumference of a circle in terms of its radius is $C = 2\pi r$. Thus, if the circumference is to be 85 inches with a maximum error of ± 0.3 inch, we write $|2\pi r - 85| \leq 0.3$ and proceed as follows:

$$|2\pi r - 85| \leq 0.3$$
$$-0.3 \leq 2\pi r - 85 \leq 0.3 \qquad \text{Property 1}$$
$$84.7 \leq 2\pi r \leq 85.3 \qquad \text{Adding 85}$$
$$\frac{84.7}{2\pi} \leq r \leq \frac{85.3}{2\pi} \qquad \text{Dividing by } 2\pi$$
$$13.48 \leq r \leq 13.58 \qquad \text{Approximating}$$

So the radius r may be between 13.48 and 13.58 inches.

Polynomial Inequalities Consider the polynomial inequality $x^2 - x - 2 < 0$. If we let $y = x^2 - x - 2$, then we are interested in the x-values for which $y < 0$. Thus, the solutions of the inequality $x^2 - x - 2 < 0$ correspond to the x-values for which the graph of $y = x^2 - x - 2$ lies

Figure 88

below the x-axis. Figure 88 shows the graph of $y = x^2 - x - 2$. Notice that the portion of the graph shown in red below the x-axis is between the two x-intercepts, which appear to be -1 and 2. Now, to be precise, we should find exact values for the x-intercepts. We do this by solving the equation $x^2 - x - 2 = 0$.

$$x^2 - x - 2 = 0$$
$$(x - 2)(x + 1) = 0$$
$$x = 2, x = -1$$

Thus, the graph of $y = x^2 - x - 2$ is below the x-axis for x-values between -1 and 2, and so the solution set of $x^2 - x - 2 < 0$ is the interval $(-1, 2)$.

More generally, if $p(x)$ denotes a polynomial expression in x, then the solution set to a polynomial inequality of the form $p(x) > 0$ (or $p(x) < 0$, $p(x) \leq 0$, or $p(x) \geq 0$) consists of intervals, the endpoints of which are solutions of $p(x) = 0$. This suggests the following strategy for solving polynomial inequalities.

Strategy for Solving Polynomial Inequalities

1. Write the inequality in **standard form** with 0 on the right-hand side and a polynomial $p(x)$ on the left-hand side (that is, $p(x) > 0$; $p(x) < 0$; $p(x) \geq 0$; or $p(x) \leq 0$).
2. Solve the equation $p(x) = 0$. If the solutions cannot be found algebraically, use a graphing calculator to estimate the x-intercepts of the graph of $y = p(x)$.
3. Determine the x-values for which the graph of $y = p(x)$ is above or below the x-axis (depending on *which* inequality symbol is being used—above for $>$ and below for $<$). Express the result using interval notation. If the inequality symbol is $\geq$ or $\leq$, the intervals will include their endpoints; otherwise not.

EXAMPLE 12

Solving a Polynomial Inequality

Solve $-x^3 - x^2 \geq -6x$.

Solution We begin by writing the inequality in standard form.

$$-x^3 - x^2 + 6x \geq 0$$

Next we find the solutions of $-x^3 - x^2 + 6x = 0$.

$$-x^3 - x^2 + 6x = 0$$
$$-x(x^2 + x - 6) = 0$$
$$x(x + 3)(x - 2) = 0$$
$$x = 0, x = -3, x = 2$$

Then we use a graphing calculator to produce a graph of $y = -x^3 - x^2 + 6x$, as shown in Figure 89. Since the inequality symbol is $\geq$, we are interested in the x-values of the portion of the graph lying *on or above* the x-axis, shown in red in Figure 89. The graph is on or above the x-axis to the left of -3 and between 0 and 2. Thus, our solution is $(-\infty, -3] \cup [0, 2]$. Note that we include the endpoints since the inequality symbol is $\geq$ and not $>$.

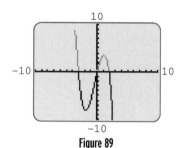

Figure 89

We recall that an alternative method for solving equations graphically is to graph the left and right sides of the equation on the same set of coordinate axes and find the x-coordinates of the points of intersection. Similarly, the inequality $-x^3 - x^2 \geq -6x$ from Example 12 can be solved by graphing $y = -x^3 - x^2$ and $y = -6x$ on the same set of coordinate axes, as in Figure 90, and identifying those intervals for which the graph of $y = -x^3 - x^2$ is on or above the graph of $y = -6x$.

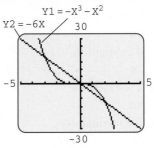

Figure 90

Now, in general, we can approximate the x-coordinates of the points of intersection of the graphs, but in this instance we know from our work in Example 12 that the graphs cross exactly at $x = -3$, $x = 0$, and $x = 2$. (Why?) Since the graph of $y = -x^3 - x^2$ is on or above the graph of $y = -6x$ to the left of 3 and between 0 and 2 (including -3, 0, and 2), the solution set is $(-\infty, -3] \cup [0, 2]$.

EXAMPLE 13

Solving a Polynomial Inequality

Solve $x^5 + 3x^4 + 3x^3 + 5x^2 - 7 < 0$.

Solution We must first solve the equation $x^5 + 3x^4 + 3x^3 + 5x^2 - 7 = 0$. In this case, a factorization of the left side is not apparent, and so we are unable to solve the equation algebraically. Instead, we use the graph of $y = x^5 + 3x^4 + 3x^3 + 5x^2 - 7$, shown in Figure 91, to *estimate* the x-intercepts. By using the calculate-zero feature or by zooming in on each of the x-intercepts and then using the trace feature, we produce the following approximations of the x-intercepts (accurate to the nearest hundredth): -2.41, -1.33, and 0.83. Since the inequality symbol is $<$, we are interested in the x-values of the portion of the graph lying *below* the x-axis, which is red in Figure 91. The graph is below the x-axis to the left of -2.41 and between -1.33 and 0.83. Thus, our solution is $(-\infty, -2.41) \cup (-1.33, 0.83)$.

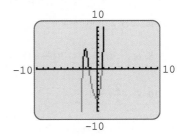

Figure 91

WARNING!

When finding x-intercepts of unfamiliar polynomials with a graphing calculator, be careful not to omit any x-intercepts that are out of the viewing window. In Example 13, it would have been prudent to have zoomed out several times to make sure there were no other x-intercepts.

Understanding and Mastery Checklists

Concepts to Understand

Linear inequalities

❧

Compound inequalities

❧

Union and intersection

❧

Absolute value inequalities

❧

Polynomial inequalities

Skills to Master

Solve linear inequalities.

❧

Solve compound inequalities.

❧

Find the union or intersection of two sets
of numbers.

❧

Express answers to compound inequalities as
an interval or as the union of intervals.

❧

Solve absolute value inequalities, and express
the solution using interval notation.

❧

Solve polynomial inequalities graphically.

Exercises 1.6

Exercises 1-10 *Solve the inequality. Write your answer using interval notation and graph the solution.*

1. $x - 8 < 5$

2. $t + 3 \geq -1$

3. $23 - 4v < 27$

4. $5x - 3 \leq 22$

5. $2z + 7 \geq 5 - 6z$

6. $2y - 2 > 4y - 5$

7. $3(x - 1) \leq 17 - (8 - 3x)$

8. $6t - 2(t + 3) < 3(t + 11)$

9. $\frac{1}{4}w + 1 < w$

10. $2 - \frac{2}{5}x \leq 1$

Exercises 11-18 *Solve the compound inequality and graph the solution.*

11. $x + 3 > 1$ and $x - 2 < 4$

12. $\frac{1}{3}y \leq -1$ or $6y + 1 \geq -5$

13. $5 \leq 8t - 3 \leq 9$

14. $3u - 2 \leq 1$ or $4u + 1 > 9$

15. $-2x + 7 \leq 1$ and $\frac{1}{2}x \leq \frac{9}{4}$

16. $3w + 2 < -4$ and $-3w - 2 < 4$

17. $2z \geq 5$ or $\frac{1}{4}(z - 1) \geq 1$

18. $-3 < 1 - \frac{1}{2}x < 1$

Exercises 19-24 *Find the solution set for the absolute value equation and check your answer.*

19. $|x| = 6$

20. $|3y| = \frac{5}{2}$

21. $|2u - 5| = 1$

22. $|1 + 3v| = \frac{1}{5}$

23. $\left|4 - \frac{1}{2}z\right| = -3$

24. $\left|\frac{2}{3}a + 4\right| = 16$

Exercises 25-36 *Solve the absolute value inequality. Express your answer in interval notation.*

25. $|2w| < 6$

26. $\left|\frac{x}{7}\right| \leq 1$

27. $\frac{1}{3}|z| > 4$

28. $|-3x| < 1$

29. $|s - 5| \le 2$

30. $|2u - 1| \ge \dfrac{5}{3}$

31. $|-2r + 1| < 8$

32. $|-8p - 3| \ge 2$

33. $\left|\dfrac{3z - 2}{5}\right| \le 1$

34. $-2\left|\dfrac{4x - 3}{5}\right| > 1$

35. $|6t - 4| + 3 < 8$

36. $\dfrac{1}{3}|3y - 7| + 8 \le 10$

Exercises 37–40 *Write the verbal statement as an absolute value inequality and then solve it.*

37. The distance between x and 1 is less than 5.

38. The distance between t and -2 is at least 4.

39. The distance between twice a number and -3 is greater than 6.

40. The distance between -3 times a number and 4 is less than 2.

Exercises 41–44 *Use a graphing calculator to help you determine the answer.*

41. Plot the graph of $y = |3x - 2| - 5$. How is the solution set of $|3x - 2| = 5$ represented on the graph? What about the solution sets of $|3x - 2| > 5$ or $|3x - 2| < 5$?

42. Plot the graph of $y = |3x - 2| + 1$. How can you tell from this graph that $|3x - 2| < -1$ has no solutions, whereas the solution set of $|3x - 2| > -1$ is $(-\infty, \infty)$?

43. Find all solutions of $|x - 1| + |x - 2| = 3|x - 3|$.

44. Find all solutions of $|x| = |x - 2|$.

Exercises 45–58 *Solve the polynomial inequality. Give exact answers wherever possible; when approximating, give answers accurate to the nearest hundredth.*

45. $(t + 3)(t - 1) \le 0$

46. $(2y - 1)(y + 2) > 0$

47. $x^2 + 4x \le 5$

48. $2x^2 + 3 > 7x$

49. $u^2 \le 8$

50. $x^3 < x^2 + 3$

51. $x^4 + x > 1$

52. $x(x + 1)(x - 3) > 0$

53. $(x - 5)^2(x + 1)^3 < 0$

54. $x(x - 2)(x + 3) < -2$

55. $(7x - 2)^4(2x + 3)^6 \le 0$

56. $3t^3 + 3t^2 > 0$

57. $2v^3 - 8v \ge 0$

58. $(x - 15)(x + 20)(x - 27) > -100$

Applications

59. Truck Rental A moving truck can be rented for $19.95 per day plus $0.39 per mile. The total cost for 1 day can be expressed as $C = 0.39x + 19.95$, where x represents the number of miles traveled. What range of miles can be traveled if the total rental expense must be between $75 and $100?

60. Scale Accuracy The manufacturer of a bathroom scale claims the scale is accurate to within $\frac{1}{2}$ pound. If Susan's true weight is 95 pounds, write an absolute value inequality that expresses the possible error in the weight W given by the scale.

61. Tire Radius If the rear tires of a certain car average 750 revolutions per minute, the distance traveled in 1 hour over the course of many trials varies between 59.80 and 60.69 miles. Find the range of values for the radius of the tires. (*Hint:* The distance traveled is the circumference of the tire, $2\pi r$, times the number of revolutions.)

62. Scale Accuracy A grocery store has a scale located in the produce department for customers to weigh the produce they buy. A sign next to the scale warns that the scale may be off by as much as 2 ounces, and so it should be used only to estimate the weight. If you buy a bag of apples that weighs 3.5 pounds (according to the inaccurate scale) and if the apples cost $0.79 per pound, how much variation in price could there be when you take it to the cash register, where there is an accurate scale?

63. Temperature Conversion On a recent trip outside of the United States, Drew begins to feel ill. The only thermometer he can find gives readings in Celsius and is accurate to within 1° Celsius. If the thermometer gives a reading of 38° Celsius, what range of Fahrenheit temperatures does this indicate? The relationship between degrees Celsius C and degrees Fahrenheit F is given by $C = \frac{5}{9}(F - 32)$.

64. Tire Spokes If a bicycle tire is to have a circumference of 85 inches, with a maximum possible error of 0.3 inch, the hub of the tire is 1.5 inches in diameter, and the rim and tire have a combined radial thickness of 1 inch (see Figure 92), find the allowable range of values for the length of a spoke.

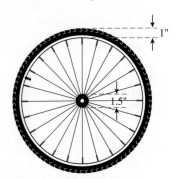

Figure 92

65. Height of a Projectile   A projectile is fired straight up with an initial velocity of 80 feet per second. Its height in feet after t seconds is $h = -16t^2 + 80t$. During what time period will the projectile be above 96 feet?

66. Volume of a Box   A rectangular package to be sent by the U.S. Postal Service can have a maximum combined length and girth (perimeter of the base) of 108 inches (see Figure 93). The volume of a box with a square base that exactly meets this restriction is $V = x^2(108 - 4x)$ cubic inches, where x is the width of the base in inches. Find all possible values of x for which the volume is positive.

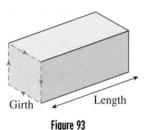

Girth   Length

Figure 93

67. Missile Flight   If a missile is shot straight up at 4000 feet per second and we assume that air resistance is negligible (is it?), then the height of the missile in feet t seconds after blast-off is given by $h = -16t^2 + 4000t$. The stratosphere, which includes the ozone layer, extends to approximately 30 miles from Earth's surface. For what period of time will the missile be outside of the stratosphere? (There are 5280 feet in 1 mile.)

68. Calling Card Comparison   The following comparison of several popular calling card plans appeared in a 1998 *USA Today* advertisement.

	AT&T	MCI	Sprint
Service charge (per call)	0¢	99¢	30¢
Cost per minute	25¢	40¢	30¢

For each of the three plans, set up and solve a linear inequality to find the interval of possible times for a single call if the total bill for the call may not exceed $5.00. Assume that fractions of minutes are billed at the corresponding fraction of the per-minute rate.

Concepts and Critical Thinking

Exercises 69-72   *Answer true or false.*

69. If $-2x < 4$, then $x < -2$.

70. The solution set to a linear inequality is either the empty set or a single interval.

71. The solution set to an absolute value inequality may be the union of two disjoint intervals.

72. If $x^2 < 4$, then we can conclude that $x < 2$.

Exercises 73-76   *Give an example of each.*

73. An absolute value equation with empty solution set

74. A difference between the solution sets of inequalities of the form $|\square| < a$ and $|\square| > a$, where $a > 0$

75. An absolute value inequality having the empty set as its solution

76. A difference between the techniques for solving linear equations and inequalities

Questions for Discussion or Essay

77. In Example 6, we considered a diet and exercise program that claimed a weight loss of $1\frac{1}{2}$ pounds each week. This implies that weight loss will occur in a linear way, which is based on the assumption that calories will continue to be burned at the same rate. Why isn't this valid? What modification could be made to the weight-loss model?

78. In what sort of real-world contexts are absolute value inequalities likely to arise? Do you see a common thread in the applications in this exercise set?

79. Many students, when solving an absolute value inequality such as $|x + 3| > 2$, will write down $-2 > x + 3 > 2$ as their first step and then arrive at $-5 > x > -1$ as an answer. Even if you know nothing about absolute value, what is it about the answer that tells you something is wrong?

80. Solution sets of rational inequalities can be approximated in much the same way as those for polynomial inequalities. For example, to find the solution set of

(1) $$\frac{x^2 + 1}{(x - 2)(2 - x)} > 0$$

we can plot the graph of

$$y = \frac{x^2 + 1}{(x - 2)(2 - x)}$$

with a graphing calculator to determine where the graph lies above the x-axis.

a. Use the graphing procedure to find the solution set of inequality (1).

b. It is tempting to solve rational inequalities by clearing fractions, as we do when solving rational equations. If this (incorrect) technique is applied to inequality (1), we would be led to conclude that $x^2 + 1 > 0$. Describe the solution set of $x^2 + 1 > 0$. Is this consistent with your result from part a?

c. Explain why multiplying both sides of a rational inequality by a variable expression can lead to erroneous results.

Projects for Enrichment

81. Estimating Distance with a Bicycle In this project, we discover how an ordinary bicycle tire can be used as a measuring device. You need a bicycle to perform this experiment.

a. Measure the circumference of a tire to the nearest centimeter by marking the distance that the bicycle travels in one revolution of the tire. Perform this experiment five times.

b. Explain why there is some discrepancy in the results of the five trials from part a.

c. Compute the average of the five results from part a.

d. Compute the difference between the largest and smallest results in part a. We will assume that the actual circumference deviates from the average obtained in part c by no more than this difference.

e. Let C be the actual (unknown) circumference of the tire. Express the range of values of C as the solution set to an absolute value inequality and as an interval.

f. Suppose that you compute the distance traveled on a lengthy trip by counting revolutions of your tire. Using the result from part c, you estimate that the trip was 10 kilometers. Express the range of values of the actual trip length as the solution to an absolute value inequality and as an interval.

g. Take the bicycle to a track on which a 100-meter course is marked. Count the smallest integer number of revolutions required to travel the length of the track; then compute the

distance by which the bicycle has overshot the starting point. For example, it might happen that 180 revolutions are required, with the bicycle ending up 1 meter past the finish line at the end of the 180th revolution.

h. Express the total distance traveled in part g in meters by adding the overshot distance to 100 meters.

i. Use the result from part h to estimate the circumference of the tire.

j. Assuming that the length of the track is 100 meters, with a maximum error of 1 centimeter, and that the error involved in computing the overshoot is negligible, find the maximum error in computing the circumference by this method. Now express the range of values of the circumference as the solution set to an absolute value inequality and as an interval.

k. Now suppose that, using the value of the circumference found in part i, you find that a certain trip was 10 kilometers long. Use the result of part j to express the range of values of the actual trip length as the solution set to an absolute value inequality and as an interval. What's the maximum error in computing the trip length by this method?

l. Compare and contrast the results from the two methods of computing the circumference.

m. What physical conditions could affect the accuracy of your results?

Section 1.7   Lines

- How can lines be used to compute invisible entries in mathematical tables?
- If you know the freezing and boiling points of water in both Celsius and Fahrenheit, then you know enough to convert from one system to the other. How?
- If children grow at a steady rate, then why aren't there more 3-inch newborns?
- How do accountants determine the value of an asset that depreciates in value?

Slope

The **slope** of a nonvertical line is the number of units the line rises for every unit of horizontal change. For example, consider the line shown in Figure 94.

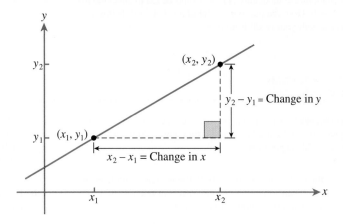

Figure 94

In passing from (x_1, y_1) to (x_2, y_2), the x-value is increased by $x_2 - x_1$ units, whereas the y-value is increased by $y_2 - y_1$. Thus, the number of units the line rises for each unit of horizontal change is given by

$$\frac{y_2 - y_1}{x_2 - x_1}$$

In other words, the slope of the line is the ratio of the difference in the y-coordinates to the difference in the x-coordinates for any two points on the line.

Definition of the Slope of a Line

The slope of the line passing through the points (x_1, y_1) and (x_2, y_2) (where $x_1 \neq x_2$) is traditionally denoted by m and is defined by

$$m = \frac{\text{Change in } y}{\text{Change in } x} = \frac{y_2 - y_1}{x_2 - x_1}$$

An essential point is that for a given line, the quantity

$$m = \frac{y_2 - y_1}{x_2 - x_1}$$

is always the same; it doesn't matter which two points (x_1, y_1) and (x_2, y_2) are chosen. This is illustrated in Figure 95, in which it can be seen that triangle PQR is similar to (that is, has the same shape as) triangle STU. A consequence of this similarity is that corresponding sides are proportional, so that

$$\frac{QR}{PR} = \frac{TU}{SU}$$

In other words, computing the slope with points P and Q would yield the same result as computing the slope with points S and T.

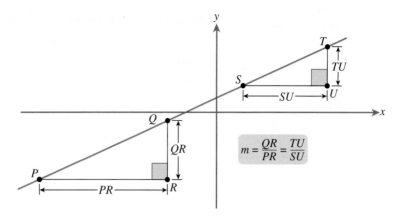

Figure 95

When computing the slope of a line given two points on the line, it doesn't matter which point is considered to be (x_1, y_1) and which is considered to be (x_2, y_2). It is essential, however, that the order of subtraction be consistent from numerator to denominator. In other words, we may compute slope as

$$m = \frac{y_2 - y_1}{x_2 - x_1} \quad \text{or} \quad m = \frac{y_1 - y_2}{x_1 - x_2} \quad \text{but } not \quad m = \frac{y_2 - y_1}{x_1 - x_2} \quad \text{or} \quad m = \frac{y_1 - y_2}{x_2 - x_1}$$

The slope of a line is a measure of its steepness; the steeper the line, the greater the absolute value of its slope. For example, a line with slope 1 will rise 1 unit for every unit that we move to the right, whereas a line with slope 5 will rise 5 units for every unit that we move to the right. Thus, a line with slope 5 is much steeper than a line with slope 1. If the slope of a line is negative, then moving 1 unit to the right will result in a *negative* change in the y-coordinate; that is, the line will descend. Figure 96 shows several lines, along with their slopes.

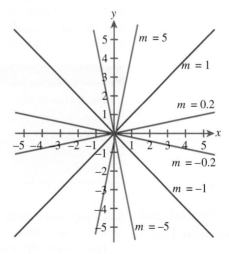

Figure 96

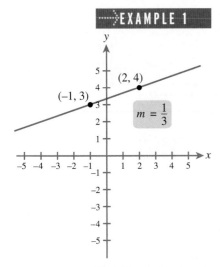

Figure 97

EXAMPLE 1

Finding the Slope of a Line Given Two Points on the Line

The points $(-1, 3)$ and $(2, 4)$ are on the line l. Find the slope of l.

Solution   We take (x_1, y_1) to be the point $(-1, 3)$ and (x_2, y_2) to be the point $(2, 4)$. Then

$$m = \frac{y_2 - y_1}{x_2 - x_1}$$
$$= \frac{4 - 3}{2 - (-1)}$$
$$= \frac{1}{3}$$

The graph of l is shown in Figure 97. Note that since m is positive, the line is rising. Because m is relatively small, the line rises gently.

EXAMPLE 2

Finding the Slope of a Horizontal Line

Find the slope of the horizontal line shown in Figure 98.

Solution   We choose two points on the line, say $(2, 3)$ and $(5, 3)$. Then we have

$$m = \frac{3 - 3}{5 - 2} = \frac{0}{3} = 0$$

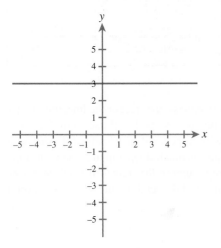

Figure 98

It isn't difficult to see that the slope of any horizontal line is zero. All y-coordinates of points on a horizontal line are the same, so the difference in the y-coordinates of two points on a horizontal line is always zero. Thus, the slope is

$$m = \frac{\text{Change in } y}{\text{Change in } x} = \frac{0}{\text{Change in } x} = 0$$

Similarly, since all x-coordinates of points on a vertical line are the same, the difference in the x-coordinates of two points on a vertical line is always zero. Thus, the slope of a vertical line is undefined. The relationship between the orientation of a line and its slope may be summarized as follows.

Relationship Between Slope and Orientation	
Description of line	**Slope**
Rising from left to right	Positive
Falling from left to right	Negative
Horizontal	Zero
Vertical	Not defined

Parallel and Perpendicular Lines

Since slope is a measure of the steepness of a line, it is not surprising that parallel lines have the same slope. Figure 99 shows three parallel lines, all with slope 2.

The relationship between the slopes of perpendicular lines is slightly more complicated. In Figure 100, a line l with slope m is shown, along with a line l' perpendicular to l. For convenience, we have assumed that the lines cross at the origin. Now the point

$(1, m)$ is on line l and, using elementary geometry, it can be shown that the point $(-m, 1)$ is on line l'. To compute the slope of l' we will use the points $(0, 0)$ and $(-m, 1)$. If we let m' denote the slope of l', then we have

$$m' = \frac{1 - 0}{-m - 0} = -\frac{1}{m}$$

Thus, the slope of l' is the negative reciprocal of the slope of l.

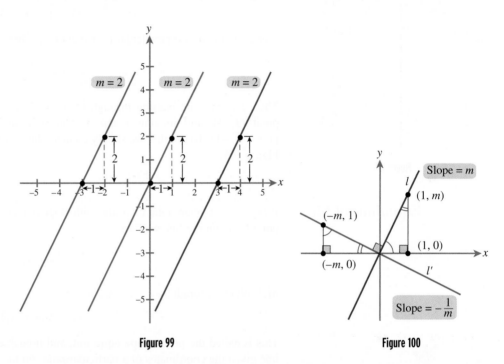

Figure 99   **Figure 100**

We summarize the relationships between parallel and perpendicular lines as follows.

Slopes of Parallel and Perpendicular Lines

1. Two nonvertical lines are parallel if and only if they have the same slope.
2. Two lines, neither of which is vertical, are perpendicular if and only if their slopes are negative reciprocals of one another—that is, if and only if the product of their slopes is -1.

⋯⋯EXAMPLE 3

Finding the Slope of a Line Parallel to a Given Line

Find the slope of a line parallel to the line passing through the points $(2, 3)$ and $(4, 7)$.

Solution   Let l denote the line passing through $(2, 3)$ and $(4, 7)$. Then l has slope

$$m = \frac{7 - 3}{4 - 2} = \frac{4}{2} = 2$$

Since parallel lines have the same slope, a line parallel to l also has slope 2.

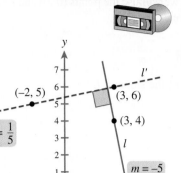

EXAMPLE 4

Figure 101

Graphing a Line Passing Through a Given Point and Perpendicular to a Given Line

Graph the line passing through the point $(3, 4)$ that is perpendicular to the line passing through $(-2, 5)$ and $(3, 6)$.

Solution   Let l' denote the line passing through $(-2, 5)$ and $(3, 6)$, and let l denote the line perpendicular to l' and passing through $(3, 4)$. The slope of l' is given by

$$m' = \frac{6 - 5}{3 - (-2)} = \frac{1}{5}$$

Since l and l' are perpendicular to one another, l has slope

$$m = -\frac{1}{m'} = -\frac{1}{(1/5)} = -5$$

Thus, l is the line passing through $(3, 4)$ with slope -5. Now, beginning with the point $(3, 4)$, and moving 1 unit to the right and 5 units down, we obtain the point $(4, -1)$. To graph l, we simply connect the points $(3, 4)$ and $(4, -1)$, as shown in Figure 101.

Equations of Lines

If (x_1, y_1) is a given point on a line with slope m (see Figure 102), then for an arbitrary point (x, y) on the line we have

$$\frac{y - y_1}{x - x_1} = m$$

Multiplying through by $x - x_1$, we obtain

$$y - y_1 = m(x - x_1)$$

This is called the **point-slope equation**, and it enables us to produce an equation of a line given the coordinates of a particular *point* on the line and the *slope* of the line.

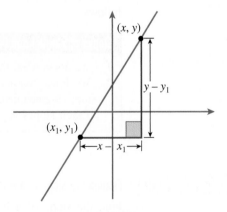

Figure 102

Point-Slope Equation of a Line

A line with slope m passing through the point (x_1, y_1) has equation $y - y_1 = m(x - x_1)$.

EXAMPLE 5

Finding the Equation of a Line

Find an equation of the line with slope -2 passing through the point $(1, -5)$. Graph the line with a graphing calculator.

Solution   Using the point-slope equation with $m = -2$, $x_1 = 1$, and $y_1 = -5$, we obtain the following equation:

$$y - y_1 = m(x - x_1)$$
$$y - (-5) = -2(x - 1)$$
$$y + 5 = -2(x - 1)$$

Solving for y, we obtain

$$y = -2(x - 1) - 5$$
$$= -2x - 3$$

The graph of $y = -2x - 3$ is shown in Figure 103.

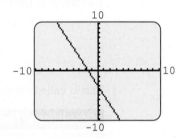

Figure 103

The following examples illustrate how the point-slope equation can be used when either the slope or a point on the line is not given directly.

EXAMPLE 6

Finding the Equation of a Line Given Two Points on the Line

Find an equation of the line passing through the points $(-2, 6)$ and $(2, 7)$.

Solution   We begin by finding the slope.

$$m = \frac{7 - 6}{2 - (-2)} = \frac{1}{4}$$

Now that we have the slope we can select either $(-2, 6)$ or $(2, 7)$ as the point (x_1, y_1). We choose $(-2, 6)$. From the point-slope equation, we have

$$y - y_1 = m(x - x_1)$$
$$y - 6 = \frac{1}{4}(x - (-2))$$
$$y - 6 = \frac{1}{4}(x + 2)$$

Note that if we had chosen $(2, 7)$ instead, we would have obtained the equation $y - 7 = \frac{1}{4}(x - 2)$. But the two equations are equivalent and can both be written in the form $y = \frac{1}{4}x + \frac{13}{2}$.

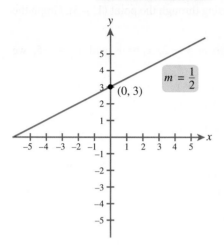

Figure 104

> **EXAMPLE 7**

Finding the Equation of a Line Given the Slope and the *y*-intercept

Find an equation of the line with *y*-intercept 3 and slope $\frac{1}{2}$, the graph of which is shown in Figure 104.

Solution   Since the *y*-intercept is 3, we know that the graph crosses the *y*-axis at the point (0, 3). Thus, we have

$$y - y_1 = m(x - x_1)$$

$$y - 3 = \frac{1}{2}(x - 0)$$

$$y = \frac{1}{2}x + 3$$

In Example 7, we saw that the point-slope equation could be used when the slope and the *y*-intercept were given. More generally, if the line *l* has slope *m* and *y*-intercept *b*, then the point (0, *b*) lies on *l*, and so the equation of *l* is

$$y - b = m(x - 0)$$
$$y - b = mx$$
$$y = mx + b$$

This is called the **slope-intercept equation**.

Slope-Intercept Equation of a Line

A line with slope *m* and *y*-intercept *b* has equation $y = mx + b$.

> **EXAMPLE 8**

Finding the Slope-Intercept Equation of a Line

Find an equation of the line with slope -4 and *y*-intercept 6.

Solution   We apply the slope-intercept equation with $m = -4$ and $b = 6$.

$$y = mx + b$$
$$y = -4x + 6$$

The following examples illustrate the utility of the slope-intercept form.

> **EXAMPLE 9**

Finding the Slope Given the Equation of a Line

Find the slope and the *y*-intercept of the line with equation $5x + 4y = 8$.

Solution   We begin by writing the equation in slope-intercept form. To do this, we solve for *y*.

$$5x + 4y = 8$$
$$4y = -5x + 8$$
$$y = -\frac{5}{4}x + 2$$

Thus, the slope is $-\frac{5}{4}$ and the *y*-intercept is 2.

EXAMPLE 10 **Finding the Equation of a Line through a Given Point and Parallel to a Given Line**

Find an equation of the line l that passes through $(-3, 5)$ and is parallel to the line with equation $2x + y = 10$.

Solution   We begin by finding the slope of the line with equation $2x + y = 10$. To do this, we write its equation in slope-intercept form.

$$2x + y = 10$$
$$y = -2x + 10$$

Thus, we can conclude that the line $2x + y = 10$ has slope -2. Since l is parallel to this line, it also has slope -2. Thus, l has slope -2 and passes through the point $(-3, 5)$. Now that we know the slope and a point on the line, we can apply the point-slope equation.

$$y - y_1 = m(x - x_1)$$
$$y - 5 = -2(x - (-3))$$
$$y - 5 = -2(x + 3)$$

If desired, we can also express the equation in slope-intercept form, as follows:

$$y - 5 = -2x - 6 \qquad \text{Distributing}$$
$$y = -2x - 1 \qquad \text{Solving for } y$$

EXAMPLE 11 **Graphing a Line Given Its Equation**

Graph the line with equation $y = 2x - 4$.

Solution   We see that the slope is 2 and the y-intercept is -4. To graph the line, we begin by plotting the y-intercept $(0, -4)$. Next we find one additional point by substituting a convenient value (such as 1) for x. With $x = 1$, we have

$$y = 2x - 4$$
$$= 2 \cdot 1 - 4$$
$$= -2$$

Thus, $(1, -2)$ is a second point on the line. Finally, we simply "connect the dots," as shown in Figure 105.

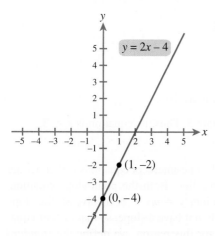

Figure 105

The slope-intercept equation can be used to derive the equation of a horizontal line. Since the slope of a horizontal line is zero, we have

$$y = mx + b$$
$$= 0x + b$$
$$= b$$

This is not a surprising result since points on a given horizontal line must have the same y-coordinates. Similarly, points on a vertical line must have the same x-coordinates. These observations lead to the following equations.

The Equations of Vertical and Horizontal Lines

Horizontal lines have equations of the form $y = C$, where C is a constant.

Vertical line have equations of the form $x = C$, where C is a constant.

EXAMPLE 12

Finding the Equation of a Horizontal Line

Find an equation of the horizontal line passing through the point $(2, 5)$.

Solution Since the line is horizontal, we know that its equation is of the form $y = C$. Since the point $(2, 5)$ is on the line and has y-coordinate 5, C must be 5. Thus, the equation of the line is $y = 5$. See Figure 106.

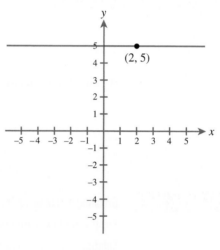

Figure 106

EXAMPLE 13

Finding the Equation of a Vertical Line

Find an equation of the vertical line shown in Figure 107.

Solution Every point on the line has x-coordinate 3. Thus, its equation is $x = 3$.

One minor deficiency of the point-slope and slope-intercept equations is that neither is sufficiently general to include every possible line. Both the point-slope equation, $y - y_1 = m(x - x_1)$, and slope-intercept equation, $y = mx + b$, involve m, the slope of the line. The difficulty is that vertical lines do not have a slope, so that neither equation is applicable for vertical lines. Primarily for this reason, we define the *standard equation of a line* as follows.

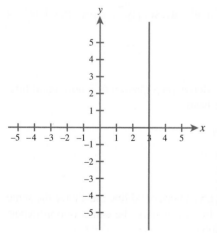

Figure 107

Standard Equation of a Line

An equation of the form $Ax + By = C$ is said to be a **standard equation of a line**. Every line has an equation of this form, and, conversely, every equation of this form is that of a line, provided A and B are not both 0. If the equation of a line is a standard equation, then it is said to be in **standard form**.

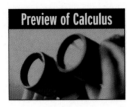

EXAMPLE 14 Finding the Standard Form for the Equation of a Line

Find an equation in standard form for the line passing through (2, 5) with slope -2.

Solution Since we are given a point on the line, (2, 5), and the slope of the line, -2, we use the point-slope equation.

$$y - 5 = -2(x - 2)$$

Now we rearrange to write the equation in standard form.

$$y - 5 = -2x + 4$$
$$2x + y = 9$$

Preview of Calculus

Local Linearity

Many ancient civilizations viewed Earth as an essentially flat plane. Though the Greeks of 2000 years ago were well aware of the roundness of Earth, belief systems based on a flat Earth survive in this, the third millennium. In spite of the fact that day changes to night and back again as Earth rotates on its axes, in spite of the fact that ships disappear mast first as they cross the horizon, and in spite of photographs of Earth from space showing a pale blue ball, modern-day Flat Earthers tenaciously cling to their belief. But why? The answer is simple: Earth really is flat—at least its surface appears flat when viewed from up close. Globally, Earth may be curved, but locally it looks like a plane. Approximating the surface of Earth as a plane may not be useful for developing models of climate, seismography, or astronomy, but it works pretty well in building construction and furniture design!

Local flatness is a characteristic of not just the surface of Earth, but of any smooth surface or curve. Take a point on the graph of a parabola or some other smooth curve, zoom in repeatedly, and eventually the curve looks like a line (see Figures 108–110). When we're close enough to something curved, the curvature disappears and what we see appears completely flat. This property, called **local linearity**, is the basis for some of the most important ideas in calculus—ideas that allow us to find the length of a curve, the instantaneous speed of a projectile, and optimal pricing strategies—ideas without which space travel, electronic computers, and most of the creations of modern-day science and technology would have been all but impossible. Even your calculator relies on local linearity to graph functions and automatically compute their zeros.

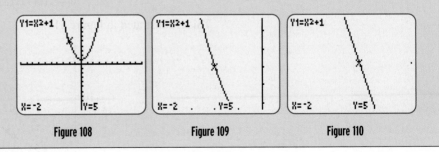

| Figure 108 | Figure 109 | Figure 110 |

Linear Interpolation and Extrapolation

It is often the case that we have no equation to express the precise relationship between two variables. Perhaps our only information is in the form of a limited number of data points. In order to use these data to make predictions, we must make certain assumptions regarding the (unknown) equation. **Linear interpolation** is the process of predicting values based on the assumption that the graph of an equation is linear "between" data points; **linear extrapolation**, on the other hand, is based on the assumption that the graph of an equation will extend in a linear fashion outside the range of the data points. These techniques are illustrated in the following example.

---->**EXAMPLE 15**   **Linear Interpolation and Extrapolation**

Table 8 gives the average weight for boys ages 10–14. Use linear interpolation or extrapolation to predict the average weight for a boy at each of the given ages.

Table 8
Average weight for boys

Age (years)	Weight (pounds)
10	71
11	79
12	89
13	102
14	114

Data source: U.S. Centers for Disease Control.

a. 11 years, 3 months
b. 14 years, 6 months
c. 30 years

Solution   Let t represent the age in years and w the weight in pounds.

a. We interpolate by assuming the relationship between w and t is linear between the data points $(11, 79)$ and $(12, 89)$. We have

$$m = \frac{89 - 79}{12 - 11} = \frac{10}{1} = 10$$

Using the point-slope equation, we have

$$w - 79 = 10(t - 11)$$

or

$$w = 10t - 31$$

Now, 3 months = 0.25 year, so we are interested in $t = 11.25$. Thus,

$$w = 10(11.25) - 31 = 81.5$$

This is illustrated in Figure 111.

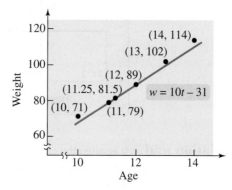

Figure 111

b. Since $t = 14.5$ (14 years, 6 months) lies outside our given data, we use linear extrapolation. We find the equation of the line passing through the last two data points, (13, 102) and (14, 114), and we assume that this relationship holds for $t = 14.5$ as well. The slope of this line is given by

$$m = \frac{114 - 102}{14 - 13} = 12$$

Thus, the line has equation

$$w - 102 = 12(t - 13)$$
$$w = 12t - 54$$

When $t = 14.5$, we have $w = 12(14.5) - 54 = 120$.

c. Again, $t = 30$ is outside the range of the given data, so we extrapolate from the last two data points. From part b, we know that the equation of the line joining these points is $w = 12t - 54$. Evaluating at $t = 30$ gives us

$$w = 12(30) - 54 = 306$$

Obviously, the average weight of 30-year-old "boys" is not 306 pounds: Extrapolation far from original data is not to be trusted.

Understanding and Mastery Checklists

Concepts to Understand	Skills to Master
Slope of a line	Estimate the slope of a line given its graph.
Parallel and perpendicular lines	Compute the slope of a line through two points.
Point-slope equation of a line	Find the slope of a line given the slope of a line parallel or perpendicular to it.
Slope-intercept equation of a line	Find an equation of a line given a point on the line and the slope of the line.
Standard equation of a line	Find an equation of a line given two points on the line.
Linear interpolation and extrapolation	Find the slope and y-intercept of a line given its equation.
	Place the equation of a line in standard form.
	Estimate the values between or outside data points using linear interpolation or extrapolation.

Exercises 1.7

Exercises 1-12 *Find the slope of the line passing through the given pair of points.*

1. $(1, 3)$ and $(2, 7)$

2. $(3, 4)$ and $(-2, 4)$

3. $(-2, 6)$ and $(3, -5)$

4. $(3, -4)$ and $(-7, -1)$

5. $(0, 32)$ and $(100, 212)$

6. $(1492, 1)$ and $(1999, 1000)$

7. $\left(\frac{1}{2}, 2\right)$ and $\left(-1, 3\frac{1}{2}\right)$

8. $\left(\frac{2}{3}, -3\right)$ and $\left(-\frac{3}{5}, -2\right)$

9. $(2, 3.5)$ and $(4.6, 7.9)$

10. $(1.02, 100)$ and $(0.98, 101)$

11. $(35, 1.5 \times 10^7)$ and $(40, 2.0 \times 10^7)$

12. $(2.3 \times 10^{-2}, 3.5)$ and $(4.5 \times 10^{-2}, 5.7)$

Exercises 13-16 *The graph of a line is given. Estimate the slope of the line.*

13.

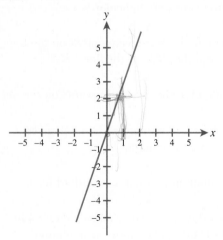

14.

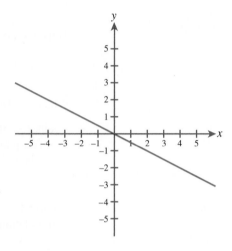

15.

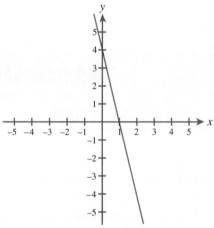

16.

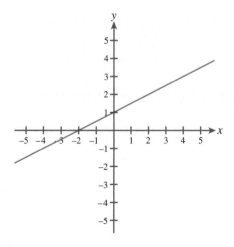

Exercises 17-20   *Find the slope of the line l with the given property.*

17. The line l is parallel to a line with slope -3.

18. The line l is perpendicular to the line passing through the points $(2, 7)$ and $(-1, 9)$.

19. The line l is perpendicular to a line with slope $\frac{3}{5}$.

20. The line l is parallel to the line passing through the points $(3, 4)$ and $(-2, 6)$.

Exercises 21-22   *Use the given information to determine the slopes of the lines l_1, l_2, l_3, and l_4.*

21. The lines l_1, l_2, l_3, and l_4, have slopes of $\frac{2}{3}$, -2, 0, and 3, but not necessarily in that order.

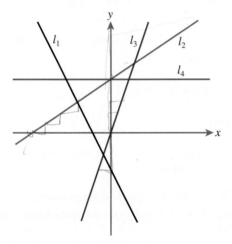

22. The lines l_1, l_2, l_3, and l_4 have slopes of 1, 4, $-\frac{1}{2}$, and -3, but not necessarily in that order.

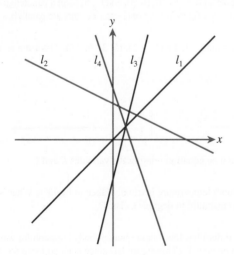

Exercises 23-24   *Find the slope of each of the lines l_1, l_2, and l_3.*

23.

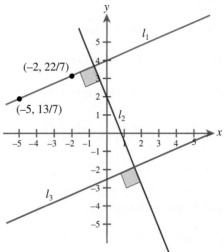

24.

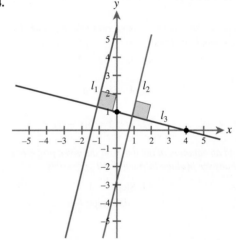

Exercises 25-34   *Find an equation of the line with the given slope passing through the indicated point P. Express your final answer in slope-intercept form, if possible.*

25. $m = 1$, $P = (2, 4)$

26. $m = 3$, $P = (-1, 5)$

27. $m = -1$, $P = (2, 8)$

28. $m = 0$, $P = (0, 1)$

29. m is undefined, $P = (3, 7)$

30. $m = 7$, $P = (-1, -2)$

31. $m = 5$, $P = \left(\frac{1}{2}, \frac{2}{3}\right)$

32. $m = -\frac{2}{5}$, $P = (-6, 2)$

33. $m = 0$, $P = (2, 1)$

34. m is undefined, $P = (-2, 1)$

Exercises 35-46 *Find an equation of the line passing through the indicated pair of points. Express your final answer in slope-intercept form, if possible.*

35. $(1, 4)$ and $(-2, 3)$ **36.** $(2, 2)$ and $(4, 7)$

37. $(2, 0)$ and $(0, 4)$ **38.** $(-3, 0)$ and $(0, 2)$

39. $(3, -14)$ and $(3, 7)$ **40.** $(19, 5)$ and $(23, 5)$

41. $(1.2, 3.4)$ and $(4.6, 5.9)$ **42.** $\left(\frac{1}{2}, \frac{3}{5}\right)$ and $\left(-\frac{1}{4}, \frac{2}{5}\right)$

43. $\left(3, \frac{7}{8}\right)$ and $\left(-2, \frac{1}{8}\right)$

44. $(1, 0.000002)$ and $(1,$ Your weight in kilograms$)$

45. $(\pi, 2 \times 10^{100})$ and $(\pi, 2 \times 10^{-10})$

46. $(1999, 1.2 \times 10^9)$ and $(2001, 1.4 \times 10^9)$

Exercises 47-52 *Find the slope and the y-intercept of the line with the given equation. Then graph the line.*

47. $2x - 3y = 8$ **48.** $3x + 6y = 9$

49. $x = 3y + 4$ **50.** $4 - 3y = 0$

51. $y - 3 = -2(x + 3)$ **52.** $y + 2 = 3(x - 7)$

Exercises 53-60 *Find an equation of the line with the given properties. Express your final answer in slope-intercept form, if possible.*

53. Slope: $\frac{2}{3}$ **54.** Slope: -4
y-intercept: 2 y-intercept: -3

55. Parallel to: $y = 4x + 2$
y-intercept: 5

56. Parallel to: The line containing $(0, 3)$ and $(4, 0)$
Point on line: $(-2, 5)$

57. Perpendicular to: $x - 3y = 8$
Point on line: $(2, -4)$

58. Perpendicular to: $4x + 8y = 7$
Point on line: $(-2, -2)$

59. Parallel to: $x = 6$
Point on line: $(3, 8)$

60. Perpendicular to: $x = 19$
y-intercept: 7

Exercises 61-64 *Write the given equation in standard form.*

61. $y = 3x + 4$ **62.** $y - 8 = -2(x - 3)$

63. $y - \frac{2}{3} = 2\left(x - \frac{1}{2}\right)$ **64.** $x = 2y + 8$

Exercises 65-68 *Find the point of intersection of the given lines.*

65. $y = 3x + 5$ and $y = -2x - 8$

66. $2x - 3y = 6$ and $3x + 4y = 12$

67. $x = 2$ and $y = 5$

68. $y = m_1 x + b_1$ and $y = m_2 x + b_2$. (Under what circumstances do these lines not intersect?)

Exercises 69-72 *Use slope to verify the given statement.*

69. The points $(1, 0)$, $(7, 4)$, $(15, 9)$ are not collinear (that is, do not lie on a line).

70. The points $(-2, -9)$, $(2, -1)$, and $(4, 3)$ are collinear (that is, lie on a line).

71. The points $(2, 3)$, $(3, 5)$, $(6, 3)$, and $(7, 5)$ form a parallelogram (that is, a four-sided figure having opposite sides that are parallel).

72. The points $(-1, 1)$, $(1, -1)$, $(5, 3)$, and $(3, 5)$ form a rectangle.

Applications

73. **Mail-Order Cost**   It costs $1000 in overhead per month to run a mail-order business. In addition, each customer order costs $5.00 to produce. Express the total monthly cost C in terms of N, the number of orders received.

74. **Temperature Conversion**   The relationship between degrees Fahrenheit (F) and Celsius (C) is linear. At atmospheric pressure, water freezes at 32°F and 0°C, whereas water boils at 212°F and 100°C.

a. Find an equation relating the variables F and C.

b. Room temperature is usually taken to be 72°F. What is room temperature in degrees Celsius?

c. For decades, the average human body temperature was assumed to be 98.6°F. (This figure has since been adjusted to 98.2°.) What is 98.6°F in degrees Celsius?

d. The temperature at the surface of the sun can reach 5500°C. Express this in degrees Fahrenheit.

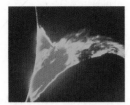

The surface of the sun during a solar flare.

e. Absolute zero is the lowest possible temperature. On the Celsius scale, absolute zero is –273°. What is absolute zero in degrees Fahrenheit?

75. Growth Rate Suppose that a child were to grow at a constant rate from conception to maturity. Thus, if h represents the height (or length) of the child in inches and t represents the number of years *since conception*, then h and t are related linearly. Assume that at $t = 0$, $h = 0$. Assume further that the child is fully grown at 16 years old, at which time she is 5′6″ tall.

a. Find an equation relating h and t.

b. Use the result of part a to estimate the length of the baby at birth.

c. Estimate the height of the child at age 12.

76. Sales Commission Each month a salesman earns a $2000 salary plus a 10% commission on all sales over $10,000. Let P be the salesman's pay and S be his monthly sales.

a. Write an equation relating S and P (assuming $S > 10,000$).

b. Graph the equation of part a.

c. Use the result of part a to find the salesman's total pay in a month when he sold $25,000.

77. Machinery Depreciation When computing the value of a piece of machinery for tax purposes, it is common practice to use linear depreciation. With linear depreciation, we assume that the value decreases at a constant rate, so that value V and time T are related linearly. Suppose that in 2003, a crane was purchased for $200,000. After 10 years, the crane will have a salvage value of only $20,000. Use linear depreciation to estimate the value of the crane in 2006.

78. Computer Depreciation Suppose that a computer system purchased new in 2001 for $1200 is worth $500 in 2004. Assuming linear depreciation—that is, assuming the value of the computer system and its age are related linearly—estimate the value of the computer in 2002.

79. Population Interpolation The U.S. population was approximately 249 million in 1990 and 281 million in 2000.

a. Use linear interpolation to estimate the population in 1997.

b. Use linear extrapolation to estimate the population in 2010 (the next census year).

80. GDP Interpolation The U.S. Gross Domestic Product (GDP) was approximately $6708 billion in 1990 and $9319 billion in 2000 (both in 1996 dollars).

a. Use linear interpolation to estimate the GDP in 1998.

b. Use linear extrapolation to predict the GDP in 2012.

Concepts and Critical Thinking

Exercises 81–86 *Answer true or false.*

81. Slope isn't defined for horizontal lines.

82. A vertical line has a slope of 0.

83. If the y-coordinates of points on a line increase as the x-coordinates decrease, then the line has negative slope.

84. Parallel lines have the same slope.

85. A line with slope 3 is steeper than a line with slope -4.

86. Two lines having the same slope and a point of intersection are identical.

Exercises 87–90 *Give an example of each.*

87. An equation of a line that has undefined slope

88. An equation of a horizontal line

89. A line that does not intersect $2x + 7y = 11$

90. A line perpendicular to $3x - y = 5$

91. Explain why it doesn't matter whether one uses

$$\frac{y_2 - y_1}{x_2 - x_1} \quad \text{or} \quad \frac{y_1 - y_2}{x_1 - x_2}$$

to compute the slope.

Questions for Discussion or Essay

92. A line is an idealized, abstract construction. But do lines really exist? To what extent does a sketch of a line (whether done by hand or with a graphing calculator) differ from the line itself? Are rays of light lines? More generally, what connections are there between the abstract spaces of mathematics (like the coordinate plane and the number line) and physical space?

93. It can easily be shown that if the lines l_1 and l_2 have equations $a_1x + b_1y = c_1$ and $a_2x + b_2y = c_2$, respectively, then either l_1 and l_2 are the same line, intersect in a single point, or are parallel. However, as you might recall from high school geometry, lines may also be **skew**—nonintersecting and yet not parallel either. How do you explain this apparent contradiction?

94. One of the most common abuses of mathematics is to assume, with very little evidence, that some particular quantity and time are related linearly. In essence, a variable will vary linearly with time if it changes by the same amount for each unit of time—that is, if its rate of change is *constant*. Which of the following variables do you think will have a relationship with time that is approximately linear? Give reasons to support your answers.

a. The world's population

b. The national debt

c. The price of a 12-ounce can of Coke

d. The world-record time for the 100-meter run

e. The number of floating-point operations per second of the world's fastest computer

f. The planet Earth's total reserves of crude oil

g. The number of AIDS patients in the United States

h. The total number of solar eclipses of which there is a human record

Projects for Enrichment

95. Distance Between a Line and a Point In this project, we derive a formula for the distance between a line l and a point P not on l. We assume that l is nonvertical. Suppose then that l has equation $y = mx + b$ and that P has coordinates (c, d), as shown in Figure 112. Clearly, the shortest path between P and l is along l', the line perpendicular to l through P. Then if Q is the point of intersection of l and l', the distance between P and l is given by PQ.

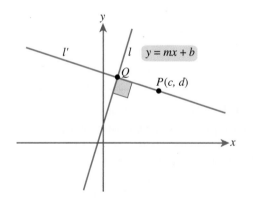

Figure 112

a. What is the slope of l'?

b. What is the equation of l'?

c. Find the coordinates of Q.

d. Find an expression for the distance between l and P. Your answer should involve only the variables m, b, c, and d.

e. Use the formula obtained in part d to find the distance between the line with equation $y = 4x + 7$ and the point $(2, 5)$.

f. Use the formula obtained in part d to find the distance between the parallel lines $y = mx + b_1$ and $y = mx + b_2$. [*Hint*: Consider the point $(0, b_1)$ on the first line.]

96. Life Tables and Linear Interpolation Although it is impossible to predict with certainty the life span of an individual, it is possible—and essential—to make meaningful predictions about the life spans of groups of people. Social security withholdings are determined, in part, by forecasts of the number of citizens who will be eligible for social security benefits decades later. Life insurance rates depend on death rates for individuals of a given age, sex, and health status. The entire health-care industry must prepare for an uncertain future by developing demographic models that include projections of death rates.

Inferences regarding death rates can often be gleaned from what is (ironically) called a life table. Table 9 is a portion of such a table. In essence, the table indicates the number of people from an original group (called a **cohort**) of 100,000 live births that we would expect to be alive at various ages. Specifically, we define l_x to be the number of people from the original cohort that we would expect to be alive at age x. Thus, for example, since $l_{70} = 68{,}248$, we would expect that out of 100,000 live births, about 68,000 of them, or 68%, would survive to age 70.

a. Use the life table to estimate the following quantities:

i. The percentage of newborns who will survive to age 80

ii. The percentage of people aged 70 who will live at least 10 more years

iii. From a group of 1000 people aged 40, the number who will die before they reach the age of 60

b. Use the life table and linear interpolation to estimate the following quantities:

i. The percentage of newborns who will survive to age 65

ii. The median lifetime—that is, the age at which exactly half of the original cohort of 100,000 people will be alive

iii. The age at which only half of a graduating high school class will survive (Assume that all students graduate at exactly 18 years of age.)

Table 9

Age x	Expected number living l_x
0	100,000
10	98,347
20	97,741
30	96,477
40	94,926
50	91,526
60	83,726
70	68,248
80	43,180
90	14,154
100	1,150

Chapter 1 Review

Exercises 1-4   *List all the terms that describe the given number. Choose from natural, whole, integer, rational, irrational, real, and complex.*

1. -2.33

2. $\dfrac{12}{3}$

3. A number that, when squared, gives -11

4. π

Exercises 5-8   *Write the set using interval notation and sketch the graph.*

5. $\{x \mid -5 \le x < -1\}$

6. $\{x \mid x > 3\}$

7. $\left\{x \mid x \le \dfrac{1}{2}\right\}$

8. $\{x \mid \pi \le x \le 2\pi\}$

Exercises 9-12   *Simplify the given expression. Write your answer without using negative exponents.*

9. $(-3)^4$

10. $\left(\dfrac{4}{5}\right)^{-2}$

11. $\dfrac{\left(a^3 b\right)^2}{a^4 b^6}$

12. $\dfrac{x^{-2} y^3}{(xy^{-1})^{-3}}$

Exercises 13-14   *Convert the given number to scientific notation.*

13. $31,400,000$

14. 0.00461

Exercises 15-20   *Simplify the given radical expression.*

15. $\sqrt[3]{a^6 b^4}$

16. $\sqrt{32 u^3 v^6}$

17. $\dfrac{9}{\sqrt{3}}$

18. $\dfrac{2}{\sqrt[3]{2}}$

19. $\dfrac{1}{x - \sqrt{2}}$

20. $\dfrac{x - 4}{\sqrt{x} - 2}$

Exercises 21-24   *Perform the indicated operation and simplify. Assume all variables represent positive numbers.*

21. $\sqrt{6s^5t} \cdot \sqrt{2st}$

22. $\dfrac{\sqrt[3]{54xy^2}}{\sqrt[3]{2xy}}$

23. $\sqrt{28} + 9\sqrt{7}$

24. $3\sqrt{2} - \sqrt{8}$

Exercises 25-26   *Convert from radical notation to exponential notation. Assume all variables represent positive numbers. Write your answer without using negative exponents.*

25. $\sqrt[4]{t^2}$

26. $\dfrac{1}{\sqrt[3]{x}}$

Exercises 27-28   *Convert from exponential notation to radical notation.*

27. $x^{2/3}$

28. $t^{-5/2}$

Exercises 29-32   *Simplify the given exponential expression. Express the answer in exponential form without negative exponents.*

29. $(x^{1/3})^{1/4}$

30. $y^{2/5}y^{3/2}$

31. $\dfrac{a^{2/3}b^{4/3}}{a^{5/3}b^{-2/3}}$

32. $\left(\dfrac{xy^{-1/2}}{x^{1/4}y}\right)^2$

Exercises 33-36   *Perform the indicated operation and simplify.*

33. $(2t^3 + 4t) - (3t^3 + t^2 - t)$

34. $(2y + 3)(y - 4)$

35. $(2 - w)(4 - 3w)$

36. $(u^2 - 3u + 7)(u - 2)$

Exercises 37-42   *Factor the given polynomial.*

37. $9u^2 - 4$

38. $8x^3 + 125$

39. $t^2 + 4t - 32$

40. $2y^3x - 54x$

41. $9x^2 + 6x + 1$

42. $2x^2 + 7x - 4$

Exercises 43-46   *Simplify the rational expression.*

43. $\dfrac{x + 3}{x^2 - 9}$

44. $\dfrac{v^2 - v - 2}{v^2 + v - 6}$

45. $\dfrac{t^3 - 1}{t^3 - t}$

46. $\dfrac{y^2x^2 + 4y^2}{x^2y^2 + 4y^2 + 4x^2 + 16}$

Exercises 47-52   *Perform the indicated operation and simplify.*

47. $\dfrac{3}{y + 2} - \dfrac{1}{y}$

48. $\dfrac{1}{x^2 + x} + \dfrac{1}{x^2 - x}$

49. $\dfrac{x^2}{x^2 - 9} \cdot \dfrac{3 - x}{x^2 + x}$

50. $\dfrac{v^3 + 8}{v^2 - 4} \div \dfrac{v^2 - 2v + 4}{v - 2}$

51. $\dfrac{\dfrac{1}{t} + \dfrac{1}{2}}{t + 2}$

52. $\dfrac{1 - \dfrac{1}{x}}{1 + \dfrac{1}{x}}$

Exercises 53-54   *Plot the given points, find the distance between them, and find the coordinates of the midpoint of the line segment connecting them.*

53. $(4, 5), (-2, -3)$ **54.** $(-1, -2), (-3, 7)$

Exercises 55-58   *Find the center and radius of the circle and sketch its graph.*

55. $x^2 + y^2 = 16$ **56.** $(x + 3)^2 + (y - 2)^2 = 25$

57. $x^2 + y^2 + 10x - 8y + 32 = 0$

58. $x^2 + y^2 - 2x - 6y + 6 = 0$

Exercises 59-60   *Find the equation of the circle with the given properties.*

59. Center (2, 5) and radius 3

60. The points (5, 5) and (−1, −3) are endpoints of a diameter.

Exercises 61-64   *Find the graph that matches the given equation. Choose from i–iv.*

i.

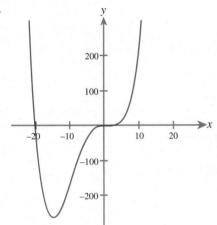

iv.

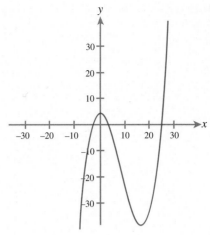

ii.

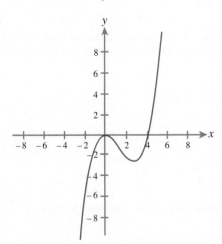

61. $y = \dfrac{x^3}{48} - \dfrac{x^2}{2} + 4$

62. $y = \dfrac{x^4}{80} + \dfrac{x^3}{5} - x^2$

63. $y = \dfrac{x^3}{4} - \dfrac{33x^2}{32}$

64. $y = 4 - 0.48x^2$

Exercises 65-68   *Use the trace feature of a graphing calculator to estimate the indicated value(s) to the nearest hundredth.*

65. $y = \dfrac{x^3}{8} - \dfrac{15x^2}{16} - \dfrac{129x}{32} + 5$; x- and y-intercepts

66. $y = \dfrac{1440}{x^2 + 36} - 2x - 32$; x- and y-intercepts

67. $y = \sqrt{-x^2 + 4x + 5}$; coordinates of the highest point on the graph

68. $y = x^2 - 3x + 1$; coordinates of the lowest point on the graph

Exercises 69-88   *Use any method to find all solutions.*

69. $3x + 4 = 9$

70. $\dfrac{2}{3}(x - 4) = \dfrac{1}{6}(x - 1)$

71. $2(x + 1) - 3(2x - 2) = 1$

72. $(x + 1)(x - 2) = (x - 3)^2$

73. $x(2x - 1)(x + 3) = 0$

74. $x^2 + 5x + 6 = 0$

iii.

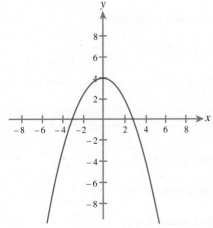

75. $x^2 + x - 12 = 0$

76. $2x^2 - 4x = 9$

77. $w^2 + 4w + 2 = 0$

78. $x^2 + x + 2 = 0$

79. $6y^2 - y - 2 = 0$

80. $x^4 - 3x^2 = 4$

81. $\dfrac{2}{x} = \dfrac{3}{x+1}$

82. $\dfrac{3x-1}{2x} = \dfrac{3x+4}{2x+1}$

83. $\dfrac{x}{2x+1} = \dfrac{x+2}{5x}$

84. $\dfrac{x^2}{x+1} = -3 + \dfrac{1}{x+1}$

85. $\sqrt{3x-5} = 1$

86. $\sqrt{2x-3} + 7 = 5$

87. $\sqrt{3a+10} - a = 2$

88. $\sqrt{x-1} + \sqrt{x+4} = 5$

Exercises 89-92 *Solve the given equation with a graphing calculator. Give answers accurate to the nearest hundredth.*

89. $x^7 - x^3 + 1 = 0$

90. $4x^3 - 2x - 1 = 0$

91. $x^4 - 3x = 1$

92. $x^3 - 2x^2 = 5x - 6$

Exercises 93-98 *Solve the given equation for y. Use a graphing calculator to plot the graph of the resulting equation and estimate the x-intercept(s) to the nearest hundredth.*

93. $2x + 3y = 12$

94. $xy - x^2 + 4 = 0$

95. $y^4 - x = 5$

96. $y^3 + x = 2$

97. $x + 4y = y^2 + 5$

98. $x^2 - 4y = 5 - y^2$

Exercises 99-106 *Solve the inequality and graph the solution.*

99. $-3x - 4 < 8$

100. $2x + 1 \geq 5$

101. $\dfrac{1}{3}(x+2) - \dfrac{1}{2}(x-2) > 0$

102. $-2 < \dfrac{2x-3}{4} \leq 6$

103. $2x - 1 \leq 5$ or $3 - 3x \leq 0$

104. $2x - 2 > -6$ and $5 - x \geq 1$

105. $|-3x + 4| < 2$

106. $|2x + 4| > 1$

Exercises 107-110 *Solve the given polynomial inequality. Find exact values whenever possible. If you must approximate, give answers correct to the nearest hundredth.*

107. $(s+1)(s+5) > 0$

108. $4t^3 > 4t^2$

109. $x^4 - 2x < 1$

110. $x^3 - 2x^2 \leq 5x - 6$

Exercises 111-114 *Find the slope of the line satisfying the given properties.*

111. Passing through $(2, 5)$ and $(-1, 4)$

112. Passing through $(2, 5)$ with y-intercept 5

113. With equation $y = 4x - 7$

114. With equation $3x + 4y = 6$

Exercises 115-124 *Find the equation of the line satisfying the indicated properties. Express your answer in slope-intercept form $y = mx + b$, if possible.*

115. Slope 3 and passing through $(1, 5)$

116. Slope undefined and passing through $(2, \pi)$

117. Passing through $(-1, 6)$ and $(3, -2)$

118. Passing through (2, 4) and (−2, 7)

119. With slope −2 and y-intercept 2

120. With y-intercept 5 and x-intercept 1

121. Parallel to $y = -3x + 1$ and passing through (7, 1)

122. Passing through (−2, 4) and perpendicular to the line containing (3, −1) and (5, −1)

123. Perpendicular to $2x - 3y = 1$ and passing through (−4, 2)

124. Parallel to $5x + 2y = 10$ and passing through (0, 0)

Exercises 125-132   *Parts a and b are connected: Part a involves an elementary concept, whereas part b involves related material from this chapter. First answer a, and then use this result to answer b.*

125. a. Expand $(2x - 3y)^2$.

   b. It is given that $2x \geq 3y$. Simplify $\sqrt{4x^2 - 12xy + 9y^2}$.

126. a. Expand $3a(2a - 3)(a - 5)$.

   b. Factor $6a^3 - 39a^2 + 45a$.

127. a. Find the midpoint of the line segment connecting (−2, 3) and (2, 11).

   b. Find a point P such that the line segment connecting (−2, 3) and (0, 7) is half the length of the line segment connecting (−2, 3) and P.

128. a. Simplify $2x(x + 1)(x - 1)$.

   b. Simplify $\dfrac{2x^3 - 2x}{x + 1}$.

129. a. Factor $x^2 - x - 6$.

   b. Solve $\dfrac{x^2 - x - 6}{x^{14} + 7x^{12} + 6x^3 + 192.35x + 8} = 0$.

130. a. Evaluate $(x + 4)(x + 5)(x + 6)$ for $x = 1$.

   b. Find three consecutive integers whose product is 210.

131. a. Expand $(x - 3)(x + 3)(x + 9)$.

   b. Find all solutions of $Z^3 + 9Z^2 - 9Z = 81$.

132. a. Evaluate $(x + 1)^2 + (x - 2)^2$ for $x = 11$.

   b. Find the dimensions of a right triangle with hypotenuse 15 in which one side is 3 units longer than the other.

133. Volume of a Pyramid   The volume of a pyramid with height h and a square base with side length x is given by the monomial $\frac{1}{3}x^2h$. Find the volume of the Great Pyramid of Egypt given that the base has length 750 feet and the height is 450 feet. The dimensions of an Olympic-size swimming pool are approximately $164' \times 75' \times 6\frac{1}{2}'$. How many Olympic-size swimming pools could be drained into the pyramid?

The Great Pyramid at Giza near Cairo.

134. Maximizing Area   Consider a rectangle with dimensions l and w. Denote the perimeter of the rectangle by P, so that $P = 2l + 2w$, and let $d = l - w$, the difference between the dimensions.

   a. Show that $\frac{1}{4}(P + 2d) = l$ and $\frac{1}{4}(P - 2d) = w$.

   b. Use the expressions given in part a to find a formula for the area of the rectangle in terms of P and d.

   c. Use your formula from part b to deduce that for a fixed perimeter P, the area is greatest when $d = 0$. For what kind of rectangle is $d = 0$?

135. CD Price   A compact disc (CD) costs $12.19 including 6% sales tax. What is the price of the CD before sales tax?

136. Interception Risk   Since many professional football players can jump and extend their hands to a height of 11 feet or more, a passed football is at some risk of interception whenever its height is 11 feet or below. If the quarterback releases the ball from a height of 6 feet with an upward velocity of 30 feet per second, then the height of the ball in feet t seconds after it is released is given by $h = -16t^2 + 30t + 6$. Find the time interval for which the football is at risk of being intercepted.

137. Fortune 500   A corporate CEO would like her company's total sales to surpass $625 million, which she estimates would be enough to be listed in the *Fortune 500*. If the company's revenue (in millions of dollars) for the sale of x million units of a product is given by $R = \frac{1}{3}x(100 - x)$, then for what interval of unit sales will the company attain her goal?

138. Rectangle Dimensions   The length of a rectangle is $\frac{1}{2}$ foot less than 6 times its width. Its area is 3 square feet. Find the dimensions of the rectangle.

139. Car Wash Rate   It takes Daniel 1 hour longer to wax the car than it does his brother Roberto. Together it takes them 50 minutes to wax the car. How long does it take each working separately?

140. Triangle Dimensions   What are the dimensions of a right triangle if one side is 3 inches less than twice the length of the other and the length of the hypotenuse is 51 inches?

Chapter 1 Test

Problems 1-8   *Answer true or false.*

1. All rational numbers are real.

2. For some real numbers x, $|x| = -x$.

3. The principal root of any positive number is positive.

4. If $(x + 1)(x - 4) = 4$, then either $x = 3$ or $x = 6$.

5. If $(x + 1)(x - 4) = 0$, then either $x = -1$ or $x = 4$.

6. If $|x - 2| = 3$, then x is a distance of 3 units from 2.

7. If $|x| > 1$, then $-1 > x > 1$.

8. If $a^2 = b^2$, then $a = b$.

Problems 9-12   *Give an example of each.*

9. An irrational number

10. A perfect square trinomial

11. A quadratic equation with 3 as its only solution

12. An absolute value inequality with no solution

Problems 13-15   *Simplify the given expression. Assume all variables represent positive numbers.*

13. $\left(\dfrac{a^{-3}b^4}{a^2b^{-1}}\right)^2$

14. $\sqrt{27x^5y^6}$

15. $\dfrac{2}{\sqrt[3]{4}}$

Problems 16-17   *Perform the indicated operation and simplify whenever possible.*

16. $(2x + 3)(3x - 2)$

17. $\dfrac{1}{x^2 + 2x} + \dfrac{x}{x^2 - 4}$

Problems 18-19   *Factor the given polynomial.*

18. $4x^2 - 9$

19. $u^3 + 8v^3$

Problems 20-24   *Solve the given equation or inequality. Give exact answers wherever possible. When approximating, give answers accurate to the nearest hundredth.*

20. $|3 - 2y| > 1$

21. $x^4 - 2x^2 + x - 1 = 0$

22. $2x^2 - 4x = 8$

23. $x^2 - 5 \le 4x$

24. $\sqrt{2x + 1} + 1 = x$

Problems 25-26   *Find an equation of the line with the given properties.*

25. Passes through the points $(-2, 3)$ and $(-2, 7)$

26. Passes through the point $(-1, 4)$ and is perpendicular to the line with equation $4x + 3y = 12$

27. The graph of an equation is shown in Figure 113, with a viewing rectangle defined by $\texttt{Xmin} = -10$, $\texttt{Xmax} = 10$, $\texttt{Ymin} = -10$, $\texttt{Ymax} = 10$. Match the viewing rectangles labeled i–iii with the range values in parts a–c.

Figure 113

i.

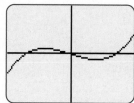

ii.

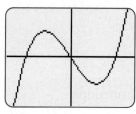

iii.

a. $\texttt{Xmin} = -2$, $\texttt{Xmax} = 2$, $\texttt{Ymin} = -2$, $\texttt{Ymax} = 2$

b. $\texttt{Xmin} = -2$, $\texttt{Xmax} = 2$, $\texttt{Ymin} = -10$, $\texttt{Ymax} = 10$

c. $\texttt{Xmin} = -10$, $\texttt{Xmax} = 10$, $\texttt{Ymin} = -2$, $\texttt{Ymax} = 2$

28. A ball thrown upward from the ground with an initial velocity of 44 feet per second has height in feet after t seconds given by the polynomial $-16t^2 + 44t$. Find the height of the ball at time $t = 2$. When does the ball hit the ground?

Chapter 2

Functions and Their Graphs

The Philippine volcano Mount Pinatubo erupted on June 15, 1991, spewing millions of tons of sulfur dioxide and ash into Earth's atmosphere. Rising, the hot gas and ash combined with water vapor to form a haze of sulfuric acid droplets, blocking sunlight and cooling Earth's atmosphere. Such naturally occurring phenomena complicate the task of measuring the effects of greenhouse gases (such as carbon dioxide) on global warming. Atmospheric scientists believe that volcanic eruptions and emissions of chemicals known as CFCs from aerosol sprays and other sources contribute to the depletion of stratospheric ozone. In this chapter, we mathematically model both carbon dioxide and atmospheric ozone levels and discuss the difficulty of modeling Earth's atmosphere.

Section 2.1 | Functions

- What was the volume of the sphere of influence of Martin Luther King's "I have a dream" speech at the turn of the century?
- How fast must a cyclist ride in order to generate enough power to replace a small power plant?
- Does the association of a U.S. president to his vice president constitute a function?
- What type of business software is designed specifically for the rapid computation of thousands of function values?
- Why is it that no matter what expression you enter into your graphing calculator, its graph will not be a circle?

It is impossible to overstate the significance of functions—both as a foundational mathematical concept and as a tool for describing relationships between real-world variables. The language of functions permeates every branch of the physical and social sciences: IQ scores are a function of nature and nurture, pressure is a function of temperature and volume, consumer demand is a function of price, population is a function of food supply, and computational power is a function of processor speed. We will see that whenever the value of one variable determines the value of another, this correspondence defines a function. In fact, many of the equations involving two variables that we saw in the first chapter of this text define functional relationships. Specifically, every time we solved for y in an equation involving both x and y, we were expressing y as a function of x. It is thus in a very real sense that this section is largely a formalization of a concept with which we already are quite familiar.

Definition of Function

Correspondences between sets of objects are very common in mathematics as well as in everyday life. Table 1 gives some examples of such correspondences.

Table 1
Correspondences between sets

To each human . . .	there corresponds . . .	a biological mother.
To each license plate number in a given state . . .	there corresponds . . .	a vehicle.
To each package weight . . .	there corresponds . . .	a parcel post delivery rate.
To each positive real number . . .	there corresponds . . .	the square of that number.

Each of the examples in Table 1 has the following form:

> To each element of a set D . . . there corresponds . . . an element of a set R.

More specifically, to each element of a set D there corresponds *exactly* one element of a set R. Correspondences with this property are called *functions*.

Definition of a Function

A **function** from a set D to a set R is a correspondence or rule that assigns to each element x of D exactly one element y of R. The set D is called the **domain** of the function and the elements x of D are the **input values**. The elements y of R that correspond to the input values are the **output values**. The set of all possible output values is called the **range** of the function.

EXAMPLE 1

Examples of Functions

Verify that each of the examples given in Table 1 is a function and identify the domain and range for each.

Solution

a. Since each human has exactly one biological mother, this correspondence is a function. The domain is the set of all humans. The range is the set of all mothers.

b. You can imagine the resulting chaos if the same license number were assigned to two different vehicles. So it is by necessity that to each license number there corresponds exactly one vehicle, and thus this correspondence is a function. The domain is the set of all current license numbers, and the range is the set of all licensed vehicles.

c. For any given package weight, there is only one parcel post rate, and so this correspondence is a function. The domain is the set of all allowable package weights (up to 70 pounds for the Postal Service). The range is the set of all parcel post rates.

d. Each positive real number has exactly one square, and so this correspondence is a function. The domain is the set of all real numbers. Since the square of a real number can never be negative, the range is the set of nonnegative real numbers.

EXAMPLE 2

An Example of a Nonfunction

Explain why the correspondence between the set D of all telephone numbers and the set R of all telephones is not a function.

Solution It is not uncommon for several telephones in a home to have the same number. Thus, there are elements of the set D that are paired with more than one element of R, and so the correspondence is not a function.

Equations and Functions

Functions involving sets of real numbers are often defined by an equation in two variables. By convention, we typically let the variable x represent the input and y the output. For example, the function described earlier that assigns to each positive real number the square of that number can be defined by the equation $y = x^2$. This equation specifies that to each input value x there corresponds the output value $y = x^2$, and so the equation defines a function. More generally, an equation of the form

$$y = \square$$

where $\square$ is an expression involving x, defines y as a function of x. The domain of such a function is the set of all values that may be assigned to the input variable x, or the **independent variable**, as it is sometimes called. The range is the set of all resulting values for the output variable y, otherwise known as the **dependent variable**. It is important to note, however, that not all equations in x and y define y as a function of x. In order for y to be a function of x, there must correspond exactly one y for each x.

EXAMPLE 3

Testing an Equation to See if It Defines a Function

Determine whether the equation $y - x^3 = 0$ defines y as a function of x. If so, find the unique y-values that correspond to the x-values given in the following table:

x	y
−2	
0	
1	
3	

Solution   Solving for y, we have $y = x^3$. This shows that for each value of x there is exactly one value for y. Thus, the equation does define y as a function of x. We complete the table as follows:

x	$y = x^3$
−2	$(-2)^3 = -8$
0	$0^3 = 0$
1	$1^3 = 1$
3	$3^3 = 27$

⋯⋯▷EXAMPLE 4   **An Equation that Does Not Define y as a Function of x**

Show that the equation $y^2 - x = 0$ does not define y as a function of x.

Solution

We first solve for y.

$$y^2 - x = 0$$
$$y^2 = x$$
$$y = \pm\sqrt{x}$$

This indicates that for each positive value of x there are *two* values for y—namely, $y = \sqrt{x}$ and $y = -\sqrt{x}$. Thus, this equation does *not* define y as a function of x. Note, however, that since the equation can be written as $x = y^2$, with x on the left and an expression involving only y on the right, $y^2 - x = 0$ *does* define x as a function of y.

The graph of the equation $y^2 - x = 0$ from Example 4 shows us very clearly why it does not define the variable y as a function of x. By rewriting the equation as $x = y^2$ and plotting points, we obtain the graph shown in Figure 1. Notice that for any positive value of a, both of the points $(a, \sqrt{a})$ and $(a, -\sqrt{a})$ are on the graph: There are two distinct y-values corresponding to the same x-value. Graphically, this means that the vertical line $x = a$ intersects the graph twice. This is not allowed if the equation is to define y as a function of x since such a function must pair each x with exactly one y, which suggests the following test.

The Vertical Line Test

An equation defines y as a function of x if and only if no vertical line crosses the graph of the equation more than once.

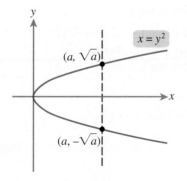

Figure 1

EXAMPLE 5

Using the Vertical Line Test

Use the vertical line test to identify which of the following equations define y as a function of x.

a. $x^2 + y^2 = 4$ **b.** $y = (x + 1)^3 - 2$

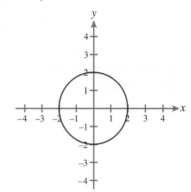 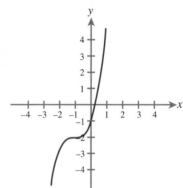

Solution

a. $x^2 + y^2 = 4$ does not define y as a function of x since some vertical lines intersect the graph in two points.

b. $y = (x + 1)^3 - 2$ does define y as a function of x since no vertical line intersects the graph in more than one point.

Function Notation

As we observed earlier, an equation of the form $y = \square$, where $\square$ is a single-valued expression involving x, defines a function. In many circumstances, particularly when more than one function is under discussion, it is convenient to attach names to functions. For example, the functions defined by the equations $y = \sqrt{x}$ and $y = x^2$ might be referred to as f and g, respectively. The output value of a function f associated with an input value x is denoted by $f(x)$, read "f of x." Thus, instead of writing $y = \sqrt{x}$, we write $f(x) = \sqrt{x}$. Functions expressed in this way are said to be written in **function notation**.

EXAMPLE 6

Using Function Notation

Write each of the following functions using function notation:

a. h is the function that assigns to each nonzero real number its reciprocal.
b. g is the function that assigns to each real number t the number $t^3 - t$.
c. P is the function that assigns to each real number one less than its square.

Solution

a. The reciprocal of a number x is the number $1/x$. Thus, $h(x) = 1/x$.
b. $g(t) = t^3 - t$
c. One less than the square of a real number x is the number $x^2 - 1$. Thus, $P(x) = x^2 - 1$.

EXAMPLE 7

Constructing a Function from a Formula

Express the area of a circle as a function of its circumference.

Solution   The area of a circle in terms of its radius is $A = \pi r^2$. The circumference of a circle in terms of its radius is $C = 2\pi r$. Solving the latter equation for r, we obtain $r = C/(2\pi)$. Substituting for r in the area formula gives

$$A = \pi r^2 = \pi \left(\frac{C}{2\pi} \right)^2 = \pi \frac{C^2}{4\pi^2} = \frac{C^2}{4\pi}$$

Thus $A = \dfrac{C^2}{4\pi}$.

Examples 6 and 7 illustrate that not all functions are named f and that the independent variable need not be x. In fact, the independent variable simply serves as a placeholder, and thus could be replaced by any symbol. For example, the function $f(x) = x^2 - 3x$ is exactly the same as the function $f(\square) = \square^2 - 3\square$. With either definition, the output is obtained by subtracting three times the input from the square of the input. Thus, in order to find the value of the function $f(x) = x^2 - 3x$ when $x = -1$, we simply replace all occurrences of x with -1 to obtain

$$f(-1) = (-1)^2 - 3(-1) = 1 + 3 = 4$$

WARNING!

> The function notation $f(-1)$ represents the output value of the function f when $x = -1$. It does not mean we are to multiply f by -1.

EXAMPLE 8

Evaluating Functions

Let $g(t) = t^3 - t$. Find the following function values.

a. $g(-2)$ **b.** $g\left(\begin{smallmatrix} \circ \\ \curlywedge \end{smallmatrix} \right)$ **c.** $g(x + 1)$

Solution

a. $g(-2) = (-2)^3 - (-2) = -8 + 2 = -6$

b. $g\left(\begin{smallmatrix} \circ \\ \curlywedge \end{smallmatrix} \right) = \left(\begin{smallmatrix} \circ \\ \curlywedge \end{smallmatrix} \right)^3 - \begin{smallmatrix} \circ \\ \curlywedge \end{smallmatrix}$

c. $g(x + 1) = (x + 1)^3 - (x + 1)$
$$= (x^3 + 3x^2 + 3x + 1) - (x + 1)$$
$$= x^3 + 3x^2 + 3x + 1 - x - 1$$
$$= x^3 + 3x^2 + 2x$$

The process of evaluating a function can be illustrated using a "function machine," such as the one suggested by Figure 2. When an input value x is fed into the machine, the machine operates on the number and produces an output value $f(x)$. If no restrictions are specifically placed on the input values, the domain of a function is assumed to be the set of all real number input values for which the function is defined. The range is then the set of all resulting output values.

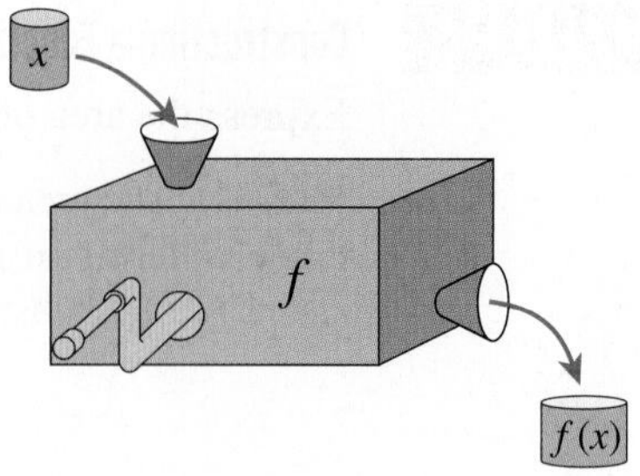

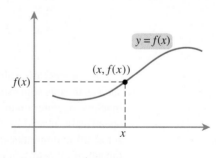

Figure 2

⋯⋯►EXAMPLE 9

Finding Domain

Find the domain of $g(t) = \dfrac{t}{t^2 - 1}$.

Solution All input values of t are valid except those that lead to a zero in the denominator. To determine where this occurs, we set $t^2 - 1 = 0$ and solve for t.

$$t^2 - 1 = 0$$
$$(t + 1)(t - 1) = 0$$
$$t = -1, t = 1$$

So the domain of g is the set of all real numbers t except -1 and 1. This set can be written in set-builder notation as $\{t \mid t \neq -1, t \neq 1\}$. In interval notation, we write $(-\infty, -1) \cup (-1, 1) \cup (1, \infty)$.

⋯⋯►EXAMPLE 10

Finding Domain

Find the domain of $h(x) = \sqrt{2x - 3}$.

Solution Since we are only considering values of x for which $h(x)$ is real, the valid input values of x are those for which $2x - 3 \geq 0$. Solving this linear inequality for x yields $x \geq \frac{3}{2}$. Thus, the domain of h is $\left[\frac{3}{2}, \infty\right)$.

Graphs of Functions

The **graph of a function** f is the set of points (x, y) such that $y = f(x)$. In other words, the graph of a function f is simply the graph of the equation $y = f(x)$. Thus, the techniques developed for graphing equations can also be applied to graphing functions.

The graph of a function depicts the correspondence between input values in the domain and output values in the range. Indeed, if (x, y) is a point on the graph of a function f, then y is the value of the function at x, as shown in Figure 3.

Figure 3

····EXAMPLE 11   **Graphing a Function**

Sketch the graph of the function $f(x) = \frac{1}{2}x - 3$.

Solution   Replacing $f(x)$ with y, we obtain $y = \frac{1}{2}x - 3$, a straight line with y-intercept -3 and slope $\frac{1}{2}$. The graph is shown in Figure 4.

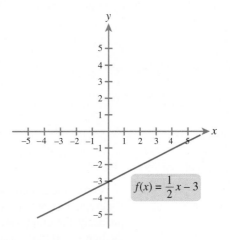

Figure 4

····EXAMPLE 12   **Graphing a Function with a Restricted Domain**

Sketch the graph of the function $f(x) = \frac{1}{x}$.

Solution   We first note that 0 is not in the domain of f. So there will be no point on the graph with x-coordinate 0. Next, we find and plot some points satisfying $y = \frac{1}{x}$, as shown in Figure 5.

x	$y = \dfrac{1}{x}$	x	$y = \dfrac{1}{x}$
1	1	$\dfrac{1}{2}$	2
2	$\dfrac{1}{2}$	$\dfrac{1}{3}$	3
3	$\dfrac{1}{3}$	$-\dfrac{1}{2}$	-2
-1	-1	$-\dfrac{1}{3}$	-3
-2	$-\dfrac{1}{2}$		
-3	$-\dfrac{1}{3}$		

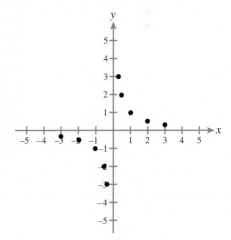

Figure 5

Notice that as the x-values get larger (in magnitude), the corresponding y-values get smaller, and so the points get closer and closer to the x-axis. Moreover, as the x-values get smaller, the y-values get larger, shooting up or down along the y-axis. We use these observations to complete the graph, as shown in Figure 6.

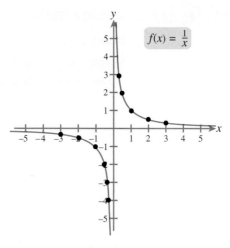

Figure 6

As we noted, there is no point on the graph with x-coordinate 0. As a consequence, the graph does not cross the line $x = 0$ but rather approaches it arbitrarily closely. More precisely, the graph rises (or falls) without bound as x gets close to 0. For this reason, the line $x = 0$ (the y-axis) is called a *vertical asymptote*. The line $y = 0$ (the x-axis), on the other hand, is a *horizontal asymptote* since the graph approaches it arbitrarily closely as x goes toward ∞ and $-\infty$. We will study vertical and horizontal asymptotes in detail in Section 3.5.

····⫶EXAMPLE 13   **Determining Domain and Range Using a Graphing Calculator**

Use a graphing calculator to plot the graph of the function $f(x) = \sqrt{x^2 - 16}$, and use the graph to determine the domain and range of f.

Solution   We enter the equation $y = \sqrt{x^2 - 16}$ and plot it with a graphing calculator, as shown in Figures 7 and 8. To find the domain, we look for the set of all possible x-coordinates of points on the graph. The view in Figure 7 suggests that no portion of the graph lies between $x = -4$ and $x = 4$. The two views together suggest that the graph extends infinitely to the right and left. Thus, we conclude that the domain is $(-\infty, -4] \cup [4, \infty)$. Note that the numbers -4 and 4 are included because both values can be substituted into the function. The range consists of the set of all possible y-coordinates of points on the graph. Since the graph lies strictly above the x-axis, and appears to extend infinitely upward, we conclude that the y-coordinates are all positive (or zero) and approach infinity. In other words, we conclude that the range is $[0, \infty)$.

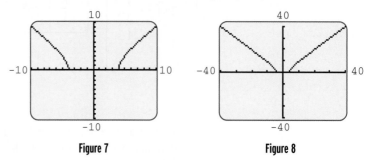

Figure 7   **Figure 8**

Functions are generally the method of choice for expressing relationships between variables in the real world, and even the not-so-real world, as the following example illustrates.

····→EXAMPLE 14   **Computing the Power Output of a Cyclist**

For a 170-pound cyclist on a 30-pound bicycle pedaling at v miles per hour, the power output in watts is given by $P(v) = 0.0178678v^3 + 2.01168v$.

a. Find the power output if the cyclist is traveling at 20 miles per hour.
b. A small power plant will produce around 500,000,000 watts of power. Find the speed required for a single cyclist to generate enough power to replace a small power plant. Answer to the nearest mile per hour.

*The human-powered Gossamer
Albatross in flight.*

Solution

a. Evaluating $P(20)$, we have

$$P(20) = 0.0178678(20)^3 + 2.01168(20) \approx 183.176 \text{ watts}$$

b. We could set $P(v) = 500,000,000$ and then solve for v using the graphical techniques developed in Chapter 1. Instead, for the sake of illustration, we plot the graph of P with a graphing calculator and search for a point with y-coordinate 500,000,000. After some experimentation, we settle on the view shown in Figure 9. Next we zoom in until we can estimate the x-coordinate of the point to the nearest mile per hour, as shown in Figure 10. Thus, a cyclist capable of traveling at 3036 miles per hour would generate enough power to replace a small power plant.

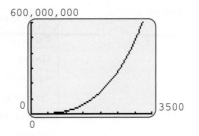

Figure 9

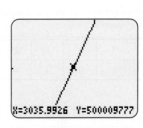

Figure 10

Understanding and Mastery Checklists

Concepts to Understand	Skills to Master
Functions ✤ Domain and range of a function ✤ Vertical line test ✤ Function notation ✤ Graphs of functions	Determine when a correspondence defines a function. ✤ Find the domain of a function. ✤ Use a graphing calculator to estimate the domain and range of a function. ✤ Given the graph of an equation, determine if the graph defines y as a function of x. ✤ Evaluate functions expressed with function notation. ✤ Express functional relationships using function notation. ✤ Sketch the graph of a function.

Exercises 2.1

Exercises 1-6 *Determine whether the correspondence from the set D to the set R defines a function.*

1.

D	−2	−1	0	1	2
R	4	1	0	1	4

2.

D	4	1	0	1	4
R	−2	−1	0	1	2

3.

U.S. Presidents and Their Vice Presidents	
President (D)	**Vice President (R)**
George Washington	John Adams
John Adams	Thomas Jefferson
Thomas Jefferson	Aaron Burr
Thomas Jefferson	George Clinton
James Madison	George Clinton
James Madison	Elbridge Gerry

4.

Car and Driver Top Ten for 2002	
Make (D)	**Model (R)**
Acura	RSX
Audi	A4
BMW	3-Series/M3
BMW	5-Series
Chevrolet	Corvette
Ford	Focus
Honda	Accord
Honda	S2000
Porsche	Boxster
Subaru	Impreza WRX

5.

Internet Growth	
Year (D)	**Internet Hosts (R)**
1991	376,000
1992	727,000
1993	1,313,000
1994	2,217,000
1995	4,852,000
1996	9,472,000
1997	16,146,000
1998	29,670,000
1999	43,230,000
2000	72,398,092
2001	109,574,429
2002	147,344,723

6.

Density	
Substance (D)	**Density (lb/ft³) (R)**
Hydrogen	0.006
Air	0.1
Styrofoam	6.2
Whale oil	48.0
Ice	57.4
Water	62.4
Aluminum	168.4
Lead	704.2
Gold	1203.8
Neutron star	4.4×10^{16}

Exercises 7-12 *Determine whether the given equation defines y as a function of x.*

7. $3x + 4y = 12$

8. $y^2 - x^2 = 0$

9. $x^2 + y^2 = 1$

10. $x^2 + x - y = 0$

11. $0 = x - y^3$

12. $x = \sqrt{y}$

Exercises 13-18 *Use the graph to determine whether the given equation defines y as a function of x.*

13. $y = -x^2 + 1$

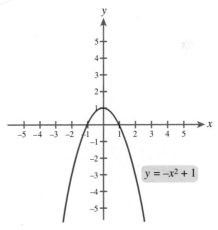

$y = -x^2 + 1$

14. $\dfrac{x^2}{16} + \dfrac{y^2}{9} = 1$

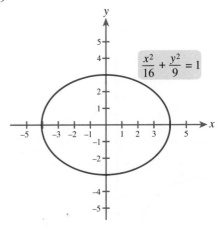

$\dfrac{x^2}{16} + \dfrac{y^2}{9} = 1$

15. $x = y^3 - 4y$

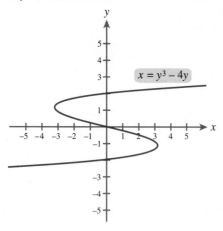

16. $x = y^3$

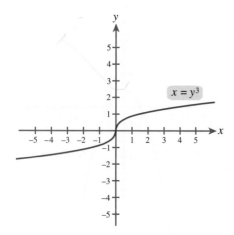

17. $xy = 4$

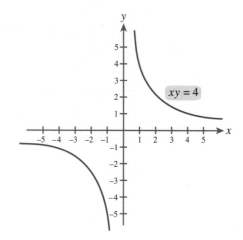

18. $|x| - |y| = 0$

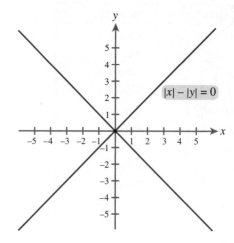

Exercises 19-28 *Evaluate the given function as indicated.*

19. $f(x) = x^2 + 3x + 2$

    a. $f(2)$

    b. $f(-3)$

    c. $f(0)$

20. $g(t) = t^3 - 4t$

    a. $g(1)$

    b. $g(-2)$

    c. $g(3)$

21. $f(x) = \dfrac{3x + 2}{4x - 5}$

    a. $f(4)$

    b. $f(5)$

    c. $f(-t)$

22. $h(y) = \dfrac{y^2 - 3}{y + 7}$

    a. $h(2)$

    b. $h(-6)$

    c. $h(x)$

23. $x(t) = 3t^2$

    a. $x(t^2)$

    b. $x(t - 1)$

    c. $x\left(\dfrac{5}{t}\right)$

24. $f(x) = \dfrac{2x + 2}{2x - 3}$

    a. $f(t)$

    b. $f(x - 1)$

    c. $f\left(\dfrac{x}{2}\right)$

25. $g(x) = (x + 1)^2$

    a. $g(x - 1)$

    b. $g(x^2)$

    c. $g\left(\dfrac{1 - a}{a}\right)$

26. $p(x) = \sqrt{9 - x^2}$

    a. $p(\Box)$

    b. $p(x + 3)$

    c. $p(x - 3)$

27. $f(x) = \dfrac{1}{x - 1}, g(x) = x^2$

    a. $f(g(3))$

    b. $f(g(t))$

    c. $g(f(x + 1))$

28. $f(x) = x^2, g(y) = 2y + 1$

    a. $f(g(2))$

    b. $f(g(t))$

    c. $g(f(n^2))$

Exercises 29-32 *Express the given function using function notation.*

29. h assigns to every real number the absolute value of the number.

30. g assigns to every real number twice the cube of the number.

31. f assigns to every real number in its domain the reciprocal of 3 more than the number.

32. p assigns to each real number in its domain the reciprocal of the square root of the number.

Exercises 33-44 *Find the domain of the function.*

33. $f(x) = x^2$

34. $g(x) = \dfrac{4}{x}$

35. $h(x) = \dfrac{3}{x - 4}$

36. $r(s) = |s|$

37. $r(t) = \dfrac{5t^2 + 3t + 1}{t^2 - 9}$

38. $p(x) = \dfrac{x + 1}{x^2 - 5x + 6}$

39. $f(x) = \sqrt{x + 1}$

40. $g(t) = \sqrt{t - 3}$

41. $f(x) = \dfrac{\sqrt{x + 2}}{x^2 - 4x + 3}$

42. $g(y) = \dfrac{2}{\sqrt{2y - 6}}$

43. $f(x) = \dfrac{2}{|x| - 3}$

44. $g(x) = \dfrac{x}{x}$

Exercises 45-48 *Find a formula f(x) for a function that satisfies the given correspondences between input and output values. There may be more than one correct answer.*

45.

x	0	1	2	3
$f(x)$	1	3	5	7

46.

x	0	1	2	3
$f(x)$	2	1	0	−1

47.

x	−1	0	1	2
$f(x)$	undefined	1	$\dfrac{1}{2}$	$\dfrac{1}{3}$

48.

x	2	5	10	17	26
$f(x)$	1	2	3	4	5

Exercises 49-52 *Find a function f for which the given set is the domain. There may be more than one correct answer.*

49. $(-\infty, 7]$

50. All real numbers except -4 and 1

51. All real numbers except 0 and 3

52. $[-2, \infty)$

Exercises 53-56 *Complete the table and use the resulting points to sketch the graph of the function.*

53.

x	$f(x) = -2x + 5$
0	
2	
4	

54.

x	$f(x) = \frac{2}{3}x - 2$
-3	
0	
3	

55.

x	$f(x) = (x + 1)^2 - 3$
-3	
-2	
-1	
0	
1	

56.

x	$f(x) = -(x - 2)^2 + 4$
0	
1	
2	
3	
4	

Exercises 57-60 *Use a graphing calculator to sketch the graph of the function, and then use the graph to help identify or approximate the domain and range of the function.*

57. a. $f(x) = x^2$

   b. $f(x) = (x + 1)^2$

   c. $f(x) = x^2 + 1$

   d. $f(x) = \dfrac{1}{x^2 + 1}$

58. a. $f(x) = \sqrt{x}$

   b. $f(x) = \sqrt{x + 4}$

   c. $f(x) = \sqrt{x} + 4$

   d. $f(x) = \dfrac{1}{\sqrt{x + 4}}$

59. a. $f(x) = \sqrt{x^2 + 1}$

   b. $f(x) = x + \sqrt{x^2 + 1}$

   c. $f(x) = x\sqrt{x^2 + 1}$

   d. $f(x) = \dfrac{x}{\sqrt{x^2 + 1}}$

60. a. $f(x) = \sqrt{9 - x^2}$

   b. $f(x) = x + \sqrt{9 - x^2}$

   c. $f(x) = x\sqrt{9 - x^2}$

   d. $f(x) = \dfrac{x}{\sqrt{9 - x^2}}$

Exercises 61-64 *The average rate of change of a function f over an interval [a, b] is defined to be*

$$\frac{f(b) - f(a)}{b - a}$$

For the given function, compute the average rate of change over the indicated interval.

61. $f(x) = 2x^2$, $[0, 2]$ **62.** $f(x) = \dfrac{1}{x + 1}$, $[1, 4]$

63. $f(x) = 7x + 5$, $[2.3, 6.8]$ **64.** $f(x) = \sqrt{x}$, $[4, 9]$

65. a. Use the quadratic formula to solve the equation
   $y = x^2 + 4x + 6$ for x.

   b. For what values of y does the equation of part a have a real-valued solution?

   c. What is the range of the function g defined by
   $g(x) = x^2 + 4x + 6$?

66. a. Find the domain of $f(x) = \dfrac{|x|}{x}$.

   b. Evaluate $f(2)$, $f(-3.76)$, and $f(10^{2345})$

   c. Find the range of f.

Applications

67. Volume of a Sphere   The volume of a sphere with radius r is given by $V = \frac{4}{3}\pi r^3$. Express the volume of a sphere as a function of its diameter.

68. Area of a Circle   The area of a circle with radius r is given by $A = \pi r^2$. Express the area of a circle as a function of its diameter.

69. Area of a Square   A square is inscribed in a circle. Express the area of the square as a function of the radius of the circle.

70. Volume of a Cube   A cube is inscribed in a sphere. Express the volume of the cube as a function of the radius of the sphere.

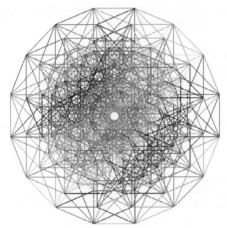

Just as a 2-dimensional cube (a square) has volume s^2, and a 3-dimensional cube has volume s^3, an 8-dimensional hypercube, as depicted here, has volume s^8.

71. Area of a Triangle   A line passing through the point $(2, 1)$ has intercepts $(a, 0)$ and $(0, b)$, which form the vertices of a right triangle, as shown in Figure 11. Express the area of the triangle as a function of a.

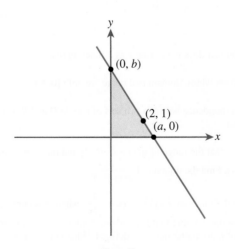

Figure 11

72. Area of a Triangle   Two perpendicular lines intersect at the point $(2, 3)$ as shown in Figure 12. Their x-intercepts form the vertices of a right triangle, as shown. Express the area of the triangle as a function of the length of its hypotenuse.

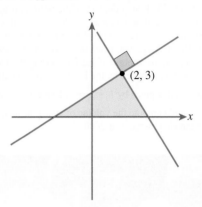

Figure 12

73. Volume of a Box   A box with a square base and no top is formed by cutting squares out of the corners of a piece of cardboard and then folding up the sides (Figure 13). If the cardboard measures $16'' \times 16''$ and the length of the edge of each cut-out square is denoted by x, express the volume of the resulting box as a function of x. What is the domain of this function? (*Hint:* Include in the domain only those values that make sense in the context of this problem.)

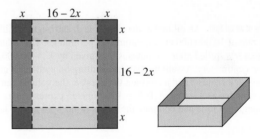

Figure 13

74. Area of a Pasture   A rancher wishes to enclose a rectangular pasture at the base of a cliff: 200 linear feet of fencing are available and no fencing is needed along the cliff side of the pasture (Figure 14). Express the area of the pasture as a function of x, the dimension perpendicular to the cliff face.

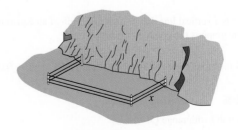

Figure 14

75. Sphere of Influence Electromagnetic waves—like those in radio and television transmissions, for example—travel at the speed of light, often in all directions. Although our radios pick up a signal almost instantaneously, the signal continues to travel through space at the speed of light. Thus, for example, Dr. Martin Luther King's famous 1963 "I have a dream" speech is even now reaching new parts of the galaxy. The radio waves from the broadcast began traveling through space at the speed of light in all directions from the point of origin. Since light travels at a rate of approximately 5.9×10^{12} miles per year, the distance the waves will have traveled after t years is given by $r = (5.9 \times 10^{12})t$. Moreover, when the waves are a distance r from the point of origin, the volume of space (ignoring the fact that some of the signal will be blocked by the Earth itself) through which they will have traveled is given by $V = \frac{4}{3}\pi r^3$. Express the volume V as a function of t, and compute the volume of space through which the speech had traveled by the year 2000.

Dr. Martin Luther King delivers his "I have a dream" speech.

76. Radius of an Oil Spill An oil tanker develops a leak and begins spilling oil at a rate of 10,000,000 cubic centimeters per minute. The volume of oil that has spilled after t minutes is thus given by $V = 10,000,000t$ cubic centimeters. Assuming that the resulting oil spill has a circular shape with radius r and a constant thickness h, the volume can also be expressed as $V = \pi r^2 h$. If $h = 1$ centimeter, express the radius as a function of time and compute the radius of the spill after 2 hours.

In spite of containment efforts, oil spills continue to cause immeasurable damage to the environment.

77. Subsidized Housing An apartment complex receives a government subsidy and, in return, must base its monthly rent on family size and monthly income. Selected income and rent amounts for a family of four are given in the following table. Plot the points and construct a function $R(x)$ that gives the monthly rent for a family of four with an income of x dollars per month.

Income ($)	800	950	1150	1400
Rent ($)	100	160	240	340

78. Cab Fare Selected cab fares are given in the following table. Plot the points and construct a function $F(x)$ that gives as output values the fare that would be charged for any given distance x.

Distance (mi)	1	3	7	13
Fare ($)	1.50	2.50	4.50	7.50

Concepts and Critical Thinking

Exercises 79-82 *Answer true or false.*

79. The domain of a function is the set of all output values.

80. No circle is the graph of $y = f(x)$, where f is a function.

81. If there is a vertical line that crosses the graph of an equation once, then the graph is the graph of a function.

82. For all functions f and real numbers x and y, $f(x + y) = f(x) + f(y)$.

Exercises 83-86 *Give an example of each.*

83. A function f with range $[5, \infty)$

84. A graph that does not define y as a function of x

85. A function whose domain and range are both $[0, \infty)$

86. A correspondence between two sets of people D and R that is not a function

87. Suppose that the range of $g(x)$ is $(-\infty, 4]$ and the domain of $g(x)$ is $(-\infty, \infty)$. Find the domain of $\dfrac{1}{g(x) - 5}$.

88. Define the function $f(x)$ by $f(x) = \dfrac{1}{q(x)}$, where q is another function. There is one number that is guaranteed *not* to be an element of the range of f, no matter how q is defined. What number is it?

89. Rational expressions are simplified by dividing out equal factors of the numerator and denominator. Thus, for example,

$$\frac{x^2 - 1}{x + 1} = \frac{(x + 1)(x - 1)}{x + 1} = x - 1$$

However, if we define

$$f(x) = \frac{x^2 - 1}{x + 1} \quad \text{and} \quad g(x) = x - 1$$

these two functions are not the same. Explain why this is so. In what sense is $\frac{x}{x} \neq 1$?

Questions for Discussion or Essay

90. A graphing calculator is capable of graphing equations of the form $y = \Box$, where $\Box$ is an expression involving x. As we have seen in this section, such equations define y as a function of x, and their graphs pass the vertical line test. Explain how a graphing calculator can be used as an aid in graphing an equation such as $x^2 + y^2 = 1$, even though its graph does not pass the vertical line test and so does not define y as a function of x.

91. In our discussion on equations and functions, we were careful to say that an equation "defines y as a function of x" rather than just saying that the equation was a function. What does it mean to define x as a function of y? Give an example of an equation that defines x as a function of y but that does not define y as a function of x. Now explain why it is not meaningful to say that an equation either is or is not a function.

92. In Example 14, we saw that a cyclist would have to travel at a rate of 3036 miles per hour in order to produce enough power to run a small power plant. If we agree that 20 miles per hour is a "normal" speed, the cyclist would have to travel 152 times faster than normal. If, instead of one cyclist traveling 152 times faster than normal, we had many cyclists traveling at normal speed, how many cyclists does your intuition tell you it should take to run the plant? Now compute how many it would take by using the facts from Example 14 that a cyclist traveling at 20 miles per hour can produce approximately 183 watts of power and a small power plant produces around 500,000,000 watts. What seems surprising about the answer? Can you explain why there is such a disparity between your intuitive answer and the actual number?

Projects for Enrichment

93. Functions of Several Variables Many of the correspondences between real-world quantities require the use of more than one variable. Consider, for example, the correspondence that assigns to each rectangle of length l and width w an area A equal to lw. The equation $A = lw$ defines the area as a function of two variables that we can write as $A(l, w) = lw$. In this case, each pair of values (l, w) is associated with *exactly one* area A. For example, a rectangle with length $l = 8$ and width $w = 5$ yields an area of $A(8, 5) = 8 \cdot 5 = 40$, and no other area is possible.

In general, we define a **function f of two variables** to be a rule or correspondence that assigns to a pair of numbers (x, y) exactly one number $f(x, y)$. The **domain** of a function of two variables is the set of all pairs (x, y) that can be used as input values. The **range** is the set of all corresponding output values. The domain of the function $f(x, y) = x^2 + y^2$ is the set of all pairs (x, y) such that x and y are real numbers. The range is the set of all positive real numbers.

a. Evaluate the following functions for the indicated values:

i. $d(r, t) = rt$; $d(55, 4)$, $d\left(20, \frac{1}{4}\right)$

ii. $f(x, y) = x^2 + 2xy + y^2$; $f(2, 3)$, $f(-1, 5)$

iii. $h(u, v) = 1/(u - v)$; $h(10, 8)$, $h(0.3, 0.4)$

iv. $g(x, y) = \sqrt{1 - x^2 - y^2}$; $g(0, 1)$, $g\left(\frac{1}{2}, \frac{1}{4}\right)$

b. Describe the domain of each of the functions in part a.

Functions of three or more variables are also possible. Consider the correspondence defined by the equation $B = P(1 + r)^t$, where B is the balance in an account after a principal of P dollars is deposited for t years at an annual rate of interest r. So B is a function of the three variables P, t, and r, and we write $B(P, r, t) = P(1 + r)^t$. Notice here that each triple of values (P, r, t) leads to *exactly one* balance B. For example, if $P = 1000$, $t = 4$, and $r = 0.05$, then

$$B(1000, 4, 0.05) = 1000(1 + 0.05)^4 \approx 1215.51$$

and this is the only balance possible for these values of P, t, and r.

c. Give a precise definition for a function of three variables and also for its domain and range.

d. Evaluate the following functions for the indicated values:

i. $V(l, w, h) = lwh$; $V(4, 6, 3)$, $V\left(5, \frac{3}{2}, 8\right)$

ii. $g(x, y, z) = \sqrt{x^2 + y^2 + z^2}$; $g(0, -4, 3)$, $g(-2, 1, 5)$

iii. $f(u, v, w) = \dfrac{uv + vw + uw}{uvw}$; $f(1, 2, 3), f\left(-4, 3, \dfrac{1}{3}\right)$

iv. $x(a, b, c) = \dfrac{-b + \sqrt{b^2 - 4ac}}{2a}$; $x(1, -7, -10)$,

$x(2, 0, -9)$

e. Describe the domain of each of the functions in part d.

94. Spreadsheet Functions A spreadsheet is well-suited for studying functions. In this project, we use a spreadsheet to investigate the definition of a function, the domain and range of a function, and the graph of a function.

a. Consider the spreadsheet fragment shown in Figure 15. Use the definition of function given in this section to explain why these data define a function. We refer to such functions as **numerical functions**. What is the domain of this numerical function? What is the range?

	A	B
1	**Input**	**Output**
2	0	1
3	1	2
4	2	5
5	3	10
6	4	17
7	5	26
8		

Figure 15

b. Set up a worksheet of your own with the labels **Input** and **Output** in cells A1 and B1, respectively. Enter the input values **0** through **3** in cells A2 through A5. Next, enter the formula $=$**sqrt(4-a2^2)** in cell B2. Figure 16 shows what you should see on your screen just before you press the Enter key after typing the formula. Copy the formula from cell B2 to cells B3 through B5. Your screen should look similar to Figure 17. Write out, using standard function notation, the symbolic definition of the function $f(x)$ whose output values agree with those given in the output column of Figure 17. What does the entry **#NUM!** in cell B5 suggest about the input value 3?

	A	B
1	**Input**	**Output**
2	0	=sqrt(4-a2^2)
3	1	
4	2	
5	3	

Figure 16

	A	B
1	**Input**	**Output**
2	0	2
3	1	1.732050808
4	2	0
5	3	#NUM!

Figure 17

c. Use your spreadsheet's "fill" feature to enter the numbers -3.0, -2.9, -2.8, $\ldots$, 2.9, 3.0 in the cell range A2:A62. Copy the formula in cell B2 to the cell range B3:B62. Based on the numerical data, what appears to be the domain of the function $f(x)$ you identified in part a? What is the range? How do the domain and range of the symbolic function $f(x)$ differ from the domain and range of the numerical function defined by the input–output columns of your spreadsheet?

d. Use your spreadsheet to produce input–output tables for the following functions. Use the given starting, stopping, and increment values for the input column. (In part c, for example, the starting value was -3.0, the stopping value was 3.0, and the increment value was 0.1.) In each case, estimate the domain and range of the symbolic function $f(x)$.

Function	Start	Stop	Increment
$f(x) = x\sqrt{4 - x^2}$	-3	3	0.1
$f(x) = \sqrt{-x^4 + 10x^2 - 9}$	-5	5	0.2
$f(x) = \dfrac{1}{\sqrt{25x + 1}}$	-1	3	0.1

e. Use algebraic techniques to find the domain and range of the function $f(x) = \dfrac{1}{\sqrt{25x + 1}}$ from part d. What does your answer tell you about the spreadsheet method for investigating domain and range?

Graphs produced by spreadsheets also define functions. Figures 18–20 show three different spreadsheet graphs of the same set of data.

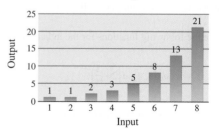

Figure 18

XY Scatter Graph

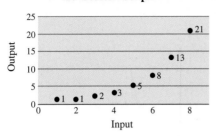

Figure 19

Line Graph

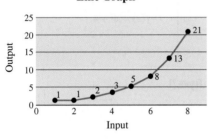

Figure 20

f. Use the definition of function given in this section to explain why these graphs define functions. We refer to such functions as **graphical functions**. Identify the domain and range of the graphical functions shown in Figures 18–20.

g. Use your spreadsheet's graphing feature to plot a column (or bar) graph, an *xy* (scatter) graph, and a line graph of the data from part c. Which type of graph most closely resembles a typical mathematical graph of a function?

Section 2.2 | **Graphs of Functions**

- Why does a soft drink can have the shape it does?
- What are the optimum dimensions of a box to be sent by mail?
- Why are functions satisfying $f(-x) = -f(x)$ said to be *odd*?
- How can a still photograph of a stretch of highway be used to estimate the speed at which the cars are traveling?
- How can the graph of a function be used to ensure that a bicycle racer receives a smooth handoff from a support vehicle?

Zeros of a Function   A **zero** of a function f is any number c such that $f(c) = 0$. It follows that if c is a zero of f, then $(c, 0)$ is an x-intercept of the graph of f. Thus, we can find the zeros of a function f algebraically by solving the equation $f(x) = 0$ or graphically by identifying the x-intercepts of the graph of $y = f(x)$.

·····➤**EXAMPLE 1**   **Finding Zeros Algebraically**

Find the zeros of the function $f(x) = x^2 - 4$ and illustrate them graphically.

Solution   To find the zeros, we set $f(x) = 0$ and solve for x.

$$f(x) = 0$$
$$x^2 - 4 = 0$$
$$(x + 2)(x - 2) = 0$$
$$x = -2, x = 2$$

In Figure 21, we see that the zeros of f, -2 and 2, correspond to the x-intercepts of the graph of $y = x^2 - 4$.

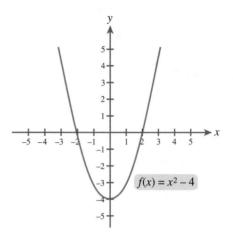

Figure 21

◦◦◦◦◦**EXAMPLE 2**

Finding Zeros Graphically

Use a graphing calculator to estimate the zeros of $f(x) = x^3 + 3x^2 - 3$ to the nearest hundredth.

Solution The graph of $y = x^3 + 3x^2 - 3$, shown in Figure 22, suggests three zeros (x-intercepts). We estimate each zero using the calculate-zero feature, obtaining $x \approx -2.53$, $x \approx -1.35$, and $x \approx 0.88$. As we will show in Chapter 3, a polynomial of degree 3 can have at most three zeros. Thus, there are no other zeros.

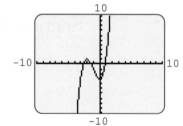

Figure 22

Translations

Many graphs are easily sketched by recognizing that they are merely shifts or **translations** of familiar graphs. For example, consider the graphs of $y = x^3$ and $y = (x - 5)^3$ shown in Figure 23. It is evident from the graphs and also from the following tables that the y-values are the same whenever the x-values for $y = (x - 5)^3$ are 5 greater than those for $y = x^3$. The net effect is that the graph of $y = (x - 5)^3$ has the same shape as the graph of $y = x^3$, but it is translated (shifted) 5 units to the right.

x	$y = x^3$		x	$y = (x - 5)^3$
-2	-8		3	-8
-1	-1		4	-1
0	0		5	0
1	1		6	1
2	8		7	8

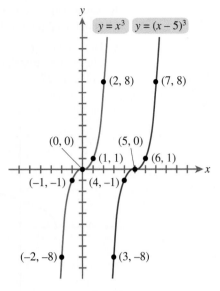

Figure 23

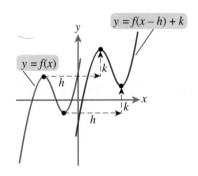

Figure 24

More generally, replacing x with $x - h$ in an equation results in a horizontal shift of the graph by h units (to the right if $h > 0$, to the left if $h < 0$). Thus, we conclude that the graph of $y = f(x - h)$ is a horizontal shift by h units of the graph of $y = f(x)$. Similarly, if we replace y by $y - k$ in an equation, then the graph will be shifted k units vertically (up if $k > 0$, down if $k < 0$). Consequently, the graph of $y - k = f(x)$ or, equivalently, $y = f(x) + k$ is a vertical shift by k units of the graph of $y = f(x)$. These observations are illustrated in Figure 24 (for $h, k > 0$) and are summarized as follows.

Translations of Functions

The graph of $y = f(x - h) + k$ is a translation h units horizontally and k units vertically of the graph of $y = f(x)$. If h is positive, then the translation is to the right; if h is negative, then the translation is to the left. If k is positive, then the translation is upward; if k is negative, then the translation is downward.

····▷ **EXAMPLE 3**

Graphing Translated Functions

Use the graph of $f(x) = \sqrt{x}$ to sketch the graph of the given function.

a. $q(x) = \sqrt{x} + 3$ **b.** $r(x) = \sqrt{x + 2}$ **c.** $s(x) = \sqrt{x - 1} - 2$

Solution

a. Since

$$q(x) = \sqrt{x} + 3$$
$$= f(x) + 3$$
$$= f(x - 0) + 3$$

we see that $q(x) = f(x - h) + k$ with $h = 0$ and $k = 3$. Thus, the graph of q is obtained by translating the graph of f 3 units upward, as shown in Figure 25.

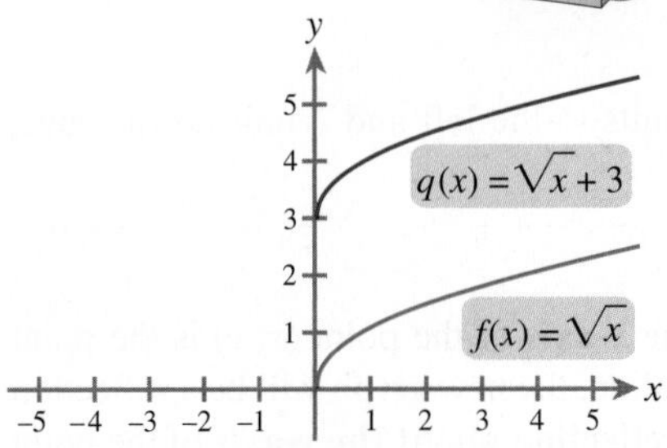

Figure 25

b. Since $r(x) = \sqrt{x + 2} = f(x + 2)$, we have $h = -2$ and $k = 0$. Thus, the graph of r is obtained by translating the graph of f 2 units to the left, as shown in Figure 26.

c. Here $s(x) = f(x - 1) - 2$, so that $h = 1$ and $k = -2$. Thus, the graph of s is obtained by translating the graph of f 1 unit to the right and 2 units down, as shown in Figure 27.

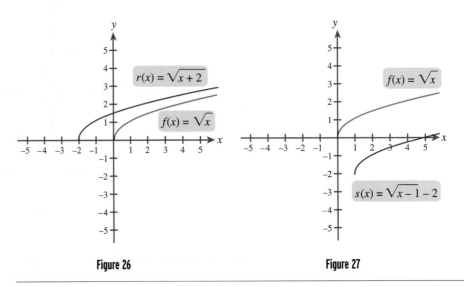

Figure 26   Figure 27

·····➢EXAMPLE 4

Constructing a Translated Function

The graph of $f(x) = x^3$ is shown in Figure 28, together with the graph of a translated function $h(x)$. Find $h(x)$.

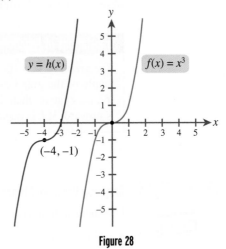

Figure 28

Solution   The graph of h is that of f shifted 4 units to the left and 1 unit down. Thus, $h(x) = f(x + 4) - 1 = (x + 4)^3 - 1$.

Reflections   As Figure 29 illustrates, the **reflection about the x-axis** of the point (x, y) is the point $(x, -y)$. Thus, if we substitute $-y$ for y in an equation, the new graph will be a reflection of the old graph about the x-axis. Similarly, the **reflection about the y-axis** of the point

(x, y) is the point $(-x, y)$, and so substitution of $-x$ for x in an equation causes a reflection about the y-axis. **Reflection about the origin** is defined by reflecting about both the x-axis and the y-axis. Thus, the reflection about the origin of the point (x, y) is the point $(-x, -y)$, and it follows that if we substitute $-x$ for x and $-y$ for y in an equation, the new graph will be the reflection of the old graph about the origin.

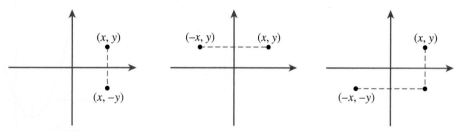

Reflection about the x-axis   Reflection about the y-axis   Reflection about the origin

Figure 29

When these principles are applied to the graph of a function $y = f(x)$, we obtain the following rules for reflections.

Reflections of Functions

Function	Graph
$h(x) = -f(x)$	Reflection about the x-axis of the graph of f
$h(x) = f(-x)$	Reflection about the y-axis of the graph of f
$h(x) = -f(-x)$	Reflection about the origin of the graph of f

EXAMPLE 5

Graphing Reflected Functions

Use the graph of $f(x) = x^2 - 2x$, shown in Figure 30, to sketch the graphs of the following functions:

a. $h(x) = (-x)^2 - 2(-x) = x^2 + 2x$

b. $h(x) = -(x^2 - 2x) = -x^2 + 2x$

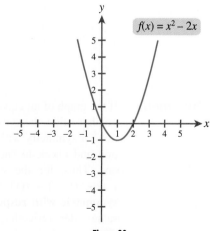

Figure 30

Solution

a. We note that $h(x) = f(-x)$ since $f(-x) = (-x)^2 - 2(-x) = x^2 + 2x$. Thus, the graph of h is the reflection of the graph of f about the y-axis, as shown in Figure 31.

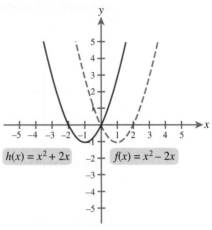

Figure 31

b. Here $h(x) = -f(x)$ since $-f(x) = -(x^2 - 2x) = -x^2 + 2x$. Thus, the graph of h is the reflection of the graph of f about the x-axis, as shown in Figure 32.

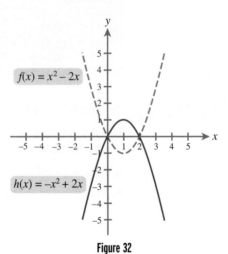

Figure 32

Symmetry   If the graph of an equation remains the same after reflecting about the y-axis, then the graph is said to be **symmetric with respect to the y-axis**. To test the graph of an equation for symmetry with respect to the y-axis, we simply replace x with $-x$ in the equation and check to see whether the resulting equation is equivalent to the original one. Thus, for the graph of a function f to be symmetric with respect to the y-axis, $f(-x) = f(x)$. Such functions are said to be **even**. Similarly, functions that are **symmetric with respect to the origin**—that is, functions whose graphs remain the same after reflecting about the origin—can be shown to satisfy the equation $f(-x) = -f(x)$ and are said to be **odd**. We summarize these facts as follows.

Even and Odd Functions

Type of function	Graphical property	Algebraic property
Even	Symmetry with respect to the y-axis	$f(-x) = f(x)$ for each x in the domain of f
Odd	Symmetry with respect to the origin	$f(-x) = -f(x)$ for each x in the domain of f

····►EXAMPLE 6   **Identifying Even and Odd Functions**

Use graphical as well as algebraic tests to determine whether the following functions are even or odd:

a. $f(x) = |x|$ **b.** $f(x) = -x^2 + 6x$ **c.** $f(x) = x^3 - 3x$

Solution

Function	Graph	Algebraic Test	Conclusion
a. $f(x) = |x|$	The graph appears to be symmetric with respect to the y-axis.	$f(-x) = |-x| = |x| = f(x)$	Even
b. $f(x) = -x^2 + 6x$	The graph is not symmetric with respect to either the y-axis or the origin.	$f(-x) = -(-x)^2 + 6(-x)$ $= -x^2 - 6x$ This is not equal to $f(x)$ or $-f(x)$.	Neither even nor odd
c. $f(x) = x^3 - 3x$	The graph appears to be symmetric with respect to the origin.	$f(-x) = (-x)^3 - 3(-x)$ $= -x^3 + 3x$ $= -f(x)$	Odd

Rule of Thumb

Notice that the function $f(x) = x^3 - 3x$ from part c of Example 6 includes only odd powers of x and was determined to be an odd function. This was not a coincidence. It can be shown (see Exercise 70) that a polynomial is even if all the powers of x are even, and it is odd if all the powers of x are odd.

EXAMPLE 7

Completing the Graph of a Function

Complete the graph of the function f with the given property.

a. f is even.   **b.** f is odd.

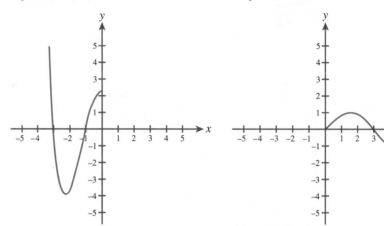

Solution

a. Since an even function is symmetric with respect to the y-axis, we simply reflect the graph across the y-axis, as shown in Figure 33.

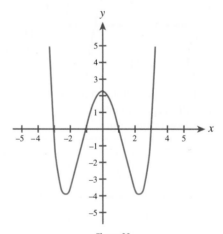

Figure 33

b. Since an odd function is symmetric with respect to the origin, we first reflect the graph across the x-axis to obtain the dashed portion shown in Figure 34, and then we reflect the *dashed* portion across the y-axis to obtain the complete graph of f in Figure 35.

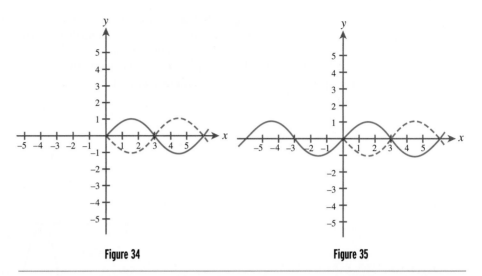

Figure 34   Figure 35

Preview of Calculus

Slope of a Curve

The slope of a line is a measure of its steepness; moreover, it tells us whether the line is rising or falling. Since the steepness of a line doesn't vary, neither does the slope: Each line has one and only one slope. By contrast, for nonlinear curves [like the graph of $f(x) = x^2 + 1$ shown in Figure 36], the steepness does vary from point to point. It follows that the slope of a nonlinear curve depends on which point on the curve we are considering.

But what do we mean by the slope of a curve at a point? As we get closer and closer to a point on a smooth curve, the curve flattens out, eventually appearing to be completely linear. We define the slope of the curve at a given point to be the slope of this line. For example, if we repeatedly zoom in on the point $(-2, 5)$ on the graph of $f(x) = x^2 + 1$, then the curve will eventually appear to be linear, as shown in Figures 37–38. By dividing the change in y by the change in x for 2 points in the viewing window, we obtain a slope of -4. Thus, we conclude that the slope of $f(x) = x^2 + 1$ at the point $(-2, 5)$ is approximately -4.

In the language of calculus, the line that the graph approaches as we zoom in on a point is called the **tangent line**, and the slope of the tangent line is the **derivative** of the function at the point. These concepts are the cornerstones of the differential calculus, one of the two main branches of calculus.

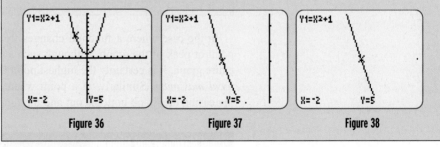

Figure 36   Figure 37   Figure 38

Increasing and Decreasing Functions

In addition to showing symmetry, graphs can be used to unveil many other important characteristics of functions. Consider, for example, the function $f(x) = x^3 - 3x^2$ graphed in

Figure 39. Notice that as we move from left to right, the graph rises until $x = 0$, falls in the interval from $x = 0$ to $x = 2$, and then rises to the right of $x = 2$. We say that the function f is *increasing* in the interval $(-\infty, 0)$, *decreasing* in the interval $(0, 2)$, and *increasing* in the interval $(2, \infty)$. The points $(0, 0)$ and $(2, -4)$, where the graph of f changes direction, are called *turning points*. More generally, we make the following definitions.

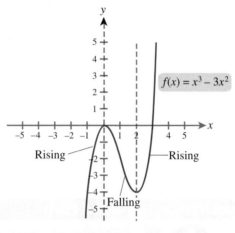

Figure 39

Increasing, Decreasing, and Turning Points

A function f is said to be **increasing** or **decreasing** on an interval I according to the following properties.

Type of behavior	Graphical property	Algebraic property
Increasing	As we move from left to right in the interval I, the graph rises.	For any x_1 and x_2 in the interval I with $x_1 < x_2$, we have $f(x_1) < f(x_2)$.
Decreasing	As we move from left to right in the interval I, the graph falls.	For any x_1 and x_2 in the interval I with $x_1 < x_2$, we have $f(x_1) > f(x_2)$.

The points where f changes from increasing to decreasing or decreasing to increasing are called **turning points**.

In the case where a function changes from increasing to decreasing, we hit a high point or peak of the graph. Although such a peak might not be the highest point on the entire graph, it is certainly the highest point nearby and so the function value is called a *local maximum*. Similarly, at a point where the function changes from decreasing to increasing the graph bottoms out and the function has a *local minimum*.

Local Maximum and Minimum

The value of f at $x = c$ is called a **local maximum** if it is the largest value of f for x near c. The value of f at $x = c$ is called a **local minimum** if it is the smallest value of f for x near c. In Figure 40, we have identified points where a function f has local maximum or minimum values.

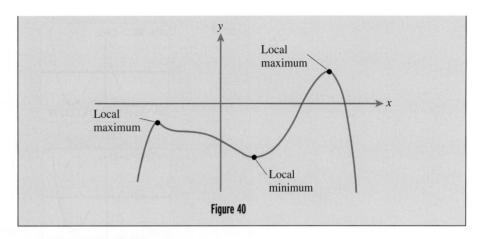

Figure 40

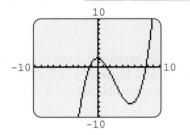

Figure 41

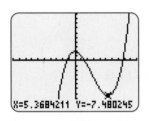

Figure 42

Using a Graphing Calculator to Find Turning Points

Use a graphing calculator to approximate the locations of the turning points for the function $f(x) = \frac{1}{8}x^3 - x^2 + 2$. Also, indicate the intervals on which f is increasing and the intervals on which it is decreasing.

Solution   The graph of $y = \frac{1}{8}x^3 - x^2 + 2$, shown in Figure 41, has two turning points: a local maximum at $x = 0$ and a local minimum at a point in the fourth quadrant. Since $f(0) = 2$, there is a local maximum of 2 at $x = 0$. Using the trace feature, as shown in Figure 42, we find that f has a local minimum of approximately -7.48 near $x = 5.37$.

Since the graph of f is rising for x less than 0 and also for values of x greater than 5.37, f is increasing on $(-\infty, 0) \cup (5.37, \infty)$. Since the graph of f is falling between $x = 0$ and $x = 5.37$, f is decreasing on $(0, 5.37)$.

In Example 8, we were able to quickly obtain rough approximations for the coordinates of turning points using the trace feature of a graphing calculator. As one might guess, more accurate approximations could be found by zooming in. Alternatively, we can accurately estimate the coordinates of turning points in a more direct and efficient manner using a built-in feature of our graphing calculator.

Calculator Keys

Turning Points

Most graphing calculators are capable of "automatically" calculating the coordinates of turning points. With one popular model, for example, a turning point can be found by plotting the graph with a view that clearly shows the turning point, selecting the **calculate-minimum** or **calculate-maximum** feature (depending on whether the turning point corresponds to a local minimum or maximum), and specifying a **left bound** (a point to the left of the turning point), a **right bound** (a point to the right of the turning point), and an **initial guess** (a rough approximation for the turning point, one that is located between the two bounds). The bounds and initial guess can be given either by positioning the trace cursor or by entering a number for the x-coordinate. Figures 43–46 illustrate the use of the calculate-minimum feature for estimating the coordinates of one turning point of $f(x) = \frac{1}{8}x^3 - x^2 + 2$ from Example 8.

(continued)

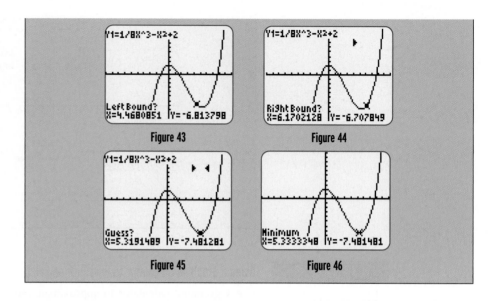

Figure 43

Figure 44

Figure 45

Figure 46

> **EXAMPLE 9**

Maximizing Volume with a Graphing Calculator

A rectangular package to be sent by the Postal Service can have a maximum combined length and girth (perimeter of the base) of 108 inches (see Figure 47). What are the dimensions of a box with a square base and with the largest possible volume that meets this restriction?

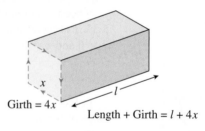

Girth = $4x$

Length + Girth = $l + 4x$

Figure 47

Solution The volume of a box with length l and a square base of side length x is

$$V = x^2 l$$

To maximize volume, we let the sum of the length and girth be the largest allowable value. Thus, $l + 4x = 108$, or

$$l = 108 - 4x$$

Substituting $l = 108 - 4x$ into $V = x^2 l$, we obtain

$$V = x^2(108 - 4x)$$

Since neither x nor l can be negative, we have $x \geq 0$ and $108 - 4x \geq 0$, from which it follows that $x \leq 27$. Thus, we are interested in the value of x between 0 and 27 that makes $V = x^2(108 - 4x)$ as large as possible. We set the window variables Xmin and Xmax to 0 and 27, respectively. After some experimentation, we settle on values for Ymin and Ymax of 0 and 12,000, respectively. A plot of $y = x^2(108 - 4x)$ with this viewing window is shown in Figure 48. If only a rough approximation is required, we could simply use the trace feature, as shown in Figure 49. But if greater accuracy is desired, we would use the calculate-maximum feature, as shown in Figure 50. In either case, our work suggests that the maximum occurs at approximately $x = 18$.

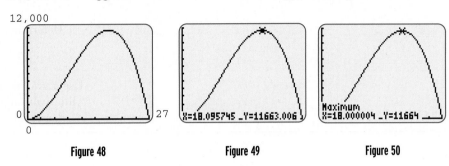

| Figure 48 | Figure 49 | Figure 50 |

Computing the corresponding values of l and V, we find that

$$l = 108 - 4x \qquad \text{and} \qquad V = x^2 l$$
$$= 108 - 4 \cdot 18 \qquad\qquad = 18^2 \cdot 36$$
$$= 36 \qquad\qquad\qquad\quad = 11{,}664$$

Thus, the box should be $18'' \times 18'' \times 36''$, which will result in a volume of 11,664 cubic inches.

Understanding and Mastery Checklists

Concepts to Understand

Zeros of functions

❧

Translations

❧

Reflections about the axes and the origin

❧

Symmetry

❧

Even and odd functions

❧

Increasing and decreasing functions

❧

Turning points

❧

Local maximum and minimum

Skills to Master

Approximate zeros graphically.

❧

Given the graph of a function, sketch the graph of its translation.

❧

Given the graph of a function, sketch the graph of its reflection.

❧

Given a formula for $f(x)$ and the graphs of f and its translation g, find an equation for $g(x)$.

❧

Given a formula for $f(x)$ and the graphs of f and its reflection g, find an equation for $g(x)$.

❧

Given a function, determine the symmetries of its graph.

❧

Use symmetry as an aid in graphing a function.

❖

Determine whether a function is even, odd, or neither.

❖

Use a graphing calculator to approximate the intervals on which a graph is increasing or decreasing, and find its turning points.

  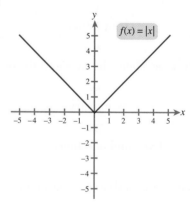

Exercises 2.2

Exercises 1-10 *Find the exact values of all the real zeros of the given function.*

1. $f(x) = 3x + 1$

2. $p(s) = 4s + 6$

3. $g(x) = x^2 - 2x - 8$

4. $x(t) = t^2 - 9$

5. $h(x) = |2x + 5|$

6. $f(x) = \sqrt{3x - 9}$

7. $f(x) = \dfrac{x^2 - 4}{x^{345} + 345x^{23} - 172{,}896}$

8. $g(x) = |2x - 8| - 2$

9. $f(a) = 2 - \dfrac{6}{a}$

10. $h(y) = 9 - \sqrt{y}$

Exercises 11-16 *Use a graphing calculator to estimate the zeros of the given function to the nearest hundredth.*

11. $f(x) = 2x^3 - 3x^2 - 20x + 2$

12. $h(x) = x^3 - 2x^2 - 14x + 30$

13. $f(x) = \dfrac{15}{x^2 + 1} - 5x - 14$

14. $g(x) = x + 8 - \dfrac{5x + 88}{x^2 + 1}$

15. $f(x) = \sqrt{2x + 5} - \sqrt{x^2 + 1}$

16. $g(x) = \sqrt{4x + 6} - x - 2$

Exercises 17-20 *A function f and its graph are given. For each translated or reflected graph, determine the corresponding function g.*

17. $f(x) = |x|$

a.

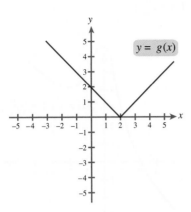

18. $f(x) = \sqrt{9 - x^2}$

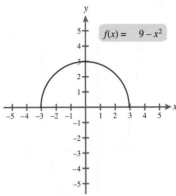

a.

b.

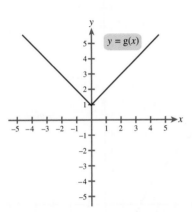

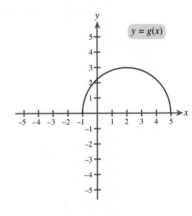

b.

c.

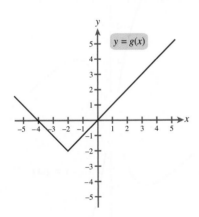

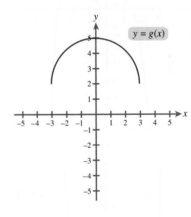

c.

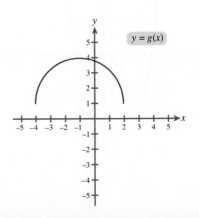

19. $f(x) = \frac{1}{4}x^4 - \frac{1}{3}x^3 - x^2$

20. $f(x) = x + 1 + \dfrac{6}{x^2 + 1}$

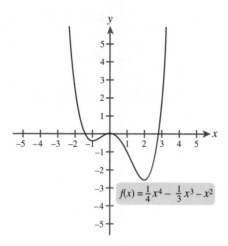

$f(x) = \frac{1}{4}x^4 - \frac{1}{3}x^3 - x^2$

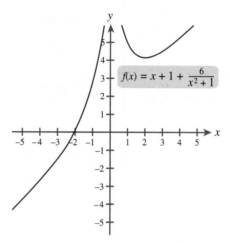

$f(x) = x + 1 + \dfrac{6}{x^2 + 1}$

a.

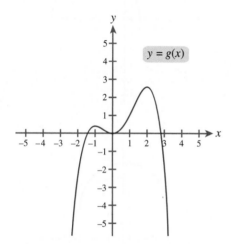

$y = g(x)$

a.

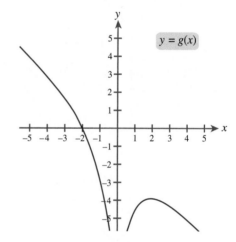

$y = g(x)$

b.

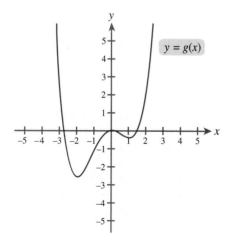

$y = g(x)$

b.

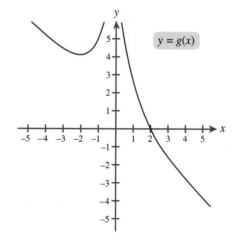

$y = g(x)$

Exercises 21-24   *Use the graph of the function f to sketch the graph of the function h.*

21.

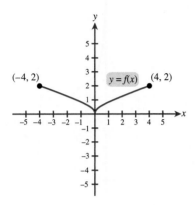

a. $h(x) = f(x - 2)$ **b.** $h(x) = f(x) + 3$

c. $h(x) = f(x + 1) - 4$ **d.** $h(x) = f(-x)$

e. $h(x) = -f(x) + 2$ **f.** $h(x) = -f(x - 1)$

22.

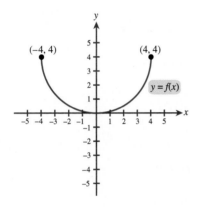

a. $h(x) = f(x) - 2$

b. $h(x) = f(x + 3)$

c. $h(x) = f(x + 1) - 4$

d. $h(x) = f(-x)$

e. $h(x) = -f(x) + 2$

f. $h(x) = -f(x - 1)$

23.

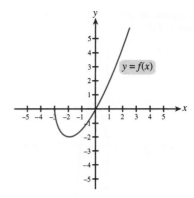

a. $h(x) = f(x - 3) + 2$ **b.** $h(x) = f(-x) - 1$

c. $h(x) = -f(x + 1)$ **d.** $h(x) = f(-x + 2) - 3$

24.

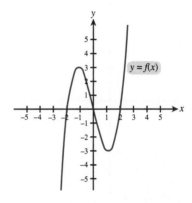

a. $h(x) = f(x + 2) - 3$ **b.** $h(x) = f(-x) - 1$

c. $h(x) = -f(x + 1)$ **d.** $h(x) = -f(x - 2) + 3$

Exercises 25-30   *Determine whether the given function is even, odd, or neither.*

25. $f(x) = x^4 - 4x^2$

26. $f(x) = 2x^3 - 8x + 1$

27. $f(x) = x\sqrt{x^2 + 1}$

28. $f(x) = \dfrac{1}{x^2 + 1}$

29. $f(x) = x^4 - \dfrac{x}{12}$

30. $f(x) = |x + 2| - |x - 2|$

Exercises 31-34 *Complete the graph of the function f by assuming that the function is (a) even and (b) odd.*

31.

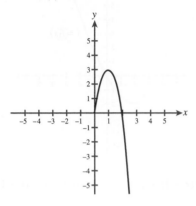

32.

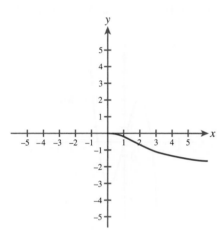

33.

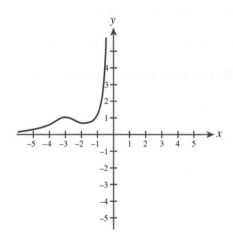

34.

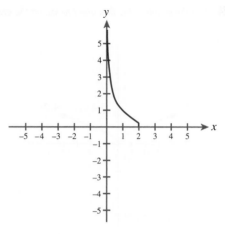

Exercises 35-42 *Use a graphing calculator to approximate the intervals on which the given function is increasing, the intervals on which it is decreasing, and the coordinates of any turning points. Classify each turning point as a local maximum or local minimum.*

35. $f(x) = x^2 - 4x + 5$

36. $g(x) = -x^2 - 6x - 5$

37. $h(x) = 2x^3 - 3x^2 - 6x + 5$

38. $p(x) = x^3 - 4x^2 + x + 1$

39. $q(x) = \dfrac{5}{x^2 - 4x + 5}$

40. $r(x) = \dfrac{-3}{x^2 - 6x + 10}$

41. $f(x) = x^4 - 3x^2 + x$

42. $g(x) = -2x^4 + 5x^2 + 7$

Applications

43. Translating a Population Function   Suppose a certain state's deer population can be approximated by the function $P(x) = x^2 + 14x + 100$, where x is the year ($x = 0$ corresponds to 1995) and $P(x)$ is the population (in hundreds) for that year. Use a horizontal shift to find a function $Q(x)$ that gives the same population approximations but for which $x = 0$ corresponds to 2000.

44. Translating a Speed Record Function   The 1-mile automobile speed records for the years 1906–1939 can be approximated by the function $f(x) = 0.29x^2 - 2.7x + 134$, where x is the year ($x = 0$ corresponds to 1906) and $f(x)$ is the speed record in miles per hour. Use a horizontal shift to find a function $g(x)$ that gives the same speed approximations but for which $x = 0$ corresponds to 1910.

45. Translating a Height Function   The height of a ball thrown from ground level with an initial velocity of 80 feet per second is given by $f(t) = -16t^2 + 80t$, where $f(t)$ is the height in feet and t is the time in seconds. Find a function $g(t)$ that gives the height of the ball at time t if it is thrown with the same initial velocity from the top of a 20-foot building. How do the graphs of f and g compare?

46. Translating a Cost Function   A clothing company has daily production costs given by $C(x) = 20{,}000 - 12x + 0.05x^2$, where $C(x)$ is the total cost in dollars and x is the number of units produced. Because of a new labor contract, the daily cost is expected to rise $2000, independent of the number of units produced. Find the new cost function. How does its graph compare to the original function?

47. Maximum Profit   An electronics company has determined that the cost for producing x security systems per month is $C(x) = 50x + 4050$ and that the revenue from the sale of x systems per month is $R(x) = 200x - 0.5x^2$. How many systems must be sold per month to break even (that is, for the profit to be zero)? Determine the number of systems that must be sold each month for the profit to be as large as possible.

48. Maximum Height   A ball is thrown up into the air from the top of a 112-foot building with an initial velocity of 96 feet per second. Its height (in feet) after t seconds is given by $s(t) = -16t^2 + 96t + 112$. Find the maximum height of the ball. When does the ball hit the ground?

49. Maximum Area   A point $P(x, y)$ lies on the line $y = -\frac{3}{4}x + 3$, as shown in Figure 51. Find the location of P so that the area of the shaded rectangle is as large as possible. (*Hint:* First find a function that expresses the area in terms of x.)

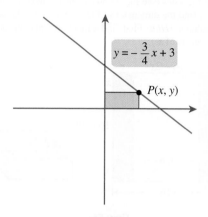

$$y = -\frac{3}{4}x + 3$$

$P(x, y)$

Figure 51

50. Minimum Area   A line passing through the point (2, 1) has intercepts $(a, 0)$ and $(0, b)$. Find the values of a and b so that the area of the shaded triangle in Figure 52 is as small as possible. (*Hint:* First find a function that expresses the area in terms of a.)

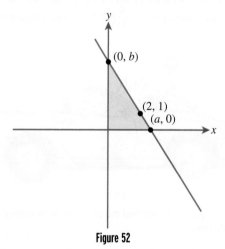

$(0, b)$

$(2, 1)$
$(a, 0)$

Figure 52

51. Maximum Area   A rancher wishes to enclose a rectangular pasture at the base of a cliff, as shown in Figure 53. Two hundred linear feet of fencing are available, and no fencing is needed along the cliff side of the pasture. Find the dimensions of the pasture that will have the largest possible area. (*Hint:* First find a function that expresses the area in terms of x.)

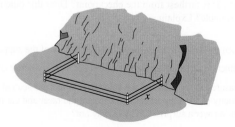

Figure 53

52. Maximum Volume   A box with a square base and no top is formed by cutting squares out of the corners of a piece of cardboard and then folding up the sides (see Figure 54). If the cardboard measures $16'' \times 16''$, find the dimensions of the box that has the largest possible volume. (*Hint:* First find a function that expresses the volume in terms of x.)

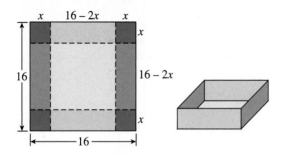

Figure 54

53. Water Bottle Handoff   A bicycle racer passes a checkpoint at a constant rate of 33 feet per second. At that instant, her support car accelerates from a dead stop in order to overtake her and hand her a water bottle. The car's distance (in feet) from the checkpoint after t seconds is given by the function $f(t) = \dfrac{4}{33}(33t^2 - t^3)$ for t from 0 to 33.

a. Plot the graph of f and find the car's maximum distance from the checkpoint.

b. Find a function $g(t)$ that gives the racer's distance from the checkpoint after t seconds.

c. Determine when the racer and support car meet. This can be done algebraically, but you may wish to plot the graphs of f and g as well. How does this time compare to the time at which the support car is farthest from the checkpoint? Does this outcome seem reasonable? Explain.

d. Calculus can be used to show that the velocity of the support car as a function of t is given by $v(t) = \dfrac{4}{11}(22t - t^2)$, where t is again measured in seconds and the velocity is in feet per second. Find the velocity of the support car at the instant the racer and car meet. Is your answer a surprise? Why or why not?

54. Traffic Flow   A model for estimating traffic flow on a certain two-lane highway is given by the function
$$f(x) = -0.000111x^3 - 0.118x^2 + 65x$$
where x denotes the number of vehicles per kilometer (traffic density) and $f(x)$ is the corresponding number of cars passing a given point on the road in 1 hour (traffic flow). For example, when the traffic density is 100 vehicles per kilometer, the traffic flow is $f(100) \approx 5209$ vehicles per hour.

a. Estimate the traffic flow when the traffic density is 150 vehicles per kilometer.

b. Estimate the traffic density when the traffic flow is measured to be 3000 vehicles per hour.

c. Estimate the average vehicle speed when the traffic flow is measured to be 3000 vehicles per hour. (*Hint:* Traffic flow is the product of traffic density and the average vehicle speed.)

d. Explain how a still photograph of a stretch of highway could be used to estimate the average vehicle speed.

e. For what traffic density will traffic come to a stop? (*Hint:* For what value of x will $f(x) = 0$?)

f. For what traffic density is the traffic flow greatest? What is the greatest possible traffic flow?

[*Source*: Neville D. Fowkes and John J. Mahony, *An Introduction to Mathematical Modeling* (Chichester, England: Wiley, 1994), p. 390.]

55. Modeling CO₂ Concentration   The atmospheric concentration of CO_2 has shown an overall upward trend in recent years, but it also oscillates quite predictably during the course of each year. For May 1999 through May 2001, CO_2 levels recorded at the Mauna Loa Observatory in Hawaii can be approximated by the function
$$f(x) = -14.3x^5 + 84x^4 - 175x^3 + 153.4x^2 - 49.4x + 371$$
where x denotes the year ($x = 0$ corresponds to May 1999) and $f(x)$ is the concentration of CO_2 in parts per million. Find the turning points of f. At approximately what time of the year is the concentration the lowest? At what time is it the highest?

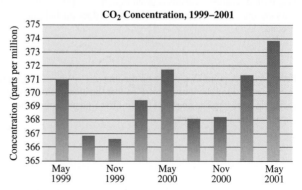

Data source: The Carbon Dioxide Information Analysis Center, U.S. Department of Energy.

56. Global Temperature Long-term trends in average global temperature during the years 1985–1997 can be modeled by the function

$$f(x) = 0.0022x^3 - 0.041x^2 + 0.24x + 61.7$$

where $f(x)$ represents the average temperature (measured in degrees Fahrenheit) and x denotes the year (with $x = 0$ corresponding to 1985). According to this model, during which years was global temperature rising? When was it falling? This model doesn't reflect short-term fluctuations brought about by atmospheric events. Which of the following occurrences might be responsible for the fluctuations suggested by the data depicted in the bar chart?

Occurrence	Date
Chernobyl meltdown	April 26, 1986
Yellowstone fire	Summer 1988
Exxon Valdez oil spill	March 24, 1989
San Francisco earthquake	October 17, 1989
Mt. Pinatubo eruption	June 15, 1991
El niño	1991–1995, 1997
Hurricane Andrew	August 1992
Peruvian tsunami	February 21, 1996

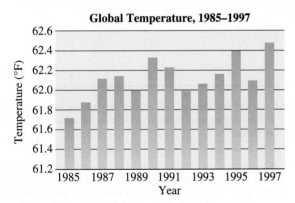

Global Temperature, 1985–1997

Data source: U.K. Meteorological Office.

57. Modeling Gold Reserves Gold reserves in the United States for the years 1975–1995 can be modeled by the function

$$f(x) = 0.0005x^4 - 0.026x^3 + 0.48x^2 - 3.9x + 275$$

where x denotes the year ($x = 0$ corresponds to 1975) and $f(x)$ is the gold reserve for that year measured in millions of fine troy ounces.

Find the intervals on which f is increasing and those on which it is decreasing. According to this model, for what year was the gold reserve lowest?

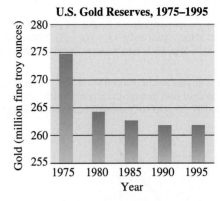

U.S. Gold Reserves, 1975–1995

Data source: International Monetary Fund Statistics.

58. Modeling Waste Production The average number of pounds of waste produced each day by each person in the United States for the years 1992–2001 can be approximated by the polynomial function

$$f(x) = 0.007x^3 - 0.085x^2 + 0.19x + 5.24$$

where x denotes the year ($x = 0$ corresponds to 1992) and $f(x)$ is the number of pounds of waste. Use the model to determine the time period(s) during which waste production was rising.

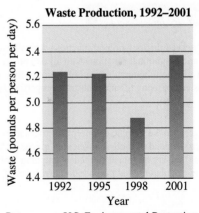

Waste Production, 1992–2001

Data source: U.S. Environmental Protection Agency.

Concepts and Critical Thinking

Exercises 59–62 *Answer true or false.*

59. If the graph of a function f lies entirely in the third quadrant, then 2 is in neither the domain nor the range of f.

60. If the graph of a function f crosses the y-axis once, then we can conclude that f has exactly one zero.

61. If a function f satisfies $f(-x) = f(x)$ for all x, then its graph will be symmetric with respect to the y-axis.

62. If h is defined by $h(x) = f(x - 2)$, then the graph of h will be a translation, 2 units to the left, of the graph of f.

Exercises 63–66 *Give an example of each.*

63. A function for which 2 and -2 are zeros

64. A function with a local maximum of 5 occurring at $x = 0$

65. An even function, the graph of which is a line

66. An odd function, the graph of which is a line

67. Suppose that f is an even function (so that the graph of f is symmetric with respect to the y-axis) and $g(x) = f(x - 3)$. What can be said about the graph of g? If f is odd, what can be said about the graph of g?

68. Show that if an odd function is defined at $x = 0$, then the graph of the function must pass through the origin. Explain why this fact is not contradicted by the graph of the odd function shown in Figure 55.

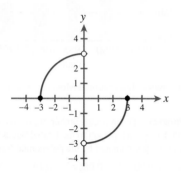

Figure 55

69. If n is even, show that $f(x) = x^n$ is an even function. If n is odd, show that f is an odd function.

70. Generalize Exercise 69 to show that if $f(x)$ is a polynomial, then f is even if all the powers of x are even, and f is odd if all the powers of x are odd.

71. An even function is symmetric with respect to the y-axis. An odd function is symmetric with respect to the origin. Why isn't there a term for a function that is symmetric with respect to the x-axis?

Questions for Discussion or Essay

72. We have observed that polynomial functions with all even powers of x are even and polynomial functions with all odd powers of x are odd. Why can't the functional notions of even and odd be defined this way instead of using the more abstract properties $f(-x) = f(x)$ and $f(-x) = -f(x)$?

73. To what extent are the graphical concepts of translation, reflection, and symmetry needed as tools for sketching graphs of functions? If you feel that they are no longer needed as graphing tools because of the graphing calculator, what purpose do they serve? If, on the other hand, you feel that they are still needed as graphing tools, when and how should they be used in conjunction with (or instead of) a graphing calculator?

74. When zooming in on turning points, it can happen that the graph will eventually appear as a horizontal line across the entire graphing calculator display. Experiment with zoom boxes of varying dimensions to see how this problem can be avoided. What is your conclusion as to the best dimensions of a zoom box for zooming in on turning points?

Projects for Enrichment

75. Designing a Soft Drink Can   An ordinary soft drink can has a volume of 355 cubic centimeters and, for the sake of simplicity, may be assumed to be a right circular cylinder. In this project, you will determine the dimensions of the cylindrical can that will provide the required volume with the least amount of aluminum.

   a. Write out the formula for computing the surface area of a cylindrical can in terms of its radius r and height h.

   b. Write out the formula for computing the volume of a cylindrical can in terms of its radius r and height h.

   c. Use the volume formula from part b and the fact that the volume must be 355 cubic centimeters to obtain an expression for h in terms of r. Substitute this expression into the surface area formula of part a. The result should be a function $S(r)$ that gives the surface area of a can of radius r with a volume of 355.

d. Use a graphing calculator to approximate the value for r that gives the minimum value for $S(r)$. This is the radius of the can with minimum surface area. Find the height that corresponds to the radius you found.

e. Approximate the actual radius and height of an ordinary soft drink can. How do the actual dimensions compare to the ones you found in part d? What explanation can you give for the discrepancy?

f. Now suppose the top of the can is twice as thick as the sides and bottom. Find the dimensions of the can that will minimize the amount of aluminum in the can.

g. Experiment with values for the thickness of the top (compared to the sides) until the dimensions of the can with a minimum amount of aluminum approximately coincide with the actual dimensions of a soft drink can. Does your value for the relative thickness of the top seem reasonable? Do you think this explains why the dimensions of a soft drink can are the way they are, or have other important factors still been ignored?

76. Stretching and Shrinking a Function We have already seen how the graph of a function $y = f(x)$ is affected by the modifications $f(x + c)$, $f(x - c)$, $f(x) + c$, and $f(x) - c$. In this project, we investigate the effect of the modifications $cf(x)$ and $f(cx)$.

a. Sketch the graph of $f(x) = x^2 + 1$.

b. Sketch the graph of each of the following functions:

   i. $h(x) = 2(x^2 + 1)$ [that is, $h(x) = 2f(x)$]

   ii. $h(x) = 4(x^2 + 1)$

   iii. $h(x) = \dfrac{1}{2}(x^2 + 1)$

   iv. $h(x) = \dfrac{1}{4}(x^2 + 1)$

c. Describe how each of the graphs in part b compare to the graph in part a. You may find it helpful to first consider what happens to several specific points on the graph of f. For example, the point $(2, 5)$ on the graph of f is "moved" away from the x-axis to the point $(2, 10)$ on the graph in (i) of part b.

d. Carefully write out a rule that describes how the graph of $h(x) = cf(x)$ can be obtained from the graph of a function f. You may find it convenient to consider two cases, one where $c > 1$ and the other where $0 < c < 1$. Explain how the case $c < 0$ can be dealt with by using your new rule together with a reflection.

e. Try out your rule by sketching the graphs of $y = cf(x)$ for $c = 3$ and $c = \frac{1}{3}$ and the graph of the function f, shown in Figure 56.

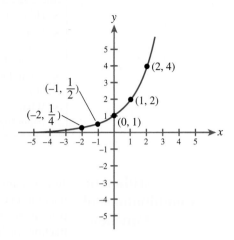

Figure 56

f. Sketch the graph of each of the following functions:

   i. $h(x) = (2x)^2 + 1$ [that is, $h(x) = f(2x)$]

   ii. $h(x) = (4x)^2 + 1$

   iii. $h(x) = \left(\dfrac{1}{2}x\right)^2 + 1$

   iv. $h(x) = \left(\dfrac{1}{4}x\right)^2 + 1$

g. Describe how each of the graphs in part f compare to the graph in part a. You may find it helpful to first consider what happens to several specific points on the graph of f. For example, the point $(2, 5)$ on the graph of f is "moved" toward the y-axis to the point $(1, 5)$ on the graph in (i) of part f. In other words, the same y-coordinate is obtained from an x-coordinate that is half the distance to the y-axis.

h. Carefully write out a rule that describes how the graph of $h(x) = f(cx)$ can be obtained from the graph of a function f. You may find it convenient to consider two cases, one where $c > 1$ and the other where $0 < c < 1$. Explain how the case $c < 0$ can be dealt with by using your new rule together with a reflection.

i. Try out your rule by sketching the graphs of $y = f(cx)$ for $c = 3$ and $c = \frac{1}{3}$ and the graph of the function f, shown in Figure 56.

Section 2.3 | Combinations of Functions

- ◑ How can the beating of a butterfly's wings in China create a tropical storm in Mexico?
- ◑ If the number of hot dogs consumed after t minutes by 113-pound hot dog–eating champion Takeru Kobayashi is $f(t)$, and his body weight after consuming n hot dogs is $g(n)$, then what does the function $g \circ f$ model?
- ◑ If f and g are functions and not just real numbers, what is meant by $f + g$, $f - g$, fg, or f/g?
- ◑ What do you get if you choose a number, square it, add 1, and then repeat this process (squaring and adding 1) 8 times?

Arithmetic Combinations of Functions

Just as numbers can be combined by the basic arithmetic operations of addition, subtraction, multiplication, and division, so too can functions. We can define the sum of two functions f and g by the formula $(f + g)(x) = f(x) + g(x)$. The operations of subtraction, multiplication, and division of functions are defined in a similar fashion, as follows.

Arithmetic Operations on Functions

Operation	Definition
Addition	$(f + g)(x) = f(x) + g(x)$
Subtraction	$(f - g)(x) = f(x) - g(x)$
Multiplication	$(fg)(x) = f(x)g(x)$
Division	$\left(\dfrac{f}{g}\right)(x) = \dfrac{f(x)}{g(x)}$

⋯⋯▷ EXAMPLE 1

Finding Combinations of Functions

Let $f(x) = x^2 - 4$ and $g(x) = x + 2$. Compute each of the following:

a. $(f + g)(x)$ **b.** $(f - g)(x)$

c. $(fg)(x)$ **d.** $\left(\dfrac{f}{g}\right)(x)$

Solution

a. $(f + g)(x) = f(x) + g(x)$
$$= (x^2 - 4) + (x + 2)$$
$$= x^2 + x - 2$$

b. $(f - g)(x) = f(x) - g(x)$
$$= (x^2 - 4) - (x + 2)$$
$$= x^2 - x - 6$$

c. $(fg)(x) = f(x)g(x)$
$$= (x^2 - 4)(x + 2)$$
$$= x^3 + 2x^2 - 4x - 8$$

d. $\left(\dfrac{f}{g}\right)(x) = \dfrac{f(x)}{g(x)}$

$= \dfrac{x^2 - 4}{x + 2}$

$= \dfrac{(x - 2)(x + 2)}{x + 2}$

It is tempting to simplify this expression to $x - 2$ by canceling the common factor $x + 2$. Note, however, that $x - 2$ is defined for all real numbers x, whereas

$$\frac{(x - 2)(x + 2)}{x + 2}$$

is undefined for $x = -2$. Thus, to be precise, we should express our answer as $(f/g)(x) = x - 2$ for $x \neq -2$.

EXAMPLE 2

Computing Combinations of Functions Defined Numerically

Suppose that f and g are defined by Table 2. Compute $f + g$, fg, and f/g.

Table 2

x	$f(x)$	$g(x)$
0	-2	1
1	0	2
2	2	5
3	4	10
4	6	17

Solution Since $(f + g)(x) = f(x) + g(x)$, we have, for example, $(f + g)(0) = f(0) + g(0) = -2 + 1 = -1$. In other words, the output values of $f + g$ are found simply by adding together the corresponding output values of f and g, as shown in Table 3. In a similar fashion, the output values of fg and f/g are found by multiplying or dividing the corresponding output values of f and g. See Table 3.

Table 3

x	$f(x)$	$g(x)$	$(f + g)(x)$	$(fg)(x)$	$(f/g)(x)$
0	-2	1	-1	$-2 \cdot 1 = -2$	$-2/1 = -2$
1	0	2	2	$0 \cdot 2 = 0$	$0/2 = 0$
2	2	5	7	$2 \cdot 5 = 10$	$2/5 = \dfrac{2}{5}$
3	4	10	14	$4 \cdot 10 = 40$	$4/10 = \dfrac{2}{5}$
4	6	17	23	$6 \cdot 17 = 102$	$6/17 = \dfrac{6}{17}$

In keeping with our convention of assuming that the domain (unless otherwise specified) is the set of real numbers for which the function "makes sense," the domain of $f + g$ is the set of real numbers x for which both $f(x)$ and $g(x)$ are defined. Thus, the domain of $f + g$ will consist of those real numbers that are in the domains of both f and g. Similarly, the domains of $f - g$ and fg consist of all real numbers that are contained in the domains of both f and g. However, since the expression $f(x)/g(x)$ is not defined when $g(x) = 0$, the domain of f/g consists of those real numbers x in the domains of both f and g such that $g(x) \neq 0$.

> **EXAMPLE 3**

Finding the Domains of Combinations of Functions

Let $f(x) = x - 2$, and $g(x) = \sqrt{x - 1}$. Find the domains of each of the following functions:

a. $(f + g)(x)$ **b.** $(f - g)(x)$

c. $(fg)(x)$ **d.** $\left(\dfrac{f}{g}\right)(x)$

e. $\left(\dfrac{g}{f}\right)(x)$

Solution   We begin by noting that $f(x) = x - 2$ is defined for all real numbers, whereas $g(x) = \sqrt{x - 1}$ is defined only for $x - 1 \geq 0$ or, equivalently, $x \geq 1$. Thus, the domain of f is $(-\infty, \infty)$ and the domain of g is $[1, \infty)$.

a–c. The domain of $f + g$, $f - g$, and fg is the set of real numbers x in the domains of both f and g. Thus, the domain of $f + g$, $f - g$, and fg is $[1, \infty)$.

d.   As already noted, both f and g are defined on the interval $[1, \infty)$. However, since $g(1) = \sqrt{0} = 0$, 1 is excluded from the domain of f/g. Thus, the domain of f/g is $(1, \infty)$.

e.   Since $f(2) = 2 - 2 = 0$, the number 2 is excluded from the domain of g/f. Thus, the domain of g/f consists of all real numbers in the interval $[1, \infty)$, with the exception of 2—that is, $[1, 2) \cup (2, \infty)$.

Composition of Functions

Functions can also be combined by *composing* or "stringing" them together—using the output of one function as the input for another. The **composition** of two functions f and g, written $f \circ g$, is defined by the formula $(f \circ g)(x) = f(g(x))$. The function $f \circ g$ takes an input value x and produces $f(g(x))$ as the corresponding output value, as shown in Figure 57.

As we see in Figure 57, $g(x)$ must be defined so that x must be in the domain of g. We also see that, because we are using $g(x)$ as an input for f, $g(x)$ must be in the domain of f. Thus, we have the following formal definition of composition.

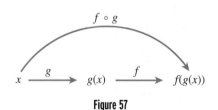

Figure 57

Definition of Composition of Functions

The composition of the functions f and g, written $f \circ g$, is defined by

$$(f \circ g)(x) = f(g(x))$$

The domain of $f \circ g$ consists of those real numbers x in the domain of g such that $g(x)$ is in the domain of f.

EXAMPLE 4

Computing the Composition of Two Functions

Given that $f(x) = x^2$ and $g(x) = 4 - 3x$, compute $(f \circ g)(2)$.

Solution

$$(f \circ g)(2) = f(g(2))$$

$$= f(-2) \qquad \text{Replacing } g(2) \text{ with } 4 - 3 \cdot 2 = -2$$

$$= (-2)^2$$

$$= 4$$

EXAMPLE 5

Computing the Composition of Two Functions

Given that $f(x) = 2x^2 + 4x + 5$ and $g(x) = 2x + 1$, compute $(f \circ g)(x)$.

Solution

$$(f \circ g)(x) = f(g(x))$$

$$= f(2x + 1) \qquad \text{Replacing } g(x) \text{ with } 2x + 1$$

$$= 2(2x + 1)^2 + 4(2x + 1) + 5 \qquad \begin{array}{l}\text{Substituting } 2x + 1 \text{ for } x \\ \text{in the formula for } f(x)\end{array}$$

$$= 2(4x^2 + 4x + 1) + 8x + 4 + 5$$

$$= 8x^2 + 16x + 11$$

Rule of Thumb

When evaluating the composition of functions, begin on the inside and work your way out. That is, at each stage, evaluate the innermost function first.

EXAMPLE 6

Computing Compositions of Functions

Let

$$f(x) = \frac{x + 1}{x - 1} \quad \text{and} \quad g(x) = x^2$$

Compute each of the following compositions:

a. $(f \circ g)(x)$ **b.** $(g \circ f)(x)$

Solution

a. $(f \circ g)(x) = f(g(x))$

$$= f(x^2)$$

$$= \frac{x^2 + 1}{x^2 - 1}$$

b. $(g \circ f)(x) = g(f(x))$

$$= g\left(\frac{x + 1}{x - 1}\right)$$

$$= \left(\frac{x + 1}{x - 1}\right)^2$$

$$= \frac{(x + 1)^2}{(x - 1)^2}$$

EXAMPLE 7

Computing Compositions of Functions

Let $f(x) = \sqrt{x}$ and $g(x) = x^2 + 1$. Compute each of the following compositions and find their domains:

a. $(f \circ g)(x)$ **b.** $(g \circ f)(x)$

Solution

a. $(f \circ g)(x) = f(g(x))$

$\qquad\qquad = f(x^2 + 1)$

$\qquad\qquad = \sqrt{x^2 + 1}$

Since $g(x)$ is defined for all real numbers and the expression under the radical, $x^2 + 1$, is always positive, the domain of $f \circ g$ is the set of all real numbers.

b. $(g \circ f)(x) = g(f(x))$

$\qquad\qquad = g(\sqrt{x})$

$\qquad\qquad = (\sqrt{x})^2 + 1 \quad$ for $x \geq 0$

$\qquad\qquad = x + 1 \quad$ for $x \geq 0$

Note that although $x + 1$ is defined for all real numbers x, $(g \circ f)(x)$ is not since the inside function, $f(x) = \sqrt{x}$, is defined only for nonnegative x. Thus, the domain of $g \circ f$ is the set of all nonnegative real numbers.

WARNING!

As the previous examples suggest, $f \circ g$ and $g \circ f$ are generally different functions. Thus, composition of functions is *noncommutative*—the order matters when composing functions.

EXAMPLE 8

Computing Compositions of Functions Defined Numerically

Suppose that f and g are defined by Table 4. Compute $(f \circ g)(x)$ and $(g \circ f)(x)$ for $x = 0, 1, 2, 3, 4$.

Solution By definition, $(f \circ g)(x) = f(g(x))$. Thus, using the appropriate values from Table 4, we have

$$(f \circ g)(0) = f(g(0)) = f(1) = 0$$
$$(f \circ g)(1) = f(g(1)) = f(2) = 2$$

But

$$(f \circ g)(2) = f(g(2)) = f(5)$$

Table 4

x	$f(x)$	$g(x)$
0	−2	1
1	0	2
2	2	5
3	4	10
4	6	17

and $f(5)$ is not defined in Table 4. Thus, $(f \circ g)(2)$ is undefined. Similarly, $(f \circ g)(3)$ and $(f \circ g)(4)$ are undefined, as shown.

$$(f \circ g)(3) = f(g(3)) = f(10) \text{ is undefined}$$
$$(f \circ g)(4) = f(g(4)) = f(17) \text{ is undefined}$$

The computations for $g \circ f$ are much the same as those for $f \circ g$.

$$(g \circ f)(0) = g(f(0)) = g(-2) \text{ is undefined}$$
$$(g \circ f)(1) = g(f(1)) = g(0) = 1$$

$$(g \circ f)(2) = g(f(2)) = g(2) = 5$$
$$(g \circ f)(3) = g(f(3)) = g(4) = 17$$
$$(g \circ f)(4) = g(f(4)) = g(6) \text{ is undefined}$$

Table 5 summarizes the values for $f \circ g$ and $g \circ f$.

Table 5

x	$(f \circ g)(x)$	$(g \circ f)(x)$
0	0	undefined
1	2	1
2	undefined	5
3	undefined	17
4	undefined	undefined

Dynamical Systems and Chaos Theory

Suppose we enter a number into our calculator and repeatedly press the x^2 key. What will happen? The result depends on the value of the original number. If the original number is less than 1, then the result becomes smaller each time we press the x^2 key, so we get closer and closer to 0. On the other hand, if the original number is greater than 1, then the result becomes larger and larger. This is an example of a **dynamical system**: A process is repeated over and over again, and the output from one step is the input for the next.

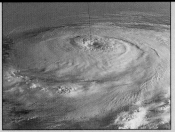

A hurricane as photographed from a satellite

Our lives are tapestries of dynamical systems. The amount that we eat influences our metabolism, which affects the amount that we eat, which influences our metabolism. . . . Our personalities determine our friends, who influence our personalities, which determine our friends. . . . The amount that we sleep, our moods, our eating, drinking, and exercise habits are all bound together in an elaborate dynamical system. Every ecosystem is, in essence, a dynamical system. On every day in every acre of tropical rain forest, an intricate dance is played out among millions of competing, cooperating, and coexisting dynamical systems. Earth's atmosphere is an enormous dynamical system, the complexity of which has thus far vexed our most brilliant atmospheric scientists. In fact, one could argue that Earth's atmosphere is the second most complex dynamical system known, second only to what is perhaps the most complex structure in the known universe, the human brain. Our very thoughts are the product of the interactions and interconnections between literally billions of neurons. Synapses are being reinforced here and weakened there, leaving in their wake the stuff of which thoughts and fears, joy and sorrow are made.

In the last few decades, computers have provided us with a lens through which dynamical systems can be viewed, the results of millions of iterations of certain processes played out on computer screens in a matter of minutes. These powerful instruments have illuminated largely unexplored mathematical territory and have given rise to the new science of **chaos**, which has caused nothing less than a paradigm shift in our understanding of the way in which dynamical systems unfold.

For an illustration of chaos, consider a billiard ball on a frictionless pool table with bumpers that absorb no energy. One would guess that the path of the billiard ball is completely determined by the initial velocity, position, and spin of the ball—the ball, without friction to slow it, will bounce around the table in a completely determined pattern. But, in fact, small, seemingly negligible factors will, after surprisingly few caroms, give rise to enormous variation. Surely one needn't account for the gravitational pull of Jupiter or the vibrations from a tree limb falling 20 miles away or the change in air pressure near the ball caused by the breeze from the hamster wheel at the house next door—but one must, if accuracy is desired after the first few caroms. Shockingly small changes in conditions can have enormous consequences.

This important aspect of chaos, extreme sensitivity to initial conditions, is often called **the butterfly effect**. The name arises from the degree to which atmospheric conditions can be chaotic; minor perturbations can become small eddies that can give rise to storms. It is said that the beating of a butterfly's wings at the right location at the right time can cause a tropical storm a thousand miles away a few days later. There is a certain degree of unpredictability in the weather that no computer, no matter how powerful, and that no measuring instruments, no matter how sensitive, will ever overcome.

Understanding and Mastery Checklists

Concepts to Understand	Skills to Master
Sum of functions ❖ Difference of functions ❖ Product of functions ❖ Quotient of functions ❖ Composition of functions	Compute the sum, difference, product, and quotient of two given functions and find their respective domains. ❖ Evaluate arithmetic combinations of functions. ❖ Given two functions f and g, compute $(f \circ g)(x)$ and $(g \circ f)(x)$. ❖ Determine the domain of a function formed by composition.

Exercises 2.3

Exercises 1–12 *Find the following combinations. Specify the domain if it is anything other than all real numbers.*

a. $(f + g)(x)$ **b.** $(f - g)(x)$

c. $(fg)(x)$ **d.** $\left(\dfrac{f}{g}\right)(x)$

1. $f(x) = x, g(x) = 5$

2. $f(x) = 3x + 2, \ g(x) = x - 7$

3. $f(x) = x^2, g(x) = 3x + 1$

4. $f(x) = 2x + 5, g(x) = x^2 + 1$

5. $f(x) = x, g(x) = x$

6. $f(x) = \dfrac{2}{x}, g(x) = 3x$

7. $f(x) = 2x - 2, g(x) = \dfrac{2}{x + 5}$

8. $f(x) = \dfrac{1}{x}, g(x) = -\dfrac{1}{x}$

9. $f(x) = \sqrt{x - 2}, g(x) = x - 4$

10. $f(x) = 3x, g(x) = \sqrt{2x - 6}$

11.

x	$f(x)$	$g(x)$
0	7	0
1	0	4
2	−2	8

12.

x	$f(x)$	$g(x)$
1999	100	undefined
2000	0	10
2001	200	5

Exercises 13–30 *Compute both $(f \circ g)(x)$ and $(g \circ f)(x)$. Specify the domain if it is anything other than all real numbers.*

13. $f(x) = 2x + 3, g(x) = 4x - 5$

14. $f(x) = 3 - 2x, g(x) = \dfrac{3 - x}{2}$

15. $f(x) = x^2, g(x) = 2x + 7$

16. $f(x) = 3x - 4$, $g(x) = x^3$

17. $f(x) = \dfrac{3}{x}$, $g(x) = \dfrac{1}{3x}$

18. $f(x) = \dfrac{4}{x + 1}$, $g(x) = 2x + 4$

19. $f(x) = x + 1$, $g(x) = x^3 - 3x$

20. $f(x) = x^2 + 3x$, $g(x) = x - 5$

21. $f(x) = 7$, $g(x) = x^2 + 3x + 1$

22. $f(x) = \dfrac{1}{x}$, $g(x) = \dfrac{1}{x}$

23. $f(x) = \sqrt{x}$, $g(x) = x^4$

24. $f(x) = |x|$, $g(x) = 4x - 1$

25. $f(x) = 3x + 5$, $g(x) = \sqrt{x - 2}$

26. $f(x) = 2x + 3$, $g(x) = \sqrt{2x - 5}$

27. $f(x) = \dfrac{2}{x^3 + 8}$, $g(x) = \sqrt[3]{\dfrac{2}{x} - 8}$

28. $f(x) = \dfrac{1}{x^2 - 9}$, $g(x) = 3$

29.

x	f(x)	g(x)
1	2	1
2	4	2
3	6	4
4	8	8

30.

x	f(x)	g(x)
0	10	5
5	15	10
10	20	15
15	10	0

Exercises 31-32   *Use the given graphs of f and g to complete the tables.*

31.

x	f(x)	g(x)	(f ∘ g)(x)	(g ∘ f)(x)
0				
1				
2				

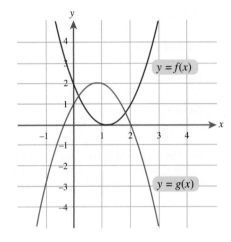

32.

x	f(x)	g(x)	(f ∘ g)(x)	(g ∘ f)(x)
0				
3				
6				
9				

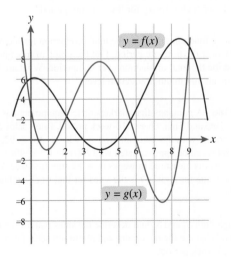

Exercises 33-38 *These exercises deal with the **iterates** of a function f, which are defined in the following way:*

$$f^1(x) = f(x)$$
$$f^2(x) = f(f(x))$$
$$f^3(x) = f(f(f(x)))$$

.
.
.

Thus, the nth iterate of a function f, denoted $f^n(x)$, is found by composing f with itself n times. For example, if $f(x) = 3x$, then $f^2(x) = f(f(x)) = f(3x) = 3(3x) = 9x$.

33. If $f(x) = x + 5$, find $f^3(x)$.

34. If $f(x) = 2x + 1$, find $f^2(x)$.

35. If $f(x) = 2x$, find $f^4(x)$.

36. If $f(x) = x^2$, find $f^3(x)$.

37. If $f(x) = x^2$, then $f^n(a)$ can be found by entering a into a calculator and pressing the x^2 key (followed by the Enter key, if necessary) n times. Use this method to find $f^{10}(0.98)$ and $f^{10}(1.02)$. What happens to $f^n(0.98)$ and $f^n(1.02)$ as n gets larger?

38. If $f(x) = x^2 + 1$, then $f^n(a)$ can be found by performing the following procedure n times: enter a into a calculator, compute the square, and add 1. Compute $f^8(a)$ for the values of $a = -1$ and $a = 0.5$. What happens to $f^n(a)$ as n gets larger?

Exercises 39-42 *For each of parts a and b, simplify the expression for the given function f.*

a. $\dfrac{f(x) - f(2)}{x - 2}$ **b.** $\dfrac{f(x + h) - f(x)}{h}$

39. $f(x) = 3x + 4$

40. $f(x) = x^2$

41. $f(x) = x^3$

42. $f(x) = \dfrac{1}{x}$

Applications

43. **Refinery Production** An oil company owns two refineries. Over a 6-month period, the number of barrels of crude oil refined at each can be approximated with the functions $B_1(t) = 75t^2 - 450t + 1800$ and $B_2(t) = -85t^2 + 510t + 900$, where t is measured in months ($t = 1$ is the first month), and the function value for a given t is the production for that month, in thousands of barrels.

    a. Graph both B_1 and B_2 and describe the production pattern for each refinery. What are the maximum and minimum production levels for each refinery for the 6-month period?

    b. Write a function that gives the total production for the oil company over the 6-month period. Plot the graph of this function, and describe the total production over this period. What are the maximum and minimum production levels?

44. **Company Sales** A company that markets office equipment has two salespeople, Pete and Tina. Over a 10-week period, Pete's sales can be approximated with the function $S_1(t) = -3t^2 + 36t + 92$, and Tina's sales can be approximated with $S_2(t) = 4t^2 - 40t + 200$. In both cases, t is measured in weeks ($t = 1$ is the first week), and the function value for a given t is the sales for that week, in hundreds of dollars.

    a. Graph both S_1 and S_2 and describe the sales pattern for Pete and Tina. What are the maximum and minimum sales for both during the 10-week period?

    b. Write a function that gives the total sales for the company for the 10-week period. Plot the graph of this function, and describe the total sales over this period. What are the maximum and minimum sales?

45. **Hot Dog Contest** Japanese professional speed-eater Takeru "The Tsunami" Kobayashi chewed through the competition at the 2002 Nathan's Famous Hot Dog–Eating Contest by consuming 50 hot dogs and buns in 12 minutes. The 113 pound Kobayashi defeated second-place winner, 400-pound Eric "Badlands" Booker, by more than 20 hot dogs, thus covering the spread of Internet gamblers. Bitter and humiliated opponents have accused him of premature regurgitation and having a second stomach.

    a. Let $f(t)$ be the total number of hot dogs that Kobayashi consumed after t minutes. Find an expression for $f(t)$. (Assume that Kobayashi eats at a constant rate.)

    b. Let $g(n)$ represent Kobayashi's weight after consuming n hot dogs. Find an expression for $g(n)$. (You will need to estimate the weight of a hot dog and bun.)

    c. Find expressions for $(f \circ g)(t)$ and $(g \circ f)(t)$.

    d. Which of the functions, $f \circ g$ or $g \circ f$, is a real-world model? What does it model?

46. Homicides and Population Growth   The population of Indianapolis, Indiana, can be modeled by the function $g(t) = 0.045t + 6.96$, where t represents the year (with $t = 0$ corresponding to 1980), and $g(t)$ the population (in hundreds of thousands). An analysis of Bureau of Justice data reveals that the number of homicides per year in a city of a given population is approximately $f(P) = 25.25P + 15.998$, where P is the population of the city (in hundreds of thousands).

a. Find an expression for $(f \circ g)(t)$ and $(g \circ f)(P)$. Which of these two functions can be interpreted as a mathematical model? What does it model?

b. Use the results of part a to predict the number of homicides in Indianapolis in the year 2020.

c. The model used in part b rests on the assumption that population is the only factor that affects the homicide rate for a city. In reality, what other factors would come into play?

Concepts and Critical Thinking

Exercises 47-50   *Answer true or false.*

47. $(f \circ g)(x) = f(x)g(x)$ for all real numbers x.

48. If $f(2) = g(2)$, then 2 is a zero of $f - g$.

49. If the domain of each of f and g is the set of all real numbers, then the domain of f/g is the set of all real numbers.

50. For any two functions f and g, $f \circ g = g \circ f$.

Exercises 51-54   *Give an example of each.*

51. A function f such that $f(x + c) = f(x)$ for all real numbers c

52. A pair of functions f and g such that $\left(\dfrac{f}{g}\right)(x) = x^2$

53. A pair of functions f and g such that $f \circ g = g \circ f$

54. A pair of functions f and g such that $(f \circ g)(x) = x$

55. Show that composition of functions is associative; that is, show that $f \circ (g \circ h) = (f \circ g) \circ h$.

56. It can be shown that
$$1 + x + x^2 + x^3 + \cdots + x^n = \frac{1 - x^{n+1}}{1 - x} \quad \text{for } x \neq 1$$

a. Find a formula for
$$1 + (x + 1) + (x + 1)^2 + (x + 1)^3 + \cdots + (x + 1)^n$$

b. Find a formula for $1 - x + x^2 - x^3 + \cdots + (-x)^n$.

57. Find a combination of rotations by 90° and reflections about the axes that yield a reflection about the line $y = x$.

58. Produce functions f and g such that $f \circ g$ is nowhere defined, even though g is defined everywhere. [*Hint*: Start by finding a function f that is defined only for $x > 0$. Then select a function g such that $g(x) < 0$.]

Questions for Discussion or Essay

59. If f and g are two linear functions, then which of the functions $f + g$, $f - g$, fg, f/g, $f \circ g$, and $g \circ f$ are guaranteed to be linear also? Justify your answer.

60. In what sense could an automobile manufacturing line be considered as the composition of many functions? Give an example of one of these factory functions and its domain and range. Find another multistep process that can be described in terms of composition of functions.

Projects for Enrichment

61. Computing Iterates of a Function Graphically   In Exercises 33–38, we defined the iterate $f^n(x)$ to be the function formed by composing f with itself n times. Suppose that for a given function f and a real number x_0, we wish to determine the behavior of $f^n(x_0)$ as n gets larger and larger. Does $f^n(x_0)$ jump around seemingly at random, does it converge to a single point, or does it bounce back and forth between two values?

To answer such questions we could simply compute $f(x_0)$, $f^2(x_0)$, $f^3(x_0)$, and so on until, if we are lucky, we discern a pattern. Computing $f^n(x_0)$ can be a time-consuming process, even with the aid of computers. There is, however, a way to use the graph of the function to estimate its iterates. For example, suppose that we are interested in computing iterates of the function $f(x) = 2x(1 - x)$

with $x_0 = 0.2$. We begin by graphing the function f and also the line $y = x$. Next we locate $x_0 = 0.2$ on the x-axis and move vertically to locate the point $(0.2, f(0.2))$ on the graph. We then move horizontally until we hit the line $y = x$. We should now be at the point $(f(0.2), f(0.2))$. We then move vertically until we hit the graph of f; we are now at $(f(0.2), f(f(0.2)))$ or, more briefly, $(f(0.2), f^2(0.2))$. We again move horizontally until we hit the line $y = x$, which leaves us at $(f^2(0.2), f^2(0.2))$. We continue in this way—moving vertically until we hit the graph of the function, then moving horizontally until we hit the line $y = x$—until we find the desired iterate. Figure 58 illustrates this procedure. It is evident from this figure that as n gets larger and larger, $f^n(0.2)$ approaches 0.5.

a. Let $f(x) = 2x(1 - x)$, as previously stated. Use a calculator to compute $f(0.2)$, $f^2(0.2)$, $f^3(0.2)$, and $f^4(0.2)$, confirming that these iterates indeed approach 0.5.

b. Use the graphical technique illustrated in Figure 58 to determine the first four iterates of f with $x_0 = 0.8$.

c. Now let $g(x) = 3.5x(1 - x)$. Use the graphical technique to determine the first 10 iterates of g with $x_0 = 0.3$. A sketch of the graph of g (along with the line $y = x$) is shown in Figure 59. Describe what happens to $g^n(x)$ as n gets larger and larger.

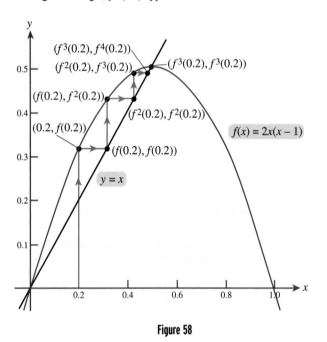

Figure 58

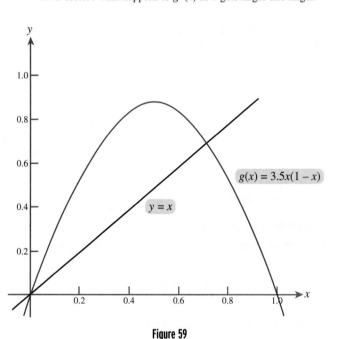

Figure 59

Section 2.4 | Inverses of Functions

- If $f(x) = 3x + 2$, why isn't $f^{-1}(x) = \dfrac{1}{3x + 2}$?

- Why is Los Angeles considered a heat island, and what do we predict its maximum temperature to be in the year 2035?

- What function gives the maximum weight that a person of a given height can be without being considered obese? What does the inverse of this function represent?

- How can we mentally compute the time it takes for an investment to double?

- If h is the function that performs a computer-generated translation from English to German and back to English again, then for what English text string x is $h(x) =$ "Row, row, rudders your boat easily down the river"?

Inverse Functions Consider the function f defined by $f(x) = 2x + 1$. It takes the input, multiplies it by 2, and then adds 1. Now generally, when we reverse a multistep process, we undo actions in the reverse order in which they were done. (When undressing, we remove our shoes before our socks.) Thus, it seems reasonable that an inverse of f would first subtract 1, and then divide by 2. In other words, we are guessing that the inverse of f is $g(x) = \dfrac{x - 1}{2}$. If g

really does undo the action of f, then applying g to $f(x)$ should return us to x. In other words, we expect that $g(f(x)) = x$. In fact,

$$g(f(x)) = g(2x + 1)$$

$$= \frac{(2x + 1) - 1}{2}$$

$$= \frac{2x}{2}$$

$$= x$$

As expected, g undid the action of f. Now let's see if f undoes the action of g by computing $f(g(x))$.

$$f(g(x)) = f\left(\frac{x - 1}{2}\right)$$

$$= 2\left(\frac{x - 1}{2}\right) + 1$$

$$= x - 1 + 1$$

$$= x$$

Our work here suggests the following, more formal definition of the inverse of a function.

Definition of f^{-1}

A function g satisfying $(g \circ f)(x) = x$ for all x in the domain of f and $(f \circ g)(x) = x$ for all x in the domain of g is said to be the **inverse** of the function f. We write $g = f^{-1}$ or, equivalently, $f = g^{-1}$.

EXAMPLE 1   **Verification of Inverse Functions**

Show that the functions

$$f(x) = \frac{3x + 2}{4} \quad \text{and} \quad g(x) = \frac{4x - 2}{3}$$

are inverses of one another.

Solution   We must show that $(f \circ g)(x) = (g \circ f)(x) = x$ for all x.

$$(f \circ g)(x) = f(g(x)) \qquad\qquad (g \circ f)(x) = g(f(x))$$

$$= f\left(\frac{4x - 2}{3}\right) \qquad\qquad\qquad = g\left(\frac{3x + 2}{4}\right)$$

$$= \frac{3\left(\dfrac{4x - 2}{3}\right) + 2}{4} \qquad\qquad = \frac{4\left(\dfrac{3x + 2}{4}\right) - 2}{3}$$

$$= \frac{4x - 2 + 2}{4} \qquad\qquad\qquad = \frac{3x + 2 - 2}{3}$$

$$= \frac{4x}{4} \qquad\qquad\qquad\qquad = \frac{3x}{3}$$

$$= x \qquad\qquad\qquad\qquad\quad = x$$

Thus, f and g are indeed inverses, and we can write, for example, $f^{-1}(x) = \frac{1}{3}(4x - 2)$.

WARNING!

> The notation $f^{-1}(x)$ does not mean $\dfrac{1}{f(x)}$.

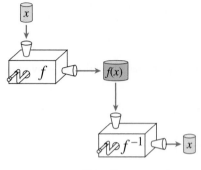

Figure 60

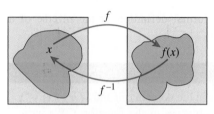

Figure 61

The inverse of a function is its opposite; it *undoes* what the function *does,* and vice versa, as illustrated in Figure 60. It is also useful to think of f and f^{-1} as mappings between sets, as depicted in Figure 61. Here we see that the domain of f is the range of f^{-1} and that the domain of f^{-1} is the range of f. In other words, the input values for f are the output values of f^{-1}, and vice versa.

The inverse of a function f can be viewed as that function which, when given an output of the function f, determines the input. However, it isn't always possible to determine the input to a function from the output, and for this reason not all functions have inverses. For example, consider the function $f(x) = x^2$. If it is given that the output of this function is 9, we cannot determine the input; it might have been either 3 or -3. Thus, the squaring process cannot be reversed: If the result of a squaring is a given positive number, then the original number cannot be specified with certainty. Thus, the function f doesn't have an inverse. In fact, the only functions that *do* have inverses are those for which distinct inputs yield distinct outputs. We define *one-to-one functions* to be such functions. More formally, we have the following definition.

Definition of One-to-One Functions

If the function f has the property that $f(a) = f(b)$ only if $a = b$, then f is said to be **one-to-one**, or 1–1. That is, f is 1–1 if, for each output value, there is exactly one input value.

A function can easily be tested to see whether or not it is 1–1 by looking at its graph. If the function is 1–1, then for each y-value, there will be only one x-value. Thus, a line with equation $y = c$ (a horizontal line) will be crossed at most once by the graph of a 1–1 function.

Horizontal Line Test

A function f is 1–1 if and only if no horizontal line crosses the graph of f more than once.

·····EXAMPLE 2

Testing to See Whether a Function is 1-1

Determine which of the following are the graphs of 1–1 functions:

a.

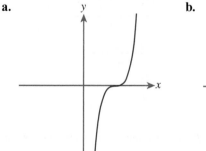

b.

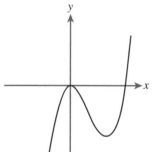

c.

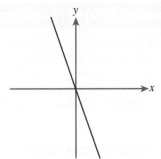

d.

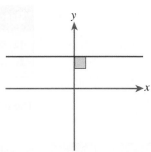

e.

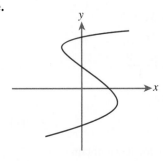

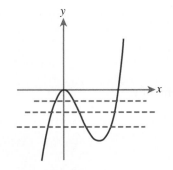

Figure 62

Solution

a. It appears that no horizontal line crosses this graph more than once; the function appears to be 1–1.

b. Several horizontal lines are shown in Figure 62 that cross the graph more than once; this function is not 1–1.

c. It appears that no horizontal line crosses this graph more than once; hence this function appears to be 1–1.

d. The graph of this function is itself a horizontal line. Therefore, there is one horizontal line (the graph itself) that intersects the graph more than once. In fact, it intersects itself infinitely many times. This function is not 1–1.

e. Trick question! As we see in Figure 63, this graph fails the vertical line test, and so it is not even the graph of a function.

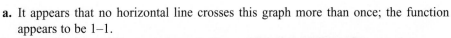

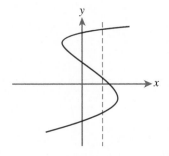

Figure 63

We have seen that if a function is 1–1, then it has an inverse. We now turn our attention to computing the inverse. Let f be a 1–1 function, and let $y = f(x)$. Since the inverse of a function returns the input to f when given the output, we can compute the inverse of f by solving the equation $y = f(x)$ for x. In fact, if we apply f^{-1} to both sides of the equation $y = f(x)$, we obtain

$$y = f(x)$$
$$f^{-1}(y) = f^{-1}(f(x)) \qquad \text{Applying } f^{-1} \text{ to both sides}$$
$$f^{-1}(y) = x \qquad \text{Simplifying by using the definition of } f^{-1}$$

Of course, once we know $f^{-1}(y)$, we can compute $f^{-1}(x)$ by simply substituting x for y in the formula for $f^{-1}(y)$. This suggests the following strategy for computing inverses.

Steps for Computing $f^{-1}(x)$ for 1-1 Functions f

1. Solve the equation $y = f(x)$ for x to obtain $x = f^{-1}(y)$.
2. Substitute x for y to find $f^{-1}(x)$.

The following examples illustrate this technique.

EXAMPLE 3

Computing the Inverse of a Linear Function

Find the inverse of the function $f(x) = 3x + 2$.

Solution We begin by solving the equation $y = 3x + 2$ for x.

$$y = 3x + 2$$
$$y - 2 = 3x \qquad \text{Subtracting 2 from both sides}$$
$$x = \frac{y - 2}{3} \qquad \text{Dividing both sides by 3 and exchanging the left and right sides}$$

Thus, we have

$$f^{-1}(y) = \frac{y - 2}{3}$$

Substituting x for y, we obtain

$$f^{-1}(x) = \frac{x - 2}{3}$$

EXAMPLE 4

Finding the Inverse of a Function Involving Radicals

Find $f^{-1}(x)$ if $f(x) = \sqrt[3]{2x - 7}$.

Solution We let $y = f(x)$ and solve for x in terms of y.

$$y = \sqrt[3]{2x - 7}$$
$$y^3 = \left(\sqrt[3]{2x - 7}\right)^3 \qquad \text{Cubing both sides}$$
$$y^3 = 2x - 7$$
$$y^3 + 7 = 2x$$
$$2x = y^3 + 7$$
$$x = \frac{y^3 + 7}{2}$$

Thus, $f^{-1}(y) = \frac{y^3 + 7}{2}$, so that $f^{-1}(x) = \frac{x^3 + 7}{2}$.

EXAMPLE 5

Finding the Inverse of a Function

Find the inverse of $f(x) = x^2$, $x \geq 0$.

Solution We let $y = x^2$ and solve for x in terms of y.

$$y = x^2$$
$$\sqrt{y} = \sqrt{x^2} \qquad \text{Taking a square root on both sides}$$

$$\sqrt{y} = x \qquad \text{Using the fact that } x \geq 0$$
$$x = \sqrt{y}$$

Thus, $f^{-1}(y) = \sqrt{y}$, so that $f^{-1}(x) = \sqrt{x}$.

Graphs of Inverse Functions

Suppose that (a, b) is on the graph of $y = f(x)$. Then $b = f(a)$, from which it follows that $a = f^{-1}(b)$. Thus, (b, a) is a point on the graph of $y = f^{-1}(x)$. As Figure 64 suggests, the point (b, a) is the reflection of the point (a, b) about the line $y = x$. Thus, the graph of $y = f^{-1}(x)$ is the reflection of the graph of $y = f(x)$ about the line $y = x$.

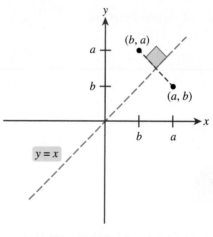

Figure 64

EXAMPLE 6

Using the Graph of f to Graph f^{-1}

The graph of a function f is given in Figure 65. Show that f^{-1} exists, and graph it.

Solution   Since the graph passes the horizontal line test, f^{-1} exists. To graph f^{-1}, we simply reflect the graph of f about the line $y = x$, as shown in Figure 66.

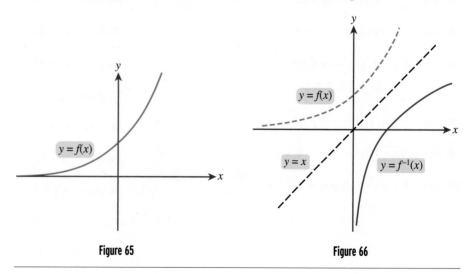

Figure 65

Figure 66

Understanding and Mastery Checklists

Concepts to Understand

Inverse functions

❖

One-to-one functions

❖

Horizontal line test

❖

Graphs of inverse functions

Skills to Master

Verify or disprove that a given pair of functions are inverses of one another.

❖

Use the horizontal line test to determine whether a function is one-to-one.

❖

Compute the inverse of a simple function.

❖

Given the graph of a function, sketch its inverse.

Exercises 2.4

Exercises 1-10   *Determine whether the functions f and g are inverses of one another by evaluating $f(g(x))$ and $g(f(x))$.*

1. $f(x) = x + 2; g(x) = x - 2$

2. $f(x) = 3x; g(x) = \dfrac{x}{3}$

3. $f(x) = 2x + 1; g(x) = \dfrac{1}{2}x - 1$

4. $f(x) = \dfrac{1}{4}x - 3; g(x) = 4x + 12$

5. $f(x) = (x + 1)^3; g(x) = \sqrt[3]{x} - 1$

6. $f(x) = \sqrt[5]{x - 2}; g(x) = (x + 2)^5$

7. $f(x) = \dfrac{x + 3}{2x - 1}; g(x) = \dfrac{x + 3}{2x - 1}$

8. $f(x) = \dfrac{1}{x} - 4; g(x) = \dfrac{1}{x + 4}$

9. $f(x) = \sqrt{x - 6}; g(x) = x^2 + 6, x \geq 0$

10. $f(x) = 4 - x^2, x \geq 0; g(x) = \sqrt{x + 4}$

Exercises 11-20   *Determine whether the function is 1–1. Use a graphing calculator as necessary.*

11. $f(x) = -2x + 5$

12. $g(x) = \dfrac{x + 3}{4}$

13. $f(x) = x^2$

14. $f(x) = \sqrt[3]{x}$

15. $h(x) = x^3 + 2x + 1$

16. $f(x) = x^3 - x^2$

17. $g(x) = \dfrac{x^4}{12} - x^3$

18. $f(x) = x^5 + 1$

19. $h(x) = |x - 3|$

20. $f(x) = \sqrt{x} + 2$

Exercises 21-34 *Determine whether the function f is 1–1. If it is, find* $f^{-1}(x)$.

21. $f(x) = \frac{1}{3}x - 1$

22. $f(x) = \frac{-x + 3}{4}$

23. $f(x) = x^4$

24. $f(x) = x^3$

25. $f(x) = x^2 + 1, \ x \geq 0$

26. $f(x) = (x + 4)^2$

27. $f(x) = \sqrt{2x + 5}$

28. $f(x) = \sqrt{4 - x^2}$

29. $f(x) = \frac{x^2}{x^2 + 1}$

30. $f(x) = \frac{x}{x + 4}$

31.

x	f(x)
0	4
1	7
2	10
3	13
4	16

32.

x	f(x)
0	1
1	−1
2	−3
3	−1
4	1

33.

x	f(x)
−2	5
−1	1
0	0
1	1
3	10

34.

x	f(x)
−5	15
0	10
5	5
10	0
15	−5

Exercises 35-40 *Determine whether* f^{-1} *exists and, if so, sketch its graph.*

35.

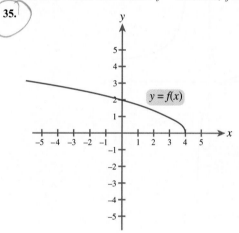

36.

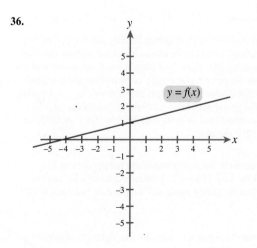

37.

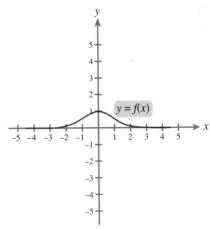

39.

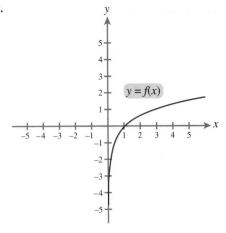

38.

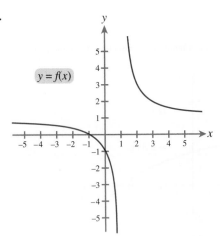

40.

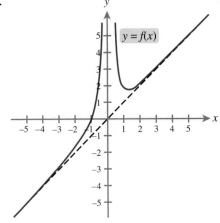

Applications

41. Heat Islands Covered with heat-absorbing asphalt and stripped of cooling vegetation, large cities can be as much as 10°F warmer than the surrounding countryside. These heat islands wreak environmental and economic havoc by pumping up energy consumption and contributing to smog. As cities expand, the problem grows worse. In Los Angeles, for example, the yearly maximum temperature (computed as a 10-year running average) rose from 98°F in 1935 to 104°F in 1985, an increase of 0.12°F per year.

a. Let $M(t)$ represent the yearly maximum temperature in Los Angeles in the year $1935 + t$. Find an expression for $M(t)$ by assuming that the yearly maximum temperature increases 0.12°F every year.

b. Use the result from part a to estimate the highest temperature in Los Angeles in 2035.

c. Find an expression for $M^{-1}(t)$.

d. Give an interpretation of $M^{-1}(t)$. In particular, what does $M^{-1}(120)$ represent?

42. The Kelvin Scale The Kelvin temperature scale is commonly used in many of the sciences, including astronomy, chemistry, and physics. Essentially, the Kelvin temperature scale is a translated Celsius scale: A degree Kelvin is the same as a degree Celsius, with 0 Kelvin defined to be absolute zero (the coldest physically possible temperature). It can be shown that if K and F represent the temperature in degrees Kelvin and Fahrenheit, respectively, then

$$K = g(F) = \frac{5F + 2300}{9}$$

a. Find absolute zero in degrees Fahrenheit.

b. Interpret g^{-1}, and then find a formula for it.

c. According to data collected by the Mars Pathfinder mission, the temperature on Mars varies from 190 to 240 on the Kelvin scale. Find the corresponding temperature range in degrees Fahrenheit.

Martian landscape as photographed by the Mars Pathfinder.

43. Rule of 72 The time t (in years) it takes an investment to double if it is earning interest at an annual rate of $i\%$ is approximated by

$$t = f(i) = \frac{72}{i}$$

This formula is usually referred to as the rule of 72.

a. Some credit cards charge interest at rates of 18% and more. If no payments are made (which is generally not an option), find the time required for an initial charge of $400 to grow to $800, assuming an interest rate of 18%.

b. Interpret f^{-1}, and then find a formula for it.

c. Use your result from part b to estimate the interest rate for an investment that doubled in value after 8 years.

44. Training Heart Rate Tolerance to cardiovascular exertion depends on a variety of factors, including the subject's overall health and general level of aerobic fitness. Still, all other things being equal, younger people can tolerate higher pulse rates than can older people, so naturally the recommended training heart rate (THR) is adjusted for age. A general rule of thumb for the THR (in beats per minute) for a subject aged t years is given by

$$\text{THR} = h(t) = \frac{3(220 - t)}{4}$$

a. Find an appropriate training heart rate for a 20-year-old.

b. Interpret h^{-1}, and then find a formula for it.

c. Use your result from part b to estimate the age of a subject whose recommended training heart rate is 120 beats per minute.

d. Some health professionals consider formulas such as the one we have given for THR to be downright dangerous. Wherein lies the danger in applying this simple formula?

45. Body Mass Index Few mathematical models have created as much widespread controversy as those used to determine appropriate body weight ranges. One such model, based on the body mass index (BMI), dictates that the maximal healthy weight (in pounds) for a person h inches tall is given by

$$w = f(h) = 0.0355264\, h^2$$

a. When the 6′2″ Arnold Schwarzenegger competed as a bodybuilder, he sometimes weighed as much as 250 pounds. According to this formula, was he obese?

Arnold Schwarzenegger posing in competition.

b. What restriction must we place on the domain of f so that f will have an inverse? With this restriction in place, compute the inverse of f and give an interpretation.

c. When Walter Hudson died in 1991, he was estimated to have weighed 1400 pounds. Use your result from part b to estimate the minimum height for a healthy 1400-pounder.

d. List a few of the many factors not taken into account by our model f that are relevant to the determination of obesity.

Concepts and Critical Thinking

Exercises 46-49   *Answer true or false.*

46. A function has an inverse if and only if it passes the vertical line test.

47. If $g(x) = \dfrac{1}{f(x)}$, then $g = f^{-1}$.

48. Only 1–1 functions have inverses.

49. If f is a 1–1 function, then so is $g(x) = f(x - 1)$.

Exercises 50-53   *Give an example of each.*

50. The graph of an equation that is not the graph of a function

51. The graph of a function that is not a 1–1 function

52. The graph of a 1–1 function having –3 as its only zero

53. A pair of functions f and g such that $f + g$ is 1–1, even though neither f nor g is 1–1

54. a. Simplify $f^{-1} \circ (f \circ g)$.

    b. Suppose that f and $f \circ g$ are given. Explain how the result of part a could be used to find g.

    c. Employ the technique you described in part b to find $g(x)$ if it is given that $(f \circ g)(x) = 6x + 21$ and $f(x) = 3x + 9$.

55. Define $(f \circ g \circ h)(x) = f(g(h(x)))$.

    a. Show that $f^{-1} \circ f \circ p = p$ for any function p.

    b. Show that $p \circ f \circ f^{-1} = p$ for any function p.

    c. Suppose that the function f has two inverses, g and h. Use the result of part a to show that $g \circ f \circ h = h$.

    d. Use the result of part b to show that $g \circ f \circ h = g$.

    e. Use parts c and d to conclude that g and h are the same function and that a function can have at most one inverse.

56. An arbitrary linear function can be written as $f(x) = ax + b$. What restrictions must be placed on a and b in order for f^{-1} to exist? Under these restrictions, compute f^{-1}.

57. Find a function f such that $f^{-1} = f$. (*Hint:* Start by looking for an operation that, when performed twice in a row, returns you to the original number.)

58. Which 1–1 functions have graphs that are symmetric with respect to the y-axis?

Questions for Discussion or Essay

59. Discuss the analogy between functions and their inverses on the one hand and numbers and their multiplicative inverses on the other.

60. What is the relationship between a function and its inverse with respect to the following properties: always positive, always negative, always increasing, always decreasing, 1–1, even, and odd? In other words, for example, if a function is always positive, then what can be said about its inverse?

61. The Web search engine Altavista offers an online translation service. Given a textual document, it can translate from English to one of eight languages, or it can translate from one of these languages to

English. If we denote the Altavista functions that translate from English to German and from German to English by f and g, respectively, then what are the domains and ranges of f and g? Would you expect f and g to be inverses of one another? Use the following facts to check.

- $(f \circ g)$("A rolling stone gathers no moss") = A role stone does not seize Moos."
- $(f \circ g)$("Row, row, row your boat gently down the stream.") = "Row, row, rudders your boat easily down the river."
- $(f \circ g)$("A scrub is a guy that thinks he's fly, And is also known as a buster") = "A Schrubben is a cord, which is it fly and admits also as a chap thinks."

Projects for Enrichment

62. Solving Functional Equations   If x, y, and z are real numbers, then the equation $xy = z$ can be solved for y by multiplying both sides by x^{-1}:

$$xy = z$$
$$x^{-1}xy = x^{-1}z$$
$$y = x^{-1}z$$

This strategy can be modified for solving certain **functional equations**, equations in which the unknown is a function. The basic idea is that instead of multiplying by the multiplicative inverse, we compose with the functional inverse. For example, if the functions f and h are known and f has an inverse, then the equation $f \circ g = h$ can be solved for g as follows:

$$f \circ g = h$$
$$f^{-1} \circ f \circ g = f^{-1} \circ h$$
$$g = f^{-1} \circ h$$

a. Find a function g such that $f \circ g = h$, where $f(x) = 2x - 4$ and $h(x) = \dfrac{2x - 10}{x - 2}$.

b. Find a function f such that $f \circ g = h$, where $g(x) = 2x - 1$ and $h(x) = \dfrac{10x - 2}{4x - 3}$.

c. Consider the functional equation $f \circ g = h$, with $f(x) = 3x + 4$ and $h(x) = \dfrac{7x + 1}{x + 1}$. The following is an *incorrect* solution of this functional equation:

$$f \circ g = h \qquad \text{The original equation}$$
$$f^{-1} \circ f \circ g = h \circ f^{-1} \qquad \text{Composing with } f^{-1} \text{ on both sides}$$
$$g = h \circ f^{-1} \qquad \text{Simplifying}$$
$$g(x) = (h \circ f^{-1})(x) \qquad \text{Evaluating both sides at } x$$

$$g(x) = h(f^{-1}(x)) \qquad \text{Using the definition of composition of functions}$$
$$g(x) = h\!\left(\dfrac{x - 4}{3}\right) \qquad \text{Substituting } \dfrac{x-4}{3} \text{ for } f^{-1}(x)$$
$$= \dfrac{7\!\left(\dfrac{x-4}{3}\right) + 1}{\left(\dfrac{x-4}{3}\right) + 1} \qquad \text{Evaluating, using the fact that } h(x) = \dfrac{7x+1}{x+1}$$
$$= \dfrac{7x - 25}{x - 1} \qquad \text{Simplifying}$$

If we check our solution by evaluating $(f \circ g)(x)$, we find that

$$(f \circ g)(x) = f\!\left(\dfrac{7x - 25}{x - 1}\right)$$
$$= 3\!\left(\dfrac{7x - 25}{x - 1}\right) + 4$$
$$= \dfrac{25x - 79}{x - 1}$$
$$\neq h(x)$$

What went wrong?

Section 2.5   Selected Functions

- How long must one wait after drinking alcoholic beverages to ensure that the blood alcohol level falls below the legal limit?
- Will a baseball hit at 144 feet per second (about 98 miles per hour) at an angle of 60° hit the ceiling of Minute Maid Park, home of the Houston Astros?
- How can a single function be used to compute federal income tax even when there are several tax brackets?
- How can you quickly mentally compute the day of the week for any date in any century?

To fully exploit the power of functions for modeling the real world, it is not sufficient to merely master general concepts such as notation, graphs, combinations, and inverses. Instead, we must also gain familiarity with a few particularly useful classes of frequently occurring functions. In this section, we discuss linear, quadratic, and piecewise-defined functions; in the next three chapters, we will tackle polynomial, rational, exponential, logarithmic, and trigonometric functions. As we begin this functional odyssey, pay special attention to the key features that distinguish one species of function from another.

Linear Functions

Recall that the graph of an equation of the form $y = mx + b$ is a line with slope m and y-intercept b. In fact, any nonvertical line can be put in this form, and it is for this reason that we define a **linear function** to be a function of the form $f(x) = mx + b$. In the case where $m = 0$, we obtain the **constant function** $f(x) = b$. Linear functions are certainly the easiest to study, and they are also among the most useful for applications.

EXAMPLE 1

Linear Cost and Revenue Functions

A sidewalk hot dog vendor has a fixed daily cost of $60 for the rental of a cart and the vendor permit. The variable costs (for the hot dogs, buns, and condiments) average 30¢ per hot dog. Each hot dog sells for $1.50.

a. Construct a linear cost function that gives total cost for the day as a function of the number of hot dogs sold.

b. Construct a linear revenue function that gives the total revenue for the day as a function of the number of hot dogs sold.

c. How many hot dogs would have to be sold in a given day in order to break even? Illustrate this graphically.

Solution

a. If x represents the number of hot dogs sold in one day, then

$$\text{Total cost} = \text{Fixed cost} + \text{Variable cost}$$
$$= 60 \qquad + (\text{Cost per hot dog})(\text{Number of hot dogs})$$
$$= 60 \qquad + (0.3)x$$

So the total cost function is $C(x) = 0.3x + 60$.

b. Again, if x represents the number of hot dogs sold in one day, then

$$\text{Total revenue} = (\text{Revenue per hot dog})(\text{Number of hot dogs})$$
$$= (1.50)x$$

So the total revenue function is $R(x) = 1.5x$.

c. To break even, the total revenue must equal the total cost. Thus,

$$R(x) = C(x)$$
$$1.5x = 0.3x + 60$$
$$1.2x = 60$$
$$x = 50$$

So 50 hot dogs must be sold in one day to break even that day. Graphically, 50 is the x-coordinate of the point of intersection of the graphs of R and C, as shown in Figures 67 and 68.

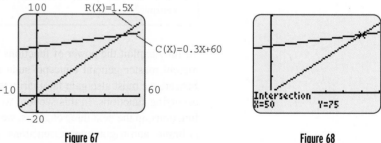

Figure 67   **Figure 68**

Quadratic Functions

After linear functions, the next simplest class of functions is the **quadratic** functions—functions of the form $f(x) = ax^2 + bx + c$, where a, b, and c are constants and $a \neq 0$. Of course, we could simply use a graphing calculator to plot the graph of a quadratic function (and, in fact, we often do just that), but learning how to graph a quadratic by hand will help develop our intuition and "feel" for this important category of functions. This task is greatly simplified by the fact that, as we will soon see, there is a sense in which all quadratic functions descend from a common ancestor: $f(x) = x^2$.

Since $x^2 \geq 0$ for all x, it follows that the smallest possible value of $f(x) = x^2$ is $f(0) = 0$. Furthermore, since f is even [that is, $f(-x) = f(x)$], the graph of f should be symmetric with respect to the y-axis. Indeed, the graph of f shown in Figure 69 reveals that $(0, 0)$ is the lowest point on the curve and that the graph is symmetric about the y-axis. You might recognize the familiar shape of this graph as being that of a parabola. The point where a parabola changes direction is called its **vertex**. The line through the vertex dividing the parabola into two equal pieces is called the **axis of symmetry**. For the graph of $f(x) = x^2$, the vertex is the origin and the axis of symmetry is the y-axis, as shown in Figure 70.

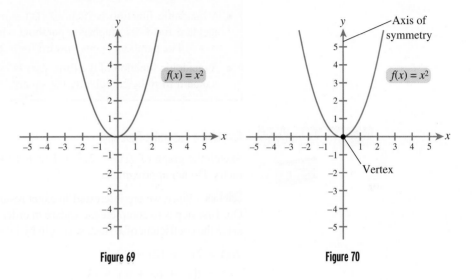

Figure 69 **Figure 70**

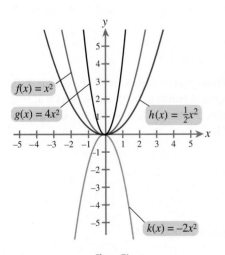

Figure 71

Like the graph of $y = x^2$, the graph of $f(x) = ax^2$, where $a \neq 0$, is a parabola, with the origin as its vertex and the y-axis as its axis of symmetry. The sign of a determines the parabola's orientation: For $a > 0$, the parabola opens upward, and for $a < 0$, the parabola opens downward. The magnitude of a affects the "pointiness" of the graph: The larger the magnitude of a, the narrower the graph of f. Figure 71 shows the graphs of several functions of the form $f(x) = ax^2$.

Recall from Section 2.2 that the graph of $g(x) = f(x - h) + k$ is a translation h units horizontally and k units vertically of the graph of f. Thus, $g(x) = a(x - h)^2 + k$ is a translation of the graph of $f(x) = ax^2$, so that the graph of g is a parabola with (h, k) as its vertex and $x = h$ as its axis of symmetry. Furthermore, the parabola will open upward for $a > 0$ and downward for $a < 0$. These features are suggested by the graphs shown in Figures 72 and 73.

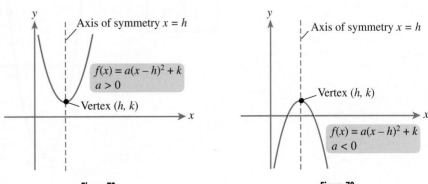

Figure 72 **Figure 73**

Now, for our final step, we note that by completing the square, any quadratic function $g(x) = ax^2 + bx + c$ can be written in the standard form $g(x) = a(x - h)^2 + k$. We summarize the key points of the preceding discussion as follows.

Quadratic Functions

- A function of the form $f(x) = ax^2 + bx + c \ (a \neq 0)$ is called a **quadratic function**. Its graph is a parabola.

- A quadratic function written as $f(x) = a(x - h)^2 + k \ (a \neq 0)$ is said to be in standard form. Its graph is a parabola with vertex (h, k) and axis of symmetry $x = h$. The parabola opens upward for $a > 0$ and downward for $a < 0$.

- A quadratic function of the form $f(x) = ax^2 + bx + c \ (a \neq 0)$ can be written in standard form by completing the square.

EXAMPLE 2

Graphing a Quadratic Function

Sketch the graph of $f(x) = 2x^2 + 12x + 13$ and identify the vertex and axis of symmetry. Do not approximate.

Solution Since we are interested in exact results, we will graph the quadratic by hand. Our first step is to complete the square in order to write $f(x)$ in standard form. Note that since the coefficient of x^2 is 2, we begin by factoring 2 from both the x^2 and the x terms.

$$f(x) = 2x^2 + 12x + 13$$

$$= 2\left(x^2 + 6x + \rule{0.6cm}{0.35cm}\right) + 13 \qquad \text{Factoring 2 from the square and first-power terms; preparing to complete the square by leaving space for a constant term}$$

$$= 2\left(x^2 + 6x + \boxed{9}\right) + 13 - 2 \cdot 9 \qquad \text{Adding } \left(\tfrac{6}{2}\right)^2 = 9 \text{ inside the parentheses to complete the square; compensating by subtracting } 2 \cdot 9 \text{ outside the parentheses}$$

$$= 2(x + 3)^2 - 5 \qquad \text{Simplifying}$$

This is a quadratic function in the standard form $f(x) = a(x - h)^2 + k$, with $a = 2$, $h = -3$, and $k = -5$. Thus, its graph opens upward (since $a > 0$), and it has $(-3, -5)$ as its vertex and $x = -3$ as its axis of symmetry. Plotting a few points and using symmetry, we obtain the sketch shown in Figure 74.

x	y
-3	-5
-2	-3
-1	3

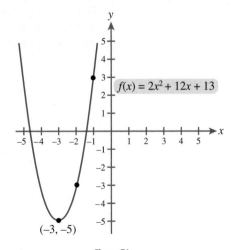

$(-3, -5)$

Figure 74

Minute Maid Park

······**EXAMPLE 3**

The Path of a Baseball

The ballpark of the Houston Astros (currently called Minute Maid Park) in Houston, Texas, features a retractable roof that rises to a maximum height of 242 feet above the playing field. If a baseball is hit from a height of 3 feet, with an initial velocity of 144 feet per second and an initial angle of 60°, and if we ignore air resistance, the height of the ball is approximated by $f(x) = -\frac{1}{324}x^2 + \sqrt{3}x + 3$, where x is the horizontal distance from home plate and all distances are in feet. Assuming the ball is hit fair and the playing field extends at least 300 feet from home plate, show that the ball will hit the dome ceiling somewhere above the field.

Solution   Because of the complexity of the coefficients of f, we use a graphing calculator to plot the graph of this quadratic function, as shown in Figure 75. By using the trace feature, we see that a height of more than 242 feet is obtained at an x-value of about 246 feet—which corresponds to a position within the playing field. It follows that, indeed, the ball will hit the ceiling of the dome somewhere above the playing field.

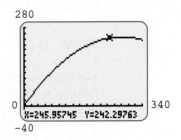

Figure 75

Piecewise-defined Functions

Occasionally, it is desirable to use two or more formulas to define a function. Consider, for example, the function f defined by

$$f(x) = \begin{cases} -x & x \leq -1 \\ 2x + 3 & x > -1 \end{cases}$$

This *piecewise* definition indicates that for an input value x less than or equal to -1, the output value is $-x$. For an input value x greater than -1, the output value is $2x + 3$. Thus, $f(-4) = -(-4) = 4$, whereas $f(1) = 2(1) + 3 = 5$. Some additional input–output pairs are given in Table 6. Notice that for each $x \leq -1$, $f(x)$ is computed using $-x$, whereas for $x > 1$, $f(x)$ is computed using $2x + 3$. The graph of f is obtained by plotting the line $y = -x$ for $x \leq -1$ and the line $y = 2x + 3$ for $x > -1$, as shown in Figure 76.

Figure 76

Table 6

x	Interval	Formula	$f(x)$
-3	$x \leq -1$	$-x$	$-(-3) = 3$
-2	$x \leq -1$	$-x$	$-(-2) = 2$
-1	$x \leq -1$	$-x$	$-(-1) = 1$
0	$x > -1$	$2x + 3$	$2(0) + 3 = 3$
1	$x > -1$	$2x + 3$	$2(1) + 3 = 5$
2	$x > -1$	$2x + 3$	$2(2) + 3 = 7$

EXAMPLE 4

Evaluating and Graphing a Piecewise-Defined Function

Evaluate the function

$$f(x) = \begin{cases} x + 1 & x < 2 \\ x^2 - 4 & x \geq 2 \end{cases}$$

at the x-values given in the following table, and then sketch its graph.

x	$f(x)$
-1	
0	
1	
2	
3	

Solution   In order to compute a function value for a given x, we simply note whether $x < 2$ or $x \geq 2$ and use the corresponding "piece" of the function definition. For example, to compute $f(-1)$, we note that $-1 < 2$, and so we use the $x + 1$ piece of the function to obtain $f(-1) = (-1) + 1 = 0$. Similarly, since $3 \geq 2$, we compute $f(3)$ using the $x^2 - 4$ piece, obtaining $f(3) = (3)^2 - 4 = 5$. The remaining output values are computed in Table 7. The graph of f is obtained by plotting the line $y = x + 1$ for $x < 2$ and the parabola $y = x^2 - 4$ for $x \geq 2$, as shown in Figure 77. Notice that an open circle is used at the point $(2, 3)$ to indicate that the point is not included on the graph, and a closed circle is used at $(2, 0)$ to indicate that this point is included.

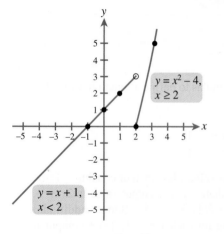

Figure 77

Table 7

x	Interval	Formula	$f(x)$
-1	$x < 2$	$x + 1$	$(-1) + 1 = 0$
0	$x < 2$	$x + 1$	$(0) + 1 = 1$
1	$x < 2$	$x + 1$	$(1) + 1 = 2$
2	$x \geq 2$	$x^2 - 4$	$(2)^2 - 4 = 0$
3	$x \geq 2$	$x^2 - 4$	$(3)^2 - 4 = 5$

WARNING!

> A piecewise-defined function is *one* function with several pieces, not several different functions. Thus, for a given input value x, the corresponding output value $f(x)$ is computed using only one of the pieces.

Piecewise-defined functions can also be plotted with the aid of a graphing calculator. Although some graphing calculators have the ability to plot such functions directly, we can simply graph all of the expressions used to define the piecewise-defined function, while recognizing that for a given x-value, only one of the graphs applies.

EXAMPLE 5   Graphing a Piecewise-Defined Function With a Graphing Calculator

Graph the following function with the aid of a graphing calculator:

$$f(x) = \begin{cases} -x^3 + 6x^2 - 9x + 4 & x < 3 \\ x - 3 & x \geq 3 \end{cases}$$

Solution   We begin by plotting the graphs of both $y = -x^3 + 6x^2 - 9x + 4$ and $y = x - 3$ on the same screen, as shown in Figure 78. Now we recognize that only the portion of the graph of $y = -x^3 + 6x^2 - 9x + 4$ for which $x < 3$ (shown in red) is actually part of the graph of f; moreover, only the portion of the graph of $y = x - 3$ for which $x \geq 3$ (shown in blue) is part of the graph of f. Figure 79 shows the graph of f.

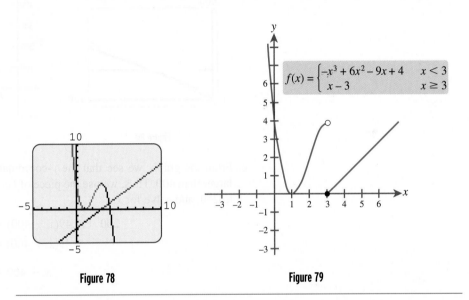

$$f(x) = \begin{cases} -x^3 + 6x^2 - 9x + 4 & x < 3 \\ x - 3 & x \geq 3 \end{cases}$$

Figure 78   **Figure 79**

EXAMPLE 6   Constructing and Graphing a Piecewise-Defined Function

A moving van rental company has two different rate schedules for a certain size of moving van. For vans used locally (a total distance of less than 100 miles), the rate is $40 plus 39¢ per mile. For long-distance rentals (a total distance of 100 miles or more), the rate is $200 plus 49¢ per mile for each mile over 400.

a. Construct a piecewise-defined function that can be used to compute the rate for any desired distance.
b. Plot the graph with the aid of a graphing calculator.
c. Determine how far one can move with $250.

Solution

a. Let x represent the total distance traveled, and let $f(x)$ be the corresponding rental rate (in dollars). If $x < 100$, the rate is $40 + 0.39(x)$. If $100 \leq x \leq 400$, the rate is a fixed $200. If $x > 400$, the rate is $200 plus $0.49 for each mile over 400, or $200 + 0.49(x - 400)$. Thus, f is defined by

$$f(x) = \begin{cases} 40 + 0.39x & 0 \leq x < 100 \\ 200 & 100 \leq x \leq 400 \\ 200 + 0.49(x - 400) & x > 400 \end{cases}$$

b. We begin by graphing all three of the formulas that are used to define f, as shown in Figure 80. Note that only the red portion of each graph is actually part of the graph of f. In Figure 81, we show the graph of f.

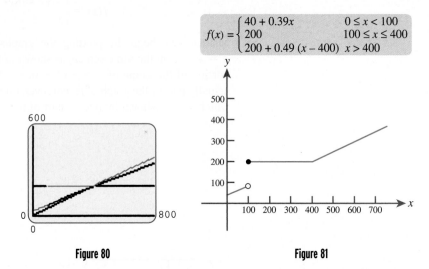

$$f(x) = \begin{cases} 40 + 0.39x & 0 \le x < 100 \\ 200 & 100 \le x \le 400 \\ 200 + 0.49\,(x - 400) & x > 400 \end{cases}$$

Figure 80   Figure 81

c. From the graph, we see that the y-coordinate attains the value 250 for an x-value larger than 400. Thus, we use the piece of f corresponding to $x > 400$, set $f(x)$ equal to 250, and solve for x.

$$200 + 0.49(x - 400) = 250$$
$$0.49(x - 400) = 50$$
$$x - 400 = \frac{50}{0.49}$$
$$x = \frac{50}{0.49} + 400$$

Thus, the total distance one can move with \$250 is $x = \frac{50}{0.49} + 400 \approx 502$ miles.

Two special examples of piecewise-defined functions are the absolute value function and the so-called step functions. Although the **absolute value function** $f(x) = |x|$ can be written as a single equation, it is equivalent to

$$f(x) = \begin{cases} -x & x < 0 \\ x & x \ge 0 \end{cases}$$

So the graph of $f(x) = |x|$ can be found by graphing $y = -x$ for $x < 0$ and $y = x$ for $x \ge 0$, as shown in Figure 82.

A **step function** is a piecewise-defined function taking only constant values over its pieces. For example, the function g defined by

$$g(x) = \begin{cases} 1 & x \le -3 \\ -1 & x > -3 \end{cases}$$

takes on constant values over each of its pieces and is, thus, a step function. Its graph is shown in Figure 83.

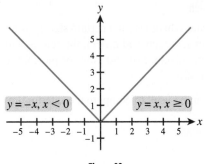

$y = -x,\ x < 0$ $y = x,\ x \ge 0$

Figure 82

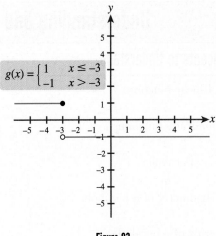

$$g(x) = \begin{cases} 1 & x \le -3 \\ -1 & x > -3 \end{cases}$$

Figure 83

Another example of a step function is the *floor function* $f(x) = \lfloor x \rfloor$, where $\lfloor x \rfloor$ is defined as follows.

The Floor Function

The **floor** of a real number x, denoted $\lfloor x \rfloor$, is the greatest integer less than or equal to x. The function $f(x) = \lfloor x \rfloor$ is called the **floor function** (or the **greatest integer function**).

Applying this definition, we see that $\lfloor 2.6 \rfloor = 2, \lfloor 8 \rfloor = 8, \lfloor -4.1 \rfloor = -5$. Note in particular that the floor function does not simply round or truncate the decimal places. More generally, we observe that for every real number x between two consecutive integers n and $n + 1$, $\lfloor x \rfloor = n$. Thus, the graph of the floor function is constant between each pair of consecutive integers, as shown in Figure 84.

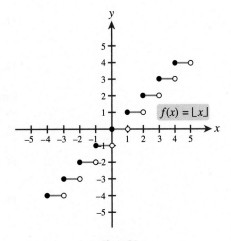

Figure 84

Understanding and Mastery Checklists

Concepts to Understand	Skills to Master

Concepts to Understand

Linear functions

✧

Quadratic functions

✧

Parabolas

✧

Axis of symmetry of a parabola

✧

Vertex of a parabola

✧

Piecewise-defined functions

✧

Step functions

✧

The floor function

Skills to Master

Recognize a linear function and sketch its graph.

✧

Sketch the graph of a quadratic function.

✧

Find the vertex of a parabola.

✧

Evaluate and graph piecewise-defined functions.

Exercises 2.5

Exercises 1-12 *Identify the type of function that is given (piecewise-defined, quadratic, linear, or none of these).*

1. $f(x) = 3x^2 + 4$

2. $g(x) = 2x^3 + 5x + 7$

3. $h(x) = \dfrac{x^2 - 5x + 3}{x + 1}$

4. $p(x) = \sqrt{2x + 5}$

5. $q(x) = 2^x$

6. $H(x) = \begin{cases} 0 & x < 0 \\ 1 & x > 0 \end{cases}$

7. $f(x) = 3x + 9$

8. $g(x) = 2$

9. $f(t) = 4^{13}t + 7$

10. $r(s) = 2s^2 - 1$

11. $f(x) = \begin{cases} x & x < 0 \\ x^2 & x > 0 \end{cases}$

12. $g(p) = 3p$

Exercises 13-16 *Plot the graph of the linear function and find the slope and y-intercept.*

13. $f(x) = 2x - 1$

14. $g(x) = -3x + 2$

15. $h(x) = \dfrac{-3x + 4}{4}$

16. $g(x) = \dfrac{x - 6}{2}$

Exercises 17-20 *Find the linear function having the given properties.*

17. The graph of h has slope $-\frac{1}{2}$ and $h(2) = 4$.

18. $f(-2) = 3$ and $f(2) = -3$.

19. The graph of g has y-intercept -4 and $g(3) = 0$.

20. The graph of f is the perpendicular bisector of the line segment connecting $(-2, 1)$ and $(4, 3)$.

Exercises 21-28 *Find the coordinates of the vertex of the graph of the given quadratic function.*

21. $f(x) = 3x^2 - 4$

22. $g(x) = x^2 + 4x + 9$

23. $f(x) = 2x^2 - 6x + 10$

24. $g(x) = ax^2 + c$

25. $m(x) = x^2 + 16x + 10$

26. $f(x) = 2^5 x^2 + 2^7 x$

27. $P(x) = 1 - 2x + x^2$

28. $h(x) = 3x + x^2 + 5$

Exercises 29-32 *Find the quadratic function having the given properties, and sketch its graph.*

29. The graph of f has vertex $(1, 3)$ and y-intercept 5.

30. The graph of h has vertex $(-2, -1)$ and $h(-1) = -3$.

31. The graph of g has axis of symmetry $x = 3$, $g(3) = -1$, and $g(2) = -2$.

32. The graph of f passes through the origin, $f(-1) = 0$, and $f(1) = 4$.

Exercises 33-38 *Evaluate the piecewise-defined function as indicated, and then sketch its graph.*

33. $f(x) = \begin{cases} -x - 4 & x \le 0 \\ 2x - 4 & x > 0 \end{cases}$
$f(-2), f(0), f(3)$

34. $g(x) = \begin{cases} 3x + 2 & x < 0 \\ -x + 2 & x \ge 0 \end{cases}$
$g(-4), g(0), g(1)$

35. $h(x) = \begin{cases} x^2 - 1 & x < 2 \\ \frac{1}{2}x + 1 & x \ge 2 \end{cases}$
$h(0), h(2), h(4)$

36. $g(x) = \begin{cases} -\frac{2}{3}x + 2 & x \le -3 \\ 13 - x^2 & x > -3 \end{cases}$
$g(-4), g(-3), g(0)$

37. $f(x) = \begin{cases} x & x < -1 \\ 1 & -1 \le x \le 1 \\ x + 2 & x > 1 \end{cases}$
$f(-3), f(-1), f(4)$

38. $h(x) = \begin{cases} -x - 1 & x < -2 \\ x + 3 & -2 \le x \le 0 \\ 3 & x > 0 \end{cases}$
$h(-3), h(0), h(6)$

Exercises 39-42 *Use the graphs of f and g to sketch the graph of h.*

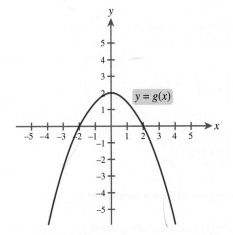

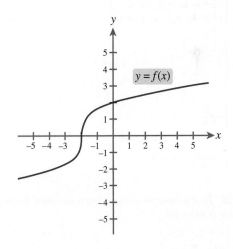

39. $h(x) = \begin{cases} f(x) & x \le -2 \\ g(x) & x > -2 \end{cases}$

40. $h(x) = \begin{cases} g(x) & x \le 0 \\ f(x) & x > 0 \end{cases}$

41. $h(x) = \begin{cases} g(x) & x \le -2 \\ x & -2 < x < 0 \\ f(x) & x \ge 0 \end{cases}$

42. $h(x) = \begin{cases} f(x) & x < -2 \\ 0 & -2 \le x \le 2 \\ g(x) & x > 2 \end{cases}$

Exercises 43-46 *Use a graphing calculator to help you graph the piecewise-defined function.*

43. $f(x) = \begin{cases} -x^2 + 2x & x \le 1 \\ (x-2)^2 & x > 1 \end{cases}$

44. $g(x) = \begin{cases} -x^2 - 4x - 3 & x < -2 \\ x^2 + 4x + 5 & x \ge -2 \end{cases}$

45. $g(x) = \begin{cases} x^3 + 2x^2 & x < -1 \\ 1 & -1 \le x \le 1 \\ x^3 - 2x^2 + 2 & x > 1 \end{cases}$

46. $g(x) = \begin{cases} \dfrac{2}{x^2 + 1} & x < -1 \\ -x & -1 \le x \le 1 \\ \dfrac{-2}{x^2 + 1} & x > 1 \end{cases}$

Exercises 47-50 *Find a value for the constant a for which the graph of the given function is unbroken.*

47. $f(x) = \begin{cases} x + 1 & x \le 1 \\ ax^2 & x > 1 \end{cases}$

48. $f(x) = \begin{cases} x + a & x \le 0 \\ \sqrt{x + 4} & x > 0 \end{cases}$

49. $f(x) = \begin{cases} \dfrac{x^2 - 1}{x - 1} & x \ne 1 \\ a & x = 1 \end{cases}$

50. $f(x) = \begin{cases} \dfrac{x - 4}{\sqrt{x} - 2} & x \ne 4 \\ a & x = 4 \end{cases}$

51. Let $f(x) = \lfloor x + 2 \rfloor$. Sketch the graph of f.

52. Let $h(x) = \lfloor x \rfloor - 2$. Sketch the graph of h.

53. For an interval $[a, b]$, define
$$f_{[a,b]}(x) = \begin{cases} 1 & a \le x \le b \\ 0 & \text{otherwise} \end{cases}$$

a. Graph $f_{[0,1]}$.

b. Graph $f_{[0,2]} + f_{[1,3]}$.

c. Graph $f_{[0,1]} + f_{[1,2]} + f_{[2,3]} + \cdots$.

54. Let $f(x) = \dfrac{|x|}{x}$.

a. Find the domain of f.

b. Evaluate the following:

     i. $f(2)$ **ii.** $f(5)$

     iii. $f(-2)$ **iv.** $f(-1)$

c. Find the range of f.   **d.** Graph f.

55. Let $g(x) = ((x))$, where $((x))$ denotes the fractional part of x, so that, for example, $((3.456)) = 0.456$, $\left(\left(2\frac{7}{8}\right)\right) = \frac{7}{8}$, and so on.

a. Evaluate the following:

     i. $g(2.8)$ **ii.** $g(4.7)$

     iii. $g\left(\dfrac{23}{5}\right)$

b. Graph g for $x \ge 0$.

c. Simplify $\lfloor x \rfloor + ((x))$ for $x \ge 0$.

56. The function "sgn," the signum function, is defined by
$$\text{sgn}(x) = \begin{cases} 1 & x > 0 \\ -1 & x < 0 \end{cases}$$

a. Graph sgn(x).   **b.** Let $f(x) = \text{sgn}(x - 2)$. Graph f.

Applications

57. Rental Profit A landlord owns 50 apartments that can be rented on a monthly basis. He has fixed overhead costs of $5000 per month, plus $80 per month for every apartment that is rented. He has determined that all of the apartments will be rented if he charges $300 per month. However, for every $10 increase in rent, two fewer apartments will be rented.

a. Find a linear cost function that expresses the landlord's total monthly cost as a function of the number of apartments rented.

b. Find a linear demand function that expresses the rental price as a function of the number of apartments rented. [*Hint*: Find two points (x, p), where x is the number of units rented and p is the corresponding monthly rent, and then find the equation of the line passing through the two points.]

c. Use the demand function from part b to find a revenue function that expresses the landlord's monthly revenue as a function of the number of apartments rented.

d. Use the cost and revenue functions to find the profit function.

e. How many apartments should be rented, and at what price, in order for the landlord to maximize his profit?

58. Ticket Profit A university has a football stadium that can hold 80,000 people. A recent analysis has determined that the cost to staff the stadium for a game is $50,000 plus 25¢ for each person in attendance. It has also been determined that all the seats will be filled if the average ticket price is $10, but 5000 fewer people will attend for each $1 increase in ticket price.

a. Find a linear cost function that expresses the university's total cost to staff the stadium as a function of the number of people in attendance.

b. Find a linear demand function that expresses the ticket price as a function of the number of people in attendance. [*Hint*: Find two points (x, p), where x is the number of people in attendance and p is the corresponding ticket price, and then find the equation of the line passing through the two points.]

c. Use the demand function from part b to find a revenue function that expresses the university's revenue as a function of the number of people in attendance.

d. Use the cost and revenue functions to find the profit function.

e. How many people must attend, and at what ticket price, for the university to maximize profit?

59. Path of a Baseball If a baseball is hit from a height of 3 feet, with an initial velocity of 100 feet per second and an initial angle of 45°, and if we ignore air resistance, the ball will follow the path given by the quadratic function $f(x) = -\frac{2}{625}x^2 + x + 3$, where x is the horizontal distance in feet from home plate and $f(x)$ is the corresponding height in feet above the playing field.

a. Find the maximum height of the ball.

b. Will the ball clear a 12-foot-high fence located 300 feet from home plate?

60. Path of a Baseball If a baseball is thrown from a height of 5 feet, with an initial velocity of 80 feet per second and an initial angle of 30°, and if we ignore air resistance, the ball will follow the path given by the quadratic function

$$f(x) = -\frac{1}{300}x^2 + \frac{\sqrt{3}}{3}x + 5$$

where x is the horizontal distance in feet from where the ball is thrown and $f(x)$ is the corresponding height in feet.

a. Find the maximum height of the ball.

b. How far will the ball land from where it was thrown?

61. Farm Acreage The number of acres of U.S. farm land during the years 1930–1997 can be approximated with the quadratic function

$$f(x) = -0.148x^2 + 8.885x + 1004.5$$

where x is the year ($x = 0$ corresponds to 1930) and $f(x)$ is the total acreage in millions of acres. At what time was the number of acres highest? How many acres were there at that time?

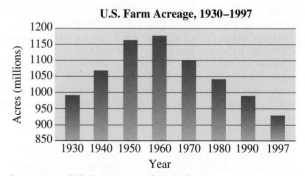

U.S. Farm Acreage, 1930–1997

Data source: U.S. Department of Agriculture.

62. Alcohol Consumption The average number of gallons of alcoholic beverages consumed per year by each adult in the United States for the years 1970–1995 can be approximated with the quadratic function

$$f(x) = -0.0024x^2 + 0.046x + 2.52$$

where x is the year ($x = 0$ corresponds to 1970) and $f(x)$ is the average number of gallons consumed by each adult in that year. For what year was the alcohol consumption the greatest and what was the average rate of consumption at that time?

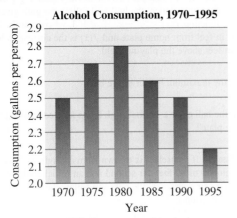

Alcohol Consumption, 1970–1995

Data source: U.S. Department of Agriculture.

63. **Telephone Rate**   Suppose the cost of a cell phone call from New York to Los Angeles is 25¢ for the first minute and 15¢ for each additional minute or fraction thereof. Then the cost of a phone call lasting x minutes can be computed using the function $C(x) = 0.25 - 0.15\lfloor 1 - x \rfloor$. Plot the graph of this function. What is the longest phone call that can be made if the cost cannot exceed $3.00?

64. **Postal Rate**   In 2002, the U.S. Postal Service rate for first-class mail was raised to 37¢ for the first ounce and 23¢ for each additional ounce or fraction thereof. The rate for an item weighing x ounces can be computed with the function $C(x) = 0.37 - 0.23\lfloor 1 - x \rfloor$. Plot the graph of this function. What is the heaviest first-class item that can be sent for $3.00 or less?

65. **Tax Rates**   Fred and Irma are trying to decide if they should file a joint tax return or if they should file separately. If they file separately, Fred has a taxable income of $60,000 and Irma has a taxable income of $86,000. If they file jointly, their combined taxable income is $146,000. The function S below computes taxes for separate returns and the function J computes taxes for joint returns. In both cases, x denotes the taxable income. Compute the total taxes they will pay if they file separately. How does this amount compare to the taxes they will pay if they file jointly?

$$S(x) = \begin{cases} 0.1x & 0 \le x \le 6000 \\ 600 + 0.15(x - 6000) & 6000 < x \le 23{,}350 \\ 3202.50 + 0.27(x - 23{,}350) & 23{,}350 < x \le 56{,}425 \\ 12{,}132.75 + 0.30(x - 56{,}425) & 56{,}425 < x \le 85{,}975 \\ 20{,}997.75 + 0.35(x - 85{,}975) & 85{,}975 < x \le 153{,}325 \\ 44{,}640.25 + 0.386(x - 153{,}325) & x > 153{,}325 \end{cases}$$

$$J(x) = \begin{cases} 0.1x & 0 \le x \le 12{,}000 \\ 1200 + 0.15(x - 12{,}000) & 12{,}000 < x \le 46{,}700 \\ 6405 + 0.27(x - 46{,}700) & 46{,}700 < x \le 112{,}850 \\ 24{,}265.50 + 0.30(x - 112{,}850) & 112{,}850 < x \le 171{,}950 \\ 41{,}995.50 + 0.35(x - 171{,}950) & 171{,}950 < x \le 307{,}050 \\ 89{,}280.50 + 0.386(x - 307{,}050) & x > 307{,}050 \end{cases}$$

66. **Tax Rates**   The tax rates for a single person filing a U.S. tax return in 2002 are given in Table 8. Construct a piecewise-defined function that expresses the total tax that must be paid as a function of taxable income.

Table 8   2002 Tax rates (single)

If taxable income is over . . .	but not over . . .	the tax is . . .	of the amount over . . .
$0	$6000	10%	$0
$6000	$27,950	$600 + 15%	$6000
$27,950	$67,700	$3892.50 + 27%	$27,950
$67,700	$141,250	$14,635 + 30%	$67,700
$141,250	$307,050	$36,690 + 35%	$141,250
$307,050		$94,720 + 38.6%	$307,050

67. **Secret Code**   At one time, a secret decoder ring could be found inside each specially marked box of Cap'n Crunch cereal. The ring consisted of concentric circles on which the alphabet was written. A message could be encoded by rotating one of the circles and replacing the letter on the inner circle with the letter on the outer circle. A message could be decoded in a similar fashion. For a certain alignment of circles, the function

$$f(x) = x + 6 - 26\left\lfloor \frac{x + 5}{26} \right\rfloor$$

can be used to decode a message by assigning each letter of the alphabet the number corresponding to its position (A $\leftrightarrow$ 1, B $\leftrightarrow$ 2, . . .) and then computing $f(x)$ for that number. Finally, replace the number with the letter in that position. Using this technique, descramble the message GUNB LIWEM.

68. **Christmas Break**   At a certain university, the spring semester begins on the first Monday after January 5. It follows (trust us!) that if Christmas falls on day n (with $n = 0$ being Sunday, $n = 1$ being Monday, and so on), then the January date on which the next spring semester will begin is given by

$$f(n) = 9 - n + 7\left\lfloor \frac{n + 3}{7} \right\rfloor$$

Also, the day of the week on which Christmas falls in the year $2000 + x$ is given by

$$g(x) = 1 + x + \left\lfloor \frac{x}{4} \right\rfloor - 7\left\lfloor \frac{1 + x + \lfloor \frac{x}{4} \rfloor}{7} \right\rfloor$$

a. On what day of the week will Christmas fall in 2014?

b. If Christmas falls on a Saturday, on what date does the next spring semester start?

c. What interpretation can be given to the function $f \circ g$? In particular, if the year is $2000 + x$, what does $(f \circ g)(x)$ represent?

d. Find an expression for $(f \circ g)(x)$. (It won't be pretty.)

e. Use your answer to part d to find the day on which the spring semester will start for the 2012–2013 academic year.

Concepts and Critical Thinking

Exercises 69-72   *Answer true or false.*

69. If x^2 appears in the formula defining a function, then the function is said to be quadratic.

70. The graph of a quadratic function is a parabola with a vertical axis of symmetry.

71. The floor function is an example of a step function.

72. A quadratic function cannot have both a maximum value and a minimum value.

Exercises 73-78   *Give an example of each.*

73. A piecewise-defined function that is linear on one piece and quadratic on another

74. A function whose graph is a parabola with vertex (1, 3)

75. Two linear functions, the graphs of which do not intersect

76. A quadratic function having no zeros

77. A quadratic function having one zero

78. A quadratic function having two zeros

Questions for Discussion or Essay

79. The function $f(x) = -\frac{1}{324}x^2 + \sqrt{3}x + 3$ in Example 3 takes into account the initial velocity, angle, and height of the ball, as well as the force of gravity. What other factors might also affect the path of the ball? How do these factors compare in significance to the ones that were taken into account? How will these factors influence such things as the maximum height of the ball and the maximum distance the ball travels before it hits the ground (assuming it doesn't hit the ceiling first)?

80. How can you tell whether a graphical image is the plot of the graph of one piecewise-defined function or of several functions?

81. Some quadratic functions have a local maximum, whereas others have a local minimum. How can you determine whether a given quadratic function has a local maximum or a local minimum without graphing the function?

82. Some people view taxes as payment for services rendered by the government: defending our borders, educating our children, protecting our environment, insuring our elderly and indigent, and so forth. Certainly the cost of a quart of milk is not dependent on one's income, nor is the cost of virtually any commodity available in the marketplace. Why then are taxes dependent on income? If we let i represent income and $T(i)$ represent the corresponding income tax, then what kind of function would T be if all individuals were taxed the same amount? What if all individuals were taxed at a flat rate? Some have suggested that anyone earning under a certain cutoff amount should not pay any tax but that anyone earning above the cutoff should be taxed at a flat rate on the money they earn above the cutoff level. What kind of function would $T(i)$ be under these circumstances? What method of tax computation do you feel would be fairest?

Projects for Enrichment

83. Driving Under the Influence   Concern over the incidence of alcohol-related motor vehicle accidents has prompted lower limits on the legal level of alcohol in the bloodstream. Unfortunately, most people are unaware of the relationship between alcohol consumption and the level of alcohol in the bloodstream. Indeed, many are unaware that even small quantities of alcohol can raise the level beyond legal limits. In this project, we investigate the relationship between alcohol consumption and the alcohol level in the bloodstream, and we also see what bearing body weight has on the relationship.

Ethanol, or grain alcohol, is the form of alcohol that is found in beer, wine, and other liquors. It is eliminated from the body by the liver via an enzymatic process at the constant rate of 12 grams per hour. We assume that each serving of beer, wine, or other liquor contains 18 grams of ethanol. We further assume that the ethanol enters the bloodstream immediately after the drink is consumed.

a. Suppose one drink is consumed at time $t = 0$. Find a linear function $A_1(t)$ that gives the number of grams of ethanol remaining in the bloodstream after t hours. How long will it take for all the ethanol to be eliminated?

b. Now suppose that one drink is consumed at time $t = 0$ and a second is consumed 1 hour later, at time $t = 1$. Find a piecewise-defined function $A_2(t)$ that gives the number of grams of ethanol remaining in the bloodstream after t hours. Note that one piece will correspond to the interval $0 \leq t < 1$ and the second will correspond to $t > 1$. Plot the graph of A_2. How long will it take for all the ethanol to be eliminated?

c. Suppose that drinks are consumed at a regular rate of 2 per hour for 2 hours (that is, at times $t = 0$, $t = \frac{1}{2}$, $t = \frac{3}{2}$, $t = 2$). Find a piecewise-defined function $A_3(t)$ that gives the number of grams of ethanol remaining after t hours, and plot the graph of A_3.

d. Repeat part c to find a function A_4 for consumption at a rate of 3 drinks per hour.

If $A(t)$ denotes the number of grams of ethanol in the bloodstream at time t, the percentage levels $L(t)$ of blood serum ethanol at time t are approximated by

$$L(t) = \frac{A(t)}{310 \times (\text{Body weight in pounds})} \quad \text{for adult males}$$

and

$$L(t) = \frac{A(t)}{250 \times (\text{Body weight in pounds})} \quad \text{for adult females}$$

e. For each of the functions A_1–A_4 in parts a through d, find the functions $L_1(t)$ through $L_4(t)$ for an adult male weighing 160 pounds and an adult female weighing 130 pounds.

f. In many states, the legal limit is 0.1%. In other words, one is legally under the influence if $L(t) \geq 0.001$. For each of the functions $L_1(t)$ through $L_4(t)$ in part e, find the time interval during which the blood serum ethanol level is at or above 0.001. It may be helpful to sketch the graphs of the functions.

g. Discuss some of the limitations of this model for computing blood serum ethanol level.

84. Day of the Week Computation To the list of human calculators—people capable of extraordinary mental feats such as rapid multiplication, extraction of cube roots, memorization of entire books—one addition should be made: you. You are about to learn a method that will enable you to compute the day of the week for any known date—the signing of the Declaration of Independence, the day you were born, next Christmas, Halloween in the year 2025, the day of your twenty-first birthday—quickly and without a calculator or even a pencil and paper.

In this project, we investigate a technique developed by mathematician John Horton Conway for computing the day of the week on which any given date falls in the standard or Gregorian calendar. It is based on the fact that in any given year, a certain collection of easily memorized dates (called reference dates) fall on the same day of the week—a day we will call Doomsday. Once Doomsday is known for a given year, the day on which any given date falls can easily be computed by comparison to a nearby reference date. Here are a few facts about the Gregorian calendar that we will need:

i. There are 365 days in an ordinary year; thus, leap years have 366 days. The extra day is February 29.

ii. Leap years occur every four years—whenever the last two digits form a number divisible by 4. Thus 1840, 1984, and 1996 were all leap years. (These are the same years in which Summer Olympics and U.S. presidential elections are held.)

iii. There is one exception to rule ii: Only every fourth century year—a year ending in "00"—is a leap year. Thus, 1700, 1800, and 1900 were not leap years, but 2000 was a leap year; 2100, 2200, and 2300 will not be leap years, but 2400 will be a leap year, and so forth.

a. Find the number of days separating each pair of dates in the tables given, and then explain why all these dates fall on the same day of the week.

Ordinary Year:

Dates	1/3, 2/28	2/28, 4/4	4/4, 5/9	5/9, 6/6
Days Apart				

Dates	6/6, 7/11	7/11, 8/8	8/8, 9/5	9/5, 10/10
Days Apart				

Dates	10/10, 11/17	11/17, 12/12
Days Apart		

Leap Year:

Dates	1/4, 2/29	2/29, 4/4	4/4, 5/9	5/9, 6/6
Days Apart				

Dates	6/6, 7/11	7/11, 8/8	8/8, 9/5	9/5, 10/10
Days Apart				

Dates	10/10, 11/17	11/17, 12/12
Days Apart		

We call the dates appearing in these tables *reference dates*. Because we will be using them extensively, let's see if we can develop a mnemonic device (memory aid) to help us remember them. First, we notice the following pattern: 4/4, 6/6, 8/8, 10/10, and 12/12 are all reference dates. Second, imagine that you have a 9 to 5 job at a 7–11 convenience store—a handy way of remembering the dates 9/5 and 7/11, and their transposes 5/9 and 11/7. We then note that no matter if we are in an ordinary or a leap year, the last day of February is a reference date. For convenience, we will think of this as the 0th day of March, or 3/0. Finally, we must remember that 1/3 is a reference date in an ordinary year, whereas 1/4 is a reference date in a leap year. Memorize the reference dates; all reference dates will fall on Doomsday for a given year.

b. At this point, we are able to find the day of the week on which any date falls provided that we know Doomsday for the year in question. For example, suppose that we are interested in finding the day on which Christmas falls in a year in which Doomsday is Saturday. Now December 12 is a reference day, so we know that December 12 is a Saturday. It follows that December 26 is a Saturday also and hence Christmas, December 25, is on a Friday. Use this technique to complete the following table.

Year	1776	1865	1998	2020
Doomsday	Thurs	Mon	Sat	
Valentine's Day (2/14)				
Independence Day (7/4)				Wed
Halloween (10/31)				
Christmas (12/25)				

c. Since 52 weeks is 364 days, but an ordinary year has 365 days and a leap year has 366 days, the day on which April 4 falls will advance 1 day in an ordinary year and 2 days in a leap year. Thus, Doomsday itself advances 1 day for an ordinary year and 2 days for a leap year. For example, if this year Doomsday is Monday, then next year it will be either Tuesday or Wednesday, depending on whether next year is an ordinary or leap year. Use this fact to complete the following table.

Year	1942	1943	1944	1945	1996	1997	1998
Doomsday		Sun					Sat

d. It is a fact that Doomsday 1900 was Wednesday. Thus, to figure Doomsday for any year in the 20th century, we begin with Wednesday and advance one day for each year and an additional day for each leap year. Thus, for example, to compute Doomsday for 1910, we advance 10 days—one for each year after 1900—and an additional 2 days to take into account the leap years of 1904 and 1908. Advancing 12 days from Wednesday brings us to a Monday. Thus, Doomsday 1910 was Monday. A similar technique can be used to find Doomsday for any year in the 21st century (advancing from Tuesday instead of Wednesday). Use this technique to find Doomsday for each of the following years.

Year	1926	1953	1986	2002	2050
Doomsday					

e. Show that the number of days that one must advance from Wednesday to compute Doomsday for the year $1900 + x$ (with $0 \le x \le 99$) is given by $x + \lfloor \frac{x}{4} \rfloor$. (Similarly, the number of days that we advance from Tuesday to find Doomsday in $2000 + x$ is given by $x + \lfloor \frac{x}{4} \rfloor$.)

The formula given in part e often involves computations that are difficult to perform without pencil and paper. We now provide an alternative method that, although more complicated, involves smaller numbers and is thus ideal for mental computation. To find Doomsday in $1900 + x$, divide x by 12 to obtain a quotient q and a remainder r. Next, divide r by 4 to obtain a quotient s—ignore the remainder of this second division. Now add the numbers q, r, and s. For example, to compute Doomsday 1981, we divide 81 by 12 to get a quotient of 6 and a remainder of 9; dividing 9 by 4 gives us a quotient of 2. We add to obtain $6 + 9 + 4 = 17$; thus, Doomsday will fall 17 days after Wednesday—that is, on a Saturday. (An explanation for why this works is just beyond the scope of this text.) Try using this technique, without pencil and paper, to find the day of the week on which you were born.

f. To find Doomsdays for other centuries, we need only know Doomsday for the century year. Doomsdays for century years are as follows: 1700, Sunday; 1800, Friday; 1900, Wednesday; 2000, Tuesday. After this, the cycle repeats: Sunday, Friday, Wednesday, Tuesday, Sunday, Friday, Wednesday, Tuesday, and so forth. Use this information to complete the following table.

Event	Date	Day of the Week
Signing of the Declaration of Independence	7/4/1776	
Assassination of Abraham Lincoln	4/14/1865	
D-day	6/6/1944	
Neil Armstrong walks on the moon	7/20/1969	
Y2K eve	12/31/1999	

Section 2.6 — Modeling with Functions and Variation

- What is the gravitational pull exerted on the moon by 600-pound sumo wrestler Konishiki?
- If a man were 48 times as tall as a grasshopper is long, and a grasshopper can jump 20 inches, how far could a man-sized grasshopper jump?
- What will the total health-care expenditures be in the year 2005?
- How much is a graduate degree worth?
- What concentration of the ozone-depleting chemical CFC-11 was present in the atmosphere in the year 2002?
- How are the gas mileage and weight of a car related?

Mathematical Models

A **mathematical model** is a mathematical description of the behavior of some aspect of the real world. Mathematical models have been formulated for such diverse phenomena as signal transmission in the human nervous system, the aerodynamics of a hummingbird, the spread of AIDS, deforestation of the Amazon Rain Forest, behavior of the jet stream, the path of Halley's comet, and the entire global economy. All the major theories of the physical sciences are, in essence, mathematical models. Examples include Einstein's general theory of relativity, which models gravity, and superstring theory, which is a model of the very fabric of space itself.

Real-world phenomena tend to be so extraordinarily complex and subject to so many random influences that any description, mathematical or otherwise, is necessarily incomplete. A mathematical model is an attempt to capture the salient features of a real-world phenomenon, but we sometimes find that in striving for simplicity, we have left out so many key features that our model is like a bad made-for-TV movie: overly simplistic and unrealistic. Nonetheless, the accuracy and utility of certain mathematical models is startling. For example, our current model of the motion of Earth, the sun, and the moon is so good that it correctly predicts the time of eclipses to within seconds, decades in advance. On the other hand, in spite of an astronomical amount of data collection, the use of the most powerful supercomputers in existence, and the efforts of some of the most brilliant minds on the planet, existing models of Earth's atmosphere are not good enough to reliably predict the weather even 3 or 4 days in advance!

You have already seen many examples of mathematical models in the text—from falling objects and stopping distance in the previous chapter to CO_2 levels and alcohol consumption in this chapter. However, in these earlier examples, we were concerned primarily with how a model could be used to solve a real-world problem. In this section, we investigate not just how models are used to solve problems, but also how models are constructed.

Variation

Many of the most important mathematical models arising in the natural sciences can be conveniently described using the language of **variation**. For example, the force acting on an object *varies directly* as the acceleration that the object is undergoing; the strength of the gravitational pull between two bodies *varies inversely* as the square of the distance between them and *varies jointly* as the masses of the bodies. The terminology of variation is summarized in the following table.

Variation Terminology		
Type of variation	**Equation**	**Terminology**
Direct variation	$y = kx$	"y varies directly as x," "y is proportional to x," or "y is directly proportional to x"
Inverse variation	$y = \dfrac{k}{x}$	"y varies inversely as x" or "y is inversely proportional to x"
Joint variation	$z = kxy$	"z varies jointly as x and y"
In all cases, the constant k is called the **constant of proportionality**.		

If y varies directly as x, then we can conclude that $y = kx$ for some constant k. Note, however, that this information alone is not enough to completely specify the relationship between the two variables. Additional data are required to determine *k,* the constant of proportionality, as illustrated in the following examples.

EXAMPLE 1

Computing the Constant of Proportionality

It is given that y varies directly as x and that $y = 3$ when $x = 2$. Find the constant of proportionality.

Solution Since y varies directly as x, we may write $y = kx$ for some constant k. Because $y = 3$ when $x = 2$, we have $3 = k \cdot 2$, from which it follows that $k = \frac{3}{2}$. Therefore, $y = \frac{3}{2}x$.

EXAMPLE 2

Computing the Constant of Proportionality

Suppose that z varies jointly with x and y, and that $z = 40$ when $x = 2$ and $y = 5$. Find the constant of proportionality.

Solution Since z varies jointly with x and y, we know that $z = kxy$ for some constant k. Substituting 2 for x, 5 for y, and 40 for z gives us

$$40 = k \cdot 2 \cdot 5$$
$$40 = 10k$$
$$k = 4$$

Therefore, $z = 4xy$.

Once the constant of proportionality is determined, the model is complete and can then be used to make estimates, as in the following examples.

EXAMPLE 3

Estimating the Time of a Trip

The travel time for a certain trip varies inversely with the average speed. If the trip takes 30 minutes at an average speed of 25 miles per hour, how long will it take at 100 miles per hour?

Solution Let t represent the time for the trip in minutes and r the average speed in miles per hour. Then, since the time varies inversely with the rate, we have.

$$t = \frac{k}{r}$$

Using the fact that $t = 30$ when $r = 25$ gives us

$$30 = \frac{k}{25}$$

and so

$$k = 25 \cdot 30 = 750$$

Thus, we have $t = 750/r$. When $r = 100$, this gives us $t = 750/100 = 7.5$ minutes.

Robert Wadlow, who reached a world record height of 8′11″ shortly before his death at the age of 22, towers above actresses Maureen O'Sullivan and Ann Morris.

-----→EXAMPLE 4

-----→EXAMPLE 5

Estimating Body Weight as a Function of Height

If all men were shaped similarly, then increasing a man's size proportionally would result in increases in all three dimensions. In this case, the weight of a man would be proportional to the cube of his height. Suppose that the weight of a man 5′10″ is 170 pounds. What would the weight of a similarly shaped 8-foot-tall man be?

Solution Let w represent the weight of a man in pounds and h his height in inches. Since weight is proportional to the cube of height, we have $w = kh^3$. Now 5′10″ is 70 inches, so we have

$$170 = k \cdot 70^3$$

$$k = \frac{170}{70^3} \approx 0.0004956$$

Thus $w = 0.0004956h^3$. Since 8 feet is 96 inches, the weight of an 8-foot-tall man would be

$$w = 0.0004956 \cdot 96^3 \approx 438.5 \text{ pounds}$$

Often it is convenient to express the relationship between several variables as a combination of two or more of the standard types of variation, as in the following examples.

Expressing the Relationship Between Several Variables

It is given that w varies jointly as the square of x and the cube of y and inversely as z. Find w when $x = 1$, $y = 2$, and $z = 10$, given that $w = 9$ when $x = 2$, $y = 3$, and $z = 36$.

Solution Combining the variation statements, we have

$$w = k\frac{x^2y^3}{z}$$

Now, substituting $w = 9$, $x = 2$, $y = 3$, and $z = 36$, we can solve for k.

$$9 = k\frac{2^2 3^3}{36}$$

$$9 = 3k$$

$$k = 3$$

Thus,

$$w = 3\frac{x^2y^3}{z}$$

Finally, substituting $x = 1$, $y = 2$, and $z = 10$, we can determine the value of w.

$$w = 3 \cdot \frac{1^2 2^3}{10}$$

$$= 3 \cdot \frac{4}{5}$$

$$= \frac{12}{5}$$

::::EXAMPLE 6

Estimating the Gravitational Pull of an Object

The gravitational pull between two objects varies jointly as their masses and inversely as the square of the distance between them. The weight of an object on Earth is, by definition, the gravitational pull between the object and Earth. Find the gravitational pull that 600-pound sumo wrestler Konishiki exerts on the moon, which weighs 1.62064×10^{23} pounds and is 234,912 miles from Earth.

Solution   We begin by finding the general equation of variation. Let d be the distance between two objects, let m_1 and m_2 be their masses (or weights), and let F be the force due to gravity. Since F varies jointly with m_1 and m_2 and inversely with the square of d, we have

$$F = k\frac{m_1 m_2}{d^2}$$

In order to find the constant k, we choose a scenario in which all of the quantities F, m_1, m_2, and d are known. Because the gravitational force on a 1-pound object on the surface of Earth is 1 pound, if we knew the weight of Earth and the distance from Earth's surface to its center, we would have enough information to evaluate k. After consulting an almanac (or the numerography inside the front cover of this text), we discover that Earth has a weight of approximately 1.3176×10^{25} pounds and that the radius of Earth is approximately 3963 miles. Thus, we have $F = 1$ pound, $d = 3963$ miles, $m_1 = 1$ pound, and $m_2 = 1.3176 \times 10^{25}$ pounds. Substituting into the equation of variation, we find

$$1 \text{ lb} = k\frac{1 \text{ lb} \cdot 1.3176 \times 10^{25} \text{ lb}}{(3963 \text{ mi})^2}$$

and solving for k, we obtain

$$k = \frac{(3963)^2 \text{ mi}^2}{1.3176 \times 10^{25} \text{ lb}} \approx 1.19197 \times 10^{-18} \frac{\text{mi}^2}{\text{lb}}$$

Now, to find the gravitational pull between Konishiki and the moon, we use this value of k and let $m_1 = 600$ pounds, $m_2 =$ the weight of the moon $\approx 1.62064 \times 10^{23}$ pounds, and $d =$ the distance from Earth to the moon $\approx 234,912$ miles. Thus, the force is given by

$$F \approx 1.19197 \times 10^{-18} \frac{\text{mi}^2}{\text{lb}} \cdot \frac{600 \text{ lb} \cdot 1.62064 \times 10^{23} \text{ lb}}{(234,912 \text{ mi})^2}$$

$$\approx 0.00210 \text{ lb}$$

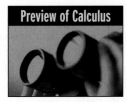

Preview of Calculus

Functions of Several Variables and Surfaces

As we have seen, if for each value of x there is exactly one value of y, then we say that y is a function of x. If the function is called f, then we can express this relationship algebraically by the equation $y = f(x)$. In many real-world contexts, however, a given quantity depends on the value of two or more variables. For example, the volume V of a (right circular) cylinder depends on both the radius r and the height h, according to the equation $V = \pi r^2 h$. In fact, complex models arising in such diverse fields as pharmacology, biology, chemistry, physics, telecommunications, industrial engineering, and climatology can involve hundreds and even thousands of variables. (Quantum physicists even consider functions involving infinitely many variables!) Whereas the graph of a function of a single variable is a curve in the plane, the graph of a function of two variables is a surface in three-dimensional space, such as that shown in Figure 85. Functions of several variables and their graphs are examined in detail in multivariate calculus.

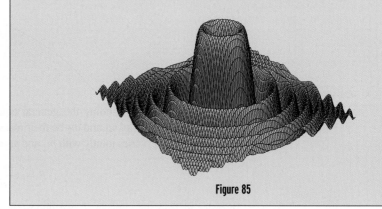

Figure 85

Data Fitting   Although some sophisticated mathematical models are constructed from underlying principles, many are formed simply by finding a function that fits a given set of data. Often the *form* of the function (for example, linear or quadratic) is either known or assumed, and the given data are then used to determine the function precisely. For example, suppose that we wish to find a linear function that is consistent with the following data.

x	y
2	5
3	7

One approach would be to find the slope of the line connecting $(2, 5)$ and $(3, 7)$ and then use the point-slope form for the equation of a line. However, we will use a method that is more general and can be applied in many other situations. We first note that since y is to be a linear function of x, we must have $y = ax + b$ for some choice of constants a and b. (In the context of data fitting, it is conventional to represent linear functions in the form $y = ax + b$ rather than $y = mx + b$.) Substituting 2 for x and 5 for y, we have

$$5 = a(2) + b$$

Now substituting 3 for x and 7 for y gives us

$$7 = a(3) + b$$

Thus, we must solve the following system of two linear equations in two unknowns:

(1) $\qquad\qquad\qquad\qquad\qquad 2a + b = 5$

(2) $\qquad\qquad\qquad\qquad\qquad 3a + b = 7$

Now a system of equations such as this could be solved in many different ways. In fact, several entire sections of Chapter 9 are devoted to different techniques for solving systems of equations. Here we simply use substitution—arguably the most straightforward technique for solving systems of two linear equations in two unknowns. Solving equation (1) for b we have

(3) $\qquad\qquad\qquad\qquad\qquad b = 5 - 2a$

Substituting from equation (3) into equation (2) we obtain

$$3a + b = 7$$
$$3a + (5 - 2a) = 7$$
$$a + 5 = 7$$
$$a = 2$$

Substituting 2 for a in equation (1) gives

$$2(2) + b = 5$$
$$4 + b = 5$$
$$b = 1$$

Thus, we have $y = ax + b = 2x + 1$. It can be easily verified that the line with equation $y = 2x + 1$ passes through the points $(2, 5)$ and $(3, 7)$. In the following two examples, we apply a similar procedure to find quadratic functions that "fit" a given set of data.

EXAMPLE 7

Finding a Parabola Passing Through Two Points

Suppose we are given that the variable y is a function of the variable x of the form $y = x^2 + bx + c$ and that we are provided with the following data.

x	y
2	4
4	8

Find values for b and c so that $y = x^2 + bx + c$ is consistent with the data.

Solution   We use the two data points to form two equations in the unknowns b and c. Using the point $(2, 4)$, we set $x = 2$ and $y = 4$ to obtain

$$4 = 2^2 + b \cdot 2 + c$$
$$4 = 4 + 2b + c$$
$$2b + c = 0$$

Similarly, the point (4, 8) gives us

$$8 = 4^2 + b \cdot 4 + c$$
$$8 = 16 + 4b + c$$
$$4b + c = -8$$

Thus, we must solve the following system of equations:

(1) $$2b + c = 0$$
(2) $$4b + c = -8$$

Solving equation (1) for c gives us $c = -2b$. Substituting into equation (2) we have

$$4b + c = -8$$
$$4b - 2b = -8$$
$$2b = -8$$
$$b = -4$$

Substituting -4 for b in equation (1) gives us $2(-4) + c = 0$, so that $c = 8$. Thus, we have $y = x^2 - 4x + 8$.

EXAMPLE 8

Finding a Parabola Passing Through Three Points

It is known that s can be expressed as a quadratic function $f(t)$.

a. Find $f(t)$ and sketch its graph given the following data.

t	s
0	13
1	7
5	23

b. Find the value of t that minimizes s.

Solution

a. An arbitrary quadratic function of t is of the form $f(t) = at^2 + bt + c$. We use the given data points to form a system of equations in the unknowns a, b, and c.

Using (0, 13): $$13 = a \cdot 0^2 + b \cdot 0 + c$$
$$c = 13$$
Using (1, 7): $$7 = a \cdot 1^2 + b \cdot 1 + c$$
$$a + b + c = 7$$
Using (5, 23): $$23 = a \cdot 5^2 + b \cdot 5 + c$$
$$25a + 5b + c = 23$$

Thus, we must solve the following system of equations:

(1) $$c = 13$$
(2) $$a + b + c = 7$$
(3) $$25a + 5b + c = 23$$

Substituting $c = 13$ in equations (2) and (3) and simplifying gives us

(4) $$a + b = -6$$
(5) $$25a + 5b = 10$$

Solving equation (4) for b gives us $b = -6 - a$. Substituting into equation (5), we have

$$25a + 5b = 10$$
$$25a + 5(-6 - a) = 10$$
$$20a - 30 = 10$$
$$20a = 40$$
$$a = 2$$

Substituting 2 for a in equation (4) gives $2 + b = -6$, and so $b = -8$. Hence, $a = 2$, $b = -8$, and $c = 13$, so that $f(t) = 2t^2 - 8t + 13$. The graph of $f(t)$ is shown in Figure 86.

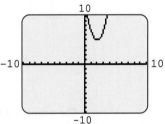

Figure 86

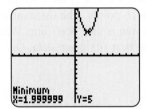

Figure 87

b. Since $f(t) = 2t^2 - 8t + 13$ is a quadratic function, its graph is a parabola, and its minimum value will occur at the vertex of the parabola. We can find the vertex using the techniques of Section 2.5, or by approximating with a graphing calculator, as shown in Figure 87, which suggests that the vertex is the point $(2, 5)$. Thus, the minimum occurs at $t = 2$ and the minimum value is 5.

In the previous examples, the form of the model was given, and just enough data was provided to determine the model precisely. In practice, the data will not fit the model precisely; the model can only approximate the data. Our task, then, is to choose the model that most nearly approximates the given data. For example, suppose that we wish to model the data given in Table 9 with a linear function. Unless we are extremely fortunate, the given points will not lie on any one line. Therefore, we must choose the line that, in some sense, best approximates the given data.

Table 9

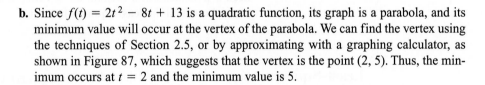

Year	1985	1990	1995	2000
U.S. heath-care expenditure (in billions of dollars)	422.6	666.2	991.4	1299.5

Data source: Centers for Disease Control.

If we let t be the number of years since 1985 (so that 1985 corresponds to $t = 0$, 1990 corresponds to $t = 5$, and so on) and let E be the total U.S. health-care expenditures in billions of dollars, then we obtain the graph of the data points (t, E) shown in

Figure 88. Now it appears that no line passes through all of these points, so the best that we can do is to find a line that "almost" passes through all of these points. Shown in Figure 89 is the line that appears to the authors to best fit the data; your eyes may tell you something different.

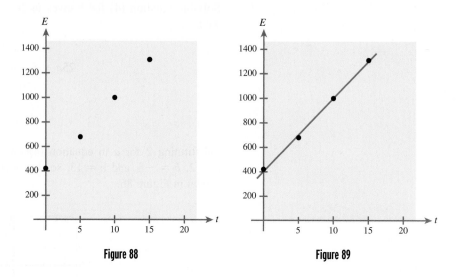

Figure 88   Figure 89

The primary advantage of this method is its simplicity; we simply drew the line that seemed to come the closest to passing through the given points. The disadvantage is obvious; there is no guarantee that the line we have chosen is the "best" line. What we need is a mathematical method for selecting a line that best fits given data. One such method is called the **least-squares best fit**.

Least-Squares Best Fit

Suppose we are given three data points (x_1, y_1), (x_2, y_2), and (x_3, y_3), as shown in Figure 90, and we would like to find the equation $y = ax + b$ of the line that in some sense most nearly fits the data. The distances d_1, d_2, and d_3 are a measure of the error involved in approximating the data points with the line. The line with the least-squares best fit is obtained by selecting a and b in such a way as to minimize the sum of the squares of the errors—that is, to minimize the quantity $d_1^2 + d_2^2 + d_3^2$. The following definition extends this idea to any number of data points.

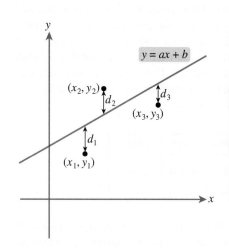

Figure 90

Least-Squares Best Fit

If data points (x_1, y_1), (x_2, y_2), . . . , (x_n, y_n) are given, then the **line with the least-squares best fit**, or **regression line**, is that line for which the sum of the squares of the errors, $d_1^2 + d_2^2 + \cdots + d_n^2$, is as small as possible. The process of finding the coefficients a and b for the line $y = ax + b$ with the least-squares best fit is called **linear regression**.

Fortunately, graphing calculators can perform linear regressions for us. The general steps are as follows.

Calculator Keys

Linear Regression

The **linear regression** feature of a graphing calculator can be used to find a linear function of the form $y = ax + b$ that has the best least-squares fit to a collection of data points (x, y). There are several steps to the process.

Step 1.   Clear any existing statistical data.
Step 2.   Enter the points (x, y) as statistical data.
Step 3.   Select the linear regression option from the list of available regression options.

After step 3 has been completed, the calculator will display values for a and b, the slope and y-intercept, respectively. Some calculators will also display a third number, r, the coefficient of correlation. Although a full discussion of the significance of r is beyond the scope of this text, we note that when r^2 is close to 1, the regression line is a good fit to the data. If r is positive, y tends to increase as x increases, whereas if r is negative, y tends to decrease as x increases. Note also that some calculators use $y = a + bx$ instead of $y = ax + b$ as the general form for the regression line. In this case, when the values for a and b are reported, a gives the y-intercept and b gives the slope. Be sure to check your calculator to see which form is used.

⋯⋯EXAMPLE 9

Performing Linear Regression with a Graphing Calculator

Find an equation of the form $y = ax + b$ with the least-squares best fit to the points $(1, 5)$, $(2, 6)$, and $(6, 13)$.

```
LinReg
  y=ax+b
  a=1.642857143
  b=3.071428571
  r²=.9943609023
  r=.997176465
```

Figure 91

Solution   We begin by clearing any existing statistical data and entering the points $(1, 5)$, $(2, 6)$, and $(6, 13)$ as statistical data. Next we select the linear regression feature. The calculator should return a screen much like the one in Figure 91, which indicates that $a \approx 1.64$, $b \approx 3.07$, and $r \approx 0.997$. Thus, the regression line has equation $y = 1.64x + 3.07$. Since the value of r is very close to 1, the line is a good fit.

⋯⋯EXAMPLE 10

Finding the Least-Squares Best Fit

a. Use the data in Table 10 to find a linear model with the least-squares best fit.

Table 10

Year	1985	1990	1995	2000
U.S. heath-care expenditure (in billions of dollars)	422.6	666.2	991.4	1299.5

b. Use the model developed in part a to predict the total cost of health care in the year 2005.

Solution

a. Let t represent the number of years after 1985 and $E(t)$ the total U.S. health-care expenditure (in billions of dollars) for year t. Then we wish to obtain a model of the

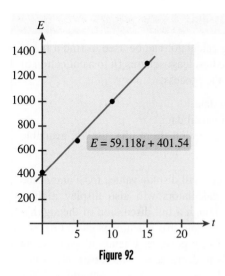

$E = 59.118t + 401.54$

Figure 92

form $E(t) = at + b$. We begin by clearing any existing statistical data from the calculator and then entering the following data points:

$$(0, 422.6), (5, 666.2), (10, 991.4), (15, 1299.5)$$

Next we employ the linear regression feature to obtain $a \approx 59.118$, $b \approx 401.54$, and $r \approx 0.998$. We conclude that the linear model with the least-squares best fit to the given data has equation $E(t) = 59.118t + 401.54$, and because the coefficient of correlation r is nearly 1, the fit is a good one. Figure 92 shows a graph of the least-squares best fit line along with a plot of the data points themselves.

b. Recall that t represents the number of years after 1985. Thus, the year 2005 corresponds to $t = 20$. Substituting 20 for t in the linear model gives

$$E = 59.118t + 401.54 = 59.118(20) + 401.54 \approx 1583.9$$

Thus, according to this model, the total of all health-care expenditures in the year 2005 will be approximately \$1584 billion.

Understanding and Mastery Checklists

Concepts to Understand	Skills to Master
Mathematical models	Determine the constant of proportionality.
❖	❖
Variation: direct, joint, and inverse	Solve a variation problem.
❖	❖
Constant of proportionality	Find a function of a given type whose graph passes through given data points.
❖	❖
Data fitting	Find the equation of a line with the least-squares best fit to a collection of data points.
❖	❖
Least-squares best fit	Perform linear regression on a collection of data points and use the result to estimate missing data.
❖	
Linear regression	

Exercises 2.6

Exercises 1-6 *Find the constant of proportionality.*

1. y varies directly as x, and $y = 8$ when $x = 2$.

2. z varies directly as y, and $z = 10$ when $y = 5$.

3. z varies inversely as x, and $z = 12$ when $x = 4$.

4. y varies inversely as t, and $y = 8$ when $t = \frac{1}{2}$.

5. w varies jointly as x and y, and $w = 15$ when $x = \frac{1}{2}$ and $y = 3$.

6. z varies jointly as s and t, and $z = 2$ when $s = 10$ and $t = 20$.

Exercises 7-12   *Solve the variation problem.*

7. w varies directly as x. Find w when $x = 4$, given that $w = 9$ when $x = 3$.

8. y varies directly as x. Find y when $x = 6$, given that $y = 10$ when $x = 3$.

9. q varies jointly as x and y. Find q when $x = 2$ and $y = 5$, given that $q = 90$ when $x = 3$ and $y = 5$.

10. A varies jointly as x and y. Find A when $x = 12$ and $y = 5$, given that $A = 18$ when $x = 3$ and $y = 10$.

11. V varies directly as the square of x and inversely as the cube of y. Find V when $x = 3$ and $y = 2$, given that $V = 2$ when $x = 4$ and $y = 3$.

12. W varies jointly as x and y and inversely as z. Find W when $x = 5$, $y = 3$, and $z = 1$, given that $W = 2$ when $x = 7$, $y = 2$, and $z = 4$.

Exercises 13-18   *Find the function of the given form whose graph passes through the indicated points.*

13. $f(x) = \dfrac{k}{x^2}; (-3, 2)$

14. $f(x) = ax + b; (2, 5), (4, 9)$

15. $f(x) = ax^2 + bx; (2, -4), (-1, 5)$

16. $g(x) = a(x + 2)(x - 3) + bx(x + 2) + cx(x - 3); (0, 1), (-2, 4),$ $(3, 10)$

17. $h(x) = ax^2 + bx + c; (0, 5), (1, 3), (2, 3)$

18. $g(x) = ax^2 + bx + c; (-1, -1), (0, 4), (1, 5)$

Exercises 19-22   *Find the equation of the line with the least-squares best fit to the given points.*

19. $(1, 3), (4, 4), (7, 6)$

20. $(-1, -4), (-2, -2), (-4, 0)$

21. $(-1, -6), (0, -5), (1, -4), (2, -1)$

22. $(1, 2), (2, 5), (3, 4), (4, 7)$

Exercises 23-26   *Find the line with the least-squares best fit to the given data points, and then use the equation of the line to predict the unknown y-value.*

23.

x	0	5	10	15	20	25
y	90	101	116	130	138	?

24.

x	0	4	8	12	16	20
y	35	45	56	69	78	?

25.

x	0	1	2	3	4	5
y	80	45	12	?	-50	-78

26.

x	0	2	4	6	8	10
y	200	130	?	-10	-60	-120

Applications

27. Board Feet   Suppose that the number of board feet of lumber in a ponderosa pine varies directly as the cube of its circumference at waist height. If a ponderosa pine with a circumference of 100 inches yields 1500 board feet of lumber, how much can be obtained from one with a circumference of 120 inches?

28. Falling Ball   The distance traveled by a ball dropped from the top of a tall building varies directly as the square of the time since its release. If a ball has fallen 64 feet after 2 seconds, how far will it have fallen after 3 seconds?

29. Kinetic Energy   The kinetic energy of a body in motion varies jointly as the mass of the object and the square of its velocity. If the kinetic energy of a 4000-pound vehicle traveling at a rate of 30 miles per

hour is 200,000 joules, find the kinetic energy of the same vehicle traveling at 60 miles per hour.

30. Potential Energy   The potential energy of an object at rest near the surface of Earth varies jointly as its mass and height above the ground. If an object with a mass of 10 kilograms is 3 meters above the ground and has a potential energy of 294 joules, find the potential energy of an object with a mass of 15 kilograms that is 12 meters above the ground.

31. Coulomb's Law   According to Coulomb's law, the electric force exerted between two charged particles at rest varies jointly as the charges of the particles and inversely as the square of the distance between the particles. Two electrons, each with a charge of 1.6×10^{-19} coulomb, are 1 centimeter apart and exert a force of 2.3×10^{-24} newton. How much force is exerted if the electrons are 2 centimeters apart?

32. Grasshopper Jump If a $1\frac{1}{2}$-inch-long grasshopper can jump to a height of 20 inches, and if jumping ability varies directly as length, how high could a 6-foot-tall human jump? Do you think jumping ability varies directly as length? Explain.

33. Ant Strength If an ant weighing 0.01 gram can lift an object weighing 0.2 gram, and if strength varies directly as body weight, how much could a 150-pound human lift? Do you think strength varies directly as body weight? Explain.

Leaf cutter ants

34. Skidding Vehicles The force needed to prevent a vehicle from skidding on a circular curve varies jointly as the weight of the car and the square of the speed and inversely as the radius of the curve. If it takes 4840 pounds of force to keep a certain car traveling at a certain speed from skidding on a curve, how much force will it take to prevent a truck that is twice as heavy but is traveling half as fast from skidding on the same curve?

35. Powerlifting Records The following table shows several weight classes of the men's world powerlifting bench-press records. Find the line with the least-squares best fit to the data, and estimate the bench-press record for the 100-kilogram weight class. The actual record is 265 kilograms, held by Laszlo Meszaros.

Weight class (kg)	52	60	75	90
Bench-press record (kg)	173	205.5	236.0	255

Data source: International Powerlifting Federation.

36. Blood Serum Ethanol Level Ethanol is a form of alcohol found in beer, wine, and other liquors. It is eliminated from the body by the liver, via an enzymatic process, at a constant rate. The following table shows the blood serum ethanol level for a 150-pound male taken at half-hour intervals following four 12-ounce beers. Find the line with the least-squares best fit to the data and predict when the level is below 0.001 (the legal limit for operating a motor vehicle in many states).

Time (minutes)	0	30	60	90
Blood serum ethanol ($\times 10^{-3}$)	1.56	1.42	1.29	1.15

37. Ozone Depletion Chlorofluorocarbons (CFCs) are a family of chemicals used as refrigerants and in the manufacture of numerous consumer products. The increasing concentration of CFCs in the environment has been shown to be a cause in the depletion of the ozone layer. The following table shows the rise in atmospheric concentration of one member of this chemical family, CFC-11, from 1980 to 1990. Find the line with the least-squares best fit to the data, with 1980 corresponding to year 0, and predict the concentration of CFC-11 in 2002. In fact, the CFC-11 in 2002 was about 262 parts per trillion. Can you offer a possible explanation for the disagreement between the prediction of the model and the recorded data for 2002?

Year	1980	1982	1984	1986	1988	1990
Concentration of CFC-11 (parts per trillion)	180	195	215	230	255	275

Data source: Carbon Dioxide Information Analysis Center.

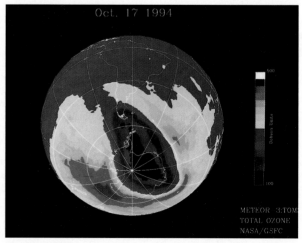

An elongated ozone hole covers much of Antarctica and the tip of South America.

38. Water Consumption The use of bottled water in the United States has shown a steady increase in recent years. The following table shows the per capita (that is, per person) consumption for certain years in the range 1990–2000. Find the line with the least-squares best fit to the data and predict the per capita consumption in 2005.

Year	1990	1992	1994	1996	1998	2000
Bottled water consumption (gallons per person per year)	8.1	8.2	9.6	11.0	11.8	13.2

Data source: USDA.

39. Prison Population The following table shows the number of inmates in federal and state prisons in 1980, 1990, and 2000. Find a quadratic model that expresses the number of prisoners as a function of the year, with 1980 corresponding to year 0. Predict the number of inmates in the year 2005.

Year	1980	1990	2000
Number of inmates (in thousands)	320	743	1316

Data source: U.S. Bureau of Justice Statistics.

40. Average Income The average yearly earnings for a year-round, full-time worker are related to education level. Use the following data from 1990 to find a quadratic model that expresses the average earnings as a function of the number of years of education beyond 8th grade. According to the model you found, what average earnings could be expected for someone with a master's degree (usually 17 years of education)? What educational level corresponds to the smallest average earnings? Does this seem reasonable?

Years of education beyond 8th grade	0	4	8
Average earnings	$16,000	$24,000	$36,000

Data source: U.S. Bureau of the Census.

Concepts and Critical Thinking

Exercises 41-44 *Answer true or false.*

41. If y varies directly as x, then doubling x will cause y to double as well.

42. The line with the least-squares best fit to a set of points must pass through at least one of the points.

43. If y varies inversely as x, then doubling x will cause y to double as well.

44. The line with the least-squares best fit is the line for which the sum of the squares of the errors is as small as possible.

Exercises 45-48 *Give an example of each.*

45. A collection of three points for which the regression line passes through all three points

46. An equation involving two variables such that one variable varies directly with respect to the other

47. A collection of three data points and a quadratic model that fits the data points with absolutely no error

48. A type of variation

Questions for Discussion or Essay

49. Explain what is meant by the statement "Three distinct nonlinear points completely determine a parabola." Is the statement true? Why or why not? What would happen if you tried to fit a function of the form $f(x) = ax^2 + bx + c$ to three points that lie on a straight line?

50. Discuss the primary hazards in the use of mathematical models for predicting real-world phenomena. What safeguards must be taken in order to minimize the risks?

51. The area of a circle varies directly as the square of the radius. If the radius of a circle is doubled, does this mean that the area will double? Explain, using an example to illustrate. In general, if a quantity y varies directly as the nth power of x, what happens to y if x is doubled? Justify your answer.

52. If the surface area of an object of a given shape varies directly as the square of its length and the volume varies directly as the cube of its length, then what can you say about the ratio of volume to surface area? How does this discussion relate to the rate at which small hamburgers grill compared to large ones? What about the rate at which large ice cubes melt compared to small ones?

53. If all men were shaped similarly, then body weight w would vary directly with the cube of height h, so that $w = kh^3$, as mentioned in Example 4. Clearly, men are not all of the same shape. But what

about the shape of the "typical" 5-foot-tall man? Is it the same as that of the "typical" 6-foot-tall man or "typical" 7-foot-tall man? Using the units of pounds for weight and feet for height, estimate appropriate constants of proportionality for 5-footers, 6-footers, and 7-footers. Explain your results.

Projects for Enrichment

54. Transformed Least-Squares Fit We have seen how the least-squares method can be used to find the line of the form $y = ax + b$ that has the least-squares best fit to a set of data points. Here we consider a slight extension that will enable us to fit an equation of the form $y = a \cdot u + b$, where u is a function of x. Consider, for example, the set of data points shown in Table 11 and graphed in Figure 93.

Table 11

x	0	1	2	3
y	-1.3	-0.8	0.7	3.2

Figure 93

From the graph, we suspect that the data points lie on the graph of an equation of the form $y = ax^2 + b$. Notice that if we define $u = x^2$, this equation has the form $y = au + b$, which is linear in u (or x^2) and y. Thus, we may be successful in looking for a linear fit for the data points that are *transformed* in Table 12 by pairing each y with x^2. Indeed, the plot of x^2 versus y in Figure 94 does look linear. So we find the line with the least-squares best fit to the data in Table 12. We find $a = 0.5$ and $b = -1.3$. Our model is thus $y = 0.5x^2 - 1.3$, and it is not hard to check that this equation fits the original data points very well.

Table 12

x^2	0	1	4	9
y	-1.3	-0.8	0.7	3.2

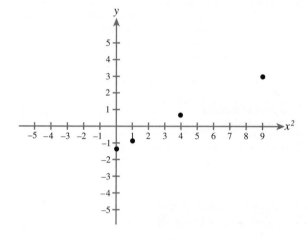

Figure 94

For each of the sets of data in parts a through d, find an equation of the given form that has the best transformed least-squares fit. Plot each equation along with the data points to show that the fit is good.

a. $y = ax^3 + b$

x	0	1	2	3
y	2.2	1.8	0.8	-2.4

b. $y = ax^4 + b$

x	1	$\sqrt{2}$	2	$\sqrt{5}$
y	-4.1	-3.9	-2.8	-1.8

c. $y = a\sqrt{x} + b$

x	1	2	3	4
y	11	13	14	16

d. $y = a\dfrac{1}{x} + b$

x	−3	−1	2	4
y	6	7	4	5

55. Car Models Table 13 provides a compilation of data on 10 different 1997 automobiles. In this project, we investigate relationships between the various characteristics given in the table. For example, we will see how gas mileage relates to engine size, power, and weight. You will need a graphing calculator or spreadsheet that is capable of performing linear regression (least-squares best fit). A spreadsheet may be useful for organizing and plotting the data.

Table 13

Make and Model	Engine Size (liters)	Power (hp)	Weight (lb)	Top Speed (mph)	Highway Mileage (mpg)
BMW 328i	2.8	190	3197	128	29
Cadillac Eldorado	4.6	275	3821	148	26
Cheverolet Camaro Z28	5.7	285	3442	149	25
Dodge Neon	2.0	132	2428	117	33
Ferrari 456 GT	5.5	436	3898	185	16
Ford Taurus LX	3.0	145	3326	112	29
Honda Civic LX	1.6	106	2387	116	35
Jaguar XJ6	4.0	245	4080	144	23
Mercedes Benz C36	3.6	276	3549	155	22
Toyota Camry LE	2.2	133	3020	116	30

a. Describe any general relationships you see in the table. For example, larger engine sizes tend to generate more horsepower.

b. Let y denote the top speed of a car and let x denote its engine size. Plot the set of points (x, y) given by Table 13. Do the points appear to lie on (or near) a single straight line? Find the least-squares best fit model of the form $y = ax + b$ for the engine size and top speed data in Table 13.

c. Repeat part b using power instead of engine size as the x variable. How do your results compare? Which appears to be the better predictor of top speed: power or engine size? (*Hint*: Compare the r values for the two models.)

d. Repeat part b using weight instead of engine size as the x variable. How do your results compare? Which of engine size, power, or weight appears to be the best predictor of top speed? Use your best predictor to estimate the top speed of a car with a 2.4-liter engine, capable of 141 hp, and weighing 2822 lb.

e. Develop and compare three linear models for predicting the highway gas mileage of a car as a function of engine size, power, and weight. Which of the three appears to be the best predictor? Use your best predictor to estimate the highway gas mileage of a car with a 2.4-liter engine, capable of 141 hp, and weighing 2822 lb.

So far, we have attempted to model the top speed and gas mileage of a car as a function of just one of the variables engine size, power, or weight. In actuality, all three of these variables, as well as perhaps others, play roles of varying importance in determining the top speed and gas mileage of a car.

f. Identify as many variables as you can think of that might affect top speed and gas mileage.

Using a technique called *multiple linear regression* we can take more than one variable into account. For the following discussion, let s denote the top speed of a car, m denote the highway gas mileage, e denote the engine size, p denote the power, and w denote the weight. Applying multiple linear regression to the data in Table 13, we obtain the following two models:

$$s = 0.27p - 2.89e + 91.1$$
$$m = -0.004w - 0.05p + 1.15e + 47$$

g. Notice that the coefficient of w is very small in the gas mileage model and that w doesn't even appear in the top speed model. What does this suggest?

h. Find the top speed and highway gas mileage for a car with a 2.4-liter engine, capable of 141 hp, and weighing 2822 lb. A Mitsubishi Galant with these characteristics has a top speed of 130 mph and highway gas mileage of 28 mpg. How do these values compare to those predicted by the models? How might this information be used?

Chapter 2 Review

Exercises 1-4 *Use the following table to determine whether the indicated correspondence from the set D to the set R defines a function.*

Los Angeles Lakers 2002–2003 Roster		
Player	**Height**	**Weight**
Kobe Bryant	6-7	215
Derek Fisher	6-1	205
Rick Fox	6-7	235
Devean George	6-8	225
A.J. Guyton	6-2	185
Robert Horry	6-10	238
Mark Madsen	6-9	245
Stanislav Medvedenko	6-10	255
Tracy Murray	6-7	228
Shaquille O'Neal	7-1	335
Jannero Pargo	6-2	170
Guy Rucker	6-9	270
Kareem Rush	6-6	215
Soumalia Samake	6-10	240
Brian Shaw	6-6	200
Nick Sheppard	6-11	200
Jefferson Sobral	6-8	210
Samaki Walker	6-9	255

1. D = the set of players, R = the set of heights

2. D = the set of heights, R = the set of weights

3. D = the set of weights, R = the set of heights

4. D = the set of players, R = the set of weights

Exercises 5-8 *Determine whether the given equation defines y as a function of x.*

5. $y - x^2 + 1 = 0$

6. $x - y^3 = 1$

7. $x^2 + y^2 = 5$

8. $x = y^4 - 2$

Exercises 9-14 *Evaluate the given function as indicated.*

9. $g(x) = 2x^2 - 3x + 1$

    a. $g(4)$

    b. $g(-3)$

10. $h(x) = \dfrac{x}{2x + 5}$

    a. $h(0)$

    b. $h(-5)$

11. $f(x) = \dfrac{x^2}{x^2 + 1}$

    a. $f(-x)$

    b. $f(x + 1)$

12. $g(x) = (x + 2)^2$

    a. $g(x - 2)$

    b. $g\left(\dfrac{1}{x}\right)$

13. $h(x) = \begin{cases} 4 - x^2 & x \le 2 \\ -x + 4 & x > 2 \end{cases}$

    a. $h(0)$

    b. $h(5)$

    c. $h(a)$ if $a > 2$

14. $f(x) = \dfrac{|x - 3|}{x - 3}$

    a. $f(-2)$

    b. $f(10)$

    c. $f(c)$ if $c < 3$

Exercises 15-18 *Find the domain of the function.*

15. $f(x) = x^3 + 1$

16. $g(x) = \sqrt{3 - x}$

17. $h(t) = \dfrac{t}{\sqrt{2t + 1}}$

18. $f(x) = \dfrac{x^2}{4 - x^2}$

Exercises 19-22   *Use a graphing calculator to estimate the domain and range.*

19. $h(x) = \sqrt{x^2 - 4}$

20. $f(x) = \dfrac{2x^2}{x^2 + 1}$

21. $f(x) = \dfrac{|1 - x|}{1 - x}$

22. $g(x) = \dfrac{\sqrt{x + 1}}{\sqrt{x} + 1}$

Exercises 23-24   *A function f and its graph are given. For each translated or reflected graph, determine the corresponding function g.*

23. $f(x) = 2x^3 - 6x + 1$

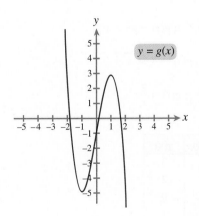

a.

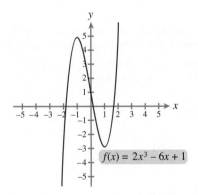

b.

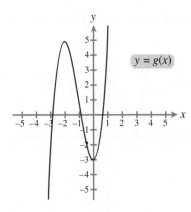

24. $f(x) = x^3 - 3x^2 + 1$

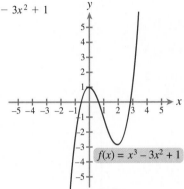

a.

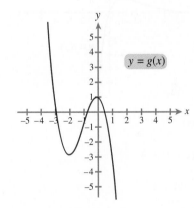

b.

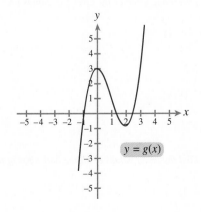

Exercises 25-26 *Use the graph of f to sketch the graph of h.*

25.

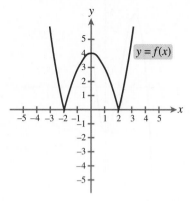

a. $h(x) = f(x + 3)$

b. $h(x) = f(x - 2) + 1$

c. $h(x) = -f(x) - 2$

26.

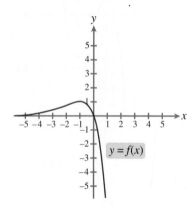

a. $h(x) = f(x) - 3$

b. $h(x) = f(x + 1) - 4$

c. $h(x) = f(-x) + 2$

Exercises 27-30 *Determine whether the given function is even, odd, or neither.*

27. $f(x) = x(4x^2 - 1)$

28. $g(x) = x^4 - x$

29. $f(x) = |x + 1|$

30. $h(x) = \sqrt{x^2 - 1}$

Exercises 31-34 *Find the exact values of all the real zeros of the given function.*

31. $h(x) = 4x - 9$

32. $f(x) = x^2 - x - \dfrac{5}{16}$

33. $g(x) = 3 - \dfrac{8}{x}$

34. $g(x) = |3x - 1| - 5$

Exercises 35-38 *Use a graphing calculator to estimate the following features of the given function to the nearest hundredth:*

a. *Zeros*

b. *Coordinates of the turning points (classify as local maximum or minimum)*

c. *Intervals on which the function is increasing or decreasing*

35. $f(x) = x^3 - 3x + 1$

36. $g(x) = 3x^4 + 4x^3 - 12x^2 + 2$

37. $h(x) = \dfrac{x^2 - 3}{2x - 4}$

38. $g(x) = |x^3 - x^2 + 1|$

Exercises 39-46 *Find the following combinations of the functions f and g. Specify the domain if it is anything other than the set of all real numbers.*

a. $(f + g)(x)$ **b.** $(fg)(x)$ **c.** $(f/g)(x)$
d. $(f \circ g)(x)$ **e.** $(g \circ f)(x)$

39. $f(x) = x^2$; $g(x) = 2x - 1$

40. $f(x) = 4x + 3$; $g(x) = \dfrac{2}{x}$

41. $f(x) = \dfrac{1}{x - 4}$; $g(x) = x^3$

42. $f(x) = \sqrt{x + 2}$; $g(x) = x^2 - 1$

43. $f(x) = 2x^2$; $g(x) = \sqrt{2x + 4}$

44. $f(x) = \dfrac{x}{x - 1}$; $g(x) = \dfrac{2}{x}$

45.

x	$f(x)$	$g(x)$
0	1	3
1	2	2
2	3	0
3	0	1

46.

x	$f(x)$	$g(x)$
-3	-2	-1
-2	0	2
-1	1	0
0	-3	3

Exercises 47-50   *Determine whether the given functions are inverses.*

47. $f(x) = x - 5$; $g(x) = x + 5$

48. $f(x) = 3x - 1$; $g(x) = \dfrac{x}{3} + 1$

49. $f(x) = x^3 - 4$; $g(x) = \sqrt[3]{x} + 4$

50. $f(x) = (x - 2)^2, x \geq 2$; $g(x) = \sqrt{x} + 2$

Exercises 51-58   *Determine whether the function is 1–1. If so, compute its inverse.*

51. $f(x) = 3x - 2$

52. $g(x) = x^2 + 5$

53. $h(x) = x^3 - x + 1$

54. $g(x) = (x + 2)^3 + 3$

55. $f(x) = \sqrt{x - 4}$

56. $h(x) = \dfrac{1}{x - 1}$

57.

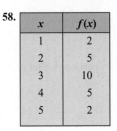

x	$f(x)$
1	2
2	5
3	10
4	17
5	26

58.

x	$f(x)$
1	2
2	5
3	10
4	5
5	2

Exercises 59-60   *Use the graph of f to sketch the graph of f^{-1}.*

59.

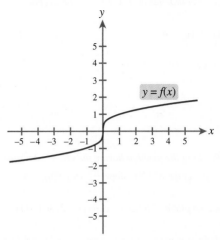

60.

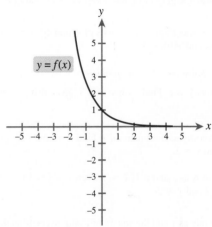

Exercises 61-66   *Identify the type of function that is given (linear, quadratic, piecewise-defined, or none of these), sketch its graph, and locate its intercepts. Use a graphing calculator as necessary.*

61. $h(x) = -(x - 3)^2 + 1$

62. $g(x) = \dfrac{1}{2}x - 4$

63. $f(x) = \dfrac{x + 2}{x - 2}$

64. $h(x) = \dfrac{1}{4}x^4 - 2x^2 + 2$

65. $g(x) = \begin{cases} 2x + 3 & x \leq 1 \\ 5 - x^2 & x > 1 \end{cases}$

66. $f(x) = \begin{cases} x^2 + 4x & x \leq -2 \\ x & x > -2 \end{cases}$

Exercises 67-70 *Sketch the graph of the given quadratic function and locate the exact coordinates of its vertex and intercepts.*

67. $f(x) = (x + 1)^2 - 4$

68. $f(x) = -2(x - 1)^2$

69. $g(x) = -x^2 + 8x - 17$

70. $h(x) = 3x^2 - 12x + 16$

Exercises 71-74 *Find the function having the given properties.*

71. f is linear, the graph of f has slope $-\frac{1}{2}$, and $f(0) = 3$.

72. g is linear, the graph of g has y-intercept -2, and $g(1) = 5$.

73. h is quadratic, the graph of h has vertex $(2, -1)$, and $h(4) = 3$.

74. f is quadratic, the graph of f has axis of symmetry $x = -1$, $f(-1) = 4$, and $f(0) = 3$.

Exercises 75-78 *Solve the variation problem.*

75. z varies directly as a. Find z when $a = 3$, given that $z = 20$ when $a = 4$.

76. y varies inversely as the square of x. Find y when $x = \frac{1}{4}$, given that $y = 3$ when $x = 3$.

77. T varies jointly as x and y. If $T = 1$ when $x = \frac{1}{2}$ and $y = \frac{1}{3}$, find T when $x = 2$ and $y = 3$.

78. P varies jointly as x and the square of y and inversely as the cube of z. If $P = 5$ when $x = 4$, $y = 5$, and $z = 2$, find P when $x = 2$, $y = 3$, and $z = 3$.

Exercises 79-80 *Find the function of the given form whose graph passes through the indicated points.*

79. $f(x) = ax^3 + b$; $(1, 2), (-1, 4)$

80. $h(x) = \sqrt{ax + b}$; $(2, 3), (10, 5)$

Exercises 81-82 *Find the equation of the line with the least-squares best fit to the given points.*

81. $(1, 2), (3, 3), (6, 5)$

82. $(-3, 5), (-1, 3), (2, 1), (5, -2)$

Exercises 83-84 *Find the line with the least-squares best fit to the given data points, and then use the equation of the line to predict the unknown y value.*

83.

x	0	10	20	30	40	50
y	42	67	88	118	135	?

84.

x	0	3	6	9	12	15
y	205	175	150	?	90	65

Exercises 85-94 *Parts a and b are connected: Part a involves a concept from earlier in the text; part b involves related material from this chapter. First answer a, and then use this result to answer b.*

85. a. Solve $x^2 + 6 = -5x$.

 b. Find all zeros of $f(x) = x^2 + 5x + 6$.

86. a. Solve $x^2 + 3x < 1$.

 b. Find the domain of $f(x) = \sqrt{x^2 + 3x - 1}$.

87. a. Solve $y = \dfrac{1}{x + 2}$ for x.

 b. Find $f^{-1}(x)$ for the function $f(x) = \dfrac{1}{x + 2}$.

88. a. Expand $(x + 2)(x^2 + 1)$.

 b. Find all zeros of $g(t) = t^3 + 2t^2 + t + 2$.

89. a. Expand $(2x - 1)^3$.

 b. Find functions f and g such that $(f \circ g)(x) = 8x^3 - 12x^2 + 6x - 1$.

90. a. Solve $y(x - 5) = 2x - 7$ for x.

 b. Let $f(x) = \dfrac{2x - 7}{x - 5}$. Find $f^{-1}(x)$.

91. a. Simplify $\dfrac{3}{x + 5} - \dfrac{4}{x - 2}$.

 b. Find the domain and all zeros of $f - g$, where $f(x) = \dfrac{3}{x + 5}$ and $g(x) = \dfrac{4}{x - 2}$.

92. a. Solve the following equation for k: $9 = k\dfrac{4}{7}$

 b. If y is directly proportional to x and $y = 9$ when $x = \frac{4}{7}$, find y when $x = 4$.

93. a. Simplify $\left(k\sqrt{x}\right)^4$.

 b. If y is proportional to the square root of x and z is proportional to the fourth power of y, how does z vary with x?

94. a. Graph $y = x^3 - x^2$ and $y = 2x - 1$ on the same set of coordinate axes.

b. Estimate the zeros of $f(x) = x^3 - x^2 - 2x + 1$.

95. Area of a Circle Express the area of a circle as a function of its circumference.

96. Area of a Rectangle A point $P(x, y)$ lies on the parabola $y = 4 - x^2$, as shown in Figure 95. Express the area of the shaded rectangle as a function of x. Plot the graph of the function and estimate the value of x that will yield the largest possible area for the rectangle.

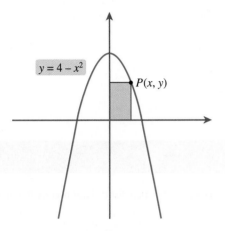

$y = 4 - x^2$

$P(x, y)$

Figure 95

97. Height of a Balloon A hot-air balloon is rising vertically from a point 200 feet from an observer on the ground (Figure 96). Express the height h of the balloon as a function of its distance d from the observer. What domain of d values makes sense for the function h?

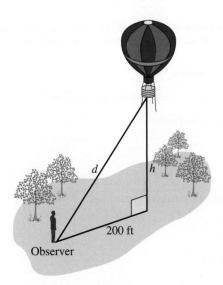

d

h

200 ft

Observer

Figure 96

98. Minimum Cost A company has determined that the total production cost for manufacturing x units is

$$C(x) = 0.01x^3 - 4.5x^2 + 475x + 36,000.$$

Estimate the intervals on which C is increasing and the intervals on which it is decreasing. Estimate the number of units that must be manufactured to minimize the cost.

99. Path of a Football If a football is kicked from ground level with an initial velocity of 64 feet per second and an initial angle of 45°, and if we ignore air resistance, the ball will follow the path given by the quadratic function $f(x) = -\frac{1}{128}x^2 + x$, where x is the horizontal distance in feet from where the ball was kicked and $f(x)$ is the corresponding height of the ball in feet.

a. Find the maximum height of the ball.

b. Assuming the ball is kicked straight, will it clear a 10-foot-high goal post 30 yards away? What is the farthest distance from which the ball could be kicked in order to clear the goal post?

100. Overnight Rate Suppose an overnight package delivery service charges $8.50 for the first pound and $2.00 for each additional pound or fraction thereof. Then the rate for a package weighing x pounds can be determined using the function $C(x) = 8.5 - 2\lfloor 1 - x \rfloor$. Plot the graph of this function. What is the heaviest package that can be sent if the cost cannot exceed $31?

101. Volunteer Model The percentage of the adult population doing volunteer work is related to educational level, as suggested by the following table. Find the line with the least-squares best fit to the data and predict the percentage of the population with 18 years of education who are involved in volunteer work.

Years of education	10	12	14	16
Percent doing volunteer work	8.3	18.8	28.1	38.4

Data source: U.S. Bureau of Labor.

102. Accidental Death Rate The death rate for work-related nonmanufacturing accidental deaths has been on the decline in recent years. The following table shows the death rates for the years 1970, 1975, 1980, 1990, and 1995. Find the line with the least-squares best fit to the data and predict the death rate for the years 2005 and 2010. Are your answers reasonable? Explain.

Year	1970	1975	1980	1990	1995
Death rate (per 100,000)	18	15	13	9	4

Data source: U.S. Bureau of the Census.

Construction of the Hoover Dam in the 1930s resulted in 114 accidental deaths.

103. Compact Disc Sales The following table shows the number of CDs sold during each of the years 1985, 1991, and 1999. Find a quadratic model that expresses the number of CDs sold as a function of the year, with 1985 corresponding to year 0. Use the model to predict the number of CDs to be sold in 2010.

Year	1985	1991	1999
CDs sold (millions)	23	339	995

Data source: Recording Industry Association of America.

Chapter 2 Test

Problems 1-8 *Answer true or false.*

1. If a horizontal line intersects the graph of an equation in more than one point, the equation does not define y as a function of x.

2. Unless otherwise specified, we assume that the domain of a function f is the set of all input values such that the expression $f(x)$ is defined.

3. If $h(x)$ is defined by $h(x) = f(x - 2) + 3$, then the graph of h can be obtained by translating the graph of f 2 units to the right and 3 units upward.

4. An even function cannot be 1–1.

5. Only functions satisfying the horizontal line test are 1–1, and only 1–1 functions have inverse functions.

6. The graph of a piecewise-defined function is really just the graph of several different functions plotted on the same set of coordinate axes.

7. If y varies directly as x, then y is also a linear function of x.

8. Linear regression is the process of finding an equation of the line that passes through a given set of data points.

Problems 9-14 *Give an example of each.*

9. A linear function and its inverse

10. A quadratic function and the coordinates of the vertex of its graph

11. A piecewise-defined function that is linear over each of its three pieces

12. A set of three data points such that the regression line passes through all of the points

13. A function, the graph of which is symmetric with respect to the origin

14. A polynomial function that is neither even nor odd

15. Determine whether the equation $(1/y) + x = 2$ defines y as a function of x. If it does, find the domain of the function.

16. Verify or disprove that
$$f(x) = \frac{3x + 1}{2} \quad \text{and} \quad g(x) = \frac{2x - 1}{3}$$
are inverses of one another.

17. Given that
$$f(x) = \begin{cases} 3x + 1 & x \le 2 \\ 4x - 5 & x > 2 \end{cases} \quad \text{and} \quad g(x) = x^2$$
compute each of the following quantities:

a. $f(3)$

b. $(f \circ g)(\sqrt{2})$

c. $(g \circ f)(3)$

18. Find the domain of the function

$$f(x) = \frac{\sqrt{x-2}}{x-3}$$

19. Use the graph of f given in Figure 97 to produce the graph of $h(x)$, where $h(x) = f(x-1) - 3$.

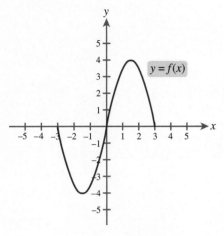

$y = f(x)$

Figure 97

20. Find the coordinates of all zeros and turning points of $f(x) = x^3 - 6x^2$ to the nearest hundredth, and classify each turning point as a local maximum or a local minimum. In addition, indicate the intervals on which f is increasing and the intervals on which it is decreasing.

21. Find the coordinates of the vertex of $f(x) = 3x^2 - 6x + 10$, and sketch the graph.

22. It is given that z varies inversely as x and directly as y. Find z when $x = 3$ and $y = 4$, given that $z = 5$ when $x = 2$ and $y = 3$.

23. It is given that $f(x) = ax^2 + bx$ and that $(2, -4)$ and $(-1, 5)$ are points on the graph of f. What is $f(3)$?

24. As of April 2003, the International Association of Athletics Federation gave the following world records for women's track and field events:

- 100 meters: Delorez Florence Griffith Joyner (USA) 10.49 seconds
- 200 meters: Delorez Florence Griffith Joyner (USA) 21.34. seconds
- 400 meters: Marita Koch (East Germany) 47.60 seconds

Use linear regression to estimate the world record time for 800 meters.

Chapter 3
Polynomial and Rational Functions

This mountain range exists only in cyberspace; it is a computer-generated *fractal landscape*. Fractals are self-similar; that is, a small portion of a fractal appears to be a scaled-down version of the whole. Compare the appearance of the mountain peak with the mountain itself and, in turn, the mountain with the entire range of mountains. Fractals are used to model many aspects of nature, including coastlines, clouds, ocean waves, and plants, all of which have self-similarity properties. In this chapter, we investigate polynomial functions, an essential tool for constructing computer-generated fractal images.

Section 3.1

Polynomial Functions

- How did fertility rates fluctuate in the 20th century?
- Why is it that polynomial functions can provide such startlingly accurate models of real-world phenomena for short time periods and yet are nearly useless for making long-term predictions?
- How can polynomials be used to generate stunning fractal images?
- How can we ever be certain that a graphing calculator is showing us all the important graphical features of a polynomial?

Polynomials can be viewed as fundamental building blocks from which other important classes of functions can be constructed. And as we will discover, because of their simplicity, variety, and the smoothness of their graphs, polynomials are often used to construct mathematical models of real-world phenomena. In Section 2.5, we explored first- and second-degree polynomials and their graphs (lines and parabolas) in some detail. Much of this chapter, and this section in particular, is devoted to extending our analysis to higher-degree polynomials. We begin with some definitions.

In Chapter 1, we defined polynomial expressions, such as $2x^3 - 5x^2 + 1$. A *polynomial function* is simply a function defined by a polynomial. More formally, we have the following definition.

Definition of a Polynomial Function

A **polynomial function of degree n** has the form

$$f(x) = a_n x^n + a_{n-1} x^{n-1} + \cdots + a_1 x + a_0$$

where n is a nonnegative integer and $a_n \neq 0$. The **coefficients** of $f(x)$ are the numbers $a_0, a_1, \ldots, a_n$.

Polynomial functions of degree 1, such as $f(x) = \frac{1}{2}x - 3$, and polynomial functions of degree 2, such as $g(x) = 2x^2 + 4x + 5$, can be graphed by hand using the techniques for lines and parabolas discussed in Chapter 2. However, it is often quite tedious to sketch the graphs of polynomial functions of degree 3 or higher. For this reason, we often rely on a graphing calculator to help us with polynomial graphs. Still, as we have seen many times, a poor choice of viewing window can lead to misleading results, and so a basic understanding of the graphical features of polynomials is essential.

End Behavior of Polynomials

The behavior of the graph of a function to the far right and to the far left (that is, for large positive and negative values of the independent variable) is called the **end behavior** of the function. Consider the polynomial function $f(x) = x^3$. For large positive values of x, $f(x)$ will be very large. For example, $f(1000) = 1000^3 = 1,000,000,000$. In fact, as x gets larger and larger, $f(x)$ becomes arbitrarily large. We say that $f(x)$ approaches infinity as x approaches infinity, and we write $f(x) \to \infty$ as $x \to \infty$. Similarly, for large negative values of x, $f(x)$ will be a very large negative number. For example, $f(-1000) = -1,000,000,000$. Thus, $f(x) \to -\infty$ as $x \to -\infty$. The graph of f provides compelling evidence of this end behavior. Notice how the graph soars upward as we move to the far right and plummets sharply as we move to the far left, as shown in Figure 1.

Let us now determine the end behavior of the polynomial function $f(x) = x^2$. When a large positive number is squared, the result is a very large positive number. For example, $1000^2 = 1,000,000$. If a large negative number is squared, the result is also a large positive number. For example, $(-1000)^2 = 1,000,000$. Thus, $f(x) \to \infty$ as $x \to \infty$ and $f(x) \to \infty$ as $x \to -\infty$. These facts can also be determined by noting that since the graph of $y = x^2$ is a parabola opening upward, the function values are approaching infinity on the far right and the far left, as shown in Figure 2.

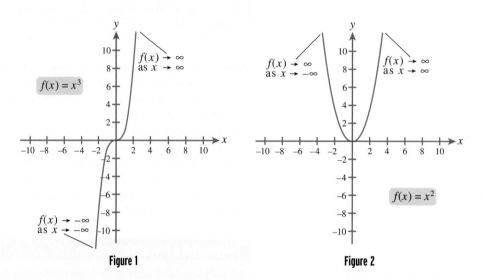

Figure 1   Figure 2

The difference in the end behaviors of x^2 and x^3 arises from the fact that x^2 is never negative, whereas x^3 *is* negative for $x < 0$. More generally, the end behavior of x^n depends on whether n is even or odd: For n odd, x^n behaves like x^3, whereas when n is even, x^n behaves like x^2.

Generalizing still further, the end behavior of any function of the form $f(x) = ax^n$ will either be identical to or opposite that of x^n, according to whether a is positive or negative. Collecting these results gives us the following strategy for determining the end behavior of monomial functions.

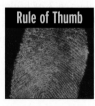

Rule of Thumb

To determine the end behavior of $f(x) = ax^n$

1. First determine the end behavior of x^n. Use the fact that x^n behaves like x^3 for n odd and like x^2 for n even. That is, for n even, the graph of x^n approaches ∞ on both ends, whereas for n odd, the graph approaches $-\infty$ on the left and ∞ on the right.
2. If $a > 0$, ax^n behaves like x^n; if $a < 0$, the end behavior of ax^n will be opposite that of x^n, in the sense that where the graph of one goes up, the other goes down.

····▷EXAMPLE 1

Determining the End Behavior of Monomials

Determine the end behavior of each of the following functions and match each to its graph:

a. $f(x) = x^4$ **b.** $g(x) = x^7$
c. $h(x) = -2x^3$ **d.** $p(x) = -3x^2$

i.

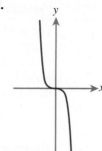

ii.

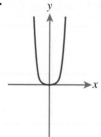

iii.

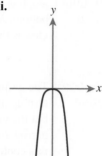

iv.

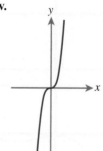

Solution

a. Since the exponent 4 is even, $f(x) = x^4$ behaves like x^2. Thus, $f(x) \to \infty$ as $x \to \infty$ and also as $x \to -\infty$. The graph is (ii).

b. Since the exponent 7 is odd, $g(x) = x^7$ behaves like x^3. Thus, $g(x) \to \infty$ as $x \to \infty$, and $g(x) \to -\infty$ as $x \to -\infty$. The graph is (iv).

c. The graph of $y = x^3$ tends toward infinity on the right and negative infinity on the left. Multiplication by -2 reverses the end behavior. Thus, $h(x) = -2x^3 \to -\infty$ as $x \to \infty$, and $h(x) \to \infty$ as $x \to -\infty$. The graph is (i).

d. The exponent is even, but multiplication by -3 changes the sign. Thus, $p(x) \to -\infty$ as $x \to \infty$ and also as $x \to -\infty$. The graph is (iii).

Thus far, we have dealt with polynomials having only one term. For a more general polynomial, it is tempting to simply graph the function using a graphing calculator and then to note whether the graph is rising or falling on the far right and on the far left of the viewing window. The difficulty is in knowing how *wide* the viewing window should be, as the following example illustrates.

EXAMPLE 2

Investigating the End Behavior Graphically

Investigate the end behavior of $f(x) = x^3 - 1{,}000{,}000x^2$ with a graphing calculator.

Solution We have chosen a very naive approach: Begin with a standard viewing window and make adjustments until we are confident that we are, in fact, witnessing the "true" end behavior of the function. The main steps are shown in Table 1; considerations of space prevent us from showing all of the missteps that we might have taken along the way.

Table 1

Plot	Range	Problem	Solution
	Xmin=−10 Xmax=10 Ymin=−10 Ymax=10	The graph is invisible because of the choice of scale. In fact, the graph is "on top of" the y-axis.	Increase the range of y-values.
	Xmin=−10 Xmax=10 Ymin=−1,000,000 Ymax=1,000,000	The range of y-values is still too small; the graph is cut off at the bottom. Also, the view is too narrow.	Widen the view by increasing the range of x-values. We will also make the range of y-values *huge*.
	Xmin=−1,000,000 Xmax=1,000,000 Ymin=−1E18 Ymax=1E18	The graph begins to turn up at the far right, but the view is too narrow to determine if this continues.	Widen the view. Again, we must increase our range of y-values to produce a usable plot.
	Xmin=−1E10 Xmax=1E10 Ymin=−1E30 Ymax=1E30		

The final graph in Table 1 suggests that $f(x) \to \infty$ as $x \to \infty$, and $f(x) \to -\infty$ as $x \to -\infty$.

Even after going through this extraordinary effort, we still have little confidence in our answer! Perhaps an even wider viewing window might show yet another turn in the graph. Again, we see the limitations of the graphing calculator. Fortunately, the end behavior of a polynomial can be determined without using a graphing calculator at all.

The end behavior of $f(x) = x^3 - 1,000,000x^2$ can be viewed as a contest between the monomials x^3 and $1,000,000x^2$. As x approaches infinity, x^3 approaches infinity, whereas $-1,000,000x^2$ approaches negative infinity. The end behavior of the sum of these two monomials—namely, $f(x)$—depends on which of them is dominant. As we will see in Exercise 54, the monomial of highest degree, which in this case is x^3, "wins." Thus, x^3 approaches infinity faster than $-1,000,000x^2$ approaches negative infinity. In fact, the end behavior of $f(x)$ is identical to that of x^3. More generally, the end behavior of a polynomial is determined by its term of highest degree.

End Behavior of Polynomials

The end behavior of a polynomial is identical to that of its term of highest degree.

EXAMPLE 3

Finding the End Behavior of a Polynomial

Find the end behavior of each of the following polynomials:

a. $f(x) = x^4 - 2x^3 + 4x - 5$

b. $g(x) = 7 + 4x - 3x^2 - 2x^3$

c. $h(x) = -7x^6 - 4x^5 + 3x + 1$

Solution

a. Since the term of highest degree is x^4, the end behavior of $f(x)$ will be identical to that of x^4 (see Figures 3 and 4). Thus, $f(x) \to \infty$ as $x \to \infty$ and $f(x) \to \infty$ as $x \to -\infty$.

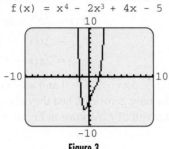

Figure 3

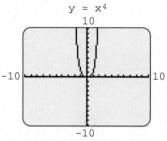

Figure 4

b. The term of highest degree is $-2x^3$, so the end behavior of $g(x)$ will be identical to that of $-2x^3$ (see Figures 5 and 6). Thus, $g(x) \to -\infty$ as $x \to \infty$ and $g(x) \to \infty$ as $x \to -\infty$.

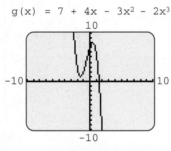

Figure 5

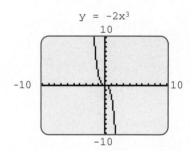

Figure 6

c. The term of highest degree is $-7x^6$, so the end behavior of $h(x)$ is identical to that of $-7x^6$ (see Figures 7 and 8). Since $-7x^6 \to -\infty$ as $x \to \infty$ and $-7x^6 \to -\infty$ as $x \to -\infty$, the same is true of $h(x)$. Thus, $h(x) \to -\infty$ as $x \to \infty$ and $h(x) \to -\infty$ as $x \to -\infty$.

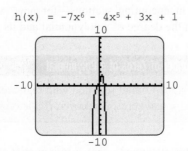

Figure 7

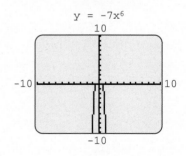

Figure 8

Zeros of Polynomials

Recall from Section 2.2 that a zero of a function f is a number c such that $f(c) = 0$. We also saw in Section 2.2 that the real zeros of a function coincide with its x-intercepts. The zeros, and hence the x-intercepts, of a polynomial function can often be found by factoring, as in the following example.

>EXAMPLE 4

Finding Zeros of a Polynomial Function

Find the real zeros of $f(x) = 2x^3 + 6x^2 - 20x$ and locate the x-intercepts of the graph of f.

Solution   We first factor $f(x)$ as completely as possible.

$$\begin{aligned} f(x) &= 2x^3 + 6x^2 - 20x \\ &= 2x(x^2 + 3x - 10) \qquad \text{Factoring out the monomial } 2x \\ &= 2x(x + 5)(x - 2) \qquad \text{Factoring} \end{aligned}$$

Setting $2x(x + 5)(x - 2) = 0$ and solving gives us $x = 0$, $x = -5$, and $x = 2$. Thus, these are the only zeros of f, and they correspond to the x-intercepts $(0, 0)$, $(-5, 0)$, and $(2, 0)$. The graph of f is shown in Figure 9.

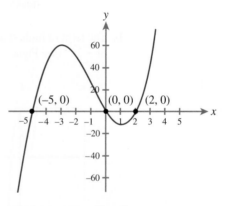

Figure 9

Notice that the zeros of the polynomial function f in Example 4 were found by setting the factors of $f(x)$ equal to zero. This suggests an important connection between the zeros and factors of a polynomial. We will investigate this connection more thoroughly in upcoming sections—for now, we simply note the following connections between the degree of a polynomial and its real zeros.

The Number of Zeros of a Polynomial

If f is a polynomial function with real coefficients and degree n ($n \geq 1$), then f has at most n real zeros. Moreover, if n is odd, then f must have at least one real zero.

The latter statement—that a polynomial with real coefficients and odd degree has at least one real zero—is evident from end behavior considerations. We ask you to discuss this in Exercise 49.

·····⟫EXAMPLE 5   **Finding the Zeros of a Polynomial with a Graphing Calculator**

Estimate the zeros of $f(x) = 0.05x^4 + 0.75x^3 - 0.2x^2 - 3x + 2$ to the nearest hundredth using a graphing calculator.

Solution   The graph shown in Figure 10 suggests that f has three zeros near the origin. By zooming in and using the trace feature, or by using the calculate-zero feature, we estimate the zeros, as shown in Figures 11–13, to be approximately -2.32, 0.73, and 1.57.

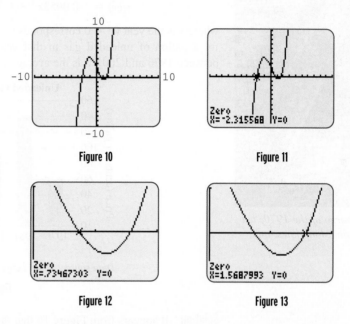

Figure 10   **Figure 11**

Figure 12   **Figure 13**

Since the term of highest degree is $0.05x^4$, we know $f(x) \to \infty$ as $x \to -\infty$, and so the graph of f must turn upward to the left of the viewing window. After zooming out, a fourth zero emerges to the left of the origin, as shown in Figure 14. Using the calculate-zero feature, we obtain $x \approx -14.99$, as in Figure 15.

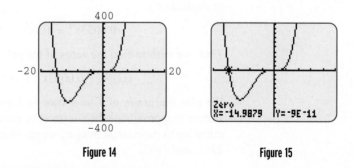

Figure 14   **Figure 15**

Thus, the zeros of f are approximately -14.99, -2.32, 0.73, and 1.57. Since a fourth-degree polynomial has at most four real zeros, there are no others.

The main reason we study zeros of polynomials is that they arise as solutions of polynomial equations. For example, the solutions of the polynomial equation $2x^3 + 6x^2 = 20x$ are the same as the solutions of $2x^3 + 6x^2 - 20x = 0$. Thus, solving $2x^3 + 6x^2 = 20x$ is equivalent to finding the zeros of the polynomial

$f(x) = 2x^3 + 6x^2 - 20x$. We saw in Example 4 that f has exactly three zeros—$x = -5$, $x = 0$, and $x = 2$—so these are the only solutions of the polynomial equation.

The Arab oil embargo of the 1970s led to gasoline rationing, long lines, and inflated prices.

·····❯EXAMPLE 6

An Application Involving Zeros

Unleaded gas prices were sampled at the end of 5-year periods beginning in 1976, as depicted in Figure 16. Based on this limited data set, the average price of unleaded gas in the U.S. for the years 1976–2001 can be modeled by the polynomial

$$g(x) = -0.0053x^4 + 0.3x^3 - 5.4x^2 + 33.8x + 60$$

where x is the year ($x = 0$ corresponds to 1976) and $g(x)$ is the average price (in cents) for a gallon of unleaded gas in that year. According to this model, for what years between 1976 and 2001 was the average price of a gallon of unleaded gas $1.00?

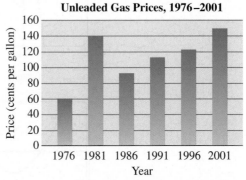

Unleaded Gas Prices, 1976–2001

Data source: U.S. Energy Information Administration.

Figure 16

Solution It appears from Figure 16 that there were three times between 1976 and 2001 when the average price was $1.00. Since we are interested in the times at which the price was $1.00 and $g(x)$ gives the price in cents, we set $g(x) = 100$ to obtain

$$-0.0053x^4 + 0.3x^3 - 5.4x^2 + 33.8x + 60 = 100$$

or equivalently,

$$-0.0053x^4 + 0.3x^3 - 5.4x^2 + 33.8x - 40 = 0$$

Thus, we wish to find the zeros of the polynomial

$$f(x) = -0.0053x^4 + 0.3x^3 - 5.4x^2 + 33.8x - 40$$

We plot the graph of f as shown in Figure 17 and use the trace or calculate-zero feature to approximate the x-intercepts, giving us $x \approx 1.52$, $x \approx 11.0$, and $x \approx 16.0$. By rounding to the nearest integer, we see that these values correspond to the years 1978, 1987, and 1992.

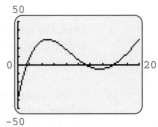

Figure 17

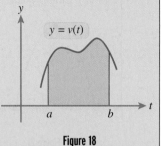

Preview of Calculus

Area Under a Curve

Suppose you wish to measure the distance of an automobile trip, but your odometer is broken. Is it possible to compute the distance using only your watch and the car's speedometer? If the velocity is held constant, then the answer is a resounding yes: Simply multiply the rate by the time to compute the distance. If the velocity varies, then the answer is still yes (at least in principle), but the computation is considerably more complex. It can be shown that if the velocity of the car is given by the function $v(t)$, then the distance traveled between times a and b is given by the area of the region bounded by the graph of $y = v(t)$, the t-axis, the line $t = a$, and the line $t = b$, as shown in Figure 18.

This then raises a second question: How can we find the area underneath a curve? If the curve is linear, then the region will be in the shape of a trapezoid, and we can easily compute its area using basic geometric formulas. But if $v(t)$ is a second-degree or higher polynomial, we will be hard-pressed to find any directly applicable geometric formulas. Now if we only required an estimate of the area, then we could sketch in a few rectangles that roughly approximate the region and compute the area by adding together the areas of each of the rectangles (Figure 19). If we wanted a more accurate approximation, then we could simply use more—but smaller—rectangles (Figure 20).

Figure 18

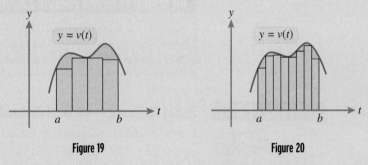

Figure 19   **Figure 20**

Calculus provides us with a technique called **integration** that enables us to find an exact value of such an area, by, in a sense, viewing the region as composed of infinitely many rectangles. But the applications of integration extend far beyond computations of area. Integrals allow us to compute the length of a curve, the balancing point of an object, the energy required to pump out a tank, or even the probability that a child's height is in a given range. Integration forms the cornerstone of the integral calculus, one of the primary branches of this far-reaching subject.

Turning Points of Polynomials

If we were to show a small child the graphs of polynomial functions and ask her to describe them, she would no doubt characterize them in terms of the number and size of the "humps"; in essence, the child would be focusing on the *turning points*—the points at which the graph changes from increasing to decreasing or decreasing to increasing. Not only are turning points the most visually prominent aspect of the graph of a polynomial function, they also play a key role in the application of polynomial models to the real world.

Surprisingly, there is a close connection between the zeros of a polynomial and its turning points. To see this, suppose c_1 and c_2 are zeros of a polynomial function f, and suppose further that f has no other zeros between c_1 and c_2. Then the graph of f must cross the x-axis at $x = c_1$ and $x = c_2$ but nowhere between them. Although the precise shape of f between c_1 and c_2 is impossible to determine from this limited information, we show several possibilities in Figures 21–23.

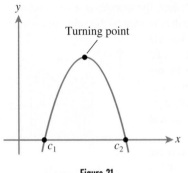

Figure 21

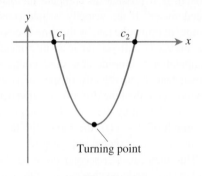

Figure 22

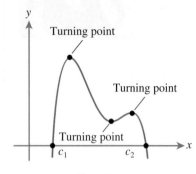

Figure 23

Notice that in either case, f has at least one turning point somewhere between c_1 and c_2. In fact, this is always the case: A polynomial always has at least one turning point between adjacent zeros. So if f has m distinct real zeros, then f must have at least $m - 1$ turning points. It can also be shown, using techniques from calculus, that a polynomial of degree n has at most $n - 1$ turning points. These facts are summarized as follows.

The Number of Turning Points of a Polynomial

Suppose f is a polynomial of degree n with m distinct real zeros, where $1 \leq m \leq n$. Then f has at least $m - 1$ and at most $n - 1$ turning points. If the degree n is even, then f has at least one turning point.

  EXAMPLE 7

Finding the Turning Points of a Polynomial with a Graphing Calculator

Use a graphing calculator to estimate the coordinates of *all* the turning points of $f(x) = \frac{1}{4}x^4 - 7x^3 + 10x^2$.

Solution   Two turning points are visible in the graph of f shown in Figure 24. After zooming in and using the trace feature, or alternatively using the calculate-maximum/minimum feature, we estimate these points to be $(0, 0)$ and $(1, 3.25)$. See Figure 25 for the latter.

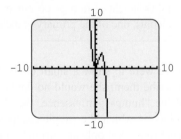

Figure 24

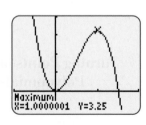

Figure 25

Now the graph of f is decreasing on the far right of the viewing window. However, because $f(x)$ will behave like its term of highest degree, $\frac{1}{4}x^4$, we know that $f(x) \to \infty$

as $x \to \infty$. Thus, the graph must turn upward somewhere to the right of the viewing window. After a process of trial and error, we obtain a view of the graph (Figure 26) that shows the third turning point. We estimate its coordinates to be approximately $(20, -12{,}000)$, as shown in Figure 27.

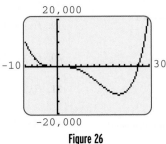

Figure 26

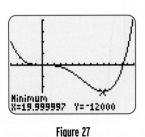

Figure 27

Since f has degree 4, there can be at most $4 - 1 = 3$ turning points. Thus, we have found *all* of the turning points of f—namely $(0, 0)$, $(1, 3.25)$, and $(20, -12{,}000)$.

Understanding and Mastery Checklists

Concepts to Understand	Skills to Master
Polynomial function	Identify a polynomial function and its degree.
✦	✦
End behavior	Determine the end behavior of a polynomial function.
✦	✦
Zero	Find the zeros of a polynomial function.
✦	✦
Turning point	Find the turning points of a polynomial function.

Exercises 3.1

Exercises 1-4 *Determine the end behavior of the given polynomial and match it with its graph.*

1. $f(x) = x^6$

2. $g(x) = -x^5$

3. $p(x) = -2x^4$

4. $h(x) = 4x^3$

i.

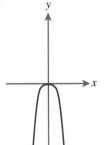

ii.

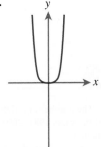

iii.

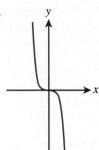

iv.

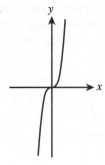

Exercises 5-12 *Determine the end behavior of the given polynomial. Verify your results with a graphing calculator.*

5. $f(x) = x^2 - 3x + 7$

6. $h(x) = -2x^3 + 6x^2 + 3x - 1$

7. $p(x) = -4x^4 + 7x - 1$

8. $g(x) = -3x^2 + 4x^4$

9. $h(x) = 10{,}000x^3 - 0.001x^5$

10. $f(x) = 2x^5 + 3x^2 - 100$

11. $g(x) = 10^7 + x^3$

12. $h(x) = -0.000001x^2 + 1{,}000{,}000x$

Exercises 13-16 *Indicate*
a. *the maximum number of real zeros of the given polynomial, and*
b. *the maximum number of turning points.*

13. $P(x) = mx + b$

14. $P(x) = ax^2 + bx + c$

15. $P(x) = ax^5 + bx^4 + cx^3 + dx^2 + ex + f$

16. $P(x) = ax^6 + bx^5 + cx^4 + dx^3 + ex^2 + fx + g$

Exercises 17-22 *Factor the polynomial and find its zeros.*

17. $f(x) = x^2 - 2x - 8$

18. $f(x) = x^2 + 8x + 15$

19. $f(x) = 3x^3 - 3x$

20. $f(x) = 4x^3 - 16x^2 + 16x$

21. $f(x) = -x^4 - 12x^3 - 36x^2$

22. $f(x) = x^4 - 25x^2$

Exercises 23-32 *Use a graphing calculator to estimate the zeros and the coordinates of the turning points of the polynomial to the nearest hundredth. Classify each turning point as a local maximum or minimum.*

23. $f(x) = x^3 - 6x^2 + 5x + 13$

24. $g(x) = x^3 + 9x^2 + 20x - 1$

25. $h(x) = \dfrac{x^3}{8} + 2x^2 + 2x + 3$

26. $g(x) = -\dfrac{x^3}{4} + 5x^2 - \dfrac{x}{4} + 2$

27. $p(x) = -\dfrac{x^4}{4} + 3x^3 + \dfrac{x^2}{2} - 9x$

28. $f(x) = \dfrac{x^4}{4} + 4x^3 + 10x^2 + 1$

29. $h(x) = 0.1x^3 - 0.7x^2 - 5.6x - 4$

30. $f(x) = 0.05x^3 + 0.55x^2 - 5.8x + 8$

31. $g(x) = 0.1x^5 - 0.01x^6 - 1$

32. $q(x) = 4x - 0.001x^4 + 9$

33. **AIDS Deaths** The number of AIDS deaths reported in the United States for each of the years 1981–2001 can be approximated with the polynomial function
$$f(x) = 1.25x^4 - 73x^3 + 980x^2 + 130$$
where x is the year ($x = 0$ corresponds to 1981) and $f(x)$ is the number of deaths reported in that year. According to this model, when did the number of deaths peak and in what year after 1990 did the number of deaths drop to 13,000? Explain why this polynomial model

works only for a limited time period by considering the end behavior of $f(x)$. If $f(1)$ represents the number of AIDS deaths in 1982 (from January 1 through December 31), then what would $f(1.5)$ represent?

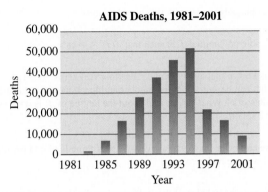

AIDS Deaths, 1981–2001

Data source: U.S. Centers for Disease Control.

34. Family Size   The average number of people per family unit in the United States for the years 1940–2000 can be modeled by the polynomial function

$$f(x) = 0.000001x^4 - 0.0001x^3 + 0.003x^2 - 0.04x + 3.76$$

where x is the year ($x = 0$ corresponds to 1940) and $f(x)$ is the average number of people per family unit during that year. Does f have any turning points? If so, approximate the coordinates. Can you give any explanations for the trends described by the function? Does it appear as though the model is valid for the year 2010? What graphical feature of polynomials can help explain why this polynomial model works only for a limited time period?

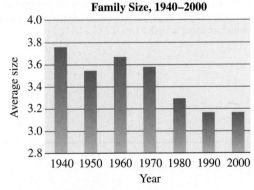

Family Size, 1940–2000

Data source: U.S. Bureau of the Census.

35. Paper Waste   The percentage of paper waste remaining after recycling during the years 1960–2000 can be modeled by the polynomial function

$$f(x) = -0.0000002x^4 + 0.0003x^3 + 0.012x^2 - 0.23x + 18.1$$

where x is the year ($x = 0$ corresponds to 1960) and $f(x)$ is the percentage of waste remaining in that year. Approximate the coordinates of any turning points. During which years between 1960 and 2000 was recycling increasing?

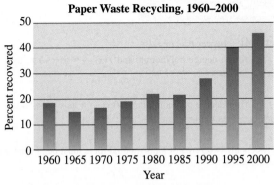

Paper Waste Recycling, 1960–2000

Data source: U.S. Environmental Protection Agency.

36. Fertility Rates   The fertility rate in the United States (the number of live births per 1000 women of childbearing age) for the years 1930–2000 can be modeled by the polynomial function

$$f(x) = 0.000044x^4 - 0.0055x^3 + 0.18x^2 - x + 89$$

where x is the year ($x = 0$ corresponds to 1930) and $f(x)$ is the fertility rate for that year. Approximate the coordinates of any turning points. Why does this model appear to be of limited utility for years after 2000?

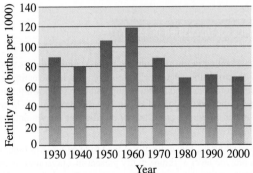

Fertility Rates, 1930–2000

Data source: U.S. Bureau of the Census.

37. Death Row Numbers   The number of U.S. prisoners awaiting execution on death row for the years 1990 to 1995 can be modeled either by

$$f(t) = -1.2t^3 + 16.1t^2 + 90.9t + 2349.3$$

or

$$g(t) = 7.2t^2 + 107.2t + 2345.8$$

where t represents the year (with $t = 0$ corresponding to 1990).

a. Describe the behaviors of the graphs of both f and g as $t \to \infty$.

b. Which of the two models is most likely to provide the best fit for the 21st century? Why?

c. When (if ever) do the two models agree?

d. Use a graphing calculator to estimate the maximum amount by which the models differ for the years 1990–1993. [*Hint:* Consider the function $|f(x) - g(x)|$.]

Concepts and Critical Thinking

Exercises 38-43   *Answer true or false.*

38. If f is a fourth-degree polynomial and $f(x) \to \infty$ as $x \to \infty$, then $f(x) \to -\infty$ as $x \to -\infty$.

39. If f is a fifth-degree polynomial and $f(x) \to \infty$ as $x \to \infty$, then $f(x) \to -\infty$ as $x \to -\infty$.

40. Between any two zeros of a polynomial there must be at least one turning point.

41. Between any two turning points of a polynomial there must be at least one zero.

42. Only the term of highest degree has any effect on the graph of a polynomial.

43. Only the term of highest degree has any effect on the end behavior of the graph of a polynomial.

Exercises 44-47   *Give an example of each.*

44. A fourth-degree polynomial with no zeros

45. A cubic polynomial with no turning points

46. A cubic polynomial with two turning points

47. A cubic polynomial $f(x)$ such that $f(x) \to \infty$ as $x \to -\infty$

48. It is known that a polynomial f has four real zeros: -3, 1, 5, and 7. What can be said about the number and location of the (turning points of f? What can be said about the degree of f; is it necessarily 4? Explain.

49. Use end behavior to explain why all polynomials of odd degree have at least one real zero.

50. Use your knowledge of the end behavior of polynomials to explain why a fourth-degree polynomial can have one or three turning points, but never two. More generally, what can you say about the possible number of turning points for polynomials of degree n, where n is even? What about polynomials of degree n, where n is odd?

Questions for Discussion or Essay

51. When zooming in to find a turning point, it often happens that the view shown by the calculator is "too flat"; that is, the entire graph appears horizontal, and thus it is impossible to pin down the x-coordinate of the turning point. Explain how this problem can be remedied either by using a zoom box or by adjusting the window variables.

52. In the real world, many quantities tend to level off over time. For example, the level of radioactivity present in a person exposed to radiation approaches zero as time goes on. Give three examples of real-world phenomena that tend to level off, and explain why polynomials don't make good models for such phenomena.

53. In Example 6, we considered a polynomial
$$g(x) = -0.0053x^4 + 0.3x^3 - 5.4x^2 + 33.8x + 60$$
that approximated the average price of unleaded gas in the United States for the years 1976–2001. For example, $g(5) \approx 128.2$ approximates the average price for 1981 (in cents). What interpretation can be given to the values of $g(x)$ if x is not an integer? For example, how should we interpret $g(5.5) \approx 127.6$?

Projects for Enrichment

54. Relating the Magnitudes of Monomials with Their Degrees   In this project, we explore the relationship between the degree of a monomial and its magnitude. Let $f(x) = 0.0001x^3$ and $g(x) = 1000x^2$.

a. Compute $f(10)$ and $g(10)$ and then $f(100)$ and $g(100)$. Explain why $g(x)$ is larger than $f(x)$ for these relatively small values of x.

b. Find a value of x (other than 0) for which $f(x) = g(x)$.

c. For what values of x will $f(x)$ be greater than $g(x)$?

d. Explain why $[f(x) - g(x)] \to \infty$ as $x \to \infty$.

e. Suppose that $P(x) = ax^n$ and $Q(x) = bx^{n+1}$, where both a and b are positive. For what values of x is $Q(x) > P(x)$?

f. Explain why the end behavior of a polynomial is determined by the term of highest degree.

55. Complex-Valued Polynomials and the Mandelbrot Set   Until now, we have considered only *real-valued* functions with domains in the set of real numbers. In other words, we assumed that every function had only real numbers as input and output values. But we can also consider functions that allow nonreal input values and yield complex output values that may or may not be real. For example, suppose f is defined as $f(x) = x^2$, and we allow both real and nonreal input values. Thus, for example, $f(i) = i^2 = -1$ and $f(1 + i) = (1 + i)^2 = 1 + 2i + i^2 = 2i$. We say that f is a **complex-valued** function or, since f is a polynomial, a complex-valued polynomial. We can also consider functions that have nonreal numbers in their definitions—$f(x) = x^2 + ix$, for example. Here we compute $f(i) = i^2 + i^2 = -2$ and $f(1 + i) = (1 + i)^2 + i(1 + i) = -1 + 3i$.

a. Evaluate each of the following functions at $x = i$:

    i. $f(x) = x^2 - x$

    ii. $f(x) = x^3 + 2x^2 - 3i$

    iii. $f(x) = 3x^4 - 2x^2 + x$

    iv. $f(x) = x^5 - 2ix^4 + 3x^2 - 5ix + 1$

b. Evaluate each of the following functions at $x = 1 + i$ and $x = a + bi$:

    i. $f(x) = x^2 + 1$ **ii.** $f(x) = x^2 + i$

    iii. $f(x) = x^2 + (1 + i)$ **iv.** $f(x) = x^2 + x$

The complex-valued polynomial $f(x) = x^2 + c$ generates a surprisingly intricate *fractal* image called the **Mandelbrot set**. To see how this happens, we must first consider the geometric representation of a complex number in the **complex plane**. The complex plane looks very much like the rectangular coordinate plane, but instead of the x-axis, we have the real axis, and instead of the y-axis, we have the imaginary axis. A complex number $a + bi$ is represented on the complex plane as the point with coordinates (a, b). The complex plane is shown in Figure 28 with several complex numbers plotted.

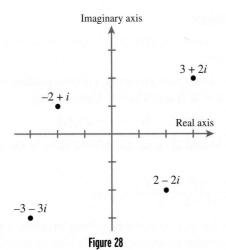

Figure 28

The Mandelbrot set is defined to be the set of numbers c in the complex plane for which the sequence $c, f(c), f(f(c)), f(f(f(c))), \ldots$ is *bounded* in the sense that if we were to plot these numbers in the complex plane, they would all be within a fixed distance of the origin. Notice that each number in the sequence (starting with the second) is found by using the previous number as the input value for f. We show below how to compute a few terms of the sequence for two different values of c.

$c = -2, f(x) = x^2 - 2$	$c = 1 + i, f(x) = x^2 + 1 + i$
$f(-2) = 2$	$f(1 + i) = 1 + 3i$
$f(2) = 2$	$f(1 + 3i) = -7 + 7i$
$f(2) = 2$	$f(-7 + 7i) = 1 - 97i$
$\vdots$	$\vdots$

Now the sequence $-2, 2, 2, \ldots$ is bounded, and so -2 is a member of the Mandelbrot set. However, the sequence $1 + i, 1 + 3i,$ $-7 + 7i, 1 - 97i, \ldots$ is not bounded; in fact, each successive point is more than twice as far from the origin as the previous point. Thus, $1 + i$ is not in the Mandelbrot set.

c. Determine whether the following numbers are in the Mandelbrot set:

    i. -1 **ii.** 2

    iii. i **iv.** $\dfrac{1}{4}i$

If it were possible to plot all the points in the complex plane that belong to the Mandelbrot set, we would obtain a graph similar to the one shown in Figure 29. A more interesting "graph" can be obtained if we assign colors to the points that are not in the Mandelbrot set. This is done on the basis of how quickly the sequence $c, f(c),$ $f(f(c)), \ldots$ moves away from the origin. One such image is shown in Figure 30. One of the fascinating properties of the Mandelbrot set (the black region) is that it contains an infinite number of miniature copies of itself. These can be found by zooming in near the boundary of the set. Also near the boundary is an endless collection of stunning images, two of which are shown in Figures 31 and 32.

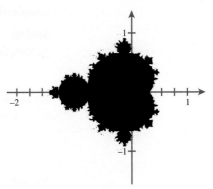

Figure 29

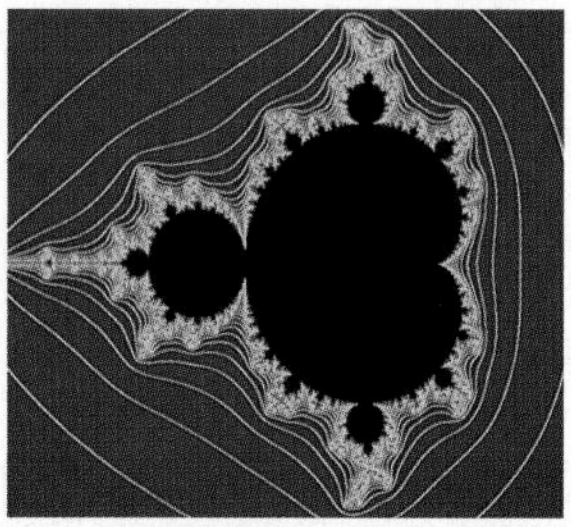

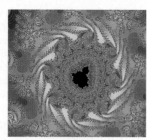

Figure 30

Figure 31

Figure 32

Section 3.2 | Division of Polynomials

- ● How can you divide two polynomials without writing down a single variable?
- ● How can you evaluate a 17th-degree polynomial function for any given value of x using only a ☒ key and a ⊕ key 17 times each?
- ● If every person with a flu virus passes it on to three other people, how many people will have had the flu 20 days after the first person became infected?

Dividing Polynomials

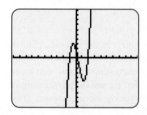

Figure 33

We have seen that the exact values of the zeros of a polynomial can be found by factoring. But what if a polynomial cannot be easily factored? Consider, for example, the polynomial $f(x) = 3x^3 - 4x^2 - 5x + 2$. It is not at all clear how to factor this polynomial, if indeed it can be factored. However, using a graphing calculator, we see that there are three zeros (Figure 33). Two of these zeros, $x = -1$ and $x = 2$, can be determined from the graph and verified by substitution into f. The third zero can at least be approximated using either the trace or calculate-zero feature. However, if we wish to find it exactly, we must take a different approach. Intuition suggests that since $x = -1$ and $x = 2$ are zeros, $f(x)$ should have factors $(x + 1)$ and $(x - 2)$ (we will see why this is the case in Section 3.3). In other words, $f(x) = (x + 1)(x - 2)q(x)$ for some polynomial $q(x)$. As we see in the following example, this information is sufficient to find the remaining zero.

####···❭EXAMPLE 1

Factoring a Polynomial Using Long Division

Given that $(x + 1)$ and $(x - 2)$ are factors of $f(x) = 3x^3 - 4x^2 - 5x + 2$, factor $f(x)$ completely and determine the zeros.

Solution   Since $(x + 1)$ and $(x - 2)$ are both factors of $f(x)$, the product of these two factors, $(x + 1)(x - 2) = x^2 - x - 2$, is also a factor. Thus, we have

$$3x^3 - 4x^2 - 5x + 2 = (x^2 - x - 2)q(x)$$

for some polynomial $q(x)$. Now to find $q(x)$, we divide both sides of the equation by $x^2 - x - 2$ to obtain

$$\frac{3x^3 - 4x^2 - 5x + 2}{x^2 - x - 2} = q(x)$$

We can simplify this expression for $q(x)$ by performing **long division**, a process for dividing two polynomials that is similar to ordinary long division of natural numbers.

As the first step, we choose a monomial that, when multiplied by the leading term of the denominator, yields the leading term of the numerator. We select $3x$ since $3x \cdot x^2 = 3x^3$. The monomial $3x$ is then multiplied through the denominator, $x^2 - x - 2$, as illustrated.

$$\begin{array}{r} 3x \\ x^2 - x - 2\overline{)3x^3 - 4x^2 - 5x + 2} \\ \underline{3x^3 - 3x^2 - 6x} \\ -x^2 + x + 2 \end{array}$$

We select $3x$ since $3x \cdot x^2 = 3x^3$.

Multiplying $x^2 - x - 2$ by **$3x$** to obtain $3x^3 - 3x^2 - 6x$

Subtracting the second line from the first

Next, we choose a monomial that, when multiplied by the leading term of the denominator, produces the leading term of the remainder just found, $-x^2 + x + 2$. In this case, we select -1 and continue as follows:

$$\begin{array}{r} 3x - 1 \\ x^2 - x - 2\overline{)3x^3 - 4x^2 - 5x + 2} \\ \underline{3x^3 - 3x^2 - 6x} \\ -x^2 + x + 2 \\ \underline{-x^2 + x + 2} \\ 0 \end{array}$$

We select -1 since $-1 \cdot x^2 = -x^2$

Multiplying $x^2 - x - 2$ by **-1** to obtain $-x^2 + x + 2$

Subtracting

The completed division shows that

$$\frac{3x^3 - 4x^2 - 5x + 2}{x^2 - x - 2} = 3x - 1$$

and so

$$\begin{aligned} 3x^3 - 4x^2 - 5x + 2 &= (x^2 - x - 2)(3x - 1) \\ &= (x + 1)(x - 2)(3x - 1) \end{aligned}$$

Setting the factors of $f(x)$ equal to 0, we find that the three zeros of f are $x = -1$, $x = 2$, and $x = \frac{1}{3}$.

Rule of Thumb

When performing long division, write the terms of each polynomial in descending order and fill in any missing terms using zero coefficients. For example, when computing

$$\frac{-4x^2 + 1 - 2x^3}{1 + 3x}$$

begin by rewriting as

$$\frac{-2x^3 - 4x^2 + 0x + 1}{3x + 1}$$

In Example 1, $x^2 - x - 2$ divided evenly into $3x^3 - 4x^2 - 5x + 2$ because $x^2 - x - 2$ was a factor of $3x^3 - 4x^2 - 5x + 2$. In general, if the denominator of a quotient is not a factor of the numerator, then the long division process will produce a remainder. More formally, we have the following.

Division Algorithm

Suppose $f(x)$ and $d(x)$ are polynomials, $d(x) \neq 0$, and the degree of $d(x)$ is less than or equal to the degree of $f(x)$. Then there are unique polynomials $q(x)$ and $r(x)$ such that

$$f(x) = d(x) \cdot q(x) + r(x) \quad \text{or} \quad \frac{f(x)}{d(x)} = q(x) + \frac{r(x)}{d(x)}$$

where either $r(x) = 0$ or the degree of $r(x)$ is less than the degree of $d(x)$. The polynomials $f(x)$, $d(x)$, $q(x)$, and $r(x)$ are referred to, respectively, as the **dividend**, **divisor**, **quotient**, and **remainder**.

EXAMPLE 2

Using Long Division to Find a Quotient and Remainder

Use long division to find the quotient and remainder when dividing $f(x) = x^3 - 5x^2 + 20$ by $d(x) = x - 3$. In other words, find $q(x)$ and $r(x)$ so that $x^3 - 5x^2 + 20 = (x - 3)q(x) + r(x)$.

Solution   Since there is no x term in the polynomial $x^3 - 5x^2 + 20$, we insert $0x$ to ensure that terms with like powers line up. We divide as follows:

$$
\begin{array}{r}
x^2 - 2x - 6 \\
x - 3 \overline{\smash{\big)}\, x^3 - 5x^2 + 0x + 20} \\
\underline{x^3 - 3x^2} \\
-2x^2 + 0x + 20 \\
\underline{-2x^2 + 6x} \\
-6x + 20 \\
\underline{-6x + 18} \\
2
\end{array}
$$

Thus, $q(x) = x^2 - 2x - 6$ and $r(x) = 2$. It follows that

$$x^3 - 5x^2 + 20 = (x - 3)(x^2 - 2x - 6) + 2$$

Equivalently, we can write

$$\underbrace{\frac{\overbrace{x^3 - 5x^2 + 20}^{\text{Dividend}}}{\underbrace{x - 3}_{\text{Divisor}}}}_{} = \overbrace{x^2 - 2x - 6}^{\text{Quotient}} + \frac{\overbrace{2}^{\text{Remainder}}}{x - 3}$$

Synthetic Division   A special process called **synthetic division** has been developed for performing divisions in the case where the divisor is of the form $x - c$, as in Example 2. Essentially, synthetic division is a compact and efficient bookkeeping method for keeping track of the steps performed in long division.

Let us begin our derivation of this procedure by finding shortcuts in the work we performed for Example 2. The following table shows several successive simplifications.

Remove all x's	Remove unnecessary numbers and "+" signs	Condense by bringing numbers up	Combine top line with bottom line
$\begin{array}{r} 1 \ -2 \ -6 \\ -3\ )\overline{1\ -5\ +0\ +20} \\ \underline{1\ -3} \\ -2\ +0\ +20 \\ \underline{-2\ +6} \\ -6\ +20 \\ \underline{-6\ +18} \\ 2 \end{array}$	$\begin{array}{r} 1 \ -2 \ -6 \\ -3\ )\overline{1\ -5\quad 0\quad 20} \\ \underline{-3} \\ -2 \\ \underline{6} \\ -6 \\ \underline{18} \\ 2 \end{array}$	$\begin{array}{r} 1 \ -2 \ -6 \\ -3\ )\overline{1\ -5\quad\ \ 0\quad 20} \\ \underline{-3\quad\ \ 6\quad 18} \\ -2\ -6\quad 2 \end{array}$	$\begin{array}{r} -3\ \underline{)\ 1\ -5\quad\ \ 0\quad 20} \\ \underline{-3\quad\ \ 6\quad 18} \\ 1\ -2\ -6\quad 2 \end{array}$

We can further simplify the procedure if we omit the "$-$" from the divisor -3. This changes the signs of the numbers in the second row and, as a consequence, allows us to add down the columns instead of subtracting. The final version of this **synthetic division** procedure is shown in Figure 34. To summarize, the numbers circled in the top row of Figure 35 are the coefficients of the dividend $x^3 - 5x^2 + 20$. The 3 on the far left is from the divisor $x - 3$. The 1 in the bottom row is carried down from the top row. Each boxed number is used to obtain the number to its right by following these steps.

$$3\ \underline{\begin{array}{rrrr} 1 & -5 & 0 & 20 \\ & 3 & -6 & -18 \end{array}}$$
$$\begin{array}{rrrr} 1 & -2 & -6 & 2 \end{array}$$

Figure 34

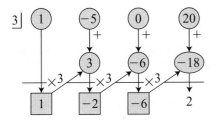

Figure 35

1. Multiply the boxed number by 3 and place the product in the circle above and to the right.

2. Add the two circled numbers to obtain the boxed number in the bottom row.

The resulting boxed numbers in the bottom row are the coefficients of the quotient $x^2 - 2x - 6$, and the unboxed 2 is the remainder. Note that the degree of the quotient $x^2 - 2x - 6$ is 1 less than the degree of the dividend $x^3 - 5x^2 + 20$.

EXAMPLE 3

Using Synthetic Division to Find a Quotient and Remainder

Use synthetic division to find the quotient and remainder when $3x^4 - 12x^2 + 8x + 4$ is divided by $x + 2$.

Solution   Since the divisor is $x + 2 = x - (-2)$, we place -2 to the far left. The top row is formed using the coefficients of the dividend $3x^4 - 12x^2 + 8x + 4$, *including a 0 for the coefficient of the missing x^3 term*. After bringing the leading coefficient 3 down to the bottom row, we proceed as in Figure 35, multiplying by -2 and then adding at each stage.

$$-2\ \underline{\begin{array}{rrrrr} 3 & 0 & -12 & 8 & 4 \\ & -6 & 12 & 0 & -16 \end{array}}$$
$$\begin{array}{rrrrr} 3 & -6 & 0 & 8 & -12 \end{array}$$

Since the divisor $x + 2$ has degree 1, and the dividend $3x^4 - 12x^2 + 8x + 4$ has degree 4, the quotient will have degree 3. Using the first four numbers in the bottom row as the coefficients, we obtain the quotient $3x^3 - 6x^2 + 0x + 8$. The remainder is -12. Thus,

$$3x^4 - 12x^2 + 8x + 4 = (x + 2)(3x^3 - 6x^2 + 8) - 12$$

EXAMPLE 4

Using Synthetic Division to Factor a Polynomial

Given that $(x + 4)$ and $(x - 1)$ are factors of the polynomial

$$f(x) = 4x^4 + 16x^3 - 3x^2 - 13x - 4$$

use synthetic division to factor $f(x)$ and then find the zeros. Verify using a graphing calculator.

Solution   We first apply synthetic division to divide $4x^4 + 16x^3 - 3x^2 - 13x - 4$ by $x + 4$.

$$
\begin{array}{r|rrrrr}
-4 & 4 & 16 & -3 & -13 & -4 \\
   &   & -16 & 0 & 12 & 4 \\
\hline
   & 4 & 0 & -3 & -1 & 0
\end{array}
$$

Since the remainder is 0, we have confirmed that $x + 4$ is indeed a factor of $f(x)$. The first four numbers in the bottom row are the coefficients of the quotient $4x^3 - 3x - 1$, and so $f(x) = (x + 4)(4x^3 - 3x - 1)$. Now if $x - 1$ is a factor of $f(x)$, it must also be a factor of $4x^3 - 3x - 1$. Thus, we use synthetic division to divide $x - 1$ into $4x^3 - 3x - 1$.

$$
\begin{array}{r|rrrr}
1 & 4 & 0 & -3 & -1 \\
  &   & 4 & 4 & 1 \\
\hline
  & 4 & 4 & 1 & 0
\end{array}
$$

Since the remainder is 0, $x - 1$ is indeed a factor. The first three numbers in the bottom row are the coefficients of $4x^2 + 4x + 1$. Because $4x^2 + 4x + 1 = (2x + 1)^2$, it follows that

$$4x^4 + 16x^3 - 3x^2 - 13x - 4 = (x + 4)(x - 1)(2x + 1)^2$$

Thus, setting each of the factors of $f(x)$ equal to zero, we determine that the zeros of f are $x = -4$, $x = 1$, and $x = -\frac{1}{2}$. These values are consistent with the x-intercepts shown in Figure 36.

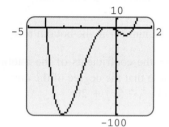

Figure 36

Notice that in all of the synthetic division examples, the remainder turned out to be zero or some other constant. The division algorithm guarantees that this will always be the case. Indeed, since the divisor $x - c$ is a polynomial of degree 1, and since the remainder's degree must be less than that of the divisor, the remainder must have degree 0 and so must be a constant. In light of this, the division algorithm for a polynomial $f(x)$ and divisor $x - c$ gives us

$$f(x) = (x - c)q(x) + k$$

Now, if we set $x = c$, we obtain $f(c) = 0 \cdot q(c) + k = k$. In other words, $f(c)$ is the remainder obtained by dividing $f(x)$ by $x - c$. This fact is known as the **Remainder Theorem**.

Remainder Theorem

If a polynomial $f(x)$ is divided by $x - c$, the remainder is $f(c)$.

EXAMPLE 5

Evaluating a Polynomial with the Remainder Theorem

Use the Remainder Theorem to evaluate $f(x) = 2x^3 - 5x^2 + x - 4$ at $x = -3$.

Solution   We use synthetic division to find the remainder when $2x^3 - 5x^2 + x - 4$ is divided by $x + 3 = x - (-3)$.

$$
\begin{array}{r|rrrr}
-3 & 2 & -5 & 1 & -4 \\
   &   & -6 & 33 & -102 \\
\hline
   & 2 & -11 & 34 & -106
\end{array}
$$

Thus, $f(-3) = -106$.

Understanding and Mastery Checklists

Concepts to Understand	Skills to Master
Long division	Use long division to find a quotient and remainder when dividing one polynomial by another.
Division algorithm	Use synthetic division to find a quotient and remainder when dividing one polynomial by $x - c$.
Synthetic division	Divide a given factor into a polynomial to find the other factors.
Remainder Theorem	Use synthetic division and the Remainder Theorem to find a function value.

Exercises 3.2

Exercises 1–12 *Use long division to find the quotient $q(x)$ and remainder $r(x)$ when $f(x)$ is divided by $d(x)$.*

1. $f(x) = 2x^2 - 5x - 12;\ d(x) = x - 4$

2. $f(x) = 3x^2 + x - 10;\ d(x) = x + 2$

3. $f(x) = 4x^3 - 3x^2 + 20x - 15;\ d(x) = x^2 + 5$

4. $f(x) = 2x^3 + 4x^2 - 1;\ d(x) = x^2 + 2x - 1$

5. $f(x) = 2x^4 - 7x^3 - 6x + 9;\ d(x) = 2x^2 - x + 3$

6. $f(x) = 9x^4 - 3x^3 + 2x - 4;\ d(x) = 3x^2 - 2$

7. $f(x) = x^2 + 1;\ d(x) = x + 1$

8. $f(x) = 2x^2 - x + 4;\ d(x) = x - 2$

9. $f(x) = 3x^3 - 4x + 2;\ d(x) = x^2 - 3$

10. $f(x) = x^4;\ d(x) = x^3 + 1$

11. $f(x) = x^5;\ d(x) = x^3 + x + 1$

12. $f(x) = 3x^4 + x^2 - 1;\ d(x) = x^2 - x + 3$

Exercises 13–24 *Use synthetic division to find the quotient $q(x)$ and remainder $r(x)$ when $f(x)$ is divided by $d(x)$.*

13. $f(x) = 2x^2 + 9x - 18;\ d(x) = x + 6$

14. $f(x) = 5x^2 - 13x - 6;\ d(x) = x - 3$

15. $f(x) = x^3 - 2x^2 - 2x + 4; d(x) = x - 2$

16. $f(x) = x^3 + 3x^2 - 9x + 5; d(x) = x + 5$

17. $f(x) = 3x^4 - 8x^3 - 10x + 3; d(x) = x - 3$

18. $f(x) = 2x^4 - 14x^3 + 5; d(x) = x - 7$

19. $f(x) = x^2 - 2; d(x) = x + 1$

20. $f(x) = x^2 + 2x; d(x) = x - 2$

21. $f(x) = x^3 + 6x^2 - 16x + 2; d(x) = x + 8$

22. $f(x) = 1 - 2x^3; d(x) = x - 4$

23. $f(x) = 1 + x^3 - 4x^4; d(x) = x - \dfrac{1}{3}$

24. $f(x) = 2x^4 + x^3 - 4x^2 + 1; d(x) = x + \dfrac{1}{2}$

Exercises 25-32   *Use the given factor(s) of the polynomial to find the remaining factors, and then identify the zeros. Verify the zeros graphically.*

25. $x + 2; f(x) = x^3 + 6x^2 + 3x - 10$

26. $x - 4; f(x) = x^3 - 3x^2 - 10x + 24$

27. $x - 3; p(x) = 2x^3 - 9x^2 + 7x + 6$

28. $x + 1; h(x) = 3x^3 + 10x^2 + x - 6$

29. $x + 4$ and $x - 1; g(x) = x^4 + 9x^3 + 22x^2 - 32$

30. $x - 6$ and $x + 3; f(x) = x^4 - 2x^3 - 27x^2 + 108$

31. $2x + 1$ and $x - 8; h(x) = 6x^4 - 47x^3 - 13x^2 + 38x + 16$

32. $5x - 1$ and $x - 1; p(x) = 10x^4 + 13x^3 - 43x^2 + 23x - 3$

Exercises 33-38   *Use synthetic division and the Remainder Theorem to find the indicated function values.*

33. $f(x) = 2x^3 - x^2 - 4x + 6$

    a. $f(3)$ **b.** $f(-2)$ **c.** $f\left(\dfrac{1}{2}\right)$

34. $g(x) = 3x^3 + 7x^2 - 18x + 4$

    a. $g(-4)$ **b.** $g(2)$ **c.** $g\left(-\dfrac{1}{3}\right)$

35. $h(x) = 3x^4 - 2x^3 - 4x + 3$

    a. $h(-1)$ **b.** $h(5)$ **c.** $h(-2.1)$

36. $h(x) = -x^4 + 3x^2 + 5x - 10$

    a. $h(3)$ **b.** $h(-1)$ **c.** $h(1.4)$

37. $g(x) = \dfrac{1}{2}x^4 - \dfrac{5}{3}x^3 + \dfrac{2}{3}x^2 - \dfrac{9}{2}x$

    a. $g(4)$ **b.** $g(-2)$

38. $f(x) = \dfrac{2}{5}x^4 - \dfrac{21}{2}x^2 + \dfrac{5}{2}x + \dfrac{3}{4}$

    a. $f(5)$ **b.** $f(-5)$

Concepts and Critical Thinking

Exercises 39-42   *Answer true or false.*

39. When one polynomial is divided by another, the degree of the remainder is always greater than that of the quotient.

40. If the remainder obtained when dividing the polynomial $f(x)$ by $x + 3$ is 8, then $f(3) = 8$.

41. If the remainder obtained when dividing the polynomial $f(x)$ by $x - 3$ is 8, then $f(3) = 8$.

42. The most efficient method for computing the quotient and remainder when dividing a fifth-degree polynomial by a fourth-degree polynomial is synthetic division.

Exercises 43-46   *Give an example of each.*

43. A third-degree polynomial having $x + 2$ as a factor

44. A polynomial $P(x)$ such that when $P(x)$ is divided by $x - 1$ the remainder is 3.

45. A fourth-degree polynomial having $x + 1$ and $x - 2$ as factors

46. An application of synthetic division

Questions for Discussion or Essay

47. Explain how the trace feature on your graphing calculator could be used to evaluate a polynomial function $f(x)$ at specific values for x. Discuss the advantages and disadvantages of this technique, as compared to the synthetic division process described in this section. Does your calculator have any built-in feature for evaluating functions? If so, how does it compare to using the trace feature or synthetic division?

48. Compare and contrast long division of polynomials with long division of real numbers.

49. a. How many additions and multiplications are required to divide $x - c$ into a third-degree polynomial? A fourth-degree polynomial? An nth-degree polynomial?

b. Explain how one can evaluate a 17th-degree polynomial function for any given value of x using only a $\boxed{\times}$ key and a $\boxed{+}$ key 17 times each.

Projects for Enrichment

50. Factors of $f(x) = x^n - a^n$ and Sums of the Form $1 + a + a^2 + \cdots + a^n$ In this project, we use synthetic division to investigate the factors of $x^n - a^n$, and then see how this helps us find sums of the form $1 + a + a^2 + \cdots + a^n$. To begin, we consider the polynomial $g(x) = x^2 - a^2$. Using the difference of squares formula, we can factor this as $g(x) = (x - a)(x + a)$. Next we consider $h(x) = x^3 - a^3$. The difference of cubes formula gives us $h(x) = (x - a)(x^2 + ax + a^2)$. Notice that both $g(x)$ and $h(x)$ have $x - a$ as a factor.

a. Complete the following synthetic division to divide $f(x) = x^4 - a^4$ by $x - a$, and then write out the factored form of $f(x)$.

$$
\begin{array}{r|cccc}
a & 1 & 0 & 0 & 0 & -a^4 \\
  &   & a & a^2 & & \\
\hline
  & 1 & a & & &
\end{array}
$$

b. Use synthetic division to divide $f(x) = x^5 - a^5$ by $x - a$, and write out the factored form of $f(x)$.

c. Complete the following synthetic division to divide $f(x) = x^n - a^n$ by $x - a$. Write out the factored form of $f(x)$.

$$
\begin{array}{r|ccccc}
 & & \overbrace{\quad\quad}^{n\,-\,1\ \text{zeros}} & & & \\
a & 1 & 0 \ 0 \ \cdots \ 0 & -a^n \\
  &   & a \ a^2 & & \\
\hline
  & 1 & a & & &
\end{array}
$$

d. By setting $x = 1$ in the factored form of $f(x)$ from part c, show that

$$1 - a^n = (1 - a)(1 + a + a^2 + \cdots + a^{n-1})$$

and thus

$$(1) \qquad \frac{1 - a^n}{1 - a} = 1 + a + a^2 + \cdots + a^{n-1}$$

Equation (1) enables us to find sums of the form $1 + a + a^2 + a^3 + \cdots + a^n$. Indeed, if we replace n with $n + 1$ in equation (1), we obtain

$$(2) \qquad 1 + a + a^2 + \cdots + a^n = \frac{1 - a^{n+1}}{1 - a}$$

Now suppose we wish to find the sum $1 + 3 + 3^2 + 3^3 + \cdots + 3^6$. In equation (2), we set $a = 3$ and $n = 6$ to obtain

$$
1 + 3 + 3^2 + 3^3 + \cdots + 3^6 = \frac{1 - 3^{6+1}}{1 - 3}
$$
$$
= \frac{1 - 2187}{-2}
$$
$$
= 1093
$$

e. Use equation (2) to find the following sums:

i. $1 + 4 + 4^2 + 4^3 + \cdots + 4^8$

ii. $1 + \dfrac{1}{2} + \left(\dfrac{1}{2}\right)^2 + \cdots + \left(\dfrac{1}{2}\right)^{10}$

iii. $1 + 6 + 36 + 216 + \cdots + 10{,}077{,}696$

iv. $1 + \dfrac{2}{3} + \dfrac{4}{9} + \cdots + \dfrac{4096}{531{,}441}$

f. A job pays \$1 in the first month, \$2 in the second month, \$4 in the third month, and so on. How much will the job pay over the course of 5 years?

g. Suppose a highly contagious flu virus is brought into the United States by a returning tourist. To model the spread of this virus, we will assume that the flu is passed to three other people on the first day, and it continues to spread in such a way that each person who gets the flu on a given day passes it to three other people the next day. If this pattern continues, how many people will have had the flu after 10 days? After 20 days?

h. Is the assumption in part g realistic? Why or why not? What factors affect the rate at which a contagious disease spreads? Is it likely that an entire population will contract a contagious disease? Explain.

51. Synthetic Division on a Spreadsheet A spreadsheet is an ideal tool for performing synthetic division. Suppose, for example, we wish to divide $f(x) = 2x^5 - 12x^4 - 13x^3 + 5$ by $x - 7$. We begin by setting up the first row of the synthetic division in cells A1 through G1, as shown in Figure 37. Some of the formulas needed for rows 2 and 3 are also shown in Figure 37. The remaining formulas for rows 2 and 3 can be entered simply by copying from cell C2 to cells D2 through G2, and from cell C3 to cells D3 through G3. The coefficients of the resulting quotient and the remainder will be computed and displayed in cells B3 through G3. In this case, the quotient is $q(x) = 2x^4 + 2x^3 + x^2 + 7x + 49$ and the remainder is 348 (see Figure 38). Higher-degree polynomials can be handled by copying the formulas to more columns.

	A	B	C	D	E	F	G
1	7	2	-12		-13	0	0	5
2			=A1*B3				
3		=B1	=C1+C2				

Figure 37

	A	B	C	D	E	F	G
1	7	2	-12	-13	0	0	5
2			14	14	7	49	343
3		2	2	1	7	49	348

Figure 38

a. Use synthetic division to find the quotient $q(x)$ and remainder $r(x)$ when $f(x)$ is divided by $d(x)$.

   i. $f(x) = 2x^5 - 19x^4 + 25x^3 + 2x^2 - 86x + 87$; $d(x) = x - 8$

   ii. $f(x) = 3x^6 + 33x^5 - 5x^4 - 43x^3 + 130x^2 - 7x + 177$; $d(x) = x + 11$

   iii. $f(x) = x^8 - 10$; $d(x) = x - 2$

b. Use synthetic division and the Remainder Theorem to find the indicated function value.

   i. $f(x) = x^5 + 3x^2 - 5x + 1$; $f(5)$

   ii. $f(x) = 3x^6 - 10x^4 + 2x^2 + 5$; $f(-3)$

   iii. $f(x) = x^8 + x^7 + x^6 + \cdots + x + 1$; $f(2)$

Section 3.3   Zeros and Factors of Polynomials

- How can a polynomial whose graph never touches the x-axis have six zeros?
- How can a graphing calculator be used as an aid in factoring polynomials?
- Is there a cubic formula for solving third-degree polynomial equations much like the quadratic formula for solving quadratic equations? If so, why haven't either of the authors memorized it?
- How can the zeros of a polynomial be used to model bungee jumping?

The Factor Theorem

We have already seen evidence of a close connection between the factors and zeros of polynomials. In Section 3.1, for example, we used the factors of a polynomial to find the exact values of its zeros. Now we complete the connection by showing that the zeros of a polynomial can be used to find its factors. Suppose that $x = c$ is a zero of a polynomial f. Then $f(c) = 0$. But according to the Remainder Theorem, $f(c)$ is the value of the remainder when $f(x)$ is divided by $x - c$. Thus, $f(x) = (x - c)q(x) + 0$, which shows that $x - c$ is a factor of $f(x)$. We have just proven the Factor Theorem.

Factor Theorem

A polynomial function f has a zero c if and only if $x - c$ is a factor of $f(x)$. In other words, $f(c) = 0$ if and only if $f(x) = (x - c)q(x)$ for some nonzero polynomial $q(x)$.

EXAMPLE 1

Using Known Zeros of a Polynomial to Find the Remaining Zeros

The polynomial function $f(x) = 5x^4 + 8x^3 - 29x^2 - 20x + 12$ has among its zeros $x = -3$ and $x = 2$. Find the remaining zeros.

Solution Since f has zeros $x = -3$ and $x = 2$, the Factor Theorem tells us that $f(x)$ has factors $x + 3$ and $x - 2$. Thus, we use synthetic division to divide $f(x)$ by $x + 3$ and then apply synthetic division again to divide the quotient by $x - 2$. [Alternatively, we could use long division to divide $f(x)$ by $(x + 3)(x - 2) = x^2 + x - 6$.]

$$
\begin{array}{r|rrrrr}
-3 & 5 & 8 & -29 & -20 & 12 \\
   &   & -15 & 21 & 24 & -12 \\
\hline
   & 5 & -7 & -8 & 4 & 0
\end{array}
\qquad
\frac{5x^4 + 8x^3 - 29x^2 - 20x + 12}{x + 3} = 5x^3 - 7x^2 - 8x + 4
$$

$$
\begin{array}{r|rrrr}
2 & 5 & -7 & -8 & 4 \\
  &   & 10 & 6 & -4 \\
\hline
  & 5 & 3 & -2 & 0
\end{array}
\qquad
\frac{5x^3 - 7x^2 - 8x + 4}{x - 2} = 5x^2 + 3x - 2
$$

From the last line of the division by $x - 2$, we see that $5x^2 + 3x - 2$ is a factor of $f(x)$. But $5x^2 + 3x - 2 = (x + 1)(5x - 2)$, and so we have

$$
\begin{aligned}
5x^4 + 8x^3 - 29x^2 - 20x + 12 &= (x + 3)(5x^3 - 7x^2 - 8x + 4) \\
&= (x + 3)(x - 2)(5x^2 + 3x - 2) \\
&= (x + 3)(x - 2)(x + 1)(5x - 2)
\end{aligned}
$$

Setting each factor equal to zero, we determine that the zeros of f are $-3, 2, -1,$ and $\frac{2}{5}$.

EXAMPLE 2

Finding a Polynomial with Given Zeros

Find a possible third-degree polynomial function f for the graph shown in Figure 39. Verify with a graphing calculator.

Solution The function graphed in Figure 39 appears to have zeros of $-3, -1,$ and $\frac{1}{2}$. According to the Factor Theorem, each zero c corresponds to a factor of the form $x - c$. So $f(x)$ must have factors $x + 3$, $x + 1$, and $x - \frac{1}{2}$. Because $f(x)$ could also have a constant factor, we have

$$
f(x) = a\left(x - \frac{1}{2}\right)(x + 3)(x + 1)
$$

To find the value of a, we note that the graph appears to have a y-intercept of $(0, 3)$, and thus we will assume that $f(0) = 3$. This gives us

$$
f(0) = 3
$$

$$
a\left(0 - \frac{1}{2}\right)(0 + 3)(0 + 1) = 3 \qquad \text{Setting } x = 0 \text{ in } f(x)
$$

$$
a\left(-\frac{3}{2}\right) = 3 \qquad \text{Multiplying out}
$$

$$
a = 3\left(-\frac{2}{3}\right) = -2
$$

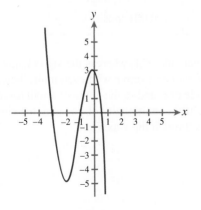

Figure 39

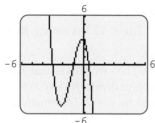

Figure 40

Finally, we multiply out the factors of $f(x)$ to obtain the standard polynomial form.

$$f(x) = -2\left(x - \frac{1}{2}\right)(x + 3)(x + 1)$$
$$= (-2x + 1)(x + 3)(x + 1) \qquad \text{Multiplying } -2 \text{ through } \left(x - \frac{1}{2}\right)$$
$$= (-2x + 1)(x^2 + 4x + 3) \qquad \text{Multiplying } (x + 3)(x + 1)$$
$$= (-2x^3 - 8x^2 - 6x) + (x^2 + 4x + 3) \qquad \text{Distributing } (-2x + 1)$$
$$= -2x^3 - 7x^2 - 2x + 3$$

The graph of f shown in Figure 40 looks very much like that shown in Figure 39.

If a certain factor $x - c$ appears more than once in the factorization of a polynomial function, then c is said to be a *multiple zero*. For example, the fifth-degree polynomial $f(x) = x^5 - 2x^4 + x^3$ factors as $f(x) = x^3(x - 1)^2$, and so $x = 0$ and $x = 1$ are multiple zeros. More specifically, since the zero $x = 0$ arises from x^3, a factor of degree 3, it is said to be a zero of multiplicity 3. Likewise, $x = 1$ is a zero of multiplicity 2 since it arises from the factor $(x - 1)^2$.

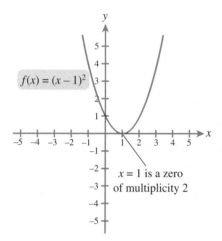

Figure 41

Multiple Zeros

If k is the largest positive integer for which $(x - c)^k$ is a factor of a polynomial $f(x)$, then c is called a **zero of multiplicity k**.

The multiplicity of a real zero of a polynomial affects the shape of the graph of the polynomial near the zero. Consider the polynomials $f(x) = (x - 1)^2$ and $g(x) = (x - 1)^3$, shown in Figures 41 and 42. Both have $x = 1$ as a zero, but the multiplicity is 2 for f and 3 for g. When we compare the graphs of f and g, we see that the graph of f touches but does not cross the x-axis, whereas the graph of g passes through the x-axis. To see why, let's consider the signs of $f(x)$ and for $g(x)$ for x-values just to the left and just to the right of 1.

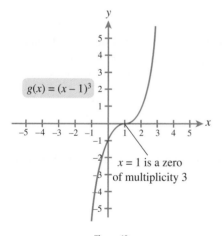

Figure 42

x	$f(x) = (x - 1)^2$ (even multiplicity)	$g(x) = (x - 1)^3$ (odd multiplicity)
0.9	$(0.9 - 1)^2 = 0.01$; positive	$(0.9 - 1)^3 = -0.001$; negative
1.1	$(1.1 - 1)^2 = 0.01$; positive	$(1.1 - 1)^3 = 0.001$; positive

Note that the sign of $f(x)$ is the same on either side of 1, whereas the sign changes for $g(x)$. More generally, if a polynomial function h has a zero c with even multiplicity, then the sign of $h(x)$ will be the same on either side of c, and so the graph of h will touch but not cross the x-axis at c. On the other hand, if the multiplicity of c is odd, the sign of $h(x)$ will change at c, and the graph will pass through the x-axis at c.

EXAMPLE 3   Finding a Polynomial with Multiple Zeros

Find the third-degree polynomial $f(x)$ with the graph shown in Figure 43. Verify with a graphing calculator.

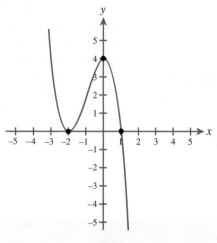

Figure 43

Solution   Since the graph turns at $x = -2$ but passes through the x-axis at $x = 1$, we conclude that $x = -2$ is a zero of even multiplicity, whereas $x = 1$ is a zero of odd multiplicity. Since $f(x)$ is a third-degree polynomial, it follows that

$$f(x) = a(x + 2)^2(x - 1)$$

for some constant a. To find the value of a, we note that f has y-intercept $(0, 4)$, and so $f(0) = 4$.

$$
\begin{aligned}
f(0) &= 4 \\
a(0 + 2)^2(0 - 1) &= 4 \qquad \text{Setting } x = 0 \text{ in } f(x) \\
a(-4) &= 4 \qquad \text{Simplifying} \\
a &= -1
\end{aligned}
$$

Thus, $f(x) = -1(x + 2)^2(x - 1)$. The graph of f shown in Figure 44 looks very much like the graph in Figure 43.

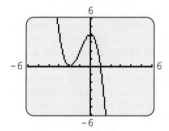

Figure 44

···▷EXAMPLE 4

Modeling Bungee Jumping with a Cubic Polynomial

Suppose you are standing on a building 25 feet above the ground. A bungee jumper leaps from a platform 64 feet above you. She passes your level after 2 seconds, again after 6 seconds on the way back up, and after 9 seconds on the way back down. Assuming the height of the jumper can be roughly approximated with a cubic polynomial, how close to the ground does she get and how high does she rebound?

Solution   We first construct a cubic polynomial $h(t)$ that gives the height of the jumper with respect to your level. The times at which the jumper passes your level—namely, $t = 2$, $t = 6$, and $t = 9$—are the zeros of the function h since the height (with respect to you) at those times is 0. Thus, we know that $h(t)$ must have factors $t - 2$, $t - 6$, and $t - 9$. We must also allow for the fact that $h(t)$ may have a constant factor, and so we settle on the form $h(t) = a(t - 2)(t - 6)(t - 9)$. Because the initial height of the jumper is 64 feet, we have $h(0) = 64$. This enables us to solve for a.

$$
\begin{aligned}
h(0) &= 64 \\
a(0 - 2)(0 - 6)(0 - 9) &= 64 \\
-108a &= 64 \\
a &= -\frac{64}{108} = -\frac{16}{27}
\end{aligned}
$$

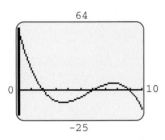

Figure 45

Thus, $h(t) = -\frac{16}{27}(t - 2)(t - 6)(t - 9)$. The graph of h is shown in Figure 45. Using the trace or calculate-maximum/minimum feature, we determine that the minimum height is 12.3 feet below your level, or 12.7 feet above the ground, and the maximum height after the rebound is 7.5 feet above you, or 32.5 feet above ground level.

However accurate this model might be for the first 9 seconds, it is invalid soon after. The cubic polynomial $h(t)$ approaches $-\infty$ as t approaches ∞, but it seems unlikely that the bungee jumper will puncture Earth's crust, pass through its mantle and molten core, knife through its crystalline core, and then be shot out the other side into space.

Morphing

When you see faces of different genders and races blend from one to another in a Michael Jackson video, when you see an automobile transform into a tiger in a commercial for Mobil, or the T-2000 robot metamorphose from liquid metal to human in the movie *Terminator II*, you are witnessing *morphing*, a spectacular graphic image manipulation technique. The most striking aspect of a morphing sequence is its fluidity; one image melts into another so seamlessly that our eyes tell us that the transformation, no matter how logic-defying, is actually taking place.

Although morphing has existed only since the 1980s, it is now commonplace in several visual media. Films, television commercials, television series, and music videos have employed morphing techniques to create startling special effects, and now software is available that enables those with personal computers to create their own morphing sequences.

A visual effect created with the aid of morphing.

Let us consider the steps required to create a morph sequence from an original to a final image. First, the original and final images are digitized; that is, the images are encoded as numerical data, in much the same way as images and music are stored on compact discs. Then the animator defines a correspondence between points on the initial and final images—for example, a point on the nose of a tiger might correspond to a point on the

beak of an eagle; a certain point on the hindquarters of the tiger might correspond to a point on the wing of the eagle; and so forth. Next the images are blended from one to the other, step by step, in such a way that the intermediate steps seem plausible. For example, when morphing a tiger into an eagle, we don't want to see a cavity form in the middle of the tiger's body that is then gradually filled in by feathers, nor do we want to see a paw *abruptly* change into a talon, or see a creature with whiskers on one side of its face and feathers on the other. The production of a smooth, plausible morph requires a combination of sophisticated mathematics, computational power, and often the artistic touch and intuition of the animator.

Mathematically, morphing is essentially an interpolation problem: A set of data points is given and intermediate values are to be determined. There are many ways in which data points can be "connected," and so there are many ways in which the morphing sequence can be generated. One of the most popular methods of interpolating is to use polynomial functions. Because polynomial functions are "smooth" (their graphs have no breaks or sharp changes of direction), the resulting morphing sequence likewise appears smooth.

Complex Zeros

We have seen many examples of polynomials with one or more zeros. Are there examples of polynomials with no zeros? If we consider only real-valued zeros—that is, those that correspond to x-intercepts—the answer is yes. A simple example is the polynomial $f(x) = x^2 + 1$. It is clear from the graph in Figure 46 that f has no x-intercepts and hence has no real zeros. However, if we set $x^2 + 1 = 0$ and solve for x within the complex number system, we see that $x = \pm i$. Thus, the imaginary numbers i and $-i$ are both zeros of f. As it turns out, if we allow complex-valued zeros, every polynomial of degree 1 or higher will have at least one zero. This was first proven in 1799 by the

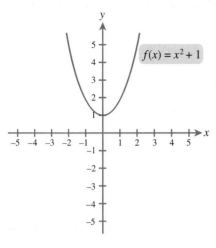

$f(x) = x^2 + 1$

Figure 46

famous German mathematician Carl Friedrich Gauss (he was 22 at the time). Because of its importance to algebra, his result has come to be called the **Fundamental Theorem of Algebra**.

Fundamental Theorem of Algebra

If $f(x)$ is a polynomial of degree 1 or higher, then f has at least one zero in the complex number system.

We can extend the Fundamental Theorem of Algebra to show that a polynomial of degree n must have exactly n zeros (counting multiplicities). To do this, we first show that a polynomial of degree n can be factored into n factors of the form $x - c$. Suppose $f(x)$ is a polynomial of degree n, with $n \geq 1$. Then, by the Fundamental Theorem of Algebra, f has a zero; call it c_1. By the Factor Theorem, this means $f(x)$ can be factored as

$$f(x) = (x - c_1)q_1(x)$$

where $q_1(x)$ is a polynomial of degree $n - 1$. Applying the Fundamental Theorem of Algebra to the polynomial $q_1(x)$, we know that it too must have a zero; call it c_2. By the Factor Theorem, $q_1(x) = (x - c_2)q_2(x)$ for some polynomial $q_2(x)$ of degree $n - 2$. Combined with the earlier factorization, we now have

$$f(x) = (x - c_1)(x - c_2)q_2(x)$$

If we continue this process, we eventually arrive at a polynomial factor $q_n(x)$ of degree 0 (a constant), and here the process stops. The result is a *complete* factorization of $f(x)$ into linear factors.

Linear Factorization Theorem

If $f(x)$ is a polynomial of degree n, with $n \geq 1$, then $f(x)$ can be expressed as a product of linear factors in the following way:

$$f(x) = a(x - c_1)(x - c_2) \cdots (x - c_n)$$

where $c_1, c_2, \ldots, c_n$ are complex numbers and a is the leading coefficient of $f(x)$.

An immediate consequence of the Linear Factorization Theorem is that an nth-degree polynomial has n zeros (counting multiplicities), each of which corresponds to a factor of the form $x - c$.

EXAMPLE 5

Finding the Complex Zeros of a Polynomial

Given that 4 is a zero of $f(x) = x^3 - 4x^2 + 4x - 16$, find all the complex zeros of f and write $f(x)$ as a product of linear factors.

Solution   Since 4 is a zero of f, $x - 4$ must be a factor of $f(x)$. We apply synthetic division to find a second factor.

$$\begin{array}{r|rrrr} 4 & 1 & -4 & 4 & -16 \\ & & 4 & 0 & 16 \\ \hline & 1 & 0 & 4 & 0 \end{array}$$

$$\frac{x^3 - 4x^2 + 4x - 16}{x - 4} = x^2 + 4$$

So $f(x) = (x - 4)(x^2 + 4)$. Now the zeros that correspond to $x^2 + 4$ can be found as follows:

$$\begin{aligned} x^2 + 4 &= 0 \\ x^2 &= -4 \\ x &= \pm\sqrt{-4} = \pm 2i \end{aligned}$$

Thus, the zeros of f are $x = 4$, $x = 2i$, and $x = -2i$, and we have

$$f(x) = (x - 4)(x - 2i)(x + 2i)$$

EXAMPLE 6

Finding the Complex Zeros of a Polynomial

Find all the complex zeros of the polynomial $f(x) = x^5 - 2x^4 - x^3 + 4x^2 - 2x - 4$ and express the polynomial as a product of linear factors.

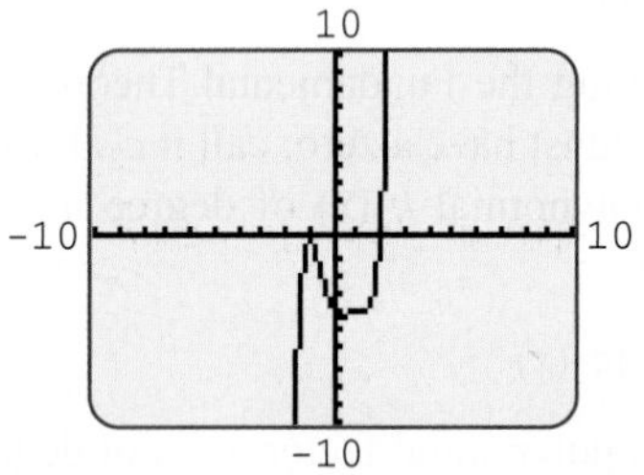

Figure 47

Solution We first plot the graph of f to see if any integer zeros can be easily obtained. It appears from the graph in Figure 47 that $x = -1$ and $x = 2$ are zeros of f. It also appears that $x = -1$ has even multiplicity, and so we deduce that $f(x)$ has factors $(x + 1)^2(x - 2)$. To verify these factors, and also to find any remaining factors, we divide $f(x)$ by $(x + 1)^2(x - 2)$. Although this division could be accomplished by performing synthetic division three times in succession (once for each of the factors), we choose to divide in one step, using long division. Now $(x + 1)^2(x - 2) = x^3 - 3x - 2$, so we divide $f(x)$ by $x^3 - 3x - 2$ as follows:

$$\begin{array}{r|l} & x^2 - 2x + 2 \\ \hline x^3 - 3x - 2 & x^5 - 2x^4 - x^3 + 4x^2 - 2x - 4 \\ & x^5 \qquad\quad - 3x^3 - 2x^2 \\ \hline & -2x^4 + 2x^3 + 6x^2 - 2x - 4 \\ & -2x^4 \qquad\quad + 6x^2 + 4x \\ \hline & 2x^3 \qquad\quad - 6x - 4 \\ & 2x^3 \qquad\quad - 6x - 4 \\ \hline & 0 \end{array}$$

Thus,

$$\begin{aligned} f(x) &= (x^3 - 3x - 2)(x^2 - 2x + 2) \\ &= (x + 1)^2(x - 2)(x^2 - 2x + 2) \end{aligned}$$

Since $x^2 - 2x + 2$ cannot be factored using integer coefficients, we find its zeros using the quadratic formula.

$$x^2 - 2x + 2 = 0$$

$$x = \frac{-(-2) \pm \sqrt{(-2)^2 - 4(1)(2)}}{2(1)} = \frac{2 \pm \sqrt{-4}}{2} = \frac{2 \pm 2i}{2} = 1 \pm i$$

Thus, $x^2 - 2x + 2 = [x - (1 + i)][x - (1 - i)]$. Putting all the factors together, we have

$$f(x) = (x + 1)^2(x - 2)[x - (1 + i)][x - (1 - i)]$$

The five zeros of f are $x = -1$ (multiplicity 2), $x = 2$, $x = 1 + i$, and $x = 1 - i$.

It is no coincidence that the two nonreal zeros in Example 6 ($x = 1 + i$ and $x = 1 - i$) differ only in the operators $+$ and $-$. Pairs of complex numbers of the form $a + bi$ and $a - bi$ are called **complex conjugates**. If a polynomial has real coefficients, all nonreal zeros will appear as complex conjugate pairs. In other words, if a polynomial $f(x)$ with real coefficients has a complex zero $a + bi$, then it must also have a zero $a - bi$. (See Exercise 63 for an outline of the proof.)

----›EXAMPLE 7

Finding a Polynomial with Known Complex Zeros

Find a polynomial $f(x)$ of least degree with real coefficients and zeros $x = -3$ and $x = 2 - 4i$.

Solution   The polynomial must have a second nonreal zero $x = 2 + 4i$. The two factors corresponding to the nonreal zeros $2 - 4i$ and $2 + 4i$ are $x - (2 - 4i)$ and $x - (2 + 4i)$. Multiplying these out yields a quadratic factor with real coefficients, as shown.

$$[x - (2 - 4i)][x - (2 + 4i)] = x^2 - (2 - 4i)x - (2 + 4i)x + (2 - 4i)(2 + 4i)$$
$$= x^2 - 4x + [2^2 - (4i)^2]$$
$$= x^2 - 4x + 4 + 16$$
$$= x^2 - 4x + 20$$

The factor corresponding to the real zero -3 is $x + 3$, and so we know that $f(x) = (x^2 - 4x + 20)(x + 3)$ has the desired zeros. Multiplying out gives $f(x) = x^3 - x^2 + 8x + 60$.

Note that if we did not require the polynomial in Example 7 to have real coefficients, a polynomial with least degree and zeros $x = -3$ and $x = 2 - 4i$ would be the second-degree polynomial

$$f(x) = (x + 3)[x - (2 - 4i)] = x^2 + (1 + 4i)x + (-6 + 12i)$$

Complex zeros need not come in conjugate pairs if the coefficients of the polynomial are allowed to be nonreal.

Understanding and Mastery Checklists

Concepts to Understand	Skills to Master
The Factor Theorem	Use given zeros of a polynomial to find the remaining zeros.
◊	◊
Multiplicity of a zero	Find a polynomial of a specified degree and with given zeros.
◊	◊
The Fundamental Theorem of Algebra	Find the complex zeros of a polynomial.
◊	
The Linear Factorization Theorem	

Exercises 3.3

Exercises 1-6   *Use the given zero(s) of f to factor $f(x)$, and then find the remaining zeros.*

1. -3; $f(x) = x^3 + 3x^2 - 4x - 12$

2. 1; $f(x) = x^3 + 5x^2 + 3x - 9$

3. 5; $f(x) = 2x^3 - 7x^2 - 14x - 5$

4. -6; $f(x) = 3x^3 + 10x^2 - 44x + 24$

5. $-\dfrac{1}{3}, -4$; $f(x) = 3x^4 + 7x^3 - 19x^2 + 5x + 4$

6. $\dfrac{3}{2}, -1$; $f(x) = 2x^4 - x^3 - 5x^2 + x + 3$

Exercises 7-12   *Use a graphing calculator to identify the integer zeros of f, and then factor $f(x)$ completely to find the remaining zeros.*

7. $f(x) = 5x^3 + 3x^2 - 12x + 4$

8. $f(x) = 3x^3 - 10x^2 - x + 12$

9. $f(x) = 6x^3 - 13x^2 + x + 2$

10. $f(x) = 6x^3 + 13x^2 + 4x - 3$

11. $f(x) = x^4 - x^3 - 5x^2 + 3x + 6$

12. $f(x) = x^4 + x^3 - 7x^2 - 5x + 10$

Exercises 13-20   *Find a polynomial function f that satisfies the given properties.*

13. Degree 2; zeros -3 and 4

14. Degree 3; zeros -1, 2, and 4

15. Degree 3; zeros -2, 1, and 5; $f(0) = 5$

16. Degree 2; zeros -4 and -1; $f(0) = 8$

17. Degree 4; zeros 0 and 2 (both with multiplicity 2); $f(-1) = 3$

18. Degree 4; zeros -1 and 1 (both with multiplicity 2); $f(2) = 1$

19. Degree 5; zeros -1 (multiplicity 3) and $\dfrac{1}{2}$ (multiplicity 2)

20. Degree 5; zeros 0 (multiplicity 1), $-\dfrac{2}{3}$ (multiplicity 2), and 1 (multiplicity 2)

Exercises 21-26   *Find the polynomial with the given degree and graph. Check your answer with a graphing calculator.*

21. Degree 2

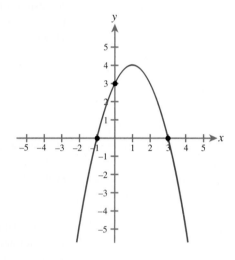

22. Degree 2

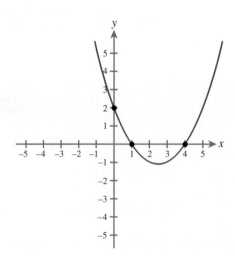

23. Degree 3

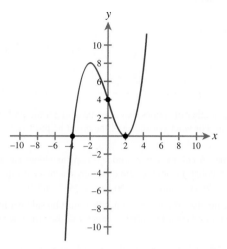

24. Degree 3

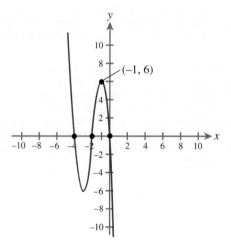

25. Degree 4

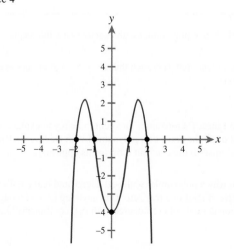

26. Degree 4

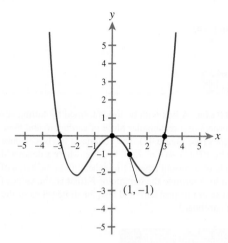

Exercises 27-30 *Use the given zero of f to factor f(x), and then find the remaining zeros.*

27. $-2i$; $f(x) = x^3 - 3x^2 + 4x - 12$

28. i; $f(x) = x^3 + 2x^2 + x + 2$

29. $1 + i$; $f(x) = x^4 - 5x^3 + 4x^2 + 2x - 8$

30. $2 - i$; $f(x) = x^4 - 3x^3 - 5x^2 + 29x - 30$

Exercises 31-38 *Find all the complex zeros of f and write f(x) as a product of linear factors. Use a graphing calculator as necessary.*

31. $f(x) = x^2 + 1$

32. $f(x) = x^2 - 4x + 13$

33. $f(x) = x^3 - 2x^2 + 5x$

34. $f(x) = x^3 + 2x^2 + 4x$

35. $f(x) = x^4 + 3x^2 - 4$

36. $f(x) = x^4 - 4x^3 + 5x^2$

37. $f(x) = x^5 - 5x^4 + 9x^3 - 5x^2$

38. $f(x) = x^5 - x^4 - x + 1$

Exercises 39-44 *Find a polynomial of least degree with real coefficients having the given zeros.*

39. -2 and $2i$

40. 0 and $2 + i$

41. i and $1 - i$

42. -1, 3, and $-4i$

43. 0, $\pm\sqrt{2}$, and $-2 + 2i$

44. 1, $-1 + 3i$, $1 - 3i$

Applications

45. Dimensions of a Box A box with no top is formed by cutting squares out of the corners of a rectangular piece of cardboard and then folding up the sides (see Figure 48). The volume of the box in cubic inches is given by $V(x) = 4x^3 - 72x^2 + 320x$, where x denotes the length of the side of a cut-out square (in inches). Find the zeros of V and use them to determine the domain of V (that is, the values for x that make sense in this problem) and also the dimensions of the original piece of cardboard.

Figure 48

46. Dimensions of a Box A box with no top is formed by cutting squares out of the corners of a rectangular piece of cardboard and then folding up the sides (see Figure 48). The volume of the box in cubic inches is given by $V(x) = 4x^3 - 48x^2 + 135x$, where x denotes the length of the side of a cut-out square (in inches). Find the zeros of V and use them to determine the domain of V (that is, the values for x that make sense in this problem) and also the dimensions of the original piece of cardboard.

47. Breaking Even A company that manufactures bicycle frames has determined that the monthly profit (in dollars) from the sale of x frames is given by

$$P(x) = -0.005x^3 + 1.5x^2 + 50x - 15,000$$

Find the number of frames that must be sold per month for the profit to be zero. For what x-values will the profit be positive?

48. Breaking Even A company that produces copy machines has a monthly profit (in dollars) from the sale of x copy machines given by

$$P(x) = -0.1x^3 + 20.5x^2 - 75x - 67,500$$

Find the number of copy machines that must be sold per month for the profit to be zero. For what x-values will the profit be positive?

49. Height of a Ball A ball is projected upward from 64 feet below ground level. It passes ground level after 1 second and again after 4 seconds on its way back down. It is known that the height of the ball with respect to ground level is given by a quadratic function $s(t)$, where t denotes the number of seconds after the object has been projected upward. Find $s(t)$. What is the maximum height of the ball?

50. Maximum Profit A roofing company has determined that because of fixed operating costs, they will lose \$10,000 in a given month if no roofs are completed. Moreover, because of the company's variable cost structure, they will break even (that is, zero profit) if they complete either 10 roofs or 30 roofs each month. Find a quadratic function $P(x)$ that expresses the company's monthly profit in terms of the number of roofs completed. How many roofs should be completed to maximize profit?

Concepts and Critical Thinking

Exercises 51-54 *Answer true or false.*

51. If -4 is a zero of a polynomial $f(x)$, then $x - 4$ is a factor of $f(x)$.

52. All polynomials of degree 1 or greater have at least one real zero.

53. All nth-degree polynomials have exactly n distinct complex zeros.

54. If $2 + 3i$ is a zero of a polynomial $f(x)$ with real coefficients, then $2 - 3i$ is a zero of $f(x)$ also.

Exercises 55-58 *Give an example of each.*

55. A second-degree polynomial with no real zeros

56. A third-degree polynomial with exactly two real zeros

57. A fourth-degree polynomial with exactly two x-intercepts

58. A cubic polynomial $f(x)$ such that $x^2 + 5x + 6$ divides evenly into $f(x)$

59. Use the Linear Factorization Theorem to help you explain why the product of all the zeros of a polynomial with leading coefficient 1 must be equal to the constant term (or its opposite).

60. Explain why a polynomial with odd degree and real coefficients must have at least one real zero. Can anything be said along this line about polynomials of even degree with real coefficients? Explain.

Questions for Discussion or Essay

61. The Factor Theorem states that there is a one-to-one correspondence between the zeros of a polynomial and its linear factors. In other words, each zero c of a polynomial $f(x)$ corresponds to a factor $x - c$, and vice versa. This does *not* mean, however, that there is only one polynomial of a given degree that corresponds to a given set of zeros. Explain why this is so. It may be helpful to think of how two different polynomials of the same degree can be constructed so that they have the same zeros. If we know all the zeros of f, what additional information is required to determine $f(x)$?

62. Many physical applications involve cyclical or oscillating behavior. For example, the motion of a pendulum, the path of a weight attached to a spring, alternating current, and average monthly temperatures all exhibit cyclical behavior. Discuss the limitations in using polynomials to model cyclical phenomena.

Projects for Enrichment

63. Complex Conjugates   The complex conjugate of a number $a + bi$ is $a - bi$. In general, the conjugate of a complex number z is denoted by $\bar{z}$. Thus, if $z = a + bi$, then $\bar{z} = a - bi$. In this project, we investigate some of the properties of complex conjugates. Throughout, we let $z = a + bi$ and $w = c + di$.

a. Show that $z + \bar{z}$ is a real number.

b. Show that $z\bar{z}$ is a real number.

c. Show that $\overline{z + w} = \bar{z} + \bar{w}$.

d. Show that $\overline{z \cdot w} = \bar{z} \cdot \bar{w}$.

e. Use part d to argue that $\overline{z^n} = \left(\bar{z}\right)^n$.

f. If a is a real number, show that $\bar{a} = a$.

g. If $f(x) = a_n x^n + a_{n-1} x^{n-1} + \cdots + a_1 x + a_0$ is a polynomial with real coefficients, show that $f(\bar{z}) = \overline{f(z)}$.

h. If $f(x) = a_n x^n + a_{n-1} x^{n-1} + \cdots + a_1 x + a_0$ is a polynomial with real coefficients and z is a zero of f, show that $\bar{z}$ is also a zero of f.

64. Cardan's Formula   The zeros of a general quadratic polynomial $f(x) = ax^2 + bx + c$ are easily found with the quadratic formula. To find exact values for the zeros of the general cubic polynomial $f(x) = ax^3 + bx^2 + cx + d$, where $a \neq 0$, we require **Cardan's Formula**, a formula that gives the solution to cubic equations in terms of their coefficients.

a. A cubic equation is said to be in **reduced form** if the square term has coefficient 0 and the cubed term has coefficient 1—that is, if it is of the form $z^3 + pz + q = 0$. Show that the equation $ax^3 + bx^2 + cx + d = 0$ can be transformed into a cubic equation in reduced form by making the substitution $x = z - \dfrac{b}{3a}$ and then dividing through by a.

b. Write the cubic equation $x^3 - 9x^2 + 9x + 62 = 0$ in reduced form.

Cardan's Formula can be applied only to cubic equations in reduced form. To express Cardan's Formula concisely, we define the following three variables:

$$s = \sqrt{\frac{q^2}{4} + \frac{p^3}{27}}, \quad u = \sqrt[3]{\frac{-q}{2} + s}, \quad \text{and} \quad v = \sqrt[3]{\frac{-q}{2} - s}$$

According to Cardan's Formula, the three solutions to the equation $z^3 + pz + q = 0$ are given by

$$z_1 = u + v,$$

$$z_2 = \frac{-(u + v) + (u - v)\sqrt{3}i}{2}$$

and

$$z_3 = \frac{-(u + v) - (u - v)\sqrt{3}i}{2}$$

c. Use Cardan's Formula to find the zeros of each of the following polynomials:

  i. $f(x) = x^3 + 63x - 316$

  ii. $g(x) = x^3 - 27x - 54$

  iii. $h(x) = x^3 - 9x^2 + 9x + 62$ (*Hint:* First write in the reduced form $z^3 + pz + q = 0$ and solve the resulting equation for z. Then use the relationship $x = z - \dfrac{b}{3a}$ to find the corresponding values of x.)

d. Use Cardan's Formula in conjunction with the Factor Theorem to find factorizations of each of the polynomials f, g, and h defined in part c.

Section 3.4   Real Zeros of Polynomials

⟐ Why is it that the polynomial $P(x) = x^{20} - $ (Your age in seconds)$x^{19} + $ (Your weight in kilograms) $x^5 + 2x + 2.7$ cannot possibly cross the positive x-axis exactly once?

⟐ You are told that a certain polynomial has an x-intercept somewhere between 1 and 100. How could you evaluate the function at 20 points and, based on this information, determine the x-intercept to within three decimal places of accuracy?

⟐ Why is it that any polynomial $P(x)$ with integer coefficients, leading coefficient 1, and satisfying $P(0) = 149$ has at most two rational zeros?

⟐ How can it be that for a given loan and repayment schedule, there are two possible interest rates?

We have already seen that graphing calculators are invaluable aids for investigating the real zeros of a polynomial. However, we have also seen that a graphing calculator has its limitations. If the zeros are not integers, we may be limited to finding approximations. More importantly, if the calculator is not used carefully, it can give misleading information about the number of zeros or the multiplicity of a zero. Two examples of these limitations are given in Table 2.

Table 2

Limitations of Finding Zeros of Polynomials Graphically

Polynomial	$f(x) = 27x^3 - 27x^2 - 18x + 8$	$f(x) = 6x^3 - 89x^2 + 216x - 144$
Graph		
Problem	The three zeros turn out to be noninteger rational numbers and cannot be obtained exactly from the graph.	The apparent zero of multiplicity 2 turns out to be two zeros of multiplicity 1. A third zero is outside the viewing window.
Strategy	Develop a test for finding rational zeros.	Develop tests for predicting the number and size of zeros.

The Rational Zero Theorem

Suppose that we are interested in finding only the rational zeros of a quadratic polynomial $f(x) = ax^2 + bx + c$, where the coefficients a, b, and c are integers. Now, because any rational number can be written as a fraction p/q in lowest terms, we will search for zeros of the form p/q, where p and q have no common factors other than 1. Since $f(p/q) = 0$, we have

$$a\left(\frac{p}{q}\right)^2 + b\left(\frac{p}{q}\right) + c = 0 \qquad \text{Definition of a zero}$$

$$ap^2 + bpq + cq^2 = 0 \qquad \text{Multiplying through by } q^2$$

$$ap^2 + bpq = -cq^2 \qquad \text{Subtracting } cq^2 \text{ on both sides}$$

$$p(ap + bq) = -cq^2 \qquad \text{Factoring out } p \text{ on the left}$$

The last step shows that p is a factor of the left-hand side, and so p must be a factor of the right-hand side as well. But since p and q have no common factors other than 1, p must be a factor of c. In a similar fashion, we can show that q must be a factor of a. This narrows our search considerably: The only possible rational zeros of $f(x) = ax^2 + bx + c$ are those of the form p/q, where p divides into the constant term c, and q divides into the leading coefficient a. The Rational Zero Theorem is a generalization of these results for polynomials of nth degree.

Rational Zero Theorem

If the polynomial $f(x) = a_n x^n + a_{n-1} x^{n-1} + \cdots + a_1 x + a_0$ has integer coefficients, then every rational zero of f has the form p/q where p and q have no common factors other than 1, p is a factor of the constant term a_0, and q is a factor of the leading coefficient a_n.

According to the Rational Zero Theorem, if we find all the factors of the constant term and the leading coefficient and form a list of the rational numbers of the form

$$\frac{\text{Factor of the constant term}}{\text{Factor of the leading coefficient}}$$

then all the rational zeros of the polynomial will be in the list. Trial and error (perhaps with the assistance of a graphing calculator) will lead us to the actual rational zeros.

EXAMPLE 1

Finding Rational Zeros

Use the Rational Zero Theorem to find the rational zeros of

$$f(x) = 27x^3 - 27x^2 - 18x + 8.$$

Solution First we list all the factors of the constant term and leading coefficient.

Factors of the constant term 8: $\quad \pm 1, \pm 2, \pm 4, \pm 8$
Factors of the leading coefficient 27: $\quad \pm 1, \pm 3, \pm 9, \pm 27$

Next we form all quotients of the factors of 8 divided by the factors of 27.

$$\text{Possible rational zeros:} \quad \pm 1, \pm\frac{1}{3}, \pm\frac{1}{9}, \pm\frac{1}{27}$$

$$\pm 2, \pm\frac{2}{3}, \pm\frac{2}{9}, \pm\frac{2}{27}$$

$$\pm 4, \pm\frac{4}{3}, \pm\frac{4}{9}, \pm\frac{4}{27}$$

$$\pm 8, \pm\frac{8}{3}, \pm\frac{8}{9}, \pm\frac{8}{27}$$

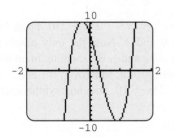

Figure 49

Since there are so many possible zeros, naive trial and error will likely be quite inefficient. Instead, we use information from the graph of f to help narrow down the list. From the graph in Figure 49, we can rule out integer zeros. Moreover, the middle zero appears to be approximately $\frac{1}{3}$, which is in our list of possible zeros. Thus, we begin by testing $\frac{1}{3}$ with synthetic division.

$$\begin{array}{r|rrrr} \frac{1}{3} & 27 & -27 & -18 & 8 \\ & & 9 & -6 & -8 \\ \hline & 27 & -18 & -24 & 0 \end{array}$$

The remainder 0 tells us that $x = \frac{1}{3}$ is indeed a zero. This fact, together with the first three numbers in the bottom row, indicates that two of the factors of $f(x)$ are $x - \frac{1}{3}$ and $27x^2 - 18x - 24$. Furthermore,

$$27x^2 - 18x - 24 = 3(9x^2 - 6x - 8) = 3(3x + 2)(3x - 4)$$

Thus, the other two zeros occur when $3(3x + 2)(3x - 4) = 0$—namely, when $x = -\frac{2}{3}$ and $x = \frac{4}{3}$. So the three zeros of f are $-\frac{2}{3}, \frac{1}{3},$ and $\frac{4}{3}$.

WARNING!

> The Rational Zero Theorem can be applied only to polynomials with integer coefficients. Also, the Rational Zero Theorem only produces *possible* zeros. In fact, many polynomials with integer coefficients have no rational zeros (see Example 6).

⋯⋗EXAMPLE 2

Finding the Zeros of a Polynomial

Find all the zeros of $f(x) = x^4 + \frac{3}{5}x^3 - \frac{27}{5}x^2 - 3x + 2$ and write $f(x)$ as a product of linear factors.

Solution   Notice that f has noninteger coefficients and so we cannot apply the Rational Zero Theorem to f. However, the zeros of f coincide with the solutions of $x^4 + \frac{3}{5}x^3 - \frac{27}{5}x^2 - 3x + 2 = 0$, and we can clear the fractions in this equation by multiplying through by 5 to obtain $5x^4 + 3x^3 - 27x^2 - 15x + 10 = 0$. Thus, we will apply the Rational Zero Theorem to the polynomial function $g(x) = 5x^4 + 3x^3 - 27x^2 - 15x + 10$ (note that f and g are *different* polynomials— they just have the same zeros).

Factors of the constant term 10: $\pm 1, \pm 2, \pm 5, \pm 10$

Factors of the leading coefficient 5: $\pm 1, \pm 5$

Possible rational zeros: $\pm 1, \pm\frac{1}{5}, \pm 2, \pm\frac{2}{5}, \pm 5, \pm 10$

These are the only possible rational zeros of g and so also of f. To narrow down the list, we plot the graph of g. It appears in Figure 50 that g has a zero at $x = -1$, and we can easily test this using synthetic division.

$$\begin{array}{r|rrrrr} -1 & 5 & 3 & -27 & -15 & 10 \\ & & -5 & 2 & 25 & -10 \\ \hline & 5 & -2 & -25 & 10 & 0 \end{array}$$

So $x = -1$ is a zero and $g(x)$ factors as $(x + 1)(5x^3 - 2x^2 - 25x + 10)$. From the graph, we see that there is another zero between 0 and 1. Now the only numbers between 0 and 1 in our list of possible rational zeros are $\frac{1}{5}$ and $\frac{2}{5}$. By zooming in on this positive zero, as shown in Figure 51, we can rule out $\frac{1}{5} = 0.2$. Moreover, if $\frac{2}{5}$ is a zero of $g(x)$, then it must also be a zero of $5x^3 - 2x^2 - 25x + 10$. Dividing synthetically, we obtain

$$\begin{array}{r|rrrr} \frac{2}{5} & 5 & -2 & -25 & 10 \\ & & 2 & 0 & -10 \\ \hline & 5 & 0 & -25 & 0 \end{array}$$

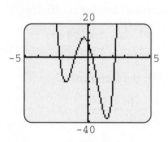

Figure 50

X=.39893617   Y=.0360215

Figure 51

So $\frac{2}{5}$ is indeed a zero. The synthetic division for $\frac{2}{5}$ also tells us that another factor of $g(x)$ is $5x^2 - 25 = 5(x^2 - 5)$. This factor has two irrational zeros—namely, $x = \pm\sqrt{5}$. Thus, the four zeros of g (and f also) are $x = -1$, $x = \frac{2}{5}$, $x = -\sqrt{5}$, and $x = \sqrt{5}$. This gives us

$$f(x) = a(x + 1)\left(x - \frac{2}{5}\right)(x + \sqrt{5})(x - \sqrt{5})$$

Since $f(x) = x^4 + \frac{3}{5}x^3 - \frac{27}{5}x^2 - 3x + 2$ has a leading coefficient of 1, it follows that $a = 1$, which gives us

$$f(x) = (x + 1)\left(x - \frac{2}{5}\right)(x + \sqrt{5})(x - \sqrt{5})$$

Descartes' Rule of Signs

Our next test gives us information about the number of real zeros of a polynomial. In Section 3.3, we saw that a polynomial of degree n will have n zeros. However, because of possible multiplicities and complex zeros, an nth-degree polynomial may have fewer than n distinct real zeros. **Descartes' Rule of Signs** gives us more information about the number of real zeros by considering the number of times that successive coefficients of the polynomial change from positive to negative or negative to positive. These changes are referred to as **variations in sign**. The following table gives several examples.

Polynomial	Coefficient signs	Variations in sign
$x^3 + 2x + 5$	$+\ +\ +$	0
$2x^3 - x^2 - 1$	$+\ -\ -$	1
$5x^3 - 3x^2 + x + 4$	$+\ -\ +\ +$	2

Descartes' Rule of Signs

Let $f(x) = a_n x^n + a_{n-1} x^{n-1} + \cdots + a_1 x + a_0$ be a polynomial with real coefficients.

1. The number of positive real zeros (counting multiplicities) is either equal to the number of variations in sign of $f(x)$ or is less than that number by an even integer.
2. The number of negative real zeros (counting multiplicities) is either equal to the number of variations in sign of $f(-x)$ or is less than that number by an even integer.

EXAMPLE 3

Using Descartes' Rule of Signs

Apply Descartes' Rule of Signs to the polynomial $f(x) = 2x^3 - x^2 - 1$.

Solution   Since $f(x)$ has only one variation in sign, f must have exactly one positive real zero. To test for negative zeros, we first simplify $f(-x)$.

$$f(-x) = 2(-x)^3 - (-x)^2 - 1 = -2x^3 - x^2 - 1$$

Since $f(-x)$ has no variations in sign, we conclude that f has no negative zeros. The graph of f in Figure 52 shows one zero near 1. Because of Descartes' Rule of Signs, we

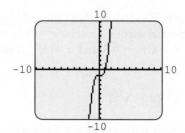

Figure 52

can be certain that there are no other real zeros. (Note that if we had just used the graph and not Descartes' Rule of Signs, we could not rule out the presence of zeros outside the viewing window.)

Since Descartes' Rule of Signs does not "pin down" the number of zeros unless there is only one variation in sign or none at all, it is usually best to complement the information given by Descartes' Rule of Signs by graphing the polynomial with a graphing calculator.

EXAMPLE 4

Determining the Number of Real Zeros of a Polynomial

Apply Descartes' Rule of Signs to $f(x) = x^3 - 2x^2 + x + 2$.

Solution   Since $f(x)$ has two variations in sign, f has either two or no positive real zeros. Moreover, since

$$f(-x) = -x^3 - 2x^2 - x + 2$$

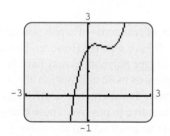

Figure 53

has only one variation in sign, f has one negative zero. From the graph of f in Figure 53, we see one negative zero, no positive zeros, and two turning points. Because f is a cubic polynomial, and cubic polynomials have at most two turning points, f has no additional turning points. Thus, the graph of f will continue to rise as x gets larger, and f has no additional real zeros.

In the previous example, Descartes' Rule of Signs was superfluous. All the pertinent information could be obtained directly from the graph. This is not always the case, as the following examples illustrate.

EXAMPLE 5

Finding the Zeros of a Polynomial

Find the zeros of $f(x) = 6x^3 - 89x^2 + 216x - 144$.

Solution   The graph of f is shown in Figure 54. At first glance it appears that f has only one zero of multiplicity 2—or perhaps two distinct zeros very close together—somewhere in the interval $[1, 2]$. However, since $f(x)$ has three variations in sign, Descartes' Rule of Signs tells us that f has either one or three positive zeros (counting multiplicities), and so a single positive zero of multiplicity 2 is impossible, as is the presence of exactly two positive zeros very close together. Thus, there must be a zero to the right of the viewing window. After some experimentation, we settle on the view shown in Figure 55. This suggests another zero at $x = 12$, which can be confirmed by synthetic division.

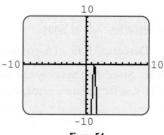

Figure 54

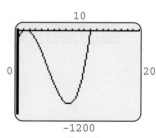

Figure 55

$$\begin{array}{r|rrrr} 12 & 6 & -89 & 216 & -144 \\ & & 72 & -204 & 144 \\ \hline & 6 & -17 & 12 & 0 \end{array}$$

Because the remainder is 0, we know that $f(x) = (x - 12)(6x^2 - 17x + 12)$. Moreover, $6x^2 - 17x + 12 = (2x - 3)(3x - 4)$, and so f has two more zeros at $x = \frac{3}{2}$ and $x = \frac{4}{3}$.

It is not uncommon for a polynomial to have no rational zeros, and in this case it is very difficult—perhaps impossible—to obtain exact values for the zeros. Fortunately, we can approximate irrational zeros with a graphing calculator.

EXAMPLE 6

Finding the Zeros of a Polynomial

Show that $f(x) = x^3 + x - 1$ has exactly one real zero, which is irrational, and use a graphing calculator to approximate it to the nearest hundredth.

Solution   One real zero is readily apparent from the x-intercept of the graph of f shown in Figure 56. However, f could have as many as three real zeros because it is a polynomial of degree 3. To investigate the number of real zeros, we apply Descartes' Rule of Signs. Since $f(x)$ has only one variation in sign, f must have one positive zero. Since $f(-x) = -x^3 - x - 1$ has no variations in sign, f has no negative zeros. Thus, f has precisely one real zero. To see if a rational value can be found, we apply the Rational Zero Theorem.

Factors of the constant term -1:	± 1
Factors of the leading coefficient 1:	± 1
Possible rational zeros:	± 1

But $f(-1) = -3$ and $f(1) = 1$, so neither 1 nor -1 is a zero. Thus, f has no rational zeros. To approximate the irrational zero, we can either zoom in and use the trace feature or apply the calculate-zero feature to obtain $x \approx 0.68$.

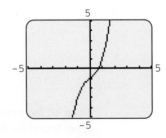

Figure 56

Upper and Lower Bounds Test   Our final test helps us determine upper and lower bounds for the zeros of a polynomial. We say that a real number b is an **upper bound** for the real zeros of a polynomial if none of the zeros is greater than b. Similarly, a number b is a **lower bound** if none of the zeros is less than b. Suppose we suspect that a number b is an upper (or lower) bound for the real zeros of a polynomial f. How can we determine this for sure? The graph of f can be misleading because of the possibility that a zero lies outside the viewing window. Instead, we can apply the following test, the proof of which is beyond the scope of this text.

Upper and Lower Bounds Test

Let $f(x) = a_n x^n + a_{n-1} x^{n-1} + \cdots + a_1 x + a_0$ be a polynomial with real coefficients and positive leading coefficient a_n.

1. If $b > 0$ and each number in the last row of the synthetic division of $f(x)$ by $x - b$ is either positive or zero, then b is an upper bound for the real zeros of f.
2. If $b < 0$ and the numbers in the last row of the synthetic division of $f(x)$ by $x - b$ alternate in sign (with 0 counting either as positive or negative), then b is a lower bound for the real zeros of f.

 EXAMPLE 7

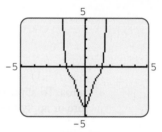

Determining the Number of Zeros of a Polynomial

Use the Upper and Lower Bounds Test to help you determine the number of zeros of the function $f(x) = x^6 - 4x^4 + 6x^2 - 4$.

Solution   The graph of f in Figure 57 shows two zeros. However, since f has degree 6, there could be as many as six zeros. Moreover, all we can tell from Descartes' Rule of Signs is that, since $f(x)$ and $f(-x)$ both have three variations in sign, f could have either one or three positive zeros and either one or three negative zeros.

Because the graph suggests that there are no zeros greater than 2, we apply the Upper and Lower Bounds Test to see if $x = 2$ is an upper bound for the real zeros of f.

Figure 57

$$
\begin{array}{r|rrrrrr}
2 & 1 & 0 & -4 & 0 & 6 & 0 & -4 \\
  &   & 2 & 4 & 0 & 0 & 12 & 24 \\
\hline
  & 1 & 2 & 0 & 0 & 6 & 12 & 20
\end{array}
$$

Since all the numbers in the last row are positive or zero, we know that $x = 2$ is an upper bound on the real zeros of f—there are no zeros to the right of the viewing window. Next we test to see if $x = -2$ is a lower bound for the real zeros of f.

$$
\begin{array}{r|rrrrrrr}
-2 & 1 & 0 & -4 & 0 & 6 & 0 & -4 \\
   &   & -2 & 4 & 0 & 0 & -12 & 24 \\
\hline
   & 1 & -2 & 0 & 0 & 6 & -12 & 20 \\
   & + & - & + & - & + & - & +
\end{array}
$$

Because the numbers in the last row alternate in sign (note how 0 is first counted as positive and then as negative), $x = -2$ is indeed a lower bound. Thus, f has only the two real zeros shown in Figure 57.

Understanding and Mastery Checklists

Concepts to Understand

The Rational Zero Theorem

٥

Descartes' Rule of Signs

٥

Upper and Lower Bounds Test

Skills to Master

Use the Rational Zero Theorem to find the rational zeros of a polynomial.

٥

Apply Descartes' Rule of Signs to a polynomial to obtain information about the number of zeros.

٥

Apply the Upper and Lower Bounds Test to a polynomial to obtain information about its zeros.

Exercises 3.4

Exercises 1-10 *Use the Rational Zero Theorem to list all the possible rational zeros of the function. Use synthetic division and/or a graphing calculator to help you determine which are actually zeros.*

1. $f(x) = x^4 - 12x^2 + 27$

2. $f(x) = x^3 + 3x^2 - 2x - 6$

3. $g(x) = x^3 - 3x^2 - 4x + 12$

4. $h(x) = x^4 - 17x^2 + 16$

5. $f(x) = 4x^4 - 4x^3 - 9x^2 + x + 2$

6. $h(x) = 9x^3 + 9x^2 - 16x + 4$

7. $g(x) = 2x^5 + 3x^4 - 2x - 3$

8. $h(x) = 25x^5 - 4x^3 + 25x^2 - 4$

9. $f(x) = x^3 - \frac{9}{2}x^2 + \frac{1}{2}x + 6$

10. $g(x) = x^4 - \frac{10}{3}x^3 - 12x^2 + \frac{58}{3}x - 5$

Exercises 11-20 *Apply Descartes' Rule of Signs to the given function. Then use a graphing calculator to determine the precise number of positive and negative zeros.*

11. $f(x) = x^3 + 4$

12. $h(x) = x^4 - 3x^2 - 1$

13. $g(x) = 2x^4 + x^2 + 3$

14. $g(x) = 5x^3 - x^2 - 1$

15. $f(x) = x^3 - 4x^2 + 7x - 2$

16. $h(x) = 3x^3 + x^2 + 2x + 6$

17. $f(x) = x^5 - 10x^4 - 11x^3 - 5$

18. $g(x) = x^4 + 11x^3 - 12x^2 + 3$

19. $h(x) = x^3 - x^2 - \frac{101}{100}x + \frac{99}{100}$

20. $f(x) = x^3 + \frac{1}{100}x^2 - \frac{301}{100}x - \frac{101}{50}$

Exercises 21-26 *Use the Upper and Lower Bounds Test to confirm the given bounds for the zeros of the function.*

21. $f(x) = x^4 - 5x^3 - 11x^2 + 33x - 18$
Lower: -4; Upper: 7

22. $f(x) = x^4 + 3x^3 - 27x^2 + 3x - 28$
Lower: -8; Upper: 5

23. $f(x) = x^4 - 10x^3 - 5$
Lower: -1; Upper: 11

24. $f(x) = x^4 + 11x^3 - 12x^2 + 6$
Lower: -13; Upper: 1

25. $f(x) = x^4 - 62x^3 + 962x^2 - 62x + 960$
Lower: -1; Upper: 62

26. $f(x) = x^4 - 3024x^2 - 3024$
Lower: -56; Upper: 55

Exercises 27-36   *Find the exact values of the real zeros of the function. Use any of the tools described in this chapter.*

27. $h(x) = x^3 + 6x^2 - x - 6$

28. $g(x) = x^3 - 13x - 12$

29. $f(x) = x^3 - 14x^2 + 25x - 12$

30. $g(x) = x^3 + 11x^2 + 2x + 22$

31. $h(x) = x^3 + \dfrac{7}{3}x^2 - \dfrac{23}{12}x + \dfrac{1}{4}$

32. $g(x) = x^3 - \dfrac{7}{12}x^2 - \dfrac{7}{8}x - \dfrac{1}{6}$

33. $h(x) = 8x^3 + x^2 - 16x - 2$

34. $f(x) = 4x^3 - 9x^2 - 6x + 2$

35. $g(x) = x^4 + x^3 - 120x^2 - 121x - 121$

36. $h(x) = x^4 + 26x^3 + 170x^2 + 26x + 169$

Exercises 37-42   *Find the exact values of the real solutions of the polynomial equation.*

37. $x^3 + x^2 - 4x - 4 = 0$

38. $x^3 + 3x^2 + 2x + 6 = 0$

39. $x^4 - 27 = 6x^2$

40. $2x^4 - 21x^2 - 5 = 5x^3 + 19x$

41. $8x^5 - 8x^3 = 1 - x^2$

42. $x^5 + 18x = 11x^3$

Exercises 43-48   *Use the Rational Zero Theorem to show that the function has no rational zeros. Use a graphing calculator to approximate any irrational zeros to the nearest hundredth.*

43. $f(x) = x^3 + 3x + 1$

44. $g(x) = x^3 - x^2 - 2$

45. $h(x) = x^4 + 2x - 2$

46. $g(x) = x^4 - x^3 - 1$

47. $h(x) = x^5 - 10x^4 - 11x^3 - 5$

48. $f(x) = x^4 + 11x^3 - 12x^2 + 3$

Applications

49. Dimensions of a Box   A box with a square base is to be constructed so that its height is 1 inch more than twice its base length. Find the dimensions of the box if the volume must be 9 cubic inches.

50. Dimensions of a Box   A box with a square base is to be constructed so that its height is 1 inch less than three times its base length. Find the dimensions of the box if the volume must be 100 cubic inches.

51. Makeshift Box   A box with no top is formed by cutting squares out of the corners of a $10'' \times 5''$ rectangular piece of cardboard and then folding up the sides (see Figure 58). What size square must be cut out of each corner if the resulting box is to have a volume of 18 cubic inches?

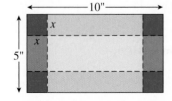

Figure 58

52. Postal Regulations   A rectangular package to be sent by the U.S. Postal Service can have a maximum combined length and girth (perimeter of the base) of 108 inches (see Figure 59). Find the dimensions of a box with a square base and volume 10,800 cubic inches if the combined length and girth is exactly 108 inches.

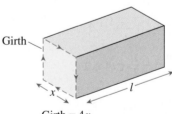

Girth $= 4x$
Length + Girth $= l + 4x$

Figure 59

53. Poverty Percentage   The percentage of U.S. families below the poverty level for the years 1960–2000 can be modeled by the polynomial

$$p(x) = -0.001x^3 + 0.07x^2 - 1.4x + 17.5$$

where x is the year (with $x = 0$ corresponding to 1960) and $p(x)$ is the percentage of families below the poverty level in that year. According to this model, in what year(s) were 10.3% of U.S. families below the poverty level?

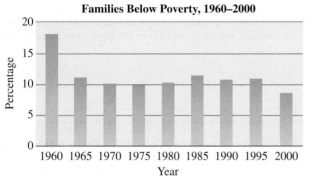

Families Below Poverty, 1960–2000

Data source: U.S. Bureau of the Census.

54. Sunday Accidents Of the total number of auto accidents that will occur on a given Sunday, the percentage that will have occurred by hour x (with $x = 0$ corresponding to midnight) can be modeled by the polynomial function

$$p(x) = -0.0015x^4 + 0.065x^3 - 0.75x^2 + 5.4x$$

for x between 0 and 24. According to this model, by what time(s) had half of the Sunday accidents occurred? During what Sunday hour do the greatest number of accidents occur?

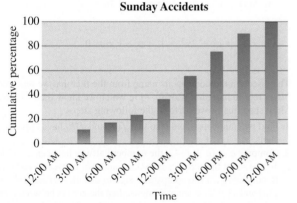

Sunday Accidents

Data source: Traffic Safety Facts 2000, U.S. Department of Transportation.

Exercises 55-58 *The **present value** of a payment P made t years from now is defined to be $P(1 + r)^{-t}$, which is the amount that, if*

invested now, would grow to P in t years. For convenience, we define $v = 1/(1 + r)$, so that the present value of a payment P made t years from now is Pv^t. The present value of a series of payments is simply the sum of the present values of each of the payments. To compute the interest rate(s) corresponding to a loan repayment schedule, we equate the present value of all payments made to that of all distributions received.

55. Installment Loan A loan of \$3000 is received at time 0, and three payments of \$1100 are made at times 1, 2, and 3.

    a. Find an expression (involving v) for the present value of the payments.

    b. Set the present value of the payments equal to \$3000 (the present value of the loan distribution) and apply Descartes' Rule of Signs to show that there is a unique positive value for v and, hence, a unique interest rate r.

    c. Use a graphing calculator to approximate the value of v, and use $v = 1/(1 + r)$ to find the corresponding interest rate r.

56. Mortgage Loan A mortgage loan of L dollars is to be paid off in n monthly installments of p dollars (assume $np > L$). Use Descartes' Rule of Signs to show that there is a unique positive value for v and, hence, a unique positive interest rate.

57. Line-of-Credit Loan A line-of-credit loan is initiated with a distribution of \$1000 at time 0, payments at times 1 and 2, a second distribution at time 3, and a final payment at time 4.

    a. Denote the unspecified payments by p_1, p_2, and p_3, and find an expression (involving v) for the present value of the payments.

    b. Denote the unspecified distribution by d, and find an expression (involving v) for the present value of the distributions.

    c. Equate the expressions found in parts a and b, and apply Descartes' Rule of Signs to determine the number of possible interest rates.

    d. Use a graphing calculator to approximate the interest rate(s) if the payments were all \$400 and the second distribution was \$500.

58. Line-of-Credit Loan A line-of-credit loan is initiated with a payment of \$25 at time 0, a distribution of \$50 at time 1, and a payment of \$26 at time 3. Show that two different positive interest rates result.

Concepts and Critical Thinking

Exercises 59-62 *Answer true or false.*

59. A polynomial of the form $f(x) = x^3 + ax^2 + bx + 1$ (where a and b are integers) can have at most 2 rational zeros.

60. According to Descartes' Rule of Signs, the number of positive zeros of a polynomial $f(x)$ is equal to the number of variations in sign of $f(x)$.

61. It is possible that $\frac{1}{3}$ is a zero of a polynomial of the form $f(x) = x^4 + ax^2 + x + 3$, where a is an integer.

62. The Upper and Lower Bounds Test enables us to narrow our search for real zeros of a polynomial.

Exercises 63-66   *Give an example of each.*

63. A polynomial that, according to Descartes' Rule of Signs, has either 1, 3, or 5 positive zeros

64. A fourth-degree polynomial that, according to Descartes' Rule of Signs, has either 2 or 0 negative zeros

65. A fifth-degree polynomial that, according to the Rational Zero Theorem, has ± 1 as its only possible rational zeros

66. A fourth-degree polynomial with five terms that, according to Descartes' Rule of Signs, has no positive zeros

67. Apply Descartes' Rule of Signs to $P(x) = x^{20} -$ (Your age in seconds)$x^{19} +$ (Your weight in kilograms)$x^5 + 2x + 2.7$.

68. Explain why any polynomial $P(x)$ with integer coefficients, leading coefficient 1, and satisfying $P(0) = 149$ has at most 2 rational zeros.

Questions for Discussion or Essay

69. According to the Rational Zero Theorem, if p/q is a zero of a polynomial $f(x)$, then p must be a factor of the constant term and q must be a factor of the leading coefficient. What can be said if the constant term is zero? What if the leading coefficient is 1? Explain how the Rational Zero Theorem can be used to show that $\sqrt{2}$ is an irrational number.

70. For what types of polynomials is Descartes' Rule of Signs most useful? Give some examples to help support your claim.

71. In light of all the other procedures we've seen for obtaining information about the zeros of a polynomial, discuss the usefulness of the Upper and Lower Bounds Test.

Projects for Enrichment

72. Proof of the Rational Zero Theorem

a. Prior to stating the Rational Zero Theorem, we showed that if p/q is a zero of $f(x) = ax^2 + bx + c$ and p and q have no common factors, then p must be a factor of c. Show that q must be a factor of a as well.

b. Generalize the technique used in the text for quadratic polynomials to show that if p and q have no common factors and p/q is a zero of
$$f(x) = a_n x^n + a_{n-1}x^{n-1} + \cdots + a_1 x + a_0$$
then p is a factor of a_0.

c. Generalize the technique used in part a to show that if p and q have no common factors and p/q is a zero of
$$f(x) = a_n x^n + a_{n-1}x^{n-1} + \cdots + a_1 x + a_0$$
then q is a factor of a_n.

73. The Bisection Method   There often is an easy way to show that a polynomial f has a zero between two numbers a and b. We simply compute $f(a)$ and $f(b)$ and check to see if one output value is negative and the other positive. If so, then there is a zero between a and b. For example, we can be certain that $f(x) = x^3 + x^2 - 4$ has a zero between 1 and 2 since $f(1) = -2 < 0$ and $f(2) = 8 > 0$.

a. Show that the given polynomial has a zero between a and b.

   i. $g(x) = x^3 + x - 3; a = 1, b = 2$

   ii. $f(x) = -x^3 + x^2 - 2x + 9; a = 2, b = 3$

   iii. $h(x) = x^4 - 9x^2 + x + 4; a = 2, b = 3$

   iv. $g(x) = x^4 - 8x + 2; a = 0, b = 1$

   v. $h(x) = x^5 + 4x^2 + 3; a = -2, b = -1$

   vi. $f(x) = x^5 + 2x^3 + 1; a = -1, b = 0$

The bisection method gets its name from the fact that we repeatedly *bisect* (cut in half) intervals $[a, b]$ in which we know there is a zero. More specifically, we perform the following steps:

1. Find two numbers a and b so that $f(a)$ and $f(b)$ have opposite signs.

2. Compute $c = (a + b)/2$, the number halfway between a and b.

3. Test to see if the zero lies between a and c or between c and b. That is, compute $f(c)$ and compare its sign to that of $f(a)$ and $f(b)$. If $f(a)$ and $f(c)$ have opposite signs, then the zero is between a and c. If $f(c)$ and $f(b)$ have opposite signs, then the zero is between c and b.

4. Rename a and b so they denote the two numbers (either the old a and c or the old c and b) between which the zero lies.

5. Repeat steps 2, 3, and 4 until the zero is approximated to the desired accuracy.

Consider the function $f(x) = x^3 + x^2 - 4$. As we saw earlier, f has a zero between $a = 1$ and $b = 2$. Thus, we set $c = 1.5$. Since $f(1) = -2$, $f(1.5) = 1.625$, and $f(2) = 8$, we know there is a zero between 1 and 1.5. So we set $a = 1$ and $b = 1.5$ and repeat the procedure.

b. Continue the process to estimate the zero of $f(x) = x^3 + x^2 - 4$ that lies between 1 and 2. Stop when a and b agree to two decimal places.

c. You are told that a certain polynomial function has an x-intercept somewhere between 1 and 100. Explain how you could evaluate the function at 20 points and, based on this information, determine the x-intercept to within three decimal places of accuracy.

d. *For students with programming experience* The bisection method is ideally suited for a computer or programmable calculator. The following pseudocode suggests how the program might be written. Note that the program requests values for a and b and also for the desired accuracy e. When the desired accuracy is reached, the program displays the approximate zero.

INPUT a Input the left endpoint.
INPUT b Input the right endpoint.
IF $f(a)*f(b) > 0$ THEN   If $f(a)$ and $f(b)$ don't have opposite
    OUTPUT "$f(a)$ and $f(b)$ signs, there may not be a zero
    must have opposite sign"   between a and b, so we should stop.
    STOP
ENDIF

INPUT e Input the desired accuracy.
WHILE $b - a > e$ DO   The program stops when $b - a \le e$.
    $(a + b)/2 \to c$ Let c be the point halfway between a and b.
    IF $f(a)*f(c) < 0$ THEN   If $f(a)$ and $f(c)$ are opposite in sign:
        $c \to b$ let c be the new right endpoint;
    ELSE   otherwise:
        $c \to a$ let c be the new left endpoint.
    ENDIF
ENDWHILE
OUTPUT $(a + b)/2$ The approximate zero is $(a + b)/2$

Write a program for your graphing calculator that implements the bisection method. Test the program on the function $f(x) = x^3 + x^2 - 4$ using $a = 1$, $b = 2$, and $e = 0.0001$. You should get 1.3146 as an approximate answer. Once the program is working correctly, use it to approximate the zeros of the functions in part a to within $e = 0.000001$.

Section 3.5   Rational Functions

- Which rational functions have graphs that resemble lines?
- How much more might a barrel of oil cost if refineries were required to reduce their discharge by 100%?
- What percentage of U.S. males between the ages of 18 and 24 are 6'3"tall to the nearest inch?
- What is the relationship between education and unemployment?
- How can the inventory costs of a small business be minimized?

A **rational function** is a function of the form

$$f(x) = \frac{p(x)}{q(x)}$$

where p and q are polynomials. Examples include

$$g(x) = \frac{2x + 3}{x^2 - 4} \quad \text{and} \quad h(x) = \frac{2}{x - 5}$$

as well as all polynomial functions, such as $f(x) = x^2 + 3x$, where the denominator is assumed to be 1. In this section, we consider only rational functions *in lowest terms*—that is, rational functions $f(x) = p(x)/q(x)$, for which $p(x)$ and $q(x)$ have no common factors. See Exercise 66 for a discussion of rational functions that are not in lowest terms.

Domain and Zeros

Since a rational function is a ratio of polynomials, its domain is found by excluding the zeros of the polynomial in the denominator, and its zeros correspond to the zeros of the polynomial in the numerator.

Domain of a Rational Function

The domain of a rational function $f(x) = p(x)/q(x)$ consists of all real numbers x such that $q(x) \neq 0$.

Zeros of a Rational Function

If $f(x) = p(x)/q(x)$ is a rational function in lowest terms, then $f(x) = 0$ if and only if $p(x) = 0$.

EXAMPLE 1

Finding the Domain and Zeros of a Rational Function

Find the domain and zeros of the rational function $f(x) = \dfrac{3x + 6}{x + 1}$.

Solution The values that must be excluded from the domain can be found by locating the zeros of the denominator:

$$x + 1 = 0$$
$$x = -1$$

Thus, the only zero of the denominator is $x = -1$, and so the domain of f is the set of all real numbers except -1. To find the zeros of f, we simply find the zeros of the numerator.

$$3x + 6 = 0$$
$$3x = -6$$
$$x = -2$$

Thus, -2 is the only zero of f.

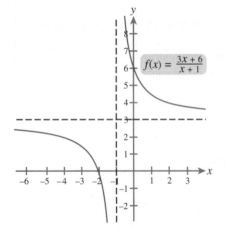

$$f(x) = \frac{3x + 6}{x + 1}$$

Figure 60

As we saw in Section 3.1, graphs of polynomial functions are always smooth, unbroken curves that, in the long run, either rise to infinity or sink to negative infinity. With rational functions, there is far more variation. The graph of a rational function can abruptly break, forming a steep, nearly vertical cliff as it approaches a *vertical asymptote*. Its end behavior might be like that of a polynomial—approaching positive or negative infinity on the far left and far right—or, alternatively, its graph might level off, gradually approaching a *horizontal asymptote*. For example, consider the graph of the function $f(x) = (3x + 6)/(x + 1)$ from Example 1, as shown in Figure 60. In particular, notice the break in the graph across the line $x = -1$ (this is a vertical asymptote). Notice also that the graph of f levels off along the horizontal line $y = 3$ as x approaches $\pm\infty$ (this is a horizontal asymptote). The remainder of this section is devoted to investigating these features of rational functions in more detail.

Calculator Keys

Rational Functions

Your graphing calculator may occasionally connect pieces of the graph of a rational function that should not be connected. An example of this phenomenon is shown in Figure 61, where the graph of the function $f(x) = (3x + 6)/(x + 1)$ is shown. As we can see in Figure 60, the graph of f should approach the line $x = -1$, but it should not cross it. Unfortunately, your calculator does not realize this—it simply plots points and connects them with line segments. If two adjacent points happen to be on opposite sides of the line $x = -1$, they will be connected. This is what occurred in Figure 61. One way to avoid this problem is to change your calculator to dot mode. In this mode, your calculator will plot points, but it will not connect them. The graphs of f in Figures 62 and 63 were done in dot mode.

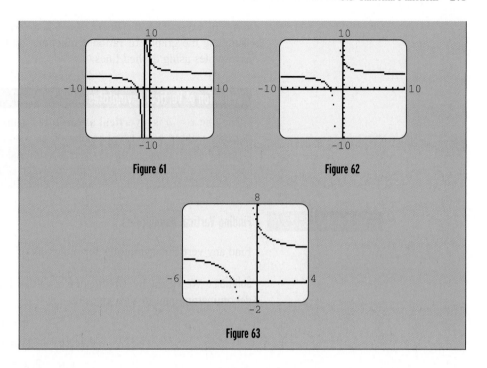

Figure 61   Figure 62

Figure 63

Vertical Asymptotes

Let's take a closer look at the breaks that sometimes occur in the graphs of rational functions. In general, the graph of a rational function will be broken at an x-value for which the function is not defined. In other words, there will be a break in the graph of the rational function $f(x) = p(x)/q(x)$ at every zero of q. For example, consider again the function

$$f(x) = \frac{3x + 6}{x + 1}$$

from Example 1. Since the denominator has a zero of -1, it follows that the graph of f will break at $x = -1$. Let's investigate the behavior of $f(x)$ near this x-value.

We begin by forming a table of function values $f(x)$ for x near -1.

x approaches -1 from the right:

x	-0.9	-0.99	-0.999
$f(x) = \dfrac{3x + 6}{x + 1}$	33	303	3003

x approaches -1 from the left:

x	-1.1	-1.01	-1.001
$f(x) = \dfrac{3x + 6}{x + 1}$	-27	-297	-2997

Our work suggests that as x approaches -1 from the right, $f(x)$ approaches ∞, whereas as x approaches -1 from the left, $f(x)$ approaches $-\infty$. It follows that the graph of f will become nearly vertical near -1, shooting up on one side and down on the other, as we have already seen in Figures 60–63. Because the graph approaches the vertical line

$x = -1$, we say that the line $x = -1$ is a *vertical asymptote* for the graph of f. When sketching the graphs of rational functions by hand, it is customary to depict vertical asymptotes using dashed lines.

Definition of Vertical Asymptotes

The line $x = a$ is a **vertical asymptote** for the graph of a rational function f if $f(x)$ grows without bound (approaches ∞ or $-\infty$) as x approaches a from the right and from the left. If $f(x) = p(x)/q(x)$ is a rational function in lowest terms, then the line $x = a$ is a vertical asymptote for the graph of f if and only if $q(a) = 0$.

EXAMPLE 2

Finding Vertical Asymptotes

Find any vertical asymptotes for the graph of $f(x) = \dfrac{6}{x^2 - 4}$.

Solution   We begin by finding the x-values for which f is not defined. Setting the denominator equal to zero gives us

$$x^2 - 4 = 0$$
$$(x + 2)(x - 2) = 0$$
$$x = -2, x = 2$$

Thus, the graph of f will have vertical asymptotes at $x = -2$ and $x = 2$. Figure 64 shows the graph of f as depicted by a graphing calculator. Using this graph as a guide, we've sketched the graph of f shown in Figure 65.

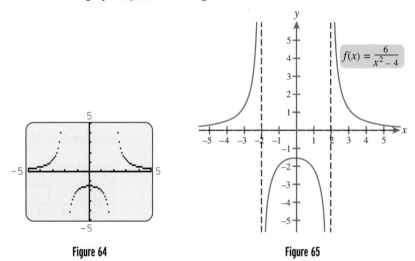

Figure 64   Figure 65

Horizontal Asymptotes   As we have already seen, the graph of a rational function may eventually level off, approaching a horizontal line called a *horizontal asymptote*. More precisely, we have the following definition.

Definition of a Horizontal Asymptote

The line $y = b$ is a **horizontal asymptote** for the graph of a rational function f if $f(x)$ approaches b as x approaches ∞ or $-\infty$.

To approximate a horizontal asymptote graphically, we select a viewing window that reveals the end behavior of the graph. Then, if the graph appears to be horizontal, we can use the trace feature to estimate the limiting value b. We illustrate this technique with the function from Example 1.

EXAMPLE 3 **Finding a Horizontal Asymptote Graphically**

Use a graphing calculator to estimate the horizontal asymptote (if any) of

$$f(x) = \frac{3x + 6}{x + 1}$$

Solution Figure 66 shows the graph of f with the standard viewing window. In order to view the end behavior of f, we set the window variables to $\texttt{Xmin}=-500$, $\texttt{Xmax}=500$, $\texttt{Ymin}=-10$, and $\texttt{Ymax}=10$, obtaining the view shown in Figure 67. As we trace the graph from left to right (Figure 68) it appears that the function values are approaching 3. Tracing from right to left, we see that $f(x)$ approaches 3 as well. Based on this evidence, it appears that $y = 3$ is a horizontal asymptote for the graph of f.

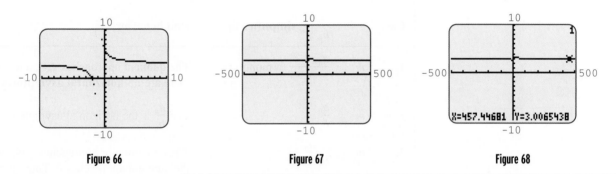

Figure 66   Figure 67   Figure 68

We now use our knowledge of the end behavior of polynomials to develop an *algebraic* method for finding horizontal asymptotes of rational functions. Since a rational function is the quotient of two polynomials, it can be written in the form

$$f(x) = \frac{a_n x^n + a_{n-1} x^{n-1} + \cdots + a_1 x + a_0}{b_m x^m + b_{m-1} x^{m-1} + \cdots + b_1 x + b_0}$$

where neither a_n nor b_m is 0. Because polynomials, in the long run, tend to behave like their terms of highest degree, we would expect that the end behavior of $f(x)$ could be found by replacing the numerator and denominator by their terms of highest degree. In fact, $f(x)$ does indeed behave like the quotient

$$\frac{a_n x^n}{b_m x^m}$$

Thus, for example, to determine the end behavior of

$$f(x) = \frac{x^2 + 1}{2x^3 - 3x + 4}$$

we need only consider the end behavior of the much simpler function

$$\frac{x^2}{2x^3} = \frac{1}{2x}$$

Now as x gets large, the denominator does also, and the quotient tends toward 0. Thus, $f(x) \to 0$ as $x \to \infty$ and as $x \to -\infty$. It follows that the line $y = 0$ is a horizontal asymptote of the graph of $f(x)$. With hindsight, this result was predictable: Since the degree of the denominator is greater than that of the numerator, the denominator grows much more rapidly than the numerator, and thus their quotient tends toward 0. More generally, the end behavior of a rational function depends on the relative degrees of numerator and denominator, as follows.

Locating Horizontal Asymptotes

The end behavior of the rational function

$$f(x) = \frac{a_n x^n + a_{n-1}x^{n-1} + \cdots + a_1 x + a_0}{b_m x^m + b_{m-1}x^{m-1} + \cdots + b_1 x + b_0}$$

is the same as that of

$$\frac{a_n x^n}{b_m x^m}$$

There are three possible cases, depending on the relatives sizes of n and m.

Case	$\dfrac{a_n x^n}{b_m x^m}$ simplifies to	End behavior
1. $n > m$	$\dfrac{a_n x^{n-m}}{b_m}$ (a monomial)	The quotient approaches $\pm\infty$, and so there are **no horizontal asymptotes**.
2. $n = m$	$\dfrac{a_n}{b_n}$	$y = \dfrac{a_n}{b_n}$ is the horizontal asymptote.
3. $n < m$	$\dfrac{a_n}{b_m x^{m-n}}$	The denominator approaches $\pm\infty$, so the quotient approaches 0. Thus, $y = 0$ (the x-axis) is the horizontal asymptote.

EXAMPLE 4

Locating Vertical and Horizontal Asymptotes

Locate any vertical and horizontal asymptotes and sketch the graph of

$$f(x) = \frac{x^2 - 2x + 2}{2x^2 - 4x}$$

Solution $f(x)$ is undefined where $2x^2 - 4x = 0$. Solving for x we have

$$2x^2 - 4x = 0$$
$$2x(x - 2) = 0$$
$$x = 0, x = 2$$

Thus, there are vertical asymptotes at both $x = 0$ and $x = 2$. To determine the end behavior of f, we simply ignore all but the leading terms in the numerator and denominator to obtain $\dfrac{x^2}{2x^2} = \dfrac{1}{2}$, so that $y = \dfrac{1}{2}$ is the horizontal asymptote. (This is Case 2 described in the preceding box.) One view of the graph is shown in Figure 69. The graph in Figure 70 includes the asymptotes.

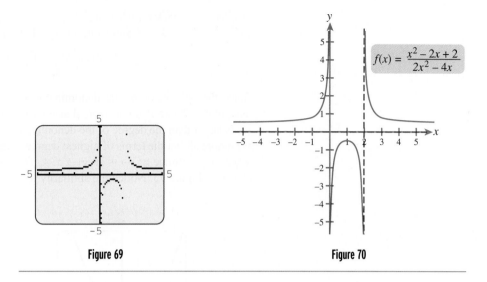

Figure 69   Figure 70

EXAMPLE 5   **Locating Vertical and Horizontal Asymptotes**

Locate any vertical and horizontal asymptotes and sketch the graph of

$$f(x) = \frac{80x}{16x^2 + 1}$$

Solution   The graph of f has no vertical asymptotes since the denominator $16x^2 + 1$ has no real zeros. The horizontal asymptote is the x-axis because the degree of the numerator is less than the degree of the denominator (Case 3). One view of the graph is shown in Figure 71. At first glance, it appears that the y-axis is a vertical asymptote. However, we know that is not the case since f has no vertical asymptotes. To see more clearly what is happening near the y-axis, we adjust the scale on the x-axis; the result is shown in Figure 72.

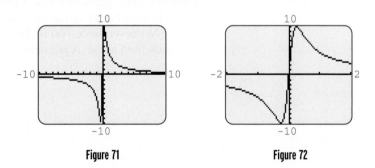

Figure 71   Figure 72

EXAMPLE 6   **A Rational Function with No Vertical or Horizontal Asymptotes**

Show that

$$f(x) = \frac{x^4}{x^2 - 2x + 2}$$

has no vertical or horizontal asymptotes. Describe the end behavior of f.

Solution To check for vertical asymptotes, we find the zeros of the denominator by setting $x^2 - 2x + 2 = 0$ and using the quadratic formula.

$$x = \frac{-(-2) \pm \sqrt{(-2)^2 - 4(1)(2)}}{2(1)} = \frac{2 \pm \sqrt{-4}}{2}$$

Since the only zeros of the denominator are nonreal, the graph of f has no vertical asymptotes. There is no horizontal asymptote either because the degree of the numerator is larger than the degree of the denominator (Case 1). To determine the end behavior, we ignore all but the terms of highest degree in the numerator and denominator to obtain $x^4/x^2 = x^2$. Since $x^2 \to \infty$ as $x \to \pm\infty$, $f(x) \to \infty$ also as $x \to \pm\infty$. The graph of f is shown in Figure 73. Notice the similarity to the graph of $y = x^2$ shown in Figure 74.

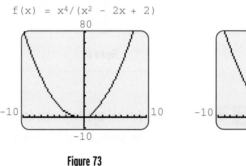

Figure 73   Figure 74

::::⟩**EXAMPLE 7**

Graphing an Unknown Rational Function

Suppose f is a rational function with vertical asymptote $x = -3$ and horizontal asymptote $y = 0$. Moreover, suppose $f(x) > 0$ for $x \neq -3$. Sketch a possible graph of f.

Solution Since $x = -3$ is a vertical asymptote, and since $f(x)$ is always positive, we know that $f(x)$ approaches ∞ as x approaches -3 from the left and the right. Because $y = 0$ is a horizontal asymptote, $f(x)$ must approach 0 (the x-axis) as x approaches ∞ or $-\infty$. Moreover, since $f(x)$ is always positive, we know that the graph must approach the x-axis from above. A *possible* sketch for f is shown in Figure 75.

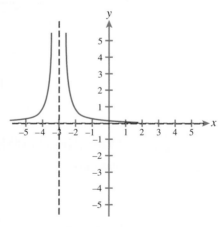

Figure 75

Inclined Asymptotes

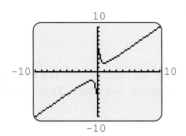

Figure 76

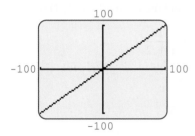

Figure 77

We have seen that the end behavior of a rational function is like that of a polynomial whenever the degree of the numerator exceeds the degree of the denominator. More specifically, such a rational function has no horizontal asymptote, and the function values approach either positive or negative infinity as $x \to \infty$ and as $x \to -\infty$. We now consider the particular case in which the degree of the numerator is exactly one more than the degree of the denominator. Consider, for example, the function

$$f(x) = \frac{x^2 + 1}{x}$$

The graph of f shown in Figure 76 certainly suggests that there is no horizontal asymptote. To investigate the end behavior further, we zoom out to obtain the view shown in Figure 77. Here we see that not only does f have end behavior like that of a polynomial, but the graph of f actually behaves like that of a linear (first-degree) polynomial. Indeed, the graph of f appears linear with a sufficiently large viewing window.

To see more clearly what's happening, let us investigate f algebraically. Dividing the numerator by x gives us $f(x) = x + 1/x$. Now for very large positive or negative values of x, $1/x$ is a very small number, and hence $f(x) = x + 1/x \approx x$. Thus, we would expect that in the long run, the graph of f would approach the graph of $y = x$. Since the line $y = x$ is neither vertical nor horizontal but rather at an incline, we say that $y = x$ is an **inclined (slant) asymptote** for the function f. Figure 78 shows the graph of f with its inclined asymptote $y = x$ depicted as a dashed line.

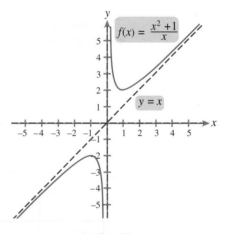

Figure 78

More generally, we find inclined asymptotes as follows.

Inclined Asymptotes

Suppose $f(x) = p(x)/q(x)$ is a rational function for which the degree of $p(x)$ is *exactly one more* than the degree of $q(x)$. Then we can use division to rewrite f in the form

$$f(x) = ax + b + \frac{r(x)}{q(x)}$$

where the degree of $r(x)$ is less than the degree of $q(x)$. The line $y = ax + b$ is an inclined asymptote for f.

EXAMPLE 8

Locating Vertical and Inclined Asymptotes

Find the vertical and inclined asymptotes for

$$f(x) = \frac{x^2 + 5x + 7}{x + 2}$$

and sketch the graph.

Solution Setting the denominator equal to 0 and solving, we see that there is a vertical asymptote at $x = -2$. To find the inclined asymptote, we must first divide $x^2 + 5x + 7$ by $x + 2$. We use synthetic division.

$$\begin{array}{r|rrr} -2 & 1 & 5 & 7 \\ & & -2 & -6 \\ \hline & 1 & 3 & 1 \end{array}$$

From the last row, we see that the quotient is $x + 3$ and the remainder is 1. Thus

$$\frac{x^2 + 5x + 7}{x + 2} = x + 3 + \frac{1}{x + 2}$$

and so

$$f(x) = x + 3 + \frac{1}{x + 2}$$

This shows that $y = x + 3$ is an inclined asymptote. Using a graphing calculator, we obtain the view shown in Figure 79. The graph in Figure 80 includes the asymptotes.

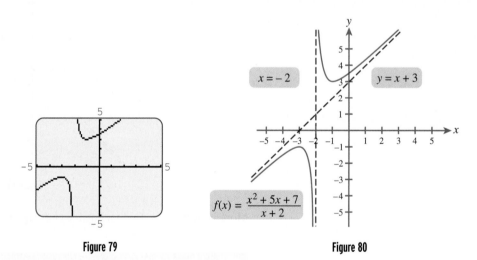

Figure 79   Figure 80

EXAMPLE 9

Minimizing Inventory Cost

An appliance retailer sells 500 microwave ovens each year. By averaging ordering costs and storage costs, the retailer has determined that if x microwaves are ordered at a time, the yearly inventory cost will be

$$C(x) = 5x + 40{,}000 + \frac{4500}{x}$$

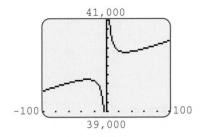

41,000

-100 ....‖....... 100

39,000

Figure 81

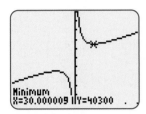

Minimum
X=30.000009   Y=40300

Figure 82

a. Estimate the number of microwaves that should be ordered each time so that the cost is as small as possible. How often should each order be placed?

b. Find the inclined asymptote of C and interpret its slope.

Solution

a. We want to choose a viewing window so that a minimum value is shown. After some experimentation, we settle on the view shown in Figure 81. Because we are only concerned with positive values of x, we look for the lowest point on the graph of C to the right of the y-axis. Using the trace or calculate-minimum feature, we estimate the minimum cost to be \$40,300 per year when 30 microwaves are ordered each time (see Figure 82). If 500 microwaves are sold each year and 30 must be ordered each time to minimize inventory cost, an order must be placed $\frac{500}{30} \approx 17$ times each year, or every 22 days.

b. By ignoring the rational term $4500/x$, we see that $y = 5x + 40{,}000$ is the inclined asymptote for the graph of C. Since the slope of this line is 5, we conclude that for very large values of x, the y-coordinates on the graph of C will increase by approximately 5 units for every unit increase in x. In other words, for very large order sizes, the inventory cost will increase by approximately \$5 for every additional unit ordered.

Understanding and Mastery Checklists

Concepts to Understand	Skills to Master
Domain and zeros of a rational function	Find the domain and zeros of a rational function.
↻	↻
Vertical asymptotes	Find the vertical asymptote(s) of a rational function.
↻	↻
Horizontal asymptotes	Find the horizontal asymptote of a rational function.
↻	↻
Inclined asymptotes	Find the inclined asymptote of a rational function.
	↻
	Sketch the graph of a rational function.

Exercises 3.5

Exercises 1–8 *Find the domain and zeros of the rational function and match it with its graph.*

1. $f(x) = \dfrac{1}{x-2}$

2. $f(x) = \dfrac{x}{x-1}$

3. $f(x) = \dfrac{x+1}{x}$

4. $f(x) = \dfrac{x-2}{x-1}$

5. $f(x) = \dfrac{1}{x + 2}$

6. $f(x) = \dfrac{1}{x^2 + x - 2}$

7. $f(x) = \dfrac{x^2 - 4}{x^2 - 1}$

8. $f(x) = \dfrac{x + 1}{x + 2}$

iv.

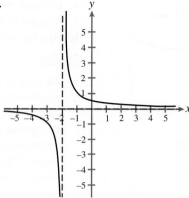

i.

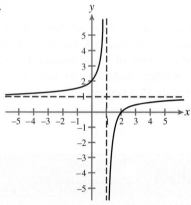

v.

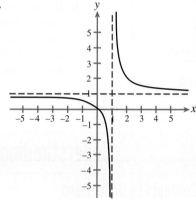

ii.

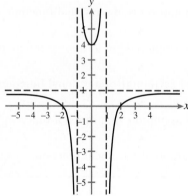

vi.

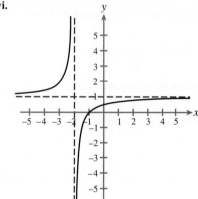

iii.

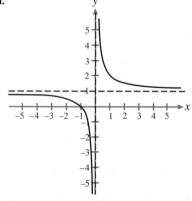

vii.

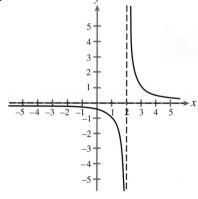

viii.

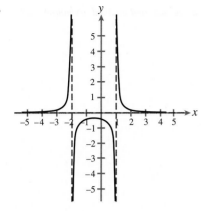

a. $g(x) = \dfrac{1}{x^2} - 3$

b. $g(x) = \dfrac{1}{(x + 2)^2}$

c. $g(x) = -\dfrac{1}{x^2}$

Exercises 9-12   *Use the graph of the function f to sketch the graph of the function g.*

9.

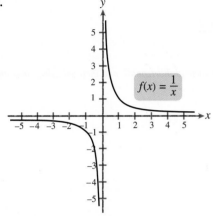

$f(x) = \dfrac{1}{x}$

a. $g(x) = \dfrac{1}{x - 3}$

b. $g(x) = \dfrac{1}{x} + 2$

c. $g(x) = -\dfrac{1}{x}$

10.

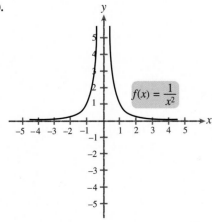

$f(x) = \dfrac{1}{x^2}$

11.

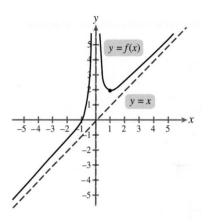

$y = f(x)$

$y = x$

a. $g(x) = f(x - 1)$

b. $g(x) = f(x) - 2$

c. $g(x) = -f(x)$

12.

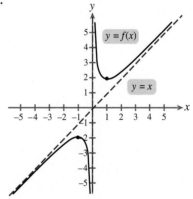

$y = f(x)$

$y = x$

a. $g(x) = f(x) + 2$

b. $g(x) = f(x + 1)$

c. $g(x) = -f(x)$

Exercises 13-16   *Construct a possible graph for the rational function with the given properties. There may be more than one correct answer.*

13. *f* has a vertical asymptote at $x = 4$, horizontal asymptote $y = 0$, and $f(x) < 0$ for all $x \neq 4$.

14. h has a vertical asymptote at $x = -2$, horizontal asymptote $y = 0$, $h(x) > 0$ if $x < -2$, and $h(x) < 0$ if $x > -2$.

15. g has a vertical asymptote at $x = -1$, horizontal asymptote $y = 2$, an x-intercept at $\left(-\frac{3}{2}, 0\right)$, and a y-intercept at $(0, 3)$.

16. h has a vertical asymptote at $x = 2$, horizontal asymptote $y = -3$, and zeros at $x = 0$ and $x = 4$.

Exercises 17-30   *Find all vertical and horizontal asymptotes and sketch the graph.*

17. $f(x) = \dfrac{1}{x + 3}$

18. $f(x) = \dfrac{1}{x - 2}$

19. $h(x) = \dfrac{x}{x + 2}$

20. $g(x) = \dfrac{x - 3}{x + 4}$

21. $f(x) = \dfrac{3x}{x - 2}$

22. $g(x) = \dfrac{x + 1}{3 - 2x}$

23. $g(x) = \dfrac{x}{x^2 - 9}$

24. $f(x) = \dfrac{x^2}{x^2 - 9}$

25. $g(x) = \dfrac{45x^2}{9x^2 + 1}$

26. $f(x) = \dfrac{1}{x^2 + 9}$

27. $h(x) = \dfrac{1}{(x + 1)^2}$

28. $f(x) = \dfrac{-x}{(x - 3)^2}$

29. $f(x) = \dfrac{x^2}{(x - 2)^2}$

30. $g(x) = \dfrac{(x - 2)^2}{x^2}$

Exercises 31-36   *Find the vertical and inclined asymptotes and sketch the graph.*

31. $g(x) = 2x + \dfrac{1}{x}$

32. $h(x) = -x + \dfrac{1}{x}$

33. $f(x) = \dfrac{x^2}{x - 1}$

34. $h(x) = \dfrac{x^2 + 1}{x + 1}$

35. $g(x) = \dfrac{-3x^2 - 6x + 1}{x + 2}$

36. $f(x) = \dfrac{4x^2 - 4x + 2}{x - 1}$

Exercises 37-46   *Find all vertical, horizontal, and inclined asymptotes and sketch the graph.*

37. $f(x) = \dfrac{2}{x^2 + 1}$

38. $h(x) = \dfrac{8}{x^2 - 2x + 3}$

39. $h(x) = \dfrac{12x + 24}{x - 15}$

40. $h(x) = \dfrac{-2}{x - 3}$

41. $g(x) = \dfrac{x^2 - 4x + 7}{x - 3}$

42. $f(x) = \dfrac{1 - 16x}{x + 12}$

43. $h(x) = \dfrac{-1}{x^2 - 9}$

44. $h(x) = \dfrac{2x^2 + 9x + 5}{x + 4}$

45. $f(x) = \dfrac{x^3}{x - 3}$

46. $g(x) = \dfrac{2x^4}{x^2 - 1}$

Exercises 47-50 *The graphs of many other functions besides rational functions have horizontal asymptotes. Use a graphing calculator to estimate the horizontal asymptote(s) of the given function.*

47. $f(x) = \dfrac{\sqrt{1 + 4x^2}}{4 + x}$

48. $f(x) = \sqrt{1 + x} - \sqrt{x}$

49. $f(x) = \dfrac{3x}{\sqrt{x^2 + 1}}$

50. $f(x) = \dfrac{\sqrt{x^4 + 2}}{x^2}$

Applications

51. Pollution Control Based on data from a 1973 study on the cost of reducing oil refinery discharge, the increase in refinery costs can be modeled by the function

$$f(x) = \dfrac{x}{1000(100 - x)}$$

where x is the desired percent reduction in discharge and $f(x)$ is the corresponding additional cost in dollars per barrel crude. Identify any vertical or horizontal asymptotes and sketch the graph of f. What percent reduction is possible with an increase of $0.003 per barrel crude? According to this model, is a 100% reduction attainable? Explain. [*Data source:* Maynard Hufschmidt et al., *Environment, Natural Systems, and Development* (Baltimore: The Johns Hopkins University Press, 1983).]

52. Nerve Excitation The minimum voltage needed to excite a nerve fiber is related to the time during which the current flows. According to data from one source, this relationship can be modeled by the function

$$f(x) = \dfrac{17}{x + 0.113} + 23.2$$

where x is the time during which current flows (in milliseconds) and $f(x)$ is the minimum current needed to excite the nerve fiber (in millivolts). Identify any vertical or horizontal asymptotes and sketch the graph of f. How long must the current flow to excite a nerve fiber with a voltage of 40 millivolts? [*Data source:* Douglas Riggs, *The Mathematical Approach to Physiological Problems* (Cambridge: M.I.T. Press, 1970).]

53. Unemployment Rate The percentage rate of unemployment in the U.S. in 2000 for people with x years of education can be modeled by the function

$$f(x) = \dfrac{77}{(x - 10)^2 + 7}$$

Identify any vertical or horizontal asymptotes and sketch the graph of f. What level of education corresponds to an unemployment rate of 4.9%? What happens to the unemployment rate as the level of education increases? According to this model, is there an education level that corresponds to an unemployment rate of 0%? Explain. (*Data source:* U.S. Bureau of Labor Statistics.)

54. Height Distribution The percentage of U.S. males between the ages of 18 and 24 years who are within a half inch of a given height can be modeled by the function

$$f(x) = \dfrac{256}{(2x - 139)^2 + 16}$$

where x is the height (in inches). Identify any vertical or horizontal asymptotes and sketch the graph of f. What height corresponds to the highest percentage? What interpretation could be given to this height? (*Data source:* U.S. National Center for Health Statistics.)

7′5″ NBA center Yao Ming with Shaquille O'Neal and Christina Aguilera

55. Bicycle Average Cost A company that manufactures bicycles has fixed costs of $100,000 and variable costs of $100 per bicycle. The total cost of producing x bicycles is $C(x) = 100{,}000 + 100x$. The average cost per bicycle is found by dividing the total cost by the number of bicycles produced. Thus, the average cost function is

$$\overline{C}(x) = \dfrac{100{,}000 + 100x}{x}$$

Sketch the graph of the average cost function. What is the horizontal asymptote and what information can be obtained from it?

56. Skate Average Cost A company that manufactures inline skates has fixed costs of $80,000 and variable costs of $50 per pair of skates, so that the total cost of producing x pairs of skates is thus $C(x) = 80{,}000 + 50x$. The average cost per pair of skates is found by dividing the total cost by the number of pairs of skates produced. Thus, the average cost function is

$$\overline{C}(x) = \dfrac{80{,}000 + 50x}{x}$$

Sketch the graph of the average cost function. What is the horizontal asymptote and what information can be obtained from it?

57. Basketball Inventory Cost A sporting goods retailer sells 580 basketballs each year. By averaging ordering costs and storage costs, the retailer has determined that if x balls are ordered at a time, the yearly inventory cost will be

$$C(x) = \dfrac{3}{2}x + 600 + \dfrac{5046}{x}$$

a. Estimate the number of balls that should be ordered each time so that the cost is as small as possible. How often should each order be placed?

b. Find the inclined asymptote of C and interpret its slope.

58. Television Inventory Cost A retail appliance store sells 1500 television sets per year. By averaging ordering costs and storage costs, the retailer has determined that if x televisions are ordered at a time, the yearly inventory cost will be

$$C(x) = 5x + 30{,}000 + \frac{28{,}200}{x}$$

a. Estimate the number of televisions that should be ordered each time so that the cost is as small as possible. How often should each order be placed?

b. Find the inclined asymptote of C and interpret its slope.

Concepts and Critical Thinking

Exercises 59-62 *Answer true or false.*

59. The graphs of all rational functions have vertical asymptotes.

60. The graph of a rational function may cross its horizontal asymptote but does not cross any of its vertical asymptotes.

61. The zeros of a rational function in lowest terms are the zeros of its numerator.

62. The vertical asymptotes of a rational function can be found by setting the numerator equal to zero.

Exercises 63-64 *Give an example of each.*

63. A rational function with horizontal asymptote $y = 3$ and vertical asymptotes $x = 1$ and $x = -1$

64. A rational function having neither vertical nor horizontal asymptotes

Questions for Discussion or Essay

65. In Exercises 55 and 56, we considered average cost functions of the form

$$\overline{C}(x) = \frac{a + bx}{x}$$

Could a function of this form have a minimum value? In other words, is it possible to find a production level x for which the average cost is as small as possible? Explain. Is this situation consistent with what you would expect for a company's average cost? Why or why not?

66. We have seen that if $f(x) = p(x)/q(x)$ is in simplest terms, with $q(a) = 0$, then the line $x = a$ is a vertical asymptote for the graph of f. What happens if f is not in simplest terms? To help you answer this question, consider some examples:

$$f(x) = \frac{x^2 - 4}{x - 2}, \quad f(x) = \frac{x}{x^3 - x}, \quad \text{and} \quad f(x) = \frac{x^2 - 2x + 1}{x^2 + x - 2}$$

If $f(x)$ is not in simplest terms and $q(a) = 0$, is it necessarily the case that $x = a$ is a vertical asymptote? If not an asymptote, what happens at $x = a$?

67. Rational functions of the form

$$f(x) = \frac{ax + b}{cx + d}, \quad c \neq 0$$

can be graphed quite easily without the use of a graphing calculator. Consider the function

$$f(x) = \frac{2x + 3}{x + 1}$$

The vertical and horizontal asymptotes divide the plane into four regions, which we have numbered in Figure 83. It can be shown that the graph must lie either entirely in regions 1 and 3 or entirely in regions 2 and 4. Why do you think this is? Once the regions have been determined, a sketch of the graph is easily completed, as shown in Figure 84.

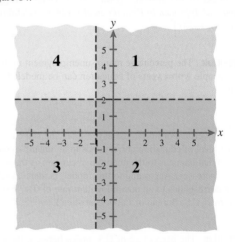

Figure 83

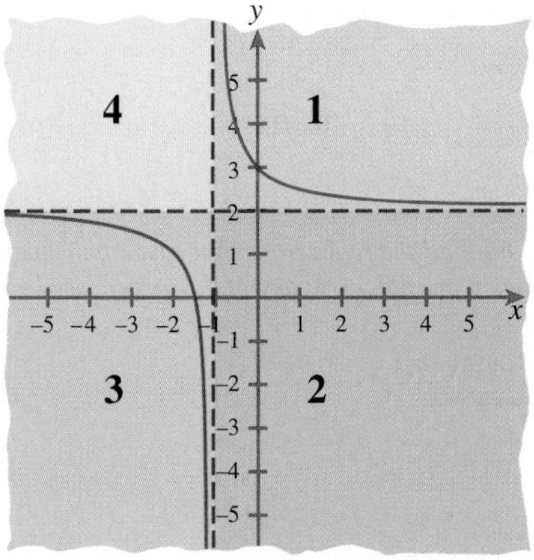

Figure 84

Describe a procedure for locating the horizontal and vertical asymptotes of the graph of a function

$$f(x) = \frac{ax + b}{cx + d}, \quad c \neq 0$$

and for deciding in which pair of regions to place the graph. Illustrate your procedure with an example.

68. In Example 7, we sketched the graph of a function f with a vertical asymptote at $x = -3$, a horizontal asymptote at $y = 0$, and with $f(x) > 0$ for $x \neq -3$. Explain why there could be more than one correct answer for this example, and give an example of another possible solution. Do you think one answer could be "more right" than another? Why or why not? What additional information would have to be given to narrow down the choice of possibilities? It has been said that one of the strengths of mathematics over some other disciplines is that an answer is either right or wrong. There is no gray area. In your experience with mathematics, how often have you found this to be the case? Do you think this "all or nothing" phenomenon is more or less likely to be the case when mathematics is applied to the real world? Explain.

Projects for Enrichment

69. Minimizing Inventory Cost A shoe retailer wishes to minimize the costs incurred in handling a certain style of shoe that must be ordered periodically and kept in stock as they are sold to customers. Since the expenses involved in ordering and storing the shoes depend on how often orders are made and how long items are stored, the retailer wishes to minimize the *total cost per year*. We consider three separate costs that are involved in the total cost to the retailer.

  i. A *fixed cost* of \$30 *per order* that is independent of the amount ordered. This cost includes such things as record keeping and other paperwork, employees' time, and so on.
  ii. A *purchase cost* of \$15 *per pair* (throughout, "pair" will mean "pair of shoes").
  iii. An *inventory holding cost* of \$2 *per pair per year.* This cost covers the expenses of keeping the shoes in the store or warehouse.

Let x denote the number of pairs ordered at a time, and assume that the shoes are sold at a constant rate of 1200 pairs per year. Because shortages are not tolerable, the time between two consecutive orders depends only on the number ordered and the rate they are sold.

  a. Find an expression for the number of reorders per year.

  b. Find an expression for the total cost per order (not including inventory cost).

  c. Use the expressions from parts a and b to find an expression for the yearly ordering costs.

  d. Find an expression for the yearly inventory cost. Since x pairs are ordered at a time, assume that the average number of pairs of shoes in inventory at any given time is $x/2$.

  e. Use the expressions from parts c and d to obtain a function for the total cost per year. Denote this total cost function by $T(x)$.

  f. Use a graphing calculator to plot the graph of $T(x)$ and approximate the value for x that yields the minimum cost. This represents the number of shoes that should be ordered at a time. How often must this quantity of shoes be ordered?

  g. What would change in this scenario if the retailer can order shoes only in quantities of 18?

Chapter 3 Review

Exercises 1-6 *Determine the end behavior of the given polynomial. Verify your results with a graphing calculator.*

1. $g(x) = -3x^6$

2. $h(x) = 2x^5$

3. $f(x) = 4x^3 - x^2 + 9x + 3$

4. $g(x) = -x^4 + 5x^2 - x + 8$

5. $h(x) = x^3 + 100 - 0.00005x^4$

6. $f(x) = (2 \times 10^4)x^2 + (2 \times 10^{-4})x^3$

Exercises 7-10 *Use a graphing calculator to estimate the zeros and the coordinates of any turning points of the given polynomial to the nearest hundredth. Classify each turning point as a local maximum or minimum.*

7. $f(x) = \dfrac{x^3}{3} - 3x^2 - 3$

8. $g(x) = \dfrac{x^4}{9} + x^2 - 4x + 2$

9. $g(x) = 0.1x^4 + x^3 - 4x^2 + x - 4$

10. $h(x) = 0.01x^5 - 0.15x^4 - 0.4x^3 + 2x^2$

Exercises 11-14 *Use long division to find the quotient q(x) and remainder r(x) when f(x) is divided by d(x).*

11. $f(x) = 2x^2 + 3x - 2; d(x) = x + 2$

12. $f(x) = 9x^3 - 3x^2 + 7x + 1; d(x) = 3x - 2$

13. $f(x) = x^4 - 5x^2 + 2; d(x) = x^2 + 4x$

14. $f(x) = x^5 + 2x^3 - x^2 - 2; d(x) = x^3 - 1$

Exercises 15-18 *Use synthetic division to find the quotient q(x) and remainder r(x) when f(x) is divided by d(x).*

15. $f(x) = 3x^3 - 17x^2 + 22x - 8; d(x) = x - 4$

16. $f(x) = 4x^3 + 8x^2 - 9x - 13; d(x) = x + 2$

17. $f(x) = -2x^4 + 4x^2 + x - 3; d(x) = x + 3$

18. $f(x) = 32x^5 - 1; d(x) = x - \dfrac{1}{2}$

Exercises 19-22 *Use the Remainder Theorem to find the indicated function values.*

19. $f(x) = 3x^3 + 11x^2 + 2x - 6$

    a. $f\left(\dfrac{3}{2}\right)$ **b.** $f(-2)$

20. $g(x) = 5x^4 - 6x^3 - 32x^2 + 7$

    a. $g(-3)$ **b.** $g(3.2)$

21. $h(x) = -2x^4 + 5x^3 + 3x$

    a. $h(5)$ **b.** $h\left(\dfrac{1}{2}\right)$

22. $f(x) = \dfrac{1}{3}x^5 + 2x^4 + \dfrac{1}{2}x + 3$

    a. $f(-6)$ **b.** $f(2)$

Exercises 23-28 *Find all the real zeros of the given polynomial function. Give exact values if possible. Use any of the tools described in this chapter.*

23. $f(x) = 2x^2 + 5x - 3$

24. $g(x) = 4x^3 - 4x^2 - 24x$

25. $h(x) = x^4 - 11x^2 + 18$

26. $g(x) = 4x^3 + 5x^2 - 18x + 9$

27. $h(x) = 6x^3 + 7x^2 - 1$

28. $f(x) = x^4 - 2x^3 - 5x^2 + 8x + 4$

Exercises 29-36 *Find a polynomial function f of least degree with real coefficients satisfying the given properties.*

29. Zeros -2, 3, and 5

30. Zeros -3, 0, and 4; $f(1) = 10$

31. Zeros 1 and -1 (both with multiplicity 2); $f(0) = 4$

32. Zeros $\dfrac{1}{2}$ (multiplicity 2) and 0 (multiplicity 3)

33. Graph appears as follows:

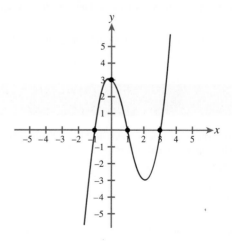

34. Graph appears as follows:

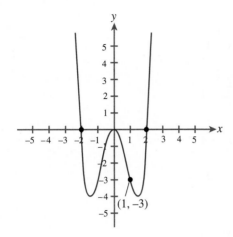

35. Zeros -3 and $4i$

36. Zeros 0, 2, and $1 - 2i$

Exercises 37-40 *Write the polynomial function as a product of linear factors, and identify all the complex zeros. Use any of the tools described in this chapter.*

37. $f(x) = x^2 - 4x + 5$

38. $g(x) = x^3 + 3x^2 + 4x + 12$

39. $h(x) = x^4 + 7x^2 - 144$

40. $f(x) = x^5 - 4x^4 + 10x^3 - 12x^2 + 5x$

Exercises 41-44 *Use the Rational Zero Theorem to list all the possible rational zeros of the polynomial. Then use synthetic division and/or a graphing calculator to help you determine which are actually zeros.*

41. $f(x) = x^3 - 2x^2 - 5x + 6$

42. $h(x) = 3x^3 - 10x^2 + 11x - 4$

43. $g(x) = 4x^4 - 8x^3 - x^2 + 8x - 3$

44. $h(x) = 9x^5 - x^3 - 9x^2 + 1$

Exercises 45-48 *Apply Descartes' Rule of Signs to the given function. Then use a graphing calculator to determine the precise number of positive and negative zeros.*

45. $h(x) = x^3 + 2x + 1$

46. $f(x) = x^4 - 4x^3 - x + 5$

47. $g(x) = -x^4 - 5x^3 + 2x^2 + 8x$

48. $h(x) = 2x^5 - 6x^3 + 3x^2 - 1$

Exercises 49-50 *Use the Upper and Lower Bounds Test to confirm the given bounds for the zeros of the function.*

49. $f(x) = x^4 - 2x^3 - 49x^2 - 2x + 20$
  Lower: -7; Upper: 9

50. $g(x) = x^4 + 5x^3 - 299x^2 + 5x - 400$
  Lower: -21; Upper: 16

Exercises 51-54 *Find the exact real solutions of the given polynomial equation.*

51. $x^3 + 4x + 12 = 7x^2$

52. $4x^3 + 56x^2 = x + 14$

53. $x^3 - \dfrac{37}{12}x^2 - \dfrac{7}{2}x - \dfrac{2}{3} = 0$

54. $3x^4 - 125x^2 + 19x = -57x^3 + 42$

Exercises 55-58 *Show that the function has no rational zeros. Use a graphing calculator to approximate all irrational zeros to the nearest hundredth.*

55. $f(x) = x^3 - 5x - 1$

56. $g(x) = x^4 + 2x^3 - x^2 - 7$

57. $g(x) = 0.05x^4 + x^3 - x^2 + x - 3$

58. $h(x) = x^5 - 15x^4 - x + 5$

Exercises 59-62 *Find the domain and zeros of the rational function and match it with its graph.*

59. $f(x) = \dfrac{1}{x - 3}$

60. $f(x) = \dfrac{x + 2}{x - 3}$

61. $f(x) = \dfrac{x + 2}{x^2 - 9}$

62. $f(x) = \dfrac{1}{x + 3}$

i.

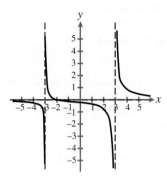

ii.

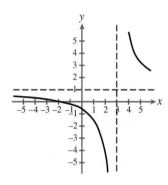

iii.

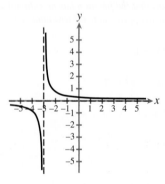

iv.

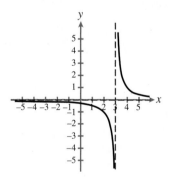

Exercises 63-72 *Find all vertical and horizontal asymptotes and sketch the graph.*

63. $f(x) = \dfrac{1}{x + 2}$

64. $h(x) = \dfrac{x + 3}{x - 1}$

65. $g(x) = \dfrac{2x + 5}{3x - 1}$

66. $h(x) = \dfrac{-1}{x^2 - 4}$

67. $f(x) = \dfrac{3 - x}{x^2}$

68. $g(x) = \dfrac{x}{x^2 + 5}$

69. $g(x) = \dfrac{2x^2}{x^2 + x + 2}$

70. $h(x) = \dfrac{x^4}{x + 2}$

71. $f(x) = \dfrac{2}{(1 - x)^2}$

72. $h(x) = \dfrac{x + 1}{(x - 3)^2}$

Exercises 73-76 *Find the vertical and inclined asymptotes and sketch the graph.*

73. $h(x) = -2x - \dfrac{1}{x}$

74. $g(x) = \dfrac{x^2 + 2x}{x - 2}$

75. $h(x) = \dfrac{3x^2 + 8x - 9}{x + 3}$

76. $f(x) = \dfrac{x^2 - 6x + 14}{2x - 8}$

Exercises 77-86 *Parts a and b are connected: Part a involves an elementary concept, whereas part b involves related material from this chapter. First answer part a, and then use this result to answer part b.*

77. a. Simplify $x - 3 + \dfrac{2}{x - 1}$.

b. Find the inclined asymptote of $f(x) = \dfrac{x^2 - 4x + 5}{x - 1}$.

78. a. Expand $(x + 2)(x - 2)(x - 1)$.

   b. Find the zeros of $f(x) = x^4 - x^3 - 4x^2 + 4x$.

79. a. Solve $x^2 + 3x - 1 = 0$.

   b. Find all zeros of $f(x) = x^3 + 3x^2 - x$.

80. a. Factor $x^3 - 5x^2 - 6x$.

   b. Find all vertical asymptotes of $\dfrac{x + 1}{x^3 - 5x^2 + 6x}$.

81. a. Simplify $(x - 1)(x + 5) + 9$.

   b. Simplify $\dfrac{x^2 + 4x + 4}{x - 1}$.

82. a. Find all complex solutions of $x^2 + 2x + 5 = 0$.

   b. Express $f(x) = x^2 + 2x + 5$ as a product of linear factors.

83. a. If a function f has the x- and y-axes as its horizontal and vertical asymptotes, respectively, what are the asymptotes for $h(x) = f(x - 3) - 2$?

   b. Find a function with vertical asymptote $x = 3$ and horizontal asymptote $y = -2$.

84. a. Compute $f(-x)$ for $f(x) = x^4 + 4x^2 + 2$.

   b. Show that f has no real zeros.

85. a. Compute $(2 - 3i) + (2 + 3i)$ and $(2 - 3i)(2 + 3i)$.

   b. Find a polynomial with zeros $2 \pm 3i$.

86. a. Find $f(0)$, $f(1)$, and $f(2)$ for $f(x) = x^3 - 2x^2 - x + 2$.

   b. Find an interval of length 1 in which the function f has a turning point.

87. Maximum Profit A company has determined that its profits can be modeled by a function of the form $f(x) = ax^2 + bx + c$, where x is the number of units sold and $f(x)$ is the corresponding profit. Find f given that the profit is zero for $x = 1000$ and $x = 5000$ and, because of fixed costs, the company loses $15,000 if no units are sold. How many units must be sold in order to maximize profit? What is the largest possible profit?

88. Oscillating Spring A spring is hung from a fixed point and an object is attached to its free end, as shown in Figure 85. If no external force is applied to the object, it will remain at rest in its *equilibrium* position. If the object is pulled down and then released, its position relative to

the equilibrium position can be modeled for a brief period of time with a polynomial function of the form $f(t) = at^4 + bt^2 + c$, where t is the time after the object is released and $f(t)$ is the directed distance away from the equilibrium position at that time (a negative value indicates that the object is above the equilibrium position). Find f if the object is pulled 6 centimeters below the equilibrium position and then passes the equilibrium position at $t = 0.11$ and again at $t = 0.27$. (*Hint:* Since f is an even function, its graph is symmetric about the y-axis.) Where is the object at $t = 0.18$ second? At what point does the model fail to provide reasonable results?

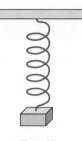

Figure 85

89. Alaskan Temperature The average monthly temperature in Juneau, Alaska, over a 12-month period can be approximated with the function

$$f(x) = 0.0134x^4 - 0.33x^3 + 2.07x^2 - 0.456x - 5.04$$

where x denotes the month ($x = 0$ corresponds to January) and $f(x)$ is the average Celsius temperature for that month. Approximate the zeros of f in the interval $[0,12)$. What do these zeros represent?

90. Mathematics Degrees The number of bachelor's degrees conferred in mathematics during the years 1970–2000 can be approximated with the polynomial function

$$f(x) = -3.1x^3 + 165x^2 - 2680x + 27,450$$

where x denotes the year ($x = 0$ corresponds to 1970) and $f(x)$ is the number of degrees for that year. Estimate the year(s) in which 14,500 bachelor's degrees in mathematics were conferred. Approximate the coordinates of any turning points.

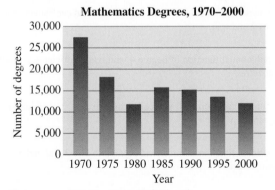

Mathematics Degrees, 1970–2000

Data source: U.S. Department of Education.

91. Calculator Cost A company that manufactures calculators has determined that the total cost for producing x calculators is given by $C(x) = 15,000 + 20x$. The average cost per calculator is thus given by

$$\overline{C}(x) = \frac{15,000 + 20x}{x}$$

Sketch the graph of $\overline{C}$. As the number of calculators increases, what value is the average cost approaching?

Chapter 3 Test

Problems 1-8 *Answer true or false.*

1. A polynomial of degree 3 with real coefficients must have at least one real zero.

2. The graph of a polynomial function of degree 4 can have at most three turning points.

3. According to Descartes' Rule of Signs, if a polynomial function has three variations in sign, then it must have three positive zeros.

4. There is exactly one polynomial of degree 3 with zeros 1, -2, and 4.

5. If a polynomial of degree 2 divides evenly into one of degree 5, the quotient will have degree 3.

6. If $2 + 3i$ is a zero of a polynomial with real coefficients, then so is $2 - 3i$.

7. A rational function must have a vertical asymptote.

8. A rational function can have at most one horizontal asymptote.

Problems 9-14 *Give an example of each.*

9. A polynomial function f for which $f(x) \to -\infty$ as $x \to \infty$

10. A polynomial function of least degree and with zeros -1, 3, and 4

11. A polynomial function of least degree with real coefficients and with zeros 2 and $-3i$

12. A polynomial function of degree 4 with real coefficients and with no real zeros

13. A rational function with vertical asymptote $x = -1$ and horizontal asymptote $y = 1$

14. A rational function with vertical asymptote $x = 2$ and inclined asymptote $y = x - 3$

Problems 15-16 *Factor the given polynomial and find the exact values of all the zeros.*

15. $f(x) = 2x^4 - 7x^3 - 4x^2$

16. $f(x) = 6x^3 + x^2 - 19x + 6$

17. Find the quotient and remainder when
$f(x) = 3x^4 - 12x^2 + 5x + 14$ is divided by $d(x) = x + 2$.

18. Given that $f(x) = 5x^4 - 2x^3 + 45x^2 - 18x$ has $3i$ as one of its zeros, find the remaining zeros and express $f(x)$ as a product of linear factors.

19. Use the Rational Zero Theorem to list all the possible rational zeros of $f(x) = 2x^3 + x^2 - 12x + 9$. Use synthetic division and/or a graphing calculator to help you determine which are actually zeros.

20. Apply Descartes' Rule of Signs to
$f(x) = 2x^4 - 6x^3 + x^2 - 8x - 7$. Then use a graphing calculator to determine the precise number of positive and negative zeros.

21. Show that $f(x) = x^3 - x - 2$ has no rational zeros, and use a graphing calculator to estimate any irrational zeros to the nearest hundredth.

22. Find any vertical or horizontal asymptotes for the function
$$f(x) = \frac{x + 1}{2x - 3}$$
and sketch its graph.

23. U.S. homicide rates from 1985 to 2000 can be modeled by the function
$$f(x) = 0.001x^4 - 0.03x^3 + 0.23x^2 - 0.29x + 8.2$$
where x is the year ($x = 0$ corresponds to 1985) and $f(x)$ is the number of homicides per 100,000 people.

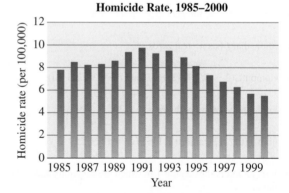

Homicide Rate, 1985–2000

Data source: Crime in the United States—2000, FBI, Uniform Crime Reports.

a. Estimate the year(s) between 1985 and 2000 in which the homicide rate was 8.7 per 100,000.

b. Approximately when during 1985–2000 did homicide rates peak? What was the rate at that time?

c. According to this model, during what period did the crime rate drop?

Chapter 4
Exponential and Logarithmic Functions

As the world population has exploded, so too has humankind's influence on Earth. This satellite photograph of North America at night shows the extent of our influence: where there is light, there is humanity. Population growth profoundly affects the consumption of natural resources, environmental quality, and the entire global economy, and thus predicting future population has become increasingly important. In this chapter, we develop mathematical models for predicting the growth of human and other populations.

Section 4.1 | Exponential Functions

- If a certain population of bacteria grows exponentially, how long would it take for it to fill a space the size of our solar system?
- How can a doctor estimate the concentration of a medication in the bloodstream?
- Can a bank compound interest every instant of every day without going bankrupt?
- In 1965, Intel founder and technovisionary, Gordon Moore, predicted that the number of transistors on a microprocessor would double every 18–24 months. How accurate has he been so far?

From the population growth of a city to the growth of single cell, from the value of a retirement account to the price of a personal computer, from drug concentration in the body to drug use in society, there are countless contexts in which mathematical models are used to predict how a given quantity will vary over time. One of the first steps in developing mathematical models of this type is to consider the growth rate of the quantity. For example, the average global temperature in 1980 was 61.9° Fahrenheit, and, since then, it has increased at an average rate of 0.029°F per year. Thus, t years after 1980, we would expect that the average global temperature would have risen $0.029t$ degree. It follows that in the year $1980 + t$, the average global temperature (in degrees Fahrenheit) can be estimated by the linear function

$$G(t) = 61.9 + 0.029t$$

In this section and much of this chapter, we model phenomena that grow not at a constant rate (like global temperature), but rather at a *constant percentage rate*. For example,

- During the 1990s the cost of living increased by about 3% per year.
- The amount of radioactive carbon 14 in the Dead Sea Scrolls is decreasing by about 0.012% per year.
- The number of Internet hosts grew by about 90% per year in the 1990s.
- The world population in 2000 was 6 billion, and it is growing at about 1.3% per year.

If a quantity $Q = Q(t)$ grows at a constant percentage rate, then $Q(t)$ is what is known as an *exponential function*. Among the many real-world phenomena that can be described using exponential functions are population growth, the spread of disease and information, radioactive decay, concentrations of drugs in the body, sales patterns, and compound interest. The first of these—population growth—will motivate our definition of an exponential function.

The Exponential Function with Base a

Suppose a biologist has been conducting a carefully controlled experiment on the growth of a bacteria culture. Unfortunately, a lab assistant misplaced all the data except that which appears in Table 1. To avoid the expense of starting the experiment over, the biologist decides to construct a function that estimates the number of bacteria at any given time.

The pattern in Table 1 suggests that the bacteria population is doubling (growing by 100%) every hour. In fact, the values in the second column are successive powers of 2. After 1 hour, the population (in thousands) is $2^1 = 2$; after 2 hours, it is $2^2 = 4$; after 3 hours, it is $2^3 = 8$; and so on. In general, if we let t represent the number of hours after 12:00 noon, then the number of bacteria after t hours is 2^t thousand. Thus, the function that the biologist is seeking would appear to be the *exponential function*

$$f(t) = 2^t$$

Table 1

Time	Number of bacteria (in thousands)
12:00 noon	1
1:00 P.M.	2
2:00 P.M.	4
3:00 P.M.	8
4:00 P.M.	16

where t denotes the number of hours after 12:00 noon, and the population $f(t)$ is measured in thousands. Note that the function f can be used to estimate the bacteria population at any time. Indeed, since 12:00 noon represents $t = 0$, times before or after 12:00 noon can be treated as negative or positive values of t, respectively. For example, 9:00 A.M. is 3 hours before 12:00 noon, and so we set $t = -3$ to obtain

$$f(-3) = 2^{-3} = \frac{1}{2^3} = \frac{1}{8} = 0.125$$

which corresponds to 125 bacteria. The population at 6:00 P.M. $(t = 6)$ is given by

$$f(6) = 2^6 = 64$$

which corresponds to 64,000 bacteria. For fractions of an hour, we use fractional exponents. Thus, for example, the bacteria population at 2:30 P.M. $\left(t = \frac{5}{2}\right)$ is given by

$$f\left(\frac{5}{2}\right) = 2^{5/2} = (\sqrt{2})^5 \approx 5.66$$

which corresponds to 5660 bacteria. Irrational input values can be justified using methods from calculus. We simply assume that expressions such as $2^{\sqrt{2}}$ are valid and can be approximated, should the need arise, using the power key on a calculator.

----》**EXAMPLE 1**

Graphing an Exponential Function

Sketch the graph of the function $f(x) = 2^x$, and determine its domain and range.

Solution   We begin by selecting a few convenient x-values and computing the corresponding function values. For example, with $x = -3$, $f(-3) = 2^{-3} = \frac{1}{8}$. Thus, $\left(-3, \frac{1}{8}\right)$ is a point on the graph. Other values are shown in Table 2. After plotting these points and connecting them with a smooth curve, we obtain the graph shown in Figure 1.

Table 2

x	$f(x) = 2^x$
-3	$\frac{1}{8}$
-2	$\frac{1}{4}$
-1	$\frac{1}{2}$
0	1
1	2
2	4
3	8

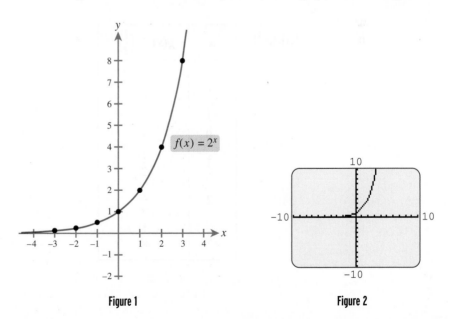

Figure 1

Figure 2

Alternatively, we could use a graphing calculator to obtain a graph similar to that shown in Figure 2. Recall that the domain of a function is the set of all possible input values (x-coordinates on the graph), and the range is the set of all possible output values (y-coordinates on the graph). Thus, our graph suggests that the domain of f is the set of all real numbers, and the range of f is the set of positive real numbers, or $(0, \infty)$.

The function $f(x) = 2^x$ is more properly called the *exponential function with base 2*. For arbitrary bases, we give the following definition.

Definition of the Exponential Function with Base a

The function $f(x) = a^x$, for $a > 0$ and $a \neq 1$, is the **exponential function with base a**.

When $a > 1$, the graph of the exponential function $f(x) = a^x$ rises sharply as x gets larger, as shown in Figure 2, for example. However, when $a < 1$, the graph falls as x gets larger, as illustrated by the following example.

EXAMPLE 2

Graphing an Exponential Function with a Base Less than 1

Sketch the graph of the exponential function $g(x) = \left(\frac{1}{2}\right)^x$, and determine its domain and range.

Solution Once again, we begin by selecting a few convenient x-values and computing the corresponding function values. For example, with $x = -3$, $g(-3) = \left(\frac{1}{2}\right)^{-3} = 2^3 = 8$. Thus, $(-3, 8)$ is a point on the graph. Other values are shown in Table 3. After plotting these points and connecting them with a smooth curve, we obtain the graph shown in Figure 3.

Table 3

x	$g(x) = \left(\frac{1}{2}\right)^x$
-3	8
-2	4
-1	2
0	1
1	$\frac{1}{2}$
2	$\frac{1}{4}$
3	$\frac{1}{8}$

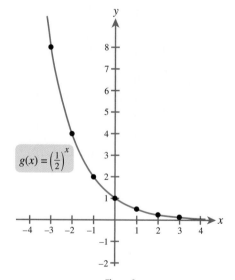

Figure 3

On the basis of the graph, we again conclude that the domain is the set of all real numbers, and the range is the set $(0, \infty)$.

The graph of $g(x) = \left(\frac{1}{2}\right)^x$ in Figure 3 appears to be a mirror image of the graph of $f(x) = 2^x$ from Figure 1. To confirm this, recall that the graph of a function g is the reflection about the y-axis of the graph of a function f if and only if $f(-x) = g(x)$. In this case, we have

$$f(x) = 2^x$$
$$f(-x) = 2^{-x}$$
$$= \frac{1}{2^x} = \left(\frac{1}{2}\right)^x = g(x)$$

Thus, the graphs of f and g are indeed reflections of one another about the y-axis. More generally, we have the following property.

The Reflective Property of Exponential Functions

For $a > 0$ and $a \neq 1$, the graph of $g(x) = \left(\frac{1}{a}\right)^x$ is the reflection about the y-axis of the graph of $f(x) = a^x$.

Figure 4 illustrates the reflective property of exponential functions.

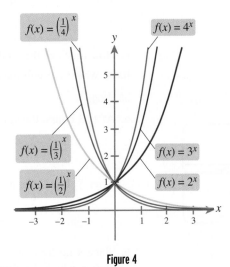

Figure 4

Several important properties of exponential functions are illustrated by the examples shown in Figure 4. First, we note that each exponential function f satisfies $f(x) > 0$ for all real numbers x. Thus, the domain of f is the set of all real numbers, and the range is the set of all *positive* real numbers. Note also that since $a^0 = 1$, all exponential functions of the form $f(x) = a^x$ have y-intercept $(0, 1)$. Also, when the base a is greater than 1 [as with $f(x) = 2^x$, $f(x) = 3^x$, or $f(x) = 4^x$], the graph rises from left to right, so that f is an increasing function. However, when the base is less than 1 [as with $f(x) = \left(\frac{1}{2}\right)^x$, $f(x) = \left(\frac{1}{3}\right)^x$, or $f(x) = \left(\frac{1}{4}\right)^x$], the graph falls from left to right, and so f is a decreasing function. We also observe that graphs of exponential functions pass the horizontal line test, and so exponential functions are 1–1. Finally, we see that the x-axis is a horizontal asymptote for the graph of each exponential function f. These properties of exponential functions are summarized as follows.

Properties of Exponential Functions

Let $f(x) = a^x$, $a > 0$, $a \neq 1$.

1. The domain of f is the set of all real numbers, and the range of f is the set of all positive real numbers.
2. The x-axis is a horizontal asymptote for the graph of f.
3. f has y-intercept $(0, 1)$ (that is, $a^0 = 1$).
4. f is increasing if $a > 1$ and decreasing if $0 < a < 1$.
5. f is 1–1 (that is, the graph of f passes the horizontal line test).

Exponential functions arise frequently in medical sciences such as pharmacology. In the next example, we see how drug levels can be modeled with exponential functions.

EXAMPLE 3

Medication in the Body

A 4-milligram dose of a certain medication is eliminated from the body in such a way that the quantity remaining at a given time is $\frac{2}{3}$ of the amount that was there 1 hour earlier. Thus, after t hours, the amount remaining in the body is given by the function

$$g(t) = 4\left(\frac{2}{3}\right)^t$$

a. How much of the medication remains after 3 hours? After $5\frac{1}{2}$ hours?

b. At what time will the amount in the body be half the original quantity?

Solution

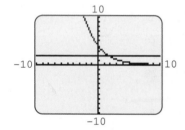

Figure 5

a. After 3 hours, the amount of medication remaining is given by

$$g(3) = 4\left(\frac{2}{3}\right)^3 = 4 \cdot \frac{8}{27} \approx 1.185 \text{ mg}$$

After $5\frac{1}{2}$ hours, the amount is

$$g\left(\frac{11}{2}\right) = 4\left(\frac{2}{3}\right)^{\frac{11}{2}} = 4\left(\frac{2}{3}\right)^{5.5} \approx 0.430 \text{ mg}$$

b. Since 4 milligrams of medication are in the body initially, we are interested in the time at which 2 milligrams of the substance remain. Thus, we are interested in the value of t for which

$$4\left(\frac{2}{3}\right)^t = 2$$

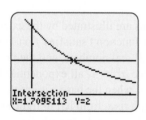

Figure 6

The solution of this equation corresponds to the x-coordinate of the point of intersection of the graphs of $y = 2$ and $y = 4(2/3)^t$ (see Figure 5). Zooming in on the point of intersection and using the calculate-intersect feature gives us a t-value of approximately 1.71, as shown in Figure 6. Thus, after 1.71 hours (roughly 1 hour and 43 minutes), half the original dose of medication remains in the body.

The 1–1 property of exponential functions is often useful for solving simple equations involving exponents. It can be stated in the following way.

The 1-1 Property of Exponential Functions

For all real numbers r and s, if $a^r = a^s$, then $r = s$.

EXAMPLE 4

Solving an Exponential Equation

Use the 1–1 property of exponential functions to solve the exponential equation

$$\frac{1}{4^x} = 64$$

Solution   First we must rewrite both sides of the equation as a power of 4—that is, in the form $4^{\square}$. This is possible because $1/4^x$ can be written as 4^{-x} and 64 can be written as 4^3. We proceed as follows:

$$\frac{1}{4^x} = 64$$

$$4^{-x} = 4^3 \qquad \text{Using the identity } \frac{1}{a^n} = a^{-n} \text{ and rewriting 64 as } 4^3$$

$$-x = 3 \qquad \text{Applying the 1–1 property}$$

$$x = -3$$

Notice that an exponential identity was needed in the preceding example when we claimed that $1/4^x = 4^{-x}$. Using techniques from calculus, the familiar exponential identities that we have already seen for integer and rational exponents can be shown to be valid for any real number exponent. For convenience, we summarize these identities again.

Exponential Identities

Let a and b be positive real numbers, and let r and s be real numbers.

1. $a^r a^s = a^{r+s}$ **5.** $\left(\dfrac{a}{b}\right)^r = \dfrac{a^r}{b^r}$

2. $\dfrac{a^r}{a^s} = a^{r-s}$ **6.** $a^0 = 1$

3. $(a^r)^s = a^{rs}$ **7.** $a^{-r} = \dfrac{1}{a^r}$

4. $(ab)^r = a^r b^r$

EXAMPLE 5

Solving an Exponential Equation

Solve the exponential equation $3^x \cdot 3^{x+1} = \frac{1}{27}$.

Solution

$$3^x \cdot 3^{x+1} = \frac{1}{27}$$

$$3^{x+x+1} = \frac{1}{27} \qquad \text{Applying the exponential identity } a^r a^s = a^{r+s}$$

$$3^{2x+1} = \frac{1}{3^3} \qquad \text{Rewriting 27 as a power of 3}$$

$$3^{2x+1} = 3^{-3}$$ Using the exponential identity $\dfrac{1}{a^n} = a^{-n}$
$$2x + 1 = -3$$ Applying the 1–1 property
$$2x = -4$$
$$x = -2$$

It should be noted that the technique shown in Examples 4 and 5 is practical only for exponential equations that can easily be written in the form

$$a^\square = a^\bigcirc$$

For example, the method does not apply for the equation $2^x = 25$. In this case, it is not possible to rewrite 25 as an integer power of 2. We will develop an alternative approach using logarithms in Section 4.4.

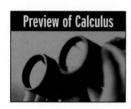

Preview of Calculus

Differential Equations

In this section, our focus has been on exponential functions, which arise whenever the percentage rate of change of a quantity is constant or, equivalently, when the rate at which a quantity grows (or shrinks) is proportional to the quantity itself. Now, if we assume that such a quantity is represented by a function $f(t)$, then we have

Rate of change of $f(t) = k \cdot f(t)$

This is an example of a **differential equation**, an equation involving the rate at which a quantity changes.

Virtually any statement about how a quantity changes over time can be recast as a differential equation. For example, a certain drug may enter the bloodstream via absorption into the gastrointestinal tract and then gradually leave the bloodstream as the blood is filtered by the liver and kidneys. This information leads us to a differential equation relating the rate at which the drug level is changing to the amount of drug in the bloodstream and the elapsed time since the drug was introduced into the system. By solving the differential equation, researchers are able to predict the amount of drug in the bloodstream as a function of time, which, in turn, allows for the determination of safe and effective dosages.

The subject of differential equations, though an important offshoot of calculus, is an entire branch of mathematics in its own right, with important applications in nearly every branch of science, engineering, and technology.

The Natural Exponential Function

The frequency with which the number e appears in mathematical formulas from diverse branches of mathematics (number theory, probability, statistics, and calculus, for example) is rivaled only by that of π. Like π, e is an irrational number, and so its decimal expansion neither repeats nor terminates. Its decimal expansion begins

$$e = 2.71828\ldots$$

Perhaps the most important function in all of mathematics is the exponential function with base e, $f(x) = e^x$, also known as the **natural exponential function**. Since e is a number between 2 and 3, the graph of the natural exponential function $f(x) = e^x$ lies between the graphs of $y = 2^x$ and $y = 3^x$, as shown in Figure 7.

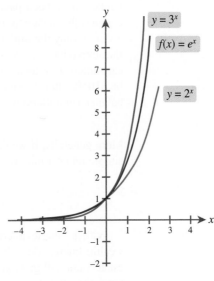

Figure 7

EXAMPLE 6

Using a Translation to Graph an Exponential Function

Use the graph of $f(x) = e^x$ to sketch the graph of $g(x) = e^{x+3}$.

Solution   Since $g(x)$ can be obtained by replacing x with $x + 3$ in the function $f(x) = e^x$, the graph of g can be found by translating the graph of f to the left 3 units, as shown in Figure 8. Note that the y-intercept $(0, 1)$ on the graph of $f(x) = e^x$ is translated 3 units to the left to obtain the corresponding point $(-3, 1)$ on the graph of $g(x) = e^{x+3}$.

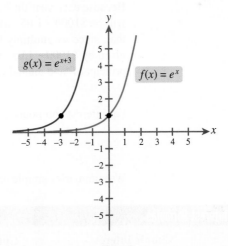

Figure 8

Simple and Compound Interest

Financial contracts have existed since the dawn of recorded history. In fact, evidence suggests that writing was invented in Mesopotamia around 3000 B.C. for the purpose of recording financial transactions. For thousands of years, beginning even before the invention of currency, lenders have been charging interest on loaned commodities, and bankers have been paying interest on deposited commodities. Up until the early 19th century, the most common form of interest was **simple interest**. With a simple interest account, only the initial amount invested (the **principal**) earns interest. For example, if the simple interest rate is 5% and the principal is $1000, then each year the account will earn interest in the amount of 5% · $1000 = $50, so that after 1 year, the balance will be $1050; after 2 years, the balance will be $1100; and so forth. Thus, after t years the balance (in dollars) will be given by

$$B = 1000 + 50t$$

More generally, if we denote the balance after t years by B, the principal by P, and the simple interest rate by r, we have

$$B = P + (\text{Interest per year})t$$
$$= P + (rP)t$$
$$= P(1 + rt)$$

With a simple interest account, the principal does all the work of creating new wealth: Interest doesn't earn new interest. It follows that the balance of a simple interest account will grow at a constant rate. For a **compound interest** account, on the other hand, the entire balance is put to work creating wealth: Both principal and interest alike earn interest. For example, if interest is 5% compounded annually, then after the first year the balance will be $1050, just as with simple interest. But in the second year, the entire balance of $1050 will earn interest, so that the interest earned in the second year will be 5% · $1050 = $52.50, giving us a balance after 2 years of $1102.50. In fact, each year we will earn interest equal to 5% of the balance at the beginning of the year. In other words, for each year we have

$$\text{New balance} = \text{Beginning balance} + \text{Interest}$$
$$= \text{Beginning balance} + 5\% \text{ of Beginning balance}$$
$$= (\text{Beginning balance})(1.05)$$

Because each year the balance is multiplied by a factor of 1.05, the balance after 1 year will be $1000 · 1.05; after 2 years, we will have $1000 · 1.05^2$; and so forth. It follows that, since we multiply by a factor of 1.05 each year, the balance after t years will be given by $1000 · 1.05^t$. More generally, the balance B after t years for an account with principal P and annual compound interest rate r is given by

$$B = P(1 + r)^t$$

For interest compounded n times per year, a similar development would yield

$$B = P\left(1 + \frac{r}{n}\right)^{nt}$$

We summarize simple and compound interest as follows.

Simple and Compound Interest Formulas

	Simple interest	Compound interest
Interest earned by ...	principal only	entire balance—both principal and interest
Balance grows ...	at a constant rate (for example, $50 per year)	at a constant percentage rate (for example, 5% per year)
Balance is given by ...	$B = P(1 + rt)$	$B = P\left(1 + \frac{r}{n}\right)^{nt}$, where n is the number of compounding periods per year

EXAMPLE 7

Computing Compound Interest

There is an apocryphal tale of a young George Washington throwing a silver dollar across the Potomac River. Suppose that instead of throwing the dollar, he had deposited it in 1776 in a bank that paid 6% interest. How much would that dollar be worth in the year 2005 if interest were compounded

a. annually?   **b.** quarterly?

c. daily?   **d.** every minute?

Solution

a. Letting $P = 1$, $r = 0.06$, $t = 229$, and $n = 1$ in the compound interest formula, we obtain

$$B = P\left(1 + \frac{r}{n}\right)^{nt} = 1\left(1 + \frac{0.06}{1}\right)^{1(229)} = (1.06)^{229} \approx 623{,}796.80$$

b. Using $n = 4$, the number of quarters in a year, we obtain

$$B = P\left(1 + \frac{r}{n}\right)^{nt} = 1\left(1 + \frac{0.06}{4}\right)^{4(229)} = (1.015)^{916} \approx 837{,}326.24$$

c. With $n = 365$, we obtain

$$B = 1\left(1 + \frac{0.06}{365}\right)^{365(229)} \approx 926{,}223.47$$

d. Letting $n = 365 \cdot 24 \cdot 60 = 525{,}600$, the number of minutes in a year, we obtain

$$B = 1\left(1 + \frac{0.06}{525{,}600}\right)^{525{,}600(229)} \approx 927{,}269.21$$

Note that the answer your calculator gives may differ because of round-off error.

From the preceding example, it is evident that the greater the number of compounding periods, the greater the balance. One might expect that the balance could be made as large as desired simply by compounding with sufficient frequency. Surprisingly, though, there is a limit; no matter how many compounding periods are used, George Washington's dollar could never accumulate to more than $927,269.94 in 229 years. To see where this amount comes from, we first examine how the natural exponential function is connected to compound interest.

Suppose you were lucky enough to find a bank that paid 100% interest, and you deposited a dollar for 1 year. By setting $P = 1$, $r = 1.00$, and $t = 1$ in the compound interest formula, and by considering larger and larger values of n (the number of compounding periods), we obtain the table of values (right).

Notice that as the number of compounding periods increases, the balance gets closer to the number e. Actually, this is no coincidence since the number e can be *defined* as the limiting value of $\left(1 + \frac{1}{n}\right)^n$. So if the bank compounded interest *continuously*, your dollar would increase in value to exactly e dollars after 1 year. For arbitrary values of P, r, and t, we have the following general formula for **continuously compounded interest**.

n	$B = \left(1 + \dfrac{1}{n}\right)^n$
1	2
10	2.59374
100	2.70481
1000	2.71692
10,000	2.71815
100,000	2.71827
1,000,000	2.71828

Continuously Compounded Interest

If a principal of P dollars is deposited in an account paying an annual rate of interest r compounded continuously, then the balance B in the account after t years is given by

$$B = Pe^{rt}$$

EXAMPLE 8

Computing Continuously Compounded Interest

Suppose that George Washington deposited his dollar in 1776 in a bank that compounded interest continuously at a rate of 6%. How much would it be worth in the year 2005?

Solution By applying the formula for continuously compounded interest with $P = 1$, $r = 0.06$, and $t = 229$, the balance would be

$$B = 1e^{(0.06)229} = e^{13.74} \approx 927{,}269.94$$

Thus, the dollar would be worth $927,269.94 in the year 2005.

Understanding and Mastery Checklists

Concepts to Understand

Exponential function with base a

The 1–1 property of exponential functions

Exponential identities

The natural exponential function

Compound interest

Continuously compounded interest

Skills to Master

Sketch the graph of an exponential function.

Use the 1–1 property of exponential functions to solve exponential equations.

Use the compound interest formula to find the balance in an account.

Use the continuously compounded interest formula to find the balance in an account.

Exercises 4.1

Exercises 1-2 *Sketch the graph of f and use translations or reflections to sketch the graph of g.*

1. $f(x) = 3^x$

   a. $g(x) = -3^x$ **b.** $g(x) = 3^{-x}$

   c. $g(x) = 3^x + 2$ **d.** $g(x) = 3^{x+2}$

2. $f(x) = \left(\frac{1}{4}\right)^x$

   a. $g(x) = -\left(\frac{1}{4}\right)^x$ **b.** $g(x) = \left(\frac{1}{4}\right)^{-x}$

c. $g(x) = \left(\frac{1}{4}\right)^x - 3$ **d.** $g(x) = \left(\frac{1}{4}\right)^{x-2}$

Exercises 3-6 *Determine the value of a for which $f(x) = a^x$ has the indicated graph.*

3.

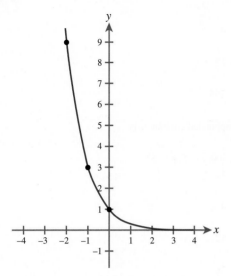

4.

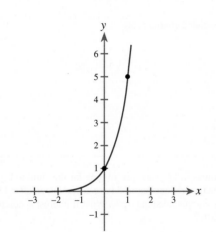

5.

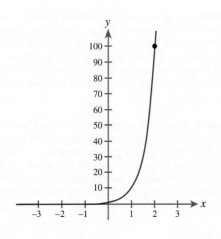

6.

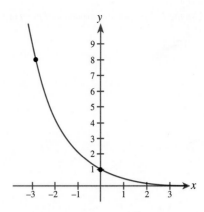

Exercises 7-10 *Use the given graph of $f(x) = e^x$ to sketch the graph of $g(x)$.*

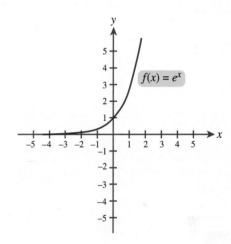

7. $g(x) = e^{-x}$

8. $g(x) = -e^x$

9. $g(x) = e^{x-2}$

10. $g(x) = e^x + 3$

Exercises 11-14 *Use a graphing calculator to help you solve the given equation. Approximate all answers to the nearest hundredth.*

11. $4e^{0.1t} = 13$

12. $2e^{-0.2t} = 1$

13. $1200e^{-0.05t} = 100$

14. $8000e^{0.07t} = 9000$

Exercises 15-24   *Use the 1–1 property of exponential functions to solve the given equation.*

15. $3^x = 81$

16. $4^{-x} = 64$

17. $2^{3x-4} = 32$

18. $5^{1-x} = 125$

19. $\dfrac{1}{3^x} = 27$

20. $2^{x-1} = \dfrac{1}{64}$

21. $4^{x-3} = 16 \cdot 4^{2x}$

22. $3^{x+2} = \dfrac{81}{3^x}$

23. $2^{x^2+x} = 4$

24. $5^{x^2-5x} = \dfrac{1}{625}$

Exercises 25-26   *Determine the balance B if a principal of P dollars is deposited in an account for t years at an annual rate of interest r compounded n times per year. Give answers to the nearest cent.*

25. $P = \$1000,\ t = 10$ years, $r = 6\%$

   a. $n = 1$

   b. $n = 4$

   c. $n = 12$

   d. $n = 365$

   e. Compounded continuously

26. $P = \$2500,\ t = 5,\ r = 8\%$

   a. $n = 1$

   b. $n = 4$

   c. $n = 12$

   d. $n = 365$

   e. Compounded continuously

Applications

27. Population Growth   A breeder who supplies pet stores with gerbils has determined that a certain gerbil population can be approximated over a 4-month period with the function

$$f(t) = 20\left(\frac{3}{2}\right)^t$$

where t is measured in months. How many gerbils were present initially? Create a table showing the gerbil population at the end of each of the first 4 months. Assuming the pattern continues beyond 4 months, predict how many gerbils will be present after 6 months and again after 5 years.

28. Population Growth   An exterminating company has developed a new chemical-free method for exterminating cockroaches. They claim that a cockroach population with initial size 10,000 will number $f(t) = (10,000)2^{-t}$ after t days. Plot this function. Predict how many cockroaches will remain after 1 week. Estimate the day on which the last cockroach will die.

29. Compound Interest   On the occasion of the birth of their first grandchild, two proud grandparents deposit $5000 in a certificate of deposit (CD) in the child's name. The CD pays an annual interest rate of 8%,

and it matures in 18 years, just in time for the child to begin college. Determine the amount that will have accumulated after 18 years if interest is compounded quarterly. What if interest is compounded continuously?

30. Compound Interest   The grandparents in Exercise 29 wish to determine whether the CD they purchased will pay for the first year of their grandchild's college education. Assuming that 1 year at their alma mater currently costs $15,000 and the amount is likely to increase an average of 5% per year, use the formula for interest compounded annually to estimate what 1 year of college will cost in 18 years.

31. Energy Consumption   Energy consumption in the United States during the years 1935–2000 can be approximated using the exponential function $Q(t) = 23.17 \cdot 1.023^t$, where t denotes the year ($t = 0$ corresponds to 1935) and $Q(t)$ is the energy consumed in that year (in quadrillions of Btu's). Estimate the energy consumption in the years 1940 and 2000, and use these values to determine the percent increase from 1940 to 2000. What does this model predict for energy consumption in 2010?

Energy Consumption, 1935–2000

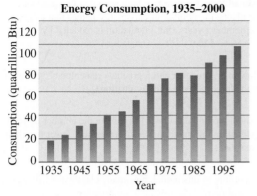

Data source: U.S. Department of Energy.

A field of windmills generates electricity near Altamont, California

A heliostat field focuses solar energy on a central receiving tower

32. Computer Transistors Moore's law, formulated in 1965 by Gordon Moore, cofounder of Intel, states that the number of transistors that can be put on a computer processor chip will approximately double every 18–24 months. In 1971, Intel introduced the 4004

processor with 2300 transistors. According to Moore's law, the number of transistors t years after 1971 can be approximated by $N = 2300 \cdot 2^{t/2}$. Use this function to estimate the number of transistors on a processor in 2000. The actual number of transistors on the Pentium 4 processor, introduced by Intel in 2000, was 42 million. How well does your estimate compare to this value? What does Moore's law predict for the number of transistors in the year 2005?

Intel Processors, 1971–2000

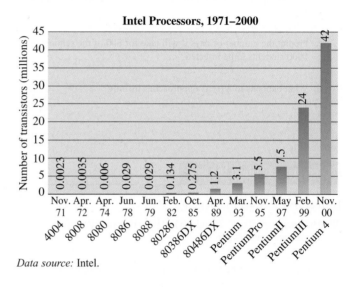

Data source: Intel.

33. Population Equation At the beginning of this section, we considered a bacteria population that was modeled by the function $f(t) = 2^t$, where t is in hours ($t = 0$ corresponds to 12:00 noon) and $f(t)$ is the population in thousands. Estimate the time at which the population reaches 25,000.

34. Intersection Points Use a graphing calculator to estimate the point(s) of intersection for the graphs of $y = 2^x$ and $y = x^2$. [*Hint:* It may be simpler to approximate the *three* zeros of $h(x) = 2^x - x^2$ than to use the calculate-intersect feature.]

35. Density Function The function $f(x) = (1/\sqrt{2\pi})e^{-x^2/2}$ is known as the *standard normal probability density function* and is important in the study of probability and statistics. Use a graphing calculator to determine the maximum value of f. What happens to the value of $f(x)$ as x approaches infinity or negative infinity?

36. Medication Concentration A certain medication is ingested and begins spreading throughout the bloodstream. The concentration (in nanograms per milliliter) of the medication in the bloodstream after t hours is given by $Q(t) = 4te^{-0.8t}$. Use a graphing calculator to plot this function. Estimate the maximum concentration of the medication and the time at which it occurs. What happens to the concentration as time increases beyond this point?

37. Contagious Virus A group of 100 students returns from a trip to Europe, where they were exposed to a highly contagious virus. The virus begins spreading among the other 900 students in their school in such a way that t days after their return, the number of students with the virus is modeled by

$$N(t) = \frac{100{,}000}{100 + 900e^{-0.15t}}$$

a. How many students have the virus after 2 days? After 5 days?

b. Use a graphing calculator to help you plot the graph of $N(t)$ for $t > 0$.

c. Predict the number of days it will take for 800 of the school's 1000 students to be infected.

d. What happens to $N(t)$ as t approaches infinity?

38. **Knowledge Retention** For a certain high school with a graduating class of 200 students, the number of classmates' names that a typical student can remember t years after graduation is modeled by $R(t) = 80 + 120e^{-0.75t}$.

a. How many students will a graduate remember 1 year after graduation? 5 years after graduation?

b. Use a graphing calculator to estimate the number of years after graduation until a graduate will remember fewer than half of her classmates.

c. According to this model, how many classmates will a graduate never forget?

Concepts and Critical Thinking

Exercises 39–42 *Answer true or false.*

39. The graphs of $f(x) = 2^x$ and $g(x) = \left(\frac{1}{2}\right)^x$ are reflections of one another about the y-axis.

40. Simple interest can be modeled with an exponential function.

41. All exponential functions are 1–1.

42. For some exponential functions f, $f(x) \to 0$ as $x \to \infty$.

Exercises 43–46 *Give an example of each.*

43. An exponential function that is decreasing over its entire domain

44. Two exponential functions such that the graph of the second is a translation, 6 units to the left, of the graph of the first

45. A point on the graph of every exponential function of the form $f(x) = a^x$

46. A number that is not in the range of any exponential function of the form $f(x) = a^x$

Questions for Discussion or Essay

47. Discuss the restrictions placed on the value of a in the definition of the exponential function $f(x) = a^x$. In particular, consider the following questions: Why does the definition exclude the value $a = 1$? Isn't $f(x) = 1^x$ a valid function? Why isn't it considered an exponential function? Why must we have $a > 0$? What difficulties do we encounter in defining the function $f(x) = (-2)^x$?

48. Discuss the limitations of the technique used in Example 4 for solving simple exponential equations. Describe a procedure for using a graphing calculator to find approximate solutions to exponential equations.

49. Although many quantities grow exponentially over relatively short periods of time, exponential models can yield bizarre predictions for longer time periods. Consider, for example, the exponential function $f(t) = 2^t$, which was used to model the bacteria population at the

beginning of this section. If the growth were to continue at this exponential pace, show that the number of bacteria at the end of 1 week would be 3.74×10^{50}. Assuming an "average" size of bacteria, this would be enough bacteria to fill a sphere with diameter that of our solar system. What is wrong with the model? What would really happen to the bacteria population over time?

50. A careful comparison of banking practices would reveal that many methods are used by banks to compound interest. Even among banks that claim to have the same compounding periods, there are differences in when the interest is actually paid into an account. Contact several local banks and inquire about their interest rates and compounding schemes for a specific type of account. Discuss some ways for comparing banks so that you can choose the one that pays "the best" interest.

Projects for Enrichment

51. Savings Plans We have seen how the balance in an account can be computed using the compound interest formula. Here we consider the effect of compound interest on deposits made at equally spaced time intervals throughout the year. To that end, suppose that an amount A is deposited k times each year in an account earning an annual rate of interest r compounded n times per year (we assume the account is opened with the first deposit of the amount A and that the deposits are equally spaced from that point on). After t years, the balance in the account will have accumulated to

(1)
$$B = \frac{A\left[\left(1 + \frac{r}{n}\right)^{nt} - 1\right]}{1 - \left(1 + \frac{r}{n}\right)^{-\frac{n}{k}}}$$

a. Use equation (1) to find the balance after 10 years if monthly deposits of $100 are placed into an account paying 5% interest compounded quarterly.

b. Use equation (1) to find the balance for the following savings plans:

   i. Monthly deposits of $50 in an account paying 6% interest compounded monthly for 20 years

   ii. Quarterly deposits of $150 in an account paying 6% interest compounded monthly for 20 years

Note that for each plan, the same total is deposited each year. Which plan yields the larger balance? Explain.

c. Solve equation (1) for A. This will yield a formula for finding the deposit amount that will yield a balance B after t years in an account where the annual rate of interest r is compounded n times per year and deposits are made k times per year.

d. Find the deposit amount necessary to achieve the following savings goals:

   i. A balance of $50,000 after 20 years with an interest rate of 5% compounded daily and deposits made monthly

   ii. A balance of $1,000,000 after 40 years with an interest rate of 8% compounded monthly and deposits made monthly

For both of parts i and ii, use the standard compound interest formula to determine the amount of a single deposit that would be necessary to achieve the indicated savings goals.

e. Use a graphing calculator to find the number of years that would be necessary to achieve the following savings goals:

   i. A balance of $10,000 in an account with an interest rate of 5.5% compounded yearly and $50 deposits made monthly

   ii. A balance of $100,000 in an account with an interest rate of 7.25% compounded monthly and $50 deposits made weekly

52. Continuously Compounded Interest Using the fact that $\left(1 + \frac{1}{m}\right)^m$ approaches e as m approaches infinity, we can derive the formula for continuously compounded interest from the compound interest formula $B = P\left(1 + \frac{r}{n}\right)^{nt}$.

a. Show that the formula for compound interest can be rewritten as
$$B = P\left[\left(1 + \frac{1}{n/r}\right)^{n/r}\right]^{rt}$$

b. Let $m = n/r$. What happens to the value of m if r remains fixed but n approaches infinity?

c. Substituting $m = n/r$ into the formula in part a, we obtain
$$B = P\left[\left(1 + \frac{1}{m}\right)^m\right]^{rt}$$

Using your observation from part b, argue that as the number of compounding periods approaches infinity, the balance gets closer in value to Pe^{rt}.

d. Surprisingly, continuous compounding does not offer a significant advantage over annual compounding at the same rate. To see this, suppose you were to find a bank that pays 5% interest compounded continuously. If you deposit P dollars, the balance after t years would be $Pe^{0.05t}$. Now suppose a different bank pays 5.13% interest compounded annually. If you deposit P dollars there, the balance after t years would be $P(1 + 0.0513)^t$. In which of these banks should you invest your money?

e. Find a value for r for which quarterly compounding at an annual rate r will be equivalent to continuous compounding at an annual rate of 5%.

Section 4.2 | # Logarithmic Functions

- According to Newton, how long does it take a 185° turkey to cool to 100°?
- What makes acid rain acidic?
- How could you cross a river with an unlimited supply of pizza boxes, a parachute, and a mathematician?
- How many decibels is the song of the blue whale, a sound with more than 6 quintillion times the energy of the faintest sound perceptible to the human ear?
- How can a pathologist use body temperature to estimate time of death?

The Logarithm Function with Base a

As Figures 9 and 10 suggest, and as we observed in the previous section, the exponential function $f(x) = a^x$ (with $a > 0$ and $a \neq 1$) passes the horizontal line test and is thus 1–1: If $a^x = a^y$, then $x = y$.

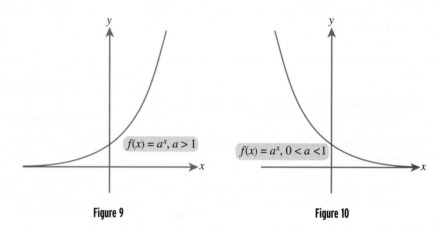

Figure 9   **Figure 10**

Because exponential functions are 1–1, they have inverses. We define *logarithms* to be the inverse functions of exponentials. Intuitively, we think of logarithmic functions as undoing exponential functions in much the same way that subtracting 7 undoes adding 7, that dividing by 3 undoes multiplication by 3, that taking a cube root undoes cubing a number, and so forth. More precisely, we have the following definition.

Definition of the Logarithm Function with Base a

The function $g(x) = \log_a x$ (read "log base a of x") for $a > 0$ and $a \neq 1$, the **logarithm function with base a**, is the inverse of the exponential function $f(x) = a^x$. Equivalently,

$$y = \log_a x \quad \text{if and only if} \quad a^y = x$$

Thus, we may think of $\log_a x$ as the power to which a must be raised in order to obtain x.

Since logarithmic functions are inverses of exponential functions, their graphs are reflections of one another about the line $y = x$. The graphs of a few exponential functions and the logarithmic functions that are their inverses are shown in Figure 11.

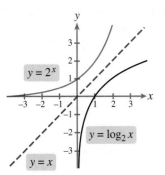
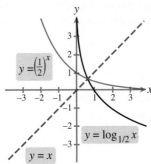
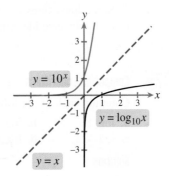
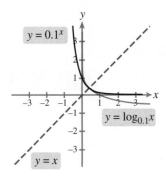

Figure 11

Several properties of logarithmic functions are suggested by the graphs in Figure 11, each of which can be verified from the definition of logarithms. For example, the graphs suggest that all logarithmic functions $g(x) = \log_a x$ have an x-intercept of 1, or that $\log_a 1 = 0$. Since $a^0 = 1$, $\log_a 1$ is indeed 0. This and other properties of logarithmic functions are summarized as follows.

Properties of Logarithmic Functions

Let $g(x) = \log_a x$, $a > 0$, $a \neq 1$.

1. The domain of g is $(0, \infty)$, and the range of g is $(-\infty, \infty)$.
2. The y-axis is a vertical asymptote of the graph of g.
3. g has x-intercept 1 (that is, $\log_a 1 = 0$).
4. g is increasing if $a > 1$ and decreasing if $0 < a < 1$.
5. g is 1–1 (that is, if $\log_a x = \log_a y$, then $x = y$).

Two additional properties of logarithms can be obtained directly from the definition of logarithmic functions as inverses of exponential functions. Since $f(x) = a^x$ and $g(x) = \log_a x$ are inverses, each "undoes" the other; that is, $f(g(x)) = x$ and $g(f(x)) = x$. This gives us

$$f(g(x)) = x \qquad g(f(x)) = x$$
$$f(\log_a x) = x \qquad g(a^x) = x$$
$$a^{\log_a x} = x \qquad \log_a a^x = x$$

The first of these properties is merely a restatement of the fact that $\log_a x$ is the power to which a must be raised in order to get x. The second property gives us the self-evident statement that "x is the power to which a must be raised in order to get a^x."

Inverse Identities

Let $a > 0$ and $a \neq 1$.

Property	Example
1. $a^{\log_a x} = x$ for all $x > 0$	$2^{\log_2 5} = 5$
2. $\log_a(a^x) = x$ for all x	$\log_3(3^{-8}) = -8$

Some logarithms can be computed directly from the definition, or, equivalently, by using the inverse identities, as illustrated by the following example.

EXAMPLE 1

Computing Logarithms

Compute each of the following quantities:

a. $\log_2 8$ **b.** $\log_{1/3} 9$

c. $\log_{1.23} 1$ **d.** $\log_4(-2)$

Solution

a. We need to find the power to which 2 must be raised in order to get 8. Since $2^3 = 8$, $\log_2 8 = 3$.

b. We are interested in finding the power to which $\frac{1}{3}$ must be raised in order to get 9. Since $\left(\frac{1}{3}\right)^{-2} = 9$, $\log_{1/3} 9 = -2$.

c. To what power must 1.23 be raised in order to get 1? Since $1.23^0 = 1$, $\log_{1.23} 1 = 0$. As previously stated, $\log_a 1 = 0$ for all possible bases a.

d. Here we are interested in solutions to $4^y = -2$. But $4^y > 0$ for all values of y, so this equation has no solution. In other words, $\log_4(-2)$ is not defined. More generally, the domain of logarithmic functions consists of all positive numbers. Thus, $\log_a x$ is defined only for $x > 0$.

Most logarithms cannot be evaluated using the technique illustrated in Example 1. Consider, for example, $\log_2 5$. Here we wish to find the power to which 2 must be raised in order to get 5. Since $2^2 = 4$ and $2^3 = 8$, we would expect that the desired power is somewhere between 2 and 3. In the following example, we consider a graphical approach to estimating logarithms such as $\log_2 5$. A more computational approach is considered in Section 4.3.

EXAMPLE 2

Estimating a Logarithm Graphically

Use a graphing calculator to estimate $\log_2 5$.

Solution   Since $\log_2 5$ is that power to which 2 must be raised in order to get 5, we are looking for the value of x for which $2^x = 5$. Thus, we plot the graph of $y = 2^x$ and $y = 5$, and use the calculate-intersect feature to estimate the x-coordinate of the point of intersection, as shown in Figure 12. We obtain $x \approx 2.32$.

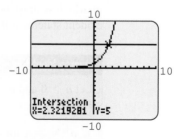

Figure 12

From the definition of logarithm, $y = \log_a x$ if and only if $a^y = x$. In other words, for every logarithmic equation, there is a corresponding exponential equation, and vice versa.

> **Equivalence of Exponential and Logarithmic Equations**
>
> The logarithmic equation $y = \log_a x$ and the exponential equation $a^y = x$ are equivalent.

EXAMPLE 3

Writing Logarithmic Equations as Exponential Equations

Write an equivalent exponential equation for each of the following logarithmic equations:

a. $\log_{10} 1000 = 3$ **b.** $\log_x 100 = 2.1$ **c.** $\log_3(x^2) = 12$

Solution

a. $10^3 = 1000$

b. $x^{2.1} = 100$

c. $3^{12} = x^2$

Rule of Thumb

When converting between exponential equations and logarithmic equations, it is helpful to keep the following rules of thumb in mind:

- The base of the logarithm is the base of the exponent (that is, the quantity being raised to the power).
- The value of the logarithm is the exponent.

EXAMPLE 4

Writing Exponential Equations as Logarithmic Equations

For each exponential equation, write an equivalent logarithmic equation.

a. $0.5^{-2} = 4$ **b.** $e^0 = 1$ **c.** $1.015^{12t} = 1,000,000$

Solution

a. $\log_{0.5} 4 = -2$

b. $\log_e 1 = 0$

c. $\log_{1.015} 1,000,000 = 12t$

The correspondence between exponential and logarithmic equations is also useful for finding inverses of functions involving logarithms or exponents, as we see in the following example.

EXAMPLE 5

Finding the Inverse of a Function Involving Logarithms

Compute the inverse of $f(x) = 4 \log_2(x + 1)$.

Solution Recall from Section 2.4 that if $y = f(x)$ defines a 1–1 function, then $x = f^{-1}(y)$. Thus, to find $f^{-1}(y)$, we solve the equation $y = f(x)$ for x.

$$y = 4 \log_2(x + 1) \qquad \text{Replacing } f(x) \text{ with } y$$

$$\frac{y}{4} = \log_2(x + 1)$$

$$2^{y/4} = x + 1 \qquad \text{Converting to exponential form}$$

$$x = 2^{y/4} - 1 \qquad \text{Solving for } x$$

Thus, we have $f^{-1}(y) = 2^{y/4} - 1$. Substituting x in place of y, we obtain

$$f^{-1}(x) = 2^{x/4} - 1$$

Natural and Common Logarithms

The most commonly used logarithmic bases—by far—are 10 and e. The relative prominence of base 10 logarithms, also known as **common logarithms**, derives from the fact that 10 is the base for our number system. Although their role in computation (see Exercises 83 and 84 in Section 4.3) is mostly of historical significance in the posttransistor era, common logarithms survive in many scientific contexts, including decibel ratings for loudness, the Richter scale for earthquakes, and the pH level for measuring acidity. If we were to encounter intelligent life from a distant galaxy, we would be shocked if we found that common logarithms enjoyed any special status there. By contrast, one could argue that base e, or **natural logarithms**, would eventually emerge as the logarithm of choice in any civilization. Arising naturally in such seemingly unrelated areas as continually compounded interest and radioactive decay, natural logarithms (and their base e) seem part of the very fabric of our universe.

Special notation has developed for expressing common and natural logarithms. In the case of common logarithms, the base is omitted, whereas with natural logarithms, the symbol "ln" is used.

Notation for Common and Natural Logarithms

$\log x = \log_{10} x$. Base 10 logarithms are called **common logarithms**.
$\ln x = \log_e x$. Base e logarithms are called **natural logarithms**.

All of the properties that apply to general logarithm functions apply to common and natural logarithms as well. For this reason, the following properties should not be viewed as new formulas, but rather as particular instances of properties that we have already encountered.

Properties of Common and Natural Logarithms

- $10^{\log x} = x$ and $e^{\ln x} = x$
- $\log(10^x) = x$ and $\ln(e^x) = x$
- $\log 1 = \ln 1 = 0$
- The domain of both $f(x) = \log x$ and $g(x) = \ln x$ is the set of positive real numbers.
- The function $f(x) = \log x$ is the inverse of $y = 10^x$, and the function $g(x) = \ln x$ is the inverse of $y = e^x$.

EXAMPLE 6

Computing Common and Natural Logarithms

Compute each of the following quantities, using a calculator as necessary:

a. $\log 1000$ **b.** $\ln \dfrac{1}{e}$ **c.** $\log(2.4 \times 10^8)$

Solution

a. Because the base isn't given, it is understood to be 10. Thus, we must find the power to which 10 is raised in order to get 1000. Since $10^3 = 1000$, $\log 1000 = 3$.

b. Here the base is e. We must find the power to which e is raised in order to get $1/e$. Since $e^{-1} = 1/e$, $\ln(1/e) = -1$.

c. This one requires a calculator. However, even without a calculator, we can make a rough estimate. Since $\log(10^x) = x$, it follows that, for example, $\log(10^8) = 8$ and $\log(10^9) = 9$. Thus, since 2.4×10^8 is between 10^8 and 10^9, its common logarithm is somewhere between 8 and 9. Using the common logarithm key on our calculator, we find that $\log(2.4 \times 10^8) \approx 8.38$.

EXAMPLE 7

Graphing a Function Involving the Natural Logarithm

Sketch the graph of $f(x) = \ln x$, and use it to sketch the graph of $g(x) = \ln x - 2$.

Solution Since $f(x) = \ln x$ is the inverse of the natural exponential function e^x, the graph of f is the reflection of the graph of $y = e^x$ about the line $y = x$, as shown in Figure 13. Note that the graph of f has the y-axis as a vertical asymptote, an x-intercept of 1, and is only defined for positive values of x. Moreover, since the base e is greater than 1, the graph of f increases from left to right. Now $g(x) = \ln x - 2 = f(x) - 2$. So the graph of g can be obtained by translating the graph of f down 2 units. The result is shown in Figure 13.

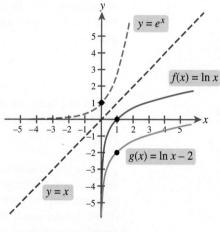

Figure 13

In the following example, we use a graphing calculator to explore Newton's Law of Cooling, a mathematical model of temperature change.

> EXAMPLE 8

An Application Involving Natural Logarithms

According to Newton's Law of Cooling, the time it takes for an object to cool from an initial temperature T_0 to a temperature T, given that the temperature of the surrounding air is a constant C, is given by

$$t = k \ln\left(\frac{T_0 - C}{T - C}\right)$$

where k is a constant. Suppose a turkey is taken out of the oven and is placed in air having a constant temperature of 70°F. If the turkey initially has an internal temperature of 185°F and $k = 12.5$ (for time measured in minutes), find the time required for the turkey to cool to the following temperatures:

a. $T = 160°F$ **b.** $T = 100°F$ **c.** $T = 80°F$

What happens to the time as T gets closer to 70°F? According to this model, will the turkey ever cool to 70°F? Use a graphing calculator to sketch the graph of t as a function of T, and then identify the domain.

Solution   We first set $C = 70$, $T_0 = 185$, and $k = 12.5$ to obtain

$$t = 12.5 \ln\left(\frac{185 - 70}{T - 70}\right) = 12.5 \ln\left(\frac{115}{T - 70}\right)$$

Now we substitute the given values for T:

a. $T = 160$: $t = 12.5 \ln\left(\dfrac{115}{160 - 70}\right) \approx 3.1$ minutes

b. $T = 100$: $t = 12.5 \ln\left(\dfrac{115}{100 - 70}\right) \approx 16.8$ minutes

c. $T = 80$: $t = 12.5 \ln\left(\dfrac{115}{80 - 70}\right) \approx 30.5$ minutes

As T gets closer to 70, t increases without bound. The value $T = 70$ would yield an undefined expression inside the logarithm. Thus, according to the model, the turkey will never reach 70°F. In practice, of course, we suspect that it will. We conclude that the model fails to give realistic results for values of T near 70. This is illustrated by the graph of

$$t = 12.5 \ln\left(\frac{115}{T - 70}\right)$$

shown in Figure 14. Here we see that t (the y-coordinate) approaches infinity as T (the x-coordinate) approaches 70 from the right. The domain of the function is $(70, \infty)$.

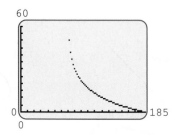

Figure 14

Logarithmic scales are often used when measuring quantities that vary over a large range of possible values. Examples include the decibel scale for measuring the loudness of sound and the Richter scale for measuring the intensity of earthquakes. In the case of the decibel scale, we establish a baseline by assigning an intensity I_0 to the faintest sound perceptible to most humans (approximately 10^{-16} watts per square centimeter). A sound with intensity I is then assigned a decibel rating of

$$D = 10 \cdot \log\left(\frac{I}{I_0}\right)$$

Some common sounds and their corresponding decibel ratings are given in Table 4.

Table 4

Sound	Decibels
Human breathing	10
Whisper	20
Office activities	50
Automobile	70
Lawn mower	85
Crowd noise at football game	100
Loud rock music	115
Jet airplane	140
Rocket launching	180

EXAMPLE 9

A blue whale flirts with onlookers.

Measuring the Loudness of Sound

The sound of a blue whale (which can be heard up to 530 miles away) can reach an intensity of 6.3×10^{18} times that of I_0. Find the number of decibels of this sound.

Solution We set $I = (6.3 \times 10^{18})I_0$ and compute D.

$$D = 10 \cdot \log\left(\frac{I}{I_0}\right)$$
$$= 10 \cdot \log\left(\frac{6.3 \times 10^{18}I_0}{I_0}\right)$$
$$= 10 \cdot \log(6.3 \times 10^{18})$$
$$\approx 10 \cdot 18.8$$
$$= 188$$

Thus, the sound of a blue whale can reach 188 decibels.

Understanding and Mastery Checklists

Concepts to Understand

Logarithm function with base a

◊

Inverse properties of logarithms and exponents

◊

Natural logarithms

◊

Common logarithms

Skills to Master

Sketch the graph of a logarithmic function.

◊

Evaluate a logarithm by applying the definition.

◊

Estimate a logarithm graphically.

◊

Compute a natural logarithm using a calculator.

◊

Compute a common logarithm using a calculator.

◊

Find the inverse of a function involving logarithms or exponents.

Exercises 4.2

Exercises 1-8   *Evaluate the expression using a calculator. Give answers accurate to five decimal places.*

1. $\ln 15.2$

2. $\log 110$

3. $\log\left(\dfrac{2}{3}\right)$

4. $\ln \sqrt{21}$

5. $\log(2\sqrt{3})$

6. $\ln\left(\dfrac{17}{5}\right)$

7. $(\ln 0.41)^3$

8. $\log(3.52 \times 10^{47})$

Exercises 9-24   *Evaluate the expression without the use of a calculator.*

9. $\log_2 16$

10. $\log_3 \dfrac{1}{9}$

11. $\log_3 0$

12. $\log_{23} 1$

13. $\ln e^2$

14. $\log 10,000$

15. $\log_6 \dfrac{1}{216}$

16. $\log_7 343$

17. $\log_{1/5} 25$

18. $\log_{3/4}\left(\dfrac{4}{3}\right)^{100}$

19. $\log_{0.1} 1000$

20. $\log_{\sqrt{2}} 16$

21. $\log(10^{100})$

22. $\ln \sqrt{e}$

23. $\ln(e^4 \cdot e^3)$

24. $\log(10^4 \cdot 10^{-3})$

Exercises 25-30   *Use a graphing calculator to estimate the value of the expression to the nearest hundredth.*

25. $\log_5 37$

26. $\log_2 12$

27. $\log_2 50$

28. $\log_{100} 8$

29. $\log_{1/2} 108$

30. $\log_{1/3} 11$

Exercises 31-38   *Write an exponential equation equivalent to the given logarithmic equation.*

31. $\log_2 \dfrac{1}{4} = -2$

32. $\log_{1/3} 9 = -2$

33. $\log 1000 = 3$

34. $\ln 7 = x$

35. $\ln(x + 1) = 2$

36. $\log x = 4$

37. $\log_x 10 = 3$

38. $\ln(3x + 1) = x$

Exercises 39-46   *Write a logarithmic equation equivalent to the given exponential equation.*

39. $3^2 = 9$

40. $10^{-1} = \dfrac{1}{10}$

41. $\left(\dfrac{1}{2}\right)^{-3} = 8$

42. $\left(\dfrac{1}{4}\right)^{2} = \dfrac{1}{16}$

43. $e^x = 5$

44. $e^{3x} = \dfrac{1}{2}$

45. $e^{-0.013t} = \dfrac{1}{2}$

46. $1.06^t = 2$

Exercises 47-52 *Use the given graph of* $f(x) = \ln x$ *to graph* $g(x)$, *and then find the domain of* g. *Verify your result with a graphing calculator.*

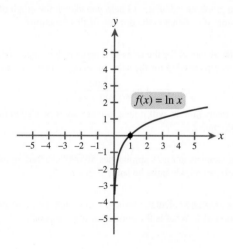

47. $g(x) = -\ln x$

48. $g(x) = \ln(-x)$

49. $g(x) = \ln(x - 3)$

50. $g(x) = 2 + \ln x$

51. $g(x) = \ln(-x) - 1$

52. $g(x) = -\ln(x + 2)$

Exercises 53-58 *Compute the inverse of each function, and graph both the function and its inverse on the same set of coordinate axes.*

53. $f(x) = \log_3 x$

54. $f(x) = 2 \ln x$

55. $f(x) = e^x - 5$

56. $f(x) = \log(x - 4)$

57. $f(x) = \log(1000x)$

58. $f(x) = 5^{x-3}$

Exercises 59-62 *Use a graphing calculator to help you sketch the graph of the given function, and find its domain.*

59. $f(x) = x \ln x$

60. $f(x) = (x^2 - 1)\ln x$

61. $f(x) = \dfrac{x}{\ln(x + 1)}$

62. $f(x) = \dfrac{\ln x}{x - 1}$

Applications

Exercises 63-66 *The following exercises deal with Newton's Law of Cooling, which can be expressed using the formula*

$$t = k \ln\left(\frac{T_0 - C}{T - C}\right)$$

Here t is the time it will take for an object to cool to a temperature T, T_0 is the initial temperature, C is the temperature of the surrounding air, and k is a constant.

63. Coffee Temperature Find the time required for a hot cup of coffee with an initial temperature of 190°F to cool to a temperature of 80°F if the temperature of the surrounding air is 70°F and $k = 20$ (for time measured in minutes).

64. Ice Tea Temperature A freshly brewed pitcher of tea, currently at a temperature of 175°F, is placed in a refrigerator that has a constant temperature of 40°F. Find the time required for the tea to reach a temperature of 45°F given that $k = 50$ (for time measured in minutes).

65. Body Temperature Inspector Magill is called to the scene of a murder. The victim is lying in a room that has a constant temperature of 75°F. The victim's body temperature is currently 81.2°F, down from the normal body temperature of 98.6°F. After considering the height and weight of the victim, Magill arrives at a value of $k = 10$ (for time measured in hours). How long has the victim been dead?

66. **Body Temperature Revisited** After further consideration (see Exercise 65), Inspector Magill decides that he is not sufficiently confident in his k value to fix the time of death. To determine k, he notes that the victim's body temperature at 3:00 P.M. Wednesday is 81.2°F and that by 4:00 A.M. Thursday it has dropped to 77.0°F. Assuming that the body was kept in a 75°F room, what value does Magill find for k, and what was the time of death?

Exercises 67-70 *The following exercises deal with the loudness of sound, which is measured using* ***decibels****. The faintest sound perceptible to most individuals is assigned an intensity of I_0, and a sound of intensity I has a decibel rating of*

$$D = 10 \cdot \log\frac{I}{I_0}$$

Find the decibel rating for the given sound.

67. A whisper with $I = 110I_0$

68. A quiet conversation with $I = 9000I_0$

69. A circular saw with intensity 10 billion times that of I_0

70. An auto horn with intensity 100 billion times that of I_0

Exercises 71-74 *The following exercises deal with the acidity of an aqueous solution, which is dependent on the solution's hydrogen ion concentration. Since these concentrations may be very small, it is convenient to measure acidity using the formula*

$$pH = -\log[H^+]$$

where $[H^+]$ is the hydrogen ion concentration in moles per liter. In general, the smaller the pH, the more acidic the solution. Find the pH for the given solution.

71. Orange juice with $[H^+] = 2.8 \times 10^{-4}$

72. Milk with $[H^+] = 3.97 \times 10^{-7}$

73. Beer with $[H^+] = 3.16 \times 10^{-5}$

74. Lemon juice with $[H^+] = 6.3 \times 10^{-3}$

75. **Mile-thick Paper** How many times must a 0.003-inch-thick piece of paper be folded to obtain a thickness of at least 1 mile?

76. **Doubling Your Money** Suppose that you have invested $1 into First Fantasy Bank's Double-Your-Money Account, in which your balance doubles every year. Use the graph of $y = 2^x$ to estimate the number of years required to obtain a balance of $1,000,000,000 (one billion dollars). Then express your answer in terms of logarithms.

77. **Typing Speed** A learning model suggests that the time in weeks that it will take to learn to type w words per minute (wpm) is given by

$$t = 5 \ln\left(\frac{100}{100 - w}\right)$$

   a. How many weeks will it take to learn to type 60 wpm? 80 wpm?

   b. What happens to the time as w approaches 100 wpm? According to this model, can 100 wpm be reached? Explain.

   c. Use a graphing calculator to help you sketch the graph of t as a function of w. What is the domain of this function?

78. **Sales Growth** A model for the sales of a new product suggests that the time in months needed for the sales to reach S units is given by

$$t = 4 \ln\left(\frac{50,000}{50,000 - S}\right)$$

   a. How many months will it take to reach sales of 15,000 units? 40,000 units?

   b. What happens to t as S approaches 50,000? According to this model, can 50,000 units be sold? Explain.

   c. Use a graphing calculator to help you sketch the graph of t as a function of S. What is the domain of this function?

Concepts and Critical Thinking

Exercises 79-84 *Answer true or false.*

79. The natural logarithm function and the exponential function with base e are inverses of one another.

80. For any $a > 0$, $a \neq 1$, $\log_a 0 = 1$.

81. The logarithm of a negative number is negative.

82. If $a > b$, then $\log a > \log b$.

83. $y = \log_a x$ is equivalent to $a^y = x$.

84. The domain of the natural logarithm function consists of all real numbers.

Exercises 85-89 *Give an example of each.*

85. A number having a negative common logarithm

86. A number whose natural and common logarithms are the same

87. A real-world context in which common logarithms are employed

88. A characteristic shared by all logarithmic functions but possessed by no exponential function

89. A function $f(x)$ satisfying $\ln x < f(x) < e^x$ for all $x > 0$

90. Graph the function

$$f(x) = \frac{\ln x}{x}$$

What value does $f(x)$ approach as x becomes larger and larger?

91. Graph the function

$$f(x) = \log\!\left(\frac{x + 1}{2x - 8}\right)$$

Find the domain of f and any vertical or horizontal asymptotes.

Questions for Discussion or Essay

92. John Napier of Scotland invented logarithms in the 16th century as a labor-saving device for doing computations. But *who* in 16th-century Europe would have had need for them? What computationally intensive activities might have occupied scientists and mathematicians of this era?

93. Although Napier is generally credited with the invention of logarithms, the Swiss watchmaker Burgi discovered logarithms independently at about the same time. The mathematical literature is filled with such instances. For example, Newton and Leibniz invented calculus independently at about the same time. Are such occurrences simply coincidences? What explains the tendency for mathematicians, working independently, to make major discoveries simultaneously?

94. In Exercises 71–74, we considered the acidity of a solution as measured by the pH level. If normal rain is considered to have a pH of approximately 5.6, and a recent rainfall is determined to have a hydrogen ion concentration of 6.3×10^{-4}, is the recent rain more or less acidic than normal rain? Which has the larger hydrogen ion concentration? If the primary causes of acid rain are sulfur dioxide and nitrogen oxides, do these compounds increase or decrease the hydrogen ion concentration of rain? In general, describe the effect on the pH level if the hydrogen ion concentration increases or decreases.

High levels of sulfur in industrial emissions are a primary cause of acid rain.

95. In Exercises 65 and 66, Inspector Magill applied Newton's Law of Cooling to estimate the time of death. What characteristics of a body (in addition to height and weight) would influence the cooling rate? Does Newton's Law of Cooling take into account all means by which heat is transferred? If so, of what relevance is the windchill factor, and how do you explain microwave ovens?

Projects for Enrichment

96. *The Leaning Tower of Pizza*   It can be shown that (in principle) Domino's Pizza boxes can be stacked on a table with each box overhanging the box below it so that the top box extends as far as you please past the edge of the table—without toppling! Unfortunately, producing a large overhang requires a tremendous number of boxes. It can be shown that the greatest possible overhang with n ($n > 1$) 12-inch boxes is given by

$$6 \cdot \left(1 + \frac{1}{2} + \frac{1}{3} + \frac{1}{4} + \frac{1}{5} + \cdots + \frac{1}{n-1}\right) \text{ inches}$$

a. Compute the greatest overhang possible with 6 pizza boxes.

b. It can be shown that

$$1 + \frac{1}{2} + \frac{1}{3} + \frac{1}{4} + \frac{1}{5} + \cdots + \frac{1}{n-1} - \ln(n-1) \approx \epsilon$$

where ϵ is a constant known as Euler's number. Estimate Euler's number by computing

$$1 + \frac{1}{2} + \frac{1}{3} + \frac{1}{4} + \frac{1}{5} + \cdots + \frac{1}{n-1} - \ln(n-1)$$

for $n = 10$.

c. Find an approximation for the maximum overhang possible with n boxes that involves Euler's number and natural logarithms. [*Hint:* If

$$S = 1 + \frac{1}{2} + \frac{1}{3} + \frac{1}{4} + \frac{1}{5} + \cdots + \frac{1}{n-1}$$

and $S - \ln(n-1) \approx \epsilon$, then what can be said of S?]

d. Using the approximation for the maximum possible overhang, derived in part c, compute the number of pizza boxes required to obtain an overhang of 10 feet.

e. Suppose that a pizza box is 2 inches thick. How high would the stack of pizza boxes in part d be? (Express your answer in light-years.)

f. Explain how a very narrow river could be crossed using only a parachute and Domino's Pizza boxes.

g. In this project, we've ignored many aspects of reality in order to obtain our result. Describe some of the factors that would complicate the unusual river-crossing technique described in part f.

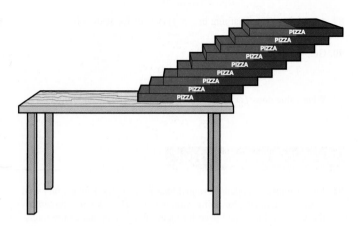

Section 4.3   Logarithmic Identities and Equations

- How powerful would the "bombquake" be that would result from simultaneously detonating all of the world's nuclear warheads?
- How can multiplication and division be performed simply by sliding pieces of paper?
- How many boom boxes would it take to cause a fatal musical overdose?
- If the entire population of China simultaneously jumped a foot off the ground, how strong would the resulting earthquake be?

Logarithmic Identities

As we have already seen, exponential and logarithmic functions are closely connected. In the previous section, we exploited this connection to recast exponential equations as logarithmic equations and vice versa, so that, for example, the exponential equation $y = a^x$ could be rewritten as the equivalent logarithmic equation $x = \log_a y$. In fact, any statement written in the language of exponentials can be translated into an equivalent statement involving logarithms. In this section, we use three familiar exponential identities to derive three new logarithmic identities. Then, armed with our graphing calculator and our newly enlarged collection of logarithmic identities, we tackle logarithmic equations.

We begin our development of logarithmic identities by considering the exponential identity

$$a^b a^c = a^{b+c}$$

If we let $u = a^b$ and $v = a^c$, so that $b = \log_a u$ and $c = \log_a v$, then we have

$$uv = a^{b+c}$$

The corresponding logarithmic equation is

$$\log_a(uv) = b + c$$

Thus, we have the logarithmic identity

$$\log_a(uv) = \log_a u + \log_a v$$

Similarly, the exponential identities

$$\frac{a^b}{a^c} = a^{b-c} \quad \text{and} \quad (a^b)^n = a^{bn}$$

yield the logarithmic identities

$$\log_a\left(\frac{u}{v}\right) = \log_a u - \log_a v \quad \text{and} \quad \log_a(u^n) = n \log_a u$$

respectively.

The following is a summary of logarithmic identities. For the sake of completeness, we have included the inverse identities developed in the previous section and have restated each of the rules for common and natural logarithms as well.

Logarithmic Identities

Let a, u, and v be positive real numbers with $a \neq 1$, and let n be any real number.

Arbitrary logarithms	Common logarithms	Natural logarithms
1. $\log_a(uv) = \log_a u + \log_a v$	$\log(uv) = \log u + \log v$	$\ln(uv) = \ln u + \ln v$
2. $\log_a\left(\frac{u}{v}\right) = \log_a u - \log_a v$	$\log\left(\frac{u}{v}\right) = \log u - \log v$	$\ln\left(\frac{u}{v}\right) = \ln u - \ln v$
3. $\log_a(u^n) = n \log_a u$	$\log(u^n) = n \log u$	$\ln(u^n) = n \ln u$
4. $\log_a a^u = u$	$\log a^u = u$	$\ln a^u = u$
5. $a^{\log_a u} = u$	$10^{\log u} = u$	$e^{\ln u} = u$

The identities just listed are relatively simple, and, indeed, most students commit them to memory relatively quickly. Strangely, the difficulty lies not in the correct application of these five identities, but with resisting the overwhelming temptation to manufacture similar-looking—but false—formulas. For example, after learning the first logarithmic identity, beginning students are often lured into believing that $\log_a(u + v) = \log_a u \log_a v$ or that $\log_a u \log_a v = \log_a u + \log_a v$, neither of which is true!

WARNING!

> Be aware of the temptation to use seemingly plausible (but bogus) formulas that, on the surface, resemble the logarithmic identities. In particular, the following expressions cannot, in general, be simplified using $\log_a$.
>
> $$\log_a(u + v)$$
> $$\frac{\log_a u}{\log_a v}$$
> $$\log_a u \log_a v$$

The following two examples illustrate the use of logarithmic identities for eliminating products, quotients, and powers within logarithmic expressions.

⋯EXAMPLE 1

Expanding with Logarithmic Identities

Expand the following expression as a sum, difference, or multiple of logarithms:

$$\log\left(\frac{x\sqrt{y}}{z}\right)$$

Solution

$$\log\!\left(\frac{x\sqrt{y}}{z}\right) = \log(x\sqrt{y}) - \log z \qquad \text{Identity 2}$$

$$= \log x + \log(y^{1/2}) - \log z \qquad \text{Identity 1}$$

$$= \log x + \frac{1}{2}\log y - \log z \qquad \text{Identity 3}$$

EXAMPLE 2

Expanding with Logarithmic Identities

Expand the following expression as a sum, difference, or multiple of logarithms:

$$\ln\!\left(\frac{3y^2}{x^5}\right)^3$$

Solution

$$\ln\!\left(\frac{3y^2}{x^5}\right)^3 = 3\ln\!\left(\frac{3y^2}{x^5}\right) \qquad \text{Identity 3}$$

$$= 3[\ln(3y^2) - \ln(x^5)] \qquad \text{Identity 2}$$

$$= 3[\ln 3 + \ln(y^2) - \ln(x^5)] \qquad \text{Identity 1}$$

$$= 3(\ln 3 + 2\ln y - 5\ln x) \qquad \text{Identity 3}$$

$$= 3\ln 3 + 6\ln y - 15\ln x$$

Logarithmic identities can be useful for simplifying complex expressions involving logarithms. As we will see shortly, the simplification of complicated expressions involving logarithms is often the first step in solving logarithmic equations.

EXAMPLE 3

Simplifying with Logarithmic Identities

Express $3\log_a(x^2) + 4\log_a x$ as a single logarithm.

Solution

$$3\log_a(x^2) + 4\log_a x = \log_a(x^2)^3 + \log_a(x^4) \qquad \text{Identity 3}$$

$$= \log_a(x^6 \cdot x^4) \qquad \text{Identity 1}$$

$$= \log_a(x^{10})$$

EXAMPLE 4

Simplifying with Logarithmic Identities

Express $2\ln x - \ln y + 6\ln z$ as a single logarithm.

Solution

$$2\ln x - \ln y + 6\ln z = \ln(x^2) - \ln y + \ln z^6 \qquad \text{Identity 3}$$

$$= \ln\!\left(\frac{x^2}{y}\right) + \ln(z^6) \qquad \text{Identity 2}$$

$$= \ln\!\left(\frac{x^2 z^6}{y}\right) \qquad \text{Identity 1}$$

Logarithmic
Equations

Equations involving one or more logarithmic expressions are called **logarithmic equations**. Many logarithmic equations can be solved by converting to exponential equations. For example, the equation

$$\ln x = 2$$

is equivalent to

$$x = e^2$$

This solution could also be found by exponentiating both sides of the original equation, as follows:

$$\ln x = 2$$
$$e^{\ln x} = e^2 \qquad \text{Exponentiating both sides}$$
$$x = e^2 \qquad \text{Using the property } a^{\log_a x} = x$$

In general, any equation of the form $\log_a \square = c$ can be solved for $\square$, either by converting to exponential form or by exponentiating both sides.

▶EXAMPLE 5

Solving a Logarithmic Equation

Solve $\log_2(x + 1) = 5$.

Solution   We convert from the base 2 logarithmic form to the base 2 exponential form.

$$\log_2(x + 1) = 5$$
$$x + 1 = 2^5$$
$$x = 32 - 1 = 31$$

Another useful technique for solving logarithmic equations is to apply the 1–1 property of logarithms: if $\log_a u = \log_a v$, then $u = v$.

▶EXAMPLE 6

Using the 1-1 Property of Logarithms to Solve an Equation

Solve $\log(x^2 + 2x - 5) = \log(x + 1)$.

Solution

$$\log(x^2 + 2x - 5) = \log(x + 1)$$
$$x^2 + 2x - 5 = x + 1 \qquad \text{Using the 1–1 property of logarithms}$$
$$x^2 + x - 6 = 0$$
$$(x - 2)(x + 3) = 0 \qquad \text{Factoring}$$
$$x = 2, x = -3$$

Check

$x = 2$		$x = -3$	
$\log(x^2 + 2x - 5)$	$\log(x + 1)$	$\log(x^2 + 2x - 5)$	$\log(x + 1)$
$\log[(2)^2 + 2(2) - 5]$	$\log(2 + 1)$	$\log[(-3)^2 + 2(-3) - 5]$	$\log[(-3) + 1]$
$\log(4 + 4 - 5)$	$\log 3$	$\log(9 - 6 - 5)$	$\log(-2)$
$\log 3$	✓		Not defined

Thus, the only solution is $x = 2$.

Since logarithms are defined only for positive real numbers, it is important to check all solutions to logarithmic equations by substituting them into the *original* equation.

Some logarithmic equations must be algebraically rearranged using logarithmic identities before they can be solved, as illustrated in the following example.

EXAMPLE 7

Solving an Equation Involving Natural Logarithms

Solve $\ln 2 + 3 \ln(x - 1) = 2 \ln 4$.

Solution   We begin by expressing each side in the form $\ln \square$.

$$\ln 2 + 3 \ln(x - 1) = 2 \ln 4$$
$$\ln 2 + \ln(x - 1)^3 = \ln(4^2) \qquad \text{Using } n \ln u = \ln(u^n)$$
$$\ln[2(x - 1)^3] = \ln 16 \qquad \text{Using } \ln(uv) = \ln u + \ln v$$
$$2(x - 1)^3 = 16 \qquad \text{Using the 1–1 property of logarithms}$$
$$(x - 1)^3 = 8$$
$$x - 1 = 2 \qquad \text{Taking the cube root of both sides}$$
$$x = 3$$

Check

$$x = 3$$

$\ln 2 + 3 \ln(x - 1)$	$2 \ln 4$
$\ln 2 + 3 \ln(3 - 1)$	$\ln(4^2)$
$\ln 2 + 3 \ln 2$	$\ln 16$
$4 \ln 2$	
$\ln(2^4)$	
$\ln 16$	✓

EXAMPLE 8

Comparing Sound Intensity

The decibel rating D of a sound with intensity I is given by

$$D = 10 \cdot \log \frac{I}{I_0}$$

where I_0 is the intensity of the faintest sound perceptible to the human ear. How much more intense is the 140-decibel sound of a jet taking off than that of a 120-decibel jackhammer?

Solution   For comparison purposes, we set $I_0 = 1$. The relative intensity of the jet at takeoff can be found as follows:

$$140 = 10 \log I$$
$$14 = \log I$$
$$I = 10^{14} \qquad \text{Converting to exponential form}$$

For the jackhammer, we have

$$120 = 10 \log I$$
$$12 = \log I$$
$$I = 10^{12} \qquad \text{Converting to exponential form}$$

Now $10^{14} = 100 \cdot 10^{12}$, and so the jet is 100 times more intense than the jackhammer.

EXAMPLE 9

A loaded ICBM missile is an ominous reminder of the nuclear threat.

The Nuclear Earthquake

An analysis of published data suggests that the total nuclear stockpile has been reduced from 25,000 megatons in 1988 to about 10,000 megatons in 2000. Based on empirical evidence, the energy E (in ergs) of an earthquake is related to the Richter scale reading R by the equation $\log E = 11.8 + 1.5R$.

a. If all the warheads available in 1988 were simultaneously detonated, how powerful would the resulting "bombquake" be? (A megaton is one million tons, and a ton of TNT carries about 3.4×10^{16} ergs of energy.)

b. How powerful would the bombquake be resulting from the detonation of all the nuclear warheads in 2000?

c. How many megatons of TNT would it take to create an earthquake of magnitude 7 on the Richter scale?

Solution

a. We first compute the energy involved in such an explosion.

$$E = (25{,}000 \text{ megatons of TNT}) \times \frac{10^6 \text{ tons of TNT}}{1 \text{ megaton of TNT}} \times \frac{3.4 \times 10^{16} \text{ ergs}}{1 \text{ ton of TNT}}$$

$$= 8.5 \times 10^{26} \text{ ergs}$$

Using the equation $\log E = 11.8 + 1.5R$, we have

$$\log(8.5 \times 10^{26}) = 11.8 + 1.5R$$
$$26.929 = 11.8 + 1.5R$$
$$1.5R = 15.129$$
$$R = 10.086$$

Thus, the world's supply of nuclear warheads contains an energy equal to that of an earthquake of magnitude 10.086 on the Richter scale! This is greater than that of any recorded earthquake.

b. We have

$$E = (10{,}000 \text{ megatons of TNT}) \times \frac{10^6 \text{ tons of TNT}}{1 \text{ megaton of TNT}} \times \frac{3.4 \times 10^{16} \text{ ergs}}{1 \text{ ton of TNT}}$$

$$= 3.4 \times 10^{26} \text{ ergs}$$

Using $\log E = 11.8 + 1.5R$, we have

$$\log(3.4 \times 10^{26}) = 11.8 + 1.5R$$
$$26.531 = 11.8 + 1.5R$$
$$1.5R = 14.731$$
$$R = 9.821$$

Thus, even after cutting more than half the world's supply of nuclear warheads, the stockpile still contains an energy equal to that of an earthquake of magnitude 9.821.

c. We set $R = 7$ and solve for E.

$$\log E = 11.8 + 1.5R$$
$$\log E = 11.8 + 1.5 \cdot 7$$
$$\log E = 22.3$$
$$E = 10^{22.3} \approx 2.0 \times 10^{22} \text{ ergs}$$

Converting to megatons of TNT, we find

$$2.0 \times 10^{22} \text{ ergs} \times \frac{1 \text{ ton of TNT}}{3.4 \times 10^{16} \text{ ergs}} \times \frac{1 \text{ megaton of TNT}}{10^6 \text{ tons of TNT}}$$
$$\approx 0.59 \text{ megaton of TNT}$$

Approximations Involving Logarithms

In Example 2 of Section 4.2, we estimated $\log_2 5$ graphically. Now we consider a more algebraic approach using the logarithmic identities discussed in this section.

EXAMPLE 10

Approximating Logarithmic Expressions

Estimate $\log_2 5$.

Solution   We begin by setting $x = \log_2 5$ and then converting to exponential form.

$$x = \log_2 5$$
$$2^x = 5 \qquad \text{Converting to exponents}$$
$$\ln 2^x = \ln 5 \qquad \text{Taking the natural logarithm of both sides}$$
$$x \ln 2 = \ln 5 \qquad \text{Using the identity } \ln u^n = n \ln u$$
$$x = \frac{\ln 5}{\ln 2} \qquad \text{Solving for } x$$

Finally, using a calculator, we obtain $x \approx 2.3219$.

The process used in Example 10 can be used to rewrite any logarithmic expression of the form $\log_a u$ as an expression involving only natural logarithms, common logarithms, or even logarithms to an arbitrary base b. In so doing, we obtain the *change-of-base formula*. This formula, though useful, need not be memorized; it is always possible to change the base by using the procedure illustrated in Example 10.

The Change-of-Base Formula

Let a and b be positive real numbers with $a \neq 1$ and $b \neq 1$. Then, for $u > 0$,

$$\log_a u = \frac{\log_b u}{\log_b a} = \frac{\log u}{\log a} = \frac{\ln u}{\ln a}$$

EXAMPLE 11

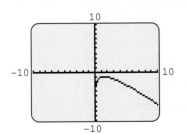

Figure 15

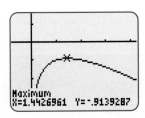

Figure 16

A Graph Involving the Base 2 Logarithm Function

Estimate (to the nearest hundredth) the coordinates of any turning points of the function $f(x) = \log_2 x - x$.

Solution   We assume that our graphing calculator does not compute base 2 logarithms directly. Thus, we must convert to either common or natural logarithms. We will use natural logarithms. From the change-of-base formula, we obtain

$$f(x) = \log_2 x - x$$
$$= \frac{\ln x}{\ln 2} - x$$

Plotting f in the standard viewing window, as in Figure 15, we see a single turning point (a local maximum). After zooming out a few times, we are reasonably certain there are no others. Figure 16 shows the view after zooming in and applying the calculate-maximum feature. The turning point has approximate coordinates $(1.44, -0.91)$.

The need for approximation also arises in situations where logarithmic equations cannot be solved algebraically. As we have seen many times in the past, approximate solutions of equations can be found by first writing the equation in the form $f(x) = 0$, then graphing the function f on a graphing calculator, and finally estimating the zeros with the trace or calculate-zero feature. However, if an equation involves nonpolynomial terms (such as $\ln x$ or $\log x$), it can be difficult to tell how many solutions the equation has. Because of this, and because of our knowledge of polynomials, it is often helpful to first put the equation in the form $g(x) = p(x)$, where $p(x)$ is a polynomial, and then plot the graphs of g and p to see how many times they intersect. Solution values can be approximated either by determining the intersection points of g and p or by finding the zeros of f.

EXAMPLE 12

Determining the Number of Solutions of a Logarithmic Equation

Determine the number of solutions of the equation $\frac{1}{18}x^3 + 4x + 1 = x^2 + \ln x$.

Solution   We first regroup terms so that all the polynomial terms are on one side of the equation.

$$\frac{1}{18}x^3 + 4x + 1 = x^2 + \ln x$$

$$\frac{1}{18}x^3 - x^2 + 4x + 1 = \ln x$$

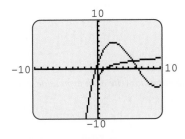

Figure 17

Next we plot the functions $p(x) = \frac{1}{18}x^3 - x^2 + 4x + 1$ and $g(x) = \ln x$, as shown in Figure 17. Because the solutions to $p(x) = g(x)$ correspond to the x-coordinates of the points of intersection of the graphs of p and g, we attempt to locate and approximate any intersection points of the two curves. One intersection point can be seen near the point $(6, 2)$. Since $p(x)$ is a third-degree polynomial with a positive leading coefficient, we know that its graph must turn around and start increasing to the right of the viewing window. Thus, a second intersection point is certain. With the slightly modified viewing window shown in Figure 18, we see that there is indeed a second intersection point. Since p can have no other turning points, there can be no other intersection points. Thus, there are exactly two solutions. We could approximate these solutions either by zooming in on the points of intersection or by using the calculate-intersect feature.

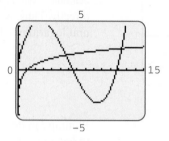

Figure 18

Understanding and Mastery Checklists

Concepts to Understand

Logarithmic identities

✧

Logarithmic equations

✧

The 1–1 property of logarithms

✧

The change-of-base formula

Skills to Master

Use the logarithmic identities to rewrite expressions involving logarithms.

✧

Solve equations involving logarithms.

✧

Approximate a logarithm base a by changing to natural or common logarithms.

Exercises 4.3

Exercises 1-8   *Use logarithmic identities to expand each expression, rewriting it as a sum, difference, or multiple of logarithms.*

1. $\ln(x^2 y)$

2. $\log_2(xy\sqrt{z})$

3. $\log_5\left(\dfrac{x^3 y^4}{z^{-3}}\right)$

4. $\log_4\left(\dfrac{15\sqrt{q}}{\sqrt{r}}\right)$

5. $\ln\left(\dfrac{e}{7x^3}\right)$

6. $\ln\left(\dfrac{xy^2}{\sqrt{z}}\right)^3$

7. $\log\sqrt{\dfrac{y\sqrt{x}}{z^2}}$

8. $\log_a\sqrt{x^3 + y^3}$

Exercises 9-18   *Use logarithmic identities to write each expression as a single logarithm.*

9. $\log x - 2\log y$

10. $2\ln x + \dfrac{1}{2}\ln y$

11. $3 \ln x + 4 \ln y - \ln z$

12. $\dfrac{1}{2} \log(2x - y) + \dfrac{1}{2} \log(x + 2y)$

13. $\dfrac{1}{3} \ln(x - z) - \dfrac{2}{3} \ln(x + z)$

14. $4[\log_5(x^2 - y^2) - \log_5(x + y)]$

15. $2 \log 4 - \log 6 + 3 \log 2$

16. $\log_3 8 - 4 \log_3(x^2)$

17. $2 \ln(x + 3) - \ln(x^2 + 5x + 6)$

18. $\log(x + y) - \dfrac{1}{2} \log(x^2 + 2xy + y^2)$

Exercises 19-44   *Solve the logarithmic equation.*

19. $\dfrac{1}{3} \log_3 x = 1$

20. $2 \log_4 x = 32$

21. $\log_x 4 = 2$

22. $\log_x \dfrac{1}{9} = 2$

23. $3 \log_5 x = 2$

24. $2 \log_2 x = 3$

25. $\log_4(2x + 1) = -2$

26. $\log_2(1 - 3x) = -1$

27. $\log_3(27^x) = -1$

28. $\log_5(5^x) = 2$

29. $2 \log_3(x - 1) + \log_3 9 = 2$

30. $\log(2x + 3) = 2 \log x$

31. $\ln(x + 7) - \ln(x + 2) = \ln(x + 1)$

32. $\log_2(3 - x) + \log_2(2 - x) = \log_2(1 - x)$

33. $\log(2x^2 + 11x + 5) - \log(2x + 1) = 2$

34. $\log_a(x + 1) + \log_a(x - 2) = \log_a(x + 6)$

35. $\log_2(27x^3 - 1) - \log_2(9x^2 + 3x + 1) = 1$
   (*Hint:* Factor $27x^3 - 1$.)

36. $\log_{1/4} x + \log_{1/4}(3x - 5) = -1$

37. $\log_a(x + 7) = \log_a(x + 1)$

38. $\log_5(x^7) = 7 \log_5 x$

39. $\log \sqrt{x^2 + 1999} = 3$

40. $\log \sqrt[3]{x^2 - 15x} = \dfrac{2}{3}$

41. $\ln(x^2) = 2 \ln x$

42. $(\ln x)^2 = 2 \ln x$

43. $\ln(\ln x) = 1$

44. $\log[\log(x + 1)] = \log 2$

Exercises 45-50   *Use the technique of Example 10 to estimate the logarithm to five decimal places.*

45. $\log_3 16$

46. $\log_4 25$

47. $\log_{12} 0.341$

48. $\log_{23} 1250$

49. $\log_\pi 10$

50. $\log_\pi e$

Exercises 51-56   *Approximate the solutions of the given equation to the nearest hundredth.*

51. $x - 2 = \ln x$

52. $\ln(x + 1) = x^2 - 1$

53. $\log_5 x = \dfrac{3}{2} x - 2$

54. $3 = x + \log_3 x$

55. $0.288 x^2 = \ln(1 - x) + 0.002 x^4 + 1$

56. $1.5 x^2 = 0.0625 x^3 + 5x + \ln x$

Applications

Exercises 57-60   *The following exercises deal with the decibel rating D of a sound with intensity I as given by $D = 10 \cdot \log(I/I_0)$, where I_0 is the faintest sound perceptible to the human ear.*

57. Hearing Loss   The threshold of pain for most individuals is 120 decibels, whereas permanent hearing damage can be caused by even brief exposure to a sound rated at 160 decibels. How many times more intense is a sound of 160 decibels than one of 120 decibels?

58. Roaring Fans   According to *The Guinness Book of World Records,* fans at Denver's Mile High Stadium shrieked a record roar of 128.7 decibels for 10 seconds during half-time of a Broncos and Patriots game on October 1, 2000. Though impressive by human standards, the Mile High Crazies are no match for the blue whale (up to 188 decibels). How many times more intense is the sound produced by a blue whale than that produced by the Bronco fans?

59. Mass Choir   Suppose that 1000 people sing a note at precisely the same volume and pitch at exactly the same moment. Assuming that the collective sound is 1000 times more intense than that of an individual singer, how much louder is the sound of the group than that of an individual?

60. Fatal Boom Box   It is said that a 200-decibel sound can be deadly. How many times more intense is a 200-decibel sound than a 100-decibel boom box? Does it follow that this number of boom boxes would result in a fatal musical overdose?

Exercises 61-64   *The following exercises deal with the equation $\log E = 11.8 + 1.5R$, which relates the energy E (in ergs) of an earthquake to the Richter scale reading R.*

61. TNT Earthquake   If 1 ton of TNT carries 3.4×10^{16} ergs of energy, how many tons of TNT would it take to equal the energy output of an earthquake registering 5 on the Richter scale?

62. Automobile Earthquake   The total number of miles driven by cars and light trucks in the United States in 1999 was approximately 2.7×10^{12}. If we assume that each vehicle was being driven at 40 miles per hour and had a 200-horsepower engine, then each expended about 1.34×10^{14} ergs per mile. What magnitude earthquake has the same energy output as the total energy used by vehicles in the United States in 1999?
(*Data source:* U.S. Department of Transportation.)

63. Comparing Earthquakes   The earthquake that shook Armenia on December 7, 1988, registered 6.8 on the Richter scale. The San Francisco earthquake of April 18, 1906, registered 8.3 on the Richter scale. How many times more powerful, as measured by energy output, was the San Francisco earthquake than that of Armenia?

64. Jumping Earthquake   It has been hypothesized that if each Chinese man, woman, and child were to jump at precisely the same instant, an earthquake of magnitude 4.3 on the Richter scale would result. Assuming that the energy of the ground movement is proportional to the number of people jumping and that there are 1.3 billion people in China and 6.25 billion people in the world, of what magnitude would the earthquake be that would result if all the people on Earth were to jump simultaneously?

Concepts and Critical Thinking

Exercises 65-72   *Answer true only if the given equation is an identity for all positive x and y; otherwise answer false.*

65. $\log xy = (\log x)(\log y)$

66. $\log(x + y) = \log x + \log y$

67. $\log xy = \log x + \log y$

68. $y \ln x = \ln yx$

69. $y \ln x = \ln x^y$

70. $\log_2 \dfrac{x}{y} = \dfrac{\log_2 x}{\log_2 y}$

71. $\log_2 \dfrac{x}{y} = \log_2 x - \log_2 y$

72. $\log_2 \dfrac{x}{y} = \log_2(x - y)$

Exercises 73-76   *Give an example of each.*

73. A reason why solution candidates to logarithmic equations must be checked

74. A number greater than 20 whose logarithm can be computed if the logarithm of 3 is known

75. Two numbers x and a such that $\log_a x = 2$

76. A logarithmic equation having no solution

77. Graph the following functions on the same coordinate system:

$$f(x) = \log x$$
$$g(x) = \log(x \cdot 10)$$
$$h(x) = \log(x \cdot 0.1)$$

What relationship do you see among the graphs of these functions? What property of logarithms explains this relationship?

78. Graph $y = \ln(e^3 x)$. Now find a number k so that the graph of $y = k + \ln x$ is the same. What properties of logarithms explain this?

Questions for Discussion or Essay

79. As was mentioned in the text, many students struggle with logarithms because there are so many seemingly plausible identities that are simply not true. For example, it is not true that for all x and y, $\log_a x \cdot \log_a y = \log_a(x + y)$. What makes this nonrule so tempting to accept as true? What other tempting nonrules involving logarithms can you develop? How would one go about showing that a nonrule (such as the one given) is false?

80. In Exercise 64, we made the assumption that the ground movement arising from a number of people jumping at the same moment would be proportional to the number of people jumping. Is this a reasonable assumption? What other factors besides the number of people jumping would affect the resulting seismic disturbance?

81. As we have seen, the decibel rating of a sound with intensity I is computed using the formula $D = 10 \log(I/I_0)$, where I_0 is the intensity of the faintest sound that is perceptible to the human ear. Sound intensity itself is measured in watts per square centimeter. For example, I_0 is, by common agreement, assigned the value 10^{-16} watts per

square centimeter, and the intensity of a whisper is approximately 10^{-13} watts per square centimeter. Why do you think the decibel is preferred as a measure of the loudness of a sound rather than the intensity? More generally, under what circumstances do you think logarithms would be useful in formulas that define measurement scales?

82. In this section, we saw how a calculator could be used to compute the logarithm to any base of any number. Prior to the advent of the hand-held calculator, such tasks were typically accomplished using a process that required extensive logarithm tables (see Exercise 84 for an example of such a table). This is just one example of how calculators have changed the way in which mathematics is taught. What are some other changes that have occurred? (A discussion with your parents or an older student may help with this question.) What are some changes that you think have yet to occur but are inevitable? Give some examples of basic skills that you think should never be replaced by a calculator.

Projects for Enrichment

83. Constructing a Slide Rule Slide rules were widely used before inexpensive electronic calculators became available. In this project, we construct a simple slide rule and use it to perform multiplications.

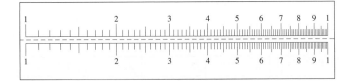

Figure 19

Begin by copying Figure 19 and separating the copied version into two pieces by cutting along the dashed line. Now secure the lower piece so that it will not move.

To multiply two numbers between 1 and 10 whose product is less than 10, say 2.2 and 3.1, simply slide the top piece over the bottom so that the 1 at the left of the top piece is directly above 2.2 on the bottom piece. Next find 3.1 on the top piece and read off the number directly below it (≈ 6.8) on the bottom piece. If the numbers are such that their product exceeds 10, say 4.2 and 5.3, place the 1 at the *right-hand end* of the top piece directly above 4.2; then find 5.3 on

the top piece and read off the number directly below it (≈ 2.2). The answer is then obtained by multiplying 2.2 by 10 to give 22.

In general, when using slide rules, we must keep track of powers of 10. For this reason, it is helpful to first convert the numbers that we are multiplying to scientific notation. For example, to compute 120×4100, we first rewrite the product as $(1.2 \times 10^2)(4.1 \times 10^3)$, then multiply 1.2 and 4.1 with the slide rule to obtain 4.9, next multiply 10^2 and 10^3 to obtain 10^5, and finally write our answer as 4.9×10^5.

a. Compute the following products using your makeshift slide rule:

   i. 2.5×6.8 **ii.** 0.0024×39

   iii. 350×180

A slide rule like that depicted in Figure 19 could be formed in the following way. First, take a piece of ordinary graph paper, and mark a number line from 0 to 1. Next, compute the common logarithms of the integers from 1 to 10. Then, for each integer, make a tick mark at the position corresponding to its logarithm, and label the tick mark with the corresponding integer. For example, the number 2 was written at the position 0.301 since $\log 2 \approx 0.301$. For increased accuracy, smaller tick marks could be made at locations corresponding to the logarithms of the multiples of 0.1—that is, the numbers 1.1, 1.2, . . . , 9.9. Thus, the location of a number on the slide rule is determined by its logarithm.

b. Find the ratio of the distance between 1 and 2 to the distance between 2 and 4 on the slide rule.

c. What number on the slide rule is the same distance from 9 as 1 is from 2?

d. Explain the process by which numbers are multiplied with a slide rule.

84. Logarithm Tables John Napier developed logarithms as a labor-saving device for doing arithmetic calculations. Today, although logarithmic functions are extremely important in modeling real-world phenomena, logarithms are rarely used as a computational aid for performing arithmetic because inexpensive electronic calculators are readily available. In this project, we investigate the methods by which our predecessors performed lengthy arithmetic computations.

I. Tables
Until fairly recently, logarithms were calculated by hand (using complicated formulas from calculus) and then tabulated in lengthy logarithm tables. Table 5 is an abbreviated version of a common (base 10) logarithm table.

To find the logarithm of a number between 1 and 10, say 3.9, we read off the entry in row 3 and column 9—namely, 0.5911. Thus, $\log 3.9 \approx 0.5911$. Logarithms of numbers larger than 9.9 or smaller than 1.0 can be found by converting to scientific notation and using the properties of logarithms. For example, we compute $\log 230$ as follows:

$$\begin{aligned} \log 230 &= \log(2.3 \times 10^2) \\ &= \log 2.3 + \log(10^2) \\ &= 0.3617 + 2 \\ &= 2.3617 \end{aligned}$$

a. Approximate each of the following logarithms using Table 5:

   i. $\log 6700$ **ii.** $\log 0.0023$ **iii.** $\log 15{,}000{,}000$

To perform calculations using logarithms, we must be able to compute **antilogarithms**. The antilogarithm of x is simply the number 10^x. If the number x is between 0 and 1, then the antilogarithm of x (10^x) can be found by locating x in Table 5 and then reading off the number whose logarithm is x. For example, to compute $10^{0.1139}$, we locate 0.1139 in the table, finding that 1.3 is the number whose logarithm is 0.1139. Thus, $10^{0.1139} \approx 1.3$. If the number x does not appear in the table, then, as an estimate, simply choose the nearest number in the table. For example, to compute $10^{0.52}$, we locate the entry in the table nearest to 0.52—namely, 0.5185—and find that $10^{0.52} \approx 10^{0.5185} \approx 3.3$. Antilogarithms of numbers larger than 1 or smaller than 0 can be found using the properties of exponents. For example, we compute $10^{4.75}$ and $10^{-2.8}$ as follows:

$$\begin{aligned} 10^{4.75} &= 10^4 10^{0.75} & 10^{-2.8} &= 10^{-3} \cdot 10^{0.2} \\ &\approx 10^4 \cdot 10^{0.7482} & &\approx 10^{-3} \cdot 10^{0.2041} \\ &\approx 10^4 \cdot 5.6 & &\approx 10^{-3} \cdot 1.6 \\ &= 56{,}000 & &\approx 0.0016 \end{aligned}$$

b. Approximate each of the following antilogarithms using the logarithms in Table 5:

   i. $10^{3.89}$ **ii.** $10^{-4.05}$

II. Multiplication and Division
The product or quotient of any two numbers can be estimated using the properties of logarithms and a logarithm table. For example, the identity $x \cdot y = 10^{\log(x \cdot y)} = 10^{\log x + \log y}$ can be used to compute 348×5700 as follows:

$$\begin{aligned} 348 \times 5700 &= 10^{\log(348 \times 5700)} \\ &= 10^{\log 348 + \log 5700} \\ &\approx 10^{2.5441 + 3.7559} && \text{Using the table to estimate} \\ & && \log 348 \text{ and } \log 5700 \\ &= 10^{6.3} \\ &= 10^{0.3} 10^6 \\ &\approx 2 \cdot 10^6 && \text{Using the table to estimate} \\ & && \text{the antilogarithm of } 0.3 \\ &= 2{,}000{,}000 \end{aligned}$$

Table 5

N	0	1	2	3	4	5	6	7	8	9
1	.0000	.0414	.0792	.1139	.1461	.1761	.2041	.2304	.2553	.2788
2	.3010	.3222	.3424	.3617	.3802	.3979	.4150	.4314	.4472	.4624
3	.4771	.4914	.5051	.5185	.5315	.5441	.5563	.5682	.5798	.5911
4	.6021	.6128	.6232	.6335	.6435	.6532	.6628	.6721	.6812	.6902
5	.6990	.7076	.7160	.7243	.7324	.7404	.7482	.7559	.7634	.7709
6	.7782	.7853	.7924	.7993	.8062	.8129	.8195	.8261	.8325	.8388
7	.8451	.8513	.8573	.8633	.8692	.8751	.8808	.8865	.8921	.8976
8	.9031	.9085	.9138	.9191	.9243	.9294	.9345	.9395	.9445	.9494
9	.9542	.9590	.9683	.9685	.9731	.9777	.9823	.9868	.9912	.9956

c. Perform each of the following operations using only addition, subtraction, and Table 5:

   i. $79{,}000 \times 62.5$ **ii.** 0.0003×426.7

d. Show that for any two positive numbers x and y, the quotient x/y is equal to the antilogarithm of the difference in their logarithms. Use this fact to compute $426/0.032$.

e. Explain how one could evaluate $3200^{2.3}$ using only a logarithm table, addition, subtraction, and multiplication. (*Hint:* Start with the fact that $3200^{2.3} = 10^{\log(3200^{2.3})}$, and then use a logarithmic identity.)

Section 4.4 | # Exponential Equations and Applications

- Could the Shroud of Turin be 2000 years old?
- Which would you rather have: $1000 in an account paying 10% interest or $2000 in an account paying 5% interest?
- If eliminating all sources of contamination drops the pollution level of a lake by 15% after 5 years, then how long will it take for the pollution level to drop by 50%?
- How can a coroner estimate the initial dose of a heart medication taken 24 hours before a fatal heart attack, even if later doses were given?

Exponential Equations

Equations such as $2^x = 2^3$, $3^x = 50$, and $2^{2x-1} = 3^x$, in which the variable appears in an exponent, are called **exponential equations**. Some exponential equations can be solved using algebraic techniques we have already seen. For example, to solve $2^x = 2^3$, we use the fact that exponential functions are 1–1 to obtain $x = 3$. Until now, we have solved equations like $3^x = 50$ and $2^{2x-1} = 3^x$ graphically, a surefire—though sometimes lengthy—method. Our focus in this section is on the development of algebraic techniques for solving exponential equations, techniques that are, in many instances, both more efficient and more accurate than graphical techniques.

The most straightforward technique for solving exponential equations is also the most effective: taking the logarithm of both sides. The central idea is to eliminate exponents using the identity $\log_a u^n = n \log_a u$. For example, consider the equation $3^x = 50$. Taking common logarithms of both sides gives us

$$3^x = 50$$
$$\log 3^x = \log 50$$
$$x \log 3 = \log 50$$
$$x = \frac{\log 50}{\log 3} \approx 3.56088$$

Note that taking the logarithm of both sides has the effect of "bringing the variable out of the exponent."

◦◦◦▸EXAMPLE 1 **Solving an Exponential Equation**

Solve $7^{3x} = 15$.

Solution

$$7^{3x} = 15$$

$\log 7^{3x} = \log 15$	Taking the common logarithm of both sides
$3x \log 7 = \log 15$	Using the identity $\log u^n = n \log u$ on both sides
$(3 \log 7)x = \log 15$	Rearranging on the left side
$x = \dfrac{\log 15}{3 \log 7}$	Solving for x
$x \approx 0.46389$	Using a calculator

EXAMPLE 2

Solving an Exponential Equation

Solve $2^{2x-1} = 3^x$. Check the result graphically.

Solution

$$2^{2x-1} = 3^x$$

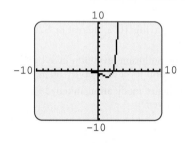

Figure 20

Figure 21

$\ln(2^{2x-1}) = \ln(3^x)$	Taking the natural logarithm of both sides
$(2x - 1) \ln 2 = x \ln 3$	Using the logarithmic identity $\ln u^n = n \ln u$
$(2 \ln 2)x - \ln 2 = (\ln 3)x$	Multiplying out on the left side
$(2 \ln 2)x - (\ln 3)x = \ln 2$	Rearranging so that all variable terms are on one side
$(2 \ln 2 - \ln 3)x = \ln 2$	Factoring out x
$x = \dfrac{\ln 2}{2 \ln 2 - \ln 3}$	Solving for x
≈ 2.40942	Approximating using a calculator

To check this solution graphically, we plot the graph of $y = 2^{2x-1} - 3^x$ (see Figure 20) and approximate the x-intercept, obtaining $x \approx 2.40942$ (see Figure 21).

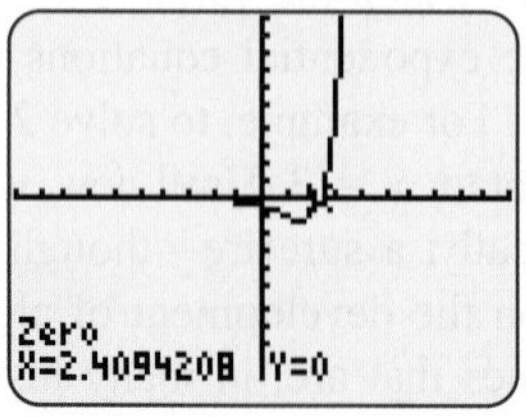

When solving exponential equations such as $2^{2x-1} = 3^x$ from Example 2, linear equations like $(2x - 1)\ln 2 = x \ln 3$ often arise. Such equations are easily solved if it is kept in mind that "$\ln 2$" and "$\ln 3$" are nothing more than real numbers like 5 or 7.

EXAMPLE 3

Compound Interest

Suppose that $1000 is deposited into an account paying 10% interest, and on the same day, $2000 is deposited into an account paying 5% interest. How long will it take for the balance in the 10% account to equal the balance in the 5% account? Assume that interest is compounded annually.

Solution   Let t be the number of years required for the balances to be equal. Using the formula for compound interest, $B = P(1 + r)^t$, the balance in the 5% account is given by $B = 2000 \cdot 1.05^t$, whereas the balance in the 10% account is given by $B = 1000 \cdot 1.1^t$. Equating the balances, we have

Balance on $2000 at 5% = Balance on $1000 at 10%

$$2000 \cdot 1.05^t = 1000 \cdot 1.1^t$$

$2 \cdot 1.05^t = 1.1^t$	Dividing both sides by 1000

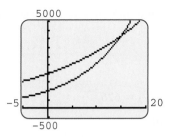

5000

-5   20

-500

Figure 22

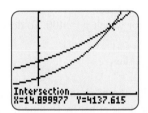

Intersection
X=14.899977 Y=4137.615

Figure 23

$$\log(2 \cdot 1.05^t) = \log(1.1^t)$$ Taking the common logarithm of both sides

$$\log 2 + \log(1.05^t) = \log(1.1^t)$$ Using $\log uv = \log u + \log v$

$$\log 2 + t \log 1.05 = t \log 1.1$$ Using $\log u^n = n \log u$

$$t \log 1.05 - t \log 1.1 = -\log 2$$ Rearranging

$$t(\log 1.05 - \log 1.1) = -\log 2$$ Factoring out the variable

$$t = \frac{-\log 2}{\log 1.05 - \log 1.1}$$ Solving for the variable

$$\approx 14.9$$ Approximating using a calculator

Thus, it will take approximately 14.9 years for the balances to be equal. We can check this solution by finding the intersection of the graphs of $y = 2000 \cdot 1.05^x$ and $y = 1000 \cdot 1.1^x$, as in Figure 22. In Figure 23, we see that the point of intersection is approximately (14.9, 4138), which confirms our solution of $t \approx 14.9$.

Many exponential equations either cannot be solved exactly using the methods discussed in this section or are very difficult to solve. In such cases, we must resort to approximating the solutions using a graphing calculator.

······**EXAMPLE 4**

The Gateway Arch

When a cable is suspended from two points, it takes the shape of a *catenary* curve. This same curve was used in the design of the Gateway Arch in St. Louis, Missouri. The centers of mass of cross sections of the Arch (which happen to be equilateral triangles) are traced out by the graph of the equation

$$y = 694 - 34(e^{0.01x} + e^{-0.01x})$$

where x denotes the horizontal position measured from the central axis of the arch and y denotes the corresponding height in feet above the ground (see Figure 24). Roughly speaking, the graph of this equation traces out the path of the trams that are used to transport people up and down the Arch.

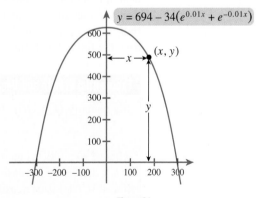

Figure 24

The Gateway Arch just before its completion in 1965.

a. How high is the Arch?

b. How high above the ground is a person riding in a tramcar when it is 50 feet from the central axis?

c. How far from the central axis is a person riding in a tramcar when the car is at a height of 400 feet?

Solution

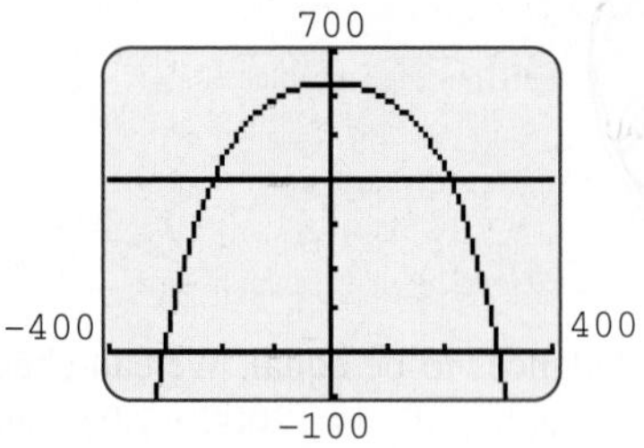

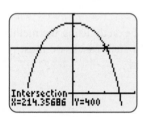

700

−400　　　400

−100

Figure 25

a. Setting $x = 0$, we have

$$y = 694 - 34(e^{0.01(0)} + e^{-0.01(0)}) = 626$$

and so the Arch is 626 feet high at its highest point.

b. We simply set $x = 50$ to obtain

$$y = 694 - 34(e^{0.01(50)} + e^{-0.01(50)}) \approx 617$$

Thus, the height above the ground is approximately 617 feet.

c. In this case, we wish to determine the value(s) of x for which $y = 400$. To do so, we must solve the equation

$$400 = 694 - 34(e^{0.01x} + e^{-0.01x})$$

Although this equation could be solved algebraically, the work is quite tedious, and so we opt for a graphical approximation. This can be done either by finding the point of intersection of $y = 400$ and $y = 694 - 34(e^{0.01x} + e^{-0.01x})$ or by writing the equation in the form $f(x) = 0$ and finding the x-intercepts of the graph of f. Choosing the former, we plot the graphs of $y = 694 - 34(e^{0.01x} + e^{-0.01x})$ and $y = 400$, as shown in Figure 25. Using the calculate-intersect feature, we obtain $x \approx 214$, as shown in Figure 26. Note that, because of symmetry, there is a second solution at $x \approx -214$. In either case, the tramcar is approximately 214 feet from the central axis when it is 400 feet above the ground.

Intersection
X=214.35686　Y=400

Figure 26

Exponential Growth and Decay

Exponential functions are used to model diverse phenomena in a variety of disciplines, from physics to pharmacology, chemistry to criminology, and medicine to meteorology. This wide-ranging applicability stems from the fact that exponential functions increase (or decrease) at a constant percentage rate, a property shared by many real-world quantities. Thus, for example, both a population growing at the constant rate of 3% per year and a radioactive sample decaying by 50% every 1000 years can be modeled by exponential functions. For this reason, if the percentage growth rate of a quantity is constant, then the quantity is said to be either **growing exponentially** or **decaying exponentially**, according to whether the quantity is increasing or decreasing. More precisely, we have the following definitions.

Exponential Growth and Decay

If a quantity $Q = Q(t)$ changes at a constant percentage rate, then

$$Q(t) = Q_0 e^{kt}$$

where $Q_0 = Q(0)$ is the initial quantity, and k is called the **growth constant**.
If $k > 0$, $Q(t)$ is increasing, and we say that Q is **increasing exponentially**.
If $k < 0$, $Q(t)$ is decreasing, and we say that Q is **decaying exponentially**.

The exponential growth or decay function $Q(t) = Q_0 e^{kt}$ can be written in the equivalent form $Q(t) = Ca^t$ (see Exercise 51). Other characterizations of exponential growth and decay include the following.

Characterizations of Exponential Growth and Decay

1. If a quantity is increasing at a constant percentage rate, then the quantity is growing exponentially.
2. If a quantity is decreasing at a constant percentage rate, then the quantity is decaying exponentially.
3. If the time it takes for a quantity to double (the doubling time) is constant, then the quantity is growing exponentially.
4. If the time it takes for a quantity to decrease to one-half its initial amount (the half-life) is constant, then the quantity is decaying exponentially.

EXAMPLE 5

Population Growth

The population of Florida was 12.9 million in 1990 and 16.0 million in 2000. Assuming that Florida's population grows at a constant percentage rate, estimate the population of Florida in 2005.

Solution Assuming exponential growth with an initial population of 12.9 million, we conclude that the population of Florida is modeled by a function of the form

$$P(t) = 12.9e^{kt}$$

where t is the number of years after 1990 and $P(t)$ is the population in millions at time t. Since the population was 16.0 million in 2000, we can find k as follows:

Population after 10 years $= 16.0$

$$P(10) = 16.0$$

$$12.9e^{k \cdot 10} = 16.0$$

$$e^{10k} = \frac{16.0}{12.9}$$

$$\ln e^{10k} = \ln\left(\frac{16.0}{12.9}\right) \qquad \text{Taking the natural logarithm of both sides}$$

$$10k = \ln\left(\frac{16.0}{12.9}\right) \qquad \text{Using the logarithmic identity } \ln e^x = x$$

$$k = \frac{\ln\left(\dfrac{16.0}{12.9}\right)}{10} \approx 0.021536$$

Thus, we have $P(t) = 12.9\, e^{0.021536t}$. Since the year 2005 corresponds to $t = 15$, we predict a population of

$$P(15) = 12.9\, e^{0.021536(15)} \approx 17.8 \text{ million}$$

Rule of Thumb

In general, when using a calculator to solve an applied problem, you should retain all intermediate results in the calculator's memory. This is particularly advisable when working with exponential functions because small round-off errors can have profound consequences (see Exercise 52). In the text, we often round intermediate results so that individual steps in the solution can be shown.

Since a radioactive substance decays at a constant percentage rate, it follows that the mass of a radioactive substance decays exponentially according to $Q(t) = Q_0 e^{kt}$, where $k < 0$. It is customary to report the rate of decay of a substance in terms of its **half-life**: the time required for one-half of the substance to decay. The half-lives of a few common isotopes are given in Table 6.

Table 6
Half-life

Element	Atomic weight	Half-life
Uranium	235	7×10^8 years
Carbon	14	5760 years
Lead	210	22.3 years
Barium	133	10.5 years
Iron	59	44.5 days
Gold	198	64.6 hours
Arsenic	76	26.5 hours
Copper	64	12.7 hours
Nitrogen	13	9.97 minutes

Data source: http://www.inus.com/doc/isotope.htm.

EXAMPLE 6

Carbon-14 Dating

Carbon-14 (C-14) decays exponentially with a half-life of approximately 5760 years. If an artifact originally had 12 grams of C-14 and now has only 4 grams remaining, how old is the artifact?

Solution   Let $Q(t)$ represent the quantity of C-14 remaining at time t. Since there were initially 12 grams of C-14, and the quantity is decaying exponentially, we may write $Q(t) = 12 e^{kt}$. We next find the constant k by using the half-life of 5760 years. Since there were originally 12 grams, there would be only 6 grams remaining after 5760 years.

The Dead Sea Scrolls, consisting mostly of thousands of fragments such as this, have been carbon dated to around the time of Christ.

$$\begin{aligned}
\text{Quantity of C-14} & \\
\text{remaining after 5760 years} &= 6 \\
Q(5760) &= 6 \\
12e^{5760k} &= 6 && \text{Setting } t = 5760 \text{ in } Q(t) \\
e^{5760k} &= \frac{1}{2} && \text{Dividing by 12} \\
\ln(e^{5760k}) &= \ln\left(\frac{1}{2}\right) && \text{Taking the natural logarithm of both sides} \\
5760k &= \ln\left(\frac{1}{2}\right) && \text{Using the logarithmic identity } \ln e^a = a \\
k &= \frac{\ln\left(\frac{1}{2}\right)}{5760} \approx -0.00012034 && \text{Solving for } k
\end{aligned}$$

Thus, $Q(t) = 12e^{-0.00012034t}$. Because there are currently 4 grams remaining, we have

$$\begin{aligned}
\text{Quantity of C-14} \\
\text{remaining after } t \text{ years} &= 4 \\
Q(t) &= 4 \\
12e^{-0.00012034t} &= 4 \\
e^{-0.00012034t} &= \frac{1}{3} \qquad \text{Dividing both sides by 12} \\
\ln(e^{-0.00012034t}) &= \ln\left(\frac{1}{3}\right) \qquad \text{Taking the natural logarithm of both sides} \\
-0.00012034t &= \ln\left(\frac{1}{3}\right) \qquad \text{Using the logarithmic identity } \ln e^a = a \\
t &= \frac{\ln\left(\frac{1}{3}\right)}{-0.00012034} \approx 9129 \text{ years} \qquad \text{Solving for } t
\end{aligned}$$

Carbon-14 Dating and the Shroud of Turin

The Shroud of Turin, a linen relic bearing the image of a man, first appeared in 14th-century France. Although the Roman Catholic Church took no official position on its authenticity, it was widely believed to be the burial Shroud of Jesus. The origins of the Shroud were questioned from the very beginning, but there was no way to disprove its authenticity. In fact, the Catholic Church allowed a team of investigators to conduct a scientific investigation on the Shroud in 1978, and no conclusive evidence of fraud was produced, although one member of the team concluded that the image on the Shroud was, in fact, a very thin watercolor.

In order to protect the possibly sacred Shroud, the Church had not allowed **carbon-14 dating** to be performed. This test, if conducted using the technology available in 1978, would have caused the destruction of an unacceptably large portion of the Shroud. By 1988, however, technological developments made it possible to conduct carbon-14 dating on the Shroud with only very minimal destruction. The results demonstrated that

Shroud of Turin

the Shroud dated from the 14th century and hence could not have been the burial shroud of Jesus.

Carbon-14 is a radioactive isotope of carbon that is present in all living organisms. After the death of an organism, the carbon-14 gradually decays to an isotope of nitrogen. As a consequence, the quantity of carbon-14 slowly decreases. By measuring the remaining carbon-14, scientists can determine how much time has passed since the organism died, and thus they can date pieces of wood, cloth, hides, bones, and other organic matter.

Since carbon-14 decays exponentially, we know that the amount remaining t years after the death of the organism is given by $Q(t) = Q_0 e^{kt}$, where Q_0 is the amount of carbon-14 prior to death. It can be shown (see Example 6) that since the half-life of carbon-14 is 5760 years, $k = -0.00012034$. Thus, $Q(t) = Q_0 e^{-0.00012034t}$.

EXAMPLE 7

Pollution in Lake Michigan

Under suitable conditions, and assuming all incoming pollution is stopped, the concentration of pollution in a lake decays exponentially over time. If it takes 5 years for the pollution concentration in Lake Michigan to decrease to 85% of its current level, how many years will it take for the pollution concentration to decrease to 50% of its current level?

Solution Let $Q(t) = Q_0 e^{kt}$ denote the concentration of pollution in Lake Michigan at time t. We must first determine the value of k. Since the concentration in 5 years will be 85% of the initial concentration Q_0, we obtain the following equation:

$$\text{Concentration after 5 years} = 85\% \text{ of } Q_0$$
$$Q(5) = 0.85Q_0$$
$$Q_0e^{k(5)} = 0.85Q_0$$

Now we solve for k.

$$Q_0e^{k(5)} = 0.85Q_0$$
$$e^{5k} = 0.85 \qquad \text{Dividing both sides by } Q_0$$
$$\ln e^{5k} = \ln 0.85 \qquad \text{Taking the natural logarithm of both sides.}$$
$$5k = \ln 0.85$$
$$k = \frac{\ln 0.85}{5} \approx -0.032504$$

Thus, we have $Q(t) = Q_0e^{-0.032504t}$. To find how long it takes for the concentration to reach 50% of its present level, we proceed as follows:

$$\text{Concentration after } t \text{ years} = 50\% \text{ of } Q_0$$
$$Q_0e^{-0.032504t} = 0.50Q_0$$
$$e^{-0.032504t} = 0.5 \qquad \text{Dividing both sides by } Q_0$$
$$\ln e^{-0.032504t} = \ln 0.5 \qquad \text{Taking the natural logarithm of both sides}$$
$$-0.032504t = \ln 0.5$$
$$t = \frac{\ln 0.5}{-0.032504} \approx 21.3 \text{ years}$$

Understanding and Mastery Checklists

Concepts to Understand	Skills to Master
Exponential equations	Solve an exponential equation algebraically.
❖	❖
Exponential growth and decay	Approximate the solutions of an exponential equation graphically.
❖	❖
Half-life	Construct an exponential function that models exponential growth or decay.
	❖
	Compute half-life and doubling time.

Exercises 4.4

Exercises 1-18   *Solve the exponential equation. When approximating, give answers to five decimal places of accuracy. You may wish to check your solution graphically.*

1. $2^x = 16$

2. $3^{-x} = \dfrac{1}{9}$

3. $25^{3x} = 125^{x+1}$

4. $3^{x^2+1} = 27$

5. $7^x = 10$

6. $5^x = 10$

7. $3^{-x} = 11$

8. $3^{x-1} = 2^x$

9. $\left(\frac{3}{5}\right)^x = 6^{2-x}$

10. $0.4^{2+x} = 1.5^{3x-1}$

11. $e^x = 2^{x+1}$

12. $5e^{-2.3x} = 4$

13. $e^{2x-1} = \pi^{2x}$

14. $e^{x/2} = 3^{x-4}$

15. $\dfrac{1}{2^{x+2}} = \dfrac{1}{8}$

16. $\left(\dfrac{2}{3}\right)^{-x+1} = \dfrac{81}{16}$

17. $1.06^t = 1000$

18. $1000e^{0.05t} = 2000$

Exercises 19-24 *Approximate the solution(s) of the given equation to the nearest hundredth.*

19. $e^{2x} - 5e^x = -6$

20. $4^x - 3 \cdot 2^x + 2 = 0$

21. $e^{-x} = x$

22. $4 - x^2 = e^{x-2}$

23. $\dfrac{1}{25}x^4 + 4 = x^2 + e^{x+1}$

24. $\dfrac{1}{8}x^3 + \dfrac{7}{8}x^2 = x + 2 - e^x$

Applications

25. Compound Interest An account is established with a principal of $200, paying 4% interest compounded semiannually. How long will it take for the balance to reach $300?

26. Tripling Money How long does it take for money to triple in an account paying 5% interest compounded annually?

27. Doubling Money An account paying interest compounded continuously at a rate r doubles in size every 12 years. Find r.

28. Competing Accounts Suppose that Bob deposits $1000 into an account paying 6% interest compounded quarterly at the same time that Janet deposits $1200 into an account paying 6% interest compounded annually. How long will it take for Bob's balance to exceed Janet's?

29. Competing Accounts Suppose that $200 was deposited on January 1, 2002, into an account that earned 5% interest compounded annually. Suppose further that $200 was deposited on January 1, 2003, into a different account that earned 6% interest compounded annually. In what month of what year will the balance in the account earning 6% interest overtake the balance in the account earning 5%?

30. Competing Accounts On January 1, 2003, $1000 is deposited into an account paying 15% simple interest, and $100 is deposited into an account paying 5% interest compounded annually. In what month of what year will the balance in the simple interest account be overtaken by that in the compound interest account?

31. Compact Disk Inflation Suppose that the average price of a compact disk was $18.00 on January 1, 2003, and that the price increases at the rate of 3% per year. Give the month and the year in which the average price of a compact disk will be $100.

32. Annual Inflation Suppose that inflation is 3.5% annually. Then goods or services that cost P dollars today will, on the average, cost $P(1.035)^t$ dollars in t years. If tuition at a certain prestigious university is $25,000 per year in 2004–05, in what year will tuition reach $40,000?

33. Consumer Price Index The Consumer Price Index (CPI) was 99.6 in 1983 and 172.2 in 2000. In other words, goods and services that cost $99.60 in 1983 cost $172.20 in 2000. Assuming exponential growth, what will the CPI be in 2010? When will it reach 300?

34. Dollar Devaluation One dollar in 2000 was worth approximately 17¢ in 1960 dollars. Assuming this is an exponential decrease, when will a dollar be worth 1¢ in 1960 dollars?

35. Population Growth At 8:00 A.M., July 4, 2003, a colony of 1000 of the smallest free-living entity, *Mycoplasma laidlawii* (average weight $\approx 10^{-16}$ grams), is established. If the population were to double every hour, at what time would the collective mass of the colony exceed that of the largest organism on Earth, the 190-ton blue whale? (*Hints:* If P represents the population of the colony and t the number of hours since the establishment of the colony, then $P = 1000 \cdot 2^t$. A ton is 2000 pounds, and 1 pound is 454 grams.)

36. Population Decline   The population of a certain endangered species of owl is declining exponentially. There are currently 500 living specimens, whereas just 10 years ago there were 10,000. If the population continues to decline exponentially, how long will it be until there is only a single owl left?

37. Population Decline   The metropolitan area encompassing Steubenville, Ohio, and Weirton, West Virginia, suffered the greatest percentage population loss in the United States between 1990 and 2000. (*Data source:* U.S. Bureau of the Census.) The population of Steubenville-Weirton was 142,523 in 1990 and 132,008 in 2000. Assuming the population is decreasing exponentially, what will the population be in the year 2010?

38. Population Growth   The fastest growing metropolitan area in the United States between the years 1990 and 2000 was Las Vegas, Nevada-Arizona. (*Data source:* U.S. Bureau of the Census.) The population was 852,737 in 1990 and 1,563,282 in 2000. Assuming exponential growth, what will the population be in the year 2010?

39. Carbon Dating   Physicists found that the remains of a certain woolly mammoth contained only 25% of its original carbon-14. Given that the half-life of carbon-14 is 5760 years, how old is the mammoth?

40. Carbon Dating   The well-preserved corpse (see photo) of a Bronze Age man was found frozen in ice in the Austrian Alps in September 1991.

If only 53.3% of the carbon-14 is found, and the half-life of carbon-14 is 5760 years, how old is the body?

41. Pollution Turnover   The rate of pollution turnover in Lake Erie is quite rapid. Suppose that if all pollution inflow into Lake Erie were stopped, the concentration of pollution would decrease to 80% of its present level in 6 months. Assuming the pollution concentration decays exponentially, estimate how many years it will take for the concentration to decrease to 10% of its present level.

42. Styrofoam Decay   Suppose that Styrofoam cups are biodegraded at an exponential rate and that after 10 years 99.5% of the cup remains. Compute the half-life of a Styrofoam cup.

Concepts and Critical Thinking

Exercises 43–46   *Answer true or false.*

43. For a given real number y, the equation $2^x = y$ will either have a unique solution or no solution at all.

44. The half-life of a substance is equal to one-half the age of the substance.

45. A quantity will grow exponentially if it grows at a constant rate.

46. A quantity will grow exponentially if it grows at a constant percentage rate.

Exercises 47–50   *Give an example of each.*

47. An exponential equation having no solution

48. An exponential equation that is more easily solved by using common logarithms than natural logarithms

49. A real-world context in which exponential growth occurs

50. A real-world context in which exponential decay occurs

51. In this text, we have chosen to represent exponential growth and decay models with functions of the form $Q = Q_0 e^{kt}$, but it is not uncommon to see them represented with functions of the form $Q = Ca^t$. Show that, in fact, these two representations are equivalent. [*Hint:* $e^{kt} = (e^k)^t$.]

52. In Example 5, round the value of k to two decimal places and compare the population you find with this new value to that found in the example. What does your result suggest about rounding the k-value in exponential growth and decay models?

Questions for Discussion or Essay

53. When a substance is decaying exponentially, the percentage of substance lost for any time period remains constant. It follows that for a radioactive substance, such as radium, a 1-gram sample will take exactly the same amount of time to decay as two 0.5-gram samples or ten 0.1-gram samples. Does ice melt exponentially? Support your answer. What does the rate of melting of ice depend on?

54. In Exercise 33, we made an assumption about the exponential growth of the Consumer Price Index and observed that the costs for goods and services increase dramatically over time. What additional information would be necessary to test this assumption? What would be the inevitable effect of exponential CPI growth on our monetary system? The yearly rise in the CPI is known more commonly as inflation. Do you think inflation is one of the certainties in life, along with death and taxes? Why or why not? Have there ever been periods of deflation?

55. Suppose that the function $f(t) = 60{,}000e^{0.2t}$ gives the population of a city at time t, and the function $h(t) = 3t + 85{,}000$ measures the number of "people units" of housing available in the city at time t. Thus, the population is growing exponentially, whereas the available housing is growing at a constant rate (that is, linearly). Plot the two functions and approximate the time at which their graphs intersect. What does this point represent? What do these functions imply about the housing situation in the city over time? Is this realistic? Explain.

Projects for Enrichment

56. **A Medical Mystery** A patient was admitted to the hospital early in the morning complaining of chest pains. An initial dose of the heart medicine digoxin was administered at 9 A.M. although, due to an oversight, the amount was not recorded. Subsequent doses in a quantity sufficient to immediately raise the digoxin level in the bloodstream an additional 0.5 ng/ml were administered at 9 P.M. that evening and again at 9 A.M. the next morning. Unfortunately, the patient's condition did not improve, and at 9:05 A.M. he had a fatal heart attack. A test run at that time showed a digoxin level of 1.5 ng/ml in his bloodstream. In order to avoid a large malpractice settlement, the hospital must prove that the level of the drug was always within the therapeutic range, which is between 0.8 and 1.6 ng/ml. It is known from many clinical observations that a given dose of digoxin decays exponentially over time and is reduced to half its original level in the body in 36 hours.

a. Determine the level of digoxin in the bloodstream after the initial dose.

b. Illustrate graphically whether the level was always within the therapeutic range.

57. **Comparing Interest Rates and Compounding Periods** Suppose you are trying to decide whether to invest some money in an account that pays 5% compounded quarterly or 4.9% compounded monthly. Naturally, one way would be to determine which would yield the highest balance at the end of a given period of time. Another way would be to convert both rates to their equivalent rates compounded annually. In other words, we would like to find the annually compounded interest rates that would yield the same balance as 5% compounded quarterly and 4.9% compounded monthly. For this task, it would be helpful to have a formula that "converts" an interest rate r compounded n times a year to its equivalent rate s compounded annually. To that end, suppose a principal P is deposited at an interest rate r compounded n times a year. Suppose that, at the same time, a principal P is deposited at an interest rate s compounded annually. For the rates r and s to be equivalent, they must yield equal balances at the end of any time t. Thus, we must have

(1)
$$P\left(1 + \frac{r}{n}\right)^{nt} = P(1 + s)^t$$

a. Solve equation (1) for s to show that

(2)
$$s = \left(1 + \frac{r}{n}\right)^n - 1$$

The quantity s in equation (2) is called the **effective annual interest rate**.

b. Compute the effective annual interest rate that corresponds to 5% compounded quarterly and 4.9% compounded monthly. Which of the two interest-earning schemes would you choose? Why?

c. Solve equation (2) for r and use the resulting equation to find the interest rate r that, when compounded daily, would be equivalent to an effective annual rate of 6%.

d. Use equation (2) and your graphing calculator to estimate the number of compounding periods that would be needed for an interest rate of 5.85% to have an effective rate of 6%.

A continuously compounded interest rate can also be converted to an effective annual interest rate. If a principal P is deposited at an interest rate r compounded continuously, the effective annual interest rate is the value s that satisfies the equation

(3)
$$Pe^{rt} = P(1 + s)^t$$

e. Solve equation (3) for s to show that

(4)
$$s = e^r - 1$$

f. Find the effective annual rate corresponding to 4.8% compounded continuously.

g. Solve equation (4) for r and use the resulting formula to find the continuously compounded interest rate r that would be equivalent to an effective annual rate of 6%.

h. What does your solution to part g say about the number of compounding periods that would be necessary for an interest rate of 5.8% to have an effective rate of 6%?

Section 4.5   Modeling with Exponential and Logarithmic Functions

- ↺ Is hunting necessary to limit the exponential growth of a deer population?
- ↺ How many Internet hosts will there be in 2005?
- ↺ How much faster do pedestrians walk in Mexico City than in New York City?
- ↺ When will the number of transistors on a microprocessor exceed the number of neurons in the human brain?
- ↺ How much slower are the reflexes of a 60-year-old than those of a 19-year-old?

Constructing Mathematical Models

Throughout the text, we have seen examples of how mathematics can be used to model real-world phenomena. However, in most of these instances, we were concerned primarily with using a given model to solve a real-world problem. With few exceptions, we allowed ourselves the luxury of assuming that a given mathematical model would provide accurate results. We now wish to consider in more detail the work that goes on "behind the scenes" in the construction and validation of a mathematical model.

The construction of a mathematical model is often preceded by detailed observations of some naturally occurring event, as was the case with the bacteria population example in Section 4.1. The actual construction may then be as simple as drawing a graph based on those observations or as complex as finding a formula that provides results similar to the observations. The goal is to provide a model that will accurately describe trends, predict future outcomes, and perhaps even provide insight into how future outcomes can be controlled or at least influenced. Unfortunately, because of the compromise that must often be made between accuracy and simplicity, even the best mathematical models have limitations. Thus, it is important to test and possibly modify models as the need arises. We discuss these issues as we study the development and testing of mathematical models.

 **EXAMPLE 1**

Modeling a Deer Population Graphically

Consider the following scenario: In 1999, a certain state's Department of Wildlife recommended a ban on deer hunting because of a dangerously low deer population. Now, 5 years later, the department wishes to reevaluate its recommendation. Table 7 gives estimates of the deer population since 1999 ($t = 0$). Construct a graphical model to illustrate any trends in the data and to predict what might happen in future years. Use the graph to find a rough approximation for the population in 2005 ($t = 6$).

Table 7

Year	0	1	2	3	4	5
Population	10,000	11,500	13,200	15,100	17,400	20,100

Solution   The graph in Figure 27 shows that the population growth is occurring in a "smooth" way. Unless a dramatic change occurs, it is likely that the deer population will continue to grow—at least for a few years—according to the pattern that has been established. By fitting a smooth curve to the data, as shown in Figure 28, we obtain a rough estimate of 24,000 deer in 2005.

Increasing deer populations and human encroachment on natural habitat force deer into urban and suburban settings.

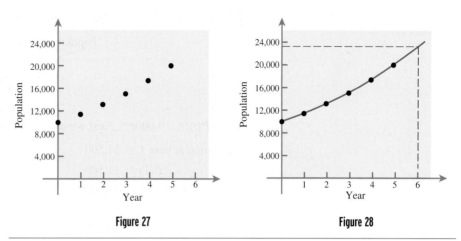

Figure 27   **Figure 28**

As useful as graphs are for displaying data, describing trends, and suggesting future behavior, they do have limitations when it comes to making precise predictions. The curve-fitting approach used in Example 1 involves some uncertainty: What does the graph look like beyond the known data points? Is it linear, parabolic, exponential, or something altogether different? It would be useful to have an explicit mathematical formula that captures the relationship between the quantities involved. Consider, for example, the data given in Table 7 and graphed in Figure 28. It appears that the growth is more rapid than linear. To verify this, we compute the percentage increase from one year to the next.

Table 8

Year	Population	Percentage increase over previous year
0	10,000	
1	11,500	$\dfrac{11{,}500 - 10{,}000}{10{,}000} = 0.15 = 15\%$
2	13,200	$\dfrac{13{,}200 - 11{,}500}{11{,}500} \approx 0.148 = 14.8\%$
3	15,100	$\dfrac{15{,}100 - 13{,}200}{13{,}200} \approx 0.144 = 14.4\%$
4	17,400	$\dfrac{17{,}400 - 15{,}100}{15{,}100} \approx 0.152 = 15.2\%$
5	20,100	$\dfrac{20{,}100 - 17{,}400}{17{,}400} \approx 0.155 = 15.5\%$

Based on Table 8, it appears that the percentage rate of growth is roughly constant (at 15% per year). This suggests exponential growth, and so we expect a function of the form $P(t) = ae^{bt}$.

EXAMPLE 2   **Constructing an Exponential Model**

Use the first two data values in Table 7 to construct a population model of the form $P(t) = ae^{bt}$, and then estimate the population after 6 years and again after 10. Here $P(t)$ denotes the population after t years.

Solution   We first use the fact that the population was 10,000 when $t = 0$.

$$\text{Population at time } 0 = 10,000$$
$$P(0) = 10,000$$
$$ae^{b(0)} = 10,000$$
$$a = 10,000$$

Thus, $P(t) = 10,000e^{bt}$. Next we use the population value 11,500 for $t = 1$.

$$\text{Population at year } 1 = 11,500$$
$$P(1) = 11,500$$
$$10,000e^{b(1)} = 11,500$$
$$e^b = \frac{11,500}{10,000} = 1.15 \qquad \text{Dividing both sides by 10,000}$$
$$b = \ln 1.15 \approx 0.1398 \qquad \text{Taking the natural logarithm of both sides}$$

Thus, $P(t) = 10,000e^{0.1398t}$. At 6 years, we have $P(6) = 10,000e^{0.1398(6)} \approx 23,100$. At 10 years, $P(10) \approx 40,500$.

Notice that the model in Example 2 was constructed using only two of the six available data points. Slightly different models would result if we were to use different combinations of data points. For example, if we were to use the points (0, 10,000), and (3, 15,000), we would obtain the exponential model $P(t) = 10,000e^{0.1374t}$. This raises an important question: Is there a way to construct an exponential model that takes all available data points into consideration? Fortunately, the answer is yes. The **exponential regression** feature of your graphing calculator is capable of finding a model of the form $P(t) = ae^{bt}$ that fits all available data points as closely as possible.

Calculator Keys

Exponential Regression

Exponential regression can be used to find an exponential function of the form $y = ae^{bx}$ (on some calculators, $y = ab^x$) that best fits a collection of data points (x, y). There are several steps to the process.

1. Clear any existing statistical data.
2. Enter the points (x, y) as statistical data.
3. Select exponential regression from the list of available regression options.

After step 3 has been completed, the calculator will compute values for a and b. Note that if your calculator finds a and b for a model of the form $y = ab^x$, you can obtain a model of the form $y = ae^{cx}$ by letting $c = \ln b$.

EXAMPLE 3

Using Exponential Regression to Fit a Model

Use exponential regression to find a model of the form $P(t) = ae^{bt}$ that best fits the data in Table 7.

Solution   Following the steps just outlined, we clear any existing statistical data in the calculator, enter the population data from Table 7 as statistical data (using time as the x-coordinate and population as the y-coordinate), and then select the exponential regression option. The computed values for a and b are $a \approx 9992$ and $b \approx 0.1391$. Thus, the best-fit model is $P(t) = 9992e^{0.1391t}$.

Testing Mathematical Models

Before a model can be used with confidence, it must be tested for accuracy. When actual data are available, we can test a mathematical model for accuracy by comparing its predictions to the actual data. With models such as the ones in Examples 2 and 3, this can be done either by computing tables of actual and predicted data values or by graphing the model along with the actual data points.

EXAMPLE 4

Testing Models

Compare the population data in Table 7 with the populations predicted by the following models:

a. $P(t) = 10,000 \, e^{0.1398t}$ **b.** $P(t) = 9992 \, e^{0.1391t}$

Solution

a. Table 9 shows how the actual data values compare with the ones predicted by $P(t) = 10,000 \, e^{0.1398t}$ for $t = 0$ through $t = 5$. Figure 29 shows the actual population data points plotted along with the graph of $P(t) = 10,000 \, e^{0.1398t}$. The data and the graph show very close agreement. In fact, because of the scale that was chosen, the graph appears to pass through all the data points.

Table 9

t	Actual values from Table 7	$P(t) = 10,000e^{0.1398t}$
0	10,000	10,000
1	11,500	11,500
2	13,200	13,226
3	15,100	15,210
4	17,400	17,493
5	20,100	20,117

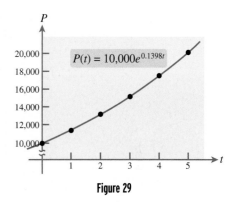

Figure 29

b. Table 10 shows how the actual data values compare with the ones predicted by $P(t) = 9992e^{0.1391t}$ for $t = 0$ through $t = 5$. In Figure 30, we have used a graphing calculator to plot both $P(t) = 9992e^{0.1391t}$ and the actual population data. The data and graph show very close agreement.

Table 10

t	Actual values from Table 7	$P(t) = 9992e^{0.1391t}$
0	10,000	9,992
1	11,500	11,483
2	13,200	13,197
3	15,100	15,166
4	17,400	17,430
5	20,100	20,031

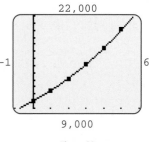

Figure 30

If a more precise measure of accuracy is desired, one can compute the *absolute error* or *deviation* between the actual data and the predicted values. For example, using the model $P(t) = 10,000e^{0.1398t}$ for the population data in Table 7, the absolute error at $t = 2$ is

$$|(\text{Actual population at } t = 2) - (\text{Predicted population at } t = 2)|$$
$$= |13,200 - 13,226| = 26$$

If we compute the absolute error for all known data values, then we can use either the sum of the errors or the largest of the errors as a single measure of the accuracy of the model.

⋯▷EXAMPLE 5

Computing Absolute Error

Determine the absolute errors between the data values given in Table 7 and those predicted by the following two models. Find the maximum absolute error and also the sum of the absolute errors. Which of the two models is more accurate?

a. $P(t) = 10,000e^{0.1398t}$ **b.** $P(t) = 9992e^{0.1391t}$

Solution

a. Using Table 9, we obtain the following absolute errors for the model $P(t) = 10,000e^{0.1398t}$.

t	Actual values from Table 7	$P(t) = 10,000e^{0.1398t}$	Absolute error
0	10,000	10,000	$\|10,000 - 10,000\| = 0$
1	11,500	11,500	$\|11,500 - 11,500\| = 0$
2	13,200	13,226	$\|13,200 - 13,226\| = 26$
3	15,100	15,210	$\|15,100 - 15,210\| = 110$
4	17,400	17,493	$\|17,400 - 17,493\| = 93$
5	20,100	20,117	$\|20,100 - 20,117\| = 17$

Thus, the maximum absolute error is 110 when $t = 3$. The sum of the errors is 246.

b. Using Table 10, we obtain the following absolute errors for $P(t) = 9992e^{0.1391t}$.

t	Actual values from Table 7	$P(t) = 9992e^{0.1391t}$	Absolute error
0	10,000	9,992	8
1	11,500	11,483	17
2	13,200	13,197	3
3	15,100	15,166	66
4	17,400	17,430	30
5	20,100	20,031	69

Thus, the maximum absolute error is 69 when $t = 5$. The sum of the errors is 193. Since the maximum absolute error and the sum of the absolute errors is smaller for the model $P(t) = 9992e^{0.1391t}$, we conclude that it is the more accurate of the two.

The absolute errors found in the preceding example are quite small, especially when the size of the population is taken into consideration. However, the acceptable error must also be weighed against the simplicity of a model. A model that is extremely accurate but difficult to use may not serve the purpose for which it was designed. In the case of our deer example, we were fortunate in finding a simple exponential model that was also extremely accurate. This is not completely unexpected. Numerous statistical studies have shown that uninhibited population growth tends to be exponential in nature. In fact, the British economist and sociologist Thomas Malthus (1766–1834) had proposed exponential population models as early as 1798. The Malthusian model, as it has come to be called, has the form $P(t) = P_0 e^{kt}$, where t denotes time, P_0 represents the **initial population**, and k is the **growth constant**.

Many mathematical models are not as easily verified as our Malthusian deer population model, however, especially when observed data are not readily available for testing purposes. For example, it would be difficult—if not impossible—to rigorously verify mathematical models that are intended to predict the effect of ozone depletion. In such cases, one must be careful about depending too heavily on the model's predictions. Even if a model has been thoroughly verified, one should be wary about placing too much confidence in its predictions.

Predicting with Mathematical Models

Once we have verified that a model is accurate and have judged it to be sufficiently simple, the next step is to use the model to make inferences about future behavior. After all, the purpose for constructing and verifying a model is not just to showcase its accuracy and simplicity, but to use it to gain insight into how the future will unfold.

EXAMPLE 6

Predicting with Exponential Models

Suppose that on the basis of historical data, the Department of Wildlife (see Example 1) estimates that the natural resources in the state can support at most 240,000 deer. Proponents of deer hunting argue that the ban on deer hunting should be lifted when the population reaches half of that limiting value in order to prevent the deaths by starvation and auto accidents that would likely occur as the population nears the limiting value. Use the population model $P(t) = 9992e^{0.1391t}$ to predict when the deer population will reach 120,000.

Solution   We want to determine the value for t at which the population predicted by $P(t) = 9992e^{0.1391t}$ will be 120,000. Thus, we must set $P(t)$ equal to 120,000 and solve for t.

$$P(t) = 120,000$$

$$9992e^{0.1391t} = 120,000$$

$$e^{0.1391t} = \frac{120,000}{9992} \qquad \text{Dividing both sides by 9992}$$

$$\ln e^{0.1391t} = \ln\left(\frac{120,000}{9992}\right) \qquad \text{Taking the natural logarithm of both sides}$$

$$0.1391t = \ln\left(\frac{120,000}{9992}\right) \qquad \text{Using the identity } \ln(e^x) = x$$

$$t = \frac{\ln\left(\dfrac{120,000}{9992}\right)}{0.1391} \approx 17.9$$

Thus, our model predicts that it will take approximately 17.9 years for the deer population to reach 120,000.

Although there has been human settlement on what is now Istanbul for thousands of years, the city has not reached its limiting population.

Notice that although the Department of Wildlife has estimated that the natural resources can support a maximum of 240,000 deer, our model actually suggests that the population growth has no bound. For example, the model predicts that in 25 years the deer population will be $9992e^{(0.1391)25}$, or approximately 324,000 and still rising. Perhaps our estimate of the limiting population is too low, but clearly there is a limiting population: If the deer population grew without bound, then eventually the entire state (and the airspace above it) would be nothing but a gigantic mound of deer. Thus, the problem with the model is the assumption that the deer population will continue to grow as it has over the first 5-year period. Due to limited natural resources, a slower population growth would be a certain, though gradual, outcome. In order to extend the range of applicability, we must modify our Malthusian exponential model to reflect this new assumption of limited growth.

Modifying Mathematical Models

The inability of a model to make long-term predictions is a problem that is common to exponential models and many other mathematical models. Thus, we must be wary of predictions that are far from the known data. We must also be on the lookout for ways in which a model can be modified to make it more accurate for a larger domain. One such modification for population models is possible using the logistic model developed by the Dutch mathematical biologist Pierre-François Verhulst (1804–1849). It has the general form

$$P(t) = \frac{L}{1 + Ce^{-kt}}$$

The logistic model improves on our earlier Malthusian model in that it imposes a limit on the size of a population. Such a limit is a natural consequence of competition for food, living space, and other natural resources.

EXAMPLE 7

A Logistic Model for Population Growth

Using techniques from calculus, we can construct the following logistic model of a deer population with a limiting value of 240,000:

$$P(t) = \frac{240,000}{1 + 23e^{-0.1398t}}$$

The graph of P versus t is shown in Figure 31. Plot the graph with a graphing calculator and predict when the population is growing the fastest and when the population will reach 90% of the limiting population of 240,000.

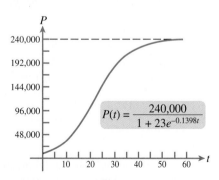

Figure 31

Solution One view of the graph of $P(t)$ is shown in Figure 32. The population is growing the fastest where the graph is rising most rapidly. In other words, the growth rate is largest where the graph is the steepest. Using the trace feature, we determine that this occurs when t (the x-coordinate) is approximately 23, as shown in Figure 33. To find when the population will reach 90% of the limiting population, or 216,000 deer, we locate the point on the graph where the y-coordinate is approximately 216,000. After zooming in several times, we see in Figure 34 that $x \approx 38$.

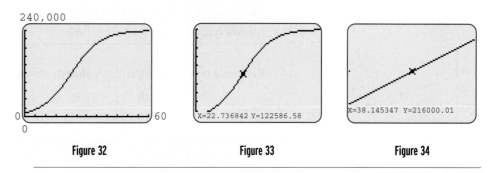

| Figure 32 | Figure 33 | Figure 34 |

Although the logistic model for population growth appears to be more realistic than our earlier exponential model, it too does not account for all of the many factors that may affect population growth. For example, the logistic model suggests continued (though slower) growth, and so it would not accurately model the deer population if hunting were again allowed. Also, the logistic model assumes a fixed limiting value, while in reality this value itself can change. We will ask you to supply some other overlooked factors in Exercise 32.

Logarithmic Models

Logarithmic models, though not as common as exponential models, are useful for such tasks as measuring sound and shock intensity and modeling human response systems. In the following example, we consider a human memory model.

·····◆**EXAMPLE 8**

A Logarithmic Memory Model

In a study on human memory, subjects were given an initial spelling exam on obscure words in the English language and then retested monthly. The average scores for the initial exam and those of the next 3 months, out of a possible score of 20, are given in the following table.

Month	0	1	2	3
Score	17	14	12	11

Construct a model of the form $y = a \log(t + 1) + b$ that approximates the average score y as a function of time t measured in months. Test the model for accuracy and use it to predict the average score after 6 months.

Solution We wish to select constants a and b so that the resulting model $y = a \log(t + 1) + b$ most nearly fits the data. Close inspection of the data, or even plotting a graph of the data, probably will not give us much insight into an appropriate choice of constants: We just don't have enough familiarity with functions of the form $f(t) = a \log(t + 1) + b$ to predict the effect on the graph of adjusting the constants a and b. We can simplify things a great deal, however, by making the substitution

Dominic O'Brien is listed in The Guinness Book of World Records *for memorizing 35 decks of playing cards (1820 cards) with only 2 errors.*

$x = \log(t + 1)$ to obtain the linear equation $y = ax + b$. To find a and b, we first compute $x = \log(t + 1)$ for $t = 0$ to $t = 3$, as shown in Table 11, and then apply the linear regression feature of a graphing calculator to the points (x, y). We obtain $a \approx -10.12$ and $b \approx 16.99$. The resulting best-fit line, $y = -10.12x + 16.99$, is graphed in Figure 35 along with the data points (x, y) from Table 11. It appears that the line is a very good fit.

Table 11

Month (t)	0	1	2	3
$x = \log(t + 1)$	0	0.30103	0.47712	0.60206
Score (y)	17	14	12	11

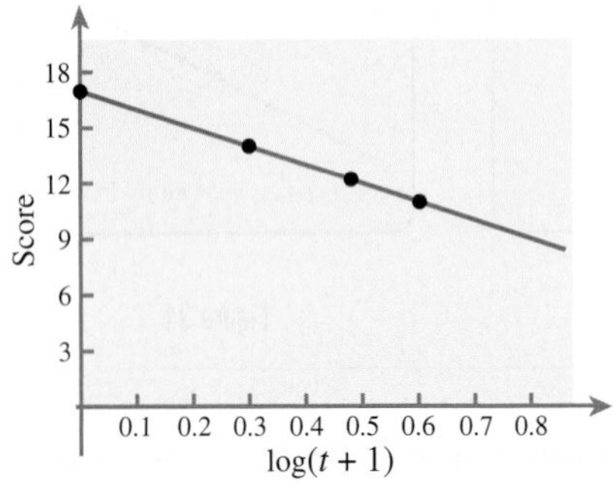

Figure 35

Returning now to the logarithmic model, we have

$$y = -10.12x + 16.99$$
$$= -10.12 \log(t + 1) + 16.99 \qquad \text{Using } x = \log(t + 1)$$

The following table compares the actual test scores for the first 3 months with those predicted by the model $y = -10.12 \log(t + 1) + 16.99$.

Month	0	1	2	3
Actual Score	17	14	12	11
Predicted	16.99	13.94	12.16	10.90
Absolute error	0.01	0.06	0.16	0.10

With a maximum absolute error of 0.16, this model is quite accurate. For $t = 6$, the model predicts an average score of $y = -10.12 \log(6 + 1) + 16.99 \approx 8.44$.

Understanding and Mastery Checklists

Concepts to Understand

Graphical model

❧

Exponential model

❧

Exponential regression

❧

Absolute error

❧

Logistic model

❧

Logarithmic model

Skills to Master

Find a graphical model for a set of data and use it to make predictions.

❧

Find an exponential model for a set of data and use it to make predictions.

❧

Use exponential regression to fit an exponential model to a set of data.

❧

Compute the absolute error between actual and predicted values.

❧

Use a logistic model to make predictions.

❧

Find a logarithmic model for a set of data and use it to make predictions.

Exercises 4.5

Applications

1. Rabbit Population   A population of rabbits has been observed over a 5-month period, and the following data have been collected.

Month	0	1	2	3	4	5
Rabbits	20	24	30	36	45	54

a. Plot the data in the table on a coordinate system with an appropriate scale.

b. Visually fit a smooth curve to the data points you plotted in part a.

c. Use the curve in part b to estimate the population after 6 months.

2. City Population   The population of a city has been observed over a 5-year period, and the following data have been collected.

Year	0	1	2	3	4	5
Population (in thousands)	125	127.5	130.1	132.7	135.4	138.2

a. Plot the data on a coordinate system with an appropriate scale.

b. Visually fit a smooth curve to the data points you plotted in part a.

c. Use the curve in part b to estimate the population after 6 years.

3. Rabbit Population   The population of rabbits in Exercise 1 has been determined to be growing exponentially. Using the fact that there are 20 rabbits initially and 24 rabbits after 1 month, find an exponential function of the form $P(t) = ae^{bt}$ that estimates the number of rabbits after t months. Compare the values given by your function for the first 5 months with the values given in the table in Exercise 1. What is the largest absolute error? What is the sum of the absolute errors?

4. City Population   The population of the city in Exercise 2 has been determined to be growing exponentially. Using the fact that there are 125,000 people initially and 127,500 after 1 year, find an exponential function of the form $P(t) = ae^{bt}$ that estimates the population of the city after t years. Compare the values given by your function for the first 5 years with the values given in the table in Exercise 2. What is the largest absolute error? What is the sum of the absolute errors?

5. Rabbit Population   Repeat Exercise 3, this time using the fact that there are 20 rabbits initially and 54 rabbits after 5 months. How does your model differ from that of Exercise 3? Is it more or less accurate?

6. City Population   Repeat Exercise 4, this time using the fact that there are 125,000 people initially and 138,200 people after 5 years. How does your model differ from that of Exercise 4? Is it more or less accurate?

7. Rabbit Population   Use the exponential regression feature of your graphing calculator to find a model of the form $P(t) = ae^{bt}$ that best fits the rabbit population data given in Exercise 1. How does the accuracy of this model compare to that of Exercise 3?

8. City Population   Use the exponential regression feature of your graphing calculator to find a model of the form $P(t) = ae^{bt}$ that best fits the population data given in Exercise 2. How does the accuracy of this model compare to that of Exercise 4?

9. Alaska Population   Population data for Alaska for the census years 1960, 1970, 1980, 1990, and 2000 are given in the table.

Year	1960	1970	1980	1990	2000
Population	226,000	303,000	402,000	550,000	627,000

Data source: U.S. Bureau of the Census.

Construct and test an exponential model of the form $P(t) = ae^{bt}$ that estimates the population t years after 1960. According to the model, what will the population be in the year 2010? When will the population first exceed 1,500,000?

10. New Hampshire Population   Population data for New Hampshire for the census years 1960, 1970, 1980, 1990, and 2000 are given in the table.

Year	1960	1970	1980	1990	2000
Population	607,000	728,000	921,000	1,109,000	1,236,000

Data source: U.S. Bureau of the Census.

Construct and test an exponential model of the form $P(t) = ae^{bt}$ that estimates the population t years after 1960. According to the model, what will the population be in the year 2010? When will the population first exceed 2,000,000?

11. Sales Decline   In an effort to cut spending, a company decides to stop all advertising and promotions for one of its products. Over the next 4 months, the following monthly sales of this product are observed.

Month (*t*)	0	1	2	3	4
Sales (in thousands) (*S*)	80	72	66	61	58

Construct and test a model of the form $S(t) = ae^{bt}$ that estimates the sales for month *t*. What does your model predict for the sales for month 6? When will the sales reach 20,000?

12. Sales Increase   Due to unacceptable sales figures, a company decides to start a new promotion for one of its products. The following increases in monthly sales are observed over a 4-month period.

Month (*t*)	0	1	2	3	4
Sales (in thousands) (*S*)	58	61	65	68	73

Construct and test a model of the form $S(t) = ae^{bt}$ that estimates the sales for month *t*. What does your model predict for the sales for month 6? During what month will the sales reach 100,000?

13. Computer Viruses   In a study on computer viruses conducted in 1991, the following data were obtained concerning the percentage of American companies that encountered one or more computer viruses during 1990 and 1991.

Quarter	0 (Oct–Dec 1990)	1 (Jan–Mar 1991)	2 (Apr–Jun 1991)	3 (Jul–Sep 1991)
Percent	8	19	26	40

Data source: Dataquest/National Computer Security Association, 11/91.

Construct and test a model of the form $p(t) = ae^{bt}$ that estimates the percentage after *t* quarters. When does the model predict that 90% of American companies will have encountered a computer virus? According to this model, what percentage of companies will have encountered a virus by 2004? Why is this model not useful for very long? What modification would you suggest?

The Falling Letters virus causes characters to fall to the bottom of the screen.

14. Internet Growth   The following table shows the approximate number of Internet hosts since 1991.

Year	1991	1993	1995	1997	1999	2001
Internet hosts (in thousands)	376	1313	5846	21,819	43,230	109,574

Data source: Internet Software Consortium, http://www.isc.org.

Construct and test a model of the form $N(t) = ae^{bt}$ that estimates the number of hosts *t* years after 1991. How many hosts are predicted for the year 2005?

15. Transistor Growth   The following table shows the approximate number of transistors in various Intel microprocessors introduced since 1971.

Year	Processor	Transistors (in thousands)
1971	4004	2
1974	8080	6
1978	8086	29
1982	286	134
1985	386	275
1989	486	1200
1993	Pentium	3100
1997	P II	7500
1999	P III	24,000
2001	P 4	42,000

Data source: Intel Technology Briefing.

Construct and test a model of the form $N(t) = ae^{bt}$ that estimates the number of transistors *t* years after 1971. How many transistors are predicted for the year 2005? The human brain is estimated to have between 10 and 100 billion neurons. According to the model, when will the number of transistors exceed 100 billion?

16. Spanish Flu Pandemic   The Spanish flu, first reported in the United States in late 1917 and carried to Europe by American troops sent to fight in World War I, is estimated to have caused the deaths of 25 to 40 million people worldwide. For the 13-week period beginning with the week of October 6, 1918, the spread of the Spanish flu among U.S. Army personnel can be approximated using the logistic model

$$N(t) = \frac{21,250}{1 + 395e^{-0.992t}}$$

where $N(t)$ is the number of deaths reported in week *t* (*t* = 0 corresponds to the week of October 6, 1918). [*Data source:* Alfred W. Crosby, *America's Forgotten Pandemic: The Influenza of 1918* (Cambridge: Cambridge University Press, 1989).]

a. During what week did the number of deaths first exceed 10,000?

b. According to this model, what was the upper bound on the number of deaths?

17. Spreading a Rumor   A rumor begins to spread around a college campus. The number of people N who have heard the rumor after t hours can be approximated using the logistic model

$$N = \frac{2000}{1 + 499e^{-0.3t}}$$

a. How many students started the rumor?

b. How many students have heard the rumor after 10 hours?

c. How long before 500 students have heard the rumor?

d. Plot the graph of N and estimate the limiting number of students who will hear the rumor.

e. During which hour did the greatest number of students hear the rumor?

18. Contagious Virus   A group of tourists returns home after a 2-week tour; all have unknowingly been infected by a contagious flu virus. Suppose that, if left untreated, the spread of the flu virus in their city can be approximated using the logistic model

$$Q = \frac{150,000}{1 + 2999e^{-0.05t}}$$

where Q is the number of people with the flu virus after t days.

a. How large was the group of tourists?

b. How many people will be infected with the virus after 14 days?

c. How long before 10,000 people are infected?

d. Plot the graph of Q and estimate the limiting number of people who will be infected.

e. During which day did the greatest number of people become infected?

19. Automobile Speed Record   The 1-mile automobile speed records for the years 1906–1983 can be approximated with the logistic model

$$S = \frac{700}{1 + 6.8e^{-0.057t}}$$

where S is the 1-mile speed record in miles per hour (mph) for year t ($t = 0$ corresponds to 1906).

Richard Noble's Thrust SSC breaks the sound barrier, traveling at more than 760 miles per hour.

a. In what year did the speed record first exceed 400 mph?

b. According to this model, what is the upper bound on the speed record?

c. The land speed record of 763 mph (faster than the speed of sound!) was set by Andy Green in 1997 in a jet-powered vehicle on the Bonneville Salt Flats near Bonneville, Utah. Is this consistent with the prediction made by the model?

20. Memory Loss   A high school French class is given a test covering newly learned vocabulary words and then retested monthly on the same material. The average scores for the first test and those of the next 3 months are given in the following table.

Month	0	1	2	3
Avg score	75.0	70.5	67.8	66.0

Using t to denote the month and y to denote the score, construct a logarithmic model of the form $y = m \log(t + 1) + b$ that best fits the data in the table. Test the model against the data in the table, and use it to predict the average score after 7 months.

21. Walking Speed   Various studies have found a correlation between the size of a city and the average walking speed of pedestrians. One such study obtained the following data.

Population	5,500	14,000	71,000	138,000	342,000
Velocity (ft/sec)	3.3	3.7	4.3	4.4	4.8

Data source: Marc and Helen Bornstein, "The Pace of Life," *Nature* 259 (19 February 1976): 557–559.

Using P to denote population and v to denote velocity, construct a logarithmic model of the form $v = m \log P + b$ that best fits the data in the table. Test the model against the data in the table, and use the model to predict the walking speed in Little Rock, Arkansas (population 183,000), New York (population 8,008,000), and Mexico City (population 18,330,000).

22. Reflex Speed   Numerous studies have shown a relationship between age and reflex speed. One such study obtained the following data.

Median age	19	27	45	60
Reflex time (sec.)	0.18	0.20	0.23	0.25

Data source: Jean Hodgkins, "Reaction Time and Speed of Movement in Males and Females of Various Ages," *Res. Quart. Amer. Assoc. Hlth. Phys. Educ. Recr.* 34, no. 3 (1963): 335–343.

Using a to denote age and t to denote reflex time, construct a logarithmic model of the form $t = m \log a + b$ that best fits the data in the table. Test the model against the data in the table, and use it to predict your reflex time.

Concepts and Critical Thinking

Exercises 23-26   *Answer true or false.*

23. If a mathematical model fits known data exactly, then we can have 100% confidence in any predictions made that are based on the model.

24. The Malthusian exponential model tends to be accurate over very long time periods.

25. Exponential regression is used to fit a model of the form $P(t) = ae^{bt}$ to a collection of data.

26. The logistic model of population growth imposes an upper limit on the population.

Exercises 27-30   *Give an example of each.*

27. A reason why populations tend not to continue to grow exponentially over long time periods

28. A set of three data points such that the graph of the function obtained by exponential regression passes through all three data points

29. A graphical feature of a logistic model that distinguishes it from an exponential model

30. A logistic function with a limiting value of one million

Questions for Discussion or Essay

31. Discuss the pros and cons of graphical models such as the one we constructed in Example 1.

32. What are some factors that have not been considered in the logistic model for the deer population discussed in this section? What modifications would have to be made in the model to account for these factors? What are the dangers in using mathematical models such as the ones considered here to justify or condemn deer hunting?

33. Describe in your own words the various steps involved in the construction, testing, and use of an exponential model.

34. Exponential models for population growth have been validated using both theoretical and experimental observations. The theoretical explanation is based on the fact that, left unchecked, populations tend to increase at a constant percentage rate. Experimental observations have confirmed that exponential population models often work quite well for limited periods of time but that eventually the growth rate levels off. Describe several factors that prevent unlimited growth of human populations.

Projects for Enrichment

35. **Comparing Models**   In this section, we considered methods for constructing and testing exponential models of the form $y = ae^{bx}$. Similar methods can be used to construct and test models of many different forms. Two specific examples are **linear models** of the form $y = ax + b$ and **power models** of the form $y = ax^b$. In some cases, where it is not known which model form is needed, it may be necessary to construct and test several different models for the same set of data. Then it is possible to select the model that works the best.

    a. As of 2003, the three largest cities in the United States were New York, Los Angeles, and Chicago. Census data for each of these cities for the years 1880–1930 are given in the following table.

Year	New York	Los Angeles	Chicago
1880	1,912,000	11,000	503,000
1890	2,507,000	50,000	1,100,000
1900	3,437,000	102,000	1,699,000
1910	4,767,000	319,000	2,185,000
1920	5,620,000	577,000	2,702,000
1930	6,930,000	1,238,000	3,376,000

    For each city, use your graphing calculator to construct and test models of the form $y = ax + b$ and $y = ae^{bx}$ to see which fits the data best. Use the models that fit the best to predict the population of the city in 1940, and compare your prediction with the actual values (New York: 7,455,000; Los Angeles: 1,504,000; Chicago: 3,397,000). What is your conclusion as to the accuracy of the models?

    b. Early in the 17th century, the German astronomer Johannes Kepler discovered three laws that govern the motion of the planets around the sun. The first states that the shape of each planet's orbit around the sun is an ellipse with the sun at one focus. The second states that a line from a planet to the sun sweeps out equal areas in equal times. Thus, referring to Figure 36, if it takes the same amount of time for a planet to travel from A to B as it does from A' to B', then the areas of ABF and $A'B'F$ are equal. Kepler's third law relates a planet's average distance from the sun to its orbital period. The data in the following table give the average distance and the orbital period for each of the nine planets.

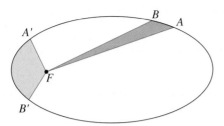

Figure 36

Planet	Average distance (miles)	Period (days)
Mercury	36,000,000	88
Venus	67,000,000	225
Earth	93,000,000	365
Mars	142,000,000	687
Jupiter	484,000,000	4329
Saturn	887,000,000	10,753
Uranus	1,784,000,000	30,660
Neptune	2,795,000,000	60,150
Pluto	3,671,000,000	90,670

Construct and test models of the form $y = ae^{bx}$ and $y = ax^b$, where y is the period and x is the distance. Which model is more accurate? State Kepler's third law.

36. Modeling the Spread of Lyme Disease   Lyme disease is a bacterial infection that is spread by the bite of certain species of ticks. The disease was first discovered in the United States in 1975, after a mysterious outbreak of arthritis near Lyme, Connecticut. Since then, cases have been reported in all states except Montana, with a cumulative total of over 99,000 cases between 1982 and 1996 (see Table 12).

Table 12
Lyme Disease

Year	Cases*
1982	491
1983	1086
1984	2604
1985	5352
1986	6739
1987	9131
1988	14,013
1989	22,816
1990	30,759
1991	40,229
1992	50,137
1993	58,394
1994	71,437
1995	83,137
1996	99,169

* Cumulative total since 1982
Data source: U.S. Centers for Disease Control.

In this project, we investigate various exponential models for the spread of Lyme disease. We begin with a standard exponential growth model.

a. Use the data for the years 1982 and 1983 from Table 12 to construct a model of the form $N(t) = Ce^{kt}$, where N is the number of cases and t is the year (with $t = 0$ corresponding to 1982). How well does the model predict the number of cases in 1990 and 1995? What do you suspect about a prediction for 2000?

b. Use exponential regression on the data for the years 1982–1985 from Table 12 to construct a model of the form $N(t) = Ce^{kt}$, where N is the number of cases and t is the year (with $t = 0$ corresponding to 1982). How well does the model predict the number of cases in 1990 and 1995? What do you suspect about a prediction for 2000?

c. Based on your findings in parts a and b, justify the need for a model $N(t)$ that has a limiting value for the size of N.

Next we consider a logistic exponential model of the form

(1) $$N(t) = \frac{L}{1 + Ce^{-kt}}$$

where L, C, and k are positive constants, and $N(t)$ denotes the number of people infected with a disease at time t. The constant L is often referred to as the limiting value of the logistic model since $N(t)$ approaches (but never exceeds) L as t increases without bound. A typical graph of a model of this form is given in Figure 37.

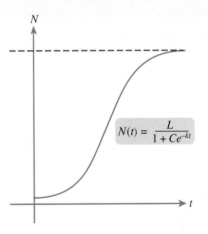

Figure 37

A detailed analysis of the data in Table 12 suggests that the number of cases had begun to level off by 1994. After some trial and error, we settle on a limiting value of 120,000 for the number of Lyme cases. We will use this value to construct a logistic model of the form given in equation (1). We first rewrite the logistic equation in a form to which we can apply an exponential regression.

d. Show that the logistic equation

$$N = \frac{120,000}{1 + Ce^{-kt}}$$

can be written in the form

$$Ce^{-kt} = \frac{120,000}{N} - 1$$

Note that this equation is of the form $ae^{bx} = y$ and is suitable for an exponential regression.

e. Use the data given in Table 12 to complete Table 13 for t ranging from 0 to 14. Note that $t = 0$ corresponds to 1982. For example, for $t = 0$ we have $N = 491$ so that

$$\frac{120,000}{N} - 1 = 243.4$$

You may find a spreadsheet to be extremely useful for this task.

Table 13

t	N	$\dfrac{120,000}{N} - 1$
0	491	243.4
1	1086	109.5
2	2604	45.1
⋮	⋮	⋮
14	99,169	0.2

f. Perform an exponential regression on the data in the columns labeled "t" and "$\mathbf{120,000/N - 1}$" in Table 13, with those columns representing the x- and y-coordinates, respectively. What are the computed values for the constants k and C that appear in the equation in part d?

g. Write out the complete logistic model for the data given in Table 12. Use the model to predict the number of cases in the years 2000 and 2005.

h. Some graphing calculators have logistic regression capabilities. If yours does, use it to find a logistic model for the data given in Table 12. How does this model compare to the one you found in part g? In particular, how does the limiting value compare?

Chapter 4 Review

Exercises 1-2 *Sketch the graph of f and use it to sketch the graph of g.*

1. $f(x) = 2^x$

   a. $g(x) = 2^{-x}$

   b. $g(x) = 2^x + 2$

   c. $g(x) = 2^{x+1}$

2. $f(x) = \left(\frac{1}{3}\right)^x$

   a. $g(x) = -\left(\frac{1}{3}\right)^x$

   b. $g(x) = \left(\frac{1}{3}\right)^x - 4$

   c. $g(x) = \left(\frac{1}{3}\right)^{x-2}$

Exercises 3-6 *Solve the given equation without using logarithms.*

3. $3^x = 27$

4. $\dfrac{1}{4^x} = 16$

5. $5^{1-2x} = 25$

6. $3^{x^2+4x} = \dfrac{1}{27}$

Exercises 7-12 *Evaluate each logarithmic expression without using a calculator.*

7. $\log_2 \dfrac{1}{4}$

8. $\log_3 81$

9. $\log 0.01$

10. $\ln e^3$

11. $\log_5 \sqrt{5}$

12. $\log_\pi 1$

Exercises 13-16 *Write an equivalent exponential equation for the given logarithmic equation.*

13. $\log_2 \dfrac{1}{16} = -4$

14. $\ln(3 - x) = 1$

15. $\log(2x + 1) = 4$

16. $\log_3 18 = x$

Exercises 17-20   *Write an equivalent logarithmic equation for the given exponential equation.*

17. $e^5 = x$

18. $3^{x-5} = 7$

19. $2^{x+3} = 5$

20. $10^{1/x} = \dfrac{1}{2}$

Exercises 21-24   *Sketch the graph of the given function and find its domain.*

21. $f(x) = \log_5 x$

22. $f(x) = 3 + \log x$

23. $f(x) = \ln(x - 2)$

24. $f(x) = -\ln(5x)$

Exercises 25-28   *Use logarithmic identities to expand each expression, rewriting it as a sum, difference, or multiple of logarithms.*

25. $\log_2(xy^2)$

26. $\log_3\!\left(\dfrac{x^2}{z}\right)$

27. $\log \sqrt[3]{x^2 y}$

28. $\ln\!\left(\dfrac{x^3 y^2}{\sqrt{z}}\right)$

Exercises 29-32   *Use logarithmic identities to rewrite each expression as a single logarithm.*

29. $2 \log x + 3 \log y$

30. $4 \log_2 p - 3 \log_2 q$

31. $\dfrac{1}{3}\ln x + \dfrac{2}{3}\ln y - \ln z$

32. $3 \log 4 - \log 6 + 2 \log 2$

Exercises 33-34   *Estimate the given logarithmic expression. Give answers accurate to five decimal places.*

33. $\log_4 10$

34. $\log_{2/5} 130$

Exercises 35-42   *Solve the logarithmic equation.*

35. $\log_2 x = -1$

36. $\log_8 x = \dfrac{1}{3}$

37. $\log_5(2x + 4) = \log_5 10$

38. $\ln(3x + 5) = \ln x$

39. $\log(5y - 2) = 1$

40. $2 \log_4 x - \log_4(x + 1) = 1 - \log_4 3$

41. $\ln(x - 1) + \ln(x + 2) = \ln 4$

42. $\log \sqrt{x + 1000} = 2$

Exercises 43-50   *Solve the exponential equation. When approximating, give answers to five decimal places of accuracy. You may wish to check your solution graphically.*

43. $5^x = 10$

44. $e^x = 2$

45. $1.05^t = 10$

46. $3^{-2x} = 9$

47. $4e^{0.1x} = 20$

48. $350 - 280e^{-0.05t} = 180$

49. $3^{x+2} = 2^{2x}$

50. $2^{2x} - 5 \cdot 2^x = 0$

Exercises 51-54   *Estimate the solution to the given equation to the nearest hundredth.*

51. $e^x = 8 - x^2$

52. $\ln x = x^3 - 3x^2$

53. $\ln(3x) + 0.02x^3 + 4x = x^2$

54. $-\dfrac{x^3}{9} + e^{x+2} = 2x^2 + 5x$

Exercises 55-66   *Parts a and b are connected: Part a involves an elementary concept, whereas part b involves related material from this chapter. First answer part a, and then use this result to answer part b.*

55. a. Simplify $\dfrac{2^x}{3^x}$.

b. Solve $\dfrac{2^x}{3^x} = 2$.

56. a. Simplify $\sqrt[3]{x^6 y^9}$.

b. Write $\ln\left(\sqrt[3]{x^6 y^9}\right)$ as a sum of logarithms.

57. a. Solve $x^2 + x - 2 = 0$.

b. Solve $3^{x^2 + x} = 9$.

58. a. Simplify $\dfrac{x^2 - y^2}{x - y}$.

b. Write $\log(x^2 - y^2) - \log(x - y)$ as a single logarithm.

59. a. Graph $y = e^{-x}$ and $y = x$ on the same set of coordinate axes.

b. How many solutions does $e^{-x} - x = 0$ have?

60. a. Simplify $\dfrac{x^2 - 5x + 6}{x - 3}$.

b. Solve $\log(x^2 - 5x + 6) - \log(x - 3) = \log 2$.

61. a. Solve $2x - 3 \le 0$.

b. Find the domain of $f(x) = \ln(2x - 3)$.

62. a. Solve $u^2 - 11u + 10 = 0$.

b. Solve $10^{2x} - 11(10^x) + 10 = 0$.

63. a. Solve $u + \dfrac{1}{u} = 2$.

b. Solve $e^x + e^{-x} = 2$.

64. a. Simplify $f(f(f^{-1}(x)))$.

b. Simplify $e^{(e^{\ln x})}$.

65. a. Graph $y = 4x - x^2$.

b. Find the domain of $g(x) = \ln(4x - x^2)$.

66. a. Compute $16\left(\dfrac{3}{2}\right)^4$.

b. How long will it take $16 to grow to $81 with 50% interest compounded annually?

67. Compound Interest Suppose that $50 is deposited into an account paying 4% interest compounded quarterly. Find the balance in the account after 3 years.

68. Compound Interest If $42 is deposited into an account paying 3% compounded continuously, what is the balance after 4 years?

69. Doubling Money How long does it take money that is deposited into an account paying 3% interest compounded monthly to double?

70. Sound Intensity The decibel rating of a sound of intensity I is given by $D = 10 \log(I/I_0)$, where I_0 is the intensity of a sound that is just perceptible. How many times more intense is a sound of 100 decibels than one of 50 decibels?

71. Population Growth A certain bacteria colony is growing exponentially. Initially there are 10 bacteria. After 1 hour, there are 25. How long will it take until the population of the colony surpasses the 1 million mark?

72. Radioactive Decay Suppose that a sample of a radioactive substance having a half-life of 12,000 years was placed in a time capsule 2000 years ago. What percentage of the original sample remains today?

73. Infectious Disease A community of astronauts living on a base on Mars is exposed to an infectious disease. The number of people who have contracted the disease t days from the initial exposure is given by

$$Q = \frac{200{,}000}{1 + 1999e^{-0.08t}}$$

a. How many astronauts have the disease after 5 days?

b. How many days will it take before 1000 astronauts have become infected?

c. Plot the graph of Q, and estimate the limiting number of astronauts who will become infected.

d. During what day did the greatest number of astronauts become infected: day 5, day 96, or day 120?

74. Memory Model It is hypothesized that if a person were introduced to 100 guests at a cocktail party, the number of people whose names the person would remember after t months would be given by $y = m \log(t + 1) + b$. Suppose one person remembered 50 people immediately after the party and 36 people 2 years after the party. How many guests will the person remember 10 years after the party? How many guests did the person remember 1 year after the party?

75. Car Depreciation An SUV purchased new for $28,000 in 2000 depreciates in value each year, as shown in the following table.

Year	2000	2001	2002	2003	2004
Value	$28,000	$24,000	$20,800	$17,800	$15,400

Construct and test an exponential model of the form $V(t) = ae^{bt}$ that estimates the value of the car t years after 2000. When will the car be worth $4000?

76. **U.S. Population Growth** U.S. census data for the years 1790–1840 are given in the table. Construct and test an exponential model of the form $P(t) = ae^{bt}$ that estimates the population t years after 1790. Use the model to predict the population in 1850. How does the predicted value compare to the actual population of 23.2 million in 1850?

Year	1790	1800	1810	1820	1830	1840
Population (in millions)	3.9	5.3	7.2	9.6	12.9	17.1

Data source: U.S. Bureau of the Census.

Chapter 4 Test

Problems 1-8 *Answer true or false.*

1. $\log_a x \cdot \log_a y = \log_a x + \log_a y$ for all $x > 0$ and $y > 0$.

2. $\log_a(x^y) = y \log_a x$ for all $x > 0$ and $y > 0$.

3. If $f(x) = a^x$, then $f^{-1}(x) = \dfrac{1}{a^x}$.

4. The logarithm of a number cannot be negative.

5. The logarithm of a negative number is not defined.

6. Natural logarithms are simply logarithms with base e.

7. It is possible that a sample of a substance undergoing exponential decay will take longer to decay from 100 grams to 50 grams than from 50 grams to 25 grams.

8. Both $f(x) = a^x$ and $g(x) = \log_a x$ are 1–1 functions.

Problems 9-12 *Give an example of each of the following.*

9. An exponential function that is always decreasing

10. A number not in the domain of $\ln(x - 5)$ but in the domain of $\ln x$

11. A real number whose common logarithm is a negative integer

12. A reason why exponential growth of populations of bacteria colonies cannot continue indefinitely

Problems 13-16 *Solve the given equation. When approximating, give answers to five decimal places of accuracy.*

13. $2^{x^2+x} = 4$

14. $\log(x + 1) = 12$

15. $3^{x-1} = 2^x$

16. $5e^{0.3x} = 10$

Problems 17-18 *Evaluate the given logarithm.*

17. $\log_2 32$

18. $\log_3 \dfrac{1}{27}$

Problems 19-21 *Graph the given function.*

19. $f(x) = e^{x-2}$

20. $g(x) = \left(\dfrac{2}{3}\right)^x$

21. $f(x) = \ln(x + 3)$

22. Express $\log\left(\dfrac{x^4 y^2}{z^{-2}}\right)$ in terms of logarithms of x, y, and z.

23. Write $3 \ln x - 4 \ln y + 5 \ln z^2$ as a single logarithm.

24. Suppose that $300 is deposited into an account paying 6% interest compounded annually. How long does it take the balance to grow to $500?

25. Suppose that a radioactive sample is decaying exponentially and that 10 grams of the sample were present on January 1, 1999, and 7 grams on January 1, 2004. In what year will the quantity of the sample first drop below 1 gram?

26. AIDS, first diagnosed in the United States in 1981, grew exponentially during much of the 1980s. The approximate number of new AIDS cases reported in each of the years 1983–1986 is given in the following table.

Year	1983	1984	1985	1986
New AIDS cases	2100	4400	8200	13,100

Data source: U.S. Centers for Disease Control.

Construct and test a model of the form $N(t) = ae^{bt}$ that estimates the number of new cases reported t years after 1983. Use the model to predict the number of new cases reported in 1987. How does the prediction compare to 21,100, the approximate number of new cases reported in 1987? In 2002 approximately 40,000 new cases were reported. How does this compare to the model? What factors may have caused the rate of new AIDS cases in the United States to drop below the rate predicted by the model?

Chapter 5
Trigonometric Functions

The science of surveying is a marriage of mathematics and measurement. Trigonometric techniques are used to complete a positional database, the framework of which is formed from measurements of angles and distances. Our ability to accurately determine position is limited only by the precision of our rulers, and modern rulers are capable of astonishing feats. The horizontal laser in the photograph is used to measure the distance from Earth to the moon (the bright spot at the left) to within 3 centimeters, whereas the vertical beam is used to determine satellite positions to within even narrower margins.

Section 5.1 | Angles and Their Measurements

- Why do figure skaters tuck in their arms while spinning?
- How fast does the tip of Big Ben's minute hand move?
- How can the linear speed of a bike be determined from the rate at which the pedals are moving?
- Could the bizarre occurrences in the Bermuda Triangle be explained by the fact that the sum of its angles is greater than 180°?
- Why doesn't light travel in straight lines?

A ray of light formed from nuclear fires deep within the sun streaks through space for 8 minutes until it caroms off the cratered lunar surface at an angle equal to that at which it struck. Thus reflected, it travels another quarter million miles (in less than 2 seconds) and then passes through the lens of your eye, an eye precisely aligned at an angle of elevation and azimuth to accommodate the incoming moonbeam. If Earth lies between you and the sun—if it is night—then the moonlight striking your retina will trigger a complex series of electrochemical reactions, and you will see a slice of the moon. How much moon you see—crescent, quarter, half, full, or something in between—depends on the angles in the triangle formed from Earth, the moon, and the sun. Understanding how this magical sequence of events unfolds requires the dual notions of distance and angle: distance for proximity and angle for direction and orientation. From the tilt of a towering skyscraper to the direction of a thundering gale, from the facets of an ideally cut diamond to the bearing of a supersonic jet, angles pervade physical reality. Indeed, for the scientist, there is a sense in which music, matter, electricity, light, and even our thoughts themselves have aspects that are angular.

Angles   In geometry, an **angle** is defined as the set of points determined by two rays or half-lines, called **sides**, that have a common endpoint, called the **vertex**. If A is a point on one side and B is a point on the other, and if O is the vertex, as shown in Figure 1, then we refer to the angle as $\angle AOB$.

In trigonometry, it is convenient to think of angles in terms of rotated rays. To form an angle, we start with two rays that are initially in the same position. One of the rays—the **initial side**—is fixed, whereas the other—the **terminal side**—is rotated to its final position, as illustrated in Figure 2. It is customary to denote angles with lowercase Greek letters, such as α, β, and θ, and to indicate the direction of rotation with curved arrows. Several examples are given in Figure 3. Note in the case of the angles α and β in Figure 3 that it is possible for two or more angles to have the same initial and terminal sides. Such angles are said to be **coterminal**.

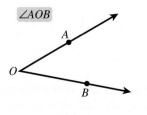

Figure 1

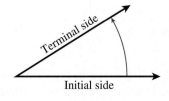

Figure 2

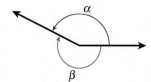

Figure 3

An angle in a rectangular coordinate system is said to be in **standard position** if its vertex is at the origin and if its initial side coincides with the positive *x*-axis. Angles formed by a counterclockwise rotation are considered **positive**, whereas those formed by a clockwise rotation are considered **negative**. See Figure 4 for an example of each.

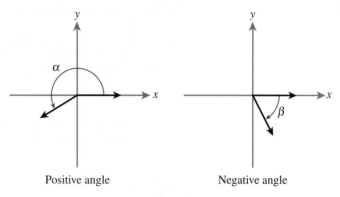

Positive angle   Negative angle

Figure 4

Degree Measure   The measure of an angle indicates the amount of rotation from the initial to the terminal side. The most common unit of angle measure is the **degree**, and it is denoted by the symbol °.

Definition of Degree Measure

An angle measure of **1 degree** (1°) is equivalent to $\frac{1}{360}$ of a complete revolution about the vertex. In other words, a measure of 360 degrees (360°) is equivalent to a full revolution.

Figure 5 shows some common angles and their corresponding degree measures.

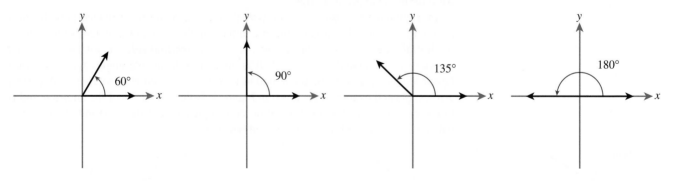

Figure 5

The angles α, β, θ, and ϕ given in Figure 5 can be placed into four categories: acute, right, obtuse, and straight, respectively. These categories are summarized as follows.

Angle Categories

An angle θ is said to be **acute**, **right**, **obtuse**, or **straight** according to the following conditions:

Type of angle	Condition
Acute	$0° < \theta < 90°$
Right	$\theta = 90°$
Obtuse	$90° < \theta < 180°$
Straight	$\theta = 180°$

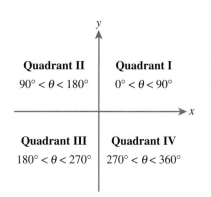

Figure 6

Angles can also be classified according to the quadrant in which the terminal side lies. Thus, angles between 0° and 90° are said to lie in quadrant I, angles between 90° and 180° are said to lie in quadrant II, and so on, as illustrated in Figure 6. The angles 0°, 90°, 180°, and 270° are called **quadrant angles**.

Recall that two angles are said to be coterminal if they have the same initial and terminal sides. In fact, if a given angle is in standard position, infinitely many coterminal angles can be found by adding or subtracting multiples of 360° to the degree measure of the given angle.

EXAMPLE 1

Finding and Sketching Coterminal Angles

Find and sketch an angle that is coterminal with the given angle.
a. $\alpha = 45°$ **b.** $\theta = 210°$

Solution   In each case, we add or subtract a multiple of 360° to obtain a coterminal angle.

a. We choose to add 360°, obtaining an angle β given by

$$\beta = 45° + 360° = 405°$$

The angles α and β are shown in Figure 7.

b. This time we subtract 360°, obtaining an angle ϕ given by

$$\phi = 210° - 360° = -150°$$

The angles θ and ϕ are shown in Figure 8.

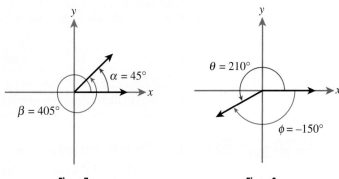

Figure 7   **Figure 8**

Two positive angles are said to be **complementary** (and are called complements of each other) if the sum of their measures is 90°. For example, angles with measures 35° and 55° are complementary since 35° + 55° = 90°. Two positive angles are said to be **supplementary** (and are called supplements of each other) if the sum of their measures is 180°. Thus, for example, angles with measures 112° and 68° are supplementary since 112° + 68° = 180°.

> **EXAMPLE 2**

Finding Complementary and Supplementary Angles

Find the complementary and supplementary angles for the angle θ having measure 27°.

Solution   To find the complement of θ, we subtract 27° from 90° to obtain

$$90° - 27° = 63°$$

To find the supplement of θ, we subtract 27° from 180° to obtain

$$180° - 27° = 153°$$

When expressing the measures of angles with a fractional degree measure, we have two distinct methods at our disposal. The first, and most widely used, is based on our base 10, or decimal system, in which a whole is divided into tenths, hundredths, and so forth. Using the decimal system, we might refer, for example, to an angle of 14.213°. The second method originated with Babylonian astronomers, who sometimes worked in the base 60, or **sexagesimal** system. They first split the circle into 360 degrees, then divided each degree into 60 *minutes,* and each minute, in turn, into 60 *seconds*. The notation and formal definitions of minutes and seconds are as follows.

Minutes and Seconds

A **minute** is defined to be $\frac{1}{60}$ of a degree and is denoted by the symbol ′. Thus,

$$1' = \left(\frac{1}{60}\right)^° \text{ or, equivalently, } 1° = 60'$$

A **second** is defined to be $\frac{1}{60}$ of a minute and is denoted by the symbol ″. Thus,

$$1'' = \left(\frac{1}{60}\right)' \text{ or, equivalently, } 1' = 60''$$

Since there are 60 seconds in a minute and 60 minutes in a degree, it follows that there are 60 · 60 = 3600 seconds in a degree. Thus, we have

$$1'' = \left(\frac{1}{3600}\right)^° \text{ or, equivalently, } 1° = 3600''$$

The notation $a° \ b' \ c''$ indicates a measure of $a° + b' + c''$, and we say the angle is written in **degree-minute-second (DMS)** form.

> **EXAMPLE 3**

Converting to Decimal Degree Notation

Convert 43° 12′ 54″ to decimal degree form.

Solution   We write 43° 12′ 54″ as the sum of three angles and then convert all units to degrees. We have

$$43° \ 12' \ 54'' = 43° + 12' + 54''$$
$$= 43° + 12' \cdot \frac{1°}{60'} + 54'' \cdot \frac{1°}{3600''}$$
$$= 43° + 0.2° + 0.015°$$
$$= 43.215°$$

EXAMPLE 4

Converting to DMS Notation

Convert 110.355° to DMS notation.

Solution Clearly there are 110 degrees. Converting the remaining 0.355° to minutes gives us

$$110.355° = 110° + 0.355°$$
$$= 110° + 0.355° \cdot \frac{60'}{1°}$$
$$= 110° + 21.3'$$

Thus, there are 110° and 21'. Converting the remaining 0.3' to seconds gives us

$$110.355° = 110° + 21.3'$$
$$= 110° + 21' + 0.3'$$
$$= 110° + 21' + 0.3' \cdot \frac{60''}{1'}$$
$$= 110° + 21' + 18''$$

Thus, 110.355° = 110° 21' 18''.

Calculator Keys

Converting Between Degree Forms

Many calculators are capable of converting between decimal degree form and DMS form. To convert from DMS form to decimal degrees with one popular model, simply enter the angle as $a°b'c''$; the value returned will be in decimal degree form. To convert from decimal degrees to DMS form, key in the decimal angle and select →DMS from the list of angle options.

Radian Measure

Degree measure is used extensively as a measurement of real-world angles that arise in applications such as surveying, astronomy, architecture, navigation, and even graphic design. But in contexts in which the behavior of a function of an angular measure is the subject of investigation, a second unit of measure—the radian—is used almost exclusively. Whereas degree measure is defined somewhat arbitrarily (1 degree is $\frac{1}{360}$ of a full circle), radian measure is defined quite naturally as the ratio of an arc of a circle to its radius.

The formal definition of radian measure involves central angles and subtended arcs. A **central angle** of a circle is an angle whose vertex is at the center. The arc **subtended** by a central angle is the arc of the circle swept out by the angle. We define the radian measure of the central angle as the ratio of the length of the subtended arc to the radius of the circle.

Definition of Radian Measure

Let s denote the length of the arc subtended by a central angle θ in a circle of radius r. The **radian measure** of θ is the ratio $\frac{s}{r}$, as shown in Figure 9. Note that for a circle of radius r, one radian is the measure of a central angle θ that subtends an arc of length r. For a circle of radius 1, the number of radians in an angle θ is the same as the length of the arc subtended by θ.

$\theta = \frac{s}{r}$ radians

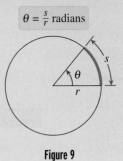

Figure 9

It is important to note that the radian measure of a central angle θ is actually independent of the size of the circle. One way to see this is to consider the effect of increasing the radius of the circle by a certain factor. Because of similarity, the arc length subtended by angle θ would increase by the same factor. Thus, the ratio of arc length to radius would remain unchanged.

EXAMPLE 5

Finding the Radian Measure of an Angle

Find the radian measure of a central angle θ that subtends an arc of length π inches on a circle of radius 4 inches.

Solution According to the definition of radian measure

$$\theta = \frac{s}{r} = \frac{\pi \text{ inches}}{4 \text{ inches}} = \frac{\pi}{4}$$

Notice that radian measure is *dimensionless*—it has no unit. However, the label "radian" is sometimes attached to emphasize that we are dealing with an angular measure.

Since the circumference of a circle is $2\pi r$, it follows that the radian measure of one revolution is $\frac{2\pi r}{r} = 2\pi$. Moreover, since the degree measure of one revolution is $360°$, it follows that $360° = 2\pi$ radians, or $180° = \pi$ radians. Thus, $1 = \frac{180°}{\pi} = \frac{\pi}{180°}$, which leads us to the following conversion rules.

Converting Degrees and Radians

To convert from degrees to radians: multiply by $\frac{\pi}{180°}$

To convert from radians to degrees: multiply by $\frac{180°}{\pi}$

······**EXAMPLE 6**

Converting from Degrees to Radians

Convert the following degree measures to radians:

a. $60°$ **b.** $220.4°$

Solution

a. $60° = 60°\left(\dfrac{\pi}{180°}\right) = \dfrac{\pi}{3}$

b. $220.4° = 220.4°\left(\dfrac{\pi}{180°}\right) = \dfrac{220.4\pi}{180} \approx 3.8467$

······**EXAMPLE 7**

Converting from Radians to Degrees

Convert the following radian measures to degrees:

a. $\dfrac{\pi}{6}$ **b.** 5

Solution

a. $\dfrac{\pi}{6} = \dfrac{\pi}{6}\left(\dfrac{180°}{\pi}\right) = 30°$ **b.** $5 = 5\left(\dfrac{180°}{\pi}\right) = \left(\dfrac{900}{\pi}\right)^{\circ} \approx 286.48°$

······**EXAMPLE 8**

Finding an Angle Complement

Find the radian measure of the complement of an angle of $\frac{\pi}{12}$ radians.

Solution Since $90° = \frac{\pi}{2}$ radians, the complement of an angle of $\frac{\pi}{12}$ radians is

$$\frac{\pi}{2} - \frac{\pi}{12} = \frac{6\pi}{12} - \frac{\pi}{12} = \frac{5\pi}{12}$$

Calculator Keys

Degrees and Radians

Many calculators are capable of dealing with angle measures in either degrees or radians and are also able to convert between the two forms. The **default angle mode** can be set to either degrees or radians, so that future calculations involving angle measure will assume the default form. With one popular model of graphing calculator, a degree measure θ can be converted to radians by setting the calculator in **radian mode** and entering $\theta°$. With the calculator set in **degree mode**, a radian measure x can be converted to degrees by entering x^{r}. The symbols ^{r} and $°$ can be found in the menu of angle options.

Applications Locations of points on the surface of Earth are often expressed in terms of *meridians of longitude* and *parallels of latitude*. A **meridian** is a circle on Earth's surface passing through the north and south geographic poles. The **prime meridian** is the meridian passing through Greenwich, England. The **longitude** of a point is the angular "distance," measured in degrees east or west, from the prime meridian to the meridian passing through the point. The **latitude** of a point is its angular "distance" north or south of

Earth's equator, measured in degrees along a meridian. In Figure 10, we illustrate the point on the surface of Earth at 45° N latitude and 90° W longitude. In Figure 11, we show how the latitude and longitude of this point can be viewed as central angles of circles centered at the center of Earth.

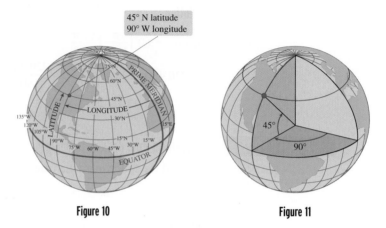

Figure 10   Figure 11

EXAMPLE 9

Locating a City

The city of Angle, Utah (really!), has latitude 38° 14′ 57″ N and longitude 111° 58′ 33″ W. Express its latitude and longitude in terms of decimal degrees, and describe its location relative to Salt Lake City, which has latitude 41.76° N and longitude 111.89° W.

Solution   We first convert from DMS notation to decimal degrees.

$$\text{Latitude: } 38° \, 14′ \, 57″ = \left(38 + \frac{14}{60} + \frac{57}{3600}\right)° \approx 38.249°$$

$$\text{Longitude: } 111° \, 58′ \, 33″ = \left(111 + \frac{58}{60} + \frac{33}{3600}\right)° \approx 111.976°$$

Since both latitudes are measured north of the equator, 38.249° N is not as far north as is 41.76° N. Thus, Angle is south of Salt Lake City. Similarly, because both longitudes are measured west from the prime meridian, Angle is slightly west of Salt Lake City.

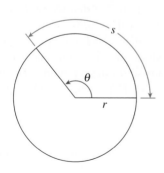

Radian measure is useful in applications involving arc length, area of circular sectors, and angular speed. For arc length, we consider a circle of radius r and a central angle of θ radians. Denote by s the length of the arc subtended by θ, as shown in Figure 12. From the definition of radian measure, we have

$$\theta = \frac{s}{r}$$

By solving for s, we obtain the following formula for arc length.

Figure 12

Arc Length

The length s of an arc on a circle of radius r subtended by a central angle of θ radians is given by

$$s = r\theta$$

EXAMPLE 10

Finding Arc Length

Find the length of an arc subtended by a central angle measuring $150°$ in a circle of radius 3 centimeters.

Solution   We first convert from degrees to radians.

$$150° = 150° \left(\frac{\pi}{180°} \right) = \frac{5\pi}{6}$$

Applying the arc length formula $s = r\theta$ with $r = 3$ and $\theta = \frac{5\pi}{6}$, we have

$$s = 3 \left(\frac{5\pi}{6} \right) = \frac{5\pi}{2} \approx 7.85$$

Thus, the length of the arc is approximately 7.85 centimeters.

The formula for arc length can be used to relate *linear* and *angular* speeds. Just as linear speed is a measure of how much an object's position changes in a given time period, the angular speed of an object rotating at a constant rate is a measure of how much the object's angle changes in a given time period. Thus, the angular speed of an object tells us how fast the object is rotating. In many real-world contexts, rates of rotation are given in revolutions per unit time period. Since 1 revolution equals 2π radians, we can easily convert from revolutions per unit time period to radians per unit time period to obtain the angular speed. For example, a second hand sweeps out 1 revolution or 2π radians every minute, so that its angular speed is given by

$$\text{Angular speed} = \frac{1 \text{ revolution}}{1 \text{ minute}} = \frac{2\pi \text{ radians}}{1 \text{ minute}} = 2\pi \text{ radians per minute}$$

More generally, the **angular speed** ω of an object rotating at a constant rate is given by

$$\omega = \frac{\text{Change in angle}}{\text{Change in time}} = \frac{\theta}{t}$$

where θ is the angle through which the object rotates in the time period t. For an object rotating at a constant rate, the **linear speed** is given by

$$v = \frac{\text{Change in position}}{\text{Change in time}} = \frac{s}{t}$$

Applying the arc length formula $s = r\theta$ gives us

$$v = \frac{s}{t}$$
$$= \frac{r\theta}{t}$$
$$= r \cdot \frac{\theta}{t}$$
$$= r\omega$$

In other words, the linear speed is found by multiplying the angular speed (expressed in radians per time period) by the radius.

····▷EXAMPLE 11   **Finding the Linear Speed of the Tip of a Helicopter Blade**

A helicopter has a 20-foot diameter main rotor that rotates at a rate of 420 revolutions per minute. Find the linear speed of the tip of each blade.

Solution   Each revolution of the rotor corresponds to a change in angle of 2π radians, and so the angular speed of the rotor is given by

$$\omega = \frac{\text{Change in angle}}{\text{Change in time}} = \frac{420 \text{ revolutions}}{1 \text{ minute}} \cdot \frac{2\pi \text{ radians}}{1 \text{ revolution}} = 840\pi \text{ radians per minute}$$

The tip of each blade is 10 feet from the center. Thus, the linear speed of the tip of each blade is

$$v = r\omega = 10(840\pi) \approx 26{,}390 \text{ feet per minute}$$

····▷EXAMPLE 12   **Finding the Angular Speed of a Bicycle Tire**

A bicycle with 27-inch-diameter tires is moving at a constant rate of 15 miles per hour. Find the angular speed of the tires in revolutions per minute.

Solution   Our strategy is to first find the radius r and the linear velocity v and then exploit the equation $v = \omega r$ to determine ω. Since the radius is half the diameter, we have

$$r = \frac{27}{2} = 13.5 \text{ inches}$$

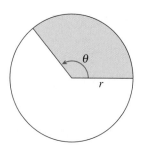

If the bicycle is rolling without slipping, then the linear speed v of a point on the edge of a tire relative to the center of the wheel is equal to the speed of the bicycle. Thus, $v = 15$ miles per hour. Since r is expressed in inches and ω is to be expressed in revolutions per minute, we convert to inches per minute, giving us

$$v = \left(15 \frac{\text{miles}}{\text{hour}}\right)\left(\frac{1 \text{ hour}}{60 \text{ minutes}}\right)\left(\frac{5280 \text{ feet}}{1 \text{ mile}}\right)\left(\frac{12 \text{ inches}}{1 \text{ foot}}\right) = 15{,}840 \text{ inches per minute}$$

From the equation $v = \omega r$, it follows that $\omega = v/r$, and so

$$\omega = \frac{v}{r} = \frac{15{,}840}{13.5} \approx 1173.3 \text{ radians per minute}$$

Since there are 2π radians in each revolution, this angular speed is equivalent to

$$\left(1173.3 \frac{\text{radians}}{\text{minute}}\right)\left(\frac{1 \text{ revolution}}{2\pi \text{ radians}}\right) \approx 186.7 \text{ revolutions per minute}$$

To compute the area of a circular sector (such as the shaded region in Figure 13), we use the fact that the area is proportional to the central angle θ. Thus, for example, if the central angle is half of a full revolution, then the area will be half that of the full circle; if the central angle is one-quarter of a full revolution, then the area will be one-fourth that of the full circle, and so forth. It follows that the ratio of the area of the sector (A) to the area of the circle (πr^2) is equal to the ratio of the central angle (θ) to a full revolution (2π). Thus, we have

$$\frac{A}{\pi r^2} = \frac{\theta}{2\pi}$$

Figure 13

and, by solving for A, we obtain the following area formula.

Area of a Circular Sector

The area of a circular sector formed by a central angle of θ radians in a circle of radius r is given by

$$A = \frac{1}{2}r^2\theta$$

·····EXAMPLE 13   **Finding the Area and Perimeter of a Circular Sector**

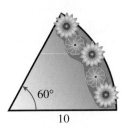

Figure 14

Anton wishes to construct a decorative flower garden in the shape of a circular sector with central angle 60° and radius 10 feet (see Figure 14). To purchase the proper amount of topsoil and border material, he wishes to determine the area and perimeter of the garden. Find both values.

Solution   Converting to radians, we have $60° = \frac{\pi}{3}$ radians. Thus, the area of the sector is

$$A = \frac{1}{2}r^2\theta$$

$$= \frac{1}{2}10^2\left(\frac{\pi}{3}\right)$$

$$\approx 52.36 \text{ square feet}$$

The perimeter P is the sum of the lengths of the two radii that form the sides of the sector and the length of the arc subtended by θ. Thus,

$$P = r + r + r\theta$$

$$= 10 + 10 + 10\left(\frac{\pi}{3}\right)$$

$$\approx 30.47 \text{ feet}$$

The Digits of Pi

The number π, the ratio of the circumference of a circle to its diameter, is irrational, and hence its decimal expansion neither repeats nor ends. It begins
3.14159265358979323846264338327950288419716939937510582097494459 23 . . .

For most practical purposes, the first few digits of π provide us with a sufficiently accurate approximation. Indeed, the 65 digits just given provide us with sufficient accuracy to estimate the circumference of a circle with a radius equal to the distance of the furthest known object (a quasar some 7.8×10^{22} miles away) to within the length of the smallest hypothesized object in the universe (a "string" approximately 10^{-33} centimeters in length). But mathematicians, who are not constrained by the boundaries of the physical universe, have computed π to what, at first glance, seems a ridiculous number of decimal places. In 2002, Yasumasa Kanada and oth-

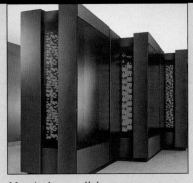

Massively parallel computer

ers at the University of Tokyo computed π to 1.24 trillion decimal places. To put this in perspective: If their decimal expansion were written on a single line with the same size type that is used in this sentence, it would wrap around Earth's equator 100 times!

Mathematicians have been competing to compute the most accurate version of π since ancient times. Although the need for increased accuracy may have been initially dictated by practical concerns, the driving force behind the quests of many modern "digit hunters" is clearly the competition itself. Also, computation of π to millions of decimal places provides an excellent means for testing the speed and accuracy of supercomputers, such as the massively parallel computer shown here.

But there is more here than meets the eye. Although we define π to be the ratio of the circumference of a circle to its diameter, it appears in many settings having nothing to do with circles. It appears in probability theory, calculus, number theory, and physics. To many mathematicians, π is not simply a human convention; π is an integral part of our universe. If we were one day to encounter intelligent beings from another galaxy, then surely they would know about π. To a digit hunter, the digits of π represent a piece of the cosmic code, a divine communiqué containing, perhaps, revelations of nature. And so they compute digits of π, looking for patterns, trying to detect structures. Are the digits truly random? Are there as many strings of consecutive digits as we would expect? These are the sorts of questions that have not yet been answered, but perhaps we will know more once we see the next trillion digits.

Understanding and Mastery Checklists

Concepts to Understand

Angle
◦
Coterminal angles
◦
Standard position
◦
Degree
◦
Degree-minute-second (DMS) form
◦
Radian
◦
Complementary angles
◦
Supplementary angles
◦
Central angle
◦
Arc length
◦
Angular speed
◦
Area of a circular sector

Skills to Master

Find coterminal angles for a given angle.
◦
Find complementary and supplementary angles for a given angle.
◦
Convert between decimal degrees and degree-minute-second form.
◦
Convert between degrees and radians.
◦
Find the length of an arc subtended by a given central angle.
◦
Find the angular or linear speed of a rotating object.
◦
Find the area of a circular sector.

Exercises 5.1

Exercises 1-10   *Match the given angle measure with the corresponding sketch.*

1. $30°$

2. $-60°$

3. $-135°$

4. $270°$

5. $540°$

6. $-360°$

7. $\dfrac{\pi}{3}$

8. $-\dfrac{\pi}{4}$

9. $-\dfrac{3\pi}{2}$ **10.** $\dfrac{5\pi}{6}$

i.
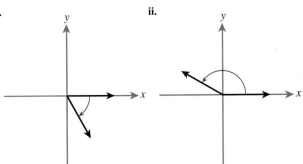

ii.

ix.
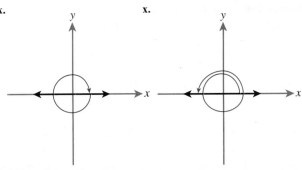

x.

iii.
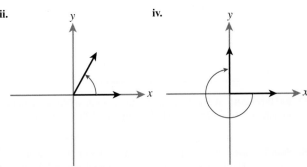

iv.

Exercises 11-18 *Find two coterminal angles (one positive and one negative) for the given angle.*

11. $30°$ **12.** $-60°$

13. $-135°$ **14.** $270°$

15. $\dfrac{\pi}{3}$ **16.** $-\dfrac{\pi}{4}$

17. $-\dfrac{3\pi}{2}$ **18.** $\dfrac{5\pi}{6}$

Exercises 19-22 *Find the complementary and supplementary angles for the given angle.*

19. $20°$ **20.** $37°$

21. $\dfrac{\pi}{6}$ **22.** $\dfrac{\pi}{4}$

v.
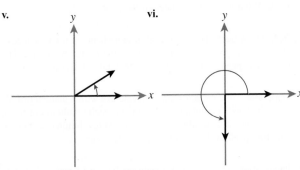

vi.

Exercises 23-26 *Convert the given angle measure to decimal form.*

23. $25°\,16'$ **24.** $142°\,50'$

25. $173°\,20'\,35''$ **26.** $30°\,28'\,42''$

Exercises 27-30 *Convert the given angle measure to DMS form.*

27. $132.4°$ **28.** $47.15°$

29. $15.625°$ **30.** $163.32°$

Exercises 31-36 *Find the exact radian measure of the given angle.*

31. $30°$ **32.** $120°$

33. $225°$ **34.** $-45°$

35. $-150°$ **36.** $300°$

Exercises 37-40 *Find the radian measure of the given angle. Round to three decimal places.*

vii.
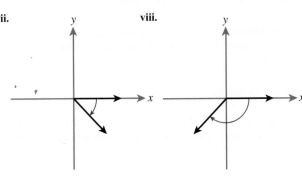

viii.

37. $15°$ **38.** $-20°$

39. $-117.4°$ **40.** $326.7°$

Exercises 41-46 *Find the exact degree measure of the given angle.*

41. $\dfrac{3\pi}{2}$ **42.** $\dfrac{11\pi}{6}$

43. $-\dfrac{4\pi}{3}$ **44.** $\dfrac{\pi}{9}$

45. $\dfrac{11\pi}{12}$ **46.** $-\dfrac{\pi}{4}$

Exercises 47-50 *Find the degree measure of the given angle. Round to three decimal places.*

47. $-\dfrac{2\pi}{7}$ **48.** $\dfrac{7\pi}{11}$

49. 2   **50.** -3

Exercises 51-54 *For the given central angle θ in a circle of radius r, find*

a. *The length of the arc subtended by θ*
b. *The area of the sector determined by θ*

51. $\theta = \dfrac{5\pi}{3}, r = 20$ meters

52. $\theta = \dfrac{3\pi}{4}, r = 16$ inches

53. $\theta = 75°, r = 4$ inches

54. $\theta = 300°, r = 2$ feet

Applications

55. Pulley Angles A bucket in a well is raised by turning a crank attached to a pulley with an 8-inch radius (see Figure 15).

    a. How far is the bucket raised if the crank is turned through an angle of 10π radians (5 revolutions)?

    b. Through what angle must the crank be turned in order to raise the bucket 9 feet? Express your answer in degrees.

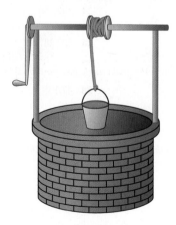

Figure 15

56. Clock Angles Find the exact radian measure of the smaller angle made by the hands of a clock when the time is

    a. 2:00   **b.** 1:30

57. Wind Turbine Speed A wind turbine rotates at a rate of 40 revolutions per minute. Find the linear speed of the tip of a 100-foot blade.

58. Mower Blade Speed A 20-inch-diameter lawn mower blade rotates at a rate of 3200 revolutions per minute. Assume that a rock is struck by the tip of the blade and is propelled away from the lawn mower with an initial speed equal to that of the linear speed of the tip of the blade. What is the initial speed of the rock?

59. Big Ben's Linear Speed The tip of each minute hand on the four faces of Big Ben is approximately 11 feet from the center of the face. Find the linear speed of the tip of a minute hand in miles per hour.

60. Bicycle Speed In a certain gear, a bicycle tire with a diameter of 27 inches makes 3 complete revolutions for each revolution of the pedals. If a cyclist is pedaling at a rate of 100 revolutions per minute, find the speed of the bicycle in miles per hour.

61. Tire Revolutions The tires on a certain car have a 14-inch radius. Find the number of revolutions per minute for each tire when the car is traveling at a rate of 90 feet per second.

62. Earth's Rotational Speed Earth rotates about its axis approximately once every 24 hours. Find the linear speed, relative to the axis of rotation, of a point

    a. On the equator

    b. With latitude 45° N (*Hint:* Find the radius r of the circle formed by all points on the surface of Earth with latitude 45° N, as shown in Figure 16.)

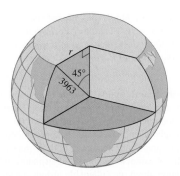

Figure 16

The apparent circular motion of celestial bodies is due to the rotation of Earth.

63. Stained Glass A piece of stained glass is to be cut in the shape of a circular sector with a radius of 12 inches and a central angle of 120°.

a. Find the area of the piece of glass.

b. Find the length of a lead strip that is to be placed around the edges of the piece of glass.

64. Estimating the Moon's Diameter For a small angle θ and a large radius r, the arc length $s = \theta r$ is approximately equal to the length of the line segment connecting the endpoints of the arc. Suppose an observer estimates a 1° angle between the lines of sight to the opposite ends of the diameter of the moon, as shown in Figure 17. Given that the distance to the moon is approximately 230,000 miles, estimate the diameter of the moon.

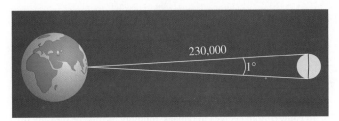

Figure 17

65. Deflection of Light As predicted by Einstein's Theory of Relativity and also observed experimentally, even light is subject to the force of gravity. Thus, when a ray of light passes by a massive heavenly body such as our sun, the ray is deflected by an angle α that is dependent on the mass m of the body and the distance r from the center of the body to the passing light ray. (See Figure 18.) In fact, the angle α satisfies

$$\alpha \approx 4.05\frac{Gm}{rc^2}$$

where $G \approx 6.67 \times 10^{-11}$ m³/(kg · sec²) is Newton's gravitational constant, $c \approx 3 \times 10^8$ m/sec is the speed of light, m is in kilograms, and r is in meters. Find the angle of deflection for a light ray passing close to the surface of the sun, in which case $m \approx 1.989 \times 10^{30}$ kilograms and $r \approx 6.95 \times 10^8$ meters. Express your answer both in radians and degrees.

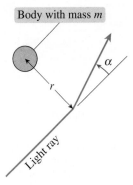

Figure 18

A solar eclipse on May 12, 1919, provides the first evidence of the gravitational bending of light.

66. Angular Momentum A figure skater begins spinning with her arms extended horizontally and with 2-kilogram weights in each hand. As she brings her hands down to her side, her angular velocity increases. This phenomenon is due to *conservation of angular momentum*. For a certain skater, the angular velocity ω satisfies

$$\omega = \left(\frac{8.7}{5.4 + 4r^2}\right)\omega_0$$

where ω_0 is the initial angular velocity of the skater, and r meters is the horizontal distance from the weights to the axis of revolution. If the skater is initially spinning at a rate of 3 revolutions per second, find the angular velocity after the weights are at a final distance of 0.15 meter from the axis of revolution.

Exercises 67-70 *If two locations on the surface of Earth have the same longitude (that is, one lies due north of the other), the distance between them can be approximated using their latitudes. Assuming that all points on the surface of Earth with the same longitude form a circle with a radius of 3963 miles, we find the length of the arc subtended by the central angle between the two latitudes, as suggested by Figure 19. Use this technique to approximate the distance between the following pairs of cities, which have roughly the same longitude. Answer to the nearest mile.*

Figure 19

67. *Distance between Phoenix and Salt Lake City*
Phoenix: 33° 30′ N latitude
Salt Lake City: 40° 45′ N latitude

68. *Distance between San Francisco and Seattle*
San Francisco: 37° 45′ N latitude
Seattle: 47° 35′ N latitude

69. *Distance between Montreal and Bogotá*
Montreal, Canada: 45° 30′ N latitude
Bogotá, Colombia: 4° 38′ N latitude

70. *Distance between St. Petersburg and Alexandria*
St. Petersburg, Russia: 59° 55′ N latitude
Alexandria, Egypt: 31° 13′ N latitude

Exercises 71-72 *It is a well-known fact that the sum of the angles of a triangle in the plane is 180°. But for a spherical triangle—that is, a triangle on the surface of a sphere, such as Earth—the sum of the angles is actually greater than 180°. In fact, it can be shown that if A, B, and C are the angle measures of a triangle on the surface of a sphere, then*

$$A + B + C = 180° + \left(\frac{T}{S}\right)720°$$

where T is the area of the triangle and S is the surface area of the sphere. In the case of triangles on the surface of Earth, S denotes the surface area of Earth. Note that the surface area of a sphere of radius r is given by $4\pi r^2$ and that the radius of Earth is 3963 miles.

71. The Bermuda Triangle The Bermuda Triangle is a region in the Atlantic Ocean bounded by imaginary lines connecting Bermuda, Puerto Rico, and Melbourne, Florida. Find the sum of the angles of the Bermuda Triangle given that its area is approximately 490,000 square miles. Express your answer in both degrees and radians. (Numerous disappearances of ships and airplanes in this area have led to speculation about unexplainable turbulence and other atmospheric disturbances. However, investigations to date have produced no scientific evidence of unusual phenomena.)

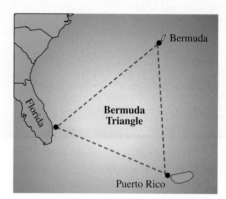

72. Sales Region A salesman covers a region roughly bounded by imaginary lines connecting Chicago, Minneapolis, and Des Moines. The spherical triangle with these three cities as vertices has angle measures 40.64°, 58.82°, and 80.67°. Estimate the area of the region.

Concepts and Critical Thinking

Exercises 73-76 *Answer true or false.*

73. If two angles are coterminal, then their measures must differ by 360 degrees.

74. One radian is the length of the arc subtended by a central angle measuring 1 degree in a circle of radius 1.

75. There are 3600 seconds in 1°.

76. To convert from degrees to radians, multiply the degree measure of an angle by $\frac{180°}{\pi}$.

Exercises 77-80 *Give an example of each.*

77. Two complementary angles

78. Two supplementary angles

79. A quadrant angle

80. The radian measure of an acute angle expressed as a decimal

81. Suppose α and β are coterminal central angles of a circle with radius r. Explain why the arc lengths subtended by α and β need not be the same. Give an example that illustrates this.

Questions for Discussion or Essay

82. In Exercise 62, we considered the linear speed of a point on the surface of Earth at the equator. If it were possible to construct a tower of sufficient height, how high would the tower have to be in order for its tip to have a linear speed of 186,000 miles per second relative to Earth's axis? According to Einstein's Theory of Relativity, speeds beyond the speed of light are not possible. Explain this apparent contradiction.

83. If two points on the surface of Earth have latitude 45° N and longitudes differing by 15°, the distance between them is approximately 734 miles. On the other hand, if two points have latitude 30° N and longitudes differing by 15°, they are approximately 899 miles apart. Explain this discrepancy.

84. The unit of length commonly referred to as a mile is more precisely called a **statute mile**. A **nautical mile**, on the other hand, is 1′ of arc length as measured on a great circle such as the equator. (A great circle is a circle described by the intersection of the surface of a sphere with a plane passing through the center of the sphere.) Why do you suppose nautical miles are so named? What is the radian measure corresponding to 1 nautical mile? Which is traveling faster, a boat with a rate of 40 knots (40 nautical miles per hour) or one with a rate of 40 statute miles per hour?

Projects for Enrichment

85. Bicycle Gearing The modern bicycle is a direct descendent of, and bears a striking resemblance to, the so-called *safety bicycle* designed by John Kemp Starley in the 1880s. Starley's 1885 Rover (see photo) was the first chain-drive bicycle with changeable front and rear sprockets. The eventual development of derailleur systems further improved upon the chain-drive concept by allowing the chain to be shifted between sprockets of different sizes. In this project, we investigate the connection between the angular speeds of the sprockets and the resulting speed of the bicycle.

Starley's 1885 Rover

Gt Bicycles' 1996 Team USA Superbike

When two sprockets of different sizes are connected by a chain, their linear speeds are forced to be the same. As a consequence, their angular speeds can be shown to be proportional. Consider the front and rear sprockets shown in Figure 20, with radii R and r, respectively. Denote the angular speed of the front sprocket (usually called a **chain ring**) by ω_R and the angular speed of the rear sprocket by ω_r. Finally, define the **gear ratio** to be the ratio $\frac{R}{r}$.

a. Show that the angular speed of the rear sprocket can be found by multiplying the angular speed of the chain ring by the gear ratio.

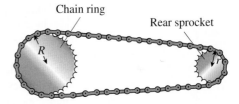

Chain ring

Rear sprocket

Figure 20

b. If an 8-inch-diameter chain ring has an angular speed of 80 revolutions per minute, find the angular speed of a 3-inch-diameter rear sprocket. Given that the rear tire has a 27-inch diameter, find the linear speed of the bicycle and the distance it has traveled during 1 revolution of the pedals.

c. Complete the following table. As before, R and r denote the radii of the chain ring and rear sprocket, v is the linear speed of the bike, and d is the distance traveled during 1 revolution of the pedals. Assume the angular speed of the chain ring is 80 revolutions per minute and the diameter of the rear tire is 27 inches.

R (in.)	r (in.)	Gear ratio	v (mph)	d (in.)
8	4			
8	3			
8	2			
6	4			
6	3			
6	2			

d. Based on your results from part c, describe how the linear speed of the bicycle depends on the sizes of the chain ring and rear sprocket. How does the gear ratio make this dependency easier to describe?

Since the spacing between each link of the chain is fixed, the diameter of a sprocket is completely determined by the number of teeth it has. Thus, the gear ratio can be found by computing the ratio of the number of teeth in the chain ring to the number of teeth in the rear sprocket. For example, a 52-tooth chain ring and a 14-tooth rear sprocket would yield a gear ratio of $\frac{52}{14} \approx 3.71$.

e. Find a multispeed bicycle with at least two chain rings and at least five rear sprockets. Determine the diameter of the rear tire and the number of teeth in each chain ring and sprocket. Complete a chart that shows all of the possible chain ring/sprocket combinations, the corresponding gear ratios, and the linear distance that would be traveled for each gear ratio for a single revolution of the pedals. What does your chart suggest about the order in which you should shift gears if you wish to progress from lowest to highest in sequential order?

Many cycling experts recommend a pedaling rate of 90 revolutions per minute. This will allow for a good balance between leg speed and power.

f. Measure the length of the pedal crankarm and use that length to determine the linear speed of your feet when pedaling at a rate of 90 revolutions per minute.

g. Add a column to your table in part e that gives the speed you would be traveling for each gear ratio if you were pedaling at a rate of 90 revolutions per minute.

Section 5.2 Trigonometric Functions of Acute Angles

- How can the length of a shadow and knowledge of trigonometry destroy a suspect's alibi?
- How can a piece of wood, a protractor, and a weight be used to estimate the height of a tall building?
- According to OSHA standards, how high should the tip of a 20-foot ladder reach?
- How can the angles of elevation to known landmarks be used to determine one's location?

Narrowly defined, trigonometry is the study of relationships between sides and angles of triangles. But the scope of trigonometry's applications extends far from its geometric point of origin. As we will see in later sections, trigonometric functions have a cyclical quality that renders them ideal for mathematically modeling a vast array of periodic phenomena—from the rise and fall of tides on the shore to the phases of the moon that lights the night sky, from complex cadences beat by our hearts and brains to the rhythmic pulsing of electrons and photons, the flow of which moves voices and images, power and information. Where there is repetition, where there is periodicity, where cycles can be found, there lives trigonometry.

There are several, equivalent ways of defining the trigonometric functions. We begin with an intuitive, easily accessible method based on ratios of side lengths associated with an acute angle of a right triangle. In the next section, we will extend our definitions to angles of any measure.

Right Triangle Definitions   The right triangle definitions of the trigonometric functions are based on an important theorem from geometry, generally attributed to the Greek mathematician Euclid. Euclid showed that the ratio of any two sides of a given triangle must be the same as the corresponding ratio of sides of any similar triangle. Thus, for a right triangle with an acute angle θ, the ratios of any two sides are dependent *only* on the angle θ and *not* on the size of the right triangle. To see more clearly what this statement means, let us consider the right triangle shown in Figure 21, where we have labeled the angle θ, the length of the side adjacent to θ as "adj," the length of the side opposite θ as "opp," and the length of the hypotenuse as "hyp." The six possible side ratios are as follows:

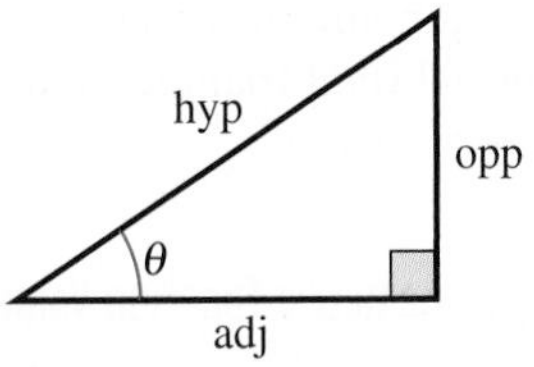

hyp

opp

θ

adj

Figure 21

$$\frac{\text{opp}}{\text{hyp}}, \frac{\text{adj}}{\text{hyp}}, \frac{\text{opp}}{\text{adj}}, \frac{\text{hyp}}{\text{opp}}, \frac{\text{hyp}}{\text{adj}}, \frac{\text{adj}}{\text{opp}}$$

According to Euclid's theorem, each of the six ratios will remain unchanged if we increase or decrease the size of the triangle, as long as we keep the measure of θ the same. Thus, each ratio is dependent only on the angle θ and hence defines a function of θ. The six functions that result, known collectively as the **trigonometric functions**, are the **sine**, **cosine**, **tangent**, **cosecant**, **secant**, and **cotangent** functions. They are usually abbreviated **sin**, **cos**, **tan**, **csc**, **sec**, and **cot**, respectively, so that, for example, $\sin\theta$ denotes the sine of the angle θ. Their formal definitions are as follows.

Right Triangle Definitions of Trigonometric Functions

Let θ be an acute angle of a right triangle, and let "opp" denote the length of the side opposite θ, "adj" the length of the side adjacent to θ, and "hyp" the length of the hypotenuse (Figure 22). We define the six trigonometric functions of θ as follows:

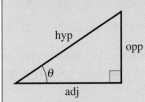

hyp

opp

θ

adj

Figure 22

$$\sin\theta = \frac{\text{opp}}{\text{hyp}} \qquad \cos\theta = \frac{\text{adj}}{\text{hyp}} \qquad \tan\theta = \frac{\text{opp}}{\text{adj}}$$

$$\csc\theta = \frac{\text{hyp}}{\text{opp}} \qquad \sec\theta = \frac{\text{hyp}}{\text{adj}} \qquad \cot\theta = \frac{\text{adj}}{\text{opp}}$$

Note that since each of the lengths opp, adj, and hyp is positive, the values of the trigonometric functions for an acute angle θ are all positive. Moreover, since the hypotenuse is the longest side, both $\sin\theta = \text{opp}/\text{hyp}$ and $\cos\theta = \text{adj}/\text{hyp}$ are less than 1 for all acute angles θ.

····**EXAMPLE 1**

Evaluating Trigonometric Functions

Find the values of the six trigonometric functions for the angle θ shown in Figure 23.

Solution   The lengths of the hypotenuse and the side opposite θ are given in Figure 23. Thus, hyp = 13 and opp = 5. The length of the side adjacent to θ can be found by applying the Pythagorean Theorem.

$$\text{adj} = \sqrt{13^2 - 5^2} = \sqrt{144} = 12$$

Using the definitions of the trigonometric functions, we obtain the following values:

$$\sin\theta = \frac{\text{opp}}{\text{hyp}} = \frac{5}{13} \qquad \cos\theta = \frac{\text{adj}}{\text{hyp}} = \frac{12}{13} \qquad \tan\theta = \frac{\text{opp}}{\text{adj}} = \frac{5}{12}$$

$$\csc\theta = \frac{\text{hyp}}{\text{opp}} = \frac{13}{5} \qquad \sec\theta = \frac{\text{hyp}}{\text{adj}} = \frac{13}{12} \qquad \cot\theta = \frac{\text{adj}}{\text{opp}} = \frac{12}{5}$$

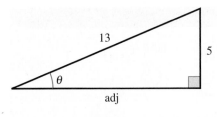

13

5

θ

adj

Figure 23

The trigonometric function values for the angles 30°, 45°, and 60° can be found using special right triangles and some facts from geometry, as illustrated in the following example.

┄┄►EXAMPLE 2

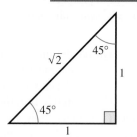

Figure 24

Finding Trigonometric Function Values for 30°, 45°, and 60°

Find the values of the sine, cosine, and tangent functions for 30°, 45°, and 60°.

Solution   For a 45° angle, we consider the isosceles right triangle shown in Figure 24 with leg length 1. By the Pythagorean Theorem, the hypotenuse has length $\sqrt{2}$. Applying the definitions of the trigonometric functions, we obtain the following values:

$$\sin 45° = \frac{\text{opp}}{\text{hyp}} = \frac{1}{\sqrt{2}} = \frac{\sqrt{2}}{2} \qquad \cos 45° = \frac{\text{adj}}{\text{hyp}} = \frac{1}{\sqrt{2}} = \frac{\sqrt{2}}{2} \qquad \tan 45° = \frac{\text{opp}}{\text{adj}} = \frac{1}{1} = 1$$

For 30° and 60° angles, we consider the equilateral triangle shown in Figure 25 with side length 2. If we construct line segment $\overline{AD}$ from the top vertex to the midpoint of the opposite side, we obtain right triangle ADB with acute angles of 30° and 60°. By the Pythagorean Theorem, the length of $\overline{AD}$ is $\sqrt{3}$. Once again, we apply the definitions of the trigonometric functions to obtain the following values:

$$\sin 30° = \frac{\text{opp}}{\text{hyp}} = \frac{1}{2} \qquad \cos 30° = \frac{\text{adj}}{\text{hyp}} = \frac{\sqrt{3}}{2} \qquad \tan 30° = \frac{\text{opp}}{\text{adj}} = \frac{1}{\sqrt{3}} = \frac{\sqrt{3}}{3}$$

$$\sin 60° = \frac{\text{opp}}{\text{hyp}} = \frac{\sqrt{3}}{2} \qquad \cos 60° = \frac{\text{adj}}{\text{hyp}} = \frac{1}{2} \qquad \tan 60° = \frac{\text{opp}}{\text{adj}} = \frac{\sqrt{3}}{1} = \sqrt{3}$$

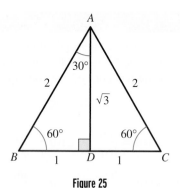

Figure 25

Table 1 summarizes the sine, cosine, and tangent values for 30°, 45°, and 60°. Because these values occur so frequently in trigonometry, we strongly recommend that they be memorized.

Table 1
Special Angles

θ	θ (radians)	$\sin \theta$	$\cos \theta$	$\tan \theta$
30°	$\dfrac{\pi}{6}$	$\dfrac{1}{2}$	$\dfrac{\sqrt{3}}{2}$	$\dfrac{\sqrt{3}}{3}$
45°	$\dfrac{\pi}{4}$	$\dfrac{\sqrt{2}}{2}$	$\dfrac{\sqrt{2}}{2}$	1
60°	$\dfrac{\pi}{3}$	$\dfrac{\sqrt{3}}{2}$	$\dfrac{1}{2}$	$\sqrt{3}$

Fundamental Identities   Many important relationships can be derived from the definitions of the trigonometric functions. For example, since the sine of an angle is the ratio opp/hyp and the cosecant is the ratio hyp/opp, we see that the sine and cosecant functions are reciprocals of each other. These relationships are referred to as **ratio identities**, and they are summarized as follows.

Ratio Identities

$$\sin \theta = \frac{1}{\csc \theta} \qquad \cos \theta = \frac{1}{\sec \theta} \qquad \tan \theta = \frac{1}{\cot \theta}$$

$$\csc \theta = \frac{1}{\sin \theta} \qquad \sec \theta = \frac{1}{\cos \theta} \qquad \cot \theta = \frac{1}{\tan \theta}$$

$$\tan \theta = \frac{\sin \theta}{\cos \theta} \qquad \cot \theta = \frac{\cos \theta}{\sin \theta}$$

·····≽EXAMPLE 3   **Evaluating Trigonometric Functions of Special Angles**

Complete the following table using ratio identities.

θ	θ (radians)	$\sin\theta$	$\cos\theta$	$\tan\theta$	$\sec\theta$	$\csc\theta$	$\cot\theta$
30°	$\dfrac{\pi}{6}$	$\dfrac{1}{2}$	$\dfrac{\sqrt{3}}{2}$	$\dfrac{\sqrt{3}}{3}$			
45°	$\dfrac{\pi}{4}$	$\dfrac{\sqrt{2}}{2}$	$\dfrac{\sqrt{2}}{2}$	1			
60°	$\dfrac{\pi}{3}$	$\dfrac{\sqrt{3}}{2}$	$\dfrac{1}{2}$	$\sqrt{3}$			

Solution   Since $\sec\theta = 1/\cos\theta$, the values in the $\sec\theta$ column can be obtained by taking the reciprocal of the entries in the $\cos\theta$ column. Similarly, the entries in the $\csc\theta$ and $\cot\theta$ columns can be obtained by exploiting the identities

$$\csc\theta = \frac{1}{\sin\theta} \quad \text{and} \quad \cot\theta = \frac{1}{\tan\theta}$$

Thus, the entries in the $\csc\theta$ and $\cot\theta$ columns are the reciprocals of the corresponding entries in the $\sin\theta$ and $\tan\theta$ columns, respectively. After simplifying, we obtain the values shown in Table 2.

Table 2
Special Angles

θ	θ (radians)	$\sin\theta$	$\cos\theta$	$\tan\theta$	$\sec\theta = \dfrac{1}{\cos\theta}$	$\csc\theta = \dfrac{1}{\sin\theta}$	$\cot\theta = \dfrac{1}{\tan\theta}$
30°	$\dfrac{\pi}{6}$	$\dfrac{1}{2}$	$\dfrac{\sqrt{3}}{2}$	$\dfrac{\sqrt{3}}{3}$	$\dfrac{2}{\sqrt{3}} = \dfrac{2\sqrt{3}}{3}$	$\dfrac{2}{1} = 2$	$\dfrac{3}{\sqrt{3}} = \sqrt{3}$
45°	$\dfrac{\pi}{4}$	$\dfrac{\sqrt{2}}{2}$	$\dfrac{\sqrt{2}}{2}$	1	$\dfrac{2}{\sqrt{2}} = \sqrt{2}$	$\dfrac{2}{\sqrt{2}} = \sqrt{2}$	$\dfrac{1}{1} = 1$
60°	$\dfrac{\pi}{3}$	$\dfrac{\sqrt{3}}{2}$	$\dfrac{1}{2}$	$\sqrt{3}$	$\dfrac{2}{1} = 2$	$\dfrac{2}{\sqrt{3}} = \dfrac{2\sqrt{3}}{3}$	$\dfrac{1}{\sqrt{3}} = \dfrac{\sqrt{3}}{3}$

For most angles, it is not possible to express trigonometric function values in rational or radical form. Instead, we must rely on calculator approximations.

Calculator Keys

Trigonometric Function Values

Approximations for the sine, cosine, and tangent of an angle θ can be found using the $\boxed{\text{SIN}}$, $\boxed{\text{COS}}$, and $\boxed{\text{TAN}}$ keys, respectively. Approximations for the cosecant, secant, and cotangent of θ can be found by applying the ratio identities and the reciprocal key $\boxed{x^{-1}}$. Note that you must be sure your calculator is set in the intended mode, either degree or radian.

·····EXAMPLE 4

Using a Calculator to Find Trigonometric Function Values

Find decimal approximations for the values of the trigonometric functions for 20°.

Solution   We set our calculator in degree mode (so that it is not necessary to enter the ° symbol), and obtain the following approximations:

$$\sin 20° \approx 0.342020 \qquad\qquad \csc 20° = \frac{1}{\sin 20°} \approx 2.923804$$

$$\cos 20° \approx 0.939693 \qquad\qquad \sec 20° = \frac{1}{\cos 20°} \approx 1.064178$$

$$\tan 20° \approx 0.363970 \qquad\qquad \cot 20° = \frac{1}{\tan 20°} \approx 2.747477$$

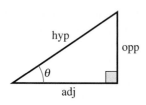

Figure 26

The next three identities are consequences of the Pythagorean Theorem and so are called the **Pythagorean identities**. For an acute angle θ in a right triangle with side lengths opp, adj, and hyp, as given in Figure 26, we have

$$(\text{opp})^2 + (\text{adj})^2 = (\text{hyp})^2$$

If we divide both sides by $(\text{hyp})^2$, we obtain

$$\left(\frac{\text{opp}}{\text{hyp}}\right)^2 + \left(\frac{\text{adj}}{\text{hyp}}\right)^2 = 1$$

or

$$(\sin\theta)^2 + (\cos\theta)^2 = 1$$

Similarly, by dividing both sides of $(\text{opp})^2 + (\text{adj})^2 = (\text{hyp})^2$ by either $(\text{opp})^2$ or $(\text{adj})^2$, we obtain $1 + (\tan\theta)^2 = (\sec\theta)^2$ and $1 + (\cot\theta)^2 = (\csc\theta)^2$, respectively.

Pythagorean Identities

$$\sin^2\theta + \cos^2\theta = 1 \qquad 1 + \tan^2\theta = \sec^2\theta \qquad 1 + \cot^2\theta = \csc^2\theta$$

Notice that we denote $(\sin\theta)^2$ by $\sin^2\theta$, and we use the same convention for the squares of the other trigonometric functions. In general, positive integer powers of the trigonometric functions are written using this convention. Thus, for example, we write $\tan^5\theta$ instead of $(\tan\theta)^5$.

·····EXAMPLE 5

Using Identities to Find Trigonometric Function Values

If θ is an acute angle for which $\tan\theta = \sqrt{15}$, find the exact values of the other five trigonometric functions of θ.

Solution   We illustrate two methods. The first involves constructing a right triangle, and the second uses trigonometric identities.

Method 1:   We begin by sketching a right triangle with an acute angle θ and with opp $= \sqrt{15}$ and adj $= 1$, as shown in Figure 27. This ensures that $\tan\theta = \sqrt{15}$. Next, we apply the Pythagorean Theorem to find the length of the hypotenuse.

$$(\text{hyp})^2 = \left(\sqrt{15}\right)^2 + 1^2$$

$$(\text{hyp})^2 = 16$$

$$\text{hyp} = \sqrt{16} = 4$$

Finally, we apply the definitions of the trigonometric functions to obtain

$$\sin\theta = \frac{\sqrt{15}}{4} \qquad\qquad \cos\theta = \frac{1}{4}$$

$$\csc\theta = \frac{4}{\sqrt{15}} = \frac{4\sqrt{15}}{15} \qquad \sec\theta = \frac{4}{1} = 4 \qquad \cot\theta = \frac{1}{\sqrt{15}} = \frac{\sqrt{15}}{15}$$

Method 2: We can find $\sec\theta$ by applying the Pythagorean identity $1 + \tan^2\theta = \sec^2\theta$.

$$\sec^2\theta = 1 + \tan^2\theta$$
$$\sec^2\theta = 1 + \left(\sqrt{15}\right)^2$$
$$\sec^2\theta = 16$$
$$\sec\theta = \sqrt{16} = 4$$

Notice that we have chosen only the positive root because all trigonometric function values are positive for acute angles. Now, using the ratio identity $\cos\theta = 1/\sec\theta$, we have

$$\cos\theta = \frac{1}{\sec\theta} = \frac{1}{4}$$

Since $\tan\theta = \sin\theta/\cos\theta$, it follows that $\tan\theta\cos\theta = \sin\theta$, and so

$$\sin\theta = \tan\theta\cos\theta = \sqrt{15}\cdot\frac{1}{4} = \frac{\sqrt{15}}{4}$$

The remaining two function values, $\csc\theta$ and $\cot\theta$, can be found using the ratio identities.

$$\csc\theta = \frac{1}{\sin\theta} = \frac{4}{\sqrt{15}} = \frac{4\sqrt{15}}{15} \qquad \cot\theta = \frac{1}{\tan\theta} = \frac{1}{\sqrt{15}} = \frac{\sqrt{15}}{15}.$$

Figure on left:

hyp = 4   opp = $\sqrt{15}$

θ

adj = 1

Figure 27

Notice in the preceding example that it was not necessary to find the angle θ in order to determine the remaining trigonometric function values for θ. However, if exact values are not needed, an alternative approach would be to use a calculator to first find a decimal approximation for the angle θ and then find decimal approximations for the other trigonometric function values for θ.

Calculator Keys

Estimating an Angle

An approximation for an acute angle whose sine, cosine, or tangent is known can be found using the $\boxed{\text{SIN}^{-1}}$, $\boxed{\text{COS}^{-1}}$, or $\boxed{\text{TAN}^{-1}}$ key (usually in conjunction with the $\boxed{\text{2nd}}$ key). For example, if it is known that θ is an acute angle with $\sin\theta = 0.4$, then θ can be approximated by entering $\boxed{\text{2nd}}\ \boxed{\text{SIN}^{-1}}\ \boxed{.}\ \boxed{4}$. If the calculator is in degree mode, the answer will be given as approximately $23.578178°$. Note that $\sin^{-1}(x)$ does not equal $1/\sin(x)$; rather, it represents the inverse of the function $\sin(x)$. The same applies to $\cos^{-1}(x)$ and $\tan^{-1}(x)$. These inverses will be considered in more detail in Section 5.6.

EXAMPLE 6

Using a Calculator to Find Trigonometric Function Values

Given that θ is an acute angle for which $\sec\theta = 8$, approximate θ and the values of the other trigonometric function values at θ.

Solution   Since most calculators do not have a $\boxed{\text{SEC}^{-1}}$ key, we first apply a ratio identity to obtain

$$\cos\theta = \frac{1}{\sec\theta} = \frac{1}{8} = 0.125$$

Now, using the $\boxed{\text{COS}^{-1}}$ key with a calculator set in degree mode, we see that $\theta \approx 82.819244°$. Finally, using the $\boxed{\text{SIN}}$, $\boxed{\text{TAN}}$, and $\boxed{\text{x}^{-1}}$ keys we obtain

$$\sin\theta \approx 0.992157 \qquad \tan\theta \approx 7.937254$$

$$\csc\theta = \frac{1}{\sin\theta} \qquad\qquad \cot\theta = \frac{1}{\tan\theta}$$

$$\approx \frac{1}{0.992157} \qquad\qquad \approx \frac{1}{7.937254}$$

$$\approx 1.007905 \qquad\qquad \approx 0.125988$$

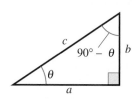

Figure 28

The final group of identities we consider here are known as the **cofunction identities**. Consider the acute angle θ shown in Figure 28. From the definition of the sine function, we have $\sin\theta = b/c$. The angle complementary to θ is $90° - \theta$, and $\cos(90° - \theta) = b/c$. Thus, $\sin\theta = \cos(90° - \theta)$. Similar identities hold for the other cofunction pairs—tangent/cotangent and secant/cosecant—as follows.

Cofunction Identities

$\sin\theta = \cos(90° - \theta)$	$\tan\theta = \cot(90° - \theta)$	$\sec\theta = \csc(90° - \theta)$
$\cos\theta = \sin(90° - \theta)$	$\cot\theta = \tan(90° - \theta)$	$\csc\theta = \sec(90° - \theta)$

⋯⋯EXAMPLE 7

Using a Cofunction Identity

It is known that $\tan\frac{\pi}{8} = \sqrt{2} - 1$. Find the exact value of $\tan\frac{3\pi}{8}$.

Solution   Using the radian form of the cofunction identity for tangent, we obtain

$$\tan\frac{3\pi}{8} = \cot\left(\frac{\pi}{2} - \frac{3\pi}{8}\right)$$

$$= \cot\frac{\pi}{8}$$

$$= \frac{1}{\tan\dfrac{\pi}{8}} \qquad \text{Using a ratio identity}$$

$$= \frac{1}{\sqrt{2} - 1}$$

Rationalizing the denominator, we obtain that $\tan\frac{3\pi}{8} = \sqrt{2} + 1$.

Applications   The early development of trigonometry was motivated by problems in astronomy, navigation, and surveying. Consider, for example, the problem of a surveyor who wishes to determine the distance between two points, labeled A and B in Figure 29, on opposite sides of a river. One approach would be to mark a third point, labeled C in Figure 29, a certain distance from point B and along a line perpendicular to AB. The surveyor could then measure the angle θ formed by the line segments BC and AC and use trigonometry to determine the distance from A to B. We illustrate this process in the following example.

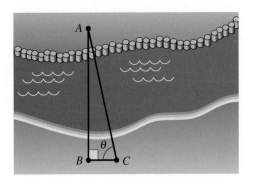

Figure 29

EXAMPLE 8 **Solving a Right Triangle**

Find all unknown side lengths and angle measures for the triangle shown in Figure 29 given that θ is $78°$ and the distance from B to C is 50 feet.

Solution   Figure 30 shows the right triangle from Figure 29 with the unknown side lengths labeled. Note that c is the length of the side opposite θ, and the side adjacent to θ has length 50. Thus, we require a trigonometric function that relates θ to its opposite and adjacent sides. We choose tangent instead of cotangent because its values can be calculated directly with a calculator. Since $\tan \theta = \text{opp}/\text{adj}$, we have

$$\tan 78° = \frac{c}{50} \qquad \text{Using the definition of tangent}$$

$$50 \tan 78° = c \qquad \text{Multiplying both sides by 50}$$

$$c = 50 \tan 78°$$

$$c \approx 50(4.7046) = 235.2 \qquad \text{Approximating with a calculator}$$

Thus, the distance c from A to B is approximately 235.2 feet. The distance b is the length of the hypotenuse, and so we apply the Pythagorean Theorem:

$$b = \sqrt{50^2 + c^2} \approx \sqrt{50^2 + 235.2^2} \approx 240.5$$

Finally, the angle at vertex A is the complement of θ and so has measure

$$90° - \theta = 90° - 78° = 12°$$

Figure 30

In many applications involving trigonometry, we consider angles formed by a horizontal line and the line of sight from a reference point on the horizontal line to an object above or below it. We refer to such an angle as an **angle of elevation** or **angle of depression** according to whether the object is above or below the horizontal, as illustrated in Figures 31 and 32.

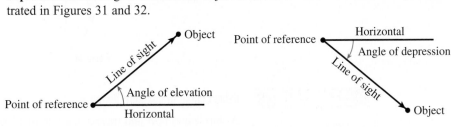

Figure 31   **Figure 32**

EXAMPLE 9

The World's Steepest Railroad Tracks

According to *The Guinness Book of World Records*, the world's steepest standard-gauge railroad tracks are between Chedde and Servoz, France. On average, a train will rise 100 feet in elevation for every 1105 feet of track length traveled. Find the average angle of elevation of the tracks.

Solution We begin by sketching the right triangle shown in Figure 33. To find the average angle of elevation θ, we note that the sine function establishes a relationship between θ and the two given distances. Thus,

$$\sin\theta = \frac{\text{opp}}{\text{hyp}} = \frac{100}{1105} \approx 0.0905$$

Now, using the $\boxed{\text{SIN}^{-1}}$ key on a calculator in degree mode, we obtain

$$\theta \approx 5.19°$$

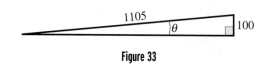

Figure 33

In surveying and navigation, directions are often specified in terms of bearings or courses. Generally speaking, a **bearing** gives a direction of observation, whereas a **course** gives a direction of movement. Historically, a bearing (or course) was given as the acute angle made with a north-south line. In Figure 34, for example, ship A has a course of N 40° W, whereas ship B has a course of S 15° W. The modern approach is to express bearing (or course) as the angle measured clockwise from the north about a north-south line. Thus, ships A and B have courses of 320° and 195°, respectively, as shown in Figure 35.

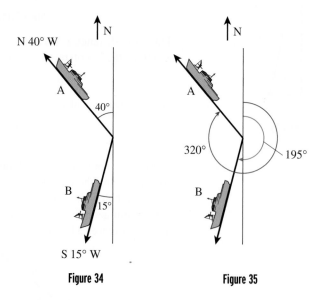

Figure 34   **Figure 35**

EXAMPLE 10

Using Course to Compute Distance

A ship leaves port and travels at a rate of 15 miles per hour on a course of N 40° W. If the shoreline is roughly north-south, how far is the ship from shore after $3\frac{1}{2}$ hours?

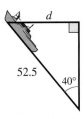

Figure 36

Solution   After $3\frac{1}{2}$ hours, the ship has traveled $(3.5)(15) = 52.5$ miles from port. From Figure 36, we have

$$\sin 40° = \frac{\text{opp}}{\text{hyp}} = \frac{d}{52.5}$$

and it follows that

$$d = 52.5 \sin 40° \approx 33.7 \text{ miles}$$

Understanding and Mastery Checklists

Concepts to Understand	Skills to Master
Right triangle definitions of trigonometric functions: sine, cosine, tangent, cosecant, secant, and cotangent	Find the trigonometric function value for an acute angle in a right triangle.
Ratio identities	Find the trigonometric function value for an acute angle given information about other trigonometric values.
Pythagorean identities	Find an exact value for a trigonometric function of a special acute angle.
Cofunction identities	Use a calculator to approximate trigonometric function values.
Angle of elevation	Solve a right triangle.
Angle of depression	
Course and bearing	

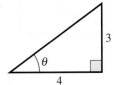

Exercises 5.2

Exercises 1–6   *Find the values of the six trigonometric functions for the given angle θ.*

1.

2.
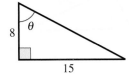

3.
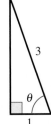

4.
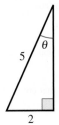

5.

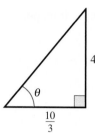

6.

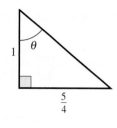

Exercises 7-12 *Use a right triangle to find the other five trigonometric function values of the acute angle θ.*

7. $\sin\theta = \dfrac{5}{13}$

8. $\cos\theta = \dfrac{3}{5}$

9. $\sec\theta = 4$

10. $\cot\theta = 3$

11. $\tan\theta = \dfrac{2}{3}$

12. $\csc\theta = \dfrac{7}{4}$

Exercises 13-20 *Use trigonometric identities to find the indicated function value for the acute angle θ.*

13. $\cos\theta$; $\sec\theta = 2$

14. $\tan\theta$; $\cot\theta = 5$

15. $\sin\theta$; $\cos\theta = \dfrac{1}{3}$

16. $\sec\theta$; $\tan\theta = \dfrac{3}{2}$

17. $\cot\theta$; $\sec\theta = \dfrac{25}{24}$

18. $\csc\theta$; $\cos\theta = \dfrac{8}{17}$

19. $\tan\theta$; $\cot(90° - \theta) = 8$

20. $\cos(90° - \theta)$; $\sin\theta = 0.25$

Exercises 21-28 *Find the exact trigonometric function value.*

21. $\sin 30°$

22. $\tan 45°$

23. $\tan\dfrac{\pi}{3}$

24. $\cos\dfrac{\pi}{6}$

25. $\sec 30°$

26. $\csc 60°$

27. $\cot\dfrac{\pi}{6}$

28. $\sec\dfrac{\pi}{4}$

Exercises 29-32 *Use trigonometric identities to find the exact function value.*

29. $\sin\dfrac{\pi}{12}$, given that $\cos\dfrac{5\pi}{12} = \dfrac{1}{4}(\sqrt{6} - \sqrt{2})$

30. $\cos 67.5°$, given that $\sin 22.5° = \dfrac{1}{2}\sqrt{2 - \sqrt{2}}$

31. $\cot 22.5°$, given that $\cot 67.5° = \sqrt{2} - 1$

32. $\tan\dfrac{\pi}{12}$, given that $\tan\dfrac{5\pi}{12} = \sqrt{3} + 2$

Exercises 33-42 *Use a calculator to find a decimal approximation for the given function, correct to four decimal places.*

33. $\sin 20°$

34. $\cos 75°$

35. $\cot 88.4°$

36. $\csc 15.8°$

37. $\sec 20° \, 42'$

38. $\tan 51° \, 11'$

39. $\cos\dfrac{\pi}{8}$

40. $\sin\dfrac{\pi}{12}$

41. $\csc 1$

42. $\cos 0.5$

Exercises 43-48 *Use a calculator to find a decimal approximation, correct to four decimal places, for the acute angle θ measured in degrees.*

43. $\cos\theta = 0.2$

44. $\sin\theta = 0.75$

45. $\tan\theta = 4.36$

46. $\cos\theta = 0.94$

47. $\csc\theta = 5.6$

48. $\sec\theta = 1.3$

Exercises 49-58 *Find all unknown side lengths and angle measures; that is, solve the given triangle.*

49.

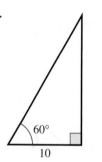

50.

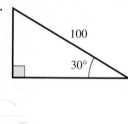

51.

52.

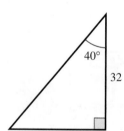

53.

54.

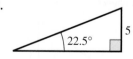

55.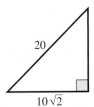

20

$10\sqrt{2}$

56.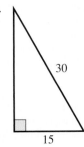

30

15

57.

134

54

58.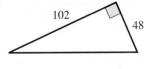

102

48

Applications

59. Height of the World's Tallest Tree   According to *The Guinness Book of World Records,* the tallest tree currently standing is the Mendocino Tree, a coast redwood at Montgomery State Reserve near Ukiah, California. At a distance of 200 feet from the base of the tree, the angle of elevation to the top of the tree is approximately 61.5°. How tall is the tree?

60. Leaning Ladder   Based on OSHA (Occupational Safety and Health Administration) standards, an extension ladder should make a 75° angle with the ground. Using this rule of thumb, find the height of the tip of a 20-foot ladder.

61. Pinpointing a Fire   A U.S. Forest Service helicopter is flying at a height of 500 feet. The pilot spots a fire in the distance with an angle of depression of 12°. Find the horizontal distance to the fire.

62. Wheelchair Ramp   According to the ADA (Americans with Disabilities Act) Accessibility Guidelines, the maximum angle of a wheelchair ramp in new construction may not exceed 4.764°. If the door to a building is 5 feet above sidewalk level, how long must the ramp be in order to extend from sidewalk to door?

63. Computing Area by Measuring One Distance   The length of the diagonal of a rectangular plot of land is 225 feet, and the angle made between the diagonal and one of the sides is 20°. Find the area of the plot.

64. Estimating a Long Jump   Mike Powell, world-record holder (as of 2003) in the long jump at approximately 29 feet 4 inches, comes upon a ravine on a hike. From a point directly opposite a rock on the other side of the ravine, he steps off a distance of 20 feet. From this vantage point, his line of sight to the rock forms an angle of 57° with the edge of the ravine (see Figure 37). Assuming the two sides of the ravine are parallel, can he make the jump to the other side?

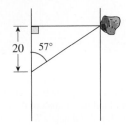

20   57°

Figure 37

Mike Powell breaks a 23-year-old long jump record in 1991.

65. **Estimating Golf Distances** On approximately level ground, the angle of elevation from a point 5 feet off the ground to the top of a 7-foot golf flag is 1.2°. How far away is the hole?

66. **Skiing Distance** From experience, a skier knows that he cannot maintain control on slopes steeper than 30°. If he is at the top of a 5000-foot slope, skis a constant speed of 40 feet per second, and by zigzagging is always able to keep his skis at an angle of 30° with the horizontal, how long will it take him to get to the bottom of the slope?

67. **Volume of a Grain Pile** The base of a conical pile of grain has a circumference of 160 feet, and the angle made between the ground and the side of the pile is 35°. Find the volume of the pile. (*Hint*: The volume of a cone with radius r and height h is $V = \frac{1}{3}\pi r^2 h$.)

68. **Locating a Car** From the 102nd-floor observation deck of the Empire State Building, an observer spots his car at an angle of depression of 5°. If the observation deck is 1250 feet high, how far is his car from the building?

69. **Hiking Course** A hiker leaves a north-south highway, setting out on a constant course of 153° clockwise from north. After 3 miles, he trips over a fallen branch and sprains his ankle. How far is he from the highway, using the shortest possible route?

70. **Windsurfer Course** A windsurfer starts from shore and travels 5 miles on a course of N 42° W. If the shoreline is roughly north-south, how far from shore is she?

71. **Supersonic Travel** Sound travels at approximately 742 miles per hour. A fighter plane is traveling at Mach 2—that is, at twice the speed of sound—with a course of N 35° E. At 7:00 P.M. the pilot is warned that the enemy border, which forms an east-west line, is 100 miles due north. If the plane continues on its current course, at what time will it cross into enemy airspace?

72. **Calculating a Course** A ship leaves port and travels on a constant course and with a constant speed. After 2 hours, the ship is 55 miles south and 20 miles west of the port. What is its course and speed?

73. **Shadow Length Mystery** Inspector Magill is investigating a bank robbery that occurred in New York on June 20, 1997. The prime suspect claims to have been in London on that date and makes a plea on national television for someone to corroborate his story. Several days later, a witness steps forward with a videotape showing the suspect in front of Big Ben in London. The date on the video shows June 20, 1997, and the time, 3:00 P.M., coincides with that shown on Big Ben. After a visit to London, Inspector Magill determines that the suspect's shadow in the video was 8.7 feet long. The suspect is 6 feet tall. If the angle of elevation of the sun in London, England, at 3:00 P.M. on June 20, 1997, was 45° 59', how does Magill determine that the suspect and the witness are lying?

Concepts and Critical Thinking

Exercises 74-77 *Answer true or false.*

74. The equation $\sin^2 x + \cos^2 x = 1$ is true for all acute angles x.

75. The equation $\tan^2 x - \sec^2 x = 1$ is true for all acute angles x.

76. The tangent of an angle is equal to the cotangent of the angle's complement.

77. If both acute angle measures of a right triangle are given, then all side lengths can be found.

Exercises 78-81 *Give an example of each.*

78. An angle θ such that $\cos \theta = \sin \frac{\pi}{7}$

79. A right triangle (with side lengths and angles labeled) that has an angle of $\frac{\pi}{3}$

80. An angle θ such that $\sin\theta < \frac{1}{2}$

81. A possible navigational course for a hot-air balloon flying directly from New York to Africa

82. What can be said about the relative measures of α and β if $\cos\alpha > \cos\beta$? Explain.

83. Explain why $\tan\theta$ can exceed 1 even though neither $\sin\theta$ nor $\cos\theta$ can.

84. An acute angle θ in a right triangle is such that $\sin\theta = a/b$. Explain why this does not necessarily imply that the side opposite θ has length a and the hypotenuse has length b.

Questions for Discussion or Essay

85. Explain why enormous errors would result if right triangle trigonometry were used to compute great distances on the surface of Earth.

86. In Exercise 66, you were asked to compute the time it would take for a skier to get to the bottom of a slope. Explain why the answer doesn't depend on the specific path the skier follows.

87. If the distance to a building is known, what additional information is required to obtain the height?

Projects for Enrichment

88. Measuring Angle of Elevation In this project, we use the following items to construct a device for measuring angles of elevation.

- A long straight stick (a yardstick is ideal)
- A protractor
- A piece of string approximately 1 foot long
- A weight that can be attached to one end of the string
- A tape measure

a. Attach the protractor to the stick so that its base is parallel to the side of the stick. Then attach one end of the string to the "center point" of the protractor, and attach the weight to the other end of the string. See Figure 38.

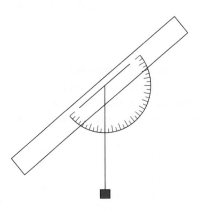

Figure 38

To estimate an object's angle of elevation, line the stick up with the object (at eye level), allow the string to hang freely, as shown in Figure 38, and record the angle indicated by the string on the protractor. The angle of elevation from your eye to the object is the complement of the angle indicated on the protractor.

b. Use your device to estimate the heights of five campus (or community) landmarks. Describe the steps you took, and include all relevant measurements. Don't forget to compensate for the eye-level height.

89. Trigonometry to the Rescue A tourist wakes up in a strange room in a dilapidated building in Washington, D.C. Fearing for his safety, the man is afraid to leave the building and instead uses his cell phone to call his good friend Inspector Magill. Unfortunately, the battery in the phone is almost drained, and so a short call is all that is possible. Thinking quickly, Magill asks the man if any landmarks are visible outside the ground-floor window. The Washington Monument is to the right and the Capitol Building is to the left. Magill then asks his friend to estimate the angles of elevation to the top of each landmark. The man responds with "7° to the top of the Washington Monument" and "5° to the tip of the Capitol Building dome." At this point, the phone connection is broken. Several hours later, the police arrive and rescue the distraught man. In the following steps, we reconstruct the method used by Magill to estimate the location of the building.

a. The Washington Monument is approximately 550 feet high. Approximately how far is the building from the Washington Monument?

b. The tip of the dome of the Capitol Building is approximately 300 feet high. Approximately how far is the building from the Capitol Building?

c. The Washington Monument and the Capitol Building are approximately 7400 feet apart. By setting up a coordinate system with the Capitol Building at the origin, find the approximate location of the building.

d. Assume the tourist's estimates of the angle of elevation were off by as much as 1°. Sketch the region of Washington, D.C., that would have to be searched by the police in this case. (*Hint:* A compass may be useful.)

Section 5.3 | Trigonometric Functions of Real Numbers

- What initial angle and velocity are needed to break the world record in the javelin throw?
- How can the average temperature in Orlando, Florida, and the blood pressure of a dog be modeled with trigonometric functions?
- How can trigonometric functions be approximated using polynomials?
- How can trigonometry be used to estimate the area of the state of Wyoming?

In the last section, we witnessed the application of trigonometry to the determination of unknown distances in a right triangle. Seen through the eyes of a child, this right triangle trigonometry is capable of performing astonishing feats of measurement magic—determining the height of a tree without climbing it or the breadth of a river without crossing it. In each instance, we first identify a right triangle and then express any unknown side length in terms of a trigonometric function of an acute angle. But not all triangles are right, and not all angles are acute. Our goal in this section is to expand the range of applications of trigonometry by extending the definitions of the trigonometric functions to arbitrary angles.

Trigonometric Functions of Any Angle

As a first step toward extending our definition of trigonometric functions, we consider the **unit circle**—that is, a circle of radius 1 centered at the origin. As Figure 39 suggests, to each angle θ there corresponds a point (x, y) on the unit circle. If θ is acute, then we can apply the right triangle definitions of the trigonometric functions to obtain

$$\sin \theta = \frac{\text{opp}}{\text{hyp}} = \frac{y}{1} = y \qquad \cos \theta = \frac{\text{adj}}{\text{hyp}} = \frac{x}{1} = x \qquad \tan \theta = \frac{\text{opp}}{\text{adj}} = \frac{y}{x}$$

$$\csc \theta = \frac{\text{hyp}}{\text{opp}} = \frac{1}{y} \qquad \sec \theta = \frac{\text{hyp}}{\text{adj}} = \frac{1}{x} \qquad \cot \theta = \frac{\text{adj}}{\text{opp}} = \frac{x}{y}$$

Alternatively, we could think of these equations as *defining* the trigonometric functions, so that, for example, $\sin \theta$ is defined to be the y-coordinate of the point on the unit circle corresponding to the angle θ; $\cos\theta$ is defined to be the x-coordinate corresponding to θ, and so forth. The advantage of this alternative definition is that now the trigonometric functions are defined for every angle θ, whether acute or not. The formal definitions of the six trigonometric functions are thus as follows.

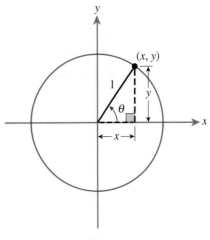

Figure 39

Unit Circle Definition of Trigonometric Functions

Let θ be an angle in standard position, with (x, y) the associated point on the unit circle. We define the six trigonometric functions of θ as follows:

$$\sin\theta = y \qquad\qquad \cos\theta = x \qquad\qquad \tan\theta = \frac{y}{x}, x \neq 0$$

$$\csc\theta = \frac{1}{y}, y \neq 0 \qquad \sec\theta = \frac{1}{x}, x \neq 0 \qquad \cot\theta = \frac{x}{y}, y \neq 0$$

A few important properties of the sine and cosine functions follow immediately from their unit circle definitions.

1. Since $(\cos\theta, \sin\theta)$ is a point on the unit circle, it follows that both $\cos\theta$ and $\sin\theta$ must lie between -1 and 1.

2. Angles that are coterminal naturally correspond to the same point on the unit circle and thus have identical trigonometric function values. In particular, since θ and $\theta + 2\pi$ are coterminal, we conclude that $\sin(\theta + 2\pi) = \sin\theta$ and $\cos(\theta + 2\pi) = \cos\theta$ for all angles θ. This repeating property of trigonometric functions is called *periodicity* and will be explored in greater detail in the next section.

3. The unit circle definition of the trigonometric functions preserves all of the fundamental identities established in the previous section. For example, $\tan\theta = \sin\theta/\cos\theta$ for all angles θ, whether acute or not.

Rule of Thumb

When finding trigonometric function values for an angle θ with associated point (x, y) on the unit circle, it is only necessary to remember that

$$\cos\theta = x \qquad \text{and} \qquad \sin\theta = y$$

The remaining function values can be found using the ratio identities.

·····>**EXAMPLE 1**

Trigonometric Function Values and the Unit Circle

For the given point on the unit circle, find an associated angle θ, and then find the trigonometric function values for that angle.

a. $(0, 1)$ **b.** $\left(-\dfrac{\sqrt{2}}{2}, \dfrac{\sqrt{2}}{2}\right)$

Solution   Because of periodicity, infinitely many angles are associated with a given point on the unit circle. We will restrict our attention to angles in the interval $[0°, 360°)$.

a. The point $(0, 1)$ corresponds to an angle of $90°$ (see Figure 40). Thus, we have

$$\cos 90° = x = 0 \quad \text{and} \quad \sin 90° = y = 1$$

The values of the other trigonometric functions can be computed using ratio identities, as follows:

$$\tan 90° = \frac{\sin 90°}{\cos 90°} = \frac{1}{0} \text{ is undefined} \qquad \cot 90° = \frac{\cos 90°}{\sin 90°} = \frac{0}{1} = 0$$

$$\csc 90° = \frac{1}{\sin 90°} = \frac{1}{1} = 1 \qquad \sec 90° = \frac{1}{\cos 90°} = \frac{1}{0} \text{ is undefined}$$

b. The point $\left(-\sqrt{2}/2, \sqrt{2}/2\right)$ lies in the second quadrant on the line $y = -x$, which makes a $45°$ angle with the negative x-axis. Thus, $\theta = 180° - 45° = 135°$ (see Figure 41). Applying the definitions to determine sine and cosine and the ratio identities to determine the values of the other trigonometric functions, we obtain

$$\sin 135° = y = \frac{\sqrt{2}}{2} \qquad \cos 135° = x = -\frac{\sqrt{2}}{2} \qquad \tan 135° = \frac{\sin 135°}{\cos 135°}$$

$$= \frac{\sqrt{2}/2}{-\sqrt{2}/2} = -1$$

$$\csc 135° = \frac{1}{\sin 135°} \qquad \sec 135° = \frac{1}{\cos 135°} \qquad \cot 135° = \frac{1}{\tan 135°}$$

$$= \frac{1}{\sqrt{2}/2} = \sqrt{2} \qquad = \frac{1}{-\sqrt{2}/2} = -\sqrt{2} \qquad = \frac{1}{-1} = -1$$

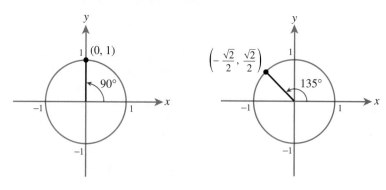

Figure 40   Figure 41

The trigonometric function values for the other quadrant angles can be found in the same way as was done for $90°$ in part a of Example 1. These values are summarized in Table 3.

θ	θ (radians)	$\sin\theta$	$\cos\theta$	$\tan\theta$
$0°$	0	0	1	0
$90°$	$\dfrac{\pi}{2}$	1	0	undefined
$180°$	π	0	-1	0
$270°$	$\dfrac{3\pi}{2}$	-1	0	undefined

Table 3
Quadrant Angles

EXAMPLE 2

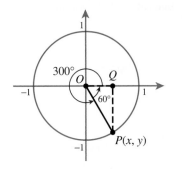

Figure 42

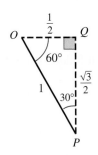

Figure 43

Finding Trigonometric Function Values for a Given Angle

Find the coordinates of the point P on the unit circle associated with $300°$, and determine the sine, cosine, and tangent of $300°$.

Solution   The point P is in the fourth quadrant, as shown in Figure 42. The vertical line passing through P intersects the x-axis at a point we denote by Q. Since $300°$ is $60°$ less than a full revolution, $\angle POQ = 60°$. Moreover, since P lies on the unit circle, the length of the hypotenuse is 1. Thus, since $\sin 60° = \sqrt{3}/2$ and $\cos 60° = 1/2$, the legs of $\triangle OQP$ have length $1/2$ and $\sqrt{3}/2$, as shown in Figure 43. Thus, P has coordinates $\left(1/2, -\sqrt{3}/2\right)$. We can now apply the definitions of sine, cosine, and tangent to obtain

$$\sin 300° = y = -\frac{\sqrt{3}}{2} \qquad \cos 300° = x = \frac{1}{2} \qquad \tan 300° = \frac{y}{x} = \frac{-\sqrt{3}/2}{1/2} = -\sqrt{3}$$

An alternate approach to finding trigonometric function values for angles larger than $90°$ is to use information about the *signs* of the trigonometric function values, together with the notion of a *reference angle*. The following comparison of the values for the sine, cosine, and tangent of $60°$ and $300°$ suggests how this alternate approach works.

$$\sin 60° = \frac{\sqrt{3}}{2} \qquad\qquad \cos 60° = \frac{1}{2} \qquad\qquad \tan 60° = \sqrt{3}$$

$$\sin 300° = -\frac{\sqrt{3}}{2} \qquad\qquad \cos 300° = \frac{1}{2} \qquad\qquad \tan 300° = -\sqrt{3}$$

Notice that trigonometric function values for $300°$ and $60°$ differ by at most a sign ($+$ or $-$). Thus, the trigonometric function values for $300°$ can be found simply by attaching the appropriate signs to the corresponding values for $60°$. We now describe these ideas more fully.

Signs and Reference Angles

The signs of the trigonometric function values for a given angle θ are determined by the quadrant in which θ lies. For example, if θ is in the second quadrant with associated point $P(x, y)$ on the unit circle, then $x < 0$ and $y > 0$. Thus,

$$\sin\theta = y > 0 \qquad \cos\theta = x < 0$$

In a similar way, we obtain the signs shown in Figure 44 for the sine and cosine functions. Notice that in each of quadrants II and IV, exactly one of the two is positive, whereas both are positive in quadrant I and both are negative in quadrant III. It follows that the tangent function is positive in quadrants I and III only. These facts lead us to Figure 45, where we indicate which of the sine, cosine, or tangent are positive in each quadrant. The signs of the other functions can be determined by applying appropriate ratio identities.

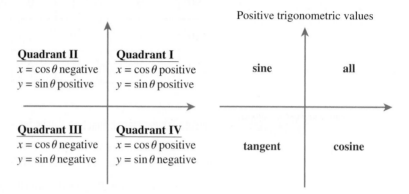

| Figure 44 | Figure 45 |

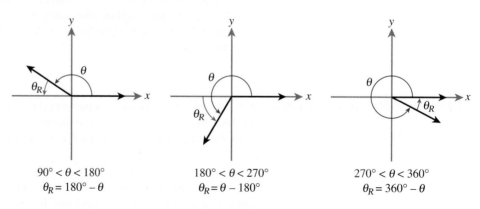

====>**EXAMPLE 3**

Finding the Sine of an Angle

Find $\sin \theta$ given that $\cos \theta = -\frac{1}{4}$ and $\tan \theta > 0$.

Solution We apply the Pythagorean identity $\sin^2 \theta + \cos^2 \theta = 1$ to obtain

$$\sin^2 \theta = 1 - \cos^2 \theta = 1 - \left(-\frac{1}{4}\right)^2 = \frac{15}{16}$$

$$\sin \theta = \pm\sqrt{\frac{15}{16}} = \pm\frac{\sqrt{15}}{4}$$

Since $\tan \theta = \sin \theta / \cos \theta$ is positive and $\cos \theta$ is negative, it follows that $\sin \theta$ is negative. Thus, we choose the negative square root to obtain

$$\sin \theta = -\frac{\sqrt{15}}{4}$$

The **reference angle** for an angle θ in standard position is the acute angle θ_R formed between the terminal side of θ and the horizontal axis. Figure 46 shows several illustrations of reference angles.

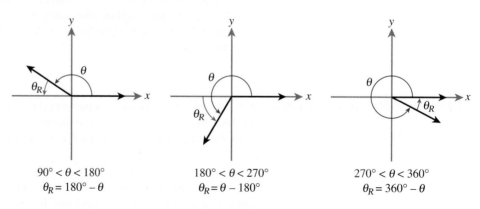

| $90° < \theta < 180°$ | $180° < \theta < 270°$ | $270° < \theta < 360°$ |
| $\theta_R = 180° - \theta$ | $\theta_R = \theta - 180°$ | $\theta_R = 360° - \theta$ |

Figure 46

----- **EXAMPLE 4**

Finding Reference Angles

Find the reference angle θ_R for the given angle θ.

a. $\theta = 240°$ **b.** $\theta = \dfrac{5\pi}{6}$

c. $\theta = -225°$ **d.** $\theta = 5$

Solution

a. We first sketch the angle $\theta = 240°$ in the third quadrant, as shown in Figure 47. Its reference angle is the acute angle made with the negative x-axis. Thus,

$$\theta_R = 240° - 180° = 60°$$

b. The angle $\theta = \dfrac{5\pi}{6}$ is in the second quadrant (Figure 48). Its reference angle is

$$\theta_R = \pi - \frac{5\pi}{6} = \frac{\pi}{6}$$

c. As we see in Figure 49, the angle $\theta = -225°$ is in the second quadrant and is coterminal with $135°$. Thus, its reference angle is

$$\theta_R = 180° - 135° = 45°$$

d. The angle $\theta = 5$ is between $\dfrac{3\pi}{2} \approx 4.71$ and $2\pi \approx 6.28$, and so is in the fourth quadrant (Figure 50). Thus, its reference angle is

$$\theta_R = 2\pi - 5 \approx 1.2832$$

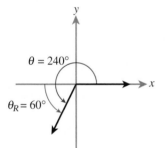

Figure 47

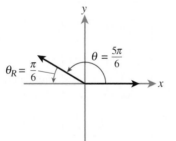

Figure 48

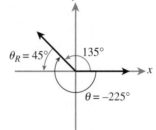

Figure 49

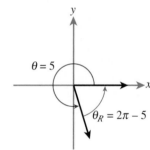

Figure 50

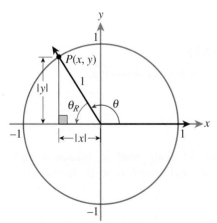

Figure 51

Reference angles can be used to evaluate trigonometric functions for nonacute angles. To see how this is done, consider the nonacute angle θ with associated point $P(x, y)$ on the unit circle shown in Figure 51. From the definitions of sine, cosine, and tangent, we have

$$\sin\theta = y, \quad \cos\theta = x, \quad \text{and } \tan\theta = \frac{y}{x}$$

Now consider the right triangle with acute angle θ_R, the reference angle for θ, as shown in Figure 51. The side opposite θ_R has length $|y|$, the side adjacent to θ_R has length $|x|$, and the hypotenuse has length 1. Thus,

$$\sin\theta_R = |y|, \quad \cos\theta_R = |x|, \quad \text{and } \tan\theta_R = \left|\frac{y}{x}\right|$$

So the sine, cosine, and tangent of θ and θ_R differ by at most a sign, which can easily be determined using quadrant information as described earlier.

We give the following procedure for evaluating trigonometric functions using reference angles.

Evaluating Trigonometric Functions Using Reference Angles

Let θ be an angle in standard position whose terminal side does not lie on a coordinate axis. The value of a trigonometric function of θ can be found as follows:

1. Sketch the angle θ to help determine its quadrant and reference angle θ_R.
2. Find the value of the trigonometric function for the associated reference angle θ_R.
3. Prefix the appropriate sign, according to the quadrant in which θ lies.

We can apply the reference angle technique to find exact trigonometric function values for any angle that has a reference angle of 30°, 45°, or 60°. For reference, Table 4 summarizes information useful for finding exact values of trigonometric functions.

Table 4
Special Angles

θ	θ (radians)	$\sin \theta$	$\cos \theta$	$\tan \theta$
0°	0	0	1	0
30°	$\dfrac{\pi}{6}$	$\dfrac{1}{2}$	$\dfrac{\sqrt{3}}{2}$	$\dfrac{\sqrt{3}}{3}$
45°	$\dfrac{\pi}{4}$	$\dfrac{\sqrt{2}}{2}$	$\dfrac{\sqrt{2}}{2}$	1
60°	$\dfrac{\pi}{3}$	$\dfrac{\sqrt{3}}{2}$	$\dfrac{1}{2}$	$\sqrt{3}$
90°	$\dfrac{\pi}{2}$	1	0	undefined
180°	π	0	-1	0
270°	$\dfrac{3\pi}{2}$	-1	0	undefined
360°	2π	0	1	0

EXAMPLE 5

Using Reference Angles to Evaluate Trigonometric Functions

Use reference angles to find the exact values for each of the following:

a. $\tan 150°$ **b.** $\sin\left(-\dfrac{2\pi}{3}\right)$ **c.** $\sec \dfrac{7\pi}{4}$

Solution

a. The angle $\theta = 150°$ is sketched in Figure 52 along with its reference angle $\theta_R = 30°$. Now $\tan 30° = \sqrt{3}/3$ (see Table 4), and since θ is in the second quadrant, $\tan \theta = \sin \theta / \cos \theta$ is negative. Thus,

$$\tan 150° = -\tan 30° = -\frac{\sqrt{3}}{3}$$

b. The angle $\theta = -\frac{2\pi}{3}$ is sketched in Figure 53 along with its reference angle $\theta_R = \frac{\pi}{3}$. Since θ is in the third quadrant, the sine of θ is negative. Thus,

$$\sin\left(-\frac{2\pi}{3}\right) = -\sin\frac{\pi}{3} = -\frac{\sqrt{3}}{2}$$

c. The angle $\theta = \frac{7\pi}{4}$ is sketched in Figure 54 along with its reference angle $\theta_R = \frac{\pi}{4}$. Since θ is in the fourth quadrant, the secant of θ, which has the same sign as the cosine of θ, is positive. Thus,

$$\sec\frac{7\pi}{4} = \sec\frac{\pi}{4} = \frac{1}{\cos\dfrac{\pi}{4}} = \frac{1}{\sqrt{2}/2} = \sqrt{2}$$

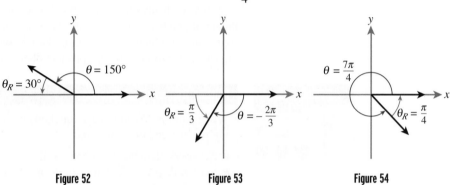

Figure 52   Figure 53   Figure 54

When exact values cannot be found, a calculator may be used to approximate trigonometric function values. As discussed in the previous section, it is important to be sure that the calculator is set in the appropriate angle measurement mode (degrees or radians).

⋯EXAMPLE 6

Approximating Trigonometric Function Values with a Calculator

Approximate the given trigonometric function value.

a. $\cos 200°$ **b.** $\cot(-3)$

Solution

a. With a calculator set in degree mode, we use the [COS] key to obtain

$$\cos 200° \approx -0.939693$$

b. With a calculator set in radian mode, we first use the [TAN] key to obtain

$$\tan(-3) \approx 0.142547$$

Next we use the reciprocal key [x⁻¹] to compute

$$\cot(-3) = \frac{1}{0.142547} \approx 7.015253$$

Trigonometric Functions of Real Numbers

So far we have viewed trigonometric functions as having angular domains (the inputs are angles). But by defining the value of a trigonometric function for a real number t to be the value of that function for an acute angle measuring t radians, we can define trigonometric functions with real number domains. This enables us to apply trigonometry to sound waves, electric current, and many other natural phenomena in which angles do not play a central role.

⋯⋯▷EXAMPLE 7

Evaluating a Trigonometric Function

Given $f(t) = \sin t$, find the indicated function value.

a. $f(0)$ **b.** $f\left(\dfrac{\pi}{6}\right)$ **c.** $f(7.1)$

Solution

a. Using Table 4, we obtain $f(0) = \sin 0 = 0$.

b. Using Table 4, we have $f\left(\dfrac{\pi}{6}\right) = \sin\dfrac{\pi}{6} = \dfrac{1}{2}$.

c. Using a calculator set in radian mode, we obtain $f(7.1) = \sin 7.1 \approx 0.728969$.

Just as with any other real-valued function, the graph of a trigonometric function f is the set of points (x, y) that satisfy the equation $y = f(x)$. We will study the graphs of the trigonometric functions in detail in the following sections; here we simply use a graphing calculator to view the graphs and obtain information about the domain and range.

Calculator Keys

Plotting Trigonometric Functions

When plotting trigonometric functions with a graphing calculator, keep the following recommendations in mind:

- Set your calculator in radian mode.
- Start with your calculator's built-in TRIG range settings (if it has them), and then zoom in or out as necessary.
- Use the SIN, COS, and TAN keys for the sine, cosine, and tangent functions, respectively. For cosecant, secant, and cotangent, use the ratios $1/\sin x$, $1/\cos x$, and $1/\tan x$, respectively.
- It is sometimes helpful to set your calculator in dot mode when plotting the tangent, cotangent, secant, or cosecant functions to avoid possible confusion over vertical asymptotes.

⋯⋯▷EXAMPLE 8

Graphing the Sine Function with a Graphing Calculator

Use a graphing calculator to estimate the domain and range of $f(x) = \sin x$.

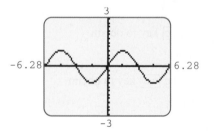

Figure 55

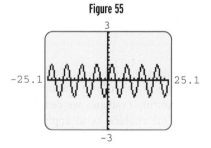

Figure 56

Solution Recall that the domain is the set of inputs of a function, and the range is the set of outputs. Graphically, the domain of f is the set of all x-values for points on the graph, whereas the range is the set of y-values. With a graphing calculator set in radian mode, and using the built-in TRIG window settings, we enter and plot $y = \sin x$ to obtain the graph shown in Figure 55. This view suggests that the y-coordinates of points on the graph are between -1 and 1 and that the function is defined for all x-values in the viewing window. To investigate further, we zoom out horizontally to obtain the graph shown in Figure 56. The wavelike repetition shown in Figure 56 suggests that the graph will continue to oscillate between -1 and 1 indefinitely. (This repeating property of the graph of f is called *periodicity* and will be discussed in greater detail in the next section.) Thus, we conclude that the domain of $f(x) = \sin x$ is the set of all real numbers and the range is $[-1, 1]$.

The results of Example 8 are easily confirmed by returning to the unit circle definition of the sine function. Since $\sin x$ is defined to be the y-coordinate of the point on the unit circle corresponding to a central angle with radian measure x, and since the

y-coordinates of points on the unit circle vary between -1 and 1, the range is indeed $[-1, 1]$. Moreover, sine is defined for all angles, and hence the domain of $f(x) = \sin x$ is indeed the set of all real numbers.

Understanding and Mastery Checklists

Concepts to Understand	Skills to Master
Unit circle	Find an angle that corresponds to a given point on the unit circle.
Unit circle definition of trigonometric functions	Find the point on the unit circle that corresponds to a given angle.
Reference angle	Find the trigonometric function value for an angle, given other trigonometric values.
Signs of trigonometric functions	Find the reference angle for a given angle.
	Use a reference angle to find the exact trigonometric function value of a special angle.
	Use the graph of a trigonometric function to find its domain and range.

Exercises 5.3

Exercises 1-6 *Find the angle θ in $[0, 2\pi)$ associated with the given point, and then find the trigonometric function values for that angle.*

1. $(-1, 0)$

2. $\left(-\dfrac{\sqrt{2}}{2}, -\dfrac{\sqrt{2}}{2}\right)$

3. $\left(\dfrac{1}{2}, -\dfrac{\sqrt{3}}{2}\right)$

4. $\left(\dfrac{\sqrt{3}}{2}, \dfrac{1}{2}\right)$

5. $\left(-\dfrac{\sqrt{2}}{2}, \dfrac{\sqrt{2}}{2}\right)$

6. $(0, -1)$

Exercises 7-10 *Find the coordinates of the point on the unit circle associated with the given angle, and use the definitions of the trigonometric functions to determine their values.*

7. $\theta = -90°$

8. $\theta = 315°$

9. $\theta = \dfrac{5\pi}{4}$

10. $\theta = 5\pi$

Exercises 11-16 *Determine the quadrant in which θ lies.*

11. $\sin\theta < 0, \cos\theta > 0$

12. $\tan\theta > 0, \sin\theta < 0$

13. $\sec\theta < 0, \tan\theta > 0$

14. $\cot\theta < 0, \sin\theta > 0$

15. $\cot\theta < 0, \csc\theta > 0$

16. $\sec\theta < 0, \csc\theta < 0$

Exercises 17-22 *Use the given information about the angle θ to find the indicated function value.*

17. $\sin\theta = \dfrac{1}{3}, \tan\theta < 0; \cos\theta = $ ____

18. $\tan\theta = -\dfrac{1}{2}, \cos\theta > 0; \sec\theta = $ ____

19. $\sec\theta = -3, \sin\theta < 0; \cot\theta = $ ____

20. $\cos \theta = 0.4$, $\cot \theta > 0$; $\csc \theta =$ _____

21. $\tan \theta = 2$, $\sec \theta < 0$; $\cos \theta =$ _____

22. $\csc \theta = -2$, $\tan \theta < 0$; $\cos \theta =$ _____

Exercises 23-30 *Find the reference angle for the given angle.*

23. $\theta = 120°$

24. $\theta = -330°$

25. $\theta = -50°$

26. $\theta = 410°$

27. $\theta = \dfrac{5\pi}{4}$

28. $\theta = \dfrac{11\pi}{6}$

29. $\theta = -2$

30. $\theta = 8$

Exercises 31-40 *Use a reference angle to find the exact value of the given trigonometric function.*

31. $\sin 210°$

32. $\tan(-120°)$

33. $\cos(-30°)$

34. $\csc 135°$

35. $\tan \dfrac{5\pi}{6}$

36. $\sec \dfrac{10\pi}{3}$

37. $\cot\left(-\dfrac{3\pi}{4}\right)$

38. $\sin\left(-\dfrac{\pi}{6}\right)$

39. $\csc(765°)$

40. $\cos(-210°)$

Exercises 41-46 *Use a calculator to find a decimal approximation, correct to four decimal places.*

41. $\sin 110°$

42. $\tan 305°$

43. $\sec 221.4°$

44. $\csc(-143.8°)$

45. $\cot(-2.3)$

46. $\cos 15.2$

Exercises 47-50 *Find the indicated function values. Give exact values whenever possible.*

47. $f(t) = \cos t$

   a. $f(0)$ **b.** $f\left(-\dfrac{\pi}{4}\right)$ **c.** $f(2)$

48. $f(t) = \tan t$

   a. $f\left(\dfrac{\pi}{4}\right)$ **b.** $f\left(\dfrac{\pi}{2}\right)$ **c.** $f(-1)$

49. $f(t) = \csc t$

   a. $f\left(-\dfrac{2\pi}{3}\right)$ **b.** $f(\pi)$ **c.** $f(5.3)$

50. $f(t) = \sec t$

   a. $f\left(\dfrac{7\pi}{6}\right)$ **b.** $f(-5\pi)$ **c.** $f(21.2)$

Exercises 51-54 *Use a graphing calculator to estimate the domain and range.*

51. $f(x) = \cos x$

52. $f(x) = \tan x$

53. $f(x) = \csc x$

54. $f(x) = \sec x$

Applications

55. Javelin Throw If we ignore air resistance, the range in feet of a projectile thrown with an initial speed of v feet per second, an initial angle θ, and from an initial height of h feet is given by

$$R = \frac{v \cos \theta}{32}\left(\sqrt{v^2 \sin^2 \theta + 64h} + v \sin \theta\right)$$

Suppose a javelin thrower releases a javelin with an initial speed of 100 feet per second from an initial height of 6 feet.

a. Find the range of the javelin using initial angles of 0°, 30°, and 90°.

b. Use a graphing calculator to estimate the angle that gives the maximum range. A world-record javelin throw of 323.1 feet was set by Jan Zelezny in 1996. Is this range possible using the given initial velocity?

Jan Zelezny prepares to throw the javelin at the 1995 world championship games in Gothenburg, Sweden.

56. Rolling Wheel   If a bicycle tire were spun in place, then a point on the tire would travel in a circular path around the axle. When the bicycle is moving, however, the path of a point on the bicycle tire is more complicated, involving both rotation around the axis and an overall forward motion. It can be shown that if a bicycle is traveling at a constant speed, then the forward speed of a fixed point on the bicycle tire is given by $f(\theta) = v + v \cos \theta$, where v is the speed of the bicycle and θ is the angle shown in Figure 57.

Figure 57

a. Find the forward speed of a point on the tire of a bicycle traveling at a rate of 20 miles per hour for $\theta = \frac{\pi}{4}$, $\theta = \frac{\pi}{3}$, and $\theta = \frac{\pi}{2}$.

b. What is the maximum forward speed of the point and when is it attained? What about the minimum speed? Assume a rate of 20 miles per hour.

57. Average Temperature   The average monthly temperature (in degrees Fahrenheit) for Orlando, Florida, can be approximated using the function

$$f(t) = 73 - 5.5 \cos\left(\frac{\pi t}{6}\right) - 13 \sin\left(\frac{\pi t}{6}\right)$$

where $t = 1$ corresponds to January 2003, $t = 2$ corresponds to February 2003, and so on.

a. Compute $f(3)$ and $f(15)$ and interpret your result.

b. Use a graphing calculator to estimate the month when the average temperature is highest and the month when it is lowest.

58. Canine Blood Pressure   The aortic blood pressure (measured in mm Hg) of a dog can be approximated using the function
$$f(t) = -6.3 \cos(42t) + 5 \sin(42t)$$
$$- 7.8 \cos(21t) + 16.2 \sin(21t) + 115$$
where t is measured in seconds. [*Data source:* Bjorn Folkow and Eric Neil, *Circulation* (New York: Oxford, 1971), p. 428.]

a. Compute $f(0)$ and $f\left(\frac{\pi}{21}\right)$ and interpret your result.

b. Use a graphing calculator to estimate the highest and lowest blood pressures.

Concepts and Critical Thinking

Exercises 59-62   *Answer true or false.*

59. An angle whose sine and cosine are both negative is in the third quadrant.

60. Infinitely many angles can have the same sine.

61. The tangent function is undefined for all integer multiples of π.

62. The reference angle of an angle θ in the second quadrant is given by $180° - \theta$.

Exercises 63-66   *Give an example of each.*

63. A nonacute angle with reference angle $10°$

64. Two angles in the interval $[0, 2\pi)$ that have the same tangent

65. An angle whose secant is undefined

66. An angle for which the sine and tangent functions have the same value

67. Use the definition of the secant function to help you explain why $\sec \frac{\pi}{2}$ is undefined.

68. If α and β are two angles for which corresponding values of the trigonometric functions agree (that is, $\sin \alpha = \sin \beta$, $\cos \alpha = \cos \beta$, and so on), does it follow that $\alpha = \beta$? Explain.

69. If α and β are two angles that have the same reference angle, is it true that $\sin \alpha = \sin \beta$? Conversely, if $\sin \alpha = \sin \beta$, does it follow that α and β have the same reference angle? Explain.

70. Recall that a function f is 1–1 if, whenever $f(a) = f(b)$, it follows that $a = b$. Equivalently, a function f is 1–1 if its graph passes the horizontal line test. Use your graphing calculator to determine whether any of the trigonometric functions are 1–1. Summarize your findings.

Questions for Discussion or Essay

71. Explain why it might be that trigonometric functions are sometimes called "circular functions."

72. In some texts, the trigonometric functions are defined using a circle with an arbitrary radius r instead of the unit circle. What modifications would need to be made in our definitions of the trigonometric functions in order to follow this approach? Of the two approaches— using the unit circle or a circle of radius r—which do you prefer and why?

Projects for Enrichment

73. Taylor Polynomials   The sine and cosine functions can be approximated with surprising accuracy using **Taylor polynomials**. In order to define these polynomials, we must first consider the notion of a **factorial**. We define $0! = 1$ and, for natural numbers $n \geq 1$, $n! = n(n-1)(n-2) \cdots (2)(1)$. Thus, for example, $4! = 4 \cdot 3 \cdot 2 \cdot 1 = 24$.

a. Compute the following factorials:

   i. 3!   **ii.** 5!

   iii. 6!

The Taylor polynomials for the sine function are defined as follows:

$$S_0(x) = x$$

$$S_1(x) = x - \frac{x^3}{3!}$$

$$S_2(x) = x - \frac{x^3}{3!} + \frac{x^5}{5!}$$

$$\vdots$$

$$S_n(x) = x - \frac{x^3}{3!} + \frac{x^5}{5!} - \cdots + (-1)^n \frac{x^{2n+1}}{(2n+1)!}$$

b. Write out the Taylor polynomials S_3 and S_4. Describe in words any patterns you see in the polynomials.

c. With your calculator set in radian mode, compute $\sin 1$, $S_0(1)$, $S_1(1)$, $S_2(1)$, $S_3(1)$, and $S_4(1)$. Note that you can save yourself a lot of work by using the fact that

$$S_1(1) = S_0(1) - \frac{1}{3!}, \quad S_2(1) = S_1(1) + \frac{1}{5!}$$

and so on. In other words, each successive polynomial value can be found by adding a positive or negative term to the previous polynomial value. What do you notice about the values of $S_n(1)$ as n gets larger?

d. Complete the following table.

x	$\sin x$	$S_0(x)$	$S_1(x)$	$S_2(x)$	$S_3(x)$	$S_4(x)$
0.1						
0.5						
2						
3						

What does this table suggest about using the Taylor polynomials $S_n(x)$ to approximate values of $\sin x$? Be specific about how the sizes of n and x affect the approximation.

e. Use your graphing calculator to plot $y = \sin x$, $y = S_0(x)$, $y = S_1(x)$, and $y = S_2(x)$ on the same set of axes. Use an x-axis range of $-2 \leq x \leq 2$ and a y-axis range of $-1 \leq y \leq 1$. How do these graphs confirm what you found in part d?

The Taylor polynomials for the cosine function are defined as follows:

$$C_0(x) = 1$$

$$C_1(x) = 1 - \frac{x^2}{2!}$$

$$C_2(x) = 1 - \frac{x^2}{2!} + \frac{x^4}{4!}$$

$$\vdots$$

$$C_n(x) = 1 - \frac{x^2}{2!} + \frac{x^4}{4!} - \cdots + (-1)^n \frac{x^{2n}}{(2n)!}$$

f. Repeat parts b through e for the Taylor polynomials $C_n(x)$.

74. Areas of Latitude/Longitude Zones   We define a **latitude zone** as a region bounded on the north and south by two parallels of latitude. Similarly, we define a **longitude zone** as a region bounded on the east and west by two meridians of longitude. A **latitude/longitude zone** is a "rectangular" region bounded on four sides by parallels of latitude and meridians of longitude. (See Figure 58.) In this project, we develop techniques for finding the area of such zones.

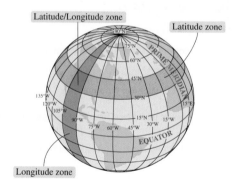

Figure 58

The area of a longitude zone bounded by longitudes with a degree difference θ is given by

(1) $$A = \frac{\theta}{360} S$$

where S is the surface area of Earth.

a. Explain why equation (1) gives the area of a longitude zone. Does the equation still hold for a zone bounded by longitudes on opposite sides of the prime meridian? Explain.

b. Find the surface area of Earth, assuming it is a sphere with radius 3963 miles.

c. Find the area of the zone bounded by the given longitudes.

   i. 45° W, 50° W

   ii. 173° W, 160° E

d. The boundaries of time zones are often quite jagged, but they roughly coincide with meridians of longitude. Estimate the area of Earth for which the time at a given instant is 3:00 P.M. Why do you think the boundaries of time zones don't coincide exactly with meridians of longitude?

To find the area of a latitude zone, we first convert each latitude to an angle between 0° and 180°. Given a latitude m, we define the **location angle** ϕ as follows:

$$\phi = \begin{cases} 90° - m & \text{for north latitude} \\ 90° + m & \text{for south latitude} \end{cases}$$

Now, for two latitudes with corresponding location angles ϕ_1 and ϕ_2, the area of the latitude zone is given by

$$A = \frac{1}{2}|\cos \phi_1 - \cos \phi_2|S$$

e. Find the area of the zone bounded by the given latitudes, and determine the percentage of Earth's surface area covered by the zone.

   i. The North Pole (90° N) and the Arctic Circle (66° 30′ N).

   ii. The Tropic of Cancer (23° 27′ N) and the Tropic of Capricorn (23° 27′ S).

The area of a latitude/longitude zone bounded by latitudes with location angles ϕ_1 and ϕ_2 and longitudes having degree difference θ is given by

(2) $$A = \frac{\theta}{720}|\cos \phi_1 - \cos \phi_2|S$$

f. Explain how equation (2) is obtained.

g. Estimate the area of the state of Colorado, which is bounded by longitudes 102° W and 109° W and latitudes 37° N and 41° N.

h. Estimate the area of the state of Wyoming, which is bounded by longitudes 104° W and 111° W and latitudes 41° N and 45° N.

i. Using parts g and h as a guide, estimate the area of the state of Utah. An atlas is a good reference for the boundaries.

j. Compute the areas of the following "1 degree square zones." Explain why they are not the same.

   i. 85° N, 86° N, 10° W, 11° W

   ii. 45° N, 46° N, 10° W, 11° W

   iii. 5° N, 6° N, 10° W, 11° W

Section 5.4   Graphs of Sine and Cosine

- How can an astronaut be weighed in a weightless environment?
- Why do soldiers march out of step whenever they cross a bridge?
- What connection exists between electrical current, music, and the ratios of the sides of a right triangle?
- Which body is most responsible for tides: the sun or the moon?

We have seen two distinct yet compatible points of view regarding the trigonometric functions—sine and cosine in particular. Initially, we defined them using the right triangle approach as functions of acute angles, and, in this capacity, they are powerful tools for the solution of geometric problems arising in surveying, navigation, and the physics of force and motion. By extending to arbitrary angles with the unit circle definition, we recast sine and cosine as functions of real numbers. But until now, we have only hinted at the riches *this* approach will yield. The sine and cosine functions are the

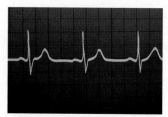

An EKG displays periodicity.

Waves emerge in the desert sand.

basis for mathematical models in disciplines as diverse as cardiology and chemistry, television transmission and tide prediction, musical synthesis and magnetic resonance. But why? What key property—what salient feature—is shared by ocean tides, television, EKGs, and also sine and cosine? The answer, in a word, is *periodicity*. In each instance, some aspect is periodic or wavelike: Ocean tides rise, fall, and rise again; televised images ride electromagnetic waves through space; and EKGs trace out the rhythmic pulsing of a beating heart.

Now imagine a ray rotating counterclockwise, sweeping out an angle t that grows larger and larger as the point $(\cos t, \sin t)$ traces and retraces the unit circle. Wherever that point is now, there it will return after each revolution. This suggests (and soon we will prove) that the graphs of both cosine and sine are likewise periodic, repeating themselves forever, one cycle after another, on and on indefinitely. But sine and cosine are more than mere examples of periodic functions; they are the fundamental building blocks from which nearly all periodic functions are constructed. Because of this, we explore the graphs of sine and cosine and the concept of periodicity in great detail in this section.

Graphs of $f(t) = \sin t$ and $g(t) = \cos t$

The graph of the sine function can be plotted easily with a graphing calculator, as we did in Section 5.3. But to deepen our understanding of the sine function and develop our graphical intuition, in this section we produce the graph of the sine function by hand, guided only by the unit circle definition and our knowledge of special angles. Recall from Section 5.3 that $\sin t$ is the y-coordinate of the point on the unit circle corresponding to the angle t, as shown in Figure 59. (See Exercise 79 for a discussion of our rationale for naming the independent variable t instead of x.)

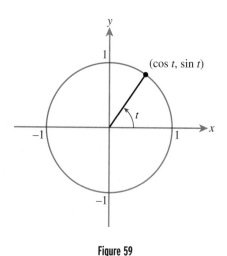

Figure 59

Thus, the graph of $f(t) = \sin t$ for t between 0 and 2π reflects the behavior of the y-coordinate of a point P on the unit circle as P rotates counterclockwise once around the circle. Beginning with $t = 0$, the y-coordinate of P first rises to 1 $\left(\text{at } t = \frac{\pi}{2}\right)$, then falls to 0 (at $t = \pi$). Continuing around the circle, y continues to fall, bottoming out at -1 $\left(t = \frac{3\pi}{2}\right)$, before rising back to 0 ($t = 2\pi$). Our circuit around the unit circle and its implications for the graph of the sine function are summarized in Table 5.

Table 5

Interval for t	Unit circle	Behavior of y-coordinate of P and graph of $f(t) = \sin t$
$\left[0, \dfrac{\pi}{2}\right]$		Increasing (from 0 to 1)
$\left[\dfrac{\pi}{2}, \pi\right]$		Decreasing (from 1 to 0)
$\left[\pi, \dfrac{3\pi}{2}\right]$		Decreasing (from 0 to -1)
$\left[\dfrac{3\pi}{2}, 2\pi\right]$		Increasing (from -1 to 0)

To pin down the shape of the graph more precisely, we determine a few additional points on the graph of $f(t) = \sin t$. In Table 6, we have chosen some angles t at which the sine function is easily computed.

Table 6

t	0	$\dfrac{\pi}{4}$	$\dfrac{\pi}{2}$	$\dfrac{3\pi}{4}$	π	$\dfrac{5\pi}{4}$	$\dfrac{3\pi}{2}$	$\dfrac{7\pi}{4}$	2π
$y = \sin t$	0	$\dfrac{\sqrt{2}}{2}$	1	$\dfrac{\sqrt{2}}{2}$	0	$-\dfrac{\sqrt{2}}{2}$	-1	$-\dfrac{\sqrt{2}}{2}$	0

By plotting these points and applying the information about where the graph increases and decreases, which is displayed in Table 5, we obtain the graph shown in Figure 60.

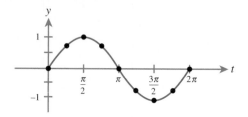

Figure 60

Now, to complete the graph, we note that the angles t and $t + 2\pi$ are coterminal; that is, the point on the unit circle corresponding to the angle t is the same as the point on the unit circle corresponding to the angle $t + 2\pi$, as shown in Figure 61.

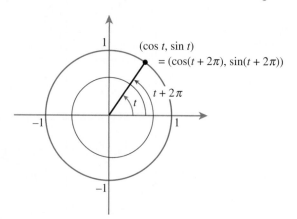

Figure 61

Thus, it follows that $\sin(t + 2\pi) = \sin t$. More generally, all angles of the form $t + 2\pi k$, where k is an integer, are coterminal with the angle t. Consequently, the graph of the sine function repeats itself every 2π units, as suggested by Figure 62.

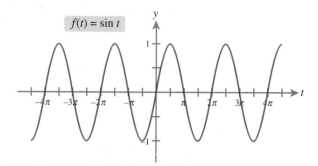

Figure 62

A similar process yields the graph of the cosine function shown in Figure 63. Notice that the graph of cosine appears to be a translation of the graph of sine, a phenomenon that we investigate in Exercise 76.

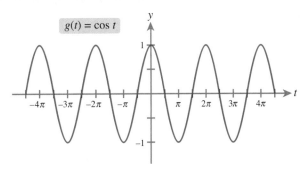

$g(t) = \cos t$

Figure 63

The graphs of both sine and cosine exhibit familiar forms of symmetry. In the case of the cosine function, the graph appears to be symmetric with respect to the *y*-axis. Recall from Section 2.2 that such a function *g* is said to be *even* and satisfies $g(-t) = g(t)$ for all *t* in its domain. Thus, we speculate that $\cos(-t) = \cos t$. Similarly, the graph of the sine function appears to be symmetric with respect to the origin, which suggests that $f(t) = \sin t$ is an *odd* function, so that $\sin(-t) = -\sin t$. To verify these observations, consider the angle *t* shown in Figure 64, with associated point (a, b) on the unit circle. The angle $-t$ has associated point $(a, -b)$. Thus, by the definitions of sine and cosine we have

$$\sin(-t) = -b = -\sin t \quad \text{and} \quad \cos(-t) = a = \cos t$$

The even and odd properties of the sine and cosine functions are summarized as follows.

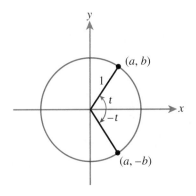

Figure 64

Even/Odd Identities

$$\sin(-t) = -\sin t \qquad \cos(-t) = \cos t$$

Periodicity

Functions like sine and cosine that have repeating graphs are said to be **periodic**. As we have already noted, it is this property that makes sine and cosine invaluable for modeling so many real-world phenomena. The following definitions are associated with periodic functions.

Periodic Functions

A function *f* is said to be **periodic** if there is a number *p* such that $f(x + p) = f(x)$ for all *x* in the domain of *f*.

The smallest such number *p* is called the **period** of *f*.

The **amplitude** of a periodic function is one-half the "height" of the graph. That is, the amplitude can be obtained by subtracting the minimum value of the function from the maximum value, and then dividing by 2:

$$\text{Amplitude} = \frac{\text{Max value} - \text{Min value}}{2}$$

A **cycle** is the graph of a single period of a periodic function.

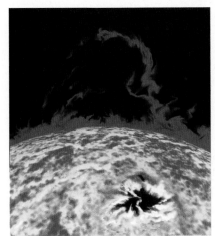

Sun spot activity varies according to an 11-year cycle.

According to the definitions just given, the functions $f(x) = \sin x$ and $g(x) = \cos x$ are periodic functions with period 2π and amplitude 1. It will prove useful to remember the appearance of one cycle of both the sine and cosine functions. Figure 65 shows the graphs of sine and cosine over the interval $[0, 2\pi]$.

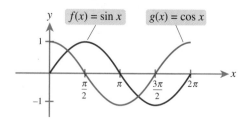

Figure 65

EXAMPLE 1

Determining the Period and Amplitude Graphically

Find the period and the amplitude of the function whose graph is shown in Figure 66.

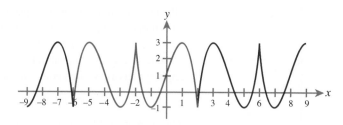

Figure 66

Solution From Figure 66, we see that it takes 8 units for the graph to begin repeating itself. Thus, the period is 8. (A single period of the function is shown in red.) Since the maximum value attained by the graph is 3 and the minimum value is -1, the amplitude is given by

$$\frac{3 - (-1)}{2} = 2$$

A classic example of a periodic phenomenon is the up-and-down motion of an oscillating spring. This is an example of *simple harmonic motion*.

EXAMPLE 2

Harmonic Motion of a Weighted Spring

Consider an object at rest, suspended by a spring, as shown in Figure 67. After pulling the object and releasing it, the object will oscillate up and down in a periodic fashion. If all forces except those exerted by the spring and gravity are ignored, then the position of the spring after t seconds is given by

$$x(t) = x_0 \cos\left(\sqrt{k/m}\ t\right) + v_0 \sqrt{m/k}\ \sin\left(\sqrt{k/m}\ t\right)$$

where $x(t)$ gives the directed distance of the object below the equilibrium (resting) position, and the other constants are defined as follows.

Constant	Definition
x_0	The initial directed distance (in meters) of the object from its rest position
k	The spring constant—a measure of the stiffness of the spring (in Newtons per meter)
m	The mass of the object in kilograms
v_0	The initial velocity of the object in meters per second

Figure 67

Find the period and amplitude of the motion resulting from releasing a 20-kilogram object from a point 3 meters below the equilibrium position with an initial velocity of 2 meters per second. Assume that the spring constant is 3.2 Newtons per meter.

Solution   From the given information, we determine that $x_0 = 3$, $k = 3.2$, $m = 20$, and $v_0 = 2$. Thus, the motion of the object is given by

$$x(t) = 3 \cos\left(\sqrt{3.2/20}\ t\right) + 2\sqrt{20/3.2}\ \sin\left(\sqrt{3.2/20}\ t\right)$$
$$= 3 \cos(0.4t) + 5 \sin(0.4t)$$

Using our graphing calculator, we obtain the graph shown in Figure 68.

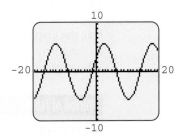

Figure 68

To find the period, we must first identify a cycle. One convenient cycle is the portion of the graph between the first two peaks on the right side of the vertical axis. Using the calculate-maximum feature, we see that the maximum value of 5.83 occurs at $t = 2.58$ and $t = 18.28$. Thus, the period is given by

$$18.28 - 2.58 = 15.7$$

The calculate-minimum feature gives us a minimum value of -5.83, and thus the amplitude is

$$\frac{\text{Max value} - \text{Min value}}{2} = \frac{5.83 - (-5.83)}{2}$$
$$= 5.83$$

In the previous examples, we relied on the graphs of the trigonometric functions to estimate the period and amplitude. We now develop algebraic techniques for computing the period and amplitude.

Stretching and Shrinking: Functions of the Form $f(x) = a \sin bx$ (or $a \cos bx$)

We begin our investigation of functions of the form $f(x) = a \sin bx$ (or $a \cos bx$) by plotting the graphs of $g(x) = \sin x$ and $h(x) = \sin 2x$, as shown in Figures 69 and 70.

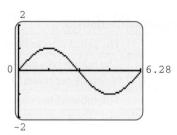

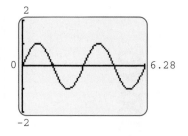

Figure 69   **Figure 70**

We have already seen that $g(x) = \sin x$ has period 2π, and this is evident in the graph in Figure 69. From the graph in Figure 70, we estimate that the period of $h(x) = \sin 2x$ is π. Indeed,

$$h(x + \pi) = \sin[2(x + \pi)]$$
$$= \sin(2x + 2\pi)$$
$$= \sin 2x$$
$$= h(x)$$

Thus, the period of $h(x) = \sin 2x$ is exactly one-half the period of $g(x) = \sin x$; in other words, the graph of h completes two cycles for every cycle of the graph of g. In general, the graph of $f(x) = \sin bx$ completes a cycle $|b|$ times as fast as $g(x) = \sin x$ and so has a period of $2\pi/|b|$. A similar result holds for the cosine function. These observations generalize to the following property of periodic functions.

Period of $f(bx)$

If $f(x)$ is a periodic function with period p, then $g(x) = f(bx)$ is a periodic function with period $\frac{p}{|b|}$.

 EXAMPLE 3

Finding a Period

Find the period of $f(x) = \cos \frac{x}{2}$.

Solution   Since $g(x) = \cos x$ has period 2π and $f(x) = \cos bx$ with $b = \frac{1}{2}$, the period of f is given by

$$\frac{2\pi}{|b|} = \frac{2\pi}{1/2} = 4\pi$$

One cycle of f is shown in Figure 71.

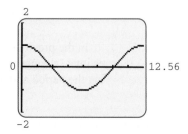

Figure 71

Next, we consider the effect of multiplying the output of the sine function by a number $a > 0$; that is, we consider the function $f(x) = a \sin x$ for $a > 0$. Essentially, because all the y-values of $y = \sin x$ have been multiplied by a, the graph of $f(x) = a \sin x$ is a times "taller" than the graph of $g(x) = \sin x$, as suggested by Figure 72. In other words, the amplitude of $f(x) = a \sin x$ is a times that of $g(x) = \sin x$. Thus, since $g(x) = \sin x$ has amplitude 1, $f(x) = a \sin x$ has amplitude a.

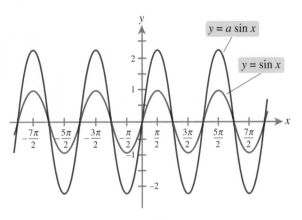

Figure 72

Finally, we consider the case that the number a in the expression $f(x) = a \sin x$ is negative. Since multiplying the output of a function by -1 has the effect of reflecting the graph of the function about the x-axis, we can obtain the graph of $f(x) = a \sin x$ for $a < 0$ by graphing $y = |a| \sin bx$ and then reflecting about the x-axis. A similar result holds for functions of the form $g(x) = a \cos x$ where a is negative. We summarize our results as follows.

Graphs of Functions of the Form $f(x) = a \sin bx$ (or $a \cos bx$)

If $f(x) = a \sin bx$ (or $a \cos bx$), then the following hold:

1. The period of f is given by $\frac{2\pi}{|b|}$.
2. The amplitude of f is given by $|a|$.
3. If $a < 0$, then the graph of f is the reflection of the graph of $y = |a| \sin bx$ (or $y = |a| \cos bx$) about the x-axis.

EXAMPLE 4

Finding Period and Amplitude

Find the period and amplitude of the function $f(x) = -2 \sin 3x$. Then sketch several cycles of the graph.

Solution   Here $a = -2$ and $b = 3$. Thus, the amplitude is given by $|-2| = 2$, and the period is given by

$$\frac{2\pi}{|b|} = \frac{2\pi}{3}$$

We begin by sketching one cycle of the sine function and then labeling the coordinate axes to reflect the fact that the period is $\frac{2\pi}{3}$ and the amplitude is 2, as shown in Figure 73.

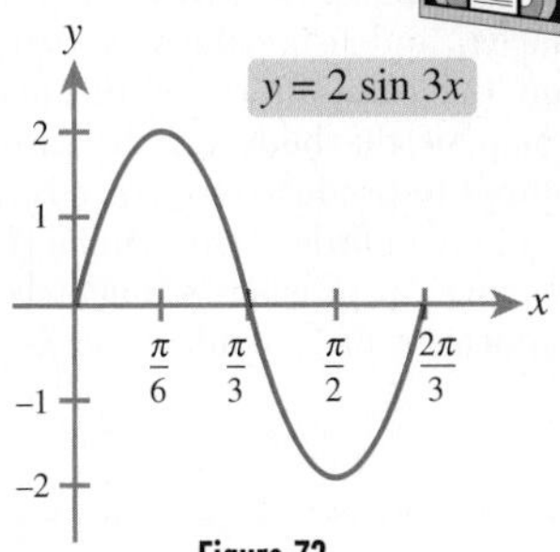

$y = 2 \sin 3x$

Figure 73

Next we repeat this basic shape, as shown by the red dashed graph in Figure 74. Finally, since the coefficient a is negative, we reflect this graph about the x-axis to obtain the graph shown in blue in Figure 74.

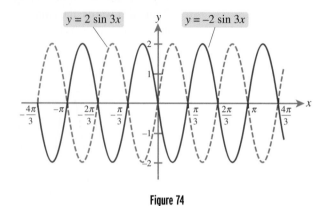

Figure 74

This graph can then be confirmed by plotting with a graphing calculator.

Resonance

We often hear political pundits speak of an idea (or movement) "resonating" with voters, by which they mean that the idea reinforces beliefs or attitudes already "in the air," so that the idea, once articulated, gathers strength as it echoes throughout the electorate. This notion of resonance as the effect created when a natural tendency is encouraged or reinforced is consistent with the scientific usage of the word.

Resonance occurs whenever an entity with a natural frequency is driven by a periodic "force" with the same frequency. For example, there is a natural frequency to the motion of a child on a swing; this is the frequency at which the child would swing if you pulled her back and released her. If you now provide her with periodic pushes, then you will, of course, affect her motion in some way. But if you push her at her natural or *resonant* frequency, then with very little effort on your part, her swings will have ever greater amplitude.

This 3D computer rendering is based in part on MRI data.

This phenomenon accounts for the shattering of crystalware occasionally performed by accomplished singers. If you strike a crystal wine glass, it will vibrate with a certain frequency that your ears will detect as sound of a given pitch. This pitch is the natural frequency of the glass. Now, sound of any frequency reaching the glass will impart a tendency for the glass to vibrate sympathetically, but sound vibrating at the *natural* frequency of the glass will, in effect, *encourage* the glass to vibrate at precisely the frequency at which the glass *likes* to vibrate. The effect can be extraordinary. The amplitude of the vibration of the glass becomes greater and greater until the glass shatters—no longer able to respond elastically to the strain imparted by the vibrations.

Legend has it that resonance is responsible for the 1831 collapse of the Broughton Suspension Bridge near Manchester, England. A column of soldiers marched across the bridge, and, unfortunately, their cadence matched the resonant frequency of the bridge, causing it to swing wildly and finally fall. It is for this reason that soldiers break step whenever they cross a footbridge. Although it is now viewed as unlikely, it had long been thought that the Tacoma Narrows Bridge at Puget Sound toppled due to vortices created by a gale wind that drove the bridge at a frequency matching the resonant frequency of the bridge. In 1959 (and again in 1960), an airplane resonated right out of the sky when an engine vibrated at the resonant frequency of one of the wings.

But the phenomenon of resonance is not entirely malevolent; in fact, it is responsible for one of the most potent medical diagnostic tools—magnetic resonance imaging (MRI). In essence, a subject is immersed in magnetic fields in such a way as to produce resonance in the atomic nuclei of the body. When resonance occurs, the nuclei release energy in the form of radio waves. By tracing these radio waves to their source nuclei, and using the fact that the resonant frequency depends on the composition of the nuclei, a detailed three-dimensional map of the body can be formed. Because the magnetic field required to produce resonance is relatively weak, an MRI procedure results in little of the cellular damage that arises from X rays. Moreover, it provides a much clearer view of the soft tissues, which appear as mere shadows on X rays.

Translations   Recall from Chapter 2 that the graph of $y = f(x - h) + k$ is a translation h units horizontally and k vertically of the graph of $y = f(x)$. Thus, the graph of $y = a \sin[b(x - h)] + k$ is a translation, h units horizontally and k units vertically, of the graph of $f(x) = a \sin bx$. A similar observation is true for the graph of $y = a \cos[b(x - h)] + k$.

⋯⋯⋗**EXAMPLE 5**   **Graphing a Translation of the Cosine Function**

Graph the function $g(x) = \cos\left(x - \frac{\pi}{2}\right) + 2$.

Solution   The graph of g is a translation, $\frac{\pi}{2}$ units to the right and 2 units up, of the graph of $f(x) = \cos x$. In Figure 75, we have graphed f (shown in red) and then shifted the graph of f $\frac{\pi}{2}$ units to the right and 2 units up to obtain the graph of g (shown in blue).

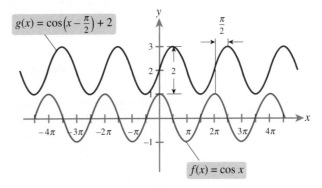

Figure 75

⋯⋯⋗**EXAMPLE 6**   **Plotting the Graph of a Function of the Form $f(x) = a \sin[b(x - h)] + k$**

Let $f(x) = 4 \sin\left(\frac{x}{2} + \frac{\pi}{2}\right) - 3$.

  a. Find the period and the amplitude of f and sketch three cycles of its graph.
  b. Find an interval of length 2π on which f is increasing and an interval of length 2π on which it is decreasing.
  c. Find the maximum and minimum values of f.

Solution

  a. We begin by rewriting the function so that it can be recognized as a translation of a function of the form $g(x) = a \sin bx$.

$$f(x) = 4 \sin\left(\frac{x}{2} + \frac{\pi}{2}\right) - 3$$

$$= 4 \sin\left[\frac{1}{2}(x + \pi)\right] - 3$$

In this form, it is clear that f is merely a translation, π units to the left and 3 units down, of the graph of $g(x) = 4 \sin(x/2)$, which has amplitude $|4| = 4$ and period

$$\frac{2\pi}{1/2} = 4\pi$$

Thus, we graph $y = 4 \sin (x/2)$ (shown in red in Figure 76) and then translate to obtain the graph of $y = f(x)$ (shown in blue). The graph of f can be confirmed using a graphing calculator, as shown in Figure 77.

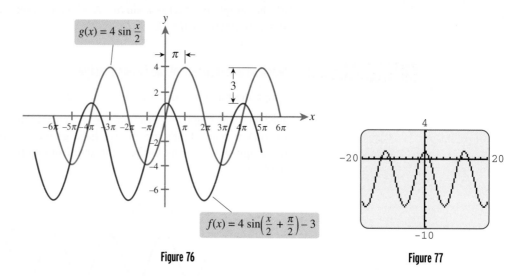

Figure 76

Figure 77

b. From the graph, it is evident that f is rising, and hence increasing, on intervals such as $(-6\pi, -4\pi)$, $(-2\pi, 0)$, and $(2\pi, 4\pi)$. Similarly, we see that it is falling, and hence decreasing, on intervals such as $(-4\pi, -2\pi)$ and $(0, 2\pi)$.

c. From the graph (or considerations of amplitude), we see that the maximum value of f is 1 and the minimum is -7.

The **phase shift** of a function of the form $f(x) = a \sin[b(x - h)] + k$ is defined to be the number h. In other words, the phase shift indicates the horizontal translation required to obtain the graph of f from the graph of $g(x) = a \sin bx$. Thus, for example, the function $f(x) = 4 \sin\left[\frac{1}{2}(x + \pi)\right] - 3$ from Example 6 has a phase shift of $-\pi$. A similar definition holds for functions of the form $f(x) = a \cos[b(x - h)] + k$. We summarize the notions of period, amplitude, and phase shift as follows.

Period, Amplitude, and Phase Shift

Let $f(x) = a \sin[b(x - h)] + k$ or $f(x) = a \cos[b(x - h)] + k$. Then the period, amplitude, and phase shift of f are as follows:

Period	$\dfrac{2\pi}{|b|}$
Amplitude	$|a|$
Phase shift	h

Tides, the periodic rise and fall of coastal waters, arise from a gravitational tug-of-war between Earth and the moon and sun. The moon and the sun attempt to pull the water toward them, which Earth steadfastly resists. In spite of the sun's great mass (about 27 million times that of the moon), its pull on Earth is dwarfed by that of the smaller, but far closer moon. In fact, a rough model of tidal variation can be obtained by considering only lunar forces.

EXAMPLE 7   ### Tidal Variation at Key West

If we assume that the moon is the only body exerting gravitational force on Earth and that it travels in a perfect circle around Earth's equator, then based on data collected at Station 216 in Key West, Florida, from 1985 to 1997, we would expect the depth of the water in centimeters at this station in 2004 to be given by

$$M(t) = 17.549 \cos(0.506t + 1.176) + 165.926$$

where t is the time in hours, with $t = 0$ corresponding to 12:00 A.M. (midnight) January 1, 2003. (*Data source:* Web site of the British Oceanographic Data Centre.)

a. Find the period, amplitude, and phase shift of the function M.
b. Determine the average water level at Station 216, and then find the maximum and minimum heights.
c. Estimate the number of tide cycles that occur in a day. (The actual length of 1 day is 23.934 hours.)
d. Estimate the first time on New Year's Day 2003 that the water level at Station 216 reached 175 centimeters.

Solution

a. Rewriting in standard form, we have

$$M(t) = 17.549 \cos(0.506t + 1.176) + 165.926$$
$$= 17.549 \cos[0.506(t + 2.324)] + 165.926$$

Here $a = 17.549$, $b = 0.506$, and $h = -2.324$. Thus, the amplitude is 17.549, the period is $2\pi/0.506 \approx 12.417$, and the phase shift is -2.324.
b. This function will oscillate above and below a baseline level of 165.926 centimeters. Since the amplitude is 17.549, we conclude that the average level is 165.926 centimeters, the maximum is $165.926 + 17.549 = 183.475$ centimeters, and the minimum is $165.926 - 17.549 = 148.377$ centimeters.
c. Since each cycle requires a period of 12.417 hours and a day has 23.934 hours, there are roughly $23.934/12.417 = 1.928$ cycles per day.
d. The graphs of $y = M(t)$ and $y = 175$ are shown in Figure 78. Using the calculate-intersect feature, we find that $M(t) = 175$ at $t \approx 8.06$, which corresponds to 8:04 A.M. on January 1, 2003.

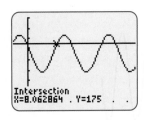

Figure 78

Understanding and Mastery Checklists

Concepts to Understand	Skills to Master

Concepts to Understand

Functions of the form $f(x) = a \sin(bx + c)$

⟡

Functions of the form $f(x) = a \cos(bx + c)$

⟡

Periodic function

⟡

Period

⟡

Amplitude

⟡

Cycle

⟡

Phase shift

Skills to Master

Sketch the graph of $f(x) = a \sin(bx + c)$.

⟡

Sketch the graph of $f(x) = a \cos(bx + c)$.

⟡

Find the period of a function involving sine or cosine.

⟡

Find the amplitude of a function involving sine or cosine.

⟡

Identify the period and amplitude from a graph.

Exercises 5.4

Exercises 1-8 *Find the period and the amplitude of the function whose graph is given.*

1.

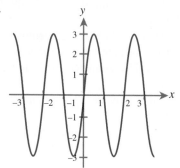

2.

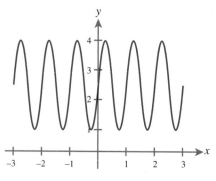

3.

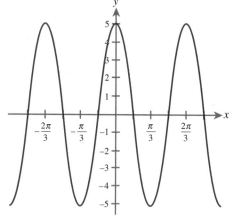

4.

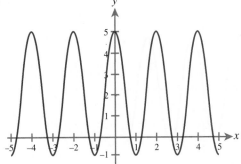

5.

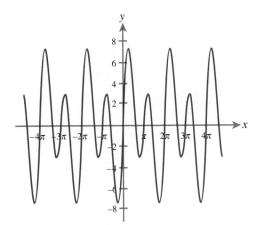

6.

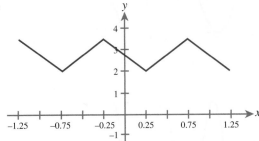

7.

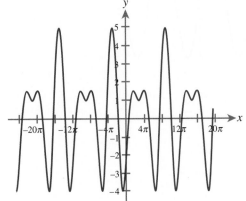

8.

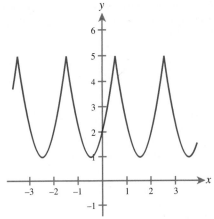

Exercises 9-14   *Find the period and the amplitude of the given function. Use a graphing calculator as necessary.*

9. $f(x) = 5 \sin 4x$

10. $g(x) = 3 \cos\left(2x - \dfrac{\pi}{2}\right)$

11. $h(x) = \sin 3x - 2 \cos x$

12. $f(x) = \cos 4x + 2 \cos x + 3$

13. $g(x) = 2^{\sin 2x} - 3^{\cos x}$

14. $h(x) = 1 + 2 \sin^2(3\pi x)$

Exercises 15-22   *Graph at least three cycles of the given function. Indicate the amplitude, period, and phase shift. Give exact values and verify your results with a graphing calculator.*

15. $h(x) = \cos 2x$

16. $g(t) = \sin \pi t$

17. $f(y) = 2 \sin 2\pi y$

18. $f(x) = -3 \cos 4x$

19. $p(t) = 3 \cos(2t + 1)$

20. $h(\theta) = 2 \sin\left(\pi\theta - \dfrac{\pi}{2}\right)$

21. $g(x) = -4 \sin\left(3x - \dfrac{\pi}{4}\right) + 4$

22. $f(t) = \dfrac{3}{2} \cos\left(\dfrac{1}{2}t + \pi\right) - 1$

Exercises 23-30 *Find the amplitude, period, phase shift, and maximum and minimum values of the function. Give exact values and verify your results with a graphing calculator.*

23. $f(x) = 3 \sin \pi x$

24. $g(x) = 4 \cos 2\pi x$

25. $p(t) = -4 \sin(2t + 1)$

26. $q(r) = 5 \sin(r - 2)$

27. $h(t) = 6 \sin\left(2t + \dfrac{\pi}{2}\right)$

28. $f(x) = -3 \sin(2x - \pi)$

29. $g(y) = 5 - 2 \cos(3y - 2)$

30. $r(t) = 5 \sin\left(2t + \dfrac{\pi}{4}\right) + 3$

Exercises 31-36 *Indicate whether the function is increasing or decreasing on the entire interval or changes direction at a point within the interval. Use a graphing calculator as necessary.*

31. $f(x) = \sin 2x;\ \left(\dfrac{\pi}{4}, \dfrac{3\pi}{4}\right)$

32. $g(x) = \cos 3x;\ \left(\pi, \dfrac{4\pi}{3}\right)$

33. $h(x) = 2 \cos(\pi x - \pi);\ \left(-\dfrac{1}{10}, \dfrac{1}{10}\right)$

34. $f(x) = -3 \sin\left(x - \dfrac{\pi}{4}\right) + 1;\ (0, 1)$

35. $g(x) = e^{\sin x};\ \left(0, \dfrac{\pi}{2}\right)$

36. $h(x) = 2 \cos^3 x - 4 \cos x;\ (3.1, 3.2)$

Exercises 37-42 *Each of the following graphs is that of a function of the form $f(x) = a \sin(bx + c)$. Find a, b, and c. Check your answer using a graphing calculator.*

37.

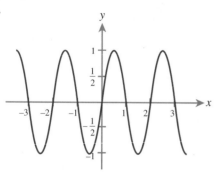

38.

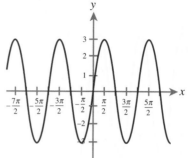

39.

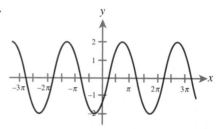

40.

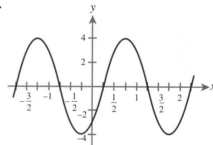

41.

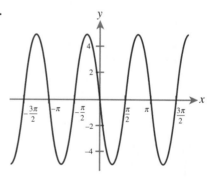

42.

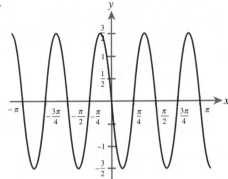

Exercises 43-48 *Use a graphing calculator to determine which of the following are identities—that is, which are true for all values of x.*

43. $\sin(-x) = \sin x$

44. $\cos(-x) = \cos x$

45. $\sin 3x - 3 \sin x = 4 \sin^3 x$

46. $\cos 4x = 4 \cos x$

47. $\sin \dfrac{x}{2} \cos \dfrac{x}{2} - \dfrac{\sin x}{2} = 0$

48. $\sin 2x = 2 \sin x \cos x$

Exercises 49-52 *Find a function of the form $f(x) = a \sin(bx + c)$ that satisfies the given properties. Check your answer using a graphing calculator.*

49. Amplitude 3, period π, and phase shift $\dfrac{\pi}{3}$

50. Amplitude 2, period 2, and phase shift 1

51. Amplitude $\sqrt{2}$, period $\dfrac{\pi}{4}$, and phase shift $\dfrac{\pi}{8}$

52. Amplitude 6, period 3, and phase shift $-\dfrac{\pi}{4}$

Exercises 53-56 *Use a graphing calculator to estimate the two smallest positive zeros of the given function.*

53. $f(x) = 4 \sin(3x + 1) + 2$

54. $g(x) = 1 - 3 \cos(2x - 5)$

55. $h(x) = \dfrac{4}{5} \cos 20x - \dfrac{1}{2}$

56. $f(x) = 0.1 - 0.2 \sin 4\pi x$

Applications

57. Musical Sound Wave A tuning fork that produces a C on the musical scale creates a wave on an oscilloscope that can be modeled by the function $f(x) = 0.002 \sin(528\pi x)$. Find the amplitude and period of this wave.

58. Electromagnetic Wave A certain electromagnetic wave can be modeled by the function $g(t) = \sin[(2 \times 10^5)\pi t]$. Find the amplitude and period of this wave.

59. Spring Motion When an object hung from a certain stretched spring (see Figure 79) is released, its displacement in centimeters from the equilibrium (resting) position t seconds after release is approximated by

$$y(t) = 5 \sin\left(4\pi t - \frac{\pi}{2}\right)$$

  a. Find the period of y.
  b. Find the maximum displacement of the object.

60. Sunset Times Sunset times in Indianapolis, Indiana, can be roughly approximated using the function

$$f(t) = 89 \sin\left(\frac{2\pi}{365}t + \frac{3\pi}{2}\right) + 409$$

where t is the day of the year and $f(t)$ is the corresponding sunset time, in minutes, after 12:00 P.M. (noon).

Figure 79

  a. Find the period and amplitude of f.

  b. Find the earliest and latest sunset times, and estimate the days on which they occur.

61. AC Voltage The voltage of a 110-volt outlet actually varies according to the function

$$V(t) = 110\sqrt{2} \sin 120\pi t$$

where t is in seconds.

a. Find the amplitude and period.

b. What is the first positive time for which the voltage is exactly 110 volts?

62. Bobbing Buoy A buoy is released at the surface of still water and begins to bob up and down. For a short period of time, its motion can be approximated with the function

$$d(t) = 10 - 10 \cos\left(\sqrt{980}\, t\right)$$

where $d(t)$ is the depth of the bottom surface of the buoy t seconds after it is released.

a. Find the period of d.

b. Find the maximum depth of the bottom surface of the buoy.

63. Ferris Wheel Motion The height in feet of a certain passenger on a Ferris wheel is given by

$$y(t) = 55 + 50 \sin\left(\frac{\pi t}{15} - \frac{\pi}{2}\right)$$

where t is the time in seconds, and $t = 0$ coincides with the time at which the wheel was set in motion.

a. Find the initial height of the passenger.

b. Find the maximum and minimum heights of the passenger and the first times at which those heights are attained.

c. How long does it take for the Ferris wheel to make one complete revolution?

64. Weighing an Astronaut The mass of an astronaut in a "weightless" environment can be determined by setting the astronaut in motion on an oscillating machine, measuring the period of the motion, and calculating the mass from the period. Suppose the astronaut's displacement in centimeters is given by the function

$$x(t) = a \cos\left(\sqrt{\frac{260}{m}}\, t\right)$$

where m is the mass of the astronaut in kilograms and t is in seconds. If the period is measured to be 0.32 second, find the mass of the astronaut. Is this result reasonable? (*Hint*: 1 kilogram = 2.2 pounds.)

65. Body Temperature Body temperature, like blood pressure and hormone levels, does not remain constant, but rather fluctuates according to the body's *circadian* rhythm. For a healthy subject with a typical sleep pattern, the temperature in degrees Fahrenheit t hours after midnight is approximated by

$$f(t) = 0.8 \cos(0.2618t + 1.5708) + 98.4$$

Find and interpret the period of f, and determine the maximum and minimum body temperatures and the times of day at which they occur.

66. Key West Tides Revisited In Example 7, we gave a model of water levels at the Key West station based on the assumptions that the moon traveled in a perfect circle in the plane of the equator and that all other factors could be ignored. An improvement to this model can be made by including a rough approximation of the sun's influence, obtained by assuming that Earth travels in a perfect circle around the sun in the plane of equator. With this refinement, the water level (in centimeters) at the Key West station is modeled by

$$M(t) = 17.549 \cos(0.506t + 1.176) +$$
$$5.144 \cos(0.524t + 1.578) + 165.926$$

where t is in hours ($t = 0$ corresponds to midnight, January 1, 2003). Plot $M(t)$ with a viewing window large enough to exhibit the behavior of the function over the course of 1000 hours, and determine the period of the Key West tides and the maximum and minimum water levels.

Concepts and Critical Thinking

Exercises 67–70 *Answer true or false.*

67. The graph of the cosine function is a translation of the graph of the sine function.

68. If $f(x) = f(x + 2\pi)$, then it follows that f has period 2π.

69. The amplitude of a function of the form $f(x) = a \sin(bx + c)$ equals the maximum value of $f(x)$.

70. The phase shift of a function of the form $f(x) = a \sin(bx + c)$ corresponds to the smallest positive zero of f.

Exercises 71–74 *Give an example of each.*

71. A function of the form $f(x) = a \sin(bx + c)$ whose period is the same as its amplitude

72. The graph of a function with y-intercept 2 and period 1

73. A function of the form $f(x) = a \sin(bx + c)$ that has π as one of its zeros

74. A function of the form $f(x) = a \sin(bx + c)$ with range $[-2, 2]$

75. Explain why the function $\sin\left(2x - \frac{\pi}{3}\right)$ does not have a phase shift of $\frac{\pi}{3}$.

76. Show that $\sin\left(x + \frac{\pi}{2}\right) = \cos x$. [*Hint*: First write $\sin\left(x + \frac{\pi}{2}\right)$ as $\sin\left(\frac{\pi}{2} - (-x)\right)$. Then apply a cofunction identity and use the fact that the cosine function is even.]

Questions for Discussion or Essay

77. Which of the following phenomena would you expect to be periodic? What would you expect the period to be? Explain.

  a. Lunar stages

  b. National debt

  c. Retail sales

  d. Number of people attending a house of worship on a given day

Give some additional examples—ones that were not considered in class or in the text—of periodic phenomena.

78. What makes for a good clock? Write a short paper describing devices that have been used throughout the ages to tell time. What characteristic do all of these clocks have in common?

79. When developing the graph of the sine function, we deliberately chose to name the independent variable t instead of x. The reason for this lies in the connection we wished to establish between the unit circle definitions of sine and cosine and their graphs. What confusion might have arisen had we used x to represent the angle instead of t?

Projects for Enrichment

80. Modeling Average Monthly Precipitation
  I. Consider the function $f(t) = A + B \cos(Ct - \theta)$, where A, B, and C are positive real numbers.

  a. Find the maximum value of f.

  b. Find the minimum value of f.

  c. Find the period of f.

  II. The average monthly precipitation for Duluth, Minnesota, is given in Figure 80.

  a. Explain why a function of the form $f(t) = A + B \cos(Ct - \theta)$ could be considered a candidate for modeling these data.

  b. Use the results of part I to find values for A, B, and C.

  c. Find θ so that the precipitation predicted by the model for each month differs from the actual rainfall by no more than 0.5 inch.

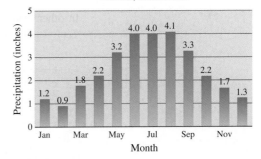

Monthly Precipitation
Duluth, Minnesota

Figure 80

Section 5.5 | ## Graphs of Other Trigonometric Functions

- ⚬ If a ball is thrown so that it lands 100 feet away, what is the length of its journey?
- ⚬ How can it be that the period of the tangent function is π when $\tan x = \sin x / \cos x$ and yet both sine and cosine have periods of 2π?
- ⚬ If the angle of elevation from a viewer to an object approaches $\frac{\pi}{2}$, what can be said about the height of the object?
- ⚬ At what angle should a 6-foot-tall outfielder release a baseball if she wishes it to travel as far as possible?

As we learned in the previous section, the graphs of sine and cosine are of particular importance since countless real-world phenomena can be modeled by functions having sinusoidal (sinelike) graphs. By contrast, the graphs of tangent, cotangent, secant, and cosecant are far less prevalent in mathematical models. In large part, this is because their graphs have **singularities**—points at which the graph approaches a vertical asymptote. It follows that these four trigonometric functions are ill-suited for modeling periodic phenomena for which there are no abrupt breaks, or "explosions." Because of their somewhat limited applicability, we will not explore the graphs of these functions in as much detail as we did those of sine and cosine. Instead, we focus on the application of general graphing principles and the location of asymptotes to generate their graphs.

Graphs of Tangent and Cotangent

Let us first consider the graph of $f(x) = \tan x$. In Table 7, we have computed values of $y = \tan x$ for selected values of x between 0 and $\frac{\pi}{2}$. Note that for values of x near (but less than) $\frac{\pi}{2}$, $\sin x$ is nearly 1 and $\cos x$ is a very small positive number. Consequently, for such x we have

$$\tan x = \frac{\sin x}{\cos x} \approx \frac{1}{\text{Very small positive number}} = \text{Very large number}$$

In other words, as x increases toward $\frac{\pi}{2}$, $\tan x$ approaches infinity. Thus, the graph of $f(x) = \tan x$ has a vertical asymptote at $x = \frac{\pi}{2}$.

Table 7

x	$\tan x = \dfrac{\sin x}{\cos x}$
0	0
$\dfrac{\pi}{4}$	1
$\dfrac{\pi}{3}$	$\sqrt{3} \approx 1.73$
1.5	≈ 14.10
1.56	≈ 92.62
1.57	≈ 1255.77
1.5707	$\approx 10{,}381.33$
$\dfrac{\pi}{2}(\approx 1.5708)$	undefined

Moreover, since the sine function is odd and the cosine function is even, we have

$$\tan(-x) = \frac{\sin(-x)}{\cos(-x)} = \frac{-\sin x}{\cos x} = -\tan x$$

Thus, the tangent function is odd, and hence its graph is symmetric with respect to the origin. In Figure 81, we have sketched a graph of $y = \tan x$ for x between 0 and $\frac{\pi}{2}$ (shown in red) and then reflected this portion of the graph about the origin to obtain the portion of the graph of the function between $-\frac{\pi}{2}$ and 0 (shown in blue).

Next, we consider the period of the tangent function. From Figure 82, we see that $\sin(x + \pi) = -\sin x$, and $\cos(x + \pi) = -\cos x$.

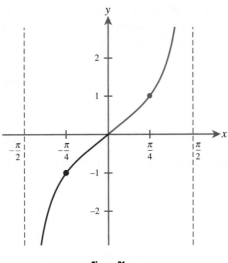

Figure 81

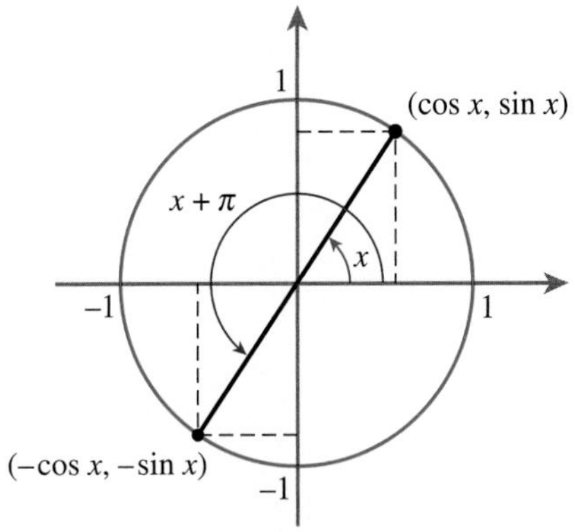

Figure 82

Thus, we have

$$\tan(x + \pi) = \frac{\sin(x + \pi)}{\cos(x + \pi)}$$

$$= \frac{-\sin x}{-\cos x}$$

$$= \frac{\sin x}{\cos x}$$

$$= \tan x$$

From this and the graph shown in Figure 81 (where we see that the period cannot be less than π), we conclude that the tangent function has period π. Figure 83 shows several cycles of the graph of $y = \tan x$.

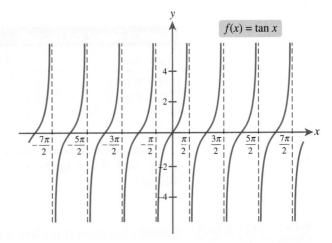

Figure 83

The tangent function is undefined (and its graph has a vertical asymptote) wherever $\cos x = 0$. From Figure 83, we see that this happens whenever x is of the form $\frac{\pi}{2} + \pi k$, where k is an integer. Similarly, since $\cot x = \cos x / \sin x$, the graph of $y = \cot x$ has vertical asymptotes wherever $\sin x = 0$—namely, at integer multiples of π. The period of the cotangent function is also π since

$$\cot(x + \pi) = \frac{1}{\tan(x + \pi)}$$

$$= \frac{1}{\tan x}$$

$$= \cot x$$

A graph showing several cycles of $f(x) = \cot x$ is depicted in Figure 84.

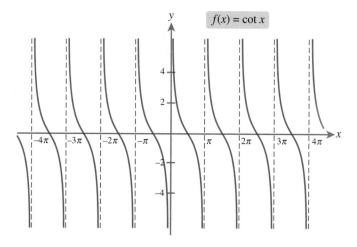

Figure 84

We summarize the most important aspects of the graphs of the tangent and cotangent functions as follows.

Graphs of Tangent and Cotangent		
Function	**Vertical asymptotes**	**Period**
$f(x) = \tan x$	$x = \frac{\pi}{2} + \pi k$, where k is an integer	π
$f(x) = \cot x$	$x = \pi k$, where k is an integer	π

We can use techniques similar to those of the previous section to graph functions of the form $f(x) = \tan(bx + c)$ or $g(x) = \cot(bx + c)$, as the following examples illustrate.

EXAMPLE 1

Graphing a Translation of the Tangent Function

Sketch a graph of the function $f(x) = \tan\left(x - \frac{\pi}{4}\right)$

Solution The graph of f is just a translation $\frac{\pi}{4}$ units to the right of the graph of $y = \tan x$. Thus, the asymptotes are at $\left(\frac{\pi}{2} + \pi k\right) + \frac{\pi}{4} = \frac{3\pi}{4} + \pi k$. The graph is shown in Figure 85.

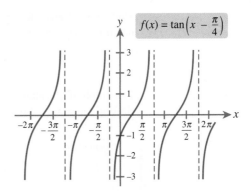

Figure 85

····◈**EXAMPLE 2**

Graphing a Function of the Form $f(x) = \tan bx$

Sketch a graph of the function $f(x) = \tan 2x$.

Solution   Recall from Section 5.4 that replacing x by bx in the formula for a periodic function has the effect of dividing the period by a factor of $|b|$. Since the period of the tangent function is π, the period of $f(x) = \tan 2x$ is then $\frac{\pi}{2}$. Thus, the graph of $f(x) = \tan 2x$ can be obtained from the graph of $y = \tan x$ by compressing by a factor of 2 in the horizontal direction. Hence, the vertical asymptotes occur not at multiples of $x = \frac{\pi}{2} + \pi k$, as with $y = \tan x$, but rather at $x = \frac{1}{2}\left(\frac{\pi}{2} + \pi k\right) = \frac{\pi}{4} + \frac{\pi}{2}k$, as shown in Figure 86.

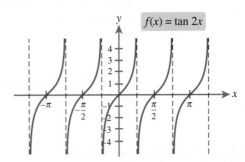

Figure 86

From the preceding example, we surmise that any function of the form $g(x) = \tan bx$ has period $\pi/|b|$. Moreover, the graph of a function of the form $f(x) = \tan[b(x - h)]$ is merely a translation, h units to the right, of the graph of $g(x) = \tan bx$. Similar results hold for functions of the form $f(x) = \cot[b(x - h)]$.

····◈**EXAMPLE 3**

Graphing a Function of the Form $f(x) = \tan(bx + c)$

Sketch a graph of $f(x) = \tan\left(\frac{x}{2} - \frac{\pi}{4}\right)$.

Solution   We have

$$f(x) = \tan\left(\frac{x}{2} - \frac{\pi}{4}\right)$$

$$= \tan\left[\frac{1}{2}\left(x - \frac{\pi}{2}\right)\right]$$

Thus, f is a translation, $\frac{\pi}{2}$ units to the right, of the graph of $g(x) = \tan(x/2)$. Now the function g has period

$$\frac{\pi}{|b|} = \frac{\pi}{1/2} = 2\pi$$

The graph of g is shown in Figure 87. The graph of f shown in Figure 88 is obtained by shifting the graph of g to the right $\frac{\pi}{2}$ units.

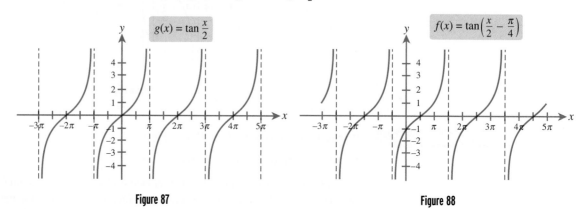

Figure 87   Figure 88

Notice that in each of the previous examples one cycle of the (translated) tangent function occurs between each consecutive pair of vertical asymptotes. This suggests another method for quickly graphing functions of the form $f(x) = \tan(bx + c)$. We find the period as before and then, since $\tan \frac{\pi}{2}$ is undefined, find an asymptote by setting

$$bx + c = \frac{\pi}{2}$$

Additional vertical asymptotes can be located using the period. Finally, the graph can be completed simply by sketching cycles of the tangent function between consecutive asymptotes.

EXAMPLE 4

Finding Asymptotes

Find the asymptotes of $f(x) = \tan(\pi x + \pi)$ and sketch three cycles of the graph. Confirm with a graphing calculator.

Solution   Using the technique just suggested, we locate one asymptote as follows:

$$\pi x + \pi = \frac{\pi}{2}$$

$$\pi x = -\frac{\pi}{2}$$

$$x = -\frac{1}{2}$$

Now the period of f is given by

$$\frac{\pi}{|b|} = \frac{\pi}{\pi} = 1$$

Thus, the asymptotes are spaced 1 unit apart, at $x = -\frac{1}{2} + k$ for any integer k. To sketch the graph, we locate asymptotes at $x = -1.5$, $x = -0.5$, $x = 0.5$, and $x = 1.5$

and plot one cycle of the tangent function between each pair of asymptotes (see Figure 89). Notice that each cycle has an x-intercept exactly halfway between adjacent asymptotes. Using a graphing calculator in dot mode, we obtain the graph shown in Figure 90.

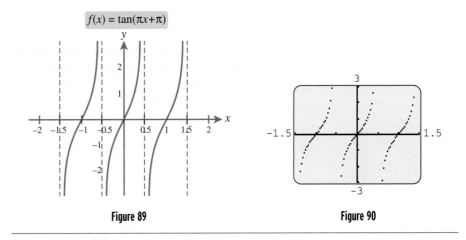

| Figure 89 | Figure 90 |

Graphs of Secant and Cosecant

Since $\sec x = 1/\cos x$, the graph of $f(x) = \sec x$, like that of $g(x) = \tan x$, has vertical asymptotes wherever $\cos x = 0$—that is, at numbers of the form $\frac{\pi}{2} + \pi k$, where k is an integer. Moreover, since $|\cos x| \leq 1$, it follows that $|\sec x| = |1/\cos x| \geq 1$. In other words, $\sec x$ is either greater than or equal to 1, or less than or equal to -1. The period of the secant function is the same as that of the cosine function—namely, 2π. In fact, we have

$$\sec(x + 2\pi) = \frac{1}{\cos(x + 2\pi)}$$

$$= \frac{1}{\cos x}$$

$$= \sec x$$

Similar arguments hold for the cosecant function; the graphs of $f(x) = \sec x$ and $g(x) = \csc x$ are shown in Figures 91 and 92.

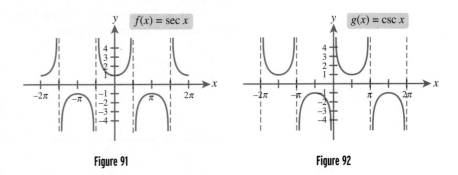

| Figure 91 | Figure 92 |

Since both the secant function and the cosecant function have periods of 2π, any function of the form $f(x) = \sec bx$ or $g(x) = \csc bx$ has period $2\pi/|b|$.

·····⫶EXAMPLE 5

Graphing a Function of the Form $f(x) = \csc bx$

Sketch a graph of the function $f(x) = \csc \dfrac{x}{2}$.

Solution   The period of f is given by

$$\frac{2\pi}{1/2} = 4\pi$$

and so the graph of f will repeat "half as fast" as the graph of $y = \csc x$. Thus, whereas the graph of $y = \csc x$ has vertical asymptotes at multiples of π, the graph of $f(x) = \csc(x/2)$ will have vertical asymptotes at multiples of 2π. The graph of f is shown in Figure 93.

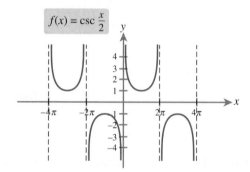

Figure 93

We could obtain the graphs of functions of the form $g(x) = \csc(bx + c)$ and $f(x) = \sec(bx + c)$ by first determining the period and then applying a translation. Alternatively, we can take advantage of the familiarity we already have with the sine and cosine functions. Suppose, for example, that the graph shown in Figure 94 is that of $h(x) = \sin(bx + c)$ for some b and c. Since $\csc(bx + c) = 1/\sin(bx + c)$, the graph of $f(x) = \csc(bx + c)$ has an asymptote wherever $\sin(bx + c) = 0$—that is, wherever the graph of h crosses the x-axis. Moreover, because cosecant is the reciprocal of sine, the graph of f peaks wherever the graph of h bottoms out, and the graph of f bottoms out wherever the graph of h peaks. Finally, if x is such that $\sin(bx + c) = 1$, then $\csc(bx + c) = 1$ also. Similarly, if $\sin(bx + c) = -1$, then $\csc(bx + c) = -1$. Thus, the graphs of f and h intersect wherever $\sin(bx + c) = \pm 1$—that is, wherever the graph of h peaks or bottoms out. Putting these facts together, we obtain the graph of $f(x) = \csc(bx + c)$ shown in Figure 95.

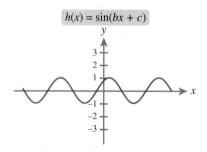

Figure 94

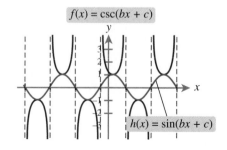

Figure 95

EXAMPLE 6

Graphing a Function of the Form $f(x) = \sec(bx + c)$

Sketch a graph of the function $f(x) = \sec\left(3x + \frac{\pi}{2}\right)$. Confirm with a graphing calculator.

Solution We begin by considering the function $h(x) = \cos\left(3x + \frac{\pi}{2}\right)$. Rewriting, we obtain

$$h(x) = \cos\left(3x + \frac{\pi}{2}\right)$$

$$= \cos\left[3\left(x + \frac{\pi}{6}\right)\right]$$

Thus, the graph of h is a translation, $\frac{\pi}{6}$ units to the left, of the graph of $g(x) = \cos 3x$, which has a period of $\frac{2\pi}{3}$. The graphs of both g and h are shown in Figure 96. Next, we isolate the graph of h and draw asymptotes through the x-intercepts, as shown in Figure 97. Finally, using the graph of h and the asymptotes as guides, we sketch the graph of f, as shown in Figure 98. A graphing calculator in dot mode produces the graph shown in Figure 99.

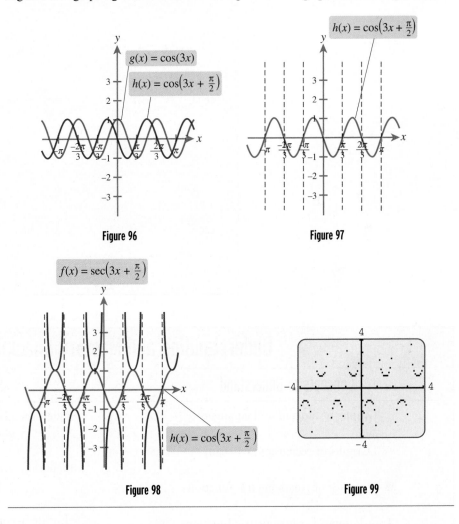

Figure 96

Figure 97

Figure 98

Figure 99

For convenience, we provide the following summary of the graphs of tangent, cotangent, secant, and cosecant.

Graphs of Tangent, Cotangent, Secant, and Cosecant

Function	Period	Vertical asymptotes	Graph of a single cycle
$\tan x$	π	$x = \dfrac{\pi}{2} + \pi k$	
$\cot x$	π	$x = \pi k$	
$\sec x$	2π	$x = \dfrac{\pi}{2} + \pi k$	
$\csc x$	2π	$x = \pi k$	

Understanding and Mastery Checklists

Concepts to Understand

Graphs of tangent and cotangent

Graphs of secant and cosecant

Asymptotes of trigonometric functions

Translations of trigonometric functions

Skills to Master

Sketch the graph of $f(x) = \tan(bx + c)$ or $g(x) = \cot(bx + c)$.

Sketch the graph of $f(x) = \sec(bx + c)$ or $g(x) = \csc(bx + c)$.

Locate asymptotes and use as an aid in graphing functions involving tangent, cotangent, secant, and cosecant.

Exercises 5.5

Exercises 1-8 *Match the given function with its graph.*

1. $\tan 2x$

2. $\cot \pi x$

3. $\sec \dfrac{\pi x}{2}$

4. $\tan\left(2x + \dfrac{\pi}{2}\right)$

5. $\cot\left(\pi x - \dfrac{\pi}{2}\right)$

6. $\sec\left(x + \dfrac{\pi}{3}\right)$

7. $\tan(2\pi x - \pi)$

8. $\csc 3x$

vii. **viii.**

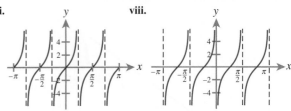

Exercises 9-22 *Graph at least three cycles of the given function. Indicate the asymptotes and determine the period.*

9. $f(x) = \tan 3x$

10. $f(x) = \tan \pi x$

11. $f(x) = \cot \dfrac{\pi x}{2}$

12. $f(x) = \cot 2x$

13. $f(x) = \tan\left(x - \dfrac{\pi}{2}\right)$

14. $f(x) = \tan\left(2x + \dfrac{\pi}{2}\right)$

15. $f(x) = \cot\left(\pi x + \dfrac{\pi}{4}\right)$

16. $f(x) = \cot\left(\dfrac{x}{2} - \dfrac{\pi}{6}\right)$

17. $f(x) = \sec 4x$

18. $f(x) = \sec\left(x - \dfrac{\pi}{2}\right)$

19. $f(x) = \sec(\pi x + \pi)$

20. $f(x) = \csc 3x$

21. $f(x) = \csc\left(2x - \dfrac{\pi}{2}\right)$

22. $f(x) = \sec\left(\dfrac{x}{4} - \dfrac{\pi}{8}\right)$

Exercises 23-26 *Use a graphing calculator to estimate the smallest positive zero of the given function.*

23. $f(x) = \tan x + \tan 2x$

24. $f(x) = \tan x - \sec x$

25. $f(x) = 1 - \tan x - \cot x$

26. $f(x) = \sec x - \sec 2x$

i.

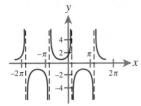

ii.

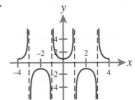

iii.

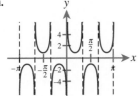

iv.

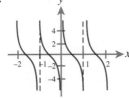

v.

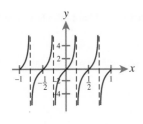

vi.

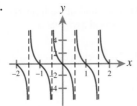

Applications

27. Space Shuttle Height A space shuttle launch is being observed from a location 5 miles away from the launch pad. The height h of the shuttle can be determined using the angle of elevation θ from the observer to the shuttle (see Figure 100).

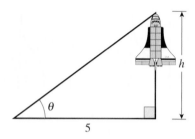

Figure 100

a. Express h as a function of the angle θ.

b. For what values of θ is the function h valid?

c. Sketch the graph of h over the interval you found in part b.

d. Interpret the behavior of the graph of h near $\theta = \frac{\pi}{2}$.

Space Shuttle

28. Tracking an Oil Tanker An oil tanker is traveling parallel to a north-south coastline at a distance of 1000 yards from shore. Let d be the number of yards that the tanker is south of the Coast Guard station, and let θ be the angle shown in Figure 101.

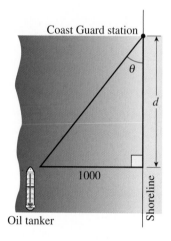

Figure 101

a. Express d as a function of the angle θ.

b. For what values of θ is the function d valid?

c. Sketch the graph of d over the interval you found in part b.

d. Interpret the behavior of the graph of d near $\theta = \pi$.

Concepts and Critical Thinking

Exercises 29–32 *Answer true or false.*

29. The graph of $f(x) = \sec(bx + c)$ has an asymptote wherever the graph of $h(x) = \cos(bx + c)$ has an x-intercept.

30. The graph of $f(x) = \cot(bx + c)$ has an asymptote wherever the graph of $h(x) = \cos(bx + c)$ has an x-intercept.

31. Functions of the form $f(x) = \tan(bx + c)$ have infinitely many vertical asymptotes.

32. No function of the form $f(x) = \sec(bx + c)$ passes through the origin.

Exercises 33–36 *Give an example of each.*

33. A trigonometric function with range $(-\infty, -1] \cup [1, \infty)$

34. A trigonometric function with no vertical asymptotes

35. A trigonometric function with vertical asymptotes at $x = 0$ and $x = 2$, but none between 0 and 2

36. A trigonometric function whose smallest positive zero is π and that has a vertical asymptote at $x = 0$

37. Explain why the secant and cosecant functions have no x-intercepts.

38. Explain why the equation $\tan x = a$ has a solution x in the interval $\left(-\frac{\pi}{2}, \frac{\pi}{2}\right)$ no matter how large (or small) the value of a.

Questions for Discussion or Essay

39. In Section 5.4, we defined the amplitude of a periodic function as one-half the "height" of its graph. According to this definition, what is the amplitude of the tangent function? Is the notion of amplitude of any use in distinguishing between tangent functions of the form $f(x) = a \tan(bx + c)$? Why or why not?

40. In this section, we described a method for obtaining the graph of a function of the form $f(x) = \sec(bx + c)$ from the graph of $g(x) = \cos(bx + c)$. Describe a method for plotting the graph of a function of the form $f(x) = \cot(bx + c)$ by using the graph of $g(x) = \tan(bx + c)$.

Projects for Enrichment

41. Projectile Distance If someone were to ask you the question "How far can you throw a baseball?" you would probably answer with an estimate of your **range**—that is, with an estimate of the horizontal distance from the point where the ball is thrown to the point where it hits the ground. However, there are two other interpretations of the word *distance* as it relates to projectile motion. One is the maximum height of the projectile and the other is the length of the projectile's path. Our goal in this project is to investigate the range, maximum height, and path length of a projectile. Specifically, we wish to see how these quantities are affected by the initial angle at which an object is propelled into the air.

If an object is thrown from ground level with an initial angle θ and an initial speed of v feet per second, its range (ignoring air resistance) is given by $(v^2 \sin 2\theta)/32$. For the following problems, assume $v = 64$ feet per second (approximately 44 miles per hour). You may also find a graphing calculator useful.

a. Find the range for initial angles of $\frac{\pi}{6}, \frac{\pi}{4}$, and $\frac{\pi}{3}$.

b. What initial angle would be required for a range of 100 feet?

c. What initial angle will give the greatest range?

The maximum height of an object thrown from ground level with an initial angle θ and an initial speed of v feet per second is given by $(v^2 \sin^2 \theta)/64$. Once again, assume $v = 64$ feet per second.

d. Find the maximum height for initial angles of $\frac{\pi}{6}, \frac{\pi}{4}$, and $\frac{\pi}{3}$.

e. What initial angle would be required for a maximum height of 40 feet?

f. What initial angle will give the greatest maximum height?

Using techniques from calculus, it can be shown that the length of a projectile's path is given by

$$\frac{v^2}{32}\left[\sin\theta - \cos^2\theta\left(\ln\left[\tan\left(\frac{\pi - 2\theta}{4}\right)\right]\right)\right]$$

Assume as before that $v = 64$ feet per second.

g. For what values of θ is this expression meaningful?

h. Find the length of the ball's path for initial angles of $\frac{\pi}{6}, \frac{\pi}{4}$, and $\frac{\pi}{3}$.

i. What initial angle would be required for a path length of 100 feet?

j. What initial angle will give the greatest path length? How does this angle compare to the angles that give the greatest range or maximum height?

k. One of our simplifying assumptions is that gravity is the only force acting on the ball. In reality, what other factors might influence the path of the ball? Under what circumstances would these other factors be significant? How accurate are these formulas for modeling the path of a golf ball? A rock? A frisbee?

Section 5.6 | Inverse Trigonometric Functions

- Where is the best seat in a movie theater?
- Why aren't road signs the most readable when we are closest to them?
- If the sine function has no inverse, then what is meant by the inverse sine function?
- How can the optimal position of a solar panel be determined in an urban area?

Most of the applications that we have encountered in this chapter have involved the computation of trigonometric function values (outputs) at specified real number inputs. In this section, we turn our attention to the inverse problem: finding the input to a trigonometric function given the output.

The Inverse Sine Function

Consider the problem of determining the angle x shown in Figure 102. We may use the right triangle definition of the sine function to obtain

$$\sin x = \frac{4}{5}$$

from which it follows that x is the angle whose sine is $\frac{4}{5}$. To find this angle, we have seen that we can use the $\boxed{\text{SIN}^{-1}}$ key on a calculator, obtaining the approximation

$$x \approx 0.9273 \text{ (or } 53.13°)$$

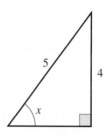

Figure 102

The calculator produces a single answer. However, as Figure 103 suggests, the equation $\sin x = \frac{4}{5}$ actually has infinitely many solutions. Thus, it is impossible to speak of *the* angle whose sine is $\frac{4}{5}$. So how does the calculator "know" which solution to give?

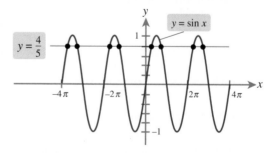

Figure 103

In the language of functions, we would say that the function $f(x) = \sin x$ is not 1–1 and so has no inverse. Recall that 1–1 functions satisfy the horizontal line test: Any horizontal line crosses the graph of a 1–1 function at most once. Figure 103 clearly shows just how badly the graph of $f(x) = \sin x$ fails to satisfy the horizontal line test. However, if we restrict the sine function to the interval $\left[-\frac{\pi}{2}, \frac{\pi}{2}\right]$, as shown in Figure 104, then the horizontal line test will indeed be satisfied. This function, obtained by restricting the sine function, is 1–1 and has an inverse, which we call arcsin x or $\sin^{-1} x$. Thus, given any number x between -1 and 1, **arcsin x** will represent the *unique* angle between $-\frac{\pi}{2}$ and $\frac{\pi}{2}$ whose sine is x. More formally, we have the following definition.

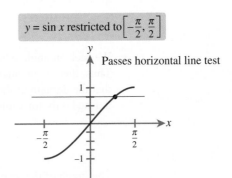

Figure 104

Definition of the Inverse Sine Function

The **inverse sine function**, denoted by **arcsin** or **$\sin^{-1}$**, is defined by

$$y = \arcsin x \quad \text{if and only if} \quad \sin y = x$$

where $-1 \le x \le 1$ and $-\frac{\pi}{2} \le y \le \frac{\pi}{2}$. The domain of $g(x) = \arcsin x$ is $[-1, 1]$ and the range is $\left[-\frac{\pi}{2}, \frac{\pi}{2}\right]$.

EXAMPLE 1

Finding Exact Values of the Inverse Sine Function

Find the exact value of the given quantity.

a. $\sin^{-1}\dfrac{1}{2}$ **b.** $\arcsin\left(-\dfrac{\sqrt{3}}{2}\right)$ **c.** $\sin^{-1}(-3)$

Solution

a. We are interested in the unique angle between $-\frac{\pi}{2}$ and $\frac{\pi}{2}$ whose sine is $\frac{1}{2}$. In other words, if we let $y = \sin^{-1}\frac{1}{2}$, then

$$\sin y = \frac{1}{2} \quad \text{and} \quad -\frac{\pi}{2} \le y \le \frac{\pi}{2}$$

Clearly $\frac{\pi}{6}$ satisfies both conditions. Thus, we conclude that

$$\sin^{-1}\frac{1}{2} = \frac{\pi}{6}$$

b. Let $y = \arcsin(-\sqrt{3}/2)$. Then, from the definition of the inverse sine function, we have

$\sin^{-1}(-\sqrt{3}/2)$ into calc

$$\sin y = -\frac{\sqrt{3}}{2} \quad \text{and} \quad -\frac{\pi}{2} \leq y \leq \frac{\pi}{2}$$

Thus, we are looking for an angle y between $-\frac{\pi}{2}$ and $\frac{\pi}{2}$ whose sine is $-\sqrt{3}/2$. Since $\sin\left(-\frac{\pi}{3}\right) = -\sqrt{3}/2$ and $-\frac{\pi}{3}$ is between $-\frac{\pi}{2}$ and $\frac{\pi}{2}$, it follows that

$$\arcsin\left(-\frac{\sqrt{3}}{2}\right) = -\frac{\pi}{3}$$

c. We seek an angle whose sine is -3. But the sine of any angle is between -1 and 1. Thus, there is no angle whose sine is -3, and so $\sin^{-1}(-3)$ is undefined. We recall that the domain of the inverse sine function is the interval $[-1, 1]$, so that $\sin^{-1}x$ is defined only for values of x from -1 to 1.

The graph of the inverse sine function can be determined by creating a table of values for the relation $y = \sin^{-1}x$, $-\frac{\pi}{2} \leq y \leq \frac{\pi}{2}$, and plotting points. Since $y = \sin^{-1}x$ is equivalent to $x = \sin y$, we select convenient y-values from $-\frac{\pi}{2}$ to $\frac{\pi}{2}$ and then find the corresponding x-values. The graph is shown in Figure 105.

$x = \sin y$	y
-1	$-\dfrac{\pi}{2}$
$-\dfrac{\sqrt{3}}{2}$	$-\dfrac{\pi}{3}$
$-\dfrac{\sqrt{2}}{2}$	$-\dfrac{\pi}{4}$
$-\dfrac{1}{2}$	$-\dfrac{\pi}{6}$
0	0
$\dfrac{1}{2}$	$\dfrac{\pi}{6}$
$\dfrac{\sqrt{2}}{2}$	$\dfrac{\pi}{4}$
$\dfrac{\sqrt{3}}{2}$	$\dfrac{\pi}{3}$
1	$\dfrac{\pi}{2}$

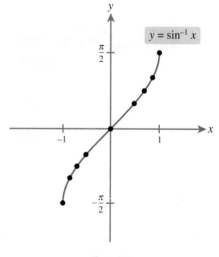

Figure 105

Although point plotting can be instructive, the graph of the inverse sine function can be obtained much more easily. Recall from Chapter 2 that the graph of the inverse of any 1–1 function f is simply the reflection of the graph of f about the line $y = x$. In the case of the inverse sine function, it is the graph of the sine function over the interval $\left[-\frac{\pi}{2}, \frac{\pi}{2}\right]$ that is to be reflected about the line $y = x$, as shown in Figure 106.

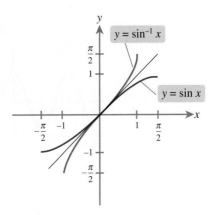

Figure 106

EXAMPLE 2   Graphing a Translation of the Inverse Sine Function

Sketch the graph of $y = \sin^{-1}(x + 1)$.

Solution   Note that the equation $y = \sin^{-1}(x + 1)$ can be obtained from the equation $y = \sin^{-1} x$ by substituting $x + 1$ in place of x. Since replacing x with $x + 1$ in an equation has the effect of translating the graph 1 unit to the left, the graph of $y = \sin^{-1}(x + 1)$ is just a translation 1 unit to the left of the graph of $y = \sin^{-1} x$, as shown in Figure 107.

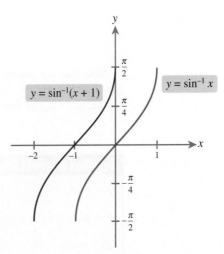

Figure 107

The Inverse Cosine and Tangent Functions

The inverse cosine and tangent functions may be defined in much the same way as the inverse sine function. Both the cosine and tangent functions fail to be 1-1 and, thus, have no inverse, as Figure 108 shows. But just as with the sine function, this situation can be remedied by restricting the domains to appropriate intervals, as shown in Figure 109.

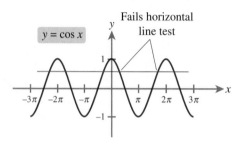
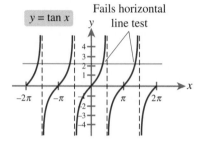

Figure 108

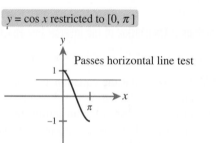
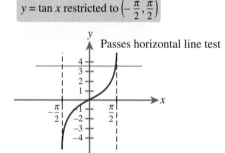

Figure 109

Note that the cosine function has been restricted to the interval $[0, \pi]$ and the tangent function to $\left(-\frac{\pi}{2}, \frac{\pi}{2}\right)$. The formal definitions and graphs of each of the inverse sine, cosine, and tangent functions are as follows.

Definition of Inverse Trigonometric Functions
$y = \arcsin x$ (or $y = \sin^{-1}x$) if and only if $\sin y = x$ and $-\dfrac{\pi}{2} \le y \le \dfrac{\pi}{2}$.
$y = \arccos x$ (or $y = \cos^{-1}x$) if and only if $\cos y = x$ and $0 \le y \le \pi$.
$y = \arctan x$ (or $y = \tan^{-1}x$) if and only if $\tan y = x$ and $-\dfrac{\pi}{2} < y < \dfrac{\pi}{2}$.

Graphs of Inverse Trigonometric Functions

Function	Domain	Range	Graph
$y = \arcsin x$	$-1 \leq x \leq 1$	$-\dfrac{\pi}{2} \leq y \leq \dfrac{\pi}{2}$	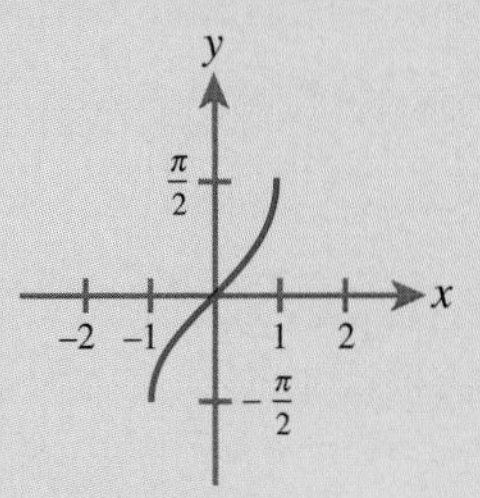
$y = \arccos x$	$-1 \leq x \leq 1$	$0 \leq y \leq \pi$	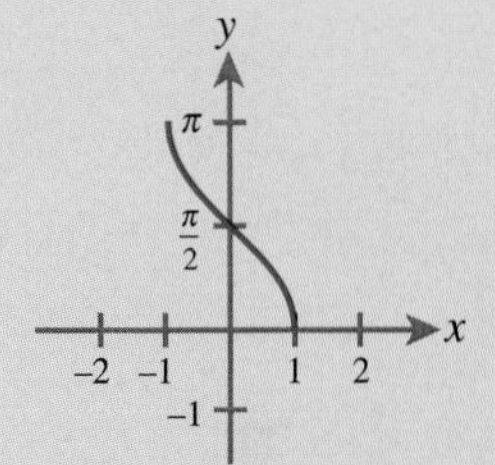
$y = \arctan x$	$-\infty < x < \infty$	$-\dfrac{\pi}{2} < y < \dfrac{\pi}{2}$	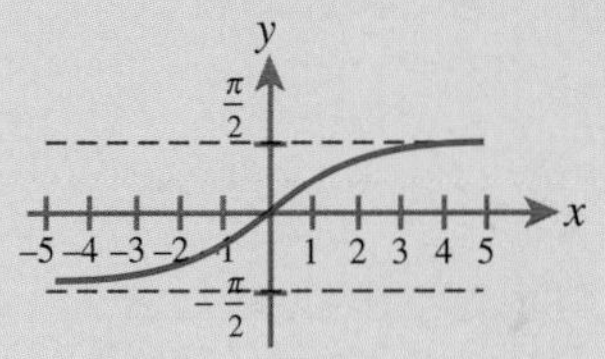

EXAMPLE 3

Finding Exact Values of the Inverse Cosine Function

Find the exact value of the given quantity.

a. $\cos^{-1} 1$ **b.** $\arccos\left(-\dfrac{\sqrt{2}}{2}\right)$

Solution

a. We are interested in an angle in the interval $[0, \pi]$ whose cosine is 1. To be precise, if we let $y = \cos^{-1} 1$, then y must satisfy the conditions

$$\cos y = 1 \quad \text{and} \quad 0 \leq y \leq \pi$$

Since 0 satisfies these conditions, we have

$$\cos^{-1} 1 = 0$$

b. Let $y = \arccos\left(-\sqrt{2}/2\right)$. Then the following conditions must hold:

$$\cos y = -\frac{\sqrt{2}}{2} \quad \text{and} \quad 0 \leq y \leq \pi$$

Since $\cos\frac{3\pi}{4} = -\frac{\sqrt{2}}{2}$, and $\frac{3\pi}{4}$ is between 0 and π, we see that $y = \frac{3\pi}{4}$. Thus,

$$\arccos\left(-\frac{\sqrt{2}}{2}\right) = \frac{3\pi}{4}$$

┈┈⟩EXAMPLE 4   Finding Values of the Inverse Tangent Function

Find the exact value of the given quantity. Confirm with a calculator.

a. $\tan^{-1}\sqrt{3}$ **b.** $\arctan(-1)$

Solution

a. Let $y = \tan^{-1}\sqrt{3}$. By the definition of the inverse tangent function, y is the unique angle between $-\frac{\pi}{2}$ and $\frac{\pi}{2}$ for which the tangent is $\sqrt{3}$. Thus, we have

$$\tan y = \sqrt{3} \text{ and } -\frac{\pi}{2} < y < \frac{\pi}{2}$$

Since $\tan\frac{\pi}{3} = \sqrt{3}$, and $\frac{\pi}{3}$ is between $-\frac{\pi}{2}$ and $\frac{\pi}{2}$, we have $y = \frac{\pi}{3}$. Thus,

$$\tan^{-1}\sqrt{3} = \frac{\pi}{3}$$

Using the $\boxed{\text{TAN}^{-1}}$ key on a calculator set in radian mode, we obtain $\tan^{-1}\sqrt{3} \approx 1.0472$, which is approximately $\frac{\pi}{3}$.

b. Since $\tan\left(-\frac{\pi}{4}\right) = -1$, and $-\frac{\pi}{4}$ is between $-\frac{\pi}{2}$ and $\frac{\pi}{2}$, we have

$$\arctan(-1) = -\frac{\pi}{4}$$

Using the $\boxed{\text{TAN}^{-1}}$ key on a calculator set in radian mode, we obtain $\tan^{-1}(-1) \approx -0.7854$, which is approximately $-\frac{\pi}{4}$.

Solving Trigonometric Equations

Although the inverse sine function gives only a single solution of the equation $\sin x = y$, other solutions can be found by considering reference angles and the unit circle, as illustrated by the following example.

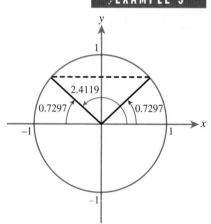

Figure 110

┈┈⟩EXAMPLE 5   Solving a Trigonometric Equation with the Inverse Sine Function

Estimate all solutions of $3 \sin x = 2$ in the interval $[0, 2\pi)$.

Solution   We begin by solving for $\sin x$.

$$3 \sin x = 2$$

$$\sin x = \frac{2}{3}$$

Since we know of no angle that has sine $\frac{2}{3}$, we use the $\boxed{\text{SIN}^{-1}}$ key on a calculator set in radian mode to obtain

$$x = \sin^{-1}\frac{2}{3} \approx 0.7297$$

As Figure 110 shows, there is also an angle between $\frac{\pi}{2}$ and π with a sine (y-coordinate) of $\frac{2}{3}$; this is $\pi - \sin^{-1}\frac{2}{3} \approx 2.4119$. Thus, our solutions are approximately 0.7297 and 2.4119.

EXAMPLE 6

Solving a Trigonometric Equation

Use inverse trigonometric functions to find all solutions of each of the following equations on the interval $[0, 2\pi)$:

a. $\cos x = \dfrac{1}{4}$ **b.** $\tan x = -5$

Solution

a. We are interested in finding all angles from 0 to 2π with a cosine of $\frac{1}{4}$. Clearly, there are two such angles, one between 0 and $\frac{\pi}{2}$ and the other between $\frac{3\pi}{2}$ and 2π, as shown in Figure 111. By definition, $\cos^{-1}\frac{1}{4}$ is an angle θ between 0 and π satisfying $\cos\theta = \frac{1}{4}$. Thus, we have labeled the solution in the first quadrant as $\cos^{-1}\frac{1}{4}$. The solution between $\frac{3\pi}{2}$ and 2π has $\cos^{-1}\frac{1}{4}$ as its reference angle and is equal to $2\pi - \cos^{-1}\frac{1}{4}$. Using a calculator, we find that our solutions are approximately 1.3181 and 4.9651.

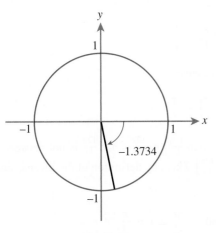

Figure 111

b. Because the tangent function is negative in both the second and fourth quadrants, the equation $\tan x = -5$ has two solutions between 0 and 2π: one between $\frac{\pi}{2}$ and π and another between $\frac{3\pi}{2}$ and 2π. Using the inverse tangent function, we have

$$\tan x = -5$$
$$x = \arctan(-5)$$
$$\approx -1.3734$$

Of course, this solution is not between 0 and 2π (see Figure 112), but it does have the same reference angle—namely, 1.3734—as the two desired solutions. The angle between $\frac{\pi}{2}$ and π with reference angle 1.3734 is $\pi - 1.3734 \approx 1.7682$ (Figure 113). The solution between $\frac{3\pi}{2}$ and 2π is given by $2\pi - 1.3734 \approx 4.9098$ (Figure 114).

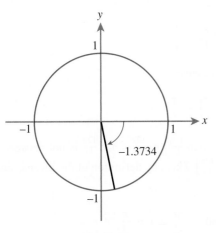

Figure 112

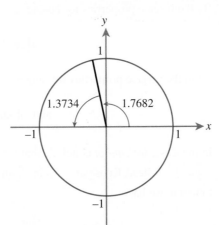

Figure 113

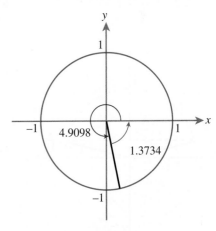

Figure 114

Composing Trigonometric and Inverse Trigonometric Functions

Recall that a function and its inverse "undo" one another. In other words, if f is a 1–1 function (so that f^{-1} exists), then for appropriate values of x,

$$f^{-1}(f(x)) = x \quad \text{and} \quad f(f^{-1}(x)) = x$$

If we let $f(x) = \sin x$ restricted to $\left[-\frac{\pi}{2}, \frac{\pi}{2}\right]$, then $f^{-1}(x) = \arcsin x$, and we have

$$\arcsin(\sin x) = x, \quad \text{provided } -\frac{\pi}{2} \le x \le \frac{\pi}{2}$$

and

$$\sin(\arcsin x) = x, \quad \text{provided } -1 \le x \le 1$$

Similar results hold for the restrictions of the cosine and tangent functions and their inverses. We summarize all such results as follows.

Sine/Inverse Sine	
Property	**Restriction**
$\sin(\sin^{-1}x) = x$	$-1 \le x \le 1$
$\sin^{-1}(\sin x) = x$	$-\frac{\pi}{2} \le x \le \frac{\pi}{2}$

Cosine/Inverse Cosine	
Property	**Restriction**
$\cos(\cos^{-1}x) = x$	$-1 \le x \le 1$
$\cos^{-1}(\cos x) = x$	$0 \le x \le \pi$

Tangent/Inverse Tangent	
Property	**Restriction**
$\tan(\tan^{-1}x) = x$	None
$\tan^{-1}(\tan x) = x$	$-\frac{\pi}{2} < x < \frac{\pi}{2}$

·····**EXAMPLE 7**

Simplifying with the Inverse Properties

Evaluate the following quantities:

a. $\cos\left(\cos^{-1}\frac{3}{4}\right)$ **b.** $\arctan\left(\tan\frac{\pi}{7}\right)$ **c.** $\sin^{-1}\left(\sin\frac{12\pi}{13}\right)$

Solution

a. By the inverse properties for cosine, we have

$$\cos\left(\cos^{-1}\frac{3}{4}\right) = \frac{3}{4}$$

b. From the inverse properties for tangent, we have

$$\arctan\left(\tan\frac{\pi}{7}\right) = \frac{\pi}{7}$$

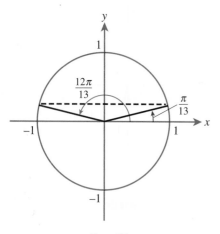

Figure 115

c. In this case, we *cannot* conclude that $\sin^{-1}\left[\sin\frac{12\pi}{13}\right] = \frac{12\pi}{13}$ since $\frac{12\pi}{13}$ is not between $-\frac{\pi}{2}$ and $\frac{\pi}{2}$. Instead, let us put $y = \sin^{-1}\left[\sin\frac{12\pi}{13}\right]$. Then, by definition of the inverse sine function, we have

$$\sin y = \sin\frac{12\pi}{13} \quad \text{and} \quad -\frac{\pi}{2} \le y \le \frac{\pi}{2}$$

Thus, we seek to find an angle y between $-\frac{\pi}{2}$ and $\frac{\pi}{2}$ whose sine is the same as that of $\frac{12\pi}{13}$. As Figure 115 indicates, $\frac{\pi}{13}$ is the desired angle. We conclude that

$$\sin^{-1}\left(\sin\frac{12\pi}{13}\right) = \frac{\pi}{13}$$

----∶EXAMPLE 8

Simplifying an Expression Involving the Inverse Sine Function

Simplify $\sin(\arccos x)$.

Solution   Let $\theta = \arccos x$. Our task then is to simplify $\sin \theta$. From the definition of the inverse cosine function, we have

$$\cos \theta = x \quad \text{and} \quad 0 \le \theta \le \pi$$

Thus, the problem can be summarized as follows: *Find a simplified expression for* $\sin \theta$ *given that* $\cos \theta = x$ *and* $0 \le \theta \le \pi$. Using the identity $\sin^2 \theta + \cos^2 \theta = 1$, we can solve for $\sin \theta$ as follows:

$$\sin^2 \theta + \cos^2 \theta = 1$$
$$\sin^2 \theta + x^2 = 1$$
$$\sin^2 \theta = 1 - x^2$$
$$\sin \theta = \pm\sqrt{1 - x^2}$$

Since θ is between 0 and π, $\sin \theta \ge 0$. Thus,

$$\sin(\arccos x) = \sin \theta = \sqrt{1 - x^2}$$

There is an alternate, more intuitive approach that works well when θ is an acute angle. We sketch a right triangle having an angle θ whose cosine is x and then use the right triangle definition of the trigonometric functions to compute $\sin x$. We begin by labeling one of the acute angles of a right triangle as θ. To ensure that $\cos \theta = x$, we label the adjacent side x and the hypotenuse 1. By the Pythagorean Theorem, the side opposite θ must be $\sqrt{1 - x^2}$. The sketch is shown in Figure 116. From the triangle, we see that

$$\sin \theta = \frac{\text{opp}}{\text{hyp}} = \frac{\sqrt{1 - x^2}}{1} = \sqrt{1 - x^2}$$

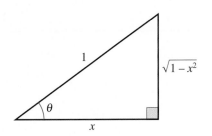

Figure 116

----∶EXAMPLE 9

Evaluating an Expression Involving the Inverse Tangent Function

Find the exact value of $\csc[\arctan(-2)]$.

Solution   Let $\theta = \arctan(-2)$. Then we are interested in evaluating $\csc \theta$. By definition of the inverse tangent function, we have

$$\tan \theta = -2 \quad \text{and} \quad -\frac{\pi}{2} < \theta < \frac{\pi}{2}$$

Since $\tan \theta < 0$, we know that θ must be between $-\frac{\pi}{2}$ and 0. Our task then can be summarized as follows: *Find* $\csc \theta$ *given that* $\tan \theta = -2$ *and* $-\frac{\pi}{2} < \theta < 0$. We accomplish this task using a reference angle.

Let θ_R be the reference angle for θ. Then $\tan \theta_R = 2$. Because cosecant is negative in the fourth quadrant, we first find $\csc \theta_R$ and then attach a negative sign. In Figure 117, we have sketched a right triangle having θ_R as one of its acute angles. In order that $\tan \theta_R = 2$, we have labeled the side opposite θ_R as 2 and the side adjacent to θ_R as 1. By the Pythagorean Theorem, the hypotenuse is $\sqrt{5}$. From the definition of the cosecant function, we have $\csc \theta_R = \sqrt{5}/2$. Thus, $\csc \theta = -\sqrt{5}/2$, and it follows that

$$\csc[\arctan(-2)] = -\frac{\sqrt{5}}{2}$$

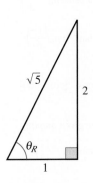

Figure 117

Understanding and Mastery Checklists

Concepts to Understand	Skills to Master
Inverse sine function	Find the exact value of an inverse trigonometric function.
◈	◈
Inverse cosine function	Use an inverse trigonometric function to approximate the solution(s) of a trigonometric equation.
◈	
Inverse tangent function	◈
◈	Simplify compositions of trigonometric functions and inverse trigonometric functions.
Trigonometric equations	
◈	
Compositions involving trigonometric functions	
◈	
Inverse properties of trigonometric functions	

Exercises 5.6

Exercises 1-16   *Find the exact value of the given quantity.*

1. arcsin 1

2. $\cos^{-1} 0$

3. $\cos^{-1}\left(-\dfrac{1}{2}\right)$

4. arctan(−1)

5. $\tan^{-1} 1$

6. $\sin^{-1}(-1)$

7. arctan $\sqrt{3}$

8. arccos $\dfrac{\sqrt{3}}{2}$

9. $\sin^{-1}\left(-\dfrac{\sqrt{2}}{2}\right)$

10. arctan$\left(-\dfrac{\sqrt{3}}{3}\right)$

11. arccos(−1)

12. $\sin^{-1}\dfrac{1}{2}$

13. $\tan^{-1} 0$

14. arccos 2

15. arcsin π

16. $\tan^{-1}(-\sqrt{3})$

Exercises 17-24   *Use inverse trigonometric functions to approximate all solutions of the given equation in the interval* $[0, 2\pi)$.

17. $\sin x = \dfrac{1}{3}$

18. $\cos x = -\dfrac{1}{5}$

19. $\tan x = -20$

20. $\sin x = \dfrac{1}{4}$

21. $\cos x = 0.7$

22. $\tan x = 0.3$

23. $\sin x = \dfrac{7}{6}$

24. $\cos x = -\pi$

Exercises 25-34   *Find the exact value of the given quantity.*

25. $\sin\left(\arcsin \dfrac{1}{3}\right)$

26. $\cos^{-1}\left(\cos \dfrac{\pi}{8}\right)$

27. $\arctan\left[\tan\left(-\dfrac{2\pi}{5}\right)\right]$

28. cos(arccos 0.2)

29. $\tan[\tan^{-1}(-7)]$

30. $\arcsin\left[\sin\left(-\dfrac{2\pi}{3}\right)\right]$

31. $\arccos\left[\cos\left(-\dfrac{3\pi}{4}\right)\right]$

32. $\tan^{-1}\left(\tan \dfrac{11\pi}{6}\right)$

33. $\sin^{-1}\left(\sin \dfrac{3\pi}{2}\right)$

34. $\cos^{-1}[\cos(-\pi)]$

Exercises 35-42   *Find the exact value of the given quantity.*

35. $\sin\left(\arctan \dfrac{4}{3}\right)$

36. $\tan\left(\sin^{-1} \dfrac{12}{13}\right)$

37. $\sec\left[\cos^{-1}\left(-\dfrac{1}{8}\right)\right]$

38. cos(tan^{-1} 3)

39. $\cot\left(\sin^{-1}\dfrac{2\sqrt{5}}{5}\right)$

40. $\csc\left[\arccos\left(-\dfrac{4}{5}\right)\right]$

41. $\cos\left[\arctan\left(-\sqrt{6}\right)\right]$

42. $\sec\left(\arcsin\dfrac{\sqrt{3}}{3}\right)$

Exercises 43-48   *Simplify the given expression. Assume $x > 0$.*

43. $\tan(\arccos x)$

44. $\sin(\tan^{-1}3x)$

45. $\sec\left(\sin^{-1}\dfrac{2}{x}\right)$

46. $\cot\left(\arctan\dfrac{1+x}{1-x}\right)$

47. $\cos\left(\arctan\dfrac{\sqrt{x^2-4}}{2}\right)$

48. $\csc\left(\cos^{-1}\sqrt{1-9x^2}\right)$

Exercises 49-54   *Sketch the graph of the given function.*

49. $y = \sin^{-1}(x-2)$

50. $y = \tan^{-1}(x) - \dfrac{\pi}{2}$

51. $y = \cos^{-1}(x) - \pi$

52. $y = \sin^{-1}(x+2)$

53. $y = \tan^{-1}(x-1) + \dfrac{\pi}{2}$

54. $y = \cos^{-1}(x+1) - \pi$

Exercises 55-58   *Use a graphing calculator to estimate the solutions of the given equation to the nearest hundredth.*

55. $\arcsin x = x + \dfrac{1}{2}$

56. $2\cos^{-1}x = x^2 + 1$

57. $\arctan x = 1 - x^2$

58. $\arcsin x = \arccos x$

Exercises 59-62   *Use a graphing calculator to help determine whether the given equation is an identity.*

59. $\arcsin^2 x + \arccos^2 x = 1$

60. $\arcsin x + \arccos x = \dfrac{\pi}{2}$

61. $\cos(2\cos^{-1}x) = 2x^2 - 1$

62. $\arctan x = \dfrac{\arcsin x}{\arccos x}$

Applications

---

63. Rising Balloon   A hot-air balloon is rising straight into the air at a constant rate of 6 feet per second.

   a. Find the height of the balloon after 8 seconds.

   b. Find the angle of elevation to the balloon after 8 seconds from an observer lying on the ground at a horizontal distance of 100 feet away from where the balloon was launched.

   c. Express the angle of elevation from the observer to the balloon as a function of time.

64. Altitude of the Sun   The altitude of the sun is defined to be the angle made between the line of sight to the sun and the horizontal.

   a. Find the altitude of the sun if the length of the shadow of a 6-foot man is 3.5 feet.

   b. If the man in part a is standing near the Washington Monument, and if the length of the monument's shadow is approximately 324 feet, find the height of the Washington Monument.

65. Air-Sea Rescue   A plane flying at a constant altitude of 2000 feet is in search of a boat in distress. Based on information from a flare sighting, the pilot anticipates that she will fly directly over the boat.

   a. Find the angle of depression from the plane to the boat when the horizontal distance between the plane and the boat is 1200 feet.

b. Express the angle of depression from the plane to the boat as a function of the horizontal distance from the plane to the boat.

c. If the pilot is scanning ahead of the plane with an angle of depression of 60°, at what horizontal distance will she spot the boat?

66. Filming a Stunt A cameraman is filming a scene in which a stunt man falls from a height of 50 feet. The camera is positioned 20 feet away from the point on the ground where the stunt man will land, and the camera is 6 feet above the ground.

a. Find the angle of elevation of the camera when the stunt man is 30 feet above the ground.

b. Express the angle of elevation of the camera as a function of the height (off the ground) of the stunt man in feet.

c. Given that the stunt man's height in feet t seconds after he begins to fall is $h = 50 - 16t^2$, express the angle of elevation of the camera as a function of t.

67. Sunshine Angle A solar panel is to be placed in an empty lot between two buildings that have heights 70 feet and 30 feet and that are 140 feet apart. Denote by θ the angle made between the lines running from the panel to the top of each building, as shown in Figure 118. A straightforward computation would show that the angle θ can be written as a function of the distance x in feet between the panel and the shorter building, as follows:

$$\theta = \pi - \tan^{-1}\left(\frac{30}{x}\right) - \tan^{-1}\left(\frac{70}{140 - x}\right)$$

a. Complete the following table of angles for the given distances.

x (feet)	20	50	100
θ (radians)			

What do you notice about the angles as the distance increases?

b. Use a graphing calculator to estimate the distance for which the angle is as large as possible. Explain why this would be the most desirable place to put the panel.

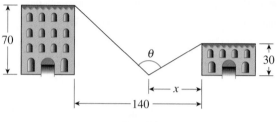

Figure 118

68. Road Sign Viewing Angle The bottom edge of a 5-foot-high freeway sign is 20 feet above the eye level of a motorist heading toward it, as shown in Figure 119. A straightforward computation would reveal that the vertical viewing angle θ can be expressed as a function of the distance x in feet from the motorist to a point directly beneath the sign, as follows:

$$\theta = \tan^{-1}\left(\frac{25}{x}\right) - \tan^{-1}\left(\frac{20}{x}\right)$$

a. Complete the following table of viewing angles for the given distances.

x (feet)	50	20	10
θ (radians)			

What do you notice about the angles as the distance decreases?

b. Use a graphing calculator to estimate the distance for which the viewing angle is as large as possible.

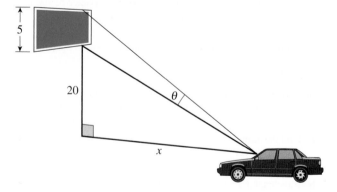

Figure 119

Concepts and Critical Thinking

Exercises 69-72 *Answer true or false.*

69. Given that $y = \arcsin x$ and $-1 \le x \le 1$, it must follow that $x = \sin y$.

70. $\cos^{-1}(\cos x) = x$ for all x.

71. $\cos(\cos^{-1} x) = x$ for all x.

72. The function $f(x) = \arctan x$ is 1–1 and so has an inverse.

Exercises 73-76 *Give an example of each.*

73. A number x for which $\sin(\arcsin x) \ne x$

74. A number x for which $\arctan(\tan x) \ne x$

75. An inverse trigonometric function that is increasing over its entire domain

76. A positive solution of $\tan x = \square$, if it is known that $\tan^{-1}\square = -\dfrac{\pi}{5}$

Questions for Discussion or Essay

77. In this section, we defined inverses for functions that are not 1–1 by first restricting the domain so that the resulting function was 1–1. In the case of $f(x) = \sin x$, we restricted the domain to the interval $\left[-\dfrac{\pi}{2}, \dfrac{\pi}{2}\right]$. Are there other intervals to which we could have restricted the sine function? Is the sine function 1–1 on the interval $\left[-\dfrac{\pi}{4}, \dfrac{\pi}{4}\right]$? What disadvantage would there be to defining the inverse sine function by "$y = \sin^{-1} x$ if and only if $\sin y = x$ and $-\dfrac{\pi}{4} \le x \le \dfrac{\pi}{4}$"?

78. The function $f(x) = x^2$ fails to be 1–1 since its graph fails the horizontal line test. How could the domain of f be restricted so that the resulting function would become 1–1? What would the inverse function be?

79. We did not define any of $\cot^{-1} x$, $\sec^{-1} x$, or $\csc^{-1} x$, but each could be defined in a fashion similar to that of $\cos^{-1} x$, $\sin^{-1} x$, or $\tan^{-1} x$. Define each of these functions.

80. Exercises 68 and 81 both relate to the concept of viewing angle. One would expect that the optimal position from which to view an object is where the viewing angle is as large as possible. But in practice, other considerations affect the optimal viewing angle. What factors (besides viewing angle) determine the best seat from which to view a theater screen? In the case of the road sign, how does the motion of the car affect the ideal position from which to read a road sign?

Projects for Enrichment

81. The Best Seat in the House When you walk into a movie theater, where do you look first for an available seat? If you're like most people, you probably look for an empty row somewhere in the middle of the theater, not too close to the front. But suppose there are no empty seats in the middle. In fact, suppose the only available seats are along the right wall of the theater. Which row would you choose? Would you be surprised to learn that there is a mathematical way of finding the "best" row? There is, provided we define the "best" row to be the one in which the viewing angle to the screen is as large as possible. Consider the diagram shown in Figure 120. The screen is 40 feet wide, and the available seats are in a column 30 feet away from the right edge of the screen. From a distance of x feet from the front of the theater, the viewing angle θ is the angle made between the lines of sight to the left and right edges of the screen. The angle α is the angle made between the column of available seats and the line of sight to the right edge of the screen. The angle β is the angle made between the column of available seats and the line of sight to the left edge of the screen.

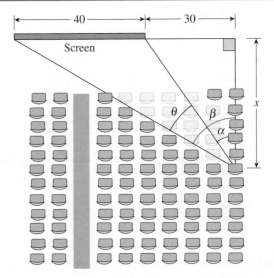

Figure 120

a. Express the angle α as a function of x.

b. Express the angle β as a function of x.

c. Express the angle θ as a function of x.

d. Use the function you found in part c to find θ for the following values of x:

   i. $x = 10$

   ii. $x = 50$

   iii. $x = 100$

e. Use a graphing calculator to plot the graph of θ as a function of x.

f. Use a graphing calculator to estimate the value for x for which θ is the given value.

   i. $\theta = 15°$ **ii.** $\theta = 30°$

   iii. $\theta = 45°$

g. Use a graphing calculator to find the distance x for which θ is as large as possible.

h. Explain how the optimum value for x would change if the distance from the right edge of the screen to the column of empty seats were smaller or larger than 30 feet.

Chapter 5 Review

Exercises 1-4 *Match the given angle measure with the corresponding sketch.*

1. $60°$

2. $-\dfrac{\pi}{6}$

3. $-\dfrac{3\pi}{2}$

4. $315°$

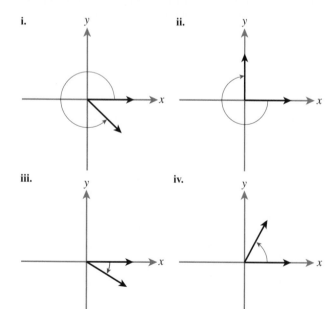

i.

ii.

iii.

iv.

Exercises 5-8 *Find two coterminal angles (one positive and one negative) for the given angle.*

5. $60°$

6. $-\dfrac{\pi}{6}$

7. $-\dfrac{3\pi}{2}$

8. $225°$

Exercises 9-12 *Convert to radian measure.*

9. $60°$

10. $315°$

11. $-210°$

12. $390°$

Exercises 13-16 *Convert to degree measure.*

13. $-\dfrac{5\pi}{6}$

14. $\dfrac{5\pi}{4}$

15. $\dfrac{\pi}{8}$

16. $-\dfrac{3\pi}{2}$

Exercises 17-18 *Find the length of the arc subtended by the central angle θ in a circle of radius r.*

17. $\theta = \dfrac{2\pi}{3}, r = 6$

18. $\theta = 225°, r = 9$

Exercises 19-22 *Find the values of the six trigonometric functions for the given angle θ.*

19.

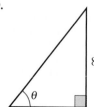

20.

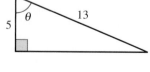

21.

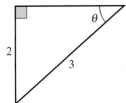

22.

Exercises 23-26   *Use a right triangle to find the other five trigonometric functions of the acute angle θ.*

23. $\cos \theta = \dfrac{15}{17}$

24. $\tan \theta = \dfrac{3}{4}$

25. $\csc \theta = 2$

26. $\sin \theta = \dfrac{1}{\sqrt{3}}$

Exercises 27-30   *Use a calculator to find a decimal approximation, correct to four decimal places, for the acute angle θ measured in degrees.*

27. $\tan \theta = 3.6$

28. $\cos \theta = 0.25$

29. $\sin \theta = 0.32$

30. $\cot \theta = 0.41$

Exercises 31-34   *Find all unknown side lengths and angle measures; that is, solve the given triangle.*

31.

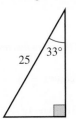

32.

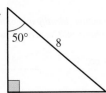

33.

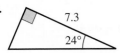

34.

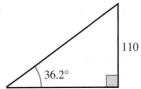

Exercises 35-38   *Find the reference angle for the given angle.*

35. $\theta = 210°$

36. $\theta = \dfrac{7\pi}{6}$

37. $\theta = -\dfrac{3\pi}{4}$

38. $\theta = -20°$

Exercises 39-44   *Use reference angles to find the exact values of the given trigonometric functions.*

39. $\sin \dfrac{2\pi}{3}$

40. $\cot\left(-\dfrac{\pi}{6}\right)$

41. $\sec(-150°)$

42. $\csc 210°$

43. $\tan \dfrac{5\pi}{4}$

44. $\cos 120°$

Exercises 45-50   *Use a calculator to find a decimal approximation for the given function, correct to four decimal places.*

45. $\sin(-200°)$

46. $\tan 95°$

47. $\sec \dfrac{7\pi}{10}$

48. $\csc 4.1$

49. $\cot(125° \, 10' \, 15'')$

50. $\cos(-32° \, 25' \, 30'')$

Exercises 51-54   *Use the given information about the angle θ to find the indicated function value.*

51. $\tan \theta = -2$, $\cos \theta < 0$; $\sin \theta =$ _____

52. $\sin \theta = \dfrac{1}{4}$, $\tan \theta < 0$; $\cos \theta =$ _____

53. $\cos \theta = -\dfrac{2}{3}$, $\sin \theta > 0$; $\tan \theta =$ _____

54. $\cot \theta = 10$, $\sec \theta > 0$; $\csc \theta =$ _____

Exercises 55-62   *Match the given function with its corresponding graph.*

55. $f(x) = \sin \pi x$

56. $f(x) = \sin(\pi x + \pi)$

57. $f(x) = \cos 2x$

58. $f(x) = \cos\left(\dfrac{2x - \pi}{2}\right)$

59. $f(x) = \tan\left(x + \dfrac{\pi}{3}\right)$

60. $f(x) = \cot\left(x - \dfrac{\pi}{3}\right)$

61. $f(x) = \sec\dfrac{x}{2}$

62. $f(x) = \sec\dfrac{\pi x}{2}$

i.
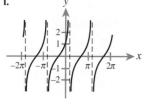

ii.
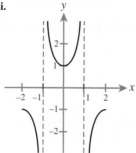

iii.
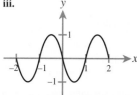

iv.
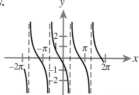

v.
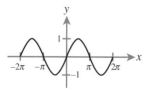

vi.
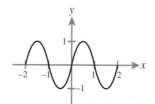

vii.
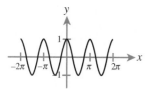

viii.
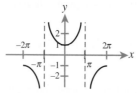

Exercises 63-66 *Find the period and the amplitude of the function with the given graph.*

63.
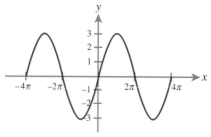

64.
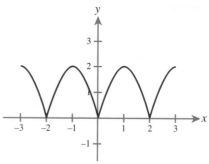

65.
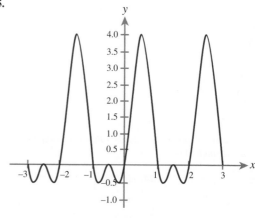

66.
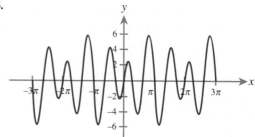

Exercises 67-70 *Find the period and the amplitude of the given function. Use a graphing calculator as necessary.*

67. $g(x) = 2\sin(3x + \pi)$

68. $f(x) = -3\cos(\pi x - 1)$

69. $h(x) = 2\sin x + \cos 2x$

70. $g(x) = \sin 3x - \cos 4x$

Exercises 71-84 *Graph three cycles of the given function. Identify any of the following features that apply: amplitude, period, phase shift, or asymptotes.*

71. $f(x) = \sin\dfrac{\pi x}{2}$

72. $g(x) = \cos 3x$

73. $h(t) = \cot 2t$

74. $g(\theta) = -\tan\dfrac{\pi\theta}{4}$

75. $f(t) = \sec 2\pi t$

76. $f(t) = \csc 4t$

77. $g(x) = 3 \cos \dfrac{x}{2} - 1$ **78.** $h(t) = 4 - 2 \sin 3t$

79. $y = \tan\left(x - \dfrac{\pi}{4}\right)$ **80.** $y = \cot(2x + \pi)$

81. $y = \csc(\pi x - \pi)$ **82.** $y = \sec\left(x + \dfrac{\pi}{6}\right)$

83. $y = -2 \cos\left(\pi x + \dfrac{\pi}{2}\right)$ **84.** $y = 4 \sin\left(\dfrac{x}{2} - \dfrac{\pi}{3}\right)$

Exercises 85-88 *The given graph is that of a function of the form $f(x) = a \sin(bx + c)$. Find a, b, and c.*

85.
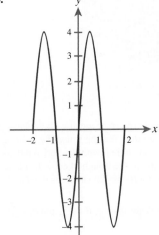

86.
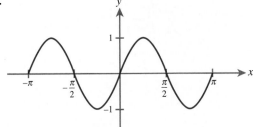

87.
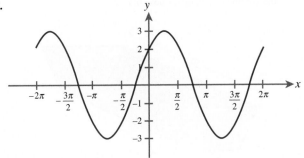

88.
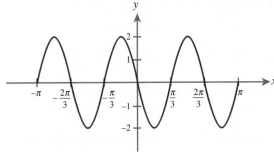

Exercises 89-90 *Find a function of the form $f(x) = a \sin(bx + c)$ that satisfies the given properties.*

89. Amplitude 4, period $\dfrac{\pi}{2}$, phase shift $\dfrac{\pi}{4}$

90. Amplitude 4, period 3, phase shift 1

Exercises 91-92 *Use a graphing calculator to estimate the smallest positive zero of the given function.*

91. $f(x) = \sin x - 2 \cos 3x$

92. $f(x) = \tan x + \sec x$

Exercises 93-98 Find the exact value of the given quantity.

93. $\arccos 1$ **94.** $\tan^{-1} 1$

95. $\sin^{-1} \dfrac{\sqrt{3}}{2}$ **96.** $\cos^{-1}\left(-\dfrac{1}{2}\right)$

97. $\arctan\left(-\dfrac{\sqrt{3}}{3}\right)$ **98.** $\arcsin(-2)$

Exercises 99-102 *Use inverse trigonometric functions to approximate all solutions of the given equation in the interval $[0, 2\pi)$.*

99. $\cos x = -\dfrac{1}{4}$ **100.** $\sin x = \dfrac{2}{3}$

101. $\sin x = \dfrac{1}{\sqrt{3}}$ **102.** $\tan x = -14$

Exercises 103-110 *Find the exact value of the given quantity.*

103. $\cos\left(\tan^{-1} \dfrac{3}{4}\right)$ **104.** $\tan\left(\sin^{-1} \dfrac{12}{13}\right)$

105. $\csc\left[\arccos\left(-\dfrac{1}{3}\right)\right]$ **106.** $\sec(\arctan \sqrt{2})$

107. $\tan(\tan^{-1} 16)$ **108.** $\sin[\arcsin(-0.1)]$

109. $\sin^{-1}(\sin 3\pi)$ **110.** $\arccos\left(\cos \dfrac{5\pi}{2}\right)$

Exercises 111-114 *Simplify the given expression. Assume x > 0.*

111. $\sec(\arcsin x)$

112. $\cot(\cos^{-1} x)$

113. $\cos\left(\tan^{-1}\dfrac{x}{3}\right)$

114. $\sin\left(\arctan\dfrac{1}{x}\right)$

Exercises 115-118 *Sketch the graph of the given function.*

115. $y = \arctan(x + 1)$

116. $y = \sin^{-1}(x - 3)$

117. $y = \arccos(x) + \dfrac{\pi}{2}$

118. $y = \tan^{-1}(x - 2) - \pi$

Exercises 119-122 *Parts a and b are connected: Part a involves an elementary concept, whereas part b involves related material from this chapter. First answer part a, and then use this result to answer part b.*

119. a. Find the missing side length in Figure 121.

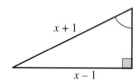

Figure 121

   b. Simplify $\sin\left[\arccos\left(\dfrac{x - 1}{x + 1}\right)\right]$.

120. a. Simplify $\sqrt{1 - \left(\dfrac{5}{13}\right)^2}$.

   b. Find the possible values of $\sin \theta$ given that $\cos \theta = \dfrac{5}{13}$.

121. a. Assuming that $f^{-1}(x)$ is defined, simplify $f(f^{-1}(x))$.

   b. Evaluate $\tan(\arctan 2001)$.

122. a. Solve $2x + \dfrac{\pi}{4} = k\pi$ for x.

   b. Graph $\csc\left(2x + \dfrac{\pi}{4}\right)$.

123. Pulley Angle A 4-inch-diameter pulley is used to reel in 20 feet of electrical cable. Through how many radians does the pulley turn, and to how many revolutions does this correspond?

124. Bicycle Speed A bicycle tire with a diameter of 20 inches makes $2\frac{1}{2}$ complete revolutions for each revolution of the pedals. If a child is pedaling at a rate of 40 revolutions per minute, find the speed of the bicycle in miles per hour.

125. World's Largest Clock Face According to *The Guinness Book of World Records*, the world's largest clock face—101 feet in diameter—is that of the floral clock in Matsubara Park, Toi, Japan. Find the linear speed, in miles per hour, of the tip of the minute hand if it is 41 feet from the center of the clock.

126. Yacht Distance The yacht *Jeffmiller* leaves port on a course of 48° and travels for 25 miles. It then travels due south for an unknown distance and returns to port by traveling due west. Find the distance traveled by the ship.

127. Viewing Distance The Colossus of Rhodes, one of the Seven Wonders of the World, was a 120-foot statue of Apollo at the entrance to the harbor of Rhodes. If an ancient observer were situated so that the angle of elevation to the top of the statue was 35°, then how far was the observer from the statue?

Colossus of Rhodes

128. Volume of a Feeding Trough A rectangular piece of metal with dimensions 3 feet by 20 feet is to be formed into a feeding trough by bending up two sides so that each forms an angle θ with the horizontal, as shown in Figure 122. The volume of the trough in cubic feet as a function of the angle θ is given by $f(\theta) = 20 \sin \theta(\cos \theta + 1)$.

   a. Find the volume of the trough for $\theta = \dfrac{\pi}{4}$, $\theta = \dfrac{\pi}{3}$, and $\theta = \dfrac{\pi}{2}$.

   b. Use a graphing calculator to estimate the value of θ for which the volume is as large as possible.

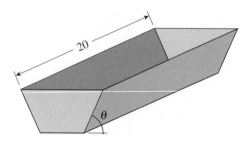

Figure 122

129. Weight Fluctuations   A certain dieter's weight has been found to "yo-yo" according to the formula

$$w(t) = 10 \cos\left(\frac{\pi}{6}t\right) + 200$$

where t represents the number of months that have elapsed since January 1.

a. Find the period of $w(t)$. Interpret this result.

b. Find the amplitude of w.

c. What is the greatest weight that the dieter attains in the course of the year? When does the dieter reach this weight?

d. What is the lightest weight that the dieter attains in the course of the year? When does the dieter reach this weight?

130. Angle of Elevation   A package dropped from a helicopter descends straight down in such a way that its height in feet t seconds after being dropped is given by $y(t) = 2000 - 16t^2$. From the perspective of an observer on the ground 100 feet from the point of impact, the angle of elevation of the package is θ. Express θ as a function of t, and find θ 10 seconds after the package has been dropped.

Chapter 5 Test

Problems 1-10   *Answer true or false.*

1. Given any angle θ, there are exactly two angles that are coterminal with θ.

2. $\sin(\arcsin x) = x$ for $-1 \le x \le 1$.

3. The function $f(x) = \arcsin x$ has period 2π.

4. If α and β are coterminal, then $\alpha = \beta$.

5. If α and β are coterminal, then $\sin \alpha = \sin \beta$.

6. The phase shift of the function $f(x) = a \sin(bx + c)$ is $-c$.

7. The period of $f(x) = a \sin(bx + c)$ depends on the value of b but not on the values of a and c.

8. If $\sin \alpha = \sin \beta$, then $\alpha = \beta$.

9. A reference angle is always acute.

10. The graphs of $f(x) = a \sin bx$ and $g(x) = -a \sin bx$ have the same amplitude.

Problems 11-14   *Give an example of each of the following.*

11. A trigonometric function with amplitude 3 and period $\frac{\pi}{2}$

12. A nonacute angle θ whose tangent is 1

13. A trigonometric function with an asymptote at π

14. The side lengths of a right triangle with an angle measuring $40°$

15. Find the length of the arc subtended by a central angle of $25°$ in a circle of radius 5.

16. Find the values of the six trigonometric functions for the angle θ shown in Figure 123.

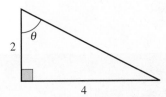

Figure 123

17. From a distance of 50 feet away from the base of a radio tower, the angle of elevation to the top of the tower is $75°$. Find the height of the tower.

18. Find the exact value of $\cos\left(\frac{7\pi}{6}\right)$.

19. Graph three cycles of the function $f(x) = 2 \sin(3x - \pi) + 1$. Find the amplitude, period, and phase shift.

20. Use a graphing calculator to estimate the period and amplitude of the function

$$f(x) = 4 \sin\left(\frac{\pi x}{2}\right) - 3 \cos(\pi x)$$

21. Graph three cycles of the function $g(x) = \tan(\pi x - \pi)$. Find the period and indicate the asymptotes.

22. Find the exact value of $\arccos\left(\cos \frac{8\pi}{5}\right)$.

23. Use inverse trigonometric functions to find all solutions of $\sin x - \frac{1}{3} = 0$ on the interval $[0, 2\pi)$.

Chapter 6
Trigonometric Identities and Equations

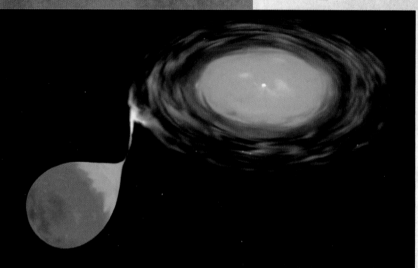

This illustration depicts a double star system at the heart of the globular cluster NGC 6624, in the constellation Sagittarius. The intense gravitational pull of the very dense neutron star (upper right) distorts its less massive white dwarf counterpart (lower left), cannibalizing its gasses across a narrow accretion bridge. The stars rotate about each other every 11 minutes, locked in an intricate celestial dance punctuated by sporadic outbursts of X rays. The script of this distant drama is encoded in these X rays and in the ultraviolet radiation that reaches the Hubble Space Telescope but can only be decoded by mathematics that is, in essence, an extension of the trigonometric concepts introduced in this chapter.

Section 6.1 | Fundamental Trigonometric Identities

- Can an equation have infinitely many solutions and yet not be an identity?
- What is the shape of the Gateway Arch in St. Louis?
- How many trigonometric identities are there?
- How do engineers determine the angle at which a turn in a highway is to be banked?

Classification of Trigonometric Equations

Equations can be classified into three categories according to the nature of the solution set. Equations such as $2(x + 1) = 2x + 2$, which are *always* true, are called **identities**; equations that are *sometimes* true, like $x + 1 = 3$, are called **conditional equations**; and those that are *never* true, like $x + 1 = x + 2$, are said to be **contradictions**.

Most identities involving nontrigonometric functions are transparently true once a few basic rules are understood. For example, the equation $2(x + 1) = 2x + 2$ is clearly an identity; it is just a special case of the distributive law. Similarly, the equation $2^{x+5} = 2^x 2^5$ is a special case of the rule $a^{x+y} = a^x a^y$. But consider the trigonometric equation

$$\frac{\sin^2 x}{\cos x} + \frac{\cos^2 x}{\sin x} = \sec x + \csc x - \cos x - \sin x$$

As it turns out, this equation is also an identity. But who among us can glance at such a complex equation and immediately recognize the left and right sides as being equivalent?

Fortunately, there is an easy way to expose identities. By plotting the graphs of both the left and right sides of the equation, we should see one of the following cases:

1. One graph: Both sides are identical; therefore, the equation is an identity.

2. Two graphs, at least one point of intersection: The equation is sometimes true; therefore, it is a conditional equation.

3. Two graphs, no points of intersection: The equation is never true; therefore, it is a contradiction.

EXAMPLE 1

Classifying a Trigonometric Equation with a Graphing Calculator

Use a graphing calculator to help decide whether the following equation is a contradiction, a conditional equation, or an identity:

$$2 \sin x = 2 - 2 \cos x$$

Solution Figure 1 shows the graph of $y = 2 \sin x$. Figure 2 depicts the graph of $y = 2 \sin x$ and $y = 2 - 2 \cos x$ on the same set of coordinate axes. Since the graphs intersect but are not identical, the equation is conditional.

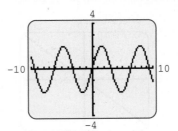

Figure 1

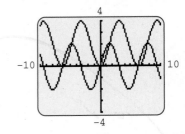

Figure 2

······>**EXAMPLE 2**

Classifying a Trigonometric Equation with a Graphing Calculator

Use a graphing calculator to help decide whether the following equation is a contradiction, a conditional equation, or an identity:

$$(\sin x + \cos x)^2 = 1 + \sin 2x$$

Solution Figure 3 shows the graph of $y = (\sin x + \cos x)^2$. In Figure 4, we have graphed $y = (\sin x + \cos x)^2$ and $y = 1 + \sin 2x$ on the same set of coordinate axes. Since the graphs are indistinguishable (there appears to be just a single graph), we conclude that the equation is most likely an identity

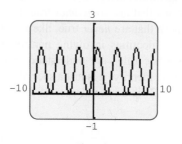

Figure 3 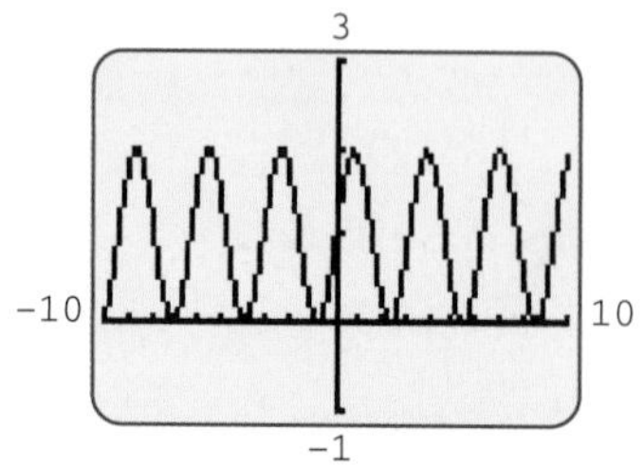   **Figure 4**

······>**EXAMPLE 3**

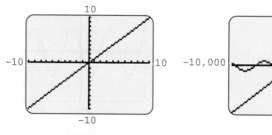

Figure 5

Classifying a Trigonometric Equation with a Graphing Calculator

Use a graphing calculator to decide whether the following equation is a contradiction, a conditional equation, or an identity:

$$2 - \sin x = \cos x$$

Solution The graphs of $y = 2 - \sin x$ and $y = \cos x$ are shown in Figure 5. Since the graphs do not appear to intersect, we deduce that the equation is a contradiction.

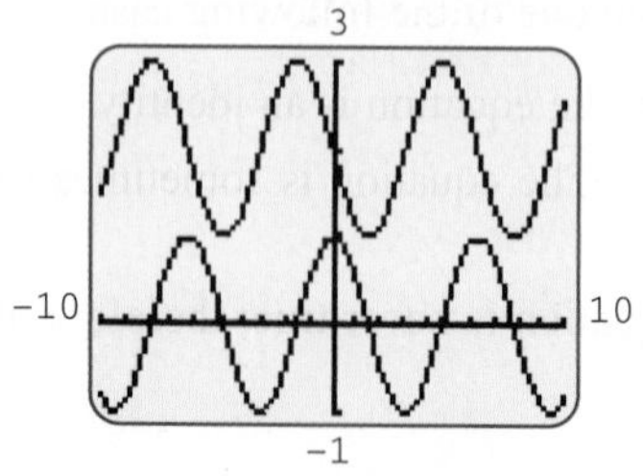

As usual, there are perils in relying on technology alone to settle mathematical questions. For example, consider the equation

$$1000 \sin \frac{x}{1000} = x$$

If we graph $y = 1000 \sin \frac{x}{1000}$ and $y = x$ on the same set of coordinate axes with the standard viewing window, we obtain the graph shown in Figure 6. From this, we would conclude that our equation is an identity. But if we change the window variables to `Xmin = -10000`, `Xmax = 10000`, `Ymin = -10000`, and `Ymax = 10000`, we obtain the graph shown in Figure 7.

Figure 6 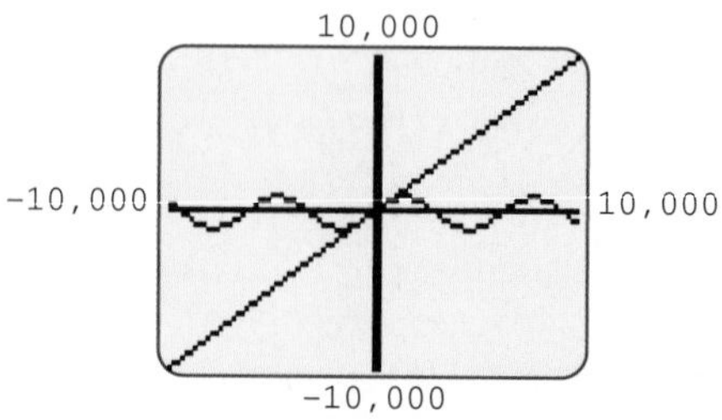   **Figure 7**

With this viewing window, it is quite clear that the equation is *not* an identity. Evidently, for relatively small values of x, $1000 \sin(x/1000)$ is nearly identical to x, which caused us to be misled by our first graph. The point is that graphical information, although providing us with valuable insight, is not enough to verify an identity. Instead, an identity is verified by showing that the two sides of an equation are equivalent using previously established identities.

Among the infinitely many trigonometric identities, a mere handful are sufficiently simple and useful to merit memorization. These fundamental identities can then be used to verify other, more complex, identities. We encountered these fundamental identities in the previous chapter, but we restate them here for convenience.

Pythagorean Identities

$$\sin^2\theta + \cos^2\theta = 1 \qquad 1 + \tan^2\theta = \sec^2\theta \qquad 1 + \cot^2\theta = \csc^2\theta$$

Ratio Identities

$$\sin\theta = \frac{1}{\csc\theta} \qquad \cos\theta = \frac{1}{\sec\theta} \qquad \tan\theta = \frac{1}{\cot\theta}$$

$$\csc\theta = \frac{1}{\sin\theta} \qquad \sec\theta = \frac{1}{\cos\theta} \qquad \cot\theta = \frac{1}{\tan\theta}$$

$$\tan\theta = \frac{\sin\theta}{\cos\theta} \qquad \cot\theta = \frac{\cos\theta}{\sin\theta}$$

Verifying Identities   To verify an identity, we must show that the equation is true for all values of the variable for which both sides are defined. To do this, we begin with one side of the equation and rewrite it either by applying a basic trigonometric identity or by algebraically rearranging. We repeat this process as many times as necessary until at last we arrive at the expression on the other side of the equation. Equivalently, we can work on both sides (independently) until we are able to show that both sides are equivalent to a third expression.

Verifying Identities

Rule 1   Wherever possible, simplify an expression rather than make it more complex. Thus, if given an equation in which one side is much more complex than the other, begin working on the complex side. If both sides are complex, simplify both sides independently.

Rule 2   If all else fails, convert both sides to sines and cosines. This is not always the most efficient way of verifying an identity, but it almost always works.

Rule 3   Look for opportunities to exploit algebraic factoring—particularly the difference of squares formula $a^2 - b^2 = (a - b)(a + b)$. For example, the expression

$$\frac{\cos^2 x - \sin^2 x}{\cos x + \sin x}$$

can be simplified by first factoring the numerator:

$$\frac{\cos^2 x - \sin^2 x}{\cos x + \sin x} = \frac{(\cos x - \sin x)(\cos x + \sin x)}{\cos x + \sin x}$$
$$= \cos x - \sin x$$

WARNING!

It is often tempting to perform an algebraic operation on both sides of the equation at the same time, such as squaring both sides. This is not a legitimate method of verification. (See Exercise 82 for an exploration of this issue.)

EXAMPLE 4

Verifying an Identity

Verify that the following equation is an identity:

$$\sin x \cot x = \cos x$$

Solution We begin with the left side since it is more complex than the right.

$$\sin x \cot x = \sin x \frac{\cos x}{\sin x} \qquad \text{Using } \cot x = \frac{\cos x}{\sin x}$$

$$= \cos x$$

EXAMPLE 5

Verifying an Identity

Verify that the following equation is an identity:

$$\frac{1 - \cos x}{\cos x}(\sec x + 1) = \tan^2 x$$

Solution We work with the left side since it is much more complex than the right.

$$\frac{1 - \cos x}{\cos x}(\sec x + 1) = \left(\frac{1}{\cos x} - \frac{\cos x}{\cos x}\right)(\sec x + 1)$$

$$= (\sec x - 1)(\sec x + 1) \qquad \text{Using } \sec x = \frac{1}{\cos x}$$

$$= \sec^2 x - 1$$

$$= (\tan^2 x + 1) - 1 \qquad \text{Using } \sec^2 x = \tan^2 x + 1$$

$$= \tan^2 x$$

Many identities can be established in several different (but equally simple) ways. This reflects the fact that, at any stage of the verification process, there is no single *right* step; there are, however, *sensible* steps for which the likelihood of success is high.

EXAMPLE 6

Verifying an Identity

Verify that the following equation is an identity:

$$1 = (1 - \cos^2 u)(1 + \cot^2 u)$$

Solution Since the right side is more complex than the left, we begin on the right. There are several ways to attack this identity—we show two.

Method 1

$$(1 - \cos^2 u)(1 + \cot^2 u) = \sin^2 u \csc^2 u \qquad \text{Using Pythagorean identities}$$

$$= \sin^2 u \frac{1}{\sin^2 u} \qquad \text{Using } \csc u = \frac{1}{\sin u}$$

$$= 1$$

Method 2

$$(1 - \cos^2 u)(1 + \cot^2 u) = \sin^2 u \left(1 + \frac{\cos^2 u}{\sin^2 u}\right)$$ Using $1 - \cos^2 u = \sin^2 u$ and $\cot u = \frac{\cos u}{\sin u}$

$$= \sin^2 u + \cos^2 u$$ Multiplying through

$$= 1$$

For cases in which the two sides of the equation are of comparable complexity and there is no obvious way to convert one to the other, converting all trigonometric functions to expressions involving sine and cosine is often the best strategy.

EXAMPLE 7   **Verifying an Identity**

Verify that the following equation is an identity:

$$\frac{\tan \theta}{1 - \sec \theta} = \frac{1}{\cot \theta - \csc \theta}$$

Solution   In this case, both sides are equally complex. We convert both sides to sines and cosines, beginning with the left side.

$$\frac{\tan \theta}{1 - \sec \theta} = \frac{\dfrac{\sin \theta}{\cos \theta}}{1 - \left(\dfrac{1}{\cos \theta}\right)}$$ Using ratio identities

$$= \frac{\dfrac{\sin \theta}{\cos \theta}}{\dfrac{\cos \theta - 1}{\cos \theta}}$$ Simplifying the denominator

$$= \left(\frac{\sin \theta}{\cos \theta}\right)\left(\frac{\cos \theta}{\cos \theta - 1}\right)$$ Inverting and multiplying

$$= \frac{\sin \theta}{\cos \theta - 1}$$

On the other hand, the right side simplifies as follows:

$$\frac{1}{\cot \theta - \csc \theta} = \frac{1}{\dfrac{\cos \theta}{\sin \theta} - \dfrac{1}{\sin \theta}}$$ Using ratio identities

$$= \frac{1}{\dfrac{\cos \theta - 1}{\sin \theta}}$$

$$= \frac{\sin \theta}{\cos \theta - 1}$$

Since both sides of the original equation simplify to the same expression, the equation is an identity.

····▷EXAMPLE 8

Verifying an Identity

Verify that the following equation is an identity:

$$\frac{\cos \alpha}{1 - \sin \alpha} = \sec \alpha + \tan \alpha$$

Solution We begin by converting the right side to sines and cosines.

$$\sec \alpha + \tan \alpha = \frac{1}{\cos \alpha} + \frac{\sin \alpha}{\cos \alpha} \qquad \text{Using ratio identities}$$

$$= \frac{1 + \sin \alpha}{\cos \alpha}$$

Unfortunately, this is not the same as the expression on the left side. Since the denominator of the desired expression is $1 - \sin \alpha$, we multiply the numerator and denominator of $(1 + \sin \alpha)/\cos \alpha$ by $1 - \sin \alpha$ as follows:

$$\left(\frac{1 + \sin \alpha}{\cos \alpha}\right)\left(\frac{1 - \sin \alpha}{1 - \sin \alpha}\right) = \frac{1 - \sin^2 \alpha}{\cos \alpha \,(1 - \sin \alpha)}$$

$$= \frac{\cos^2 \alpha}{\cos \alpha \,(1 - \sin \alpha)} \qquad \text{Using } 1 - \sin^2 \alpha = \cos^2 \alpha$$

$$= \frac{\cos \alpha}{1 - \sin \alpha}$$

Thus,

$$\sec \alpha + \tan \alpha = \frac{1 + \sin \alpha}{\cos \alpha} = \frac{\cos \alpha}{1 - \sin \alpha}$$

Mathematical Applications of Trigonometric Identities

There are many contexts in which one might wish to simplify trigonometric expressions. For example, suppose we are given the function

$$f(x) = \frac{\cos^2 x}{1 - \sin x}$$

and we wish to determine its period, its maximum and minimum values, and its amplitude. Of course, these quantities could be *estimated* by inspecting the graph of f, but by rewriting the function as

$$f(x) = 1 + \sin x$$

we see that the graph of f is nothing more than a vertical translation (1 unit upward) of the graph of $y = \sin x$. As such, we would conclude that f has period 2π, maximum value 2, minimum value 0, and amplitude 1. The process by which the expression $\cos^2 x/(1 - \sin x)$ is simplified to obtain $1 + \sin x$ is the subject of the following example.

····▷EXAMPLE 9

Simplifying a Trigonometric Expression

Simplify the following expression:

$$\frac{\cos^2 x}{1 - \sin x}$$

Solution   We notice that the numerator involves $\cos^2 x$, which can easily be rewritten in terms of $\sin x$ using the Pythagorean identity $\sin^2 x + \cos^2 x = 1$. From there, we simplify algebraically.

$$\frac{\cos^2 x}{1 - \sin x} = \frac{1 - \sin^2 x}{1 - \sin x}$$

$$= \frac{(1 - \sin x)(1 + \sin x)}{1 - \sin x} \qquad \text{Factoring a difference of squares}$$

$$= 1 + \sin x$$

Note that, strictly speaking, $f(x) = \cos^2 x/(1 - \sin x)$ and $g(x) = 1 + \sin x$ are not identical since their domains are different. For example, $f(x)$ is not defined for $x = \frac{\pi}{2}$, but $g(x)$ is defined for all x.

Facility with the basic trigonometric identities enables us to express any trigonometric function in terms of any other, although there may be an unknown sign.

EXAMPLE 10

Expressing One Trigonometric Function in Terms of Another

Write $\tan x$ in terms of $\csc x$. Assume that $0 \le x < \frac{\pi}{2}$.

Solution   It would be nice if one of the fundamental identities directly related the tangent and cosecant functions, but none do. However, there *are* fundamental identities that connect the tangent and cotangent functions and the cosecant and cotangent functions. Specifically, we have

(1) $$\tan x = \frac{1}{\cot x}$$

and

(2) $$\cot^2 x + 1 = \csc^2 x$$

In essence, equations (1) and (2) allow us to use cotangent as an intermediary between tangent and cosecant. We begin by solving equation (2) for $\cot x$.

$$\cot^2 x + 1 = \csc^2 x$$
$$\cot^2 x = \csc^2 x - 1$$
$$\cot x = \pm\sqrt{\csc^2 x - 1}$$

Since $0 \le x < \frac{\pi}{2}$ and the cotangent function is positive in the first quadrant, we have

$$\cot x = +\sqrt{\csc^2 x - 1}$$

Now, from equation (1), we have

$$\tan x = \frac{1}{\cot x}$$

$$= \frac{1}{\sqrt{\csc^2 x - 1}}$$

Understanding and Mastery Checklists

Concepts to Understand

Identity

❖

Conditional equation

❖

Contradiction

❖

Pythagorean identities

❖

Ratio identities

Skills to Master

Classify an equation as an identity, a contradiction, or a conditional equation.

❖

Verify an identity.

❖

Express a given trigonometric function in terms of another.

Exercises 6.1

Exercises 1-12 *Use a graphing calculator to determine whether the given equation is most likely an identity, a contradiction, or a conditional equation.*

1. $2 \cos x = \sin x$

2. $\cos x = 1$

3. $\sin^2 x + \cos^2 x = 1.45$

4. $\sec x \cot x = \csc x$

5. $\sec^2 x - \csc^2 x = \tan^2 x - \cot^2 x$

6. $\sin 2x + \sin^2 x = 2$

7. $\cos 2x + \cos x = 3$

8. $\cos 3x + \sin 3x = 3x$

9. $2 \cot 2x = \cot x - \tan x$

10. $\sin x = 1.45 - \cos x$

11. $\cos x = 1 - \dfrac{x^2}{2} + \dfrac{x^4}{4}$

12. $\sin\left(x + \dfrac{\pi}{6}\right) = \dfrac{\sqrt{3}}{2} \sin x + \dfrac{1}{2} \cos x$

Exercises 13-22 *First classify the equation as an identity, conditional equation, or contradiction. If the equation is an identity, verify it. If it is a conditional equation, estimate the smallest positive solution.*

13. $2 \sin^2 x - 1 = 1 - 2 \cos^2 x$

14. $\cos^3 x + \sin^3 x = 1$

15. $\sin(x - \pi) = \sin x - \pi$

16. $\dfrac{1}{\cos v} - \cos v = \dfrac{\sin^2 v}{\cos v}$

17. $\cos x \sin x = \cos x + \sin x$

18. $\sin\left(x - \dfrac{\pi}{2}\right) = \sin x - \sin \dfrac{\pi}{2}$

19. $\tan x \cot x - 1 = 0$

20. $\cos x = 1.1 - \dfrac{x^2}{2} + \dfrac{x^4}{24}$

21. $\sin 4x = 4 \sin x$

22. $\dfrac{\cot \theta}{\csc \theta + 1} = \dfrac{\csc \theta - 1}{\cot \theta}$

Exercises 23-58 *Verify the given identity algebraically.*

23. $\dfrac{\tan x}{\sin x} = \sec x$

24. $\dfrac{\cot x}{\cos x} = \csc x$

25. $\dfrac{\sec x}{\csc x} = \tan x$

26. $\dfrac{\tan x}{\cot x} = \sec^2 x - 1$

27. $\sin^2 x = 1 - \cos^2 x$

28. $(1 + \cos s)(1 - \cos s) = \sin^2 s$

29. $\sec^2 x - \tan^2 x = 1$

30. $\tan^2 \omega + 8 = \sec^2 \omega + 7$

31. $\tan^2 \beta - \sin^2 \beta = \tan^2 \beta \sin^2 \beta$

32. $\cos x - \sin x = \dfrac{1 - 2\sin^2 x}{\cos x + \sin x}$

33. $\cos t + \sin t = \dfrac{\tan t + 1}{\sec t}$

34. $\tan x + \cot x = \sec x \csc x$

35. $\sec^2 x + \csc^2 x = \sec^2 x \csc^2 x$

36. $\dfrac{\cos u}{\sec u} + \dfrac{\sin u}{\csc u} = 1$

37. $\sin u + \cos u \cot u = \csc u$

38. $\sec^2 \theta = \sec \theta \csc \theta \tan \theta$

39. $(\sin^2 \alpha - 1)(\cot^2 \alpha + 1) = 1 - \csc^2 \alpha$

40. $\dfrac{1}{\csc^2 x} + \dfrac{1}{\sec^2 x} = 1$

41. $\dfrac{\csc x + \cot x}{\csc x - \cot x} = (\csc x + \cot x)^2$

42. $\dfrac{\sin \theta \tan \theta}{1 - \cos \theta} = \sec \theta + 1$

43. $\dfrac{1}{1 - \sin x} = \sec^2 x + \sec x \tan x$

44. $\dfrac{1 - \cos x}{1 + \cos x} = 2\csc^2 x - 2\cot x \csc x - 1$

45. $\dfrac{\sin^2 x}{\cos x} + \dfrac{\cos^2 x}{\sin x} = \sec x + \csc x - \cos x - \sin x$

46. $\dfrac{\tan^4 x - 1}{\tan^2 x - 1} = \sec^2 x$

47. $\dfrac{\sin y \tan y}{\tan y - \sin y} = \dfrac{\tan y + \sin y}{\sin y \tan y}$

48. $\dfrac{1}{\tan x + 1} + \dfrac{1}{\cot x + 1} = 1$

49. $\sec^4 z - 2\tan^2 z \sec^2 z + \tan^4 z = 1$

50. $\dfrac{\sin w}{\sec w + 1} + \dfrac{\sin w}{\sec w - 1} = 2\cot w$

51. $(\csc \phi - \cot \phi)^2 = \dfrac{1 - \cos \phi}{1 + \cos \phi}$

52. $\sin^3 v + \cos^3 v = (1 - \sin v \cos v)(\sin v + \cos v)$

53. $|\cos x| = \sqrt{1 - \sin^2 x}$

54. $|\tan x| = \sqrt{\sec^2 x - 1}$

55. $\ln|\cot s| = -\ln|\tan s|$

56. $\ln|\tan x| = \ln|\sin x| - \ln|\cos x|$

57. $\ln|\sec w + \tan w| = -\ln|\sec w - \tan w|$

58. $\ln|\csc \theta + \cot \theta| = -\ln|\csc \theta - \cot \theta|$

Exercises 59-64 *Use fundamental trigonometric identities to write the given expression as a single trigonometric function.*

59. $\dfrac{\sin^2 x + \cos^2 x}{\cos x}$

60. $\dfrac{\tan^2 x + 1}{\sec x}$

61. $\dfrac{\tan x \cot x}{\sin x}$

62. $\cos x \sec^2 x \cot x$

63. $\dfrac{\sin^2 x - \cos^2 x}{\sin x + \cos x} + \dfrac{1}{\sec x}$

64. $(\csc x - 1)(\csc x + 1)\tan x$

Exercises 65-72 *Use fundamental trigonometric identities to rewrite the first function in terms of the second.*

65. $\tan x;\ \cot x$

66. $\sec x;\ \cos x$

67. $\cos x;\ \sin x$

68. $\sec x;\ \tan x$

69. $\cot x;\ \cos x$

70. $\tan x;\ \sin x$

71. $\sec x;\ \sin x$

72. $\cot x;\ \cos x$

Applications

73. Banked Curves For vehicles to travel safely around curves at high speeds, it is often necessary to bank the curves (see Figure 8). By considering frictional and centripetal forces, it can be shown that the proper angle θ for banking a curve with radius r feet so a vehicle can travel safely around the curve at a velocity of v feet per second is given by the equation

$$\sin \theta = \frac{v^2}{32r} \cos \theta$$

a. Solve this equation for v and simplify as much as possible.

b. What is the safest speed that a vehicle can travel around a curve with a radius of 2400 feet and banked at an angle of $5°$?

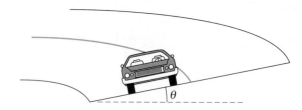

Figure 8

Concepts and Critical Thinking

Exercises 74-77   *Answer true or false.*

74. No equation of the form $\sin x = a$ has exactly 2 solutions.

75. If the graphs of f and g are identical, then the equation $f(x) = g(x)$ is an identity.

76. If an equation has infinitely many solutions, then it is an identity.

77. If an equation is an identity, then it has infinitely many solutions.

Exercises 78-81   *Give an example of each.*

78. A trigonometric identity

79. A conditional trigonometric equation

80. A trigonometric equation with no solutions

81. An expression E involving only the secant and cosecant functions such that $E = \tan \theta$ is an identity

82. In the text, it was stated that identities may be verified either by simplifying one side of an equation until it looks like the other or simplifying both sides separately until they are equal to a common expression. What is wrong with the following "proof" of the "identity" $\sin x = -\sqrt{1 - \cos^2 x}$?

$$\sin x = -\sqrt{1 - \cos^2 x}$$
$$\sin^2 x = 1 - \cos^2 x$$
$$\sin^2 x + \cos^2 x = 1$$
$$1 = 1$$

Evaluate both sides of the equation

$$\sin x = -\sqrt{1 - \cos^2 x}$$

at $x = \frac{\pi}{2}$. Is this equation an identity? Explain.

83. Suppose that the tangent of an angle θ is known. Which of the other five trigonometric functions can be determined at θ? What if you also know that $\sin \theta$ is positive?

Questions for Discussion or Essay

84. Explain how any polynomial in $\sin x$ and $\cos x$, such as $\cos^5 x + \cos^4 x \sin^3 x + 3 \cos x \sin^2 x - 4 \sin^3 x$, can be rewritten so that $\sin x$ never appears to a power higher than 1.

85. In the text, it was stated that most *algebraic* identities are transparently true, such as $x + 2 = 2 + x$, whereas most trigonometric identities require effort to verify. We have seen, however, a category of nontrigonometric functions for which there were several nontrivial identities. To what type of function are we referring? List three identities involving these functions.

86. Consider the identity $\tan x \cos x = \sin x$. This identity is true for virtually all values of x. However, there are values of x for which the expression on the left side of the equal sign is not equivalent to the expression on the right-hand side. For which values of x does this identity not hold? Is the equation still worthy of being called an identity if it doesn't hold for certain values of x?

Projects for Enrichment

87. Hyperbolic Functions   The trigonometric functions are often called **circular functions** because of the fact that cos θ and sin θ can be defined to be the x- and y-coordinates, respectively, of the point on the unit circle corresponding to the angle θ. If we begin with a hyperbola (instead of a circle), we obtain the **hyperbolic functions**: cosh x, sinh x, tanh x, sech x, csch x, and coth x. Notice that the notation for the hyperbolic functions is quite similar to that of the trigonometric functions.

The hyperbolic functions arise in many physical contexts. In particular, a cable of uniform density strung between supports (such as a power or telephone line) hangs in the shape of a **catenary**, the graph of $y = \cosh x$. Moreover, the St. Louis Arch is in the shape of an "upside-down" catenary. In this project, we investigate the algebraic properties of hyperbolic functions and the hyperbolic analogs of the trigonometric identities. The hyperbolic functions are defined as follows:

$$\cosh x = \frac{e^x + e^{-x}}{2} \qquad \operatorname{sech} x = \frac{2e^x}{e^{2x} + 1}$$

$$\sinh x = \frac{e^x - e^{-x}}{2} \qquad \operatorname{csch} x = \frac{2e^x}{e^{2x} - 1}$$

$$\tanh x = \frac{e^{2x} - 1}{e^{2x} + 1} \qquad \coth x = \frac{e^{2x} + 1}{e^{2x} - 1}$$

St. Louis Gateway Arch

Cable hanging in the shape of a catenary

a. Use the definitions of the hyperbolic functions to verify the following identities:

   i. $\tanh x = \dfrac{\sinh x}{\cosh x}$ **ii.** $\coth x = \dfrac{1}{\tanh x}$

   iii. $\operatorname{sech} x = \dfrac{1}{\cosh x}$ **iv.** $\operatorname{csch} x = \dfrac{1}{\sinh x}$

   v. $\cosh^2 x - \sinh^2 x = 1$ **vi.** $\operatorname{sech}^2 x + \tanh^2 x = 1$

   vii. $\coth^2 x - \operatorname{csch}^2 x = 1$ **viii.** $\cosh(-x) = \cosh x$

   ix. $\sinh(-x) = -\sinh x$

b. Use the identities from part a to verify the following identities:

   i. $\dfrac{\operatorname{csch} x}{\operatorname{sech} x} = \coth x$

   ii. $\cosh^2 x + \sinh^2 x = 2\cosh^2 x - 1$

   iii. $\dfrac{\tanh x}{\cosh x - \sinh x \tanh x} = \sinh x$

   iv. $\cosh x + \sinh x = \dfrac{1}{\cosh x - \sinh x}$

c. Write the corresponding trigonometric identity for each hyperbolic identity given in part a.

d. Show that the point $(\cosh x, \sinh x)$ is on the hyperbola $x^2 - y^2 = 1$.

Section 6.2 | # Sum and Difference Identities

- ◈ What note do our ears hear when two instruments play a note of the same frequency?
- ◈ Is the sine of the sum of two angles equal to the sum of the sines of the angles?
- ◈ What connection is there between the ratios of sides in a triangle and the chemical composition of distant galaxies?

Sum and Difference Identities for Sine

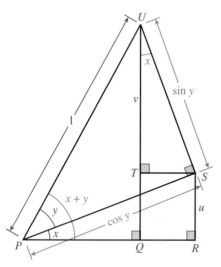

Figure 9

Suppose that the sine and cosine of the angles x and y are known. Is this sufficient information to compute $\sin(x + y)$? That is, can we somehow express $\sin(x + y)$ in terms of $\cos x$, $\sin x$, $\cos y$, and $\sin y$? A natural (but dead wrong!) guess would be that $\sin(x + y)$ simplifies to $\sin x + \sin y$. Indeed, we need simply note that $\sin\left(\frac{\pi}{2} + \frac{\pi}{2}\right) = 0$ but $\sin\frac{\pi}{2} + \sin\frac{\pi}{2} = 2$.

There is, in fact, an identity for $\sin(x + y)$, as well as for $\sin(x - y)$, $\cos(x + y)$, and $\cos(x - y)$. We begin by using right triangles to derive the identity for $\sin(x + y)$ for the special case where x, y, and $x + y$ are all acute angles. (A more traditional approach for proving these identities, one that is somewhat less intuitive but works for all angles x and y, is considered in Exercise 68. In Section 7.4, we consider a third approach using complex numbers.)

In Figure 9, we have adjoined two right triangles, one with an angle of x and another with an angle of y, in such a way as to form an angle of $x + y$. The angle measures and labels shown in blue are by assumption, whereas those in red follow from elementary geometry and right triangle trigonometry. Thus, for example, it can be shown that $\angle TUS = x$ using elementary geometry, whereas $\angle RPS = x$ by design. The fact that $US = \sin y$ and $PS = \cos y$ follows from applying the definitions of sine and cosine, respectively, to the right triangle PSU.

Let us assume that the diagram in Figure 9 is accurate. We are interested in an expression for $\sin(x + y)$, so we focus on $\triangle PQU$, a right triangle with one angle of measure $x + y$. From the definition of sine, we have that

$$\sin(x + y) = \frac{QU}{PU}$$

$$= \frac{u + v}{1}$$

$$= u + v$$

Now, if we can find expressions for u and v in terms of sines and cosines of x and y, we will be done. We begin with u. From $\triangle RPS$, we see that

$$\sin x = \frac{u}{\cos y}$$

and so it follows that $u = \sin x \cos y$. To find an expression for v, we consider $\triangle UTS$. We have

$$\cos x = \frac{v}{\sin y}$$

Solving for v gives us $v = \sin y \cos x$. Collecting our results, we have

$$\sin(x + y) = u + v$$

$$= \sin x \cos y + \sin y \cos x$$

Thus, we have shown that $\sin(x + y) = \sin x \cos y + \sin y \cos x$ for all acute angles x, y, and $x + y$. An approach involving the distance formula can be used to show that the formula holds for all angles x and y (see Exercise 68).

We can obtain the difference identity for sine by substituting $-y$ for y in the sum identity as follows:

$$\sin[x + (-y)] = \sin x \cos(-y) + \sin(-y) \cos x$$

Now, using the even/odd identities for sine and cosine [that is, $\sin(-x) = -\sin x$ and $\cos(-x) = \cos x$], we arrive at

$$\sin(x - y) = \sin x \cos y - \sin y \cos x$$

We summarize the sum and difference identities as follows.

Sum and Difference Identities for Sine

$$\sin(x + y) = \sin x \cos y + \sin y \cos x$$

$$\sin(x - y) = \sin x \cos y - \sin y \cos x$$

Thus, if the sines and cosines of two angles are known, the sines of the sums and differences can be computed.

EXAMPLE 1   **Applying the Difference Identity for Sine**

Given that $\sin u = \frac{3}{5}$ and $\cos u = \frac{4}{5}$, compute $\sin\left(u - \frac{3\pi}{2}\right)$.

Solution   According to the difference identity for sine, we have

$$\sin\left(u - \frac{3\pi}{2}\right) = \sin u \cos \frac{3\pi}{2} - \sin \frac{3\pi}{2} \cos u$$

$$= \left(\frac{3}{5}\right) \cdot 0 - (-1)\left(\frac{4}{5}\right)$$

$$= \frac{4}{5}$$

EXAMPLE 2   **Applying the Sum Identity for Sine**

Find an exact value for $\sin 75°$.

Solution   We begin by searching for two special angles, the sum or difference of which is 75°. After some trial and error, we happen upon the relation $75° = 30° + 45°$. Thus, we use the sum identity for sine.

$$\sin 75° = \sin(30° + 45°)$$

$$= \sin 30° \cos 45° + \sin 45° \cos 30°$$

$$= \left(\frac{1}{2}\right)\left(\frac{\sqrt{2}}{2}\right) + \left(\frac{\sqrt{2}}{2}\right)\left(\frac{\sqrt{3}}{2}\right)$$

$$= \frac{\sqrt{2} + \sqrt{6}}{4}$$

In the previous examples, the sines and cosines of both angles were either given or known. However, if we know the sine (or cosine) of an angle and its quadrant, we can determine the cosine (or sine) using the identity $\sin^2 x + \cos^2 x = 1$.

···· **EXAMPLE 3**   **Applying the Sum and Difference Identities**

Find $\sin(u + v)$ given that $\sin u = \frac{12}{13}$ and $\cos v = \frac{1}{2}$, where both u and v are acute angles.

Solution   The sum identity for $\sin(u + v)$ involves $\cos u$ and $\sin v$, neither of which is given. However, each can easily be computed using the identity $\sin^2 u + \cos^2 u = 1$. In the case of $\cos u$, we have

$$\sin^2 u + \cos^2 u = 1$$

$$\left(\frac{12}{13}\right)^2 + \cos^2 u = 1$$

$$\frac{144}{169} + \cos^2 u = 1$$

$$\cos^2 u = \frac{25}{169}$$

$$\cos u = \pm \frac{5}{13}$$

Since u is in the first quadrant, we conclude that $\cos u = \frac{5}{13}$. We compute $\sin v$ in a similar fashion.

$$\sin^2 v + \cos^2 v = 1$$

$$\sin^2 v + \frac{1}{4} = 1$$

$$\sin^2 v = \frac{3}{4}$$

$$\sin v = \pm \frac{\sqrt{3}}{2}$$

Again, since v is in the first quadrant, we conclude that $\sin v = \sqrt{3}/2$. Thus, we have

$$\sin(u + v) = \sin u \cos v + \sin v \cos u$$

$$= \left(\frac{12}{13}\right)\left(\frac{1}{2}\right) + \left(\frac{\sqrt{3}}{2}\right)\left(\frac{5}{13}\right)$$

$$= \frac{12 + 5\sqrt{3}}{26}$$

Sum and Difference Identities for Cosine

The difference identity for sine and the cofunction identity $\cos u = \sin\left(\frac{\pi}{2} - u\right)$ can be used to find the sum identity for cosine. In particular, with $u = x + y$, we have

$$\cos(x + y) = \sin\left[\frac{\pi}{2} - (x + y)\right]$$

$$= \sin\left[\left(\frac{\pi}{2} - x\right) - y\right]$$

$$= \sin\left(\frac{\pi}{2} - x\right)\cos y - \sin y \cos\left(\frac{\pi}{2} - x\right)$$

$$= \cos x \cos y - \sin y \sin x$$

We can now find the difference identity for cosine by substituting $-y$ for y and again applying the even/odd identities for sine and cosine to obtain

$$\cos[x + (-y)] = \cos x \cos(-y) - \sin(-y) \sin x$$

$$\cos(x - y) = \cos x \cos y + \sin y \sin x$$

We summarize these results as follows.

Sum and Difference Identities for Cosine

$$\cos(x + y) = \cos x \cos y - \sin x \sin y$$

$$\cos(x - y) = \cos x \cos y + \sin x \sin y$$

EXAMPLE 4

Applying the Sum and Difference Identities for Cosine

Given that $\cos u = \frac{3}{5}$, $\sin u = -\frac{4}{5}$, $\cos v = \frac{4}{5}$, and $\sin v = \frac{3}{5}$, compute each of the following quantities:

a. $\cos(u + v)$ **b.** $\cos(u - v)$

Solution

a. From the sum identity for cosine, we have

$$\cos(u + v) = \cos u \cos v - \sin u \sin v$$

$$= \left(\frac{3}{5}\right)\left(\frac{4}{5}\right) - \left(-\frac{4}{5}\right)\left(\frac{3}{5}\right)$$

$$= \frac{12}{25} + \frac{12}{25}$$

$$= \frac{24}{25}$$

b. From the difference identity for cosine, we have

$$\cos(u - v) = \cos u \cos v + \sin u \sin v$$

$$= \left(\frac{3}{5}\right)\left(\frac{4}{5}\right) + \left(-\frac{4}{5}\right)\left(\frac{3}{5}\right)$$

$$= \frac{12}{25} - \frac{12}{25}$$

$$= 0$$

On occasion, complex expressions can be simplified using the sum or difference identities.

EXAMPLE 5

Applying the Sum Identity for Cosine

Simplify $\cos 59° \cos 31° - \sin 59° \sin 31°$.

Solution   This can be simplified using the sum identity for cosine—namely,

$$\cos(u + v) = \cos u \cos v - \sin u \sin v$$

with $u = 59°$ and $v = 31°$.

$$\cos 59° \cos 31° - \sin 59° \sin 31° = \cos(59° + 31°)$$
$$= \cos(90°)$$
$$= 0$$

Countless identities involving special angles are easily verified using sum or difference identities.

EXAMPLE 6   **Verifying a Trigonometric Identity**

Verify that the following equation is an identity:

$$\cos(x - \pi) = -\cos x$$

Solution   From the difference identity for cosine, we have

$$\cos(x - \pi) = \cos x \cos \pi + \sin x \sin \pi$$
$$= \cos x \cdot (-1) + \sin x \cdot 0$$
$$= -\cos x$$

Sum and Difference Identities for Tangent

Since the tangent of an angle is simply the ratio of its sine to its cosine—that is,

$$\tan x = \frac{\sin x}{\cos x}$$

the sum identity for tangent can be derived from those of sine and cosine. In particular, we have

$$\tan(x + y) = \frac{\sin(x + y)}{\cos(x + y)}$$
$$= \frac{\sin x \cos y + \sin y \cos x}{\cos x \cos y - \sin x \sin y}$$

Dividing numerator and denominator by $\cos x \cos y$, we obtain

$$\tan(x + y) = \frac{\dfrac{\sin x \cos y}{\cos x \cos y} + \dfrac{\sin y \cos x}{\cos x \cos y}}{\dfrac{\cos x \cos y}{\cos x \cos y} - \dfrac{\sin x \sin y}{\cos x \cos y}}$$

$$= \frac{\dfrac{\sin x}{\cos x} + \dfrac{\sin y}{\cos y}}{1 - \left(\dfrac{\sin x}{\cos x}\right)\left(\dfrac{\sin y}{\cos y}\right)}$$

$$= \frac{\tan x + \tan y}{1 - \tan x \tan y}$$

Using the fact that the tangent function is odd—that is, that $\tan(-x) = -\tan x$—we obtain the difference identity for tangent.

$$\tan(x - y) = \tan[x + (-y)]$$

$$= \frac{\tan x + \tan(-y)}{1 - \tan x \tan(-y)}$$

$$= \frac{\tan x - \tan y}{1 + \tan x \tan y}$$

We summarize the sum and difference identities for tangent as follows.

Sum and Difference Identities for Tangent

$$\tan(x + y) = \frac{\tan x + \tan y}{1 - \tan x \tan y}$$

$$\tan(x - y) = \frac{\tan x - \tan y}{1 + \tan x \tan y}$$

EXAMPLE 7   **Applying the Sum and Difference Identities for Tangent**

Given that $\tan u = 2$ and $\tan v = \frac{1}{4}$, compute the following quantities:

a. $\tan(u + v)$ **b.** $\tan(u - v)$

Solution

a. $\tan(u + v) = \dfrac{\tan u + \tan v}{1 - \tan u \tan v}$

$$= \frac{2 + \dfrac{1}{4}}{1 - 2 \cdot \dfrac{1}{4}}$$

$$= \frac{9/4}{1/2}$$

$$= \frac{9}{2}$$

b. $\tan(u - v) = \dfrac{\tan u - \tan v}{1 + \tan u \tan v}$

$$= \frac{2 - \dfrac{1}{4}}{1 + 2 \cdot \dfrac{1}{4}}$$

$$= \frac{7/4}{3/2}$$

$$= \frac{7}{6}$$

Spectral Analysis

A musical score encodes information regarding several well-defined aspects of the music, including pitch, amplitude, and duration. But the score contains no information regarding one of the most interesting (and difficult to define) musical qualities, **timbre**. Roughly, timbre is that quality that allows us to distinguish among different musical instruments. Mathematically, music arises from periodic sound waves. The graph or **waveform** associated with a pure tone, for example, is simply that of a function of the form $y(t) = a \cos wt$. On the other hand, the waveform associated with the same note played on a rich-sounding instrument (like an oboe or a violin) will have the same period but will be much more complex. In essence, it is the intricacy of the waveform that determines the richness of the timbre.

Now we can produce fairly complex waveforms by taking sums of cosine functions with different periods. For example, consider the graph of the function

$$f(t) = 4 \cos t + 3 \cos 3t + 2 \cos 5t + 5 \cos 7t$$

shown in the accompanying figure. *Which* periods are selected determines the **spectrum** of a waveform.

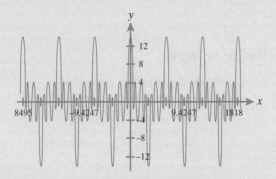

Spectral analysis is the process by which one identifies the constituent tones ($4 \cos t$, $3 \cos 3t$, and so on) of an underlying function given a limited amount of data. Since each musical instrument has a unique spectral "signature," knowledge of this signature enables us to electronically reconstruct or **synthesize** music that sounds like a particular instrument. But the sound being analyzed need not be music in order for spectral analysis to be fruitful. For example, spectral analysis is a powerful tool in speech recognition.

Moreover, spectral analysis is performed on many different kinds of data—the waveform needn't correspond to a sound at all. Important information can be gleaned and mathematical models constructed by examining the spectra associated with such diverse phenomena as earthquakes, tides, and sunspots. One of the most intriguing applications of spectral analysis is **image processing**. By analyzing the spectrum of a digital photograph, important features become apparent. For example, it is sometimes possible to enhance the resolution of a satellite photograph.

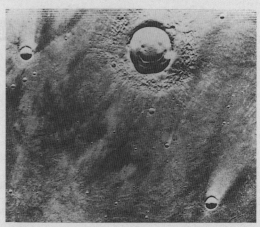

Image enhancement sharpens the detail of the "Happy face crater" on Mars, as viewed by the Viking Orbiter 2 spacecraft.

Perhaps the most astounding application of spectral analysis is **spectroscopy**, in which the spectrum of electromagnetic radiation (light, infrared radiation, and so on) is determined; this enables us to determine the chemical composition of unknown substances. Spectroscopy techniques have even been used to determine the chemical composition of distant galaxies.

Spectral analysis provides us with a compelling example of the unreasonable effectiveness of mathematics. How mysterious and wonderful it is that a subject arising from the study of triangles should prove to be a powerful tool for exploring music, understanding earthquakes, and analyzing starlight.

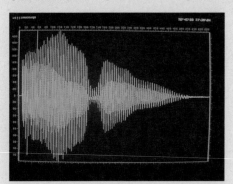

A voice print as it appears on a computer monitor.

Understanding and Mastery Checklists

Concepts to Understand	Skills to Master
Sum and difference identities for sine	Use sum and difference identities to compute exact trigonometric function values.
Sum and difference identities for cosine	Use sum and difference identities to simplify trigonometric expressions.
Sum and difference identities for tangent	Use sum and difference identities to verify trigonometric identities.

Exercises 6.2

Exercises 1-8   *Use the given information to compute the exact values of the indicated trigonometric functions.*

1. $\cos u = \dfrac{4}{5}$, $\sin u = -\dfrac{3}{5}$; $\cos v = \dfrac{5}{13}$, $\sin v = \dfrac{12}{13}$

   a. $\cos(u - v)$ **b.** $\cos(u + v)$

   c. $\sin(u - v)$ **d.** $\sin(u + v)$

2. $\cos u = -\dfrac{\sqrt{5}}{3}$, $\sin u = \dfrac{2}{3}$; $\cos v = \dfrac{3}{4}$, $\sin v = -\dfrac{\sqrt{7}}{4}$

   a. $\cos(u - v)$ **b.** $\cos(u + v)$

   c. $\sin(u - v)$ **d.** $\sin(u + v)$

3. $\cos u = \dfrac{3}{5}$, $0 \le u < \dfrac{\pi}{2}$; $\sin v = \dfrac{12}{13}$, $0 \le v < \dfrac{\pi}{2}$

   a. $\cos(u - v)$ **b.** $\cos(u + v)$

   c. $\sin(u - v)$ **d.** $\sin(u + v)$

4. $\cos u = -\dfrac{3}{5}$, $\dfrac{\pi}{2} \le u < \pi$; $\cos v = \dfrac{5}{13}$, $\dfrac{3\pi}{2} \le v < 2\pi$

   a. $\cos(u - v)$ **b.** $\cos(u + v)$

   c. $\sin(u - v)$ **d.** $\sin(u + v)$

5. $\sin r = -\dfrac{4}{5}$, $\pi \le r < \dfrac{3\pi}{2}$; $\sin s = \dfrac{3}{5}$, $0 \le s < \dfrac{\pi}{2}$

   a. $\cos(r - s)$ **b.** $\cos(r + s)$

   c. $\sin(r - s)$ **d.** $\sin(r + s)$

6. $\sin \alpha = \dfrac{15}{17}$, $\dfrac{\pi}{2} \le \alpha < \pi$; $\cos \beta = -\dfrac{3}{5}$, $\dfrac{\pi}{2} \le \beta < \pi$

   a. $\cos(\alpha - \beta)$ **b.** $\cos(\alpha + \beta)$

   c. $\sin(\alpha - \beta)$ **d.** $\sin(\alpha + \beta)$

7. $\sec v = \dfrac{5}{4}$, $0 \le v < \dfrac{\pi}{2}$; $\csc w = -\dfrac{13}{12}$, $\pi \le w < \dfrac{3\pi}{2}$

   a. $\cos(v - w)$ **b.** $\cos(v + w)$

   c. $\sin(v - w)$ **d.** $\sin(v + w)$

8. $\csc \theta = -\dfrac{17}{8}$, $\dfrac{3\pi}{2} \le \theta < 2\pi$; $\sec \phi = -\dfrac{5}{4}$, $\dfrac{\pi}{2} \le \phi < \pi$

   a. $\cos(\theta - \phi)$ **b.** $\cos(\theta + \phi)$

   c. $\sin(\theta - \phi)$ **d.** $\sin(\theta + \phi)$

Exercises 9-18 *Rewrite the given expression as a trigonometric function of a single angle and find the exact value, if possible.*

9. $\sin 62° \cos 32° - \sin 32° \cos 62°$

10. $\cos 100° \cos 10° + \sin 100° \sin 10°$

11. $\cos \dfrac{\pi}{8} \cos \dfrac{5\pi}{8} - \sin \dfrac{\pi}{8} \sin \dfrac{5\pi}{8}$

12. $\dfrac{\tan \dfrac{\pi}{5} + \tan \dfrac{\pi}{20}}{1 - \tan \dfrac{\pi}{5} \tan \dfrac{\pi}{20}}$

13. $\sin \dfrac{\pi}{5} \cos \dfrac{\pi}{15} + \sin \dfrac{\pi}{15} \cos \dfrac{\pi}{5}$

14. $\sin 152° \cos 12° - \sin 12° \cos 152°$

15. $\cos y \cos 2y - \sin y \sin 2y$

16. $\sin \dfrac{u}{2} \cos \dfrac{u}{2} + \sin \dfrac{u}{2} \cos \dfrac{u}{2}$

17. $\dfrac{\tan 4a - \tan 2a}{1 + \tan 4a \tan 2a}$

18. $\cos 4x \cos x + \sin 4x \sin x$

Exercises 19-22 *Find the exact value of the given trigonometric function. (Hint: Represent the given angle as a sum or difference of special angles.)*

19. $\cos 105°$

20. $\sin \dfrac{\pi}{12}$

21. $\tan \dfrac{11\pi}{12}$

22. $\cos 75°$

Exercises 23-36 *Simplify the given trigonometric expression.*

23. $\cos(x + \pi)$

24. $\sin(x + \pi)$

25. $\sin\left(\dfrac{\pi}{2} + x\right)$

26. $\cos\left(x - \dfrac{\pi}{4}\right)$

27. $\cos\left(\dfrac{\pi}{3} - x\right)$

28. $\sin\left(\dfrac{\pi}{6} - x\right)$

29. $\cos\left(x + \dfrac{2\pi}{3}\right)$

30. $\sin\left(x + \dfrac{\pi}{4}\right)$

31. $\tan\left(x - \dfrac{\pi}{4}\right)$

32. $\tan(x + \pi)$

33. $\sin\left(\arcsin \dfrac{3}{5} + \arccos \dfrac{3}{5}\right)$

34. $\cos\left(\arcsin \dfrac{4}{5} + \dfrac{\pi}{4}\right)$

35. $\tan\left(\dfrac{\pi}{4} - \arctan \dfrac{25}{24}\right)$

36. $\sin\left(\arcsin \dfrac{5}{13} - \arccos \dfrac{3}{5}\right)$

Exercises 37-50 *Verify the given identity using the sum and difference identities.*

37. $\cos\left(\dfrac{\pi}{2} - x\right) = \sin x$

38. $\sin\left(\dfrac{\pi}{2} - x\right) = \cos x$

39. $\sec(u + v) = \dfrac{\sec v \csc u}{\cot u - \tan v}$

40. $\cot(u + v) = \dfrac{\cot u \cot v - 1}{\cot u + \cot v}$

41. $\cot(u - v) = \dfrac{\cot u \cot v + 1}{\cot v - \cot u}$

42. $\csc(u + v) = \dfrac{\csc u \sec v}{1 + \tan v \cot u}$

43. $\cos(-x) = \cos x$ (*Hint:* $-x = 0 - x$.)

44. $\sin(-x) = -\sin x$

45. $\sin u \sin v = \dfrac{1}{2}[\cos(u - v) - \cos(u + v)]$

46. $\cos u \cos v = \dfrac{1}{2}[\cos(u + v) + \cos(u - v)]$

47. $\sin u \cos v = \dfrac{1}{2}[\sin(u + v) + \sin(u - v)]$

48. $\cos 2x = \cos^2 x - \sin^2 x$ (*Hint:* $2x = x + x$.)

49. $\sin 2x = 2 \sin x \cos x$

50. $\tan 2x = \dfrac{2 \tan x}{1 - \tan^2 x}$

Concepts and Critical Thinking

Exercises 51-54 *Answer true or false.*

51. $\sin\left(\dfrac{\pi}{2} + x\right) = \cos x$ for all x.

52. $\sin\left(\dfrac{\pi}{2} - x\right) = \cos x$ for all x.

53. $\cos\left(\dfrac{\pi}{2} + x\right) = \sin x$ for all x.

54. $\cos\left(\dfrac{\pi}{2} - x\right) = \sin x$ for all x.

Exercises 55-58 *Give an example of each.*

55. An angle θ that is not a multiple of 30° for which a sum identity can be used to determine $\sin \theta$

56. An angle θ that is not a multiple of 30° for which a difference identity can be used to determine $\sin \theta$

57. Two angles x and y for which $\cos(x + y) = \cos x + \cos y$

58. Two angles x and y for which $\cos(x + y) \neq \cos x + \cos y$

59. Suppose that u and v are angles in the first quadrant. Show that $u + v$ is also in the first quadrant if and only if $\tan u \tan v < 1$. [*Hint:* What can be said about $\tan(u + v)$ when $u + v$ is in the first quadrant?]

60. Consider the triangle with vertices $O(0, 0)$, $P(a, b)$, and $Q(c, d)$, as shown in Figure 10. Show that $\triangle POQ$ is a right triangle if $ac + bd = 0$. [*Hint:* Begin with the fact that $\cos \theta = \cos(\alpha - \beta)$.]

61. Produce formulas for $\cos(x + y + z)$ and $\sin(x + y + z)$ in terms of $\cos x$, $\cos y$, $\cos z$, $\sin x$, $\sin y$, and $\sin z$.

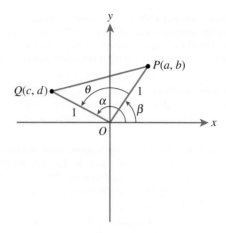

Figure 10

Questions for Discussion or Essay

62. Show that $\sin(\pi + 0) = \sin \pi + \sin 0$, but $\sin\left(\frac{\pi}{2} + \pi\right) \neq \sin \frac{\pi}{2} + \sin \pi$. What can be said in general about the statement $\sin(x + y) = \sin x + \sin y$? In general, what can you say about the use of examples to prove or disprove a statement in mathematics?

63. Use the sum identities to show that $\cos(x + \pi) = -\cos x$ and $\sin(x + \pi) = \sin x$. Explain how these could be proven without using the sum identities. (*Hint:* Sketch the angles x and $x + \pi$ on the unit circle.)

64. Prior to this section, for how many acute angles did you know exact trigonometric function values? Using the techniques of this section, for how many more can you find exact values?

65. Explain how angle addition identities can be used to compute $\sec(x \pm y)$, $\csc(x \pm y)$, and $\cot(x \pm y)$.

Projects for Enrichment

66. Product-Sum Identities   In this section, we have seen that the sine and cosine of the sum or difference of two angles can be written in terms of products involving the sine or cosine of the two angles. In this project, we use these sum and difference identities to develop additional identities involving sums and products of sine and cosine. These identities are often referred to as the **product–sum identities**. We first consider the product–sum identities involving cosine.

a. Recall that for any two angles x and y,
$$\cos(x - y) = \cos x \cos y + \sin x \sin y$$
$$\cos(x + y) = \cos x \cos y - \sin x \sin y$$
By adding these two equations together, show that
$$\cos x \cos y = \frac{1}{2}\cos(x - y) + \frac{1}{2}\cos(x + y)$$

b. Let $a = x - y$ and $b = x + y$. Solve this system of equations to find expressions for x and y in terms of a and b. Then make appropriate substitutions into the product–sum identity in part a to show that

$$\cos a + \cos b = 2\cos\left(\frac{a + b}{2}\right)\cos\left(\frac{a - b}{2}\right)$$

c. Follow steps similar to those outlined in parts a and b to show that the following product–sum identities hold. (*Hint:* In the first step, subtract the sum and difference identities for cosine.)
$$\sin x \sin y = \frac{1}{2}\cos(x - y) - \frac{1}{2}\cos(x + y)$$
$$\cos a - \cos b = -2\sin\left(\frac{a + b}{2}\right)\sin\left(\frac{a - b}{2}\right)$$

d. Use the sum and difference identities for sine to show that the following product–sum identities hold:
$$\sin x \cos y = \frac{1}{2}\sin(x - y) + \frac{1}{2}\sin(x + y)$$
$$\sin a + \sin b = 2\sin\left(\frac{a + b}{2}\right)\cos\left(\frac{a - b}{2}\right)$$
$$\sin a - \sin b = 2\cos\left(\frac{a + b}{2}\right)\sin\left(\frac{a - b}{2}\right)$$

67. Music Sound is created whenever a disturbance causes changes in air pressure at frequencies in the audible range (20–20,000 Hz). Musical tones are created when the oscillations of air pressure are periodic. As a consequence, musical tones can be modeled with functions of the form

$$f(t) = A\cos(\omega t + C)$$

where A is the amplitude and ω is the frequency. Changes in the amplitude correspond to changes in the loudness of the sound, and changes in the frequency correspond to changes in pitch.

In this project, we investigate what happens when two musical tones of the same frequency are out of phase. To that end, consider two musical tones with corresponding functions $f(t) = A_1\cos(\omega t)$ and $g(t) = A_2\cos(\omega t + \theta)$. The resulting sound is given by

$$h(t) = f(t) + g(t)$$
$$= A_1\cos(\omega t) + A_2\cos(\omega t + \theta)$$

We would like to show that this sound is again a musical tone. In other words, we would like to show that $h(t)$ can be written in the form

$$h(t) = A\cos(\omega t + C)$$

As a first step, we consider two specific functions f and g.

a. Let $f(t) = 3\cos(\pi t)$ and $g(t) = 4\cos\left(\pi t + \frac{\pi}{2}\right)$. Plot $h(t) = f(t) + g(t)$ and estimate its amplitude and phase shift. Then produce a function of the form $k(t) = A\cos(\pi t + C)$ whose graph is nearly identical to that of h.

Next we consider the general case.

b. Use the sum identity for cosine to rewrite $A_1\cos(\omega t) + A_2\cos(\omega t + \theta)$ in terms of sines and cosines of ωt and θ.

c. Use the sum identity for cosine to rewrite $A\cos(\omega t + C)$ in terms of sines and cosines of ωt and C.

d. Define

$$A = \sqrt{A_1^2 + A_2^2 + 2A_1A_2\cos\theta}$$

and let C be an angle for which

$$\cos C = \frac{A_1 + A_2\cos\theta}{A}$$

Show that

$$\sin^2 C = \frac{A_2^2(1 - \cos^2\theta)}{A^2}$$

and, assuming $A_2 > 0$ and θ is in the first quadrant,

$$\sin C = \frac{A_2\sin\theta}{A}$$

e. Combine your results from parts b, c, and d to show that

$$A_1\cos(\omega t) + A_2\cos(\omega t + \theta) = A\cos(\omega t + C)$$

Now we not only know that the sound resulting from two musical tones is again a musical tone, but we can also determine how its amplitude depends on the original tones.

f. Suppose two tones with the same frequency and amplitude are in phase. What is the amplitude of the resulting tone? What if the tones are in phase with different amplitudes? (*Hint:* Consider what happens in part c if $\theta = 0$.)

g. Suppose two tones with the same frequency and amplitude are out of phase by π radians. What is the amplitude of the resulting tone? Does this seem reasonable? What potential uses do you see for this outcome? How might it lead to the notion of "antinoise"?

68. Deriving the Sum and Difference Identities In this project, we derive the sum and difference identities for sine and cosine using an approach that involves the distance formula. Consider two angles u and v, as suggested by Figure 11. For convenience, we assume that $0 < v < u < 2\pi$.

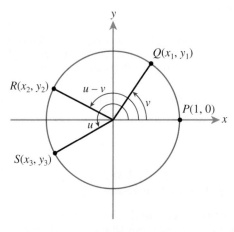

Figure 11

a. Find an expression for the distance between P and R.

b. Find an expression for the distance between Q and S.

c. From geometry, we know that the arcs RP and SQ have central angles with the same measure (namely, $u - v$), and hence the line segments RP and SQ have the same length. Use this fact and the distances you found in parts a and b to show that

$$x_2 = x_3x_1 + y_3y_1$$

d. Now show that

$$\cos(u - v) = \cos u \cos v + \sin u \sin v$$

e. Write $\cos(u + v)$ as $\cos[u - (-v)]$ and use the result of part d to show that

$$\cos(u + v) = \cos u \cos v - \sin u \sin v$$

f. Use a cofunction identity to show that

$$\sin(u + v) = \cos\left[\left(\frac{\pi}{2} - u\right) - v\right]$$

and then use the result from part d to show that

$$\sin(u + v) = \sin u \cos v + \cos u \sin v$$

g. Write $\sin(u - v)$ as $\sin[u + (-v)]$ and use the result from part f to show that

$$\sin(u - v) = \sin u \cos v - \cos u \sin v$$

Section 6.3 | # Double-Angle and Half-Angle Identities

- Why does a straw in a glass of water look bent?
- How can cubic equations be solved using trigonometry?
- How can the exact value of $\cos \frac{\pi}{8}$ be determined?
- If one line makes twice the angle with the horizontal as a second line, is it twice as steep?

The Double-Angle Identities

A simple application of the sum identities for sine, cosine, and tangent gives us the **double-angle identities**, which enable us to express any trigonometric function of twice an angle in terms of trigonometric functions of the angle itself. For example, the sum identity for sine tells us that $\sin(x + y) = \sin x \cos y + \sin y \cos x$. Replacing y with x gives us the double-angle identity for sine.

$$\sin(x + y) = \sin x \cos y + \sin y \cos x$$
$$\sin(x + x) = \sin x \cos x + \sin x \cos x$$
$$\sin 2x = 2 \sin x \cos x$$

A double-angle identity for the cosine function can be derived in a similar fashion. This time, we begin with the sum identity for cosine, as follows:

$$\cos(x + y) = \cos x \cos y - \sin x \sin y$$
$$\cos(x + x) = \cos x \cos x - \sin x \sin x$$
$$\cos 2x = \cos^2 x - \sin^2 x$$
$$= (1 - \sin^2 x) - \sin^2 x \quad \textbf{or} \quad = \cos^2 x - (1 - \cos^2 x)$$
$$= 1 - 2\sin^2 x \qquad\qquad\qquad = 2\cos^2 x - 1$$

Note that this establishes *three* identities for $\cos 2x$.

The double-angle identity for tangent also derives from the corresponding sum identity.

$$\tan(x + y) = \frac{\tan x + \tan y}{1 - \tan x \tan y}$$

$$\tan(x + x) = \frac{\tan x + \tan x}{1 - \tan x \tan x}$$

$$\tan 2x = \frac{2 \tan x}{1 - \tan^2 x}$$

Double-Angle Identities for Sine, Cosine, and Tangent

$$\sin 2x = 2 \sin x \cos x$$

$$\cos 2x = \cos^2 x - \sin^2 x = 2\cos^2 x - 1 = 1 - 2\sin^2 x$$

$$\tan 2x = \frac{2 \tan x}{1 - \tan^2 x}$$

The double-angle identities allow us to find the trigonometric function values of twice an angle, provided we have the appropriate trigonometric values for the angle itself. We illustrate the use of these identities in the following examples.

····**EXAMPLE 1**

Using the Double-Angle Identity for Sine

Compute $\sin 2x$, given that $\sin x = -\frac{4}{5}$ and $\cos x = \frac{3}{5}$.

Solution

$$\sin 2x = 2 \sin x \cos x$$
$$= 2\left(-\frac{4}{5}\right)\frac{3}{5}$$
$$= -\frac{24}{25}$$

····**EXAMPLE 2**

Computing the Cosine of Twice an Angle

Find $\cos 2\theta$, given that $\cos \theta = \frac{2}{5}$.

Solution   Since $\cos \theta$ is given but $\sin \theta$ is unknown, we select the form of the double-angle identity for cosine involving only the cosine function.

$$\cos 2\theta = 2 \cos^2 \theta - 1$$
$$= 2\left(\frac{2}{5}\right)^2 - 1$$
$$= \frac{8}{25} - 1$$
$$= -\frac{17}{25}$$

····**EXAMPLE 3**

Computing the Tangent of a Double-Angle

Evaluate $\tan 2x$, given that $\tan x = 2$.

Solution

$$\tan 2x = \frac{2 \tan x}{1 - \tan^2 x}$$
$$= \frac{4}{1 - 4}$$
$$= -\frac{4}{3}$$

····**EXAMPLE 4**

An Application of the Double-Angle Identity for Sine

Given that $\sin \theta = \frac{3}{5}$ and that $\frac{\pi}{2} < \theta < \pi$, compute $\sin 2\theta$.

Solution   To use the double-angle identity for sine, we must find $\cos \theta$.

$$\sin^2 \theta + \cos^2 \theta = 1$$
$$\left(\frac{3}{5}\right)^2 + \cos^2 \theta = 1$$
$$\frac{9}{25} + \cos^2 \theta = 1$$

$$\cos^2 \theta = \frac{16}{25}$$

$$\cos \theta = \pm \frac{4}{5}$$

Since θ is in the second quadrant and the cosine function is negative there, we have

$$\cos \theta = -\frac{4}{5}$$

Now we use the double-angle identity for sine.

$$\sin 2\theta = 2 \sin \theta \cos \theta$$

$$= 2\left(\frac{3}{5}\right)\left(-\frac{4}{5}\right)$$

$$= -\frac{24}{25}$$

Infinitely many new identities can be verified with the aid of the double-angle identities for sine, cosine, and tangent. In particular, any trigonometric function of nx (where n is an integer) can be expressed in terms of trigonometric functions of x.

EXAMPLE 5

Deriving an Identity for sin 3x

Express $\sin 3x$ in terms of $\sin x$ and $\cos x$.

Solution   We begin by recognizing that $3x = 2x + x$ and applying the sum identity for sine.

$$\sin 3x = \sin(2x + x)$$

$$= \sin 2x \cos x + \sin x \cos 2x \qquad \text{Applying the sum identity for sine}$$

$$= (2 \sin x \cos x)\cos x + \sin x(\cos^2 x - \sin^2 x) \qquad \text{Applying the double-angle identity for sine and cosine}$$

$$= 2 \sin x \cos^2 x + \sin x \cos^2 x - \sin^3 x$$

$$= 3 \sin x \cos^2 x - \sin^3 x$$

Thus far, we have used double-angle identities to convert trigonometric functions of $2x$ to equivalent expressions involving trigonometric functions of x. In practice, however, we are more likely to translate in the opposite direction—to convert trigonometric functions of x to equivalent, but simpler or more revealing, expressions involving $2x$.

EXAMPLE 6

Simplifying an Expression with Double-Angle Identities

Find the period of $f(x) = (\cos x + \sin x)^2$.

Solution   Since both the sine and cosine functions have periods of 2π and $f(x)$ is constructed from these functions, we might predict that f also has a period of 2π. Indeed, it is certainly the case that $f(x + 2\pi) = f(x)$. What is in question is whether 2π is the *smallest* number p satisfying $f(x + p) = f(x)$. Let's see if we can rewrite $f(x)$ so that its period is transparent. We begin by expanding.

$$f(x) = (\cos x + \sin x)^2$$
$$= \cos^2 x + 2 \cos x \sin x + \sin^2 x$$
$$= 1 + 2 \cos x \sin x \qquad \text{Applying the Pythagorean identity—} \sin^2 x + \cos^2 x = 1$$

Now, since $\sin 2x = 2 \sin x \cos x$, we have

$$f(x) = 1 + \sin 2x$$

From our work in Section 5.4, we see that the period of f is given by

$$\frac{2\pi}{2} = \pi$$

The Half-Angle Identities

Because each of the double-angle identities relates trigonometric functions of a given angle to trigonometric functions of *twice* that angle, we would intuitively expect there to be similar identities relating trigonometric functions of an angle to trigonometric functions of *half* the given angle. This is indeed the case. In fact, we use double-angle identities to derive the half-angle identities for sine and cosine.

Specifically, consider the following double-angle identities for cosine.

(1) $$\cos 2u = 2 \cos^2 u - 1$$

(2) $$\cos 2u = 1 - 2 \sin^2 u$$

If we solve equation (1) for $\cos^2 u$, we obtain

(3) $$\cos^2 u = \frac{1 + \cos 2u}{2}$$

Similarly, solving equation (2) for $\sin^2 u$ gives us

(4) $$\sin^2 u = \frac{1 - \cos 2u}{2}$$

Now equations (3) and (4), in addition to being key steps in the derivation of the half-angle identities for sine and cosine, are useful identities in their own right.

Formulas for $\cos^2 u$ and $\sin^2 u$

$$\cos^2 u = \frac{1 + \cos 2u}{2}$$
$$\sin^2 u = \frac{1 - \cos 2u}{2}$$

To complete the derivation of the half-angle identities for sine and cosine, we take square roots and substitute $x/2$ for u.

$$\cos^2 u = \frac{1 + \cos 2u}{2}$$
$$\cos u = \pm \sqrt{\frac{1 + \cos 2u}{2}}$$
$$\cos \frac{x}{2} = \pm \sqrt{\frac{1 + \cos x}{2}}$$

$$\sin^2 u = \frac{1 - \cos 2u}{2}$$
$$\sin u = \pm \sqrt{\frac{1 - \cos 2u}{2}}$$
$$\sin \frac{x}{2} = \pm \sqrt{\frac{1 - \cos x}{2}}$$

Note that the sign ($+$ or $-$) can be determined if the quadrant of $x/2$ is known.

An identity for $\tan(x/2)$ can be obtained by dividing the expressions for $\sin(x/2)$ and $\cos(x/2)$.

$$\tan\frac{x}{2} = \frac{\sin\dfrac{x}{2}}{\cos\dfrac{x}{2}}$$

$$= \pm\frac{\sqrt{\dfrac{1-\cos x}{2}}}{\sqrt{\dfrac{1+\cos x}{2}}}$$

$$= \pm\sqrt{\frac{1-\cos x}{1+\cos x}}$$

Fortunately, two alternative expressions for $\tan(x/2)$ can be derived (see Exercise 60), neither of which involves the awkward $\pm$ symbol.

$$\tan\frac{x}{2} = \frac{1-\cos x}{\sin x} = \frac{\sin x}{1+\cos x}$$

For convenience, we summarize the half-angle identities.

Half-Angle Identities for Sine, Cosine, and Tangent

$$\sin\frac{x}{2} = \pm\sqrt{\frac{1-\cos x}{2}}$$

$$\cos\frac{x}{2} = \pm\sqrt{\frac{1+\cos x}{2}}$$

$$\tan\frac{x}{2} = \frac{1-\cos x}{\sin x} = \frac{\sin x}{1+\cos x}$$

EXAMPLE 7

Evaluating a Trigonometric Function Using a Half-Angle Identity

Verify the identity $\sec^2\dfrac{x}{2} = \dfrac{2}{1+\cos x}$.

Solution   Since $\sec^2 u = 1/\cos^2 u$, we are able to apply the half-angle identity for cosine as follows:

$$\sec^2\frac{x}{2} = \frac{1}{\left(\cos\dfrac{x}{2}\right)^2}$$

$$= \frac{1}{\left(\pm\sqrt{\dfrac{1+\cos x}{2}}\right)^2} \qquad \text{Using the half-angle identity for cosine}$$

$$= \frac{1}{\dfrac{1+\cos x}{2}}$$

$$= \frac{2}{1+\cos x}$$

> **EXAMPLE 8**   **Using the Half-Angle Identity for Sine**

Compute $\sin(\theta/2)$ given that $\cos\theta = \frac{1}{4}$ and $\frac{3\pi}{2} < \theta < 2\pi$.

Solution   One approach would be to use the $\boxed{\text{COS}^{-1}}$ key on a calculator to estimate the reference angle for θ, then determine the angle θ from the given quadrant information, and finally estimate $\sin(\theta/2)$ using the $\boxed{\text{SIN}}$ key. Instead, we illustrate how the half-angle identity for sine can be used, thus eliminating the need to actually find θ. From the half-angle identity for sine, we have

$$\sin\frac{\theta}{2} = \pm\sqrt{\frac{1 - \cos\theta}{2}}$$

$$= \pm\sqrt{\frac{1 - \frac{1}{4}}{2}}$$

$$= \pm\sqrt{\frac{\left(\frac{3}{4}\right)}{2}}$$

$$= \pm\sqrt{\frac{3}{8}}$$

$$= \pm\frac{\sqrt{3}}{\sqrt{8}} \cdot \frac{\sqrt{2}}{\sqrt{2}}$$

$$= \pm\frac{\sqrt{6}}{4}$$

To determine the appropriate sign, we determine the quadrant of $\frac{\theta}{2}$ as follows:

$$\frac{3\pi}{2} < \theta < 2\pi$$

$$\frac{3\pi}{4} < \frac{\theta}{2} < \pi \qquad \text{Dividing through by 2}$$

Since $\theta/2$ is in the second quadrant and sine is positive there, we have

$$\sin\frac{\theta}{2} = \frac{\sqrt{6}}{4}$$

> **EXAMPLE 9**   **Applying the Half-Angle Identity for Tangent**

Find the exact value of $\tan\frac{\pi}{8}$.

Solution   Using a half-angle identity for tangent $\left(\text{with } x = \frac{\pi}{4}\right)$, we have

$$\tan \frac{\pi}{8} = \tan\left(\frac{1}{2} \cdot \frac{\pi}{4}\right)$$

$$= \frac{1 - \cos \dfrac{\pi}{4}}{\sin \dfrac{\pi}{4}}$$

$$= \frac{1 - \dfrac{\sqrt{2}}{2}}{\dfrac{\sqrt{2}}{2}}$$

$$= \frac{2}{\sqrt{2}} - 1$$

$$= \sqrt{2} - 1$$

Understanding and Mastery Checklists

Concepts to Understand	Skills to Master
Double-angle identities for sine, cosine, and tangent	Use half-angle identities to compute exact trigonometric function values.
❖	❖
Half-angle identities for sine, cosine, and tangent	Use double-angle and half-angle identities to verify trigonometric identities.

Exercises 6.3

Exercises 1-8 *Use the given information to compute*

a. $\sin 2\theta$ **b.** $\cos 2\theta$ **c.** $\tan 2\theta$

1. $\sin \theta = \dfrac{3}{5}, \cos \theta = -\dfrac{4}{5}$ **2.** $\sin \theta = -\dfrac{24}{25}, \cos \theta = \dfrac{7}{25}$

3. $\sin \theta = -\dfrac{3}{5}, \pi \leq \theta < \dfrac{3\pi}{2}$ **4.** $\cos \theta = \dfrac{5}{13}, 0 \leq \theta < \dfrac{\pi}{2}$

5. $\tan \theta = \dfrac{8}{15}, 0 \leq \theta < \dfrac{\pi}{2}$ **6.** $\csc \theta = -\dfrac{5}{3}, \dfrac{3\pi}{2} \leq \theta < 2\pi$

7. $\cot \theta = -\dfrac{12}{5}, \dfrac{\pi}{2} \leq \theta < \pi$ **8.** $\sin \theta = \dfrac{1}{2}, 0 \leq \theta < \dfrac{\pi}{2}$

Exercises 9-16 *Use the given information to compute*

a. $\sin \dfrac{x}{2}$ **b.** $\cos \dfrac{x}{2}$ **c.** $\tan \dfrac{x}{2}$

9. $\cos x = \dfrac{7}{25}, 0 \leq x < \dfrac{\pi}{2}$ **10.** $\cos x = -\dfrac{7}{25}, \pi \leq x < \dfrac{3\pi}{2}$

11. $\cos x = -\dfrac{1}{9}, \dfrac{\pi}{2} \leq x < \pi$ **12.** $\cos x = \dfrac{119}{169}, 0 \leq x < \dfrac{\pi}{2}$

13. $\sec x = \dfrac{9}{7}, \dfrac{3\pi}{2} \leq x < 2\pi$ **14.** $\sec x = -8, \dfrac{\pi}{2} \leq x < \pi$

15. $\sin x = -\dfrac{\sqrt{15}}{8}, \dfrac{3\pi}{2} \leq x < 2\pi$ **16.** $\sin x = -\dfrac{4\sqrt{5}}{9}, \pi \leq x < \dfrac{3\pi}{2}$

Exercises 17-20 *Find the exact value of the given quantity.*

17. $\sin 15°$

18. $\cos \dfrac{\pi}{8}$

19. $\sec \dfrac{\pi}{8}$

20. $\cot 165°$

Exercises 21-30 *Verify that the given equation is an identity.*

21. $\tan 2x = \dfrac{2 \cot x}{\cot^2 x - 1}$

22. $\sec 2x = \dfrac{1}{\cos^4 x - \sin^4 x}$

23. $\sin x \cos x \csc 2x = \dfrac{1}{2}$

24. $2 \sin x \sin 2x = \cos x - \cos 3x$

25. $\tan 2x = \dfrac{2 \sin x}{2 \cos x - \sec x}$

26. $2 \sin^2 x + \cos 2x = 1$

27. $2 \tan x \cot 2x = 2 - \sec^2 x$

28. $\dfrac{1}{4} \sin 4x = \sin x \cos x - 2 \sin^3 x \cos x$

29. $2 \csc 30x = \csc 15x \sec 15x$

30. $\dfrac{\cos\left(2x + \dfrac{\pi}{2}\right)}{\cos x} = -2 \sin x$

Exercises 31-34 *Find an equivalent expression involving trigonometric functions of 2x. (Answers may vary.)*

31. $4 \sin^2 x \cos^2 x$

32. $\cos^2 x - \sin x \cos x - \sin^2 x$

33. $\dfrac{1 - \tan^2 x}{\tan x}$

34. $\dfrac{1}{1 - 2 \sin^2 x}$

Exercises 35-42 *Verify that the given equation is an identity.*

35. $\sin\left(2x - \dfrac{\pi}{3}\right) + \dfrac{\sqrt{3}}{2} = \sin x\left(\cos x + \sqrt{3} \sin x\right)$

36. $\tan\left(-\dfrac{x}{2}\right) = \cot x - \csc x$

37. $\cot \dfrac{x}{2} = \csc x + \cot x$

38. $\sin x \sec \dfrac{x}{2} = \pm \sqrt{2 - 2 \cos x}$

39. $\tan \dfrac{3x}{2} = \dfrac{3 \sin x - 4 \sin^3 x}{4 \cos^3 x - 3 \cos x + 1}$

40. $4 \cos^4 \dfrac{x}{2} - 2 \cos^2 x = \sin^2 x + 2 \cos x$

41. $\tan\left(\dfrac{x}{2} + \dfrac{\pi}{4}\right) = \tan x + \sec x$ $\left[Hint: \dfrac{x}{2} + \dfrac{\pi}{4} = \dfrac{1}{2}\left(x + \dfrac{\pi}{2}\right)\right]$

42. $\tan\left(\dfrac{x}{2} - \dfrac{\pi}{4}\right) = \tan x - \sec x$

Applications

43. TV Area The area of the TV screen shown in Figure 12, in terms of its diagonal d and the angle θ made between the bottom of the screen and the diagonal, is given by $A = d^2 \sin \theta \cos \theta$.

   a. Express the area in terms of a single trigonometric function.

   b. Find the area of a 27-inch TV given that $\theta = 40°$.

Figure 12

44. Projectile Range If an object is thrown from ground level with an initial angle θ and an initial speed of v feet per second, its range in feet is given by

$$\dfrac{v^2 \sin \theta \cos \theta}{16}$$

   a. Express the range in terms of a single trigonometric function.

   b. Find the range of an object thrown with an initial angle of 30° and an initial speed of 50 feet per second.

45. Refraction of Light When light waves pass from one medium to another of a different density, they bend. This phenomenon is called *refraction*. Figure 13 illustrates light passing from air to a liquid of greater density. The result is that the *angle of incidence* α is greater than the *angle of refraction* β. In general, these angles are related by the formula

$$\dfrac{c_1}{c_2} = \dfrac{\sin \alpha}{\sin \beta}$$

where c_1 and c_2 denote the speed of light in air and the liquid, respectively. If the angle of incidence is twice the angle of refraction—that is, if $\alpha = 2\beta$—show that $c_2 = \dfrac{1}{2} c_1 \sec \beta$.

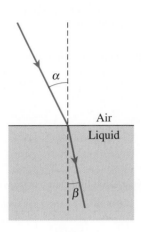

Figure 13

*Refraction of light causes the
apparent break in the straw.*

Concepts and Critical Thinking

Exercises 46-49   *Answer true or false.*

46. The cosine of twice an angle is always twice the cosine of the angle.

47. $\sin^2 2x + \cos^2 2x = 1$ for all x.

48. $\sin 4x = 2 \sin 2x \cos 2x$ for all x.

49. $\sin 4x = 4 \sin x \cos x$ for all x.

Exercises 50-53   *Give an example of each.*

50. An angle whose exact sine, cosine, or tangent can be computed using a half-angle identity

51. An angle x for which $\sin 2x \neq 2 \sin x$

52. A double-angle identity for cosecant

53. A half-angle identity for secant

54. Figure 14 suggests a proof of the identity $\sin 2x = 2 \sin x \cos x$. Can you find it? (*Hint:* Compute the area of the largest triangle in Figure 14 two different ways.)

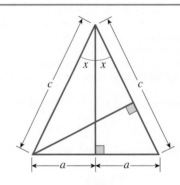

Figure 14

55. If x and y are coterminal, does it follow that $\sin(x/2) = \sin(y/2)$? Explain.

56. If $0 < \theta < 2\pi$ and $\cos(\theta/2) > 0$, what can be said about θ? Explain.

57. Explain how the identity for $\sin 2x$ can be obtained from the identity for $\cos 2x$.

Questions for Discussion or Essay

58. Explain why there are infinitely many acute angles for which exact values of the trigonometric functions can be found.

59. Let θ_m denote the angle made between the line with equation $y = mx + b$ and the positive x-axis. Describe the connection between m and θ_m. (*Hint:* Use the tangent function.) When does doubling θ_m result in doubling m?

Projects for Enrichment

60. Half-Angle Identity for Tangent In this project, we complete the derivation of the half-angle identities for tangent.

a. Verify the identity

$$\tan u = \frac{2(1 - \cos^2 u)}{\sin 2u}$$

b. Use the identity of part a to show that

$$\tan \frac{x}{2} = \frac{1 - \cos x}{\sin x}$$

c. Verify the identity

$$\frac{1 - \cos x}{\sin x} = \frac{\sin x}{1 + \cos x}$$

to show that

$$\tan \frac{x}{2} = \frac{\sin x}{1 + \cos x}$$

61. Cubics and Cosines A cubic equation is one that can be written in the form $x^3 + rx^2 + sx + t = 0$. By making the substitution $x = y - (r/3)$, it is possible to write such an equation in the form $y^3 - py + q = 0$. Thus, if we can develop a technique for solving equations of the form $y^3 - py + q = 0$, we will be able to solve any cubic equation. We will concentrate our efforts on the case where $(q^2/4) - (p^3/27) < 0$, since a technique was considered in Exercise 99 of Section 1.5 for the case where $(q^2/4) - (p^3/27) > 0$.

To begin, we note that if $(q^2/4) - (p^3/27) < 0$, then

$$\frac{q^2}{4} < \frac{p^3}{27}$$

Multiplying both sides by $27/p^3$ and taking square roots gives us

$$\left| q \sqrt{\frac{27}{4p^3}} \right| < 1$$

Thus, there is an angle ϕ for which

$$\cos \phi = -q \sqrt{\frac{27}{4p^3}}$$

Define y by

$$y = \frac{2\sqrt{3p}}{3} \cos \frac{\phi}{3}$$

The following steps show that y is a solution to the equation $y^3 - py + q = 0$.

a. Find an expression for $\cos 3\theta$ that involves only $\cos \theta$. [*Hint:* $\cos 3\theta = \cos(2\theta + \theta)$.]

b. Use your result from part a to show that

$$\cos^3 \frac{\phi}{3} = \frac{1}{4}\left(\cos \phi + 3 \cos \frac{\phi}{3}\right)$$

c. Use the fact that

$$y = \frac{2\sqrt{3p}}{3} \cos \frac{\phi}{3}$$

together with your result from part b to show that

$$y^3 = \frac{8\sqrt{3p^3}}{9}\left[\frac{1}{4}\left(\cos \phi + 3 \cos \frac{\phi}{3}\right)\right]$$

d. Use the fact that

$$\cos \phi = -q \sqrt{\frac{27}{4p^3}}$$

and your result from part c to show that $y^3 - py + q = 0$.

Using similar steps, it can be shown that the other two solutions to the cubic are given by

$$y = \frac{2\sqrt{3p}}{3} \cos\left(\frac{\phi}{3} + 120°\right)$$

$$y = \frac{2\sqrt{3p}}{3} \cos\left(\frac{\phi}{3} + 240°\right)$$

e. Solve the given cubic equation.

  i. $y^3 - 9y + 9 = 0$

  ii. $x^3 - 3x - 1 = 0$

  iii. $\frac{1}{8}t^3 - 6t + 4 = 0$

  iv. $x^3 + 3x^2 - 9x - 3 = 0$
  (*Hint:* Substitute $x = y - 1$ and simplify.)

Section 6.4 | Conditional Trigonometric Equations

- How can the distance between any two points on Earth be determined?
- How can trigonometry be used to predict the flight of a golf ball?
- At what angle must a ball be thrown so that its range equals its maximum height?
- What does the sound of a violin look like?

In the previous three sections, we have been verifying trigonometric identities; that is, we have been showing that certain equations are *always* satisfied, no matter what the value of the variable. In this section, we discuss techniques for solving **conditional trigonometric equations**, equations involving trigonometric functions that are true only for certain values of the variable.

By and large, trigonometric equations are solved in much the same way that algebraic equations are solved. We can solve them analytically (by hand) using the properties of equality, or we can solve them graphically. There are, however, two aspects of solving trigonometric equations that distinguish them from ordinary algebraic equations. First, since the trigonometric functions are periodic, most trigonometric equations have infinitely many solutions. For example, solutions to the equation $\sin x = \frac{1}{3}$ correspond to points of intersection of the graphs of $y = \sin x$ and $y = \frac{1}{3}$, as shown in Figure 15. Clearly there are infinitely many solutions.

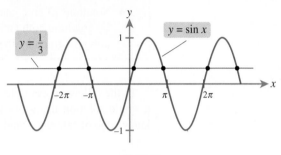

Figure 15

Second, if a trigonometric equation involves more than one trigonometric function, then we may need to employ one or more trigonometric identities to obtain exact solutions. (When approximating solutions graphically, it isn't necessary to employ trigonometric identities.)

Solving trigonometric equations requires us to synthesize much of what we have already learned about trigonometric functions. For convenience, the facts that are used most extensively in this section are summarized as follows.

Tools for Solving Trigonometric Equations

Zero-product Property

- If $ab = 0$, then $a = 0$ or $b = 0$.

Signs of Trigonometric Functions

- The sign of both $\cos \theta$ and $\sin \theta$ can be determined by considering the sign of the x- and y-coordinates of the point on the unit circle corresponding to the angle θ.

- The sign of any other trigonometric function can be determined by expressing the function in terms of sine and cosine.

Reference Angles

- $\sin x = \pm \sin y$ implies that x and y have the same reference angle. Similar statements hold for the other trigonometric functions.

(continued)

Special Angles

- The sines and cosines of the angles $\frac{\pi}{6}, \frac{\pi}{4}$, and $\frac{\pi}{3}$ are known. All other trigonometric function values for these angles can then be determined from basic identities.

Periodicity

- The sine, cosine, secant, and cosecant functions have period 2π; the tangent and cotangent functions have period π.

Identities

- All of the identities included in the first three sections of this chapter can be used to simplify or rewrite trigonometric expressions.

Graphical Techniques

- Solutions of an equation of the form $f(x) = g(x)$ can be approximated either by graphing both f and g and estimating their points of intersection or (preferably) by graphing the function $f - g$ and estimating its zeros.

In the following example, we solve a simple trigonometric equation by first finding a single solution using our knowledge of the special angles and then exploiting our knowledge of reference angles and periodicity to find the other solutions.

EXAMPLE 1

Solving a Linear Equation in sin x

a. Find all solutions of $\sin x = \frac{1}{2}$ in the interval $[0, 2\pi)$.

b. Find all solutions of the equation $\sin x = \frac{1}{2}$.

Solution

a. Since $\sin \frac{\pi}{6} = \frac{1}{2}$, we know that $\frac{\pi}{6}$ is a solution and that any other solution has a reference angle of $\frac{\pi}{6}$. Since $\sin x$ is positive for angles x in the second quadrant, the angle in the second quadrant with reference angle $\frac{\pi}{6}$ is a solution. As Figure 16 shows, this angle is $\frac{5\pi}{6}$. The figure also shows that no other angle in $[0, 2\pi)$ has sine (y-coordinate) $\frac{1}{2}$. Thus, the solution set is $\left\{\frac{\pi}{6}, \frac{5\pi}{6}\right\}$.

b. Since the sine function is periodic with period 2π, the following equations are true for all integers k:

$$\sin\left(\frac{\pi}{6} + 2\pi k\right) = \sin \frac{\pi}{6} = \frac{1}{2}$$

$$\sin\left(\frac{5\pi}{6} + 2\pi k\right) = \sin \frac{5\pi}{6} = \frac{1}{2}$$

Thus, the solution set consists of all numbers of the form $\frac{\pi}{6} + 2\pi k$ or $\frac{5\pi}{6} + 2\pi k$, where k is an integer.

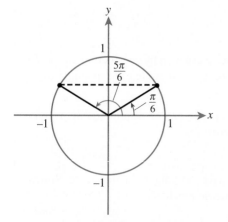

Figure 16

In some instances, the trigonometric functions appearing in an equation have arguments (inputs) other than just x. In the following two examples, the trigonometric functions involve double angles. In such cases, we first find the double angle, and then we simply divide by 2 to find the unknown angle.

······**EXAMPLE 2**   **Solving a Trigonometric Equation Involving Multiple Angles**

Find all solutions of $\cot 2x - 1 = 0$.

Solution   Solving for $\cot 2x$, we obtain

$$\cot 2x = 1$$

If we let $\theta = 2x$, then our equation becomes

$$\cot \theta = 1$$

Because the period of the cotangent function is π, we confine our initial search to values of θ in the interval $[0, \pi)$. One solution is given by $\frac{\pi}{4}$, and since the cotangent function is negative in the second quadrant, $\frac{\pi}{4}$ is the only solution in the interval $[0, \pi)$. Moreover, since the cotangent function has period π, all solutions of $\cot \theta = 1$ are of the form

$$\theta = \frac{\pi}{4} + \pi k, \quad \text{for } k \text{ an integer}$$

Since $\theta = 2x$, we have

$$2x = \frac{\pi}{4} + \pi k, \quad \text{for } k \text{ an integer}$$

Dividing by 2, we obtain

$$x = \frac{\pi}{8} + \frac{\pi}{2}k, \quad \text{for } k \text{ an integer}$$

Checking, we see that for any integer k

$$\cot\left[2\left(\frac{\pi}{8} + \frac{\pi}{2}k\right)\right] - 1 = \cot\left(\frac{\pi}{4} + \pi k\right) - 1 = \cot\left(\frac{\pi}{4}\right) - 1 = 0$$

······**EXAMPLE 3**   **Golf Ball Loft**

A projectile launched from ground level at an initial speed of v_0 feet per second with an initial angle θ has a range (ignoring air resistance) of

$$R = \frac{v_0^2 \sin 2\theta}{32}$$

Find the initial angle θ if a golf ball is struck with an initial speed of 200 feet per second and first hits the ground 150 yards from the tee.

Solution   Setting $R = 450$ (150 yards is 450 feet) and $v_0 = 200$, we obtain the equation

$$450 = \frac{200^2 \sin 2\theta}{32}$$

Solving for $\sin 2\theta$, we have

$$\sin 2\theta = \frac{32 \cdot 450}{200^2} = 0.36$$

Now, one solution of the equation $\sin 2\theta = 0.36$ can be obtained using the inverse sine function, giving us

$$2\theta = \sin^{-1}(0.36) \approx 21.1°$$

Golf swing

496 Chapter 6 Trigonometric Identities and Equations

It follows that any other angle 2θ must have a reference angle of approximately $21.1°$. Because the sine function is positive in the first and second quadrants, we are interested in an angle between $90°$ and $180°$ with reference angle $21.1°$. This angle is $180° - 21.1° = 158.9°$. Thus, we have

$$2\theta \approx 21.1° \quad \text{or} \quad 2\theta \approx 158.9°$$

and so

$$\theta \approx 10.55° \quad \text{or} \quad \theta \approx 79.45°$$

EXAMPLE 4

Solving a Trigonometric Equation in Factored Form

Find all solutions of $(\sin x - 2)(\csc x - 2) = 0$ on the interval $[0, 2\pi)$.

Solution   By the zero-product property, we have that either $\sin x - 2 = 0$ or $\csc x - 2 = 0$. The first equation gives us $\sin x = 2$, which is impossible since $-1 \le \sin x \le 1$. The second equation can be solved as follows:

$$\csc x - 2 = 0$$
$$\csc x = 2$$
$$\frac{1}{\sin x} = 2$$
$$\sin x = \frac{1}{2}$$

Since $\sin \frac{\pi}{6} = \frac{1}{2}$, all other solutions of $\sin x = \frac{1}{2}$ must have $\frac{\pi}{6}$ as their reference angle. Because the sine function is positive in the first and second quadrants, we are looking for the angle between $\frac{\pi}{2}$ and π with reference angle $\frac{\pi}{6}$—namely, $\frac{5\pi}{6}$. Thus, our two solutions are $x = \frac{\pi}{6}$ and $x = \frac{5\pi}{6}$. Checking these in the original equation, we see that

$$\left(\sin \frac{\pi}{6} - 2\right)\left(\csc \frac{\pi}{6} - 2\right) = \left(\frac{1}{2} - 2\right)(2 - 2) = 0$$

and

$$\left(\sin \frac{5\pi}{6} - 2\right)\left(\csc \frac{5\pi}{6} - 2\right) = \left(\frac{1}{2} - 2\right)(2 - 2) = 0$$

EXAMPLE 5

Solving a Quadratic Equation in Cosine

Find all solutions of $3\cos^2 x = 2\cos x + 1$ in the interval $[0, 2\pi)$.

Solution   We begin by recognizing that this is a *quadratic* equation in $\cos x$. Our strategy is to solve this quadratic equation for $\cos x$ and then finally to solve for x.

$$3\cos^2 x = 2\cos x + 1$$
$$3\cos^2 x - 2\cos x - 1 = 0$$
$$(3\cos x + 1)(\cos x - 1) = 0$$
$$\cos x = -\frac{1}{3} \quad \text{or} \quad \cos x = 1$$

Now, one solution to the equation $\cos x = -\frac{1}{3}$ can be obtained by taking the inverse cosine of both sides.

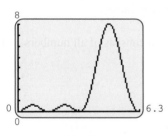

Figure 17

$$\cos x = -\frac{1}{3}$$

$$x = \cos^{-1}\left(-\frac{1}{3}\right)$$

$$\approx 1.9106$$

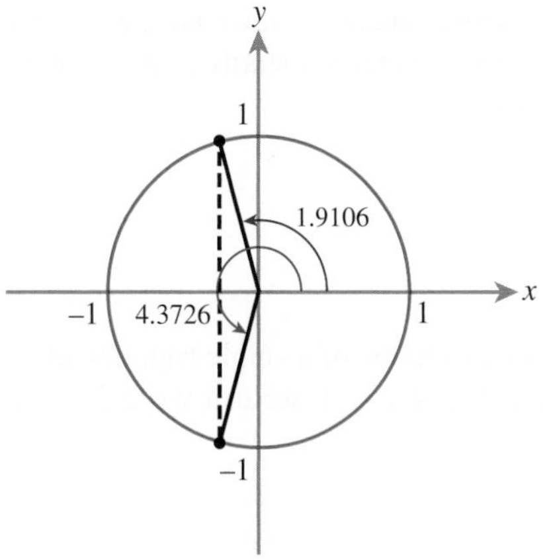

From Figure 17, we see that $2\pi - 1.9106 \approx 4.3726$ has the same cosine as 1.9106. Thus, the equation $\cos x = -\frac{1}{3}$ gives two solutions: $x \approx 1.9106$ and $x \approx 4.3726$. The equation $\cos x = 1$ gives the solution $x = 0$. Therefore, the original equation has a total of three solutions in $[0, 2\pi)$: $x = 0$, $x \approx 1.9106$, and $x \approx 4.3726$. Each of these can easily be checked in the original equation.

Of course, if we are interested only in approximate solutions to trigonometric equations, it is often much easier to employ graphical techniques. It is important to recognize that only a few of the infinitely many solutions will be visible from the graph. All other solutions can be determined by considering periodicity.

EXAMPLE 6   **Estimating Solutions of a Trigonometric Equation Graphically**

a. Estimate all solutions of $\sin 3x - 2 \cos 2x = 3 \sin x - 2$ in the interval $[0, 2\pi)$.

b. Estimate all solutions of $\sin 3x - 2 \cos 2x = 3 \sin x - 2$.

Solution

a. We begin by subtracting $3 \sin x - 2$ from both sides in order to write the equation in the standard form $f(x) = 0$.

$$\sin 3x - 2 \cos 2x = 3 \sin x - 2$$

$$\sin 3x - 2 \cos 2x - 3 \sin x + 2 = 0$$

Next, we plot the graph of $f(x) = \sin 3x - 2 \cos 2x - 3 \sin x + 2$ using a graphing calculator. Figure 18 suggests that f has as many as four distinct zeros in the interval $[0, 2\pi)$. However, upon closer inspection, the rightmost zero appears to be about 2π (6.28. . .), which is just outside our interval. A quick check confirms that 2π is indeed a zero of f, and so we consider only the other three zeros. Zooming in to estimate the values of the other three zeros, we obtain the following approximations:

$$x \approx 0, \quad x \approx 1.57, \quad \text{and } x \approx 3.14$$

These numbers are suspiciously close to the values 0, $\frac{\pi}{2}$, and π, respectively. Substitution of 0, $\frac{\pi}{2}$, and π into the equation confirms that they are, indeed, solutions.

b. In Figure 19, we have zoomed out to estimate the period of f. Close inspection reveals that the graph of f repeats itself approximately every 6.28 units. Again, this value is suspiciously close to 2π, and it is easily shown that $f(x + 2\pi) = f(x)$. Thus, the period is 2π. It follows that since 0, $\frac{\pi}{2}$, and π are solutions, so too are all numbers of the following forms (where k is any integer):

$$x = 0 + 2\pi k = 2\pi k$$

$$x = \frac{\pi}{2} + 2\pi k$$

$$x = \pi + 2\pi k$$

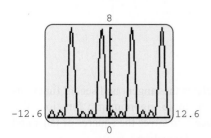

Figure 18

Figure 19

If an equation involves more than one trigonometric function but cannot be written in factored form, then it may be necessary to use a trigonometric identity to produce an equation involving just a single trigonometric function.

EXAMPLE 7

Solving an Equation Involving Sine and Cosine

Find all solutions of $2 \cos^2 x - \sin x = 1$.

Solution   Our first goal is to write the left-hand side in terms of a single trigonometric function. By using the Pythagorean identity $\sin^2 x + \cos^2 x = 1$, we can write $2 \cos^2 x$ (and hence the entire left side) in terms of $\sin x$.

$$2 \cos^2 x - \sin x = 1$$
$$2(1 - \sin^2 x) - \sin x = 1 \qquad \text{Using } \cos^2 x + \sin^2 x = 1$$
$$2 - 2 \sin^2 x - \sin x = 1 \qquad \text{Expanding}$$
$$2 \sin^2 x + \sin x - 1 = 0 \qquad \text{Simplifying}$$
$$(2 \sin x - 1)(\sin x + 1) = 0 \qquad \text{Factoring}$$

$$\sin x = \frac{1}{2} \quad \text{or} \quad \sin x = -1$$

The solution set to the original equation is formed by pooling together (taking the union of) the solution sets to each of these equations. The only solutions of $\sin x = \frac{1}{2}$ between 0 and 2π are $x = \frac{\pi}{6}$ and $x = \frac{5\pi}{6}$. Thus, the solution set of $\sin x = \frac{1}{2}$ consists of $\frac{\pi}{6}$ and $\frac{5\pi}{6}$, together with any additional angles obtained by adding multiples of 2π. Thus, we have

$$x = \frac{\pi}{6} + 2\pi k \quad \text{and} \quad x = \frac{5\pi}{6} + 2\pi k, \quad \text{where } k \text{ is any integer}$$

Now $\frac{3\pi}{2}$ is the only angle x between 0 and 2π for which $\sin x = -1$. Thus, the solution set to $\sin x = -1$ is given by

$$x = \frac{3\pi}{2} + 2\pi k, \quad \text{where } k \text{ is any integer}$$

It follows that the solution set to the original equation consists of all numbers x of one of the following forms (where k is any integer):

$$x = \frac{\pi}{6} + 2\pi k$$

$$x = \frac{5\pi}{6} + 2\pi k$$

$$x = \frac{3\pi}{2} + 2\pi k$$

EXAMPLE 8

Solving an Equation Involving Tangent and Secant

Find all solutions of $\tan x + 1 = \sec x$.

Solution   Here we need to find an identity that relates the tangent and secant functions so that we can rewrite the equation in terms of a single trigonometric function. The identity $\tan^2 x + 1 = \sec^2 x$ will do nicely, but since it involves the squares of $\tan x$ and $\sec x$, we begin by squaring both sides of the original equation.

$$\tan x + 1 = \sec x$$

$$(\tan x + 1)^2 = \sec^2 x \qquad \text{Squaring both sides}$$

$$\tan^2 x + 2 \tan x + 1 = \tan^2 x + 1 \qquad \text{Using } \sec^2 x = \tan^2 x + 1$$

$$2 \tan x = 0$$

$$\tan x = 0$$

Since $\tan x = \sin x / \cos x$, $\tan x = 0$ whenever $\sin x = 0$, which occurs at multiples of π. Thus, the solution *candidates* of $\tan x + 1 = \sec x$ have the form $x = \pi k$, where k is an integer. Note that in this example it is especially important to check our answer because squaring both sides of the equation may have introduced extraneous solutions. Indeed, if k is odd and $x = \pi k$, the left-hand side of the original equation simplifies to $\tan \pi k + 1 = 0 + 1 = 1$, whereas the right-hand side simplifies to $\sec \pi k = -1$. Thus, $x = \pi k$ is not a solution if k is odd. Consequently, the solutions are limited to those of the form $x = k\pi$, where k is even or, equivalently, $x = 2k\pi$ for any integer k.

·····EXAMPLE 9

Solving a Trigonometric Equation Involving Multiple Angles

Find the solutions of $\cos 2x = 2 \cos x$, where x is in the interval $[0, 2\pi)$.

Solution   This problem is made more difficult by the presence of the double angle $2x$. Thus, we eliminate it by using one of the double-angle identities for cosine.

$$\cos 2x = 2 \cos x$$

$$2 \cos^2 x - 1 = 2 \cos x \qquad \text{Using } \cos 2x = \cos^2 x - 1$$

$$2 \cos^2 x - 2 \cos x - 1 = 0$$

This last equation is quadratic in $\cos x$ but does not factor easily. Substituting u for $\cos x$, we obtain

$$2u^2 - 2u - 1 = 0$$

Applying the quadratic formula gives us

$$u = \frac{2 \pm \sqrt{(-2)^2 - 4(2)(-1)}}{2(2)}$$

$$= \frac{2 \pm \sqrt{12}}{4}$$

$$= \frac{1 \pm \sqrt{3}}{2}$$

$$\approx 1.366 \quad \text{or} \quad -0.366$$

It follows that $\cos x \approx 1.366$ or $\cos x \approx -0.366$. But it is impossible for $\cos x$ to be greater than 1, so we ignore $\cos x \approx 1.366$. Thus, our only solutions are those x in the interval $[0, 2\pi)$ for which $\cos x = -0.366$. Using the $\boxed{\cos^{-1}}$ key on a calculator, we obtain a solution in the second quadrant of $x \approx 1.9455$. A second solution, in the third quadrant with reference angle approximately $\pi - 1.9455$, is given by $x \approx \pi + (\pi - 1.9455) \approx 4.3377$. These two solutions can easily be checked in the original equation.

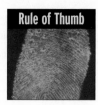

Rule of Thumb

When solving trigonometric equations, keep the following suggestions in mind:

1. Convert to a single trigonometric function if possible.
2. Look for opportunities to factor or apply the quadratic formula.
3. Check your answers in the original equation.

As we have seen throughout this and the previous chapter, trigonometric functions are an essential tool in modeling periodic and even quasi-periodic (almost periodic) functions. In the following example, trigonometric functions are used to model the path of a weight hanging from a spring. The periodic nature of the trigonometric functions mirrors the fact that the weight "bobs up and down." The motion of the spring is not truly periodic, however, since the spring gradually comes to rest.

··········>**EXAMPLE 10**   **Damped Harmonic Motion**

When a weight hung from a certain stretched spring is released (see Figure 20), its directed distance (in inches) above the equilibrium (or resting) position t seconds after the spring is released is approximated by

$$y(t) = -e^{-t}(6 \cos 6t + \sin 6t)$$

a. Find the initial position of the weight.
b. Estimate the time at which the weight first crosses the equilibrium position.
c. Estimate the highest point to which the weight rises.

Solution

a. At $t = 0$, we have

$$\begin{aligned} y(0) &= -e^{0}(6 \cos 0 + \sin 0) \\ &= -1(6 + 0) \\ &= -6 \end{aligned}$$

Evidently, the spring is released 6 inches *below* its equilibrium point.

b. The equilibrium point is defined to be where $y(t) = 0$. Thus, we are interested in the smallest positive value of t such that $y(t) = 0$. Although this equation could be solved by analytic methods, we solve it graphically. In Figure 21, we have shown the graph of $y(t)$ with the cursor near the first positive intercept. By using the calculate-zero feature, we confirm that the graph of $y(t)$ first crosses the horizontal axis at $t \approx 0.29$.

c. Figure 22 shows the graph of $y(t)$ near its highest point. By applying the calculate-maximum feature, we find that the maximum value is indeed 3.55 inches above the equilibrium position.

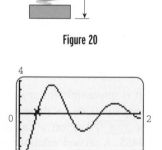

Equilibrium

y

Figure 20

X=.29473684   Y=.14709067

Figure 21

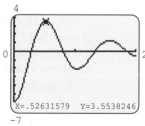

X=.52631579   Y=3.5538246

Figure 22

Understanding and Mastery Checklists

Concepts to Understand

Solutions of conditional trigonometric equations

✦

Properties of equality, as discussed in Chapter 1

✦

Signs of trigonometric functions,
as discussed in Chapter 5

✦

Reference angles, as discussed in Chapter 5

✦

Special angles, as discussed in Chapter 5

✦

Periodicity, as discussed in Chapter 5

✦

Fundamental trigonometric identities,
as discussed in Sections 6.1–6.3

✦

Graphical techniques for solving
trigonometric equations

Skills to Master

Find solutions to a trigonometric equation
in the interval $[0, 2\pi)$.

✦

Find all solutions to a trigonometric equation.

✦

Graphically estimate solutions to a
trigonometric equation.

Exercises 6.4

Exercises 1-4   *Determine whether the given number is a solution of the equation.*

1. $\cos x = \dfrac{\sqrt{3}}{2}$; $x = -\dfrac{\pi}{6}$ **2.** $\tan x = -1$; $x = \dfrac{9\pi}{4}$

3. $\csc x + 1 = 4\sin x$; $x = -\dfrac{5\pi}{3}$ **4.** $\sin x\left(\cos x + \dfrac{1}{2}\right) = 0$; $x = \dfrac{2\pi}{3}$

Exercises 5-20   *Find all solutions to the given equation in the interval $[0, 2\pi)$. Give exact answers where possible. When approximating, give answers to the nearest hundredth.*

5. $\cos x = 0$ **6.** $\sin x = \dfrac{1}{2}$

7. $\tan x - 2 = 0$ **8.** $\cot x + \sqrt{3} = 0$

9. $3\sin x + \sqrt{3} = \sin x$ **10.** $4\cos x = \cos x + 1$

11. $(\csc x - 2)(\sec x - 2) = 0$ **12.** $(\tan x + 1)(\cot x - 3) = 0$

13. $\sin x \cos x = 0$ **14.** $\tan x\left(\sec x - \sqrt{2}\right) = 0$

15. $\cos 2x = 1$ **16.** $\tan 3x = 5$

17. $\csc(2x + 1) = 3$ **18.** $\sin\left(\dfrac{x}{2} - \dfrac{\pi}{2}\right) = \dfrac{\sqrt{2}}{2}$

19. $\cot \dfrac{x}{2} = -1$ **20.** $\sin \dfrac{x}{3} = 0$

Exercises 21-34   *Find all solutions to the given equation.*

21. $\sin x = 0$ **22.** $\sqrt{3}\tan x = -1$

23. $2\cos x - 1 = 0$ **24.** $\sqrt{2}\csc x - 2 = 0$

25. $(\sin x + 1)(\tan x - 1) = 0$

26. $(2\sec x - 2)\left(2\cos x + \sqrt{3}\right) = 0$

27. $\cot x \csc x - 2\cot x = 0$

28. $2\sin x \cos x - \sqrt{2}\cos x = 0$

29. $\cos^2 x = 1$

30. $2\cos^2 x - \cos x = 1$

31. $\tan^3 x - \tan x = 0$
$$\frac{\pi}{4} + \frac{\pi}{2}k$$
33. $2\sin^2 x = \cos x + 1$

32. $2\sec x \sin^2 x - \sec x = 0$

34. $\sec x - \cos x = 0$

Exercises 35-54 *Find all solutions to the given equation in the interval* $[0, 2\pi)$. *Give exact answers where possible. When approximating, give answers to the nearest hundredth.*

35. $\sec x = -\sqrt{2}\tan x$

36. $\csc x \tan x + 2 = 0$

37. $\sin x - \cos x = 0$

38. $\csc x(\cos x + 1) = 2\cos x + \csc x$

39. $3\tan^2 x - \sin x \sec^2 x = 0$

40. $\sqrt{3}(\cot x - \tan x) = 2$

41. $\cos 2x = \cos^2 x + 1$

42. $\csc x \sin 2x + \cos x = \frac{3}{2}$

43. $\cot x = \tan 2x$

44. $\cos 2x = \sin^2 x$

45. $10\sin^2 x + \sin x - 6 = 0$

46. $\tan^2 x + \tan x - 1 = 0$

47. $9\cos^2 x - 3\cos x - 2 = 0$

48. $2\csc^2 x + \csc x - 2 = 0$

49. $\sec^2 x - \tan x = 3$

50. $\cos^2 x = -\sin x$

51. $\cos^2 x + \sin^2 x = \frac{1}{2}$

52. $\tan^2 x = \sec x - 1$

53. $\cos x \sin x = \frac{1}{4}$

54. $\tan x \cos^2 x = -\frac{\sqrt{3}}{4}$

Exercises 55-60 *Approximate all solutions in the interval* $[0, 2\pi)$ *to the nearest hundredth. Use a graphing calculator as necessary.*

55. $\sin^3 x - 3\sin x + 1 = 0$

56. $\cos^3 x = \sin^2 x$

57. $x = \cos x$

58. $x^2 = \sin x$

59. $\cos^3 x = \sin x$

60. $\tan^2 x - \sec^2 x = 0$

Exercises 61-64 *For the given function:*

a. *Estimate the period p and all zeros in the interval* $[0, p)$. *Give all answers to the nearest hundredth.*

b. *Estimate all real zeros.*

61. $f(x) = \tan^3(\pi k) - \sec^2(\pi x)$

62. $f(x) = \cos^3 x - \sin^2 x$

63. $f(x) = \ln\left(2 + \cos\frac{\pi x}{6}\right) - \frac{1}{2}$

64. $f(x) = \sin 2\pi x + \cos 3\pi x + \frac{3}{2}$

Applications

65. Ferris Wheel Height The height in feet of a certain passenger on a Ferris wheel is given by
$$y(t) = 55 + 50\sin\left(\frac{\pi t}{15} - \frac{\pi}{2}\right)$$
where t is the time in seconds and $t = 0$ coincides with the time at which the wheel was set in motion.

a. Find the first two times for which the height of the passenger is 80 feet.

b. How long does it take for the Ferris wheel to complete one rotation?

66. Spring Displacement A weight is attached to an elastic spring that is suspended from a ceiling. If the weight is pulled 1 inch below its rest position and released, its displacement in inches after t seconds is given by
$$x(t) = 2\cos\left(5\pi t + \frac{\pi}{3}\right)$$
Find the first two times for which the displacement is 1.5 inches.

67. Electric Current The current (in amperes) in an electrical circuit t seconds after the circuit is closed is given by the function
$$E(t) = -12\cos 4t$$
Find the first two times for which the current is 8 amperes.

68. Violin Sound The waveform of a certain violin note is modeled by the function
$$y(t) = 165\sin(\theta + 5.86) + 60\sin(\theta + 1.15) + 27\sin(\theta + 0.19)$$
Find the first two times for which $y(t) = 50$. [*Data source*: D.C. Miller, *The Science of Musical Sounds* (New York: Macmillan, 1916).]

Exercises 69-72 *If we ignore air resistance, then an object thrown from ground level with an initial angle* θ *and an initial speed of* v *feet per second has range (maximum horizontal distance)* $R = (v^2\sin 2\theta)/32$ *feet and maximum height* $H = (v^2\sin^2\theta)/64$ *feet.*

69. Football Range A quarterback spots a receiver in the end zone, 60 yards away. If the quarterback throws the ball with an initial velocity of 80 feet per second, what initial angle is required for the ball to just reach the receiver? (*Hint*: Assume the receiver catches the ball at the same height as it was thrown and then apply the range formula.)

70. Soccer Ball Height A soccer ball is kicked with an initial velocity of 50 feet per second and reaches a maximum height of 30 feet. At what initial angle was the ball kicked?

71. Range Equals Height An object is projected into the air from ground level in such a way that its range equals its maximum height. With what initial angle was it projected?

72. Tennis Cannon Error A certain tennis cannon can be adjusted to project tennis balls at any angle between 0 and $\frac{\pi}{2}$. However, due to design limitations, the angle indicated on the cannon may be in error by a small amount each time it is adjusted. As a consequence, for any given angle setting, the range of a tennis ball may vary.

a. For an angle setting of $\frac{\pi}{4}$, denote the *angular error* by α and show that the difference in ranges for a tennis ball projected at angles $\frac{\pi}{4} + \alpha$ and $\frac{\pi}{4}$ is given by

$$\frac{v^2}{32}(\cos 2\alpha - 1)$$

We refer to this expression as the *range error*.

b. Calculate the angular error if the cannon is set at $\frac{\pi}{4}$, the initial velocity is 40 feet per second, and the range error is measured at -2 feet.

Concepts and Critical Thinking

Exercises 73–76 *Answer true or false.*

73. If $-1 \le c \le 1$, the equation $\sin x = c$ has exactly one solution in the interval $[0, \pi)$.

74. If $c \ge 1$, the equation $\sec x = c$ has a solution.

75. The equation $\tan x = c$ has a solution for any c.

76. The equation $\cos x = x$ has exactly one solution.

Exercises 77–80 *Give an example of each.*

77. A number c for which the equation $\sin x = c$ has no solution

78. A number c for which the equation $\sec x = c$ has no solution

79. An integer n for which $\tan nx = 1$ has exactly 4 solutions in the interval $[0, 2\pi)$

80. An interval in which $\cos 2x = 1$ has exactly 4 solutions

81. If a function f is periodic with period $\frac{\pi}{3}$ and $f(x) = \frac{1}{2}$ has two solutions in the interval $\left[0, \frac{\pi}{3}\right]$, how many solutions are in the interval $\left[\frac{4\pi}{3}, \frac{7\pi}{3}\right]$?

Questions for Discussion or Essay

82. In Example 3, we considered a formula that gave the range of a golf ball in terms of the initial angle and initial velocity with which the ball was struck. Why do you think it is that dimpled golf balls fly further than undimpled balls, even when struck with the same initial angle and velocity?

83. A periodic function has either no zeros or infinitely many. Explain why this is so.

Projects for Enrichment

84. Measuring Distance on Earth Distances between points on the surface of Earth can be computed using latitude and longitude. First, we define the **location angles** ϕ and θ for a point with latitude and longitude $m°$ and $n°$, respectively.

$$\phi = \begin{cases} 90° - m° & \text{for north latitude} \\ 90° + m° & \text{for south latitude} \end{cases}$$

$$\theta = \begin{cases} 90° - n° & \text{for west longitude} \\ 90° + n° & \text{for east longitude} \end{cases}$$

See Figure 23 for a geometric interpretation of ϕ and θ.

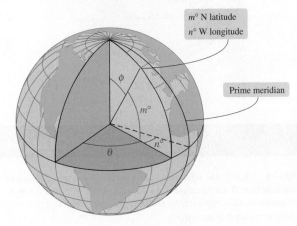

Figure 23

a. Find the location angles ϕ and θ for the following cities.

  i. New York: 40° 42′ 51″ N, 74° 00′ 23″ W

  ii. Los Angeles: 34° 03′ 08″ N, 118° 14′ 34″ W

  iii. Berlin: 52° 32′ 00″ N, 13° 25′ 00″ E

  iv. Sydney: 33° 52′ 00″ S, 151° 12′ 00″ E

  v. Rio de Janeiro: 22° 53′ 43″ S, 43° 13′ 22″ W

Figure 23

Now consider two points P_1 and P_2 with location angles ϕ_1, θ_1 and ϕ_2, θ_2, respectively. Let Φ denote the angle made between the line segments from the center of Earth to each of P_1 and P_2, as shown in Figure 24.

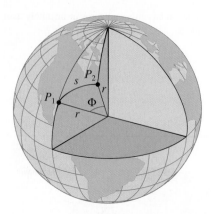

Figure 24

Note that the distance s between P_1 and P_2 is the length of the arc subtended by the angle Φ in a circle of radius r, where r is the radius of Earth. Thus, when Φ is measured in radians, $s = \Phi r = \Phi(3963)$ miles. Moreover, it can be shown that Φ satisfies

$$\cos \Phi = \sin \phi_1 \sin \phi_2 \cos \theta_1 \cos \theta_2$$
$$+ \sin \phi_1 \sin \phi_2 \sin \theta_1 \sin \theta_2 + \cos \phi_1 \cos \phi_2$$

b. Show that the preceding equation can be rewritten as
$$\cos \Phi = \sin \phi_1 \sin \phi_2[\cos(\theta_1 - \theta_2)] + \cos \phi_1 \cos \phi_2$$

To find the distance s, we simply find $\cos \Phi$, then find Φ using a calculator set in radian mode, and finally multiply by the radius of Earth.

c. Estimate the distances between New York and each of the other cities in part a.

d. A city in North America with latitude 42° 14′ 32″ N is approximately 1600 miles from New York. Find its longitude and locate the city in an atlas or a Web site such as the Geographical Names Information System home page, located (as of 2003) at http://geonames.usgs.gov/index.html. (*Hint*: In the equation given in part b, substitute the location angles for New York for ϕ_1 and θ_1, the location angle for a point with latitude 42° 14′ 32″ N for ϕ_2, and the angle subtended by an arc with length 1600 miles in a circle of radius 3963 miles for Φ. Solve the resulting equation for θ_2 and then find the longitude.)

e. Tangent, Oregon, has latitude 44° 32′ 24″ N and longitude 123° 06′ 24″ W. A city approximately 170 miles away has longitude 123° 15′ 48″ W. Find its latitude and locate the city in an atlas or on a Web site. (*Hint*: Make substitutions similar to those suggested in the hint in part d. You will need to use a graphing calculator to estimate the value of ϕ_2 that satisfies the resulting equation.)

f. Discuss the effect of elevation on this technique for estimating distances between points on Earth.

Chapter 6 Review

Exercises 1-6 *Classify the equation as an identity, conditional equation, or contradiction. If the equation is an identity, verify it. If it is a conditional equation, then estimate the smallest positive solution. Use a graphing calculator as necessary.*

1. $2 \sin x + \cos x = 2$

2. $\sec x \cot x \csc x = \cot^2 x + 1$

3. $\cos\left(\dfrac{\pi}{2} - x\right) = \sin x + 1$

4. $(\cos x - 2)(\sin x - 3) = 0$

5. $\dfrac{1}{1 - \cos x} = \csc^2 x + \csc x \cot x$

6. $\sin^2 x + 1 = 2 \cos^2 x$

Exercises 7-16 *Verify the given identity.*

7. $\dfrac{\cot x}{\tan x} = \csc^2 x - 1$

8. $\dfrac{\csc A}{\sec A} = \cot A$

9. $\cot^2 y + 5 = \csc^2 y + 4$

10. $\sin x - \cos x = \dfrac{1 - 2\cos^2 x}{\cos x + \sin x}$

11. $\cos\theta + \sin\theta\tan\theta = \sec\theta$

12. $\dfrac{1}{1 - \sin u} = \sec^2 u + \sec u\tan u$

13. $\dfrac{\sec\alpha + \tan\alpha}{\sec\alpha - \tan\alpha} = (\sec\alpha + \tan\alpha)^2$

14. $\dfrac{\cot^4 y - 1}{\cot^2 y - 1} = \csc^2 y$

15. $|\sin x| = \sqrt{1 - \cos^2 x}$

16. $\ln|\sec\theta| = -\ln|\cos\theta|$

Exercises 17-18 *Use fundamental trigonometric identities to rewrite the first function in terms of the second.*

17. $\cos x;\ \sin x$

18. $\tan x;\ \sin x$

Exercises 19-20 *Use the given information to compute the exact values of the indicated trigonometric functions.*

19. $\cos u = \dfrac{4}{5},\ 0 \le u < \dfrac{\pi}{2};\ \cos v = \dfrac{8}{17},\ 0 \le v < \dfrac{\pi}{2}$

    a. $\cos(u - v)$ **b.** $\cos(u + v)$

    c. $\sin(u - v)$ **d.** $\sin(u + v)$

20. $\sin\alpha = -\dfrac{24}{25},\ \pi \le \alpha < \dfrac{3\pi}{2};\ \cos\beta = \dfrac{3}{5},\ \dfrac{3\pi}{2} \le \beta < 2\pi$

    a. $\cos(\alpha - \beta)$ **b.** $\cos(\alpha + \beta)$

    c. $\sin(\alpha - \beta)$ **d.** $\sin(\alpha + \beta)$

Exercises 21-24 *Rewrite the given expression as a trigonometric function of a single angle.*

21. $\cos 10° \cos 50° - \sin 10° \sin 50°$

22. $\sin\dfrac{3\pi}{8}\cos\dfrac{\pi}{8} - \cos\dfrac{3\pi}{8}\sin\dfrac{\pi}{8}$

23. $\sin x \cos 2x + \cos x \sin 2x$

24. $\dfrac{\tan\dfrac{y}{2} + \tan\dfrac{3y}{2}}{1 - \tan\dfrac{y}{2}\tan\dfrac{3y}{2}}$

Exercises 25-28 *Simplify the given trigonometric expression.*

25. $\sin\!\left(x - \dfrac{3\pi}{2}\right)$

26. $\tan(\pi - x)$

27. $\cos\!\left(\dfrac{\pi}{4} - x\right)$

28. $\sin\!\left(\arcsin\dfrac{4}{5} - \arccos\dfrac{12}{13}\right)$

Exercises 29-30 *Verify the given identity using the sum and difference identities.*

29. $\dfrac{1}{\cot x - \cot y} = \dfrac{\sin x \sin y}{\sin(y - x)}$

30. $\cos(\alpha + \beta)\cos(\alpha - \beta) = \cos^2\alpha - \sin^2\beta$

Exercises 31-32 *Use the given information to compute*

    a. $\sin 2\theta$ **b.** $\cos 2\theta$ **c.** $\tan 2\theta$

31. $\tan\theta = \dfrac{3}{4},\ \pi \le \theta < \dfrac{3\pi}{2}$

32. $\cos\theta = -\dfrac{5}{12},\ 0 \le \theta < \pi$

Exercises 33-34 *Use the given information to compute*

    a. $\sin\dfrac{x}{2}$ **b.** $\cos\dfrac{x}{2}$ **c.** $\tan\dfrac{x}{2}$

33. $\sin x = -\dfrac{3}{5},\ \dfrac{3\pi}{2} \le x < 2\pi$

34. $\tan x = 2,\ 0 \le x < \dfrac{\pi}{2}$

Exercises 35-38 *Find the exact value of the given quantity.*

35. $\cos\dfrac{\pi}{12}$

36. $\sin 165°$

37. $\tan 75°$

38. $\cot\dfrac{\pi}{8}$

Exercises 39-44 *Verify that the given equation is an identity.*

39. $2\sin x = 4\sin\dfrac{x}{2}\cos\dfrac{x}{2}$

40. $\cos 2x = \cos^4 x - \sin^4 x$

41. $\cos 4x = 8\cos^4 x - 8\cos^2 x + 1$

42. $\sin 3x = \sin x(3 - 4\sin^2 x)$

43. $\tan x + \cot x = 2\csc 2x$

44. $\cot x = \dfrac{1 + \sin 2x + \cos 2x}{1 + \sin 2x - \cos 2x}$

Exercises 45-58   *Find all solutions to the given equation in the interval* $[0, 2\pi)$. *Give exact answers where possible. When approximating, give answers to the nearest hundredth.*

45. $\cos x = -1$

46. $\csc x = \sqrt{2}$

47. $(2 \sin x - 1)\left(\tan x - \sqrt{3}\right) = 0$

48. $\sin x(\cos x - \pi) = 0$

49. $2 \sin x \cos x - \sin x = 0$

50. $2 \tan^2 x \cos x - \tan^2 x = 0$

51. $2 \cos x + \sec x - 3 = 0$

52. $2 \cos^2 x - 3 \sin x = 3$

53. $\sin 2x + \sin x = 0$

54. $\tan 2x - 2 \cos x = 0$

55. $6 \sin^2 x - 2 \sin x - 1 = 0$

56. $\sec^2 x - 2 \sec x - 4 = 0$

57. $\cos 3x = -1$

58. $2 \sin 2x = 1$

Exercises 59-66   *Find all solutions to the given equation.*

59. $\cos x = \dfrac{1}{2}$

60. $\cot x = -\sqrt{3}$

61. $\left(2 \sin x - \sqrt{3}\right)\cos x = 0$

62. $\sin^2 x \cos x - \cos x = 0$

63. $\cot^3 x + \cot x = 0$

64. $\cot^2 x + \csc^2 x = 0$

65. $\sin 2x = 1$

66. $\tan 2x = -\sqrt{3}$

Exercises 67-68   *Approximate all solutions in the interval* $[0, 2\pi)$ *to the nearest hundredth. Use a graphing calculator as necessary.*

67. $\sin 2x + \cos 3x = 1$ **68.** $\sin^3 x = 3 \sin x + 1$

Exercises 69-70   *For the given function:*

  a. *Estimate the period p and all zeros in the interval* $[0, p)$. *Give all answers to the nearest hundredth.*

  b. *Estimate all real zeros.*

69. $f(x) = 2 \sin(3\pi x) - 3 \sin(2\pi x)$

70. $f(x) = \sin(\sin(\pi x))$

Exercises 71-72   *The range R in feet of a projectile launched from ground level with an initial speed of v feet per second and at an angle θ is given by*

$$R = \frac{v^2 \sin 2\theta}{32}$$

71. Baseball Range   Suppose a baseball is thrown with an initial speed of $v = 60$ feet per second and that $R = 100$ feet. Find the initial angle θ.

72. Projectile Range   Suppose a projectile fired at 32 feet per second with an initial angle θ has a range of R and that, if the angle θ is increased by $\frac{\pi}{3}$, then the resulting range will be $R + 16$. Find the initial angle θ and then find R.

73. Electric Current   The current in amperes in a certain electrical circuit at time t seconds is given by

$$i(t) = 20 \cos\left(60t - \frac{\pi}{3}\right)$$

Find all times at which the current is 10 amperes.

74. Telephone Line Distance   A temporary telephone line runs from a point P on the ground to a point Q at the base of a building and then up the side of the building to a point R, as shown in Figure 25. To run the line directly from P to R would require 100 feet, 20 feet less than with the present arrangement. Find $\angle QPR$ and the distances PQ and QR.

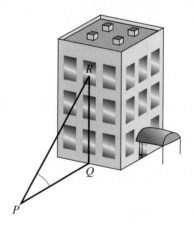

Figure 25

Chapter 6 Test

Problems 1-8 *Answer true or false.*

1. The equation $\sin(x + y) = \sin x + \sin y$ is true for all values of x and y.

2. The sine of twice an angle always equals twice the sine of the angle.

3. The equation $\sin 2x = 2 \sin x$ has infinitely many solutions.

4. Any equation that has infinitely many solutions is an identity.

5. The equation
$$1 - \sin^2(e^x) = \cos^2(e^x)$$
is true for all x.

6. There is at least one value of a for which the equation $\sin x = a$ has exactly two solutions.

7. The equation $\sin x = x$ has infinitely many solutions.

8. If x is known to be an angle terminating in the first quadrant, then we can conclude that $x/2$ terminates in the first quadrant also.

Problems 9-11 *Give an example of numbers a and b such that the equation* $\cos^a x + \sin^a x = b$ *is*

9. An identity

10. A conditional equation

11. A contradiction

Problems 12-14 *Verify the given identity.*

12. $\cos x \csc x \tan x = 1$

13. $\cos x = \sec x(1 - \sin^2 x)$

14. $\dfrac{\cos 2x}{\cos x - \sin x} = \dfrac{\csc x + \sec x}{\sec x \csc x}$

15. Simplify $\tan\left(x - \frac{\pi}{4}\right)$.

16. Find the exact value of $\sin \frac{\pi}{12}$.

17. Find all solutions of the equation $2 \tan x - 4 = 2$ in the interval $[0, 2\pi)$. Give exact answers where possible. When approximating, give answers to the nearest hundredth.

18. Find all solutions of the equation $9 \sin x + 2 \cos^2 x = 6$.

19. Use a graphing calculator to determine if the equation $2 \sin^2 x \cos x = 3$ is an identity, a contradiction, or a conditional equation. If the latter, estimate the solutions on the interval $[0, 2\pi)$.

20. A child is swinging in such a way that the height of the swing t seconds after being pushed is approximated by
$$h = 6 - 3 \cos \frac{\pi t}{2}$$
Find the first time at which the swing is 4.5 feet off the ground.

Chapter 7

Applications of Trigonometry

Stonehenge, an arrangement of enormous stone blocks on the Salisbury Plain in central southern England, is the ruin of a single stone structure, parts of which are over 4000 years old. Although the civilization that erected it and their purpose for doing so may be permanently lost in the mists of time, the alignment and orientation of the stones suggests a familiarity with the seasonal changes in position of the sun and moon. In fact, some have even suggested that Stonehenge served as an astronomical computer, capable of predicting eclipses and other astronomical events. Little is known about the mathematics employed by the builders of Stonehenge, but clearly the notions of angle and distance—cornerstones of trigonometry—were familiar to its designers.

Section 7.1 | The Law of Sines

    ◦ How can you estimate the height of a mountain from inside a car?

    ◦ What is the least amount of information necessary for finding all the side lengths and angle measures of a triangle?

    ◦ How can the height of the Eiffel Tower be estimated from a plane flying overhead?

    ◦ How can a device constructed from ordinary household objects be used to estimate distances to remote objects?

In Chapter 5, we solved right triangles by applying trigonometric function definitions, the Pythagorean Theorem, and the fact that the sum of the angles in a triangle is 180°. Now in *any* triangle, the sum of the angles is 180°, but the trigonometric function definitions and the Pythagorean Theorem do not apply to **oblique** (nonright) triangles. In this section, we develop and apply the **Law of Sines**, a relationship among the side lengths and angle measures that plays a role in the solution of oblique triangles similar to that played by the trigonometric function definitions for right triangles. In the next section, we formulate a generalization of the Pythagorean Theorem known as the **Law of Cosines**. Thus, we will have three tools for the solution of oblique triangles: the sum of the angles result, the Law of Sines, and the Law of Cosines.

Terminology and Notation

Solving a triangle consists of determining all unknown side lengths and angle measures—six pieces of information in all. Generally speaking, if we know three measurements (including at least one side length), then we can determine the other three by applying either the Law of Sines or the Law of Cosines and the sum of the angles result. The solution strategy for a given triangle depends on *which* pieces of information are provided. Essentially, there are four distinct cases:

1. Two angles and a side are given. We refer to this case as **ASA** (angle side angle) or **AAS** (angle angle side), depending on whether the given side is between the angles.

2. Two sides and an angle opposite one of the sides are given. We refer to this case as **SSA** (side side angle).

3. Three sides are given. This case is referred to as **SSS** (side side side).

4. Two sides and the angle between are given. This case is known as **SAS** (side angle side).

As we will soon discover, the first two cases can be solved using the Law of Sines, whereas the latter two are most easily solved with the Law of Cosines.

    Throughout this section and the next, we will systematically label triangles. First, we refer to angles and their vertices by the same name, generally a capital letter. Thus, the triangle with vertices named A, B, and C shown in Figure 1 has angles named A, B, and C also. Second, we refer to the length of the side opposite an angle by the lowercase version of the angle name, and so a is the length of the side opposite angle A, b is the length of the side opposite angle B, and c is the length of the side opposite angle C, as shown in Figure 1.

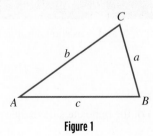

Figure 1

Solving Triangles with the Law of Sines

We develop the Law of Sines, a rule that applies to arbitrary triangles, by building from our right triangle definition of the sine function. Consider an arbitrary triangle ABC like either of those depicted in Figure 2. To introduce right triangles, we sketch the altitude h extending from the vertex C to the point D on the line containing A and B. In both cases, two right triangles are formed: ADC and BDC.

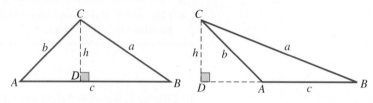

Figure 2

The Law of Sines results from computing h in two different ways. Applying the definition of the sine function to triangle ADC gives us

$$\sin A = \frac{h}{b} \quad \text{or} \quad h = b \sin A$$

On the other hand, the definition of the sine function applied to triangle BDC yields

$$\sin B = \frac{h}{a} \quad \text{or} \quad h = a \sin B$$

Equating these two expressions for h gives us $b \sin A = a \sin B$ or, equivalently,

(1)
$$\frac{a}{\sin A} = \frac{b}{\sin B}$$

A similar argument, developed by letting h denote the length of the altitude from vertex A to side BC, would yield

(2)
$$\frac{b}{\sin B} = \frac{c}{\sin C}$$

Together, equations (1) and (2) give us the Law of Sines.

Law of Sines

If a triangle with vertices A, B, and C has side lengths a, b, and c, as shown in Figure 1, then

$$\frac{a}{\sin A} = \frac{b}{\sin B} = \frac{c}{\sin C}$$

In words, no matter which angle we choose, the ratio of the opposite side to the sine of the angle is the same.

····**EXAMPLE 1**

Figure 3

Solving an AAS Triangle

Find all unknown side lengths and angle measures for the triangle shown in Figure 3.

Solution Since the angles must add up to 180°,

$$A = 180° - B - C = 180° - 107° - 32° = 41°$$

To find a, we use the fact that c, A, and C are known and apply the Law of Sines.

$$\frac{a}{\sin A} = \frac{c}{\sin C} \qquad \text{Applying the Law of Sines}$$

$$\frac{a}{\sin 41°} = \frac{17}{\sin 32°} \qquad \text{Substituting for } A, c, \text{ and } C$$

$$a = \frac{17}{\sin 32°}(\sin 41°) \approx 21.05 \qquad \text{Solving for } a$$

Similarly, we find b using the known values for c, B, and C.

$$\frac{b}{\sin B} = \frac{c}{\sin C}$$

$$\frac{b}{\sin 107°} = \frac{17}{\sin 32°}$$

$$b = \frac{17}{\sin 32°}(\sin 107°) \approx 30.68$$

Rule of Thumb

As a simple check of your work, be sure that the longest side is opposite the largest angle and the shortest side is opposite the smallest angle. Also, if there is an obtuse angle in the triangle, it must be opposite the longest side.

····**EXAMPLE 2**

Using an ASA Triangle to Find the Height of a Mountain

A motorist traveling on a straight and level highway at a rate of 60 miles per hour sees a mountain peak directly ahead at an angle of elevation of 10°. Five minutes later, the angle of elevation to the peak is 15°. Find the height of the mountain relative to the road below.

Solution At a rate of 60 miles per hour, the motorist will travel 5 miles in 5 minutes. Thus, if the first elevation reading of 10° is recorded at point A and the second reading of 15° is recorded at point B, the distance between A and B is 5 miles (see Figure 4). We wish to find the height h of the mountain relative to the road.

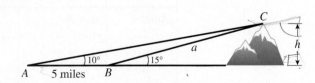

Figure 4

As a first step, we apply the Law of Sines to the triangle ABC shown in Figure 4 to find the side length a. Notice that

$$\angle ABC = 180° - 15° = 165°$$

and so

$$\angle ACB = 180° - (10° + 165°) = 5°$$

Thus, from the Law of Sines applied to triangle ABC we have

$$\frac{a}{\sin 10°} = \frac{5}{\sin 5°}$$

$$a = \frac{5}{\sin 5°}(\sin 10°) \approx 9.96 \text{ miles}$$

Now, by applying the definition of sine to the right triangle with hypotenuse length a and side length h, we obtain

$$\sin 15° = \frac{h}{a}$$

or

$$h = a \sin 15° \approx 9.96 \sin 15° \approx 2.58 \text{ miles}$$

Converting from miles to feet, we have

$$2.58 \text{ miles} \times \frac{5280 \text{ feet}}{1 \text{ mile}} \approx 13,600 \text{ feet}$$

When two sides of a triangle and an angle opposite one of the sides are known (the SSA case), finding the remaining parts of the triangle may not be as straightforward as it was in the previous examples. Three different outcomes are possible, depending on the given side lengths and angle measure: (1) There may be no triangle satisfying the given conditions; (2) there may be a unique triangle satisfying the given conditions; or (3) there may be two distinct triangles satisfying the given conditions. Table 1 illustrates the possibilities, given angle A and sides a and b.

Table 1
SSA Possibilities

Angle A	Condition ($h = b \sin A$)	Triangles possible	Example
Acute	$a < h$	0	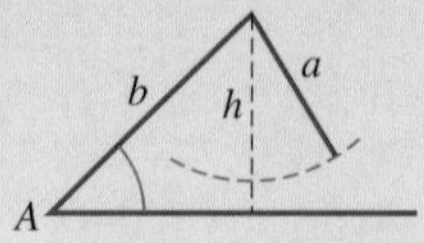
Acute	$a = h$	1	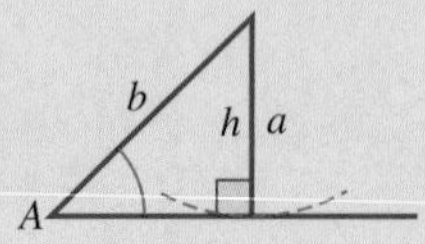

Acute	$h < a < b$	2	
Acute	$a \geq b$	1	
Obtuse	$a \leq b$	0	
Obtuse	$a > b$	1	

Note that it is not necessary to memorize the contents of Table 1. As we see in the following examples, the number of possible triangles becomes apparent as the solutions develop.

EXAMPLE 3

Solving an SSA Triangle with a Unique Solution

Find the unknown side length and angle measures for the triangle ABC, given that $a = 90$, $c = 100$, and $C = 65°$, as shown in Figure 5.

Solution   Using the Law of Sines and the known values for a, c, and C, we obtain

$$\frac{a}{\sin A} = \frac{c}{\sin C}$$

$$\frac{90}{\sin A} = \frac{100}{\sin 65°}$$

$$\sin A = \frac{90 \sin 65°}{100} \approx 0.81568 \qquad \text{Solving for } \sin A$$

Since the sine function is positive in the first and second quadrants, there are two angles between 0° and 180° whose sine is 0.81568. Using the $\boxed{\text{SIN}^{-1}}$ key on a calculator set in degree mode, we find that the acute angle with sine 0.81568 is approximately 54.65°. An angle in the second quadrant with sine 0.81568 is approximated by $180° - 54.65° = 125.35°$. However, this could not be the angle A in the triangle ABC

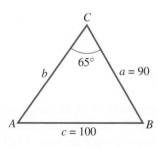

Figure 5

since the sum of 125.35° and $C = 65°$ exceeds 180°. (Note that these extra calculations could have been avoided by observing that, since a is not the longest side, A must be acute.) Thus, $A \approx 54.65°$ and

$$B = 180° - A - C \approx 180° - 54.65° - 65° = 60.35°$$

To find b, we apply the Law of Sines once more, using the known values for c, B, and C.

$$\frac{b}{\sin B} = \frac{c}{\sin C}$$

$$b = \frac{c \sin B}{\sin C} \approx \frac{100 \sin 60.35°}{\sin 65°} \approx 95.89$$

EXAMPLE 4

Solving an SSA Triangle with No Solutions

Show that there is no triangle ABC with $a = 0.3$, $b = 0.6$, and $A = 42°$.

Solution   From the Law of Sines, we know that if a triangle exists with the given side lengths and angle measure, then

$$\frac{a}{\sin A} = \frac{b}{\sin B}$$

$$\frac{0.3}{\sin 42°} = \frac{0.6}{\sin B}$$

$$\sin B = \frac{0.6 \sin 42°}{0.3} \approx 1.338$$

However, there is no angle B whose sine is greater than 1. Thus, there is no triangle with the given side lengths and angle measure. In Figure 6, we see that side a is too short to reach side c, no matter what value is chosen for angle C.

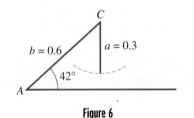

Figure 6

EXAMPLE 5

Solving an SSA Triangle with Two Solutions

Find all triangles ABC satisfying $a = 5$, $b = 3$, and $B = 31°$.

Solution   We begin by applying the Law of Sines to find angle A.

$$\frac{a}{\sin A} = \frac{b}{\sin B}$$

$$\frac{5}{\sin A} = \frac{3}{\sin 31°}$$

$$\sin A = \frac{5 \sin 31°}{3} \approx 0.8584$$

We find the acute angle having sine 0.8584 to be approximately 59.1°. The angle in the second quadrant is approximately $180° - 59.1° = 120.9°$. In this case, unlike in Example 3, the second possible angle yields a second triangle since the sum of 120.9° and $B = 31°$ is less than 180°. Thus, we consider two cases, the first with $A \approx 59.1°$ and the second with $A \approx 120.9°$.

Case 1: For $A \approx 59.1°$, we have

$$C = 180° - A - B \approx 180° - 59.1° - 31° = 89.9°$$

The remaining side, c, can now be found using the Law of Sines $\left(\dfrac{c}{\sin C} = \dfrac{b}{\sin B}\right)$.

$$c = \frac{b \sin C}{\sin B} \approx \frac{3 \sin 89.9°}{\sin 31°} \approx 5.82$$

This triangle is shown in Figure 7.

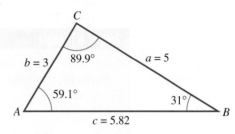

Figure 7

Case 2: For $A \approx 120.9°$, we have

$$C = 180° - A - B \approx 180° - 120.9° - 31° = 28.1°$$

and, from the Law of Sines,

$$c = \frac{b \sin C}{\sin B} \approx \frac{3 \sin 28.1°}{\sin 31°} \approx 2.74$$

This triangle is shown in Figure 8.

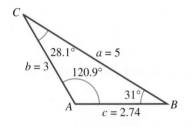

Figure 8

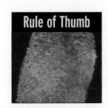

Rule of Thumb

Always check for a second triangle when solving an SSA triangle.

Area of Triangles   In proving the Law of Sines, we used the fact that the altitude h of either triangle shown in Figure 9 is given by

$$h = b \sin A$$

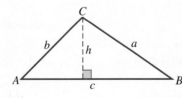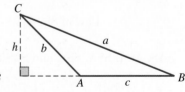

Figure 9

Thus, the area of each triangle is given by

$$\text{Area} = \frac{1}{2}(\text{Base})(\text{Height}) = \frac{1}{2}ch = \frac{1}{2}c(b \sin A) = \frac{1}{2}bc \sin A$$

Similarly, we can show that

$$\text{Area} = \frac{1}{2}ab \sin C = \frac{1}{2}ac \sin B$$

Note that in each case, the area involves two sides of the triangle and the angle between them. This leads us to the following general formula for the area of a triangle.

Area of a Triangle

Let s_1 and s_2 denote the lengths of two sides of a triangle, and let θ denote the angle between the two sides. Then the area of the triangle is given by

$$\text{Area} = \frac{1}{2}s_1s_2 \sin \theta$$

EXAMPLE 6

Finding the Area of an Oblique Triangle

Find the area of the triangle shown in Figure 10.

Solution   Since we are given two sides and the angle between, we apply the area formula with $s_1 = 75$, $s_2 = 50$, and $\theta = 130°$.

$$\text{Area} = \frac{1}{2}s_1s_2 \sin \theta$$

$$= \frac{1}{2}(75)(50) \sin 130°$$

$$\approx 1436.3 \text{ square feet}$$

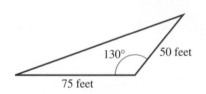

130°   50 feet

75 feet

Figure 10

Understanding and Mastery Checklists

Concepts to Understand	Skills to Master
Oblique triangle	Solve an AAS triangle using the Law of Sines.
Law of Sines	Solve an ASA triangle using the Law of Sines.
AAS, ASA, SSA, SSS, and SAS triangles	Solve an SSA triangle using the Law of Sines.
Area formula for an oblique triangle	Find the area of an oblique triangle.

Exercises 7.1

Exercises 1–8   *Find all unknown side lengths and angle measures; that is, solve the given triangle.*

1.

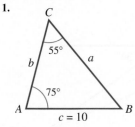

2.

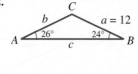

3.

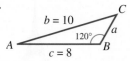

4.

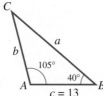

5.

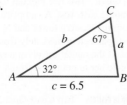

6.

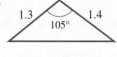

7.

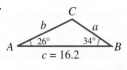

8.

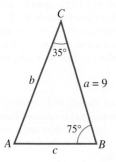

9. $c = 18, A = 17°, C = 86°$ **10.** $b = 7.5, c = 3.2, B = 140°$

11. $b = 20, c = 10, C = 30°$ **12.** $b = 80, A = 28°, C = 74°$

13. $a = 42, B = 7°, C = 51°$ **14.** $a = 15, A = 68°, B = 43°$

15. $a = 23, c = 16, A = 122°$ **16.** $a = 14, b = 18, A = 115°$

17. $b = 2, c = 9, B = 38°$ **18.** $c = 2.4, A = 116°, B = 32°$

19. $a = 14, c = 12, C = 40°$ **20.** $a = 6.3, b = 5.8, B = 55°$

21. $a = 12.5, c = 14, A = 140°$ **22.** $b = 150, B = 23°, C = 17°$

23. $b = 70, A = 62°, C = 53°$ **24.** $b = 9, c = 4, B = 34°$

25. $a = 300, A = 43°, B = 117°$ **26.** $a = 5, b = 40, A = 26°$

27. $b = 0.13, c = 0.9, C = 52°$ **28.** $a = 1, b = \sqrt{2}, A = 45°$

Exercises 29–34   *Find the area of the given triangle.*

29.

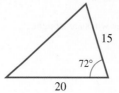

30.

31.

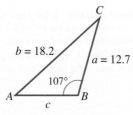

32.

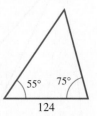

Exercises 9–28   *Solve the given triangle. Assume angles A, B, and C and sides a, b, and c are labeled as shown in Figure 11.*

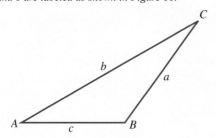

Figure 11

33.

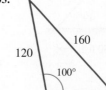

34.

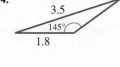

Applications

35. Length of a Power Line A power company would like to run a high-voltage power line across a canyon. It must extend from point A on the south side of the canyon to point C on the north side, as shown in Figure 12. A surveyor has marked point B on the south side of the canyon 100 feet from point A and has determined that $\angle CAB = 42°$ and $\angle ABC = 110°$. What is the minimum length of power line that will be needed?

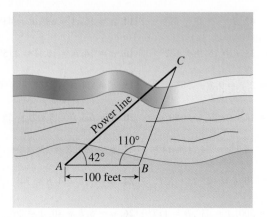

Figure 12

36. Height of a Radio Tower Two guy wires for a radio tower make angles with the horizontal of 58° and 49°, as shown in Figure 13. If the ground anchors for each wire are 150 feet apart, find

a. the length of each wire, assuming the wires are straight

b. the height of the tower

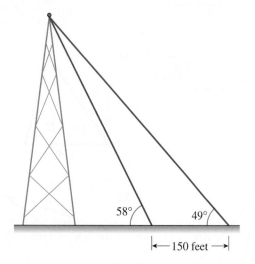

Figure 13

37. Height of the Eiffel Tower An airplane is flying at a constant altitude of 3650 feet and at a constant speed of 660 feet per second (approximately 450 miles per hour) on a path that will take it directly over the Eiffel Tower, as shown in Figure 14. At a certain point, the angle of depression to the top of the tower is 22°. Four seconds later, the angle of depression to the top of the tower is 34°. Estimate the height of the Eiffel Tower.

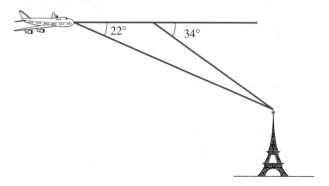

Figure 14

38. Fire Lookout Two fire lookout towers are located 1 mile apart on a line running east-west. A campfire is spotted at a bearing of N 38° E from the western tower and N 29° W from the other. A forest ranger is dispatched immediately from the tower nearest the fire to put it out. If the ranger travels at a rate of 3 miles per hour, how long will it take her to get to the fire?

39. Ship in Distress Two Coast Guard stations are located 10 miles apart on a coastline that runs north-south. A distress signal is received from a ship with a bearing of N 43° E from the southern station and a bearing of S 56° E from the northern station. How far is the ship from each station? How far is the ship from shore?

40. Great Pyramid Ramp It has been speculated that the Egyptian pyramids were built with the aid of long ramps. Given that the sides of the Great Pyramid at Giza make an angle of 51.9° with the ground, what would be the length of a ramp with a 7° angle of incline that reaches 100 feet along the face of one side of the pyramid?

41. Triangular Plot   A plot of land is bounded on two sides by rivers, as shown in Figure 15. Find the area of the plot given that the rivers meet at an angle of 35° and the lengths of the sides bounded by the rivers are 1300 feet and 650 feet. The owner would like to sell the plot for $2000 per acre. How much is the asking price, given that 1 acre is 43,560 square feet?

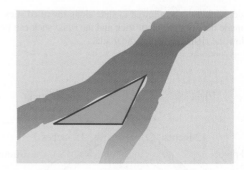

Figure 15

Concepts and Critical Thinking

Exercises 42–45   *Answer true or false.*

42. If two angles and a side length of a triangle are given, then the Law of Sines can be used to solve the triangle.

43. If two side lengths and an angle of a triangle are given, then the Law of Sines can be used to solve the triangle.

44. When the Law of Sines is used to solve a triangle, the solution is always unique.

45. For a given triangle, the ratio of a side length to the sine of the angle opposite is constant.

Exercises 46–49   *Sketch a triangle that is an example of the given case by labeling the appropriate angles and/or sides with their values.*

46. An ASA triangle

47. An AAS triangle

48. An SSA triangle

49. An SAS triangle

50. A Rule of Thumb in this section states that the largest side of a triangle is opposite the largest angle. Use the Law of Sines to explain why this is so.

Questions for Discussion or Essay

51. Exercise 35 asks for the *minimum* length of power line that is needed to span a canyon. Why might more actually be needed?

52. At the beginning of this section, we listed four categories of triangles that can be solved using the Law of Sines and the Law of Cosines. These were labeled AAS (or ASA), SSA, SSS, and SAS. What do you suppose is meant by an AAA triangle, and why is it not listed as a fifth case?

53. Explain why the Law of Sines is not effective for computing distances between cities, even if all necessary angles are known.

54. The Law of Sines was presented in this section as a tool for solving oblique (nonright) triangles. What happens if we attempt to use the Law of Sines to solve a right triangle?

Projects for Enrichment

55. Constructing a Device to Measure Angles   In this project, we construct a device to measure line-of-sight angles and then use it to estimate various distances and heights. An example of such a device is shown in Figure 16. Depending on the materials and tools you have available, yours may not look or operate exactly the same as this one, but it should be based on the same principles. Our apparatus is made from

three sticks. The two shaded sticks are both 12 inches from pivot point to pivot point. The ruled stick is somewhat longer than 24 inches. The leftmost shaded stick, which we refer to as the *pivot stick*, is anchored to the ruled stick in such a way that it is able to pivot. The second shaded stick, the *slider stick*, is anchored to one end of the pivot stick so that it also can pivot, but the other end of the slider stick must be allowed to slide along the ruled stick. The angle made between the pivot stick and the ruled stick can be read off below the right end of the slider stick.

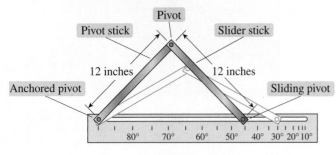

Figure 16

To determine the placement of the degree measure marks on the ruled stick, consider the isosceles triangle shown in Figure 17.

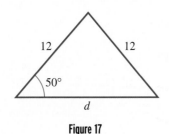

Figure 17

a. Find the length d shown in Figure 17. Explain how this value will help you locate the grid mark for 50° on the ruled stick.

b. Find the distances that correspond to each of the degree measure grid marks on the ruled stick.

c. Construct a device similar to the one shown in Figure 16. Mark the ruled stick with degree measures in increments of 5° from 85° down to 5°.

d. Choose a tall, well-known landmark on your campus. Measure an appropriate distance away from the base of the landmark. From that point, use your device to estimate the angle of elevation to the top of the landmark. This can be done by keeping the ruled stick horizontal and "lining up" the pivot stick with the top of the landmark. Now estimate the height of the landmark. Explain your steps in detail.

e. Choose a location on or near your campus where the distance between two points would be difficult or impossible to measure using standard tools (such as a tape measure). Identify these points as A and B, where A is the point at which you are positioned. Mark a point C a reasonable distance from point A and measure that distance. From point A, use your device to estimate $\angle BAC$. From point C, estimate $\angle BCA$. Now use the Law of Sines to estimate the distance between A and B. Explain your steps in detail.

Section 7.2 The Law of Cosines

- How can you find the area of a triangle if the three side lengths are known?
- If the hypotenuse and two legs of a right triangle are related by the equation $c^2 = a^2 + b^2$, what can be said about the relationship between sides of an oblique triangle?
- How can two people talking on the telephone pinpoint the location of a lightning flash?
- How far is it from the pitcher's mound to first base?

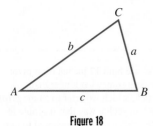

Figure 18

In the previous section, we used the Law of Sines to solve oblique triangles of the form AAS, ASA, and SSA. The remaining two cases—SAS and SSS—are more easily solved with the **Law of Cosines**. As in the previous section, we denote the angles of an oblique triangle by A, B, and C and the lengths of the corresponding opposite sides by a, b, and c, as shown in Figure 18.

The Law of Cosines can be derived using the distance formula. We begin by placing the angle A in standard position with vertex B at the point $(c, 0)$ (see Figure 19). If we denote the coordinates of vertex C by (x, y) and use the definitions of sine and cosine, we see that

$$\cos A = \frac{x}{b} \quad \text{and} \quad \sin A = \frac{y}{b}$$

Thus,

$$x = b \cos A \quad \text{and} \quad y = b \sin A$$

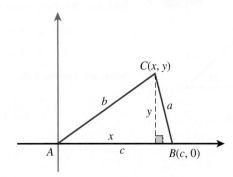

Figure 19

Note that these expressions for x and y are valid even if A is obtuse. Now by applying the formula for the distance between vertices B and C, we obtain

$$a = \sqrt{(x - c)^2 + (y - 0)^2}$$
$$a^2 = (b \cos A - c)^2 + (b \sin A)^2 \qquad \text{Squaring both sides and substituting for } x \text{ and } y$$

$$= b^2 \cos^2 A - 2bc \cos A + c^2 + b^2 \sin^2 A$$
$$= b^2(\cos^2 A + \sin^2 A) + c^2 - 2bc \cos A$$
$$= b^2 + c^2 - 2bc \cos A \qquad \text{Using the identity } \sin^2 A + \cos^2 A = 1$$

Thus, $a^2 = b^2 + c^2 - 2bc \cos A$, which is our first version of the Law of Cosines. The other two versions can be derived in a similar fashion by successively placing angles B and C in standard position.

Law of Cosines

If a triangle with vertices A, B, and C has side lengths a, b, and c, as shown in Figure 18, then the following equations hold:

$$a^2 = b^2 + c^2 - 2bc \cos A$$
$$b^2 = a^2 + c^2 - 2ac \cos B$$
$$c^2 = a^2 + b^2 - 2ab \cos C$$

Note that each of these equations has the form

$$(\text{Side opposite the angle})^2 = (\text{Side 1})^2 + (\text{Side 2})^2$$
$$- 2(\text{Side 1})(\text{Side 2}) \cos(\text{The angle})$$

Each version of the Law of Cosines involves four quantities. If any three of the four quantities are given, we can solve for the fourth. In particular, if two sides and the included angle are known (SAS), we can solve for the third side, or if the three sides are known (SSS), we can solve for any angle. We illustrate these ideas in the following examples.

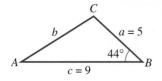

Solving an SAS Triangle

Find any unknown side lengths and angle measures for the triangle in Figure 20.

Solution   We first apply the Law of Cosines to find b.

$$b^2 = a^2 + c^2 - 2ac \cos B = 5^2 + 9^2 - 2(5)(9) \cos 44° \approx 41.26$$
$$b \approx \sqrt{41.26} \approx 6.42$$

Next, we use the Law of Sines to find A.

$$\frac{a}{\sin A} = \frac{b}{\sin B}$$

$$\frac{5}{\sin A} = \frac{6.42}{\sin 44°}$$

$$\sin A = \frac{5 \sin 44°}{6.42} \approx 0.5410 \qquad \text{Solving for } \sin A$$

Since a is not the longest side, A must be acute. Thus, $A \approx \sin^{-1} 0.5410 \approx 32.8°$, and it follows that

$$C \approx 180° - 32.8° - 44° = 103.2°$$

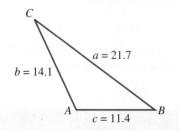

Figure 20

Solving an SSS Triangle

Find the angles for the triangle shown in Figure 21.

Solution   We first find A using the Law of Cosines.

$$a^2 = b^2 + c^2 - 2bc \cos A$$

$$\cos A = \frac{b^2 + c^2 - a^2}{2bc} \qquad \text{Solving for } \cos A$$

$$= \frac{14.1^2 + 11.4^2 - 21.7^2}{2(14.1)(11.4)}$$

$$\approx -0.4421$$

Figure 21

Thus, $\cos A \approx -0.4421$, and it follows that $A \approx \cos^{-1}(-0.4421) \approx 116.2°$. Next we find B using the Law of Sines.

$$\frac{a}{\sin A} = \frac{b}{\sin B}$$

$$\frac{21.7}{\sin 116.2°} = \frac{14.1}{\sin B}$$

$$\sin B = \frac{14.1 \sin 116.2°}{21.7} \approx 0.5830 \qquad \text{Solving for } \sin B$$

Since B must be acute, we have $B \approx \sin^{-1} 0.5830 \approx 35.7°$. Finally, we have

$$C \approx 180° - 116.2° - 35.7° = 28.1°$$

Although SSA triangles are usually solved most efficiently with the Law of Sines, they can also be solved with the Law of Cosines, as illustrated by the following example.

····▷**EXAMPLE 3**

Solving an SSA Triangle Using the Law of Cosines

Use the Law of Cosines to find the unknown side length and angle measures for any triangle ABC satisfying $b = 78$, $c = 152$, and $B = 26°$.

Solution Since b is the length of the side opposite angle B, we have

$$b^2 = a^2 + c^2 - 2ac \cos B$$
$$78^2 = a^2 + 152^2 - 2a(152) \cos 26°$$
$$6084 = a^2 + 23{,}104 - 273.23a$$
$$0 = a^2 - 273.23a + 17{,}020$$

Now the last equation is quadratic in a, so we apply the quadratic formula to obtain

$$a = \frac{273.23 \pm \sqrt{273.23^2 - 4(1)17{,}020}}{2} \approx \frac{273.23 \pm 81.08}{2}$$

or

$$a \approx 177.2, \, a \approx 96.1$$

Thus, there are two triangles satisfying the given conditions, one with $a \approx 177.2$ and the other with $a \approx 96.1$. We consider both cases.

Case 1: Using $a \approx 177.2$ and the Law of Cosines, we have

$$a^2 = b^2 + c^2 - 2bc \cos A$$
$$\cos A = \frac{b^2 + c^2 - a^2}{2bc}$$
$$\approx \frac{78^2 + 152^2 - 177.2^2}{2(78)(152)}$$
$$\approx -0.0933$$
$$A \approx \cos^{-1}(-0.0933) \approx 95.4°$$

Thus, $C = 180° - A - B \approx 180° - 95.4° - 26° = 58.6°$. The resulting triangle is shown in Figure 22.

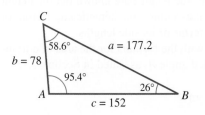

Figure 22

Case 2: For $a \approx 96.1$, we have

$$a^2 = b^2 + c^2 - 2bc \cos A$$
$$\cos A = \frac{b^2 + c^2 - a^2}{2bc}$$
$$\approx \frac{78^2 + 152^2 - 96.1^2}{2(78)(152)}$$
$$\approx 0.8415$$
$$A \approx \cos^{-1}(0.8415) \approx 32.7°$$

Thus, $C = 180° - A - B \approx 180° - 32.7° - 26° = 121.3°$. The resulting triangle is shown in Figure 23.

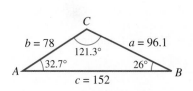

Figure 23

····▷**EXAMPLE 4**

Computing Distance

Two ships start from the same point and sail in different directions, one on a course of 40° clockwise from north at a rate of 6 miles per hour and the other on a course of 150°

Figure 24

clockwise from north at a rate of 8 miles per hour. How far apart are the ships after $2\frac{1}{2}$ hours?

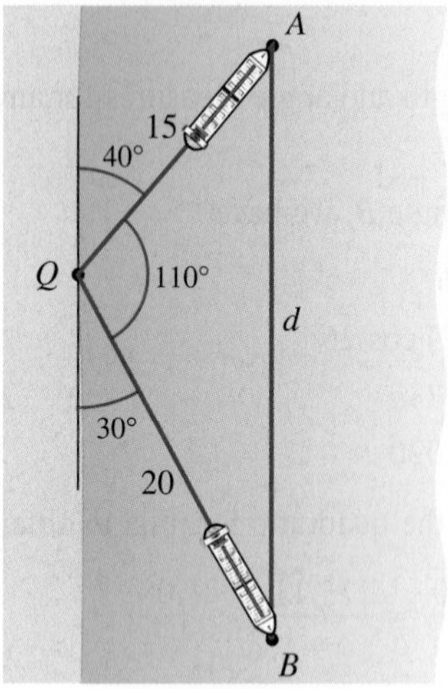

Solution   We first draw a sketch showing the courses and relative positions of the ships (see Figure 24). The position of the ship with a course of 40° is labeled A in Figure 24, whereas the position of the ship with a course of 150° is labeled B. Thus, $\angle AQB = 150° - 40° = 110°$. After $2\frac{1}{2}$ hours, the ship at point A is $(2.5)(6) = 15$ miles from Q, and the ship at B is $(2.5)(8) = 20$ miles from Q. The distance d between the two ships can be computed using the Law of Cosines. We have

$$d^2 = 15^2 + 20^2 - 2(15)(20)\cos 110° \approx 830.2$$

and so

$$d \approx \sqrt{830.2} \approx 28.8 \text{ miles}$$

Thus, the ships are approximately 28.8 miles apart after $2\frac{1}{2}$ hours.

Heron's Formula

As noted in Section 7.1, the area of a triangle with base b and height h is given by

$$A = \frac{1}{2}bh$$

But how can we find the area of a triangle if the side lengths are known but the height is not? *Heron's Formula*—named after the first-century mathematician, Heron of Alexandria—expresses the area of a triangle in terms of its side lengths.

We begin our derivation of Heron's Formula with the formula for the area of a triangle in terms of two sides b and c and the included angle A, as stated in Section 7.1.

$$\text{Area } = \frac{1}{2}bc \sin A$$

Squaring both sides and rewriting gives us

$$(\text{Area})^2 = \frac{1}{4}b^2c^2 \sin^2 A \qquad \text{Squaring both sides}$$

$$= \frac{1}{4}b^2c^2(1 - \cos^2 A) \qquad \text{Using the identity } \sin^2 A = 1 - \cos^2 A$$

$$= \frac{1}{4}b^2c^2 (1 + \cos A)(1 - \cos A) \qquad \text{Factoring the difference of squares}$$

From the Law of Cosines, we have $a^2 = b^2 + c^2 - 2bc \cos A$, and hence

$$\cos A = \frac{b^2 + c^2 - a^2}{2bc}$$

Thus,

$$(\text{Area})^2 = \frac{1}{4}b^2c^2\left(1 + \frac{b^2 + c^2 - a^2}{2bc}\right)\left(1 - \frac{b^2 + c^2 - a^2}{2bc}\right) \qquad \text{Substituting for } \cos A$$

$$= \frac{1}{4}b^2c^2\left[\frac{(b^2 + 2bc + c^2) - a^2}{2bc}\right]\left[\frac{a^2 - (b^2 - 2bc + c^2)}{2bc}\right] \qquad \text{Simplifying}$$

$$= \frac{1}{16}\left[(b + c)^2 - a^2\right]\left[a^2 - (b - c)^2\right]$$

Factoring and simplifying

$$= \frac{1}{16}[(b + c) + a][(b + c) - a][a - (b - c)][a + (b - c)]$$

Factoring the difference of squares

$$= \left(\frac{a + b + c}{2}\right)\left(\frac{a + b + c}{2} - a\right)\left(\frac{a + b + c}{2} - b\right)\left(\frac{a + b + c}{2} - c\right)$$

Rewriting

By substituting $s = \frac{1}{2}(a + b + c)$ and taking square roots on both sides, we obtain

$$\text{Area} = \sqrt{s(s - a)(s - b)(s - c)}$$

Heron's Formula

The area of a triangle with side lengths a, b, and c is given by

$$\text{Area} = \sqrt{s(s - a)(s - b)(s - c)}$$

where s is the **semiperimeter** given by

$$s = \frac{a + b + c}{2}$$

EXAMPLE 5

Computing Area with Heron's Formula

Find the area of the triangle shown in Figure 25.

Solution   The semiperimeter is

$$s = \frac{12 + 26 + 18}{2} = \frac{56}{2} = 28$$

Heron's Formula gives us

$$\begin{aligned}
\text{Area} &= \sqrt{s(s - 12)(s - 26)(s - 18)} \\
&= \sqrt{28(28 - 12)(28 - 26)(28 - 18)} \\
&= \sqrt{8960} \\
&\approx 94.7 \text{ square inches}
\end{aligned}$$

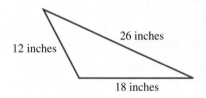

12 inches · 26 inches · 18 inches

Figure 25

EXAMPLE 6

Finding the Area of a Quadrilateral

A plot of land at the end of a cul-de-sac in a new subdivision has the shape of the quadrilateral shown in Figure 26. Find its area.

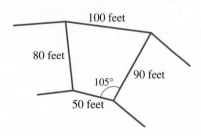

100 feet · 80 feet · 90 feet · 105° · 50 feet

Figure 26

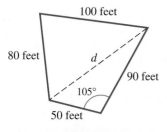

Figure 27

Solution   By constructing a diagonal between two opposite vertices, we can compute the area as the sum of the areas of two triangles. Let d denote the length of the diagonal from the lower left vertex to the upper right vertex, as shown in Figure 27. Using the Law of Cosines, we have

$$d^2 = 50^2 + 90^2 - 2(50)(90) \cos 105°$$

and so

$$d = \sqrt{50^2 + 90^2 - 2(50)(90) \cos 105°} \approx 113.7 \text{ feet}$$

Now we apply Heron's Formula to the triangle with side lengths 50, 90, and 113.7 feet to obtain

$$s = \frac{50 + 90 + 113.7}{2} = 126.85$$

and

$$\text{Area} = \sqrt{s(s - 50)(s - 90)(s - 113.7)}$$
$$= \sqrt{126.85(76.85)(36.85)(13.15)}$$
$$\approx 2173.4 \text{ square feet}$$

Similarly, for the triangle with side lengths 80, 100, and 113.7 feet, we have

$$s = \frac{80 + 100 + 113.7}{2} = 146.85$$

and

$$\text{Area} = \sqrt{s(s - 80)(s - 100)(s - 113.7)}$$
$$= \sqrt{146.85(66.85)(46.85)(33.15)}$$
$$\approx 3904.7 \text{ square feet}$$

Finally, adding the two areas together, we obtain the total area of

$$2173.4 + 3904.7 = 6078.1 \text{ square feet}$$

Understanding and Mastery Checklists

Concepts to Understand	Skills to Master
Law of Cosines	Solve an SAS, SSS, or SSA triangle using the Law of Cosines.
⬥	⬥
Heron's Formula	Solve an arbitrary triangle using an appropriate method.
	⬥
	Use Heron's Formula to find the area of an SSS, SSA, or SAS triangle.

Exercises 7.2

Exercises 1-8   *Use the Law of Cosines to solve the given triangle. Assume angles A, B, and C and sides a, b, and c are labeled as shown in Figure 28.*

27. $b = 1.8, B = 104°, C = 39°$ **28.** $a = 140, b = 65, c = 73$

Exercises 29-40   *Find x by any means.*

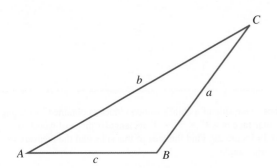

Figure 28

1. $a = 12, b = 9, C = 59°$ **2.** $a = 70, c = 85, B = 125°$

3. $a = 5, b = 8, c = 12$ **4.** $a = 12.5, b = 13.4, c = 14.7$

5. $b = 26, c = 4, A = 35°$ **6.** $a = 23, b = 17, A = 118°$

7. $b = 15, c = 20, B = 63°$ **8.** $a = 23, b = 19, C = 70°$

Exercises 9-28   *Solve the given triangle by any means. Assume angles A, B, and C and sides a, b, and c are labeled as shown in Figure 28.*

9. $b = 130, c = 180, A = 120°$ **10.** $a = 4.2, c = 3.8, B = 38°$

11. $a = 30, b = 35, c = 50$ **12.** $b = 28, c = 32, C = 165°$

13. $a = 23, B = 65°, C = 44°$ **14.** $a = 13, b = 15, c = 23$

15. $a = 26, b = 14, A = 115°$ **16.** $a = 0.12, c = 0.25, B = 130°$

17. $a = 1000, b = 2000, C = 27°$ **18.** $b = 36, A = 25°, C = 42°$

19. $a = 240, b = 175, c = 300$ **20.** $a = 56, c = 8, C = 15°$

21. $b = 14, c = 9, B = 28°$ **22.** $a = 1.7, b = 2.8, c = 3.5$

23. $a = 7, b = 25, A = 28°$ **24.** $b = 600, c = 500, B = 75°$

25. $a = 35, b = 70, c = 14$ **26.** $a = 80, A = 37°, B = 79°$

29.

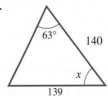

30.

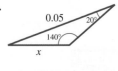

31.

32.

33.

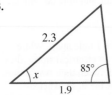

34.

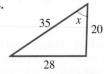

35.

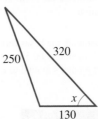

36.

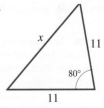

37.

38.

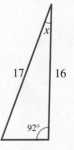

39.
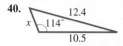

40.

12.4

x 114°

10.5

Exercises 41-46 *Use Heron's Formula to find the area of the given triangle. Assume angles A, B, and C and sides a, b, and c are labeled as shown in Figure 28.*

41. $a = 12, b = 14, c = 20$ **42.** $a = 10, b = 8, c = 5$

43. $b = 18, c = 15, B = 75°$ **44.** $a = 0.8, c = 1.6, C = 150°$

45. $a = 1.4, b = 2.8, C = 16°$ **46.** $b = 120, c = 200, A = 138°$

Applications

47. Ship Distance Two ships leave port at the same time. One travels in the direction N 35° E at a rate of 25 miles per hour. The other travels in the direction S 70° E at a rate of 30 miles per hour. How far apart are the ships after 90 minutes?

48. Hiking Distance A hiker leaves a certain point and walks in the direction 60° (clockwise from north) for 7 miles and then in the direction 320° for 5 miles. If the hiker wishes to head directly back toward the starting point, in what direction should she head, and how far will she have to walk?

49. Locating Lightning Margaret and Elizabeth live 1 mile apart. While talking to each other on the phone during an electrical storm, they both notice a bolt of lightning in the sky between their two homes. Margaret hears the thunder 5 seconds after the flash, whereas Elizabeth hears the thunder 4 seconds after the flash. Assuming that the speed of sound is 1088 feet per second, describe the location of the lightning bolt relative to Margaret's position.

50. Baseball Diamond In a major league baseball diamond, the bases form a square with side length 90 feet. The pitcher's mound is 60.5 feet from home plate. Find the distance from the pitcher's mound to each of the other three bases.

51. Block Face One side of a child's wooden block is obtained by cutting a triangular face in a 3″ × 5″ × 4″ rectangular piece of wood, as shown in Figure 29. Find the area of this side, and the measure of each vertex angle.

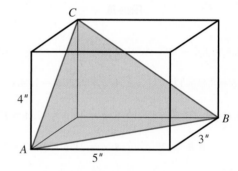

Figure 29

52. Isosceles Triangle Area An isosceles triangle is to be constructed with a perimeter of 24 inches. Let x denote the lengths of the two equal legs.

a. Use Heron's Formula to express the area of the triangle as a function of x. For what values of x is this function defined?

b. Use a graphing calculator to plot the graph of the function you found in part a. For what value of x is the area largest? What do you notice about the resulting triangle?

53. Minimizing Train Distances A passenger train traveling due west at a rate of 80 miles per hour passes a junction at 3:00 P.M. At that same instant, a freight train traveling 90 miles per hour is 100 miles away from the junction on a set of tracks that head straight toward the junction in the direction 140° (clockwise from north).

a. Find the distance between the trains at 3:30 P.M.

b. Find expressions for $p(t)$ and $f(t)$, the distances from the junction to the passenger train and freight train, respectively, t hours after 3:00 P.M.

c. Use $p(t)$, $f(t)$, and the Law of Cosines to express the distance d between the trains as a function of t.

d. Use a graphing calculator to estimate the time at which the trains are closest to each other.

Concepts and Critical Thinking

Exercises 54–57   *Answer true or false.*

54. If the three side lengths of a triangle are given, then the Law of Cosines can be used to solve the triangle.

55. If the three angles of a triangle are given, then the Law of Cosines can be used to solve the triangle.

56. The Law of Cosines can be used to solve an SSA triangle.

57. Heron's Formula is used to find the perimeter of a triangle.

Exercises 58–61   *Give an example of each.*

58. A triangle more easily solved using the Law of Sines than the Law of Cosines

59. A triangle more easily solved using the Law of Cosines than the Law of Sines

60. Three positive numbers that couldn't possibly be the side lengths of any triangle

61. The side lengths of a triangle having a semiperimeter of 12

62. When finding an unknown angle using the Law of Sines, one must eventually solve an equation of the form $\sin \theta = k$. When finding an unknown angle using the Law of Cosines, one must eventually solve an equation of the form $\cos \theta = k$. Use these facts to explain why the Law of Sines can involve ambiguity whereas the Law of Cosines never does.

63. The Law of Cosines states that $a^2 = b^2 + c^2 - 2bc \cos A$. Explain why this implies that $a^2 \geq b^2 + c^2 + 2bc$. Use this result to establish the **triangle inequality**, which states that the sum of any two sides of a triangle exceeds the length of the other side.

64. Explain why none of the factors under the radical in Heron's Formula can be negative.

Questions for Discussion or Essay

65. We have seen that an SSA triangle can be solved using either the Law of Sines or the Law of Cosines. Discuss the advantages and disadvantages of the two methods for solving SSA triangles. Which method do you prefer? Why?

66. The Law of Cosines has been described as a generalization of the Pythagorean Theorem. Explain why this is so.

67. By now you should be familiar with three formulas for finding the area of a triangle: one involving the base and height, another involving two sides and the angle between, and the third involving three sides. Describe these three formulas in detail, discuss the advantages and disadvantages of each, and give real-world examples of circumstances in which each might be used.

68. Regular *n*-gons   A **regular *n*-gon** is a polygon with *n* sides of equal
length. Some familiar examples are the regular 3-gon, also known as
an equilateral triangle; the regular 4-gon, also known as a square;
and the regular 5-gon, also known as a regular pentagon. In this
project, we consider the area and perimeter of a regular *n*-gon that is
inscribed in a circle of radius *r*. Our goal is to "discover" the area
and circumference formulas for a circle, in much the same way as
it was discovered by the ancient Greeks. We begin with a regular
pentagon inscribed in a circle of radius 1, as shown in Figure 30.

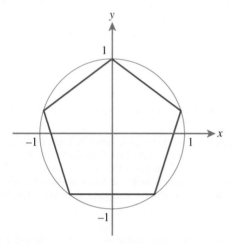

Figure 30

By constructing line segments from the center of the circle to each of
the 5 vertices of the pentagon, we can divide the pentagon into 5 triangular regions of equal area (see Figure 31). The *central angle* of
each of these 5 triangles is $\frac{360°}{5} = 72°$, and each of the sides adjacent
to the central angles has length 1, the length of the radius.

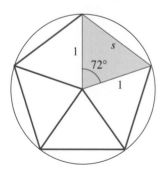

Figure 31

a. Use the area formula from Section 7.1 to find the area of the
shaded triangle in Figure 31. Then find the area of the pentagon.

b. Use the Law of Cosines to find *s*, the length of one side of the
pentagon. Then find the perimeter of the pentagon.

Now suppose a pentagon is inscribed in a circle of radius *r*. If we
divide the pentagon into five triangles, then the central angles measure 72°, as before, but the sides adjacent to these angles have length *r*.

c. Repeat parts a and b to find the area and perimeter of a pentagon
inscribed in a circle of radius *r*.

Finally, we consider the more general case of a regular *n*-gon
inscribed in a circle of radius *r*. Proceeding as before, we divide the
n-gon into *n* triangles of equal area. The central angle of each triangle measures $\frac{360°}{n}$, and the sides adjacent to these angles have length *r*
(see Figure 32).

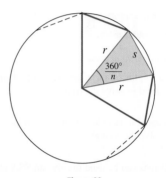

Figure 32

d. Find a formula for the area of the shaded triangle in Figure 32.
Then find a formula for the area of the regular *n*-gon.

e. Find a formula for the side length *s* in Figure 32. Then find a formula for the perimeter of the regular *n*-gon.

As the ancient Greeks observed many centuries ago, for very large
values of *n*, the area and perimeter of a regular *n*-gon closely approximate the area and circumference of a circumscribed circle. Thus, the
area and circumference of a circle of radius *r* can be approximated
using the formulas developed in parts d and e.

f. Complete the following table for a regular *n*-gon inscribed in a
circle of radius *r*.

n	10	100	1000
Area			
Perimeter			

What connection do you see between the values in this table and
the area and circumference of a circle of radius *r*?

g. Describe how the area and perimeter formulas for a regular *n*-gon
can be used to approximate π.

Section 7.3   Trigonometric Form of Complex Numbers

- If a real number can be represented graphically as a point on a number line, what is the graphical representation of a complex number?
- How can angles be useful for finding the product or quotient of complex numbers?
- What famous mathematical hypothesis has eluded proof for over 130 years, in spite of overwhelming computer evidence?

From its humble infancy as a fictitious device tolerated only for its utility in solving polynomial equations, the complex number system has matured into one of the great branches of mathematics, with profound applications in electrical circuit theory, hydrodynamics, electrodynamics, and even atomic physics. In this section, we develop a geometric interpretation of the complex number system and apply trigonometric techniques to forge links between the algebra of complex numbers and the geometry of the plane.

The Complex Plane

Recall that the complex number system consists of all numbers that can be written in the form $a + bi$, where a and b are real numbers and $i = \sqrt{-1}$. Thus, each complex number can be associated with an ordered pair of real numbers (a, b), which in turn can be represented graphically on a rectangular coordinate system known as the **complex plane**. The complex plane looks very much like the rectangular coordinate system we are already familiar with. However, instead of an x-axis and y-axis, the complex plane has a **real axis** and an **imaginary axis**. A complex number $a + bi$ is represented graphically as a point on the complex plane located above (or below) the number a on the real axis and to the right (or left) of the number b on the imaginary axis. For example, the number $3 + 2i$ is located above 3 on the real axis and to the right of 2 on the imaginary axis. Figure 33 shows $3 + 2i$ and several other points in the complex plane.

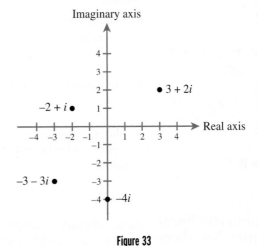

Figure 33

We defined the absolute value of a real number x to be the distance from x to 0 on the real number line. Similarly, we define the absolute value or **modulus** of a complex number $a + bi$ to be the distance from $a + bi$ to 0 in the complex plane. Thus, a complex number with modulus r lies on the circle of radius r centered at 0. The algebraic definition of modulus can be obtained by applying the Pythagorean Theorem to the triangle shown in Figure 34.

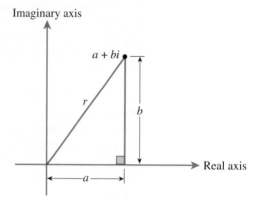

Figure 34

Definition of Modulus

The **modulus** of the complex number $a + bi$ is given by
$$r = |a + bi| = \sqrt{a^2 + b^2}$$

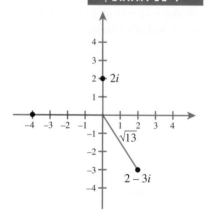

Figure 35

·····>EXAMPLE 1

Finding the Modulus of a Complex Number

Represent the given complex number graphically, and find its modulus.

a. $2 - 3i$ **b.** $2i$ **c.** -4

Solution

a. The graphical representation of $2 - 3i = 2 + (-3)i$ is shown in Figure 35. Its modulus is
$$|2 - 3i| = \sqrt{2^2 + (-3)^2} = \sqrt{4 + 9} = \sqrt{13}$$

b. The number $2i$ is shown on the imaginary axis in Figure 35. Its modulus is
$$|2i| = \sqrt{2^2} = \sqrt{4} = 2$$

c. The number -4 is shown on the real axis in Figure 35. Its modulus is $|-4| = \sqrt{(-4)^2} = 4$. Note that the modulus of a real number is simply its absolute value.

Trigonometric Form of Complex Numbers

As we have seen, a complex number z with real part a and imaginary part b can be written as $z = a + bi$. This method of expressing complex numbers works well for addition and subtraction but can be quite cumbersome for multiplication, division, and exponentiation. By contrast, expressing complex numbers in **trigonometric form** facilitates the computation of certain products, quotients, and powers.

Now the distance from $z = a + bi$ to 0 is simply the modulus $r = \sqrt{a^2 + b^2}$, and the direction from 0 to z is customarily expressed as the **argument**—the angle θ from

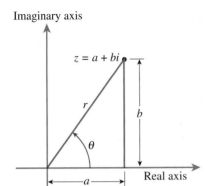

Imaginary axis

$z = a + bi$

r

b

θ

a

Real axis

Figure 36

the positive real axis to the line segment from 0 to z, as shown in Figure 36. By convention, we define counterclockwise to be the positive direction, just as we did when defining the trigonometric functions using the unit circle. Consequently, the trigonometric relations suggested by the acute triangle in Figure 36 hold for all arguments θ. Thus, we have

(1) $$\cos \theta = \frac{a}{r}$$

(2) $$\sin \theta = \frac{b}{r}$$

(3) $$\tan \theta = \frac{b}{a}$$

Equations (1) and (2) give us

$$a = r \cos \theta \quad \text{and} \quad b = r \sin \theta$$

Thus,

$$a + bi = r \cos \theta + (r \sin \theta)i = r(\cos \theta + i \sin \theta)$$

The expression $r(\cos \theta + i \sin \theta)$ is known as the trigonometric form of the complex number $a + bi$. Equation (3) will prove useful for finding θ when a and b are known. We summarize these results as follows.

Trigonometric Form of a Complex Number

The **trigonometric form** of a complex number $z = a + bi$ is

$$z = r(\cos \theta + i \sin \theta)$$

where $r = \sqrt{a^2 + b^2}$ is the modulus of z, and θ is the argument of z, the angle from the positive axis to the line segment from 0 to z in the complex plane. It follows that $a = r \cos \theta$, $b = r \sin \theta$, and $\tan \theta = \frac{b}{a}$.

Note that because there are infinitely many angles coterminal with a given angle θ, the argument, and hence the trigonometric form, of a complex number is not unique. Typically, θ is chosen to be an angle in the interval $[0, 2\pi)$.

EXAMPLE 2

Writing Complex Numbers in Trigonometric Form

Write the given complex number in trigonometric form.

a. $-5 + 5\sqrt{3}i$ **b.** $4 - i$

Solution

a. The modulus of $-5 + 5\sqrt{3}i$ is

$$r = \sqrt{(-5)^2 + \left(5\sqrt{3}\right)^2} = \sqrt{100} = 10$$

The argument can be found by noting that

$$\tan \theta = \frac{b}{a} = \frac{5\sqrt{3}}{-5} = -\sqrt{3}$$

Since $-5 + 5\sqrt{3}i$ is in the second quadrant and $\tan \theta = -\sqrt{3}$, we know that $\theta = 120°$ or, in radians,

$$\theta = \frac{2\pi}{3}$$

Thus, the trigonometric form of $-5 + 5\sqrt{3}i$ is

$$10\left(\cos \frac{2\pi}{3} + i \sin \frac{2\pi}{3}\right)$$

The graphical representation is given in Figure 37.

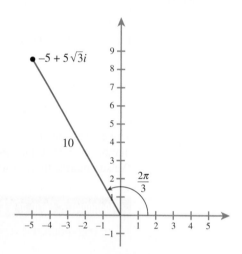

Figure 37

b. The modulus of $4 - i$ is

$$r = \sqrt{4^2 + (-1)^2} = \sqrt{17}$$

The argument θ is an angle in the fourth quadrant with

$$\tan \theta = \frac{b}{a} = -\frac{1}{4}$$

and so we take

$$\theta = \tan^{-1}\left(-\frac{1}{4}\right) \approx -14°$$

or, alternatively, $\theta = 346°$. Thus, the trigonometric form of $4 - i$ is approximately

$$\sqrt{17}(\cos 346° + i \sin 346°)$$

The graphical representation is given in Figure 38.

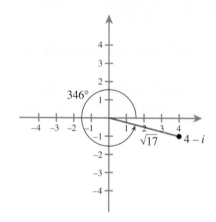

Figure 38

EXAMPLE 3   **Writing a Complex Number in Standard Form**

Write the number $2\left(\cos \frac{3\pi}{2} + i \sin \frac{3\pi}{2}\right)$ in standard form.

Solution   Evaluating, we have

$$2\left(\cos\frac{3\pi}{2} + i\sin\frac{3\pi}{2}\right) = 2[0 + (-1)i] = -2i$$

The graphical representation is given in Figure 39.

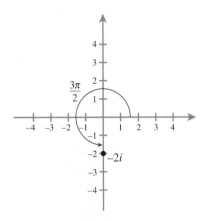

Figure 39

The Riemann Hypothesis

In 1859, the German mathematician Bernhard Riemann published an 8-page paper that is considered to be one of the most important mathematical papers ever published. In this work, Riemann stated six conjectures, or unproven assumptions, and went on to show how these assumptions could be used to prove the famous Prime Number Theorem, which roughly states that for large integers N there are approximately $N \log N$ prime numbers less than N. Since then, five of the six conjectures have been proven. The sixth has become known as the Riemann Hypothesis, and it is widely regarded as the most important unsolved problem in pure mathematics. Ironically, the Prime Number Theorem itself was proven in 1896 by the French mathematician Jacques Hadamard using an alternative approach.

The Riemann Hypothesis involves the complex zeros of the **zeta function**, defined for complex numbers z having real part greater than 1 by

$$\zeta(z) = 1 + \frac{1}{2^z} + \frac{1}{3^z} + \frac{1}{4^z} + \cdots$$

and extended to a function defined for all complex numbers by an advanced technique called analytic continuation. Simply put, the Riemann Hypothesis asserts that every zero of the zeta function (with certain trivial exceptions) lies on the vertical line in the

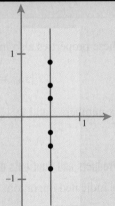

complex plane passing through $\frac{1}{2}$ on the real axis, as suggested by the accompanying figure.

It has been proven that the zeta function has an infinite number of zeros and that all of the important ones lie in the shaded strip shown in the figure. Moreover, it has been shown (using a worldwide grid of computers) that the first 100 billion solutions lie on the line $z = \frac{1}{2}$. However, in spite of this evidence, no one has proven that *all* solutions lie on this line.

Why is a proof of this hypothesis so desirable that it *"would instantaneously catapult its fortunate creator to the summit of the mathematical world"?** It is widely believed that a proof of the Riemann Hypothesis will give us important information about the distribution of prime numbers, an issue of both theoretical and practical consequence. But the great status of this problem derives, in part, from the brilliance of the mathematicians who have tried, but failed, to prove it. And just in case the promise of certain international acclaim and celebrity do not provide you with sufficient motivation to tackle this problem, the Clay Institute of Mathematics (www.claymath.org) has sweetened the pot: Solve the problem to their satisfaction and they will give you a check for $1 million.

*Douglas M. Campbell and John C. Higgins, *Mathematics: People, Problems, Results* (Wadsworth: Belmont, CA), 1984, p. 150.

Multiplication and Division of Complex Numbers

The trigonometric form of a complex number is particularly useful when computing products, quotients, powers, and roots of complex numbers. We consider products and quotients in this section and powers and roots in the next. Let z_1 and z_2 be complex numbers with trigonometric forms

$$z_1 = r_1(\cos \theta_1 + i \sin \theta_1) \quad \text{and} \quad z_2 = r_2(\cos \theta_2 + i \sin \theta_2)$$

Then the product of z_1 and z_2 is

$$
\begin{aligned}
z_1 z_2 &= r_1 r_2 (\cos \theta_1 + i \sin \theta_1)(\cos \theta_2 + i \sin \theta_2) \\
&= r_1 r_2 (\cos \theta_1 \cos \theta_2 + i \sin \theta_1 \cos \theta_2 + i \cos \theta_1 \sin \theta_2 + i^2 \sin \theta_1 \sin \theta_2) \\
&= r_1 r_2 [(\cos \theta_1 \cos \theta_2 - \sin \theta_1 \sin \theta_2) + i(\sin \theta_1 \cos \theta_2 + \cos \theta_1 \sin \theta_2)]
\end{aligned}
$$

Finally, applying the sum identities for sine and cosine, we have

$$z_1 z_2 = r_1 r_2 [\cos(\theta_1 + \theta_2) + i \sin(\theta_1 + \theta_2)]$$

In words, the product of z_1 and z_2 is the complex number whose modulus is the product of the moduli of z_1 and z_2 and whose argument is the sum of the arguments of z_1 and z_2. Using a similar procedure, we can show that the quotient of z_1 and z_2 is the complex number whose modulus is the quotient of the moduli of z_1 and z_2 and whose argument is the difference of the arguments of z_1 and z_2 (see Exercise 66). These facts are summarized as follows.

Products and Quotients in Trigonometric Form

If $z_1 = r_1(\cos \theta_1 + i \sin \theta_1)$ and $z_2 = r_2(\cos \theta_2 + i \sin \theta_2)$, then

$$z_1 z_2 = r_1 r_2 [\cos(\theta_1 + \theta_2) + i \sin(\theta_1 + \theta_2)]$$

$$\frac{z_1}{z_2} = \frac{r_1}{r_2}[\cos(\theta_1 - \theta_2) + i \sin(\theta_1 - \theta_2)], z_2 \neq 0$$

Note that these properties also imply that

$$|zw| = |z||w| \quad \text{and} \quad \left|\frac{z}{w}\right| = \frac{|z|}{|w|}$$

for complex numbers z and w.

EXAMPLE 4

Computing Products and Quotients of Complex Numbers

Perform the indicated operation.

a. $4(\cos 18° + i \sin 18°) \cdot 3(\cos 47° + i \sin 47°)$ **b.** $\dfrac{3i}{2 - 2i}$

Solution

a. According to the rule for computing products, we multiply the moduli and add the arguments to obtain

$$
\begin{aligned}
4(\cos 18° + i \sin 18°) \cdot 3(\cos 47° + i \sin 47°) &= 4 \cdot 3[\cos(18° + 47°) + i \sin(18° + 47°)] \\
&= 12(\cos 65° + i \sin 65°)
\end{aligned}
$$

b. We compute the quotient in two different ways, first using the algebraic method discussed in Section 1.3 and then by converting to trigonometric form.

Method 1: $\dfrac{3i}{2-2i} = \dfrac{3i}{2-2i} \cdot \dfrac{2+2i}{2+2i}$ Multiplying by 1 ($2 + 2i$ is the conjugate of $2 - 2i$)

$= \dfrac{6i + 6i^2}{2^2 - (2i)^2}$ Using the difference of squares formula

$= \dfrac{6i - 6}{4 - (-4)}$ Using the fact that $i^2 = -1$

$= \dfrac{-6 + 6i}{8}$

$= -\dfrac{3}{4} + \dfrac{3}{4}i$

Method 2:   We first convert to trigonometric form. The numerator $3i$ is on the positive imaginary axis, so its modulus is 3 and its argument is $90°$ or $\frac{\pi}{2}$. Thus,

$$3i = 3\left(\cos\frac{\pi}{2} + i\sin\frac{\pi}{2}\right)$$

The denominator $2 - 2i$ is in the fourth quadrant with modulus $\sqrt{2^2 + (-2)^2} = 2\sqrt{2}$ and argument $315°$ or $\frac{7\pi}{4}$. So

$$2 - 2i = 2\sqrt{2}\left(\cos\frac{7\pi}{4} + i\sin\frac{7\pi}{4}\right)$$

Thus,

$$\frac{3i}{2 - 2i} = \frac{3\left(\cos\dfrac{\pi}{2} + i\sin\dfrac{\pi}{2}\right)}{2\sqrt{2}\left(\cos\dfrac{7\pi}{4} + i\sin\dfrac{7\pi}{4}\right)}$$

$= \dfrac{3}{2\sqrt{2}}\left[\cos\left(\dfrac{\pi}{2} - \dfrac{7\pi}{4}\right) + i\sin\left(\dfrac{\pi}{2} - \dfrac{7\pi}{4}\right)\right]$ Dividing moduli and subtracting arguments

$= \dfrac{3\sqrt{2}}{4}\left[\cos\left(-\dfrac{5\pi}{4}\right) + i\sin\left(-\dfrac{5\pi}{4}\right)\right]$ Simplifying

$= \dfrac{3\sqrt{2}}{4}\left(-\dfrac{\sqrt{2}}{2} + \dfrac{\sqrt{2}}{2}i\right)$ Evaluating

$= -\dfrac{3}{4} + \dfrac{3}{4}i$ Distributing

Calculator Keys

Complex Number Computations

Many calculators are capable of performing computations involving complex numbers. One popular brand of graphing calculator, for example, can compute the modulus and argument, perform arithmetic operations, convert between standard and trigonometric forms, and find powers and roots (as discussed in the next section). Our emphasis in this chapter is on understanding *why* and *how* computational techniques work, rather than on the results themselves. Still, you may find it useful to check the results of your hand computations with a calculator.

Understanding and Mastery Checklists

Concepts to Understand	Skills to Master
Complex plane	Plot a complex number in the complex plane.
Real axis	Find the absolute value of a complex number.
Imaginary axis	Convert between trigonometric form and the standard complex number form $a + bi$.
Absolute value (modulus) of a complex number	Perform multiplication and division on complex numbers in trigonometric form.
Argument of a complex number	
Trigonometric form of a complex number	
Products and quotients in trigonometric form	

Exercises 7.3

Exercises 1-10   *Find the absolute value of the given complex number.*

1. $3 + 4i$

2. $-8 - 6i$

3. $1 + 3i$

4. $-1 - i$

5. $1 - \sqrt{3}i$

6. $\sqrt{7} + 3i$

7. -15

8. πi

9. $-\dfrac{3}{5} + \dfrac{4}{5}i$

10. $\dfrac{5}{13} + \dfrac{2}{13}i$

Exercises 11-20   *Represent the complex number graphically and write it in trigonometric form.*

11. $-4i$

12. 3.5

13. -6

14. $3i$

15. $1 + i$

16. $2\sqrt{3} + 2i$

17. $1 - \sqrt{3}i$

18. $8 - 8i$

19. $-2 - 3i$

20. $3 + i$

Exercises 21-34   *Represent the complex number graphically and write it in the form $a + bi$.*

21. $4(\cos 90° + i \sin 90°)$

22. $3\left(\cos \dfrac{11\pi}{6} + i \sin \dfrac{11\pi}{6}\right)$

23. $2(\cos 120° + i \sin 120°)$

24. $0.5\left(\cos \dfrac{3\pi}{2} + i \sin \dfrac{3\pi}{2}\right)$

25. $9\left(\cos \dfrac{3\pi}{4} + i \sin \dfrac{3\pi}{4}\right)$

26. $5[\cos(-270°) + i \sin(-270°)]$

27. $1[\cos(-\pi) + i \sin(-\pi)]$

28. $\pi(\cos 240° + i \sin 240°)$

29. $6(\cos 380° + i \sin 380°)$

30. $10(\cos 4 + i \sin 4)$

31. $\sqrt{2}(\cos 100° + i \sin 100°)$

32. $\sqrt{3}\left(\cos \dfrac{11\pi}{3} + i \sin \dfrac{11\pi}{3}\right)$

33. $\dfrac{3}{2}(\cos 2 + i \sin 2)$

34. $7(\cos 340° + i \sin 340°)$

Exercises 35-48   *Perform the indicated operation. Express the answer in the form used in the statement of the problem.*

35. $2(\cos 30° + i \sin 30°) \cdot 3(\cos 40° + i \sin 40°)$

36. $(3 - i)(4 + 2i)$

37. $\dfrac{7 - i}{1 + 2i}$

38. $\dfrac{16(\cos 135° + i \sin 135°)}{2(\cos 15° + i \sin 15°)}$

39. $\dfrac{\sqrt{5}\left(\cos \dfrac{\pi}{4} + i \sin \dfrac{\pi}{4}\right)}{2\sqrt{5}\left(\cos \dfrac{\pi}{3} + i \sin \dfrac{\pi}{3}\right)}$

40. $\sqrt{2}(\cos 83° + i \sin 83°) \cdot 2\sqrt{2}(\cos 97° + i \sin 97°)$

41. $(-5 + 3i)(1 - i)$

42. $\dfrac{13}{2 + 3i}$

43. $\dfrac{3 + 5i}{2 - 4i}$

44. $(2 + 4i)(2 - 4i)$

45. $5\left(\cos \dfrac{3\pi}{4} + i \sin \dfrac{3\pi}{4}\right) \cdot 3\left(\cos \dfrac{\pi}{2} + i \sin \dfrac{\pi}{2}\right)$

46. $\dfrac{27(\cos 20° + i \sin 20°)}{3(\cos 110° + i \sin 110°)}$

47. $\dfrac{4.2(\cos 3\pi + i \sin 3\pi)}{0.6\left(\cos \dfrac{\pi}{4} + i \sin \dfrac{\pi}{4}\right)}$

48. $0.8(\cos 2 + i \sin 2) \cdot 1.2(\cos 4 + i \sin 4)$

Exercises 49-56 *Perform the indicated operation by first converting to trigonometric form. Check your answer by performing the operation without converting.*

49. $(2 - 2i)(-3 + 3i)$

50. $\dfrac{6}{1 - \sqrt{3}i}$

51. $\dfrac{-8 - 8i}{2 + 2i}$

52. $3i(4 + 4i)$

53. $\left(\sqrt{3} + i\right)\left(2 - 2\sqrt{3}i\right)$

54. $\dfrac{-4 + 4\sqrt{3}i}{-1 + i}$

55. $\dfrac{2i}{4 - 3i}$

56. $(2 - i)(2 + i)$

Concepts and Critical Thinking

Exercises 57-60 *Answer true or false.*

57. The absolute value of a complex number $a + bi$ is $|a| + |b|$.

58. The modulus of a complex number $a + bi$ is $\sqrt{a^2 + b^2}$.

59. The product of two complex numbers $z_1 = r_1(\cos \theta_1 + i \sin \theta_1)$ and $z_2 = r_2(\cos \theta_2 + i \sin \theta_2)$ is $z_1 z_2 = r_1 r_2[\cos(\theta_1 \theta_2) + i \sin(\theta_1 \theta_2)]$.

60. The trigonometric form of a complex number is unique.

Exercises 61-64 *Give an example of each.*

61. A complex number of modulus 5 that is neither real nor imaginary

62. A complex number with modulus $\dfrac{\pi}{4}$

63. Two nonreal complex numbers whose product is a positive real number

64. Two complex numbers that are neither real nor imaginary but whose product is imaginary

65. Show that if $r > 0$, then $|r(\cos \theta + i \sin \theta)| = r$.

66. Show that if $z_1 = r_1(\cos \theta_1 + i \sin \theta_1)$ and $z_2 = r_2(\cos \theta_2 + i \sin \theta_2)$, then

$$\frac{z_1}{z_2} = \frac{r_1}{r_2}[\cos(\theta_1 - \theta_2) + i \sin(\theta_1 - \theta_2)], z_2 \neq 0$$

(*Hint:* Multiply the numerator and denominator of $\dfrac{z_1}{z_2}$ by $\cos \theta_2 - i \sin \theta_2$.)

67. Show that if $z = r(\cos \theta + i \sin \theta)$, then

$$\frac{1}{z} = \frac{1}{r}(\cos \theta - i \sin \theta)$$

(*Hint:* Write the numerator 1 in trigonometric form.)

68. Show that if $z = r(\cos \theta + i \sin \theta)$, then

$$-z = r[\cos(\theta + \pi) + i \sin(\theta + \pi)]$$

69. Recall that the complex conjugate of a complex number $z = a + bi$ is the complex number $\bar{z} = a - bi$. Show that the modulus of $\bar{z}$ equals the modulus of z and that the argument of $\bar{z}$ may be taken to be the negative of the argument of z. In other words, show that if $z = r(\cos \theta + i \sin \theta)$, then

$$\bar{z} = r[\cos(-\theta) + i \sin(-\theta)]$$

Questions for Discussion or Essay

70. Write the complex numbers

$$2\left(\cos\frac{2\pi}{3} + i\sin\frac{2\pi}{3}\right) \quad \text{and} \quad 2\left(\cos\frac{8\pi}{3} + i\sin\frac{8\pi}{3}\right)$$

in standard form. What does this say about the trigonometric form of a complex number? What if we restricted the argument θ to be an angle between 0 and 2π?

71. Describe the set of complex numbers z for which $|z| = 2$. In general, what can be said about the set $\{z \mid z$ is complex and $|z| = r\}$ for any $r > 0$?

72. Describe the set of complex numbers z for which the argument of z is 45°. In general, what can be said about the set $\{z \mid z$ is complex and the argument of z is $\theta\}$ for any $0 \le \theta < 2\pi$?

Projects for Enrichment

73. Quaternions Beginning with the intuition that multiplication could be defined in four-dimensional space much as it is defined in the two-dimensional realm of the complex numbers, the Irish mathematician William Hamilton developed the quaternions, a new number system having extraordinary applications throughout mathematics and physics. After struggling for years to define them in such a way that they would enjoy all of the properties of the complex number system, Hamilton finally (in 1843) achieved success by relaxing the condition that multiplication be commutative.

A quaternion is an expression of the form $a + bi + cj + dk$, where a, b, c, and d are real numbers. We will see that i, j, and k have properties very much like the imaginary number i. Equality, addition, and multiplication for quaternions are defined as follows:

(1) $(a_1 + a_2i + a_3j + a_4k) = (b_1 + b_2i + b_3j + b_4k)$
    if and only if $a_n = b_n$ for $n = 1, 2, 3, 4$

(2) $(a_1 + a_2i + a_3j + a_4k) + (b_1 + b_2i + b_3j + b_4k)$
    $= (a_1 + b_1) + (a_2 + b_2)i$
    $+ (a_3 + b_3)j + (a_4 + b_4)k$

(3) Multiplication is associative, distributive, and satisfies the following equations:
$$i^2 = j^2 = k^2 = -1$$
$$ij = k, jk = i, ki = j$$
$$ji = -k, kj = -i, ik = -j$$

a. Show that quaternion multiplication is not commutative.

b. Find a formula for the product of two general quaternions; that is, compute
$$(a_1 + a_2i + a_3j + a_4k)(b_1 + b_2i + b_3j + b_4k)$$

c. Use the result from part b to confirm that multiplication is both associative and distributive.

The products in (3) are easy to remember if you use the sequence
$$i, j, k, i, j, k$$
The product from left to right of two adjacent elements is the next one to the right. The product from right to left of two adjacent elements is the negative of the next one to the left.

d. Show that $ijk = jki = kij = -1$, whereas $kji = ikj = jik = 1$.

e. The modulus of the quaternion $z = a_1 + a_2i + a_3j + a_4k$ is defined by
$$|z| = \sqrt{a_1^2 + a_2^2 + a_3^2 + a_4^2}$$

Show that for any pair of quaternions z and w,
$$|zw| = |z| \cdot |w|$$

Section 7.4 | Powers and Roots of Complex Numbers

- How can you draw a pentagon, a hexagon, or even a dodecagon on a graphing calculator?
- If i is the square root of -1, then what is the square root of i?
- How can complex numbers be used to prove trigonometric identities?

Powers of Complex Numbers

In the previous section, we saw that when multiplying two complex numbers, we need only multiply moduli and add arguments to find the modulus and argument of their product. Not surprisingly, we can extend this result to products consisting of any number of complex factors: We simply multiply all moduli and add all arguments to find the modulus and argument of the product. Consider

$$z^n = z \cdot z \cdot \cdots \cdot z$$
$$= r(\cos \theta + i \sin \theta) \cdot r(\cos \theta + i \sin \theta) \cdot \cdots \cdot r(\cos \theta + i \sin \theta)$$

where there are n factors altogether. Multiplying moduli and adding arguments gives us a modulus of $r \cdot r \cdot \cdots \cdot r = r^n$ and an argument of $\theta + \theta + \cdots + \theta = n\theta$. We have thus established the following theorem, named after the French mathematician Abraham de Moivre.

De Moivre's Theorem

If $z = r(\cos \theta + i \sin \theta)$ is a complex number and n is a positive integer, then

$$z^n = [r(\cos \theta + i \sin \theta)]^n = r^n(\cos n\theta + i \sin n\theta)$$

EXAMPLE 1

Finding Powers of Complex Numbers

Use de Moivre's Theorem to expand the given power.

a. $[3(\cos 15° + i \sin 15°)]^5$　　　　**b.** $\left(\sqrt{3} - i\right)^6$

Solution

a. Applying de Moivre's Theorem, we obtain

$$[3(\cos 15° + i \sin 15°)]^5 = 3^5[\cos(5 \cdot 15°) + i \sin(5 \cdot 15°)]$$
$$= 243(\cos 75° + i \sin 75°)$$

Checking with our calculator, we see that

$$[3(\cos 15° + i \sin 15°)]^5 \approx 62.893 + 234.720i$$

and

$$243(\cos 75° + i \sin 75°) \approx 62.893 + 234.720i$$

b. We first convert $\sqrt{3} - i$ to trigonometric form. The modulus and argument are computed as follows:

$$r = \sqrt{\left(\sqrt{3}\right)^2 + (-1)^2} = 2$$

$$\tan \theta = \frac{-1}{\sqrt{3}} = -\frac{\sqrt{3}}{3}, \theta = \frac{11\pi}{6}$$　　Using the fact that $\sqrt{3} - i$ is in the fourth quadrant

Thus,

$$\sqrt{3} - i = 2\left(\cos \frac{11\pi}{6} + i \sin \frac{11\pi}{6}\right)$$

Applying de Moivre's Theorem, we have

$$\left(\sqrt{3} - i\right)^6 = \left[2\left(\cos \frac{11\pi}{6} + i \sin \frac{11\pi}{6}\right)\right]^6$$
$$= 2^6\left[\cos\left(6 \cdot \frac{11\pi}{6}\right) + i \sin\left(6 \cdot \frac{11\pi}{6}\right)\right]$$
$$= 64(\cos 11\pi + i \sin 11\pi)$$
$$= 64(-1 + 0i)$$
$$= -64$$

Roots of Complex Numbers

Roots of complex numbers are defined identically to roots of real numbers. Thus, for example, $2i$ is a cube root of $-8i$ since $(2i)^3 = -8i$. More generally, we make the following definition.

Definition of *n*th Root of a Complex Number

For complex numbers z and w, we say that w is an ***n*th root** of z if and only if

$$w^n = z$$

Since $w^n = z$ is equivalent to $w^n - z = 0$, nth roots of z correspond to zeros of the polynomial $P(x) = x^n - z$. Using techniques from Chapter 3, we can show that this polynomial has n distinct complex zeros provided $z \ne 0$, and thus every nonzero complex number has n distinct nth roots. Some examples are given in Table 2.

Table 2

Complex number	n	*n*th roots
25	2	$5, -5$
$-64i$	3	$4i, 2\sqrt{3} - 2i, -2\sqrt{3} - 2i$
16	4	$2, -2, 2i, -2i$

Note that each nth root in Table 2 can be verified by computing its nth power, using de Moivre's Theorem if necessary. For example, after converting $2\sqrt{3} - 2i$ to trigonometric form (we omit the details), we can apply de Moivre's Theorem to verify that it is a third root of $-64i$ as follows:

$$(2\sqrt{3} - 2i)^3 = \left[4\left(\cos\frac{11\pi}{6} + i\sin\frac{11\pi}{6}\right)\right]^3$$

$$= 4^3\left[\cos\left(3 \cdot \frac{11\pi}{6}\right) + i\sin\left(3 \cdot \frac{11\pi}{6}\right)\right]$$

$$= 64\left(\cos\frac{11\pi}{2} + i\sin\frac{11\pi}{2}\right)$$

$$= 64(0 - i)$$

$$= -64i$$

More generally, if

$$z = r(\cos\theta + i\sin\theta) \quad \text{and} \quad w = \sqrt[n]{r}\left(\cos\frac{\theta}{n} + i\sin\frac{\theta}{n}\right)$$

then de Moivre's Theorem gives us

$$w^n = \left[\sqrt[n]{r}\left(\cos\frac{\theta}{n} + i\sin\frac{\theta}{n}\right)\right]^n = r(\cos\theta + i\sin\theta) = z$$

So $w = \sqrt[n]{r}\left(\cos\frac{\theta}{n} + i\sin\frac{\theta}{n}\right)$ is an nth root of z. But how do we find the other $n - 1$ roots? The key is to recognize that the trigonometric form of a complex number is not unique. Indeed, adding any integer multiple of 2π to θ will yield another form for z. Thus, for any integer k,

$$z = r[\cos(\theta + 2\pi k) + i\sin(\theta + 2\pi k)]$$

and so

$$\sqrt[n]{r}\left(\cos\frac{\theta + 2\pi k}{n} + i \sin\frac{\theta + 2\pi k}{n}\right)$$

is an nth root of z. For $k = 0, 1, 2, \ldots, n - 1$, we obtain distinct roots; other integer values for k will lead to nth roots that have already been obtained. These observations can be summarized as follows.

Formula for nth Roots of a Complex Number

For any positive integer n, the distinct nth roots of the complex number $z = r(\cos\theta + i \sin\theta)$ are given by

$$\sqrt[n]{r}\left(\cos\frac{\theta + 2\pi k}{n} + i \sin\frac{\theta + 2\pi k}{n}\right)$$

where $k = 0, 1, 2, \ldots, n - 1$.

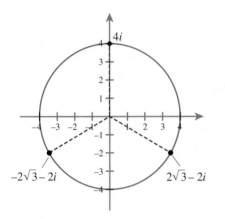

Figure 40

The nth roots of a complex number have a nice geometric interpretation. For example, consider the three cube roots of $-64i$ from Table 2—namely, $4i$, $2\sqrt{3} - 2i$, and $-2\sqrt{3} - 2i$. The graphical representations of these roots are shown in Figure 40. Notice that the points are equally spaced around a circle of radius 4 centered at the origin. In general, the nth roots of a complex number $z = r(\cos\theta + i \sin\theta)$ are equally spaced around a circle of radius $\sqrt[n]{r}$. This is a consequence of the fact that the nth roots all have modulus $\sqrt[n]{r}$ and their arguments differ by $\frac{2\pi}{n}$.

······**EXAMPLE 2**   **Finding the nth Roots of a Real Number**

Find the three cube roots of 1 and represent them graphically.

Solution   We first write 1 in trigonometric form as

$$1 = 1(\cos 0 + i \sin 0)$$

Next, we apply the nth-root formula with $r = 1$, $\theta = 0$, and $n = 3$ to obtain roots of the form

$$\sqrt[3]{1}\left(\cos\frac{0 + 2\pi k}{3} + i \sin\frac{0 + 2\pi k}{3}\right) = \cos\frac{2\pi k}{3} + i \sin\frac{2\pi k}{3}$$

Finally, setting $k = 0, 1, 2$, we compute the three cube roots as follows:

$$\text{For } k = 0: \quad \cos 0 + i \sin 0 = 1$$

$$\text{For } k = 1: \quad \cos\frac{2\pi}{3} + i \sin\frac{2\pi}{3} = -\frac{1}{2} + \frac{\sqrt{3}}{2}i$$

$$\text{For } k = 2: \quad \cos\frac{4\pi}{3} + i \sin\frac{4\pi}{3} = -\frac{1}{2} - \frac{\sqrt{3}}{2}i$$

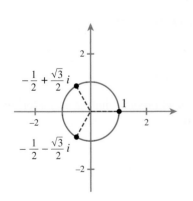

Figure 41

The graphical representations of these roots are shown in Figure 41.

When the nth roots of 1 arise in advanced areas of mathematics, they are usually referred to as the **nth roots of unity**. They have the useful property that they are equally spaced around the unit circle, starting at 1 on the real axis.

EXAMPLE 3

Finding the *n*th Roots of a Complex Number

Find the five fifth roots of $16\sqrt{2}(1 - i)$ and represent them graphically.

Solution   Since $16\sqrt{2}(1 - i)$ has modulus $r = 32$ and argument θ in the fourth quadrant satisfying $\tan\theta = -1$, its trigonometric form is

$$32(\cos 315° + i \sin 315°)$$

From the *n*th-root formula with $k = 0$, we see that one of the fifth roots is

$$\sqrt[5]{32}\left(\cos\frac{315°}{5} + i \sin\frac{315°}{5}\right) = 2(\cos 63° + i \sin 63°)$$

Now the five roots are equally spaced around a circle centered at the origin, so the arguments of "adjacent" roots must differ by $\frac{360°}{5} = 72°$. Thus, we simply add 72° to the argument of the first root to find the second, add 72° to the argument of the second to find the third, and so on. The five roots follow, and their graphical representations are shown in Figure 42.

$$2(\cos 63° + i \sin 63°) \approx 0.9080 + 1.7820i$$
$$2(\cos 135° + i \sin 135°) \approx -1.4142 + 1.4142i$$
$$2(\cos 207° + i \sin 207°) \approx -1.7820 - 0.9080i$$
$$2(\cos 279° + i \sin 279°) \approx 0.3129 - 1.9754i$$
$$2(\cos 351° + i \sin 351°) \approx 1.9754 - 0.3129i$$

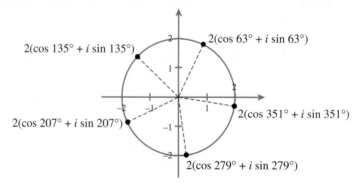

Figure 42

As we demonstrate in the following example, the *n*th-root formula is useful for finding solutions of polynomial equations of the form

$$x^n = c$$

where c is a complex number.

EXAMPLE 4

Finding Solutions of Polynomial Equations

Find all solutions of the equation $16x^4 + 1 = 0$ and represent them graphically.

Solution   The equation $16x^4 + 1 = 0$ is equivalent to

$$x^4 = -\frac{1}{16}$$

and so we must find the fourth roots of $-\frac{1}{16}$. We write $-\frac{1}{16}$ in trigonometric form as

$$\frac{1}{16}(\cos \pi + i \sin \pi)$$

and then apply the nth-root formula with $r = \frac{1}{16}$, $\theta = \pi$, $n = 4$, and $k = 0$ to obtain the root

$$\sqrt[4]{\frac{1}{16}}\left(\cos \frac{\pi}{4} + i \sin \frac{\pi}{4}\right) = \frac{1}{2}\left(\frac{\sqrt{2}}{2} + \frac{\sqrt{2}}{2}i\right) = \frac{\sqrt{2}}{4}(1 + i)$$

Since the four roots must be equally spaced around a circle centered at the origin, their arguments must be $\frac{2\pi}{4} = \frac{\pi}{2}$ radians apart. Thus, the other three roots are

$$\frac{1}{2}\left(\cos \frac{3\pi}{4} + i \sin \frac{3\pi}{4}\right) = \frac{1}{2}\left(-\frac{\sqrt{2}}{2} + \frac{\sqrt{2}}{2}i\right) = \frac{\sqrt{2}}{4}(-1 + i)$$

$$\frac{1}{2}\left(\cos \frac{5\pi}{4} + i \sin \frac{5\pi}{4}\right) = \frac{1}{2}\left(-\frac{\sqrt{2}}{2} - \frac{\sqrt{2}}{2}i\right) = \frac{\sqrt{2}}{4}(-1 - i)$$

$$\frac{1}{2}\left(\cos \frac{7\pi}{4} + i \sin \frac{7\pi}{4}\right) = \frac{1}{2}\left(\frac{\sqrt{2}}{2} - \frac{\sqrt{2}}{2}i\right) = \frac{\sqrt{2}}{4}(1 - i)$$

The graphical representations of the roots are shown in Figure 43.

Figure 43

Understanding and Mastery Checklists

Concepts to Understand

de Moivre's Theorem

❖

nth root of a complex number

Skills to Master

Use de Moivre's Theorem to find a positive integer power of a complex number.

❖

Find all nth roots of a complex number.

❖

Find all complex solutions of a polynomial equation of the form $x^n = c$.

Exercises 7.4

Exercises 1–10 *Use de Moivre's Theorem to expand the given power. Express the answer in the form used in the statement of the problem.*

1. $[2(\cos 30° + i \sin 30°)]^5$

2. $\left(\cos \frac{\pi}{6} + i \sin \frac{\pi}{6}\right)^{12}$

3. $\left[\sqrt{2}(\cos 10° + i \sin 10°)\right]^8$

4. $\left[\sqrt[5]{7}\left(\cos \frac{\pi}{5} + i \sin \frac{\pi}{5}\right)\right]^5$

5. $(1 + i)^{16}$

6. $\left(\frac{1}{2} - \frac{\sqrt{3}}{2}i\right)^{15}$

7. $\left(-\sqrt{3} + i\right)^7$

8. $\left(\sqrt{2} - \sqrt{2}i\right)^5$

9. $(1 + 2i)^6$

10. $(3 - i)^{12}$

Exercises 11-22 *Find the indicated roots, express them in the form a + bi, and represent them graphically.*

11. Square roots of $\frac{1}{25}(\cos 60° + i \sin 60°)$

12. Square roots of $9(\cos 180° + i \sin 180°)$

13. Cube roots of $1000\left(\cos \frac{3\pi}{2} + i \sin \frac{3\pi}{2}\right)$

14. Fourth roots of $\frac{1}{16}\left(\cos \frac{2\pi}{3} + i \sin \frac{2\pi}{3}\right)$

15. Square roots of $36i$ **16.** Square roots of -49

17. Cube roots of 27 **18.** Cube roots of $-8i$

19. Sixth roots of 1 **20.** Sixth roots of -64

21. Cube roots of $-\frac{\sqrt{2}}{16} + \frac{\sqrt{2}}{16}i$

22. Fourth roots of $-\frac{1}{2} - \frac{\sqrt{3}}{2}i$

Exercises 23-28 *Find the indicated roots. Express your answer in the form a + bi, with three decimal places of accuracy.*

23. Square roots of $3 - i$ **24.** Square roots of $-2 + 4i$

25. Fourth roots of i **26.** Fourth roots of $1 + \sqrt{3}i$

27. Cube roots of $2 + 5i$ **28.** Cube roots of $-1 - 6i$

Exercises 29-34 *Find all solutions of the equation and represent them graphically.*

29. $x^4 + 1 = 0$ **30.** $8x^3 - 1 = 0$

31. $x^3 + i = 0$ **32.** $x^2 - 16i = 0$

33. $x^2 - \left(2 + 2\sqrt{3}i\right) = 0$ **34.** $x^4 + 16 - 16i = 0$

35. Write $f(x) = x^6 + 64$ as a product of linear factors.

36. Write $f(x) = x^6 - 1$ as a product of linear factors.

37. Show that the cube roots of unity add up to zero.

38. Show that the fourth roots of unity add up to zero.

Concepts and Critical Thinking

Exercises 39-42 *Answer true or false.*

39. According to de Moivre's Theorem,
$[r(\cos \theta + i \sin \theta)]^n = r^n(\cos \theta^n + i \sin \theta^n)$ for any positive integer n.

40. Every complex number has three distinct third roots.

41. Every nonzero complex number has three distinct third roots.

42. An nth root of unity is a number $a + bi$ such that $(a + bi)^n = 1$.

Exercises 43-46 *Give an example of each.*

43. A positive integer n such that $(1 + i)^n$ is real and negative

44. A positive integer n such that $(1 + i)^n$ is real and positive

45. A positive integer n such that $(1 + i)^n$ is imaginary

46. A positive integer n such that $(1 + i)^n$ is neither real nor imaginary

Questions for Discussion or Essay

47. In this section, we considered positive integer powers of complex numbers. How should we define z^{-n} if z is a complex number and $n > 0$?

48. Describe a process by which the sixth roots of unity can be represented geometrically without applying the nth-root formula. Can your process be extended to locate the nth roots of unity geometrically for any n? Explain.

49. It is a consequence of Euler's Formula (see Exercise 51) that $e^{i\pi} + 1 = 0$. This has been described as one of the most beautiful formulas in mathematics, for it establishes a connection among five of the most important numbers in mathematics. Give the five numbers, and explain why each is significant.

Projects for Enrichment

50. Regular *n*-gons on a Graphing Calculator In this project, we devise a technique for drawing a regular *n*-gon (that is, a polygon with *n* sides of equal length) on a graphing calculator. We first consider a regular hexagon (6-gon), also known as a hexagon. Our strategy is to find the sixth roots of unity and then use these six points as the vertices of a regular hexagon.

a. Show that the sixth roots of unity correspond to the six points in Figure 44.

If we connect consecutive pairs of points as we move around the unit circle, we obtain the polygon shown in Figure 45.

b. How do we know that the polygon shown in Figure 45 is a regular hexagon?

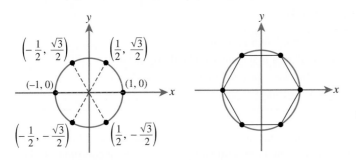

Figure 44   **Figure 45**

Next we use the line-drawing feature of a graphing calculator to draw line segments between consecutive pairs of points. With one popular brand of graphing calculator, a line segment can be drawn between two points (A, B) and (C, D) by entering `Line(A, B, C, D)`. Thus, `Line(1, 0, .5, √(3)/2)` draws a line segment between the points $(1, 0)$ and $\left(\frac{1}{2}, \frac{\sqrt{3}}{2}\right)$. Entering similar `Line` commands for the other five pairs of points produces the hexagon shown in Figure 46. Note that it is necessary to use a square viewing window to avoid distorting the hexagon.

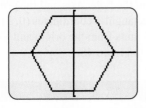

Figure 46

c. Find the eighth roots of unity and use them to draw a regular octagon. Write out the commands you used with your calculator to plot the eight line segments.

d. Repeat part c to draw a regular pentagon. Note that you will need to approximate most of the coordinates of the points.

e. *For students with programming experience* If you have access to a programmable calculator, write a program similar to the following that requests a value for *n* and then proceeds to draw a regular *n*-gon. Be sure your calculator is set in radian mode and select a square viewing window.

```
Prgm1:NGON
:Disp "NUMBER OF        Input number of sides.
 SIDES"
:Input N
:ClrDraw
:1 → A                  First vertex is (1, 0).
:0 → B
:1 → K                  Start with K = 1.
:Lbl 1                  Begin loop.
:2πK/N → T              Compute angle T for next root.
:cos T → C              Use T to find the
:sin T → D              next vertex (C, D).
:Line(A, B, C, D)       Draw a line between (A, B) and (C, D).
:C → A                  Rename vertex (C, D) as (A, B).
:D → B
:IS > (K,N)             Increment K and stop if K > N.
:Goto 1                 End loop.
```

51. Euler's Formula and Applications In this project, we give Euler's Formula for complex numbers, one of the most useful and profound formulas in all of mathematics. Then we use it to prove de Moivre's Theorem and the sum and difference identities for sine and cosine.

The great Swiss mathematician Leonhard Euler developed an ingenious technique for defining complex exponentials—expressions of the form a^b, where b is complex. At the heart of his technique is Euler's Formula, which states that for real numbers θ,

$$e^{i\theta} = \cos\theta + i\sin\theta$$

a. Apply Euler's Formula to compute $e^{i\theta}$ for each of the following values of θ:

   i. $\dfrac{\pi}{2}$ **ii.** $\dfrac{\pi}{4}$

   iii. π **iv.** $-\dfrac{\pi}{2}$

Suppose that a complex number z has polar form:

$$z = r(\cos\theta + i\sin\theta)$$

Then, according to Euler's Formula, z has the following exponential form:

$$z = re^{i\theta}$$

b. Write each of the following complex numbers in the form $z = re^{i\theta}$:

   i. $1 + i$ **ii.** $3i$

   iii. $1 - i$ **iv.** -4

c. De Moivre's Theorem states that

$$[r(\cos \theta + i \sin \theta)]^n = r^n(\cos n\theta + i \sin n\theta)$$

i. Express $\cos \theta + i \sin \theta$ in exponential form using Euler's Formula.

ii. Express $\cos n\theta + i \sin n\theta$ in exponential form using Euler's Formula.

iii. Prove de Moivre's Theorem using your results from parts i and ii and exponential properties.

d. Let $z_1 = e^{i\theta_1}$ and $z_2 = e^{i\theta_2}$.

i. Explain why $z_1z_2 = e^{i(\theta_1 + \theta_2)}$. Then use Euler's Formula to write $e^{i(\theta_1 + \theta_2)}$ in terms of the sine and cosine of $\theta_1 + \theta_2$.

ii. Use Euler's Formula to write both z_1 and z_2 in the form $a + bi$. Then multiply these expressions to obtain the product z_1z_2 in terms of sines and cosines of θ_1 and θ_2.

iii. Use the results of parts i and ii to obtain the sum identities for sine and cosine.

iv. Repeat the process just described to the quotient z_1/z_2 to derive the difference identities for sine and cosine.

Section 7.5 | Polar Coordinates

- In what sense is a spiral the graph of a function?
- What mathematical function has a graph shaped like flower petals?
- How can a single point in the plane have infinitely many coordinates?
- What kind of symmetry do roses have?

Thus far, we have specified points in the plane using rectangular coordinates. With this system, the location of a point is determined by its x- and y-coordinates: The x-coordinate tells us how far over, and the y-coordinate tells us how far up or down we must move from the origin in order to reach the point. In the *polar coordinate system*, by contrast, the position of a point is specified by r- and θ-coordinates: r tells us how far out from the origin we must travel, and θ tells us in what direction. Not surprisingly, polar coordinates and the trigonometric form of complex numbers described in the previous section are closely related.

Polar Coordinate System

The **polar coordinate system** is defined in terms of a fixed point O, called the **origin** or **pole**, and a half-line or ray with endpoint O, called the **polar axis**. To each point P in the plane, we assign **polar coordinates** $(r; \theta)$, where r is the directed distance from O to P, and θ is the directed angle between the polar axis and the segment $\overline{OP}$ (Figure 47). Thus, for example, the point P with rectangular coordinates $(0, 3)$ can be expressed as $\left(3; \frac{\pi}{2}\right)$ in polar coordinates since P lies 3 units from the origin and $\frac{\pi}{2}$ is the angle made between the polar axis and $\overline{OP}$, as shown in Figure 48.

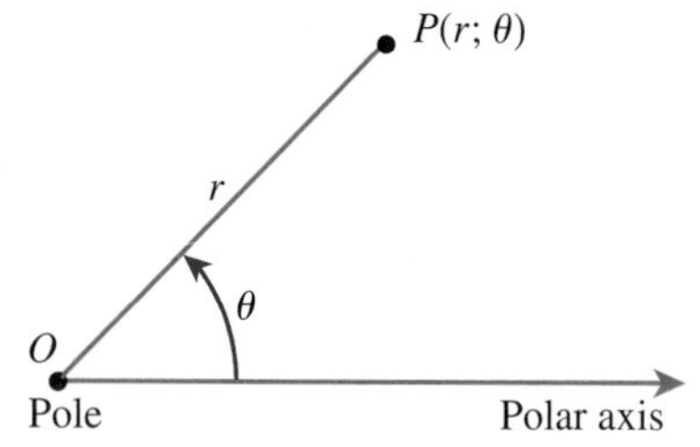

Figure 47

Polar Coordinates Notation

The point P lying a directed distance of r units from O such that θ is the directed angle from the polar axis to $\overline{OP}$ is said to have **polar coordinates** $(r; \theta)$. The semicolon separating r and θ is used to indicate that polar (as opposed to rectangular) coordinates are being used. One word of caution: Although we consistently employ semicolon notation in this text, it is not a standard convention. In most other texts, the polar coordinates r and θ are separated merely by a comma, as with rectangular coordinates.

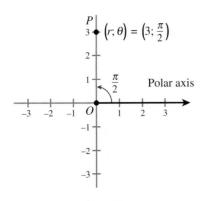

Figure 48

There is an important distinction between rectangular and polar coordinates. Although there is only one way of specifying a point in rectangular coordinates, there are infinitely many ways of specifying a point in polar coordinates. This ambiguity arises from two different sources. First of all, the angle θ associated with a point P is defined as a directed angle from the polar axis to the segment $\overline{OP}$, which essentially says that θ is an angle that, in standard position, has terminal side $\overline{OP}$. But as we have already seen, infinitely many angles are coterminal with a given angle. In the case of the point $P(0, 3)$ in Figure 48, not only can θ be taken to be $\frac{\pi}{2}$, it can also be taken to be $\frac{\pi}{2} + 2\pi, \frac{\pi}{2} + 4\pi, -\frac{3\pi}{2}$, and so on. Nowhere is this ambiguity more evident than at the pole itself, which has polar coordinates $(0; \theta)$ for any value of θ. In other words, at the pole, θ can be anything!

A second source of ambiguity follows from the fact that r is a *directed* distance (as opposed to a "distance"), and so may be negative. For negative values of r, the distance is measured along the line segment formed by extending the terminal side of θ to the opposite side of O. (This is like describing a point 2 miles to the north as being -2 miles to the south.) Figure 49 shows the point P described in several different ways.

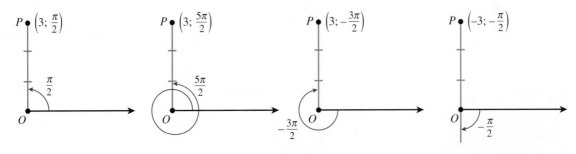

Figure 49

When working with the polar coordinate system, it is sometimes convenient to use a grid of concentric circles with the pole at the common center, the polar axis directed to the right, and line segments radiating from the pole at regular angular intervals, as shown in Figure 50.

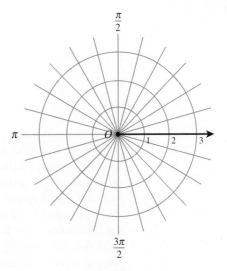

Figure 50

·····≻**EXAMPLE 1**

Plotting Points in Polar Coordinates

Plot the indicated point.

a. $\left(2; \dfrac{\pi}{6}\right)$ **b.** $\left(3; -\dfrac{2\pi}{3}\right)$ **c.** $\left(-\dfrac{5}{2}; \dfrac{3\pi}{4}\right)$

Solution   The points are shown in Figures 51–53.

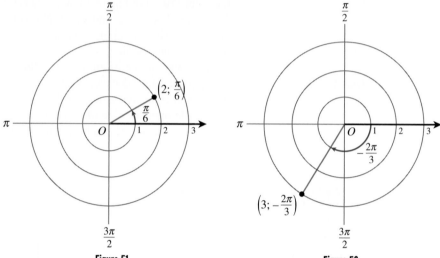

Figure 51   Figure 52

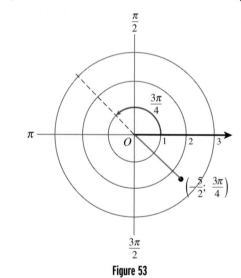

Figure 53

A **polar equation** is an equation relating r and θ. The **graph of a polar equation** consists of the collection of all points $(r; \theta)$ such that r and θ satisfy the equation. A **polar function** is defined by a polar equation of the form $r = f(\theta)$. Just as the simplest and most useful categories of rectangular functions have been given special names and studied in great detail (polynomial, rational, exponential, logarithmic, and so forth), so too have many polar functions been classified and explored in great detail. There are polar functions bearing names such as lemniscate, cardioid, limaçon, spiral, and rose. We will avoid an in-depth exploration of these exotic curves; instead, our focus will be on general techniques that apply to all polar functions.

EXAMPLE 2

Sketching the Graph of a Polar Equation

Sketch the graph of $r = 4$.

Solution Here θ is unrestricted, and so we are interested in the collection of points for which r, the distance to the origin, is 4. This is just a circle centered at the origin with radius 4, as shown in Figure 54.

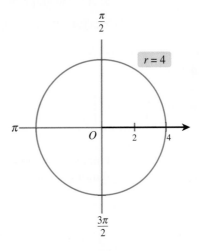

Figure 54

EXAMPLE 3

Sketching the Graph of a Polar Equation

Sketch the graph of $\theta = \frac{\pi}{3}$.

Solution In this case, r is unrestricted and θ is fixed at $\frac{\pi}{3}$. Thus, we are looking for the set of points for which θ is $\frac{\pi}{3}$. By considering both positive and negative values of r, we obtain the line shown in Figure 55.

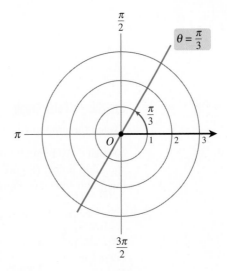

Figure 55

In each of the previous two examples, we were able to sketch the graph by using a straightforward analysis of the polar equation. However, in cases where the equation is complex and unfamiliar, this may not be possible. In such cases, we may be able to shed some light on the graph by plotting a few points, as we do in the following examples.

⋯⋯⋗EXAMPLE 4   **Sketching the Graph of a Spiral**

Sketch the graph of $r = \theta$ for $\theta \geq 0$.

Solution   We begin by making a table of values.

θ	0	$\dfrac{\pi}{4}$	$\dfrac{\pi}{2}$	π	$\dfrac{3\pi}{2}$	2π	3π	4π
r	0	$\dfrac{\pi}{4} \approx 0.79$	$\dfrac{\pi}{2} \approx 1.57$	$\pi \approx 3.14$	$\dfrac{3\pi}{2} \approx 4.71$	$2\pi \approx 6.28$	$3\pi \approx 9.42$	$4\pi \approx 12.57$

Plotting these points and connecting them with a smooth curve, we obtain the spiral shown in Figure 56.

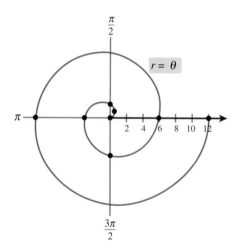

Figure 56

⋯⋯⋗EXAMPLE 5   **Sketching a Polar Graph**

Sketch the graph of $r = 2 \sin \theta$.

Solution   We begin by making a table of values for $0 \leq \theta \leq \pi$ and plotting the corresponding points (Figure 57). By connecting the points with a smooth curve, we obtain the graph shown in Figure 58.

θ	0	$\dfrac{\pi}{6}$	$\dfrac{\pi}{4}$	$\dfrac{\pi}{3}$	$\dfrac{\pi}{2}$	$\dfrac{2\pi}{3}$	$\dfrac{3\pi}{4}$	$\dfrac{5\pi}{6}$	π
$r = 2 \sin \theta$	0	1	$\sqrt{2}$	$\sqrt{3}$	2	$\sqrt{3}$	$\sqrt{2}$	1	0

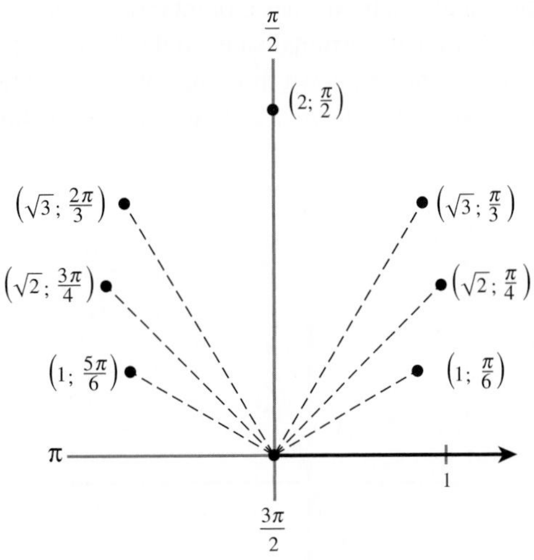

Figure 57

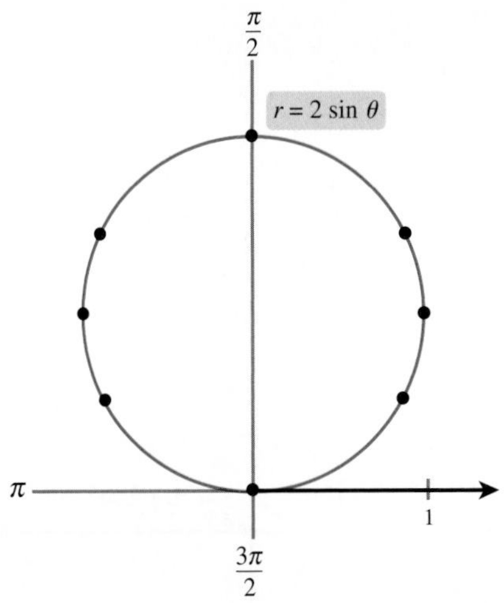

Figure 58

Note that the graph appears to be a circle (see Exercise 49 for the first step in the confirmation of this). Moreover, if $\theta > \pi$ or $\theta < 0$, the corresponding value for r would simply yield a point on the existing graph. For example, if $\theta = \frac{7\pi}{6}$, then $r = -1$, and the point $\left(-1; \frac{7\pi}{6}\right)$ is equivalent to $\left(1; \frac{\pi}{6}\right)$.

Some graphs can be fairly difficult to plot, in part because of the complications arising from negative r values. The next example illustrates the process by which a complicated polar graph can be plotted.

EXAMPLE 6

Sketching the Graph of a Function $r(\theta)$

Sketch the graph of $r = 3 \cos 2\theta$.

Solution Because of the complexity of producing a plot of this graph, we form it in several stages. We begin by plotting points for values of θ between 0 and $\frac{\pi}{4}$; connecting the points with a smooth curve gives the graph shown in Figure 59.

θ	0	$\dfrac{\pi}{12}$	$\dfrac{\pi}{8}$	$\dfrac{\pi}{6}$	$\dfrac{\pi}{4}$
$r = 3 \cos 2\theta$	3	$\dfrac{3\sqrt{3}}{2} \approx 2.60$	$\dfrac{3\sqrt{2}}{2} \approx 2.12$	$\dfrac{3}{2} = 1.5$	0

Next, we consider points for which θ lies between $\frac{\pi}{4}$ and $\frac{\pi}{2}$.

θ	$\dfrac{\pi}{4}$	$\dfrac{\pi}{3}$	$\dfrac{3\pi}{8}$	$\dfrac{5\pi}{12}$	$\dfrac{\pi}{2}$
$r = 3 \cos 2\theta$	0	$-\dfrac{3}{2} = -1.5$	$-\dfrac{3\sqrt{2}}{2} \approx -2.12$	$-\dfrac{3\sqrt{3}}{2} \approx -2.60$	-3

Notice that each of these points has a negative r value. Thus, even though the angles θ all have terminal sides in the first quadrant, the corresponding points lie in the third quadrant, as shown in Figure 60. If this process is continued to include all values of θ between 0 and 2π, we obtain the graph shown in Figure 61, called a four-petaled rose.

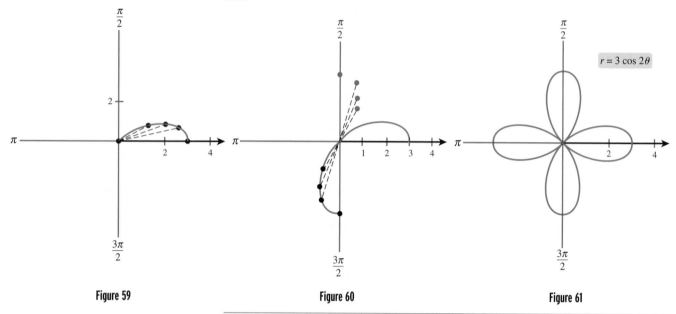

Figure 59   Figure 60   Figure 61

Most graphing calculators have the capability of plotting polar curves. The general procedure for plotting such curves is as follows.

Calculator Keys

Graphing Polar Curves

Polar curves of the form $r = f(\theta)$ can be plotted on most graphing calculators by setting the graph mode to **polar**, entering the expression $f(\theta)$, selecting an appropriate interval for θ (the interval $0 \le \theta \le 2\pi$ works well for many polar graphs), selecting an appropriate viewing window (your calculator's "square" setting is often the best choice), and then plotting the graph. Note that your calculator must be in radian mode, and you may need to adjust the interval for θ in order to see more of the graph and its key features.

·····➤**EXAMPLE 7**

Plotting a Polar Curve with a Graphing Calculator

Use a graphing calculator to plot the graphs of $r = 2\cos\theta$ and $r = 1 + 2\sin\theta$. Estimate the points of intersection.

Solution   We first set the graph mode to polar and then enter $r = 2\cos\theta$ and $r = 1 + 2\sin\theta$. Next, we select $[0, 2\pi]$ as an interval for θ. Finally, we select our calculator's "square" viewing window and zoom in to obtain the graph shown in Figure 62. The graph of $r = 2\cos\theta$ is a circle of radius 1 centered at $(1, 0)$. The graph of $r = 1 + 2\sin\theta$ is known as a limaçon with an inner loop.

One of the intersection points is clearly at the pole. The other two can be estimated using the trace feature, as shown in Figures 63 and 64.

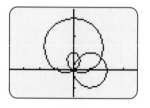

Figure 62

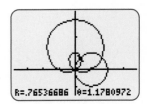

| Figure 63 | Figure 64 |

Conversion   At times, it is desirable to convert from one coordinate system to another. The following relationships between rectangular and polar coordinates are easily obtained by considering Figure 65 for the case that θ is acute and r is positive. With little effort, it can be shown that these relationships hold for all θ.

Figure 65

Coordinate Relationships

The rectangular coordinates (x, y) and polar coordinates $(r; \theta)$ of a point P are related as follows:

(1) $$x = r \cos \theta, \quad y = r \sin \theta$$

(2) $$r^2 = x^2 + y^2, \quad \tan \theta = \frac{y}{x}, \quad x \neq 0$$

The equations (1) enable us to find the rectangular coordinates of a point if the polar coordinates are known.

EXAMPLE 8

Converting from Polar to Rectangular Coordinates

Convert the given point from polar to rectangular coordinates.

a. $\left(2; \dfrac{\pi}{6}\right)$ **b.** $\left(3; -\dfrac{2\pi}{3}\right)$

Solution

a. We have $r = 2$ and $\theta = \frac{\pi}{6}$. Thus

$$x = 2 \cos \frac{\pi}{6} \qquad y = 2 \sin \frac{\pi}{6}$$

$$= 2 \cdot \frac{\sqrt{3}}{2} \qquad = 2 \cdot \frac{1}{2}$$

$$= \sqrt{3} \qquad = 1$$

and so the rectangular coordinates are $\left(\sqrt{3}, 1\right)$.

b. With $r = 3$ and $\theta = -\frac{2\pi}{3}$, we have

$$x = 3 \cos\left(-\frac{2\pi}{3}\right) \qquad y = 3 \sin\left(-\frac{2\pi}{3}\right)$$

$$= 3 \cdot \left(-\frac{1}{2}\right) \qquad = 3 \cdot \left(-\frac{\sqrt{3}}{2}\right)$$

$$= -\frac{3}{2} \qquad = -\frac{3\sqrt{3}}{2}$$

Thus, the rectangular coordinates are $\left(-\frac{3}{2}, -\frac{3\sqrt{3}}{2}\right)$.

It is slightly more difficult to convert from rectangular to polar coordinates because of the nonuniqueness of polar coordinates. The value for r can be determined using the equation $r^2 = x^2 + y^2$. An angle θ can be found by using the equation $\tan \theta = \frac{y}{x}$ and knowledge of the quadrant in which θ lies.

⋯⋯⟩EXAMPLE 9

Converting from Rectangular to Polar Coordinates

Convert the given point from rectangular to polar coordinates.

a. $\left(\frac{1}{3}, \frac{1}{3}\right)$ **b.** $\left(-2, 2\sqrt{3}\right)$ **c.** $(-1, -3)$

Solution

a. We begin by finding r. For convenience, we take r to be positive.

$$r = \sqrt{x^2 + y^2}$$
$$= \sqrt{\left(\frac{1}{3}\right)^2 + \left(\frac{1}{3}\right)^2}$$
$$= \sqrt{\frac{2}{9}} = \frac{\sqrt{2}}{3}$$

From the fact that $\tan \theta = \frac{y}{x}$, we have

$$\tan \theta = \frac{1/3}{1/3} = 1$$

Thus, θ is an angle in the first quadrant with tangent equal to 1, and so we take $\theta = \frac{\pi}{4}$. The point in polar coordinates is $\left(\frac{\sqrt{2}}{3}; \frac{\pi}{4}\right)$.

b. We find r and $\tan \theta$ as in part a.

$$r = \sqrt{(-2)^2 + \left(2\sqrt{3}\right)^2}$$
$$= \sqrt{4 + 12}$$
$$= 4$$
$$\tan \theta = \frac{2\sqrt{3}}{-2}$$
$$= -\sqrt{3}$$

Since $\tan \frac{\pi}{3} = \sqrt{3}$, we conclude that θ has a reference angle of $\frac{\pi}{3}$. Because the point is in the second quadrant, we have $\theta = \frac{2\pi}{3}$. Thus, the point is $\left(4; \frac{2\pi}{3}\right)$ in polar coordinates.

c. Once again, r and $\tan \theta$ are easily found.

$$r = \sqrt{(-1)^2 + (-3)^2}$$
$$= \sqrt{1 + 9}$$
$$= \sqrt{10}$$
$$\tan \theta = \frac{-3}{-1}$$
$$= 3$$

Thus, θ is an angle in the third quadrant with reference angle $\arctan 3 \approx 1.2490$. It follows that

$$\theta = \pi + \arctan 3 \approx 4.3906$$

Thus, $(-1, -3) \approx \left(\sqrt{10}; 4.3906\right)$.

It is often the case that an equation expressed in terms of one coordinate system is much more complex than when expressed in terms of the other system. In the following examples, we see how it is possible to convert equations from one system to the other.

EXAMPLE 10

Converting an Equation from Rectangular to Polar Form

Find a polar equation for the line $y = mx + b$.

Solution   Replacing x with $r \cos \theta$ and y with $r \sin \theta$, we have

$$r \sin \theta = m(r \cos \theta) + b$$

Solving for r gives us

$$r \sin \theta - mr \cos \theta = b$$
$$r(\sin \theta - m \cos \theta) = b$$
$$r = \frac{b}{\sin \theta - m \cos \theta}$$

EXAMPLE 11
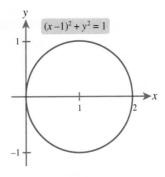

Converting an Equation from Polar to Rectangular Form

Convert the polar equation $r = 2 \cos \theta$ to an equivalent equation in x and y.

Solution   We exploit the relation $x = r \cos \theta$ by multiplying both sides of $r = 2 \cos \theta$ by r.

$$r = 2 \cos \theta$$
$$r^2 = 2r \cos \theta$$

Thus,

$$x^2 + y^2 = 2x$$

This, by the way, is the equation of a circle of radius 1 centered at $(1, 0)$. To see this, we complete the square as follows:

$$x^2 + y^2 = 2x$$
$$x^2 - 2x + y^2 = 0$$
$$(x^2 - 2x + 1) + y^2 = 1$$
$$(x - 1)^2 + y^2 = 1^2$$

The graph is shown in Figure 66.

Figure 66

Understanding and Mastery Checklists

Concepts to Understand	Skills to Master

Concepts to Understand

Polar coordinate system
⟡
Polar equation
⟡
Polar function
⟡
Graph of a polar equation
⟡
Relationship between polar and
rectangular coordinates

Skills to Master

Plot a point on the polar coordinate system.
⟡
Convert between polar and rectangular coordinates.
⟡
Plot the graph of a polar equation.
⟡
Convert between polar and rectangular equations.
⟡
Use a graphing calculator to plot a polar equation.

Exercises 7.5

Exercises 1–10 *Plot the point given in polar coordinates and find the corresponding rectangular coordinates.*

1. $\left(2; -\dfrac{\pi}{2}\right)$

2. $\left(3; \dfrac{3\pi}{2}\right)$

3. $\left(1; \dfrac{3\pi}{4}\right)$

4. $\left(-4; \dfrac{\pi}{4}\right)$

5. $\left(-2; \dfrac{\pi}{6}\right)$

6. $\left(3; -\dfrac{\pi}{3}\right)$

7. $\left(\dfrac{3}{2}; -\dfrac{5\pi}{6}\right)$

8. $\left(2\sqrt{2}; \dfrac{7\pi}{4}\right)$

9. $(-3; -\pi)$

10. $\left(-\dfrac{5}{2}; -\dfrac{7\pi}{2}\right)$

Exercises 11–20 *Plot the point given in rectangular coordinates and find two sets of polar coordinates with $0 \leq \theta < 2\pi$.*

11. $(4, 0)$

12. $(0, -3)$

13. $(2, 2)$

14. $(-3, 3)$

15. $\left(1, -\sqrt{3}\right)$

16. $\left(-2\sqrt{3}, 2\right)$

17. $(-4, -3)$

18. $(12, -5)$

19. $(3, 1)$

20. $(-2, 4)$

Exercises 21–34 *Find and plot at least four points satisfying the given polar equation, and then sketch the graph of the equation.*

21. $r = 3$

22. $r = -2$

23. $\theta = \dfrac{2\pi}{3}$

24. $\theta = \dfrac{5\pi}{6}$

25. $r = 2\theta, \theta \geq 0$

26. $r = 1 + \theta, \theta \geq 0$

27. $r = 3 \sin \theta$

28. $r = -2 \cos \theta$

29. $r = 2 - 4 \cos \theta$

30. $r = 2 \sin \theta + 2$

31. $r = \sin 2\theta$

32. $r = \cos 4\theta$

33. $r = 2 \cos 3\theta$

34. $r = 3 \sin \dfrac{\theta}{2}$

Exercises 35–42 *Convert the given rectangular equation to polar form.*

35. $y = 3$

36. $x = -1$

37. $y + x = 0$

38. $y = \sqrt{3}x$

39. $x^2 + y^2 = 9$

40. $x^2 - y^2 = 1$

41. $\ln y - \ln x = 1$

42. $e^{x^2}e^{y^2} = 2$

Exercises 43-52 *Convert the given polar equation to rectangular form.*

43. $r = 7$

44. $\theta = \dfrac{\pi}{4}$

45. $\theta = \dfrac{2\pi}{3}$

46. $r = -3$

47. $r \cos \theta = 1$

48. $r \sin \theta = 4r \cos \theta + 2$

49. $r = 2 \sin \theta$

50. $r = 2 \cos \theta + 3 \sin \theta$

51. $r = \cos 2\theta$

52. $r = \sin 2\theta$

Exercises 53-58 *Use a graphing calculator to plot the graphs of the given polar equations. Estimate the coordinates of the point(s) of intersection.*

53. $r = 2$
$r = 4 \sin \theta$

54. $r = 2 \sin \theta$
$r = 4 \cos \theta$

55. $r = 2 \cos \theta$
$r = 1 - \sin \theta$

56. $r = 3$
$r = 2 + 3 \cos \theta$

57. $r = \sin 2\theta$
$r = \sin \theta$

58. $r = \cos 3\theta$
$r = \sin \theta$

Concepts and Critical Thinking

Exercises 59-62 *Answer true or false.*

59. The rectangular coordinates of a point are unique.

60. The polar coordinates of a point are unique.

61. If a point has rectangular coordinates (x, y) and polar coordinates $(r; \theta)$, then $r = x \cos \theta$ and $r = y \sin \theta$.

62. Every point in the rectangular coordinate system can be represented with polar coordinates.

Exercises 63-66 *Give an example of each.*

63. A polar equation whose graph is a circle centered at the origin

64. A polar equation whose graph is a line

65. A polar equation whose graph is a spiral

66. A polar equation whose graph is a circle centered away from the origin

67. The graph $r = -\sin^2 \theta$ is shown in Figure 67. Does the point with polar coordinates $\left(1; \dfrac{\pi}{2}\right)$ appear to be on the graph? Compute r for $\theta = \dfrac{\pi}{2}$. Is the point $\left(1; \dfrac{\pi}{2}\right)$, in fact, on the graph of $r = -\sin^2 \theta$? Explain.

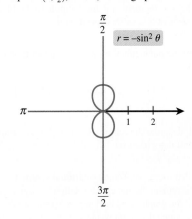

Figure 67

Questions for Discussion or Essay

68. Compare and contrast the trigonometric form of complex numbers discussed in Section 7.3 with the polar coordinate representation of points in the plane.

69. The graph of $r = 4 \sin 5\theta$ is given in Figure 68. Explain how a graphing calculator can be used to find the point farthest from the origin, without graphing in polar coordinates.

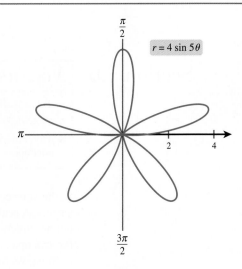

Figure 68

Projects for Enrichment

70. Rose Curves In this project, we explore **rose curves**, the graphs of polar equations of the form $r = a \sin n\theta$ or $r = a \cos n\theta$, with $n \geq 2$. Because of the great number of polar graphs involved in this project, a graphing calculator with polar capabilities would be especially useful.

a. Graph each of the following rose curves:

    i. $r = \cos 2\theta$ **ii.** $r = 3 \cos 5\theta$

    iii. $r = 2 \sin 3\theta$ **iv.** $r = 2 \sin 4\theta$

b. Indicate the number of petals for each of the rose curves graphed in part a. In general, how many petals does the graph of $a \cos n\theta$ (or $a \sin n\theta$) have?

c. For each of the rose curves graphed in part a, find the angle between the line segments joining the pole to the tips of adjacent petals. How does this value depend on n?

d. For each of the polar functions $r = f(\theta)$ given in part a, sketch the graphs of

$$r = f\left(\theta - \frac{\pi}{6}\right) \quad \text{and} \quad r = f\left(\theta + \frac{\pi}{4}\right)$$

In general, describe the relationship between the graphs of $r = f(\theta)$ and $r = f(\theta - \alpha)$.

e. It is clear that rose curves enjoy **rotational symmetry**—that is, their graphs remain the same after rotation through certain angles. For each of the graphs produced in part a, estimate the smallest angle α for which the graph would appear the same after rotating by α.

f. Find a polar function $r = f(\theta)$ corresponding to the given graph.

    i.

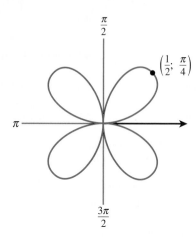

    ii.

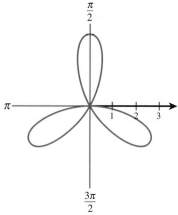

Section 7.6   Vectors

- What is the difference between speed and velocity?
- How does a steady wind affect the path of a supersonic plane?
- How much force is required to pull a 5000-pound Ford Expedition up a 10° incline?

In the sciences, physical quantities can be separated into two categories: scalars and vectors. A **scalar quantity**—such as time, distance, mass, or volume—has a magnitude but no direction. Thus, a scalar quantity can be specified by a single numerical value. For example, we might refer to a distance of 250 feet or to a mass of 5 kilograms. By contrast, **vector quantities**—such as force, velocity, and acceleration—have both magnitude and direction, and thus cannot be specified by a single numerical value. For example, the velocity of a jet might be described as 500 miles per hour on a course of 40° measured clockwise from north, or the force a dog applies to its leash could be 50 pounds at an angle of 30° with the horizontal. In each of these cases, we require two numerical values to specify magnitude and direction.

Vector Definitions

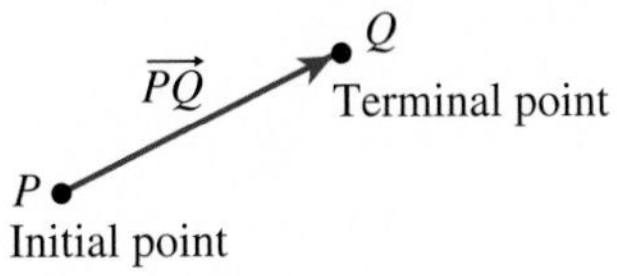

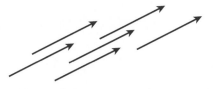

$\overrightarrow{PQ}$ Q

P
Initial point   Terminal point

Figure 69

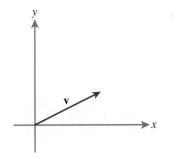

Figure 70

A vector quantity can be represented by a **directed line segment**, such as the one shown in Figure 69. The directed line segment $\overrightarrow{PQ}$ in Figure 69 has an **initial point** P and a **terminal point** Q. The arrow indicates that the vector quantity acts in the direction from P to Q, and the length of the segment is the magnitude of the vector quantity. Two directed line segments are said to be **equivalent** if they have the same magnitude and direction. In Figure 70, we see a collection of directed line segments that are equivalent to $\overrightarrow{PQ}$. We use the term **vector** to refer to the common magnitude and direction shared by a set of equivalent directed line segments. In other words, a vector with a given magnitude and direction may be represented by infinitely many directed line segments, each with a different initial point. We denote vectors by lowercase, boldface letters, such as **u**, **v**, and **w**. For convenience, we often draw vectors in **standard position** with their initial point at the origin. In Figure 71, we illustrate the standard position of the vector **v** corresponding to $\overrightarrow{PQ}$.

When a vector is represented in standard position, its magnitude and direction are completely determined by the terminal point. Thus, if a vector **v** has initial point $(0, 0)$ and terminal point (v_1, v_2), we define the **component form** of the vector **v** by

$$\mathbf{v} = \langle v_1, v_2 \rangle$$

The x- and y-coordinates of the terminal point are referred to as the x- and y-**components** of the vector **v**. A vector **w** with initial point $P(x_1, y_1)$ and terminal point $Q(x_2, y_2)$ can be placed in standard position by translating $\overrightarrow{PQ}$ so that its initial point is at the origin. In so doing, the terminal point must be translated x_1 units horizontally and y_1 units vertically. Thus, we obtain the component form

$$\mathbf{w} = \langle x_2 - x_1, y_2 - y_1 \rangle$$

The component form of a vector is unique in the sense that if $\mathbf{u} = \langle u_1, u_2 \rangle$ and $\mathbf{v} = \langle v_1, v_2 \rangle$, then $\mathbf{u} = \mathbf{v}$ if and only if $u_1 = v_1$ and $u_2 = v_2$.

Figure 71

(graph with vector **v** in standard position)

EXAMPLE 1

Sketching Vectors

Consider the points $P(-1, -2)$ and $Q(-3, 1)$.

a. Sketch the directed line segment $\overrightarrow{PQ}$.

b. Find the component form of the vector **v** represented by $\overrightarrow{PQ}$.

c. Sketch the vector **v** in standard position.

d. Sketch the directed line segment $\overrightarrow{QP}$ and its associated vector **u** in standard position.

Solution

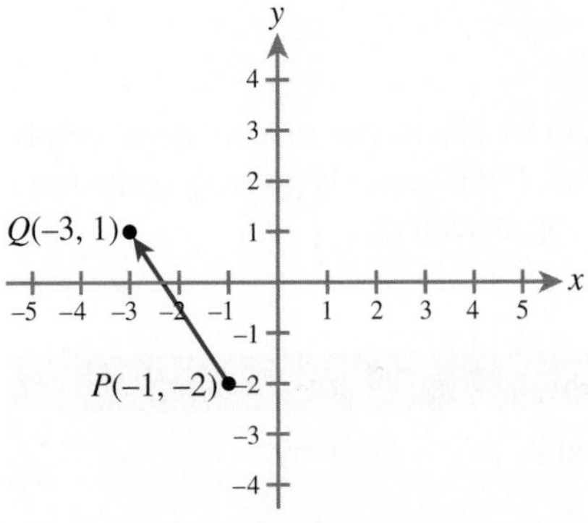

a. We first locate the points $P(-1, -2)$ and $Q(-3, 1)$, and then draw a directed line segment from the initial point P to the terminal point Q. The resulting vector $\overrightarrow{PQ}$ is shown in Figure 72.

b. To find the component form of **v**, we subtract the x- and y-coordinates of P from the x- and y-coordinates of Q.

$$\mathbf{v} = \langle -3 - (-1), 1 - (-2) \rangle = \langle -2, 3 \rangle$$

Figure 72

(graph with points $Q(-3, 1)$ and $P(-1, -2)$ and directed line segment)

c. The vector $\mathbf{v} = \langle -2, 3 \rangle$ is shown in Figure 73 with initial point $(0, 0)$ and terminal point $(-2, 3)$.

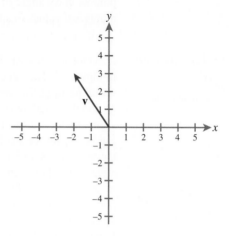

Figure 73

d. The directed line segment $\overrightarrow{QP}$ is shown in Figure 74. The component form of the associated vector $\mathbf{u}$ is given by

$$\mathbf{u} = \langle -1 - (-3), -2 - 1 \rangle = \langle 2, -3 \rangle$$

and the vector $\mathbf{u}$ is shown in Figure 74 in standard position.

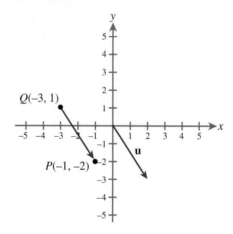

Figure 74

Since the magnitude of a vector corresponds to its length, the magnitude is simply the distance between the initial and terminal points. If the vector is given in component form, the distance formula leads us to the following statement.

Magnitude of a Vector

The magnitude of a vector $\mathbf{v} = \langle v_1, v_2 \rangle$, denoted by $\|\mathbf{v}\|$, is given by

$$\|\mathbf{v}\| = \sqrt{v_1^2 + v_2^2}$$

⋯⋯⋯⃗ **EXAMPLE 2**   **Finding the Magnitude of a Vector**

Sketch the vector $\mathbf{v} = \langle 4, -5 \rangle$ and find its magnitude.

Solution   The vector $\mathbf{v} = \langle 4, -5 \rangle$ is shown in Figure 75 with initial point $(0, 0)$ and terminal point $(4, -5)$. Its magnitude is

$$\|\mathbf{v}\| = \sqrt{4^2 + (-5)^2} = \sqrt{16 + 25} = \sqrt{41}$$

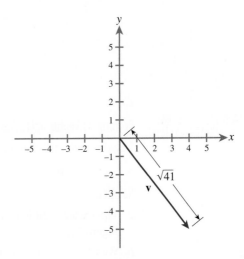

Figure 75

Notice that $\|\mathbf{v}\| > 0$ for any vector $\mathbf{v} = \langle v_1, v_2 \rangle$ as long as either v_1 or v_2 is nonzero. In the case where $\mathbf{v} = \langle 0, 0 \rangle$, $\|\mathbf{v}\| = 0$. We refer to the vector $\langle 0, 0 \rangle$ as the **zero vector** and, in keeping with our convention of writing vectors in boldface, we write

$$\mathbf{0} = \langle 0, 0 \rangle$$

Vector Operations   The two most common vector operations are addition and scalar multiplication. Geometrically, the sum of two vectors $\mathbf{u}$ and $\mathbf{v}$ can be obtained by translating $\mathbf{v}$ so that its initial point lies on the terminal point of $\mathbf{u}$. The sum $\mathbf{u} + \mathbf{v}$ is a vector whose initial point is the initial point of $\mathbf{u}$ and whose terminal point is the terminal point of $\mathbf{v}$. See Figure 76 for an illustration of this.

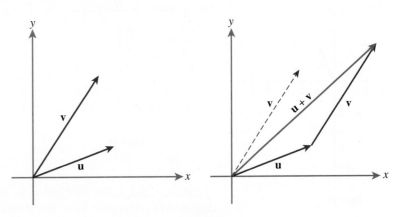

Figure 76

Scalar multiplication refers to the product of a scalar and a vector. Geometrically, the product of a scalar k and a vector $\mathbf{v}$ is a vector whose length is $|k|$ times that of $\mathbf{v}$. If k is positive, $k\mathbf{v}$ has the same direction as $\mathbf{v}$, whereas if k is negative, $k\mathbf{v}$ has the opposite direction of $\mathbf{v}$. Figure 77 gives some examples of scalar products.

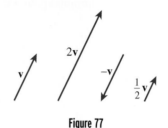

Figure 77

EXAMPLE 3

Performing Vector Operations Graphically

Use the vectors shown in Figure 78 to sketch the indicated vector.

a. $-2\mathbf{u}$ **b.** $\mathbf{u} + \mathbf{v}$ **c.** $\mathbf{u} - 2\mathbf{v}$

Solution

a. The vector $-2\mathbf{u}$ is twice the length of $\mathbf{u}$ but has the opposite direction, as shown in Figure 79.

b. We first translate $\mathbf{v}$ so that its initial point lies on the terminal point of $\mathbf{u}$. The vector $\mathbf{u} + \mathbf{v}$ has the same initial point as $\mathbf{u}$ and the same terminal point as $\mathbf{v}$, as shown in red in Figure 80.

c. Notice that $\mathbf{u} - 2\mathbf{v} = \mathbf{u} + (-2\mathbf{v})$. Thus, we first sketch the vector $-2\mathbf{v}$ with its initial point at the terminal point of $\mathbf{u}$. The vector $\mathbf{u} + (-2\mathbf{v})$ has the same initial point as $\mathbf{u}$, and the same terminal point as $-2\mathbf{v}$, as shown in green in Figure 81.

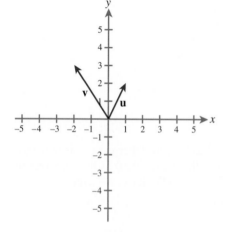

Figure 78

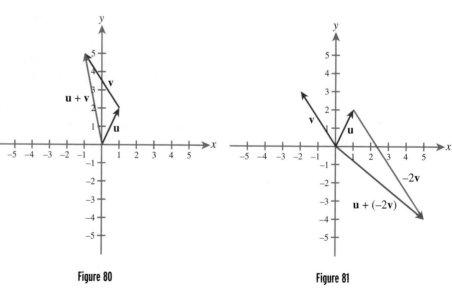

Figure 80

Figure 81

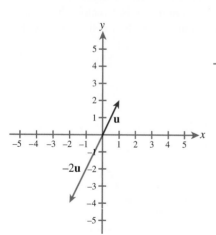

Figure 79

Addition and scalar multiplication are particularly easy to perform for vectors expressed in component form. The following definitions are easily shown to be consistent with the geometric definitions given earlier.

Definitions of Vector Addition and Scalar Multiplication

If $\mathbf{u} = \langle u_1, u_2 \rangle$ and $\mathbf{v} = \langle v_1, v_2 \rangle$, then **vector addition** and **scalar multiplication** are defined as follows:

$$\mathbf{u} + \mathbf{v} = \langle u_1 + v_1, u_2 + v_2 \rangle$$
$$k\mathbf{u} = \langle ku_1, ku_2 \rangle$$

EXAMPLE 4

Performing Vector Operations Algebraically

Perform the indicated operation on the vectors $\mathbf{u} = \langle 3, 5 \rangle$ and $\mathbf{v} = \langle 2, -4 \rangle$.

a. $\mathbf{u} + \mathbf{v}$ **b.** $-3\mathbf{u}$ **c.** $2\mathbf{u} - \dfrac{1}{2}\mathbf{v}$

Solution

a. $\mathbf{u} + \mathbf{v} = \langle 3, 5 \rangle + \langle 2, -4 \rangle = \langle 3 + 2, 5 + (-4) \rangle = \langle 5, 1 \rangle$

b. $-3\mathbf{u} = -3\langle 3, 5 \rangle = \langle (-3)3, (-3)5 \rangle = \langle -9, -15 \rangle$

c. $2\mathbf{u} - \dfrac{1}{2}\mathbf{v} = 2\langle 3, 5 \rangle - \dfrac{1}{2}\langle 2, -4 \rangle = \langle 6, 10 \rangle - \langle 1, -2 \rangle = \langle 5, 12 \rangle$

Vector addition and scalar multiplication have many of the same properties as the operations of addition and multiplication for real numbers.

Properties of Vector Addition and Scalar Multiplication

Let $\mathbf{u}$, $\mathbf{v}$, and $\mathbf{w}$ be vectors, and let r and s be scalars. Then the following properties hold:

1. $\mathbf{u} + \mathbf{v} = \mathbf{v} + \mathbf{u}$
2. $(\mathbf{u} + \mathbf{v}) + \mathbf{w} = \mathbf{u} + (\mathbf{v} + \mathbf{w})$
3. $\mathbf{u} + \mathbf{0} = \mathbf{u}$
4. $\mathbf{u} + (-\mathbf{u}) = \mathbf{0}$
5. $r(\mathbf{u} + \mathbf{v}) = r\mathbf{u} + r\mathbf{v}$
6. $(r + s)\mathbf{u} = r\mathbf{u} + s\mathbf{u}$
7. $r(s\mathbf{u}) = (rs)\mathbf{u}$
8. $1\mathbf{u} = \mathbf{u}$
9. $\|r\mathbf{u}\| = |r|\|\mathbf{u}\|$

EXAMPLE 5

Verifying a Property of Vector Addition

Verify Property 1; that is, given $\mathbf{u} = \langle u_1, u_2 \rangle$ and $\mathbf{v} = \langle v_1, v_2 \rangle$, show that $\mathbf{u} + \mathbf{v} = \mathbf{v} + \mathbf{u}$.

Solution

$$
\begin{aligned}
\mathbf{u} + \mathbf{v} &= \langle u_1, u_2 \rangle + \langle v_1, v_2 \rangle \\
&= \langle u_1 + v_1, u_2 + v_2 \rangle && \text{Using the definition of vector addition} \\
&= \langle v_1 + u_1, v_2 + u_2 \rangle && \text{Applying the commutative property of addition} \\
&= \langle v_1, v_2 \rangle + \langle u_1, u_2 \rangle && \text{Using the definition of vector addition} \\
&= \mathbf{v} + \mathbf{u}
\end{aligned}
$$

Unit Vectors

If one is only interested in the direction of a vector and not its magnitude, it is usually more convenient to work with vectors of length 1. Such vectors are called unit vectors. Thus, a **unit vector** is any vector **u** for which $\|\mathbf{u}\| = 1$. When a nonzero vector is given that is not a unit vector, it is possible to form a unit vector having the same direction as the given vector. This can be done by "dividing" the given vector by its length, in the following sense.

Forming a Unit Vector

If **v** is a nonzero vector, then the vector **u** given by

$$\mathbf{u} = \frac{1}{\|\mathbf{v}\|}\mathbf{v}$$

is a unit vector in the same direction as **v**.

The fact that $\mathbf{u} = \frac{1}{\|\mathbf{v}\|}\mathbf{v}$ is a unit vector can be verified using the scalar multiplication property $\|r\mathbf{v}\| = |r|\|\mathbf{v}\|$ as follows:

$$\left\|\frac{1}{\|\mathbf{v}\|}\mathbf{v}\right\| = \left|\frac{1}{\|\mathbf{v}\|}\right|\|\mathbf{v}\| = \frac{1}{\|\mathbf{v}\|}\|\mathbf{v}\| = 1$$

EXAMPLE 6

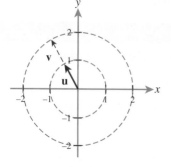

Figure 82

Finding a Unit Vector in a Given Direction

Find and sketch the unit vector **u** in the direction of $\mathbf{v} = \langle -1, \sqrt{3} \rangle$.

Solution   The magnitude of **v** is

$$\|\mathbf{v}\| = \sqrt{(-1)^2 + \left(\sqrt{3}\right)^2} = \sqrt{1+3} = \sqrt{4} = 2$$

To find the unit vector **u** in the direction of **v**, we multiply **v** by $\frac{1}{\|\mathbf{v}\|}$.

$$\mathbf{u} = \frac{1}{\|\mathbf{v}\|}\mathbf{v} = \frac{1}{2}\langle -1, \sqrt{3} \rangle = \left\langle -\frac{1}{2}, \frac{\sqrt{3}}{2} \right\rangle$$

Since **u** has magnitude 1, its terminal point must lie on the unit circle. A sketch is shown in Figure 82.

Two unit vectors of special significance are the so-called standard unit vectors **i** and **j**. As illustrated in Figure 83, these two vectors point in the directions of the positive x- and y-axes.

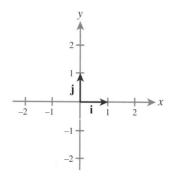

Figure 83

Standard Unit Vectors

The **standard unit vectors** are denoted **i** and **j** and are defined by

$$\mathbf{i} = \langle 1, 0 \rangle \quad \text{and} \quad \mathbf{j} = \langle 0, 1 \rangle$$

The standard unit vectors **i** and **j** can be viewed as the building blocks from which all other vectors (in the plane) can be formed. In particular, a vector $\langle a, b \rangle$ can be written as

$$\begin{aligned}
\langle a, b \rangle &= \langle a, 0 \rangle + \langle 0, b \rangle \\
&= a\langle 1, 0 \rangle + b\langle 0, 1 \rangle \\
&= a\mathbf{i} + b\mathbf{j}
\end{aligned}$$

The expression "$a\mathbf{i} + b\mathbf{j}$" is called a **linear combination** of $\mathbf{i}$ and $\mathbf{j}$. Notice that the coefficients of $\mathbf{i}$ and $\mathbf{j}$ are nothing more than the x- and y-components of the vector.

-----▷**EXAMPLE 7**

Linear Combinations of Standard Unit Vectors

Express the vector $\mathbf{v} = \langle 4, -2 \rangle$ as a linear combination of the standard unit vectors $\mathbf{i}$ and $\mathbf{j}$.

Solution   Applying the observation just made, we write

$$\mathbf{v} = \langle 4, -2 \rangle = 4\mathbf{i} - 2\mathbf{j}$$

Vectors can also be represented in what is known as *trigonometric form*. We begin by considering a unit vector $\mathbf{u}$. Let θ denote the angle measured counterclockwise from the positive x-axis to $\mathbf{u}$, as shown in Figure 84. The angle θ is referred to as the **direction angle** for the vector $\mathbf{u}$. Since $\mathbf{u}$ is a unit vector, its terminal point, (u_1, u_2) say, lies on the unit circle. From the definitions of sine and cosine, we know that $u_1 = \cos\theta$ and $u_2 = \sin\theta$. Thus,

$$\mathbf{u} = \cos\theta\,\mathbf{i} + \sin\theta\,\mathbf{j}$$

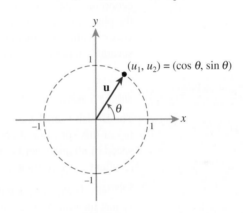

Figure 84

Now suppose $\mathbf{v}$ is any other nonzero vector with direction angle θ. Then $\mathbf{v}$ has the same direction as $\mathbf{u}$ but is $\|\mathbf{v}\|$ times as long. That is, $\mathbf{v} = \|\mathbf{v}\|\,\mathbf{u} = \|\mathbf{v}\|(\cos\theta\,\mathbf{i} + \sin\theta\,\mathbf{j})$. Based on these observations, we define the trigonometric form of a vector $\mathbf{v}$ as follows.

Trigonometric Form of a Vector

Let $\mathbf{v}$ be a nonzero vector with direction angle θ. Then the **trigonometric form** for $\mathbf{v}$ is

$$\mathbf{v} = \|\mathbf{v}\|(\cos\theta\,\mathbf{i} + \sin\theta\,\mathbf{j})$$

Notice how closely the trigonometric form of a vector parallels the trigonometric form of a complex number. In fact, if we let $r = \|\mathbf{v}\|$, then the x- and y-components of the vector $\mathbf{v}$ are $r\cos\theta$ and $r\sin\theta$, respectively, which are precisely the forms given in Section 7.4 for the real and imaginary parts of a complex number.

In practice, we determine the direction angle from the fact that if $\mathbf{v} = v_1\mathbf{i} + v_2\mathbf{j} = \|\mathbf{v}\|(\cos\theta\,\mathbf{i} + \sin\theta\,\mathbf{j})$, then

$$\frac{v_2}{v_1} = \frac{\|\mathbf{v}\|\sin\theta}{\|\mathbf{v}\|\cos\theta} = \frac{\sin\theta}{\cos\theta} = \tan\theta$$

-----⋮◇**EXAMPLE 8**

Finding the Trigonometric Form of a Vector

Find the trigonometric form of the vector $\mathbf{v} = -3\mathbf{i} + 3\mathbf{j}$.

Solution   The magnitude of $\mathbf{v}$ is

$$\|\mathbf{v}\| = \sqrt{(-3)^2 + 3^2} = \sqrt{18} = 3\sqrt{2}$$

The direction angle θ is in the second quadrant and satisfies

$$\tan \theta = \frac{3}{-3} = -1$$

Thus, $\theta = \frac{3\pi}{4}$. The trigonometric form of $\mathbf{v}$ is

$$\mathbf{v} = \|\mathbf{v}\|(\cos \theta\, \mathbf{i} + \sin \theta\, \mathbf{j}) = 3\sqrt{2}\left(\cos \frac{3\pi}{4}\, \mathbf{i} + \sin \frac{3\pi}{4}\, \mathbf{j}\right)$$

Applications   Entire textbooks have been devoted to the applications of vectors in physics, mathematics, and engineering. Vector techniques are even applied in social sciences such as economics and psychology. Still, the seminal notion of vector arose from considering the physical quantities velocity and force. In each of the following examples, note how vector addition provides a mechanism for accounting for the combined effects of several competing influences.

-----⋮◇**EXAMPLE 9**

Resultant Velocity Vector

The British-French Concorde could fly at a speed of 1320 miles per hour. If a plane flying at this speed on a course of 120° (clockwise from north) encounters wind with a speed of 80 miles per hour in the direction 60° (clockwise from north), what will be the resultant velocity (that is, the resultant speed and direction) of the plane?

Solution   If we let $\mathbf{v}_1$ denote the velocity vector of the plane in the absence of wind, then $\mathbf{v}_1$ has magnitude 1320 and direction angle 330° as measured counterclockwise from the positive x-axis. Thus,

$$\mathbf{v}_1 = 1320(\cos 330°\, \mathbf{i} + \sin 330°\, \mathbf{j}) = 660\sqrt{3}\mathbf{i} - 660\mathbf{j}$$

Similarly, if $\mathbf{v}_2$ denotes the wind velocity vector, then $\mathbf{v}_2$ has magnitude 80 and direction angle 30°, yielding

$$\mathbf{v}_2 = 80(\cos 30°\, \mathbf{i} + \sin 30°\, \mathbf{j}) = 40\sqrt{3}\mathbf{i} + 40\mathbf{j}$$

The vectors $\mathbf{v}_1$ and $\mathbf{v}_2$ are shown in Figure 85. The resultant velocity vector $\mathbf{v}$ is the sum of $\mathbf{v}_1$ and $\mathbf{v}_2$, as shown in Figure 86. Algebraically, we have

$$\begin{aligned}
\mathbf{v} &= \mathbf{v}_1 + \mathbf{v}_2 \\
&= \left(660\sqrt{3}\mathbf{i} - 660\mathbf{j}\right) + \left(40\sqrt{3}\mathbf{i} + 40\mathbf{j}\right) \\
&= 700\sqrt{3}\mathbf{i} - 620\mathbf{j}
\end{aligned}$$

The resulting speed of the plane is

$$\|\mathbf{v}\| = \sqrt{\left(700\sqrt{3}\right)^2 + (-620)^2} \approx 1361.8 \text{ miles per hour}$$

The resulting direction angle is in the fourth quadrant and satisfies

$$\tan \theta = \frac{-620}{700\sqrt{3}}$$

British-French Concorde

Thus $\theta \approx -27.1°$, which corresponds to a course of $117.1°$.

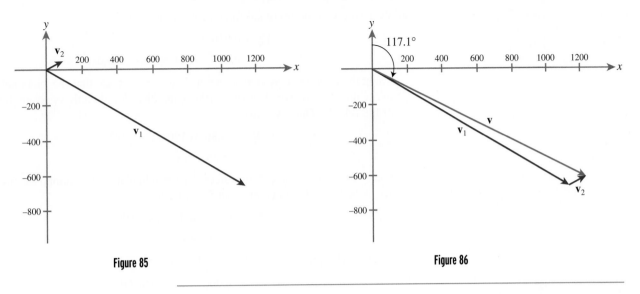

Figure 85

Figure 86

Newton's second law states that an object's acceleration is proportional to the net force acting on it. As a consequence, if an object is at rest or moving at a constant velocity, the net force acting on it must be zero. In other words, the vector sum of the external forces acting on the object must be zero. We apply this principle in the following example.

EXAMPLE 10

Towing an Expedition

An event in a strongman competition consists of tugging a 5000-pound Ford Expedition up a 10° incline, as shown in Figure 87. In addition to the pull of the strongman, there are three other forces acting on the vehicle as it moves up the ramp at a steady speed: friction **F**, which resists the vehicle's motion; the normal force **N**, an upward force directed perpendicular to the ramp (without which the vehicle would create an all-terrain sinkhole); and, of course, gravity **G**. The magnitude of the frictional force on the rolling vehicle is known to be 0.016 times the magnitude of the normal force, which is measured at 4924 pounds.

a. Express each of the forces **F**, **N**, and **G** in component form.
b. Find the magnitude of the force **S** applied by the strongman.

Figure 87

Solution

a. We begin with the frictional force **F**. We know that

$$\|\mathbf{F}\| = 0.016\|\mathbf{N}\|$$
$$= 0.016 \cdot 4924 \approx 78.8 \text{ pounds}$$

Using the coordinate system shown in Figure 88, we see that the frictional force is in the direction of $10° + 180° = 190°$—the direction directly opposite the motion of the vehicle. Thus, we have

$$\mathbf{F} = 78.8\langle \cos 190°, \sin 190° \rangle$$
$$\approx \langle -78, -14 \rangle$$

Since the normal force **N** is directed perpendicular to the ramp, its direction is $10° + 90° = 100°$. In component form, we have

$$\mathbf{N} = 4924\langle \cos 100°, \sin 100° \rangle$$
$$\approx \langle -855, 4850 \rangle$$

Since gravity pulls directly downward, we have

$$\mathbf{G} = 5000\langle \cos 270°, \sin 270° \rangle$$
$$= \langle 0, -5000 \rangle$$

b. If the vehicle is moving up the ramp at a constant speed, then the resultant of all forces acting on it must be **0**. Thus, if we let **S** denote the force applied by the yoked behemoth, then we have

$$\mathbf{F} + \mathbf{N} + \mathbf{G} + \mathbf{S} = \mathbf{0}$$

Solving for **S**, we obtain

$$\mathbf{S} = -(\mathbf{F} + \mathbf{N} + \mathbf{G})$$
$$= -(\langle -78, -14 \rangle + \langle -855, 4850 \rangle + \langle 0, -5000 \rangle)$$
$$= -\langle -933, -164 \rangle = \langle 933, 164 \rangle$$

Thus, the strongman is pulling with a force of

$$\|\mathbf{S}\| = \sqrt{933^2 + 164^2} \approx 947 \text{ pounds}$$

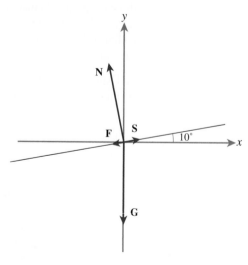

Figure 88

Understanding and Mastery Checklists

Concepts to Understand

Scalar

⋄

Directed line segment

⋄

Vector

⋄

Standard position

⋄

Component form

⋄

Magnitude of a vector

⋄

Vector addition

⋄

Scalar multiplication

⋄

Properties of vector addition and scalar multiplication

⋄

Unit vector

⋄

Standard unit vectors

⋄

Linear combination

⋄

Direction angle

⋄

Trigonometric form of a vector

Skills to Master

Find the component form of a vector.

⋄

Sketch a vector in standard position.

⋄

Find the magnitude of a vector.

⋄

Graphically illustrate vector addition
and scalar multiplication.

⋄

Algebraically perform vector addition
and scalar multiplication.

⋄

Find the unit vector in the direction
of a given vector.

⋄

Find the component form of a vector given its
magnitude and direction angle.

⋄

Find the trigonometric form of a vector given
in component form.

Exercises 7.6

Exercises 1-6 (a) *Sketch the directed line segment* $\overrightarrow{PQ}$, *(b) find the component form of the vector* **v** *represented by* $\overrightarrow{PQ}$, *and* (c) *sketch the vector* **v** *in standard position.*

1. $P(2, 5), Q(3, 7)$

2. $P(-1, 0), Q(-3, -2)$

3. $P(-3, 1), Q(2, -2)$

4. $P\left(\frac{1}{4}, \frac{1}{2}\right), Q\left(\frac{1}{2}, 0\right)$

5. $P\left(0, -\frac{1}{2}\right), Q\left(-\frac{1}{4}, 1\right)$

6. $P(6, -3), Q(11, 1)$

Exercises 7-14 *Find the magnitude of the vector.*

7. $\langle 4, 0 \rangle$

8. $\langle 3, -4 \rangle$

9. $-12\mathbf{i} - 5\mathbf{j}$

10. $4\mathbf{i} - 2\sqrt{5}\mathbf{j}$

11. $\langle 3, 5 \rangle$

12. $\langle -2, 1 \rangle$

13. $\sqrt{5}\mathbf{i} - \sqrt{7}\mathbf{j}$

14. $-\frac{4}{9}\mathbf{i} + \frac{1}{3}\mathbf{j}$

Exercises 15-20 *Use Figure 89 to sketch the graph of the indicated vector.*

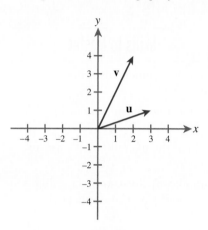

Figure 89

15. $-\mathbf{u}$ **16.** $2\mathbf{v}$

17. $\mathbf{u} + \mathbf{v}$ **18.** $2\mathbf{u} - \mathbf{v}$

19. $3\mathbf{u} - 2\mathbf{v}$ **20.** $-\mathbf{u} + 3\mathbf{v}$

Exercises 21-26 *Use Figure 90 to sketch the graph of the indicated vector.*

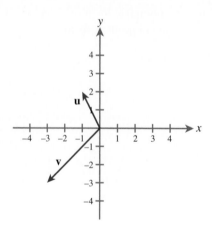

Figure 90

21. $2\mathbf{u}$ **22.** $-\dfrac{1}{3}\mathbf{v}$

23. $\mathbf{u} + \mathbf{v}$ **24.** $\mathbf{v} - \mathbf{u}$

25. $3\mathbf{u} - \mathbf{v}$ **26.** $-2\mathbf{u} + \mathbf{v}$

Exercises 27-32 *Perform the indicated operations. Express your answer in the form in which the vectors are given.*

27. $\mathbf{u} = \langle 1, 3 \rangle, \mathbf{v} = \langle 2, 5 \rangle$
   a. $-\mathbf{u}$ **b.** $\mathbf{u} + \mathbf{v}$ **c.** $-2\mathbf{u} + \mathbf{v}$

28. $\mathbf{u} = \langle -2, 2 \rangle, \mathbf{v} = \langle 3, 0 \rangle$
   a. $-\dfrac{1}{2}\mathbf{v}$ **b.** $3\mathbf{u} + \mathbf{v}$ **c.** $\dfrac{1}{2}\mathbf{u} - \dfrac{1}{3}\mathbf{v}$

29. $\mathbf{u} = 2\mathbf{i} + \mathbf{j}, \mathbf{v} = 3\mathbf{i} - \mathbf{j}$
   a. $\sqrt{2}\mathbf{v}$ **b.** $-2\mathbf{u} - \mathbf{v}$ **c.** $5\mathbf{u} + 7\mathbf{v}$

30. $\mathbf{u} = -\dfrac{1}{2}\mathbf{i} + 2\mathbf{j}, \mathbf{v} = -\mathbf{i} - \dfrac{1}{2}\mathbf{j}$
   a. $-\mathbf{u}$ **b.** $-2\mathbf{u} + 4\mathbf{v}$ **c.** $\mathbf{u} - \mathbf{v}$

31. $\mathbf{u} = \dfrac{1}{5}\mathbf{i} - \dfrac{1}{5}\mathbf{j}, \mathbf{v} = \mathbf{j}$
   a. $5\mathbf{u}$ **b.** $10\mathbf{u} - \mathbf{v}$ **c.** $\mathbf{u} + \dfrac{1}{5}\mathbf{v}$

32. $\mathbf{u} = \langle 2, 3 \rangle, \mathbf{v} = 3\langle -1, -2 \rangle$
   a. $-\dfrac{1}{3}\mathbf{v}$ **b.** $\mathbf{u} - \mathbf{v}$ **c.** $\dfrac{1}{2}(\mathbf{u} - \mathbf{v})$

Exercises 33-38 *Find the unit vector in the direction of the given vector.*

33. $\sqrt{3}\mathbf{i}$ **34.** $\langle 0, 4 \rangle$

35. $\langle -3, 4 \rangle$ **36.** $5\mathbf{i} - 12\mathbf{j}$

37. $-2\mathbf{i} - 3\mathbf{j}$ **38.** $\left\langle \dfrac{1}{2}, -\dfrac{1}{2} \right\rangle$

Exercises 39-46 *Find the component form of the vector* $\mathbf{v}$. *Assume that* θ *denotes the angle made with the positive x-axis.*

39. $\|\mathbf{v}\| = 1, \theta = \dfrac{\pi}{3}$ **40.** $\|\mathbf{v}\| = 2, \theta = 45°$

41. $\|\mathbf{v}\| = \dfrac{1}{2}, \theta = -120°$ **42.** $\|\mathbf{v}\| = 3, \theta = \pi$

43. $\|\mathbf{v}\| = 5, \theta = -\dfrac{\pi}{2}$ **44.** $\|\mathbf{v}\| = \dfrac{1}{3}, \theta = 60°$

45. $\|\mathbf{v}\| = 2, \theta = 110°$ **46.** $\|\mathbf{v}\| = 4, \theta = \dfrac{\pi}{5}$

Exercises 47-52 *Express the given vector* $\mathbf{v}$ *in the standard trigonometric form* $r(\cos\theta\mathbf{i} + \sin\theta\mathbf{j})$, *where* $r = \|\mathbf{v}\|$ *and* θ *is the angle made with the positive x-axis.*

47. $7\mathbf{i}$ **48.** $-3\mathbf{j}$

49. $\langle 2\sqrt{3}, -2 \rangle$ **50.** $\langle 2, 2 \rangle$

51. $-\dfrac{1}{8}\mathbf{i} - \dfrac{1}{8}\mathbf{j}$ **52.** $-4\mathbf{i} + 4\sqrt{3}\mathbf{j}$

Exercises 53-60   *Verify the given property of vector addition and/or scalar multiplication. Assume that* $\mathbf{u} = \langle u_1, u_2 \rangle$, $\mathbf{v} = \langle v_1, v_2 \rangle$, *and* $\mathbf{w} = \langle w_1, w_2 \rangle$, *and r and s are scalars.*

53. $(\mathbf{u} + \mathbf{v}) + \mathbf{w} = \mathbf{u} + (\mathbf{v} + \mathbf{w})$

54. $\mathbf{u} + \mathbf{0} = \mathbf{u}$

55. $\mathbf{u} + (-\mathbf{u}) = \mathbf{0}$

56. $r(\mathbf{u} + \mathbf{v}) = r\mathbf{u} + r\mathbf{v}$

57. $(r + s)\mathbf{u} = r\mathbf{u} + s\mathbf{u}$

58. $r(s\mathbf{u}) = (rs)\mathbf{u}$

59. $1\mathbf{u} = \mathbf{u}$

60. $\|r\mathbf{u}\| = |r|\|\mathbf{u}\|$

Applications

61. Airplane Velocity   An airplane with a speed in still air of 250 miles per hour is attempting to fly due north but is being blown off course by a 50-mile-per-hour wind blowing in from the northeast (toward the direction S 45° W). What is the resulting velocity (speed and direction) of the plane?

62. Cooperative Canine   A woman is walking her three dogs, holding all three leashes in one hand. The first dog pulls north with a force of 30 pounds, and the second pulls east with a force of 25 pounds. The third dog, empathizing with the plight of its mistress, pulls with the appropriate force and direction to exactly counteract the effects of its colleagues. In what direction and with how much force does the third dog pull?

63. Gulf Stream Velocity   The Gulf Stream is a warm ocean current of the northern Atlantic Ocean off North America. It has an average surface velocity of 4 miles per hour, and peak surface velocities have been known to exceed 5.5 miles per hour. Suppose an ocean liner with a velocity of 12 miles per hour on a course of 30° (clockwise from north) enters the Gulf Stream at a point where the velocity of the Gulf Stream is 4 miles per hour with a bearing of 45°. Find the resultant velocity (speed and direction) of the liner.

64. Jet Stream Velocity   The jet stream is a wind current that generally moves in a west-to-east direction at speeds often exceeding 250 miles per hour and at altitudes above 50,000 feet. Suppose an F-15 fighter with a velocity of 1200 miles per hour on a course of 145° (clockwise from north) encounters a jet stream current with a velocity of 200 miles per hour and a bearing of 80°. Find the resultant velocity (speed and direction) of the fighter.

F-15 Fighter

65. Sled Dog Team   Two sled dogs are pulling on the same sled, the first with 100 pounds of force and the second with 120 pounds of force. The first dog is pulling in the correct direction, but the second is aiming 20° off course. Find the magnitude of the net force acting on the sled. How far off course will the sled be?

66. Frictional Force   A 200-pound block is at rest on a smooth ramp that makes an angle of 30° with the horizontal. The external forces acting on the block are the force due to gravity $\mathbf{W}$, which has magnitude 200 and acts downward; the normal force $\mathbf{N}$ of the ramp pushing upward on the block with magnitude $100\sqrt{3}$ and perpendicular to the ramp; and the frictional force $\mathbf{F}$, which acts parallel to the ramp, as shown in Figure 91. Find the magnitude of the frictional force. (*Hint*: Set up a coordinate system with the origin at the block.)

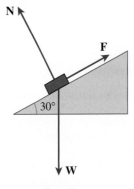

Figure 91

67. Suspended Crate   A crate is suspended by two ropes, as shown in Figure 92. The external forces acting on the crate are the force due to gravity **W**, which acts downward; the pull F_1 of the rope making an angle of 45° with the horizontal; and the pull F_2 of the rope making an angle of 30° with the horizontal. If F_1 and F_2 have magnitudes 448 and 366, respectively, find the weight of the crate; that is, find the magnitude of **W**. (*Hint:* Set up a coordinate system with the origin at the crate.)

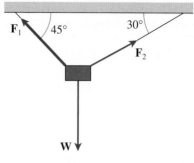

Figure 92

Concepts and Critical Thinking

Exercises 68–71   *Answer true or false.*

68. For all vectors **v**, $\|-\mathbf{v}\| = -\|\mathbf{v}\|$.

69. For all vectors **v**, $-2\mathbf{v}$ is twice as long as **v**.

70. The magnitude of the sum of two vectors is equal to the sum of the magnitudes of the vectors.

71. If $\mathbf{v} \neq \mathbf{0}$ and $\mathbf{u} = \left(\dfrac{1}{\|\mathbf{v}\|}\right)\mathbf{v}$, then $\|\mathbf{u}\| = 1$.

Exercises 72–75   *Give an example of each.*

72. A vector $\mathbf{v} = \langle a, b \rangle$, with $a \neq 0$ and $b \neq 0$ and such that $\|\mathbf{v}\| = 1$

73. A vector $\mathbf{v} = \langle a, b \rangle$, with $a \neq 0$ and $b \neq 0$ and such that $\|\mathbf{v}\| = 2$

74. Two vectors **u** and **v** such that $\|\mathbf{u} + \mathbf{v}\| < \|\mathbf{u}\| + \|\mathbf{v}\|$

75. Two vectors **u** and **v** such that $\|\mathbf{u} + \mathbf{v}\| = \|\mathbf{u}\| + \|\mathbf{v}\|$

Questions for Discussion or Essay

76. We've defined vectors as quantities having both magnitude and direction. But we've also seen that vectors (in the plane) can be represented as an ordered pair of numbers $\langle a, b \rangle$. In fact, a loose definition of a vector is "a quantity requiring two or more numbers for a full description." According to this definition, which of the following human traits are best quantified by vectors and which by scalars? Support your answers.

    a. SAT performance **b.** Height   **c.** Beauty

    d. Weight   **e.** Strength   **f.** Blood pressure

    g. Age

77. We have seen trigonometric form in three different contexts: polar coordinates, complex numbers, and now vectors. Discuss the similarities and differences among the three.

78. In Example 9, we assumed that the velocity vector of the Concorde would be the vector sum of the prevailing wind and the velocity the plane would have in the absence of wind. What factors determine how the flight of an airplane depends on air currents? Is it legitimate to assume that the air blows at a constant speed in a constant direction? How fast would the airplane travel if there were no air whatsoever?

Projects for Enrichment

79. Dot Product and Work   In this section, we introduced two operations on vectors—namely, vector addition and scalar multiplication. We now consider a third operation on vectors, the dot product. The **dot product** of two vectors $\mathbf{u} = \langle u_1, u_2 \rangle$, *and* $\mathbf{v} = \langle v_1, v_2 \rangle$ is denoted by $\mathbf{u} \cdot \mathbf{v}$ and is defined by

$$\mathbf{u} \cdot \mathbf{v} = u_1 v_1 + u_2 v_2$$

Thus, for example, the dot product of $\mathbf{u} = \langle -3, 2 \rangle$ and $\mathbf{v} = \langle 1, -4 \rangle$ is

$$\mathbf{u} \cdot \mathbf{v} = (-3)1 + 2(-4) = -11$$

Note that the dot product of two vectors is a scalar, not a vector.

    a. Find the dot product of the given pair of vectors.

        i. $\mathbf{u} = \langle 2, -1 \rangle$, $\mathbf{v} = \langle 5, 7 \rangle$

        ii. $\mathbf{u} = \langle -4, 3 \rangle$, $\mathbf{v} = \langle 2, -2 \rangle$

        iii. $\mathbf{u} = \langle 3, -2 \rangle$, $\mathbf{v} = \langle 6, 9 \rangle$

The dot product has a number of important applications, one of which is finding the angle between vectors. The angle between two nonzero vectors **u** and **v** is defined to be the angle θ with $0 \leq \theta \leq \pi$, which is formed when the vectors are in standard position, as shown in Figure 93. Using the Law of Cosines, we can show that θ satisfies

(1)
$$\cos \theta = \frac{\mathbf{u} \cdot \mathbf{v}}{\|\mathbf{u}\| \|\mathbf{v}\|}$$

To see this, consider the two vectors $\mathbf{u} = \langle u_1, u_2 \rangle$ and $\mathbf{v} = \langle v_1, v_2 \rangle$ shown in Figure 93. The vector $\mathbf{u} - \mathbf{v}$ is the vector whose initial point is the terminal point of **v** and whose terminal point is the terminal point of **u**. Now consider the triangle ABC formed by **u**, **v**, and $\mathbf{u} - \mathbf{v}$, as shown in Figure 94. Note that the side lengths are given by $\|\mathbf{u} - \mathbf{v}\|$, $\|\mathbf{v}\|$, and $\|\mathbf{u}\|$, and θ is the angle opposite the side with length $\|\mathbf{u} - \mathbf{v}\|$. Thus, we may apply the Law of Cosines to obtain

$$\|\mathbf{u} - \mathbf{v}\|^2 = \|\mathbf{u}\|^2 + \|\mathbf{v}\|^2 - 2\|\mathbf{u}\| \|\mathbf{v}\| \cos \theta$$

b. Express $\|\mathbf{u}\|^2$, $\|\mathbf{v}\|^2$, and $\|\mathbf{u} - \mathbf{v}\|^2$ in terms of u_1, u_2, v_1, and v_2, and solve for $\cos \theta$. Then simplify to obtain equation (1).

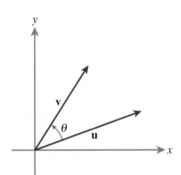

Figure 93

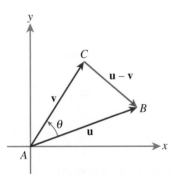

Figure 94

To find the angle θ between the two vectors $\mathbf{u} = \langle -3, 2 \rangle$ and $\mathbf{v} = \langle 1, -4 \rangle$, we proceed as follows:

$$\cos \theta = \frac{\mathbf{u} \cdot \mathbf{v}}{\|\mathbf{u}\| \|\mathbf{v}\|} = \frac{-11}{\sqrt{13}\sqrt{17}} = \frac{-11}{\sqrt{221}}$$

$$\theta = \cos^{-1}\left(\frac{-11}{\sqrt{221}}\right) \approx 137.73°$$

c. Find the angle between each pair of vectors.

   i. $\mathbf{u} = \langle 2, -1 \rangle$, $\mathbf{v} = \langle 5, 7 \rangle$

   ii. $\mathbf{u} = \langle -4, 3 \rangle$, $\mathbf{v} = \langle 2, -2 \rangle$

   iii. $\mathbf{u} = \langle 3, -2 \rangle$, $\mathbf{v} = \langle 6, 9 \rangle$

d. What geometric relationship holds for the vectors in part c, iii? State a general rule that relates the dot product to this relationship.

Another important application of the dot product is in computing work done by a constant force. When an object is moved a distance of d feet by a constant force of magnitude F applied in the direction of motion, the work done by the force is defined to be the product Fd. For example, the work done in lifting a 50-pound box of books to a height of 3 feet is

$$W = Fd = (50 \text{ pounds})(3 \text{ feet}) = 150 \text{ foot-pounds}$$

More generally, the work done by a constant force **F** in moving an object from some point P to some point Q, regardless of whether **F** is applied in the direction of motion, is given by

$$W = \mathbf{F} \cdot \overrightarrow{PQ}$$

For example, suppose you are pulling an object along flat ground by a rope attached to the front of the object. If you are pulling with a force of 80 pounds at an angle of 30° with the horizontal, then the force is given by

$$\mathbf{F} = 80(\cos 30°\mathbf{i} + \sin 30°\mathbf{j}) = 40\sqrt{3}\mathbf{i} + 40\mathbf{j}$$

If you move the object a horizontal distance of 10 feet, say from $P(0, 0)$ to $Q(10, 0)$, the *displacement vector* is

$$\overrightarrow{PQ} = 10\mathbf{i}$$

The work is found by computing the dot product of the force vector and the displacement vector

$$\begin{aligned} W &= \mathbf{F} \cdot \overrightarrow{PQ} \\ &= \left(40\sqrt{3}\mathbf{i} + 40\mathbf{j}\right) \cdot (10\mathbf{i}) \\ &= \left(40\sqrt{3}\right)(10) \\ &\approx 693 \text{ foot-pounds} \end{aligned}$$

e. Find the work done by a child pulling a wagon 100 feet along flat ground with a constant force of 50 pounds if the wagon handle makes an angle of 45° with the horizontal.

f. Two lumberjacks are pulling a log along flat horizontal ground by ropes attached to the front of the log. One is pulling with a force of 80 pounds at an angle of 30° with the horizontal; the other is pulling with a force of 100 pounds at an angle of 45°. Find the work done in moving the log a distance of 50 feet. (*Hint:* First find the resultant force.)

Chapter 7 Review

Exercises 1-4 *Use the Law of Sines to solve the given triangle. Assume angles A, B, and C and sides a, b, and c are labeled as shown in Figure 95.*

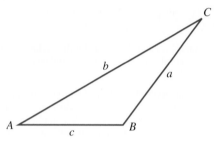

Figure 95

1. $c = 12, A = 47°, C = 52°$

2. $a = 25, B = 39°, C = 112°$

3. $a = 1.2, b = 1, B = 42°$

4. $a = 100, b = 55, A = 140°$

Exercises 5-8 *Use the Law of Cosines to solve the given triangle. Assume angles A, B, and C and sides a, b, and c are labeled as shown in Figure 95.*

5. $b = 19, c = 15, A = 10°$

6. $a = 5.8, b = 3.4, C = 64°$

7. $a = 5, b = 8, c = 9$

8. $a = 8, c = 12, A = 50°$

Exercises 9-12 *Solve the given triangle by any means. Assume angles A, B, and C and sides a, b, and c are labeled as shown in Figure 95.*

9. $a = 335, b = 260, c = 540$

10. $a = 6, A = 133°, B = 24°$

11. $a = 2, b = 7, A = 41°$

12. $b = 18, c = 16, C = 30°$

Exercises 13-16 *Solve the given triangle by any means.*

13.

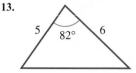

14.

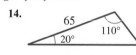

15.

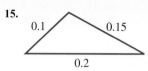

16.

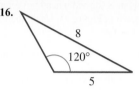

Exercises 17-20 *Find the area of the given triangle by any means.*

17.

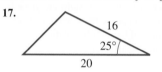

18.

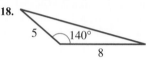

19.

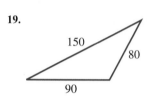

20.

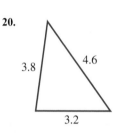

Exercises 21-24 *Find the modulus of the given complex number.*

21. $4 - 3i$

22. $\sqrt{2} + \sqrt{2}i$

23. $-9 + i$

24. $\frac{1}{2} - \frac{3}{2}i$

Exercises 25-28 *Represent the complex number graphically and write it in trigonometric form.*

25. $-2 - 2i$

26. $\sqrt{3} - i$

27. $\frac{3}{2}i$

28. $-2 + i$

Exercises 29-32 *Represent the complex number graphically and write it in the form a + bi.*

29. $2(\cos 330° + i \sin 330°)$

30. $3\left(\cos \frac{5\pi}{4} + i \sin \frac{5\pi}{4}\right)$

31. $\sqrt{3}\left(\cos \frac{\pi}{3} + i \sin \frac{\pi}{3}\right)$

32. $\frac{5}{2}(\cos 270° + i \sin 270°)$

Exercises 33-36 *Perform the indicated operation. Express the answer in trigonometric form.*

33. $5\left(\cos \frac{\pi}{3} + i \sin \frac{\pi}{3}\right) \cdot 2\left(\cos \frac{\pi}{4} + i \sin \frac{\pi}{4}\right)$

34. $\sqrt{3}(\cos 51° + i \sin 51°) \cdot \sqrt{27}(\cos 102° + i \sin 102°)$

35. $\dfrac{12(\cos 300° + i \sin 300°)}{4(\cos 100° + i \sin 100°)}$

36. $\dfrac{2.4(\cos \pi + i \sin \pi)}{0.8\left(\cos \frac{\pi}{6} + i \sin \frac{\pi}{6}\right)}$

Exercises 37-40 *Perform the indicated operation by first converting to trigonometric form. Check your answer by performing the operation without converting.*

37. $7i(-3 + 3i)$

38. $(2 - 2i)(\sqrt{3} + i)$

39. $\dfrac{-5 + 5\sqrt{3}i}{\sqrt{3} + i}$

40. $\dfrac{-3}{2 + 3i}$

Exercises 41-44 *Use de Moivre's Theorem to expand the given power. Express the answer in the same form used in the statement of the problem.*

41. $[5(\cos 10° + i \sin 10°)]^4$

42. $\left[\sqrt{2} \left(\cos \dfrac{\pi}{4} + i \sin \dfrac{\pi}{4} \right) \right]^{10}$

43. $\left(1 - \sqrt{3}i\right)^5$

44. $(1 - i)^8$

Exercises 45-48 *Find the indicated roots, express them in the form $a + bi$, and represent them graphically.*

45. Square roots of $4(\cos 120° + i \sin 120°)$

46. Third roots of $8i$

47. Fourth roots of 81

48. Sixth roots of -1

Exercises 49-50 *Find the indicated roots. Express your answer in the form $a + bi$, with three decimal places of accuracy.*

49. Square roots of $-3 + 4i$

50. Cube roots of $4 - 4i$

Exercises 51-52 *Find all solutions of the equation and represent them graphically.*

51. $x^3 + 1 = 0$

52. $x^4 - i = 0$

Exercises 53-54 *Plot the point given in polar coordinates and find the corresponding rectangular coordinates.*

53. $(2; \pi)$

54. $\left(-4; \dfrac{5\pi}{3} \right)$

Exercises 55-56 *Plot the point given in rectangular coordinates and find two sets of polar coordinates with $0 \le \theta < 2\pi$.*

55. $(-3, 3)$

56. $\left(4\sqrt{3}, 4 \right)$

Exercises 57-62 *Find and plot at least four points satisfying the given polar equation, and then sketch the graph of the equation.*

57. $r = 5$

58. $\theta = -\dfrac{\pi}{4}$

59. $r = 3\theta, \theta \ge 0$

60. $r = \cos \theta$

61. $r = 3 \sin 2\theta$

62. $r = 2 \cos \theta - 1$

Exercises 63-66 *Convert the given rectangular equation to polar form.*

63. $y = 2x$

64. $x = 5$

65. $x^2 + y^2 = 1$

66. $\dfrac{x^2}{4} - \dfrac{y^2}{9} = 1$

Exercises 67-70 *Convert the given polar equation to rectangular form.*

67. $r = -1$

68. $\theta = \dfrac{\pi}{2}$

69. $r = 3 \sin \theta$

70. $r^2 = \tan \theta$

Exercises 71-72 *Use a graphing calculator to plot the graph of the given polar equations. Estimate the coordinates of the point(s) of intersection.*

71. $r = 3$
      $r = 4 \cos \theta$

72. $r = 3 \sin \theta$
      $r = 1 - 4 \cos \theta$

Exercises 73-74 *(a) Sketch the directed line segment $\overrightarrow{PQ}$, (b) find the component form of the vector $\mathbf{v}$ represented by $\overrightarrow{PQ}$, and (c) sketch the vector $\mathbf{v}$ in standard position.*

73. $P(4, -2), Q(-1, -1)$

74. $P\left(-\dfrac{7}{2}, 3 \right), Q\left(-\dfrac{3}{2}, 1 \right)$

Exercises 75-76 *Find the magnitude of the vector.*

75. $\langle -6, 8 \rangle$

76. $4\mathbf{i} - \mathbf{j}$

Exercises 77-78 *Use Figure 96 to sketch the graph of the indicated vector.*

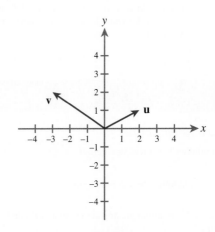

Figure 96

77. $\mathbf{u} - \mathbf{v}$

78. $3\mathbf{u} + \mathbf{v}$

Exercises 79-80 *Perform the indicated operations. Express your answer in the same form in which the vectors are given.*

79. $\mathbf{u} = \langle -4, 2 \rangle, \mathbf{v} = \langle -2, 3 \rangle$

   a. $-2\mathbf{u}$ **b.** $\mathbf{u} - 2\mathbf{v}$ **c.** $\dfrac{1}{2}\mathbf{u} + \dfrac{1}{2}\mathbf{v}$

80. $\mathbf{u} = \mathbf{i} - 3\mathbf{j}, \mathbf{v} = 2\mathbf{i} + 5\mathbf{j}$

   a. $\dfrac{1}{3}\mathbf{u}$ **b.** $\mathbf{v} - \mathbf{u}$ **c.** $-2\mathbf{u} - \mathbf{v}$

Exercises 81-82 *Find the unit vector in the direction of the given vector.*

81. $12\mathbf{i} + 5\mathbf{j}$

82. $\langle \sqrt{3}, -1 \rangle$

Exercises 83-84 *Find the component form of the vector* **v**. *Assume that θ denotes the angle made with the positive x-axis.*

83. $\|\mathbf{v}\| = 2, \theta = \dfrac{\pi}{6}$

84. $\|\mathbf{v}\| = \dfrac{1}{2}, \theta = 150°$

Exercises 85-86 *Write the given vector* **v** *in the standard trigonometric form* $r(\cos\theta\mathbf{i} + \sin\theta\mathbf{j})$, *where* $r = \|\mathbf{v}\|$ *and θ is the angle made with the positive x-axis.*

85. 4**j**

86. $-2\mathbf{i} - 2\sqrt{3}\mathbf{j}$

87. Estimating Travel Time A boater spots a lighthouse at a bearing of N 38° E. He then sails due east for 10 miles, at which time the lighthouse is at a bearing of N 47° W. If his boat is capable of 30 miles per hour, how long will it take him to reach the lighthouse if he heads directly for it?

88. Estimating Distance A child is looking through a telescope on the observation deck of a skyscraper. She spots her home, which is 18 miles from the skyscraper, and then swings the telescope 22.4° and sights her school, which is 14 miles from the skyscraper. How far is the girl's school from her home?

89. Estimating Speed A nature photographer is traveling on an overhead tram that moves parallel to the ground at a constant speed. She spots a motionless elk directly below the tram cable in front of her at an angle of depression of 30°. Two minutes later, the (still motionless) elk lies in front of her at an angle of depression of 34°, and, at this point, she estimates the distance to the elk to be 2 miles. How fast is the tram moving?

90. Finding a Pyramid Angle A man standing 200 feet from the base of a pyramid spots his friend who has traveled 130 feet up the side of the pyramid. He estimates that his friend is 300 feet away. What angle does the face of the pyramid make with the ground?

Chapter 7 Test

Problems 1-10 *Answer true or false.*

1. Every real number has a real square root.

2. Every complex number has a complex square root.

3. If the three angles of a triangle are known, then the side lengths can be determined.

4. If the three side lengths of a triangle are known, then the angles can be determined.

5. The magnitude of a vector is itself a vector.

6. Trigonometric form is more convenient than standard form when adding complex numbers.

7. The length of the sum of two vectors is the sum of the lengths of the vectors.

8. For any nonzero complex number w, there are exactly three complex numbers z for which $z^3 = w$.

9. If the point (x, y) in rectangular coordinates corresponds to the point $(r; \theta)$ in polar coordinates, then it follows that $\theta = \arctan\frac{y}{x}$.

10. The graph of $r = \theta$ crosses the graph of $\theta = \frac{\pi}{4}$ once.

Problems 11-14 *Give an example of each.*

11. Side lengths of a triangle having angles 60° and 70°

12. A cube root of i

13. Vectors **u** and **v** such that $\|\mathbf{u}\| = \|\mathbf{v}\| = 2$ and $\mathbf{u} + \mathbf{v} = \mathbf{0}$

14. A complex number with argument $\theta = \dfrac{3\pi}{4}$

Problems 15-16 *Solve the given triangle. Assume angles A, B, and C and sides a, b, and c are labeled as shown in Figure 97.*

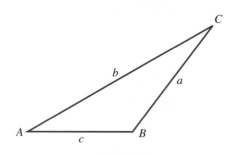

Figure 97

15. $b = 12, c = 8, A = 55°$

16. $a = 3, c = 4, C = 62°$

17. Let $z = -1 - i$.

    a. Represent z graphically and write it in trigonometric form.

    b. Compute z^6.

    c. Approximate the square roots of z to three decimal places.

18. Let $\mathbf{u} = \langle 2, 4 \rangle$ and $\mathbf{v}$ be a vector with magnitude $\sqrt{2}$ and direction angle $\theta = \frac{\pi}{4}$.

    a. Express $\mathbf{v}$ in the form $a\mathbf{i} + b\mathbf{j}$.

    b. Compute $2\mathbf{u} - 3\mathbf{v}$.

19. Find and plot at least four points satisfying $r = \cos 2\theta$, and then sketch the graph of the equation.

20. A ship leaves a certain port at 9:00 A.M. and sails on a course of 30° (clockwise from north) at a rate of 15 miles per hour. A second ship leaves the same port at 9:30 A.M and sails on a course of 100° at a rate of 12 miles per hour. How far apart are the two ships at 11:00 A.M?

Chapter 8

Relations and Conic Sections

The Arecibo radio telescope located near Arecibo, Puerto Rico, is the largest single telescope of any kind. Electromagnetic transmissions from celestial objects are reflected from the 1000-foot-diameter dish to the antenna suspended above it. Radio telescopes, although employed for such exotic purposes as identifying black holes and quasars, determining the chemical composition of distant galaxies, and even searching for extraterrestrial life, are based on the same principle as the automobile headlight: the reflective properties of conic sections.

8.1
Relations and Their Graphs

8.2
Parabolas

8.3
Ellipses

8.4
Hyperbolas and Classification
of Conic Sections

8.5
Rotation of Axes and General
Conic Sections

8.6
Parametric Equations

Section 8.1 | Relations and Their Graphs

- How can symmetry be used in data compression?
- What does the message "ti bib dod" mean to Inspector Magill?
- How can you construct a face with two left sides?
- In what sense can one relation be the "mirror image" of another?
- What types of symmetry are possessed by the human body?

Throughout the text, we have considered equations of many different varieties and in many different contexts. We have studied equations of lines, such as $2x + y = 3$; equations of circles, such as $x^2 + y^2 = 9$; and equations that define functions, such as $y = e^{3x}$. We have also encountered equations arising from real-world contexts; examples include $A = \pi r^2$, $B = P(1 + r)^t$, and $F = \frac{9}{5}C + 32$. All of these equations are examples of **relations**.

An equation involving only the variables x and y, such as $2x + y = 3$, $x^2 + y^2 = 9$, or $y = e^{3x}$, defines a **relation in x and y**. In the case where the equation defines a function, we have many tools and techniques at our disposal for studying the graph of the equation, not the least of which is the graphing calculator. You may recall, however, that not all equations in x and y define functions. For example, the graph of $x^2 + y^2 = 9$ is a circle, which fails the vertical line test. It follows that the graph of $x^2 + y^2 = 9$ could not be the graph of a function $y = f(x)$.

In this section, we generalize some of the graphing techniques developed in Chapter 2 for functions and apply them to more general relations.

Translations

Recall from Section 2.2 that the graph of $y = f(x - h) + k$ or, equivalently, $y - k = f(x - h)$, is a translation, h units horizontally and k units vertically, of the graph of $y = f(x)$. In other words, replacing x with $x - h$ and y with $y - k$ in the equation $y = f(x)$ has the effect of translating the graph h units horizontally and k units vertically. This same result holds for relations in general. Consider, for example, the graphs of $x^2 + y^2 = 9$ and $(x - 1)^2 + (y - 2)^2 = 9$ shown in Figures 1 and 2. Both are circles of radius 3, but the first is centered at the origin and the second at $(1, 2)$. Thus, the second graph is a translation of the first, 1 unit to the right and 2 units up.

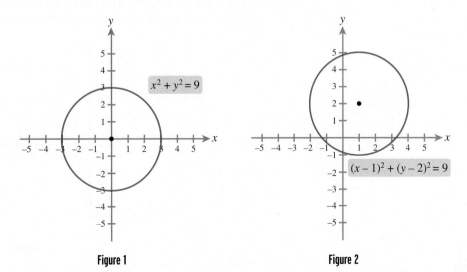

Figure 1   Figure 2

In general, substituting $x - h$ for x results in a translation of the graph h units to the right. For example, if we substitute $x - 2$ for x, the graph will be translated 2 units to the right; if we substitute $x + 3$ for x, the graph will be translated -3 units to the right (3 units to the left), and so on. Similarly, substituting $y - k$ for y translates the graph k units upward, whereas substitution of $y + k$ for y translates the graph k units downward.

Translations

Substitution	Resulting translation
$x - h$ in place of x	h units to the right
$x + h$ in place of x	h units to the left
$y - k$ in place of y	k units up
$y + k$ in place of y	k units down

⋯⋯⋗EXAMPLE 1

Translating a Parabola

The graph of $y = x^2$ is given in Figure 3. Use it to graph $y + 4 = (x - 7)^2$.

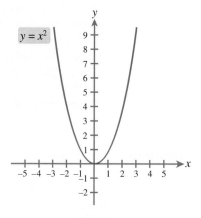

$y = x^2$

Figure 3

Solution The relation $y + 4 = (x - 7)^2$ is the result of substituting $x - 7$ for x and $y + 4$ for y. Thus, its graph can be obtained by translating the graph of $y = x^2$ to the right 7 units and down 4 units, as shown in Figure 4.

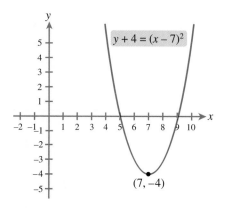

$y + 4 = (x - 7)^2$

$(7, -4)$

Figure 4

·····⋗**EXAMPLE 2**   **Graphing the Translation of a Relation**

Use the graph of $x^2/4 + y^2/9 = 1$ shown in Figure 5 to graph

$$\frac{(x+2)^2}{4} + \frac{(y-1)^2}{9} = 1$$

Solution The relation $\dfrac{(x+2)^2}{4} + \dfrac{(y-1)^2}{9} = 1$ is obtained from $\dfrac{x^2}{4} + \dfrac{y^2}{9} = 1$ by substituting $x + 2$ for x and $y - 1$ for y. Thus, the desired graph can be found by translating the graph of $\dfrac{x^2}{4} + \dfrac{y^2}{9} = 1$, as shown in Figure 6, 2 units to the left and 1 unit up.

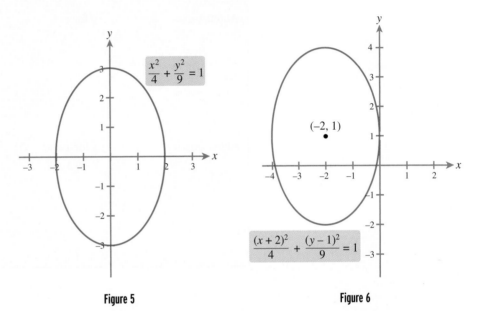

Figure 5   **Figure 6**

Reflections From our work in Section 2.2, we know that the graph of $y = f(-x)$ can be obtained by reflecting the graph of $y = f(x)$ about the y-axis. The same is true for graphs of relations. If we substitute $-x$ for x, the graph is reflected about the y-axis. Similarly, substitution of $-y$ for y, will cause a reflection about the x-axis. Reflection about the origin is defined by reflecting about both the x- and y-axes, and it is achieved by substituting both $-x$ for x and $-y$ for y in a relation. These observations can be confirmed by considering the changes that occur in the coordinates of a point (x, y) upon reflection, as suggested by Figure 7.

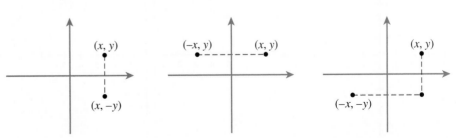

Reflection about the x-axis   **Reflection about the y-axis**   **Reflection about the origin**

Figure 7

Reflections

Type of reflection	Substitution	Graphical illustration
About y-axis	$-x$ for x	
About x-axis	$-y$ for y	
About origin	$-x$ for x and $-y$ for y	

EXAMPLE 3

Using Reflections to Find Relations

The graph of $x^2 + y^2 - 4x - 6y + 9 = 0$ is given in Figure 8. Parts a, b, and c each show a reflection of this graph. Use this information to find the equation corresponding to each of the reflected graphs.

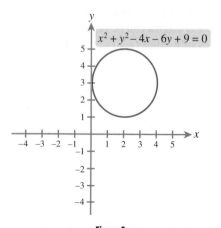

$x^2 + y^2 - 4x - 6y + 9 = 0$

Figure 8

a.

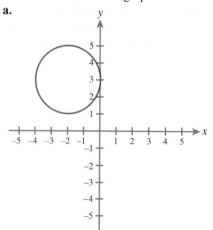

b.

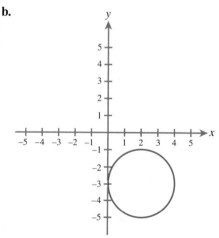

c.

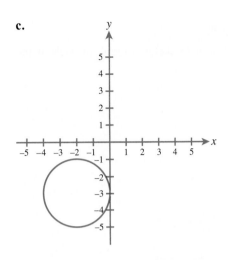

Solution

a. This is a reflection about the y-axis of the original graph. Thus, its equation is obtained by replacing x with $-x$.

$$x^2 + y^2 - 4x - 6y + 9 = 0 \qquad \text{The original relation}$$
$$(-x)^2 + y^2 - 4(-x) - 6y + 9 = 0 \qquad \text{Replacing } x \text{ with } -x$$
$$x^2 + y^2 + 4x - 6y + 9 = 0 \qquad \text{Simplifying}$$

b. This is a reflection about the x-axis of the original graph. Thus, its equation is obtained by replacing y with $-y$.

$$x^2 + y^2 - 4x - 6y + 9 = 0 \qquad \text{The original relation}$$
$$x^2 + (-y)^2 - 4x - 6(-y) + 9 = 0 \qquad \text{Replacing } y \text{ with } -y$$
$$x^2 + y^2 - 4x + 6y + 9 = 0 \qquad \text{Simplifying}$$

c. This graph has been obtained by reflecting the original graph about the origin. We find its equation by replacing x with $-x$ and y with $-y$.

$$x^2 + y^2 - 4x - 6y + 9 = 0 \qquad \text{The original relation}$$
$$(-x)^2 + (-y)^2 - 4(-x) - 6(-y) + 9 = 0 \qquad \text{Replacing } x \text{ with } -x \text{ and } y \text{ with } -y$$
$$x^2 + y^2 + 4x + 6y + 9 = 0 \qquad \text{Simplifying}$$

Certain graphs can be obtained by recognizing them as reflections of known graphs, as illustrated in the following examples.

EXAMPLE 4

Graphing a Reflection

The graph of $y = x^2 + 3$ is given in Figure 9. Use it to graph $y = -x^2 - 3$.

Solution We can rewrite $y = -x^2 - 3$ as $-y = x^2 + 3$. This relation can be obtained from $y = x^2 + 3$ by substituting $-y$ for y. Thus, its graph is a reflection about the x-axis of the graph of $y = x^2 + 3$; that is, we simply flip the graph of $y = x^2 + 3$ upside down, as shown in Figure 10.

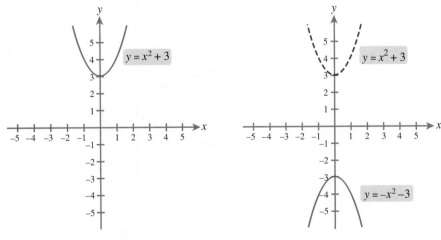

Figure 9   **Figure 10**

············>EXAMPLE 5

Graphing a Combination of Reflections and Translations

The graph of $x = y^2$ is given in Figure 11. Use it to sketch a graph of each of the following:

a. $x = -y^2$

b. $x - 3 = -(y + 2)^2$

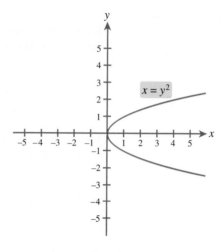

Figure 11

Solution

a. The equation $x = -y^2$ is equivalent to $-x = y^2$, and so it can be obtained from the equation $x = y^2$ by replacing x with $-x$. Thus, the graph of $x = -y^2$ can be obtained from the graph of $x = y^2$ by reflecting about the y-axis, as shown in Figure 12.

b. The equation $x - 3 = -(y + 2)^2$ can be obtained from $x = -y^2$, the equation in part a, by substituting $x - 3$ in place of x and $y + 2$ in place of y. Thus, we translate the graph from part a, as shown in Figure 13, 3 units to the right and 2 units down.

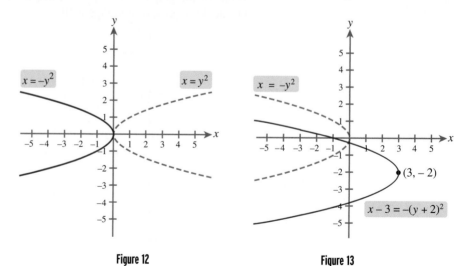

Figure 12

Figure 13

Symmetry   As we have seen, the reflection about the y-axis of the graph of a relation can be obtained by substituting $-x$ for x. Now suppose that we wish to reflect the graph of $y = x^2$ about the y-axis. If we substitute $-x$ for x, we obtain $y = (-x)^2 = x^2$. Thus, the reflection about the y-axis of the graph of $y = x^2$ is itself! Whenever the graph of a relation remains the same after undergoing some transformation (such as a reflection), we say that the graph is **symmetric** with respect to that transformation. Thus, we say that the graph of $y = x^2$ is symmetric with respect to reflection about the y-axis, or simply **symmetric with respect to the y-axis**.

The symmetry of the graph of $y = x^2$ with respect to the y-axis can be seen graphically as well. As Figure 14 shows, reflecting the graph of $y = x^2$ about the y-axis leaves the graph unchanged. This is because the right-hand portion of the graph is just a mirror image of the left-hand portion.

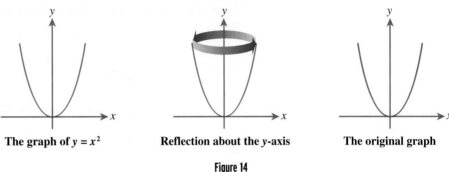

The graph of $y = x^2$ Reflection about the y-axis   The original graph

Figure 14

More generally, the graph of a relation will be symmetric with respect to the y-axis if reflecting about the y-axis leaves the graph unchanged. Algebraically, this means that the relation resulting from substituting $-x$ for x is equivalent to the original relation. In other words, $(-x, y)$ is on the graph of the relation whenever (x, y) is on the graph. Geometrically, this means that the left- and right-hand sides of the graph are mirror images of one another. Figure 15 shows several graphs that are symmetric with respect to the y-axis.

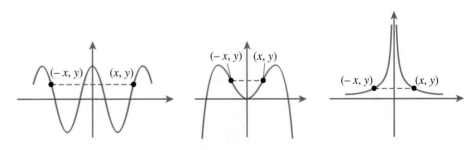

Figure 15   Graphs symmetric with respect to the y-axis

> **EXAMPLE 6**   **Testing for Symmetry with Respect to the y-axis**
>
> Without graphing, indicate whether the relation $x^2 + 2y^2 = 4y - x^4$ is symmetric with respect to the y-axis.

Solution We replace x by $-x$ to test for symmetry with respect to the y-axis.

$$x^2 + 2y^2 = 4y - x^4 \qquad \text{The original relation}$$
$$(-x)^2 + 2y^2 = 4y - (-x)^4 \qquad \text{Substituting } -x \text{ for } x$$
$$x^2 + 2y^2 = 4y - x^4 \qquad \text{Simplifying}$$

Since the resulting relation is equivalent to the original, the graph of $x^2 + 2y^2 = 4y - x^4$ is symmetric with respect to the y-axis.

In a similar fashion, the graph of a relation is said to be **symmetric with respect to the x-axis** if the graph is unchanged after reflecting about the x-axis. Algebraically, this means that substituting $-y$ for y in the relation yields an equivalent relation. Figure 16 shows the graphs of several relations that are symmetric with respect to the x-axis.

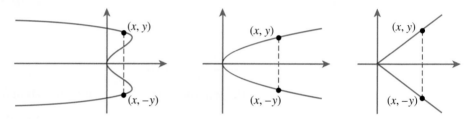

Figure 16 Graphs symmetric with respect to the x-axis

If the graph of a relation stays the same after reflection about the origin, then it is said to be **symmetric with respect to the origin**. Thus, the graph of a relation will be symmetric with respect to the origin if substitution of $-x$ for x *and* $-y$ for y yields an equivalent equation. Figure 17 shows the graphs of some relations that are symmetric with respect to the origin.

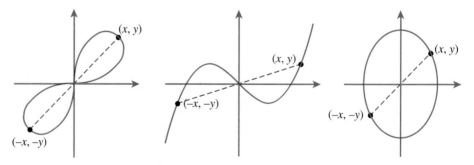

Figure 17 Graphs symmetric with respect to the origin

Symmetry can be a powerful tool for graphing relations. If we are aware of the symmetries that a graph possesses, then we can use one portion of the graph to construct the entire graph. The following examples illustrate this technique.

·····→EXAMPLE 7   **Using Symmetry to Complete the Graph of a Relation**

A portion of the graph of $\dfrac{x^2}{16} + \dfrac{y^2}{9} = 1$ is shown in Figure 18. Use symmetry to complete the graph.

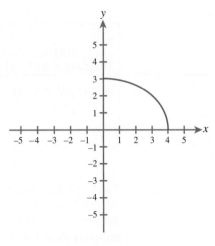

Figure 18

Solution   We begin by noting that if either x is replaced by $-x$ or y is replaced by $-y$, we will obtain an equivalent relation. Thus, the graph must be symmetric with respect to the x-axis and the y-axis. Since the graph is symmetric with respect to the y-axis, the left-hand portion and the right-hand portion will be mirror images of one another about the y-axis. Reflecting the given portion of the graph about the y-axis, we obtain Figure 19. Finally, since the graph is symmetric with respect to the x-axis, the bottom portion of the graph will be a mirror image of the top portion. Reflecting about the x-axis gives us the final graph in Figure 20.

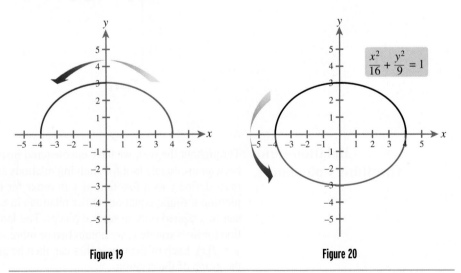

Figure 19   **Figure 20**

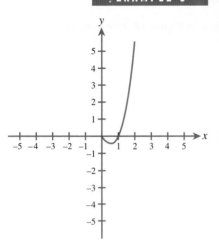

Figure 21

EXAMPLE 8

Using Symmetry to Complete the Graph of a Relation

A portion of the graph of $y = x^3 - x$ is shown in Figure 21. Use symmetry to complete the graph.

Solution We test $y = x^3 - x$ for the three types of symmetry as follows.

y-axis (Replace x with –x)	x-axis (Replace y with –y)	Origin (Replace x with –x and y with –y)
$y = (-x)^3 - (-x)$	$(-y) = x^3 - x$	$(-y) = (-x)^3 - (-x)$
$y = -x^3 + x$	$y = -x^3 + x$	$-y = -x^3 + x$
		$y = x^3 - x$

Since only the test for symmetry with respect to the origin yields the same equation, the relation is symmetric with respect to the origin but not to the y-axis or x-axis. To obtain the graph, we reflect about the origin by first reflecting about the y-axis to obtain the red curve shown in Figure 22 and then reflecting the red curve about the x-axis to obtain the final graph, shown in Figure 23.

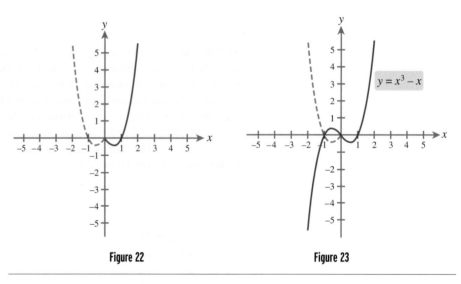

Figure 22 **Figure 23**

Relations and Graphing Calculators

Throughout the text, we have encountered instances in which a graphing calculator has been an invaluable tool for studying relations and their graphs. Unfortunately, a relation must define y as a function of x in order for its graph to be obtained by entering and plotting a single equation. Other relations in x and y either cannot be graphed at all or can be graphed only in several pieces. The latter case arises when, in solving an equation for the variable y, we obtain two or more solutions—called **branches**—of the form $y = f(x)$. Each of these branches can then be graphed, and the union of the graphs gives the graph of the relation.

··········⟩**EXAMPLE 9**   **Plotting a Relation with a Graphing Calculator**

Use a graphing calculator to plot the graph of $4x^2 + 9y^2 = 36$, and determine the highest and lowest points on the graph.

Solution   We begin by solving for y.

$$4x^2 + 9y^2 = 36$$
$$9y^2 = 36 - 4x^2$$
$$y^2 = 4 - \frac{4}{9}x^2$$
$$y = \pm\sqrt{4 - \frac{4}{9}x^2}$$

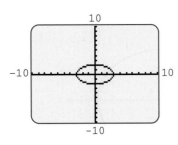

Figure 24

Thus, there are two branches, $y = \sqrt{4 - \frac{4}{9}x^2}$ and $y = -\sqrt{4 - \frac{4}{9}x^2}$. The combined graph of these branches forms the ellipse shown in Figure 24. Note that the top half is the graph of $y = \sqrt{4 - \frac{4}{9}x^2}$ and the bottom half is that of $y = -\sqrt{4 - \frac{4}{9}x^2}$. The highest and lowest points lie on the y-axis at the points $(0, 2)$ and $(0, -2)$, respectively. We will study ellipses in detail in Section 8.3.

Understanding and Mastery Checklists

Concepts to Understand	Skills to Master
Relation	Given the graph of a relation, sketch the graph of its translation.
✧	✧
Translation	Given the graph of a relation, sketch the graph of its reflection.
✧	✧
Reflection	Given a relation and its graph, together with a translation of its graph, find an equation for the translated graph.
✧	✧
Symmetry	Given a relation and its graph, together with a reflection of its graph, find an equation for the reflected graph.
	✧
	Determine the symmetries of a relation.
	✧
	Use symmetry as an aid in graphing a relation.

Exercises 8.1

Exercises 1–4 *A relation and its graph are given. Use translation to sketch the graph of the indicated relation.*

1. $y = x^3$

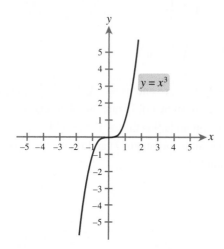

a. $y = (x + 2)^3$

b. $y + 1 = (x - 3)^3$

c. $y = (x + 4)^3 - 2$

2. $x = y^2$

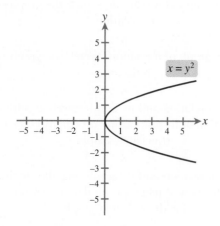

a. $x = (y - 3)^2$

b. $x - 4 = (y + 1)^2$

c. $x = (y - 5)^2 + 2$

3. $\dfrac{x^2}{16} + \dfrac{y^2}{9} = 1$

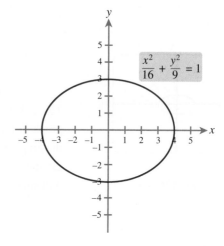

a. $\dfrac{x^2}{16} + \dfrac{(y - 1)^2}{9} = 1$

b. $\dfrac{(x - 4)^2}{16} + \dfrac{(y + 2)^2}{9} = 1$

4. $\dfrac{y^2}{4} - x^2 = 1$

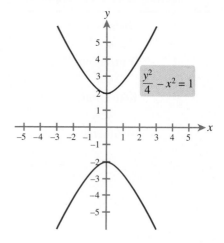

a. $\dfrac{y^2}{4} - (x - 2)^2 = 1$

b. $\dfrac{(y + 1)^2}{4} - (x + 3)^2 = 1$

Exercises 5-8 *A relation and its graph are given, together with a translated graph. Find an equation for the translated graph.*

5. Original relation: $y = \sqrt{x}$

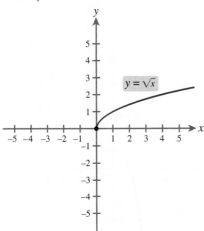

Translated graph:

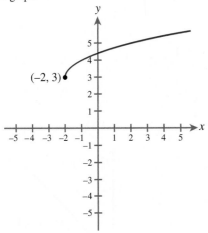

6. Original relation: $x = \sqrt[3]{y}$

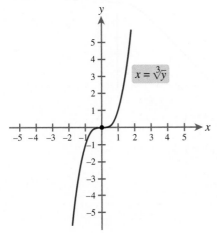

Translated graph:

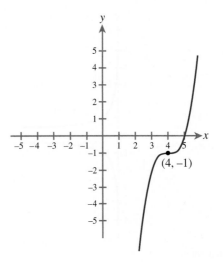

7. Original relation: $xy = 4$

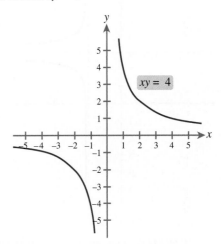

Translated graph:

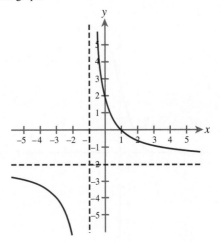

8. Original relation: $y^2x = 1$

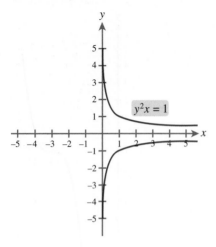

Translated graph:

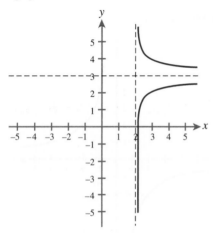

Exercises 9-12   *A relation and its graph are given. Use reflection to sketch the graph of the indicated relation.*

9. $x = \sqrt{y} - 2$

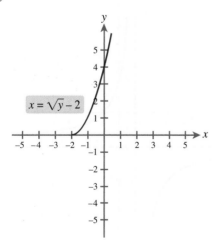

a. $-x = \sqrt{y} - 2$

b. $x = \sqrt{-y} - 2$

10. $y = x^3 + 1$

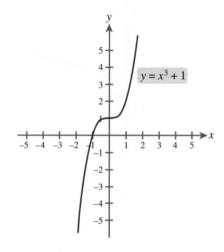

a. $-y = x^3 + 1$

b. $y = -x^3 + 1$

11. $x^3 - y^3 = 1$

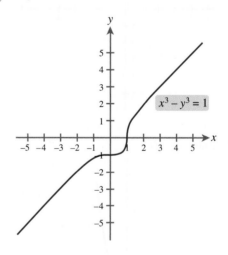

a. $y^3 - x^3 = 1$

b. $x^3 + y^3 = 1$

12. $\sqrt[3]{x} + \sqrt[3]{y} = \frac{1}{2}$

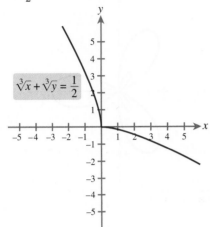

a. $\sqrt[3]{x} - \sqrt[3]{y} = \frac{1}{2}$ **b.** $\sqrt[3]{y} = \frac{1}{2} + \sqrt[3]{x}$

Exercises 13-14   *A relation and its graph are given, together with a reflected graph. Find an equation for the reflected graph.*

13. Original relation: $\sqrt{-x} + \sqrt{y} = 2$

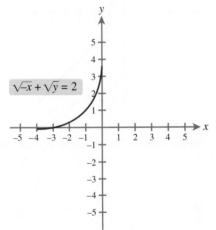

Reflected graph:

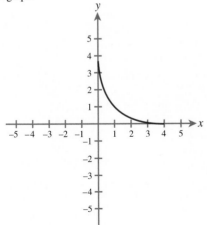

14. Original relation: $y + 1 = \frac{x + 5}{x^2 - 8x + 17}$

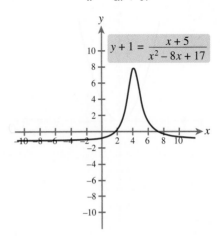

Reflected graph:

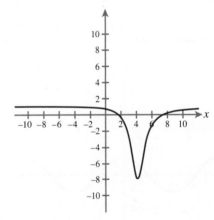

Exercises 15-16   *A relation and its graph are given. Use translation and/or reflection to sketch the graph of the indicated relation.*

15. $x = y^3$

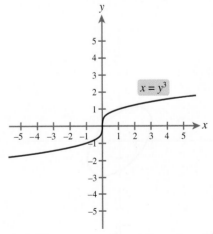

a. $x = -(y + 2)^3$ **b.** $x + 3 = -(y - 4)^3$

16. $y = x^2$

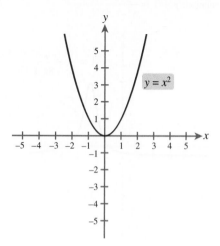

a. $y = -(x - 1)^2$ **b.** $y - 1 = -(x + 3)^2$

Exercises 17-20   *Check the given graph for symmetry with respect to the y-axis, the x-axis, and the origin.*

17.

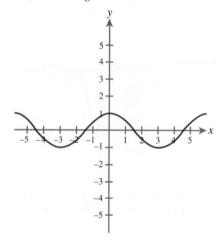

18.

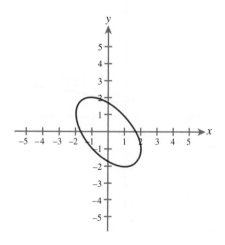

19.

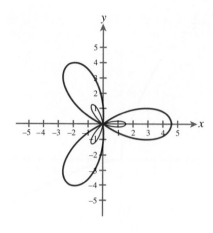

20.

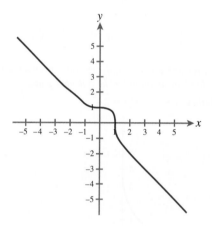

Exercises 21-28   *Test for symmetry with respect to the x-axis, the y-axis, and the origin.*

21. $x = y^2$

22. $x^2 + y^3 = 10$

23. $y = \dfrac{1}{x}$

24. $y^2 = x^4 + x^2 + 2$

25. $x^3 - y^3 = 1$

26. $y = |x|$

27. $y = \sin x$

28. $x = \cos y$

Exercises 29-36   *A portion of the graph of a relation is shown. Test for symmetry with respect to the x-axis, the y-axis, and the origin, and then use symmetry to complete the graph.*

29. $y = x^4$

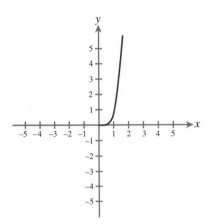

30. $xy = 5$

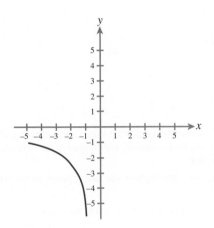

31. $10y = x^5$

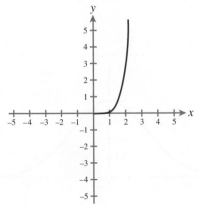

32. $y^2 = x - 2$

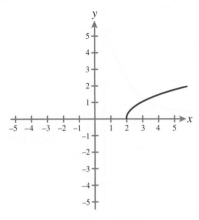

33. $y^2 - x^2 = 4$

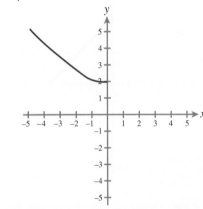

34. $x^6 + y^6 = 729$

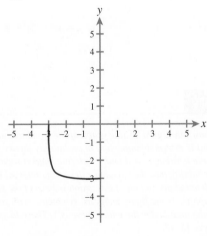

35. $x^2y + 2y = 10$

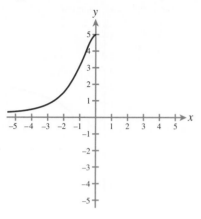

36. $|x| + |y| = 4$

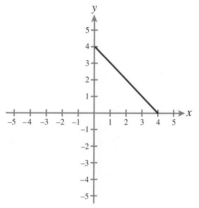

Exercises 37-40 *Use a graphing calculator to determine how the relations in parts a through c have been obtained from the given relation using translation and/or reflection.*

37. $y = x^2 + x - 2$

   a. $y = x^2 + x - 4$ **b.** $y = x^2 - 5x + 4$

   c. $y = -x^2 - x$

38. $y = x^3 - x$

   a. $y = x^3 - 9x^2 + 26x - 24$

   b. $y = x^3 - x + 4$

   c. $y = -x^3 + x + 3$

39. $y = \dfrac{5}{x^2 + 1}$

   a. $y = \dfrac{3x^2 + 8}{x^2 + 1}$

   b. $y = \dfrac{-5}{x^2 + 2x + 2}$

   c. $y = \dfrac{2x^2 + 8x + 15}{x^2 + 4x + 5}$

40. $y = \sqrt[3]{x^2 + 1}$

   a. $y = \sqrt[3]{x^2 - 4x + 5}$

   b. $y = \sqrt[3]{-x^2 - 1} - 2$

   c. $y = \sqrt[3]{x^2 + 2x + 2} + 2$

Exercises 41-44 *Solve the relation for y, and use a graphing calculator to plot the graph and to estimate the coordinates of the indicated point(s) to the nearest integer.*

41. $16x^2 + 25y^2 = 400$; highest and lowest points on the graph

42. $9y^2 - 16x^2 = 144$; point(s) closest to the origin

43. $y^2 + x^3 + x - 5 = 0$; x- and y-intercepts

44. $\dfrac{y^2}{x^4 - x^3 + 3} = 1$; x- and y-intercepts

Applications

The graph shown in Figure 25 (the graph of the standard normal probability function) is extremely important in probability theory. The area under the curve between $x = a$ and $x = b$ (the shaded region in Figure 25) is the probability that the variable x lies in the interval (a, b). Although a detailed discussion of this graph is beyond the scope of this book, the graph is, as the figure suggests, symmetric with respect to the y-axis. Also, the area under the entire curve is 1. These facts may be useful in Exercises 45–48.

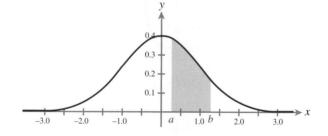

Figure 25

45. Find the area under the curve to the right of $x = 0$.

46. Find the area under the curve to the left of $x = 0$.

47. Given that the area under the curve between $x = 0$ and $x = 1$ is approximately 0.3413, compute the areas under the curve over the following intervals:

   a. $(1, \infty)$

   b. $(-\infty, -1)$

   c. $(-1, 1)$

   d. $(-\infty, 1)$

48. Given that the area under the curve to the left of $x = 2$ is approximately 0.9772, compute the area under the curve over the following intervals:

   a. $(2, \infty)$

   b. $(-2, 2)$

   c. $(-\infty, -2)$

   d. $(0, 2)$

Concepts and Critical Thinking

Exercises 49-54 *Answer true or false.*

49. If the graph of the relation obtained by replacing x with $-x$ is the same as the original graph, then we say that the graph is symmetric with respect to the y-axis.

50. Substitution of both $-x$ for x and $-y$ for y will result in a reflection about the x-axis.

51. Substituting $x + 3$ for x translates the graph 3 units to the right.

52. If a graph is symmetric with respect to both the x- and y-axes, then it will also be symmetric with respect to the origin.

53. If a graph is symmetric with respect to the origin, then it will also be symmetric with respect to both the x- and y-axes.

54. No line is symmetric with respect to both the x- and y-axes.

Exercises 55-58 *Give an example of each.*

55. Two relations such that the graph of one is the translation of the other, 5 units to the right

56. The graph of a relation that remains the same after translating 2 units to the right

57. The graph of a relation that is symmetric with respect to the line $y = x$

58. The graph of a relation that is also the graph of a function and is symmetric with respect to the origin

59. Explain why no function of the form $y = f(x)$ can have symmetry with respect to the x-axis.

Questions for Discussion or Essay

60. Suppose we wanted to electronically transmit a graph over a data line. Explain how the symmetry of a graph could be used to condense the transmission.

61. A face with two left sides can be constructed from the photo in Figure 26 by taking the mirror image of the left side of the face and attaching it on top of the right side. The result is shown in Figure 27. The same photo is used to create a face with two right sides in Figure 28. What do these two-faced photos suggest about symmetry in the human body? Give some additional evidence that supports your conclusion.

Figure 26   Figure 27   Figure 28

62. Most forms of life possess symmetries. For example, the human body is roughly *bilaterally symmetric,* starfish are *rotationally symmetric,* and sand dollars are *radially symmetric.* Perhaps the most significant and the least obvious symmetry is *symmetry of scale.* An object is said to possess a symmetry of scale if a small piece of the object resembles the object as a whole. For example, if the branching patterns of limbs on the trunk, branches on a limb, leaves on a branch, and veins on a leaf are similar, then we would say that the tree possessed a certain symmetry of scale. The human circulatory, respiratory, and nervous systems all possess a symmetry of scale. Discuss the meaning of symmetry of scale in the context of each of these systems.

Rotational symmetry in a starfish

Bilateral symmetry in a Burchell's zebra

Symmetry of scale in a tree fern

63. Inspector Magill has just been called to the scene of a homicide. The victim is lying in the middle of a carpeted floor, and the only objects near him are a pen and a hand-held mirror. A strange message is written on his forehead. It reads "ti bib dod." Almost immediately Magill knows the first name of the murderer. Explain how the notion of reflection might have helped Magill decipher the message. Who is the murderer?

Radial symmetry in organ pipe coral

Projects for Enrichment

64. Inverses of Relations The **inverse of a relation** in x and y is obtained by switching the variables in the equation that defines the relation. For example, the inverse of the relation defined by $x^3 + 3y^2 = 12$ is given by $y^3 + 3x^2 = 12$.

a. Find the inverses of the following relations:

  i. $y = x^2 + 2x - 3$ **ii.** $\dfrac{3x - 4y}{x^2 + 3y^3} = 2$

  iii. $3x^2 + 5xy + 3y^2 = 10$

In Figure 29, we have plotted the points $A(-4, 2)$, $B(1, 3)$, and $C(4, 5)$. We then plotted the points $A'(2, -4)$, $B'(3, 1)$, and $C'(5, 4)$ obtained from A, B, and C, respectively, by exchanging the x- and y-coordinates. It is easily seen that A', B', and C' are obtained from A, B, and C by reflecting about the line $y = x$. Thus, whenever we exchange the x- and y-coordinates, we reflect the point (or points) about the line $y = x$. Since the inverse of a relation is obtained by exchanging the x- and y-coordinates, the graph of a relation and its inverse are reflections about the line $y = x$. Shown in Figure 30 is the graph of a relation and its inverse.

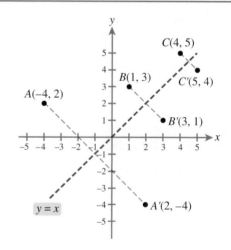

Figure 29

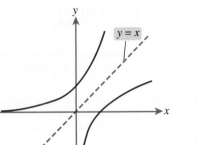

Figure 30

b. For each of the following graphs, sketch the graph of the inverse:

i.

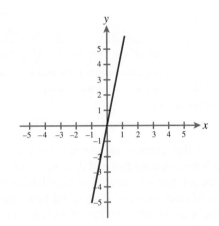

ii.

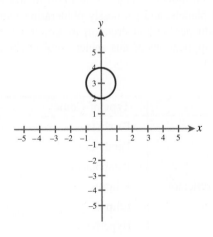

iii.

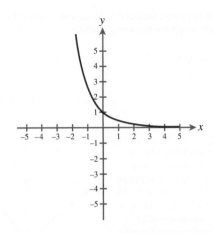

Many relations cannot be graphed directly using a graphing calculator because they cannot easily be solved for y. If, however, the relation can be solved for x, a graphing calculator *can* produce a graph of its inverse. This graph can then be reflected about the line $y = x$ to obtain a graph of the original relation. Consider, for example, the relation $x - y^3 + 3y = 0$. Although it is difficult to solve this relation for y, solving for x gives us $x = y^3 - 3y$. Now the inverse of this relation is $y = x^3 - 3x$, which can be plotted using a graphing calculator. With a viewing window defined by $\texttt{Xmin} = -15$, $\texttt{Xmax} = 15$, $\texttt{Ymin} = -10$, and $\texttt{Ymax} = 10$, we obtain the plot shown in Figure 31. Finally, we reflect about the line $y = x$ to obtain the graph of $x = y^3 - 3y$, as shown in Figure 32.

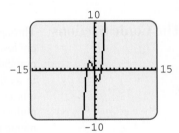

Figure 31

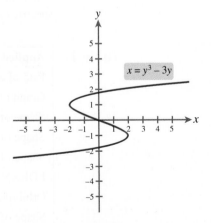

Figure 32

c. Use a graphing calculator and your understanding of inverse relations to graph the following relations:

i. $x = \sqrt{36 - y^2}$ **ii.** $x = \sqrt{y^2 - 36}$

iii. $x = \dfrac{4}{y^2 + 1}$ **iv.** $x - y^3 + 4y^2 = 0$

65. Rotation by 90° In this project, we explore rotations by 90° and their connections with reflections. Pictured in Figure 33 is an arbitrary point P with coordinates (x, y), and a point P' with coordinates (x', y'), the result of rotating the point P 90° counterclockwise.

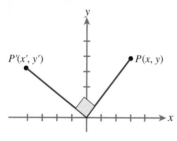

Figure 33

a. Use elementary geometry, trial and error, or any other technique to find a simple formula for x' and y' in terms of x and y.

b. Use the formula obtained in part a to find the coordinates of the point obtained by rotating $(4, 6)$ 90° counterclockwise.

c. Find a formula for the coordinates of the points obtained by rotating (x, y) 270° counterclockwise by applying three consecutive 90° counterclockwise rotations. Then give a formula for a 90° *clockwise* rotation.

d. Use the formula obtained in part a to show that a 90° counterclockwise rotation followed by a reflection about the y-axis is equivalent to a reflection about the line $y = x$.

e. Show that reflecting about the y-axis followed by a 90° counterclockwise rotation is equivalent to reflecting about the line $y = x$ and then reflecting about the origin. Compare your results to those from part d.

Section 8.2 | Parabolas

- How can parabolas help a Hollywood stunt driver or a cliff diver in Acapulco?
- What's the maximum height attained by human cannonball Emanual Zacchini?
- How might cones be connected with the disappearance of the dinosaur?
- Where on a satellite dish is the receiver placed?

The Conic Sections

Throughout much of the remainder of this chapter we consider the conic sections: circles, parabolas, ellipses, and hyperbolas. First studied and developed for purely theoretical reasons, the conic sections are yet another example of what has been referred to as the "unreasonable effectiveness of mathematics." Who could have guessed that this collection of abstract curves, the intellectual offspring of ancient Greeks more philosophers than mathematicians, would provide the mathematical weaponry for communicating across continents, beaming moving images from Earth to space and back again, guiding spacecraft to distant planets, and painlessly obliterating kidney stones with little more than a well-placed whisper? From astronomy to economics, mass communication to geolocation, modern applications of this ancient subject abound. A few specific examples are cited in Table 1.

Table 1

Applied Context	Type of Conic
Path of a projectile	Parabola
Graph of a revenue function	Parabola
Shape of a satellite dish	Parabola
Shape of a shock wave lithotripsy reflector	Ellipse
Shape of a whispering room	Ellipse
LORAN navigation	Hyperbola
Orbit of a comet	Parabola, ellipse, or hyperbola
Shape of a telescope mirror	Parabola, ellipse, or hyperbola

There are several ways to study conic sections. A purely geometric approach was developed by the ancient Greeks around 350 B.C. They described the conic sections as curves that are formed by the intersection of a plane with a double cone, as shown in Figure 34.

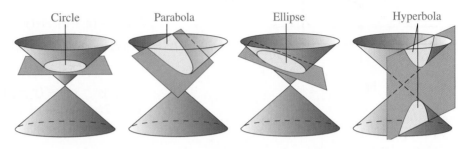

Circle   Parabola   Ellipse   Hyperbola

Figure 34

An algebraic approach to the study of conic sections was made possible after the development of analytic geometry in the 17th century. Using this approach, the conic sections are described as relations of the form

$$Ax^2 + Bxy + Cy^2 + Dx + Ey + F = 0$$

A third approach describes each conic as a *locus* or set of points satisfying certain geometric properties. This is the approach we will use to develop the standard equations of the conics. We begin with parabolas.

Parabolas

As suggested by Figure 34, a parabola is a curve that is formed when a cone is cut with a plane parallel to the side of the cone. It can be shown that any parabola formed in this manner can also be formed by applying the following definition.

Definition of a Parabola

A **parabola** is the set of all points (x, y) that are equidistant from a fixed line (called the **directrix**) and a fixed point (called the **focus**) not on the line, as shown in Figure 35. The point midway between the focus and directrix is called the **vertex**, and the line passing through the focus and vertex is called the **axis of symmetry**.

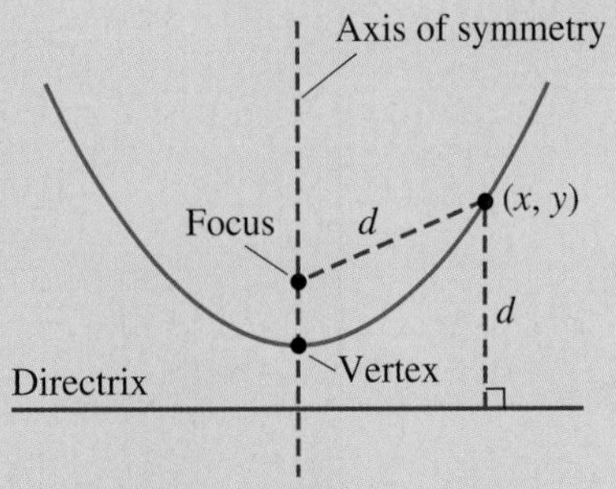

Axis of symmetry

Focus d (x, y)

d

Directrix   Vertex

Figure 35

Let us initially consider a parabola with focus $(0, p)$ and directrix $y = -p$, as shown in Figure 36. Notice that in this case the vertex is at the origin and the axis of symmetry is vertical. Moreover, if $p > 0$, the parabola opens upward, whereas if $p < 0$, the parabola opens downward. Now, according to the definition, for any point (x, y) on the

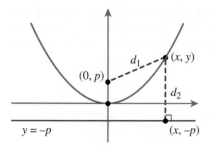

Figure 36

parabola, the distance d_1 from $(0, p)$ to (x, y) must be the same as the distance d_2 from (x, y) to the point $(x, -p)$ on $y = -p$. Applying the distance formula, we have

$$d_1^2 = (x - 0)^2 + (y - p)^2 = x^2 + (y - p)^2$$

and

$$d_2^2 = (x - x)^2 + (y - (-p))^2 = (y + p)^2$$

We now set $d_1^2 = d_2^2$ and proceed as follows:

$$d_1^2 = d_2^2$$
$$x^2 + (y - p)^2 = (y + p)^2$$
$$x^2 + y^2 - 2yp + p^2 = y^2 + 2yp + p^2 \qquad \text{Multiplying out}$$
$$x^2 = 4yp \qquad \text{Simplifying}$$
$$y = \frac{1}{4p}x^2 \qquad \text{Solving for } y$$

Finally, we set $a = 1/(4p)$ to obtain the equation $y = ax^2$. Thus, a parabola with vertex $(0, 0)$ and a *vertical* axis of symmetry has an equation of the form $y = ax^2$. A similar process could be used to show that a parabola with vertex $(0, 0)$ and a *horizontal axis* of symmetry has an equation of the form $x = ay^2$. In both cases, we have $a = 1/(4p)$ or, equivalently, $p = 1/(4a)$, and so the coefficient a gives us information about the location of the focus and directrix. We summarize these observations as follows.

Parabolas with Vertex at the Origin

The standard forms for the equation of a parabola with vertex $(0, 0)$ are as follows:

Equation	Description
$y = ax^2$	Vertical axis of symmetry Opens upward if $a > 0$, downward if $a < 0$ See Figure 37.
$x = ay^2$	Horizontal axis of symmetry Opens to the right if $a > 0$, to the left if $a < 0$ See Figure 38.

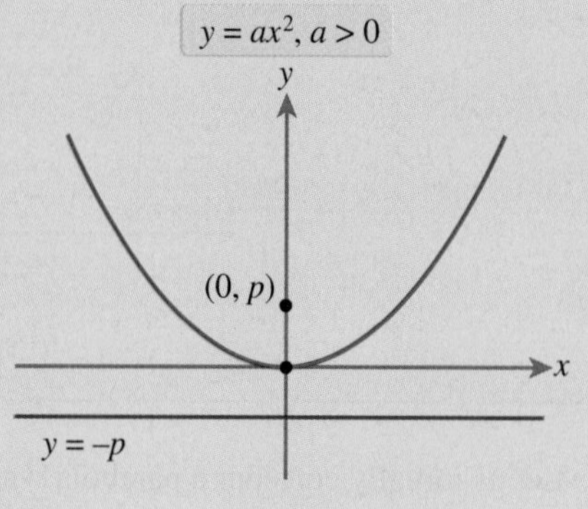

$y = ax^2, a > 0$

Figure 37

$x = ay^2, a > 0$

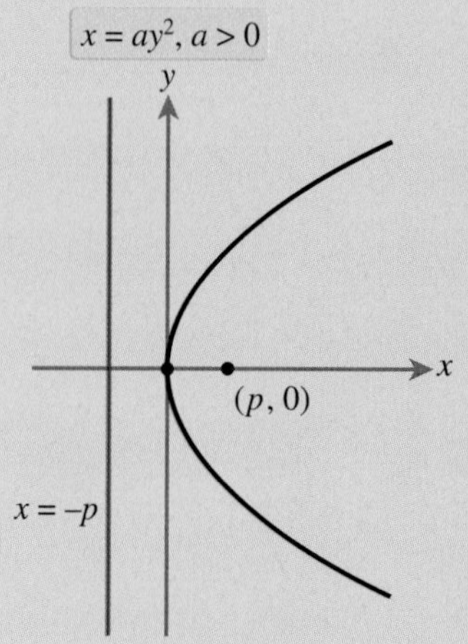

Figure 38

The focus is on the axis of symmetry in the *interior* of the parabola, at a directed distance of

$$p = \frac{1}{4a}$$

units from the vertex. The directrix is perpendicular to the axis of symmetry and is a directed distance of $-p$ units from the vertex on the side of the parabola opposite the focus.

EXAMPLE 1

Finding the Focus and Directrix of a Parabola with Vertex (0, 0)

Sketch a graph of the parabola with equation $y = \frac{1}{8}x^2$, and find its focus and directrix.

Solution   We first note that the equation is of the form $y = ax^2$, and so the parabola has a vertical axis of symmetry and vertex (0, 0). Moreover, since $a = \frac{1}{8} > 0$, the parabola opens upward. To add some detail, we locate two points on the graph by choosing 4 and -4 as convenient x-coordinates and computing $y = \frac{1}{8}(\pm 4)^2 = 2$. The graph is shown in Figure 39. From the shape of the graph, we expect the focus to be above the vertex and the directrix to be below. To locate both precisely, we compute the value of p.

$$p = \frac{1}{4a}$$

$$= \frac{1}{4\left(\frac{1}{8}\right)}$$

$$= 2$$

Thus, the focus is 2 units above the vertex, at the point (0, 2), and the directrix is 2 units below the vertex, with equation $y = -2$ (see Figure 40).

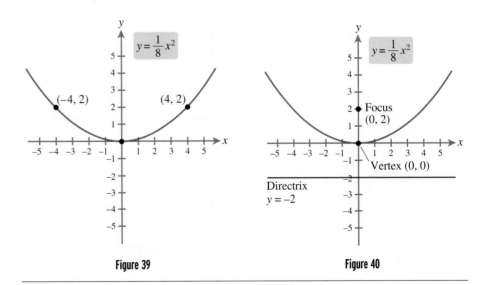

Figure 39   Figure 40

·····≥EXAMPLE 2

Finding the Focus and Directrix of a Parabola with Vertex (0, 0)

Sketch a graph of the parabola with equation $5y^2 + 24x = 0$, and find its focus and directrix.

Solution We begin by writing the equation in standard form.

$$5y^2 + 24x = 0$$
$$24x = -5y^2$$
$$x = -\frac{5}{24}y^2$$

This is the equation of a parabola with a horizontal axis of symmetry and vertex $(0, 0)$. Since $a = -\frac{5}{24} < 0$, the parabola opens to the left. To add detail to the graph, we find two points by taking $y = \pm 4$ and computing $x = -\frac{5}{24}(\pm 4)^2 = -\frac{10}{3}$. The graph is shown in Figure 41. Based on the graph, we know the focus will lie to the left of the vertex, and the directrix will lie to the right. Computing p, we obtain

$$p = \frac{1}{4a}$$
$$= \frac{1}{4\left(-\frac{5}{24}\right)}$$
$$= -\frac{6}{5}$$

Thus, the focus is $\frac{6}{5}$ units to the left of the vertex, at the point $\left(-\frac{6}{5}, 0\right)$. The directrix, lying $\frac{6}{5}$ units to the right of the vertex, has equation $x = \frac{6}{5}$. Refer to Figure 42.

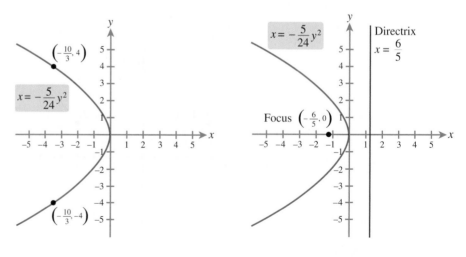

Figure 41

Figure 42

Cometary Collisions

From ancient times, astronomers and amateur stargazers alike have been fascinated with comets. One reason for this must certainly be the fact that many comets have been visible to the naked eye. Indeed, comet sightings have been recorded as early as the 15th century B.C., long before the invention of the telescope. Ironically, early comet sightings were almost always associated with catastrophe: the assassination of Julius Caesar in 44 B.C., the fall of Jerusalem to the Romans in A.D. 70, and the Norman conquest of England in A.D. 1066, to name a few. In more modern times, as comets came to be understood as astronomic phenomena, such superstitious beliefs began to wane. However, one fear has remained even to this day—that a comet may one day collide with Earth.

Devastation at Tunguska, Siberia

How likely is such a collision? Most astronomers would likely argue that the chance of a collision with any specific comet is virtually zero. However, at least one astronomer, Brian Marsden, has suggested that there is "a small but not negligible chance" that the comet Swift-Tuttle will collide with Earth on August 14, 2126. Other astronomers, whether or not they agree with Marsden, would likely admit that if the known number of comets is taken into account, a collision at some time in the future is inevitable. As evidence of this likelihood, one need only look to the collision of the comet Shoemaker-Levy 9 with Jupiter in July 1994. Further evidence can perhaps be found closer to home. Some scientists have theorized that a cometary collision caused the extinction of the dinosaur. There are even some who suggest a more recent collision. In 1908, a powerful explosion leveled thousands of acres of trees in the Tunguska River basin in central Siberia. The accompanying aftershock was detected by seismic recording equipment over 500 miles away. Eyewitnesses described a bright blue fireball, trailing smoke, that some experts have concluded was a small fragment from the comet Encke. Because no conclusive evidence of a crater was found, it has been suggested that the fragment exploded several miles above the surface of Earth.

The threat of a comet or asteroid collision has been taken seriously by astronomers and government officials, and it has led to the formation of NASA's Near-Earth Object (NEO) program. The purpose of this program is to detect, catalogue, and monitor potentially troublesome comets and asteroids. One essential ingredient of an early detection system is a thorough understanding of the path of a comet. Here is where conic sections come into play. Over 650 comets have been observed in our solar system, and each has been identified as having a path that is either elliptical, parabolic, or hyperbolic, with the sun as a focus. Comets with elliptical orbits reappear at regular (though perhaps quite lengthy) time intervals. Certainly the most famous comet with an elliptical orbit is Halley's comet, which has a period of approximately 76 years. It last appeared in 1986, and it will appear again in 2061. The Swift-Tuttle comet referenced earlier also has an elliptical orbit. It last passed Earth in 1992, and it will pass again in 2126. Comets with parabolic or hyperbolic paths are seen only once, and so it is likely that many have been missed and many have yet to be seen. The threat of a collision is greatest for those comets that have parabolic or hyperbolic paths since they are more difficult to detect and track.

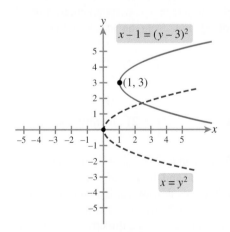

Figure 43

So far, we have considered only parabolas with vertex at the origin. Parabolas with vertical and horizontal axes and vertex *away* from the origin have standard equations that can be obtained using translations. Recall from Section 8.1 that substituting $x - h$ for x and $y - k$ for y has the effect of translating the graph h units horizontally and k units vertically. For example, the graph of the relation $x - 1 = (y - 3)^2$ can be obtained from the graph of $x = y^2$ by translating 1 unit to the right and 3 units up, as shown in Figure 43. From this, we see that the graph of $x - 1 = (y - 3)^2$ is a parabola with a horizontal axis of symmetry and vertex $(1, 3)$. More generally, we have the following standard equations.

Parabolas with Vertex (h, k)

The standard forms for the equation of a parabola with vertex (h, k) are as follows:

Equation	Description
$y - k = a(x - h)^2$	Vertical axis of symmetry Opens upward if $a > 0$, downward if $a < 0$ See Figure 44.
$x - h = a(y - k)^2$	Horizontal axis of symmetry Opens to the right if $a > 0$, to the left if $a < 0$ See Figure 45.

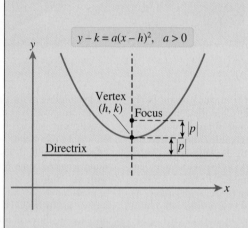

$y - k = a(x - h)^2, \quad a > 0$

Figure 44

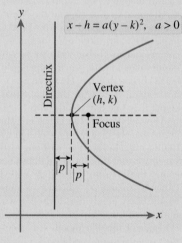

$x - h = a(y - k)^2, \quad a > 0$

Figure 45

As with parabolas having vertex (0, 0), the focus and directrix can be found using

$$p = \frac{1}{4a}$$

Specifically, the focus is on the axis of symmetry in the interior of the parabola, at a directed distance of p units from the vertex. The directrix is perpendicular to the axis of symmetry and is a directed distance of $-p$ units from the vertex on the side of the parabola opposite the focus.

······►EXAMPLE 3

Finding the Directrix and Focus of a Parabola

Sketch a graph of the parabola with equation $y - 1 = -\frac{1}{4}(x + 2)^2$. Identify the vertex, focus, and directrix.

Solution   The equation is already in standard form, with $h = -2$, $k = 1$, and the x term being squared. Thus, we know that the vertex is $(-2, 1)$ and the axis of symmetry is vertical. Moreover, since $a = -\frac{1}{4} < 0$, the parabola opens downward. By plotting points as necessary, we obtain the graph shown in Figure 46. Since $p = 1/(4a) = -1$, the focus is located 1 unit below the vertex at the point $(-2, 0)$. The directrix is located 1 unit above the vertex and has equation $y = 2$. Refer to Figure 47.

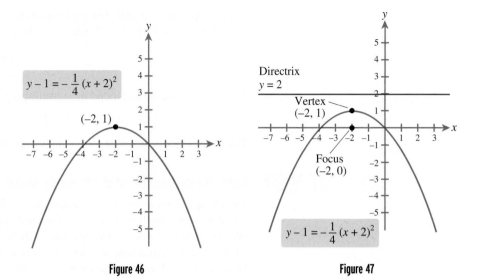

Figure 46   **Figure 47**

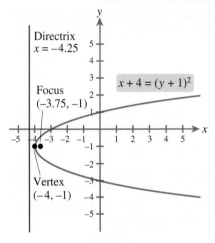

Figure 48

·····❯EXAMPLE 4

Graphing a Parabola

Sketch a graph of the parabola with equation $x = y^2 + 2y - 3$. Identify the vertex, focus, and directrix.

Solution   We must first write the equation in standard form. We do this by completing the square on the y terms.

$$x = y^2 + 2y - 3$$
$$x + 3 = y^2 + 2y \qquad \text{Adding 3 to both sides to eliminate the constant on the right}$$
$$x + 3 + 1 = y^2 + 2y + 1 \qquad \text{Adding 1 to both sides to complete the square on the right}$$
$$x + 4 = (y + 1)^2 \qquad \text{Writing the right-hand side as a perfect square}$$

This is the equation of a parabola with horizontal axis of symmetry and vertex $(-4, -1)$. Moreover, since $a = 1 > 0$, the parabola opens to the right. Since $p = 1/(4a) = \frac{1}{4}$, the focus is located $\frac{1}{4}$ unit to the right of the vertex, at the point $(-3.75, -1)$. The directrix is located $\frac{1}{4}$ unit to the left of the vertex and has equation $x = -4.25$. See Figure 48.

When equations of parabolas are given in the general form $y = ax^2 + bx + c$ or $x = ay^2 + by + c$, as was the case in Example 4, it is sometimes convenient to have a simple rule for finding the vertex. By completing the square on the general equation, we obtain the following rule.

Finding the Vertex of a Parabola

Parabola form	Axis of symmetry	Vertex
$y = ax^2 + bx + c$	Vertical	$\left(-\frac{b}{2a}, k\right)$, where k is found by setting $x = -\frac{b}{2a}$
$x = ay^2 + by + c$	Horizontal	$\left(h, -\frac{b}{2a}\right)$, where h is found by setting $y = -\frac{b}{2a}$

Consider, for example, the parabola with equation $x = y^2 + 2y - 3$ from Example 4. Since $a = 1$ and $b = 2$, the y-coordinate of the vertex is given by

$$y = -\frac{b}{2a} = -\frac{2}{2(1)} = -1$$

The x-coordinate is then found by substituting $y = -1$ into $x = y^2 + 2y - 3$, yielding

$$x = (-1)^2 + 2(-1) - 3 = 1 - 2 - 3 = -4$$

Thus, the vertex is the point $(-4, -1)$. We use this technique in the following example.

EXAMPLE 5

Finding the Maximum Height of a Human Cannonball

The muzzle velocity of the renowned human cannonball Emanual Zacchini was estimated to be as much as 128 feet per second (nearly 90 miles per hour). If we assume that this muzzle velocity was possible with the cannon pointed straight up, and if we assume that Zacchini was initially 5 feet above the ground, then his height in feet after t seconds would have been given by $s = -16t^2 + 128t + 5$. Find Zacchini's maximum height given these assumptions.

Solution   In the coordinate system having height s marked along the vertical axis and time t on the horizontal, the graph of the equation $s = -16t^2 + 128t + 5$ is a parabola opening downward. Thus, the maximum value of the height s will occur at the vertex. We find the t-coordinate of the vertex—that is, the time at which the maximum height is attained—as follows:

$$t = -\frac{b}{2a} = -\frac{128}{-32} = 4 \text{ seconds}$$

The maximum height is then found by setting $t = 4$ in the equation $s = -16t^2 + 128t + 5$.

$$s = -16(4)^2 + 128(4) + 5 = 261 \text{ feet}$$

Thus, the maximum height is 261 feet. Note that the coordinates of the vertex can easily be verified using a graphing calculator. The graph of $s = -16t^2 + 128t + 5$ is shown in Figure 49; the approximate coordinates of the vertex, as found using the calculate-maximum feature, are reported as $(4, 261)$ in Figure 50.

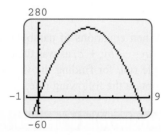

Figure 49

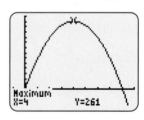

Figure 50

In all the examples to this point, we obtained information about the graphical features of a parabola from its equation. Now we consider how the equation of a parabola can be determined from its graphical features.

====⋗**EXAMPLE 6**

Finding the Equation of a Parabola

Find the standard equation of the parabola that has vertex (2, 3) and focus (2, 0).

Solution   Since the vertex and focus have the same x-coordinate, the parabola must have a vertical axis of symmetry. Combining this with the fact that the vertex is (2, 3), we know that the equation must have the form

$$y - 3 = a(x - 2)^2$$

Since the focus is 3 units below the vertex, we also know that $p = -3$, and so $a = 1/(4p) = -\frac{1}{12}$. Thus, the equation of the parabola is

$$y - 3 = -\frac{1}{12}(x - 2)^2$$

====⋗**EXAMPLE 7**

Finding the Equation of a Parabola from Its Graph

Find a possible equation for the parabola shown in Figure 51. Verify with a graphing calculator.

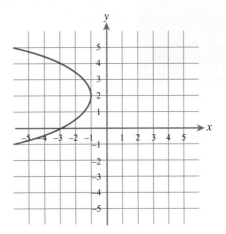

Figure 51

Solution   Figure 51 suggests a horizontal axis of symmetry and a vertex at $(-1, 2)$. These assumptions lead to an equation of the form

$$x + 1 = a(y - 2)^2$$

Since there appears to be an x-intercept at -3, we set $x = -3$ and $y = 0$ to obtain

$$-3 + 1 = a(0 - 2)^2$$
$$-2 = 4a$$
$$a = -\frac{1}{2}$$

The resulting equation is

$$x + 1 = -\frac{1}{2}(y - 2)^2$$

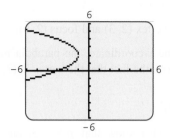

Figure 52

To plot the graph on a graphing calculator, we must solve for y.

$$x + 1 = -\frac{1}{2}(y - 2)^2$$

$$-2x - 2 = (y - 2)^2$$

$$\pm\sqrt{-2x - 2} = y - 2$$

$$2 \pm \sqrt{-2x - 2} = y$$

Graphing the branches $y = 2 + \sqrt{-2x - 2}$ and $y = 2 - \sqrt{-2x - 2}$, we obtain the parabola shown in Figure 52. It appears to be the same as that shown in Figure 51.

Reflective Property of Parabolas

The Very Large Array of radio antennas in New Mexico

If sound or light is emitted from the focus of a conic section, certain important reflective properties can be observed. In the case of a parabola, the reflected sound or light will follow a path parallel to the axis of the parabola, as shown in Figure 53. This property of parabolas is utilized in the construction of search lights. The reflector for the search light is formed by revolving a parabola around its axis, and the light source is placed at the focus of the parabola. The result is a light beam with little dissipation.

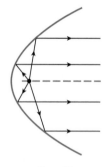

Figure 53

The reflective property of parabolas is also used in telescopes, satellite dishes, microwave antennas, and solar energy devices. When the parabolic reflector of these devices is positioned so that the incoming light rays, sound waves, or radio waves are (approximately) parallel to the axis, they will be reflected toward the focus. This results in an image or signal that is "focused" at a single point.

·····➢EXAMPLE 8

Determining Spotlight Dimensions

Find the depth of a parabolic reflector in a spotlight with a 6-inch-wide beam given that the light source is located 1 inch away from the vertex of the reflector.

Solution We begin by placing a cross section of the parabolic reflector on an xy-coordinate system in such a way that its vertex is at the origin and its focus (that is, the light source) is 1 unit to the right of the vertex, at the point $(1, 0)$. See Figure 54.

Now this parabola opens to the right and has vertex $(0, 0)$, so its equation must be of the form $x = ay^2$. Moreover, since the focus is 1 unit to the right of the vertex, we must have $p = 1$. It follows that $a = 1/(4p) = \frac{1}{4}$, so that the parabola has equation $x = \frac{1}{4}y^2$. Since the light beam is 6 inches wide, the point P in Figure 54, on the "edge" of the reflector, must have y-coordinate 3. Setting $y = 3$ in the equation $x = \frac{1}{4}y^2$ gives $x = \frac{9}{4}$, which implies that the reflector is $\frac{9}{4} \approx 2.25$ inches deep.

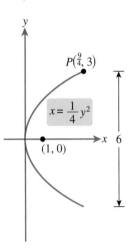

Figure 54

Understanding and Mastery Checklists

Concepts to Understand	Skills to Master
Conic section	Sketch the graph of a parabola.
Parabola	Determine the vertex, focus, and directrix of a parabola.
Vertex of a parabola	Find the equation of a parabola given information about its graph.
Focus of a parabola	
Directrix of a parabola	
Axis of symmetry of a parabola	
Reflective property of a parabola	

Exercises 1-12 *Find the vertex, focus, and directrix for the given parabola, and sketch its graph.*

1. $y = 2x^2$

2. $x = -4y^2$

3. $y^2 - 3x = 0$

4. $x^2 + 5y = 0$

5. $4y^2 + 7x = 0$

6. $2x^2 - 3y = 0$

7. $y - 3 = 3(x + 5)^2$

8. $x + 4 = -(y - 2)^2$

9. $x^2 - 4x - 4y + 8 = 0$

10. $y^2 + x - 6y + 10 = 0$

11. $x = 2y^2 + 6y + 2$

12. $y = -x^2 - x + 1$

Exercises 13-22 *Find an equation for the parabola with the given properties.*

13. Vertex at $(0, 0)$ and focus at $(2, 0)$

14. Vertex at $(0, 0)$ and focus at $(0, 4)$

15. Focus $(0, -4)$ and directrix $y = 4$

16. Vertex at $(0, 0)$ and directrix $x = 1$

17. Vertex at $(-2, 0)$, passing through $(2, 4)$, and a horizontal axis of symmetry

18. Vertex at $(-1, 3)$, y-intercept 5, and vertical axis of symmetry

19. Focus at $(4, -1)$ and directrix $y = 3$

20. Vertex at $(-2, -3)$ and focus $(-6, -3)$

21. The following graph:

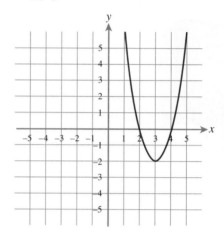

22. The following graph:

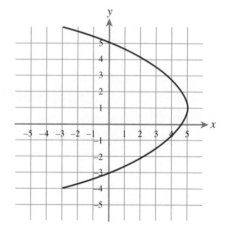

Exercises 23-30 *Solve the equation for y, and plot the resulting equation on a graphing calculator. Identify the coordinates of the vertex.*

23. $x^2 - 4x - y = 0$

24. $-x^2 + 5x + y = 0$

25. $x^2 + 8x - 2y + 22 = 0$

26. $x^2 - 6x + 3y + 12 = 0$

27. $-y^2 + 4x + 8 = 0$

28. $y^2 + 5x - 5 = 0$

29. $y^2 - 2y + x + 2 = 0$

30. $4x + y^2 - 6y = 0$

Applications

31. Path of a Ball A ball is thrown at an angle of 45° and with an initial velocity of 80 feet per second. Its path is parabolic and satisfies the equation

$$y = -\frac{x^2}{200} + x$$

where y is the height in feet when the ball is a horizontal distance of x feet from where it was thrown. Find the maximum height of the ball and the total horizontal distance it travels before hitting the ground.

32. Stunt Car The stunt coordinator for a movie is planning a stunt with a car speeding off the top floor of a parking garage. It can be shown that the path of the falling car will be parabolic with an equation of the form

$$y = h - \frac{16}{v^2}x^2$$

where h is the height of the garage in feet, v is the initial speed of the car in feet per second, and y is the height of the car in feet when its horizontal distance away from the edge of the garage is x feet. If the height of the garage is 100 feet and the car must land 200 feet from the base of the garage, determine the required speed of the car.

33. Cliff Diving The highest regularly performed dives are off La Quebrada in Acapulco, Mexico. (*Data source: The Guinness Book of World Records.*) The takeoff point is 115 feet above the water and, because of the presence of rocks at the base, the divers must land 30 feet out from the takeoff point (see Figure 55).

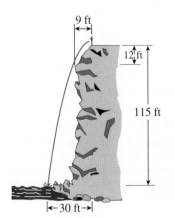

Figure 55

a. Assuming the path of a diver is a parabola with the vertex located at the takeoff point, choose a coordinate system with the origin at water level directly below the takeoff point, and find an equation for the path.

b. Suppose that 12 feet below the takeoff point, a large rock juts out 9 feet from the vertical line passing through the takeoff point. By what horizontal distance will the diver clear the rock?

34. Suspension Bridge When a cable is suspended between two towers of a suspension bridge in such a way that it bears a uniform load, the shape of the cable will be parabolic.

a. Find the equation of the shape of the cable if the towers are 200 feet apart and 40 feet high and the cable touches the road midway between the towers (see Figure 56).

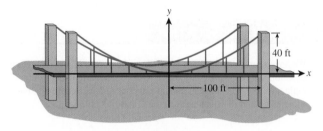

Figure 56

b. Use the equation from part a to determine the total length of cable needed for the vertical sections of cable spaced at 20-foot intervals between the towers.

35. Radio Telescope Suppose that the cross-sectional view of a radio telescope is parabolic and has, with an appropriately positioned coordinate system, an equation of the form $y = ax^2$ (Figure 57). If the dish is 1000 feet across the top and 167 feet deep at the center, where must the receiver be located?

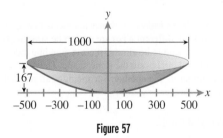

Figure 57

Concepts and Critical Thinking

Exercises 36-39 *Answer true or false.*

36. The focus of a parabola is a point on the graph of the parabola that is the same distance away from the vertex as is the directrix.

37. The graph of $x = ay^2 + by + c$ is a parabola with horizontal axis of symmetry.

38. The vertex of a parabola lies on its axis of symmetry.

39. The directrix of a parabola is a point lying on the parabola, halfway between the focus and the vertex.

Exercises 40-43 *Give an example of each.*

40. A parabola with a vertical axis of symmetry and vertex $(2, 1)$

41. A parabola with axis of symmetry $y = 4$

42. An application of the reflective property of parabolas

43. An equation of a parabola with its focus at the origin

Questions for Discussion or Essay

44. We have seen three methods for finding the vertex of a parabola with equation $y = ax^2 + bx + c$: completing the square, using the formula $x = -b/(2a)$, and graphical estimation. Discuss the relative advantages and disadvantages of these three techniques.

45. Describe a procedure for plotting the graph of a parabola of the form $x = ay^2 + by + c$ with a graphing calculator. Is this procedure any easier than sketching the graph by hand? Why or why not?

46. In the examples and exercises of this section, we've seen many problems in which the equation of a parabola was to be determined from a few characteristics, such as (1) the focus, (2) the directrix, (3) the vertex, (4) the axis of symmetry, and (5) a point other than the vertex on the parabola. In Example 6, we determined the equation of a parabola having a given vertex and focus. Are any other pairs of characteristics from among the five listed sufficient to determine the equation of the parabola? What are they? Are any three of these five sufficient to determine the equation? Explain your answers.

47. The parabolic model of projectile motion results from the assumption that the gravitational field is uniform: everywhere pointing in the same direction with the same strength. Since Earth's gravitational field points toward the center of Earth and weakens with altitude, the path of a projectile thrown from Earth's surface will *not* be a parabola, as we have assumed throughout this section, but rather an ellipse. How do the authors get away with committing such a seemingly profound error?

Projects for Enrichment

48. **Similarity Among Circles and Parabolas** Geometric objects that are the same shape but possibly different sizes are said to be **similar**. For example, two triangles are said to be similar if corresponding sides are proportional. Clearly, all circles are the same shape. Thus, we say that all circles are similar to one another. Surprisingly, all parabolas are similar to one another as well.

   I. Similarity of Circles

   a. Using graph paper, graph the circle with equation $x^2 + y^2 = 1$.

   b. Beginning with the equation $x^2 + y^2 = 1$, substitute $\frac{x}{2}$ for x, and $\frac{y}{2}$ for y, and simplify the resulting equation.

   c. Now graph $x^2 + y^2 = 1$ and the equation obtained in part b on the same set of coordinate axes. What effect did the substitution have on the size and shape of the resulting graph?

   d. How would the graphs of $x^2 + y^2 = 1$ and $\left(\frac{x}{a}\right)^2 + \left(\frac{y}{a}\right)^2 = 1$ compare in size and shape?

   e. Substituting $\frac{x}{a}$ for x and $\frac{y}{a}$ for y is equivalent to changing the scale of the graph by a factor of a. We refer to such substitutions as **scale substitutions**. Show that the circle with equation $x^2 + y^2 = r^2$ is similar to the circle with equation $x^2 + y^2 = 1$ by showing that one can be obtained from the other by a scale substitution.

   f. In part e we showed that all circles centered at the origin are similar to the circle centered at the origin with radius 1, and hence to each other. Explain why we can conclude from this that all circles, whether centered at the origin or not, are similar to one another.

II. Similarity of Parabolas
  a. Find the general equation of a parabola with a vertical axis of symmetry and vertex at the origin.

  b. Show that any such parabola can be obtained from $y = x^2$ by a scale substitution.

  c. Conclude that all parabolas (no matter what axis of symmetry or vertex they might have) are similar to one another.

  d. Explain the following statement: "All parabolas are the same shape; it is as if we are viewing the same parabola from different distances."

49. The Path of a Baseball If a baseball is hit toward center field from a height of 3 feet, at an initial angle of 45°, and with an initial velocity of 114 feet per second, will it clear a 12-foot-high center field fence 400 feet away? To answer this question, we must know something about the path of the baseball. The location of the ball at a given point in time can be expressed as an ordered pair (x, y), where x is the horizontal distance of the ball from home plate in feet and y is its height in feet. Since both x and y depend upon time, we must give equations for both. If we ignore air resistance, it can be shown that after t seconds, the position of the ball will be given by

(1) $$x = 57\sqrt{2}t$$

(2) $$y = -16t^2 + 57\sqrt{2}t + 3$$

For example, when $t = 0$, then $x = 0$ and $y = 3$, indicating that the ball leaves home plate at a height of 3 feet. When $t = 1$, $x = 57\sqrt{2} \approx 80.6$ feet, and $y = -16 + 57\sqrt{2} + 3 \approx 67.6$ feet. Thus, after 1 second, the ball is 80.6 feet from home plate (as measured from a point on the ground beneath the ball) and is 67.6 feet high. [Note that equations such as (1) and (2) are called **parametric equations** and will be explored in detail in Section 8.6.]

a. Complete the following table, plot the points (x, y), and connect the points with a smooth curve. What shape does the graph have?

t	0	1	2	3	4	5
x						
y						

We wish to find a relation in x and y whose graph gives the path of the ball.

b. Solve equation (1) for t and substitute into equation (2) to obtain an equation of the form $y = ax^2 + bx + c$.

c. Sketch the graph of the equation you found in part b. What shape is the path of the ball?

d. How far is the ball from home plate when it reaches its maximum height? What is the maximum height?

e. What is the height of the ball when it is 400 feet from home plate? Will the ball clear the 12-foot-high center field fence (Figure 58)?

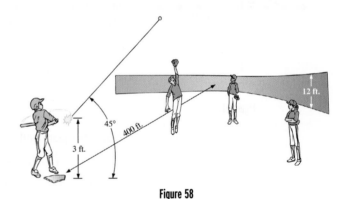

Figure 58

Section 8.3 Ellipses

- How can an ellipse be constructed with only a pencil, a piece of string, and two thumbtacks?
- It is virtually impossible to shine a flashlight in a room without seeing an ellipse. Where is it?
- How do whispering galleries work?
- If the center of Earth's orbit is *not* the center of the sun, then what is it?

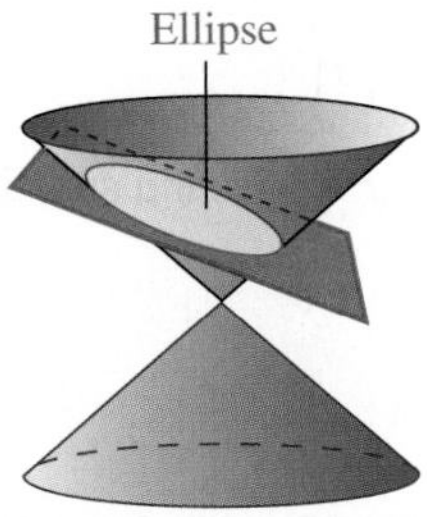

Ellipse

Figure 59

In Section 8.2, we described the conic sections as curves formed when a double cone is cut by a plane. If the plane is not parallel to the central axis or the side of the cone, we obtain an ellipse, as shown in Figure 59. Using geometric techniques, it can be shown that the following characterization of an ellipse is equivalent.

Definition of an Ellipse

An **ellipse** is the set of all points (x, y) such that the sum of the distances from two fixed points called **foci** (singular **focus**) is constant, as shown in Figure 60. The line through the foci intersects the ellipse at two points, called the **vertices**. The line segment connecting the two vertices is called the **major axis**, and the midpoint of the major axis is called the **center**. The line segment perpendicular to the major axis, passing through the center, and connecting two points on the ellipse is called the **minor axis**.

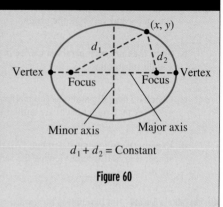

$$d_1 + d_2 = \text{Constant}$$

Figure 60

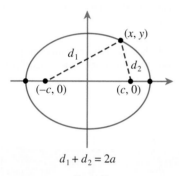

$$d_1 + d_2 = 2a$$

Figure 61

The standard equation of an ellipse centered at the origin is actually quite simple, but deriving the result from the definition takes a little work. We begin by considering an ellipse with a horizontal major axis and foci at $(-c, 0)$ and $(c, 0)$, as shown in Figure 61. According to the definition, for any point (x, y) on the ellipse, the sum of the distances to the two foci must be constant. To simplify later computations, we will denote this constant sum by $2a$. Using the distance formula for each of the two distances d_1 and d_2, we have

$$d_1 = \sqrt{(x + c)^2 + y^2} \quad \text{and} \quad d_2 = \sqrt{(x - c)^2 + y^2}$$

Setting $d_1 + d_2 = 2a$, we proceed as follows:

$$\sqrt{(x + c)^2 + y^2} + \sqrt{(x - c)^2 + y^2} = 2a$$
$$\sqrt{(x + c)^2 + y^2} = 2a - \sqrt{(x - c)^2 + y^2}$$

Now we square both sides and collect terms.

$$(x + c)^2 + y^2 = 4a^2 - 4a\sqrt{(x - c)^2 + y^2} + (x - c)^2 + y^2$$
$$x^2 + 2cx + c^2 + y^2 = 4a^2 - 4a\sqrt{(x - c)^2 + y^2} + x^2 - 2cx + c^2 + y^2$$
$$4a\sqrt{(x - c)^2 + y^2} = 4a^2 - 4cx$$
$$a\sqrt{(x - c)^2 + y^2} = a^2 - cx$$

Once again we square both sides and collect terms.

$$a^2\left[(x - c)^2 + y^2\right] = \left(a^2 - cx\right)^2$$
$$a^2\left(x^2 - 2cx + c^2 + y^2\right) = a^4 - 2a^2cx + c^2x^2$$
$$a^2x^2 - 2a^2cx + a^2c^2 + a^2y^2 = a^4 - 2a^2cx + c^2x^2$$
$$a^2x^2 - c^2x^2 + a^2y^2 = a^4 - a^2c^2$$
$$\left(a^2 - c^2\right)x^2 + a^2y^2 = a^2\left(a^2 - c^2\right)$$

Finally, since $a^2 - c^2 > 0$ (see Exercise 47), we set $b^2 = a^2 - c^2$ and substitute.

$$b^2x^2 + a^2y^2 = a^2b^2$$
$$\frac{x^2}{a^2} + \frac{y^2}{b^2} = 1$$

The standard equation of an ellipse with a vertical major axis can be derived in a similar manner.

Ellipses Centered at the Origin

The standard form for the equation of an ellipse centered at $(0, 0)$ is

$$\frac{x^2}{a^2} + \frac{y^2}{b^2} = 1, \quad a \neq b$$

If $a > b$, the major axis is horizontal with length $2a$ (see Figure 62), whereas if $b > a$, the major axis is vertical with length $2b$ (see Figure 63). The vertices are the endpoints of the major axis. The foci lie on the major axis a distance of c units from the center, where $c = \sqrt{\left| a^2 - b^2 \right|}$. Note that if $a = b$, the equation is that of a circle with radius a.

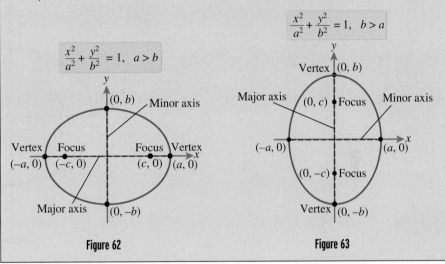

Figure 62

Figure 63

The graph of an ellipse centered at the origin can most easily be obtained by first locating the x- and y-intercepts and then connecting the intercepts with a smooth curve. To find the x-intercepts, we set $y = 0$ and solve for x. This yields $x = \pm a$. The y-intercepts are found by setting $x = 0$ and solving for y. We obtain $y = \pm b$. It will be readily apparent from the orientation of the graph whether the ellipse has a horizontal or vertical major axis.

EXAMPLE 1

An Ellipse Centered at the Origin

Sketch the graph of the ellipse with equation

$$\frac{x^2}{25} + \frac{y^2}{16} = 1$$

Identify the vertices and foci.

Solution We first rewrite in standard form as

$$\frac{x^2}{5^2} + \frac{y^2}{4^2} = 1$$

Since $a = 5$, the x-intercepts are $(-5, 0)$ and $(5, 0)$. Similarly, since $b = 4$, the y-intercepts are $(0, -4)$ and $(0, 4)$. Connecting these points with a smooth curve leads to the graph shown in Figure 64. It is evident from the graph that the major axis is horizontal, and so the vertices are $(-5, 0)$ and $(5, 0)$. The foci are on the major axis a distance of c units from the origin, where

$$c = \sqrt{|a^2 - b^2|}$$
$$= \sqrt{|25 - 16|}$$
$$= 3$$

Thus, the foci are $(-3, 0)$ and $(3, 0)$.

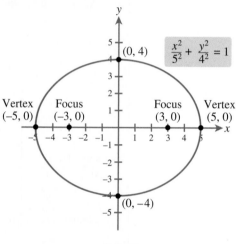

Figure 64

EXAMPLE 2

An Ellipse Centered at the Origin

Sketch the graph of the ellipse with equation $9x^2 + 4y^2 - 36 = 0$. Identify the vertices and foci.

Solution   We first write the equation in standard form.

$$9x^2 + 4y^2 - 36 = 0$$
$$9x^2 + 4y^2 = 36$$
$$\frac{9x^2}{36} + \frac{4y^2}{36} = 1$$
$$\frac{x^2}{4} + \frac{y^2}{9} = 1$$
$$\frac{x^2}{2^2} + \frac{y^2}{3^2} = 1$$

Thus, $a = 2$ and $b = 3$, giving us x-intercepts of ± 2 and y-intercepts of ± 3. The graph is shown in Figure 65. Since the major axis is vertical, the vertices have coordinates $(0, -3)$ and $(0, 3)$. To find the foci, we compute c as follows:

$$c = \sqrt{|a^2 - b^2|}$$
$$= \sqrt{|4 - 9|}$$
$$= \sqrt{5}$$

Thus, the foci are $\sqrt{5}$ units from the center on the major axis, at the points $\left(0, -\sqrt{5}\right)$ and $\left(0, \sqrt{5}\right)$.

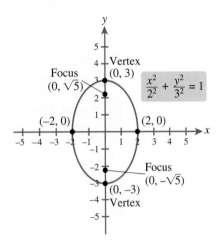

Figure 65

Just as with parabolas with vertex away from the origin, ellipses centered away from the origin have standard equations that can be obtained using translations. For example, the graph of

$$\frac{(x-1)^2}{25} + \frac{(y-3)^2}{16} = 1$$

can be obtained from the graph of

$$\frac{x^2}{25} + \frac{y^2}{16} = 1$$

by translating 1 unit to the right and 3 units up, as shown in Figures 66 and 67. From this, we see that the graph of $(x-1)^2/25 + (y-3)^2/16 = 1$ is an ellipse with a horizontal major axis and center $(1, 3)$. More generally, we have the following.

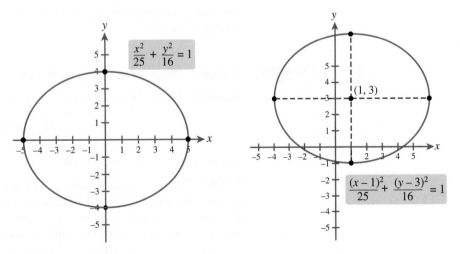

Figure 66   **Figure 67**

Ellipses Centered at (h, k)

The standard form for the equation of an ellipse centered at (h, k) is

$$\frac{(x - h)^2}{a^2} + \frac{(y - k)^2}{b^2} = 1, \quad a \neq b$$

If $a > b$, the major axis is horizontal with length $2a$ (see Figure 68), whereas if $b > a$, the major axis is vertical with length $2b$ (see Figure 69). The vertices are the endpoints of the major axis. The foci lie on the major axis a distance of c units from the center, where $c = \sqrt{|a^2 - b^2|}$. Note that if $a = b$, the equation is that of a circle centered at (h, k) with radius a.

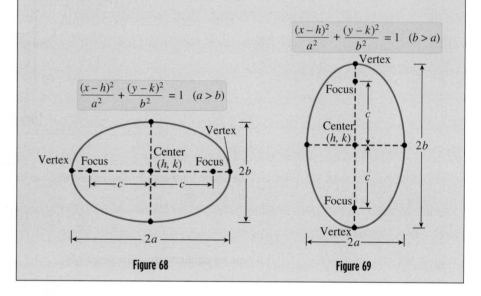

Figure 68   **Figure 69**

The graph of an ellipse centered at (h, k) can be obtained by viewing (h, k) as the origin of a new coordinate system and then applying the process used for an ellipse centered at the origin.

EXAMPLE 3

Graphing an Ellipse and Identifying Key Features

Sketch the graph of the ellipse with equation

$$\frac{(x - 2)^2}{9} + \frac{(y + 3)^2}{25} = 1$$

Identify the center, vertices, and foci.

Solution This ellipse has center $(2, -3)$. Since $a = \sqrt{9} = 3$, we plot points on the ellipse 3 units to the left and right of the center. Similarly, since $b = \sqrt{25} = 5$, we plot points on the ellipse 5 units above and below the center. Connecting these points with a smooth curve yields the graph in Figure 70. We see that the major axis is vertical and connects the vertices $(2, 2)$ and $(2, -8)$. The foci are on the major axis a distance of

$$c = \sqrt{|9 - 25|}$$
$$= \sqrt{16}$$
$$= 4$$

units from the center, at the points $(2, -7)$ and $(2, 1)$.

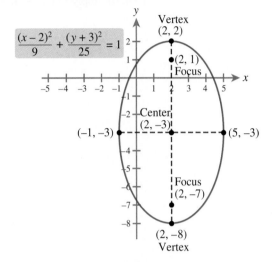

Figure 70

···◆·EXAMPLE 4

Graphing an Ellipse and Identifying Key Features

Sketch the graph of the ellipse with equation $9x^2 + y^2 + 36x + 27 = 0$. Identify the center, vertices, and foci.

Solution We must first put the equation in standard form by completing the square.

$$9x^2 + y^2 + 36x + 27 = 0$$
$$(9x^2 + 36x + \blacksquare) + y^2 = -27$$
$$9(x^2 + 4x + \blacksquare) + y^2 = -27$$
$$9(x^2 + 4x + \boxed{4}) + y^2 = -27 + 9(4) \qquad \text{Adding } 9(4) = 36 \text{ to both sides}$$
$$9(x + 2)^2 + y^2 = 9$$
$$\frac{(x + 2)^2}{1} + \frac{y^2}{9} = 1 \qquad \text{Dividing both sides by 9}$$

The center of the ellipse is $(-2, 0)$. Since $a = \sqrt{1} = 1$ and $b = \sqrt{9} = 3$, we plot points on the ellipse 1 unit to the left and right of the center and 3 units above and below the center. The graph is shown in Figure 71. The major axis is vertical, and the vertices have coordinates $(-2, 3)$ and $(-2, -3)$. The foci are on the major axis a distance of

$$c = \sqrt{|1 - 9|}$$
$$= \sqrt{8}$$
$$= 2\sqrt{2}$$

units from the center, at the points $\left(-2, -2\sqrt{2}\right)$ and $\left(-2, 2\sqrt{2}\right)$.

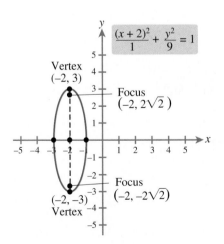

Figure 71

▶EXAMPLE 5

Finding the Equation of an Ellipse

Find the equation of the ellipse with foci $(\pm 4, 0)$ and major axis length of 12.

Solution   Since the foci are on the x-axis at equal distances from the origin, the ellipse is centered at the origin and has a horizontal major axis (see Figure 72). Moreover, since the major axis has length 12, we have $2a = 12$ or $a = 6$. The coordinates of the foci, $(\pm 4, 0)$, imply that $c = 4$. Since $b < a$, $c = \sqrt{a^2 - b^2}$ and so

$$4 = \sqrt{6^2 - b^2}$$
$$16 = 36 - b^2$$
$$b^2 = 20$$

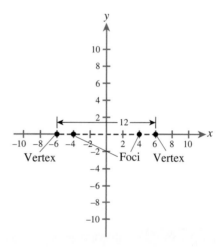

Figure 72

Thus, the equation of the ellipse is

$$\frac{x^2}{36} + \frac{y^2}{20} = 1$$

and its graph is shown in Figure 73.

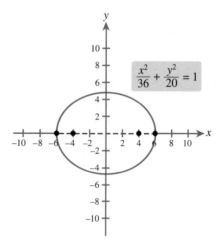

Figure 73

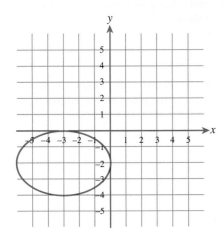

Figure 74

------EXAMPLE 6

Finding the Equation of an Ellipse from Its Graph

Find a possible equation for the ellipse shown in Figure 74. Verify with a graphing calculator.

Solution The graph in Figure 74 appears to be centered at $(-3, -2)$, so the equation has the form

$$\frac{(x + 3)^2}{a^2} + \frac{(y + 2)^2}{b^2} = 1$$

Since the horizontal axis has length 6, and the vertical axis has length 4, it follows that $a = 3$ and $b = 2$. This gives us the equation

$$\frac{(x + 3)^2}{3^2} + \frac{(y + 2)^2}{2^2} = 1$$

To graph this equation on a graphing calculator, we must first solve for y.

$$\frac{(x + 3)^2}{9} + \frac{(y + 2)^2}{4} = 1$$

$$\frac{(y + 2)^2}{4} = 1 - \frac{(x + 3)^2}{9}$$

$$(y + 2)^2 = 4 - \frac{4}{9}(x + 3)^2 \qquad \text{Multiplying both sides by 4}$$

$$y + 2 = \pm\sqrt{4 - \frac{4}{9}(x + 3)^2} \qquad \text{Taking the square root of both sides}$$

$$y = -2 \pm \sqrt{4 - \frac{4}{9}(x + 3)^2}$$

Graphing the branches $y = -2 + \sqrt{4 - \frac{4}{9}(x + 3)^2}$ and $y = -2 - \sqrt{4 - \frac{4}{9}(x + 3)^2}$, we obtain the ellipse shown in Figure 75. It appears to be the same as that shown in Figure 74.

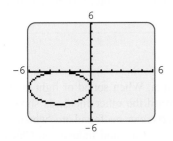

Figure 75

------EXAMPLE 7

An Application Involving an Ellipse

A train tunnel through a mountain is semielliptical, with height 18 feet at the center and width 24 feet at the base (see Figure 76). A train is hauling a load of flatcars that are piled high with freight. If the flatcars are 10 feet wide and the load reaches a height of 15 feet above the ground, will the flatcars clear the opening of the tunnel?

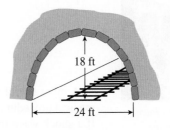

18 ft

24 ft

Figure 76

Solution To determine the clearance, we must know the height of the tunnel at the edge of the flatcar. We first construct a coordinate system with the x-axis on the ground and the origin at the center of the tracks. Next we find the equation of the ellipse that forms the tunnel opening. Using the dimensions of the tunnel opening, we know that the ellipse has a vertical major axis of length 2(18) and a horizontal minor axis of length 24. Thus $a = 12$ and $b = 18$. The equation is as follows:

$$\frac{x^2}{12^2} + \frac{y^2}{18^2} = 1$$

The edge of a 10-foot-wide flatcar corresponds to $x = 5$, and so the height of the tunnel at the edge of the flatcar is given by the y-value of the point on the ellipse when $x = 5$. Thus, we set $x = 5$ and solve for y.

$$\frac{5^2}{12^2} + \frac{y^2}{18^2} = 1$$

$$\frac{y^2}{324} = 1 - \frac{25}{144}$$

$$y^2 = 324\left(1 - \frac{25}{144}\right)$$

$$y = \pm\sqrt{324\left(1 - \frac{25}{144}\right)} \approx \pm 16.36$$

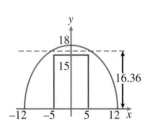

Figure 77

Thus, the height of the tunnel above the edge of the flatcar is approximately 16.36 feet, which gives 1.36 feet of clearance. See Figure 77.

Reflective Property of Ellipses

The reflective property of an ellipse involves both foci. When sound or light is emitted from one focus of an ellipse, it is reflected toward the other focus, as shown in Figure 78. An interesting application of this property can be found in "whispering rooms," such as the ones located at the Museum of Science and Industry in Chicago or the Capitol Building in Washington, D.C. These rooms have walls (or ceilings) shaped like an ellipse. If you stand at one of the foci and whisper something to a friend standing at the other focus, you will be heard clearly by your friend but nobody else.

John Quincy Adams is said to have taken advantage of the reflective property of ellipses to overhear strategies of his opponents in the U.S. Capitol Building's Statuary Hall.

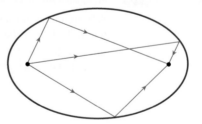

Figure 78

A more useful application of the reflective property of an ellipse can be found in a medical procedure called *shock wave lithotripsy*, used in the treatment of kidney stones. If a shock wave transmitter and an elliptic reflector are placed in such a way that the transmitter and kidney stone are located at the two foci of the reflector, the shock waves will be directed at the kidney stone, causing it to shatter, while leaving the rest of the body unharmed.

Understanding and Mastery Checklists

Concepts to Understand	Skills to Master

Concepts to Understand

Ellipse

❖

Foci of an ellipse

❖

Vertices of an ellipse

❖

Center of an ellipse

❖

Major axis of an ellipse

❖

Minor axis of an ellipse

❖

Reflective property of an ellipse

Skills to Master

Sketch the graph of an ellipse.

❖

Identify the center, vertices, and foci of an ellipse.

❖

Find the equation of an ellipse given
information about its graph.

Exercises 8.3

Exercises 1-14   *Find the center, vertices, and foci for the given ellipse, and sketch its graph.*

1. $\dfrac{x^2}{9} + \dfrac{y^2}{25} = 1$

2. $\dfrac{x^2}{16} + \dfrac{y^2}{9} = 1$

3. $4x^2 + 9y^2 = 36$

4. $16x^2 + 9y^2 = 144$

5. $9x^2 + 4y^2 = 1$

6. $16x^2 + 25y^2 = 1$

7. $5x^2 + 8y^2 = 40$

8. $3x^2 + 2y^2 = 6$

9. $\dfrac{(x+3)^2}{4} + \dfrac{(y-1)^2}{16} = 1$

10. $\dfrac{(x-2)^2}{25} + \dfrac{(y+2)^2}{9} = 1$

11. $9x^2 + 4y^2 - 18x - 24y + 9 = 0$

12. $25x^2 + 16y^2 - 200x - 32y + 16 = 0$

13. $x^2 + 36y^2 + 4x - 72y + 4 = 0$

14. $16x^2 + y^2 + 4y - 12 = 0$

Exercises 15-26   *Find an equation for the ellipse with the given properties.*

15. Center (0, 0), horizontal major axis of length 8, and vertical minor axis of length 6

16. Vertices (0, ±4) and minor axis of length 2

17. Vertices $\left(0, \pm\sqrt{6}\right)$ and foci $\left(0, \pm\sqrt{2}\right)$

18. Foci $\left(\pm\sqrt{2}, 0\right)$ and major axis of length 4

19. Vertices (±6, 0) and passes through the point (−4, 2)

20. Foci (0, ±2) and passes through the point (3, 2)

21. Center (2, −1), vertical major axis of length 6, horizontal minor axis of length 4

22. Vertices (−2, 6) and (−2, 0) and minor axis of length 1

23. Vertices (−4, −2) and (6, −2) and foci (−2, −2) and (4, −2)

24. Foci $(0, 4)$ and $(6, 4)$ and major axis of length $\sqrt{10}$

25. The following graph:

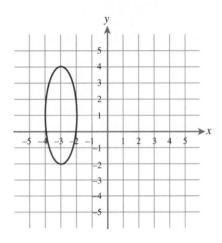

26. The following graph:

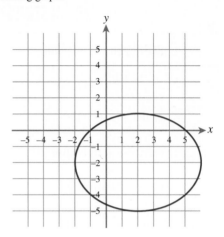

Exercises 27-30 *Solve for y, and plot the resulting equation(s) with a graphing calculator. Identify the coordinates of the vertices.*

27. $x^2 + 4y^2 + 4x = 0$ **28.** $4x^2 + y^2 - 16x = 0$

29. $4x^2 + y^2 - 2y - 3 = 0$ **30.** $x^2 + 4y^2 + 16y = 0$

Applications

31. Earth's Orbit   Earth's orbit is elliptical with major axis length 186 million miles and minor axis length 185.8 million miles. Write an equation for the path of the orbit, assuming the center is located at the origin. It may surprise you to find that the sun is not at the center of the orbit. The sun is, in fact, located at a focus, a distance of 4.3 million miles from the center. Sketch a graph of the orbit of Earth and locate the sun on the major axis.

32. Mars's Orbit   The orbit of Mars is elliptical with major axis length 283.5 million miles and minor axis length 278.5 million miles. Write an equation for the path of the orbit, assuming the center is located at the origin. The sun is located at a focus, a distance of 26.5 million miles from the center. Sketch a graph of the orbit of Mars and locate the sun on the major axis.

33. Railroad Bridge   The arch of a railroad bridge over a 2-lane highway has the shape of a semiellipse (see Figure 79). The distance from the base of the arch on one side of the road to the base on the other is 50 feet, and the height of the arch at the center is 20 feet.

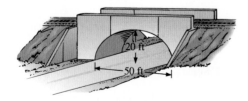

Figure 79

a. Find the equation of the ellipse that gives the shape of the arch. Assume that the *x*-axis is on the ground and runs perpendicular to the centerline of the highway, the origin is on the centerline, and the *y*-axis is vertical.

b. What is the height of the arch 5 feet from the base?

c. Could a tractor-trailer with a height of 14 feet and a width of 10 feet pass under the bridge without going over the centerline of the road?

34. Whispering Room   An elliptical whispering room is to be constructed with a maximum width of 30 feet and a maximum length of 50 feet. Locate the positions at which two people should stand if they wish to whisper to each other from opposite sides of the room without others hearing.

Exercises 35-36   *An Astronomical Unit (A.U.) refers to the mean distance from Earth to the sun. When dealing with elliptical orbits of celestial bodies with the sun located at one focus, the* **perihelion** *distance is the distance from the sun to the closest vertex, whereas the* **aphelion** *distance is the distance from the sun to the most distant vertex.*

35. Comet Encke   Some scientists have speculated that a piece of comet Encke fell to Earth in 1908, causing massive destruction in a remote region of Siberia. (See Cometary Collisions box on page 607 for more details.) The orbit of comet Encke is an ellipse, with the sun located at one focus. The major axis has length 4.4 A.U., and the minor axis has length 2.2 A.U. Find the perihelion and aphelion distances for comet Encke.

36. Halley's Comet The orbit of Halley's comet is an ellipse, with the sun located at one focus. The major axis has length 36.2 A.U., and the minor axis has length 9.1 A.U. Find the perihelion and aphelion distances for Halley's comet.

Halley's comet in space

37. Anderson Elliptical Window The Anderson Window company manufactures a glass window in the shape of a semiellipse (see Figure 80), with a major axis of length 168.3 centimeters. The area of the glass is advertised as 3995 cm².

a. The area of an ellipse with equation $\dfrac{x^2}{a^2} + \dfrac{y^2}{b^2} = 1$ is given by $A = \pi ab$. Find the equation of the ellipse that forms the boundary of the Anderson window, assuming that the origin is at the center and the major axis lies on the x-axis.

b. Find the height of the window at a point 30 centimeters from the center.

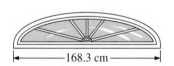

|←————— 168.3 cm —————→|

Figure 80

38. Comet Path Length There is no simple formula for the circumference, or more properly, arc length, of an ellipse. However, a crude approximation can be made by pretending that the ellipse is made up of line segments connecting points on the ellipse.

a. Find a point P on the ellipse $\dfrac{x^2}{16} + \dfrac{y^2}{9} = 1$ roughly halfway between $(4, 0)$ and $(0, 3)$.

b. Find the distance between P and $(4, 0)$ and the distance between P and $(0, 3)$.

c. Use the results from part b to estimate the arc length of the portion of the ellipse between $(4, 0)$ and $(0, 3)$. Is this result greater than or less than the actual arc length?

d. Use the result from part c to estimate the arc length of the entire ellipse. Is this result greater or less than the actual arc length?

e. How can this technique be modified to produce a more precise estimate of arc length?

f. Comet Hale-Bopp, which was visible to the naked eye during much of the spring of 1997, travels in an elliptical orbit through the solar system. Estimate the length of its orbit given that its major axis is 3.5×10^{10} miles and its minor axis is 1.0×10^9 miles.

Concepts and Critical Thinking

Exercises 39–42 *Answer true or false.*

39. An ellipse centered at the origin with horizontal major axis is symmetric with respect to the origin.

40. It is possible for the plot of an ellipse to be entirely contained within the viewing rectangle of a graphing calculator.

41. The reflective property of ellipses states that light emanating from one focus will be reflected through the other focus.

42. If the foci of an ellipse are very close together, then the ellipse will appear nearly circular.

Exercises 43–46 *Give an example of each.*

43. The equation of an ellipse centered at $(0, 3)$ with major axis of length 4

44. The equations of two ellipses such that one is a 90° rotation of the other

45. A real-world application of the reflective property of ellipses

46. The equation of an ellipse with foci 2 units apart

47. If an ellipse has a horizontal major axis of length $2a$ and foci at $(\pm c, 0)$, explain why $a^2 - c^2 > 0$.

Questions for Discussion or Essay

48. An elliptical "pool" table with only one pocket is being used for a trick-shooting demonstration. The pool shark carefully places a ball on the table but hits it in a seemingly arbitrary direction. Time after time, the ball bounces off the cushion directly into the pocket. Explain how this could happen.

49. If we use only the information given in Exercise 36, is it possible to determine how close Halley's comet comes to Earth? Explain.

50. Light from an ordinary flashlight is contained within a cone. Explain how an ellipse is formed by shining a flashlight inside a room.

Projects for Enrichment

51. Drawing Ellipses In this project, we investigate how to draw ellipses of varying sizes. For materials, locate a large piece of cardboard, blank paper, two thumbtacks, several long pieces of string, and a pencil. Lay a piece of paper on top of the cardboard. Cut a piece of string a little longer than 4 inches and attach each end around a thumb tack so that 4 inches of string is between the tacks. Stick the tacks into the cardboard 3 inches apart so there is some slack in the string. Place the tip of the pencil as shown in Figure 81 and, keeping the string taut and the tip of the pencil on the paper, carefully move the pencil along the string. The pencil will trace out the path of an ellipse with foci located at the thumbtacks. The length of the major axis is equal to the total length of the string between the tacks.

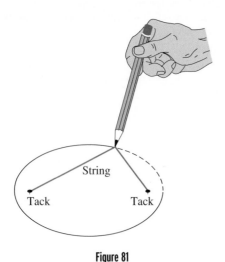

String

Tack   Tack

Figure 81

a. Use the fact that the string is 4 inches long and the tacks are 3 inches apart to find the equation of the ellipse you drew. Assume the x-axis passes through the foci and the origin is midway between them.

b. Draw an ellipse with foci a distance of 3 inches apart and major axis length of 5 inches.

c. Draw an ellipse with foci 3 inches apart and major axis length of 6 inches.

d. What general observation can you make about the shape of the ellipse if the distance between the foci is held constant but the length of the major axis (that is, the length of the string) is increased?

e. Compute the ratio of the distance between the foci to the length of the major axis for each of the three ellipses you have constructed. This ratio is called the *eccentricity* of the ellipse. What is the largest possible value of the eccentricity? Why? What is the connection between the eccentricity and the shape of the ellipse? What is the smallest the eccentricity could be? What would the shape of the ellipse be then?

Section 8.4 | Hyperbolas and Classification of Conic Sections

- What is the path of a sonic boom?
- How can hyperbolas help explain shadows on a wall?
- How can hyperbolas help navigate a ship?
- Why are the shapes of conic sections often found in the mirrors of reflective telescopes?
- How can hyperbolas be used to help Inspector Magill locate a stolen car?

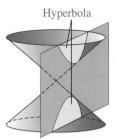

Hyperbola

Figure 82

In Section 8.2, we described the conic sections as curves formed when a double cone is cut by a plane. If the plane is parallel to the axis of the cone, we obtain a hyperbola, as shown in Figure 82. Using geometric techniques, it can be shown that the following characterization of a hyperbola is equivalent.

Definition of a Hyperbola

A **hyperbola** is the set of all points (x, y) such that the difference of the distances from two fixed points (called the **foci**) is constant, as shown in Figure 83. The line through the foci intersects the hyperbola at two points, called the **vertices**. The line segment connecting the two vertices is called the **transverse axis**, and the midpoint of the transverse axis is called the **center**.

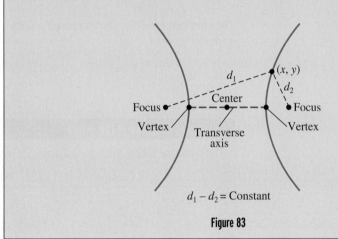

Figure 83

The standard form for the equation of a hyperbola with a horizontal transverse axis and centered at the origin can be derived in much the same way as that for the ellipse. Although the derivation is lengthy, the end result is quite simple. We begin by placing the foci at $(-c, 0)$ and $(c, 0)$, as shown in Figure 84. According to the definition, for any point (x, y) on the hyperbola, the difference of the distances from the two foci must be constant. To simplify later computations, we denote this constant difference by $2a$. Using the distance formula for each of the two distances d_1 and d_2, we have

$$d_1 = \sqrt{(x + c)^2 + y^2} \quad \text{and} \quad d_2 = \sqrt{(x - c)^2 + y^2}$$

Setting $d_1 - d_2 = 2a$, we proceed as follows:

$$\sqrt{(x + c)^2 + y^2} - \sqrt{(x - c)^2 + y^2} = 2a$$
$$\sqrt{(x + c)^2 + y^2} = 2a + \sqrt{(x - c)^2 + y^2}$$

Now we square both sides and collect terms.

$$(x + c)^2 + y^2 = 4a^2 + 4a\sqrt{(x - c)^2 + y^2} + (x - c)^2 + y^2$$
$$x^2 + 2cx + c^2 + y^2 = 4a^2 + 4a\sqrt{(x - c)^2 + y^2} + x^2 - 2cx + c^2 + y^2$$
$$4cx - 4a^2 = 4a\sqrt{(x - c)^2 + y^2}$$
$$cx - a^2 = a\sqrt{(x - c)^2 + y^2}$$

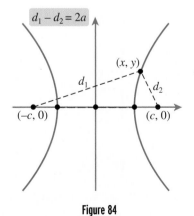

Figure 84

Once again, we square both sides and collect terms.

$$\left(cx - a^2\right)^2 = a^2\left[(x - c)^2 + y^2\right]$$
$$c^2x^2 - 2a^2cx + a^4 = a^2\left(x^2 - 2cx + c^2 + y^2\right)$$
$$c^2x^2 - 2a^2cx + a^4 = a^2x^2 - 2a^2cx + a^2c^2 + a^2y^2$$
$$c^2x^2 - a^2x^2 - a^2y^2 = a^2c^2 - a^4$$
$$\left(c^2 - a^2\right)x^2 - a^2y^2 = a^2\left(c^2 - a^2\right)$$

Finally, since $c^2 - a^2 > 0$ (see Exercise 55), we set $b^2 = c^2 - a^2$ and substitute.

$$b^2x^2 - a^2y^2 = a^2b^2$$
$$\frac{x^2}{a^2} - \frac{y^2}{b^2} = 1$$

The standard equation for a hyperbola with a vertical transverse axis can be derived in a similar manner.

Hyperbolas Centered at the Origin

The standard forms for the equation of a hyperbola centered at $(0, 0)$ are as follows:

Equation	Description
$\dfrac{x^2}{a^2} - \dfrac{y^2}{b^2} = 1$	Horizontal transverse axis
$\dfrac{y^2}{b^2} - \dfrac{x^2}{a^2} = 1$	Vertical transverse axis

The vertices are the endpoints of the transverse axis: $(\pm a, 0)$ in the case of a horizontal transverse axis (see Figure 85) and $(0, \pm b)$ for a vertical transverse axis (see Figure 86). The foci are a distance of $c = \sqrt{a^2 + b^2}$ units from the center on the same line as the vertices.

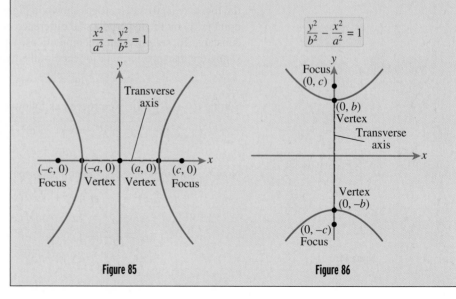

Figure 85   Figure 86

········**EXAMPLE 1**

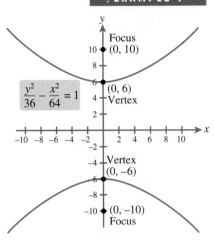

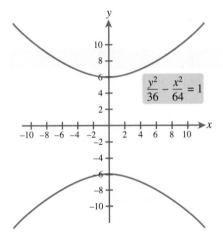

Figure 87

A Hyperbola Centered at the Origin

Determine the orientation and find the vertices and foci of the hyperbola with equation

$$\frac{y^2}{36} - \frac{x^2}{64} = 1$$

Solution   This equation is in the standard form of a hyperbola with vertical transverse axis. Thus, since $b = \sqrt{36} = 6$, the vertices are $(0, -6)$ and $(0, 6)$. To find the foci, we compute c.

$$c = \sqrt{a^2 + b^2} = \sqrt{36 + 64} = 10$$

So the foci are 10 units away from the center, on the same line as the vertices, at the points $(0, -10)$ and $(0, 10)$. The graph is shown in Figure 87.

One of the interesting features of the graph of a hyperbola is that as we zoom out, it begins to look more and more like a pair of intersecting lines. Figures 88–90 show several views of the graph of the hyperbola $\frac{y^2}{36} - \frac{x^2}{64} = 1$ from Example 1.

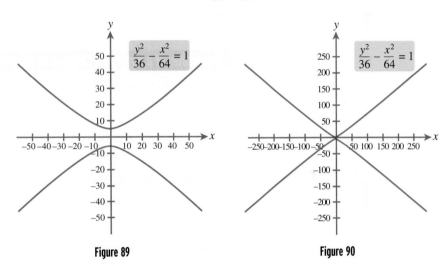

Figure 88   **Figure 89**   **Figure 90**

The lines that the graph of a hyperbola approaches are called its **asymptotes**, and they serve as useful aids in sketching the graph of the hyperbola. To determine the equations of the lines, let us look more closely at a hyperbola with a vertical axis. Solving the equation

$$\frac{y^2}{b^2} - \frac{x^2}{a^2} = 1$$

for y^2 we obtain

$$y^2 = \frac{b^2}{a^2}x^2 + b^2$$

Now, for very large values of x, the first term will dwarf the second. Thus, for large values of x we have

$$y^2 \approx \frac{b^2}{a^2}x^2 \quad \text{or} \quad y \approx \pm\frac{b}{a}x$$

Thus, the asymptotes for a hyperbola centered at the origin with a vertical axis are $y = \pm\frac{b}{a}x$. In a similar fashion, we can see that a hyperbola with a horizontal axis has asymptotes $y = \pm\frac{b}{a}x$. In either case, the asymptotes can be located by constructing a rectangle with sides passing through the points $(-a, 0)$, $(a, 0)$, $(0, b)$, and $(0, -b)$. The extended diagonals of this rectangle are the asymptotes. See Figures 91 and 92 for details.

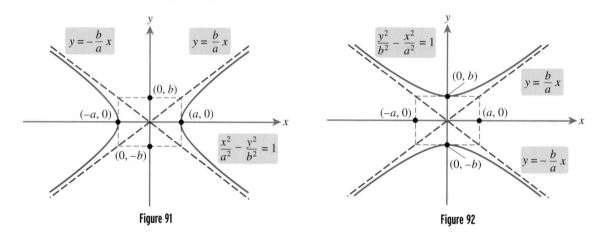

Figure 91   Figure 92

Asymptotes of a Hyperbola Centered at the Origin

The asymptotes for a hyperbola centered at the origin are

$$y = \frac{b}{a}x \quad \text{and} \quad y = -\frac{b}{a}x$$

·····>EXAMPLE 2

A Hyperbola Centered at the Origin

Find the vertices, foci, and asymptotes for the hyperbola with equation

$$9x^2 - 16y^2 = 144$$

and sketch its graph.

Solution   We first divide through by 144 to rewrite the equation in standard form.

$$9x^2 - 16y^2 = 144$$

$$\frac{9x^2}{144} - \frac{16y^2}{144} = 1$$

$$\frac{x^2}{16} - \frac{y^2}{9} = 1$$

From the standard form of the equation, we know that the hyperbola has a horizontal transverse axis with $a = \sqrt{16} = 4$ and $b = \sqrt{9} = 3$. Thus, the vertices are $(-4, 0)$ and $(4, 0)$ and the transverse axis is the line segment connecting these two points. The distance from the center to each focus is $c = \sqrt{16 + 9} = 5$, and so the foci are located at $(-5, 0)$ and $(5, 0)$. To find the asymptotes, we construct the rectangle whose sides pass through the vertices $(-4, 0)$ and $(4, 0)$ on the x-axis and the points $(0, -3)$ and $(0, 3)$ on the y-axis. The asymptotes are then formed by extending the diagonals of the rectangle. The equations of the asymptotes are $y = \frac{3}{4}x$ and $y = -\frac{3}{4}x$. The graph is shown in Figure 93.

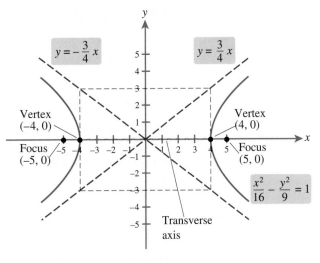

Figure 93

As with both parabolas and ellipses, the standard equations of hyperbolas centered away from the origin can be found by applying translations. These equations are summarized as follows.

Hyperbolas Centered at (h, k)

The standard forms for the equation of a hyperbola centered at (h, k) are as follows:

Equation	Description
$\dfrac{(x-h)^2}{a^2} - \dfrac{(y-k)^2}{b^2} = 1$	Horizontal transverse axis
$\dfrac{(y-k)^2}{b^2} - \dfrac{(x-h)^2}{a^2} = 1$	Vertical transverse axis

The vertices are the endpoints of the transverse axis: a units from the center in the case of a horizontal transverse axis (see Figure 94) and b units from the center in the case of a vertical transverse axis (see Figure 95). The foci are a distance of $c = \sqrt{a^2 + b^2}$ units from the center on the same line as the vertices.

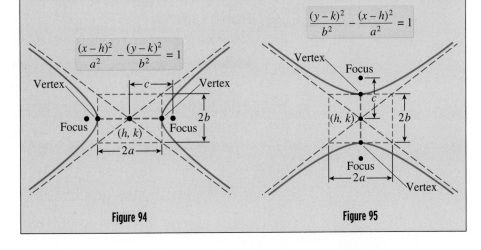

Figure 94   **Figure 95**

⋯⋯⋯⟩EXAMPLE 3

Graphing a Hyperbola and Identifying Key Features

Find the center, vertices, foci, and asymptotes for the hyperbola with equation

$$\frac{(x-2)^2}{4} - \frac{(y-1)^2}{16} = 1$$

and sketch its graph.

Solution   The center of the hyperbola is the point $(2, 1)$. The axis is horizontal, with $a = 2$ and $b = 4$. Thus, the vertices are 2 units to the left and to the right of the center, at $(0, 1)$ and $(4, 1)$. The transverse axis is the line segment connecting the two vertices. The foci are located $c = \sqrt{4 + 16} = \sqrt{20} = 2\sqrt{5}$ units from the center on the same line as the vertices, at the points $\left(2 - 2\sqrt{5}, 1\right)$ and $\left(2 + 2\sqrt{5}, 1\right)$. To locate the asymptotes, we construct a rectangle with horizontal sides 4 units above and below the center and vertical sides 2 units to the left and right of the center. The asymptotes are the extended diagonals of this rectangle. They have slope $m = \pm(b/a) = \pm 2$, and they pass through the center $(2, 1)$. Their equations are thus $y - 1 = \pm 2(x - 2)$. The graph is shown in Figure 96.

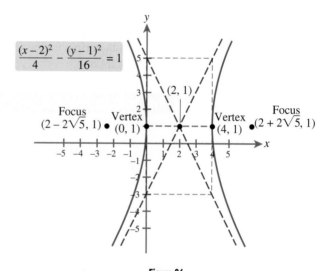

Figure 96

⋯⋯⋯⟩EXAMPLE 4

Graphing a Hyperbola

Sketch the graph of the hyperbola with equation $16x^2 - y^2 + 64x - 2y + 67 = 0$.

Solution   We first complete the square in x and y to write the equation in standard form.

$$16x^2 - y^2 + 64x - 2y + 67 = 0$$

$$(16x^2 + 64x + \blacksquare) + (-y^2 - 2y + \blacksquare) = -67$$

$$16(x^2 + 4x + \blacksquare) - (y^2 + 2y + \blacksquare) = -67$$

$$16(x^2 + 4x + \boxed{4}) - (y^2 + 2y + \boxed{1}) = -67 + 64 - 1 \qquad \text{Adding } 16(4) \text{ and } -(1) \text{ to both sides}$$

$$16(x + 2)^2 - (y + 1)^2 = -4$$

$$-4(x + 2)^2 + \frac{(y + 1)^2}{4} = 1 \qquad \text{Dividing both sides by } -4$$

$$\frac{(y + 1)^2}{4} - 4(x + 2)^2 = 1$$

$$\frac{(y + 1)^2}{4} - \frac{(x + 2)^2}{1/4} = 1$$

$$\frac{(y + 1)^2}{2^2} - \frac{(x + 2)^2}{(1/2)^2} = 1$$

The center of the hyperbola is at $(-2, -1)$. The axis is vertical with $a = \frac{1}{2}$ and $b = 2$. The vertices are 2 units above and below the center, at $(-2, -3)$ and $(-2, 1)$. To sketch the hyperbola, we construct a rectangle with center $(-2, -1)$, horizontal sides 2 units above and below the center, and vertical sides $\frac{1}{2}$ unit to the left and to the right of the center. The asymptotes are formed by extending the diagonals, as shown in Figure 97.

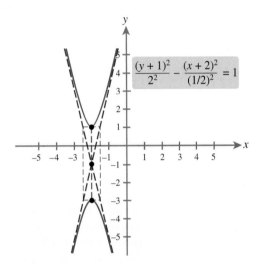

$$\frac{(y + 1)^2}{2^2} - \frac{(x + 2)^2}{(1/2)^2} = 1$$

Figure 97

····➤**EXAMPLE 5**

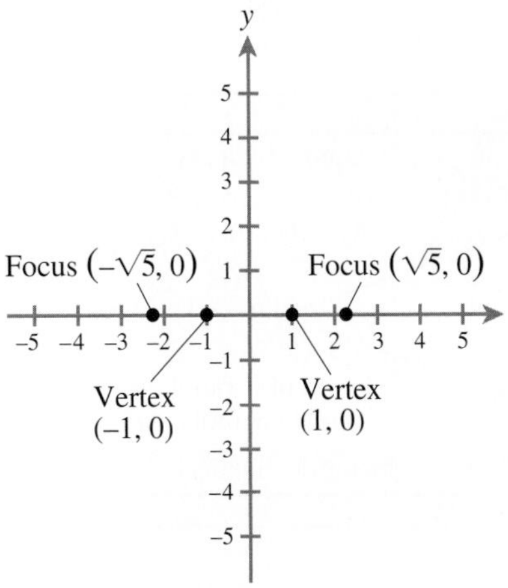

Finding an Equation of a Hyperbola

Find an equation of the hyperbola with vertices $(\pm1, 0)$ and foci $\left(\pm\sqrt{5}, 0\right)$.

Solution   We begin by plotting the vertices and foci, as shown in Figure 98. It is clear from the locations of these points that the center is at the origin (halfway between the vertices) and the transverse axis is horizontal (since the vertices are on the x-axis). Thus, the standard form of the equation is

$$\frac{x^2}{a^2} - \frac{y^2}{b^2} = 1$$

The vertices of a hyperbola with this equation form are a units from the center, and so it must follow that $a = 1$. Furthermore, since the foci are located $\sqrt{a^2 + b^2}$ units from the center, we have

$$\sqrt{a^2 + b^2} = \sqrt{5}$$

$$\sqrt{1 + b^2} = \sqrt{5}$$

$$b^2 = 4$$

Focus $\left(-\sqrt{5}, 0\right)$ Focus $\left(\sqrt{5}, 0\right)$

Vertex $(-1, 0)$ Vertex $(1, 0)$

Figure 98

Thus, the hyperbola has equation

$$\frac{x^2}{1} - \frac{y^2}{4} = 1$$

The graph is shown in Figure 99.

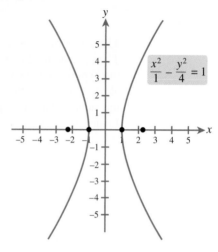

Figure 99

Reflective Property of Hyperbolas

Just as with ellipses, the reflective property of hyperbolas involves its foci. If an incoming light ray or sound wave is aimed at one focus of a hyperbola, it is reflected toward the other focus, as shown in Figure 100. Hyperbolic reflectors are often used as auxiliary mirrors in telescopes to direct the image to the eyepiece of the telescope, where it can be magnified. Two schematic diagrams of telescopes are shown in Figures 101 and 102. Notice that in both the main reflector is parabolic. Also in both, the hyperbolic reflector is positioned so that its focus coincides with that of the parabolic mirror. In Figure 101, the eyepiece of the telescope is located at the other focus of the hyperbolic reflector. In Figure 102, an elliptic reflector is added, with one of its foci at the second focus of the hyperbolic reflector and its other focus at the eyepiece.

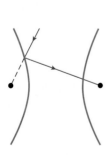

Figure 100

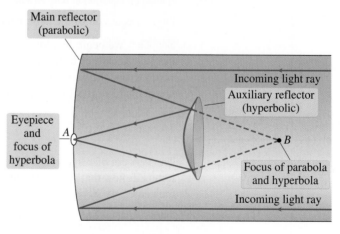

Main reflector (parabolic)

Incoming light ray

Auxiliary reflector (hyperbolic)

Eyepiece and focus of hyperbola

A

B

Focus of parabola and hyperbola

Incoming light ray

Figure 101

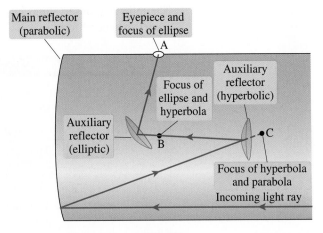

Figure 102

Classification of Conic Sections

Table 2 gives some examples of conic sections from earlier in the text.

Table 2

Equation	Conic Section
$x^2 + y^2 - 2x + 4y - 20 = 0$	Circle
$x = y^2 + 2y - 3$	Parabola
$9x^2 + y^2 + 36x + 27 = 0$	Ellipse
$16x^2 - y^2 + 64x - 2y + 67 = 0$	Hyperbola

In every case, the equation of the conic section can be written in the form $Ax^2 + Cy^2 + Dx + Ey + F = 0$. In fact, all conic sections we have seen thus far have equations of this form. More generally, we have the following.

The General Equation of Conic Sections with Vertical or Horizontal Axes

The general equation of a circle, or any parabola, ellipse, or hyperbola with a vertical or horizontal axis is $Ax^2 + Cy^2 + Dx + Ey + F = 0$.

Note that not all equations of the form $Ax^2 + Cy^2 + Dx + Ey + F = 0$ correspond to the familiar conic sections. Consider, for example, the equations

$$x^2 - y^2 = 0 \quad \text{and} \quad x^2 + y^2 + 1 = 0$$

The first equation can be rewritten as $x = \pm y$, and so it yields the equations of two lines. The second can be rewritten as $x^2 + y^2 = -1$, which has no solutions and hence no graph. These two equations are examples of **degenerate** conic sections. Throughout the remainder of the section, we consider only the general equations of **nondegenerate** conic sections—namely, circles, parabolas, ellipses, and hyperbolas. In these cases, it is possible to recognize the type of conic by inspecting the coefficients of x^2 and y^2.

The following observations can be verified by rewriting the standard equations of the conics in general form.

Classifying Conic Sections

A nondegenerate conic section of the form $Ax^2 + Cy^2 + Dx + Ey + F = 0$ can be classified as follows:

Condition	Verbal description	Resulting conic
$A = C$	The squared terms have equal coefficients.	Circle
$AC = 0$	There is only one squared term.	Parabola
$AC > 0, A \neq C$	The coefficients of the squared terms are unequal but have the same sign.	Ellipse
$AC < 0$	The coefficients of the squared terms have opposite signs.	Hyperbola

EXAMPLE 6

Classifying Conic Sections

Classify each of the following nondegenerate conics as a circle, a parabola, an ellipse, or a hyperbola:

a. $4x^2 - 9y^2 - 8x - 36y - 68 = 0$

b. $y^2 + 8x + 6y + 25 = 0$

c. $9x^2 - 36x + 31 = -4y^2 - 8y$

Solution

a. Since both squared terms are present and the coefficients have opposite signs, this is a hyperbola.

b. Since only one squared term is present, this is a parabola.

c. We first rewrite the equation in general form as

$$9x^2 + 4y^2 - 36x + 8y + 31 = 0$$

Since the coefficients of the squared terms have the same sign but are not equal, this is an ellipse.

If the equation of a conic section is given in general form, its graph can be obtained either by writing the equation in standard form and using the techniques described earlier or by solving the equation for y and using a graphing calculator.

EXAMPLE 7

Graphing a Conic Section Using a Graphing Calculator

Identify the conic $-x^2 + y^2 + 3x + 4y - 5 = 0$ and obtain its plot with a graphing calculator. In the case of a parabola, ellipse, or hyperbola, approximate the coordinates of any vertices.

Solution Since the coefficients of x^2 and y^2 have opposite signs, this is a hyperbola. We solve for y by completing the square in y.

$$-x^2 + y^2 + 3x + 4y - 5 = 0$$
$$(y^2 + 4y +) = x^2 - 3x + 5$$
$$(y^2 + 4y + 4) = x^2 - 3x + 5 + 4$$
$$(y + 2)^2 = x^2 - 3x + 9$$
$$y + 2 = \pm\sqrt{x^2 - 3x + 9}$$
$$y = -2 \pm \sqrt{x^2 - 3x + 9}$$

Thus, the two branches of the hyperbola are

$$y = -2 + \sqrt{x^2 - 3x + 9} \quad \text{and} \quad y = -2 - \sqrt{x^2 - 3x + 9}$$

After entering both equations and graphing, we obtain the graph shown in Figure 103. The coordinates of the vertices, found using either the trace or calculate-maximum/ minimum feature, are approximately $(1.5, 0.6)$ and $(1.5, -4.6)$.

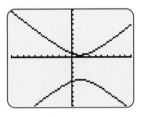

Figure 103

Understanding and Mastery Checklists

Concepts to Understand	Skills to Master
Hyperbola	Sketch the graph of a hyperbola.
Foci of a hyperbola	Identify the center, vertices, and foci of a hyperbola.
Vertices of a hyperbola	Find the asymptotes of a hyperbola.
Center of a hyperbola	Find the equation of a hyperbola given information about its graph.
Transverse axis of a hyperbola	Identify a nondegenerate conic section from its equation.
Asymptotes of a hyperbola	Graph a conic section with a graphing calculator.
Reflective property of a hyperbola	Identify a conic section from its graph.
General equation of a conic section	

Exercises 8.4

Exercises 1-12   *Find the center, vertices, foci, and asymptotes for the given hyperbola, and sketch its graph.*

1. $\dfrac{x^2}{9} - \dfrac{y^2}{16} = 1$ **2.** $\dfrac{y^2}{25} - \dfrac{x^2}{16} = 1$

3. $4y^2 - 9x^2 = 36$ **4.** $25x^2 - 16y^2 = 400$

5. $12x^2 - 3y^2 = -24$ **6.** $5x^2 - 3y^2 = 15$

7. $\dfrac{(x-1)^2}{4} - \dfrac{(y-4)^2}{9} = 1$ **8.** $\dfrac{(y+3)^2}{16} - \dfrac{(x+2)^2}{25} = 1$

9. $4x^2 - 9y^2 - 16x + 54y - 101 = 0$

10. $25x^2 - 16y^2 - 50x + 160y + 25 = 0$

11. $4x^2 - 25y^2 - 32x + 164 = 0$

12. $x^2 - 4y^2 + 4x - 24y - 36 = 0$

Exercises 13-18   *Find an equation for the hyperbola with the given properties.*

13. Vertices $(0, \pm 4)$ and foci $(0, \pm 5)$

14. Vertices $(\pm 6, 0)$ and asymptotes $y = \pm \dfrac{4}{3}x$

15. Foci $\left(0, \pm 2\sqrt{5}\right)$ and asymptotes $y = \pm \dfrac{1}{2}x$

16. Vertices $(\pm 3, 0)$ and passes through the point $(5, 4)$

17. Vertices $(-1, 3)$ and $(5, 3)$ and foci $(-3, 3)$ and $(7, 3)$

18. Vertices $(5, 2)$ and $(5, -6)$ and foci $(5, 3)$ and $(5, -7)$

Exercises 19-26   *Classify the given nondegenerate conic section as a circle, parabola, ellipse, or hyperbola.*

19. $2x^2 - x - 3y + 4 = 0$ **20.** $3x^2 + 2y^2 = 7x - 7y - 5$

21. $4x^2 + 4y^2 = 2x - 6y + 10$ **22.** $4y^2 - 9x + 23y - 9 = 0$

23. $-5x^2 - 11x = y^2$ **24.** $x^2 - 2y^2 + 5 = 0$

25. $-2x^2 + 4y^2 - 15x + 8 = 0$ **26.** $-3x^2 + 2 = 3y^2 - 5y$

Exercises 27-36   *Solve the equation for y, and plot the resulting equation(s) with a graphing calculator. Identify the conic section and approximate the coordinates of any vertices.*

27. $-x^2 + 5x + y = 0$ **28.** $x^2 - 6x + 3y + 12 = 0$

29. $5x^2 + 8y^2 = 40$ **30.** $11y^2 - 3x^2 = 33$

31. $y^2 - 3x^2 + 4x + 9 = 0$ **32.** $y^2 - 4y - x = 0$

33. $2x^2 + y^2 - 5x - 8 = 0$ **34.** $2x^2 + y^2 + 3x - 6y + 4 = 0$

35. $3x^2 - y^2 + 8x - 8y - 7 = 0$ **36.** $y^2 + 8y - 2x + 22 = 0$

Exercises 37-44   *Plot both branches with a graphing calculator. Identify the resulting conic section and find the general form of its equation.*

37. $y = \pm\sqrt{x+3}$ **38.** $y = \pm\sqrt{8x - x^2}$

39. $y = 2 \pm \sqrt{16 - 6x - x^2}$ **40.** $y = -4 \pm \sqrt{2x - 5}$

41. $y = \pm\sqrt{\dfrac{x^2}{3} - 2}$ **42.** $y = \pm\sqrt{9 - 3x^2}$

43. $y = 1 \pm \sqrt{2 - 8x - 2x^2}$ **44.** $y = -3 \pm \sqrt{x^2 - 10x}$

Applications

45. Long-Range Navigation   The long-range navigation (LORAN) system determines the locations of ships by timing radio signals. If signals are simultaneously transmitted to a ship from two stations, they will arrive at the ship at slightly different times. By assuming the ship lies on a hyperbola with foci at the transmitting stations, this time difference can be used to determine the position of the ship. Suppose transmitting stations A and B are located 200 miles apart on a coastline and a ship is located at a point P, as shown in Figure 104.

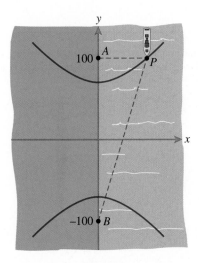

Figure 104

a. If the ship receives the signal from station A 800 microseconds (1 microsecond $= 10^{-6}$ second) before the signal from station B, find the equation of the hyperbola shown in Figure 104. Assume the signals travel at a rate of 0.186 mile per microsecond. [*Hint:* If the equation is

$$\frac{y^2}{b^2} - \frac{x^2}{a^2} = 1$$

then the difference of the distances from any point (x, y) on the hyperbola to the foci is $2b$.]

b. If it is known that the ship is due east of station A, determine the location of the ship.

46. Sonic Boom Supersonic Transports (SSTs) are aircraft capable of traveling faster than the speed of sound. When traveling at supersonic speeds, they create a conical shock wave more commonly known as a "sonic boom." The region on the ground that is affected by the sonic boom has a boundary that is hyperbolic, as can be seen in Figure 105. When a certain SST travels at Mach 2 (twice the speed of sound) and at an altitude of 65,000 feet, the vertex of the hyperbolic boundary is approximately 24 miles from the point on the ground directly beneath the nose of the plane, as shown in Figure 105. If we were to set up a coordinate system with the origin on the ground directly beneath the nose of the plane, the x-axis lying on the ground parallel to the path of the plane, and the y-axis also lying on the ground, then the asymptotes for the hyperbolic boundary would have equations $y = \pm\frac{1}{2}x$. Find the equation of the hyperbolic boundary. The region affected by the sonic boom extends as far as 55 miles behind the point on the ground below the plane. What is the width of the affected region when $x = -55$?

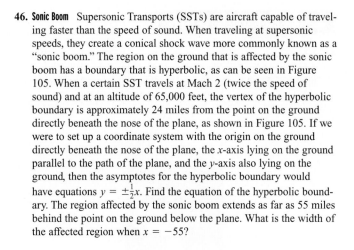
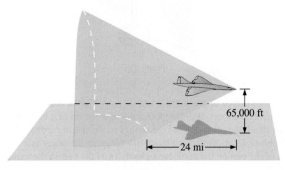

Figure 105

Exercises 47-50 *Answer true or false.*

47. The points $(\pm a, 0)$ and $(0, \pm b)$ lie on the graph of the hyperbola with equation

$$\frac{x^2}{a^2} - \frac{y^2}{b^2} = 1$$

48. Hyperbolas never intersect their asymptotes.

49. A hyperbola is the set of all points such that the sum of the distances from two fixed points (called the foci) is constant.

50. The foci of a hyperbola lie on the line determined by the vertices.

Exercises 51-54 *Give an example of each.*

51. A hyperbola for which the transverse axis is horizontal

52. A hyperbola with asymptotes $y = \pm 4x$

53. An application of the reflective property of hyperbolas

54. A characteristic of the graph of a hyperbola not shared by the graphs of the other conic sections

55. If a hyperbola has a horizontal transverse axis of length $2a$ and foci at $(\pm c, 0)$, explain why $c^2 - a^2 > 0$.

Questions for Discussion or Essay

56. A lamp with a shade is placed close to a wall in a dark room. When the lamp is turned on, the outline of the light on the wall will look similar to that shown in Figure 106. Consult Figure 82 on page 631 to help you explain why the outline forms a hyperbola.

Figure 106

57. A rough sketch of a hyperbola can be drawn in the following way. Attach two strings of different lengths near the tip of a ballpoint pen, and attach the other ends of the strings to a piece of cardboard using thumb tacks, as shown in Figure 107. Keeping the string taut and the tip of the pen on the cardboard, slowly twirl the pen so the string begins to wind around the tip. As the lengths of string shorten, the pen will follow a hyperbolic path. Explain why this is so.

Figure 107

58. In Figure 102 on page 639, a reflective telescope is shown that uses a parabolic main mirror and both elliptical and hyperbolic auxiliary mirrors to route the incoming light to the eyepiece. Can a telescope be constructed with a parabolic main mirror and only one auxiliary mirror so that it still directs the incoming light to the eyepiece? Explain, using a sketch if necessary.

Projects for Enrichment

59. Degenerate Conics Recall that the geometric definitions of the conic sections are based on the intersection of a plane with a double cone. The degenerate conics—a line, a pair of intersecting lines, and a point—arise geometrically when the plane passes through the vertex of the cone. The equations of the degenerate conics are all special cases of the general equation $Ax^2 + Cy^2 + Dx + Ey + F = 0$. For example, if A and C are both zero, the resulting equation $Dx + Ey + F = 0$ is a line.

a. Explain why each of the following equations has the indicated description:

  i. $2x^2 + 3y^2 = 0$; a point

  ii. $4x^2 - 9y^2 = 0$; a pair of intersecting lines

  iii. $x^2 + 2y^2 + 1 = 0$; no graph

b. State some general conditions under which equations of the form $Ax^2 + Cy^2 + F = 0$ will be

  i. a point   **ii.** a pair of intersecting lines

c. Complete the square on the equation $Ax^2 + Cy^2 + Dx + Ey + F = 0$ to determine a test for characterizing the equation as

  i. a point   **ii.** a pair of intersecting lines

  [*Hint:* Your test should involve the expression $D^2/(4A) + E^2/(4C) - F$, and it should take into account the signs of A and C.]

d. Use the test you developed in part c to determine which, if any, of the following equations are degenerate conics:

  i. $2x^2 - 3y^2 - 4x + 6y - 1 = 0$

  ii. $2x^2 + y^2 - 4x + 2y + 10 = 0$

  iii. $2x^2 + 3y^2 - 4x + 6y + 5 = 0$

60. The Case of the Stolen Car A victim of a carjacking calls Inspector Magill's office at 9:00 A.M. seeking help in recovering a stolen car. The following telephone conversation takes place.

Victim: This morning on my way to work, I was forced out of my car by two masked men. I overheard one say to the other that they would take the car to "the usual place" and store it there until noon. My car is equipped with an electronic homing device; unfortunately, it isn't working quite right and only emits a signal every 5 minutes. Is there any way to locate the source of the signal before the car is moved again?

Magill: Yes, but I will need time to round up three electronic receivers that can determine the precise time at which the signal is received from the homing device. I will also need three mobile phones and the help of two assistants.

Victim: You've got it!!

At 10:15, the victim and a friend arrive at Magill's office with the three receivers. After synchronizing the timing devices in the receivers, Magill takes the two friends to positions labeled A and B in Figure 108, 10 miles apart, and he drives to a third position C, 10 miles from B. His instructions to the two friends are simply to record the exact time at which the first signal after 11:00 is received from the homing device and then call him on the mobile phone. By 11:30, the car has been recovered and the criminals have been apprehended. We will reconstruct the method used by Magill to locate the car.

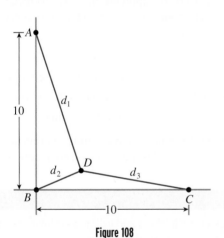

Figure 108

The times recorded by the receivers gave the following information:
- It took 3.22×10^{-5} second longer to receive the signal at point A than at point B.
- It took 2.15×10^{-5} second longer to receive the signal at point C than at point B

a. If D denotes the position of the car and d_1, d_2, and d_3 denote the distances (in miles) to A, B, and C, respectively, determine the differences $d_1 - d_2$ and $d_3 - d_2$ (round to the nearest integer).

Assume that the signal travels at the speed of light, which is 186,000 miles per second. Recall that a hyperbola is the set of all points such that the difference of the distances from the foci is constant. Point D must thus be one of the points on the hyperbola for which the difference of the distances from A and B is $d_1 - d_2$, and it must also be a point on the hyperbola for which the difference of the distances from C and B is $d_3 - d_2$. In other words, if we treat A and B as the foci of one hyperbola and B and C as the foci of a second hyperbola, point D will be a point of intersection. We simply need to find the equations of the two hyperbolas to determine their points of intersection. We will demonstrate how to find the one with foci at B and C and leave the other for you. We first assign a coordinate system to Figure 108, choosing $\overrightarrow{BC}$ as the positive x-axis, $\overrightarrow{BA}$ as the positive y-axis, and the origin at point B. Since B and C are 10 miles apart, the center of the hyperbola is at $(5, 0)$. Thus, we are looking for an equation of the form

$$\frac{(x - 5)^2}{a^2} - \frac{y^2}{b^2} = 1$$

Since the foci are 5 miles from the center, $c = 5$. The distance $d_3 - d_2$ must be twice the distance from the center to the vertices, so $2a = d_3 - d_2$. Finally, $b = \sqrt{c^2 - a^2} = \sqrt{25 - a^2}$.

b. Use the results from part a to show that the equation of the hyperbola with vertices B and C is

$$\frac{(x - 5)^2}{4} - \frac{y^2}{21} = 1$$

c. Show that the equation of the second hyperbola is

$$\frac{(y - 5)^2}{9} - \frac{x^2}{16} = 1$$

d. Estimate the coordinates of the intersection points of these two hyperbolas and determine the location of the car.

Section 8.5 | **Rotation of Axes and General Conic Sections**

    What is the equation of a conic section with rotated axes?

    How can we find the intersection points of the path of a comet and Earth's orbit around the sun?

    How can computers identify circular objects in photographs?

    Given an equation of a comet's path, how can the length of its major and minor axes be found?

In the previous three sections, we investigated conic sections with vertical and horizontal axes. These conics all had general equations of the form

$$Ax^2 + Cy^2 + Dx + Ey + F = 0$$

In this section, we consider conics with axes that are neither vertical nor horizontal. Such conics have general equations of the form

$$Ax^2 + Bxy + Cy^2 + Dx + Ey + F = 0$$

with $B \neq 0$. To graph a conic of this type, we rotate the xy-coordinate system until one of the coordinate axes is parallel to the axis of the conic. By this process, we create the **uv-coordinate system**, within which the conic will have a vertical or horizontal axis. In the uv-system, the equation of the conic will have the form

$$au^2 + cv^2 + du + ev + f = 0$$

and so we are able to sketch its graph using the techniques from the previous sections.

Rotated Coordinate Systems

We begin by considering a point P with coordinates (x, y) in the standard xy-coordinate system, and a uv-coordinate system obtained by rotating the xy-system counterclockwise by an angle θ. Our goal is to express the uv-coordinates of the point P in terms of x and y. To do this, we temporarily convert to polar coordinates. Let us suppose that the standard polar coordinates of P are $(r; \alpha)$. In the rotated system, the polar coordinates of P are $(r; \alpha - \theta)$ (see Figure 109). By applying the polar-to-rectangular and rectangular-to-polar conversion formulas from Section 7.5 and the difference identities for sine and cosine from Section 6.2, we have

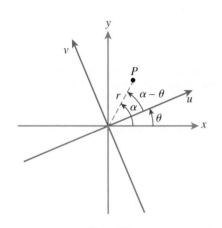

Figure 109

$$
\begin{aligned}
u &= r\cos(\alpha - \theta) & v &= r\sin(\alpha - \theta) \\
&= r(\cos\alpha\cos\theta + \sin\alpha\sin\theta) & &= r(\sin\alpha\cos\theta - \cos\alpha\sin\theta) \\
&= (r\cos\alpha)\cos\theta + (r\sin\alpha)\sin\theta & &= (r\sin\alpha)\cos\theta - (r\cos\alpha)\sin\theta \\
&= x\cos\theta + y\sin\theta & &= y\cos\theta - x\sin\theta
\end{aligned}
$$

In a similar manner, we can derive formulas for converting coordinates from the uv-system to the xy-system. Both sets of formulas are summarized as follows.

Rotation Conversion Formulas

Let P be a point with coordinates (x, y) in the standard xy-coordinate system and coordinates (u, v) in the uv-coordinate system obtained by rotating the xy-system counterclockwise by an angle θ. Then x, y, u, and v are related by the equations

$$u = x\cos\theta + y\sin\theta \qquad v = y\cos\theta - x\sin\theta$$

and

$$x = u\cos\theta - v\sin\theta \qquad y = u\sin\theta + v\cos\theta$$

EXAMPLE 1

Converting a Point from the xy-System to the uv-System

The xy-coordinates of a point are $(-2, 2)$. Find its coordinates in the uv-system obtained by a counterclockwise rotation of $\frac{\pi}{6}$.

Solution   Using the conversion formulas, we obtain

$$u = x \cos \theta + y \sin \theta$$

$$= -2 \cos \frac{\pi}{6} + 2 \sin \frac{\pi}{6}$$

$$= -\sqrt{3} + 1$$

and

$$v = y \cos \theta - x \sin \theta$$

$$= 2 \cos \frac{\pi}{6} - (-2) \sin \frac{\pi}{6}$$

$$= \sqrt{3} + 1$$

Thus, the uv-coordinates are $\left(1 - \sqrt{3}, 1 + \sqrt{3}\right)$.

EXAMPLE 2

Converting an Equation from the xy-System to the uv-System

A uv-coordinate system is obtained by a counterclockwise rotation of $\frac{\pi}{4}$. Find a relation in u and v that is equivalent to $y = 2x$.

Solution   We make the substitutions

$$x = u \cos \theta - v \sin \theta \quad \text{and} \quad y = u \sin \theta + v \cos \theta$$

with $\theta = \dfrac{\pi}{4}$ to obtain

$$y = 2x$$

$$u \sin \frac{\pi}{4} + v \cos \frac{\pi}{4} = 2\left(u \cos \frac{\pi}{4} - v \sin \frac{\pi}{4}\right)$$

$$\frac{\sqrt{2}}{2}u + \frac{\sqrt{2}}{2}v = 2\left(\frac{\sqrt{2}}{2}u - \frac{\sqrt{2}}{2}v\right)$$

$$u + v = 2u - 2v$$

$$v = \frac{1}{3}u$$

EXAMPLE 3

Circles in Rotated Coordinate Systems

A circle centered at the origin with radius r has equation $x^2 + y^2 = r^2$ in the standard xy-coordinate system. Show that in a uv-coordinate system obtained by a rotation of an angle θ, the equation $x^2 + y^2 = r^2$ is equivalent to $u^2 + v^2 = r^2$.

Solution   Substituting $x = u \cos \theta - v \sin \theta$ and $y = u \sin \theta + v \cos \theta$ into the equation $x^2 + y^2 = r^2$, we have

$$x^2 + y^2 = r^2$$

$$(u \cos \theta - v \sin \theta)^2 + (u \sin \theta + v \cos \theta)^2 = r^2$$

$$(u^2 \cos^2\theta - 2uv \sin \theta \cos \theta + v^2 \sin^2\theta) + (u^2 \sin^2\theta + 2uv \sin \theta \cos \theta + v^2 \cos^2\theta) = r^2$$

$$u^2(\cos^2\theta + \sin^2\theta) + v^2(\sin^2\theta + \cos^2\theta) = r^2$$

$$u^2 + v^2 = r^2$$

Figure 110

······>EXAMPLE 4

Inclined Soccer

When a goalie kicks a soccer ball from level ground, the ball describes a curve given by $y = x - \frac{1}{100}x^2$. Suppose she kicks the ball so that it follows this curve, but she is standing on a hill with a 30° incline. How far up the hill will the ball land?

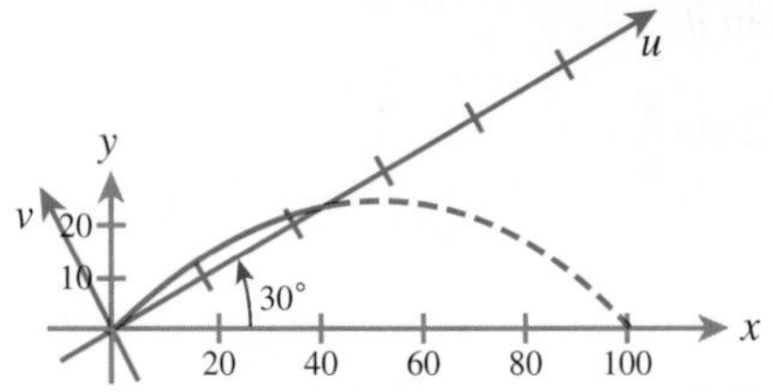

Solution We begin by defining the uv-coordinate system with the u-coordinate expressing the directed distance up the hillside, as shown in Figure 110. The coordinate transformation equations are then given by

$$x = u\cos 30° - v\sin 30° \qquad\qquad y = u\sin 30° + v\cos 30°$$

$$= \frac{\sqrt{3}}{2}u - \frac{1}{2}v \qquad\qquad\qquad = \frac{1}{2}u + \frac{\sqrt{3}}{2}v$$

Substituting, we have

$$y = x - \frac{1}{100}x^2$$

$$\frac{1}{2}u + \frac{\sqrt{3}}{2}v = \frac{\sqrt{3}}{2}u - \frac{1}{2}v - \frac{1}{100}\left(\frac{\sqrt{3}}{2}u - \frac{1}{2}v\right)^2$$

The ball will hit the hillside when $v = 0$. This gives us

$$\frac{1}{2}u = \frac{\sqrt{3}}{2}u - \frac{1}{100}\left(\frac{\sqrt{3}}{2}u\right)^2$$

$$\frac{1 - \sqrt{3}}{2}u + \frac{3}{400}u^2 = 0$$

$$u\left(\frac{1 - \sqrt{3}}{2} + \frac{3}{400}u\right) = 0$$

$$u = 0, u = \frac{200(\sqrt{3} - 1)}{3} \approx 48.8$$

Thus, the ball will land approximately 48.8 feet up the hill.

Rotated Conic Sections

Using the techniques illustrated in Examples 2 and 3, we can convert any conic equation of the form

$$Ax^2 + Bxy + Cy^2 + Dx + Ey + F = 0$$

to an equation of the form

$$au^2 + buv + cv^2 + du + ev + f = 0$$

by making the substitutions

$$x = u\cos\theta - v\sin\theta \quad\text{and}\quad y = u\sin\theta + v\cos\theta$$

Our goal now is to choose the angle of rotation θ so that $b = 0$—that is, so that there is no term involving the product uv. Without going into all the details, it can be shown that

$$b = -A\sin 2\theta + B\cos 2\theta + C\sin 2\theta$$

Thus, to ensure that $b = 0$, we must have

$$(A - C)\sin 2\theta = B\cos 2\theta$$

or, equivalently,

$$\cot 2\theta = \frac{A - C}{B}$$

In other words, if we rotate the xy-system by an angle θ satisfying $\cot 2\theta = (A - C)/B$, the resulting equation in the uv-coordinate system will have the form

$$au^2 + cv^2 + du + ev + f = 0$$

as desired.

Rotating a Conic Section

The general conic equation

$$Ax^2 + Bxy + Cy^2 + Dx + Ey + F = 0$$

where $B \neq 0$, is equivalent to an equation of the form

$$au^2 + cv^2 + du + ev + f = 0$$

in the uv-coordinate system obtained by rotating the xy-coordinate system through an angle θ satisfying

$$\cot 2\theta = \frac{A - C}{B}$$

EXAMPLE 5

Graphing a Rotated Conic

Sketch the graph of the conic section $x^2 + 2xy + y^2 + \sqrt{2}(x - y) = 0$.

Solution We begin by finding the angle of rotation for the uv-coordinate system. Since $A = 1$, $B = 2$, and $C = 1$, we want an angle θ satisfying

$$\cot 2\theta = \frac{A - C}{B} = 0$$

Thus, $2\theta = \frac{\pi}{2}$, and it follows that $\theta = \frac{\pi}{4}$. Next we turn to our conversion formulas to find expressions for x and y in terms of u and v.

$$x = u \cos \theta - v \sin \theta \qquad\qquad y = u \sin \theta + v \cos \theta$$

$$= u \cos \frac{\pi}{4} - v \sin \frac{\pi}{4} \qquad\qquad = u \sin \frac{\pi}{4} + v \cos \frac{\pi}{4}$$

$$= u \frac{\sqrt{2}}{2} - v \frac{\sqrt{2}}{2} \qquad\qquad\quad = u \frac{\sqrt{2}}{2} + v \frac{\sqrt{2}}{2}$$

$$= \frac{u - v}{\sqrt{2}} \qquad\qquad\qquad\quad = \frac{u + v}{\sqrt{2}}$$

Substituting into the original conic equation yields

$$x^2 + 2xy + y^2 + \sqrt{2}(x - y) = 0$$

$$\left(\frac{u - v}{\sqrt{2}}\right)^2 + 2\left(\frac{u - v}{\sqrt{2}}\right)\left(\frac{u + v}{\sqrt{2}}\right) + \left(\frac{u + v}{\sqrt{2}}\right)^2 + \sqrt{2}\left[\left(\frac{u - v}{\sqrt{2}}\right) - \left(\frac{u + v}{\sqrt{2}}\right)\right] = 0$$

$$\frac{1}{2}(u^2 - 2uv + v^2) + (u^2 - v^2) + \frac{1}{2}(u^2 + 2uv + v^2) - 2v = 0$$

$$2u^2 - 2v = 0$$

$$v = u^2$$

In the *uv*-system, this is the equation of a parabola with vertex at the origin. The graph is shown in Figure 111.

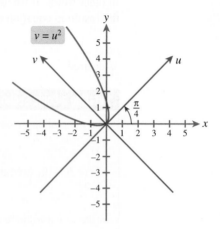

Figure 111

······**EXAMPLE 6**

Graphing a Rotated Conic

Sketch the graph of the conic section $13x^2 - 6\sqrt{3}xy + 7y^2 - 16 = 0$.

Solution   To find the angle of rotation θ, we proceed as follows:

$$\cot 2\theta = \frac{A - C}{B}$$

$$\cot 2\theta = \frac{13 - 7}{-6\sqrt{3}}$$

$$\cot 2\theta = -\frac{1}{\sqrt{3}}$$

$$2\theta = \frac{2\pi}{3} \qquad \text{Since } \cot\frac{2\pi}{3} = -\frac{1}{\sqrt{3}}$$

$$\theta = \frac{\pi}{3}$$

Setting $\theta = \frac{\pi}{3}$ in the conversion formulas, we obtain the following expressions for x and y in terms of u and v.

$$x = u\cos\theta - v\sin\theta \qquad\qquad y = u\sin\theta + v\cos\theta$$

$$= u\cos\frac{\pi}{3} - v\sin\frac{\pi}{3} \qquad\qquad = u\sin\frac{\pi}{3} + v\cos\frac{\pi}{3}$$

$$= u\frac{1}{2} - v\frac{\sqrt{3}}{2} \qquad\qquad = u\frac{\sqrt{3}}{2} + v\frac{1}{2}$$

$$= \frac{u - \sqrt{3}v}{2} \qquad\qquad = \frac{\sqrt{3}u + v}{2}$$

Substituting these expressions into the original conic equation, we have

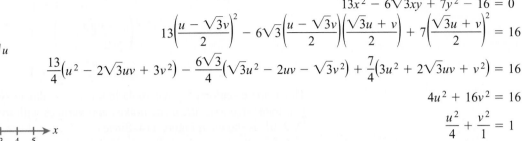

$$13x^2 - 6\sqrt{3}xy + 7y^2 - 16 = 0$$

$$13\left(\frac{u - \sqrt{3}v}{2}\right)^2 - 6\sqrt{3}\left(\frac{u - \sqrt{3}v}{2}\right)\left(\frac{\sqrt{3}u + v}{2}\right) + 7\left(\frac{\sqrt{3}u + v}{2}\right)^2 = 16$$

$$\frac{13}{4}(u^2 - 2\sqrt{3}uv + 3v^2) - \frac{6\sqrt{3}}{4}(\sqrt{3}u^2 - 2uv - \sqrt{3}v^2) + \frac{7}{4}(3u^2 + 2\sqrt{3}uv + v^2) = 16$$

$$4u^2 + 16v^2 = 16$$

$$\frac{u^2}{4} + \frac{v^2}{1} = 1$$

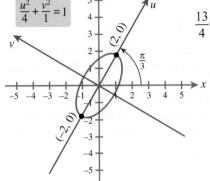

Figure 112

In the uv-system, this is the equation of an ellipse with center $(0, 0)$ and vertices $(-2, 0)$ and $(2, 0)$. In the xy-system, the vertices have coordinates

$$x = \frac{u - \sqrt{3}v}{2} = \frac{-2}{2} = -1 \quad \text{and} \quad y = \frac{\sqrt{3}u + v}{2} = \frac{-2\sqrt{3}}{2} = -\sqrt{3}$$

The graph is shown in Figure 112.

·····▷**EXAMPLE 7**

Finding the Key Features of a Rotated Hyperbola

The graph of the hyperbola $y = \frac{1}{x}$ is shown in Figure 113. Find its vertices, foci, and asymptotes.

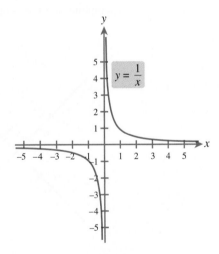

Figure 113

Solution   We first rewrite the equation to obtain $xy = 1$ or $xy - 1 = 0$. This is a general conic equation with $A = 0$, $B = 1$, and $C = 0$. Thus, we want an angle θ satisfying $\cot 2\theta = (A - C)/B = 0$. It follows that $2\theta = \frac{\pi}{2}$ or $\theta = \frac{\pi}{4}$. Since this is the same angle as in Example 5, we use the same substitutions for x and y—namely,

$$x = \frac{u - v}{\sqrt{2}} \quad \text{and} \quad y = \frac{u + v}{\sqrt{2}}$$

to obtain

$$xy - 1 = 0$$

$$\left(\frac{u - v}{\sqrt{2}}\right)\left(\frac{u + v}{\sqrt{2}}\right) = 1$$

$$\frac{u^2}{2} - \frac{v^2}{2} = 1$$

This is the equation of a hyperbola in the uv-coordinate system with center at the origin, transverse axis along the u-axis, and vertices with uv-coordinates $\left(-\sqrt{2}, 0\right)$ and $\left(\sqrt{2}, 0\right)$, as shown in Figure 114. Since $c = \sqrt{a^2 + b^2} = \sqrt{2 + 2} = 2$, the foci are 2 units from the center along the transverse axis, at the points with uv-coordinates $(-2, 0)$ and $(2, 0)$. To find the xy-coordinates of the vertices and foci, we apply the conversion formulas once again. For example, the xy-coordinates of the vertex with uv-coordinates $\left(-\sqrt{2}, 0\right)$ are found as follows:

$$x = \frac{u - v}{\sqrt{2}} \qquad \text{and} \qquad y = \frac{u + v}{\sqrt{2}}$$

$$= \frac{-\sqrt{2} - 0}{\sqrt{2}} \qquad\qquad\qquad = \frac{-\sqrt{2} - 0}{\sqrt{2}}$$

$$= -1 \qquad\qquad\qquad\qquad = -1$$

Thus, one vertex is at the point $(-1, -1)$. Similar computations would show that the other vertex is at $(1, 1)$ and the foci are at $\left(-\sqrt{2}, -\sqrt{2}\right)$ and $\left(\sqrt{2}, \sqrt{2}\right)$. See Figure 115 for a view of the graph with xy-coordinates labeled. It is evident from the graph that the asymptotes are the x- and y-axes.

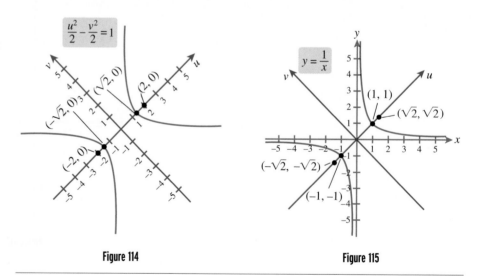

Figure 114   Figure 115

Classification of General Conic Sections

As we have now seen, any equation of the form

(1) $$Ax^2 + Bxy + Cy^2 + Dx + Ey + F = 0$$

can be converted to one of the form

(2) $$au^2 + cv^2 + du + ev + f = 0$$

by a cleverly chosen rotation of the xy-coordinate system. Because of our work with equations of form (2) in earlier sections of this chapter, we now know that any equation of form (1) that is not degenerate—that is, its graph is not a single point, a line, or a pair of lines—must be that of a parabola, circle, ellipse, or hyperbola. In fact, we can also conclude that any parabola, circle, ellipse, or hyperbola must have an equation of form (1). For this reason, we say that equation (1) is the **general equation of a conic section**.

In Section 8.4, we developed an algebraic test for classifying conic sections with vertical or horizontal axes of symmetry. This simple test is based on the fact that for all conic sections of a given type with equation

$$ax^2 + cy^2 + dx + ey + f = 0$$

the sign of the product ac is the same. In particular, for parabolas $ac = 0$, for ellipses $ac > 0$, and for hyperbolas, $ac < 0$. Of course, this test applies only to equations of conic sections with vertical or horizontal axes; it does not apply to the general equation given in (1) for which the axes may be rotated. Such equations could, in principle, be classified by first changing to an equation of the form given in (2) and then classifying as usual according to the sign of ac. But as you have already seen, converting coordinate systems is a lengthy and difficult process. Fortunately, it can be shown that for each conic section of a given type, the sign of $B^2 - 4AC$ is the same. This enables us to classify a general conic section without converting to a new coordinate system. Specifically, we have the following test.

Classifying Conic Sections

A nondegenerate conic section of the form $Ax^2 + Bxy + Cy^2 + Dx + Ey + F = 0$ can be classified as follows:

Condition	Resulting conic
$B^2 - 4AC = 0$	Parabola
$B^2 - 4AC < 0$	Ellipse (or circle if $B = 0$ and $A = C$)
$B^2 - 4AC > 0$	Hyperbola

The expression $B^2 - 4AC$ is called the **discriminant**.

EXAMPLE 8

Classifying Conic Sections

Classify each of the following nondegenerate conics as a circle, a parabola, an ellipse, or a hyperbola:

a. $x^2 - 2xy + 3y^2 - 5 = 0$

b. $2x^2 + 3xy = -y^2 + x - 4y - 1$

Solution

a. Here we have $A = 1$, $B = -2$, and $C = 3$. Thus,

$$B^2 - 4AC = (-2)^2 - 4(1)(3) = 4 - 12 = -8 < 0$$

and it follows that the equation is that of an ellipse (not a circle since $B \neq 0$).

b. We first rewrite the equation in general form as

$$2x^2 + 3xy + y^2 - x + 4y + 1 = 0$$

Now we have

$$B^2 - 4AC = 3^2 - 4(2)(1) = 9 - 8 = 1 > 0$$

Thus, the equation is that of a hyperbola.

Understanding and Mastery Checklists

Concepts to Understand

Rotated coordinate system

❖

Angle of rotation

❖

Rotated conic section

Skills to Master

Find the new coordinates of a point after rotating the axes.

❖

Find the xy-coordinates of a point given its coordinates in a rotated system.

❖

Convert a relation in x and y to one in a rotated coordinate system.

❖

Sketch the graph of a rotated conic.

❖

Find the foci, vertices, and (if applicable) asymptotes of a rotated conic section.

Exercises 8.5

Exercises 1-6   *The xy-coordinates of a point are given. Find the coordinates in the uv-system obtained by rotating the standard xy-coordinate system by the given angle θ.*

1. $(3, 0)$; $\theta = \dfrac{\pi}{4}$

2. $(0, -2)$; $\theta = \dfrac{\pi}{3}$

3. $(1, -1)$; $\theta = -\dfrac{\pi}{2}$

4. $(4, 4)$; $\theta = \dfrac{3\pi}{4}$

5. $(3, 5)$; $\theta = \pi$

6. $(7, 2)$; $\theta = \dfrac{3\pi}{2}$

Exercises 7-12   *A uv-coordinate system is obtained by rotating the standard xy-coordinate system by an angle θ. Find the xy-coordinates of the point whose uv-coordinates are given.*

7. $\theta = \dfrac{3\pi}{4}$; $(u, v) = (2, 0)$

8. $\theta = \dfrac{\pi}{4}$; $(u, v) = (0, -3)$

9. $\theta = \dfrac{3\pi}{2}$; $(u, v) = (-4, 4)$

10. $\theta = \dfrac{\pi}{2}$; $(u, v) = (2, 1)$

11. $\theta = -\dfrac{\pi}{6}$; $(u, v) = \left(\sqrt{3}, 1\right)$

12. $\theta = \dfrac{2\pi}{3}$; $(u, v) = \left(1, -\sqrt{3}\right)$

Exercises 13-18   *A uv-coordinate system is obtained by a rotation through an angle θ. Find a relation in u and v that is equivalent to the given equation.*

13. $\theta = \dfrac{\pi}{2}$; $x = -e^y$

14. $\theta = \dfrac{\pi}{2}$; $(x - 1)^2 + y^2 = 4$

15. $\theta = \dfrac{\pi}{4}$; $x - 5y = -2$

16. $\theta = \dfrac{\pi}{3}$; $2x - 3y = 6$

17. $\theta = \dfrac{\pi}{6}$; $x^2 - y^2 = 1$

18. $\theta = \dfrac{\pi}{4}$; $xy = 0$

Exercises 19-26   *Sketch the graph of the conic section.*

19. $13x^2 + 10xy + 13y^2 - 72 = 0$

20. $3x^2 - 10xy + 3y^2 - 8 = 0$

21. $11x^2 - 10\sqrt{3}xy + y^2 + 16 = 0$

22. $3x^2 + 2\sqrt{3}xy + y^2 - 2x + 2\sqrt{3}y = 0$

23. $x^2 - 2\sqrt{3}xy + 3y^2 - \sqrt{3}x - y = 0$

24. $3x^2 - 2\sqrt{3}xy + y^2 + 8x + 8\sqrt{3}y = 0$

25. $7x^2 - 4\sqrt{3}xy + 3y^2 - 9 = 0$

26. $x^2 + 2xy + y^2 + 4\sqrt{2}x - 4\sqrt{2}y = 0$

Exercises 27-30   *Sketch the graph of the conic section and find any vertices, foci, or asymptotes.*

27. $2xy - 1 = 0$

28. $17x^2 + 16xy + 17y^2 - 225 = 0$

29. $5x^2 + 2\sqrt{3}xy + 7y^2 - 8 = 0$

30. $x^2 - 2\sqrt{3}xy + 3y^2 + 8\sqrt{3}x + 8y = 0$

Exercises 31-36   *Classify the given nondegenerate conic section as a circle, parabola, ellipse, or hyperbola.*

31. $xy + x - 2 = 0$ **32.** $x^2 - 2xy + y^2 - 3x + y = 0$

33. $-x^2 + 3xy - 4y^2 + x - 4y - 1 = 0$

34. $3x^2 - xy - 2y^2 + x - y + 1 = 0$

35. $\dfrac{(x-2)^2}{4} + (y+1)^2 = 3 - xy$

36. $(x + y)(x + 2y) = xy + 1$

Applications

37. Image Processing   Suppose that a photograph of a soft drink can is taken and that data-fitting techniques reveal the presence of an elliptical border with equation
$$13x^2 - 6\sqrt{3}xy + 7y^2 - 16 = 0$$
Since circular objects appear elliptical in photographs, the elliptical edge is most likely the lip of the can. Graph the ellipse and then use this as a guide to sketch the can.

38. Cometary Orbit   Based on a series of measurements, a certain comet's position with respect to a coordinate system with the sun at the origin and with distances measured in A.U. is given by
$$17x^2 + 16xy + 17y^2 - 100\sqrt{2}x - 100\sqrt{2}y + 175 = 0$$
Find the lengths of the major and minor axes.

Concepts and Critical Thinking

Exercises 39-42   *Answer true or false.*

39. If the equation of an ellipse is given by $Ax^2 + Bxy + Cy^2 + Dx + Ey + F = 0$ and $B \neq 0$, then the major and minor axes are not parallel to the coordinate axes.

40. For any ellipse centered at the origin, we can define a *uv*-coordinate system such that the ellipse has equation
$$\frac{u^2}{a^2} + \frac{v^2}{b^2} = 1$$

41. For some values of A, C, D, E, and F, the graph of $Ax^2 + xy + Cy^2 + Dx + Ey + F = 0$ is a circle.

42. It is sometimes necessary to rotate the coordinate axes by 90° in order to eliminate the *xy* term from the equation of a conic.

Exercises 43-46   *Give an example of each.*

43. Two points in the plane such that one is a 45° rotation of the other

44. The equation of a hyperbola with foci lying on the line $y = x$

45. The equation of a conic section requiring rotation by an angle θ to eliminate the *xy* term, where $\cot 2\theta = \frac{2}{3}$

46. The equation of a conic section such that no matter what coordinate system is chosen there is no *uv* term

47. For conic sections with vertical or horizontal axes of symmetry, we now have two different tests for classification: the test described in Section 8.4 and the general test given in this section. Of course, we would hope that the two tests would give consistent results. Show that this is indeed the case.

Questions for Discussion or Essay

48. In most applied settings involving a single conic section, it isn't necessary to employ a rotated coordinate system. Instead, we simply define our coordinate system so that it is aligned with the conic section from the very beginning. Under what circumstances are we forced to use a rotated coordinate system?

49. If a point is translated and then rotated through an angle θ about the origin, does it end up in the same place as it would if it were first rotated through the angle θ and then translated? If so, explain why. If not, give an example that illustrates the difference.

50. In Example 4, we found the distance up a 30° incline that a kicked soccer ball landed by finding the equation of the parabola with respect to a rotated coordinate system. Explain how the point at which the ball landed on the hill could have been found graphically without rotating the coordinate system. Once the landing point is found, how would we find the distance up the hill to the landing point?

Projects for Enrichment

51. Comet Collision The threat of a comet or asteroid collision has created heated debate among scientists and government officials, inspired blockbuster movies, and initiated numerous efforts to detect and monitor Near-Earth Objects (NEOs) (see the Cometary Collisions box on page 607). In this project, we consider an algebraic approach for determining the possible locations of a collision between Earth and a comet. We use the following assumptions:

- Earth's orbit is elliptical, with the sun at one focus and axes of lengths 185.8 and 186 million miles.
- An *xy*-coordinate system is defined in the plane of Earth's orbit, with the sun at the origin, the *x*-axis along the major axis of Earth's orbit, and the *y*-axis through the sun and parallel to the minor axis of Earth's orbit, as shown in Figure 116. (Note that, for the sake of clarity, we have exaggerated the elliptical shape of Earth's orbit. A properly scaled graph would appear nearly circular.)
- A certain comet has an elliptical orbit in the same plane as Earth's orbit, with the sun at one focus. The major axis has length 560 million miles and is rotated 45° counterclockwise from that of Earth's major axis. The minor axis has length 330 million miles. See Figure 117.

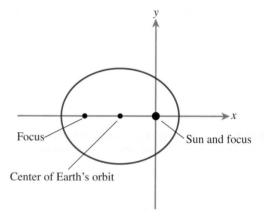

Figure 116

This painting, created for NASA by Don Davis, depicts a colossal impact on Earth.

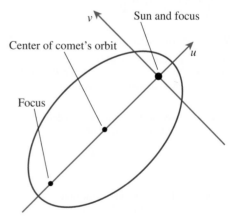

Figure 117

Our first step is to find equations for the elliptical orbits of Earth and the comet.

a. Find the distance between the center of Earth's orbit and the sun.

b. Find an equation for Earth's orbit. (*Hint:* It may be helpful to begin by assuming the origin is the center of Earth's orbit, finding the resulting equation, and then translating so that the origin is at the sun.)

c. The *uv*-coordinate system shown in Figure 117 is obtained by rotating the *xy*-coordinate system 45°counterclockwise. Find the distance between the center of the comet's orbit and the sun, and find the equation for the comet's orbit in the *uv*-system.

d. Substitute for *u* and *v* in the equation you found in part c to find an equation for the comet's orbit in terms of *x* and *y*.

Now we have two equations of ellipses in the *xy*-system. The intersection points of the graphs of these equations represent points where a collision between Earth and the comet could occur. We next find these points.

e. Solve the equation you found in part b for *y* and substitute into the equation from part d. Solve the resulting equations for *x* to find the *x*-coordinates of the points of intersection of the two elliptical paths. Finally, find the corresponding *y*-coordinates.

f. Sketch the graphs of the two orbits on the same coordinate system and identify the coordinates of the points of intersection.

g. Does the fact that the paths of the orbits of Earth and the comet intersect imply that the two will collide? Explain.

Section 8.6   Parametric Equations

> ↻ What's the equation of the letter A?
> ↻ What path is traced by a point on the outside of a quarter rolling around the edge of another quarter?
> ↻ On what date was Earth closest to the sun in 2003?
> ↻ How can the path of a ball on a billiard table be drawn with a graphing calculator?
> ↻ What sort of path is traced out by a rider on a double Ferris wheel?
> ↻ Why are elliptical crosstrainers so named?

We are now able to represent a wide array of curves as algebraic equations in x and y. For example, a circle with radius 2 centered at the origin has equation $x^2 + y^2 = 4$, a line with slope 1 and y-intercept 3 has equation $y = x + 3$, and so forth. But there are many curves—particularly paths arising from complex motion—for which it is more natural to view x and y as functions of a third variable t, called a **parameter**. The curve is then defined by the **parametric equations** $x = f(t)$ and $y = g(t)$. The **graph** of the parametric equations $x = f(t)$ and $y = g(t)$ consists of all points of the form $(f(t), g(t))$. The **orientation** of the graph is determined by the direction of motion as t increases and, when desired, it is indicated on the graph with arrows.

EXAMPLE 1   **Graphing Parametric Equations**

Graph the relation defined by the following parametric equations:

$$x = t + 1$$
$$y = 2t - 1$$

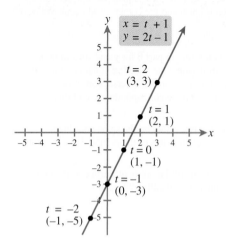

Figure 118

Solution This set of parametric equations is relatively simple, so we just plot a few points by selecting convenient values of t and computing the corresponding x- and y-values.

t	$x = t + 1$	$y = 2t - 1$
-2	-1	-5
-1	0	-3
0	1	-1
1	2	1
2	3	3

Connecting the points $(-1, -5)$, $(0, -3)$, and so forth gives us the graph shown in Figure 118. Note that the arrow indicates the orientation—that is, the direction of motion as t increases. Note also that the graph appears to be a line with slope 2 and y-intercept -3, and, if so, would have equation $y = 2x - 3$. We verify this in Example 3.

It is quite possible for different sets of parametric equations to have identical graphs. In fact, just as there are infinitely many possible journeys along a single road, so too are there infinitely many parameterizations for a given curve, as suggested by the following examples.

EXAMPLE 2   **Graphing Parametric Equations**

Graph the given parametric equations.

a. $x = 3t$, $y = 6t$

b. $x = -t + 3$, $y = -2t + 6$

c. $x = t^2$, $y = 2t^2$

Solution

a. We begin by selecting some convenient t-values and computing the corresponding x- and y-values.

t	$x = 3t$	$y = 6t$
-2	-6	-12
-1	-3	-6
0	0	0
1	3	6
2	6	12

Plotting the points, we obtain the graph shown in Figure 119. We note that since the y-coordinate ($6t$) is twice the x-coordinate ($3t$), the graph is simply the line $y = 2x$.

b. This time we note from the outset that the y-value ($-2t + 6$) is twice the x-value ($-t + 3$). Thus, just as in part a, the parametric equations define the line $y = 2x$. However, since the x- and y-values both decrease as t increases, the orientation is in the direction opposite that in part a. The resulting graph is shown in Figure 120.

c. Again, the y-coordinate is twice the x-coordinate. But in this case, both $x = t^2$ and $y = 2t^2$ are nonnegative for all values of t. Thus, the graph consists of just the portion of the line $y = 2x$ for which $x \geq 0$ and $y \geq 0$, as shown in Figure 121.

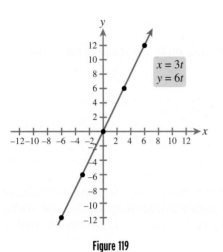

$x = 3t$
$y = 6t$

Figure 119

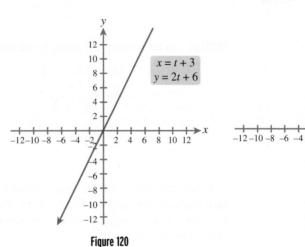

$x = t + 3$
$y = 2t + 6$

Figure 120

$x = t^2$
$y = 2t^2$

Figure 121

Eliminating the Parameter

As we have seen in the previous examples, it is sometimes possible to find an equation involving only x and y that is equivalent to a given set of parametric equations. We refer to the process of producing an equation in x and y from a set of parametric equations as **eliminating the parameter**. This process can be simple, tricky, or downright impossible (sometimes the parameter cannot be eliminated at all). The most straightforward technique is substitution—solving one equation for t and substituting into the second equation, as demonstrated by the following example.

▶EXAMPLE 3

Eliminating the Parameter by Substitution

Eliminate the parameter.

$$x = t + 1$$
$$y = 2t - 1$$

Solution   We solve the first equation for t and substitute into the second, as follows:

$x = t + 1$	The parametric equation for x
$t = x - 1$	Solving for t
$y = 2t - 1$	The parametric equation for y
$y = 2(x - 1) - 1$	Substituting $t = x - 1$
$y = 2x - 3$	Simplifying

Another standard trick for eliminating the parameter is to multiply the equations by appropriate constants and then add in such a way that the parameter is eliminated. Not surprisingly, this technique is called **elimination**.

·····>EXAMPLE 4

Elimination

Eliminate the parameter.

$$x = -t^3 + 1$$
$$y = 3t^3$$

Solution To eliminate t, we multiply the first equation by 3 and add to the second.

$$3x = -3t^3 + 3 \qquad \text{Multiplying the first equation by 3}$$
$$\underline{+ \quad y = 3t^3}$$
$$3x + y = 0t^3 + 3 \qquad \text{Adding to eliminate } t$$
$$3x + y = 3$$
$$y = -3x + 3 \qquad \text{Simplifying}$$

Parametric equations for conic sections often involve trigonometric functions. In particular, ellipses and circles are generally parameterized using the sine and cosine functions. Trigonometric identities, especially $\cos^2 t + \sin^2 t = 1$, are often helpful in eliminating the parameter.

·····>EXAMPLE 5

Eliminating the Parameter

Eliminate the parameter t and graph the relation defined by the following parametric equations:

$$x = 3 \cos t - 2$$
$$y = 5 \sin t + 1$$

Solution We first solve the equations for $\cos t$ and $\sin t$.

$$x = 3 \cos t - 2 \qquad\qquad y = 5 \sin t + 1$$
$$\cos t = \frac{x + 2}{3} \qquad\qquad \sin t = \frac{y - 1}{5}$$

Next, we apply the identity $\cos^2 t + \sin^2 t = 1$.

$$\cos^2 t + \sin^2 t = 1$$
$$\left(\frac{x + 2}{3}\right)^2 + \left(\frac{y - 1}{5}\right)^2 = 1 \qquad \text{Substituting for } \cos t \text{ and } \sin t$$
$$\frac{(x + 2)^2}{3^2} + \frac{(y - 1)^2}{5^2} = 1 \qquad \text{Simplifying}$$

This is the equation of an ellipse centered at $(-2, 1)$ with vertical major axis of length $2 \cdot 5 = 10$ and minor axis of length $2 \cdot 3 = 6$. The graph is shown in Figure 122.

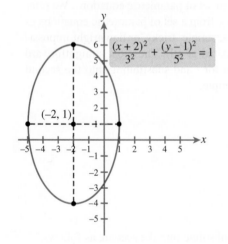

$$\frac{(x + 2)^2}{3^2} + \frac{(y - 1)^2}{5^2} = 1$$

$(-2, 1)$

Figure 122

Parametric Plotting with a Graphing Calculator

Many complex curves—particularly those arising in real-world contexts—are most naturally defined by parametric equations. In a few special cases, we can graph such curves by hand. Generally, however, we require a graphing tool. An overview of graphing parametric equations with a graphing calculator is given in the accompanying box, but in essence, our calculators require three pieces of information to produce parametric plots: $x(t)$, $y(t)$, and the interval of t-values.

Calculator Keys

Graphing Parametric Equations

To graph parametric equations with a graphing calculator, it is necessary to change the calculator to **parametric** graphing mode (as opposed to **function** graphing mode). Once this is done, the parametric expressions $x(t)$ and $y(t)$ can be entered in much the same way as one would enter an expression $f(x)$ when plotting graphs of functions. For example, with one popular brand of graphing calculator, the expressions `3cos(T)`-2 and `5sin(T)`$+1$ from Example 5 would be entered as X_{1T} and Y_{1T}, respectively. The next step is to choose initial settings for the window variables `Xmin`, `Xmax`, `Ymin`, and `Ymax` and also for the variables `Tmin` and `Tmax` that define the interval of t-values to be used for the plot. In many cases, a good place to start is with the calculator's standard settings. For example, when using the standard settings `Tmin=0`, `Tmax=6.28318` (that is, 2π), `Xmin=`-10, `Xmax=10`, `Ymin=`-10, and `Ymax=10` with the parametric equations from Example 5, we obtain the graph shown in Figure 123. For some plots, you may need to experiment with `Tmin` and `Tmax`.

Just as with graphs of functions, the trace feature can be used to move the cursor along graphs of parametric equations. One difference, however, is that the current value of t is displayed along with the coordinates of the point $(x(t), y(t))$. Figure 124 shows the cursor at the point $(-2, 6)$, which corresponds to $t \approx 1.5707963$ (that is, $\frac{\pi}{2}$).

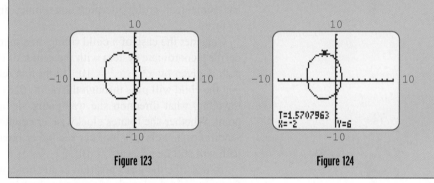

Figure 123

Figure 124

·····›**EXAMPLE 6**

Parametric Equations of a Point on a Wheel

Consider the motion of a point on a wheel. If the wheel is held up in the air and spun, then a point on the outside of the wheel will trace out a circle. On the other hand, if the wheel is rolling along in a straight line on the ground, then the actual motion of the point will be complex, as it simultaneously rotates about the axle and moves forward along the ground. The resulting curve is called a **cycloid**. In particular, if a point P on the edge of a 2-foot-diameter bicycle tire begins at the origin and the wheel is rolled in the positive x direction, then it can be shown that the path of P is given by

$$x = t - \sin t$$
$$y = 1 - \cos t$$

where $t = 0$ corresponds to the initial position of P. Use a graphing calculator to graph the cycloid and approximate the height of P after the bicyclist has traveled 5 feet.

Solution We first set the graphing calculator to parametric mode and enter the parametric equations. Next, we select a large interval for t, say [0, 20]. After adjusting the win-

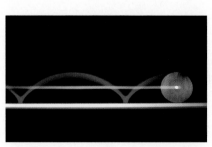

A point on the rim of a wheel traces out a cycloid, whereas the center travels in a straight line, as shown in this time-lapse photo.

dow variables, we obtain the view shown in Figure 125. To find the height of P after the cyclist has traveled 5 feet, we simply use the trace feature to locate the point on the graph for which $x \approx 5$. We conclude from Figure 126 that the point is about 1.5 feet off the ground.

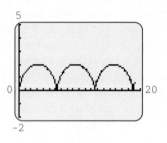

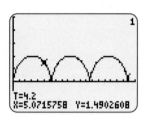

Figure 125   Figure 126

Applications   Thus far, we have attached no special meaning to the parameter t; the parameter has served simply as a device for defining a relation between x and y. In practice, however, parametric equations generally arise in contexts in which the parameter has some real-world interpretation—time, angle of rotation, distance traveled, and so forth.

Consider the case of a child riding on a merry-go-round with a 10-foot radius. If we define a coordinate system with the center of the carousel as the origin, then the child's path is given by $x^2 + y^2 = 10^2$. From this relation we could determine, for example, that the child will pass through the point (6, 8), but we wouldn't know how fast she was traveling, what direction she was going, or how many times she passed through this point. Whether she rotates clockwise or counterclockwise, goes around once or a thousand times, travels at 1 foot per second or breaks the sound barrier, the equation of her path will still be $x^2 + y^2 = 10^2$.

To describe the girl's position at any time, we could give her x- and y-coordinates as functions of t, the length of time that she has been on the ride. Her position would then be given by the parametric equations

$$x = f(t)$$
$$y = g(t)$$

In the next example, we illustrate how parametric equations can be used to describe two, very different 8-second carousel rides.

EXAMPLE 7   **Carousel Motion**

Describe the motion of a girl traveling on a merry-go-round according to the given set of parametric equations. Assume that t represents the elapsed time for the ride (in seconds) and x and y are in feet.

a. $x = 10 \cos \frac{1}{2}\pi t$ **b.** $x = 10 \cos 50\pi t$

$\qquad\qquad\qquad\qquad\qquad\qquad\qquad y = -10 \sin 50\pi t$

$\quad\ \ y = 10 \sin \frac{1}{2}\pi t$

Solution

a. We begin by finding the girl's location at conveniently chosen time intervals.

t	$x = 10 \cos \dfrac{1}{2}\pi t$	$y = 10 \sin \dfrac{1}{2}\pi t$
0	10	0
1	0	10
2	−10	0
3	0	−10
4	10	0
⋮	⋮	⋮
8	10	0

Plotting these points, as shown in Figure 127, we see that the girl begins at the point $(10, 0)$ and travels counterclockwise at the steady rate of one revolution every 4 seconds. In the indicated 8-second time interval, she makes a total of two revolutions.

b. Since the period of $f(x) = a \cos bt$ is given by $2\pi/|b|$, the x- and y-values will repeat themselves every $2\pi/|50\pi| = 0.04$ second. Thus, we suspect that the girl is moving at the rate of one revolution every 0.04 second. To determine the direction of her motion, we'll sample her position at 0.01-second time intervals.

t	$x = 10 \cos 50\pi t$	$y = -10 \sin 50\pi t$
0	10	0
0.01	0	−10
0.02	−10	0
0.03	0	10
0.04	10	0
⋮	⋮	⋮

Plotting these points, as shown in Figure 128, we see that the girl begins at $(10, 0)$, and travels clockwise at the rate of one revolution for every 0.04 second (about Mach 1.6) for 8 seconds, making a total of 200 revolutions in all.

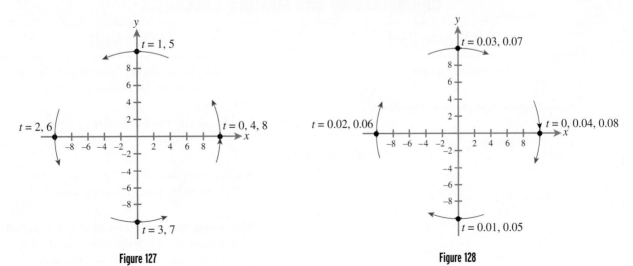

Figure 127   Figure 128

**EXAMPLE 8**

Parametric Nightmare

One of the authors has a recurrent nightmare in which he is running a race around an elliptical track. If we let x and y represent directed distances (in meters) from the center of the track and t represent elapsed time in minutes, then his motion around the track is approximated by the parametric equations

$$x(t) = 30 \cos(4 \arctan t)$$
$$y(t) = 20 \sin(4 \arctan t)$$

What's so nightmarish about this race?

Solution We use a graphing calculator to plot this curve over greater and greater time intervals. Entering the equations for x and y, setting `Tmin=0` and `Tmax=1`, and selecting an appropriate viewing window, we obtain the plot shown in Figure 129, which suggests that the author is halfway around the track after 1 minute. Perhaps the race will be over after just 2 minutes! With `Tmax=2` (Figure 130), we see that the slumbering math guy has not quite completed his lap. Figures 131 and 132 reveal that even after 10 minutes, even after 30 minutes, he has still not finished the race. In fact, it can be shown (see Exercise 40) that no matter how much time elapses, he'll never quite finish the race—the classic running-but-going-nowhere nightmare.

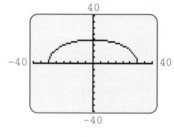

Figure 129

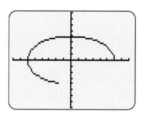

Figure 130

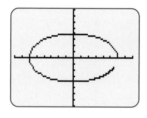

Figure 131

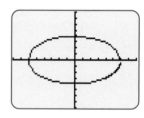

Figure 132

Understanding and Mastery Checklists

Concepts to Understand

Parametric equations

❖

Graph of parametric equations

❖

Eliminating the parameter

Skills to Master

Sketch the graph of a pair of parametric equations by plotting points.

❖

Eliminate the parameter of a pair of parametric equations.

❖

Describe the path of an object whose motion is given by parametric equations.

❖

Use a graphing calculator to graph a pair of parametric equations and estimate the coordinates of points on the curve.

Exercises 8.6

Exercises 1-2 *Complete the table and use it to sketch the graph of the parametric equations.*

1. $x = t - 3$
  $y = 2t + 1$

t	x	y
-1		
0		
1		
2		

2. $x = 3 - 2t$
  $y = 2t - 1$

t	x	y
-1		
0		
1		
2		

Exercises 3-8 *Match the parametric equations with the appropriate description. Choose from:*

 i. *Path of a point on a basketball, shot from the free-throw line with a slight backspin*
 ii. *Path of a reflector strapped to a cyclist's ankle*
 iii. *The letter A*
 iv. *Path of a point on a quarter rolling around the outside of a second quarter*
 v. *A spiral*
 vi. *Path of a ray of light inside a mirrored room*

3. $x = \dfrac{10}{t+1} \cos t$

  $y = \dfrac{10}{t+1} \sin t$

  `Tmin=0, Tmax=50`

4. $x = 0.75 \sin\left(\dfrac{8\pi}{3}t\right) + 26.4t$

  $y = 0.75 \cos\left(\dfrac{8\pi}{3}t\right) + 1.25$

  `Tmin=0, Tmax=10`

5. $x = 8 - 8\left| t - 2\left\lfloor \dfrac{1}{2}t \right\rfloor - 1 \right|$

  $y = 8 - 8\left| \dfrac{7}{11}t - 2\left\lfloor \dfrac{7}{22}t \right\rfloor - 1 \right|$

  `Tmin=0, Tmax=11`
  (*Hint:* Recall that $\lfloor\ \rfloor$ denotes the floor function, as described in Section 2.5.)

6. $x = \dfrac{4}{9}|t - 9| - |t - 6| - |t - 5| + 4|t - 4| - 3|t - 3| + \dfrac{13}{9}|t|$

  $y = 2|t - 6| - 2|t - 5| + 2|t - 3| - 4|t - 2| + 2|t|$
  `Tmin=0, Tmax=9`

7. $x = \dfrac{5}{12} \cos(4\pi t) + \dfrac{30\sqrt{34}}{17}t$

  $y = \dfrac{5}{12} \sin(4\pi t) - 16t^2 + \dfrac{76\sqrt{34}}{17}t + 6$

  `Tmin=0, Tmax=1.5`

8. $x = 2 \sin t + \sin(2t)$
  $y = 2 \cos t + \cos(2t)$
  `Tmin=0, Tmax=2π`

Exercises 9-16 *Eliminate the parameter and then graph.*

9. $x = 3 + t$
  $y = -2t + 2$

10. $x = 3t^3 - 4$
  $y = 6t^3 + 1$

11. $x = t + 1$
  $y = t^2 + 2t + 1$

12. $x = t^2$
  $y = t - 1$

13. $x = 4 \cos t$
  $y = 4 \sin t$

14. $x = 2 \cos t$
  $y = 3 \sin t$

15. $x = 1 - 2 \sin t$
  $y = 2 + 3 \cos t$

16. $x = 2 + 3 \sin t$
  $y = -3 - 3 \cos t$

Exercises 17-22 *For the given parametric equations:*

a. *Eliminate the parameter and graph the resulting equation in x and y.*
b. *Use a graphing calculator to plot the original parametric equations over $-10 \le t \le 10$.*
c. *Explain any differences between the graphs in parts a and b.*

17. $x = t^2 - 2$
  $y = t^2$

18. $x = -t^2$
  $y = t^4$

19. $x = e^t$
  $y = e^{2t}$

20. $x = \ln t^2$
  $y = \ln t$

21. $x = \cos t$
  $y = 4 \cos t$

22. $x = \sin t$
  $y = 2 \sin^2 t$

Exercises 23-26 *Use a graphing calculator to plot the graph and estimate the indicated point(s).*

23. $x = t^3 - 3t^2 + 2t + 2$
  $y = -t^2 + 2t + 3$
  Crossing point (where the curve intersects itself)

24. $x = \cos t + \cos(2t) + 3$
  $y = \sin t + \sin(2t) - 4$
  Crossing point (where the curve intersects itself)

25. $x = 2 \cos^3 t$
  $y = 4 \sin^3 t$
  Intercepts

26. $x = t^2 - t - 6$
  $y = t^2 - 4t - 5$
  Intercepts

Applications

27. Elliptical Crosstrainer   The motion of the heel of the foot when using an elliptical crosstrainer is approximated by the parametric equations

$$x = 6.1 \cos t - 0.95 \sin t - 0.19$$
$$y = 2.1 \cos t + 2.7 \sin t + 7.9$$

where y gives the height of the heel above the floor in inches. Estimate the maximum and minimum heights.

28. Earth Orbit   Earth travels around the sun along an elliptical path approximated by the parametric equations

$$x = 1.496 \cos\left(0.172t^2 - 6.459t - 3.071\right) - 0.025$$
$$y = -1.49579 \sin\left(0.172t^2 - 6.459t - 3.071\right)$$

where t is measured in years ($t = 0$ corresponds to the beginning of 2003) and x and y are measured in 10^{11} meters. Find the dates on which Earth is closest to (perihelion) and furthest from (aphelion) the sun.

29. Double Ferris Wheel   The position of a rider on a double Ferris wheel (see the photo) is approximated by the parametric equations

$$x = 20 \sin\left(\frac{\pi t}{10}\right) + 10 \sin\left(\frac{2\pi t}{5}\right)$$

$$y = 35 - 20 \cos\left(\frac{\pi t}{10}\right) - 10 \cos\left(\frac{2\pi t}{5}\right)$$

where x represents the horizontal distance in feet from the base, y is the height in feet, and t is elapsed time in seconds from the start of the ride.

a. Find the initial position of the rider.

b. How long does it take for the rider to return to the starting position?

c. Find the greatest height and the first two times at which this height is attained.

A pair of double Ferris wheels

30. Bouncing Golf Ball   The position of a golf ball t seconds after it has been struck can be approximated by

$$x = 200\left(\frac{t - 2 - \lfloor t \rfloor}{2\lfloor t \rfloor} + 2\right)$$
$$y = 25\left(t - \lfloor t \rfloor\right)\left(\lfloor t \rfloor - t + 1\right)2^{4 - \lfloor t \rfloor}$$

where x represents the horizontal distance traveled and y is the height, with both measured in feet. Recall that $\lfloor \ \rfloor$ denotes the floor function, as described in Section 2.5.

a. What horizontal distance does the ball travel?

b. What is the ball's maximum height?

c. What is the maximum height attained by the ball after its third bounce?

Concepts and Critical Thinking

Exercises 31-34   *Answer true or false.*

31. If t seconds after the cue ball has been struck the location of the 8-ball is given by $x = f_1(t)$ and $y = g_1(t)$, and the location of the 7-ball is given by $x = f_2(t)$ and $y = g_2(t)$, and the graphs of these two sets of parametric equations intersect, then the 8- and 7-balls must collide.

32. If t seconds after the cue ball has been struck the location of the 8-ball is given by $x = f_1(t)$ and $y = g_1(t)$, and the location of the 7-ball is given by $x = f_2(t)$ and $y = g_2(t)$, and for some value of t $f_1(t) = f_2(t)$ and $g_1(t) = g_2(t)$, then the 8- and 7-balls collide.

33. No portion of the curve with parametric equations $x = t^2 + 1$ and $y = g(t)$ lies in the second quadrant.

34. Each line in the plane can be parameterized in exactly one way.

Exercises 35-38   *Give an example of each.*

35. Parametric equations for the portion of the graph of $y = x^2$ lying in the first quadrant

36. Two different sets of parametric equations for the line $y = 2x$

37. Parametric equations for a curve that is not the graph of a function $y = f(x)$

38. A parameterization for the curve $y = f(x)$

Questions for Discussion or Essay

39. Any curve that lies entirely in a single plane can be plotted—at least in principle—with a graphing calculator. However, many important and interesting curves are *not* contained in a single plane. For example, the helix—the curve formed by a single strand of DNA or a partially stretched spring—is not contained in any one plane and thus cannot be easily plotted with a graphing calculator. Give at least four examples of real-world curves that, like the helix, are not contained in a single plane.

40. In Example 8, we considered the parametric equations

$$x(t) = 30 \cos(4 \arctan t)$$

$$y(t) = 20 \sin(4 \arctan t)$$

Describe the end behavior of $x(t)$ and $y(t)$ by separately graphing each function. As $t \to \infty$, what point does $(x(t), y(t))$ approach? Does it ever reach this point?

41. In parts a through e, parametric equations for an employee's daily commute from home to work for a 5-day workweek are given, but in no particular order. In addition, the employee's account of each day's commute is included. Use the trace feature of your graphing calculator to find the day of the week corresponding to the commutes described in parts a through e.

a. $x = -4 + t, y = 5; 0 \le t \le 8$

b. $x = -4 + 0.1t, y = 5; 0 \le t \le 80$

c. $x = \frac{1}{2}(|t| - |t - 4| + |t - 8| - |t - 12|), y = 5; 0 \le t \le 12$

d. $x = 4 \sin\left(\frac{\pi}{8}t - \frac{\pi}{2}\right), y = 5; 0 \le t \le 24$

e. $x = -4 + t + \sin(\pi t), y = 5; 0 \le t \le 8$

Monday: I stopped off to get gas about halfway to work, but other than that, I went pretty fast.

Tuesday: I sailed to work—there was almost no traffic.

Wednesday: My transmission acted up on me. The car kept lurching into reverse, and I would find myself hurtling backwards for several hundred yards before I could get her back in drive again.

Thursday: It was such a nice day that I jogged to work.

Friday: I got to work in good time, but then realized I had forgotten some important papers, and so I had to go back home to get them.

In general, what type of information about a journey can be conveyed by parametric equations that cannot be conveyed by an equation in x and y?

Projects for Enrichment

42. Cue Ball Trails In this project, we investigate the path made by a cue ball on a billiard table. Consider a billiard table with dimension $l \times w$, which we represent as a rectangle in the first quadrant with vertices $(0, 0)$, $(l, 0)$, (l, w), and $(0, w)$, as shown in Figure 133.

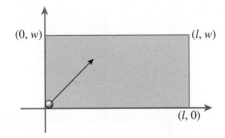

Figure 133

It can be shown that if a ball starts at the origin and initially moves along a linear path with slope m, its path is given by the parametric equations

$$x(t) = l - \left| t - 2l \left\lfloor \frac{t}{2l} \right\rfloor - l \right|$$

$$y(t) = w - \left| mt - 2w \left\lfloor \frac{mt}{2w} \right\rfloor - w \right|$$

where $\lfloor \ \rfloor$ denotes the floor function, as described in Section 2.5. We first consider a billiard table of dimension $8' \times 4'$.

a. Let $l = 8$, $w = 4$, and $m = 3$, and use a graphing calculator to plot the graph of the parametric equations over the following intervals. (*Hints:* Set your viewing window to show only the rectangle determined by $0 \le x \le 8$ and $0 \le y \le 4$. Also, when you enter the parametric equations into your calculator, you may want to use variable names such as L, W, and M and then store the values 8, 4, and 3 into these variables before you graph. These values can then easily be changed for future parts of the project.)

 i. $0 \le t \le 4$ **ii.** $0 \le t \le 8$

 iii. $0 \le t \le 12$

b. With $m = 3$, how many times does the ball "bounce" off the side of the table before the path starts to repeat?

c. Redo parts a and b with $m = 4$, $m = 5$, and $m = 6$. In general, if the slope is a positive integer k, how many times will the ball bounce before the path repeats?

d. For the given slope, experiment with the interval for t to determine how many bounces are necessary before repetition.

 i. $m = \frac{1}{3}$ **ii.** $m = \frac{1}{4}$ **iii.** $m = \frac{2}{3}$ **iv.** $m = \frac{2}{5}$

For noninteger rational slopes, it is very difficult to determine a general rule relating the slope to the number of bounces before a path begins to repeat on an 8′ × 4′ table. However, the situation is not as hopeless for square tables. Thus, we now consider a table with dimension 4′ × 4′.

e. Set $l = 4$, $w = 4$, and $m = \frac{1}{3}$, and plot the graph of the parametric equations over the following intervals. You may want to experiment with viewing window settings.

    i. $0 \le t \le 1$ **ii.** $0 \le t \le 2$

    iii. $0 \le t \le 10$ **iv.** $0 \le t \le 100$

f. With $m = \frac{1}{3}$, how many times does the ball "bounce" off the side of the table before the path starts to repeat?

g. Repeat parts e and f with $m = \frac{1}{4}$, $m = \frac{2}{3}$, $m = \frac{2}{5}$, and $m = \frac{3}{7}$. For a slope of $m = \frac{u}{v}$ (in lowest terms), how many bounces are there before the path repeats?

Until now we have considered only rational slopes. In our final step, we consider an irrational slope. (Of course, even irrational numbers are approximated by finitely many digits in your calculator.)

h. Set $m = \sqrt{2}$, and experiment with different intervals for t. What happens in this case? In general, what do you think can be said about path repetition in the case of irrational slopes?

Chapter 8 Review

Exercises 1-4 *A relation and its graph are given. Use translation and/or reflection to sketch the graph of the indicated relation.*

1. $y = x^2$

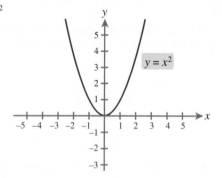

    a. $y = (x - 2)^2$ **b.** $y = x^2 + 3$

    c. $y - 4 = (x - 2)^2$ **d.** $y = -x^2$

2. $x = y^3$

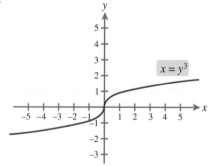

    a. $x = (y - 1)^3$ **b.** $x = y^3 + 2$

    c. $-x = y^3$ **d.** $-x = (y - 1)^3$

3. $xy^2 + x = 4$

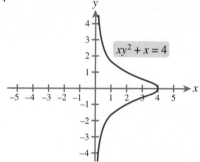

    a. $x(y + 2)^2 + x = 4$ **b.** $xy^2 + x = -4$

    c. $(x - 1)(y - 3)^2 + x - 1 = 4$

4. $x^2 + y^3 = 16$

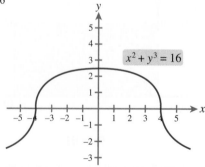

    a. $(x - 2)^2 + y^3 = 16$ **b.** $(x - 2)^2 + (y - 1)^3 = 16$

    c. $x^2 - y^3 = 16$

Exercises 5-6 *A relation and its graph are given, together with a translated graph. Find an equation for the translated graph.*

5. Original relation: $x^2 + y^4 = 16$

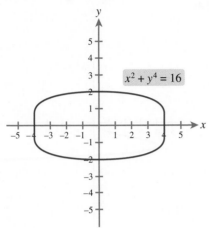

$x^2 + y^4 = 16$

Translated graph:

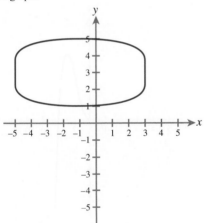

6. Original relation: $y = x^3$

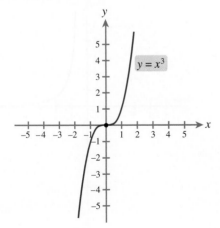

$y = x^3$

Translated graph:

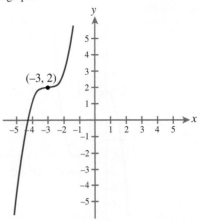

$(-3, 2)$

Exercises 7-8 *A relation and its graph are given, together with a reflected graph. Find an equation for the reflected graph.*

7. Original relation: $y = x^4 - 4x^2 + 2$

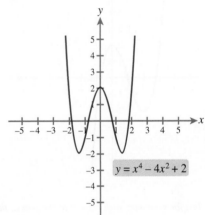

$y = x^4 - 4x^2 + 2$

Reflected graph:

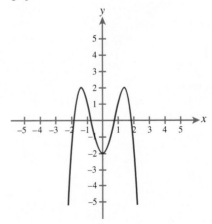

8. Original relation: $y = x^3 - 3x^2 + 2$

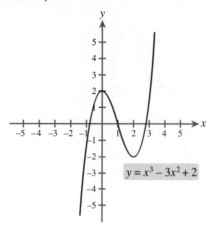

Reflected graph:

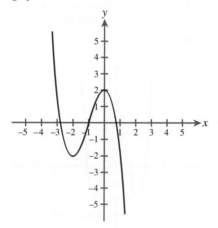

Exercises 9-12   *Test for symmetry with respect to the x-axis, the y-axis, and the origin.*

9. $y^2 - 2x = y^4$

10. $|x| + |y| = 1$

11. $x - y = 1$

12. $y \cos x = \sin x$

Exercises 13-16   *A portion of the graph of a relation is shown. Test for symmetry with respect to the x-axis, the y-axis, and the origin, and then use symmetry to complete the graph.*

13. $x = y^4 - 4y^2$

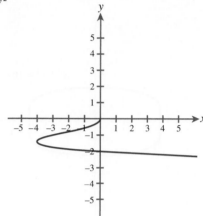

14. $y = 4x^2 - x^4$

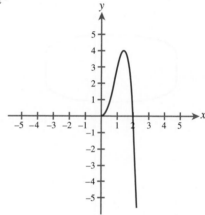

15. $xy + x^3y = 5$

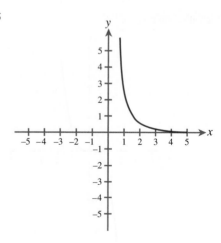

16. $9x^2 + y^4 = 81$

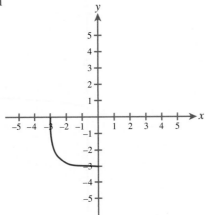

Exercises 17-22   *Find the vertex, focus, and directrix for the given parabola, and sketch its graph.*

17. $x = 2y^2$

18. $x^2 + 3y = 0$

19. $y + 4 = -2(x - 3)^2$

20. $2y - 4 = (x + 1)^2$

21. $x = 3y^2 + 6y + 7$

22. $y = 2x^2 + 6x + 1$

Exercises 23-26   *Find an equation for the parabola with the given properties.*

23. Vertex at the origin and focus at $\left(0, \dfrac{1}{4}\right)$

24. Focus at $(-1, 0)$ and directrix $x = 1$

25. Vertex at $(-1, 2)$ and directrix $y = 6$

26. Horizontal axis, vertex at $(4, 3)$, and passing through $(2, 4)$

Exercises 27-32   *Find the center, vertices, and foci for the given ellipse, and sketch its graph.*

27. $\dfrac{x^2}{49} + \dfrac{y^2}{64} = 1$

28. $2x^2 + 3y^2 = 1$

29. $9(x - 3)^2 + 25(y - 2)^2 = 225$

30. $\dfrac{(x + 2)^2}{64} + \dfrac{(y - 3)^2}{100} = 1$

31. $x^2 + 4y^2 - 2x + 24y + 21 = 0$

32. $16x^2 + 9y^2 + 128x + 18y + 121 = 0$

Exercises 33-36   *Find an equation for the ellipse with the given properties.*

33. Horizontal major axis of length 6, vertical minor axis of length 4, and center at the origin

34. Vertices at $(\pm 4, 0)$ and foci at $\left(\sqrt{5}, 0\right)$ and $\left(-\sqrt{5}, 0\right)$

35. Foci at $(0, \pm 4)$ and minor axis of length 6

36. Vertices at $(2, \pm 4)$ and minor axis of length 6

Exercises 37-42   *Find the center, vertices, foci, and asymptotes for the given hyperbola, and sketch its graph.*

37. $\dfrac{x^2}{25} - \dfrac{y^2}{9} = 1$

38. $8y^2 - 12x^2 = 1$

39. $4(y - 1)^2 - 9(x + 5)^2 = 36$

40. $\dfrac{(x + 1)^2}{4} - \dfrac{(y + 2)^2}{16} = 1$

41. $25x^2 - 16y^2 - 150x + 64y - 239 = 0$

42. $x^2 - 4y^2 - 6x + 32y = 51$

Exercises 43-44   *Find an equation for the hyperbola with the given properties.*

43. Vertices at $(\pm 5, 0)$ and foci at $\left(\pm\sqrt{34}, 0\right)$

44. Vertices at $\left(0, \pm\sqrt{5}\right)$ and passing through $(4, 5)$

Exercises 45-48   *Solve the equation for y and plot the resulting equation(s) on a graphing calculator. Identify the conic section, and approximate the coordinates of any vertices.*

45. $y^2 + 4x = 6$

46. $4y - 8 = -x^2 + 2x$

47. $2x^2 - 7x + y^2 = 6$

48. $y^2 + 6y = x^2 + 2$

Exercises 49-52   *Use a graphing calculator to plot the pair of equations on the same coordinate system. Identify the resulting conic section and find its general equation.*

49. $y = \pm\sqrt{16 - x^2}$

50. $y = \pm\sqrt{x^2 - 16}$

51. $y = 2 \pm \sqrt{x - 4}$

52. $y = 4 \pm \sqrt{16 - 4x^2}$

Exercises 53-54   *The xy-coordinates of a point are given. Find the coordinates in the uv-system obtained by counterclockwise rotation of the standard xy-coordinate system by the given angle θ.*

53. $(2, 2)$; $\theta = \dfrac{\pi}{4}$

54. $(-3, 0)$; $\theta = -\dfrac{\pi}{6}$

Exercises 55-56   *A uv-coordinate system is obtained by rotating the standard xy-coordinate system counterclockwise by an angle θ. Find the xy-coordinates of the point whose uv-coordinates are given.*

55. $\theta = \dfrac{\pi}{2}$; $(2, -1)$

56. $\theta = \dfrac{\pi}{3}$; $\left(2, -2\sqrt{3}\right)$

Exercises 57-58   *A uv-coordinate system is obtained by a counterclockwise rotation through an angle θ. Find a relation in u and v that is equivalent to the given equation.*

57. $\theta = \dfrac{\pi}{4}$; $y = 2x^2$

58. $\theta = \dfrac{\pi}{6}$; $x + 4y = 8$

Exercises 59-62 *Sketch the graph of the conic section.*

59. $x^2 + 2xy + y^2 + \dfrac{\sqrt{2}}{2}x - \dfrac{\sqrt{2}}{2}y = 0$

60. $5x^2 - 6xy + 5y^2 - 8 = 0$

61. $x^2 - 2\sqrt{3}xy - y^2 + 2 = 0$

62. $x^2 - 2\sqrt{3}xy + 3y^2 - 2\sqrt{3}x - 2y = 0$

Exercises 63-70 *Classify the given nondegenerate conic section as a circle, parabola, ellipse, or hyperbola.*

63. $4x^2 + y^2 - 1 = 0$ **64.** $4x^2 - 8x - y^2 + 8y - 16 = 0$

65. $y + 3x^2 + 2x = 0$

66. $-x^2 - y^2 + 2x + 4y = 0$

67. $x + 4xy + 2y^2 - 1 = 0$

68. $3x^2 - 6xy + 3y^2 + x - 1 = 0$

69. $(x + y)^2 = 2xy + 5$

70. $(x + y)^2 = xy + 5$

Exercises 71-76 *Eliminate the parameter and graph. Check your result with a graphing calculator.*

71. $x = 2t - 5$
    $y = -3t + 2$

72. $x = t^2$
    $y = 2t^2 + 1$

73. $x = t^3 + 1$
    $y = 2t^3 + 2$

74. $x = t - 2$
    $y = t^2$

75. $x = 2\cos t$
    $y = 2\sin t$

76. $x = 1 + 2\cos t$
    $y = -2 + 3\sin t$

Exercises 77-78 *Use a graphing calculator to plot the graph and estimate the indicated point(s).*

77. $x = 2 - \sin t \cos(3t)$
    $y = 3 + \cos t \sin(3t)$
    Crossing point (where the curve intersects itself)

78. $x = t^2 - 1$
    $y = t^2 + t - 6$
    Intercepts

Exercises 79-86 *Parts a and b are connected: Part a involves an elementary concept, whereas part b involves related material from this chapter. First answer part a, and then use this result to answer part b.*

79. a. Classify $f(x) = \dfrac{x^2 + 2}{x^2 + 4}$ as even, odd, or neither.

    b. One portion of the graph of $y(x^2 + 4) = x^2 + 2$ is shown in Figure 134. Use symmetry to complete the graph.

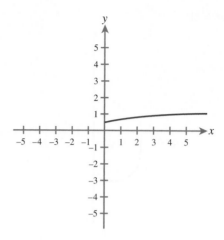

Figure 134

80. a. Expand $(x + 3)^2 + 2(y - 1)^2$.

    b. Find the center of the ellipse with equation
    $x^2 + 6x + 2y^2 - 4y + 11 = 1$.

81. a. Estimate the coordinates of the vertex of $y = -x^2 + 3x + 1$.

    b. Estimate the coordinates of the vertex of $x = -y^2 + 3y + 1$.

82. a. Find the distance between $(4, 5)$ and $(0, 2)$.

    b. Find a point in the first quadrant lying on the parabola having $(0, 2)$ as its focus and the x-axis as its directrix.

83. a. Verify the identity $\sec^2 t - \tan^2 t = 1$.

    b. Eliminate the parameter, and sketch the graph of $x = \sec t$, $y = \tan t$.

84. a. Compute $\sin\left[\tan^{-1}\left(\dfrac{3}{4}\right)\right]$ and $\cos\left[\tan^{-1}\left(\dfrac{3}{4}\right)\right]$.

    b. The xy-coordinates of a point are $(3, 4)$. Find its coordinates in the uv-system obtained by rotating the standard xy-coordinate system by $\theta = \tan^{-1}\left(\dfrac{3}{4}\right)$.

85. a. Find the midpoint and the slope of the line segment having endpoints $(-1, 0)$ and $(3, 4)$.

    b. The vertices of an ellipse are $(-1, 0)$ and $(3, 4)$, and one endpoint of the minor axis is $(2, 1)$. Find the other endpoint of the minor axis.

86. a. Eliminate the radical in the equation $y = \sqrt{1 - qx^2} + 2$ to rewrite it in the form $Ax^2 + Cy^2 + Dx + Ey + F = 0$.

    b. Find the values of q for which the graph of $f(x) = \sqrt{1 - qx^2} + 2$ is a part of (i) a parabola, (ii) an ellipse, or (iii) a hyperbola.

87. Pressure and Altitude The graph of the equation relating pressure P (measured in pounds per square inch) and altitude above sea level h (measured in feet) is given in Figure 135. Sketch the graph of the equation relating pressure and depth below sea level d.

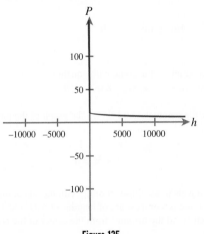

Figure 135

88. Whispering Room A whispering room is in the shape of an ellipse that is twice as long as it is wide. If the foci are 1.86 feet from the vertices, find the distance from each focus to the center of the room.

89. Parametric Stepping The location of one pedal of an aerobic stair-stepper is given by the parametric equations $x = \frac{3}{2}(1 - \cos \pi t)$ and $y = 6(1 - \cos \pi t)$, where t is in seconds and x and y are in inches. How long does it take to complete one step?

90. Home Run Path If air resistance is ignored, the path of a baseball hit with an initial angle of $45°$ and an initial velocity of 120 feet per second is given by

$$450y = -x^2 + 450x + 1350$$

where x is the horizontal distance from home plate (in feet) and y is the height (in feet).

a. Find the maximum height of the ball.

b. How far away from home plate does the ball land?

Chapter 8 Test

Problems 1–8 *Answer true or false.*

1. If a graph is reflected first about the x-axis and then about the y-axis, the result is the same as if it were reflected first about the y-axis and then about the x-axis.

2. The graph of a relation can be symmetric with respect to both the x- and y-axes.

3. The distance from the focus to the vertex of a parabola is equal to the distance from the vertex to the directrix.

4. A point on a hyperbola is equidistant from the two foci of the hyperbola.

5. It is possible to find an angle θ such that in the uv-coordinate system formed by a rotation by the angle θ, the equation $y = x^2 + 3x + 4$ becomes $2u^2 + 3v^2 = 9$.

6. If $f(a) = g(a) = 0$, then the graph of $x = f(t)$ and $y = g(t)$ passes through the origin.

7. All conic sections with vertical or horizontal axes have equations of the form $Ax^2 + Cy^2 + Dx + Ey + F = 0$.

8. The foci of an ellipse always lie on the minor axis.

Problems 9–13 *Give an example of each.*

9. A relation whose graph is symmetric with respect to the x-axis

10. Two sets of parametric equations with the same graph

11. The equation of a parabola that opens to the left

12. The equation of a conic section with two vertices on the x-axis

13. An application of a reflective property of a conic section

14. The graph of $8x = y^4$ is given in Figure 136. Sketch the graphs of the relations in parts a and b that are formed by translating and/or reflecting the graph of $8x = y^4$.

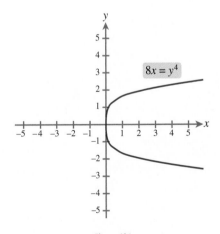

Figure 136

a. $8(x - 1) = (y + 3)^4$

b. $8x = -y^4$

15. A portion of the graph of $x^3 - y^2 = 0$ is shown in Figure 137. Test the relation for symmetry and then use symmetry, to complete the graph.

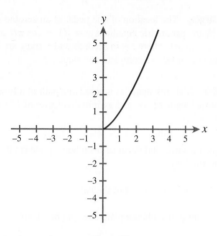

Figure 137

Problems 16-18   *Identify and sketch the graph of the given conic section. Find any of the following features that apply: center, vertices, foci, and asymptotes.*

16. $x = y^2 + 4y$

17. $4x^2 - 9y^2 = 36$

18. $4x^2 + y^2 - 16x + 6y + 9 = 0$

19. Sketch the graph of the conic with equation
$$3x^2 + 2\sqrt{3}xy + y^2 + 8x - 8\sqrt{3}y = 0.$$

20. Eliminate the parameter and sketch the graph of the parametric equations
$$x = t + 1$$
$$y = 3t - 1$$

21. A comet travels in an elliptical orbit, with the sun at one focus. The major and minor axes are of lengths 12 A.U. and 10 A.U., respectively. Find the distance from the center of the comet's orbit to the sun.

Chapter 9

Systems of Equations and Inequalities

Hurricane Andrew struck southern Florida and Louisiana in August 1992, causing over $20 billion in damage. With enough destructive force to topple homes, uproot trees, and change the course of rivers, hurricanes provide meteorologists with a strong incentive to accurately model Earth's weather system: The difference between 2 hours' notice and 2 days' notice of an impending hurricane can be the difference between life and death. Mathematical models of Earth's atmosphere are incredibly complex, often involving the solution of systems consisting of thousands of linear equations. In this chapter, we explore several methods of solving such systems.

Section 9.1 | # Systems of Equations

- How can chemical equations be balanced using systems of equations?
- How can hog lard and turkey breast be mixed to form a 90% fat-free mixture?
- If one runner defeats another by 10 meters in the first heat of a 50-meter race, and then for the second heat begins 10 meters behind the starting line to handicap himself, who will win the second heat?

Throughout the text, we have solved a variety of equations—linear equations, quadratic equations, equations with rational expressions, equations involving radicals, exponential and logarithmic equations, trigonometric equations, and so on. In this section, we deal with **systems of equations**, which consist of more than one equation and involve more than one variable. Solving a system of equations entails finding values of each of the variables so that all the equations are satisfied. For example, consider the following system of two equations in two unknowns:

$$2x + 3y = 8$$
$$6x - 3y = 0$$

It can be shown that $x = 1$, $y = 2$ is a solution of this system. In fact, the first equation is satisfied since $2 \cdot 1 + 3 \cdot 2 = 8$, and the second equation is satisfied since $6 \cdot 1 - 3 \cdot 2 = 0$. It is customary to express the solution as $(1, 2)$, which indicates that $x = 1$, $y = 2$ is a solution of the system.

In this section, we consider linear and nonlinear systems of equations. Linear equations involve only first-degree terms. For example, $2x + 3y = 5$ is linear, whereas $3x^2 + 4y^3 = 7$ and $x - 4xy = 5$ are nonlinear since the terms $3x^2$, $4y^3$, and $4xy$ all have degree larger than 1. A **linear system of equations** is a system in which every equation is linear. A **nonlinear system of equations** includes at least one equation that is nonlinear. For example,

$$2x + 3y = 8$$
$$6x - 3y = 0$$

is a linear system of equations, whereas

$$x^2 + y^2 = 1$$
$$2x + 3y = 5$$

is a nonlinear system of equations.

⋯⟩EXAMPLE 1 | ### Verifying a Solution of a System of Nonlinear Equations

In parts a and b, check to see whether the given ordered pair satisfies the system of equations

$$3x^2 - 5y = 7$$
$$2x + y = 10$$

a. $(2, 1)$ **b.** $(3, 4)$

Solution

a. Substituting $x = 2$ and $y = 1$ into the first equation, we have

$$3(2)^2 - 5(1) = 12 - 5 = 7$$

Substitution into the second equation gives

$$2(2) + 1 = 5 \neq 10$$

Since the second equation doesn't hold, (2, 1) is not a solution.

b. Substituting $x = 3$ and $y = 4$ into the first equation, we have

$$3(3)^2 - 5(4) = 27 - 20 = 7$$

Substitution into the second equation gives

$$2(3) + 4 = 6 + 4 = 10$$

Since (3, 4) satisfies both equations, it is a solution of the system.

Solving Systems Graphically   Recall that the graph of an equation in x and y is the set of points (x, y) such that x and y satisfy the equation. It follows that if (x, y) is a solution to a system of equations, then the point (x, y) must lie on the graph of each of the equations that make up the system. Thus, the solution set of a system of two equations in x and y consists of the intersection points of their graphs. This suggests a technique for approximating the solution set to a system of equations: Graph each of the equations comprising the system, and then estimate the points of intersection of the resulting curves. We illustrate this method in the following examples.

EXAMPLE 2

Solving a System Graphically

Solve the system

$$x - y = 4$$
$$x^3 + y = 7$$

Solution   To plot the graphs, we must first solve each equation for y. This gives us

$$y = x - 4$$
$$y = 7 - x^3$$

Figure 1 shows the result of plotting these equations in the standard viewing window. The graphs appear to intersect near $(2, -2)$. After zooming in a few times, we obtain the view shown in Figure 2, from which we conclude that the coordinates of the point are approximated by $(2.07, -1.93)$. By zooming out repeatedly, we are convinced that there are no other points of intersection. Thus, the only solution to the system is $x \approx 2.07$, $y \approx -1.93$. If more accurate estimates are needed, we could use the calculate-intersect feature of our graphing calculator.

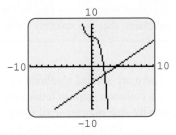

Figure 1

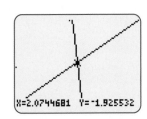

Figure 2

**EXAMPLE 3**

Solving a System Graphically

Solve the system

$$2x - y = 7$$
$$x^2 + y^2 = 9$$

Solution   Since the first equation is linear and the second is the equation of a circle, we are looking for point(s) of intersection of a line and a circle. Since a line can cross a circle once, twice, or not at all, there may be 0, 1, or 2 solutions to this system. Solving the first equation for y gives us $y = 2x - 7$. The second equation gives us

$$x^2 + y^2 = 9$$
$$y^2 = 9 - x^2$$
$$y = \pm\sqrt{9 - x^2}$$

Figure 3 shows the graphs of $y = 2x - 7$, $y = \sqrt{9 - x^2}$, and $y = -\sqrt{9 - x^2}$ (note the use of the "square" window settings, as described in Section 1.4). With this view, it is difficult to tell if there is an intersection point. The view shown in Figure 4, on the other hand, suggests that the line $2x - y = 7$ does not intersect the circle $x^2 + y^2 = 9$. Thus, we conclude that the system has no solution.

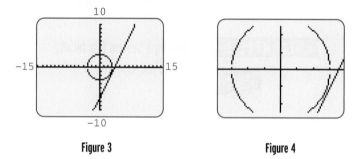

Figure 3   Figure 4

Some systems of equations have several or even infinitely many solutions. On the other hand, as the previous example illustrates, some systems have no solutions whatsoever. In this case, the equations of the system impose inconsistent conditions: It is impossible to satisfy both equations at the same time. More formally, we have the following definitions.

Consistent and Inconsistent Systems

A system with at least one real solution is said to be **consistent**. If there are no real solutions, then we say that the system is **inconsistent**.

There are several algebraic methods for solving systems of equations, all of which have a common goal: to reduce the number of equations and the number of variables in a step-by-step manner until there is only one equation in one unknown. The resulting equation is then solved for the lone remaining variable, and its value is used to find the values of the previously eliminated variables.

Solving Systems of Equations by Substitution

The most straightforward method (although not always the shortest) for solving systems of equations is **substitution**: solving one of the equations of a system for a variable and then substituting the resulting expression into the other equations that comprise the system. This technique is illustrated in the following example.

EXAMPLE 4

Solving a System of Equations by Substitution

Solve the following system of equations:

(1) $$3x + y = 10$$
(2) $$2x - 3y = -8$$

Solution   We begin by solving equation (1) for y.

$$3x + y = 10$$
(3) $$y = 10 - 3x$$

Next we substitute $10 - 3x$ for y in equation (2).

$$2x - 3y = -8 \qquad \text{Equation (2)}$$
$$2x - 3(10 - 3x) = -8 \qquad \text{Substituting } 10 - 3x \text{ for } y$$
$$2x - 30 + 9x = -8$$
$$11x - 30 = -8$$
$$11x = -8 + 30$$
$$11x = 22$$
$$x = 2$$

Finally, we use equation (3) and the fact that $x = 2$ to find y.

$$y = 10 - 3x \qquad \text{Equation (3)}$$
$$y = 10 - 3(2) \qquad \text{Substituting 2 for } x$$
$$y = 4$$

Thus, the solution is given by $(x, y) = (2, 4)$. We now verify that $(2, 4)$ is indeed a solution of the system by substitution into the original system.

Check

Equation: $3x + y = 10$		Equation: $2x - 3y = -8$	
Solution: $x = 2, y = 4$		Solution: $x = 2, y = 4$	
$3x + y$	10	$2x - 3y$	-8
$3(2) + 4$		$2(2) - 3(4)$	
$6 + 4$		$4 - 12$	
10	✔	-8	✔

Note that in the previous example, we chose to solve the first equation for y. In general, our choice of which variable to solve for (and in which equation) is determined by convenience.

In all cases, solutions to systems of equations should be checked by substitution into the original system. For the remainder of the examples in this section, we shall omit the details of the verification process.

EXAMPLE 5

Solving a System of Nonlinear Equations by Substitution

Solve the following system of equations:

(1) $$x - y = 1$$

(2) $$x^2 + 2 = 6y$$

Solution   In this case, we solve equation (1) for x and substitute into equation (2).

$$x - y = 1$$

(3) $$x = y + 1$$

We now substitute $y + 1$ for x in equation (2).

$$x^2 + 2 = 6y \qquad \text{Equation (2)}$$
$$(y + 1)^2 + 2 = 6y \qquad \text{Substituting } y + 1 \text{ for } x$$
$$y^2 + 2y + 1 + 2 = 6y \qquad \text{Expanding}$$
$$y^2 - 4y + 3 = 0 \qquad \text{Collecting all terms on the left}$$
$$(y - 1)(y - 3) = 0 \qquad \text{Factoring}$$
$$y = 1, y = 3$$

Now we use equation (3), $x = y + 1$, to find the corresponding x-values.

$$\text{When } y = 1: \qquad x = 1 + 1 = 2$$
$$\text{When } y = 3: \qquad x = 3 + 1 = 4$$

Thus, the two solutions are (2, 1) and (4, 3). Note that in this case *both* solutions need to be verified.

EXAMPLE 6

Determining the Weights of Disks Using Systems of Equations

Suppose that you are at a bench-press contest and have just witnessed the lightweight champion press a bar weighted with 2 large disks and 1 small disk on each side, and the weight is announced as 275 pounds. Then, the eventual middleweight champion presses a bar with 3 large disks and 1 small disk on each side, and the weight is announced as 365 pounds. The heavyweight champion lifts a bar weighted with 4 large disks and 1 small disk, but you cannot hear the announced weight. Assuming the bar weighs 45 pounds, how much did the heavyweight lift?

Solution   We first organize the given information in a table.

Table 1

Weight class	Large disks (per side)	Small disks (per side)	Announced Weight
Light	2	1	275
Middle	3	1	365
Heavy	4	1	?

Let x represent the weight of the larger disk and y the weight of the smaller disk. Since the lightweight champion's bar had a total of 4 large disks and 2 small disks, and the bar weighs 45 pounds, we have

$$4x + 2y + 45 = 275, \text{ or}$$

(1) $$4x + 2y = 230$$

Similarly, from the middleweight's lift we obtain

$$6x + 2y + 45 = 365, \text{ or}$$

(2) $$6x + 2y = 320$$

Now solving equation (1) for y, we have

$$4x + 2y = 230$$
$$2y = 230 - 4x$$
(3) $$y = 115 - 2x$$

Substituting into equation (2) gives us

$$6x + 2(115 - 2x) = 320$$
$$2x + 230 = 320$$
$$2x = 90$$
$$x = 45$$

Now substituting 45 for x in equation (3), we obtain

$$y = 115 - 2(45)$$
$$= 25$$

Thus, the large disks weigh 45 pounds and the small disks weigh 25 pounds. Since the heavyweight raised a total of 8 large disks and 2 small disks (plus the bar), he lifted

$$8(45) + 2(25) + 45 = 455 \text{ pounds}$$

Solving Systems of Equations by Elimination

A second method for solving systems of equations, **elimination**, consists of adding multiples of the equations that constitute the system in such a way that a variable is eliminated. For example, consider the following system of equations:

$$2x + y = 6$$
$$3x - y = 9$$

Adding the two equations, we obtain

$$\begin{array}{r} 2x + y = 6 \\ + \ 3x - y = 9 \\ \hline 5x \quad\ = 15 \end{array}$$

This gives us $x = 3$, which can then be substituted into either of the original equations to find y. We substitute into the first equation:

$$2x + y = 6$$
$$2(3) + y = 6$$
$$y = 0$$

Thus, the solution is $(3, 0)$. The technique of elimination is further illustrated in the following examples.

EXAMPLE 7

Solving a System of Linear Equations by Elimination

Solve the following system of equations by elimination:

(1) $$4x + 3y = 5$$
(2) $$3x + y = 4$$

Solution Adding these equations together will eliminate neither x nor y. In fact, addition gives

$$
\begin{array}{r}
4x + 3y = 5 \\
+\ 3x +\ \ y = 4 \\
\hline
7x + 4y = 9
\end{array}
$$

and we're no better off than before. However, if we first multiply the second equation by -3, we obtain

$$(-3)(3x + y) = (-3)4$$

(3)
$$-9x - 3y = -12$$

Since equation (3) includes the term $-3y$, and equation (1) includes the term $3y$, adding equations (1) and (3) will eliminate y.

$$
\begin{array}{rl}
4x + 3y = 5 & \text{Equation (1)} \\
+\ -9x - 3y = -12 & \text{Equation (3)} \\
\hline
-5x \qquad\quad = -7 &
\end{array}
$$

Thus $x = \frac{7}{5}$. We can now substitute into equation (2) to find y.

$$3x + y = 4$$

$$3\left(\frac{7}{5}\right) + y = 4$$

$$\frac{21}{5} + y = 4$$

$$y = 4 - \frac{21}{5}$$

$$= \frac{20}{5} - \frac{21}{5}$$

$$= -\frac{1}{5}$$

Thus, our solution is $(x, y) = \left(\frac{7}{5}, -\frac{1}{5}\right)$.

EXAMPLE 8

Solving a System of Linear Equations by Elimination

Solve the following system of equations by elimination:

$$3x - 5y = 5$$
$$4x - 7y = 6$$

Solution Although either variable could be eliminated, we will eliminate x. In order to obtain equations in which the coefficients of x are opposites, we will multiply the first equation by 4 and the second equation by -3:

Original equation	Multiply by . . .	Result
$3x - 5y = 5$	4	$12x - 20y = 20$
$4x - 7y = 6$	-3	$\dfrac{+\ -12x + 21y = -18}{y = 2}$

Substituting 2 for y in $3x - 5y = 5$ gives

$$3x - 5(2) = 5$$
$$3x - 10 = 5$$
$$3x = 15$$
$$x = 5$$

Thus, our solution is $(x, y) = (5, 2)$.

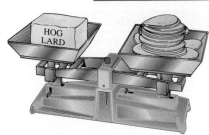

·····➤ EXAMPLE 9

Solving a Mixture Problem Using Elimination

A foolish dieter, following the instructions of his nutritionist to the letter, is insistent that fat constitute exactly 10% of his diet. On a certain day, he must select from a menu consisting strictly of 92% fat-free turkey bologna and hog lard, which is virtually 100% fat. How much of each should he consume if he wishes to eat 10 pounds of food?

Solution   Let x represent the number of pounds of turkey bologna eaten and y the number of pounds of hog lard. Then we have

Pounds of fat from turkey bologna $+$ Pounds of fat from lard $=$ Pounds of fat in mixture

$$8\% \cdot x + 100\% \cdot y = 10\% \cdot 10$$
$$0.08x + y = 1$$

On the other hand, we have

Pounds of turkey bologna $+$ Pounds of lard $=$ 10 pounds

$$x + y = 10$$

Thus, we must solve the system

(1) $$0.08x + y = 1$$
(2) $$x + y = 10$$

Multiplying equation (1) by -1 and adding to equation (2) gives us

$$\begin{array}{r} -0.08x - y = -1 \\ + \quad x + y = 10 \\ \hline 0.92x \qquad = 9 \end{array}$$

Solving for x gives us $x = \frac{9}{0.92} \approx 9.8$. Substituting $x = 9.8$ into equation (2) gives

$$x + y = 10$$
$$9.8 + y = 10$$
$$y = 0.2$$

Thus, 9.8 pounds of turkey bologna mixed with 0.2 pound of lard will yield 10 pounds of a mixture that is 90% fat free.

So far, we have illustrated the technique of elimination for solving linear systems of equations, but it can also be an effective technique for solving nonlinear systems, as illustrated in the following example.

EXAMPLE 10

Solving a Nonlinear System of Equations

Find all points of intersection of the curves $x^2 + y = 2$ and $2x^2 + y^2 = 3$ shown in Figure 5.

Solution We begin by noting that intersection points of the graphs correspond to solutions of the system

(1) $x^2 + y = 2$
(2) $2x^2 + y^2 = 3$

We eliminate the variable x by multiplying equation (1) by -2 and adding the result to equation (2).

Original equation	Multiply by ...	Result
$x^2 + y = 2$	-2	$-2x^2 - 2y = -4$
$2x^2 + y^2 = 3$	1	$+\ \ 2x^2 + y^2 = 3$
		$\overline{\qquad y^2 - 2y = -1}$

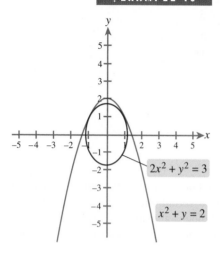

Figure 5

Solving for y, we obtain

$$y^2 - 2y = -1$$
$$y^2 - 2y + 1 = 0$$
$$(y - 1)^2 = 0$$
$$y = 1$$

Substituting $y = 1$ into equation (1) gives us

$$x^2 + 1 = 2$$
$$x^2 = 1$$
$$x = \pm 1$$

Thus, the points of intersection are $(1, 1)$ and $(-1, 1)$, as shown in Figure 6.

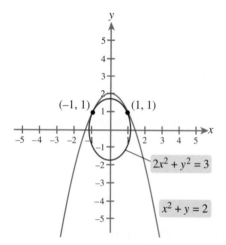

Figure 6

Elimination can be extended to systems of three equations in three unknowns, as the following example illustrates.

EXAMPLE 11

A System of Three Equations in Three Unknowns

Solve the following system of equations:

(1) $x + y + z = 4$
(2) $2x + y - z = 9$
(3) $x - 3y - 2z = -1$

Solution Our strategy is as follows. First we eliminate a variable from a pair of equations. Then we eliminate the same variable from another pair of equations. This gives us a system of two equations in two unknowns, which we then solve. Specifically, we eliminate z by adding equation (1) to equation (2) and by adding twice equation (1) to equation (3).

$$\begin{array}{lrl} & x + y + z = 4 & \text{Adding equations (1) and (2) together to eliminate } z \\ & + \; 2x + y - z = 9 & \\ \hline (4) & 3x + 2y \quad = 13 & \end{array}$$

$$\begin{array}{lrl} & 2x + 2y + 2z = 8 & \text{Multiplying equation (1) by 2 . . .} \\ & + \; x - 3y - 2z = -1 & \text{and adding to equation (3) to eliminate } z \\ \hline (5) & 3x - y \quad = 7 & \end{array}$$

Now we solve the system consisting of equations (4) and (5). We can eliminate x from this system by multiplying equation (4) by -1 and adding to equation (5).

$$\begin{array}{r} -3x - 2y = -13 \\ + \; 3x - y = 7 \\ \hline -3y = -6 \end{array}$$

It follows that $y = 2$. Substituting this value into equation (4) gives us

$$\begin{aligned} 3x + 2y &= 13 \\ 3x + 2 \cdot 2 &= 13 \\ 3x &= 9 \end{aligned}$$

Hence $x = 3$. We now substitute the values of x and y into equation (1) to find z.

$$\begin{aligned} x + y + z &= 4 \\ 3 + 2 + z &= 4 \\ z &= -1 \end{aligned}$$

Thus, our solution is $x = 3$, $y = 2$, and $z = -1$.

Understanding and Mastery Checklists

Concepts to Understand

- Linear and nonlinear systems of equations
- Intersection of graphs
- Consistent and inconsistent systems
- Substitution
- Elimination

Skills to Master

- Classify systems of equations as linear or nonlinear.
- Solve systems of equations graphically.
- Classify systems of equations as consistent or inconsistent.
- Solve systems of equations by substitution.
- Solve systems of equations by elimination.
- Apply systems techniques in applied settings.

Exercises 1-36 *Solve the system of equations. Some systems may be inconsistent.*

1. $2x + y = 10$
$3x - y = 5$

2. $3s + t = 5$
$2s - 2t = 14$

3. $-3x + y = 7$
$3x + 4y = -2$

4. $2x + 3y = -7$
$x + 5y = -14$

5. $2q + 3r = 3$
$4q + 6r = 0$

6. $3x - 4y = 8$
$9x - 12y = 2$

7. $3m - n = 2$
$-4m + 2n = 2$

8. $3x + 5y = 2$
$5x + y = 18$

9. $x - 5y = 5$
$4x + 5y = 5$

10. $-2x - y = -2$
$2x + y = -1$

11. $2z - 7w = -2$
$10z - 35w = -8$

12. $2x + 5y = 5$
$4x - 5y = 4$

13. $\dfrac{10}{3}a + 2b = -6$
$a + \dfrac{1}{5}b = 1$

14. $6a - 2b = \dfrac{7}{2}$
$3a + \dfrac{1}{2}b = \dfrac{5}{2}$

15. $2x + 3y = 9.8$
$3x - 5y = -8.1$

16. $2w - 5z = 12.9$
$3w + 4z = 4.4$

17. $0.5x - 0.2y = 1.36$
$1.5x + 0.4y = 5.78$

18. $1.2A + 4.6B = 18.5$
$-2.4A + 3.0B = 5.7$

19. $(2 \times 10^{-6})x + 10^{-6}y = 33$
$4x - y = 3 \times 10^7$

20. $(4 \times 10^4)k + (1.2 \times 10^4)m = 2 \times 10^4$
$(2 \times 10^4)k + (5 \times 10^4)m = -2.1 \times 10^5$

21. $x + y = 8$
$xy = 15$

22. $2x - 3y = 5$
$\dfrac{y}{x} = -1$

23. $2p^2 + 3q^2 = 30$
$p + q = 1$

24. $\dfrac{c}{d} - c = 4$
$c - 8d = 0$

25. $\dfrac{2}{a} + \dfrac{3}{b} = 0$
$\dfrac{4}{a} + \dfrac{12}{b} = -1$

26. $\dfrac{6}{p} + \dfrac{6}{q} = 33$
$-\dfrac{3}{p} + \dfrac{4}{q} = -6$

27. $x^2 + y^2 = 9$
$5x - y = -3$

28. $2(u + v)^3 - v = -11$
$(u + v)^3 - 3v = 7$

29. $x^4 + y^4 = 97$
$x^4 - y^4 = 65$

30. $\dfrac{x^2}{4} + \dfrac{y^2}{9} = 2$
$3x + y = 3$

31. $3x + y + 2z = 11$
$x + 3y = 6$
$3y + z = 0$

32. $2x + y - z = -3$
$x + 3y = 6$
$x + 3z = 3$

33. $x + y + z = 2$
$2y - 3z = -7$
$3y + 2z = 9$

34. $a + b + c = 2$
$2a - 3b + c = 1$
$3a - 4b + 2c = 4$

35. $u + v + w = 1$
$2u + 2v + w = 2$
$3u + v - w = 5$

36. $2A - 2B + 3C = -4$
$2A + 2B - C = 8$
$3A - 4B + 5C = -9$

Exercises 37-42 *Find the point(s) of intersection of the given pair of equations.*

37. $2x - y = 1$; $3x + y = 9$

38. $x - 4y = 8$; $-3x - 2y = 4$

39. $y + x^2 = 4$; $y - x = 2$

40. $y = x^2 + 2x$; $y = -2x - 3$

41. $x^2 + y^2 = 2$; $x = y^2$

42. $x^2 + y^2 = 25$; $x + y = 1$

Exercises 43-48 *Use a graphing calculator to estimate the coordinates of the point(s) of intersection of the given pair of equations to the nearest hundredth.*

43. $y = x^3 - x + 2$
$y = x^2 + x + 1$

44. $y = x^2 - 3x + 7$
$y = 3x^2 - 11x + 15$

45. $y = 15 - x^2$
$y = 3\sqrt{4 - x^2} + 4$

46. $y = \sqrt{x^2 - 4x + 4}$
$y = x - 2$

47. $y = 3|x - 2|$
$y = 6 - 3|x - 4|$

48. $y = x$
$y = \dfrac{x^3 + x + 1}{x^2 + 1}$

Applications

49. Counting Coins A street musician counts the coins in his hat and discovers that he has four more nickels than dimes and that he has a total of 80¢ in nickels and dimes. How many nickels and how many dimes does the street musician have?

50. Counting Coins A budding numismatist (coin collector) has twice as many silver dollars as quarters; the total face value of the silver dollars and the quarters is $4.50. How many of each does she have?

51. Low-Fat Mixture A sundae is being made from 98% fat-free frozen yogurt and 80% fat-free dessert topping. How many ounces of each should be used if we wish to have a 10-ounce, 90% fat-free sundae?

52. Achieving Racial Balance In a certain school district, 40% of the students are African American, but only 10% of the 1000 teachers are. Since the student–teacher ratio is also unacceptably high, the school board decides that it will correct both problems by doubling the number of teachers, and hiring in such a way that the racial makeup of the resulting faculty reflects that of the student body. The personnel director determines that all of the new hires will be from two institutions: a state university of which 60% of the students are African American and an historically black college, of which 93.3% of the students are African American. How many new teachers should be hired from each of the schools?

53. Age Comparison Mark is 11 times older than his son Benjamin. In 27 years, Benjamin will be $\frac{1}{2}$ the age of his father. How old are they?

54. Age Comparison Two years ago, Juanita was 2.5 times older than her younger sister, LaShonda. Ten years from now, Juanita will be 1.5 times older than LaShonda. How old are they now?

55. Ticket Sales Some floor tickets for the Halloween 2002 Rolling Stones concert in Los Angeles (part of the "Licks" tour) sold for over $2000, and it was nearly impossible to find tickets (even with an obscured view) for under $100. Suppose a corporation purchased two blocks of tickets to be given to a select group of 400 clients— a block of $200 tickets in an upper level of the Staples Center and a block of lower level/center tickets for $500. If the corporation paid a total of $95,000, then how many of each kind of ticket were purchased?

56. Painting Rate It takes Susan 1 hour less to paint a room than it does David. Together, they take 2 hours. How long does it take each of them working alone?

57. Raising a Whale Two powerlifters (whom we'll call Hans and Franz) are raising money to protect the blue whale, of which there may be fewer than 10,000, by participating in a fund-raiser called "Raising a Whale." Their goal is to lift a total weight equal to that of the largest blue whale ever captured, approximately 209 tons. (*Data source: The Guinness Book of World Records.*) If Hans were lifting alone, he could "raise a whale" in 10 fewer hours than it would take Franz working alone. Together, they can complete the task in 15 hours. How long does it take each of them working alone?

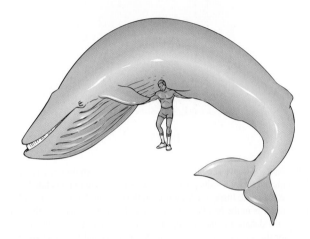

58. Car Speed Car A leaves an intersection at 2:00 P.M. traveling north at a constant speed. Car B leaves the same intersection at 4:00 P.M. and travels east at a constant speed. At 6:00 P.M. the cars are 200 miles apart, and the total distance traveled by the cars is 280 miles. How fast is each car traveling?

59. Sprint Times Two brothers are competing in a 50-meter race. The first time they race, the older brother wins by 10 meters. The older brother offers to start the next race 10 meters behind the starting line in order to make the race fair. Although both runners run at the same speed as before, the older brother still wins, this time by 0.5 second. What is the 50-meter time for both boys? (Assume each is running at a constant speed.)

60. Cattle Speed Two ancient cows of equal speeds, each well versed in the rudiments of mathematics, are standing side by side on a 120-foot railroad bridge that runs north-south. The cows hear the whistle of a train heading toward them from the north when the train is 1000 feet away from the bridge. The train is traveling at a speed of 53 feet per second. The impish bovines run at top speed in opposite directions to torment the engineer. Just as planned, they each are just grazed by the train as they clear the bridge. How fast can the cattle run, and how far were they initially from the north end of the bridge?

Get ready!

61. Radar Search   Two radar stations are located 50 miles apart along a coastline at the points marked A and B in Figure 7. A ship is in distress at a point C, which is 20 miles from A and 40 miles from B. Find the location of the ship relative to A. (*Hint:* Use point A as the origin of a coordinate system and find the points of intersection of the circle centered at A with radius 20 and the circle centered at B with radius 40.)

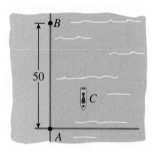

Figure 7

62. Kidnap Caper   A suitcase containing $250,000 in ransom money has just been picked up by a kidnapper. Little does he realize that his fate has just been sealed; Inspector Magill is on the case, and a miniature radio transmitter is in the handle of the suitcase. Using two receivers that can detect the distance to the signal, Magill and an assistant are able to monitor the movement of the signal. When the signal finally stops moving, it is 2 miles away from Magill's position and 3 miles away from his assistant's position. If Magill is 4 miles west of his assistant, use intersecting circles to determine the location(s) where police should be sent.

63. Unknown Constants   It is given that the variables y and x are related by the formula

$$y = a|x| + b|x - 1| + c|x - 2|$$

It is known that when $x = 0$, $y = 5$; when $x = 1$, $y = 6$; and when $x = 2$, $y = 1$. Find a, b, and c.

64. Volume of a Box   An open box of volume 108 cubic inches is formed by cutting out squares of side length x inches from the corners of an $l'' \times w''$ square sheet of cardboard and then folding along the dashed lines, as shown in Figure 8. It so happens that the cut-out squares can be taped together to form a lid for the box, as shown in Figure 8. Find l, w, and x.

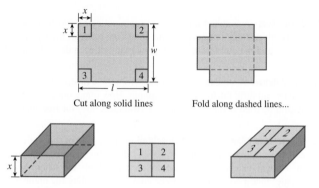

Cut along solid lines   Fold along dashed lines...

...to make an open box.   Assemble cut-out squares to make a lid...   ...that fits on the open box. The resulting closed box has a volume of 108 in³.

Figure 8

Concepts and Critical Thinking

Exercises 65–68   *Answer true or false.*

65. Replacing the first of two equations in a system by the sum of the two equations results in a system with the same solution set.

66. The system

$$x + y = 5$$
$$2x + 2y = 5$$

is inconsistent.

67. If even a single equation in a system is linear, then the entire system is considered to be linear.

68. If even a single equation in a system is nonlinear, then the entire system is considered to be nonlinear.

Exercises 69–72   *Give an example of each.*

69. A nonlinear system of equations

70. A linear system of equations with infinitely many solutions

71. A system of two linear equations for which the only solution is $x = 2$, $y = 3$

72. An inconsistent system in which one of the equations is $2x + 3y = 5$

73. One of the most common mistakes made by beginning algebra students when solving applied problems is that of not carefully defining variables. Consider the following problem:

There are four more than twice as many nickels as dimes. Together the nickels and dimes are worth $1.00. How many of each are there?

One key—but often overlooked—restriction is that any variable introduced should represent a number; many students make the mistake of defining variables to be inanimate objects, barnyard animals, and people. For example, when solving this coin problem, it is not uncommon for students to begin by writing something like, "Let N = nickels, and let D = dimes." The problem is that "nickels" is not a number. Thus, the student is forced to assume that N represents some number in some way associated with "nickels." But there is more than one number associated with "nickels"! Explain how this difficulty leads to the following incorrect solution:

Since there are 4 more than twice as many nickels as dimes, we have $N = 2D + 4$. Since the nickels and dimes together are worth 100¢, we have $N + D = 100$. Thus, we must solve the system

(1) $$N = 2D + 4$$
(2) $$N + D = 100$$

Substituting equation (1) into equation (2), we have

$$(2D + 4) + D = 100$$
$$3D + 4 = 100$$
$$3D = 96$$
$$D = 32$$

Thus, $N = 2D + 4 = 2 \cdot 32 + 4 = 68$.

Questions for Discussion or Essay

74. Explain the connection between the systems of equations in the first column and the corresponding set of statements in the second column.

System	Statements
A. $x + y = 2$	Edward is the father of Mark.
$2x + 2y = 4$	Mark is the son of Edward.
B. $x + y = 3$	I am a crook.
$x + y = 5$	I am not a crook.
C. $2x + 3y = 8$	The author was a Kennedy.
$3x - y = 1$	The author served as president.

75. Some applied problems require natural number solutions. For example, in a problem involving the number of coins, it would be unacceptable to have a final answer such as "2.3 quarters, 4.7 dimes, and $-\sqrt{2}$ nickels." How do you suppose the authors construct systems of equations in such a way that the answers are guaranteed to be of the proper form? For example, can you construct a system of equations involving the variables x and y such that the solution is $x = 2$ and $y = -5$?

76. Explain how a system of four equations in four variables could be solved by generalizing the techniques of this section.

Projects for Enrichment

77. Balancing Chemical Equations Chemical equations are a shorthand way of describing chemical changes. The left side of the chemical equation consists of the **reactants**, the chemicals that react with each other. The right side of the chemical equation consists of the **products**, the results of the chemical reaction. The left and right sides are separated by an arrow that is typically read as "yields." For example, the reaction consisting of hydrogen (in the form of hydrogen gas H_2) and oxygen (in the form of oxygen gas O_2) combining to form water would be represented as

$$H_2 + O_2 \rightarrow H_2O$$

This equation has not been balanced, however, and as such does not conform to the law of conservation of matter—there are two oxygen atoms on the left but only one on the right. The following chemical equation is a balanced version of this reaction:

$$2H_2 + O_2 \rightarrow 2H_2O$$

More generally, we must ensure that the number of atoms on each side of the yield sign are the same by affixing integer coefficients in front of the reactants and products. For example, consider the unbalanced chemical equation $Al_2O_3 + F_2 \rightarrow AlF_3 + O_2F_2$. To balance the equation, we begin by placing unknown coefficients in front of each of the reactants and products.

$$wAl_2O_3 + xF_2 \rightarrow yAlF_3 + zO_2F_2$$

Now, by equating the number of atoms of aluminum, fluorine, and oxygen, we obtain the following system of equations:

$2w = y$	Equating the number of aluminum atoms on each side
$2x = 3y + 2z$	Equating the number of fluorine atoms on each side
$3w = 2z$	Equating the number of oxygen atoms on each side

Unfortunately, there are not enough equations to determine the values of w, x, y, and z since there are four variables but only three equations. For the sake of convenience, we will assume that $w = 1$. This eliminates one of the variables, giving us the following system of three equations in three unknowns:

(1) $$2 = y$$
(2) $$2x = 3y + 2z$$
(3) $$3 = 2z$$

From equation (1) we have $y = 2$. From equation (3) we have $z = \frac{3}{2}$. Substituting these values into equation (2) gives us

$$2x = 3(2) + 2\left(\frac{3}{2}\right)$$

$$2x = 6 + 3$$

$$x = \frac{9}{2}$$

Thus, our balanced chemical equation now reads

$$Al_2O_3 + \frac{9}{2}F_2 \rightarrow 2AlF_3 + \frac{3}{2}O_2F_2$$

However, we would like our coefficients to be integers. Since multiplying both sides by the same constant will give an equivalent balanced equation, we simply clear fractions by multiplying both sides by the least common denominator of all fractions appearing in the equation. In this case, we multiply by 2 to obtain

$$2Al_2O_3 + 9F_2 \rightarrow 4AlF_3 + 3O_2F_2$$

Balance the following chemical equations by setting up and solving a system of equations:

a. $Na + Cl_2 \rightarrow NaCl$

b. $Ag + H_2S + O_2 \rightarrow Ag_2S + H_2O$

c. $Cu + HNO_3 \rightarrow Cu(NO_3)_2 + H_2O + NO$

d. $Ca_3(PO_4)_2 + H_3PO_4 \rightarrow Ca(H_2PO_4)_2$

Chemical explosions such as this one (created with 24 kilograms of gunpowder) result from the rapid release of gases and heat in a chemical reaction.

78. Partial Fraction Decomposition Consider the following sum of rational expressions:

$$\frac{3}{x} - \frac{2}{x^2} + \frac{4}{x + 1}$$

We can easily add these terms together by finding a common denominator in the usual way. That is, since the least common denominator is $x^2(x + 1)$, we have

$$\frac{3(x)(x + 1) - 2(x + 1) + 4(x^2)}{x^2(x + 1)} = \frac{3x^2 + 3x - 2x - 2 + 4x^2}{x^2(x + 1)}$$

$$= \frac{7x^2 + x - 2}{x^2(x + 1)}$$

Partial fraction decomposition is the reverse of this procedure. For example, the partial fraction decomposition of

$$\frac{7x^2 + x - 2}{x^2(x + 1)} \quad \text{is} \quad \frac{3}{x} - \frac{2}{x^2} + \frac{4}{x + 1}$$

We will confine ourselves to rational expressions of the form $P(x)/Q(x)$, where $P(x)$ and $Q(x)$ are polynomials with the degree of $P(x)$ less than the degree of $Q(x)$. We will further assume that $Q(x)$ can be factored as the product of linear factors; that is, $Q(x)$ can be factored as $(x - a)^s(x - b)^t \cdots (x - c)^u$. It can be shown (it is called the Partial Fraction Decomposition Theorem) that with these restrictions, $P(x)/Q(x)$ can be written in the form

$$\frac{A_1}{x - a} + \frac{A_2}{(x - a)^2} + \cdots + \frac{A_s}{(x - a)^s} +$$

$$\frac{B_1}{x - b} + \frac{B_2}{(x - b)^2} + \cdots + \frac{B_t}{(x - b)^t} + \cdots +$$

$$\frac{C_1}{x - c} + \frac{C_2}{(x - c)^2} + \cdots + \frac{C_u}{(x - c)^u}$$

for some choice of the coefficients $A_1, A_2, \ldots, A_s, B_1, B_2, \ldots, B_t, \ldots, C_1, C_2, \ldots, C_u$. This is not as confusing as it sounds. For example, according to the Partial Fraction Decomposition Theorem, the rational expression

$$\frac{3x^2 + 4x + 7}{x^3(x - 2)^2}$$

can be written in the form

$$\frac{A}{x} + \frac{B}{x^2} + \frac{C}{x^3} + \frac{D}{x - 2} + \frac{E}{(x - 2)^2}$$

with one term for each of the powers of x from 1 to 3, and one term for each of the powers of $x - 2$ from 1 to 2.

We illustrate the method of finding the partial fraction decomposition with the rational expression

$$\frac{7x^2 + x - 2}{x^2(x + 1)}$$

since we already know what the answer should be. We begin by noting that, indeed, the degree of the numerator (2) is less than the degree of the denominator (3) and that the denominator is the product of linear factors. Next, we use the Partial Fraction Decomposition Theorem to conclude that

$$\frac{7x^2 + x - 2}{x^2(x + 1)} = \frac{A}{x} + \frac{B}{x^2} + \frac{C}{x + 1}$$

for some choice of constants A, B, and C. Clearing all denominators by multiplying both sides through by $x^2(x + 1)$ gives us

$$7x^2 + x - 2 = Ax(x + 1) + B(x + 1) + Cx^2$$

$$7x^2 + x - 2 = Ax^2 + Ax + Bx + B + Cx^2$$

Collecting like terms on the right-hand side gives us

$$7x^2 + x - 2 = (A + C)x^2 + (A + B)x + B$$

Now, equating like powers of x on the left- and right-hand sides gives us the following system of equations:

(1) $\qquad\qquad\qquad 7 = A + C$
(2) $\qquad\qquad\qquad 1 = A + B$
(3) $\qquad\qquad\qquad -2 = B$

This system is especially easy to solve. From equation (3) we have $B = -2$. Substituting into equation (2), we get $1 = A + -2$, so that $A = 3$. Upon substituting $A = 3$ into equation (1), we get $C = 4$. Thus, as expected,

$$\frac{7x^2 + x - 2}{x^2(x + 1)} = \frac{3}{x} + \frac{-2}{x^2} + \frac{4}{x + 1}$$

Find the partial fraction decompositions for each of the following rational expressions.

a. $\dfrac{9x^2 + 4x + 1}{x(x + 1)^2}$ **b.** $\dfrac{x^3 + 8x^2 + 11x + 6}{x(x + 1)^3}$

c. $\dfrac{x^3 + 10x^2 - 12x + 5}{x^2(x - 1)^2}$

Section 9.2 | Matrices

- When is $AB \neq BA$?
- How can $AB = 0$ if neither A nor B is 0?
- How can you find the coordinates of a point after it undergoes a rotation?
- If 10% of nonsmokers take up smoking every year, whereas 20% of smokers quit smoking each year, then in a town of 2000 people, how many people will smoke after 32 years, and why doesn't the answer depend on the current number of smokers?

Matrices can be an invaluable tool for solving equations. In this section, we investigate the fundamental properties of matrices, and in later sections, we will see how matrices can be used to solve systems of equations.

Fundamentals

A matrix (plural: *matrices*) is a rectangular array used to store and manipulate data. For example, the linear system of equations

$$x + y + z = 4$$
$$2x + y - z = 9$$
$$x - 3y - 2z = -1$$

from Example 11 of Section 9.1 can be encoded as the 3×4 matrix

$$\begin{bmatrix} 1 & 1 & 1 & 4 \\ 2 & 1 & -1 & 9 \\ 1 & -3 & -2 & -1 \end{bmatrix}$$

The performance of a company whose domestic division has quarterly earnings (in millions of dollars) of 2.3, 2.7, 3.2, and 4.1, and whose international division has quarterly earnings of 1.8, 1.2, 2.0, and 1.7 can be expressed using the 2×4 matrix

$$\begin{bmatrix} 2.3 & 2.7 & 3.2 & 4.1 \\ 1.8 & 1.2 & 2.0 & 1.7 \end{bmatrix}$$

We give a formal definition of a matrix as follows.

Matrix Definitions

- A **matrix of order $m \times n$** is a rectangular array consisting of entries in m (horizontal) **rows** and n (vertical) **columns**, as shown in Figure 9.

$$\underbrace{\left.\begin{bmatrix} a_{11} & a_{12} & a_{13} \cdots a_{1n} \\ a_{21} & a_{22} & a_{23} \cdots a_{2n} \\ a_{31} & a_{32} & a_{33} \cdots a_{3n} \\ \vdots & \vdots & \vdots \quad\;\; \vdots \\ a_{m1} & a_{m2} & a_{m3} \cdots a_{mn} \end{bmatrix}\right\} m \text{ rows}}_{n \text{ columns}}$$

Figure 9

- The $m \times n$ matrix in which every entry is zero is called the **$m \times n$ zero matrix**.
- Two $m \times n$ matrices are said to be equal if corresponding entries are equal.

Note that the entries of the matrix in Figure 9 are written using a double subscript. The first subscript indicates the row of the entry, and the second indicates the column of the entry. Rows are numbered from top to bottom, and columns are numbered from left to right. Thus, the entry in the ith row from the top and the jth column from the left is denoted by a_{ij}, as shown in Figure 10.

$$\text{Row} \quad \begin{matrix} & 1 & 2 & 3 & & j & & n \\ 1 & a_{11} & a_{12} & a_{13} & \cdots & a_{1j} & \cdots & a_{1n} \\ 2 & a_{21} & a_{22} & a_{23} & \cdots & a_{2j} & \cdots & a_{2n} \\ 3 & a_{31} & a_{32} & a_{33} & \cdots & a_{3j} & \cdots & a_{3n} \\ & \vdots & \vdots & \vdots & & \vdots & & \vdots \\ i & a_{i1} & a_{i2} & a_{i3} & & a_{ij} & & a_{in} \\ & \vdots & \vdots & \vdots & & \vdots & & \vdots \\ m & a_{m1} & a_{m2} & a_{m3} & \cdots & a_{mj} & \cdots & a_{mn} \end{matrix} \quad (\text{Column})$$

Entry a_{ij} is in the ith row and the jth column.

Figure 10

The matrix for which the entry in row i and column j is a_{ij} is often denoted by $[a_{ij}]$. This is an especially convenient way of representing matrices with unknown or arbitrary entries. Thus, for example, we can write

$$[a_{ij}] = \begin{bmatrix} a_{11} & a_{12} & a_{13} & \cdots & a_{1n} \\ a_{21} & a_{22} & a_{23} & \cdots & a_{2n} \\ a_{31} & a_{32} & a_{33} & \cdots & a_{3n} \\ \vdots & \vdots & \vdots & & \vdots \\ a_{m1} & a_{m2} & a_{m3} & \cdots & a_{mn} \end{bmatrix} \quad \text{and} \quad [b_{ij}] = \begin{bmatrix} b_{11} & b_{12} & b_{13} & \cdots & b_{1n} \\ b_{21} & b_{22} & b_{23} & \cdots & b_{2n} \\ b_{31} & b_{32} & b_{33} & \cdots & b_{3n} \\ \vdots & \vdots & \vdots & & \vdots \\ b_{m1} & b_{m2} & b_{m3} & \cdots & b_{mn} \end{bmatrix}$$

┈┈**EXAMPLE 1**   **Identifying Entries Using Subscript Notation**

Let

$$A = \begin{bmatrix} 5 & 6 & \frac{1}{2} \\ -2 & 3 & -7 \end{bmatrix}$$

a. What is the order of A?

b. If $A = [a_{ij}]$, identify a_{21} and a_{13}.

Solution

a. Since A has 2 rows and 3 columns, it is of order 2×3.

b. The entry a_{21} is in the second row and the first column. Thus, $a_{21} = -2$. The entry a_{13} is in the first row and the third column, and so $a_{13} = \frac{1}{2}$.

If matrices only served as a convenient way of tabulating data, then they would be nothing more than glorified tables. It is the *algebraic* properties of matrices that set them apart from simple tables and make them especially useful as a tool for solving linear systems of equations. Under certain conditions, matrices can be added, subtracted, multiplied, and even divided.

Addition and Subtraction of Matrices

The definitions of addition and subtraction of matrices are very intuitive; we simply add or subtract corresponding entries. Thus, if

$$A = \begin{bmatrix} 2 & -4 \\ 3 & -5 \end{bmatrix} \quad \text{and} \quad B = \begin{bmatrix} -1 & 7 \\ 2 & 3 \end{bmatrix}$$

then $A + B$ and $A - B$ are computed as follows:

$$A + B = \begin{bmatrix} 2 & -4 \\ 3 & -5 \end{bmatrix} + \begin{bmatrix} -1 & 7 \\ 2 & 3 \end{bmatrix} \qquad A - B = \begin{bmatrix} 2 & -4 \\ 3 & -5 \end{bmatrix} - \begin{bmatrix} -1 & 7 \\ 2 & 3 \end{bmatrix}$$

$$= \begin{bmatrix} 2 + (-1) & -4 + 7 \\ 3 + 2 & -5 + 3 \end{bmatrix} \qquad = \begin{bmatrix} 2 - (-1) & -4 - 7 \\ 3 - 2 & -5 - 3 \end{bmatrix}$$

$$= \begin{bmatrix} 1 & 3 \\ 5 & -2 \end{bmatrix} \qquad = \begin{bmatrix} 3 & -11 \\ 1 & -8 \end{bmatrix}$$

More formally, we have the following definition.

Matrix Addition and Subtraction

For $m \times n$ matrices $A = [a_{ij}]$ and $B = [b_{ij}]$,

$A + B$ is the $m \times n$ matrix $[c_{ij}]$, where $c_{ij} = a_{ij} + b_{ij}$

$A - B$ is the $m \times n$ matrix $[d_{ij}]$, where $d_{ij} = a_{ij} - b_{ij}$

We cannot define matrix addition or subtraction unless the matrices are the same order.

·····**EXAMPLE 2**

Adding and Subtracting Matrices

Perform the following matrix operations:

a. $\begin{bmatrix} 0 & 4 & 5 \\ 1 & 6 & 7 \end{bmatrix} + \begin{bmatrix} -2 & 2 & 4 \\ -2 & 5 & 4 \end{bmatrix}$

b. $\begin{bmatrix} 2 & 3 \\ 4 & 5 \end{bmatrix} - \begin{bmatrix} 1 & 2 \\ 3 & 2 \end{bmatrix}$

c. $\begin{bmatrix} 1 & 2 \\ 4 & 5 \\ 3 & 6 \end{bmatrix} + \begin{bmatrix} 0 & 2 \\ 5 & 2 \end{bmatrix}$

Solution

a. $\begin{bmatrix} 0 & 4 & 5 \\ 1 & 6 & 7 \end{bmatrix} + \begin{bmatrix} -2 & 2 & 4 \\ -2 & 5 & 4 \end{bmatrix} = \begin{bmatrix} 0 + (-2) & 4 + 2 & 5 + 4 \\ 1 + (-2) & 6 + 5 & 7 + 4 \end{bmatrix}$

$$= \begin{bmatrix} -2 & 6 & 9 \\ -1 & 11 & 11 \end{bmatrix}$$

b. $\begin{bmatrix} 2 & 3 \\ 4 & 5 \end{bmatrix} - \begin{bmatrix} 1 & 2 \\ 3 & 2 \end{bmatrix} = \begin{bmatrix} 2 - 1 & 3 - 2 \\ 4 - 3 & 5 - 2 \end{bmatrix}$

$$= \begin{bmatrix} 1 & 1 \\ 1 & 3 \end{bmatrix}$$

c. Since the first matrix is of order 3×2 and the second is of order 2×2, the matrices are of different orders, and thus their sum is not defined.

Matrices are often useful in applications where large quantities of data are involved. In the following example, we see how matrices can be used to organize data, and also how matrix addition can be used to "combine" the information stored in matrices.

·····**EXAMPLE 3**

An Application of Matrix Addition

A telecommunications company has two divisions, one that deals primarily with computer networking and the second, with satellite data transmission. The costs and revenues (in millions of dollars) for the two divisions for the four quarters of 2003 are as follows: The networking division had costs of 1.5, 0.5, 1.2, and 1.1 million for the first through fourth quarters, respectively, and revenues of 1.8, 0.7, 1.4, and 1.2 million, respectively. The satellite transmission division had costs of 2.8, 1.4, 1.5, and 1.7 million for the first through fourth quarters, respectively, and revenues of 3.3, 1.8, 2.0, and 2.1 million, respectively. Organize this information in matrix form and use matrix addition to find the company's total cost and total revenue for each of the four quarters of 2003.

Solution We define two 4×2 matrices, one for each of the divisions. The 4 rows correspond to the 4 quarters of 2003 and the 2 columns correspond to cost and revenue (in millions of dollars). Thus, if N and S denote the matrices for the networking and satellite divisions, respectively, then

$$N = \begin{bmatrix} 1.5 & 1.8 \\ 0.5 & 0.7 \\ 1.2 & 1.4 \\ 1.1 & 1.2 \end{bmatrix} \quad \text{and} \quad S = \begin{bmatrix} 2.8 & 3.3 \\ 1.4 & 1.8 \\ 1.5 & 2.0 \\ 1.7 & 2.1 \end{bmatrix}$$

The company's total cost and revenue can be obtained by adding the matrices N and S.

$$N + S = \begin{bmatrix} 1.5 + 2.8 & 1.8 + 3.3 \\ 0.5 + 1.4 & 0.7 + 1.8 \\ 1.2 + 1.5 & 1.4 + 2.0 \\ 1.1 + 1.7 & 1.2 + 2.1 \end{bmatrix} = \begin{bmatrix} 4.3 & 5.1 \\ 1.9 & 2.5 \\ 2.7 & 3.4 \\ 2.8 & 3.3 \end{bmatrix}$$

Thus, the company's total cost and revenue for each of the quarters are as follows.

Quarter	Total cost	Total revenue
1	$4,300,000	$5,100,000
2	$1,900,000	$2,500,000
3	$2,700,000	$3,400,000
4	$2,800,000	$3,300,000

Scalar Multiplication

In the context of matrices, it is customary to refer to ordinary real or complex numbers as **scalars**. To multiply a matrix by a scalar, we simply multiply each entry of the matrix by the scalar. For example,

$$2\begin{bmatrix} 3 & 5 \\ 7 & 8 \end{bmatrix} = \begin{bmatrix} 2 \cdot 3 & 2 \cdot 5 \\ 2 \cdot 7 & 2 \cdot 8 \end{bmatrix}$$

$$= \begin{bmatrix} 6 & 10 \\ 14 & 16 \end{bmatrix}$$

More formally, we define multiplication by a scalar as follows.

Scalar Multiplication

If c is a real or complex number and $A = \begin{bmatrix} a_{ij} \end{bmatrix}$, then $cA = \begin{bmatrix} b_{ij} \end{bmatrix}$, where $b_{ij} = ca_{ij}$.

EXAMPLE 4   Performing Scalar Multiplication

Compute $-\dfrac{1}{2}\begin{bmatrix} 2 & -4 \\ 8 & 5 \\ -6 & 7 \end{bmatrix}$.

Solution   We simply multiply each entry of the matrix by $-\dfrac{1}{2}$.

$$-\frac{1}{2}\begin{bmatrix} 2 & -4 \\ 8 & 5 \\ -6 & 7 \end{bmatrix} = \begin{bmatrix} \left(-\frac{1}{2}\right)2 & \left(-\frac{1}{2}\right)(-4) \\ \left(-\frac{1}{2}\right)8 & \left(-\frac{1}{2}\right)5 \\ \left(-\frac{1}{2}\right)(-6) & \left(-\frac{1}{2}\right)7 \end{bmatrix} = \begin{bmatrix} -1 & 2 \\ -4 & -\frac{5}{2} \\ 3 & -\frac{7}{2} \end{bmatrix}$$

EXAMPLE 5   **Combining Scalar Multiplication With Subtraction**

Let

$$A = \begin{bmatrix} 1 & 4 \\ 5 & 3 \end{bmatrix} \quad \text{and} \quad B = \begin{bmatrix} 3 & 6 \\ 4 & 2 \end{bmatrix}$$

Compute $3A - 2B$.

Solution

$$3A - 2B = 3\begin{bmatrix} 1 & 4 \\ 5 & 3 \end{bmatrix} - 2\begin{bmatrix} 3 & 6 \\ 4 & 2 \end{bmatrix}$$

$$= \begin{bmatrix} 3 & 12 \\ 15 & 9 \end{bmatrix} - \begin{bmatrix} 6 & 12 \\ 8 & 4 \end{bmatrix} \qquad \text{Performing the scalar multiplications}$$

$$= \begin{bmatrix} -3 & 0 \\ 7 & 5 \end{bmatrix} \qquad\qquad \text{Subtracting corresponding entries}$$

EXAMPLE 6   **An Application of Scalar Multiplication and Addition**

Suppose that financial forecasters for the telecommunications company from Example 3 predict that all quarterly revenues and costs will increase by 5% in 2004 for the networking division and by 8% for the satellite division. Estimate the total revenue and cost for each quarter of 2004.

Solution   Increasing a quantity by 5% is equivalent to multiplying the quantity by 1.05. Thus, the quarterly networking costs and revenues for 2004 are given by

$$1.05N = 1.05\begin{bmatrix} 1.5 & 1.8 \\ 0.5 & 0.7 \\ 1.2 & 1.4 \\ 1.1 & 1.2 \end{bmatrix}$$

$$= \begin{bmatrix} (1.05)1.5 & (1.05)1.8 \\ (1.05)0.5 & (1.05)0.7 \\ (1.05)1.2 & (1.05)1.4 \\ (1.05)1.1 & (1.05)1.2 \end{bmatrix}$$

$$= \begin{bmatrix} 1.575 & 1.890 \\ 0.525 & 0.735 \\ 1.260 & 1.470 \\ 1.155 & 1.260 \end{bmatrix}$$

Similarly, an 8% increase is equivalent to multiplying a quantity by 1.08. Thus, the satellite costs and revenues are given by

$$1.08S = 1.08\begin{bmatrix} 2.8 & 3.3 \\ 1.4 & 1.8 \\ 1.5 & 2.0 \\ 1.7 & 2.1 \end{bmatrix}$$

$$= \begin{bmatrix} (1.08)2.8 & (1.08)3.3 \\ (1.08)1.4 & (1.08)1.8 \\ (1.08)1.5 & (1.08)2.0 \\ (1.08)1.7 & (1.08)2.1 \end{bmatrix}$$

$$
= \begin{bmatrix} 3.024 & 3.564 \\ 1.512 & 1.944 \\ 1.620 & 2.160 \\ 1.836 & 2.268 \end{bmatrix}
$$

The company's total cost and revenue can be obtained by adding $1.05N$ and $1.08S$.

$$
1.05N + 1.08S = \begin{bmatrix} 1.575 & 1.890 \\ 0.525 & 0.735 \\ 1.260 & 1.470 \\ 1.155 & 1.260 \end{bmatrix} + \begin{bmatrix} 3.024 & 3.564 \\ 1.512 & 1.944 \\ 1.620 & 2.160 \\ 1.836 & 2.268 \end{bmatrix} = \begin{bmatrix} 4.599 & 5.454 \\ 2.037 & 2.679 \\ 2.880 & 3.630 \\ 2.991 & 3.528 \end{bmatrix}
$$

Thus, the company's total cost and revenue for each quarter of 2004 would be as follows.

Quarter	Total cost	Total revenue
1	$4,599,000	$5,454,000
2	$2,037,000	$2,679,000
3	$2,880,000	$3,630,000
4	$2,991,000	$3,528,000

Matrix Multiplication

As we have seen, to add or subtract two matrices, we simply add or subtract corresponding entries. The product of two matrices, however, is *not* found by multiplying corresponding entries. Instead, each entry of the product matrix is the "product" of a *row* of the first matrix and a *column* of the second matrix.

Multiplication of a Row and a Column

If

$$
R = [a_1 \quad a_2 \quad a_3 \quad \cdots \quad a_n] \text{ and } C = \begin{bmatrix} b_1 \\ b_2 \\ \vdots \\ b_n \end{bmatrix}
$$

then $RC = a_1 b_1 + a_2 b_2 + \cdots + a_n b_n$. Note that the product of a row and a column is a scalar.

WARNING!

A row and column must have the same number of entries in order to be multiplied.

EXAMPLE 7

Multiplying a Row by a Column

Let

$$
A = \begin{bmatrix} 2 & -5 & 4 \\ -1 & 7 & 5 \end{bmatrix} \text{ and } B = \begin{bmatrix} 1 & 2 & -3 & 5 \\ 3 & -2 & 1 & 5 \\ 5 & 4 & 0 & -7 \end{bmatrix}
$$

Compute the product of the first row of A with the third column of B.

Solution   We must compute

$$[2 \quad -5 \quad 4]\begin{bmatrix} -3 \\ 1 \\ 0 \end{bmatrix}$$

According to the definition, we multiply corresponding entries and then add. Thus, we have

$$[2 \quad -5 \quad 4]\begin{bmatrix} -3 \\ 1 \\ 0 \end{bmatrix} = 2(-3) + (-5)1 + 4 \cdot 0 = -11$$

Note that in the preceding example, matrix A has the same number of columns as matrix B has rows. This ensures that the rows of A have the same number of entries as the columns of B. In fact, the matrix product AB is defined only if the number of columns of A is the same as the number of rows of B. If this is the case, then the entry in row i and column j of AB is simply the product of the ith row of A with the jth column of B.

$m \times n$
n columns

$n \times p$
p columns

$m \times p$
p columns

m rows $\times$ n rows $=$ m rows

Figure 11

Multiplication of Matrices

If A has the same number of columns as B has rows, then $AB = \begin{bmatrix} c_{ij} \end{bmatrix}$, where

$$c_{ij} = [\text{row } i \text{ of } A]\begin{bmatrix} \text{column} \\ j \\ \text{of} \\ B \end{bmatrix}$$

It follows that if the order of A is $m \times n$ and the order of B is $n \times p$, then AB has order $m \times p$.

Figure 11 depicts the relationships between the orders of the factors and the resulting product matrix.

Rule of Thumb

When multiplying matrices, first write their orders side by side in the order in which the product is to be computed. If the inner numbers are the same, then the product is defined, and the order of the product is given by the outer numbers. If the inner numbers are different, then the product is not defined.

The order of the product is $m \times p$.

$\boxed{m} \times \textcircled{n} \qquad \textcircled{n} \times \boxed{p}$

These numbers must be the same.

EXAMPLE 8

Multiplication of Matrices

Multiply

$$\begin{bmatrix} 1 & 2 & 3 \\ -2 & 0 & 5 \end{bmatrix}\begin{bmatrix} 2 & 1 \\ -3 & 4 \\ 2 & 1 \end{bmatrix}$$

Solution   Since the first factor is a 2 × 3 matrix and the second is a 3 × 2 matrix, the product will be a 2 × 2 matrix. Let us denote the product by $[p_{ij}]$, so that we have

$$\begin{bmatrix} 1 & 2 & 3 \\ -2 & 0 & 5 \end{bmatrix} \begin{bmatrix} 2 & 1 \\ -3 & 4 \\ 2 & 1 \end{bmatrix} = \begin{bmatrix} p_{11} & p_{12} \\ p_{21} & p_{22} \end{bmatrix}$$

Table 2 illustrates the process of computing the product $[p_{ij}]$. Note that the product of row i from the first matrix and column j from the second gives p_{ij}.

Table 2

Big picture	Computations	Update
$\begin{bmatrix} 1 & 2 & 3 \\ -2 & 0 & 5 \end{bmatrix} \begin{bmatrix} 2 & 1 \\ -3 & 4 \\ 2 & 1 \end{bmatrix} = \begin{bmatrix} p_{11} & p_{12} \\ p_{21} & p_{22} \end{bmatrix}$ Row 1 Column 1	$\begin{bmatrix} 1 & 2 & 3 \end{bmatrix} \begin{bmatrix} 2 \\ -3 \\ 2 \end{bmatrix} = 1 \cdot 2 + 2(-3) + 3 \cdot 2$ $= 2 - 6 + 6$ $= 2$	$\begin{bmatrix} 2 & p_{12} \\ p_{21} & p_{22} \end{bmatrix}$
$\begin{bmatrix} 1 & 2 & 3 \\ -2 & 0 & 5 \end{bmatrix} \begin{bmatrix} 2 & 1 \\ -3 & 4 \\ 2 & 1 \end{bmatrix} = \begin{bmatrix} 2 & p_{12} \\ p_{21} & p_{22} \end{bmatrix}$ Row 1 Column 2	$\begin{bmatrix} 1 & 2 & 3 \end{bmatrix} \begin{bmatrix} 1 \\ 4 \\ 1 \end{bmatrix} = 1 \cdot 1 + 2 \cdot 4 + 3 \cdot 1$ $= 1 + 8 + 3$ $= 12$	$\begin{bmatrix} 2 & 12 \\ p_{21} & p_{22} \end{bmatrix}$
$\begin{bmatrix} 1 & 2 & 3 \\ -2 & 0 & 5 \end{bmatrix} \begin{bmatrix} 2 & 1 \\ -3 & 4 \\ 2 & 1 \end{bmatrix} = \begin{bmatrix} 2 & 12 \\ p_{21} & p_{22} \end{bmatrix}$ Row 2 Column 1	$\begin{bmatrix} -2 & 0 & 5 \end{bmatrix} \begin{bmatrix} 2 \\ -3 \\ 2 \end{bmatrix} = -2 \cdot 2 + 0 \cdot (-3) + 5 \cdot 2$ $= -4 + 0 + 10$ $= 6$	$\begin{bmatrix} 2 & 12 \\ 6 & p_{22} \end{bmatrix}$
$\begin{bmatrix} 1 & 2 & 3 \\ -2 & 0 & 5 \end{bmatrix} \begin{bmatrix} 2 & 1 \\ -3 & 4 \\ 2 & 1 \end{bmatrix} = \begin{bmatrix} 2 & 12 \\ 6 & p_{22} \end{bmatrix}$ Row 2 Column 2	$\begin{bmatrix} -2 & 0 & 5 \end{bmatrix} \begin{bmatrix} 1 \\ 4 \\ 1 \end{bmatrix} = -2 \cdot 1 + 0 \cdot 4 + 5 \cdot 1$ $= -2 + 0 + 5$ $= 3$	$\begin{bmatrix} 2 & 12 \\ 6 & 3 \end{bmatrix}$

Thus, the product is $\begin{bmatrix} 2 & 12 \\ 6 & 3 \end{bmatrix}$.

The order in which matrices are multiplied is crucial. For example, suppose that A is a 2 × 4 matrix and B is a 4 × 3 matrix. Then AB is a 2 × 3 matrix, whereas BA isn't even defined! Even if both AB and BA are defined, they are generally not equal.

EXAMPLE 9   **Multiplying 2 × 2 Matrices in Both Orders**

Let

$$A = \begin{bmatrix} -1 & 2 \\ 3 & 4 \end{bmatrix} \quad \text{and} \quad B = \begin{bmatrix} 2 & -1 \\ 3 & 2 \end{bmatrix}$$

Show that $AB \neq BA$.

Solution

$$AB = \begin{bmatrix} -1 & 2 \\ 3 & 4 \end{bmatrix}\begin{bmatrix} 2 & -1 \\ 3 & 2 \end{bmatrix}$$

$$= \begin{bmatrix} (-1)2 + 2 \cdot 3 & (-1)(-1) + 2 \cdot 2 \\ 3 \cdot 2 + 4 \cdot 3 & 3(-1) + 4 \cdot 2 \end{bmatrix}$$

$$= \begin{bmatrix} 4 & 5 \\ 18 & 5 \end{bmatrix}$$

$$BA = \begin{bmatrix} 2 & -1 \\ 3 & 2 \end{bmatrix}\begin{bmatrix} -1 & 2 \\ 3 & 4 \end{bmatrix}$$

$$= \begin{bmatrix} 2(-1) + (-1)3 & 2 \cdot 2 + (-1)4 \\ 3(-1) + 2 \cdot 3 & 3 \cdot 2 + 2 \cdot 4 \end{bmatrix}$$

$$= \begin{bmatrix} -5 & 0 \\ 3 & 14 \end{bmatrix}$$

Thus, $AB \neq BA$.

At this point, it is natural to question our motivation for defining matrix multiplication in such a seemingly unnatural and convoluted way. Why don't we simply multiply corresponding entries? For that matter, why bother to define multiplication of matrices at all? In general, mathematical tools are forged from the fires of utility: They are constructed to solve real-world problems or assist in mathematical investigations and are kept only when they continue to be useful. Matrix multiplication, born as a purely algebraic tool for solving systems of linear equations, has proven to be a potent weapon for attacking problems in virtually all branches of the mathematical, physical, and social sciences, including geometry, statistics, chemistry, biology, physics, astronomy, economics, and psychology. The common thread in each area of application is an underlying linear relationship between a collection of variables, which returns us to the birthplace of matrix multiplication—linear systems.

EXAMPLE 10   **Expressing a System of Equations in Matrix Form**

Show that the system of equations

$$2x + 3y + 4z = 5$$
$$3x - 5y + z = 6$$
$$4x + 7y - z = 3$$

is equivalent to the matrix equation

$$\begin{bmatrix} 2 & 3 & 4 \\ 3 & -5 & 1 \\ 4 & 7 & -1 \end{bmatrix}\begin{bmatrix} x \\ y \\ z \end{bmatrix} = \begin{bmatrix} 5 \\ 6 \\ 3 \end{bmatrix}$$

Solution   We begin by multiplying the matrices on the left side of the equation.

$$\begin{bmatrix} 2 & 3 & 4 \\ 3 & -5 & 1 \\ 4 & 7 & -1 \end{bmatrix}\begin{bmatrix} x \\ y \\ z \end{bmatrix} = \begin{bmatrix} 5 \\ 6 \\ 3 \end{bmatrix}$$

$$\begin{bmatrix} 2x + 3y + 4z \\ 3x - 5y + z \\ 4x + 7y - z \end{bmatrix} = \begin{bmatrix} 5 \\ 6 \\ 3 \end{bmatrix}$$

Since two matrices are equal if and only if corresponding entries are equal, we have

$$2x + 3y + 4z = 5$$
$$3x - 5y + z = 6$$
$$4x + 7y - z = 3$$

Calculator Keys

Matrix Operations

The operations of matrix addition, subtraction, and multiplication, as well as scalar multiplication, can be performed by your graphing calculator. First, for each matrix involved in the computation, it is necessary to specify the order and then key in the entries. Once this is done, the matrix expression can be entered and evaluated. For details, consult your calculator's manual.

In the following example, we see an illustration of matrix multiplication's applicability to contexts seemingly unrelated to systems of equations. As you work through this example, note how the early investment in expressing a quantity with matrix multiplication pays off when recomputing the quantity with new data.

EXAMPLE 11

An Application of Matrix Multiplication

Lachelle has $500 withheld out of each month's paycheck and placed in a 401k retirement plan. The plan offers two fund choices: a high-risk growth fund and a low-risk income fund. Lachelle has chosen to place $350 in the growth fund and $150 in the income fund. The manager of each fund determines how investments are distributed among three categories of stocks: blue chip, biotech, and international. The distributions are summarized in Table 3.

Table 3

	To blue chip stocks	To biotech stocks	To international stocks
Growth fund	30%	50%	20%
Income fund	50%	35%	15%

(For example, the growth fund manager distributes 30% of all investments to blue chip stocks.) Express the distribution of Lachelle's $500 monthly withholding as a 2×3 matrix, and use matrix multiplication to find the earnings of each fund for a single month given that the rate of return is

a. 8% for blue chip stocks, 10% for biotech stocks, and 9% for international stocks

b. 9% for blue chip stocks, 13% for biotech stocks, and 7% for international stocks

Solution We define a 2×3 matrix A representing Lachelle's investment portfolio, with each row corresponding to a fund (growth and income, in that order) and each column corresponding to a stock option (blue chip, biotech, and international, in that order). Thus, we have

$$A = \begin{bmatrix} 30\% \text{ of } 350 & 50\% \text{ of } 350 & 20\% \text{ of } 350 \\ 50\% \text{ of } 150 & 35\% \text{ of } 150 & 15\% \text{ of } 150 \end{bmatrix} = \begin{bmatrix} 105 & 175 & 70 \\ 75 & 52.5 & 22.5 \end{bmatrix}$$

a. Multiplying the amounts invested by the corresponding rates of return, we see that the total earnings of Lachelle's growth fund account is given by

$$8\% \cdot 105 + 10\% \cdot 175 + 9\% \cdot 70$$

Similarly, her income account generates earnings of

$$8\% \cdot 75 + 10\% \cdot 52.5 + 9\% \cdot 22.5$$

In matrix form, we could express her earnings for the two funds as

$$\begin{bmatrix} 0.08 \cdot 105 + 0.10 \cdot 175 + 0.09 \cdot 70 \\ 0.08 \cdot 75 + 0.10 \cdot 52.5 + 0.09 \cdot 22.5 \end{bmatrix}$$

or equivalently,

$$\begin{bmatrix} 105 & 175 & 70 \\ 75 & 52.5 & 22.5 \end{bmatrix} \begin{bmatrix} 0.08 \\ 0.10 \\ 0.09 \end{bmatrix}$$

Note that this is the product of the matrix A and the column matrix consisting of the rates of return. Evaluating the product with a graphing calculator, we obtain

$$\begin{bmatrix} 32.2 \\ 13.275 \end{bmatrix}$$

Thus, on Lachelle's investment of $500, the growth fund earns $32.20 and the income fund earns $13.28.

b. In this case, the total earnings are given by

$$\begin{bmatrix} 105 & 175 & 70 \\ 75 & 52.5 & 22.5 \end{bmatrix} \begin{bmatrix} 0.09 \\ 0.13 \\ 0.07 \end{bmatrix} = \begin{bmatrix} 37.1 \\ 15.15 \end{bmatrix}$$

Thus, the growth fund earns $37.10 and the income fund earns $15.15.

We now summarize the fundamental algebraic properties of matrices. Proofs of some of these properties for 2×2 matrices will be developed in Exercises 73–76.

Properties of Matrix Operations

For scalars x and y, and matrices A, B, and C (of appropriate orders):

Algebraic description	Verbal description
1. $A + B = B + A$	Matrix addition is commutative.
2. $(A + B) + C = A + (B + C)$	Matrix addition is associative.
3. $x(yA) = (xy)A$	Scalar multiplication is associative.
4. $A(BC) = (AB)C$	Matrix multiplication is associative.
5. $x(AB) = (xA)B = A(xB)$	Scalar/matrix multiplication is associative.
6. $x(A + B) = xA + xB$	Scalar multiplication is distributive.
7. $A(B + C) = AB + AC$	Matrix multiplication is distributive (on the left).
8. $(A + B)C = AC + BC$	Matrix multiplication is distributive (on the right).

Note that, in general, $AB \neq BA$, so matrix multiplication is *not* commutative.

Understanding and Mastery Checklists

Concepts to Understand	Skills to Master
Matrix of order $m \times n$	Determine the order of a matrix.
❖	❖
Subscript notation for matrices	Identify an entry of a matrix by its row and column.
❖	❖
Matrix addition and subtraction	Add or subtract matrices.
❖	❖
Scalar multiplication	Find a scalar multiple of a matrix.
❖	❖
Matrix multiplication	Determine whether a matrix product can be computed.
❖	❖
Matrix form for a system of equations	Compute the product of two matrices.
❖	❖
Properties of matrix operations	Express a system of equations in matrix form.

Exercises 9.2

Exercises 1-4 *Identify the order of the given matrix.*

1. $\begin{bmatrix} 1 & 2 & 3 \\ 4 & 5 & 6 \end{bmatrix}$

2. $\begin{bmatrix} 1 & 0 \\ -1 & 0 \\ 0 & 1 \end{bmatrix}$

3. $\begin{bmatrix} a \\ b \\ c \\ d \end{bmatrix}$

4. $[i \ j \ k]$

Exercises 5-18 *Perform the indicated matrix operation, if possible.*

5. $2\begin{bmatrix} 1 & 0 \\ 3 & 5 \end{bmatrix}$

6. $-3\begin{bmatrix} 1 & 5 & 3 \\ 4 & 0 & 2 \end{bmatrix}$

7. $\begin{bmatrix} 2 & 3 \\ 6 & 4 \end{bmatrix} + \begin{bmatrix} -4 & 8 \\ 0 & 2 \end{bmatrix}$

8. $[0 \ 1 \ 2 \ 3] + [4 \ 3 \ 1 \ 0]$

9. $\begin{bmatrix} 1 & -1 \\ 0 & 3 \end{bmatrix} + \begin{bmatrix} 2 \\ 5 \end{bmatrix}$

10. $\begin{bmatrix} 2 & -3 & 5 \\ -1 & 7 & 0 \end{bmatrix} - \begin{bmatrix} 3 & -5 & 2 \\ 1 & 0 & 4 \end{bmatrix}$

11. $\begin{bmatrix} -3 \\ 1 \\ 6 \end{bmatrix} + 2\begin{bmatrix} 4 \\ -1 \\ 0 \end{bmatrix}$

12. $\begin{bmatrix} 0 & 3 \\ 6 & 7 \end{bmatrix} + \begin{bmatrix} -2 & 4 \\ 0 & 1 \end{bmatrix} - \begin{bmatrix} 2 & -5 \\ 3 & -5 \end{bmatrix}$

13. $\begin{bmatrix} a & b \\ c & d \end{bmatrix} + \begin{bmatrix} 3a & 2b \\ c & 5d \end{bmatrix}$

14. $\begin{bmatrix} 1 & 0 & -1 \\ 3 & 2 & 1 \end{bmatrix} - \begin{bmatrix} 4 & 6 \\ 6 & 4 \end{bmatrix}$

15. $2\begin{bmatrix} 1 & 2 & 0 \\ 2 & 6 & 9 \\ 3 & -1 & -4 \end{bmatrix} + \begin{bmatrix} -1 & 3 & 4 \\ 3 & -3 & 0 \\ 1 & 4 & 5 \end{bmatrix}$

16. $3\begin{bmatrix} 2 & 3 & 0 \\ 1 & -1 & 1 \\ -1 & -2 & 2 \end{bmatrix} - 2\begin{bmatrix} 0 & -6 & -1 \\ 2 & -4 & 3 \\ -1 & 2 & -2 \end{bmatrix}$

17. $\begin{bmatrix} 2 & a & 1 \\ 0 & 1 & 2 \end{bmatrix} - a\begin{bmatrix} 1 & 1 & 1 \\ 1 & 1 & 0 \end{bmatrix}$

18. $\begin{bmatrix} 1 & x \\ x^2 & 0 \\ 2 & 1 \end{bmatrix} + \begin{bmatrix} x & 2 \\ -x^2 & 3 \\ x^2 & -2 \end{bmatrix}$

Exercises 19-22 *Simplify the given quantity for*
$$A = \begin{bmatrix} 1 & 2 \\ 3 & 5 \end{bmatrix} \quad and \quad B = \begin{bmatrix} 2 & 1 \\ 4 & 9 \end{bmatrix}$$

19. $A + B$

20. $A - B$

21. $2A - 5B$

22. $3A + 2B$

Exercises 23-26 *Solve the given equation for the unknown matrix X. (Hint: Use matrix operations to solve for X, much as you would for an ordinary linear equation.)*

23. $2X = \begin{bmatrix} 4 & 6 \\ 8 & 2 \end{bmatrix}$

24. $3X + \begin{bmatrix} 1 & 0 \\ 0 & 1 \end{bmatrix} = \begin{bmatrix} 10 & 12 \\ 3 & 16 \end{bmatrix}$

25. $2X + \begin{bmatrix} 1 & 2 & 3 \\ 6 & 5 & 4 \end{bmatrix} = \begin{bmatrix} 0 & 0 & 0 \\ 0 & 0 & 0 \end{bmatrix}$

26. $-2\left(X + \begin{bmatrix} 2 & 6 \\ 3 & 3 \end{bmatrix}\right) = -4X + \begin{bmatrix} 0 & 0 \\ 2 & 4 \end{bmatrix}$

Exercises 27-32 *Let*

$$A = \begin{bmatrix} 1 & 2 & 4 \\ 2 & 3 & 0 \end{bmatrix}, \ B = \begin{bmatrix} 1 & 2 \\ 3 & 4 \end{bmatrix}, \ \text{and } C = \begin{bmatrix} 3 & 7 & 9 \\ 4 & 5 & 1 \\ 2 & 6 & 4 \end{bmatrix}$$

Indicate whether the given product is defined. If it is, give the order of the product matrix. You need not compute the product.

27. AB

28. AC

29. BA

30. BC

31. CA

32. CB

Exercises 33-54 *Compute the product, if possible.*

33. $\begin{bmatrix} 3 & 5 \\ 4 & 2 \end{bmatrix}\begin{bmatrix} 3 \\ -3 \end{bmatrix}$

34. $\begin{bmatrix} a & b \\ c & d \end{bmatrix}\begin{bmatrix} x \\ y \end{bmatrix}$

35. $\begin{bmatrix} 1 & 0 \\ 0 & 1 \end{bmatrix}\begin{bmatrix} 1 \\ 2 \\ 3 \end{bmatrix}$

36. $\begin{bmatrix} 2 & 3 \\ 4 & 6 \end{bmatrix}\begin{bmatrix} 1 & 2 \\ 0 & 1 \end{bmatrix}$

37. $\begin{bmatrix} 1 & 2 \\ 0 & 1 \end{bmatrix}\begin{bmatrix} 2 & 3 \\ 4 & 6 \end{bmatrix}$

38. $\begin{bmatrix} 1 & -2 \\ -5 & 7 \end{bmatrix}\begin{bmatrix} a & b \\ c & d \end{bmatrix}$

39. $\begin{bmatrix} 2 & 1 \\ 3 & 2 \end{bmatrix}\begin{bmatrix} 2 & -1 \\ -3 & 2 \end{bmatrix}$

40. $\begin{bmatrix} 1 & -1 & 0 \\ 2 & 0 & 3 \end{bmatrix}\begin{bmatrix} 4 & -1 \\ 1 & 0 \end{bmatrix}$

41. $\begin{bmatrix} 2 & 0 \\ 1 & 4 \\ 2 & 1 \end{bmatrix}\begin{bmatrix} 3 & 5 \\ 1 & 7 \end{bmatrix}$

42. $\begin{bmatrix} 0 & 2 & 0 \\ 0 & 0 & 3 \\ 0 & 0 & 0 \end{bmatrix}\begin{bmatrix} 0 & 2 & 0 \\ 0 & 0 & 3 \\ 0 & 0 & 0 \end{bmatrix}$

43. $\begin{bmatrix} 1.8 & 3.5 & 4.6 \\ 4.8 & 1.7 & 3.2 \\ 1.7 & 2.5 & 0.7 \end{bmatrix}\begin{bmatrix} 2.6 \\ 3.4 \\ 4.9 \end{bmatrix}$

44. $\begin{bmatrix} 2.75 & 3.01 \\ 4.07 & 0.07 \end{bmatrix}\begin{bmatrix} 1.7 & 3.4 \\ 6.8 & 5.9 \end{bmatrix}$

45. $\begin{bmatrix} 1 & 0 \\ x & 1 \end{bmatrix}\begin{bmatrix} a & b \\ c & d \end{bmatrix}$

46. $\begin{bmatrix} x & 0 \\ 0 & 1 \end{bmatrix}\begin{bmatrix} a & b \\ c & d \end{bmatrix}$

47. $\begin{bmatrix} a \\ b \end{bmatrix}\begin{bmatrix} c \\ d \end{bmatrix}$

48. $[a \ \ b \ \ c]\begin{bmatrix} i \\ j \\ k \end{bmatrix}$

49. $\begin{bmatrix} 0 & \frac{1}{b} \\ \frac{1}{a} & 0 \end{bmatrix}\begin{bmatrix} 0 & a \\ b & 0 \end{bmatrix}$

50. $[3 \ \ 6]\begin{bmatrix} \frac{1}{3} & \frac{1}{6} \end{bmatrix}$

51. $\begin{bmatrix} 1 & 0 & 0 \\ 0 & 1 & 0 \\ 0 & 0 & 1 \end{bmatrix}\begin{bmatrix} p & q \\ r & s \\ t & u \end{bmatrix}$

52. $\begin{bmatrix} a & 0 & 0 & 0 \\ 0 & b & 0 & 0 \\ 0 & 0 & c & 0 \\ 0 & 0 & 0 & d \end{bmatrix}\begin{bmatrix} p & 0 & 0 & 0 \\ 0 & q & 0 & 0 \\ 0 & 0 & r & 0 \\ 0 & 0 & 0 & s \end{bmatrix}$

53. $\begin{bmatrix} 0 & 0 & 1 \\ 0 & 1 & 0 \\ 1 & 0 & 0 \end{bmatrix}\begin{bmatrix} a_{11} & a_{12} & a_{13} \\ a_{21} & a_{22} & a_{23} \\ a_{31} & a_{32} & a_{33} \end{bmatrix}$

54. $\begin{bmatrix} a_{11} & a_{12} & a_{13} \\ a_{21} & a_{22} & a_{23} \\ a_{31} & a_{32} & a_{33} \end{bmatrix}\begin{bmatrix} 0 & 0 & 1 \\ 0 & 1 & 0 \\ 1 & 0 & 0 \end{bmatrix}$

Exercises 55-60 *Express the given system of equations as a matrix equation.*

55. $\begin{aligned} 2x - 3y &= 5 \\ -x + y &= 6 \end{aligned}$

56. $\begin{aligned} a - b + c &= 0 \\ 2a + 3b &= 0 \end{aligned}$

57. $\begin{aligned} r + s &= 1 \\ 2t - r &= 3 \\ r + s + t &= 4 \end{aligned}$

58. $\begin{aligned} p + q + r &= 0 \\ p + 3r &= -1 \\ q - r - 2 &= 0 \end{aligned}$

59. $\begin{aligned} x_1 + x_2 + x_3 + x_4 &= 4 \\ x_2 + x_3 + x_4 &= 3 \\ x_3 + x_4 &= 2 \end{aligned}$

60. $\begin{aligned} a_1 - a_4 &= 1 \\ a_2 - a_1 &= 0 \\ a_3 - a_2 &= 0 \\ a_4 - a_3 &= -1 \end{aligned}$

Exercises 61-64 *If A is an n × m matrix, then the **transpose** of A, denoted A^T, is the matrix whose rows are formed from the columns of A. More precisely, if $A = [a_{ij}]$, then $A^T = [b_{ij}]$, where $b_{ij} = a_{ji}$. For example, if*

$$A = \begin{bmatrix} 2 & 4 \\ 6 & 8 \end{bmatrix}, \text{then } A^T = \begin{bmatrix} 2 & 6 \\ 4 & 8 \end{bmatrix}$$

Compute the transposes of the given matrices.

61. $\begin{bmatrix} 1 & 2 & 3 \\ 4 & 5 & 6 \end{bmatrix}$

62. $\begin{bmatrix} a & b & c \\ d & e & f \\ g & h & i \end{bmatrix}$

63. $\begin{bmatrix} 1 & 2 & 3 \\ 2 & 5 & 6 \\ 3 & 6 & 4 \end{bmatrix}$

64. $\begin{bmatrix} x & y \\ 2x & 3y \\ -x & 4y \end{bmatrix}$

Exercises 65-68 *Perform the given operation. Note that for a square matrix A, A^2 is defined as $A \times A$. Similarly, $A^3 = A \times A \times A$, and so on.*

65. $\begin{bmatrix} 1 & 2 \\ 0 & 1 \end{bmatrix}^2$

66. $\begin{bmatrix} 0 & 1 \\ 0 & 0 \end{bmatrix}^3$

67. $\begin{bmatrix} 1 & 0 \\ 0 & 0 \end{bmatrix}^4$

68. $\begin{bmatrix} a & b \\ c & d \end{bmatrix}^2$

Exercises 69-72 *Compute the indicated powers, given that*

$$A = \begin{bmatrix} 1 & 2 & 3 \\ 0 & 1 & 4 \\ 0 & 0 & 1 \end{bmatrix}$$

69. A^2

70. A^3

71. A^{32} (*Hint:* If you enter 3 in your calculator and then press $\boxed{x^2}$, you will obtain 3^2. If you press $\boxed{x^2}$ again, you will obtain $(3^2)^2 = 3^4$. If you press $\boxed{x^2}$ once again, you obtain 3^8, and so on.)

72. A^{33} (*Hint:* $A^{33} = A^{32}A$.)

Exercises 73-76 *Confirm the given property for 2 × 2 matrices. Do so by letting*

$$A = \begin{bmatrix} a_1 & a_2 \\ a_3 & a_4 \end{bmatrix}, \quad B = \begin{bmatrix} b_1 & b_2 \\ b_3 & b_4 \end{bmatrix}, \quad and\ C = \begin{bmatrix} c_1 & c_2 \\ c_3 & c_4 \end{bmatrix}$$

and directly verifying the given identity.

73. $A + B = B + A$

74. $(xA)B = x(AB)$, where x is a scalar

75. $(A + B)C = AC + BC$

76. $x(A + B) = xA + xB$, where x is a scalar

Applications

77. Health-Care Claims All employees of a computer software company, whether hourly workers or salaried personnel, select one of two health-care plans: major medical or comprehensive. Claims totals for 2003 are shown in Table 4. Express the breakdown of claims filed by hourly workers as a 2 × 2 matrix, and do likewise for claims filed by salaried personnel. Find and interpret the sum of these two matrices.

Table 4

	Hourly		Salaried	
	Employees	Dependents	Employees	Dependents
Major medical	$230,000	$320,000	$125,000	$250,000
Comprehensive	$280,000	$300,000	$400,000	$500,000

78. Birth Distributions A city has two major hospitals: St. Mary's and Parkview. St. Mary's recorded births of 200 boys and 210 girls for the first half of 1978 and 185 boys and 177 girls for the second half. Parkview Hospital recorded 320 girls and 307 boys for the first half and 300 girls and 295 boys for the second half. Express the births for each 6-month period as a 2 × 2 matrix, and compute and interpret the sum.

79. Point Breakdown The team statistician of the Dallas Mavericks recorded the following statistics from the team's last game against the Orlando

Magic. The Mavericks made 5 three-point shots, 40 two-point field goals, and 18 free throws. The Magic made 4 three-point shots, 37 two-point field goals, and 26 free throws. Express the shot distribution for the game as a 2 × 3 matrix S. Find a 3 × 1 matrix T such that the final score is given by the matrix product ST. What is the final score?

80. Congressional Districts Since the number of congressional districts is fixed at 435 and representation in the House of Representatives is proportional to population, it is occasionally necessary to redefine congressional districts. Population growth in a certain state has lagged behind that of the rest of the nation; thus, the state will be combining portions of two districts into one. The first district is 40% Republican and 30% Democrat, whereas the second is 25% Republican and 55% Democrat. There are 200,000 registered voters in the first district and 150,000 in the second. Express the political party percentage distribution of the two districts as a 2 × 2 matrix, and then find the total number of Democrats and Republicans in the new district using matrix multiplication, assuming that the new district will be formed from:

a. 200,000 voters from the first district and 150,000 from the second

b. 175,000 voters from each district

Concepts and Critical Thinking

Exercises 81-89 *Answer true or false.*

81. The sum $A + B$ is defined for any pair of matrices A and B.

82. The product AB is defined only for matrices A and B of the same order.

83. If A and B are of the same order, then $A + B$ is defined.

84. If A and B are of the same order, then AB is defined.

85. A 5 × 3 matrix has 5 rows and 3 columns

86. Matrix addition is commutative; that is, for any pair of matrices A and B for which $A + B$ is defined, $A + B = B + A$.

87. Matrix multiplication is commutative; that is, for any pair of matrices A and B for which AB is defined, $AB = BA$.

88. Multiplication of matrices is accomplished by multiplying corresponding entries.

89. Addition of matrices is accomplished by adding corresponding entries.

Exercises 90-95   *Give an example of each.*

90. A 2×3 matrix

91. Two matrices A and B such that AB is defined but BA is not defined

92. A matrix A such that A^2 is not defined

93. Two matrices A and B such that $A + B$ is defined but neither AB nor BA is defined

94. Two matrices A and B such that AB is defined but $A + B$ is not defined

95. A 3×2 matrix such that all of its columns are identical but no two of its rows are the same

96. For real numbers a and b, the zero product property holds: If $ab = 0$, then either $a = 0$ or $b = 0$. Show that the zero product property does *not* hold for matrices by finding a 1×2 matrix A and a 2×1 matrix B such that $AB = 0$ but neither A nor B consists entirely of zero entries.

Questions for Discussion or Essay

97. How does a matrix differ from a simple table of numbers?

98. Most of the matrix products in this section can be determined using a graphing calculator. Why bother to learn how to multiply matrices by hand? For that matter, is there any point in learning how to multiply numbers by hand?

99. We are free to define the product of matrices in any way we want. So why have we chosen to define them in such a seemingly bizarre way? Why not simply define the product of the matrices $[a_{ij}]$ and $[b_{ij}]$ to be $[c_{ij}]$, where $c_{ij} = a_{ij}b_{ij}$?

100. It's reasonable to wonder how much we really gain by having the graphing calculator perform matrix operations such as matrix multiplication. To multiply a pair of 2×2 matrices by hand, a total of 12 arithmetic operations (addition or multiplication) must be performed. On the other hand, $4 + 4 = 8$ entries must be entered into the calculator. Develop a formula for the number of operations (additions and multiplications count as one each) required to multiply a pair of $n \times n$ matrices. (*Hint:* The product of two $n \times n$ matrices is an $n \times n$ matrix with n^2 entries. So compute the number of multiplications and additions required to determine each entry of the product matrix, and then multiply this number by n^2.) How many entries must be entered into a calculator when multiplying two $n \times n$ matrices? In the case of multiplication of a pair of 6×6 matrices, is the calculator worthwhile?

Projects for Enrichment

101. Linear Transformations: An Application of Matrix Multiplication   If the line segment $\overline{OP}$ in Figure 12 is rotated by an angle α about the origin, the point $P(x, y)$ is transformed to a new point, $P'(x', y')$. We can express x' and y' in terms of x and y using matrix multiplication.

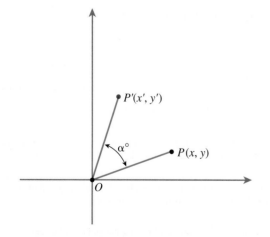

Figure 12

The point P in the Cartesian plane with coordinates (x, y) can also be represented as the coordinate matrix

$$\begin{bmatrix} x \\ y \end{bmatrix}$$

The coordinates of P' can be obtained by multiplying the coordinate matrix by a certain 2×2 matrix A. Thus, if the coordinates of P' are (x', y'), then we have

$$A\begin{bmatrix} x \\ y \end{bmatrix} = \begin{bmatrix} x' \\ y' \end{bmatrix}$$

To find the appropriate matrix A, we use the following rule:

- The first column of A represents the rotated coordinates of the point $(1, 0)$.
- The second column of A represents the rotated coordinates of the point $(0, 1)$.

For example, consider a rotation by $90°$ counterclockwise. The point $(1, 0)$ will be rotated to the point $(0, 1)$, and so the first column of A is

$$\begin{bmatrix} 0 \\ 1 \end{bmatrix}$$

The point $(0, 1)$ will be rotated to the point $(-1, 0)$, so the second column of A is

$$\begin{bmatrix} -1 \\ 0 \end{bmatrix}$$

Thus,

$$A = \begin{bmatrix} 0 & -1 \\ 1 & 0 \end{bmatrix}$$

a. Where does the point $(2, 3)$ end up if it is rotated counter-clockwise by $90°$?

b. Where does the point $(-1, 2)$ end up if it is rotated counter-clockwise by $90°$?

c. Define A to be the $90°$ counterclockwise rotation matrix just discussed. Compute A^4. Explain your answer.

d. Find the matrix B for a $45°$ counterclockwise rotation. [*Hint:* The point $(1, 0)$ will be rotated onto a point 1 unit from the origin on the line $y = x$; the point $(0, 1)$ will be rotated onto a point 1 unit from the origin on the line $y = -x$.]

e. Show directly that $B^2 = A$. Explain why this is logical.

f. What is the smallest whole number n so that $B^n = \begin{bmatrix} 1 & 0 \\ 0 & 1 \end{bmatrix}$? Explain your answer.

102. Transition Matrices: An Application of Matrix Multiplication Suppose that in a certain town, 10% of nonsmokers take up smoking each year, whereas 20% of smokers quit each year. If, in a given year, there are x smokers and y nonsmokers, then the smoking distribution of the population can be represented as

$$\begin{bmatrix} x \\ y \end{bmatrix}$$

The next year, the smokers will consist of the 80% of smokers who did *not* quit plus the 10% of former nonsmokers who take up smoking. Thus, the number of smokers in the next year will be $0.8x + 0.1y$. Similarly, the number of nonsmokers will be $0.2x + 0.9y$. Thus, after 1 year, the distribution of the population is given by

$$\begin{bmatrix} 0.8x + 0.1y \\ 0.2x + 0.9y \end{bmatrix}$$

a. Find a 2×2 matrix A such that

$$A\begin{bmatrix} x \\ y \end{bmatrix} = \begin{bmatrix} 0.8x + 0.1y \\ 0.2x + 0.9y \end{bmatrix}$$

This is called a **transition matrix**.

b. Suppose that the town has a smoking distribution of

$$\begin{bmatrix} 600 \\ 1400 \end{bmatrix}$$

Compute the smoking distribution of the town after 4 years. (*Hint:* The smoking distribution after 1 year is given by

$$A\begin{bmatrix} 600 \\ 1400 \end{bmatrix}, \quad \text{after 2 years by} \quad A\left(A\begin{bmatrix} 600 \\ 1400 \end{bmatrix}\right)$$

and so on.)

c. Find the smoking distribution for the town after 32 years, assuming its initial distribution is

$$\begin{bmatrix} 600 \\ 1400 \end{bmatrix}$$

Now find the smoking distribution after 64 years with the same initial distribution. What relationship do you find between your answers, and how can this be explained?

d. Compare the smoking distribution of two towns after 64 years, both with initial populations of 2000 people, but one having 100 smokers and the other having 1900 smokers. Explain the relationship between your results.

e. We are making numerous assumptions about the population when we employ transition matrices. For instance, we are assuming that there is no migration either into or out of the town. What other assumptions are we making, and how reasonable are they?

f. We have assumed that the smoking transition matrix for every year is the same. Of course, this is unlikely to be true. What sorts of trends would we be likely to find if we checked historical records? How would a smoking transition matrix for 1950 compare to one for the year 2000?

Section 9.3 Linear Systems and Matrices

- How can diets be balanced by solving systems of equations?
- How can an athlete choose an appropriate combination of activities as part of a cross-training routine?
- Why are there never exactly two solutions to a linear system of equations?
- How can we *quickly* find the equation of the parabola passing through three given points?

In Section 9.1, we investigated systems of equations in 3 variables or fewer and introduced the notion of a *linear* system. Throughout much of the rest of this chapter, we develop a handful of systematic methods—algorithms—that can be applied to solve linear systems of arbitrary size. In each case, the first step to solving a linear system is to translate the system into matrices, the language of linear systems.

Linear Systems of Equations

Recall from Section 9.1 that a system of equations is said to be linear if each equation is linear. Thus, a linear system of two equations in x and y is of the form

$$ax + by = c$$
$$dx + ey = f$$

For example, the system

$$3x - 4y = 12$$
$$2x + 9y = 1$$

is linear since both equations are linear, whereas the system

$$2x^2 - 3y = 5$$
$$x + 47y = 9$$

is nonlinear since one of the equations (the first in this case) is not linear.

We can consider linear systems consisting of any number of equations in any number of variables. For example, each of the following is a linear system of equations:

$$
\begin{aligned}
2x - 3y + 4z &= 10 \\
x - y + z &= 3 \\
3x + 5y - 2z &= 9
\end{aligned}
\qquad
\begin{aligned}
2p - 3q + r - 5s &= 3 \\
p + q + 2r + 3s &= -4
\end{aligned}
$$

(3 linear equations in 3 variables)   (2 linear equations in 4 variables)

More precisely, we have the following definition.

Definition of a Linear System of Equations

A **linear system of m equations in n variables $x_1, x_2, x_3, \ldots, x_n$** is a system that can be expressed in the form

$$
\begin{aligned}
a_{11}x_1 + a_{12}x_2 + a_{13}x_3 + \cdots + a_{1n}x_n &= b_1 \\
a_{21}x_1 + a_{22}x_2 + a_{23}x_3 + \cdots + a_{2n}x_n &= b_2 \\
a_{31}x_1 + a_{32}x_2 + a_{33}x_3 + \cdots + a_{3n}x_n &= b_3 \\
&\vdots \\
a_{m1}x_1 + a_{m2}x_2 + a_{m3}x_3 + \cdots + a_{mn}x_n &= b_m
\end{aligned}
$$

where the a_{ij}'s and b_i's are real numbers. The a_{ij}'s are called **coefficients**.

Solution Sets of Linear Equations

If a point (x, y) is a solution of the linear system

$$ax + by = c$$
$$dx + ey = f$$

then (x, y) must satisfy each equation. Moreover, because the graph of each equation is a line, a solution point (x, y) must lie on both lines and, hence, must be a point of intersection. Now a pair of lines may intersect in a single point, infinitely many points (if the lines are the same), or no points at all (if the lines are parallel). Thus, we see that the number of solutions to a system of two linear equations in two variables is either 0, 1, or infinity. Table 5 illustrates the range of possibilities for the solution set of a system of two equations in two variables.

Table 5
Solution possibilities

System	Graph	Points of intersection
$2x - y = 1$ $x - 2y = -4$		Just one point of intersection: $(2, 3)$
$2x - y = 1$ $2x - y = -3$		Zero points of intersection; the lines are parallel. The equations are inconsistent.
$x + 2y = 3$ $-2x - 4y = -6$		Infinitely many points of intersection; the equations are of the same line.

The results in Table 5 hold true for general linear systems regardless of the number of equations or variables.

Number of Solutions to a Linear System of Equations

A linear system of equations has 0, 1, or infinitely many solutions.

EXAMPLE 1

A Linear System with Infinitely Many Solutions

Solve the system of equations

$$x - y = 6$$
$$2x - 2y = 12$$

Solution   Since the second equation can be obtained by multiplying both sides of the first equation by 2, there is, in essence, only one equation to be satisfied. Thus, the solution set consists of all pairs (x, y) such that $x - y = 6$. If we solve this equation for y, we obtain $y = x - 6$, so that the solution set to the system could also be described as the set of pairs $(x, x - 6)$. So, for example, when $x = 1$, we obtain the point $(1, -5)$ as a solution; when $x = -3$, we obtain $(-3, -9)$, and so on.

Representing Systems with Matrices

The notation used in defining linear systems is suggestive of matrix notation. In fact, we have already seen (see Example 10 of Section 9.2) that matrices are a convenient way of expressing linear systems. As we will soon see, matrices are also an invaluable tool for *solving* linear systems of equations. The following definitions are central to our discussion.

Definition of Coefficient Matrix and Augmented Matrix

The **coefficient matrix** of the linear system of equations

$$a_{11}x_1 + a_{12}x_2 + a_{13}x_3 + \cdots + a_{1n}x_n = b_1$$
$$a_{21}x_1 + a_{22}x_2 + a_{23}x_3 + \cdots + a_{2n}x_n = b_2$$
$$a_{31}x_1 + a_{32}x_2 + a_{33}x_3 + \cdots + a_{3n}x_n = b_3$$
$$\vdots \qquad \vdots \qquad \vdots \qquad \qquad \vdots \qquad \vdots$$
$$a_{m1}x_1 + a_{m2}x_2 + a_{m3}x_3 + \cdots + a_{mn}x_n = b_m$$

is the matrix

$$[a_{ij}] = \begin{bmatrix} a_{11} & a_{12} & a_{13} & \cdots & a_{1n} \\ a_{21} & a_{22} & a_{23} & \cdots & a_{2n} \\ a_{31} & a_{32} & a_{33} & \cdots & a_{3n} \\ \vdots & \vdots & \vdots & & \vdots \\ a_{m1} & a_{m2} & a_{m3} & \cdots & a_{mn} \end{bmatrix}$$

The matrix formed by adjoining a column consisting of the b_i's to the coefficient matrix $[a_{ij}]$ is called the **augmented matrix** of the system. The augmented matrix is often written with a vertical bar separating the column of the b_i's from the coefficient matrix, as follows:

Augmented matrix

$$\left[\begin{array}{ccccc|c} a_{11} & a_{12} & a_{13} & \cdots & a_{1n} & b_1 \\ a_{21} & a_{22} & a_{23} & \cdots & a_{2n} & b_2 \\ a_{31} & a_{32} & a_{33} & \cdots & a_{3n} & b_3 \\ \vdots & \vdots & \vdots & & \vdots & \vdots \\ a_{m1} & a_{m2} & a_{m3} & \cdots & a_{mn} & b_m \end{array} \right]$$

····▷**EXAMPLE 2**

Finding a Coefficient and Augmented Matrix

Find the coefficient and augmented matrix for the following linear system:

$$2x - 4y + 2z = 6$$
$$3y + 5x = 8$$

Solution   We begin by writing the variables in the same order in each equation of the system.

$$2x - 4y + 2z = 6$$
$$5x + 3y \qquad = 8$$

The coefficient matrix is then

$$\begin{bmatrix} 2 & -4 & 2 \\ 5 & 3 & 0 \end{bmatrix}$$

The augmented matrix is formed by "augmenting" the coefficient matrix with a column formed from the constants to the right of the equal sign. Thus, the augmented matrix is given by

$$\left[\begin{array}{ccc|c} 2 & -4 & 2 & 6 \\ 5 & 3 & 0 & 8 \end{array}\right]$$

Conversely, given an augmented matrix (and an ordering of the variables), we can reproduce the original system of equations, as the following examples illustrate.

····▷**EXAMPLE 3**

Producing a System of Equations from an Augmented Matrix

Find the linear system of equations in the variables x and y that corresponds to the following augmented matrix. (Assume that the entries correspond to the order x, y.)

$$\left[\begin{array}{cc|c} 2 & 5 & 1 \\ 1 & 3 & 0 \end{array}\right]$$

Solution

$$2x + 5y = 1$$
$$x + 3y = 0$$

····▷**EXAMPLE 4**

Finding and Solving a System of Equations

Find and solve the linear system of equations corresponding to the following augmented matrix. (Assume that the variables are x, y, and z, respectively.)

$$\left[\begin{array}{ccc|c} 2 & 2 & -1 & -5 \\ 0 & 1 & 4 & 13 \\ 0 & 0 & 3 & 9 \end{array}\right]$$

Solution   The augmented matrix corresponds to the following system of equations:

(1) $$2x + 2y - z = -5$$
(2) $$y + 4z = 13$$
(3) $$3z = 9$$

Solving equation (3) for z we have

$$\begin{aligned} 3z &= 9 \\ z &= 3 \end{aligned}$$

Substituting $z = 3$ into equation (2), we obtain

$$\begin{aligned} y + 4z &= 13 \\ y + 4(3) &= 13 \\ y + 12 &= 13 \\ y &= 1 \end{aligned}$$

Now we can find x by substituting the known values of y and z into equation (1).

$$\begin{aligned} 2x + 2y - z &= -5 \\ 2x + 2(1) - 3 &= -5 \\ 2x - 1 &= -5 \\ 2x &= -4 \\ x &= -2 \end{aligned}$$

Thus, the solution is given by $x = -2$, $y = 1$, and $z = 3$.

Upper Triangular Systems and Back-Substitution

In Example 4, we saw a system of equations that was especially easy to solve. The technique we employed is called **back-substitution**. In effect, we simply solve the last equation for the last variable and then substitute this value into the second to the last equation to solve for the second to the last variable, continuing in this fashion until we have solved for all the variables. Of course, back-substitution only works for systems of a very special form. The system must have as many equations as unknowns. Moreover, the last equation must involve only the last variable, the second to the last equation must involve only the last two variables, and so on. Such systems have coefficient matrices that are said to be *upper triangular*.

Back-Substitution and Upper Triangular Matrices

- The **main diagonal** of a square matrix (a matrix with the same number of rows as columns) runs from the upper left corner to the lower right corner, as shown in Figure 13.
- Square matrices with zeros below the main diagonal are called **upper triangular**. See Figure 14.
- A system of equations for which the coefficient matrix is upper triangular is said to be an **upper triangular system**, and it can be solved using **back-substitution**.

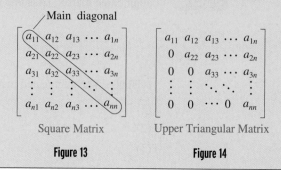

Square Matrix

Figure 13

$$\begin{bmatrix} a_{11} & a_{12} & a_{13} & \cdots & a_{1n} \\ 0 & a_{22} & a_{23} & \cdots & a_{2n} \\ 0 & 0 & a_{33} & \cdots & a_{3n} \\ \vdots & \vdots & & \ddots & \vdots \\ 0 & 0 & \cdots & 0 & a_{nn} \end{bmatrix}$$

Upper Triangular Matrix

Figure 14

⋯⋯⟩EXAMPLE 5   **Planning a Meal**

Table 6 gives nutritional information for a serving of lean ham, potatoes, and green beans.

Table 6

	Ham	Potatoes	Green beans
Calories	245	145	35
Carbohydrates (in grams)	0	33	8
Vitamin A (in I.U.)	0	0	780

Suppose you wish to prepare a meal consisting of ham, potatoes, and green beans with 650 calories, 80 grams of carbohydrates, and 1200 I.U. of vitamin A. How many servings of each type of food must be used?

Solution   Let x be the number of servings of ham, y the number of servings of potatoes, and z the number of servings of green beans. Then we have

$$\underset{\text{Calories from ham}}{} + \underset{\text{Calories from potatoes}}{} + \underset{\text{Calories from green beans}}{} = 650$$

$$245x + 145y + 35z = 650$$

$$\underset{\text{Carbohydrates from ham}}{} + \underset{\text{Carbohydrates from potatoes}}{} + \underset{\text{Carbohydrates from green beans}}{} = 80$$

$$0x + 33y + 8z = 80$$

$$\underset{\text{Vitamin A from ham}}{} + \underset{\text{Vitamin A from potatoes}}{} + \underset{\text{Vitamin A from green beans}}{} = 1200$$

$$0x + 0y + 780z = 1200$$

Combining the three equations, we obtain the following system of equations:

(1) $245x + 145y + 35z = 650$

(2) $33y + 8z = 80$

(3) $780z = 1200$

Note that this is an upper triangular system that can be solved by back-substitution. Solving equation (3) for z gives

$$780z = 1200$$

$$z \approx 1.5$$

Substituting $z = 1.5$ into equation (2) and solving for y, we have

$$33y + 8z = 80$$

$$33y + 8(1.5) = 80$$

$$33y = 68$$

$$y \approx 2.1$$

Substituting $z = 1.5$ and $y = 2.1$ into equation (1) and solving for x, we obtain

$$245x + 145y + 35z = 650$$

$$245x + 145(2.1) + 3.5(1.5) = 650$$

$$245x = 293$$

$$x \approx 1.2$$

Thus, the meal should be made with approximately 1.2 servings of ham, 2.1 servings of potatoes, and 1.5 servings of green beans.

Understanding and Mastery Checklists

Concepts to Understand	Skills to Master
Linear system of equations	Determine whether a system of equations is linear.
◆	◆
Solution possibilities for linear systems of equations	Determine whether a linear system has 0, 1, or infinitely many solutions.
◆	◆
Coefficient matrix of a linear system of equations	Find the augmented matrix corresponding to a linear system.
◆	◆
Augmented matrix of a linear system of equations	Solve a system of equations using back-substitution.
◆	
Back-substitution	
◆	
Square matrix	
◆	
Main diagonal	
◆	
Upper triangular matrix	

Exercises 9.3

Exercises 1-6 *Indicate whether the given system of equations is linear.*

1. $2x - 3y - 2z = \dfrac{1}{10}$
$x + y + z = 2$

2. $x^2 - y = 0$
$x - y = 1$

3. $\sqrt{x} - 3y = 7$
$2x + 5y = 9$

4. $\dfrac{1}{x} + y = 0$
$\dfrac{3}{x} - z = 2$

5. $x - 2y = 3$
$x + 3yz = 1$

6. $\sqrt{34}y - \dfrac{1}{7}x = z$
$x + 2y - 3z = 0$

Exercises 7-12 *Express the given system of equations as an augmented matrix.*

7. $3x + 4y = 10$
$2x - 7y = 4$

8. $-2x + 3y = 2$
$x - 6y = 0$

9. $x = 3$
$y = 4$

10. $2x - 3y + 4z = 2$
$x + 6y + 3z = 5$
$5x - 7y + 2z = 1$

11. $x - y + 2z = 1$
$2y + 6z = 3$

12. $x + 2y - 3z = 4$
$y + 3z = 0$
$z = 6$

Exercises 13-18 *Determine whether the given system of equations has 0, 1, or infinitely many solutions.*

13. $2x - 3y = 4$
$4x - 6y = 7$

14. $3x + 2y = 4$
$2x + 3y = 1$

15. $x - y = 7$
$2x - 2y = 14$

16. $x + \frac{1}{2}y = 3$

$3x + \frac{3}{2}y = 7$

17. $5x - 6y = 1$
$x - y = 3$

18. $4x + 2y = 6$
$20x + 10y = 30$

Exercises 19-24 *Solve the given system using back-substitution.*

19. $2x + 3y = 7$
$2y = 2$

20. $3x - 6y = 4$
$\frac{1}{2}y = 3$

21. $2x + y - z = 12$
$y + z = 8$
$3z = 3$

22. $2a + 3b + 4c = 2$
$2b - c = 5$
$2c = -2$

23. $q - 2r + 3s + t = 3$
$4r + s - t = 18$
$2s + 2t = 4$
$3t = -6$

24. $a + b + c + d = 2$
$b + c + d = 2$
$c + d = 1$
$d = 1$

Exercises 25-26 *An upper triangular augmented matrix is given. Find the solution to the corresponding system using back-substitution. Assume that the variable names are x and y, in that order.*

25. $\begin{bmatrix} 2 & -3 & | & -5 \\ 0 & 4 & | & 12 \end{bmatrix}$

26. $\begin{bmatrix} -1 & -2 & | & -11 \\ 0 & \frac{1}{3} & | & 3 \end{bmatrix}$

Exercises 27-30 *An upper triangular augmented matrix is given. Find the solution to the corresponding system using back-substitution. Assume that the variable names are x, y, and z, in that order.*

27. $\begin{bmatrix} 1 & 3 & 6 & | & 0 \\ 0 & 1 & 2 & | & 2 \\ 0 & 0 & 1 & | & -4 \end{bmatrix}$

28. $\begin{bmatrix} 1 & 2 & -3 & | & 1 \\ 0 & 1 & 3 & | & 5 \\ 0 & 0 & 1 & | & 0 \end{bmatrix}$

29. $\begin{bmatrix} 3 & -2 & 4 & | & 3 \\ 0 & 1 & -3 & | & 1 \\ 0 & 0 & 2 & | & -2 \end{bmatrix}$

30. $\begin{bmatrix} 4 & -6 & 5 & | & 4 \\ 0 & 2 & -3 & | & 0 \\ 0 & 0 & 3 & | & 3 \end{bmatrix}$

Applications

31. Income Tax A family has income totaling $58,000 from three different sources: salary, payments from municipal bonds, and payments from U.S. Treasury bonds. The municipal and U.S. Treasury bonds are free from state income tax, whereas salary is subject to a 6% state income tax. The U.S. Treasury bonds are free from federal tax, but both the salary and the income from the municipal bonds are subject to 20% federal tax. If the family's state tax totals $2400 and their federal tax totals $10,000, compute the income from each of the three sources.

32. Sports Drink An athlete wishes to concoct a 165-calorie 16-ounce sports beverage by mixing together orange juice (which has 13.75 calories per ounce) and water. How many ounces of each ingredient should be used?

33. Power Breakfast A breakfast consisting of milk, bananas, and whole wheat toast is to have 1200 units of vitamin A, 1.5 units of vitamin B_6, and 1.35 units of vitamin B_{12}. Table 7 shows the number of units of vitamins A, B_6, and B_{12} that are provided by a single serving of each of the three breakfast items. How many servings of each breakfast item should be consumed?

Table 7

	Milk	Bananas	Whole wheat toast
Vitamin A	500	226	0
Vitamin B_6	0.1	0.6	0.04
Vitamin B_{12}	0.9	0	0

34. Cross Training A 170-pound athlete cross-trains during the off season by cycling (at 10 miles per hour), playing volleyball, and weightlifting. He determines that he needs to burn off an extra 6000 calories per week through exercise, and that he needs to ride 80 miles per week. Because of his full-time job, he will be unable to devote more than 20 hours per week to exercise. His trainer tells him that he will burn approximately 552 calories per hour while cycling at 10 miles per hour and 234 calories per hour while playing volleyball, whereas the calories burned through weightlifting are negligible. Determine the number of hours that the athlete should spend in each of the three activities.

Concepts and Critical Thinking

Exercises 35-38 *Answer true or false.*

35. A system of equations is said to be linear if at least one of the equations is linear.

36. A system of equations is said to be nonlinear if at least one of the equations is nonlinear.

37. The augmented matrix $\begin{bmatrix} 1 & 0 & 0 & | & 3 \\ 2 & 1 & 0 & | & 2 \\ 1 & 1 & 0 & | & 3 \end{bmatrix}$ corresponds to the system
$$x = 3z$$
$$2x + y = 2z$$
$$x + y = 3z$$

38. Some linear systems have exactly two solutions.

Exercises 39-43 *Give an example of each.*

39. A nonlinear system of equations

40. A system of three equations in three unknowns such that the corresponding augmented matrix is upper triangular

41. An equation that, when paired with $2x + 3y = 5$, forms a system of equations with infinitely many solutions

42. An equation that, when paired with $2x + 3y = 5$, forms a system of equations with no solutions

43. An equation that, when paired with $2x + 3y = 5$, forms a system of equations with exactly one solution

44. Suppose that you were asked to write an exam for this class and you wished to produce a linear system of three equations with infinitely many solutions. If the first two equations are $2x + 3y - z = 5$ and $x + 2y + z = 3$, what might the third equation be? What if you wanted a system with no solution?

Questions for Discussion or Essay

45. It turns out that the graph of a linear equation in three variables is a plane in three-dimensional space. Explain why three planes will meet in 0, 1, or infinitely many points. Give an example from the room you are in now that suggests three planes meeting in a single point; also describe a situation in which three planes meet in a line. Sketch both.

46. Give an example of a system of two equations in two unknowns that has exactly two solutions. Explain why this doesn't contradict the statement that the number of solutions to a linear system of equations is 0, 1, or infinity.

47. Suppose you are given a system of three linear equations and that the equations are given in no particular order. In how many different ways can the system be represented using augmented matrices? Without checking all of these different matrix representations one by one, how can you tell just by looking at the equations whether one of the matrix representations will be upper triangular?

Projects for Enrichment

48. **Linear Systems with More Than One Solution**   In the text, it was stated that the solution set to a linear system of equations has 0, 1, or infinitely many solutions. In this project, we prove this statement using matrices.

a. Suppose we are given a linear system of m equations in the n variables $x_1, x_2, x_3, \ldots, x_n$:
$$a_{11}x_1 + a_{12}x_2 + a_{13}x_3 + \cdots + a_{1n}x_n = b_1$$
$$a_{21}x_1 + a_{22}x_2 + a_{23}x_3 + \cdots + a_{2n}x_n = b_2$$
$$a_{31}x_1 + a_{32}x_2 + a_{33}x_3 + \cdots + a_{3n}x_n = b_3$$
$$\vdots \qquad \vdots \qquad \vdots \qquad \vdots \qquad \vdots$$
$$a_{m1}x_1 + a_{m2}x_2 + a_{m3}x_3 + \cdots + a_{mn}x_n = b_m$$

i. Show that the system is equivalent to the matrix equation $AX = B$, where
$$A = \begin{bmatrix} a_{11} & a_{12} & a_{13} & \cdots & a_{1n} \\ a_{21} & a_{22} & a_{23} & \cdots & a_{2n} \\ a_{31} & a_{32} & a_{33} & \cdots & a_{3n} \\ \vdots & \vdots & \vdots & & \vdots \\ a_{m1} & a_{m2} & a_{m3} & \cdots & a_{mn} \end{bmatrix},$$
$$B = \begin{bmatrix} b_1 \\ b_2 \\ b_3 \\ \vdots \\ b_m \end{bmatrix}, \quad \text{and} \quad X = \begin{bmatrix} x_1 \\ x_2 \\ x_3 \\ \vdots \\ x_n \end{bmatrix}$$

ii. Use a property of matrices to show that if Y and Z are both solutions to $AX = B$, then $Y - Z$ is a solution to $AX = 0$.

iii. Now show that if Y and Z are solutions to $AX = B$, then so is $Y + t(Y - Z)$, where t is any real number.

b. For the given system of equations, two solutions are provided. Use the results from part a to give three more solutions.

i. $2x + 3y - z = 12$
$$x - y + z = 0$$
$$3x + 2y = 12$$
Solution 1: $x = 4, y = 0, z = -4$
Solution 2: $x = 2, y = 3, z = 1$

Hint: In matrix form these solutions correspond to

$$Y = \begin{bmatrix} 4 \\ 0 \\ -4 \end{bmatrix} \quad \text{and} \quad Z = \begin{bmatrix} 2 \\ 3 \\ 1 \end{bmatrix}$$

ii. $p + q + 3r = 3$
$$2q + 4r = 2$$
Solution 1: $p = 2, q = 1, r = 0$
Solution 2: $p = 1, q = -1, r = 1$

c. **i.** Use the result of part a to show that if a linear system of equations has two solutions, then it has infinitely many solutions.

ii. Use the result of part c-i to explain why a linear system of equations has 0, 1, or infinitely many solutions.

49. Interpolating with Polynomials It can be shown that given any collection of n points, there is a polynomial of degree $n - 1$ or lower that passes through the points. Suppose we wish to find the equation of the quadratic polynomial passing through three points, say $(1, 4)$, $(2, 6)$, and $(4, 22)$. Since the polynomial is quadratic, it is of the form $p(x) = ax^2 + bx + c$, and we have $p(1) = 4$, $p(2) = 6$, and $p(4) = 22$. From the fact that $p(1) = 4$, we have
$$p(1) = a \cdot 1^2 + b \cdot 1 + c = 4$$
which gives us the equation
$$a + b + c = 4$$
In a similar fashion, from $p(2) = 6$ and $p(4) = 22$, we obtain the equations
$$4a + 2b + c = 6$$
$$16a + 4b + c = 22$$

thus giving us the linear system
$$a + b + c = 4$$
$$4a + 2b + c = 6$$
$$16a + 4b + c = 22$$
This system could be solved using substitution, elimination, or any of the other techniques that are described in this chapter. In this project, we discover an alternative way to find a polynomial passing through a given set of points that involves solving a system of equations by back-substitution.

a. We find the equation of the quadratic polynomial passing through the points $(1, 4)$, $(2, 6)$, and $(4, 22)$. We begin by assuming that the polynomial is of the form
$$p(x) = a + b(x - 1) + c(x - 1)(x - 2)$$
Note that the red and blue numbers in the polynomial are two of the x-coordinates of the given points.

i. Set up a linear system involving the three variables a, b, and c.

ii. Solve the system obtained in part i by back-substitution.

iii. Now multiply out all products appearing in the expression for $p(x)$ in order to write $p(x)$ in standard form.

b. Use the technique developed in part a to find the quadratic polynomial passing through the points (x_1, y_1), (x_2, y_2), and (x_3, y_3).

c. Extend the approach developed in part a to find the cubic polynomial passing through

i. $(-2, 25)$, $(0, 1)$, $(1, 5)$, and $(2, 19)$

ii. $(0, 0)$, $(1, 0)$, $(4, 12)$, and $(2, -2)$

d. The U.S. Census Bureau has recorded the following data for the population of Arizona:

1970: 1,775,399
1980: 2,716,546
1990: 3,665,339
2000: 5,130,632

Let t represent the year (with $t = 0$ corresponding to 1970) and P the population of Arizona. Use the approach developed in part c to express P as a cubic polynomial in the variable t. According to this model, when will the population of Arizona reach 40 million? What are the limitations of this model?

Section 9.4

Gaussian and Gauss-Jordan Elimination

◦ What is the sum of the squares of the first 10,000 natural numbers, $1^2 + 2^2 + 3^2 + \cdots + 10{,}000^2$?

◦ A baseball is hit so it is at a height of 9 feet when it passes over the pitcher and clears the 10-foot center field wall by 4 feet. Is the ball rising or falling as it leaves the field?

◦ How can a system of equations be solved without ever writing down a single variable?

Gaussian Elimination

In Section 9.1, we presented two methods for solving systems of equations: substitution and elimination. Substitution works well when the number of equations is small or if the equations have an especially simple form. Indeed, we saw in the previous section that if the coefficient matrix of a system of equations is upper triangular, then the system can be solved easily using back-substitution. In this section, we see how systematic elimination can be used to convert augmented matrices to a form that is closely related to upper triangular—namely, **row-echelon form**.

Row-Echelon Form

A matrix is said to be in row-echelon form if the following conditions are satisfied:

- The first nonzero entry of each row is a 1. It is called the **leading 1**.
- Below each leading 1 are only 0s.
- Each leading 1 is to the right of the leading 1 in the row above.
- Any rows consisting entirely of 0s are at the bottom of the matrix.

⋯▷EXAMPLE 1

Identifying Row-Echelon Form

Indicate whether the given matrix is in row-echelon form.

a. $\begin{bmatrix} 1 & 4 & 6 \\ 0 & 1 & 1 \\ 0 & 0 & 1 \end{bmatrix}$
b. $\begin{bmatrix} 1 & 2 & 3 \\ 0 & 1 & 5 \\ 0 & 6 & 2 \end{bmatrix}$

c. $\left[\begin{array}{cccc|c} 1 & 0 & 3 & 4 & 5 \\ 0 & 2 & 0 & 1 & 4 \end{array}\right]$
d. $\left[\begin{array}{cccc|c} 1 & 0 & 2 & & 5 \\ 0 & 1 & 4 & & 2 \\ 0 & 0 & 0 & & 0 \end{array}\right]$

Solution

a. This matrix is in row-echelon form.
b. This matrix is not in row-echelon form because there is a nonzero number (6) in the third row and second column, directly below the leading 1 of the second row. Also, the first nonzero entry of the third row is 6 and not 1.
c. This augmented matrix is not in row-echelon form because the first nonzero entry of the second row is not 1.
d. This augmented matrix is in row-echelon form. Note that there is one row consisting entirely of 0s, and it is the last row.

Systems of equations represented by augmented matrices in row-echelon form can often be solved quite easily using back-substitution.

⋯⋯▷EXAMPLE 2 | **Solving a System Corresponding to a Matrix in Row-Echelon Form**

Solve the system of equations corresponding to the following augmented matrix in row-echelon form. (Assume the variables are x, y, and z, in that order.)

$$\begin{bmatrix} 1 & 3 & 2 & | & 11 \\ 0 & 1 & 4 & | & 6 \\ 0 & 0 & 1 & | & 1 \end{bmatrix}$$

Solution Expressing the augmented matrix as a system of equations, we have

(1) $\qquad\qquad x + 3y + 2z = 11$

(2) $\qquad\qquad\qquad\quad y + 4z = 6$

(3) $\qquad\qquad\qquad\qquad\quad z = 1$

Using back-substitution, we substitute $z = 1$ into equation (2).

$$y + 4z = 6$$
$$y + 4(1) = 6$$
$$y = 2$$

Now that y and z have been determined, we can find x by substituting their values into equation (1).

$$x + 3y + 2z = 11$$
$$x + 3(2) + 2(1) = 11$$
$$x = 3$$

Thus, the solution is given by $x = 3$, $y = 2$, and $z = 1$.

In Example 2, we were fortunate to have an augmented matrix that was already in row-echelon form. If the augmented matrix is *not* in row-echelon form, we can convert it to row-echelon form using a systematic procedure known as **Gaussian elimination**. At each stage of Gaussian elimination, one of three **elementary row operations** is performed on the augmented matrix. These operations are (1) interchanging two rows of the matrix, (2) multiplying a row by a constant, and (3) adding a multiple of one row to another. Performing an elementary row operation on an augmented matrix results in an equivalent system of equations. For example, interchanging two rows of an augmented matrix corresponds to interchanging two equations in the corresponding system.

Elementary Row Operations		
Operation	**Notation**	**Effect on system**
Interchanging rows i and j	R_{ij}	Equivalent. (Two equations change places.)
Multiplying row i by c, where $c \neq 0$	cR_i	Equivalent. (An equation is multiplied by a nonzero constant.)
Adding c times row i to row j	$cR_i + R_j$	Equivalent. (A multiple of an equation is added to another.)

Note that the operation $cR_i + R_j$ changes the jth row, *but not the ith row*. In other words, the jth row is replaced by $cR_i + R_j$, but the ith row is left alone. Also, elementary row operations may be performed on any matrix, augmented or not.

EXAMPLE 3

Performing Elementary Row Operations

For each matrix, perform the indicated elementary row operation.

a. $\begin{bmatrix} 1 & 3 \\ 2 & 5 \end{bmatrix}, -2R_1 + R_2$ **b.** $\begin{bmatrix} 3 & 6 & 5 & | & 3 \\ 4 & 2 & 1 & | & 7 \end{bmatrix}, -\frac{1}{3}R_1$ **c.** $\begin{bmatrix} 0 & 1 & 2 & | & 3 \\ 3 & 6 & 1 & | & 5 \\ 1 & 2 & 0 & | & 4 \end{bmatrix}, R_{13}$

Solution

a. Multiplying the first row by -2 and adding to the second row gives us

$$\begin{bmatrix} 1 & 3 \\ 2 + (-2)1 & 5 + (-2)3 \end{bmatrix} = \begin{bmatrix} 1 & 3 \\ 0 & -1 \end{bmatrix}$$

b. Multiplying the first row by $-\frac{1}{3}$ gives us

$$\begin{bmatrix} \left(-\frac{1}{3}\right)3 & \left(-\frac{1}{3}\right)6 & \left(-\frac{1}{3}\right)5 & | & \left(-\frac{1}{3}\right)3 \\ 4 & 2 & 1 & | & 7 \end{bmatrix} = \begin{bmatrix} -1 & -2 & -\frac{5}{3} & | & -1 \\ 4 & 2 & 1 & | & 7 \end{bmatrix}$$

c. Exchanging rows 1 and 3 gives us

$$\begin{bmatrix} 1 & 2 & 0 & | & 4 \\ 3 & 6 & 1 & | & 5 \\ 0 & 1 & 2 & | & 3 \end{bmatrix}$$

Gaussian elimination proceeds column by column from left to right. In each column, a leading 1 is obtained in the diagonal position, and then the entries under the leading 1 are cleared out, one by one, using elementary row operations. The following example illustrates Gaussian elimination.

EXAMPLE 4

Solving a System of Equations by Gaussian Elimination

Solve the following system of equations using Gaussian elimination.

$$2y - 3z = 12$$
$$2x + 7y - 11z = 27$$
$$x + 2y - z = 0$$

Solution We begin by writing the system as an augmented matrix.

$$\begin{bmatrix} 0 & 2 & -3 & | & 12 \\ 2 & 7 & -11 & | & 27 \\ 1 & 2 & -1 & | & 0 \end{bmatrix}$$

Next we perform Gaussian elimination to convert to row-echelon form, as shown in Table 8. For emphasis, at each stage, we shade the rows that are modified.

Table 8

Beginning matrix	Operation	Explanation	Resulting matrix
$\begin{bmatrix} 0 & 2 & -3 & \vert & 12 \\ 2 & 7 & -11 & \vert & 27 \\ 1 & 2 & -1 & \vert & 0 \end{bmatrix}$	R_{13}	Interchange rows 1 and 3 to obtain a leading 1 in row 1, column 1.	$\begin{bmatrix} 1 & 2 & -1 & \vert & 0 \\ 2 & 7 & -11 & \vert & 27 \\ 0 & 2 & -3 & \vert & 12 \end{bmatrix}$
$\begin{bmatrix} 1 & 2 & -1 & \vert & 0 \\ 2 & 7 & -11 & \vert & 27 \\ 0 & 2 & -3 & \vert & 12 \end{bmatrix}$	$-2R_1 + R_2$	Add -2 times the first row to the second row to obtain a 0 in row 2, column 1.	$\begin{bmatrix} 1 & 2 & -1 & \vert & 0 \\ 0 & 3 & -9 & \vert & 27 \\ 0 & 2 & -3 & \vert & 12 \end{bmatrix}$

The first column is now completed; we begin on the second.

Beginning matrix	Operation	Explanation	Resulting matrix
$\begin{bmatrix} 1 & 2 & -1 & \vert & 0 \\ 0 & 3 & -9 & \vert & 27 \\ 0 & 2 & -3 & \vert & 12 \end{bmatrix}$	$\frac{1}{3}R_2$	Multiply row 2 by $\frac{1}{3}$ in order to obtain a leading 1 in row 2, column 2.	$\begin{bmatrix} 1 & 2 & -1 & \vert & 0 \\ 0 & 1 & -3 & \vert & 9 \\ 0 & 2 & -3 & \vert & 12 \end{bmatrix}$
$\begin{bmatrix} 1 & 2 & -1 & \vert & 0 \\ 0 & 1 & -3 & \vert & 9 \\ 0 & 2 & -3 & \vert & 12 \end{bmatrix}$	$-2R_2 + R_3$	Add -2 times the second row to the third row to obtain a 0 in row 3, column 2.	$\begin{bmatrix} 1 & 2 & -1 & \vert & 0 \\ 0 & 1 & -3 & \vert & 9 \\ 0 & 0 & 3 & \vert & -6 \end{bmatrix}$

The second column is now completed; on to the third.

Beginning matrix	Operation	Explanation	Resulting matrix
$\begin{bmatrix} 1 & 2 & -1 & \vert & 0 \\ 0 & 1 & -3 & \vert & 9 \\ 0 & 0 & 3 & \vert & -6 \end{bmatrix}$	$\frac{1}{3}R_3$	Multiply the third row by $\frac{1}{3}$ in order to obtain a leading 1 in row 3, column 3.	$\begin{bmatrix} 1 & 2 & -1 & \vert & 0 \\ 0 & 1 & -3 & \vert & 9 \\ 0 & 0 & 1 & \vert & -2 \end{bmatrix}$

The matrix is now in row-echelon form. Translating back into equation form, we have

(1) $$x + 2y - z = 0$$
(2) $$y - 3z = 9$$
(3) $$z = -2$$

We now use back-substitution with $z = -2$ to find y.

$$y - 3z = 9$$
$$y - 3(-2) = 9$$
$$y = 3$$

Finally, we use back-substitution with $z = -2$ and $y = 3$ to find x.

$$x + 2y - z = 0$$
$$x + 2(3) - (-2) = 0$$
$$x = -8$$

Thus, our solution is given by $x = -8$, $y = 3$, $z = -2$.

Although the process of Gaussian elimination always yields a matrix in row-echelon form, this matrix is not unique. The final row-echelon matrix depends on choices made during Gaussian elimination.

Recall from the previous section that a linear system of equations will have 0, 1, or infinitely many solutions. Thus far, we have used Gaussian elimination to solve systems of equations with exactly one solution; however, it is also effective for expressing the solution sets of systems with infinitely many solutions and for exposing systems with no solution, as the following examples illustrate.

EXAMPLE 5

Solving a Linear System with Infinitely Many Solutions

Describe the solution set of the system

$$x + 3y - 5z = 2$$
$$x - y + z = 4$$
$$x + y - 2z = 3$$

Solution   We first express the system as an augmented matrix and then use Gaussian elimination to convert the matrix to row-echelon form. In moving from one matrix to the next, we shade the row that is modified and indicate the elementary row operation that is used.

$$\begin{bmatrix} 1 & 3 & -5 & | & 2 \\ 1 & -1 & 1 & | & 4 \\ 1 & 1 & -2 & | & 3 \end{bmatrix} \xrightarrow{\ -R_1 + R_2\ } \begin{bmatrix} 1 & 3 & -5 & | & 2 \\ 0 & -4 & 6 & | & 2 \\ 1 & 1 & -2 & | & 3 \end{bmatrix}$$

$$\begin{bmatrix} 1 & 3 & -5 & | & 2 \\ 0 & -4 & 6 & | & 2 \\ 1 & 1 & -2 & | & 3 \end{bmatrix} \xrightarrow{\ -R_1 + R_3\ } \begin{bmatrix} 1 & 3 & -5 & | & 2 \\ 0 & -4 & 6 & | & 2 \\ 0 & -2 & 3 & | & 1 \end{bmatrix}$$

$$\begin{bmatrix} 1 & 3 & -5 & | & 2 \\ 0 & -4 & 6 & | & 2 \\ 0 & -2 & 3 & | & 1 \end{bmatrix} \xrightarrow{\ -\frac{1}{4}R_2\ } \begin{bmatrix} 1 & 3 & -5 & | & 2 \\ 0 & 1 & -\frac{3}{2} & | & -\frac{1}{2} \\ 0 & -2 & 3 & | & 1 \end{bmatrix}$$

$$\begin{bmatrix} 1 & 3 & -5 & | & 2 \\ 0 & 1 & -\frac{3}{2} & | & -\frac{1}{2} \\ 0 & -2 & 3 & | & 1 \end{bmatrix} \xrightarrow{\ 2R_2 + R_3\ } \begin{bmatrix} 1 & 3 & -5 & | & 2 \\ 0 & 1 & -\frac{3}{2} & | & -\frac{1}{2} \\ 0 & 0 & 0 & | & 0 \end{bmatrix}$$

Converting to equations, we have

(1) $$x + 3y - 5z = 2$$

(2) $$y - \frac{3}{2}z = -\frac{1}{2}$$

(3) $$0 = 0$$

Note that equation (3) gives us no further restrictions on x, y, or z. We use equations (1) and (2) to express x and y in terms of z. Solving equation (2) for y gives us

$$y - \frac{3}{2}z = -\frac{1}{2}$$

$$y = \frac{3}{2}z - \frac{1}{2}$$

Substituting $y = \frac{3}{2}z - \frac{1}{2}$ into equation (1) and solving for x, we obtain

$$x + 3\left(\frac{3}{2}z - \frac{1}{2}\right) - 5z = 2$$

$$x - \frac{1}{2}z - \frac{3}{2} = 2$$

$$x = \frac{1}{2}z + \frac{7}{2}$$

Thus, the solution set consists of triples of numbers (x, y, z), where $x = \frac{1}{2}z + \frac{7}{2}$, $y = \frac{3}{2}z - \frac{1}{2}$, and z is arbitrary.

········**EXAMPLE 6**   **A Linear System with No Solution**

Solve the linear system of equations

$$a - b + 3c - 2d = 1$$
$$-2a + 2b - 6c + 4d = 4$$

Solution   Using Gaussian elimination, we have

$$\begin{bmatrix} 1 & -1 & 3 & -2 & | & 1 \\ -2 & 2 & -6 & 4 & | & 4 \end{bmatrix} \quad \xrightarrow{2R_1 + R_2} \quad \begin{bmatrix} 1 & -1 & 3 & -2 & | & 1 \\ 0 & 0 & 0 & 0 & | & 6 \end{bmatrix}$$

The second row of this matrix corresponds to the equation $0 = 6$, which is a contradiction. Thus, the system has no solution.

Once an augmented matrix has been converted to row-echelon form using Gaussian elimination, the corresponding system of equations can easily be characterized as consistent or inconsistent, as described by the following rule of thumb.

Rule of Thumb

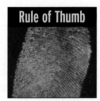

Suppose that a system has been expressed as a matrix in row-echelon form. The system is inconsistent (has no solution) if there is a row with all zeros to the left of the partition and a nonzero number to the right. Otherwise, the system is consistent.

Gauss-Jordan Elimination and Reduced Row-Echelon Form

We have seen that a system of equations can be solved by first converting the augmented matrix to row-echelon form using Gaussian elimination and then converting back to equation form and solving using back-substitution. It is possible, however, to avoid reverting to equations altogether by performing additional matrix operations that mirror the process of back-substitution, thus producing an especially simple augmented matrix from which solutions are simply "read off." Such matrices are said to be in reduced row-echelon form.

Reduced Row-Echelon Form

A matrix is said to be in **reduced row-echelon form** if the following two conditions are met:

1. The matrix is in row-echelon form.
2. There are only 0s directly above each leading 1.

········**EXAMPLE 7**   **Identifying Matrices in Reduced Row-Echelon Form**

Determine whether each of the following matrices is in reduced row-echelon form:

a. $\begin{bmatrix} 1 & 0 & 0 & | & 3 \\ 0 & 1 & 0 & | & 2 \\ 0 & 0 & 1 & | & 5 \end{bmatrix}$ **b.** $\begin{bmatrix} 1 & 0 & 0 & | & 1 \\ 0 & 1 & 2 & | & 6 \\ 0 & 0 & 1 & | & 2 \end{bmatrix}$ **c.** $\begin{bmatrix} 6 & 0 & 0 & | & 3 \\ 0 & 1 & 0 & | & 5 \\ 0 & 0 & 1 & | & 1 \end{bmatrix}$

Solution

a. This matrix is in reduced row-echelon form.

b. This matrix is in row-echelon form but not *reduced* row-echelon form. The culprit is the 2 in the second row, third column: It would need to be 0 for the matrix to be in reduced row-echelon form.

c. This matrix is not even in row-echelon form, let alone reduced row-echelon form. The culprit is the 6 in the first row, first column. If it were a 1, then this matrix would be in reduced row-echelon form.

It is easy to see why reduced row-echelon form is desirable. For example, suppose that a system of equations is represented by the following augmented matrix, which is in reduced row-echelon form (with columns corresponding to the variables x, y, and z, in that order):

$$\left[\begin{array}{ccc|c} 1 & 0 & 0 & 7 \\ 0 & 1 & 0 & -2 \\ 0 & 0 & 1 & 3 \end{array}\right]$$

Converting this augmented matrix into equation form gives

$$x = 7$$
$$y = -2$$
$$z = 3$$

The system is solved, and no back-substitution is required!

Gauss-Jordan elimination is a systematic procedure for converting a matrix into reduced row-echelon form using elementary row operations. First, Gaussian elimination is used to produce a matrix in row-echelon form. Next, elementary row operations are used to eliminate the nonzero entries *above* each leading 1, in much the same way that nonzero entries below leading 1s are eliminated when converting the matrix to row-echelon form. It can be shown that the resulting matrix is unique: No matter what decisions are made during the course of Gauss-Jordan elimination, the final reduced row-echelon matrix will be the same.

⋯⟩EXAMPLE 8 **Solving a System of Equations Using Gauss-Jordan Elimination**

Solve the following system using Gauss-Jordan elimination:

$$2x - 4y - 10z = -12$$
$$2x - 2y - 2z = 2$$
$$-3x + 4y + 11z = 8$$

Solution We first form the augmented matrix; then we begin applying elementary row operations to produce a reduced row-echelon matrix.

$$\left[\begin{array}{ccc|c} 2 & -4 & -10 & -12 \\ 2 & -2 & -2 & 2 \\ -3 & 4 & 11 & 8 \end{array}\right] \xrightarrow{\frac{1}{2}R_1} \left[\begin{array}{ccc|c} 1 & -2 & -5 & -6 \\ 2 & -2 & -2 & 2 \\ -3 & 4 & 11 & 8 \end{array}\right]$$

$$\left[\begin{array}{ccc|c} 1 & -2 & -5 & -6 \\ 2 & -2 & -2 & 2 \\ -3 & 4 & 11 & 8 \end{array}\right] \xrightarrow{-2R_1 + R_2} \left[\begin{array}{ccc|c} 1 & -2 & -5 & -6 \\ 0 & 2 & 8 & 14 \\ -3 & 4 & 11 & 8 \end{array}\right]$$

$$\left[\begin{array}{ccc|c} 1 & -2 & -5 & -6 \\ 0 & 2 & 8 & 14 \\ -3 & 4 & 11 & 8 \end{array}\right] \xrightarrow{3R_1 + R_3} \left[\begin{array}{ccc|c} 1 & -2 & -5 & -6 \\ 0 & 2 & 8 & 14 \\ 0 & -2 & -4 & -10 \end{array}\right]$$

$$\begin{bmatrix} 1 & -2 & -5 & -6 \\ 0 & 2 & 8 & 14 \\ 0 & -2 & -4 & -10 \end{bmatrix} \xrightarrow{\ \frac{1}{2}R_2\ } \begin{bmatrix} 1 & -2 & -5 & -6 \\ 0 & 1 & 4 & 7 \\ 0 & -2 & -4 & -10 \end{bmatrix}$$

$$\begin{bmatrix} 1 & -2 & -5 & -6 \\ 0 & 1 & 4 & 7 \\ 0 & -2 & -4 & -10 \end{bmatrix} \xrightarrow{\ 2R_2 + R_3\ } \begin{bmatrix} 1 & -2 & -5 & -6 \\ 0 & 1 & 4 & 7 \\ 0 & 0 & 4 & 4 \end{bmatrix}$$

$$\begin{bmatrix} 1 & -2 & -5 & -6 \\ 0 & 1 & 4 & 7 \\ 0 & 0 & 4 & 4 \end{bmatrix} \xrightarrow{\ \frac{1}{4}R_3\ } \begin{bmatrix} 1 & -2 & -5 & -6 \\ 0 & 1 & 4 & 7 \\ 0 & 0 & 1 & 1 \end{bmatrix}$$

$$\begin{bmatrix} 1 & -2 & -5 & -6 \\ 0 & 1 & 4 & 7 \\ 0 & 0 & 1 & 1 \end{bmatrix} \xrightarrow{\ 2R_2 + R_1\ } \begin{bmatrix} 1 & 0 & 3 & 8 \\ 0 & 1 & 4 & 7 \\ 0 & 0 & 1 & 1 \end{bmatrix}$$

$$\begin{bmatrix} 1 & 0 & 3 & 8 \\ 0 & 1 & 4 & 7 \\ 0 & 0 & 1 & 1 \end{bmatrix} \xrightarrow{\ -3R_3 + R_1\ } \begin{bmatrix} 1 & 0 & 0 & 5 \\ 0 & 1 & 4 & 7 \\ 0 & 0 & 1 & 1 \end{bmatrix}$$

$$\begin{bmatrix} 1 & 0 & 0 & 5 \\ 0 & 1 & 4 & 7 \\ 0 & 0 & 1 & 1 \end{bmatrix} \xrightarrow{\ -4R_3 + R_2\ } \begin{bmatrix} 1 & 0 & 0 & 5 \\ 0 & 1 & 0 & 3 \\ 0 & 0 & 1 & 1 \end{bmatrix}$$

It follows that $x = 5$, $y = 3$, and $z = 1$.

As noted earlier, one of the main benefits of Gauss-Jordan elimination is that it does not require back-substitution. However, if we were to judge solely on the basis of efficiency, Gaussian elimination with back-substitution would almost always come out ahead of Gauss-Jordan elimination. For most systems of equations, the total number of arithmetic operations required by Gauss-Jordan elimination will be much larger than the number required by the combination of Gaussian elimination with back-substitution. (See Exercise 75.)

Understanding and Mastery Checklists

Concepts to Understand

Row-echelon form of a matrix

❖

Elementary row operation

❖

Gaussian elimination

❖

Reduced row-echelon form

❖

Gauss-Jordan elimination

Skills to Master

Recognize when a matrix is in row-echelon or reduced row-echelon form.

❖

Apply an elementary row operation to a matrix.

❖

Use Gaussian elimination to find the row-echelon form of a matrix.

❖

Use Gauss-Jordan elimination to find the reduced row-echelon form of a matrix.

❖

Solve a linear system of equations by applying either Gaussian or Gauss-Jordan elimination.

Exercises 1-6   *Determine whether the given matrix is in row-echelon form. If not, support your answer.*

1. $\begin{bmatrix} 1 & 3 \\ 0 & 1 \end{bmatrix}$

2. $\begin{bmatrix} 1 & 4 & 0 \\ 0 & 2 & 1 \end{bmatrix}$

3. $\begin{bmatrix} 1 & 2 & 3 \\ 0 & 1 & 2 \\ 1 & 0 & 3 \end{bmatrix}$

4. $\begin{bmatrix} 1 & 0 & 2 & | & 2 \\ 0 & 1 & 1 & | & 3 \\ 0 & 0 & 1 & | & 4 \end{bmatrix}$

5. $\begin{bmatrix} 1 & 2 & 0 & -1 & | & 0 \\ 0 & 1 & -3 & 1 & | & 1 \\ 0 & 1 & 2 & 0 & | & 0 \end{bmatrix}$

6. $\begin{bmatrix} 1 & 0 & 2 & 0 \\ 0 & 0 & 0 & 0 \\ 0 & 1 & 4 & 0 \end{bmatrix}$

Exercises 7-10   *Perform the indicated row operation.*

7. $\begin{bmatrix} 2 & 4 \\ 3 & 0 \end{bmatrix}; \frac{1}{2}R_1$

8. $\begin{bmatrix} 1 & 0 & 3 & | & 1 \\ 0 & 0 & 2 & | & -1 \\ 0 & 1 & 6 & | & 3 \end{bmatrix}; R_{23}$

9. $\begin{bmatrix} 1 & 4 & | & 6 \\ 2 & -1 & | & 3 \end{bmatrix}; -2R_1 + R_2$

10. $\begin{bmatrix} 1 & 3 & 7 & | & 1 \\ 0 & 1 & 2 & | & 3 \\ 0 & 0 & 1 & | & -1 \end{bmatrix}; -3R_2 + R_1$

Exercises 11-22   *Use Gaussian elimination to write the augmented matrix in row-echelon form. (There may be more than one correct answer.)*

11. $\begin{bmatrix} 4 & 8 & | & 12 \\ 3 & 9 & | & 21 \end{bmatrix}$

12. $\begin{bmatrix} \frac{2}{3} & -\frac{2}{3} & | & \frac{4}{3} \\ -4 & 2 & | & -18 \end{bmatrix}$

13. $\begin{bmatrix} 0 & 2 & -4 & | & 8 \\ 3 & 0 & 9 & | & 12 \end{bmatrix}$

14. $\begin{bmatrix} 2 & 0 & 1 & | & 0 \\ 2 & 1 & 4 & | & 1 \\ 3 & 1 & 8 & | & 1 \end{bmatrix}$

15. $\begin{bmatrix} 2 & 6 & 8 & | & -2 \\ 2 & 4 & 12 & | & 5 \\ 1 & 7 & -6 & | & -21 \end{bmatrix}$

16. $\begin{bmatrix} 1 & 2 & 1 & | & 0 \\ -2 & 1 & 3 & | & 4 \\ 1 & 7 & 6 & | & 4 \end{bmatrix}$

17. $\begin{bmatrix} 1 & -3 & -1 & | & 2 \\ -2 & 6 & 4 & | & 3 \\ 2 & -1 & 2 & | & 5 \end{bmatrix}$

18. $\begin{bmatrix} 2 & 4 & 2 & | & 4 \\ 3 & 4 & 1 & | & 6 \end{bmatrix}$

19. $\begin{bmatrix} 1 & 3 & 0 & | & 4 \\ \frac{1}{3} & -1 & -4 & | & \frac{16}{3} \end{bmatrix}$

20. $\begin{bmatrix} 2 & 4 & 6 & 2 & | & 6 \\ -2 & -3 & -7 & 0 & | & 0 \\ 4 & 6 & 14 & -2 & | & -4 \end{bmatrix}$

21. $\begin{bmatrix} 1 & 0 & 1 & 2 & | & 1 \\ -1 & 1 & -1 & -1 & | & -1 \\ 0 & 2 & 1 & 3 & | & 0 \\ 1 & 0 & 2 & 4 & | & 3 \end{bmatrix}$

22. $\begin{bmatrix} 1 & 1 & 0 & 0 & | & 2 \\ 2 & 3 & 0 & 0 & | & 4 \\ 3 & 3 & 1 & 0 & | & 7 \\ 4 & 4 & 0 & 4 & | & 8 \end{bmatrix}$

Exercises 23-32   *Solve the system of equations by first expressing it as an augmented matrix and then applying Gaussian elimination and back-substitution.*

23. $2x - 6y = -2$
    $3x + 5y = 11$

24. $2u + 4v = -6$
    $5u - 3v = 24$

25. $2p + q = \dfrac{5}{3}$

    $3p - 6q = -\dfrac{5}{2}$

26. $4y + 6z = -\dfrac{8}{5}$

    $3y - 5z = \dfrac{13}{5}$

27. $x + y + 2z = 7$
    $x + 2y + 3z = 10$
    $x - 4y + z = 4$

28. $x - y + z = -2$
    $x + y - z = 0$
    $-x + y + z = -4$

29. $2p + r = 2$
    $2p + q + 4r = 9$
    $3p + q + 8r = 17$

30. $2u - 4v + 2w = 18$
    $3u - 5v + 2w = 23$
    $4u + 9v + 6w = -13$

31. $w + x + y + 2z = 1$
    $-w + x - y - z = -2$
    $2x + y + 3z = -1$
    $w + x + 2y + 5z = 1$

32. $2q + 4r + 6s - 2t = 8$
    $r - s + t = 2$
    $r + s - t = 3$
    $-r + s + t = -5$

Exercises 33-38   *Use Gauss-Jordan elimination to express the given matrix in reduced row-echelon form.*

33. $\begin{bmatrix} 1 & 2 & | & 3 \\ 0 & 1 & | & 2 \end{bmatrix}$

34. $\begin{bmatrix} 1 & 3 & | & -2 \\ 0 & 1 & | & 5 \end{bmatrix}$

35. $\begin{bmatrix} 2 & 4 & | & -8 \\ 3 & 3 & | & -3 \end{bmatrix}$

36. $\begin{bmatrix} 2 & 3 & | & 0 \\ 2 & 6 & | & -2 \end{bmatrix}$

37. $\begin{bmatrix} 1 & 2 & 2 & | & 2 \\ 0 & 1 & 3 & | & -2 \\ 0 & 0 & 1 & | & -1 \end{bmatrix}$

38. $\begin{bmatrix} 1 & 3 & 1 & | & \frac{7}{2} \\ 0 & 1 & -\frac{1}{6} & | & 0 \\ 0 & 0 & 1 & | & 2 \end{bmatrix}$

Exercises 39-46 *Solve the system of equations by expressing it in matrix form and using Gauss-Jordan elimination to obtain a matrix in reduced row-echelon form.*

39. $2x - 3y = 12$
$x + 2y = -1$

40. $3x - y = 1$
$6x + 3y = 7$

41. $4x + 6y = 0$
$x - y = \dfrac{5}{6}$

42. $2x - 4y = -\dfrac{13}{2}$
$4x + 2y = 7$

43. $x - 2y + z = 9$
$y - 3z = -10$
$z = 3$

44. $x + 3y - 2z = -9$
$y + 2z = 10$
$z = 5$

45. $x + 2y + 4z = -10$
$2x + 3y + 6z = -15$
$x - y + z = -7$

46. $x - 3y + z = -12$
$x + y + z = 0$
$2x - y + z = -8$

Exercises 47-54 *Express the system as an augmented matrix and solve using Gaussian elimination. There may be 0, 1, or infinitely many solutions.*

47. $2x + y = 5$
$4x + 2y = 10$

48. $2x - 3y = 3$
$-4x + 6y = 5$

49. $-x + 3y = -7$
$3x - 2y = 7$

50. $3x - y = 1$
$\dfrac{3}{2}x - \dfrac{1}{2}y = \dfrac{1}{2}$

51. $-3x - y + z = -1$
$x + 4y - z = 3$
$-5x + 2y + z = 2$

52. $x + 2y + z = 5$
$x - y + 3z = 6$
$2x + y + 4z = 11$

53. $2x + 4y + 2z = 0$
$3x - y + z = 1$
$x - 5y - z = 1$

54. $x - 2y + 3z = 0$
$2x + y - z = 1$
$-x + z = 2$

Applications

55. Fitting a Parabola Find the equation of the parabola that passes through the points $(1, 5)$, $(2, 7)$, and $(3, 13)$.

56. Unknown Constants It is given that the variables y and x are related by the formula
$$y = a|x| + b|x - 1| + c|x - 2|$$
If $y = 5$ when $x = 0$, $y = 6$ when $x = 1$, and $y = 1$ when $x = 2$, find a, b, and c.

57. Counting Coins Your Uncle Bob pulls 15 coins (nickels, dimes, and quarters) from behind your ear. The total value of the coins is $2.10, and there are two more quarters than nickels. How many of each coin does Uncle Bob have?

58. River Beautification Kyung-Bai, Antoine, and Vanessa are cleaning a river bank of trash. The time required for Vanessa to clear a 100-yard stretch is the average of the time required for Kyung-Bai and Antoine. The sum of their times is 27 hours, and the time required for Antoine to clear a 100-yard stretch is 7 less than the sum of the corresponding times for Kyung-Bai and Vanessa.

a. Find the time required for each of them to clear a 100-yard stretch of the river bank.

b. How long would it take them to clear a 100-yard stretch if they work together?

59. Barbell Weights   A certain weight room includes a set of preweighted barbells, with the total weight printed on some of the barbells. A barbell with 2 large disks and 1 small disk on each side weighs 70 pounds; a barbell with 1 small disk on each side weighs 30 pounds; and a barbell with 3 large disks on each side weighs 75 pounds. Find the weight of the bar and each of the disks. Also, compute the weight of an (unmarked) barbell with 4 large disks and 2 small disks on each side.

60. Unknown Power   A weightlifter is bench-pressing a total of 305 pounds using a 45-pound bar on which there are only 45-, 25-, and 5-pound weights. He uses a total of 12 weights, and the number of 5-pound weights is equal to the sum of the number of 25- and 45-pound weights. How many weights of each type does the weightlifter use?

61. Ancient Faiths   When Martin Luther broke from the Roman Catholic Church in 1517, Buddhism and Hinduism were a combined 5059 years old. When Joseph Smith founded the Mormon Church in 1827, Islam and Buddhism were a combined 3557 years old. Hinduism is 975 years older than Buddhism. Find the years in which Islam, Buddhism, and Hinduism were founded.

62. Nut Mixture   How can 30 pounds of a mixture that is 40% pecans, 40% almonds, and 20% cashews be formed if only the following cans of mixed nuts are available?

	Percent pecans	Percent almonds	Percent cashews
1-pound can	50	30	20
2-pound can	30	60	10
5-pound can	40	30	30

63. School Integration   In a certain community, the middle schools fall into three categories:
(1) Schools of about 500 students, in which roughly half the students are white and half the students are black
(2) Schools with about 450 students that are approximately half black and half Latino
(3) Schools with about 400 students that are 80% white and 20% black

A new high school with a capacity of 3750 students is being built. It is desired that the ethnic composition of the school be consistent with that of the community as a whole, which is 46.8% black, 35.2% white, and 18% Latino. How many middle schools of each type should feed into the new high school in order to achieve the desired ethnic composition?

64. Ancestral Heritage   A woman is $\frac{19}{32}$ Cherokee and $\frac{13}{32}$ Irish. Her mother is half Cherokee. Her maternal grandfather has the same percentage of Cherokee blood as does her paternal grandfather. Her maternal grandmother has half as much Cherokee blood as does her paternal grandmother. Compute the fractions of Cherokee and Irish blood of her father and all four of her grandparents. [*Hint:* Let the four variables be the fraction of Cherokee blood of each of the woman's grandparents. If a person's father is a% Cherokee and mother is b% Cherokee, then the person's percentage of Cherokee blood is given by $(a + b)/2$.]

Concepts and Critical Thinking

Exercises 65-68   *Answer true or false.*

65. Any matrix in which each row begins with a 1 is in row-echelon form.

66. Interchanging two rows is an example of an elementary row operation.

67. Multiplying each entry in a column by 5 is an example of an elementary row operation.

68. If a matrix in row-echelon form has all zeros in the first row, then every entry of the matrix must be a zero.

Exercises 69-72   *Give an example of each.*

69. A matrix in row-echelon but not reduced row-echelon form

70. An elementary row operation

71. A matrix consisting only of 1s and 0s that is not in row-echelon form

72. An upper triangular matrix that is not in row-echelon form

73. Given is the final row of an augmented matrix that has been converted to reduced row-echelon form using Gauss-Jordan elimination. What conclusion can be drawn as to whether the corresponding system is consistent or inconsistent? If the system is consistent, how many solutions does it have?

a. $[0 \ 0 \ \cdots \ 0 \ 1 \ | \ 0]$

b. $[0 \ 0 \ \cdots \ 0 \ 0 \ | \ 0]$

c. $[0 \ 0 \ \cdots \ 0 \ 0 \ | \ 1]$

Questions for Discussion or Essay

74. What similarities do you see between the techniques of Gaussian elimination and synthetic division?

75. If we define an arithmetic step to be the addition, subtraction, or multiplication of two numbers, then what is the maximum number of arithmetic steps required to solve a system of three equations in three unknowns by Gaussian elimination with back-substitution? What about for Gauss-Jordan elimination? Support your answers.

76. One of the elementary row operations is multiplying a row by a nonzero constant. This operation corresponds to multiplying both sides of an equation by the constant. Why do we insist that the constant be nonzero? After all, if we multiply both sides of an equation by 0, we obtain the equation $0 = 0$, which is certainly true.

77. Gaussian elimination and Gauss-Jordan elimination are examples of algorithms, step-by-step methods for solving a problem for which every step is prescribed. Name two other mathematical algorithms and describe two algorithms from everyday life. Does the technique of elimination as described in Section 9.1 qualify as an algorithm? Why or why not?

Projects for Enrichment

78. The Sum of the First *n* Squares   In this project, we develop a formula for the sum S of the squares of the first n natural numbers, $S = 1^2 + 2^2 + 3^2 + \cdots + n^2$. We will take advantage of the fact that, in general, if $T = 1^k + 2^k + \cdots + n^k$, where k and n are natural numbers, then T is a polynomial of degree $k + 1$ in the variable n with constant term 0. In particular, S can be expressed as a third-degree polynomial in n with constant term 0; that is, $S = an^3 + bn^2 + cn$.

  a. Evaluate $S = 1^2 + 2^2 + \cdots + n^2$ for $n = 1, 2,$ and 3.

  b. Evaluate the expression $an^3 + bn^2 + cn$ for $n = 1, 2,$ and 3.

  c. Use the results of parts a and b to form a system of three equations in the three variables a, b, and c.

  d. Use Gaussian elimination to solve the system of part c.

  e. Express S as a third-degree polynomial in n.

  f. Compute the sum of the squares of the first 10,000 natural numbers.

79. Line Drive Home Runs   Suppose a batter hits a line drive that passes over the pitcher (60 feet, 6 inches from home plate) at a height of 9 feet above the field and then clears the 10-foot center field wall (330 feet from home plate) by 4 feet. A broadcaster announces that the ball "was still rising when it cleared the center field wall."

In this project, we investigate the possible locations at which the ball reaches its zenith, and, in particular, we determine if it is possible that the ball was still rising when it cleared the wall.

  a. Assuming there is no wind resistance, the ball travels in a parabolic path. Thus, if we let y represent the height of the ball and x the horizontal distance from home plate, then $y = ax^2 + bx + c$ for some constants a, b, and c. Set up a system of equations that reflects the facts given in the introductory paragraph of this project.

  b. Solve the system from part a. Describe the solution set in terms of the variable c. Note that c represents the height of the ball when it was struck by the batter.

  c. The vertex of the parabola occurs when $x = -b/(2a)$. Find an expression for the location of the vertex in terms of the variable c.

  d. Compute the location of the vertex for values of c ranging from 2 to 6, in increments of 1.

  e. Find the value of c required so that the ball's peak occurs as it crosses the fence. Is this reasonable?

  f. Could the broadcaster have been correct?

Barry Bonds belts his record-tying 70th home run on his way to the record of 73.

Section 9.5

Inverses of Square Matrices

- How can linear systems of equations be solved using a graphing calculator?
- How can you graph the letter M on a graphing calculator?
- How can the physical laws governing the universe be discovered using matrices?
- We can divide by any real number except 0. For which matrices is division defined?

We've seen that linear systems of equations can be expressed as matrix equations of the form $AX = B$, and we have developed techniques such as Gaussian elimination for solving such systems. But suppose that instead we naively attempted to solve for X by "dividing" both sides by A, in much the same way that we would divide both sides by a when solving an ordinary algebraic equation of the form $ax = b$. Immediately we are confronted with a difficulty: What does it mean to divide one matrix by another? In the case of a real number, dividing by a is equivalent to multiplying by a^{-1}. But what does A^{-1} mean when A is a matrix? Can we define A^{-1} for any nonzero matrix A, or are there some matrices A for which A^{-1} doesn't exist? In this section, we clarify the notion of the inverse of a matrix and apply it to solve certain equations of the form $AX = B$.

The Identity Matrix and Inverse Matrices

The $n \times n$ (square) matrix with 1s on the main diagonal and 0s elsewhere (see Figure 15) is called the nth-order **identity matrix** and is denoted by I_n, or simply I if the order is apparent.

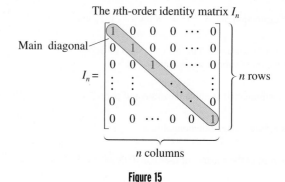

The nth-order identity matrix I_n

$$I_n = \begin{bmatrix} 1 & 0 & 0 & 0 & \cdots & 0 \\ 0 & 1 & 0 & 0 & \cdots & 0 \\ 0 & 0 & 1 & 0 & \cdots & 0 \\ \vdots & \vdots & & \ddots & & \vdots \\ 0 & 0 & & & \ddots & 0 \\ 0 & 0 & \cdots & 0 & 0 & 1 \end{bmatrix}$$

Main diagonal

n rows

n columns

Figure 15

The following example suggests why I is called the *identity* matrix.

EXAMPLE 1

Multiplication of a Matrix By *I*

Let

$$A = \begin{bmatrix} 1 & 2 & 3 \\ 4 & 5 & 6 \\ 7 & 8 & 9 \end{bmatrix}$$

Compute AI and IA.

Solution

$$AI = \begin{bmatrix} 1 & 2 & 3 \\ 4 & 5 & 6 \\ 7 & 8 & 9 \end{bmatrix} \begin{bmatrix} 1 & 0 & 0 \\ 0 & 1 & 0 \\ 0 & 0 & 1 \end{bmatrix}$$

$$= \begin{bmatrix} 1 \cdot 1 + 2 \cdot 0 + 3 \cdot 0 & 1 \cdot 0 + 2 \cdot 1 + 3 \cdot 0 & 1 \cdot 0 + 2 \cdot 0 + 3 \cdot 1 \\ 4 \cdot 1 + 5 \cdot 0 + 6 \cdot 0 & 4 \cdot 0 + 5 \cdot 1 + 6 \cdot 0 & 4 \cdot 0 + 5 \cdot 0 + 6 \cdot 1 \\ 7 \cdot 1 + 8 \cdot 0 + 9 \cdot 0 & 7 \cdot 0 + 8 \cdot 1 + 9 \cdot 0 & 7 \cdot 0 + 8 \cdot 0 + 9 \cdot 1 \end{bmatrix}$$

$$= \begin{bmatrix} 1 & 2 & 3 \\ 4 & 5 & 6 \\ 7 & 8 & 9 \end{bmatrix}$$

Thus, $AI = A$.

$$IA = \begin{bmatrix} 1 & 0 & 0 \\ 0 & 1 & 0 \\ 0 & 0 & 1 \end{bmatrix} \begin{bmatrix} 1 & 2 & 3 \\ 4 & 5 & 6 \\ 7 & 8 & 9 \end{bmatrix}$$

$$= \begin{bmatrix} 1 \cdot 1 + 0 \cdot 4 + 0 \cdot 7 & 1 \cdot 2 + 0 \cdot 5 + 0 \cdot 8 & 1 \cdot 3 + 0 \cdot 6 + 0 \cdot 9 \\ 0 \cdot 1 + 1 \cdot 4 + 0 \cdot 7 & 0 \cdot 2 + 1 \cdot 5 + 0 \cdot 8 & 0 \cdot 3 + 1 \cdot 6 + 0 \cdot 9 \\ 0 \cdot 1 + 0 \cdot 4 + 1 \cdot 7 & 0 \cdot 2 + 0 \cdot 5 + 1 \cdot 8 & 0 \cdot 3 + 0 \cdot 6 + 1 \cdot 9 \end{bmatrix}$$

$$= \begin{bmatrix} 1 & 2 & 3 \\ 4 & 5 & 6 \\ 7 & 8 & 9 \end{bmatrix}$$

Thus, $IA = A$.

In the previous example, we saw that the product of a third-order square matrix and I was the given third-order matrix. In fact, this is true for all square matrices. We leave the proof to Exercise 47.

The Identity Property of I_n

If A is an $n \times n$ matrix, then $AI_n = I_nA = A$.

The matrix identity I_n is similar to the multiplicative identity 1 in that multiplying a matrix by I_n leaves the matrix unchanged, just as multiplication by 1 leaves a number unchanged. Now, the multiplicative inverse of a number x is, by definition, a *number* that when multiplied by x gives the multiplicative identity 1; it is commonly denoted by x^{-1}. What then is the multiplicative inverse of an $n \times n$ matrix A? Of course, it should be a matrix that, when multiplied by A (on either side), gives the multiplicative identity I_n. We make the following definitions.

Definition of the Multiplicative Inverse of a Matrix

Let A be an $n \times n$ matrix. The **inverse** of A, if it exists, is an $n \times n$ matrix B satisfying $AB = I_n$ and $BA = I_n$. We say that A and B are inverses of one another and write $B = A^{-1}$. In this case, we say that the matrix A is **invertible**. If A^{-1} doesn't exist, then we say that A is **noninvertible**.

EXAMPLE 2

Confirming Inverses

Show that A and B are inverses of one another, where

$$A = \begin{bmatrix} 2 & 0 \\ -4 & 1 \end{bmatrix} \quad \text{and} \quad B = \begin{bmatrix} \frac{1}{2} & 0 \\ 2 & 1 \end{bmatrix}$$

Solution   We must show that

$$AB = BA = \begin{bmatrix} 1 & 0 \\ 0 & 1 \end{bmatrix}$$

Thus, we proceed as follows:

$$AB = \begin{bmatrix} 2 & 0 \\ -4 & 1 \end{bmatrix} \begin{bmatrix} \frac{1}{2} & 0 \\ 2 & 1 \end{bmatrix}$$

$$= \begin{bmatrix} 2 \cdot \frac{1}{2} + 0 \cdot 2 & 2 \cdot 0 + 0 \cdot 1 \\ (-4)\frac{1}{2} + 1 \cdot 2 & (-4)0 + 1 \cdot 1 \end{bmatrix}$$

$$= \begin{bmatrix} 1 & 0 \\ 0 & 1 \end{bmatrix}$$

$$BA = \begin{bmatrix} \frac{1}{2} & 0 \\ 2 & 1 \end{bmatrix} \begin{bmatrix} 2 & 0 \\ -4 & 1 \end{bmatrix}$$

$$= \begin{bmatrix} \frac{1}{2} \cdot 2 + 0(-4) & \frac{1}{2} \cdot 0 + 0 \cdot 1 \\ 2 \cdot 2 + 1(-4) & 2 \cdot 0 + 1 \cdot 1 \end{bmatrix}$$

$$= \begin{bmatrix} 1 & 0 \\ 0 & 1 \end{bmatrix}$$

Rule of Thumb

Using more advanced techniques, it can be shown that if $AB = I$, then $BA = I$ also. Thus, it is not necessary to compute both AB and BA in order to show that A and B are inverses.

Computing Inverses

We will now see how the inverse of a 2×2 matrix can be computed by solving a pair of systems of equations using Gauss-Jordan elimination. Let us consider the problem of computing the inverse of the matrix A given by

$$A = \begin{bmatrix} 1 & 3 \\ 2 & 5 \end{bmatrix}$$

If we denote A^{-1} by

$$A^{-1} = \begin{bmatrix} a & b \\ c & d \end{bmatrix}$$

then we have

$$AA^{-1} = I$$

$$\begin{bmatrix} 1 & 3 \\ 2 & 5 \end{bmatrix} \begin{bmatrix} a & b \\ c & d \end{bmatrix} = \begin{bmatrix} 1 & 0 \\ 0 & 1 \end{bmatrix}$$

$$\begin{bmatrix} a + 3c & b + 3d \\ 2a + 5c & 2b + 5d \end{bmatrix} = \begin{bmatrix} 1 & 0 \\ 0 & 1 \end{bmatrix}$$

Equating corresponding entries gives two systems of equations—one involving a and c and the other involving b and d. These systems and the corresponding augmented matrices are as follows:

	System	Augmented matrix
1	$a + 3c = 1$ $2a + 5c = 0$	$\begin{bmatrix} 1 & 3 & \vert & 1 \\ 2 & 5 & \vert & 0 \end{bmatrix}$
2	$b + 3d = 0$ $2b + 5d = 1$	$\begin{bmatrix} 1 & 3 & \vert & 0 \\ 2 & 5 & \vert & 1 \end{bmatrix}$

Each of these systems can be solved using Gauss-Jordan elimination.

System 1:

$$\begin{bmatrix} 1 & 3 & \vert & 1 \\ 2 & 5 & \vert & 0 \end{bmatrix} \xrightarrow{-2R_1 + R_2} \begin{bmatrix} 1 & 3 & \vert & 1 \\ 0 & -1 & \vert & -2 \end{bmatrix} \xrightarrow{-R_2} \begin{bmatrix} 1 & 3 & \vert & 1 \\ 0 & 1 & \vert & 2 \end{bmatrix} \xrightarrow{-3R_2 + R_1} \begin{bmatrix} 1 & 0 & \vert & -5 \\ 0 & 1 & \vert & 2 \end{bmatrix}$$

Thus, $a = -5$ and $c = 2$.

System 2:

$$\begin{bmatrix} 1 & 3 & \vert & 0 \\ 2 & 5 & \vert & 1 \end{bmatrix} \xrightarrow{-2R_1 + R_2} \begin{bmatrix} 1 & 3 & \vert & 0 \\ 0 & -1 & \vert & 1 \end{bmatrix} \xrightarrow{-R_2} \begin{bmatrix} 1 & 3 & \vert & 0 \\ 0 & 1 & \vert & -1 \end{bmatrix} \xrightarrow{-3R_2 + R_1} \begin{bmatrix} 1 & 0 & \vert & 3 \\ 0 & 1 & \vert & -1 \end{bmatrix}$$

Thus, $b = 3$ and $d = -1$.

Using the values of a, b, c, and d, we find that

$$A^{-1} = \begin{bmatrix} -5 & 3 \\ 2 & -1 \end{bmatrix}$$

Note that since systems 1 and 2 have the same coefficient matrix, the elementary row operations used at each stage of the Gauss-Jordan elimination are also the same. This suggests that we solve both systems simultaneously by applying Gauss-Jordan elimination to the following augmented matrix:

$$\begin{bmatrix} 1 & 3 & \vert & 1 & 0 \\ 2 & 5 & \vert & 0 & 1 \end{bmatrix}$$

When this matrix is in reduced row-echelon form, the first two columns will form the identity matrix I_2, the solution to system 1 will be given by the entries in the third column, and the solution to system 2 will be given by the entries in the fourth column. In other words, at the conclusion of the Gauss-Jordan elimination process, the matrix will be of the form

$$\begin{bmatrix} 1 & 0 & \vert & a & b \\ 0 & 1 & \vert & c & d \end{bmatrix}$$

with the columns to the right of the partition forming A^{-1}. We now compute A^{-1} with this procedure.

$$\begin{bmatrix} 1 & 3 & \vert & 1 & 0 \\ 2 & 5 & \vert & 0 & 1 \end{bmatrix} \xrightarrow{-2R_1 + R_2} \begin{bmatrix} 1 & 3 & \vert & 1 & 0 \\ 0 & -1 & \vert & -2 & 1 \end{bmatrix} \xrightarrow{-R_2} \begin{bmatrix} 1 & 3 & \vert & 1 & 0 \\ 0 & 1 & \vert & 2 & -1 \end{bmatrix} \xrightarrow{-3R_2 + R_1} \begin{bmatrix} 1 & 0 & \vert & -5 & 3 \\ 0 & 1 & \vert & 2 & -1 \end{bmatrix}$$

So

$$A^{-1} = \begin{bmatrix} -5 & 3 \\ 2 & -1 \end{bmatrix}$$

The preceding technique generalizes to square matrices of arbitrary order, as follows.

> ### Computing Inverses with Gauss-Jordan Elimination
>
> 1. Form the augmented matrix whose first n columns constitute A and whose last n columns form I_n; symbolically, this is written $[A|I_n]$.
> 2. Use Gauss-Jordan elimination to write this matrix in reduced row-echelon form.
>   a. If the first n columns of the reduced row-echelon matrix form the identity I_n, then the last n columns will form A^{-1}. Symbolically, the reduced row-echelon matrix is of the form $[I_n|A^{-1}]$.
>   b. If the first n columns do not form the identity matrix I_n, then A is noninvertible.

·····❭EXAMPLE 3

Computing Inverses of Matrices

Compute the inverse of the given matrix and confirm using matrix multiplication.

a. $A = \begin{bmatrix} 1 & 0 & 1 \\ 1 & -1 & 0 \\ 0 & 2 & 1 \end{bmatrix}$ **b.** $B = \begin{bmatrix} 1 & 0 \\ 0 & 0 \end{bmatrix}$

Solution

a. We form the augmented matrix $[A|I_3]$ and perform Gauss-Jordan elimination.

$$\left[\begin{array}{ccc|ccc} 1 & 0 & 1 & 1 & 0 & 0 \\ 1 & -1 & 0 & 0 & 1 & 0 \\ 0 & 2 & 1 & 0 & 0 & 1 \end{array}\right] \xrightarrow{-R_1 + R_2} \left[\begin{array}{ccc|ccc} 1 & 0 & 1 & 1 & 0 & 0 \\ 0 & -1 & -1 & -1 & 1 & 0 \\ 0 & 2 & 1 & 0 & 0 & 1 \end{array}\right]$$

$$\left[\begin{array}{ccc|ccc} 1 & 0 & 1 & 1 & 0 & 0 \\ 0 & -1 & -1 & -1 & 1 & 0 \\ 0 & 2 & 1 & 0 & 0 & 1 \end{array}\right] \xrightarrow{-R_2} \left[\begin{array}{ccc|ccc} 1 & 0 & 1 & 1 & 0 & 0 \\ 0 & 1 & 1 & 1 & -1 & 0 \\ 0 & 2 & 1 & 0 & 0 & 1 \end{array}\right]$$

$$\left[\begin{array}{ccc|ccc} 1 & 0 & 1 & 1 & 0 & 0 \\ 0 & 1 & 1 & 1 & -1 & 0 \\ 0 & 2 & 1 & 0 & 0 & 1 \end{array}\right] \xrightarrow{-2R_2 + R_3} \left[\begin{array}{ccc|ccc} 1 & 0 & 1 & 1 & 0 & 0 \\ 0 & 1 & 1 & 1 & -1 & 0 \\ 0 & 0 & -1 & -2 & 2 & 1 \end{array}\right]$$

$$\left[\begin{array}{ccc|ccc} 1 & 0 & 1 & 1 & 0 & 0 \\ 0 & 1 & 1 & 1 & -1 & 0 \\ 0 & 0 & -1 & -2 & 2 & 1 \end{array}\right] \xrightarrow{-R_3} \left[\begin{array}{ccc|ccc} 1 & 0 & 1 & 1 & 0 & 0 \\ 0 & 1 & 1 & 1 & -1 & 0 \\ 0 & 0 & 1 & 2 & -2 & -1 \end{array}\right]$$

$$\left[\begin{array}{ccc|ccc} 1 & 0 & 1 & 1 & 0 & 0 \\ 0 & 1 & 1 & 1 & -1 & 0 \\ 0 & 0 & 1 & 2 & -2 & -1 \end{array}\right] \xrightarrow{-R_3 + R_1} \left[\begin{array}{ccc|ccc} 1 & 0 & 0 & -1 & 2 & 1 \\ 0 & 1 & 1 & 1 & -1 & 0 \\ 0 & 0 & 1 & 2 & -2 & -1 \end{array}\right]$$

$$\left[\begin{array}{ccc|ccc} 1 & 0 & 0 & -1 & 2 & 1 \\ 0 & 1 & 1 & 1 & -1 & 0 \\ 0 & 0 & 1 & 2 & -2 & -1 \end{array}\right] \xrightarrow{-R_3 + R_2} \left[\begin{array}{ccc|ccc} 1 & 0 & 0 & -1 & 2 & 1 \\ 0 & 1 & 0 & -1 & 1 & 1 \\ 0 & 0 & 1 & 2 & -2 & -1 \end{array}\right]$$

Thus

$$A^{-1} = \begin{bmatrix} -1 & 2 & 1 \\ -1 & 1 & 1 \\ 2 & -2 & -1 \end{bmatrix}$$

We can check this result by multiplying A^{-1} by the original matrix A; if the product is I, then our result is correct.

$$A^{-1}A = \begin{bmatrix} -1 & 2 & 1 \\ -1 & 1 & 1 \\ 2 & -2 & -1 \end{bmatrix}\begin{bmatrix} 1 & 0 & 1 \\ 1 & -1 & 0 \\ 0 & 2 & 1 \end{bmatrix}$$

$$= \begin{bmatrix} (-1)1 + 2\cdot 1 + 1\cdot 0 & (-1)0 + 2(-1) + 1\cdot 2 & (-1)1 + 2\cdot 0 + 1\cdot 1 \\ (-1)1 + 1\cdot 1 + 1\cdot 0 & (-1)0 + 1(-1) + 1\cdot 2 & (-1)1 + 1\cdot 0 + 1\cdot 1 \\ 2\cdot 1 + (-2)1 + (-1)0 & 2\cdot 0 + (-2)(-1) + (-1)2 & 2\cdot 1 + (-2)0 + (-1)1 \end{bmatrix}$$

$$= \begin{bmatrix} 1 & 0 & 0 \\ 0 & 1 & 0 \\ 0 & 0 & 1 \end{bmatrix}$$

b. We begin by forming the augmented matrix

$$\left[\begin{array}{cc|cc} 1 & 0 & 1 & 0 \\ 0 & 0 & 0 & 1 \end{array}\right]$$

Since this matrix is already in reduced row-echelon form, and the first two columns do *not* constitute the identity matrix, B is noninvertible.

Calculator Keys

Inverse Matrices

If a matrix with numerical entries is invertible, its inverse can be computed with most graphing calculators. As with other matrix operations, you must first specify the order of the matrix and then key in its entries. Once this is done, the inverse can be computed using the $\boxed{x^{-1}}$ key.

Solving Linear Systems Using Inverses

Although inverse matrices arise in many contexts, the most important application of the inverse for our purposes is as a tool for solving linear systems of equations. Consider the following system:

$$2x + 3y + z = 6$$
$$4x - y + z = -3$$
$$x + y + \frac{1}{2}z = 1$$

In matrix form we have

$$\begin{bmatrix} 2 & 3 & 1 \\ 4 & -1 & 1 \\ 1 & 1 & \frac{1}{2} \end{bmatrix}\begin{bmatrix} x \\ y \\ z \end{bmatrix} = \begin{bmatrix} 6 \\ -3 \\ 1 \end{bmatrix}$$

If we define

$$A = \begin{bmatrix} 2 & 3 & 1 \\ 4 & -1 & 1 \\ 1 & 1 & \frac{1}{2} \end{bmatrix}, \quad X = \begin{bmatrix} x \\ y \\ z \end{bmatrix}, \quad \text{and } B = \begin{bmatrix} 6 \\ -3 \\ 1 \end{bmatrix}$$

then we have the matrix equation $AX = B$. Now, if A is invertible, we can solve this equation for the matrix X as follows:

$AX = B$	The original matrix equation
$A^{-1}(AX) = A^{-1}B$	Multiplying both sides **on the left** by A^{-1}
$(A^{-1}A)X = A^{-1}B$	The associative law of matrix multiplication
$IX = A^{-1}B$	The definition of the multiplicative inverse
$X = A^{-1}B$	Using a property of the identity matrix. Note that the product on the right-hand side is defined since A^{-1} is a 3×3 matrix and B is a 3×1 matrix.

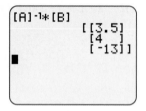

Figure 16

This technique is ideal for use with a graphing calculator because the solution involves only the operations of matrix inversion and multiplication, both of which are easily performed on a graphing calculator. After entering the matrices A and B into the calculator and computing $A^{-1}B$, the calculator returns the answer shown in Figure 16. Thus, the solution to the original system is $x = 3.5$, $y = 4$, and $z = -13$.

WARNING!

> The inverse matrix technique can be applied only to systems with an invertible coefficient matrix. If the coefficient matrix is not invertible, then some other technique, such as Gaussian or Gauss-Jordan elimination, must be employed to describe the solution set.

·····**EXAMPLE 4**   **Finding the Equation of a Parabola Passing Through Three Given Points**

The graph of a quadratic function f includes the points $(-2, 25)$, $(1, 4)$, and $(3, 20)$. Find an expression for $f(x)$.

Solution   Since f is a quadratic function, we know that $f(x) = ax^2 + bx + c$ for some choice of constants a, b, and c. Since $(-2, 25)$ is a point on the graph of f, we know that $f(-2) = 25$; thus, we have

$$f(-2) = a(-2)^2 + b(-2) + c = 25$$
$$4a - 2b + c = 25$$

Similarly, since $(1, 4)$ and $(3, 20)$ are points on the graph of f, we obtain the equations

$$f(1) = a + b + c = 4$$
$$f(3) = 9a + 3b + c = 20$$

Thus, we must solve the following system of equations:

$$4a - 2b + c = 25$$
$$a + b + c = 4$$
$$9a + 3b + c = 20$$

In matrix form, we have

$$\begin{bmatrix} 4 & -2 & 1 \\ 1 & 1 & 1 \\ 9 & 3 & 1 \end{bmatrix} \begin{bmatrix} a \\ b \\ c \end{bmatrix} = \begin{bmatrix} 25 \\ 4 \\ 20 \end{bmatrix}$$

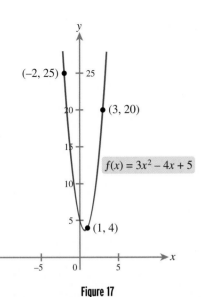

$f(x) = 3x^2 - 4x + 5$

Figure 17

Solving for the unknown matrix we have

$$\begin{bmatrix} a \\ b \\ c \end{bmatrix} = \begin{bmatrix} 4 & -2 & 1 \\ 1 & 1 & 1 \\ 9 & 3 & 1 \end{bmatrix}^{-1} \begin{bmatrix} 25 \\ 4 \\ 20 \end{bmatrix}$$

Using a graphing calculator to evaluate the expression on the right-hand side, we obtain

$$\begin{bmatrix} a \\ b \\ c \end{bmatrix} = \begin{bmatrix} 3 \\ -4 \\ 5 \end{bmatrix}$$

Thus, $a = 3$, $b = -4$, and $c = 5$. It follows that the quadratic function f is defined by

$$f(x) = 3x^2 - 4x + 5$$

The graph of f is shown in Figure 17.

Understanding and Mastery Checklists

Concepts to Understand	Skills to Master
Square matrix	Use matrix multiplication to determine if two matrices are inverses of each other.
Main diagonal of a square matrix	Use Gauss-Jordan elimination to find the inverse of a matrix or to show that it is not invertible.
Identity matrix	Use a graphing calculator to find the inverse of a matrix.
Inverse of a square matrix	Solve a linear system of equations using the inverse of the coefficient matrix.
Matrix product form of a linear system of equations	

Exercises 9.5

Exercises 1-4   *Use matrix multiplication to determine if A and B are inverses of one another.*

1. $A = \begin{bmatrix} 1 & -1 \\ 2 & -3 \end{bmatrix}$, $B = \begin{bmatrix} 3 & -1 \\ 2 & -1 \end{bmatrix}$

2. $A = \begin{bmatrix} 1 & 2 \\ -2 & 3 \end{bmatrix}$, $B = \begin{bmatrix} \frac{3}{7} & 0 \\ \frac{2}{7} & 0 \end{bmatrix}$

3. $A = \begin{bmatrix} 1 & 0 & -1 \\ 0 & 2 & 1 \\ 1 & 1 & 0 \end{bmatrix}$, $B = \begin{bmatrix} -1 & -1 & 2 \\ 1 & 1 & -1 \\ -2 & -1 & 3 \end{bmatrix}$

4. $A = \begin{bmatrix} 1 & 2 & 0 \\ 0 & -2 & -1 \\ 1 & -1 & -1 \end{bmatrix}$, $B = \begin{bmatrix} -1 & -2 & 2 \\ 1 & 1 & -1 \\ -2 & -3 & 2 \end{bmatrix}$

Exercises 5-26 *Find the inverse of the given matrix, if it exists. Use a graphing calculator as needed.*

5. $\begin{bmatrix} 1 & 0 \\ 2 & 1 \end{bmatrix}$

6. $\begin{bmatrix} \frac{1}{2} & \frac{1}{3} \\ 2 & 0 \end{bmatrix}$

7. $\begin{bmatrix} 0.2 & 0.4 \\ 0.5 & 0.5 \end{bmatrix}$

8. $\begin{bmatrix} 0.125 & 0.578 \\ 3.24 & 1.67 \end{bmatrix}$

9. $\begin{bmatrix} 1 & 2 \\ 2 & 4 \end{bmatrix}$

10. $\begin{bmatrix} a & 0 \\ 0 & b \end{bmatrix} (a \neq 0, b \neq 0)$

11. $\begin{bmatrix} 0 & a \\ b & 0 \end{bmatrix} (a \neq 0, b \neq 0)$

12. $\begin{bmatrix} x & y \\ 4x & 4y \end{bmatrix}$

13. $\begin{bmatrix} x & x+1 \\ x & x-1 \end{bmatrix} (x \neq 0)$

14. $\begin{bmatrix} 1 & x \\ x^2 & x^3 \end{bmatrix}$

15. $\begin{bmatrix} -\frac{1}{5} & 0 & 1 \\ \frac{2}{5} & -1 & -1 \\ -\frac{1}{5} & 1 & 0 \end{bmatrix}$

16. $\begin{bmatrix} 1 & 2 & 0 \\ 0 & 1 & 4 \\ 0 & 0 & 1 \end{bmatrix}$

17. $\begin{bmatrix} 1 & 4 & -2 \\ 0 & 2 & -1 \\ -3 & 0 & 2 \end{bmatrix}$

18. $\begin{bmatrix} 1 & 2 & 3 \\ 4 & 5 & 6 \\ 7 & 8 & 9 \end{bmatrix}$

19. $\begin{bmatrix} -\frac{1}{12} & \frac{7}{12} & -\frac{1}{3} \\ -\frac{1}{12} & -\frac{5}{12} & \frac{2}{3} \\ \frac{5}{12} & \frac{1}{12} & -\frac{1}{3} \end{bmatrix}$

20. $\begin{bmatrix} 1.2 & -2.3 & 4.7 \\ 0 & 1.9 & -8.1 \\ 3.1 & 0 & 0 \end{bmatrix}$

21. $\begin{bmatrix} 1 & x & 0 \\ 0 & 1 & x \\ 0 & 0 & 1 \end{bmatrix}$

22. $\begin{bmatrix} 0 & x & x \\ x & 0 & x \\ x & x & 0 \end{bmatrix}$

23. $\begin{bmatrix} 1 & 0 & 0 & 1 \\ 0 & 1 & 0 & 1 \\ 0 & 0 & 1 & 1 \\ 1 & 1 & 1 & 3 \end{bmatrix}$

24. $\begin{bmatrix} 1 & 2 & 3 & 4 \\ 2 & 3 & 4 & 1 \\ 3 & 4 & 1 & 2 \\ 4 & 1 & 2 & 3 \end{bmatrix}$

25. $\begin{bmatrix} a & 0 & 0 & 0 & 0 \\ -a & a & 0 & 0 & 0 \\ 0 & -a & a & 0 & 0 \\ 0 & 0 & -a & a & 0 \\ 0 & 0 & 0 & -a & a \end{bmatrix}$

26. $\begin{bmatrix} a & a & a & a & a \\ 0 & a & a & a & a \\ 0 & 0 & a & a & a \\ 0 & 0 & 0 & a & a \\ 0 & 0 & 0 & 0 & a \end{bmatrix}$

Exercises 27-32 *Three systems of equations are given. Solve the systems by first expressing them in matrix form as $AX = B$ and then evaluating $X = A^{-1}B$. Note that the coefficient matrix A is the same for each of parts a–c.*

27. a. $\begin{aligned} x - y &= -1 \\ x + 2y &= 12 \end{aligned}$

    b. $\begin{aligned} x - y &= -4 \\ x + 2y &= -1 \end{aligned}$

    c. $\begin{aligned} x - y &= 1 \\ x + 2y &= -\frac{1}{2} \end{aligned}$

28. a. $\begin{aligned} -2x + 6y &= 6 \\ 3x + 5y &= 5 \end{aligned}$

    b. $\begin{aligned} -2x + 6y &= 8.4 \\ 3x + 5y &= 8.4 \end{aligned}$

    c. $\begin{aligned} -2x + 6y &= -2.5 \\ 3x + 5y &= 7.25 \end{aligned}$

29. a. $\begin{aligned} 1.5x - 4y &= -12.5 \\ 2.3x + 6y &= 41.5 \end{aligned}$

    b. $\begin{aligned} 1.5x - 4y &= 2 \\ 2.3x + 6y &= 15.2 \end{aligned}$

    c. $\begin{aligned} 1.5x - 4y &= -15 \\ 2.3x + 6y &= 13.4 \end{aligned}$

30. a. $\begin{aligned} \frac{2}{5}x + \frac{3}{5}y &= 8 \\ x - y &= -5 \end{aligned}$

    b. $\begin{aligned} \frac{2}{5}x + \frac{3}{5}y &= 2 \\ x - y &= -5 \end{aligned}$

    c. $\begin{aligned} \frac{2}{5}x + \frac{3}{5}y &= 1.4 \\ x - y &= -1.5 \end{aligned}$

31. a. $\begin{aligned} x + 2y - 4z &= -8 \\ 3x - y + z &= 7 \\ 2x + 2y - z &= 6 \end{aligned}$

    b. $\begin{aligned} x + 2y - 4z &= -6 \\ 3x - y + z &= -5 \\ 2x + 2y - z &= -5 \end{aligned}$

    c. $\begin{aligned} x + 2y - 4z &= -5 \\ 3x - y + z &= 15 \\ 2x + 2y - z &= 15 \end{aligned}$

32. a. $\begin{aligned} x + y + z &= 2 \\ x + 2y - z &= 1 \\ y + z &= -1 \end{aligned}$

    b. $\begin{aligned} x + y + z &= 6 \\ x + 2y - z &= 2 \\ y + z &= 5 \end{aligned}$

    c. $\begin{aligned} x + y + z &= \frac{7}{20} \\ x + 2y - z &= \frac{4}{5} \\ y + z &= \frac{3}{20} \end{aligned}$

Applications

33. Running Regimen To safely increase aerobic capacity and burn fat, a man has developed a routine that consists of walking at 20 minutes per mile, jogging at 11 minutes per mile, and running at 8 minutes per mile. Given that walking burns 5.8 calories per minute, jogging burns 9.1 calories per minute, and running burns 14.1 calories per minute, how many minutes should he spend at each pace if he wants to

a. burn 1200 calories, work out for 2 hours, and travel 11 miles?

b. burn 750 calories, work out for 1.5 hours, and travel 7 miles?

34. Investment Options An investment company offers three different funds, each of which invests in highly volatile growth stocks, income-generating bonds, and conservative but stable blue chip stocks, but in differing proportions, as shown in Table 9.

Table 9

Fund content	Growth fund	Income fund	Security fund
Growth stocks	60%	20%	30%
Bonds	20%	50%	20%
Blue chips	20%	30%	50%

Determine the amount that should be invested in each of the three funds for an individual investor who wants to invest

a. $2000 in growth stocks, $1000 in bonds, and $1000 in blue chips.

b. $1500 in growth stocks, $1500 in bonds, and $1500 in blue chips.

35. Fitting a Cubic Find the equation of the cubic function $f(x) = ax^3 + bx^2 + cx + d$ passing through the points (1, 5), (3, 43), (5, 201), and (7, 575).

36. Unknown Interest Rates Suppose that a bank has three different money market accounts (growth, domestic, and international) paying annually compounded interest at the rates i, j, and k, respectively. Alberto, Brenda, and Carlos each invested $1000 in money market accounts at this bank, as indicated by Table 10.

Table 10

Investor	Number of years in growth fund	Number of years in domestic fund	Number of years in international fund
Alberto	4	1	0
Brenda	0	1	4
Carlos	2	2	1

a. Find an expression involving i, j, and k for each person's balance at the end of 5 years. [*Hint:* If, for example, a principal P is deposited into an account paying interest at a rate l for 4 years and then is switched to an account paying m for 3 years, at the end of 7 years the balance will be $P(1 + l)^4(1 + m)^3$.]

b. At the end of the 5 years, Alberto had $1240.05, Brenda had $1442.11, and Carlos had $1312.50. Set up a system of three equations in the three variables i, j, and k.

c. Take natural logarithms of both sides of each equation in part b. Let $I = \ln(1 + i)$, $J = \ln(1 + j)$, and $K = \ln(1 + k)$. Use properties of logarithms to write the resulting system as a *linear* system of three equations in the variables I, J, and K.

d. Use the techniques of this section to solve the linear system from part c for the variables I, J, and K.

e. Use the results of part d and the definitions of I, J, and K given in part c to find i, j, and k.

Concepts and Critical Thinking

Exercises 37-40 *Answer true or false.*

37. Only square matrices can have inverses.

38. All invertible matrices can be row-reduced to the identity matrix.

39. Every square matrix has an inverse.

40. If A is an invertible matrix, then the solution to the matrix equation $AX = B$ is given by $X = BA^{-1}$.

Exercises 41-44 *Give an example of each.*

41. A noninvertible 1×1 matrix

42. A noninvertible 2×2 matrix with 1s on the main diagonal

43. A system of equations that cannot be solved using the technique of this section

44. A scalar multiple of the 3 $\times$ 3 identity matrix and its inverse

45. Let A and B be invertible matrices. Define $C = AB$ and $D = B^{-1}A^{-1}$. Show that C and D are inverses of one another. (*Hint:* Use the definitions of C and D to show that $CD = I$ and $DC = I$.)

46. Show that a matrix with either a row or a column of all zeros is not invertible.

47. Show that if A is an $n \times n$ matrix, then $I_n A = A$.

48. Show that $(aI)^{-1} = \frac{1}{a}I$ for any scalar a.

49. Let

$$A = \begin{bmatrix} a & 0 & 0 & 0 \\ 0 & b & 0 & 0 \\ 0 & 0 & c & 0 \\ 0 & 0 & 0 & d \end{bmatrix}$$

Under what conditions is A invertible? If A is invertible, what is its inverse?

50. Let A be an invertible matrix such that $A^2 = A^4$. Show that $A = A^{-1}$.

51. The **trace** of a square matrix is defined to be the sum of the elements on the main diagonal. Compute the trace for each of the following matrices:

a. $\begin{bmatrix} 2 & -1 \\ 3 & -4 \end{bmatrix}$

b. $\begin{bmatrix} 1 & 2 & 4 \\ 6 & 9 & 8 \\ 2 & 4 & 8 \end{bmatrix}$

c. aI_n, where a is a scalar

52. To find the coordinates of the point (x, y) after a counterclockwise rotation of 45°, we multiply the matrix

$$\begin{bmatrix} x \\ y \end{bmatrix}$$

by

$$\begin{bmatrix} \frac{\sqrt{2}}{2} & -\frac{\sqrt{2}}{2} \\ \frac{\sqrt{2}}{2} & \frac{\sqrt{2}}{2} \end{bmatrix}$$

By what would we multiply

$$\begin{bmatrix} x \\ y \end{bmatrix}$$

in order to find the coordinates of the point (x, y) after undergoing a *clockwise* rotation of 45°?

53. What is I^{-1}?

54. Consider the following system of equations:

$$x^2 y^3 z^4 = 10$$
$$x^3 y^2 z^5 = 9$$
$$x^4 y^6 z^6 = 5$$

a. Is the system linear?

b. Take the natural logarithm of both sides of each equation. The resulting system is linear in the variables $\ln x$, $\ln y$, and $\ln z$. Solve the system for these variables.

c. Find x, y, and z.

Questions for Discussion or Essay

55. Up to this point, we have considered several techniques for solving linear systems of equations. Describe each technique, and discuss the circumstances under which each is preferable. Consider such issues as the size of the system, the number of nonzero coefficients, the use of a calculator, and whether other systems are to be solved that have the same coefficient matrix.

56. Explain why a nonsquare matrix could not have an inverse.

57. If A is noninvertible but we attempt to compute A^{-1} by using Gauss-Jordan elimination as outlined in the text, how will we discover that A^{-1} does not exist?

58. What similarities do you find between the inverse of a matrix and the inverse of a function? A function fails to have an inverse whenever it is not 1–1. What would it mean for a matrix to be 1–1? If f and g have inverses, then it can be shown that $(f \circ g)^{-1} = g^{-1} \circ f^{-1}$. What would be the corresponding rule for matrices?

Projects for Enrichment

59. **Dimensions** Physical laws express relationships among quantities such as mass, length, time, velocity, acceleration, force, energy, work, and many others. Each of these quantities is assigned a unit of measurement, and the physical laws relating them are valid only if the units are consistent. When we are not concerned with the particular system of measurement—SI (metric) or U.S. Customary—we can instead consider *dimensions*. To each of the three basic quantities—mass, length, and time—we associate the dimensions M, L, and T, respectively. Other quantities have dimensions that are products involving the three basic dimensions. For example, since velocity is equal to distance (dimension L) divided by time (dimension T), the dimension of velocity is $\frac{L}{T}$ or LT^{-1}.

a. Find the dimension of the following quantities. Express your answers in the form $M^a L^b T^c$. Assume that all constants have no dimension.

 i. Acceleration (= velocity ÷ time)

 ii. Force (= mass × acceleration)

 iii. Kinetic energy ($= \frac{1}{2} \times$ mass × velocity^2)

 iv. Work (= force × distance)

Equations that express physical laws must be *dimensionally compatible* in the sense that all terms must have the same dimension. Consider the equation $s = -\frac{1}{2}gt^2 + v_0 t + s_0$, which gives the height of an object at time t if it is thrown into the air with initial velocity v_0 and from an initial height s_0. Both s and s_0 have dimension L. Since g is the acceleration due to gravity and $-\frac{1}{2}$ is a dimensionless constant, the term $-\frac{1}{2}gt^2$ has dimension $(LT^{-2})T^2 = L$. Likewise, $v_0 t$ has dimension $(LT^{-1})T = L$. So each term has dimension L and the equation is dimensionally compatible.

b. Determine whether the following equations are dimensionally compatible. Assume m denotes mass; s, distance; t, time; v, velocity; a, acceleration; F, force; and W, work.

 i. $F = mv + v^2$

 ii. $v^2 = v_0^2 + 2as$

 iii. $E = \frac{1}{2}mv^2 + mgs$ (E denotes total energy, which has dimension $ML^2 T^{-2}$)

 iv. $W = msv + Fs$

Some physical laws can be derived from a basic form and dimensional considerations. Let us consider the velocity of a falling object.

It is reasonable to expect that velocity depends on the acceleration due to gravity and the total time the object has been falling. If we assume (perhaps on the basis of observations) that velocity varies directly as some power of the acceleration due to gravity and some power of time, we obtain the possible model $v = kg^a t^b$, where v is velocity, g is the acceleration due to gravity, t is time, and k, a, and b are constants. Now the dimensions of v, g, and t are LT^{-1}, LT^{-2}, and T, respectively, and the constant k is dimensionless. So the equation $v = kg^a t^b$ is dimensionally compatible only if

$$LT^{-1} = (LT^{-2})^a T^b$$

which implies

$$LT^{-1} = L^a T^{-2a+b}$$

Equating exponents, we obtain the system of equations

$$a = 1$$
$$-2a + b = -1$$

Solving the system, we find $a = 1$ and $b = 1$. Thus $v = kgt$ is dimensionally compatible.

c. Use the same process to find models of the given form.

 i. $T = kr^a g^b$, where T is the period of a pendulum with length r, k is a constant, and g is the acceleration due to gravity.

 ii. $a = kv^i r^j$, where a is the acceleration of a particle traveling with velocity v in a circular path of radius r, and k is a constant.

 iii. $F = km^a v^b r^c$, where F is the centrifugal force of a particle with mass m traveling in a circular path with radius r and velocity v, and k is a constant.

 iv. $F = k\mu^a v^b r^c$, where F is the force opposing a ball of radius r as it is falling with velocity v through air having viscosity coefficient μ (dimension $ML^{-1}T^{-1}$), and k is a constant.

 v. $V = kG^a m^b R^c$, where V is the velocity required for an object to escape the gravitational field of a planet with mass m and radius R, G is the universal gravitational constant with dimension $M^{-1}L^3 T^{-2}$, and k is a constant.

60. **Drawing Letters with a Graphing Calculator** In this project, we learn how matrix techniques can be used to draw letters on the display of a graphing calculator. In particular, we will illustrate how the letter M can be drawn on a graphing calculator. To do so, we use parametric equations. Recall from Section 8.6 that instead of defining y as a function of x, we define both x and y as functions of the parameter t. For example, suppose we define

$$x = t^2 \quad \text{and} \quad y = t^3$$

In other words, to each real number t there corresponds the pair of numbers (x, y) of the form (t^2, t^3). Thus, if $t = -2$, the corresponding pair of numbers consists of $x = (-2)^2 = 4$ and $y = (-2)^3 = -8$, or the pair $(4, -8)$. Some additional values of this correspondence are shown in Table 11. If we plot the points given in the second column of Table 11 and connect the points with a smooth curve, we obtain the graph shown in Figure 18.

Table 11

t	(t^2, t^3)
-2	$(4, -8)$
-1	$(1, -1)$
0	$(0, 0)$
1	$(1, 1)$
2	$(4, 8)$

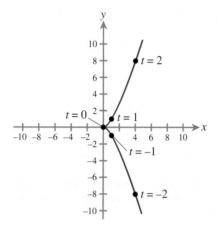

Figure 18

We now find functions $x(t)$ and $y(t)$ such that the graph of $(x(t), y(t))$ for t in an appropriate interval will resemble the letter M. Shown in Figure 19 is a depiction of the letter M superimposed on the coordinate axes.

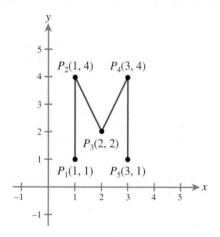

Figure 19

Let us suppose that a bug begins crawling at time 0 from point P_1 to point P_2, arriving at point P_2 at time $t = 1$; it continues to point P_3, arriving at time $t = 2$; and so on, finally arriving at point P_5 at time $t = 4$. Table 12 gives the x- and y-coordinates of the bug at times t from 0 to 4.

Table 12

t	x	y
0	1	1
1	1	4
2	2	2
3	3	4
4	3	1

a. With x as the vertical axis and t as the horizontal axis, plot the tabulated values of x versus t. Connect the points with line segments.

b. It can be shown that the figure obtained in part a is the graph of a function of the form

$$x(t) = a|t| + b|t - 1| + c|t - 2| + d|t - 3| + e|t - 4|$$

Find the constants a, b, c, d, and e by solving the system of equations arising from the fact that the graph must pass through the five indicated points.

c. Now, using the same process as that outlined in parts a and b, find constants f, g, h, i, and j so that

$$y(t) = f|t| + g|t - 1| + h|t - 2| + i|t - 3| + j|t - 4|$$

and the graph of $y(t)$ passes through the points (t, y) given in Table 12.

d. Using the parametric function capabilities of your graphing calculator, graph $(x(t), y(t))$ for t from 0 to 4.

e. Repeat the procedure outlined in steps a through d to draw the letter R.

Section 9.6   Determinants and Cramer's Rule

- How can you quickly determine whether a linear system having the same number of equations as unknowns has a unique solution?
- How can you quickly compute the value of a single variable in a system of equations without solving the entire system?
- How can matrices be used to determine whether three given points lie on a line?
- How can the area of a quadrilateral be computed from the coordinates of its vertices?

Determinants of 2 × 2 Matrices

As mentioned in the previous section, not all square matrices have inverses. We now describe a test for determining *which* matrices have inverses. We begin by considering a matrix that we happen to know is noninvertible—namely,

$$A = \begin{bmatrix} 1 & 3 \\ 4 & 12 \end{bmatrix}$$

Let's see if we can gain insight into the "cause" of invertibility by naively attempting to compute A^{-1}. Our first step gives us

$$\begin{bmatrix} 1 & 3 & | & 1 & 0 \\ 4 & 12 & | & 0 & 1 \end{bmatrix} \xrightarrow{-4R_1 + R_2} \begin{bmatrix} 1 & 3 & | & 1 & 0 \\ 0 & 0 & | & -4 & 1 \end{bmatrix}$$

Already we are in trouble; there is no longer any possibility that the first two columns of this augmented matrix will form the identity matrix. This problem occurs because the second row of A is 4 times the first row of A, so the operation $-4R_1 + R_2$ "wipes out" the second row. In fact, this problem arises whenever one row is a multiple of another. We will show in Exercise 61 that one row of the matrix

$$\begin{bmatrix} a & b \\ c & d \end{bmatrix}$$

is a multiple of the other if and only if $ad - bc = 0$. Thus, the 2 × 2 matrix

$$\begin{bmatrix} a & b \\ c & d \end{bmatrix}$$

is *not* invertible whenever $ad - bc = 0$. Moreover, it can be shown that the matrix *is* invertible whenever $ad - bc \neq 0$. Since the value of $ad - bc$ *determines* the invertibility of A, this quantity is called the **determinant** of A and is denoted by $\det(A)$ or $|A|$. We summarize determinants of 2 × 2 matrices as follows.

Determinants of 2 × 2 Matrices

- The **determinant** of the matrix

$$A = \begin{bmatrix} a & b \\ c & d \end{bmatrix}$$

is the quantity $ad - bc$, which is denoted by $\det(A)$, $|A|$, or $\begin{vmatrix} a & b \\ c & d \end{vmatrix}$.

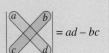

Figure 20

- The determinant of a 2 × 2 matrix is the difference of the products of the diagonal entries, as suggested by Figure 20.
- A 2 × 2 matrix is invertible if and only if its determinant is nonzero.

In the following example, we see how the determinant can be used to determine whether a given matrix is invertible.

EXAMPLE 1   Using Determinants to Test for Invertibility

Use determinants to decide which of the following matrices are invertible:

a. $\begin{bmatrix} 4 & -2 \\ 3 & 5 \end{bmatrix}$ **b.** $\begin{bmatrix} 3 & 3 \\ 2 & 2 \end{bmatrix}$ **c.** $\begin{bmatrix} 1 & -1 \\ -1 & 0 \end{bmatrix}$

Solution

a. The determinant is given by

$$\begin{vmatrix} 4 & -2 \\ 3 & 5 \end{vmatrix} = 4 \cdot 5 - (-2)3 = 26 \neq 0$$

Thus, the matrix is invertible.

b. In this case, we have

$$\begin{vmatrix} 3 & 3 \\ 2 & 2 \end{vmatrix} = 3 \cdot 2 - 3 \cdot 2 = 0$$

and so the matrix is noninvertible.

c. The determinant is given by

$$\begin{vmatrix} 1 & -1 \\ -1 & 0 \end{vmatrix} = 1 \cdot 0 - (-1)(-1) = -1 \neq 0$$

Thus, the matrix is invertible.

Evaluating Determinants with Minors

We have seen that the determinant of a 2×2 matrix indicates whether the matrix is invertible. More generally, we define determinants of higher-order square matrices so that the matrix is invertible if and only if its determinant is nonzero.

We define the determinant of a 3×3 matrix as a combination of 2×2 determinants; determinants of 4×4 matrices are expressed as a combination of 3×3 determinants; and so on. We describe the process of computing 3×3 determinants in some detail, but we leave the computation of higher-order determinants to the exercises.

The **minor of an entry** of a 3×3 matrix is defined as the determinant of the 2×2 matrix remaining after both the row and column of the entry are crossed out. For example, if

$$A = \begin{bmatrix} a_{11} & a_{12} & a_{13} \\ a_{21} & a_{22} & a_{23} \\ a_{31} & a_{32} & a_{33} \end{bmatrix}$$

then the minor of a_{11} is the determinant of the 2×2 matrix

$$\begin{bmatrix} a_{22} & a_{23} \\ a_{32} & a_{33} \end{bmatrix}$$

remaining after both the first row and first column of A are crossed out. Table 13 shows how the minors of all of the entries of the first row are computed.

Table 13
Computing minors

The minor of	Cross out row 1 and column 1	... to obtain
a_{11}	$\begin{vmatrix} a_{11} & a_{12} & a_{13} \\ a_{21} & a_{22} & a_{23} \\ a_{31} & a_{32} & a_{33} \end{vmatrix}$	$\begin{vmatrix} a_{22} & a_{23} \\ a_{32} & a_{33} \end{vmatrix} = a_{22}a_{33} - a_{23}a_{32}$
The minor of	Cross out row 1 and column 2	... to obtain
a_{12}	$\begin{vmatrix} a_{11} & a_{12} & a_{13} \\ a_{21} & a_{22} & a_{23} \\ a_{31} & a_{32} & a_{33} \end{vmatrix}$	$\begin{vmatrix} a_{21} & a_{23} \\ a_{31} & a_{33} \end{vmatrix} = a_{21}a_{33} - a_{23}a_{31}$
The minor of	Cross out row 1 and column 3	... to obtain
a_{13}	$\begin{vmatrix} a_{11} & a_{12} & a_{13} \\ a_{21} & a_{22} & a_{23} \\ a_{31} & a_{32} & a_{33} \end{vmatrix}$	$\begin{vmatrix} a_{21} & a_{22} \\ a_{31} & a_{32} \end{vmatrix} = a_{21}a_{32} - a_{22}a_{31}$

EXAMPLE 2

Computing the Minor of an Entry

Find the minor of a_{23}, where

$$[a_{ij}] = \begin{bmatrix} -2 & 0 & 5 \\ 3 & 1 & -4 \\ 2 & -6 & -1 \end{bmatrix}$$

Solution The minor of a_{23} is the determinant of the 2×2 matrix that results from crossing out row 2 and column 3.

$$\begin{bmatrix} -2 & 0 & 5 \\ 3 & 1 & -4 \\ 2 & -6 & -1 \end{bmatrix}$$

Thus, the minor of a_{23} is given by

$$\begin{vmatrix} -2 & 0 \\ 2 & -6 \end{vmatrix} = (-2)(-6) - 2 \cdot 0 = 12$$

The determinant of a 3×3 matrix can be obtained by multiplying the entries of the first row by their corresponding minors and then either adding or subtracting, as described in the following box. Determinants of higher-order matrices are defined in a similar fashion.

Definition of 3 × 3 Determinant

The determinant of

$$\begin{bmatrix} a_{11} & a_{12} & a_{13} \\ a_{21} & a_{22} & a_{23} \\ a_{31} & a_{32} & a_{33} \end{bmatrix}$$

is given by

$$\begin{vmatrix} a_{11} & a_{12} & a_{13} \\ a_{21} & a_{22} & a_{23} \\ a_{31} & a_{32} & a_{33} \end{vmatrix} = a_{11}(\text{Minor of } a_{11}) - a_{12}(\text{Minor of } a_{12}) + a_{13}(\text{Minor of } a_{13})$$

$$= a_{11}\begin{vmatrix} a_{22} & a_{23} \\ a_{32} & a_{33} \end{vmatrix} - a_{12}\begin{vmatrix} a_{21} & a_{23} \\ a_{31} & a_{33} \end{vmatrix} + a_{13}\begin{vmatrix} a_{21} & a_{22} \\ a_{31} & a_{32} \end{vmatrix}$$

This technique of computing the determinant is called **expansion about the first row**.

EXAMPLE 3

Computing a 3 × 3 Determinant by Expansion About the First Row

Find the determinant of the matrix

$$\begin{bmatrix} 1 & 4 & 3 \\ 0 & 2 & 1 \\ -2 & 6 & 0 \end{bmatrix}$$

Solution   We multiply the entries of the first row by their corresponding minors and then add or subtract according to the definition.

$$\begin{vmatrix} 1 & 4 & 3 \\ 0 & 2 & 1 \\ -2 & 6 & 0 \end{vmatrix} = 1 \cdot \begin{vmatrix} 2 & 1 \\ 6 & 0 \end{vmatrix} - 4 \cdot \begin{vmatrix} 0 & 1 \\ -2 & 0 \end{vmatrix} + 3 \cdot \begin{vmatrix} 0 & 2 \\ -2 & 6 \end{vmatrix}$$

$$= 1(0 - 6) - 4[0 - (-2)] + 3[0 - (-4)]$$
$$= -6 - 8 + 12$$
$$= -2$$

Expansion about the first row is not the only technique for evaluating 3 × 3 determinants. In fact, expansion can be done about any row or column. To do so, we simply multiply the entries of the given row or column by their corresponding minors and then either add or subtract, depending on the position of the entry. Specifically, the sign of each term is given by the corresponding entry in the following **matrix of signs**:

$$\begin{bmatrix} + & - & + \\ - & + & - \\ + & - & + \end{bmatrix}$$

For example, to evaluate the determinant of the matrix

$$\begin{bmatrix} 1 & 4 & 3 \\ 0 & 2 & 1 \\ -2 & 6 & 0 \end{bmatrix}$$

from Example 3 by expanding about the second column, we would begin by forming the products of each of the entries of the second column with their corresponding minors, as shown.

$$4 \cdot \begin{vmatrix} 0 & 1 \\ -2 & 0 \end{vmatrix} \qquad 2 \cdot \begin{vmatrix} 1 & 3 \\ -2 & 0 \end{vmatrix} \qquad 6 \cdot \begin{vmatrix} 1 & 3 \\ 0 & 1 \end{vmatrix}$$

Next we affix the appropriate sign to each term using the matrix of signs, as suggested by Figure 21. Thus, we have

$$\begin{vmatrix} 1 & 4 & 3 \\ 0 & 2 & 1 \\ -2 & 6 & 0 \end{vmatrix} = -4 \cdot \begin{vmatrix} 0 & 1 \\ -2 & 0 \end{vmatrix} + 2 \cdot \begin{vmatrix} 1 & 3 \\ -2 & 0 \end{vmatrix} - 6 \cdot \begin{vmatrix} 1 & 3 \\ 0 & 1 \end{vmatrix}$$

$$= -4 \cdot 2 + 2 \cdot 6 - 6 \cdot 1$$
$$= -2$$

$$\begin{bmatrix} + & - & + \\ - & + & - \\ + & - & + \end{bmatrix} \quad \begin{vmatrix} 1 & 4 & 3 \\ 0 & 2 & 1 \\ -2 & 6 & 0 \end{vmatrix}$$

Figure 21

Rule of Thumb

Since we may evaluate the determinant by expanding about any row or column, choose a row or column containing as many zeros as possible.

EXAMPLE 4

Evaluating a 3 × 3 Determinant by Expansion

Compute

$$\begin{vmatrix} 2 & 1 & 3 \\ 1 & 0 & 0 \\ 4 & 5 & 6 \end{vmatrix}$$

Solution   Although both the second and third columns contain a zero, the second row contains two zeros. We thus expand about the second row. Because the second row of the matrix of signs is $- \ + \ -$, we have

$$\begin{vmatrix} 2 & 1 & 3 \\ 1 & 0 & 0 \\ 4 & 5 & 6 \end{vmatrix} = (-1) \cdot \begin{vmatrix} 1 & 3 \\ 5 & 6 \end{vmatrix} + 0 \cdot \begin{vmatrix} 2 & 3 \\ 4 & 6 \end{vmatrix} - 0 \cdot \begin{vmatrix} 2 & 1 \\ 4 & 5 \end{vmatrix}$$

$$= (-1)(-9) + 0 - 0$$

$$= 9$$

Note that by choosing the second row, we had only one 2×2 determinant to evaluate.

Calculator Keys

Computing Determinants

If a square matrix has numerical entries, its determinant can be computed with a graphing calculator. As with other matrix operations, you must first specify the order of the matrix and then key in its entries. The determinant can then be computed by selecting the determinant option from the menu of available matrix operations.

A Special Rule for Computing Determinants of 3 × 3 Matrices

The technique of expanding a determinant about a row or a column works for square matrices of any order. We now describe a shortcut for computing 3×3 determinants; unfortunately, the process does not generalize to determinants of any other order. To compute the determinant

$$\begin{vmatrix} a_1 & b_1 & c_1 \\ a_2 & b_2 & c_2 \\ a_3 & b_3 & c_3 \end{vmatrix}$$

we begin by forming fourth and fifth columns consisting of copies of columns 1 and 2, respectively, as shown in Figure 22. Next, following the arrows going downward, we form three products to which we affix a + sign and, following the arrows going upward, we form three products to which we affix a − sign. Finally, computing the sum of these products, we obtain the value of the determinant.

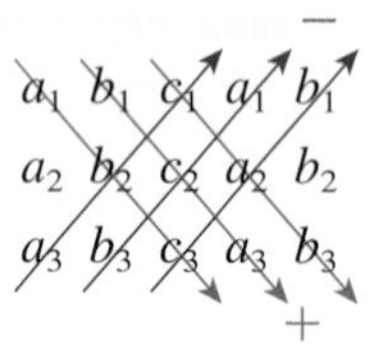

$$= a_1b_2c_3 + b_1c_2a_3 + c_1a_2b_3 - a_3b_2c_1 - b_3c_2a_1 - c_3a_2b_1$$

Figure 22

·····>EXAMPLE 5

Computing a 3 × 3 Determinant Using a Shortcut

Evaluate

$$\begin{vmatrix} 1 & -2 & 3 \\ 2 & 1 & -1 \\ 1 & 0 & 4 \end{vmatrix}$$

Solution We first form fourth and fifth columns consisting of copies of columns 1 and 2, respectively, and then apply the shortcut rule.

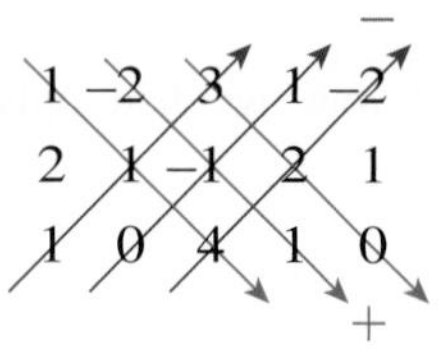

$$= 1 \cdot 1 \cdot 4 + (-2)(-1)1 + 3 \cdot 2 \cdot 0 - 1 \cdot 1 \cdot 3 - 0(-1)1 - 4 \cdot 2(-2)$$

$$= 4 + 2 + 0 - 3 - 0 + 16$$
$$= 19$$

Cramer's Rule for Solving Systems of Equations

In the preceding section, we learned how certain systems of linear equations could be solved using inverse matrices. Of course, this method works only if the system is "square" (having the same number of equations as variables) and the coefficient matrix is invertible. It so happens that if precisely these same conditions are met, the solution to such a system of equations can be expressed using determinants according to **Cramer's Rule**.

Cramer's Rule

Consider a system of n equations in the n variables $x_1, x_2, \ldots, x_n$, expressible in matrix form as $AX = B$, where A is an invertible matrix. Let A_i be the matrix obtained by replacing the ith column of A with the $n \times 1$ matrix B. Then the solution to the system is given by

$$x_1 = \frac{|A_1|}{|A|}, x_2 = \frac{|A_2|}{|A|}, \ldots, x_n = \frac{|A_n|}{|A|}$$

EXAMPLE 6

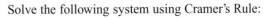

Solving a System with Cramer's Rule

Solve the following system using Cramer's Rule:

$$x - y + 2z = 3$$
$$x + y - 3z = -11$$
$$2x + 3y + z = 9$$

Solution This system can be expressed in matrix form as $AX = B$, with

$$A = \begin{bmatrix} 1 & -1 & 2 \\ 1 & 1 & -3 \\ 2 & 3 & 1 \end{bmatrix}, \quad X = \begin{bmatrix} x \\ y \\ z \end{bmatrix}, \quad \text{and } B = \begin{bmatrix} 3 \\ -11 \\ 9 \end{bmatrix}$$

Now the matrices A_1, A_2, and A_3 are each defined by replacing the appropriate column of A with the matrix B. Thus,

$$A_1 = \begin{bmatrix} 3 & -1 & 2 \\ -11 & 1 & -3 \\ 9 & 3 & 1 \end{bmatrix}, \quad A_2 = \begin{bmatrix} 1 & 3 & 2 \\ 1 & -11 & -3 \\ 2 & 9 & 1 \end{bmatrix}, \quad \text{and } A_3 = \begin{bmatrix} 1 & -1 & 3 \\ 1 & 1 & -11 \\ 2 & 3 & 9 \end{bmatrix}$$

According to Cramer's Rule, we have

$$x = \frac{|A_1|}{|A|}, \quad y = \frac{|A_2|}{|A|}, \quad \text{and } z = \frac{|A_3|}{|A|}$$

Evaluation of the determinants, either by hand or with the aid of a graphing calculator, gives

$$|A| = 19, \quad |A_1| = -38, \quad |A_2| = 57, \quad \text{and } |A_3| = 76$$

Thus

$$x = \frac{-38}{19} = -2, \quad y = \frac{57}{19} = 3, \quad \text{and } z = \frac{76}{19} = 4$$

It should be noted that Cramer's Rule is usually the least efficient of the methods we have seen for solving linear systems of equations. In Example 6, for example, it was necessary to compute four determinants, and this can be a tedious task, even with the aid of a graphing calculator—and even with matrices as small as 3×3. The inverse matrix technique discussed in Section 9.5 would have been much more efficient, requiring us simply to enter the two matrices A and B into our calculator and then compute $A^{-1}B$. When solving linear systems without the aid of a graphing calculator, Gaussian elimination with back-substitution would likely be most efficient.

Understanding and Mastery Checklists

Concepts to Understand

Determinant of a 2 × 2 matrix

❖

Minor of an entry of a matrix

❖

Determinant of a 3 × 3 matrix

❖

Matrix of signs

❖

Expansion about a row or column

❖

Cramer's Rule for solving linear systems of equations

Skills to Master

Compute a 2 × 2 determinant.

❖

Find the minor of a matrix entry.

❖

Compute a 3 × 3 determinant by expansion about a row or column.

❖

Compute a 3 × 3 determinant using the special rule.

❖

Compute a determinant using a graphing calculator.

❖

Use a determinant to determine whether a matrix is invertible.

❖

Solve a linear system of equations using Cramer's Rule.

Exercises 9.6

Exercises 1–22 *Evaluate the given determinant. (Try to select the most efficient technique.)*

1. $\begin{vmatrix} 2 & 3 \\ 4 & 5 \end{vmatrix}$

2. $\begin{vmatrix} 3 & 1 \\ -2 & 0 \end{vmatrix}$

3. $\begin{vmatrix} 1 & 6 \\ 2 & 12 \end{vmatrix}$

4. $\begin{vmatrix} 2 & 3 \\ -2 & 3 \end{vmatrix}$

5. $\begin{vmatrix} 3 & 7 \\ 4 & 1 \end{vmatrix}$

6. $\begin{vmatrix} -2 & 4 \\ -4 & 8 \end{vmatrix}$

7. $\begin{vmatrix} x & x-1 \\ x & x \end{vmatrix}$

8. $\begin{vmatrix} a & 2a \\ b & 2b \end{vmatrix}$

9. $\begin{vmatrix} 2 & 1 & 1 \\ 0 & 1 & 4 \\ 0 & 2 & 3 \end{vmatrix}$

10. $\begin{vmatrix} 1 & 0 & 4 \\ 3 & 5 & -1 \\ 2 & 0 & 6 \end{vmatrix}$

11. $\begin{vmatrix} -4 & 2 & 1 \\ -2 & 1 & 0 \\ 3 & -1 & 5 \end{vmatrix}$

12. $\begin{vmatrix} 3 & -1 & 6 \\ 2 & -4 & 1 \\ -1 & 7 & 0 \end{vmatrix}$

13. $\begin{vmatrix} 1.2 & 3.9 & 2.5 \\ 3.7 & 4.1 & 3.6 \\ 1.0 & 2.5 & 7.1 \end{vmatrix}$

14. $\begin{vmatrix} \frac{2}{3} & 4 & \frac{37}{55} \\ 1 & \frac{1}{2} & 0 \\ 0 & 0 & 3 \end{vmatrix}$

15. $\begin{vmatrix} a & b & c \\ 0 & 1 & d \\ 0 & 0 & 1 \end{vmatrix}$

16. $\begin{vmatrix} x-1 & 2 & 1 \\ 0 & x-2 & 2 \\ 0 & 2 & x-3 \end{vmatrix}$

17. $\begin{vmatrix} i & j & k \\ a_1 & a_2 & a_3 \\ b_1 & b_2 & b_3 \end{vmatrix}$

18. $\begin{vmatrix} a & a & a \\ 1 & 1 & 1 \\ x & y & z \end{vmatrix}$

19. $\begin{vmatrix} 1 & 2 & 3 & 4 \\ 2 & 2 & 5 & 7 \\ 8 & 7 & 0 & 2 \\ 8 & 3 & 9 & 0 \end{vmatrix}$

20. $\begin{vmatrix} 2 & 3 & 5 & 7 \\ 11 & 13 & 17 & 19 \\ 23 & 29 & 31 & 37 \\ 41 & 43 & 47 & 53 \end{vmatrix}$

21. $\begin{vmatrix} a & 1 & 1 & 1 & 1 \\ 0 & a & 1 & 1 & 1 \\ 0 & 0 & a & 1 & 1 \\ 0 & 0 & 0 & a & 1 \\ 0 & 0 & 0 & 0 & a \end{vmatrix}$

22. I_{100} (the 100×100 identity matrix)

Exercises 23-30   *Use determinants to test the given matrices for invertibility.*

23. $\begin{bmatrix} -3 & 5 \\ 1 & 0 \end{bmatrix}$ **24.** $\begin{bmatrix} 4 & -1 \\ -8 & 2 \end{bmatrix}$

25. $\begin{bmatrix} 2 & 5 \\ 4 & 10 \end{bmatrix}$ **26.** $\begin{bmatrix} 11 & 2 \\ 2 & 11 \end{bmatrix}$

27. $\begin{bmatrix} 2 & 0 & 5 \\ -1 & 3 & 0 \\ -1 & 9 & 5 \end{bmatrix}$ **28.** $\begin{bmatrix} 4 & -3 & 1 \\ -2 & 0 & 6 \\ 1 & 0 & 2 \end{bmatrix}$

29. $\begin{bmatrix} -12 & 1 & 14 \\ 1 & -16 & 2 \\ 15 & 0 & 0 \end{bmatrix}$ **30.** $\begin{bmatrix} 0 & 10 & -8 \\ 11 & -1 & 3 \\ 22 & 8 & -2 \end{bmatrix}$

Exercises 31-36   *Find the value(s) of x for which the given matrix is not invertible.*

31. $\begin{bmatrix} x & -1 \\ 1 & 2 \end{bmatrix}$ **32.** $\begin{bmatrix} 2 & x \\ 4 & -3 \end{bmatrix}$

33. $\begin{bmatrix} -3 & x \\ x & -3 \end{bmatrix}$ **34.** $\begin{bmatrix} x & 2 \\ 8 & x \end{bmatrix}$

35. $\begin{bmatrix} x & 0 & 1 \\ 1 & x & 0 \\ x & 1 & 1 \end{bmatrix}$ **36.** $\begin{bmatrix} 0 & x & 1 \\ x & 1 & 0 \\ 1 & 0 & x \end{bmatrix}$

Exercises 37-46   *Solve the system using Cramer's Rule.*

37. $x + 2y = 19$
$3x - 7y = -8$

38. $3x - 8y = -2$
$5x + 3y = 13$

39. $2x + z = 6$
$x - y = 4$
$-y + z = 7$

40. $x - y + 2z = 0$
$4x + y = 11$
$y - 3z = 5$

41. $x + 2y - z = -3$
$3x - 8y - 5z = -3$
$2x + 2y + z = 10$

42. $6x - 3y + 9z = -7.65$
$x - 2y + 11z = -4.15$
$3y - 4z = 9.5$

43. $x + 2y + 3z + 4w = 1$
$y - z + w = 0$
$2z + 3w = 0$
$z - w = 0$

44. $p + q + r + s + t = 1$
$q + r + s + t = 0$
$r + s + t = 1$
$s + t = 0$
$s = 1$

45. $ax + by = c$
$dx + ey = f$
(Solve for x and y. Assume $ae - bd \neq 0$.)

46. $a_1 x + a_2 y + a_3 z = 0$
$b_1 x + b_2 y + b_3 z = 0$
$c_1 x + c_2 y + c_3 z = 0$
(Assume there is a unique solution.)

Applications

Exercises 47-48   *Assume that production cost is a quadratic function of the number of units produced. Thus, if C denotes the cost and x the number of units produced, then $C = ax^2 + bx + c$ for certain values of a, b, and c. The value of c gives the fixed cost—the cost of production if no units are produced. Use the given data to set up a system of equations involving a, b, and c, and use Cramer's Rule to find the company's fixed costs.*

47.

x (units produced)	10	50	100
C (cost in dollars)	$8000	$5000	$10,000

48.

x (units produced)	150	200	300
C (cost in dollars)	$20,000	$25,000	$40,000

49. Hospital Breakfast   A hospital patient is served a breakfast consisting of milk, oatmeal, and English muffins. The total mass, caloric value, and fat content were recorded as 427 grams, 311 calories, and 4.4 grams, respectively. It is later discovered that the patient is lactose intolerant, and so it becomes necessary to determine how many grams of milk were consumed. Use the information given in Table 14, together with Cramer's Rule, to determine the number of grams of milk consumed.

Table 14

	Oatmeal	Milk	English muffin
Calories per gram	0.54	0.42	2.37
Grams of fat per gram	0.0097	0.011	0.01

50. Unknown Investment   A total of $10,000 is invested in three different funds that pay dividends and charge load fees according to Table 15.

Table 15

	First fund	Second fund	Third fund
Dividend	5%	10%	8%
Loan fee	1%	1.5%	0.5%

The total paid in dividends is $680, and the total in load fees is $110. Due to a bookkeeping error, the amount invested in the first fund is in question. Use Cramer's Rule to determine the amount.

Concepts and Critical Thinking

Exercises 51-54 *Answer true or false.*

51. The determinant of a matrix is itself a matrix.

52. Only square matrices with nonzero determinants are invertible.

53. If a is a real number, then the matrix $\begin{bmatrix} a & -1 \\ 1 & a \end{bmatrix}$ is invertible no matter what the value of a.

54. Cramer's Rule is a method for solving linear systems of equations by evaluating determinants.

Exercises 55-58 *Give an example of each.*

55. A 3×3 matrix such that the easiest method of computing the determinant is probably expansion about the second column

56. An application of determinants

57. A 3×3 matrix with determinant 125 (*Hint:* $5^3 = 125$.)

58. A matrix for which the determinant isn't defined

59. Show that if A is a 3×3 matrix such that the matrix equation $AX = 0$ has a nonzero solution X, then $\det(A) = 0$. [*Hint:* Suppose $\det(A) \neq 0$. Then A is invertible. Apply A^{-1} to both sides of the equation $AX = 0$ to find a contradiction.]

60. Let
$$A = \begin{bmatrix} a & b \\ c & d \end{bmatrix}$$
Compute both $\det(A)$ and $\det(kA)$, where k is a scalar. If A is a 3×3 matrix, what is your guess as to how $\det(kA)$ compares to $\det(A)$?

61. Complete the following steps to show that one row of the matrix
$$\begin{bmatrix} a & b \\ c & d \end{bmatrix}$$
is a multiple of the other if and only if $ad - bc = 0$.

a. If $c = ka$ and $d = kb$ (that is, row 2 is a multiple of row 1), show that $ad - bc = 0$.

b. If $ad - bc = 0$, show that $a/c = b/d$ and so $a = kc$ and $b = kd$ for some k. Thus, row 1 is a multiple of row 2.

62. It can be shown that $\det(AB) = \det(A)\det(B)$ for square matrices A and B. Use this fact to show that $\det(A^{-1}) = 1/\det(A)$.

Questions for Discussion or Essay

63. Since solving linear systems of equations with inverses or Cramer's Rule is relatively easy and can be performed by most graphing calculators, is there ever a reason to use Gauss-Jordan elimination? Explain. Are there any linear systems of equations that can be solved by Gauss-Jordan elimination but not by inverses or Cramer's Rule?

64. Suppose that 10 people are to split a jackpot and that the shares correspond to the solution of a certain system of 10 equations in 10 unknowns. If you were given the task of computing each person's share, what method would you use? If you were one of the 10 winners, what method would you use to compute *your* share? Explain your answers.

65. With Cramer's Rule, the solution values to a system of linear equations are expressed as quotients of determinants. Of course, Cramer's Rule will not provide us with a solution if a determinant appearing in the denominator is 0. If that should happen, would it make sense to attempt to solve the system using the inverse of the coefficient matrix?

66. What is the maximum total number of arithmetic operations (additions, subtractions, and multiplications) required to compute the determinant of a 3×3 matrix by expansion? What is the minimum? What about a 4×4 matrix?

Projects for Enrichment

67. Performing Elementary Row Operations by Matrix Multiplication In this project, we see that for 3×3 matrices, each of the elementary row operations (switching rows, adding a multiple of one row to another, and multiplying a row by a constant) can be represented as multiplication by an appropriate matrix.

a. Define
$$S_{12} = \begin{bmatrix} 0 & 1 & 0 \\ 1 & 0 & 0 \\ 0 & 0 & 1 \end{bmatrix}$$

Show that if B is *any* 3×3 matrix, then $S_{12}B$ is the matrix obtained by switching the first and second rows of B.

b. Find matrices S_{13} and S_{23} such that if B is an arbitrary 3×3 matrix, then $S_{13}B$ and $S_{23}B$ are the matrices obtained by switching the first and third rows of B and the second and third rows of B, respectively.

c. Find $M_1(x)$, $M_2(x)$, and $M_3(x)$ such that if B is an arbitrary 3×3 matrix, then $M_i(x)B$ is the matrix formed by multiplying row i of B by x.

d. Define $A_{13}(x)$ to be the matrix

$$\begin{bmatrix} 1 & 0 & 0 \\ 0 & 1 & 0 \\ x & 0 & 1 \end{bmatrix}$$

Show that if B is an arbitrary 3×3 matrix, then $A_{13}(x)B$ is the matrix obtained by adding x times row 1 of B to row 3 of B. Find expressions for $A_{12}(x)$ and $A_{23}(x)$.

e. Compute the determinants of all the matrices S_{ij}, $M_i(x)$, and $A_{ij}(x)$.

f. It is a well-known fact that for square matrices C and D, $\det(CD) = \det(C)\det(D)$. Use this fact and the results of part e to show that performing an elementary row operation on a matrix does not change its invertibility.

68. Determinants and Plane Geometry In this project, we explore several geometric applications of determinants. The area of a triangle in the coordinate plane with vertices (x_1, y_1), (x_2, y_2), and (x_3, y_3) can be shown to be the absolute value of the quantity

$$\frac{1}{2}\begin{vmatrix} x_1 & y_1 & 1 \\ x_2 & y_2 & 1 \\ x_3 & y_3 & 1 \end{vmatrix}$$

a. Find the area of the triangle having vertices with coordinates $(0, 0)$, $(1, 2)$, and $(3, 5)$.

b. Use the determinant formula for the area of a triangle to develop a formula for the area of quadrilateral $PQRS$ shown in Figure 23.

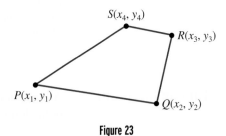

Figure 23

c. Use the formula obtained in part b to find the area of the quadrilateral with vertices $(2, 3)$, $(4, 7)$, $(1, 8)$, and $(0, 5)$.

d. If, for a given triple of points (x_1, y_1), (x_2, y_2), and (x_3, y_3), the area of the resulting triangle is 0, then what does this say about the three points?

e. Suppose that (x, y) is a point on the line determined by (x_1, y_1) and (x_2, y_2). Use the fact that the area of the triangle determined by (x, y), (x_1, y_1), and (x_2, y_2) is 0 to find a determinant form of the equation of the line.

Section 9.7 Systems of Inequalities

- How can an athlete select a fast-food snack and still stay within the guidelines of a rigid training diet?
- How is it possible to ensure that a low-fat, low-calorie diet will adequately satisfy important nutritional requirements?
- If the graph of $ax + by = c$ is a line, what does the graph of $ax + by \le c$ look like?
- What can a system of inequalities tell a farmer about land utilization, or an appliance store about inventory levels, or a retiring couple about investment options?

In the first six sections of this chapter, we developed a handful of techniques for solving systems of linear equations. We now turn our attention to a closely related topic: systems of linear inequalities. Notationally, a system of linear inequalities appears nearly identical to a system of linear equations—the only difference is that the equal sign of a linear system of equations is replaced by one of the four standard inequality symbols. But the solution sets of systems of linear inequalities are considerably more complex than their equation counterparts. For example, the solution set to a linear system of equations in two variables is either empty, a single point, or a line in the plane. By contrast, the solution set to a linear system of inequalities could be virtually any planar region whose edges are linear, including the inside of a triangle, the outside of a hexagon, quadrants, half-planes, and so on. In fact, one of the primary applications of systems of linear inequalities is to mathematically define a region in the plane. In applied settings, this region might correspond to a subset of interest: men over 35 who weigh at least 160 or students with IQs below 120 with a combined SAT score of 1000.

Linear Inequalities

A **linear inequality** in the variables x and y is an inequality that can be written in one of the forms

$$ax + by \leq c, \quad ax + by < c, \quad ax + by \geq c, \quad ax + by > c$$

where a, b, and c are real numbers. For example, $2x - 3y > 6$ is a linear inequality in x and y. A point (x, y) is a **solution** of an inequality in x and y if it satisfies the inequality. For example, $(4, -1)$ is a solution of $2x - 3y > 6$ since $2(4) - 3(-1) = 8 + 3 = 11$ and $11 > 6$. The **solution set** of an inequality is the set of all solutions.

EXAMPLE 1

Checking Solutions of Linear Inequalities

Determine whether the following ordered pairs are solutions of the linear inequality $2x - 3y > 6$:

a. $(3, 0)$ **b.** $(-3, -5)$

Solution

a. $2(3) - 3(0) = 6$ and $6 \not> 6$. Thus, $(3, 0)$ is not a solution.

b. $2(-3) - 3(-5) = -6 - (-15) = 9$ and $9 > 6$. Thus, $(-3, -5)$ is a solution.

The **graph** of an inequality in the variables x and y is the set of all points (x, y) in the coordinate plane that are solutions of the inequality. We can obtain the graph of a linear inequality by first graphing the corresponding linear equation $ax + by = c$. This line divides the coordinate plane into two **half-planes**, exactly one of which contains solutions of the inequality.

Consider, for example, the inequality $2x - 3y > 6$. The graph of the corresponding linear equation $2x - 3y = 6$ is shown in Figure 24. Notice that the graph has been drawn with a dashed line. This is because the **strict inequality** $>$ in the expression $2x - 3y > 6$ indicates that none of the points on the line $2x - 3y = 6$ are actually part of the solution set. If the inequality had been $\geq$ or $\leq$, then the line itself would have been part of the solution set, and so a solid line would have been used. In any case, the line $2x - 3y = 6$ divides the coordinate plane into two half-planes (shaded tan and blue in Figure 24), exactly one of which contains solutions of the inequality. To determine which half-plane is the correct one, we simply test a point on one side of the line to see if the inequality is satisfied. We select $(0, 0)$ as our test point and substitute $x = 0$, $y = 0$ into the original equation.

$$2(0) - 3(0) \overset{?}{>} 6$$
$$0 \not> 6$$

Thus, $(0, 0)$ does not satisfy the inequality, and so we know that the solutions of the inequality are on the other side of the line. This is the region we have shaded in Figure 25.

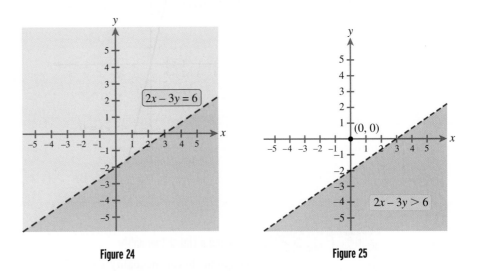

Figure 24   Figure 25

The general procedure for graphing linear inequalities is summarized as follows.

Graphing a Linear Inequality

1. Replace the inequality symbol with an $=$ and sketch the graph of the resulting line. Use a dashed line for the strict inequalities $<$ or $>$ and a solid line for $\leq$ or $\geq$.
2. Test a point on one side of the line. If the point satisfies the original inequality, then shade the half-plane containing that point. If the point does not satisfy the inequality, shade the half-plane on the other side of the line.

EXAMPLE 2

Graphing a Linear Inequality

Graph the linear inequality $4x + y \leq 0$.

Solution   First, we replace the inequality symbol with $=$ to obtain the equation $4x + y = 0$. Next, we rewrite the equation as $y = -4x$, which is the equation of a line passing through the origin with slope -4, as shown in Figure 26. Note that a solid line is used since the inequality symbol is $\leq$. To determine which side of the line to shade, we test a point on one side of the line. We choose $(0, -1)$, and substitute $x = 0$, $y = -1$ into the original inequality.

$$4(0) + (-1) \stackrel{?}{\leq} 0$$
$$-1 \leq 0$$

Since $(0, -1)$ satisfies the inequality, we shade the region containing $(0, -1)$ as shown in Figure 27.

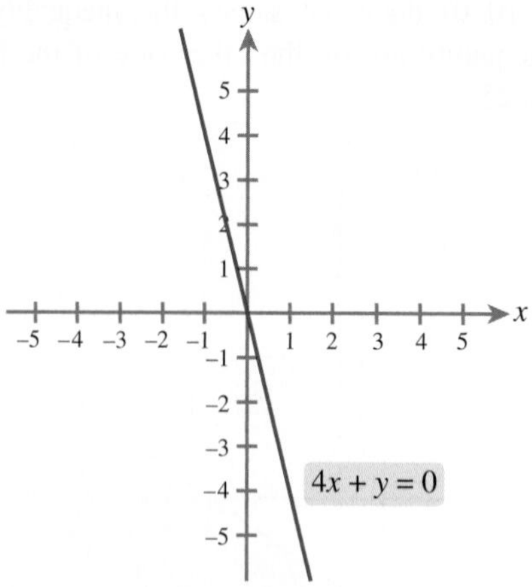

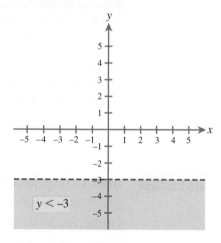

Figure 26

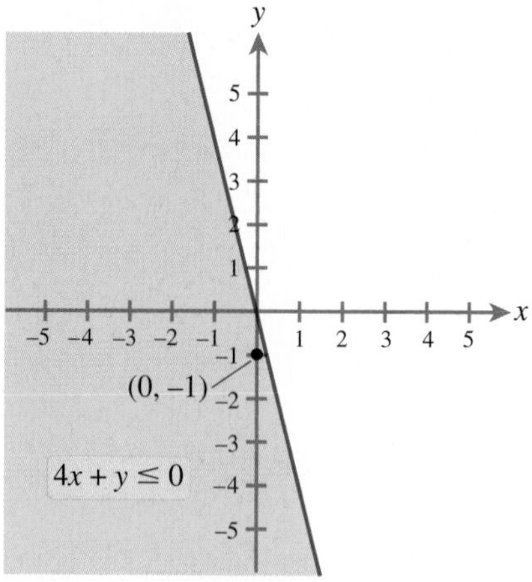

Figure 27

⋯⋯EXAMPLE 3

Graphing a Linear Inequality

Graph the linear inequality $y < -3$.

Solution The graph of $y = -3$ is a horizontal line with y-intercept -3. The points that satisfy the inequality $y < -3$ are those having y-coordinate less than -3; that is, the points that satisfy the inequality are those that lie below the line, as shown in Figure 28. Note that the line $y = 3$ is dashed since we are dealing with a strict inequality.

Figure 28

Systems of Linear Inequalities

A **system of linear inequalities** consists of a collection of two or more linear inequalities, such as

$$x + 2y > 2$$
$$-3x + 4y \leq 12$$

A **solution** of a system of inequalities in x and y is a point (x, y) that satisfies *all* of the inequalities in the system. We can graph the solution set of a system of inequalities by first sketching the graph of each individual inequality on the same coordinate system and then determining the region common to the graphs of all the inequalities. In other words, we find the intersection of the graphs of all the inequalities in the system.

EXAMPLE 4

Graphing a System of Linear Inequalities

Graph the system of inequalities

$$x + 2y > 2$$
$$-3x + 4y \leq 12$$

Solution We first graph $x + 2y > 2$. This is done by first plotting the graph of $x + 2y = 2$ as a dashed line. Next, we test the point $(0, 0)$ to see whether it satisfies the inequality. Since $0 + 2(0) = 0 \not> 2$, we shade the side of the line that does not contain $(0, 0)$. The result is shown in Figure 29. Next, on the same coordinate system and using a similar procedure, we graph $-3x + 4y \leq 12$. The two shaded regions are shown in Figure 30. The graph of the system is the region that is common to the graphs of both inequalities, as shown in Figure 31.

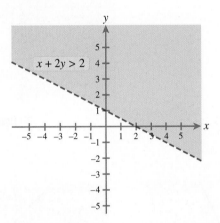

Figure 29

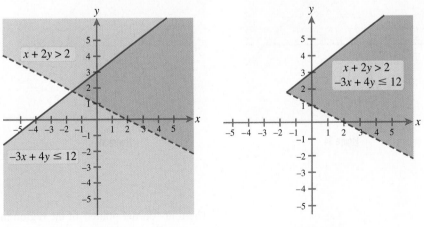

Figure 30

Figure 31

·····❯EXAMPLE 5

Graphing a System of Linear Inequalities

Graph the system of inequalities

$$
\begin{aligned}
x + y &\le 4 \\
-2x + y &\le 1 \\
y &\ge -1 \\
x &\le 2
\end{aligned}
$$

Solution Figures 32–35 illustrate the sequence of steps we took in graphing each of the inequalities on the same coordinate system. The graph of the system is the region that is common to the graphs of all four inequalities, as shown in Figure 36.

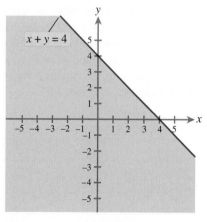

Figure 32

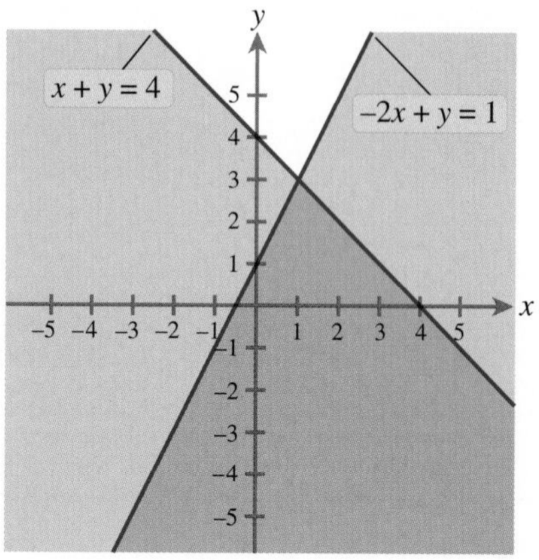

Figure 33

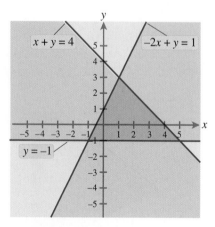

Figure 34

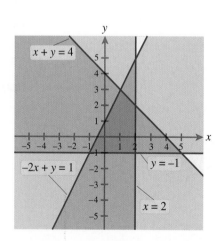

Figure 35

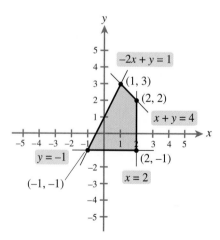

Figure 36

The four "corner points" of the region in Example 5—namely, $(1, 3)$, $(2, 2)$, $(2, -1)$, and $(-1, -1)$—are called **vertices**. In general, the vertices of a system of linear inequalities are found by locating the intersection points of appropriate pairs of lines.

Since the exact coordinates of the points of intersection may not be obvious from the graph, it is usually necessary to solve systems of linear equations. Thus, for example, the vertex $(1, 3)$ in Example 5 could be found by solving the equations $-2x + y = 1$ and $x + y = 4$ simultaneously, using any of the methods of this chapter: We choose the elimination technique from Section 9.1.

$$
\begin{aligned}
-(-2x + y) &= -1 \qquad &\text{Multiplying the first equation by } -1 \text{ and} \\
+ \qquad x + y &= 4 \qquad &\text{adding to the second} \\
\hline
3x &= 3 \\
x &= 1
\end{aligned}
$$

$$1 + y = 4 \qquad \text{Substituting } x = 1 \text{ into the second equation}$$
$$y = 3$$

⋯▷EXAMPLE 6

An Application of Systems of Linear Inequalities

Andrea is training for a bike race and so wants to keep her weight down while maintaining a high intake of protein and carbohydrates. On the way home from a particularly hard workout, she decides to treat herself to a chocolate milk shake and an order of fries at her favorite fast-food restaurant. Each gram of chocolate milk shake and fries provides the following nutritional content.

	Milk shake	Fries
Calories	1.5	3
Grams of protein	0.04	0.04
Grams of carbohydrates	0.2	0.4

For this snack, she must not exceed 405 calories but would like to have at least 6 grams of protein and 40 grams of carbohydrates. Set up and graph a system of linear inequalities that describes how many grams of milk shake and fries she can have. Locate and interpret the vertices of the solution set.

Solution   We assign the following variables to represent the number of grams of milk shake and fries.

$$x = \text{Number of grams of milk shake}$$
$$y = \text{Number of grams of fries}$$

To meet the described requirements, the following inequalities must be satisfied:

$$\text{Calories from milk shake} + \text{Calories from fries} \le 405$$
$$1.5x + 3y \le 405$$

$$\text{Protein from milk shake} + \text{Protein from fries} \ge 6$$
$$0.04x + 0.04y \ge 6$$

$$\text{Carbohydrates from milk shake} + \text{Carbohydrates from fries} \ge 40$$
$$0.2x + 0.4y \ge 40$$

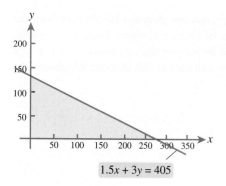

$$1.5x + 3y = 405$$

Figure 37

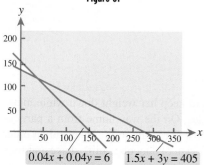

$$0.04x + 0.04y = 6 \qquad 1.5x + 3y = 405$$

Figure 38

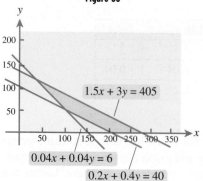

$$0.04x + 0.04y = 6$$
$$0.2x + 0.4y = 40$$

Figure 39

Since x and y cannot be negative, we must also have $x \geq 0$ and $y \geq 0$. Thus, the system of inequalities is

$$1.5x + 3y \leq 405$$
$$0.04x + 0.04y \geq 6$$
$$0.2x + 0.4y \geq 40$$
$$x \geq 0$$
$$y \geq 0$$

The last two inequalities tell us that we need to be concerned only with the first quadrant when we graph the first three inequalities. Refer to Figures 37–39 for these graphs. The resulting region is shaded in Figure 40.

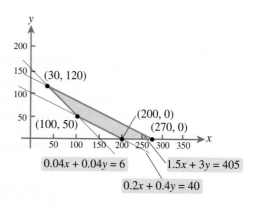

$$0.04x + 0.04y = 6 \qquad 1.5x + 3y = 405$$
$$0.2x + 0.4y = 40$$

Figure 40

Any point (x, y) in the shaded region represents an acceptable combination of milk shake and fries for Andrea. The vertices $(200, 0)$ and $(270, 0)$ are the x-intercepts of the lines $0.2x + 0.4y = 40$ and $1.5x + 3y = 405$. The other two vertices, $(100, 50)$ and $(30, 120)$, are found by solving the following systems of equations. The details of these solutions are omitted.

$(100, 50)$: $0.04x + 0.04y = 6$ $(30, 120)$: $0.04x + 0.04y = 6$
            $0.2x + 0.4y = 40$ $1.5x + 3y = 405$

From the vertex $(100, 50)$, we conclude that one possible combination would be 100 grams of milk shake and 50 grams of fries. From the vertex $(30, 120)$, we conclude that Andrea can have 120 grams of fries if she drinks only 30 grams of milk shake. The values of the other two vertices, $(200, 0)$ and $(270, 0)$, suggest that if she skips the fries altogether, she can drink anywhere from 200 to 270 grams of milk shake.

Nonlinear Inequalities

The graph of a nonlinear inequality can be obtained in much the same way as the graph of a linear inequality. We first replace the inequality symbol with an $=$, and then we sketch the graph of the resulting equation. For example, to graph $4 - x^2 \geq y$, we first graph $4 - x^2 = y$, as shown in Figure 41. As before, we use a dashed curve for the strict inequalities $<$ or $>$ and a solid curve for $\leq$ or $\geq$.

The graph of the equation normally divides the coordinate plane into two or more regions. In each region, either all of the points are solutions of the inequality or none of the points are solutions. Thus, we test a point in *each* of the regions to see which are solutions.

Test point	$4 - x^2 \overset{?}{\geq} y$	Conclusion
$(0, 5)$	$4 - (0)^2 \overset{?}{\geq} 5$ $4 \not\geq 5$	The region above the parabola is not in the solution set.
$(0, 0)$	$4 - (0)^2 \overset{?}{\geq} 0$ $4 \geq 0$	The region inside the parabola is in the solution set.

We conclude that the solution set consists of the region on and below the parabola $y = 4 - x^2$, as shown in Figure 42.

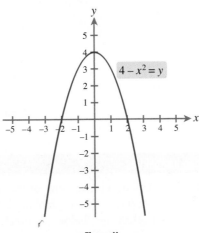

Figure 41

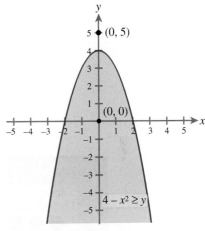

Figure 42

-----> EXAMPLE 7

Graphing a Nonlinear Inequality

Graph the inequality $\dfrac{x^2}{9} - \dfrac{y^2}{4} > 1$.

Solution   We first plot the hyperbola

$$\frac{x^2}{9} - \frac{y^2}{4} = 1$$

with dashed curves, as shown in Figure 43. The hyperbola divides the plane into three regions, and so we pick a point in each to see which satisfy the inequality.

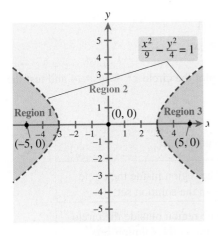

Figure 43

Test point	$\dfrac{x^2}{9} - \dfrac{y^2}{4} \overset{?}{>} 1$	Conclusion
$(-5, 0)$	$\dfrac{(-5)^2}{9} - \dfrac{0^2}{4} \overset{?}{>} 1$ $\dfrac{25}{9} > 1$	The region inside the left branch is in the solution set.
$(0, 0)$	$0 \not> 1$	The region between the branches is not in the solution set.
$(5, 0)$	$\dfrac{(5)^2}{9} - \dfrac{(0)^2}{4} \overset{?}{>} 1$ $\dfrac{25}{9} > 1$	The region inside the right branch is in the solution set.

We conclude that the solution set is inside the two branches of the hyperbola, as shaded in Figure 44.

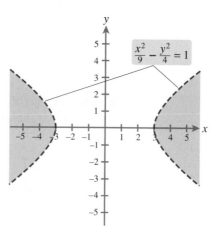

Figure 44

Graphing Inequalities

To plot the graph of an inequality of the form $y \leq f(x)$ or $y \geq f(x)$ with our graphing calculator, we first select the "below-the-graph" or the "above-the-graph" shading option, respectively. For example, the graph of $y \geq x^2 - 4x + 3$, shown in Figure 45, was obtained by graphing $y = x^2 - 4x + 3$ with the "above-the-graph" shading option.

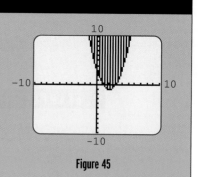

Figure 45

EXAMPLE 8

Graphing a System of Nonlinear Inequalities

Graph the system of inequalities

$$x^2 + y^2 \leq 4$$
$$y \geq |x|$$

Solution   For the inequality $x^2 + y^2 \leq 4$, we plot the circle $x^2 + y^2 = 4$ and test the points $(0, 0)$ and $(4, 0)$.

Test point	$x^2 + y^2 \overset{?}{\leq} 4$	Conclusion
$(0, 0)$	$(0)^2 + (0)^2 \overset{?}{\leq} 4$ $0 \leq 4$	The region inside the circle is in the solution set.
$(4, 0)$	$(4)^2 + (0)^2 \overset{?}{\leq} 4$ $16 \nleq 4$	The region outside the circle is not in the solution set.

The resulting region is shaded in Figure 46. Similarly, for the inequality $y \geq |x|$, we plot $y = |x|$ (on the same coordinate system) and test the points $(0, -1)$ and $(0, 1)$.

Test point	$y \overset{?}{\geq} \lvert x \rvert$	Conclusion
$(0, -1)$	$-1 \overset{?}{\geq} \lvert 0 \rvert$ $-1 \ngeq 0$	The region below $y = \lvert x \rvert$ is not in the solution set.
$(0, 1)$	$1 \overset{?}{\geq} \lvert 0 \rvert$ $1 \geq 0$	The region above $y = \lvert x \rvert$ is in the solution set.

The intersection of the two regions is shaded in dark blue in Figure 47.

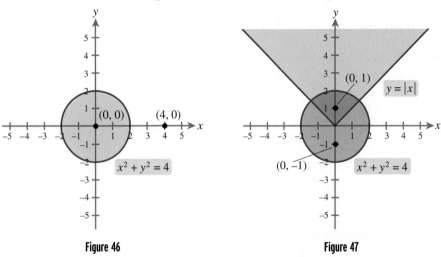

Figure 46   Figure 47

Understanding and Mastery Checklists

Concepts to Understand

Linear (or nonlinear) inequality in two variables

⬧

Solution of a linear (or nonlinear) inequality in two variables

⬧

Graph of a linear (or nonlinear) inequality in two variables

⬧

System of linear (or nonlinear) inequalities

⬧

Solution of a system of linear (or nonlinear) inequalities

⬧

Graph of a system of linear (or nonlinear) inequalities

Skills to Master

Determine whether a point is a solution of a system of inequalities.

⬧

Graph an inequality in two variables.

⬧

Graph a system of inequalities in two variables.

Exercises 9.7

Exercises 1-6 *Determine which of the ordered pairs are solutions of the inequality.*

1. $x - 2y \geq 5$;
(0,0), (4, 1), (−1, −3)

2. $y < 3x + 2$;
(0, 2), (−1, 1), (5, 5)

3. $4x + 3y > 12$;
(1, 4), (3, 0), (−2, −1)

4. $6x \geq 3y - 2$;
(−1, −5), (1, 2), (0, −3)

5. $y + x^2 \leq 4$;
(0, 2), (−1, 3), (5, −1)

6. $x^2 + y^2 > 25$;
(1, −2), (−4, 3), (5, 5)

Exercises 7-12 *Match the inequality with its graph.*

7. $3x - 2y \geq 6$

8. $3x + 2y \leq 6$

9. $x < 3$

10. $y > 3$

11. $y - x^2 \geq 0$

12. $x + y^2 \leq 0$

iii.

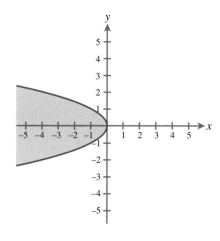

iv.

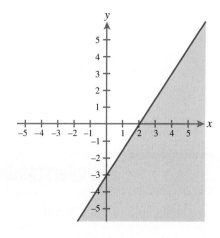

i.

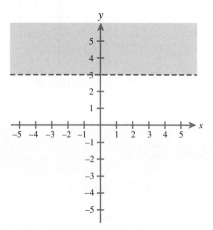

ii.

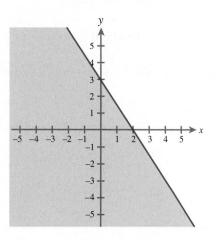

v.

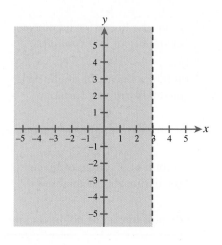

vi.

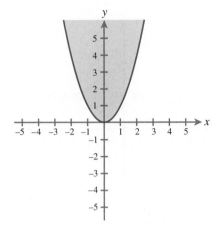

Exercises 13-28 *Graph the inequality.*

13. $x + y < 8$

14. $x - y \geq 2$

15. $6x - 2y \leq 12$

16. $5x + 4y > -20$

17. $x \geq -5$

18. $y < 4$

19. $x > -3y$

20. $2x - y \leq 0$

21. $y < x^2 + 1$

22. $x^2 - y \geq 4$

23. $y^2 \leq 1 - x^2$

24. $x^2 + y^2 > 9$

25. $y \geq \dfrac{1}{x}$

26. $x - y^2 < 0$

27. $y > e^x$

28. $y - \sin x \leq 0$

Exercises 29-32 *Match the graph display with the inequality that produced it.*

29. $y \leq -x^2 - 6x - 7$

30. $y \geq x^2 - 6x + 7$

31. $y \leq x^3 - 9x$

32. $y \geq -x^3 + 9x$

i.

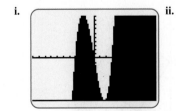

ii.

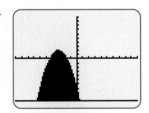

iii.

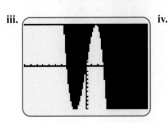

iv.

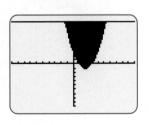

Exercises 33-54 *Graph the system of inequalities.*

33. $x + y \leq 5$
    $x - y \leq 1$

34. $2x + 3y > 6$
    $x - 3y < 3$

35. $\quad y < 2x$
    $x + 2y < 5$

36. $x - 2y \geq -6$
    $x - 2y \leq 3$

37. $\quad 4x + y \leq 6$
    $-4x - y \leq 4$

38. $5x - 2y < 5$
    $\quad 2y > 5x - 20$

39. $\quad x + y \leq 2$
    $-x + y \geq 2$
    $\quad x \geq -2$

40. $x - 2y \leq 4$
    $\quad y \geq -2x$
    $\quad y \leq 4$

41. $-3x + 2y < 6$
    $-x + 3y > 2$
    $\quad 2x + y < 3$

42. $\quad x - 2y > -4$
    $-3x + 4y > -18$
    $\quad x - 3y > 6$

43. $2x + 5y \leq 10$
    $\quad x + y \geq 3$
    $\quad x \geq 0$
    $\quad y \geq 0$

44. $2x + 3y \leq 15$
    $3x + y \leq 12$
    $\quad x \geq 0$
    $\quad y \geq 0$

45. $2x + y \geq 8$
    $2x - y \leq 8$
    $\quad x \geq 3$
    $\quad y \leq 4$

46. $y \geq 3x - 5$
    $y \leq 2x$
    $x \leq 3$
    $y \geq 1$

47. $-2x + y \geq 1$
    $\quad y \leq 4 - x^2$

48. $x - y^2 > 0$
    $x + y < 2$

49. $xy < 1$
    $\quad y < x$

50. $x^2 + y^2 \leq 9$
    $\quad x - y \leq 0$

51. $x^2 + y^2 \leq 16$
    $x^2 + y^2 \geq 4$

52. $y \leq \sqrt{x}$
    $y \geq \dfrac{1}{4}(x^2 - 2x)$

53. $y > x^3 - x$
    $y < 6x$
    $x \geq 0$

54. $y \leq e^x$
    $y \geq 1$
    $x \geq 0$

Exercises 55-64 *Find an inequality or system of inequalities with the indicated region as the solution set.*

55.

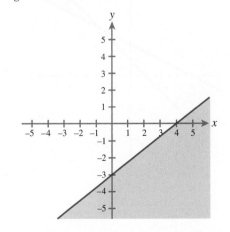

56.

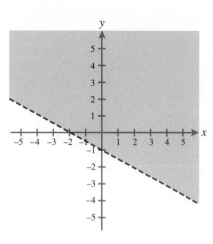

57.

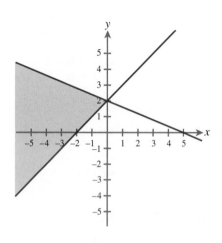

58.

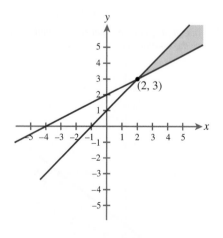

59.

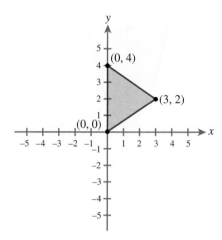

60.

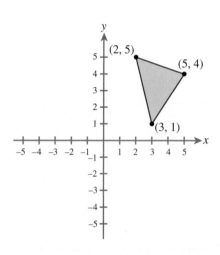

61.

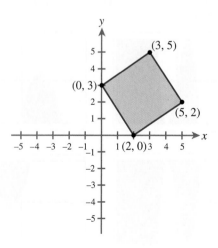

62.

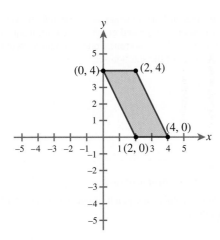

64.

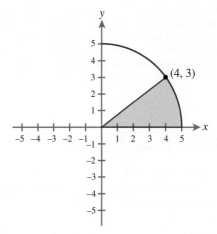

63.

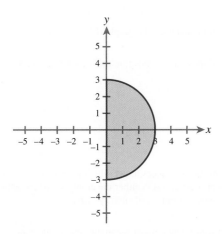

Applications

65. Retirement Planning Juan and Rosa are nearing retirement and want to invest up to $24,000 of their savings in two different mutual funds, Redbook Growth and Redbook Small Cap. Their financial adviser suggests they put at least $6000 in each fund, but because Redbook Small Cap has more risk, the amount they invest in it should be no more than half the amount invested in Redbook Growth. Set up and graph a system of inequalities that describes the various amounts they can invest in each fund. Locate the vertices of the solution set.

66. Low-fat Diet Aleshia is on a low-fat, low-calorie diet, but she prefers 2% milk over skim milk and won't give up her mozzarella cheese stick snacks. Each serving of 2% milk and mozzarella cheese contains the following.

	2% Milk	Mozzarella cheese
Calories	120	80
Grams of saturated fat	3	3
Grams of calcium	300	200

To stay within the guidelines of her diet, she must not exceed 480 calories and 15 grams of saturated fats from milk and cheese. Her daily goal for calcium from milk and cheese is at least 600 grams. Set up and graph a system of inequalities that describes how many servings of milk and cheese she can have. Locate the vertices of the solution set.

67. Land Allocation Roy has 1000 acres of land available to raise corn and soybeans although he may leave some unplanted. Each acre of corn costs $100 and requires 2 hours of labor. Each acre of soybeans costs $80 and requires 1 hour of labor. Roy does not wish his costs for the two crops to exceed $88,000, and he has at most 1600 hours of labor available for the two crops. Also, to feed his own cattle, he must plant at least 200 acres of corn. Set up and graph a system of inequalities that describes how many acres of corn and soybeans Roy should plant. Locate the vertices of the solution set.

68. Inventory Control A TV and appliance store carries a large inventory of TVs and refrigerators. Each TV requires 10 cubic feet of storage space and costs the store $500. Each refrigerator requires 50 cubic feet of storage space and costs the store $1000. The store has at most 1500 cubic feet of warehouse space and $48,000 of inventory capital available for TVs and refrigerators. Finally, because of demand, it is necessary to stock at least 8 refrigerators and at least twice as many TVs as refrigerators. Set up and graph a system of inequalities that describes how many TVs and refrigerators should be kept in stock. Locate the vertices of the solution set.

Concepts and Critical Thinking

Exercises 69-72   *Answer true or false.*

69. It is possible that the inside of a circle could be the solution set to a system of linear inequalities.

70. The solution set to a system of inequalities is the set of all points satisfying at least one of the inequalities in the system.

71. The solution set to a linear inequality is a half-plane.

72. The solution set of $mx + b < y$ is the half-plane lying entirely above the line $y = mx + b$.

Exercises 73-76   *Give an example of each.*

73. A system of inequalities, the graph of which consists of all points in the first quadrant

74. A system of inequalities, the graph of which is the inside of a semicircle

75. A system of inequalities, the graph of which is a right triangle

76. A system of inequalities, the graph of which is a square in the fourth quadrant

Questions for Discussion or Essay

77. When graphing systems of linear inequalities, it is helpful to be aware of the number of regions that are possible when one or more lines are drawn on the coordinate plane. For example, if one line is drawn on the coordinate plane, it divides the plane into two regions. If two lines are drawn, they divide the plane into either three or four regions, depending on whether the lines intersect. Discuss the connection between the number of regions formed by two lines and the solution sets of a system of two linear inequalities. What observations and connections can be made concerning the number of regions formed by three lines and the solutions of systems of three linear inequalities?

78. Describe the circumstances under which the solution set of a system of two linear inequalities is either the empty set or the entire coordinate plane. Give examples of such systems.

79. Explain in detail why a linear inequality involving $<$ or $>$ is graphed with a dashed line, whereas one involving $\leq$ or $\geq$ is graphed with a solid line.

Projects for Enrichment

80. Systems of Inequalities and Flight Paths   Inspector Magill has been called in to settle a dispute between an airline company and a group of environmentalists. The airline company had been given a court order that prohibits planes from flying over a region that is the sole breeding ground for an endangered bird species. The environmentalists claim that certain flights are still flying over this region, and so Magill must determine whether this is indeed the case. His first task is to describe the region algebraically. By setting up a coordinate system with the origin at one of the vertices of the region, Magill has determined that the other two vertices are at (20, 80) and (120, 30), as shown in Figure 48.

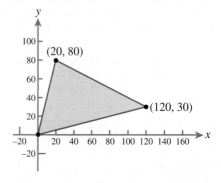

Figure 48

a. Find a system of three linear inequalities whose graph is the same as that shown in Figure 48.

Three daily flights are being contested by the environmentalists. The first is a direct flight from city A located at $(-22, 35)$ to city B located at $(110, -58)$.

b. Find the equation of the line connecting cities A and B, and sketch its graph on the same coordinate system as the triangular region. Does the flight violate the court order?

The second flight is a direct flight between city A and city C located at $(140, 197)$. A plot of the line connecting these two points is shown in Figure 49. Since it is difficult to determine graphically if the line passes through the region, Magill decides to check it algebraically.

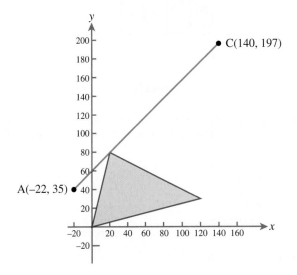

Figure 49

c. Find the equation of the line connecting cities A and C. Write the equation in the form $y = mx + b$.

d. Substitute the expression $mx + b$ found in part c for y in each of the three inequalities found in part a. After simplifying, the following three inequalities should be obtained:

$$x \geq -76$$
$$x \geq 19$$
$$x \leq 22$$

e. For the flight to pass through the region, there must be a value for x that satisfies each of the three inequalities given in part d. Does the flight pass through the region?

f. The third flight is between city B and city C. Use the technique described in parts c through e to determine if the third flight violates the court order.

81. Systems of Inequalities and Area   In this project, we consider the areas of closed regions determined by systems of inequalities. Consider, for example, the following linear system, whose solution is the triangular region shown in Figure 50.

$$x + y \leq 6$$
$$3x + y \geq 8$$
$$y \geq 2$$

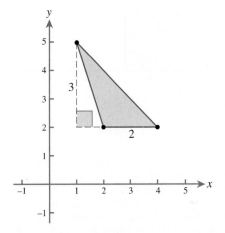

Figure 50

The area of the triangle in Figure 50 can be computed using the formula.

$$A = \frac{1}{2}bh = \frac{1}{2}(2)(3) = 3$$

a. Find the area of the region bounded by the given system of inequalities.

i. $x + y \leq 7$
$\quad$ $x + 2y \geq 8$
$\quad$ $x \geq 2$

ii. $x + 2y \leq 11$
$\quad$ $2x + 3y \leq 1$
$\quad$ $4x + y \geq 9$

iii. $x - 2y \geq -6$
$\quad$ $x - 2y \leq 0$
$\quad$ $2x - y \leq 3$
$\quad$ $2x - y \geq 0$

iv. $x^2 + y^2 \leq 4$
$\quad$ $x \geq 0$
$\quad$ $y \geq 0$

The region in part iv is unusual in that one of the boundary curves is not linear, and yet it is still possible to find the area using standard geometric formulas. In most cases, if one or more of the inequalities is nonlinear, it is not possible to find the area using standard formulas. Consider the following nonlinear system, which is graphed in Figure 51:

$$y \leq x^2 + 1$$
$$x \leq 2$$
$$x \geq 0$$
$$y \geq 0$$

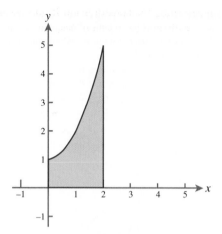

Figure 51

Because the top boundary curve is nonlinear, there is no standard formula for computing the area of the region. In fact, without the aid of calculus, we cannot compute the area exactly. Thus, we must settle for an approximation. In Figure 52, we have "inscribed" four rectangles of equal width in the region. The height of each rectangle is given by the y-coordinate of a point on the curve $y = x^2 + 1$. Thus, the first rectangle has height 1, the second has height $\frac{5}{4}$, and so on. Since each rectangle has width $\frac{1}{2}$, the area of each rectangle can be computed as follows:

$$A_1 = \frac{1}{2} \cdot 1 = \frac{1}{2} \qquad A_2 = \frac{1}{2} \cdot \frac{5}{4} \cdot = \frac{5}{8}$$

$$A_3 = \frac{1}{2} \cdot 2 = 1 \qquad A_4 = \frac{1}{2} \cdot \frac{13}{4} = \frac{13}{8}$$

The sum of the areas is

$$A = \frac{1}{2} + \frac{5}{8} + 1 + \frac{13}{8} = \frac{30}{8} = \frac{15}{4}$$

which is an approximation for the area of the shaded region in Figure 51. More accurate approximations could be found by inscribing more rectangles with smaller width.

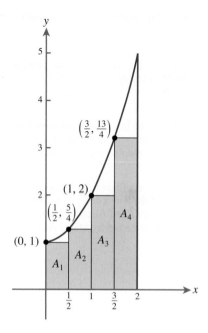

Figure 52

b. Approximate the area of the region bounded by the given system of inequalities. Use the specified number of inscribed rectangles.

i. $y \le x^2 + 1$
   $x \le 2$
   $x \ge 0$
   $y \ge 0$
   8 rectangles

ii. $y \le \dfrac{1}{x}$
   $x \ge 1$
   $x \le 4$
   $y \ge 0$
   6 rectangles

iii. $y \le x^2 - 2x$
   $y \ge 0$
   8 rectangles (2 with zero height)

Section 9.8 | **Linear Programming**

⊙ How can a business maximize its revenue without hiring additional labor?
⊙ What combination of stocks and bonds will yield the highest income for a fixed investment?
⊙ What postwar mathematical discovery helped the U.S. Air Force allocate supplies more efficiently?
⊙ How can a political party attract the largest number of TV viewers?

A common problem in business, industry, agriculture, and many other areas is that of optimizing the use of limited resources. Some examples are a factory that wishes to maximize its revenues within the constraints of limited manpower and/or raw materials, a hospital food service that wishes to minimize its costs within the constraints of necessary nutritional requirements, or a university that wishes to maximize the number of students it can house in a new dormitory within the constraints of limited available land and funding. Linear programming is a mathematical method that can be used to solve such problems. We illustrate its basic ideas with an example.

A Case Study   Willow Woods Furniture Factory makes tables and bookcases. In an effort to fine-tune production and increase profits, data have been collected on the labor requirements and profit margins for each piece of furniture. These data are summarized in Table 16. For example, each table requires 6 hours of woodworking, 3 hours of finishing, and yields a profit of $160. Without hiring additional labor, the factory has 48 hours available each day for woodworking and 30 hours each day for finishing. Because of incoming orders, the factory must make at least 2 tables each day. Naturally, the owners of the factory would like to know how many tables and how many bookcases should be made each day to maximize profit. We now begin a step-by-step process that will lead us to the answer of this question.

Table 16

	Per table	Per bookcase	Available
Woodworking time (hours)	6	4	48
Finishing time (hours)	3	4	30
Profit (dollars)	160	200	—

Step 1:   Understanding the Problem

Among the various combinations of tables and bookcases, some will bring in more profit than others. For example, if the factory sells 5 tables and 3 bookcases, the profit is $160(5) + 200(3) = \$1400$, whereas if it sells 4 tables and 6 bookcases, the profit is $160(4) + 200(6) = \$1840$. However, some combinations of tables and bookcases may exceed the constraints imposed by the available labor. The 5–3 combination would require $6(5) + 4(3) = 42$ hours of woodworking and $3(5) + 4(3) = 27$ hours of finishing, whereas the 4–6 combination would require $6(4) + 4(6) = 48$ hours of woodworking and $3(4) + 4(6) = 36$ hours of finishing. Thus, although the 4–6 combination brings in more profit, it is not allowable since it requires too many hours of finishing. We are looking for the combination that will provide the highest profit within the constraints of available labor.

Step 2:   Assigning Variables and Identifying the Expression that Is to Be Optimized

The unknowns are the number of tables and bookcases to be made. Thus, we let

$$x = \text{Number of tables made each day}$$
$$y = \text{Number of bookcases made each day}$$

We wish to maximize the profit from the sale of x tables and y bookcases. Because each table sells for a profit of $160, and each bookcase sells for a profit of $200, the total profit P is given by

$$P = 160x + 200y$$

This equation defines the **objective function**, the quantity to be maximized.

Step 3: Writing Out Inequalities that Correspond to Constraints

Production at the furniture factory is constrained by the available labor for woodworking and finishing and also by the perceived demand for tables. These constraints can be expressed algebraically using linear inequalities, as follows:

Woodworking constraint: Each table requires 6 hours, and each bookcase requires 4 hours. The total hours may not exceed 48.

Hours of woodworking for tables + Hours of woodworking for bookcases ≤ 48
$$6x \qquad + \qquad 4y \qquad\qquad \leq 48$$

Finishing constraint: Each table requires 3 hours, and each bookcase requires 4 hours. The total hours may not exceed 30.

Hours of finishing for tables + Hours of finishing for bookcases ≤ 30
$$3x \qquad + \qquad 4y \qquad\qquad \leq 30$$

Demand constraint: At least 2 tables must be made each day.

$$x \geq 2$$

Implied constraint: The number of tables and bookcases may not be negative.

$$x \geq 0$$
$$y \geq 0$$

Step 4: Graphing the Resulting System of Inequalities

The system of inequalities obtained in step 3 is

$$6x + 4y \leq 48$$
$$3x + 4y \leq 30$$
$$x \geq 2$$
$$y \geq 0$$

Notice that we have not included the constraint $x \geq 0$ since it is accounted for in the constraint $x \geq 2$. Moreover, the constraint $y \geq 0$ indicates that our region must be above the x-axis. Thus, we can restrict our attention to the first quadrant. The region that is common to the graphs of all four inequalities is shaded in dark blue in Figure 53.

The darkly shaded region in Figure 53 is called the **feasible set**, and it constitutes the set of points that satisfy all of the constraints. In the context of our problem, the feasible set corresponds to all the table–bookcase combinations that are within the labor and demand constraints. For example, the point (5, 3) is in the feasible set. It corresponds to 5 tables and 3 bookcases. As we observed in step 1, this combination is within the constraints imposed by the available labor. But many other combinations are also within the constraints. What we need is a way to determine which of the table–bookcase combinations in the feasible set is the one that will yield the greatest profit. For this task, we note that if there is an optimum solution, *it must occur at one of the vertices of the feasible set.* This will be discussed more thoroughly later in the section. For now, we simply apply the result by locating and testing each of the vertices to see which yields the maximum profit.

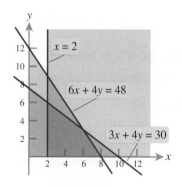

Figure 53

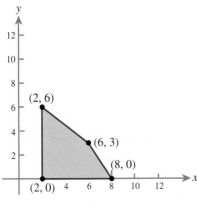

Figure 54

Step 5: Locating the Vertices of the Feasible Set

The four vertices are shown in Figure 54. The two on the x-axis are simply the x-intercepts of the lines $x = 2$ and $6x + 4y = 48$, namely $(2, 0)$ and $(8, 0)$, respectively. A third vertex is the intersection of the lines $x = 2$ and $3x + 4y = 30$. Setting $x = 2$ in the equation $3x + 4y = 30$ and solving for y gives $y = 6$. Thus, the vertex is $(2, 6)$. The fourth vertex is the intersection of the lines $6x + 4y = 48$ and $3x + 4y = 30$. Solving these simultaneously yields the solution $(6, 3)$.

Step 6: Evaluating the Objective Function and Identifying the Optimum Solution

We compute the profit at each vertex using the objective function $P = 160x + 200y$.

$$\text{At } (2, 0): P = 160(2) + 200(0) = 320$$
$$\text{At } (8, 0): P = 160(8) + 200(0) = 1280$$
$$\text{At } (6, 3): P = 160(6) + 200(3) = 1560 \quad \text{Maximum profit}$$
$$\text{At } (2, 6): P = 160(2) + 200(6) = 1520$$

So a maximum (daily) profit of $1560 can be achieved by making 6 tables and 3 bookcases each day.

The General Linear Programming Problem

The six steps we followed to determine the maximum profit for the Willow Woods Furniture Factory can be used as guidelines for solving a variety of optimization problems. We summarize the steps as follows.

Solving an Optimization Problem

1. Analyze the problem carefully. If necessary, try some specific examples to help you understand the problem.
2. Assign variables to the unknowns and write out the objective function, the quantity that is to be maximized or minimized.
3. Write out the inequalities that correspond to the constraints in the problem.
4. Graph the system of inequalities to obtain the feasible set.
5. Locate the vertices of the feasible set.
6. Evaluate the objective function at each vertex, and identify the optimum solution.

EXAMPLE 1

A Nutrition Problem

Al is on a diet. The daily fruit portion of his diet must have as few calories as possible and yet provide at least 750 units of vitamin A, 0.72 unit of vitamin B_6, and 60 units of vitamin C. His fruit bowl contains a few small bananas and oranges. Use the nutritional information given in Table 17 to determine which combination of bananas and oranges will yield the fewest calories and yet meet his nutritional needs.

Table 17

	Banana	Orange
Calories	100	60
Units of vitamin A	250	250
Units of vitamin B$_6$	0.6	0.06
Units of vitamin C	12	60

Solution The unknowns are the number of bananas and oranges that Al should eat. Thus, we let

$$x = \text{Number of bananas}$$
$$y = \text{Number of oranges}$$

We wish to minimize the number of calories. Since each banana has 100 calories and each orange has 60 calories, the function to be minimized is

$$K = 100x + 60y$$

The nutritional requirements lead to the following inequalities:

Vitamin A: Units from bananas + Units from oranges ≥ 750
$$250x + 250y \geq 750$$
Vitamin B$_6$: Units from bananas + Units from oranges ≥ 0.72
$$0.6x + 0.06y \geq 0.72$$
Vitamin C: Units from bananas + Units from oranges ≥ 60
$$12x + 60y \geq 60$$

Since x and y must be nonnegative, we also have $x \geq 0$ and $y \geq 0$. Thus, we have the following constraints.

$$
\begin{aligned}
\text{Constraints:} \quad 250x + 250y &\geq 750 \\
0.6x + 0.06y &\geq 0.72 \\
12x + 60y &\geq 60 \\
x &\geq 0 \\
y &\geq 0
\end{aligned}
$$

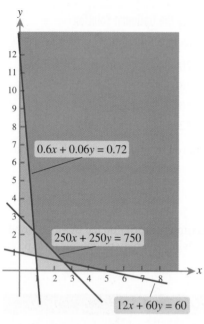

$0.6x + 0.06y = 0.72$

$250x + 250y = 750$

$12x + 60y = 60$

Figure 55

The graph of this system of inequalities is shown in Figure 55. The portion shown in dark blue is the feasible set, and the vertices shown in Figure 56 are found by locating intercepts and the intersection points of appropriate pairs of lines.

To determine the minimum number of calories, we evaluate the objective function $K = 100x + 60y$ at each vertex.

$$
\begin{aligned}
\text{At } (0, 12): \quad & K = 100(0) + 60(12) = 720 \\
\text{At } (1, 2): \quad & K = 100(1) + 60(2) = 220 \qquad \text{Minimum} \\
\text{At } \left(\tfrac{5}{2}, \tfrac{1}{2}\right): \quad & K = 100\left(\tfrac{5}{2}\right) + 60\left(\tfrac{1}{2}\right) = 280 \\
\text{At } (5, 0): \quad & K = 100(5) + 60(0) = 500
\end{aligned}
$$

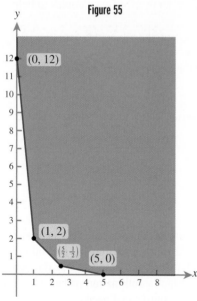

(0, 12)

(1, 2)

$\left(\tfrac{5}{2}, \tfrac{1}{2}\right)$ (5, 0)

Figure 56

Thus, to keep the fruit calories as low as possible and still obtain the required nutrients, Al should eat one banana and two oranges during the course of the day.

The two optimization problems we have considered so far have had several features in common. Each involved two variables, and each required us to maximize or minimize a linear objective function subject to two or more linear constraints. We refer to such problems as **linear programming problems** in two variables. In general, they are of the following form.

Linear Programming Problem in Two Variables

Maximize or minimize an objective function $P = Ax + By + C$ subject to constraints of the form $ax + by \leq c$ or $ax + by \geq c$.

As we have seen, the solution of a linear programming problem can be found by testing the objective function at each of the vertices of the feasible set. This fact is known as the Fundamental Theorem of Linear Programming.

The Fundamental Theorem of Linear Programming

An optimal solution to a linear programming problem, if it exists, will occur at a vertex of the feasible set.

To see why the optimal solution occurs at a vertex, let us return briefly to the furniture factory problem. Recall that we wished to maximize the objective function $P = 160x + 200y$ on the feasible set shown in Figure 57. If we approach the problem naively, without any knowledge of the Fundamental Theorem of Linear Programming, we might choose to consider certain specified profit values and the corresponding ordered pairs (x, y) that would yield that profit. For example, to obtain a profit of $1000, we must have $1000 = 160x + 200y$. This is the equation of the line graphed in Figure 58.

Since the line $1000 = 160x + 200y$ passes through the feasible set, we can see that there are points (x, y) that satisfy the constraints of the problem and also yield a profit of $1000. To determine if there are points that yield a higher profit and still satisfy the constraints, we next consider profits of $1200 and $1500. The points that satisfy the constraints and yield profits of $1200 and $1500, respectively, are the points of intersection of the lines $1200 = 160x + 200y$ and $1500 = 160x + 200y$ with the feasible set, as shown in Figure 59.

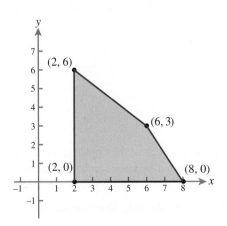

Figure 57

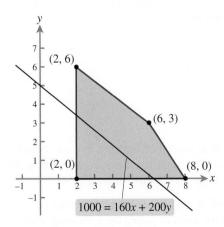

Figure 58

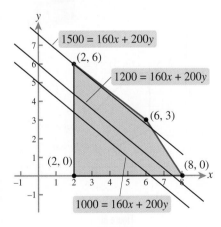

Figure 59

Notice that the lines are all parallel, and they progress farther away from the origin as the profit values increase. This suggests that we look for the line that is parallel to the other three, intersects the feasible set, and is as far from the origin as possible. This line has equation $1560 = 160x + 200y$ and intersects the feasible set at the vertex $(6, 3)$, as shown in Figure 60. Thus, we see that the greatest profit, \$1560, is attained at the vertex $(6, 3)$.

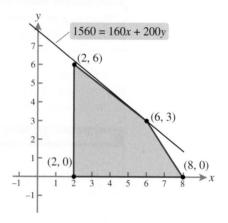

Figure 60

In general, specific values of the objective function lead to families of parallel lines. If the objective function has a maximum value, it will correspond to the line that intersects the feasible set and is furthest from, or nearest to, the origin, depending on the particular objective function $P = Ax + By + C$. This line must necessarily pass through a vertex. Likewise, if there is a minimum, it corresponds to the line that is closest to, or farthest from, the origin, again depending on the particular function $P = Ax + By + C$, and it must also pass through a vertex.

EXAMPLE 2

Solving a Linear Programming Problem

Find the maximum and minimum values of $P = 3x + 6y$ subject to the following constraints:

$$x + y \le 8$$
$$2x - y \le 7$$
$$x + 2y \ge 6$$
$$x \ge 2$$
$$y \le 5$$

Sketch several of the lines that correspond to specific values of the objective function, and show that the lines corresponding to the maximum and minimum values pass through vertices of the feasible set.

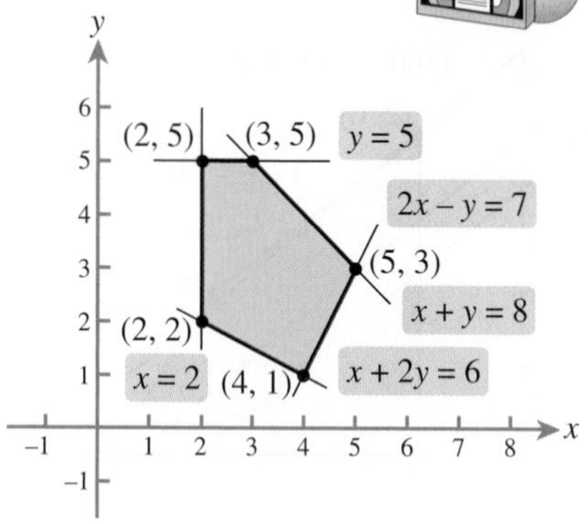

Solution   The feasible set is shown in Figure 61. Testing the objective function at each of the vertices yields the following results:

Figure 61

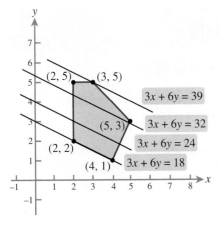

Figure 62

At (2, 5): $P = 3(2) + 6(5) = 36$

At (3, 5): $P = 3(3) + 6(5) = 39$ Maximum

At (5, 3): $P = 3(5) + 6(3) = 33$

At (4, 1): $P = 3(4) + 6(1) = 18$ Minimum

At (2, 2): $P = 3(2) + 6(2) = 18$ Minimum

Thus, the maximum value of P is 39, which occurs at (3, 5), and the minimum value of P is 18, which occurs at both (4, 1) and (2, 2). Figure 62 shows the lines corresponding to the objective function values 18, 24, 32, and 39. Note that the lines corresponding to the maximum $P = 39$ and minimum $P = 18$ pass through the vertices just found.

As can be seen in Example 2, the objective function can attain its maximum or minimum value at more than one vertex of a feasible set. Note also that the objective function need not have both a maximum and a minimum on a given feasible set. An example of this situation can be seen in Example 1, where the objective function has a minimum but not a maximum. See Exercise 38 for further discussion of this issue.

Advanced Techniques of Linear Programming

In this text, we investigate extremely simple linear programming problems involving several constraints and only two variables. We are using the geometric method: We graph the constraint inequalities and then evaluate the objective function at each of the vertices to determine where the maximum or minimum value is obtained. But what would we do if there were thousands of variables and tens of thousands of constraints? We are certainly not going to be able to graph the region in a thousand-dimensional space; indeed, it would be difficult to solve even a three-dimensional problem in this fashion. Clearly we need an **algorithm**, or step-by-step method, that doesn't require geometric visualization.

Motivated by problems arising in the context of military planning and support for the U.S. Department of the Air Force, George Dantzig in 1947 developed the **simplex method**, a systematic technique for solving linear programming problems. The simplex method has found wide application in communications, airline scheduling, and inventory control, and has hence become an integral part of management, industrial engineering, and operations research training programs throughout the world.

In the two-dimensional problems that we consider in this section, the feasible set is a region in the plane bounded by line segments, and optimal solutions exist at vertices. In the sort of multidimensional problems to which the simplex method is applied, the feasible set is a "region" in a multidimensional space bounded by portions of **hyperplanes**. If we visualize the feasible set

Narendra K. Karmarkar

as a multifaceted gemstone, then the hyperplanes that form the boundary of the feasible set are the facets of the gemstone. Just as in the two-dimensional case, optimal solutions occur at the vertices (the corners of the gemstone). The simplex method begins with a vertex in the feasible set and then at the next step selects an adjacent vertex for which the objective function is greater (in the case of a maximization problem). In this manner, step by step, the simplex method propels us ever closer to an optimal solution.

In spite of the widespread success that the simplex method has enjoyed for half a century, problems have arisen, particularly in the communications industries, for which the simplex method has proven to be inadequate. For this reason, there is an ongoing search for more efficient methods for solving high-dimensional linear programming problems. Perhaps the most significant breakthrough in this area was made in the 1980s by Narendra K. Karmarkar of Bell Laboratories (pictured here). Karmarkar's technique involves cutting a swath through the interior of the region rather than caroming from vertex to vertex on the surface. At each stage, another interior point, closer to the optimal solution than the last, is reached. Although the improvement in computation time is often modest for low-dimensional linear programming problems, the improvement can be quite dramatic when the dimension is large. The demand for more efficient techniques for solving linear programming problems will likely lead to the development of ever faster algorithms.

Understanding and Mastery Checklists

Concepts to Understand

Objective function

◦

Constraint

◦

Feasible set

◦

Vertices of the feasible set

◦

Linear programming problem in two variables

◦

The Fundamental Theorem
of Linear Programming

Skills to Master

Solve a linear programming problem
in two variables.

◦

Formulate a linear programming problem
from a real-world scenario.

Exercises 9.8

Exercises 1-6 *Find the maximum and minimum values of the objective function over the given feasible set.*

1. Objective function: $P = 8x + 10y$
Feasible set:

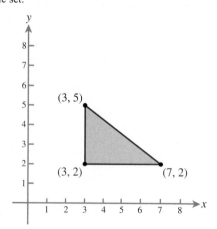

2. Objective function: $P = 10x + 4y$
Feasible set:

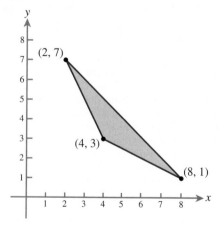

3. Objective function: $P = 4x + 5y$
Feasible set:

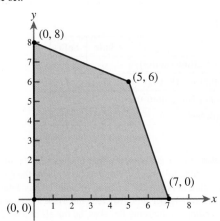

4. Objective function: $P = 6x - 2y$
Feasible set:

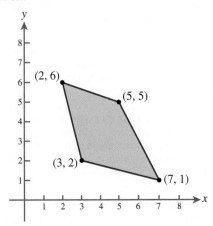

5. Objective function: $P = 4x - 4y$
Feasible set: See Exercise 3.

6. Objective function: $P = 2x + 8y$
Feasible set: See Exercise 4.

Exercises 7-16 *Solve the linear programming problem.*

7. Maximize: $P = 10x + 15y$
Subject to: $3x + 4y \leq 24$
$x + 2y \leq 10$
$x \geq 0$
$y \geq 0$

8. Maximize: $P = 14x + 30y$
Subject to: $3x + 5y \leq 30$
$2x + 4y \leq 22$
$x \geq 0$
$y \geq 0$

9. Minimize: $P = 15x + 40y$
Subject to: $2x + 5y \geq 20$
$3x + 2y \geq 19$
$x \geq 0$
$y \geq 0$

10. Minimize: $P = 30x + 10y$
Subject to: $2x + y \geq 8$
$4x + y \geq 12$
$x \geq 0$
$y \geq 0$

11. Maximize: $P = 10x + 12y$
Subject to: $x + 2y \leq 17$
$2x + 3y \leq 30$
$x \geq 3$
$y \geq 2$

12. Maximize: $P = 12x + 8y$
Subject to: $2x + y \leq 18$
$x + 2y \leq 21$
$x \geq 1$
$y \geq 4$

13. Minimize: $P = 20x + 24y$
Subject to: $2x + 3y \leq 12$
$x + y \geq 5$
$x + 4y \geq 8$
$x \geq 0$
$y \geq 0$

14. Minimize: $P = 10x + 4y$
Subject to: $x + 3y \geq 16$
$x + y \leq 10$
$2x + y \geq 12$
$x \geq 0$
$y \geq 0$

15. Maximize: $P = 2x + 10y$
Subject to: $x + y \geq 9$
$3x - 2y \leq 24$
$x + 4y \leq 36$
$x \geq 4$
$y \geq 3$

16. Minimize: $P = 10x + 8y$
Subject to: $x + 2y \geq 12$
$x - 4 \leq 4$
$3x + y \geq 16$
$x \leq 8$
$y \leq 10$

Applications

17. Clothing Profits A clothing company makes two styles of tailored suits. The labor requirements and profit margins for each style are given in Table 18. If 56 hours are available per day for cutting and 72 hours per day for sewing, how many suits of each style should be made to maximize profit?

Table 18

	Style A	Style B
Cutting (hours)	2	4
Sewing (hours)	3	2
Profit (dollars)	35	40

18. Livestock Allocation A farmer raises cattle and pigs. The investment requirements, labor requirements, and profit margins for each cow and pig are given in Table 19. If the farmer has 680 hours of labor and $52,000 of investment capital available, how many cows and pigs should be raised to maximize profit?

Table 19

	Cow	Pig
Investment (dollars)	480	1400
Labor (hours)	10	9
Profit (dollars)	120	400

19. Refinery Production An oil company owns two refineries. The daily production limits and operating costs for each refinery are given in Table 20. An order is received for 1000 barrels of high-grade oil, 1000 barrels of medium-grade oil, and 1800 barrels of low-grade oil. How many days should each refinery be operated so that the order can be filled at the least cost?

Table 20

	Refinery 1	Refinery 2
High-grade oil (barrels)	100	200
Medium-grade oil (barrels)	200	100
Low-grade oil (barrels)	300	200
Operating cost (dollars)	10,000	9000

20. Furniture Profit A furniture company makes chairs and sofas. The labor requirements and profit margins for each chair and sofa are given in Table 21. The total labor available each day for carpentry, finishing, and upholstery is 96 hours, 18 hours, and 72 hours, respectively. How many chairs and sofas should be made each day to maximize profit?

Table 21

	Chair	Sofa
Carpentry time (hours)	6	3
Finishing time (hours)	1	1
Upholstery time (hours)	2	6
Profit (dollars)	80	70

21. Vehicle Allocation Steve can borrow either his mother's car or his father's truck to commute to school, as long as he replaces the gas he uses. After averaging his costs for a period of time, he has determined that the car costs him 4¢ per mile, whereas the truck averages 6¢ per mile. Depending on the route he takes, he travels at least 300 miles each month but no more than 350. For insurance purposes, he must drive his father's truck at least twice as far as his mother's car. However, his father does not want him to use the truck for more than 250 miles each month. How many miles should Steve drive each vehicle in order to minimize his monthly cost? What is the least Steve will have to spend each month?

22. Dormitory Construction A university is making plans to build a new dormitory. Early estimates show that each single room will cost $20,000 and will require approximately 1200 cubic feet of space. Each double room will cost $25,000 and will require an average of 1800 cubic feet of space. The university can spend at most $2,200,000. Available space and zoning restrictions limit the total room space to 150,000 cubic feet. Finally, there must be at least as many double rooms as single rooms. How many of each type of room should the dormitory have in order to maximize the number of students it can hold?

23. Campaign Strategy A political party is planning a half-hour television show for its incumbent candidates for state governor and U.S. Senate. Based on a preshow survey, it is believed that 40,000 viewers will watch the show for each minute the senator is on and 60,000 viewers will watch for each minute the governor is on. The senator is a party "elder statesman" and so demands to be on the air at least one and a half times as long as the governor. The governor will not participate if her time is less than 10 minutes. And, of course, the sum of their speaking times may not exceed 30 minutes. Determine the time that should be allotted to each of the two candidates to maximize the number of viewers.

24. Power Lunch Sam's lunches consist primarily of peanut butter sandwiches and hamburgers from his favorite fast-food restaurant. Each sandwich has 1.5 grams of saturated fats, 40 grams of carbohydrates, 9 grams of protein, and 2 grams of iron. Each hamburger has 4 grams of saturated fats, 30 grams of carbohydrates, 12 grams of protein, and 4 grams of iron. Determine how many sandwiches and hamburgers Sam must eat to minimize the amount of saturated fats but still obtain at least 110 grams of carbohydrates, 30 grams of protein, and 7 grams of iron.

25. **Hospital Meal**   A hospital food service is planning a meal that includes Swiss steak and peas. Each ounce of Swiss steak costs 9¢ and has 150 units of vitamin A, 0.06 unit of vitamin B_6, and 3 units of vitamin C. Each ounce of peas costs 4¢ and has 120 units of vitamin A, 0.02 unit of vitamin B_6, and 9 units of vitamin C. From the meat and vegetable portion of the meal, each patient must receive 1260 units of vitamin A, 0.35 unit of vitamin B_6, and 45 units of vitamin C. How many ounces of Swiss steak and peas should be served to minimize the cost?

26. **Troop Maneuvers**   The leader of a U.N. peacekeeping unit wishes to lead his troops, by foot, to a hot point many miles away in as few days as possible. To save time, he decides that he will have his troops jog at a rate of 5 miles per hour for a portion of each day but no more than 2 hours total. The rest of the 8-hour day, the troops will walk at a rate of 4 miles per hour. Since the food supply is limited, it is determined that each soldier can consume no more than 3200 calories during the daily hike. (Caloric consumption is given in the following table.) How many hours of each day should the troops jog and how many should they run in order to travel as far as possible within these constraints?

	Walking	Running
Speed (mph)	4	5
Calories per hour	360	680

27. **Staffing Conferences**   A high-tech company with two offices—one in Bozeman, MT, with 15 employees and another in New Haven, CT, with 20—wishes to send each of its employees to one of two conventions: the COMDEX computer trade show to be held in the Orange County Convention Center in Orlando, FL, and a management training seminar to take place at the Fawcett Center on the campus of Ohio State University in Columbus, Ohio. Management has concluded that at least 10 workers should attend COMDEX and at least 15 should travel to Columbus. Furthermore, the company president, who confiscates all frequent flyer mileage earned by his employees, insists that a minimum of 32,000 frequent flyer miles be accumulated to facilitate his upcoming around-the-world tour. One frequent flyer mile is given for each mile of the round-trip flights to Orlando only (see Table 22). How many employees from each office should be sent in order to meet these constraints while minimizing the total airfare (see Table 23)?

Table 22   Round-trip distances for frequent flier miles

	New Haven	Bozeman
Orlando	2000	4000

Table 23   Round-trip air fare

From: To:	New Haven	Bozeman
Orlando	$500	$750
Columbus	$300	$450

Concepts and Critical Thinking

Exercises 28-31   *Answer true or false.*

28. Every point in the feasible set of a linear programming problem satisfies all of the constraints.

29. If the objective function of a linear programming problem has a maximum, then the maximum occurs at one of the vertices of the feasible set.

30. If the objective function of a linear programming problem has a minimum, then the minimum occurs at one of the vertices of the feasible set.

31. According to the Fundamental Theorem of Linear Programming, the minimum value of the function $R = x^2 + y^2$ subject to the constraints $-1 \le x \le 1$ and $-1 \le y \le 1$ will be obtained at one of the four points $(1, 1)$, $(-1, 1)$, $(-1, -1)$, or $(1, -1)$.

Exercises 32-35   *Give an example of each.*

32. A linear programming problem for which the feasible set is a half-plane

33. A linear programming problem for which the optimal solution is the origin

34. A linear programming problem for which the feasible set is an infinitely long horizontal strip with a width of 2 units

35. A linear programming problem for which the feasible set has just one vertex

Questions for Discussion or Essay

36. A friend relates the following story: ". . . so we're taking this test on Chapter 9, and there are only a couple minutes left in the hour, when 'Ivan the Terrible' announces that there is a typo in the linear programming problem I've just spent the last 10 minutes finishing. Since there isn't enough time to redo the problem, I erase everything and write a note explaining the steps I would follow if given enough time. Can you believe I only got a couple of points for the problem?" The friend shows you the test and the location of the typo. You immediately realize that, although your friend has the steps memorized, he doesn't really understand the linear programming process. Only a few simple calculations would have been necessary to correct the problem. Was the typo in the objective function or one of the constraints? Explain.

37. In examples such as the furniture factory problem, it is tempting to assume that the maximum value of the objective function will occur at the "obvious" vertex—that is, the one where the available resources have been completely utilized. Review the furniture factory problem, and explain why the solution (6, 3) is the "obvious" one. Then show that it is possible, by changing only one of the coefficients in the objective function, to make the maximum occur at the vertex (2, 6). At this vertex, which labor resource has not been fully utilized? In practice, do you think it's possible for a company to maximize profit without using all available resources? Explain.

38. In Example 1, we minimized the objective function $K = 100x + 60y$ over the feasible set shown Figure 63. The values of the objective function at the vertices are given in Table 24.

Table 24

Vertex	$K = 100x + 60y$
(0, 12)	720
(1, 2)	220
$\left(\frac{5}{2}, \frac{1}{2}\right)$	280
(5, 0)	500

Thus, we found the minimum to be 220 at the vertex (1, 2). Why can we not assume that the maximum is 720 at the vertex (0, 12)? What can be said about the maximum of the objective function over this feasible set? Does this contradict the Fundamental Theorem of Linear Programming? Explain.

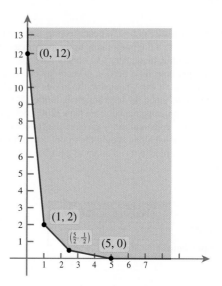

Figure 63

Projects for Enrichment

39. Linear Programming with Three Variables Linear programming problems can also involve more than two variables. In general, a linear programming problem in n variables would involve maximizing or minimizing an objective function $P = A_1 x_1 + A_2 x_2 + \cdots + A_n x_n$ subject to constraints of the form $a_1 x_1 + a_2 x_2 + \cdots + a_n x_n \leq b$ or $a_1 x_1 + a_2 x_2 + \cdots + a_n x_n \geq b$. Unfortunately, such problems usually cannot be solved geometrically unless it is possible to express $n - 2$ of the variables in terms of the remaining two. In the following example, we start with three variables but are able to replace one of the variables with an expression involving the other two.

Example: Suppose that $18,000 is to be invested in some combination of government bonds paying 6%, municipal bonds paying 8%, and stocks paying 10%. After analyzing the risks, it is decided that at most $15,000 is to be invested in government and municipal bonds, at least $10,000 in government bonds, and at most $6000 in stocks. What combination of investments will maximize the yearly income?

Solution: We initially assign the following variables:

x = Amount invested in government bonds
y = Amount invested in municipal bonds
z = Amount invested in stocks

The yearly income from investing these amounts at the given rates is
$$I = 0.06x + 0.08y + 0.10z$$
We wish to maximize this objective function subject to the following constraints:

$x + y \leq 15,000$	At most \$15,000 in government and municipal bonds
$x \geq 10,000$	At least \$10,000 in government bonds
$z \leq 6000$	At most \$6000 in stocks
$x \geq 0$	
$y \geq 0$	
$z \geq 0$	

To solve this problem geometrically, we must reduce it to only two variables. Since the total amount invested is \$18,000, we know that
$$x + y + z = 18,000$$
so that
$$z = 18,000 - x - y$$
Substituting this expression into the objective function and the constraints gives
$$I = 0.06x + 0.08y + 0.10(18,000 - x - y)$$
$$= 1800 - 0.04x - 0.02y$$
and

$$x + y \leq 15,000$$
$$x \geq 10,000$$
$18,000 - x - y \leq 6000$ Or equivalently, $x + y \geq 12,000$
$$x \geq 0$$
$$y \geq 0$$
$18,000 - x - y \geq 0$ Or equivalently, $x + y \leq 18,000$

The feasible set corresponding to these constraints is shown in Figure 64.

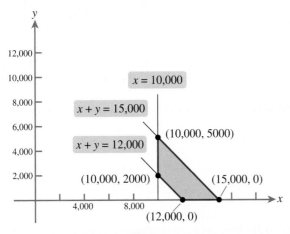

Figure 64

We now test the objective function at each of the vertices.
At (12,000,0):
$$I = 1800 - 0.04(12,000) - 0.02(0) = 1320$$

At (15,000,0):
$$I = 1800 - 0.04(15,000) - 0.02(0) = 1200$$
At (10,000, 2000):
$I = 1800 - 0.04(10,000) - 0.02(2000) = 1360$ Maximum
At (10,000, 5000):
$$I = 1800 - 0.04(10,000) - 0.02(5000) = 1300$$
So the maximum income is \$1360 when $x = 10,000$, and $y = 2000$. For these values of x and y, we obtain
$$z = 18,000 - 10,000 - 2000 = 6000$$
Thus, \$10,000 should be invested in government bonds, \$2000 in municipal bonds, and \$6000 in stocks to achieve a maximum income of \$1360.

Use this method to solve the following linear programming problems involving three variables:

a. Michael has \$20,000 to invest in some combination of mutual funds, stocks, and bonds. The projected annual rates of return are 14% for mutual funds, 16% for stocks, and 10% for bonds. He does not wish to invest more than \$12,000 in mutual funds and stocks but would like to invest at least \$4000 in stocks. Moreover, he does not wish to invest more than \$10,000 in bonds. What combination of investments will maximize Michael's income?

b. Eliza wishes to have \$2000 from her monthly paycheck deposited automatically in some combination of three different accounts at her credit union. She would like to maximize the amount of earnings from each new deposit after 1 month. The savings account pays $\frac{1}{3}$% per month, the checking account pays $\frac{1}{2}$% per month on the average balance but has a fee of \$5 per month, and the money market account pays $\frac{2}{3}$% per month. Due to bills that must be paid during the course of the month, she must put at least \$1000 into the checking account (assume that the average balance is half of what she deposits). She would like to deposit at least \$800 into savings and the money market account although the money market account has restrictions on withdrawals, and so she does not wish to deposit more than \$500 there. How should the deposit be distributed among the three accounts to maximize earnings?

c. A farmer wishes to plant exactly 1000 acres in some combination of corn, soybeans, and wheat. Each acre of corn costs \$100 and requires 2 hours of labor. Each acre of soybeans costs \$80 and requires 1 hour of labor. Each acre of wheat costs \$70 and requires 1.5 hours of labor. The total cost for the three crops may not exceed \$85,000, and the labor may not exceed 1600 hours. The profits from each acre of corn, soybeans, and wheat are \$170, \$110, and \$90, respectively. How many acres of each should be planted to maximize profit?

Chapter 9 Review

Exercises 1-6 *Solve the given system of equations using either the substitution or elimination method.*

1. $x + 5y = 13$
    $2x + 3y = 12$

2. $3x - y = -7$
    $4x + y = -7$

3. $2x - 4y = 5$
    $-x + 2y = 6$

4. $x + 2y^2 = 6$
    $x + y^2 = 5$

5. $x + 2y + z = 11$
    $y - z = -1$
    $x + z = 5$

6. $x - 3y + 2z = 3$
    $-2x + y + 4z = 7$
    $3x - y + 2z = -3$

Exercises 7-8 *Find the point(s) of intersection of the given pair of equations.*

7. $2x + 3y = 6, 4x - y = 4$

8. $x^2 - y = 1, x^2 + y^2 - 2y = 7$

Exercises 9-10 *Use a graphing calculator to estimate the coordinates of the point(s) of intersection of the given pair of equations to the nearest hundredth.*

9. $y = 2x^3 - 3x^2 - 11x$
    $y = x + 7$

10. $y = -2x^4 + 8x^3 - 8x^2 - 1$
    $y = 2x^4 - 4x^2$

Exercises 11-14 *Perform the indicated matrix operation.*

11. $3 \begin{bmatrix} 2 & -1 \\ -5 & 3 \end{bmatrix}$

12. $-2 \begin{bmatrix} -3 & 0 \\ 4 & -1 \\ 0 & -2 \end{bmatrix}$

13. $\begin{bmatrix} 3 & 0 & -2 \\ 5 & -1 & 3 \end{bmatrix} + \begin{bmatrix} 0 & -1 & 4 \\ -2 & 6 & 2 \end{bmatrix}$

14. $\begin{bmatrix} 1 & 0 \\ 3 & -8 \end{bmatrix} - \begin{bmatrix} 0 & 5 \\ -1 & -2 \end{bmatrix}$

Exercises 15-22 *Let*

$$A = \begin{bmatrix} 1 & 0 \\ -2 & 3 \end{bmatrix}, \quad B = \begin{bmatrix} 2 & -1 \\ 4 & 3 \end{bmatrix},$$

$$C = \begin{bmatrix} 4 & 0 \\ -3 & 1 \\ 0 & 2 \end{bmatrix}, \quad and \; D = \begin{bmatrix} 3 & 1 & 0 \\ -2 & 0 & 4 \\ 1 & -2 & 1 \end{bmatrix}.$$

Determine whether the given expression is defined. If so, evaluate it. If not, explain why not.

15. $A + 2B$

16. $2C - D$

17. CD

18. BA

19. DC

20. $C(A - B)$

21. C^2

22. A^2

Exercises 23-32 *Compute the given product, if possible.*

23. $\begin{bmatrix} -2 & 3 \\ 1 & 4 \end{bmatrix} \begin{bmatrix} 3 \\ -1 \end{bmatrix}$

24. $\begin{bmatrix} 3 & 0 \\ 2 & -4 \end{bmatrix} \begin{bmatrix} 2 & -1 \\ -4 & 6 \end{bmatrix}$

25. $\begin{bmatrix} 2 & 5 \\ 3 & 7 \end{bmatrix} \begin{bmatrix} 4 & 0 \end{bmatrix}$

26. $\begin{bmatrix} 5 & -3 \\ 0 & 4 \end{bmatrix} \begin{bmatrix} 2 & 1 & -1 \\ -1 & 3 & 0 \end{bmatrix}$

27. $\begin{bmatrix} 3 & 4 \end{bmatrix} \begin{bmatrix} 2 \\ -1 \end{bmatrix}$

28. $\begin{bmatrix} 2 & 3 & 5 \\ 7 & 11 & 13 \end{bmatrix} \begin{bmatrix} 2 & 4 \\ 8 & 16 \end{bmatrix}$

29. $\begin{bmatrix} 0 & 4 \\ -1 & 2 \\ 3 & 0 \end{bmatrix} \begin{bmatrix} 2 & 3 \\ -3 & 1 \end{bmatrix}$

30. $\begin{bmatrix} 1 & 1 & 0 \\ 0 & 1 & 1 \end{bmatrix} \begin{bmatrix} 1 & 2 & 3 \\ 0 & 1 & 2 \\ 0 & 0 & 1 \end{bmatrix}$

31. $\begin{bmatrix} 1 & -1 & 2 \\ 3 & 0 & -1 \\ 4 & 2 & 0 \end{bmatrix} \begin{bmatrix} b \\ c \\ a \end{bmatrix}$

32. $\begin{bmatrix} 1 & 2 & 4 \\ 2 & 4 & 1 \\ 4 & 1 & 2 \end{bmatrix} \begin{bmatrix} -1 & 0 & 2 \\ 0 & 2 & -1 \\ 2 & -1 & 0 \end{bmatrix}$

Exercises 33-34 *Write the given system as an augmented matrix.*

33. $5x + y = 3$
    $x - 3y = 4$

34. $x + 2y - 2z = 3$
    $2x + y = -1$
    $3y + 2z = 6$

Exercises 35-36 *Write out a system of equations in x, y, and z, in that order, that corresponds to the given augmented matrix, and then solve the system using back-substitution.*

35. $\begin{bmatrix} 1 & -4 & 1 & | & 3 \\ 0 & 1 & -2 & | & 5 \\ 0 & 0 & 1 & | & -2 \end{bmatrix}$

36. $\begin{bmatrix} 1 & -2 & -1 & | & 4 \\ 0 & 1 & \frac{1}{2} & | & 2 \\ 0 & 0 & 1 & | & 3 \end{bmatrix}$

Exercises 37-40 *Use Gaussian elimination to write the augmented matrix in row-echelon form. There may be more than one correct answer.*

37. $\begin{bmatrix} -1 & 3 & | & -2 \\ 4 & -9 & | & 1 \end{bmatrix}$

38. $\begin{bmatrix} 2 & 4 & -2 & | & 6 \\ 3 & 5 & -1 & | & 4 \\ -2 & -4 & 9 & | & 6 \end{bmatrix}$

39. $\begin{bmatrix} 2 & 4 & -3 & | & 1 \\ 3 & 0 & \frac{9}{2} & | & -\frac{3}{2} \end{bmatrix}$

40. $\begin{bmatrix} 3 & 10 & 4 & | & 0 \\ 1 & 3 & 1 & | & 1 \\ -1 & -6 & -3 & | & -2 \end{bmatrix}$

Exercises 41-48 *Express the system as an augmented matrix and solve using Gaussian elimination. There may be 0, 1, or infinitely many solutions.*

41. $\begin{aligned} -x + 2y &= 7 \\ 2x - 3y &= -9 \end{aligned}$

42. $\begin{aligned} 3x + 6y &= 1 \\ 2x + 2y &= 1 \end{aligned}$

43. $\begin{aligned} -2x + 3y &= 6 \\ 4x - 6y &= -12 \end{aligned}$

44. $\begin{aligned} 2x + 4y - 2z &= 9 \\ y - z &= 1 \\ -x - 3y + z &= -6 \end{aligned}$

45. $\begin{aligned} x - y + z &= -4 \\ 3x - 2y - z &= -4 \\ x - 2z &= 1 \end{aligned}$

46. $\begin{aligned} x + 2y + z &= 3 \\ 2y + 3z &= 2 \\ -x + 2z &= 1 \end{aligned}$

47. $\begin{aligned} x - y &= 4 \\ 2x - z &= 5 \\ y - z &= 1 \end{aligned}$

48. $\begin{aligned} x + y &= 0 \\ -2x - y + z &= -9 \\ 3x + y - z &= 8 \end{aligned}$

Exercises 49-50 *Use Gauss-Jordan elimination to write the augmented matrix in reduced row-echelon form.*

49. $\begin{bmatrix} 1 & 2 & | & 4 \\ 2 & 3 & | & 5 \end{bmatrix}$

50. $\begin{bmatrix} 1 & -1 & 2 & | & 8 \\ 0 & 1 & -4 & | & -9 \\ 0 & 0 & 1 & | & 3 \end{bmatrix}$

Exercises 51-54 *Solve the system of equations by first writing it in matrix form and then using Gauss-Jordan elimination.*

51. $\begin{aligned} 3x + 5y &= -5 \\ -x - 2y &= 3 \end{aligned}$

52. $\begin{aligned} x - 4y &= -0.5 \\ -2x + 9y &= 1.25 \end{aligned}$

53. $\begin{aligned} x - y + 3z &= 9 \\ y - 2z &= -3 \\ z &= 4 \end{aligned}$

54. $\begin{aligned} x - y + z &= 0 \\ -4x + 5y - 6z &= 7 \\ 2x - y + z &= 2 \end{aligned}$

Exercises 55-62 *Find the inverse of the given matrix, if it exists. Use a graphing calculator as needed.*

55. $\begin{bmatrix} 2 & 1 \\ -3 & -1 \end{bmatrix}$

56. $\begin{bmatrix} 0 & \frac{1}{2} \\ 8 & -\frac{1}{4} \end{bmatrix}$

57. $\begin{bmatrix} a & b \\ b & a \end{bmatrix}$

58. $\begin{bmatrix} 1 & x \\ x & x^2 \end{bmatrix}$

59. $\begin{bmatrix} 1 & -2 & 3 \\ 0 & 1 & -2 \\ 0 & 0 & 1 \end{bmatrix}$

60. $\begin{bmatrix} 0 & 4 & 8 \\ 4 & 0 & 4 \\ 8 & 4 & 0 \end{bmatrix}$

61. $\begin{bmatrix} 0 & 0 & a \\ 0 & 0 & 0 \\ a & 0 & 0 \end{bmatrix}$

62. $\begin{bmatrix} 1 & -1 & 0 & 0 \\ -1 & 1 & -1 & 0 \\ 0 & -1 & 1 & -1 \\ 0 & 0 & -1 & 1 \end{bmatrix}$

Exercises 63-66 *Solve the system of equations by first expressing it in matrix form as $AX = B$ and then evaluating $X = A^{-1}B$.*

63. a. $\begin{aligned} x + y &= 3 \\ 2x + y &= -4 \end{aligned}$ **b.** $\begin{aligned} x + y &= 2 \\ 2x + y &= 0 \end{aligned}$

64. a. $\begin{aligned} 3x - 2y &= 5 \\ 4x - y &= -10 \end{aligned}$ **b.** $\begin{aligned} 3x - 2y &= -2 \\ 4x - y &= 3 \end{aligned}$

65. a. $\begin{aligned} \frac{1}{2}x + \frac{3}{2}y &= 3 \\ -\frac{1}{4}x + \frac{5}{4}y &= -4 \end{aligned}$ **b.** $\begin{aligned} \frac{1}{2}x + \frac{3}{2}y &= 0 \\ -\frac{1}{4}x + \frac{5}{4}y &= -3 \end{aligned}$

66. a. $\begin{aligned} x + y - z &= 3 \\ -4x - 3y + 6z &= -3 \\ -x - 2y &= 9 \end{aligned}$

b. $\begin{aligned} x + y - z &= 1 \\ -4x - 3y + 6z &= 0 \\ -x - 2y &= 2 \end{aligned}$

Exercises 67-74 *Evaluate the given determinant.*

67. $\begin{vmatrix} 4 & 6 \\ 3 & 2 \end{vmatrix}$

68. $\begin{vmatrix} -2 & 1 \\ 3 & 5 \end{vmatrix}$

69. $\begin{vmatrix} \sqrt{5} & 2 \\ 2 & \sqrt{5} \end{vmatrix}$

70. $\begin{vmatrix} 3 & -1 & 0 \\ -1 & 3 & 0 \\ -2 & -2 & 3 \end{vmatrix}$

71. $\begin{vmatrix} x & 0 & 1 \\ 0 & x & 0 \\ 1 & 0 & x \end{vmatrix}$

72. $\begin{vmatrix} 1 & 2 & 3 \\ 1 & 1 & 1 \\ 3 & 2 & 1 \end{vmatrix}$

73. $\begin{vmatrix} 1 & -1 & 0 & 0 \\ 3 & 4 & 0 & 0 \\ 0 & 0 & 2 & -3 \\ 0 & 0 & 1 & 1 \end{vmatrix}$

74. $\begin{vmatrix} a & b & 0 & 0 & 0 \\ b & a & 0 & 0 & 0 \\ 0 & 0 & 1 & 0 & 0 \\ 0 & 0 & 0 & 1 & 0 \\ 0 & 0 & 0 & 0 & 1 \end{vmatrix}$

Exercises 75-78   *Find the value(s) of x for which the given matrix is not invertible.*

75. $\begin{bmatrix} 2 & 2 \\ 5 & x \end{bmatrix}$

76. $\begin{bmatrix} 2 & x \\ x & 8 \end{bmatrix}$

77. $\begin{bmatrix} 1 & 2x & -4 \\ 0 & x & -2 \\ 0 & 3 & x-5 \end{bmatrix}$

78. $\begin{bmatrix} 1 & 0 & 0 \\ 1 & x & x+1 \\ 5 & -4 & x \end{bmatrix}$

Exercises 79-82   *Solve the system of equations using Cramer's Rule.*

79. $3x + y = 0$
   $5x + 3y = -1$

80. $2x - 3y = 16$
   $5x + 2y = 2$

81. $x + y = 0$
   $x + 2y - z = -6$
   $-4x - 3y = -1$

82. $x + y + z - 2w = -6$
   $-2x - y + 3w = 5$
   $2x - z - 3w = -5$
   $x - w = -2$

Exercises 83-90   *Graph the inequality.*

83. $x + y \geq 4$

84. $2x - y < -3$

85. $x < 2y$

86. $y \leq 3x - 4$

87. $x \geq -2$

88. $y \geq x^2 - 1$

89. $x^2 + y^2 < 4$

90. $x - y^2 \leq 0$

Exercises 91-100   *Graph the system of inequalities.*

91. $x - y \leq 6$
   $x + y \leq 4$

92. $3x + y > 3$
   $x < 3y$

93. $4x - 7y \geq 14$
   $3y \leq x + 6$

94. $2x + y \leq 4$
   $y \geq 2x - 1$
   $y \geq -2$

95. $-x + 4y > 8$
   $-x + y > 0$
   $-x + 2y > 4$

96. $-4x + 3y \leq 24$
   $-3x + 5y \leq 30$
   $x \geq 2$
   $y \geq 1$

97. $y \geq x^2 - 2$
   $y \leq x$

98. $y + x^2 \leq 4$
   $y \geq x^2$

99. $x^2 + y^2 \leq 25$
   $x \geq 3$

100. $y \geq 2^x$
   $y \leq 4$
   $x \geq 0$

Exercises 101-102   *Find the maximum and minimum value(s) of the objective function over the given feasible set.*

101. $P = 5x + 8y$

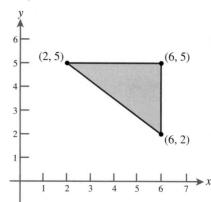

102. $P = 4x + 20y$

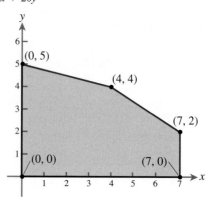

Exercises 103–106 *Solve the linear programming problem.*

103. Maximize: $P = 12x + 10y$
Subject to: $x + y \le 6$
$2x + y \le 10$
$x \ge 0$
$y \ge 0$

104. Minimize: $P = 6x + 12y$
Subject to: $5x + 4y \ge 35$
$5x + 3y \ge 30$
$x \ge 0$
$y \ge 0$

105. Minimize: $P = 8x + 10y$
Subject to: $2x + y \ge 10$
$x + y \ge 7$
$2x + 3y \ge 16$
$x \ge 2$
$y \ge 0$

106. Maximize: $P = 10x + 15y$
Subject to: $3x + 2y \le 21$
$x + y \le 8$
$y \le 2x$
$x \ge 0$
$y \ge 2$

107. Triangle Dimensions What are the dimensions of a right triangle if one side is 3 inches less than twice the length of the other and the length of the hypotenuse is 51 inches?

108. Counting Coins A woman has 11 coins in her pocket, all of which are either nickels or dimes. If the value of the coins is 75¢, how many of each type of coin does she have?

109. Estate Management An estate manager invested $100,000 in three funds, a money market fund paying 4% interest per year and charging an annual maintenance fee of 0.2%, a mutual fund (carrying substantial risk) averaging 12% per year with an annual load of 0.5%, and a savings account with no fees paying 3% interest per year. In one year, the investments earned a total of $6550, and $255 was charged in fees. How much was invested in each account?

110. Balanced Breakfast An 800-calorie breakfast is to consist of link sausages, pancakes, and orange juice and should contain 25 grams of protein and 90 grams of carbohydrates. Use the nutritional information given in Table 25 to determine how many servings of each breakfast item should be consumed.

Table 25

	Pancakes	Orange juice	Sausage links
Calories (per serving)	164	55	124
Carbohydrates (grams per serving)	23.7	13.2	0
Protein (grams per serving)	5.3	0.9	4.7

111. Fitting a Cubic Find the y-intercept of the graph of the cubic function $f(x)$ passing through the points $(-1, 5)$, $(1, 5)$, $(2, 10)$, and $(3, 31)$.

112. Unknown Coins A coin purse contains 10 coins (quarters, dimes, and nickels) with a total value of $1.60. There are as many quarters as there are nickels and dimes together. How many of each kind of coin are there?

113. Land Allocation A farmer intends to grow corn and hay. Between available cash and a bank line of credit, he has $50,000 available to invest in the two crops. Moreover, he and his son have a total of 1200 hours of labor available to devote to these two crops. Use the information given in Table 26 to determine how many acres of each crop should be planted to maximize revenue. How many acres of each should be planted to maximize profit?

Table 26

	Corn	Hay
Investment (dollars per acre)	125	100
Labor (hours per acre)	2	6
Revenue (dollars per acre)	245	210

114. Food Service Planning A university food service is planning a meal that includes broiled chicken breast and a baked potato. The chicken and potato portion of the meal must provide at least 31 units of protein, 80 units of vitamin B_6, and 16 units of iron. Use the information given in Table 27 to determine how many ounces of chicken and baked potato should be served to minimize the cost?

Table 27

	Chicken breast	Baked potato
Protein (units per ounce)	9	1
Vitamin B_6 (units per ounce)	18	6
Iron (units per ounce)	3	2
Cost (cents per ounce)	20	3

115. Investment Options Damon has $10,000 to invest in selected mutual funds and municipal bonds. For tax reasons, his accountant has advised that he invest at least twice as much in municipal bonds as in mutual funds. However, his financial adviser suggests that he invest at least $2500 in mutual funds. If the mutual funds average a 12% return and the municipal bonds average a 10% return, how much should he invest in each to maximize his earnings?

116. Studying for Finals Gina's final exams in college algebra and sociology are scheduled for the same day. She has 12 hours available to study for the two exams. Past experience suggests that each hour of studying for college algebra requires $\frac{1}{2}$ cup of coffee, whereas each hour of studying for sociology requires $\frac{1}{4}$ cup of coffee. Based on previous exam scores, each hour of studying for college algebra results in 15 points, and each hour of studying for sociology results in 10 points. If she must score at least 60 points on the sociology exam, and she only has 4 cups of coffee left, how many hours should she study for each exam to maximize the total score?

Chapter 9 Test

Problems 1-8 *Answer true or false.*

1. Matrices must be of the same order if they are to be added.

2. Matrices must be of the same order if they are to be multiplied.

3. All matrices have inverses.

4. If $\det(A) = 0$, then the system $AX = B$ can be solved by Cramer's Rule.

5. All upper triangular square matrices with 1s on the main diagonal are in row-echelon form.

6. Some linear systems of equations have exactly two solutions.

7. The solution set of a linear inequality is a half-plane.

8. According to the Fundamental Theorem of Linear Programming, the objective function will attain its maximum value at the vertex of the feasible region that is farthest from the origin.

Problems 9-13 *Give an example of each.*

9. A matrix of order 2×3

10. A noninvertible 2×2 matrix

11. An augmented matrix in row-echelon form but not reduced row-echelon form

12. A nonlinear system of equations

13. An elementary row operation

14. Solve the following system of equations using either the substitution or elimination method:
$$3x - 2y = -1$$
$$2x + 3y = 21$$

15. Let
$$A = \begin{bmatrix} 2 & -1 & 0 \\ 3 & 4 & -2 \\ 0 & -3 & 1 \end{bmatrix} \quad \text{and} \quad B = \begin{bmatrix} 0 & -2 & 5 \\ 6 & 3 & 1 \\ 4 & 0 & -1 \end{bmatrix}$$
Compute each of the following:

a. AB

b. $2A + B$

c. $A - B$

16. Express the following system as an equation involving matrices, and then solve it using an inverse matrix:
$$3x - 5y + z = 2$$
$$2x - 8y = -2$$
$$3x - z = 0$$

17. Consider the augmented matrix
$$\begin{bmatrix} 2 & -2 & 2 & | & 0 \\ 0 & 1 & -3 & | & -2 \\ 0 & 0 & 4 & | & 8 \end{bmatrix}$$

a. Give an equivalent system of equations and solve by back-substitution.

b. Write the matrix in reduced row-echelon form.

18. Solve the following system of equations by Gaussian elimination:
$$x - y - z = 4$$
$$3x + y + z = 4$$
$$x + 2y - z = 13$$

19. Use Cramer's Rule to solve the following system for b:
$$3a + b - c = 1$$
$$2a = -4$$
$$a + c = 2$$

20. In 1988, municipalities generated a total of 179.6 million tons of solid waste. Of this total, 58% was paper and yard waste; we'll classify the other 42% as miscellaneous waste. The total amount of paper waste was only 4.4 million tons less than the amount of miscellaneous waste. Find the amount of paper, yard, and miscellaneous waste. (*Data source:* Environmental Protection Agency.)

21. Find the maximum value of $P = 3x - 5y$ subject to the constraints $x \leq 0, y \geq 0, y \leq x + 6$, and $y \leq \frac{1}{4}x + 3$. Graph the feasible set and clearly indicate all relevant vertices.

22. Find a system of inequalities such that the solution set is the interior of the triangle with vertices $(0, 0)$, $(2, 3)$, and $(-1, 6)$.

Chapter 10

Integer Functions and Probability

Through personal experience, we develop a sense of the frequency of many everyday events and hence of the probability of their occurrence. But personal experience doesn't provide us with intuition about events that occur only rarely, such as being injured in a roller coaster accident, winning a lottery, or being struck by lightning. Surprisingly, our chances of being struck by lightning are greater than our chances of winning certain state lotteries and much greater than being injured in a roller coaster accident. In Section 4 of this chapter, we will investigate such probabilities in detail.

Section 10.1 Sequences

- What do flowers, pine cones, and sunflowers have in common with the family tree of a male honeybee?
- How is it possible to predict your salary 15 years into the future?
- If one checker is placed on the corner square of a checkerboard and the number of checkers in each successive square is doubled, how high will the pile of checkers in the 64th square reach?
- Why is 10 not necessarily the next number in the sequence 1, 4, 7, . . . ?

On the surface, a sequence is little more than a glorified list—of numbers, letters, or virtually anything. But the simplicity of this definition belies the significance of this fundamental notion and the role it plays in our lives. Sounds and images, documents and data are converted to binary sequences—strings of 1s and 0s—then chopped into pieces, pushed through conduits of copper, glass, and air and reassembled at distant points on the vast network that is the World Wide Web. A digital cellular telephone reconstructs voices from sequences riding radio waves through space. Plastic and magnetic media, from CDs and DVDs to floppy disks and magnetic tape, archive our music, our cinema, and much of the collective knowledge of the human race in sequential form. Even our genetic code, the detailed blueprint for life itself replicated in nearly all of the 50 trillion cells that make up the human body, is, in essence, nothing more than a simple sequence.

In this section, we focus on the notation and terminology associated with sequences and on an intuitive development of central concepts. Though our tour will be little more than a high-altitude survey of the vast sequential landscape, we will take a closer pass at two areas of particular interest—arithmetic and geometric sequences.

Number Sequences

In mathematics, the term *sequence* refers to an ordered list, often consisting of real numbers. The following lists are all examples of sequences of real numbers.

Sequence a: $-2, 0, 5, 8$

Sequence b: $1, 4, 9, 16, 25, \ldots$

Sequence c: $\dfrac{1}{2}, \dfrac{1}{4}, \dfrac{1}{8}, \dfrac{1}{16}, \ldots$

Sequence d: $-1, 0, 1, 0, -1, 0, 1, \ldots$

The numbers that make up a sequence are called **terms**. For example, the terms of a are $-2, 0, 5,$ and 8. Sequences like a, which have a final term, are called **finite**, whereas never-ending sequences like b, c, and d are said to be **infinite**. Sequences are ordered in the sense that they have a first term, a second term, a third term, and so on. This allows us to refer to specific terms of a sequence simply by attaching subscripts to the name of the sequence. For example, the fourth term of sequence b can be denoted by b_4, and so $b_4 = 16$. More generally, the nth term of sequence b can be denoted by b_n, and we observe that, since each term is the square of its position, $b_n = n^2$. We often define sequences in this way, by providing an expression for the nth term of the sequence.

----->EXAMPLE 1 **Finding Terms of a Sequence**

List the first four terms of the sequence with the given nth term.

a. $a_n = 2n + 3$ **b.** $z_n = \dfrac{n}{n + 1}$ **c.** $r_n = \dfrac{(-1)^n}{2^n - 1}$

Solution In each case, we substitute $n = 1$, $n = 2$, $n = 3$, and $n = 4$.

a. $a_1 = 2(1) + 3 = 5$

$a_2 = 2(2) + 3 = 7$

$a_3 = 2(3) + 3 = 9$

$a_4 = 2(4) + 3 = 11$

b. $z_1 = \dfrac{1}{1 + 1} = \dfrac{1}{2}$

$z_2 = \dfrac{2}{2 + 1} = \dfrac{2}{3}$

$z_3 = \dfrac{3}{3 + 1} = \dfrac{3}{4}$

$z_4 = \dfrac{4}{4 + 1} = \dfrac{4}{5}$

c. $r_1 = \dfrac{(-1)^1}{2^1 - 1} = \dfrac{-1}{1} = -1$

$r_2 = \dfrac{(-1)^2}{2^2 - 1} = \dfrac{1}{3}$

$r_3 = \dfrac{(-1)^3}{2^3 - 1} = \dfrac{-1}{7}$

$r_4 = \dfrac{(-1)^4}{2^4 - 1} = \dfrac{1}{15}$

The Information Revolution

We are currently witnessing nothing less than the most revolutionary change in the way information is stored and transmitted since Gutenberg invented the printing press. Compact disc players play back music free of hiss and distortion even after thousands of plays. Single CD-ROM discs contain entire encyclopedias. Visual images are transmitted across oceans via the World Wide Web. Entire libraries are accessible from personal computers, and interactive games are played among thousands of players living thousands of miles apart.

A visualization of data traffic over the National Science Foundation network

Although this information revolution depends on technological developments such as computer chips, fiber-optic cable, and compact discs, it would have been impossible without mathematical innovations in the ways information can be encoded and decoded. By **digitizing** data—that is, converting information to a sequence of numbers, usually 1s and 0s—we are able to convert information into a language that computers can understand and manipulate, and we are able to bring to bear powerful mathematical techniques.

The accuracy of data transmission and storage is limited by the presence of **noise**—random errors with any number of possible causes: tiny flaws in the structure of a cable or a disc, electromagnetic spikes from sunspots, or even human error. Techniques for combating noise are the province of the branch of mathematics known as **error-correcting codes**, the basic principle of which is to introduce a redundancy in the way a message is encoded so that even with a small amount of noise, the correct message can still be determined. A simple (and too inefficient to be useful) example of an error-correcting code would be to simply repeat a digit 5 times. Thus, to transmit the digit 1, we would send "1 1 1 1 1"; to transmit 0, we would send "0 0 0 0 0." If the sequence received were "1 1 1 1 0," we would still be fairly certain that the intended message was 1 and not 0.

An important key to sustaining progress in the information revolution is to continually increase the efficiency with which we store and send data. To this end, mathematicians have developed powerful **data compression** techniques that, in essence, allow us to store long sequences as shorter ones. As a simple example of a data compression technique, suppose we are dealing with sequences of 1s and 0s consisting of long runs in which a digit is repeated. Rather than send the original message, we could compress it by indicating the length of each run. Thus, the sequence consisting of 20 1s followed by 15 0s might be encoded as 20 15 (or in base 2: 10100 1111).

The exponential increase in access to information does not come without cost. The same automatic teller machine that allows us to perform financial transactions can be used by thieves to steal from our accounts; the cellular telephone system that enables us to call home from remote locations enables others to overhear our conversations; and the same medical records that can be downloaded by an emergency room physician also can be used by an unscrupulous insurance company to deny us coverage. Controlling who has access to what information poses one of the greatest challenges of the information revolution. **Cryptography** is a branch of mathematics that addresses how to transmit messages and have them remain secret, except to the intended receiver. Surprisingly, many encryption schemes involve factoring enormous numbers, and, for this reason, number theory, long viewed as pure mathematics with little direct applicability to the real world, is now an essential component of the information revolution.

It is no coincidence that the procedure for computing terms of a sequence is very much like that for evaluating a function. Indeed, a sequence can be viewed as a function whose domain is a set of integers.

Definition of Sequence

A **sequence** a is a function whose domain is a set of integers. The function values $a(n)$ are called the **terms** of the sequence, and they are denoted a_n. Unless indicated otherwise, a_n is assumed to be defined for $n = 1, 2, 3, \ldots$.

⋯EXAMPLE 2

Finding the nth Term of a Sequence

Find an expression for the nth term of a sequence whose first few terms are given.

a. $3, 6, 9, 12, 15, \ldots$ **b.** $1, \dfrac{1}{2}, \dfrac{1}{4}, \dfrac{1}{8}, \dfrac{1}{16}, \ldots$ **c.** $1, 4, 9, 16, \ldots$

Solution

a. Each term of the sequence is a multiple of 3. Specifically, $a_1 = 3(1)$, $a_2 = 3(2)$, $a_3 = 3(3)$, and so on. Thus, we define $a_n = 3n$.

b. Here each denominator is a power of 2. Thus, we first rewrite the sequence as

$$\frac{1}{2^0}, \frac{1}{2^1}, \frac{1}{2^2}, \frac{1}{2^3}, \frac{1}{2^4}, \ldots$$

Since the power of 2 in the denominator is one less than the position of the term,
$a_n = \dfrac{1}{2^{n-1}}$.

c. The terms of this sequence are the squares of consecutive positive integers. Thus, the general nth term has the form $a_n = n^2$.

The first two sequences given in the previous example are representative of two important types of sequences. The sequence $3, 6, 9, 12, 15, \ldots$ has the property that each term after the first is 3 more than the one preceding it. It is an example of an *arithmetic sequence*. The sequence $1, \frac{1}{2}, \frac{1}{4}, \frac{1}{8}, \frac{1}{16}, \ldots$ has the property that each term after the first is half of the one preceding it. It is an example of a *geometric sequence*. We now investigate arithmetic and geometric sequences in more detail.

Arithmetic Sequences

The distinctive feature of an arithmetic sequence is that each term after the first is obtained from the preceding term by adding a fixed number. In other words, each pair of consecutive terms has a common difference.

Definition of an Arithmetic Sequence

A sequence $a_1, a_2, a_3, \ldots a_n, \ldots$ is **arithmetic** if there is a number d, called the **common difference**, such that

$$a_2 - a_1 = d, a_3 - a_2 = d, \ldots, a_n - a_{n-1} = d, \ldots$$

Equivalently,

$$a_2 = a_1 + d, a_3 = a_2 + d, \ldots, a_n = a_{n-1} + d, \ldots$$

EXAMPLE 3

Examples of Arithmetic Sequences

Compute the common difference d for each of the following arithmetic sequences:

a. $-10, -3, 4, 11, 18, \ldots$ **b.** $2, \dfrac{7}{4}, \dfrac{3}{2}, \dfrac{5}{4}, 1, \ldots$

Solution Since we are given that the sequences are arithmetic, in each case we can compute the common difference using any pair of consecutive terms. We use the first two terms.

a. $d = -3 - (-10) = 7$

b. $d = \dfrac{7}{4} - 2 = -\dfrac{1}{4}$

An arithmetic sequence is completely determined by the first term a_1 and the common difference d. Algebraically, we have

$$a_1$$
$$a_2 = a_1 + d$$
$$a_3 = a_2 + d = (a_1 + d) + d = a_1 + 2d$$
$$a_4 = a_3 + d = (a_1 + 2d) + d = a_1 + 3d$$
$$\vdots$$

Thus, each term can be found by adding a multiple of d to a_1. Moreover, the multiple of d is one less than the subscript of the term. This leads us to the following formula for the nth term of the sequence.

The nth Term of an Arithmetic Sequence

The nth term of an arithmetic sequence with first term a_1 and common difference d is given by

$$a_n = a_1 + (n - 1)d$$

EXAMPLE 4

Finding the nth Term of an Arithmetic Sequence

Find an expression for the nth term of the arithmetic sequence $8, -1, -10, \ldots$

Solution The common difference is $d = -1 - 8 = -9$. Using $a_1 = 8$, we apply the nth-term formula to find a_n.

$$a_n = a_1 + (n - 1)d$$
$$= 8 + (n - 1)(-9)$$
$$= -9n + 17$$

EXAMPLE 5

Finding the nth Term of an Arithmetic Sequence

The first Saturday of 2007 falls on January 6. Find an expression for the day of the year on which the nth Saturday falls, and use this result to find the date of the tenth Saturday of 2007.

Solution   Let a_n represent the day of the year on which the nth Saturday falls. From the given information, we know that $a_1 = 6$. Since Saturdays occur 7 days apart, it follows that $d = 7$. Applying the nth-term formula for an arithmetic sequence gives us

$$a_n = 6 + 7(n - 1)$$

To find the date of the tenth Saturday, we let $n = 10$, which gives us

$$a_{10} = 6 + 7(10 - 1) = 69$$

Now there are 31 days in January and 28 days in February (in a non-leap year like 2007), so the 69th day of the year falls in March on day

$$69 - (31 + 28) = 10$$

Thus, the tenth Saturday of 2007 falls on March 10.

⋯⋯▷EXAMPLE 6

Finding a Specified Term of an Arithmetic Sequence

The fifth and eleventh terms of an arithmetic sequence are 28 and 64, respectively. Find the twentieth term.

Solution   By substituting $n = 5$ and $a_5 = 28$ into the nth-term formula, we obtain

$$a_n = a_1 + (n - 1)d$$
$$a_5 = a_1 + (5 - 1)d$$
(1) $$28 = a_1 + 4d$$

Similarly, substituting $n = 11$ and $a_{11} = 64$ gives

(2) $$64 = a_1 + 10d$$

Subtracting equation (1) from equation (2) yields

$$
\begin{array}{rl}
64 = & a_1 + 10d \\
- \ 28 = & a_1 + 4d \\
\hline
36 = & 6d \\
d = & 6
\end{array}
$$

Substituting $d = 6$ into equation (1) gives

$$28 = a_1 + 4(6)$$
$$a_1 = 4$$

Now that a_1 and d are known, we apply the nth-term formula once more with $n = 20$ to see that

$$a_{20} = 4 + (20 - 1)(6) = 118$$

Geometric Sequences

We have defined an arithmetic sequence to be a sequence whose consecutive terms have a common difference. A *geometric sequence* is a sequence whose consecutive terms have a common ratio.

Definition of a Geometric Sequence

A sequence $a_1, a_2, a_3, \ldots a_n, \ldots$ is **geometric** if there is a number $r \neq 0$, called the **common ratio**, such that

$$\frac{a_2}{a_1} = r, \frac{a_3}{a_2} = r, \ldots, \frac{a_n}{a_{n-1}} = r, \ldots$$

Equivalently,

$$a_2 = a_1 r, a_3 = a_2 r, \ldots, a_n = a_{n-1} r, \ldots$$

EXAMPLE 7

Examples of Geometric Sequences

Find the common ratio r for each of the following geometric sequences:

a. $6, 18, 54, 162, 486, \ldots$ **b.** $-\dfrac{4}{3}, \dfrac{8}{9}, -\dfrac{16}{27}, \dfrac{32}{81}, -\dfrac{64}{243}, \ldots$

Solution Since we are given that the sequences are geometric, in each case we can compute the common ratio using any pair of consecutive terms. We use the first two terms.

a. $r = \dfrac{18}{6} = 3$ **b.** $r = \dfrac{\dfrac{8}{9}}{-\dfrac{4}{3}} = \dfrac{8}{9} \cdot \left(-\dfrac{3}{4}\right) = -\dfrac{2}{3}$

EXAMPLE 8

Identifying Arithmetic and Geometric Sequences

Determine whether the given sequence appears to be arithmetic, geometric, or neither.

a. $1, 3, 7, 13, \ldots$ **b.** $16, 24, 36, 54, \ldots$

Solution

a. To test for an arithmetic sequence, we compute successive differences.

$$a_2 - a_1 = 3 - 1 = 2$$
$$a_3 - a_2 = 7 - 3 = 4$$

Already we see that there is no common difference, and so the sequence is not arithmetic. To test for a geometric sequence, we compute successive ratios.

$$\frac{a_2}{a_1} = \frac{3}{1} = 3$$

$$\frac{a_3}{a_2} = \frac{7}{3}$$

Since there is no common ratio, the sequence is not geometric either.

b. Computing differences as we did in part a, we see $a_2 - a_1 = 8$, whereas $a_3 - a_2 = 12$. Thus, the sequence is not arithmetic. The successive ratios simplify as follows:

$$\frac{a_2}{a_1} = \frac{24}{16} = \frac{3}{2}$$

$$\frac{a_3}{a_2} = \frac{36}{24} = \frac{3}{2}$$

$$\frac{a_4}{a_3} = \frac{54}{36} = \frac{3}{2}$$

Since there appears to be a common ratio of $\frac{3}{2}$, we conclude that the sequence is geometric.

A formula for the nth term of a geometric sequence can be found using a method similar to that used for arithmetic sequences. We begin with the first term a_1 and use the fact that each successive term is found by multiplying the preceding one by the common ratio r.

$$a_1$$
$$a_2 = a_1 r$$
$$a_3 = a_2 r = (a_1 r)r = a_1 r^2$$
$$a_4 = a_3 r = (a_1 r^2)r = a_1 r^3$$
$$\vdots$$

Thus, each term can be found by multiplying a_1 by an appropriate power of r. In fact, we see that the power is one less than the subscript of the term. Thus, we have the following formula.

The nth Term of a Geometric Sequence

The nth term of a geometric sequence with first term a_1 and common ratio r is

$$a_n = a_1 r^{n-1}$$

EXAMPLE 9   Using the nth-Term Formula for Geometric Sequences

Find the 25th term of the geometric sequence having common ratio $r = \frac{1}{4}$ and first term $a_1 = 3$.

Solution   Applying the nth-term formula for geometric sequences, we obtain

$$a_n = a_1 r^{n-1}$$
$$a_{25} = 3\left(\frac{1}{4}\right)^{25-1}$$
$$= \frac{3}{4^{24}}$$

EXAMPLE 10   A Sequence of Checker Stacks

A checker is placed on a corner square of a checkerboard. In each successive square, the number of checkers is doubled, as shown in Figure 1. How many checkers will be on the 64th square? If a checker is approximately 5 millimeters thick, how high will the stack of checkers reach? (*Note:* Figure 2 suggests the approximate height of the stack in the 32nd square.)

Figure 1

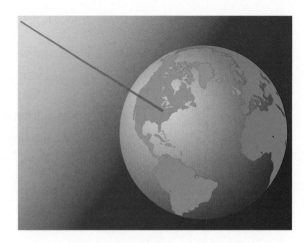

Figure 2

Solution   The number of checkers on each successive square is given by a term of the sequence 1, 2, 4, 8, 16, ... This is a geometric sequence with first term 1 and common ratio $r = 2$. Thus, by applying the nth-term formula for geometric sequences, we see that $a_n = (1)2^{n-1}$. The 64th term of this sequence is

$$a_{64} = 2^{63} \approx 9.22 \times 10^{18}$$

So there are approximately 9,220,000,000,000,000,000 checkers on the 64th square. If each checker is 5 millimeters thick, the pile of checkers would reach

$$(9.22 \times 10^{18}) \, (5 \text{ millimeters}) \left(\frac{1 \text{ kilometer}}{10^6 \text{ millimeters}} \right) = 4.61 \times 10^{13} \text{ kilometers}$$

which is approximately the distance to Proxima Centauri, the closest star (other than the sun) to Earth.

Recursively Defined Sequences

Arithmetic and geometric sequences have the property that each term after the first can be computed from the preceding term. In the case of arithmetic sequences, we simply add the common difference d, so that $a_n = a_{n-1} + d$. For geometric sequences, we multiply by the common ratio r, obtaining $a_n = a_{n-1}r$. Such formulas for the nth term of a sequence are said to be **recursive**. Generally speaking, a **recursively defined sequence** is one for which the first few terms are given and every successive term can be computed using a formula involving one or more of the preceding terms.

·····>EXAMPLE 11

Finding Terms of a Recursively Defined Sequence

Find the first four terms of the sequence defined recursively by $a_1 = 3$ and $a_n = 2a_{n-1} + 5$ for $n > 1$.

Solution

$$a_1 = 3$$
$$a_2 = 2a_1 + 5 = 2(3) + 5 = 11$$
$$a_3 = 2a_2 + 5 = 2(11) + 5 = 27$$
$$a_4 = 2a_3 + 5 = 2(27) + 5 = 59$$

·····>EXAMPLE 12

Finding Terms of a Recursively Defined Sequence

Find the first six terms of the sequence defined recursively by $a_1 = 1$, $a_2 = 1$, and $a_n = a_{n-1} + a_{n-2}$ for $n > 2$.

Solution

$$a_1 = 1$$
$$a_2 = 1$$
$$a_3 = a_2 + a_1 = 1 + 1 = 2$$
$$a_4 = a_3 + a_2 = 2 + 1 = 3$$
$$a_5 = a_4 + a_3 = 3 + 2 = 5$$
$$a_6 = a_5 + a_4 = 5 + 3 = 8$$

The recursively defined sequence 1, 1, 2, 3, 5, 8, 13, 21, 34, 55, 89, . . . of the previous example is known as the **Fibonacci sequence**. It is named after the Italian mathematician Leonardo of Pisa, whose nickname was Fibonacci. The sequence, which first appeared in Fibonacci's book *Liber Abaci* ("The Book of the Abacus") in 1202, can be described as the sequence for which the first two terms are 1 and each successive term is found by adding together the previous two terms. Despite its simplicity, the Fibonacci sequence has some fascinating mathematical properties. We will investigate some of these properties in Section 10.6.

The Fibonacci numbers occur with remarkable frequency in nature. Many species of flowers have petal arrangements consisting of 3, 5, 8, 13, 21, 34, 55, or 89 petals. The bracts of a pinecone spiral in sets of 8 and 13 rows. A sunflower has 34 petals, and its florets spiral out from the center in sets of 55 rows and 89 rows. The scales of a pineapple spiral in sets of 8, 13, and 21 rows.

Daisies, pine cone, sunflower, pineapple

The Fibonacci sequence is even present in the family tree of a male honeybee. As first observed by Johann Dzierzon in 1845, the male honeybee hatches from an unfertilized egg and so has a mother but no father (*Source:* Colin Butler, *The World of the Honeybee,* New York: Macmillan, 1955). Female honeybees, on the other hand, hatch from fertilized eggs and so have a mother and a father. Thus, if we trace the family tree of the male honeybee, we see that each male "node" of the tree has only one parent branch (a mother), whereas each female "node" has two branches (a mother and a father). The number of nodes at each level of the family tree corresponds to the Fibonacci numbers, as shown in Figure 3.

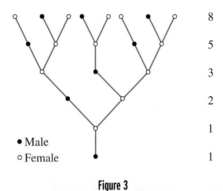

Figure 3

Understanding and Mastery Checklists

Concepts to Understand	Skills to Master
Sequence	Find the first few terms of a sequence given a formula for the nth term.
Term of a sequence	Find an expression for the nth term of a sequence given the first few terms.
Sequence notation	Determine whether a sequence is arithmetic and, if so, find its common difference.
Arithmetic sequence	Determine whether a sequence is geometric and, if so, find its common ratio.
Common difference of an arithmetic sequence	Find a specified term of an arithmetic or geometric sequence using partial information about the sequence.
nth-term formula for an arithmetic sequence	Find the first few terms of a recursively defined sequence.
Geometric sequence	
Common ratio of a geometric sequence	
nth-term formula for a geometric sequence	
Recursively defined sequence	
Fibonacci sequence	

Exercises 10.1

Exercises 1-16   *List the first four terms of the sequence with the given nth term.*

1. $a_n = 3n + 1$

2. $u_n = 4 - n$

3. $z_n = (-4)^n$

4. $c_n = -\dfrac{1}{2^n}$

5. $r_n = \dfrac{1}{n^2}$

6. $b_n = \dfrac{n-1}{n+1}$

7. $c_n = 1 + (-1)^n$

8. $x_n = (-1)^n n^2$

9. $v_n = \dfrac{(-1)^{n-1}}{n}$

10. $r_n = \dfrac{(-1)^{n+1}}{3^n}$

11. $a_n = \sqrt{n(n+1)}$

12. $y_n = \sqrt{n+1} - \sqrt{n}$

13. $b_n = \left(1 + \dfrac{1}{n}\right)^n$

14. $r_n = \dfrac{n^2}{2^n}$

15. $s_n = 1 + 2 + \cdots + n$

16. $a_n = 1 \cdot 2 \cdot \cdots \cdot n$

Exercises 17-24   *Find an expression for the nth term of a sequence whose first few terms are given.*

17. $2, 4, 6, 8, \ldots$

18. $3, 9, 27, 81, \ldots$

19. $-\dfrac{1}{2}, \dfrac{1}{4}, -\dfrac{1}{8}, \dfrac{1}{16}, \ldots$

20. $3, 7, 11, 15, \ldots$

21. $\dfrac{1}{2}, \dfrac{2}{3}, \dfrac{3}{4}, \dfrac{4}{5}, \ldots$

22. $2 \cdot 3, 3 \cdot 4, 4 \cdot 5, 5 \cdot 6, \ldots$

23. $1 - \dfrac{1}{1}, 1 - \dfrac{1}{2}, 1 - \dfrac{1}{3}, 1 - \dfrac{1}{4}, \ldots$

24. $1 + \dfrac{1}{2}, 1 - \dfrac{1}{4}, 1 + \dfrac{1}{8}, 1 - \dfrac{1}{16}, \ldots$

Exercises 25-34   *Indicate whether the sequence appears to be arithmetic, geometric, or neither. If you claim it is arithmetic, find the common difference. If you think it is geometric, find the common ratio.*

25. $5, 8, 11, 14, \ldots$

26. $1, 3, 6, 10, \ldots$

27. $3, 12, 60, 360, \ldots$

28. $3, -6, 12, -24, \ldots$

29. $5, 1, -3, -7, \ldots$

30. $3, 2, \dfrac{4}{3}, \dfrac{8}{9}, \ldots$

31. $\dfrac{1}{2}, \dfrac{3}{4}, \dfrac{7}{8}, \dfrac{15}{16}, \ldots$

32. $\dfrac{4}{3}, \dfrac{8}{3}, 4, \dfrac{16}{3}, \ldots$

33. $0.2, 0.02, 0.002, 0.0002, \ldots$

34. $1, 4, 9, 16, \ldots$

Exercises 35-40   *Find an expression for the nth term of the given arithmetic sequence.*

35. $7, 13, 19, 25, \ldots$

36. $8, 4, 0, -4, \ldots$

37. $100, 85, 70, 55, \ldots$

38. $2, 14, 26, 38, \ldots$

39. $-2, -\dfrac{3}{4}, \dfrac{1}{2}, \dfrac{7}{4}, \ldots$

40. $\dfrac{1}{2}, -\dfrac{1}{6}, -\dfrac{5}{6}, -\dfrac{3}{2}, \ldots$

Exercises 41-46   *Use the given information about the arithmetic sequence to find the indicated term.*

41. $a_1 = 3, d = 8; a_{10} = \underline{\quad}$

42. $a_1 = 22, d = -4; a_{18} = \underline{\quad}$

43. $a_1 = 10, a_5 = -10; a_{15} = \underline{\quad}$

44. $a_1 = 9, a_4 = 30; a_{21} = \underline{\quad}$

45. $a_8 = 38, a_{17} = 92; a_{31} = \underline{\quad}$

46. $a_{12} = -16, a_{20} = -40; a_{26} = \underline{\quad}$

Exercises 47-52   *Find an expression for the nth term of the given geometric sequence.*

47. $1, 3, 9, 27, \ldots$

48. $5, -10, 20, -40, \ldots$

49. $32, -16, 8, -4, \ldots$

50. $\dfrac{4}{3}, \dfrac{2}{3}, \dfrac{1}{3}, \dfrac{1}{6}, \ldots$

51. $4, 5, \dfrac{25}{4}, \dfrac{125}{16}, \ldots$

52. $-4, \dfrac{8}{3}, -\dfrac{16}{9}, \dfrac{32}{27}, \ldots$

Exercises 53-56   *Use the given information about the geometric sequence to find the indicated term.*

53. $a_1 = 2, r = 3; a_8 = \underline{\quad}$

54. $a_1 = 6, r = -\dfrac{1}{2}; a_{10} = \underline{\quad}$

55. $a_1 = 9, a_3 = 1; a_7 = \underline{\quad}$

56. $a_1 = \dfrac{1}{8}, a_3 = 2; a_6 = $ _____

Exercises 57-64 *Find the first five terms of the given recursively defined sequence.*

57. $a_1 = 2$ and $a_n = 1 - 2a_{n-1}$ for $n \geq 2$

58. $a_1 = 1$ and $a_n = 3a_{n-1} + 1$ for $n \geq 2$

59. $a_1 = 1$ and $a_n = na_{n-1}$ for $n \geq 2$

60. $a_1 = 1$ and $a_n = n + a_{n-1}$ for $n \geq 2$

61. $a_1 = 2$ and $a_n = \sqrt{a_{n-1}}$ for $n \geq 2$

62 $a_1 = 2$ and $a_n = \sqrt{(a_{n-1})^2 + 1}$ for $n \geq 2$

63. $a_1 = 1, a_2 = 2$, and $a_n = a_{n-1}a_{n-2}$ for $n \geq 3$

64. $a_1 = 1, a_2 = 2$, and $a_n = \dfrac{a_{n-1}}{a_{n-2}}$ for $n \geq 3$

Applications

65. Projecting Income Juanita is considering a new job that offers a starting base salary of $28,500, a guaranteed $1200 raise each year, and an annual bonus equal to 3% of the base salary. Determine her potential income (salary plus bonus) for each of the first 3 years and show that these values form an arithmetic sequence. What would her income be in the seventh year?

66. Projecting Income Vincent is considering a new job that offers a starting salary of $27,000 and a guaranteed $1500 raise each year. The benefit package (health, retirement, and so on) is 11% of the annual salary. Determine the total value (salary plus benefits) for each of the first 3 years and show that these totals form an arithmetic sequence. What would the total be in the eighth year?

67. Membership Increase An organization claims that its membership has increased 20% each year since it was formed with 10 charter members. How many members are there at the beginning of the 12th year?

68. Salary Increase If you start a job with an annual salary of $30,000 and a guaranteed 5% raise each year, what will your salary be at the beginning of your 15th year?

69. Accumulated Value Suppose $1000 is deposited in an account in which interest is compounded annually. The balances at the beginning of each of the first 4 years are $1000.00, $1055.00, $1113.03, and $1174.25. Assuming this pattern continues, write out a formula for the nth term of this sequence. What will the balance be at the beginning of the tenth year? What is the annual interest rate?

70. Amortization Schedule The first four entries on the amortization schedule for a $120,000 mortgage indicate monthly interest charges of $800.00, $799.46, $798.92, and $798.38. Write out a formula for the nth term of this sequence. How much interest will be charged in the 24th month?

71. Breeding Rabbits A pair of baby rabbits (one male and one female) is placed inside an enclosed area for the purposes of breeding. During the first month, the pair does not produce any new rabbits, but each month thereafter it produces a pair (one male and one female) of rabbits. Assuming each new pair follows the same reproductive pattern, complete the first six rows of the following table and find the number of rabbits at the end of 12 months. (*Hint:* Because of the 1-month maturation period, the number of pairs of rabbits born in a given month equals the number of pairs of rabbits at the beginning of the previous month.)

Month	Pairs at the beginning of the month	Pairs born during the month	Pairs at the end of the month
1	1	0	1
2	1	1	2
3	2	1	3
4	3		

72. Consumer Price Index The Consumer Price Index, or CPI, measures the prices of consumer goods and services and is a measure of the pace of U.S. inflation. If the CPI is 150 in one year and 161 in the next, then it takes roughly $161 in the second year to purchase goods and services that could have been purchased for $150 in the first year. It follows that if i_n is the inflation rate for year n and C_n is the CPI for year n, then $C_n = C_{n-1}(1 + i_n)$.

a. Use the formula for C_n to complete the following table.

Year	Average CPI	Percent change
1996	156.9	2.95%
1997		2.29%
1998		1.56%
1999		2.21%
2000		3.36%
2001		2.84%

b. The last year in which there was double-digit inflation was 1981, with an inflation rate of 10.1% and a CPI of 90.9. Find the CPI for 1980.

Concepts and Critical Thinking

Exercises 73-76 *Answer true or false.*

73. The sets $\{1, 2, 3\}$ and $\{3, 2, 1\}$ are identical.

74. The sequences 1, 2, 3 and 3, 2, 1 are identical.

75. The difference between consecutive terms of an arithmetic sequence is a constant.

76. The difference between consecutive terms of a geometric sequence is a constant.

Exercises 77-80 *Give an example of each.*

77. A recursively defined sequence

78. An arithmetic sequence with common difference 2

79. A geometric sequence with common ratio 3

80. A sequence that is both geometric and arithmetic

Questions for Discussion or Essay

81. Suppose the instructions for a set of algebra problems reads "Find an expression for the nth term of the sequence whose first few terms are given." Explain why these are not valid instructions. To help you see the problem, compute the first four terms of the following sequences, and answer the question "Why is 10 not necessarily the next number in the sequence 1, 4, 7, . . . ?"

$$a_n = 3n - 2$$
$$a_n = n^3 - 6n^2 + 14n - 8$$

82. The nth term of an arithmetic or geometric sequence can be defined either *explicitly* or *recursively*. For example, the nth term of the arithmetic sequence 4, 10, 16, 22, . . . can be defined explicitly by $a_n = 6n - 2$ for $n = 1, 2, 3, \ldots$ or recursively by $a_1 = 4$ and $a_n = a_{n-1} + 6$ for $n = 2, 3, \ldots$. Find both explicit and recursive definitions for the geometric sequence 5, 15, 45, ... Discuss the advantages and disadvantages of each definition.

83. In many of the application problems involving geometric sequences, we assumed that certain percentage rates of increase or decrease (for example, membership increase, salary increase, interest rate, population growth or decline) were constant from one year to the next, and this allowed us to make projections well into the future. Which, if any, of these assumptions are valid? Justify your answer. If a rate cannot be assumed to be constant, are the projections of any use? Explain. Give some examples from the "real world" that involve rates of increase or decrease that could be assumed to be constant.

Projects for Enrichment

84. Limiting Behavior of Sequences Although a formal definition of the limit of a sequence is beyond the scope of this text, we will say that a sequence $a_1, a_2, a_3, \ldots$ has a limit L if the terms a_n approach L (and only L) as n increases without bound. In this case we write $a_n \to L$ as $n \to \infty$. As an example, consider the sequence $a_n = \frac{1}{n}$ whose first few terms are $1, \frac{1}{2}, \frac{1}{3}, \frac{1}{4}, \ldots$. Clearly the terms of this sequence are getting closer and closer to zero. Thus, we conclude that $a_n \to 0$ as $n \to \infty$. It is sometimes possible to predict the limit of a sequence in this way, by inspecting the first "few" terms of the sequence.

   a. Write out a sufficient number of terms for the given sequence and predict the limit.

     i. $a_n = \dfrac{1}{n^2}$ **ii.** $a_n = \left(\dfrac{1}{2}\right)^n$

     iii. $a_n = \dfrac{n}{n+1}$ **iv.** $a_n = \dfrac{2n}{2^n - 1}$

Note that the method just described can also be very unreliable. A sequence may converge to a limit so slowly that it is difficult to determine the limit.

   b. Let

$$a_n = \frac{\ln n}{\sqrt{n}}$$

   Find a_n for $n = 10$, $n = 20$, $n = 50$, $n = 100$, and $n = 1000$. What do you think the limit is?

Not all sequences have a limit. For example, the terms of the arithmetic sequence 5, 8, 11, 14, . . . increase without bound and so do not approach any number. The terms of the sequence 1, −1, 1, −1, 1, . . . simply alternate between 1 and −1 and so do not approach a *single* number. For a sequence to have a limit, its terms must approach a unique finite number.

   c. Determine whether the given sequence has a limit. If so, state the limit. If not, explain why not.

i. $a_n = \dfrac{1}{2}n$

ii. $a_n = \dfrac{1}{n} + 5$

iii. $a_n = (-1)^n + 1$

iv. $a_n = \dfrac{(-1)^n}{n}$

v. $a_n = \left(\dfrac{3}{2}\right)^n$

vi. $a_n = e^{-n}$

Sometimes a graphing calculator can be a useful tool for studying limits of sequences. Consider the sequence

$$a_n = \dfrac{n}{\sqrt{2n^2 + 1}}$$

If we define the function

$$f(x) = \dfrac{x}{\sqrt{2x^2 + 1}}$$

then $f(n) = a_n$ for all n. To investigate the limit of a_n, we plot the graph of f and use the trace feature to see what value $f(x)$ approaches as x gets very large (see Figure 4). The cursor shows that $f(16) \approx 0.706$ and so $a_{16} \approx 0.706$. Since the graph does not appear to be rising further, we conclude that the limit is approximately 0.706. Using calculus, we could show that the exact limit is $\sqrt{2}/2$.

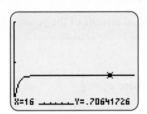

X=16 .... Y=.70641726

Figure 4

d. Use a graphing calculator to estimate the limit of the given sequence.

i. $a_n = \dfrac{2n}{3n + 1}$

ii. $a_n = \dfrac{4n^2 - 2n}{7n^2 + 3}$

iii. $a_n = \left(1 + \dfrac{1}{n}\right)^n$

e. Experiment with different viewing windows on a graphing calculator to investigate the limit of the sequence $a_n = \dfrac{0.001n^2 + n}{n + 1000}$. What can you conclude? What potential hazards can you see in using a graphing calculator to estimate limits?

85. Population Growth   Suppose a certain city has population P_0 and a constant rate of growth of k people per thousand per year. In other words, for each group of 1000 people in the population, there will be an increase of k people after 1 year. We see in this project that constant growth rates lead to geometric sequences.

a. In 2000, Las Vegas, Nevada, had a population of approximately 478,434 and a growth rate of approximately 58 people per thousand per year. Estimate the population in 2001 and 2002.

b. Argue more generally that if there are P_n people at the beginning of year n, then there will be

$$P_n + k\left(\dfrac{P_n}{1000}\right)$$

people at the end of year $n + 1$.

c. Rewrite the expression in part b to show that if P_n denotes the number of people at the beginning of year n, then $P_{n+1} = (1 + 0.001k)P_n$.

d. Use $P_0 = 478,434$ (the population in 2000) and $k = 58$ to estimate the population of Las Vegas, Nevada, in each of the years 2001–2005. Explain why this sequence is a geometric sequence. Estimate the population of Las Vegas in the year 2010.

e. In 2000, St. Louis, Missouri, had a population of approximately 348,189 and a growth rate of -26 people per thousand. Estimate the population in each of the years 2001–2005. Do these values form a geometric sequence? Why or why not? What do you notice about the population of St. Louis? How should we interpret a negative growth rate?

Section 10.2   Series

- How would a 10-year-old child prodigy add up the first 100 positive integers?
- Would you buy a $9600 car on the following terms: no money down and 48 monthly payments of $200 plus 2% per month interest on the remaining balance?
- If you had a choice between a job with a starting salary of $32,000, a yearly raise of $3000, and a 4% year-end bonus, and one with a starting salary of $32,000 and a 6% yearly raise, which would you choose?
- How is it that a Greek philosopher's "proof" of the impossibility of motion took centuries to refute?
- If you were to win the grand prize in a sweepstakes, should you take the $1,000,000 lump-sum payment or the annuity option of $100,000 each year for the next 20 years?

Series and Summation Notation

The story is told that when the German mathematician Karl Friederich Gauss was only 10 years old, he amazed his teacher by adding together the first 100 positive integers in just seconds. Historians speculate that Gauss accomplished this feat by observing that the numbers in the sum $1 + 2 + 3 + \cdots + 98 + 99 + 100$ can be grouped in pairs that add up to 101, as follows:

$$\underbrace{1 + 2 + 3 + \cdots + 49 + \overbrace{50 + 51}^{101} + 52 + \cdots + 98 + 99 + 100}_{\underbrace{\qquad\qquad\qquad\qquad}_{101}}$$

Because there are 50 such pairs, the total is simply $50(101) = 5050$. The sum $1 + 2 + 3 + \cdots + 100$ is an example of a **series**. In general, a series is a sum of a sequence of numbers, as indicated by the following definition.

Definition of Series

Given an infinite sequence $a_1, a_2, a_3, \ldots$, the sum

$$a_1 + a_2 + a_3 + \cdots + a_n$$

is called a **finite series**, whereas the sum

$$a_1 + a_2 + a_3 + \cdots$$

is called an **infinite series**. Each a_j is called a **term** of the series.

EXAMPLE 1

Evaluating a Series

Evaluate the series $a_1 + a_2 + \cdots + a_6$, where $a_n = \frac{1}{n}$.

Solution

$$
\begin{aligned}
a_1 + a_2 + \cdots + a_6 &= \frac{1}{1} + \frac{1}{2} + \frac{1}{3} + \frac{1}{4} + \frac{1}{5} + \frac{1}{6} \\
&= \frac{60 + 30 + 20 + 15 + 12 + 10}{60} \\
&= \frac{147}{60} \\
&= \frac{49}{20}
\end{aligned}
$$

Since series involving many terms can be awkward to write, we introduce a convenient shorthand notation called **summation notation**.

Summation Notation

$$\sum_{k=l}^{n} a_k = a_l + a_{l+1} + \cdots + a_n$$

$$\sum_{k=l}^{\infty} a_k = a_l + a_{l+1} + a_{l+2} + \cdots$$

The symbol Σ is the capital Greek letter **sigma**. The variable k is called the **index variable**. The expression $k = l$ below the sigma defines the starting value of the index variable, otherwise known as the **lower limit**. The ending value of the index variable—the **upper limit**—is given above the sigma.

EXAMPLE 2

Evaluating Series

Evaluate the given series.

a. $\displaystyle\sum_{k=1}^{5} 2^k$ **b.** $\displaystyle\sum_{i=1}^{6} i^2$

c. $\displaystyle\sum_{n=2}^{4} \left(-\frac{1}{10}\right)^n$

Solution

a. With a lower limit of 1 and an upper limit of 5, we obtain

$$\sum_{k=1}^{5} 2^k = 2^1 + 2^2 + 2^3 + 2^4 + 2^5$$

$$= 2 + 4 + 8 + 16 + 32$$

$$= 62$$

b. Note that the choice of index variable is immaterial. We start with $i = 1$ and end with $i = 6$.

$$\sum_{i=1}^{6} i^2 = 1^2 + 2^2 + 3^2 + 4^2 + 5^2 + 6^2$$

$$= 1 + 4 + 9 + 16 + 25 + 36$$

$$= 91$$

c. Here we have a lower limit of 2 and an upper limit of 4. Thus,

$$\sum_{n=2}^{4} \left(-\frac{1}{10}\right)^n = \left(-\frac{1}{10}\right)^2 + \left(-\frac{1}{10}\right)^3 + \left(-\frac{1}{10}\right)^4$$

$$= \frac{1}{100} - \frac{1}{1000} + \frac{1}{10,000}$$

$$= \frac{100 - 10 + 1}{10,000}$$

$$= \frac{91}{10,000}$$

EXAMPLE 3

Writing a Series With Summation Notation

Express the given series using summation notation.

a. $1 + 4 + 9 + 16 + 25 + 36 + 49 + 64 + 81$

b. $\dfrac{1}{3} + \dfrac{1}{9} + \dfrac{1}{27} + \dfrac{1}{81} + \dfrac{1}{243} + \dfrac{1}{729}$

Solution

a. The terms of this series are the squares of the numbers 1 through 9. Thus, the general kth term has the form $a_k = k^2$. The first term corresponds to $k = 1$ and the last term corresponds to $k = 9$. Thus, the lower and upper limits are 1 and 9, respectively. In summation notation we have

$$\sum_{k=1}^{9} k^2$$

b. The denominators of the terms of this series are successive powers of 3, starting with 3^1 and ending with $3^6 = 729$. Thus, the general kth term has the form $a_k = 1/3^k$. The first term corresponds to $k = 1$, and so the lower limit is 1. The last term corresponds to $k = 6$, and so the upper limit is 6. In summation notation we have

$$\sum_{k=1}^{6} \frac{1}{3^k}$$

So far, our emphasis has been on using and understanding summation notation rather than on techniques for evaluating series. We now turn our attention to methods for evaluating a few types of finite series. Infinite series will be considered in Exercise 79.

Arithmetic Series   Recall from Section 10.1 that an arithmetic sequence is one whose successive terms have a common difference. The sum of the terms of an arithmetic sequence is called an **arithmetic series**. We would like to develop a formula for evaluating *finite* arithmetic series. To that end, suppose $a_1, a_2, a_3, \ldots, a_n$ is an arithmetic sequence with common difference d, and let the sum S be given by

$$S = a_1 + a_2 + \cdots + a_{n-1} + a_n$$

Since $a_k = a_1 + (k - 1)d$, it follows that

(1) $\qquad S = a_1 + [a_1 + d] + \cdots + [a_1 + (n - 2)d] + [a_1 + (n - 1)d]$

By reversing the order of the terms of equation (1), we also have

(2) $\qquad S = [a_1 + (n - 1)d] + [a_1 + (n - 2)d] + \cdots + [a_1 + d] + a_1$

Adding equations (1) and (2) leads to the following:

$$
\begin{array}{rl}
S = a_1 & + [a_1 + d] \quad\quad + \cdots + [a_1 + (n - 2)d] \; + [a_1 + (n - 1)d] \\
+ \quad S = [a_1 + (n - 1)d] & + [a_1 + (n - 2)d] + \cdots + [a_1 + d] \quad\quad + a_1 \\
\hline
2S = [2a_1 + (n - 1)d] & + [2a_1 + (n - 1)d] + \cdots + [2a_1 + (n - 1)d] + [2a_1 + (n - 1)d]
\end{array}
$$

$$\underbrace{\qquad\qquad\qquad\qquad\qquad\qquad\qquad\qquad\qquad\qquad\qquad\qquad}_{n \text{ terms}}$$

$$
\begin{aligned}
2S &= n[2a_1 + (n - 1)d] \\
2S &= n[a_1 + a_1 + (n - 1)d] \\
2S &= n(a_1 + a_n) \qquad\qquad \text{Substituting } a_n \text{ for } a_1 + (n - 1)d \\
S &= \frac{n}{2}(a_1 + a_n) \qquad\quad \text{Dividing both sides by 2}
\end{aligned}
$$

Thus, we have the following formula.

Formula for Finite Arithmetic Series

If $a_1, a_2, \ldots, a_n$ is an arithmetic sequence, then

$$a_1 + a_2 + \cdots + a_n = \frac{n}{2}(a_1 + a_n)$$

In words, the sum of a finite arithmetic sequence is equal to the number of terms times the average of the first and last terms.

The formula for finite arithmetic series suggests another approach to the problem solved by the 10-year-old Gauss.

EXAMPLE 4   Using the Formula for Finite Arithmetic Series

Find the sum of the first 100 positive integers.

Solution   The series $1 + 2 + 3 + \cdots + 100$ is a finite arithmetic series. Applying the formula for finite arithmetic series with $n = 100$, $a_1 = 1$, and $a_{100} = 100$, we obtain

$$1 + 2 + 3 + \cdots + 100 = \frac{100}{2}(1 + 100) = 5050$$

EXAMPLE 5   Evaluating an Arithmetic Series

Compute $\sum_{j=1}^{40}(3j + 5)$.

Solution   Substituting values of j and evaluating gives us

$$\sum_{j=1}^{40}(3j + 5) = 8 + 11 + 14 + \cdots + 125$$

We note that since consecutive terms differ by 3, the series is arithmetic. Since the first term of the series is 8 and the last (40th) term is 125, we have

$$\sum_{j=1}^{40}(3j + 5) = \frac{40}{2}(8 + 125)$$

$$= 2660$$

EXAMPLE 6   An Unusual Financing Arrangement

A1 Used Cars is offering special financing. With no money down and approved credit, you can drive away in a $9600 car and pay the balance in 48 monthly payments of only $200, plus 2% interest per month on the remaining balance. How much interest will you pay over 4 years?

Solution   The 48 interest payments can be viewed as the terms of an arithmetic sequence, as shown in the following table.

Payment number	Balance before $200 payment	Interest charged	Total (payment plus interest)	Balance after $200 payment
1	$9600	(0.02)9600 = $192	$200 + $192 = $392	$9400
2	$9400	(0.02)9400 = $188	$200 + $188 = $388	$9200
3	$9200	(0.02)9200 = $184	$200 + $184 = $384	$9000
⋮	⋮	⋮	⋮	⋮
48	$200	(0.02)200 = $4	$200 + $4 = $204	$0

Applying the formula for finite arithmetic series to the interest charges, we obtain

$$192 + 188 + 184 + \cdots + 4 = \frac{48}{2}(192 + 4) = 4704$$

Thus, the total interest paid over 4 years would be $4704. An unusual feature of this financing scheme is that the monthly payments decrease over the life of the loan since the amount of interest decreases each month.

Geometric Series

In Section 10.1, we defined a geometric sequence as one whose successive terms have a common ratio. The sum of the terms of a geometric sequence is called a **geometric series**. Just as we did with arithmetic series, we would like to develop a formula for evaluating geometric series. Thus, let $a_1, a_2, \ldots, a_n$ be a geometric sequence with common ratio r, and let the sum S be given by

$$S = a_1 + a_2 + \cdots + a_{n-1} + a_n$$

Since $a_k = a_1 r^{k-1}$, we have

(3) $$S = a_1 + a_1 r + a_1 r^2 + \cdots + a_1 r^{n-2} + a_1 r^{n-1}$$

If we multiply both sides of equation (3) by r, we obtain

(4) $$rS = a_1 r + a_1 r^2 + a_1 r^3 + \cdots + a_1 r^{n-1} + a_1 r^n$$

Subtracting equation (4) from equation (3) gives us

$$
\begin{aligned}
S &= a_1 + a_1 r + a_1 r^2 + \cdots + a_1 r^{n-1} \\
- \quad rS &= \qquad\; a_1 r + a_1 r^2 + \cdots + a_1 r^{n-1} + a_1 r^n \\
\hline
S - rS &= a_1 - a_1 r^n
\end{aligned}
$$

$$S(1 - r) = a_1(1 - r^n)$$

$$S = \frac{a_1(1 - r^n)}{1 - r} \qquad \text{Assuming } r \neq 1$$

Thus, we have the following formula.

Formula for Finite Geometric Series

If $a_1, a_2, a_3, \ldots, a_n$ is a geometric sequence with common ratio r (so that $a_k = a_1 r^{k-1}$), and $r \neq 1$, then

$$a_1 + a_2 + \cdots + a_n = \frac{a_1(1 - r^n)}{1 - r}$$

⋯ EXAMPLE 7   Finding the Sum of a Geometric Sequence

Find the sum of the first 12 terms of the geometric sequence $5, 2, 0.8, 0.32, \ldots$.

Solution   The formula for finite geometric series involves only the first term, the common ratio, and the number of terms. The first term is $a_1 = 5$. The common ratio is

$$r = \frac{a_2}{a_1} = \frac{2}{5} = 0.4$$

The number of terms is $n = 12$. Thus,

$$a_1 + a_2 + \cdots + a_{12} = \frac{a_1(1 - r^n)}{1 - r}$$

$$= \frac{5[1 - (0.4)^{12}]}{1 - 0.4}$$

$$\approx 8.33319$$

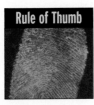

Rule of Thumb

As suggested by Example 7, it is helpful to think of a_1 and n in the expression

$$\frac{a_1(1 - r^n)}{1 - r}$$

as simply the first term and the number of terms, respectively.

EXAMPLE 8

Evaluating a Geometric Series

Evaluate $\sum_{n=3}^{10} 4\left(-\frac{1}{3}\right)^n$.

Solution The first term of the series is $4\left(-\frac{1}{3}\right)^3 = -\frac{4}{27}$. Since the starting index is 3 and the ending index is 10, there are 8 terms in the sum. Finally, the common ratio is $-\frac{1}{3}$. Thus,

$$\sum_{n=3}^{10} 4\left(-\frac{1}{3}\right)^n = \frac{(\text{First term})\left[1 - \left(-\frac{1}{3}\right)^{(\text{Number of terms})}\right]}{1 - \left(-\frac{1}{3}\right)}$$

$$= \frac{-\frac{4}{27}\left[1 - \left(-\frac{1}{3}\right)^8\right]}{1 + \frac{1}{3}}$$

$$\approx -0.111094$$

EXAMPLE 9

Comparing Salary Contracts

Suppose you are starting a new job that offers a choice between two contracts. The first has a starting base salary of $32,000, yearly increases of $3000 in the base salary, and a year-end bonus equal to 4% of the base salary for that year. The second has a starting salary of $32,000 and a 6% raise each year. Which job would pay the highest total salary over a 26-year period?

Solution The following table shows the total income for both contract options for the first 4 years.

Year	Option 1: Bonus and fixed raise			Option 2: Percentage raise
	Base salary	Bonus of 4%	Total	Salary (including 6% raise)
1	32,000	(0.04)32,000 = 1280	33,280	32,000
2	35,000	(0.04)35,000 = 1400	36,400	32,000 + (0.06)32,000 = 32,000(1.06) = 33,920
3	38,000	(0.04)38,000 = 1520	39,520	33,920(1.06) = 35,955.20
4	41,000	(0.04)41,000 = 1640	42,640	35,955.20(1.06) = 38,112.51

The salary totals for Option 1 form an arithmetic sequence with common difference $d = 36,400 - 33,280 = 3120$. Using the formula for the nth term of an arithmetic sequence, the salary for year 26 will be $a_{26} = 33,280 + (26 - 1)3120 = 111,280$.

Now we apply the formula for finite arithmetic series to determine the sum of the salaries for 26 years.

$$33{,}280 + 36{,}400 + \cdots + 111{,}280 = \frac{26}{2}(33{,}280 + 111{,}280)$$

$$= 1{,}879{,}280$$

The salaries for Option 2 form a geometric sequence with common ratio $r = 1.06$. Thus, using the formula for finite geometric series, we obtain the following salary total:

$$32{,}000 + 33{,}920 + \cdots + (\text{Salary for year 26}) = \frac{32{,}000(1 - 1.06^{26})}{1 - 1.06}$$

$$\approx 1{,}893{,}004.25$$

So even though Option 1 starts out with larger salaries, Option 2 results in a slightly higher total.

Understanding and Mastery Checklists

Concepts to Understand	Skills to Master
Series	Evaluate a finite series given in summation form.
Term of a series	Given the first few terms of a series, express the series in summation notation.
Summation notation: index, lower limit, upper limit	Evaluate a finite arithmetic series.
Arithmetic series	Evaluate a finite geometric series.
Formula for a finite arithmetic series	
Geometric series	
Formula for a finite geometric series	

Exercises 10.2

Exercises 1-4 *Find the sum of the first five terms of the sequence.*

1. $a_n = 4n + 3$

2. $a_n = n^2$

3. $a_n = \dfrac{n}{n + 1}$

4. $a_n = \dfrac{4}{10^n}$

Exercises 5-14 *Find the indicated sum.*

5. $\displaystyle\sum_{k=1}^{6}(3k - 2)$

6. $\displaystyle\sum_{i=1}^{4}(1 - 5i)$

7. $\displaystyle\sum_{n=1}^{5} -3$

8. $\displaystyle\sum_{k=1}^{4} 8$

9. $\displaystyle\sum_{j=1}^{4} 2j^2$ **10.** $\displaystyle\sum_{k=2}^{5} k^3$

11. $\displaystyle\sum_{k=3}^{6} \frac{1}{k}$ **12.** $\displaystyle\sum_{n=0}^{4} \frac{1}{n+1}$

13. $\displaystyle\sum_{i=0}^{4} 5\left(\frac{1}{2}\right)^i$ **14.** $\displaystyle\sum_{k=1}^{5} 3(-2)^k$

Exercises 15–24 *Express the series using summation notation.*

15. $\dfrac{1}{1^3} + \dfrac{1}{2^3} + \dfrac{1}{3^3} + \dfrac{1}{4^3} + \dfrac{1}{5^3} + \dfrac{1}{6^3}$

16. $\dfrac{1}{4^1} + \dfrac{1}{4^2} + \dfrac{1}{4^3} + \dfrac{1}{4^4} + \dfrac{1}{4^5}$

17. $[5(1) + 2] + [5(2) + 2] + [5(3) + 2] + \cdots + [5(12) + 2]$

18. $[7(1)^2 - 1] + [7(2)^2 - 1] + [7(3)^2 - 1] + \cdots + [7(15)^2 - 1]$

19. $3 + 9 + 27 + \cdots + 2187$

20. $2 - 4 + 8 - 16 + \cdots + 128$

21. $\dfrac{1}{1} - \dfrac{1}{4} + \dfrac{1}{9} - \dfrac{1}{16} + \cdots - \dfrac{1}{256}$

22. $\dfrac{1}{2} + \dfrac{2}{3} + \dfrac{3}{4} + \cdots + \dfrac{20}{21}$

23. $\dfrac{1}{2} + \dfrac{3}{4} + \dfrac{7}{8} + \cdots + \dfrac{63}{64}$

24. $\dfrac{1}{1 \cdot 2} + \dfrac{1}{2 \cdot 3} + \dfrac{1}{3 \cdot 4} + \cdots + \dfrac{1}{15 \cdot 16}$

Exercises 25–36 *Evaluate the given arithmetic series.*

25. $6 + 14 + 22 + \cdots + 158$ (20 terms)

26. $10 + 16 + 22 + \cdots + 154$ (25 terms)

27. $2 + 2.5 + 3 + \cdots + 12.5$

28. $47 + 44 + 41 + \cdots + 2$

29. $a_1 + a_2 + \cdots + a_{15}$, where $a_1 = 12$ and $a_2 = 8$

30. $a_1 + a_2 + \cdots + a_{18}$, where $a_1 = -3$ and $a_2 = 4$

31. $\displaystyle\sum_{n=1}^{50} (3n - 2)$ **32.** $\displaystyle\sum_{k=1}^{60} (8k + 2)$

33. $\displaystyle\sum_{k=1}^{30} (200 - 6k)$ **34.** $\displaystyle\sum_{n=1}^{80} (10 - 0.01n)$

35. $\displaystyle\sum_{i=10}^{25} \frac{5i + 3}{2}$ **36.** $\displaystyle\sum_{k=5}^{35} \left(\frac{1}{2}k + 3\right)$

Exercises 37–48 *Evaluate the given geometric series.*

37. The sum of the first 10 terms of the sequence $5, 15, 45, \ldots$

38. The sum of the first 12 terms of the sequence $20, 4, 0.8, \ldots$

39. $a_1 + a_2 + \cdots + a_{15}$, where $a_1 = 3$ and $a_2 = 1$

40. $a_1 + a_2 + \cdots + a_{10}$, where $a_1 = -2$ and $a_2 = 4$

41. $4 - 0.4 + 0.04 - 0.004 + \cdots - 0.0000004$

42. $\dfrac{1}{25} + \dfrac{1}{5} + 1 + \cdots + 5^7$

43. $\displaystyle\sum_{k=1}^{8} 3(4^k)$

44. $\displaystyle\sum_{n=1}^{11} 10(0.6)^n$

45. $\displaystyle\sum_{n=0}^{9} 2\left(-\frac{1}{6}\right)^n$

46. $\displaystyle\sum_{k=0}^{8} \frac{1}{2}(-2)^k$

47. $\displaystyle\sum_{k=5}^{16} -3(0.2)^k$

48. $\displaystyle\sum_{i=3}^{12} 9\left(\frac{2}{3}\right)^i$

Exercises 49–52 *Find the sum of the indicated integers.*

49. The first 1000 positive integers

50. The first 200 positive even integers

51. The first 100 positive odd integers

52. The first 100 positive multiples of 3

Applications

53. Pipe Pile   A pile of sewer pipes has 12 layers. The first layer has 30 pipes, the second has 29, the third has 28, and so on, as shown in Figure 5. How many pipes are in the pile?

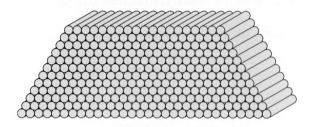

Figure 5

54. Brick Laying   A bricklayer estimates that in an 8-hour shift she can complete 15 rows of bricks in a portion of a wall near the peak of a roof, as shown in Figure 6. The first row has 24 bricks, the second has 23 bricks, and so on. How many bricks will she need that day?

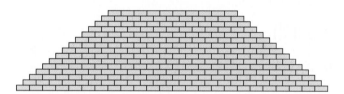

Figure 6

55. Auditorium Seats   An auditorium has 25 rows of seats. The first row has 12 seats, the second row has 14 seats, the third row has 16 seats, and so on, as shown in Figure 7. How many seats are in the auditorium?

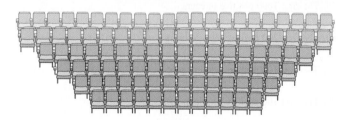

Figure 7

56. Building Height   An object is dropped from the top of the First Interstate World Center in Los Angeles. During the first second, it falls 16 feet; during the next second, it falls 48 feet; during the third second, it falls 80 feet; and so on. The object hits the ground after approximately 8 seconds. How high is the World Center?

57. Cell Phone Charges   The amount of money spent by consumers on cell phone charges in the United States for the years 1990–1999 can be approximated with the model $a_n = 4.85n + 5.86$, where n denotes the year (with $n = 0$ corresponding to 1990) and a_n is the expenditure for that year in billions of dollars. (*Data source:* U.S. Census

Bureau, Statistical Abstract of the United States.) Find the total of all cell phone charges for 1990 through 1999.

58. Government Expenditures   The annual expenditures by the federal government for the years 1990–2001 can be approximated with the model $a_n = 0.0518n + 1.258$, where n is the year (with $n = 0$ corresponding to 1990) and a_n is the outlay for that year in trillions of dollars. (*Data source:* U.S. Office of Management and Budget.) Find the total of all expenditures for 1990–2001.

59. Accumulated Value   Jamal deposits $100 at the beginning of each month in an account that pays 5% interest compounded monthly. Evaluate the following geometric series to find the balance in his account at the end of 4 years:

$$B = 100\left(1 + \frac{0.05}{12}\right)^1 + 100\left(1 + \frac{0.05}{12}\right)^2$$
$$+ 100\left(1 + \frac{0.05}{12}\right)^3 + \cdots + 100\left(1 + \frac{0.05}{12}\right)^{48}$$

60. Credit Card Debt   Todd has run up a $2000 debt on a credit card that charges interest on the unpaid balance at a rate of 1.5% per month. He resolves to stop using the card and, 1 month later, begins making monthly payments of $50. His balance after the 60th payment can be found using the following formula. What is his balance?

$$B = 2000(1.015)^{60} - [50 + 50(1.015)^1$$
$$+ 50(1.015)^2 + \cdots + 50(1.015)^{59}]$$

61. Internet Advertising   Advertising revenues from the Internet exploded at the end of the twentieth century. For the years 1995–2000, annual Internet-based advertising revenues can be approximated with the model $a_n = 84\left(\frac{8}{3}\right)^n$, where n denotes the year (with $n = 0$ corresponding to 1995) and a_n is the advertising revenue in millions of dollars. (*Data source:* Interactive Advertising Bureau.) Estimate the total Internet advertising revenue for the period 1995–2000.

62. Mail Delivery   The number of pieces of mail delivered by the U.S. Postal Service from 1971 to 2001 can be approximated with the model $a_n = 83(1.03)^n$, where n denotes the year (with $n = 0$ corresponding to 1971) and a_n is the number of pieces of mail delivered in billions. (*Data source:* U.S. Postal Service.) Find the total number of pieces of mail delivered from 1971 through 2001.

63. Loan Repayment   Molly has just borrowed $2400 from her Uncle Dave and agrees to pay it back in 24 monthly payments of $100, plus 0.5% interest per month on the remaining balance. How much interest will Molly have to pay?

64. Rumor Mill   Suppose a disgruntled accountant in a large corporation begins his workday by telling four coworkers that the president of the corporation has been embezzling. After 1 hour, each of these four has spread the rumor to four of their friends. After the second hour, each of those has told four more, and so on. How many people have heard the rumor after 8 hours? If this pattern continues through the second day, how many will have heard the rumor after 16 hours?

65. Contract Value   Suppose a college basketball star signs a 15-year contract to play professional basketball. If the first year's salary is $1.6 million and the salary for each of the following 14 years is 10% more than the preceding year's, how much is the total contract worth?

66. Job Options   If you plan to stay with a job for 30 years, would you choose a job with a starting salary of $28,000, a yearly bonus of 3%, and a guaranteed annual raise of $4000, or one with a starting salary of $36,000 and a guaranteed raise of 4% per year? Explain.

Concepts and Critical Thinking

Exercises 67–70   *Answer true or false.*

67. $\sum_{j=1}^{5} j$ is equivalent to $5j$.

68. $\sum_{j=1}^{5} k$ is equivalent to $5k$.

69. If successive terms in a series have a common ratio, then we say that the series is geometric.

70. A finite arithmetic series can be evaluated by multiplying the number of terms in the series by the average of the first and last terms.

Exercises 71–75   *Give an example of each.*

71. An arithmetic series

72. A geometric series

73. An arithmetic series with 5 terms that evaluates to 60

74. A series with m terms that evaluates to $3m$

75. A geometric series that evaluates to 0 and consists of 10 terms, none of which is zero

Questions for Discussion or Essay

76. Expand and compare the terms of the following series:

a. $\sum_{k=1}^{5} \frac{1}{k}$ **b.** $\sum_{k=3}^{7} \frac{1}{k-2}$

Find a third way of writing the series, this time using $k = 0$ as the starting index. What can you conclude about the way in which a given series can be represented using summation notation? Explain.

77. In Example 9, we saw that a percentage raise eventually beat a fixed raise with a bonus. Do you think this is always the case? How much effect does the length of time have on the situation? What are some

factors that might lead you to choose the fixed-raise option even if the percentage option ends a little better? How would these factors be affected by the length of time you plan on keeping the job?

78. Financing terms such as those described in Example 6 are uncommon because federal law requires a lender to provide the loan's A.P.R. (annual percentage rate), as well as the total interest charge. This allows a borrower to make intelligent choices between loans with different terms. Consult a local newspaper or inquire with a local lending institution to see if you find the A.P.R. and interest charge information to be readily offered. Discuss some other factors that you would consider when choosing among different loan offers.

Projects for Enrichment

79. Zeno's Paradox and Infinite Geometric Series   Around 460 B.C., the Greek philosopher Zeno of Elea posed a paradox that has intrigued philosophers and mathematicians for centuries. He stated that motion is paradoxical in that it is seemingly not possible to travel a finite distance in a finite amount of time. In order to do so, one would first have to travel half of the distance, then half of the remaining distance, then half of the remaining distance, and so on, as suggested in Figure 8.

Figure 8

Thus, one would have to cover an infinite number of distances, which is clearly not possible in a finite amount of time. Or is it? Suppose the distance between A and B is 1 mile and you start walking from A to B at a rate of 1 mile per hour. At that rate, you would cover the first half mile in $\frac{1}{2}$ hour, the next fourth mile in $\frac{1}{4}$ hour, the next eighth mile in $\frac{1}{8}$ hour, and so on. Thus, your total time in traveling from A to B would be given by the infinite sum

$$\frac{1}{2} + \frac{1}{4} + \frac{1}{8} + \cdots$$

Let us define S_n to be the sum of the first n terms of this infinite series. Then

$$S_n = \frac{1}{2} + \frac{1}{4} + \cdots + \left(\frac{1}{2}\right)^n = \sum_{k=1}^{n}\left(\frac{1}{2}\right)^k$$

a. Find S_1, S_2, S_3, and S_4 and explain what each represents.

b. Use the formula for finite geometric series to show that

$$S_n = 1 - \frac{1}{2^n}$$

c. Compute S_5, S_{10}, S_{20}, and S_{40}. What do these values approach? Since each value is the sum of more and more terms of the infinite sum $\frac{1}{2} + \frac{1}{4} + \frac{1}{8} + \cdots$, what would be a reasonable guess for the value of the infinite sum? According to this value, how much time will it take to travel from A to B? Does this agree with the amount of time suggested by the standard formula $d = rt$?

Using calculus, a tool developed centuries after Zeno, it is possible to show that if $|r| < 1$, then the sum S of the infinite geometric series $a + ar + ar^2 + ar^3 + \cdots$ is given by

$$S = \frac{a}{1 - r}$$

d. Use this formula to verify the value you found in part c.

e. Find the sum of the following infinite geometric series:

i. $\dfrac{1}{4} + \dfrac{1}{16} + \dfrac{1}{64} + \cdots$

ii. $\dfrac{2}{3} + \dfrac{4}{9} + \dfrac{8}{27} + \cdots$

iii. $\dfrac{1}{10} - \dfrac{1}{100} + \dfrac{1}{1000} - \dfrac{1}{10,000} + \cdots$

80. Geometric Series and Annuities An **annuity** is a sequence of regular payments or receipts, all subject to some underlying interest rate. Some examples of annuities include mortgage payments, installment purchases, premium payments on insurance policies, payroll deductions for pension plans, and lease payments. We limit our discussion to annuities for which payment is made at the end of the payment period. As a simple example, suppose you deposit $100 at the *end* of each year for the next 10 years at an interest rate of 5% compounded annually. The **future value** of a given payment is simply its value at the end of the last year, and this can be found using the standard formula for compound interest. Thus, the first payment, which earns 5% interest for 9 years, has a future value of $100(1 + 0.05)^9$. The second has a future value of $100(1 + 0.05)^8$, the third has a future value of $100(1 + 0.05)^7$, and so on up to the tenth, which has a future value of 100. The future value of the annuity is the sum of the future values of all the payments. If we denote this sum by S and write the terms of the sum in reverse order, we have

$$S = 100 + 100(1 + 0.05)^1$$
$$+ 100(1 + 0.05)^2 + \cdots + 100(1 + 0.05)^9$$

a. Use the formula for finite geometric series to find the future value of the annuity just described.

More generally, the future value of an annuity with periodic payments R made at the end of each of n periods with interest rate i is given by

$$S = R + R(1 + i)^1 + R(1 + i)^2 + \cdots + R(1 + i)^{n-1}$$

b. Use the formula for finite geometric series to show that

$$S = R\left[\frac{(1 + i)^n - 1}{i}\right]$$

c. Find the future value for the following annuities:

i. Eliza deposits $3000 at the end of each year for 20 years into a pension fund that earns 6% compounded annually.

ii. Charlie deposits $500 at the end of each quarter for 5 years into a bank account that earns interest at a rate of 4% compounded quarterly. Note that the periodic interest rate is $i = \frac{0.04}{4} = 0.01$ and the number of periods is $n = 4(5) = 20$.

We next consider the notion of **present value**. As an example, suppose a parent wishes to invest a certain amount of money now, at 8% compounded annually, in order to provide a child with a yearly allowance of $1000 at the end of each year for the next 10 years. We can determine this amount by considering the present value of a single payment—that is, the amount that must be invested now to make a given payment. Using the formula for compound interest, the amount A_{10} that must be invested now to make the payment at the end of the tenth year is found by solving $1000 = A_{10}(1 + 0.08)^{10}$. Thus, $A_{10} = 1000/(1 + 0.08)^{10}$. Similarly, $A_9 = 1000/(1 + 0.08)^9$, $A_8 = 1000/(1 + 0.08)^8$, and so on down to $A_1 = 1000/(1 + 0.08)$. The present value of the annuity—that is, the entire amount that must be invested—is simply the sum of the present values of each payment. Thus,

$$A = A_1 + A_2 + \cdots + A_{10}$$
$$= 1000\left(\frac{1}{1 + 0.08}\right)^1 + 1000\left(\frac{1}{1 + 0.08}\right)^2 + \cdots + 1000\left(\frac{1}{1 + 0.08}\right)^{10}$$

This is also a geometric series.

d. Use the formula for finite geometric series to find the present value of the annuity just described.

More generally, if an annuity has n periodic payments of R dollars and a periodic interest rate i, then its present value is given by

$$A = R\left(\frac{1}{1 + i}\right)^1 + R\left(\frac{1}{1 + i}\right)^2 + \cdots + R\left(\frac{1}{1 + i}\right)^n$$

e. Use the formula for finite geometric series to show that

$$A = R\left[\frac{1 - (1 + i)^{-n}}{i}\right]$$

f. Use the concept of present value to solve the following:

    i. An investment contract promises to pay $5000 at the end of each of the next 10 years. If you wish to earn 9% compounded annually, how much should you pay for the contract now?

ii. Suppose you have just won the grand prize in a sweepstakes and must choose between receiving $1,000,000 now or $100,000 at the end of each of the next 20 years. Assuming an annual yield of 10% on the annuity option, which option should you choose? Explain. What additional factors might influence your decision?

Section 10.3

Permutations and Combinations

- Why should a combination lock really be called a permutation lock?
- Why is it not a good idea for a TV program manager to select a program schedule by simply trying all possibilities?
- How many credit cards does it take to make a single dollar bill?
- How many winning combinations are possible in a state lottery?

As young children, we mastered the simplest and often most effective counting technique: Establish a one-to-one correspondence between the objects in a set and a subset of the natural numbers by pointing to the first object, announcing "one," pointing to the second object and saying "two," and so forth. As effective as this technique is for small sets, it fails miserably for larger sets. If you were capable of counting at the rate of one per second, then determining the number of possible 5-card poker hands (working around the clock) would take a solid month; counting the number of possible 9-digit social security numbers would take 30 years; and if you're planning on enumerating the ways 15 people can form a line, don't make any other plans—you're going to be booked solid for the next 40,000 years.

More sophisticated counting techniques are the province of combinatorics, a branch of mathematics essential to the computation of probability and, hence, having widespread application throughout the physical and natural sciences. Though we only scratch the surface of this deep and broad subject, the three counting tools developed in this section— the Fundamental Counting Principle, permutations, and combinations—are sufficient to perform, in a matter of minutes, all the tasks referred to in the preceding paragraph.

The Fundamental Counting Principle

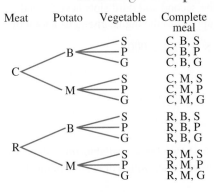

Figure 9

Suppose you are on a committee that is planning a banquet. The committee has selected a caterer that offers a choice of two meats (chicken breast or roast beef), two potatoes (baked or mashed), and three vegetables (spinach, peas, or green beans). Your task is to prepare a list of all the meals consisting of one meat, one potato, and one vegetable. In the process of creating this list, you would likely discover the need for a systematic method for finding all the possibilities without any duplication. One such method involves the use of a **tree diagram**.

We first adopt the convention that each food item is denoted by the first letter of its name (that is, C for chicken breast, R for roast beef, and so on). We next list the two choices for meat in a column labeled "Meat," as shown in Figure 9. Since each meat choice has two choices for potato, we draw two lines, called **decision branches**, from each meat to the two possible choices for potato in the column labeled "Potato." Similarly, since each potato has three possible choices for vegetable, we draw three branches from each potato to the three vegetables in the column labeled "Vegetable." All the complete meals can be found by following the different paths from meat to vegetable. There are 12 possible meals.

If it is only necessary to determine the *number* of possible meals, without actually listing them, there is a much easier way. Since each of the two meats can be paired with one of the two potatoes, there are $2 \cdot 2 = 4$ distinct meat and potato pairs. For each of these 4 meat and potato pairs, there are 3 possibilities for a vegetable, for a total of $4 \cdot 3 = 12$ meals. Thus, given a meal selection task involving 2 choices for meat, 2 choices for potato, and 3 choices of vegetable, there are $2 \cdot 2 \cdot 3 = 12$ possible meals. This observation generalizes into the **Fundamental Counting Principle**.

Fundamental Counting Principle

If a task consists of k parts, the first of which can be completed in n_1 ways, the second of which can be completed in n_2 ways, and so on up to the kth, which can be completed in n_k ways, then the total number of ways in which the task can be completed is given by the product

$$n_1 n_2 \cdots n_k$$

EXAMPLE 1

Counting License Plates

The state of Michigan has license plates of the form LLL DDD, where L denotes a letter and D denotes a digit.

a. How many different license plates of this form are possible?
b. What if no letter or digit can be repeated?

Solution

a. It is convenient to first draw six blank slots to represent the letters and digits that must be selected.

——— ——— ——— ——— ——— ———

Each of the first three slots must be filled with a letter, and so each has 26 possibilities. Each of the last three slots must be filled with one of the digits 0–9, and so each has 10 possibilities. We represent these choices as follows:

$$\underline{26} \quad \underline{26} \quad \underline{26} \quad \underline{10} \quad \underline{10} \quad \underline{10}$$

Applying the Fundamental Counting Principle, we see that there are

$$26 \cdot 26 \cdot 26 \cdot 10 \cdot 10 \cdot 10 = 17{,}576{,}000$$

possible license plates.

b. Once again, the first slot must be filled with a letter, and so there are 26 possibilities. However, since repetitions are not allowed, the second and third slots have 25 and 24 possibilities, respectively. Similar reasoning applies to the digits, giving us a total of

$$26 \cdot 25 \cdot 24 \cdot 10 \cdot 9 \cdot 8 = 11{,}232{,}000$$

possible license plates.

EXAMPLE 2

Arranging TV Shows

A TV program manager must choose among 6 shows for 4 available time slots. He decides to try each possible arrangement for 1 week to see which draws the highest ratings. How many weeks will this take?

Solution   We first draw 4 blank lines to represent the 4 time slots that must be filled. Initially, there are 6 choices for filling the first time slot, and so we place a 6 in the first blank. If we assume that the same show may not be used twice, there are only 5 choices remaining for the second time slot. Continuing in this fashion, there are 4 choices for the third slot, and 3 for the fourth.

$$\underline{\quad 6 \quad} \quad \underline{\quad 5 \quad} \quad \underline{\quad 4 \quad} \quad \underline{\quad 3 \quad}$$

By viewing the selection process as a 4-part task, each part of which can be completed in 6, 5, 4, and 3 ways, respectively, the Fundamental Counting Principle tells us that the total number of possibilities is

$$6 \cdot 5 \cdot 4 \cdot 3 = 360$$

Thus, it would take 360 weeks, or almost 7 years, to try all the possible arrangements.

Permutations

In Example 2, the *ordering* of the TV shows was an essential part of the problem. That is, if the same four shows were arranged in a different order, we would consider it to be a completely different arrangement. We say that each of the ordered arrangements is a *permutation* of 6 shows taken 4 at a time. In general, we define a permutation as follows.

Definition of Permutation

A **permutation** of n elements taken r at a time is an ordered arrangement, without repetitions, of r of the n elements. The number of permutations of n elements taken r at a time is denoted by $_nP_r$.

To compute $_nP_r$, we consider an ordered arrangement of r out of n elements without repetitions. Following the lead of Examples 1 and 2, we draw r blank slots to represent the r elements that must be chosen.

$$\underbrace{\underline{\quad\quad} \;\; \underline{\quad\quad} \;\; \underline{\quad\quad} \;\; \cdots \;\; \underline{\quad\quad}}_{r \text{ blank slots}}$$

Since there are initially n elements to choose from, the first slot can be filled in n ways. Since repetitions are not allowed, the second slot can be filled in $n - 1$ ways. Continuing in this fashion, the third slot can be filled in $n - 2$ ways, and so on up to the rth slot, which can be filled in $n - (r - 1)$ or $n - r + 1$ ways.

$$\underbrace{\underline{\quad n \quad} \quad \underline{n - 1} \;\; \underline{n - 2} \;\; \cdots \;\; \underline{n - r + 1}}_{r \text{ slots}}$$

By applying the Fundamental Counting Principle, we obtain the following formula.

Permutation Formula

For integers n and r with $1 \leq r \leq n$,

$$_nP_r = n(n - 1)(n - 2) \cdots (n - r + 1)$$

For convenience, we define $_nP_0 = 1$.

Rule of Thumb

You do not need to memorize the permutation formula just given. Instead, just remember that the product starts with n and has r factors, each of which is one less than the preceding factor.

EXAMPLE 3

Evaluating Permutations

Evaluate the following permutations:

a. $_7P_3$ **b.** $_5P_4$

Solution

a. To compute $_7P_3$, we start with 7 and write out 3 factors. Thus,

$$_7P_3 = \underbrace{7 \cdot 6 \cdot 5}_{3 \text{ factors}} = 210$$

b. We start with 5 and write out 4 factors to obtain

$$_5P_4 = \underbrace{5 \cdot 4 \cdot 3 \cdot 2}_{4 \text{ factors}} = 120$$

Calculator Keys

Permutations

Most graphing calculators are capable of computing permutations. To compute $_nP_r$, first key in the number n, then select nPr from the menu of mathematical functions, and, finally, enter the number r.

EXAMPLE 4

Using Permutations to Count Batting Orders

The manager of a coed softball team must assign a batting order by selecting 10 players from a team consisting of 14 players: 8 men and 6 women. Find the number of batting orders that are possible if

a. there are no restrictions on the number of men and women
b. league rules require that 3 women must bat first

Solution

a. If there are no restrictions, then the manager may simply select and arrange 10 out of the 14 players. Thus, there are $_{14}P_{10}$ different batting orders. Using a calculator, we obtain $_{14}P_{10} = 3,632,428,800$.

b. The number of ways to arrange 3 out of 6 women in the first 3 positions is $_6P_3 = 120$. Since no restrictions have been specified for the remaining 7 positions, the manager may select them from the 11 people that remain. This may be done in $_{11}P_7 = 1,663,200$ ways. By the Fundamental Counting Principle, the total number of ways to arrange the batting order is

$$_6P_3 \cdot {_{11}P_7} = 120 \cdot 1,663,200 = 199,584,000$$

Notice that if we let $r = n$ in the permutation formula, we obtain

$$_nP_n = n(n-1)(n-2)\cdots(1)$$

which is the product of the first n positive integers. This product occurs so frequently that it is convenient to define a special notation for it: **factorial notation**.

Factorial Notation

For any natural number n, $n! = n \cdot (n-1) \cdot \cdots \cdot 3 \cdot 2 \cdot 1$. For convenience, we define $0! = 1$. Note that $n!$ is read as "n factorial."

EXAMPLE 5

Evaluating Factorials

Evaluate the given expression.

a. $5!$ **b.** $\dfrac{20!}{18!}$

Solution

a. $5! = 5 \cdot 4 \cdot 3 \cdot 2 \cdot 1 = 120$

b. $\dfrac{20!}{18!} = \dfrac{20 \cdot 19 \cdot \cancel{18} \cdot \cancel{17} \cdot \cancel{16} \cdot \cdots \cdot 2 \cdot \cancel{1}}{\cancel{18} \cdot \cancel{17} \cdot \cdots \cdot 2 \cdot \cancel{1}} = 20 \cdot 19 = 380$

Since $_nP_r = n(n-1)(n-2)\cdots(n-r+1)$, we see that $_nP_r$ contains the first r factors of $n!$. Thus, it is perhaps no surprise that the permutation formula can be rewritten using factorials. We simply multiply $_nP_r$ by a quotient whose numerator and denominator contain the factors of $n!$ that are missing from $_nP_r$.

$$_nP_r = \frac{n(n-1)(n-2)\cdots(n-r+1)}{1} \cdot \frac{(n-r)!}{(n-r)!}$$

$$= \frac{n(n-1)(n-2)\cdots(n-r+1)(n-r)(n-r-1)\cdots(3)(2)(1)}{(n-r)!}$$

$$= \frac{n!}{(n-r)!}$$

This verifies the following factorial formula for $_nP_r$.

Factorial Formula for $_nP_r$

For integers n and r with $0 \le r \le n$,

$$_nP_r = \frac{n!}{(n-r)!}$$

The factorial symbol cannot be distributed through parentheses. Thus, the expression $(n-r)!$ in the factorial formula for $_nP_r$ *cannot* be rewritten as $n! - r!$.

·····›**EXAMPLE 6**

Applying the Factorial Formula for Permutations

Use the factorial formula for $_nP_r$ to evaluate $_7P_3$.

Solution

$$_7P_3 = \frac{7!}{(7-3)!} = \frac{7!}{4!} = \frac{7 \cdot 6 \cdot 5 \cdot \cancel{4} \cdot \cancel{3} \cdot \cancel{2} \cdot \cancel{1}}{\cancel{4} \cdot \cancel{3} \cdot \cancel{2} \cdot \cancel{1}} = 7 \cdot 6 \cdot 5 = 210$$

Combinations

We have seen that a permutation takes into account the order in which elements are selected. In many cases, however, the order of selection is not important. Suppose, for example, you are taking a course that requires you to read and report on three out of four unrelated journal articles. In this case, you would probably not be concerned about the order in which the articles are read. Thus, the list of possible reading combinations would not include all the different orderings of the same three articles. If we denote the four articles by the letters A, B, C, and D, the possible reading combinations are

$$\{A, B, C\}, \{A, B, D\}, \{A, C, D\}, \text{ and } \{B, C, D\}$$

Notice that these four combinations are simply the 3-element subsets of $\{A, B, C, D\}$. More generally, a combination is defined in the following way.

Definition of Combination

A **combination** of n elements taken r at a time is an r-element subset of a set of n elements. The number of combinations of n elements taken r at a time is denoted by $_nC_r$ or $\binom{n}{r}$.

Note that a permutation takes order into account, whereas a combination does not. However, $_nC_r$, the *number* of combinations of n elements taken r at a time, is closely related to $_nP_r$, the *number* of permutations of n elements taken r at a time. We first note that each combination of r elements can be rearranged in $r!$ distinct ways. Thus, by the Fundamental Counting Principle, there are $_nC_r \cdot r!$ ordered arrangements of r elements chosen from n. Thus,

$$_nC_r \cdot r! = {_nP_r}$$

and so

$$_nC_r = \frac{_nP_r}{r!}$$

Combination Formulas

For integers n and r with $0 \le r \le n$,

$$_nC_r = \frac{_nP_r}{r!} = \frac{n(n-1)(n-2)\cdots(n-r+1)}{r!}$$

or

$$_nC_r = \frac{n!}{r!(n-r)!}$$

EXAMPLE 7

Evaluating Combinations

Evaluate the following combinations:

a. $_5C_3$

b. $\binom{8}{6}$

Solution

a. $_5C_3 = \dfrac{_5P_3}{3!} = \dfrac{5 \cdot 4 \cdot 3}{3 \cdot 2 \cdot 1} = 10$

b. Recall that $\binom{8}{6}$ is an alternate notation for $_8C_6$. Thus,

$$\binom{8}{6} = \frac{8!}{6!(8-6)!} = \frac{8 \cdot 7 \cdot \cancel{6} \cdot \cancel{5} \cdot \cancel{4} \cdot \cancel{3} \cdot \cancel{2} \cdot \cancel{1}}{(\cancel{6} \cdot \cancel{5} \cdot \cancel{4} \cdot \cancel{3} \cdot \cancel{2} \cdot \cancel{1}) \cdot (2 \cdot 1)} = \frac{8 \cdot 7}{2 \cdot 1} = 28$$

EXAMPLE 8

Using Combinations to Count Senate Subcommittees

The U.S. Senate, consisting of 100 senators, wishes to form a 5-member subcommittee to study the selection process for Senate subcommittees.

a. How many different subcommittees are possible?
b. How many different subcommittees are possible if the Senate consists of 56 Republicans and 44 Democrats and each committee must have 3 Republicans and 2 Democrats?

Solution

a. The order in which members are selected is unimportant. Thus, we wish to find the number of combinations of 100 senators taken 5 at a time.

$$_{100}C_5 = \frac{_{100}P_5}{5!} = \frac{100 \cdot 99 \cdot 98 \cdot 97 \cdot 96}{5 \cdot 4 \cdot 3 \cdot 2 \cdot 1} = 75,287,520$$

So there are 75,287,520 possible subcommittees.
b. The number of ways of selecting 3 Republicans out of 56 without regard to order is $_{56}C_3$. The number of ways of selecting 2 Democrats out of 44 without regard to order is $_{44}C_2$. By the Fundamental Counting Principle, the total number of sub-committees is

$$_{56}C_3 \cdot {}_{44}C_2 = \frac{56 \cdot 55 \cdot 54}{3 \cdot 2 \cdot 1} \cdot \frac{44 \cdot 43}{2 \cdot 1} = 26,223,120$$

Calculator Keys

Combinations

Combinations can be computed with a calculator in much the same way as permutations. To compute $_nC_r$, first key in the number n, then select nCr from the menu of mathematical functions, and, finally, enter the number r.

Counting Card Hands

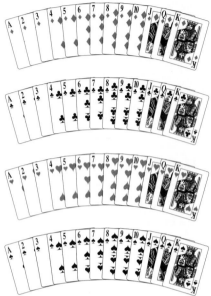

A standard 52-card deck of playing cards consists of two black suits, clubs and spades, and two red suits, hearts and diamonds. Each suit has 13 cards of different face values—A (ace), 2, 3, 4, 5, 6, 7, 8, 9, 10, J (jack), Q (queen), and K (king). In certain varieties of poker, each player is dealt a hand of 5 cards.

a. Find the number of 5-card poker hands that are possible.
b. A *full house* consists of a pair (two cards of the same face value from different suits) and three of a kind (three cards of the same face value from different suits), such as K-K-6-6-6. How many different full houses are possible?

Solution

a. Since the order in which the cards are dealt is not important, the number of ways of being dealt 5 cards out of 52 is $_{52}C_5$. Using a calculator, we obtain $_{52}C_5 = 2,598,960$.
b. A pair can be formed from 2 out of 4 cards of the same face value; there are $_4C_2$ ways in which this can be done. Since there are 13 different face values in the deck, we have

$$\text{The number of possible pairs} = 13 \cdot {}_4C_2$$

The three of a kind must consist of 3 out of 4 cards of the same face value but a different face value from that of the pair. Since 12 other face values remain, we see that

$$\text{The number of possible three of a kinds} = 12 \cdot {}_4C_3$$

Thus, applying the Fundamental Counting Principle, we see that the number of possible full houses is

$$(13 \cdot {}_4C_2) \cdot (12 \cdot {}_4C_3) = 3744$$

Understanding and Mastery Checklists

Concepts to Understand	Skills to Master
Tree diagram	Compute factorials.
Fundamental Counting Principle	Compute permutations.
Permutation definition and notation	Compute combinations.
Factorial notation	Solve a counting problem using a tree diagram.
Combination definition and notation	Solve a counting problem using the Fundamental Counting Principle.
Factorial formulas for permutations and combinations	Solve a counting problem involving permutations.
	Solve a counting problem involving combinations.

Exercises 10.3

Exercises 1-16 *Evaluate the given expression.*

1. 6!

2. (9 − 5)!

3. $\dfrac{10!}{7!}$

4. $\dfrac{8!}{2! \cdot 6!}$

5. $_6P_2$

6. $_9P_1$

7. $_5P_5$

8. $_8P_4$

9. $_6C_4$

10. $\dbinom{7}{2}$

11. $\dbinom{8}{1}$

12. $_4C_4$

13. $\dbinom{5}{0}$

14. $\dbinom{9}{8}$

15. $_8C_6$

16. $_{10}C_3$

Applications

Exercises 17-20 *Draw a tree diagram to illustrate the number of choices that are possible.*

17. The math and science requirement at a certain university includes a math course (College Algebra or Calculus I), a physics course (Elementary Physics or University Physics I), and a biology course (General Biology or Zoology). How many different course combinations will satisfy the math and science requirement?

18. The option packages for a popular pickup truck include a choice of cab style (regular or extended), bed size (short or long), and drive train (two-wheel or four-wheel). How many different option packages are possible?

19. A computer store offers three sizes of monitors (17″, 19″, or 21″), two styles of computer case (mini-tower or tower), and two types of keyboards (standard or cordless). How many different computer configurations are possible?

20. A restaurant offers a single-topping pizza special with a choice of appetizer (bread sticks or garlic bread), type of crust (thin, hand-tossed, or pan), and meat topping (pepperoni or sausage). How many different meals are possible?

Exercises 21-24 *Use the Fundamental Counting Principle to determine the number of possibilities.*

21. How many 3-digit numbers can be formed using the digits {1, 2, 3, 4, 5}

   a. if repetitions are allowed?

   b. if repetitions are not allowed?

22. How many 4-letter "words" can be formed using the letters {a, e, i, o, u, y}

   a. if repetitions are allowed?

   b. if repetitions are not allowed?

23. The state of California has license plates of the form NLLLDDD, where N denotes one of the digits 1–4, each L denotes a letter, and each D denotes one of the digits 0–9. How many license plates of this form are possible?

24. The state of Indiana has license plates of the form NN L DDDD, where NN is a number from 1 to 99 (depending upon the county in which the plate was issued), L is one of the letters A–Z (corresponding to the license branch that issued the plate), and each D is one of the digits 0–9. How many license plates of this form are possible?

Exercises 25-28 *Express your answer using permutation notation and evaluate.*

25. In a survey of TV viewer preferences, respondents are asked to rank their favorite 3 out of 8 sitcoms. How many different survey responses are possible?

26. A club consisting of 12 members wishes to elect a president, vice president, secretary, and treasurer. In how many ways can this be done if no member may hold more than one office?

27. The manager of a Little League baseball team intends to select the starting 9 players by assigning a batting order. In how many ways can this be done if there are 13 players to choose from and

   a. there are no restrictions on the order of selection?

   b. the two children of the team's sponsor must bat first and second?

28. Four boys and six girls are attending a birthday party. In how many ways can they be lined up for ice cream

  a. if there are no restrictions on the order?

  b. if the girls are served before the boys?

Exercises 29-32 *Express your answer using combination notation and evaluate.*

29. A local election ballot allows each voter to select 3 out of 8 school board candidates. In how many ways can this be done?

30. A club consisting of 12 members wishes to appoint a 4-person committee. In how many ways can this be done?

31. The manager of a high school volleyball team must select 6 first-string players from a team of 12 girls. In how many ways can this be done if

  a. there are no restrictions?

  b. exactly 2 of the 6 first-string players must be chosen from the 5 seniors on the team?

32. A 5-card poker hand is dealt from a standard 52-card deck. How many hands are possible if

  a. at least 3 cards have the same face value?

  b. exactly 3 cards have the same face value?

Exercises 33-44 *Express your answer using permutation or combination notation, if possible, and evaluate.*

33. A student has 4 remaining time slots in her schedule. Ten different classes are available, each of which has a section open at those times. In how many ways can the time slots be filled?

34. An employment agency is asked to fill secretarial openings in 6 different companies. A total of 20 people have applied. In how many ways can the positions be filled?

35. A certain state lottery is played by selecting 5 different numbers from 1 through 40. In how many ways can this be done, assuming the order of selection is not important?

36. A certain state lottery is played by selecting 6 different numbers from 1 through 50. In how many ways can this be done, assuming the order of selection is not important?

37. The dial on a combination lock is marked with the numbers 1 through 40. The lock may be opened by selecting the correct sequence of 3 numbers. How many different combinations are possible

  a. if repetitions of numbers are not allowed?

  b. if repetitions of numbers are allowed?

38. The combination lock on a briefcase consists of three dials that are each marked with the digits 0–9. The lock may be opened by selecting the correct digit on each of the three dials. How many different combinations are possible

  a. if repetition of digits is not allowed?

  b. if repetition of digits is allowed?

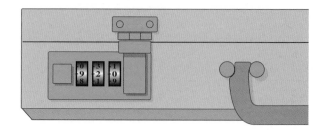

39. On a history exam, a student is asked to select 4 out of 6 essay questions. In how many ways can this be done? What if a student is also asked to select 10 out of 15 short-answer questions?

40. On a mathematics placement exam, a student is asked to select 3 out of 5 application problems. In how many ways can this be done? What if a student is also asked to select 6 out of 10 computational problems?

41. In a classroom of 30 students, how long would it take for every student to shake hands with each of the other students, assuming it takes 2 seconds for each shake and only 2 students can shake hands at a time? What about a class of 60 students?

42. If each of the 8 teams in a softball league must play every other team twice, what is the total number of games that must be played?

43. A 5-card poker hand is dealt from a standard 52-card deck. How many hands are possible if all 5 cards are of the same suit? (Such a hand is called a *flush.*)

44. A 5-card poker hand is dealt from a standard 52-card deck. How many hands are possible if 4 of the 5 cards have the same face value? (Such a hand is called *four of a kind.*)

45. Photomosaic The accompanying image is a photomosaic formed by placing Mastercards in a 23-by-49 array. Photomosaics are constructed with the aid of a powerful computer program written by Robert Silvers, MIT media lab graduate and founder of Photomosaics. The process begins with digitizing the larger image (Washington's portrait) and a collection of images (actual credit cards in this case) that will serve as tiles. Next, the program systematically selects and arranges the tiles according to color, tone, shapes within the tiling images and countless other graphical features. For this image, Silvers had images of 5000 cards at his disposal.

886428876850012173215652315719219054671701173279407926533574061150337286664005674192 36
748373355661746279198699272785426524421142444233995527639375650192264699133328341275127
367839859221860048918939762572769315859304692252489450665503536285307459622242225422522
652833045605341753656666405670549055118613279478160061936243431454580129295532954408592
028918392195136493176984330717492166949632766770879289835249249387014826608727247086706
022066325852338951830942816270488550572315161043974391204847808282991658266958280224 3
724581893036666219173519826480898269127725741589901656355874995435195434651239188031 97
540390991535025567111002888422205941182756557013899294428870481483809824994939078517 08
599850535850623405577735578870071837625137755048325158323422658180331036497785236612 39
124075028751754622608219071541950524350719527851740219444428544649520789464590838137 63
058202318734485366921132939992610199537463307429858815579653030322220820364738254708 62
122731219698647064437792070203832585458112516092246103269294729799909671990111942532 61
074839735565863472164173443458610091287214278477816546070993630424447024539169084328 19
389040735519190662840090771233461002127822071682231844300188885413244613125080805368 58
874185825193476453860752425857800747657358348137293176633761501330178119235819184332 23
692926105713269427047103930736564456329823922955831993479045493033235530366675059414 861
484574977448084410344597134051508871032078535764933246340853554407031512037141923036721
383266878322868066983355940403340666873762528032153706292317011895129780728799091209 42
478075456730482513479238653728791478899879258398849346764389636836213306540923087477 25
737537402201661659954707865264166187961101666759389403135520649494515558757637199055 966
821968716286105371870837893368791269342530250016610906064607035694005907334221511138 58
841163190976151679865456089378376196786359009779205081253596899274486831385457657775 88
908225290984676527325690761703424417173403300304168299593598036694909419100417321570 8
950993783322875695449126524872573548513057499669706664396456088091485179477175095192 77
301614714235596706179913428341044384192422678446016722231460499183554710685964131377 59
234053324634123645175339388318894893220547319097481312307088612962134631277631470652 25
204825799389160699499192801845117027755522516604387820254634454036869945181595571752 4
752699759617307816217231249439687172194591827260730033996456591596832427800143748706 27
858383940614599964483421180697652767091752663523132164786806579492783402636341979123 29
601579361863592159735993128004620557582613259575061918674053302241147796184571232240 123
885879928178919284629413131844000257040216042573584684365862758437437247338454891165 8
957233135189342534641484958165949425719527109358687220054160308952770280624220009659 843
522737264130155032379144873439507255351598970697836194986585865119478358432999677937 15
050866246435216501041853135891433115029734326708472532891668916497716209866605346774 697
207284769269491721780651786614298818438837740582808371037208339873232310509127659466 97
864050805953926551128807481242945547771415678601847545783015250694540375272920370062 03
691934518393460299952477497804056326902596964623450689928594399201825296546688406050 06
574759893687788752084879644457639736576684056369815826908154065789518987619519838736 993
155809886879790814711426908007264656338192220528199933332924580165812308004877479835 814
324300972019921097503976084912261679107718695555170408887276367050108689115091049771 24
677778238800711931092963320893961757247163461104486733478609907418686441560204924415 90
596427321459153708313254999425678994281347173363000077444558784478214577806433441560 55
523772693581959842097235303565070476275632552989486176569610456571198946485357153313 88
317786650956874966607136015905368238763481778792016534946164719041140385179751520346 25
806316569645388672475381700874033336156160000000000000000000000000000000000000000000 0
000000000000000000000000000000000000000000000000000000000000000000000000000000000000 0
000000000000000000000000000000000000000000000000000000000000000000000000000000000000 0
0000000000000000000000000000000000000000000000000000

a. Find an expression for the number of possible mosaics, given that the cards used to form the mosaic are selected from a collection of 5000 Mastercards. Assume no duplicates are allowed. Note that your calculator will not be able to simplify this expression. It is approximately 8.86×10^{4108}, a number with enough digits to fill this page. It is reproduced here in a much smaller font.

b. Find an expression for the number of possible ways of forming the mosaic if duplicates are allowed. (This number, 5.49×10^{4168}, will also be too large for your calculator to approximate.)

Concepts and Critical Thinking

Exercises 46–49   *Answer true or false.*

46. In general, combinations arise when order is important, whereas permutations occur when the order is immaterial.

47. $0! = 0$

48. For all nonnegative integers n and r with $n \geq r$, $_nP_r \geq {_nC_r}$.

49. $31! - 13! = 18!$

Exercises 50–53   *Give an example of each.*

50. Two numbers n and r such that $_nP_r = {_nC_r}$

51. A distinction between combinations and permutations

52. An alternative way of expressing $_nC_r$

53. A number n such that $n! = n$

Questions for Discussion or Essay

54. Explain why a combination lock should really be called a permutation lock. What are some other instances in which the word *combination* should perhaps be replaced with the word *permutation*? Explain your examples. To get you started, think about a TV cooking show referring to "the right combination of ingredients" or a coach referring to "a winning combination of players."

55. In Exercise 35, you were asked to determine the number of ways of selecting 5 numbers out of 40 in a state lottery. The answer to that exercise represents the number of possible winning combinations. What does this say about your chances of winning a lottery? How many different lottery combinations would you have to play in order to give yourself a 50/50 chance of winning? Explain.

56. Suppose 10 sprinters are competing in the 100-meter dash at a track meet. If the race is a qualifying round, the top 3 finishers advance to the next round. If the race is the final round, the top 3 finishers are awarded first, second, and third place. For each of these two scenarios, determine whether a permutation or combination would be used to count the number of ways of selecting 3 out of 10 sprinters. Explain in detail why you chose the method you did.

57. A basketball coach, notorious for changing his starting lineup, states that he will try every combination until he finds one that works. Assuming the team has 12 players and that a season consists of 30 games, how many seasons will it take to try every combination? In practice, would a coach really try every possible starting lineup?

Projects for Enrichment

58. Combinatorial Identities   The combination formula

$$\binom{n}{r} = \frac{n!}{r!(n-r)!}$$

leads to many useful and sometimes surprising identities. We consider a few in this project, most of which can be verified by direct simplification. One example is the identity

$$\binom{n}{n} = 1$$

This can be verified by observing that

$$\binom{n}{n} = \frac{n!}{n!(n-n)!} = \frac{n!}{n!0!} = 1$$

The interpretation of this identity is that there is only one way to select n elements (without regard to order) out of a set of n elements.

a. Simplify the following combinations and give a verbal interpretation of the result:

   i. $\binom{n}{1}$ **ii.** $\binom{n}{0}$

   iii. $\binom{n}{n-1}$

b. Simplify the following combinations:

   i. $\binom{n}{2}$ **ii.** $\binom{n}{n-2}$

You should have found that the two combinations in part b simplify to the same expression. In general,

$$\binom{n}{r} = \binom{n}{n-r}$$

for any $r \leq n$. In other words, a set of n elements has the same number of r-element subsets as it has $(n-r)$-element subsets.

c. Simplify $\binom{n}{n-r}$ to show that

$$\binom{n}{r} = \binom{n}{n-r}$$

Interpret this identity.

d. The following identities are often useful when working with complicated calculations involving combinations. Verify each.

i. $\binom{n}{m}\binom{m}{r} = \binom{n}{r}\binom{n-r}{m-r}$

   (*Hint:* Simplify both sides of the equation.)

ii. $\binom{n}{r-1} + \binom{n}{r} = \binom{n+1}{r}$

   (*Hint:* Find a common denominator for the expression to the left of the equal sign.)

iii. $\binom{n-1}{r} = \frac{r+1}{n}\binom{n}{r+1}$

59. Counting Poker Hands   In Example 9, we computed the number of possible 5-card poker hands that could be dealt from a standard 52-card deck, and we also computed the number of possible full houses. Table 1 lists all the 5-card poker hands, together with a partial list of the number of possible such hands.

Table 1   5-Card Poker Hands

Hand	Number Possible
Royal flush	4
Other straight flush	36
Four of a kind	
Full house	3,744
Flush	
Straight	10,200
Three of a kind	54,912
Two pairs	
One pair	
Nothing	1,302,540
Total	2,598,960

a. Look up and write out a description of each type of hand.

b. Show how combinations can be used to arrive at the number of possible hands for each type.

c. Although there are many varieties of 5-card poker, the general idea is that 5 cards are dealt to all players and there is one or more rounds of betting. After all betting is finished, the player with the hand that appears highest on the list wins all the money. Why does it make sense that the hand appearing highest on the list in Table 1 should win?

d. Do you think it's necessary for a poker player to know how many hands of each type are possible? Why or why not?

60. **Seating Arrangements**   Three couples—Andy and Becky, Calvin and Dawn, and Eddie and Fran—are all good friends and decide to go to a concert together. They have reserved 6 adjoining seats next to an aisle. Determine the number of seating arrangements that are possible given the following conditions:

a. There are no restrictions.

b. The men must sit in the 3 seats closest to the aisle.

c. Each couple must sit together. (*Hint:* First find the number of pairs of seats where each couple can sit, and then the number of arrangements for each of the three couples.)

d. The women must sit together. (*Hint:* First find the number of groups of 3 seats where the women can sit, and then find the number of arrangements of the three women in those seats, and, finally, find the number of arrangements of the men in the remaining 3 seats.)

e. Andy and Becky had a fight and refuse to sit next to each other. (*Hint:* First find the number of ways in which Andy and Becky can be separated, and then find the number of ways of arranging the remaining four people.)

After the concert, the three couples decide to go out for a late dinner. Determine the number of ways in which they can be seated around a circular table, given the following conditions. Keep in mind that we are only concerned with their positions relative to each other, not with the particular seat each person occupies.

f. There are no restrictions.

g. Each couple must sit together.

h. The women must sit together.

i. Andy and Becky had a fight and refuse to sit next to each other.

Section 10.4   Probability

- How can it be that approximately 60% of the people testing positive for drug usage are actually not drug users, even though the test is 97% accurate?
- What is the probability that, in a class of 30 students, at least 2 will have the same birthday?
- How likely is it that a fatal accident will involve a driver between the ages of 18 and 21 who had been drinking?
- Which is more likely, getting struck by lightning or winning a state lottery?
- Did the world's smartest human give bad advice to game show contestants?

Our language is replete with terms used to indicate the degree of certainty of uncertain events. We refer to outcomes as being "improbable," "unlikely," "a sure thing," "more likely than not," or a "longshot." **Probability** is the branch of mathematics in which degrees of certainty are quantified. Virtually every day of our lives we encounter uncertainties that can be expressed as probabilities. We are told that there is a 30% chance of thunderstorm activity, that there is a one in a million chance of winning a lottery, that one in four women will develop breast cancer, and that three out of four dentists recommend a certain sugarless gum. In addition to these everyday occurrences, probabilities arise in, and indeed are central to, such diverse areas as medicine, insurance, sociology, psychology, physics, and engineering.

As with the topics discussed earlier in the chapter, we provide here only a brief overview of the fundamental notions and broad applicability of mathematical probability. A more thorough treatment is left to later course work.

Fundamentals   In mathematical probability, a real number between 0 and 1 is associated with a possible outcome of an experiment: The greater the probability, the more likely it is that the outcome occurs. Events that have a probability of 1 are virtually certain to occur, whereas events with probability 0 virtually never occur. An event with a probability of 0.5 will occur about half the time. We begin with the following definitions.

Probability Terminology and Notation

A process of observation or measurement is referred to as an **experiment**. The **sample space** of an experiment is the set of all possible outcomes. The number of possible outcomes in a sample space S is denoted $n(S)$. An **event** is any subset of the sample space. The number of outcomes in an event E is denoted $n(E)$.

····▷EXAMPLE 1

An Experiment Involving a Die

An experiment consists of rolling an ordinary 6-sided die.

a. Find the sample space S, and determine $n(S)$.

b. Describe the event E that an even number is rolled, and determine $n(E)$.

Solution

a. Since an outcome of the experiment consists of rolling one of the numbers 1–6, the sample space S can be written as $\{1, 2, 3, 4, 5, 6\}$; $n(S)$ is the number of outcomes in S and so $n(S) = 6$.

b. The event E that an even number is rolled is the set $\{2, 4, 6\}$. Since $n(E)$ is the number of outcomes in E, $n(E) = 3$.

····▷EXAMPLE 2

A Coin-Tossing Experiment

An experiment consists of tossing a coin two times.

a. Find the sample space S, and determine $n(S)$.

b. Find $n(E)$, where E is the event that a tail is tossed both times.

Solution

a. Let H denote the outcome that a head is tossed and let T denote the outcome that a tail is tossed. Thus, for example, the outcome consisting of tossing a head and then a tail can be written as HT. The sample space S is the set $\{HH, HT, TH, TT\}$, and $n(S) = 4$.

b. The event E is the set of outcomes for which a tail is tossed both times, or $\{TT\}$. Thus $n(E) = 1$.

····▷EXAMPLE 3

Family Composition

A couple has two children.

a. Find the sample space S consisting of the possible outcomes with respect to the sexes of the children.

b. Describe the event E that the children are of different sexes.

Solution

a. Let M represent a male birth and let F represent a female birth. Thus, for example, MF represents the outcome that the first child is male and the second child is female. The sample space S can then be expressed as the set $\{MM, MF, FM, FF\}$.
b. The two outcomes in which the children are of different sexes are MF and FM. Thus, E can be expressed as the set $\{MF, FM\}$.

When all outcomes are equally likely, we can find the probability of an event simply by dividing the number of outcomes that make up the event by the total number of outcomes in the sample space. For example, the probability of rolling a 2 on a fair die (one for which all outcomes are equally likely) is $\frac{1}{6}$, since one of six outcomes corresponds to rolling a 2. Similarly, the probability of obtaining a head with one toss of a fair coin is $\frac{1}{2}$. More generally, we have the following definition of the probability of an event.

Probability of an Event

If E is an event in the sample space S and if all the outcomes of S are equally likely, then the **probability of E** is denoted by $P(E)$ and is defined by

$$P(E) = \frac{n(E)}{n(S)}$$

EXAMPLE 4

Computing a Probability Involving a Die

If an ordinary 6-sided die is rolled, what is the probability that an even number is rolled?

Solution From Example 1, we have $S = \{1, 2, 3, 4, 5, 6\}$ and $E = \{2, 4, 6\}$. Thus,

$$P(E) = \frac{n(E)}{n(S)} = \frac{3}{6} = \frac{1}{2}$$

EXAMPLE 5

Computing a Probability Involving Two Coins

If a coin is tossed two times, what is the probability that a tail is tossed both times?

Solution As we saw in Example 2, $S = \{HH, HT, TH, TT\}$ and $E = \{TT\}$. Thus,

$$P(E) = \frac{n(E)}{n(S)} = \frac{1}{4}$$

EXAMPLE 6

Computing a Probability Involving Two Dice

Two ordinary 6-sided dice are rolled. What is the probability that the total of the dice is 7?

Solution There are six possible outcomes for the first die and six possible outcomes for the second die. Thus, by the Fundamental Counting Principle, there are 36 possible outcomes in the sample space S. These outcomes are shown as ordered pairs in

Table 2. Now for each possible outcome of the first die, there is exactly one outcome for the second die that will result in a total of 7. Thus, there are six ways of rolling a total of 7. These appear in the blue diagonal of Table 2.

Table 2

Possible outcomes

		Second die				
	1	**2**	**3**	**4**	**5**	**6**
1	(1, 1)	(1, 2)	(1, 3)	(1, 4)	(1, 5)	(1, 6)
2	(2, 1)	(2, 2)	(2, 3)	(2, 4)	(2, 5)	(2, 6)
3	(3, 1)	(3, 2)	(3, 3)	(3, 4)	(3, 5)	(3, 6)
4	(4, 1)	(4, 2)	(4, 3)	(4, 4)	(4, 5)	(4, 6)
5	(5, 1)	(5, 2)	(5, 3)	(5, 4)	(5, 5)	(5, 6)
6	(6, 1)	(6, 2)	(6, 3)	(6, 4)	(6, 5)	(6, 6)

(First die)

So if E is the event that a 7 is rolled, then $n(E) = 6$. Since $n(S) = 36$, we have

$$P(E) = \frac{n(E)}{n(S)} = \frac{6}{36} = \frac{1}{6}$$

EXAMPLE 7

Computing Probabilities Involving Three Coins

Three coins are tossed. Find the probabilities of the following events:

a. All three coins come up the same.
b. Exactly two heads are tossed.
c. At least one tail is tossed.

Solution   We first find $n(S)$. There are two possible outcomes for each of the three coins—namely, heads or tails. Thus, by the Fundamental Counting Principle, there are $2 \cdot 2 \cdot 2 = 8$ possible outcomes in the sample space S. Thus, $n(S) = 8$.

a. Let E be the event that all three coins come up the same. Since the event E consists of two outcomes, *HHH* and *TTT*, $n(E) = 2$. Thus,

$$P(E) = \frac{n(E)}{n(S)} = \frac{2}{8} = \frac{1}{4}$$

b. Let F be the event that exactly two heads are tossed. Then F consists of the outcomes *HHT*, *HTH*, and *THH*. So $n(F) = 3$ and we have

$$P(F) = \frac{n(F)}{n(S)} = \frac{3}{8}$$

c. Let G be the event that at least one tail is tossed. Of the eight outcomes in S, the only one that doesn't have at least one tail is *HHH*. Thus, $n(G) = 7$ and we have

$$P(G) = \frac{n(G)}{n(S)} = \frac{7}{8}$$

----:EXAMPLE 8

Playing the Lottery

A certain state lottery is designed so that a player selects six different numbers from 1 to 40 (inclusive). If those six numbers match the six that are randomly drawn by the lottery (without regard to order), the player wins. Find the probability that a player wins the lottery after having selected 10 different combinations of six numbers.

Solution The sample space S consists of all possible combinations of six numbers chosen from the numbers 1 to 40. Thus $n(S) = {}_{40}C_6 = 3,838,380$. Let E be the event that one of the player's 10 combinations is a winner. Then $n(E) = 10$. Thus,

$$P(E) = \frac{n(E)}{n(S)} = \frac{10}{3,838,380} \approx 0.0000026$$

----:EXAMPLE 9

A 5-Card Poker Hand

A **straight flush** is a poker hand consisting of five cards in sequence from the same suit, such as A-2-3-4-5 or 8-9-10-J-Q. Find the probability of being dealt a straight flush from a standard 52-card deck.

Solution The sample space S consists of all possible 5-card hands from a deck of 52 cards. Thus,

$$n(S) = {}_{52}C_5 = 2,598,960$$

Let E be the event that a straight flush is dealt. Then $n(E)$ is the number of straight flushes that are possible in a 52-card deck. If we assume that the ace can be either low or high, then the possible straights in a given suit are A-2-3-4-5, 2-3-4-5-6, ..., 10-J-Q-K-A. Thus, there are 10 possible straights in each of the four suits, for a total of 40 possible straights. So $n(E) = 40$ and we have

$$P(E) = \frac{n(E)}{n(S)} = \frac{40}{2,598,960} \approx 0.000015$$

----:EXAMPLE 10

Reliability of Drug Testing

The president of a university with an enrollment of 30,000 is contemplating mandatory cocaine testing for all students. The test is 97% accurate in the sense that 97% of all students who are cocaine users will test positive and 97% of students who are not cocaine users will test negative. If 2% of the student body are, in fact, cocaine users, what is the probability that someone who tests positive is a cocaine user?

Solution Consider an experiment consisting of selecting a student from the group that tested positive. Then the sample space S consists of the set of all students testing positive, whether they are cocaine users or not. Next we let E be the event that a person tests positive and is indeed a cocaine user. To find $P(E)$, we must determine $n(S)$ and $n(E)$. Since 2% of the 30,000 university students are cocaine users, there are $(0.02)30,000 = 600$ cocaine users and $30,000 - 600 = 29,400$ nonusers. Now the test will come out positive for approximately 97% of the users, but it will also come out positive for about 3% of the nonusers. Thus,

$$n(S) = \text{The number of students who test positive}$$
$$= \text{The number of } users \text{ who test positive}$$
$$+ \text{ The number of } nonusers \text{ who test positive}$$
$$\approx 97\% \text{ of } 600 + 3\% \text{ of } 29{,}400$$
$$= (0.97)600 + (0.03)29{,}400$$
$$= 1464$$

and

$$n(E) = \text{The number of students who test positive and are cocaine users}$$
$$\approx (0.97)600$$
$$= 582$$

Thus, we have

$$P(E) = \frac{n(E)}{n(S)} = \frac{582}{1464} \approx 0.4$$

So the probability that someone who tests positive for cocaine use is actually a user is less than $\frac{1}{2}$! We will ask you to discuss this surprising result in Exercise 66.

Intersections, Unions, and Complements of Events

In many cases, we are interested in finding probabilities that involve two or more events from the same sample space or that indirectly involve a given event. For these tasks, we make the following definitions.

Definitions of Intersection, Union, and Complement

Let E and F be events from the same sample space S. Then

1. The **intersection** of E and F, denoted $\boldsymbol{E \cap F}$, is the event that both E and F occur.
2. The **union** of E and F, denoted $\boldsymbol{E \cup F}$, is the event that either E or F (or both) occur.
3. The **complement** of E, denoted $\boldsymbol{E'}$ and read "E complement," is the event that E does *not* occur.

EXAMPLE 11

Finding Intersections, Unions, and Complements

Suppose that a number from 1 to 10 (inclusive) is selected at random. Let E be the event that the number chosen is a multiple of 3, and let F be the event that the number chosen is even. Find expressions for each of the following events, and then compute the probability of the event.

a. $E \cap F$ **b.** $E \cup F$

c. E' **d.** F'

Solution

a. $E \cap F$ consists of all numbers from 1 to 10 that are both even *and* multiples of 3. Thus, $E \cap F = \{6\}$, and $n(E \cap F) = 1$. Since $n(S) = 10$, we have

$$P(E \cap F) = \frac{n(E \cap F)}{n(S)} = \frac{1}{10}$$

b. $E \cup F$ consists of all numbers from 1 to 10 that are even, multiples of 3, or both. Thus, $E \cup F = \{2, 3, 4, 6, 8, 9, 10\}$. Hence $n(E \cup F) = 7$, and $P(E \cup F) = \frac{7}{10}$.

c. E' is the set of numbers from 1 to 10 that are *not* multiples of 3. Thus, $E' = \{1, 2, 4, 5, 7, 8, 10\}$ and $n(E') = 7$. Hence $P(E') = \frac{7}{10}$.

d. F' is the set of numbers from 1 to 10 that are not in F—that is, those numbers that are not even. Hence F' consists of the odd numbers between 1 and 10, and so $F' = \{1, 3, 5, 7, 9\}$. Thus, $n(F') = 5$, and $P(F') = \frac{5}{10} = \frac{1}{2}$.

If E and F are events from the same sample space, and E and F have no outcomes in common, then $E \cap F = \emptyset$, the empty set, and we say that the events are **mutually exclusive**. Consider, for example, an experiment in which a die is rolled. If E is the event that an odd number is rolled and F is the event that an even number is rolled, then E and F are mutually exclusive since no outcome belongs to both events.

When two events E and F are mutually exclusive, the number of outcomes that belong to one of the two events is simply the sum of the number of outcomes in each event. In other words, for mutually exclusive events E and F

$$n(E \cup F) = n(E) + n(F)$$

Dividing by $n(S)$ gives us

$$\frac{n(E \cup F)}{n(S)} = \frac{n(E)}{n(S)} + \frac{n(F)}{n(S)}$$

$$P(E \cup F) = P(E) + P(F)$$

As an example, consider the event E that a 3 is obtained by a single throw of a die and the event F that a 5 is thrown. Since these events are mutually exclusive (we can't throw both a 3 and a 5 on the same toss), the probability of $E \cup F$ is simply the sum of the probabilities of E and F. Since $P(E) = \frac{1}{6}$ and $P(F) = \frac{1}{6}$, we have

$$P(E \cup F) = P(E) + P(F) = \frac{1}{6} + \frac{1}{6} = \frac{1}{3}$$

EXAMPLE 12

Winning the World Series

Suppose that a preseason poll of sports writers suggests that the probability of the Chicago Cubs winning the World Series is 0.06 and the probability of the New York Mets winning is 0.13. Find the probability that one of the two teams will win.

Solution Let E be the event that the Cubs win, and let F be the event that the Mets win. We would like to determine $P(E \cup F)$, the probability that E or F occurs. Since it isn't possible for both teams to win the World Series in a given year, the events E and F are mutually exclusive. Thus,

$$P(E \cup F) = P(E) + P(F) = 0.06 + 0.13 = 0.19$$

Now let us consider two events that are *not* mutually exclusive. Suppose that 2 coins are tossed. Then the sample space S can be represented as the set $\{HH, HT, TH, TT\}$. Let E be the event that the first coin comes up heads, and let F be the event that the second coin comes up heads. Thus, $E = \{HH, HT\}$ and $F = \{TH, HH\}$, so that

$n(E) = n(F) = 2$. The event $E \cup F$ consists of those outcomes for which either the first coin or the second coin is a head. That is, $E \cup F = \{HH, HT, TH\}$, and thus $n(E \cup F) = 3$. But $n(E) + n(F) = 2 + 2 = 4$, so that $n(E \cup F) \neq n(E) + n(F)$. The reason for the discrepancy is that the outcome HH belongs to both E and F so that when we add $n(E)$ and $n(F)$, we have counted this outcome twice.

In general, the sum $n(E) + n(F)$ counts the outcomes belonging to $E \cap F$ twice. To correct for this double counting, we must subtract the number of elements in the set $E \cap F$. Thus, we have

$$n(E \cup F) = n(E) + n(F) - n(E \cap F)$$

Figure 10 illustrates this formula with a **Venn diagram**—a collection of overlapping circles used to illustrate relationships among sets. In this diagram, the outcomes of the event E are represented by the squares and triangles lying inside the circle on the left. The outcomes of F consist of the solid circles and triangles lying inside the circle on the right. As the diagram suggests, the event $E \cup F$ consists of those outcomes lying in both E and F—in this case, the triangles.

Not surprisingly, a similar formula holds for finding the probability of the union of two events that may not be mutually exclusive. It is stated as follows.

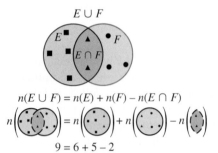

$E \cup F$

$n(E \cup F) = n(E) + n(F) - n(E \cap F)$

$9 = 6 + 5 - 2$

Figure 10

The Probability of the Union of Two Events

Let E and F be events in the same sample space. Then the probability of E or F occurring is given by

$$P(E \cup F) = P(E) + P(F) - P(E \cap F)$$

Note that this formula applies even if E and F are mutually exclusive since, in that case, $E \cap F = \emptyset$, and so $P(E \cap F) = 0$.

EXAMPLE 13

Computing the Probability of the Union of Two Events

Suppose that a single card is selected at random from a standard 52-card deck. Find the probability that the card selected is either a face card (jack, queen, or king) or a heart.

Solution Let E be the event that the card selected is a face card, and let F be the event that the card selected is a heart. Then $E \cup F$ represents the event that the card selected is a face card or a heart. Thus, we are interested in $P(E \cup F)$. Since there are three face cards in each of the four suits, there are 12 face cards in all. Thus,

$$P(E) = \frac{n(E)}{n(S)} = \frac{12}{52} = \frac{3}{13}$$

On the other hand, $\frac{1}{4}$ of the cards in the deck are hearts, so that

$$P(F) = \frac{1}{4}$$

The event $E \cap F$ consists of those outcomes in which a face card of hearts is drawn. Since there are three hearts that are face cards, we have

$$P(E \cap F) = \frac{n(E \cap F)}{n(S)} = \frac{3}{52}$$

We can now compute $P(E \cup F)$ as follows:

$$P(E \cup F) = P(E) + P(F) - P(E \cap F)$$
$$= \frac{3}{13} + \frac{1}{4} - \frac{3}{52}$$
$$= \frac{12 + 13 - 3}{52}$$
$$= \frac{22}{52} = \frac{11}{26}$$

····∴EXAMPLE 14

Probabilities Involving Fatal Car Accidents

According to the U.S. Federal Highway Administration, of the drivers involved in fatal motor vehicle accidents in 1989,

- 14.1% were 18–21 years old
- 31.9% had been drinking
- 5.3% were 18–21 years old and had been drinking

If a driver involved in a fatal motor vehicle accident is selected at random, find the probability of each of the following events:

a. The driver was 18–21 and had been drinking.
b. The driver had not been drinking.
c. The driver had been drinking but was not 18–21.

Solution   Let S be the sample space consisting of drivers involved in fatal accidents. Let E be the event that the selected driver was 18–21, and let F be the event that the driver had been drinking. According to the given data, 5.3% of the drivers in S are in $E \cap F$. Moreover, since 14.1% of the drivers are in E, there must be 14.1% $-$ 5.3% $=$ 8.8% of the drivers in E that aren't also in $E \cap F$. Similarly, there are 31.9% $-$ 5.3% $=$ 26.6% of the drivers in F that aren't also in $E \cap F$. It is convenient to summarize these percentages in a Venn diagram such as the one in Figure 11. Note that 59.3% of drivers involved in fatal accidents had not been drinking and were not 18–21 years old. Note also that the percentages themselves represent probabilities.

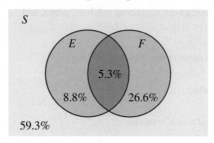

Figure 11

a. The event that the driver was 18–21 and had been drinking is $E \cap F$, and
$$P(E \cap F) = 5.3\% = 0.053$$

b. The event that the driver had not been drinking is F', and
$$P(F') = 59.3\% + 8.8\% = 0.593 + 0.088 = 0.681$$

c. The event that the driver had been drinking but was not 18–21 is $F \cap E'$. In the Venn diagram, this is the region inside F but outside E. Thus,
$$P(F \cap E') = 26.6\% = 0.266$$

Understanding and Mastery Checklists

Concepts to Understand

Experiment

⚬

Sample space

⚬

Event

⚬

Probability of an event

⚬

Intersection of events

⚬

Union of events

⚬

Complement of an event

⚬

Mutually exclusive events

Skills to Master

Find the sample space of an experiment.

⚬

Compute the probability of an event.

⚬

Describe the intersection of two events and find its probability.

⚬

Describe the union of two events and find its probability.

⚬

Describe the complement of an event and find its probability.

Exercises 10.4

Exercises 1-4   *Find the sample space S for the given experiment and determine n(S).*

1. A single ball is drawn from an urn containing red, blue, green, and white balls.

2. A button is pressed in an elevator in a 10-story building.

3. A coin is tossed three times.

4. A 6-sided die is rolled and a coin is tossed.

Exercises 5-8   *A standard 6-sided die is rolled. Find the probability of the given event.*

5. Rolling a 4

6. Rolling a number larger than 4

7. Rolling a number less than 7

8. Rolling an even number

Exercises 9-12   *A ball is selected at random from an urn containing 3 red balls, 4 blue balls, and 5 white balls. Find the probability of the given event.*

9. Selecting a red ball

10. Selecting a red or white ball

11. Selecting a ball that is not white

12. Selecting a yellow ball

Exercises 13-18   *A single card is drawn from a standard 52-card deck. Compute the probability of the given event.*

13. Drawing the ace of spades

14. Drawing a spade

15. Drawing an ace

16. Drawing an ace or a spade

17. Drawing a red face card

18. Drawing a 3 or a 4

Exercises 19-22   *Assume that the probability of a male birth is 0.5.*

19. A family has three children. Compute the probability that the children are of the same sex.

20. Compute the probability that in a family with three children, exactly one of the children is female.

21. Compute the probability that in a family with three children, exactly two of the children are male.

22. Compute the probability that in a family of three children, there is no girl with an older brother.

Exercises 23-26   *Balls are selected (without replacement) from an urn containing 5 red balls and 4 blue balls.*

23. If two balls are drawn, find the probability that both are red.

24. If two balls are drawn, find the probability that they are the same color.

25. If two balls are drawn, find the probability that they are different colors.

26. If three balls are drawn, find the probability that two of the balls are blue and the other is red.

Exercises 27-30   *A pair of standard 6-sided dice are rolled. Find the probability of the given event.*

27. Rolling the same number on each die

28. Rolling "snake-eyes" (a 1 on each die)

29. Rolling a total of 10

30. Rolling a 6 on at least one of the dice

Exercises 31-34   *Compute the probability of being dealt the given 5-card poker hand from a standard 52-card deck.*

31. Four aces

32. Four of a kind

33. A flush (all cards of the same suit)

34. Three of a kind (three cards of the same face value and two cards of different face values)

Exercises 35-40   *A fair coin is tossed. Compute the probability of the given event.*

35. Tossing 4 heads in a row

36. Having a streak of 4 heads in a row if the coin is being tossed 5 times

37. Obtaining your first head on the third toss

38. Obtaining your second head on the fourth toss

39. Tossing exactly 3 heads out of the first 5

40. Alternating heads/tails or tails/heads in the first five tosses

Exercises 41-44   *An experiment is described and two events E and F are given. Find P(E) and P(F), describe the events $E \cap F$, $E \cup F$, E', and F', and find the probability of each.*

41. A single card is drawn from a standard 52-card deck; E is the event that a red card is drawn, and F is the event that a face card is drawn.

42. A 6-sided die is rolled; E is the event that an odd number is rolled, and F is the event that a number less than 5 is rolled.

43. Two 6-sided dice are rolled; E is the event that the sum is even, and F is the event that the sum is less than 8.

44. Two integers from 1 to 10 are chosen at random (repetitions are allowed); E is the event that the sum is odd, and F is the event that the sum is greater than 10.

Applications

45. NCAA Basketball Championship   Suppose that at the beginning of the NCAA basketball tournament, it has been determined that the probability that Duke will win the championship is 0.2 and the probability that Michigan will win is 0.15. Find the probability that one of these two teams will win the championship.

46. Math Grades   Suppose you have determined that the probability of getting an A in this class is 0.25 and the probability of getting a B is 0.6. Find the probability that you will get an A or a B.

47. Lottery Chances   A state lottery is designed so that a player selects five different numbers from 1 to 40. What is the probability that a single choice of five numbers will win the lottery? What is the probability of winning if 20 combinations of numbers are chosen? How many combinations must be chosen so that the probability of winning is $\frac{1}{2}$?

48. Lottery Chances   A state lottery is designed so that a player selects six different numbers from 1 to 50. What is the probability that a single choice of six numbers will win? What is the probability of winning if 100 combinations of numbers are chosen? How many combinations must be chosen so that the probability of winning is $\frac{1}{2}$?

49. Grade Averages   The distribution of high school grade averages for college freshman in the United States in 2000 is given in Figure 12. (*Data source: The American Freshman: National Norms*, University of California, Los Angeles.) If a college freshman were selected at random, find the probability that the student

a. had a C+ or lower average

b. did not have an A− to A+ average

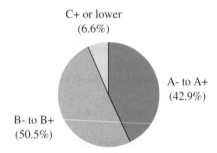

C+ or lower
(6.6%)

A− to A+
(42.9%)

B− to B+
(50.5%)

Figure 12

50. College Enrollment   The distribution of college enrollment among U.S. students in 2000 who enrolled in college in the first 12 months after graduation or receipt of a GED is given in Figure 13. (*Data source: U.S. National Center for Education Statistics.*) If a college student were selected at random, find the probability that the student

a. was Hispanic

b. was not white

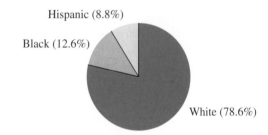

Hispanic (8.8%)

Black (12.6%)

White (78.6%)

Figure 13

51. Polygraph Testing   A large fast-food chain with 10,000 employees has been experiencing a rash of employee theft and so begins requiring employees to take a polygraph test (that is, a lie detector test). The test is 90% accurate in the sense that 90% of all employees who are stealing will fail the test and 90% of employees who are not stealing will pass the test. If 5% of the employees are in fact stealing, what is the probability that someone who fails the test is stealing?

52. Smoking Distributions   In 1998, 11.9% of the adult population of the United States was black, 24% of the adult population smoked, and 3.0% of the adult population were black smokers. If an adult American is selected at random, find the probability of each of the following events:

a. The person selected is black or smokes.

b. The person selected is nonblack.

c. The person selected does not smoke.

d. The person selected is black but does not smoke.

e. The person selected is a nonblack smoker.

53. Hispanic Californians   According to the U.S. census of 2000, approximately 12% of the population lives in California. Also, 32.4% of all Californians are Hispanic. Nationwide, 12.5% of the population is Hispanic. If an American is selected at random, find the probability of each of the following events:

a. The person is a Californian or Hispanic.

b. The person is a Californian Hispanic.

c. The person is a non-Californian.

d. The person is a Californian but is not Hispanic.

e. The person is Hispanic but is not a Californian.

54. Extracurricular Activities   At a certain university, extracurricular activities consist entirely of athletics, Greek organizations (fraternities and sororities), and clubs. You are given the following information regarding the makeup of the student body:
- 10% participate in athletics.
- 40% belong to Greek organizations.
- 45% are club members.
- 25% are involved in both Greek organizations and clubs but not athletics.
- 2% are involved in athletics, Greek organizations, and clubs.
- 15% are non-Greek, nonathlete, club members.
- 4% are Greek athletes who do not belong to any clubs.

Assume that a student is selected at random from the student body of this university. Compute the probabilities of each of the following events:

a. The student is involved in an extracurricular activity.

b. The student is involved in exactly one extracurricular activity.

c. The student is involved in a club and is an athlete but is not a member of a Greek organization.

d. The student is involved in exactly two extracurricular activities.

Concepts and Critical Thinking

Exercises 55-58 *Answer true or false.*

55. In general, for events E and F, $P(E \cup F) = P(E) + P(F)$.

56. There is no event that has a probability of 1.7.

57. If E and F are mutually exclusive events, then
$P(E \cup F) = P(E) + P(F)$.

58. If a coin is loaded in such a way that heads come up about 7 times for every 10 times that tails come up, then the probability of heads is about $\frac{7}{10}$.

Exercises 59-62 *Give an example of each.*

59. A real-world event for which the probability is $\frac{1}{7}$

60. A number that isn't the probability of any event

61. Two events E and F such that $P(E \cup F) = P(E) + P(F)$

62. A real-world event for which the probability is 1

Questions for Discussion or Essay

63. Estimates for the probability of getting struck by lightning in a given year range as high as 0.000007. How do you think such estimates are obtained? Do you think such estimates are valid? Why or why not? In Example 8, we saw that the probability of winning a certain state lottery after having played 10 combinations of numbers was 0.0000026, which is lower than the estimate for getting struck by lightning. Can we conclude that one is more likely to get struck by lightning than to win a state lottery? Explain.

64. A weathercaster announces that there is a 70% chance of rain on Saturday and a 30% chance of rain on Sunday and concludes that there is a 100% chance of rain for the weekend. Is his reasoning correct? If not, what would be a more reasonable estimate for the chance of rain over the weekend?

65. In the column "Ask Marilyn" in *Parade* magazine, Marilyn Vos Savant (who is listed in *The Guinness Book of World Records* as having the highest IQ) posed the following problem. Suppose you are a contestant on a game show. You are asked to select one of three doors. Behind one of the doors is a new car and behind each of the other two doors are goats. After choosing your door, the host opens one of the remaining doors to reveal a goat. She then gives you the choice of staying with your original selection or switching to the other door that is still closed. Ms. Vos Savant claimed that you improve your probability of winning if you switch. She was promptly

flooded with letters from irate readers (many of them educators) who argued that the probability of winning is 0.5 whether you choose to switch or not. They reasoned that there are two doors remaining—behind one is a car, behind the other is a goat—and so the probability of correctly guessing the door with the car behind it is $\frac{1}{2} = 0.5$. Surprisingly, this is not the case. You would in fact improve your probability of winning to $\frac{2}{3}$ by switching to the other door. Explain why this seems counterintuitive. How would you design an experiment to test the strategy? How relevant is the fact that the host of the game show knows which door the car is behind and so will always show you a door with a goat behind it? Suppose there were 10 doors and that after your selection, the host showed you what was behind all but two, the one you chose and one other. Would you switch doors? What if there were 1000 doors and again the host showed you what was behind all but two, the one you chose and one other? What light does this shed on the original problem?

66. In Example 10, we considered a drug test that claimed to be 97% accurate and discovered that the probability that a student who tested positive was actually a user was only 0.4. In other words, it was likely that 60% of the students who tested positive were not actually users. On the other hand, the test will fail to detect only 3% of the users. What do these observations suggest about the use of such tests or the interpretation of the results?

Projects for Enrichment

67. Assessing Psychic Phenomena   A group of 10,000 people, all of whom claim to be psychic, are assembled on New Year's Eve 2007. Each is asked to make four predictions regarding the new year:
 (1) The day in January of 2008 on which the highest temperature is achieved in Tempe, Arizona
 (2) The party (Democratic or Republican) that wins the presidential election of 2008
 (3) The state in which the percentage change in population is the greatest
 (4) Whether the economy will strengthen or weaken over the course of the year 2008

a. Find the probability of correctly guessing the outcome to the above predictions, assuming all outcomes are equally likely.

b. Find the probability of a given individual being correct on all four predictions, if he or she is guessing at random.

c. Find the probability of an individual making at least one incorrect prediction.

d. Find the probability of none of the 10,000 alleged psychics being correct on all four correct predictions.

e. Find the probability that at least one of the 10,000 alleged psychics will make all correct predictions.

f. Is it fair to say that any individual who makes four correct predictions is psychic?

g. What is your opinion as to the credibility of psychic predictions? Can mathematics be used to prove or disprove the possibility of psychic phenomena? Explain.

Nostradamus, a French astrologer and physician of the 16th century, became famous for his prediction of the death of King Henry II.

68. Coincidental Birthdays In this project, we investigate the probability that two people in a group have the same birthday. For convenience, we ignore leap years.

a. In how many ways can two people have *different* birthdays? (*Hint:* Let us call the two people Alfonso and Belinda. If we assign Alfonso any of the 365 days of the year, then that leaves only 364 possible days for Belinda's birthday. Use the Fundamental Counting Principle.)

b. Find the number of ways in which two people can have birthdays.

c. Find the probability of two people having different birthdays.

d. Find an expression for the number of ways in which n people can have different birthdays. Use the permutation symbol $_nP_r$.

e. Find an expression for the probability that among a group of n people, no two have the same birthday.

f. Use the result of part e to find the probability that among a group of 30 people, there are at least 2 people with a common birthday.

g. By trial and error, determine the minimum number of people required to ensure that the probability of at least one common birthday is greater than 0.5.

Section 10.5 The Binomial Theorem

- How many subsets does a set with 50 elements have?
- How can $\sqrt{10}$ be approximated to 10 decimal places using only the four basic arithmetic operations?
- How can you compute 11^7 *by hand* without ever multiplying?
- What is Pascal's triangle and how can it be used to expand $(a + b)^5$?

Expanding Binomials Polynomials with just two terms, such as $a + b$, are called *binomials*. In this section, we develop the **Binomial Theorem**, a general formula for the expansion of $(a + b)^n$. We are already familiar with some examples of the binomial formula. Several are listed here.

$$(a + b)^0 = 1$$
$$(a + b)^1 = 1a + 1b$$
$$(a + b)^2 = 1a^2 + 2ab + 1b^2$$
$$(a + b)^3 = 1a^3 + 3a^2b + 3ab^2 + 1b^3$$

From the above pattern, we would expect the expansion of $(a + b)^4$ to be of the form

$$(a + b)^4 = \Box a^4 + \Box a^3b + \Box a^2b^2 + \Box ab^3 + \Box b^4$$

where the entries in each of the boxes are constants called **binomial coefficients**.

We can determine the binomial coefficients using basic counting principles. Suppose that we were to evaluate $(a + b)^4$ directly by multiplication using the formula

$$(a + b)^4 = (a + b)(a + b)(a + b)(a + b)$$

The terms of the product can be formed by repeatedly selecting either an a or a b from each of the four factors of $(a + b)$. For example, the term a^4 is obtained by selecting a from each of the four factors. Since this can be done in only one way, the coefficient of a^4 is 1. The terms of the form a^3b arise by selecting b from one of the four factors, and a from the remaining three. This corresponds to choosing a subset of size 1 from a set with 4 elements. In Section 10.3, we saw that this can be done in

$$_4C_1 = \binom{4}{1} = \frac{4!}{1!3!} = 4$$

ways. We illustrate the four choices as follows:

Four ways of obtaining a^3b

$$(a + \boxed{b})(\boxed{a} + b)(\boxed{a} + b)(\boxed{a} + b)$$
$$(\boxed{a} + b)(a + \boxed{b})(\boxed{a} + b)(\boxed{a} + b)$$
$$(\boxed{a} + b)(\boxed{a} + b)(a + \boxed{b})(\boxed{a} + b)$$
$$(\boxed{a} + b)(\boxed{a} + b)(\boxed{a} + b)(a + \boxed{b})$$

Products of the form a^2b^2 arise by selecting a's from two of the four factors and b's from the other two factors. Thus, the coefficient of a^2b^2 is $\binom{4}{2}$. Similarly, the coefficients of ab^3 and b^4 are $\binom{4}{3}$ and $\binom{4}{4}$, respectively. Thus, we have

$$(a + b)^4 = \binom{4}{0}a^4 + \binom{4}{1}a^3b + \binom{4}{2}a^2b^2 + \binom{4}{3}ab^3 + \binom{4}{4}b^4$$

$$= \frac{4!}{0!4!}a^4 + \frac{4!}{1!3!}a^3b + \frac{4}{2!2!}a^2b^2 + \frac{4!}{3!1!}ab^3 + \frac{4!}{4!0!}b^4$$

$$= a^4 + 4a^3b + 6a^2b^2 + 4ab^3 + b^4$$

More generally, the coefficient of $a^{n-k}b^k$ in the expansion of $(a + b)^n$ is $\binom{n}{k}$. Thus, we have the following formula.

Binomial Theorem

For n a positive integer,

$$(a + b)^n = \binom{n}{0}a^n + \binom{n}{1}a^{n-1}b + \binom{n}{2}a^{n-2}b^2$$

$$+ \cdots + \binom{n}{n-1}ab^{n-1} + \binom{n}{n}b^n$$

Thus, the coefficient of $a^{n-k}b^k$ in the expansion of $(a + b)^n$ is $\binom{n}{k}$, where

$$\binom{n}{k} = {_nC_k} = \frac{n!}{k!(n-k)!}$$

EXAMPLE 1   **Finding a Coefficient Using the Binomial Theorem**

Find the coefficient of x^4y^2 in the expansion of $(x + y)^6$.

Solution Applying the Binomial Theorem with $n = 6$ and $k = 2$, we find that the coefficient of $x^4 y^2$ is given by

$$\binom{6}{2} = \frac{6!}{2!(6-2)!}$$

$$= \frac{6!}{2!4!}$$

$$= \frac{6 \cdot 5 \cdot 4 \cdot 3 \cdot 2 \cdot 1}{(2 \cdot 1)(4 \cdot 3 \cdot 2 \cdot 1)}$$

$$= \frac{6 \cdot 5}{2}$$

$$= 15$$

⟶ EXAMPLE 2

Finding a Given Coefficient with the Binomial Theorem

Find the coefficient of $x^2 y^3$ in the expansion of $(x - y)^5$.

Solution We begin by expressing $(x - y)^5$ as $[x + (-y)]^5$. Then, according to the Binomial Theorem, the coefficient of $x^2(-y)^3$ is

$$\binom{5}{3} = \frac{5!}{3!2!} = \frac{5 \cdot 4 \cdot 3 \cdot 2 \cdot 1}{(3 \cdot 2 \cdot 1)(2 \cdot 1)} = \frac{5 \cdot 4}{2} = 10$$

Thus the term in question is $10x^2(-y)^3 = -10x^2 y^3$, and so the coefficient of $x^2 y^3$ is -10.

⟶ EXAMPLE 3

Expanding with the Binomial Theorem

Expand $(x + 2)^5$.

Solution According to the Binomial Theorem,

$$(x + 2)^5 = \binom{5}{0}x^5 + \binom{5}{1}x^4 \cdot 2 + \binom{5}{2}x^3 \cdot 2^2 + \binom{5}{3}x^2 \cdot 2^3 + \binom{5}{4}x \cdot 2^4 + \binom{5}{5}2^5$$

$$= x^5 + \frac{5!}{1!4!}x^4 \cdot 2 + \frac{5!}{2!3!}x^3 \cdot 2^2 + \frac{5!}{3!2!}x^2 \cdot 2^3 + \frac{5!}{4!1!}x \cdot 2^4 + 2^5$$

$$= x^5 + 5x^4 \cdot 2 + 10x^3 \cdot 4 + 10x^2 \cdot 8 + 5x \cdot 16 + 32$$

$$= x^5 + 10x^4 + 40x^3 + 80x^2 + 80x + 32$$

Techniques for Evaluating Binomial Coefficients

Expanding powers of binomials using the Binomial Theorem can be quite tedious; fortunately, we can take advantage of several properties of binomial coefficients to reduce our work considerably. For example, if we evaluate the binomial coefficient $\binom{7}{3}$ using only factorials, we have

$$\binom{7}{3} = \frac{7!}{3!(7-3)!} = \frac{7!}{3!4!} = \frac{5040}{6 \cdot 24} = 35$$

which is a fairly tedious calculation. However, we can use the formula for combinations

$$\binom{n}{r} = {}_nC_r = \frac{{}_nP_r}{r!}$$

introduced in Section 10.3 to obtain

$$\binom{7}{3} = \frac{_7P_3}{3!}$$

$$= \frac{7 \cdot 6 \cdot 5}{3 \cdot 2 \cdot 1}$$

$$= 35$$

More generally, we have the following formula.

Alternative Form for Binomial Coefficients

The binomial coefficient $\binom{n}{k}$ can be computed as follows:

$$\binom{n}{k} = \frac{n(n-1)(n-2)\cdots(n-k+1)}{k(k-1)(k-2)\cdots(2)(1)}$$

Note that there are k factors in both the numerator and denominator.

EXAMPLE 4

Using the Alternative Form to Evaluate Binomial Coefficients

Evaluate each of the following binomial coefficients using the alternative form:

a. $\binom{10}{3}$ **b.** $\binom{8}{4}$

Solution

a. $\binom{10}{3} = \frac{10 \cdot 9 \cdot 8}{3 \cdot 2 \cdot 1}$

$= \frac{720}{6}$

$= 120$

b. $\binom{8}{4} = \frac{8 \cdot 7 \cdot 6 \cdot 5}{4 \cdot 3 \cdot 2 \cdot 1}$

$= 7 \cdot 2 \cdot 5$

$= 70$

The computation of binomial coefficients can be further simplified by noting that

$$\binom{n}{n-k} = \frac{n!}{(n-k)![n-(n-k)]!}$$

$$= \frac{n!}{(n-k)!k!}$$

$$= \binom{n}{k}$$

We have thus established the following property of binomial coefficients.

Symmetry Property of Binomial Coefficients

For integers n and k with $0 \le k \le n$,

$$\binom{n}{k} = \binom{n}{n-k}$$

Thus, the coefficient of $a^k b^{n-k}$ in the expansion of $(a+b)^n$ is the same as the coefficient of $a^{n-k} b^k$.

It follows from the symmetry property that the first binomial coefficient is the same as the last, the second is the same as the second to the last, and so forth. As a consequence, the work required to compute the binomial coefficients of $(a + b)^n$ is essentially cut in half, as illustrated by the following example.

EXAMPLE 5

Using the Symmetry Property to Simplify Binomial Expansions

Find the binomial expansion of $(a + b)^6$.

Solution   According to the Binomial Theorem, we have

$$(a + b)^6 = \binom{6}{0}a^6 + \binom{6}{1}a^5b + \binom{6}{2}a^4b^2 + \binom{6}{3}a^3b^3 + \binom{6}{4}a^2b^4 + \binom{6}{5}ab^5 + \binom{6}{6}b^6$$

Evaluating the "first half" of these coefficients, we have

$$\binom{6}{0} = 1, \quad \binom{6}{1} = \frac{6}{1} = 6, \quad \binom{6}{2} = \frac{6 \cdot 5}{2 \cdot 1} = 15, \quad \text{and} \quad \binom{6}{3} = \frac{6 \cdot 5 \cdot 4}{3 \cdot 2 \cdot 1} = 20$$

By the symmetry property, we have

$$\binom{6}{4} = \binom{6}{2} = 15, \quad \binom{6}{5} = \binom{6}{1} = 6, \quad \text{and} \quad \binom{6}{6} = \binom{6}{0} = 1$$

Thus,

$$(a + b)^6 = a^6 + 6a^5b + 15a^4b^2 + 20a^3b^3 + 15a^2b^4 + 6ab^5 + b^6$$

Computing Binomial Coefficients

A graphing calculator can also be used to compute binomial coefficients, but note that, for most calculators, the combination notation nCr is used. To compute a binomial coefficient $\binom{n}{r}$, first key in the number n, then select nCr from the menu of mathematical functions, and finally enter the number r.

Another convenient method for evaluating binomial coefficients arises from the observation that the binomial coefficients form a triangular array of numbers known as *Pascal's triangle*. Consider the expansions of $(a + b)^n$ for $n = 1, 2, 3$, and 4, shown in Figure 14, with the binomial coefficients boxed.

$$
\begin{aligned}
(a + b)^0 &= \boxed{1} \\
(a + b)^1 &= \boxed{1}a + \boxed{1}b \\
(a + b)^2 &= \boxed{1}a^2 + \boxed{2}ab + \boxed{1}b^2 \\
(a + b)^3 &= \boxed{1}a^3 + \boxed{3}a^2b + \boxed{3}ab^2 + \boxed{1}b^3 \\
(a + b)^4 &= \boxed{1}a^4 + \boxed{4}a^3b + \boxed{6}a^2b^2 + \boxed{4}ab^3 + \boxed{1}b^4
\end{aligned}
$$

Figure 14

If we write *only* the binomial coefficients, we have the array known as Pascal's triangle, shown in Figure 15.

$$1$$

$$1 \qquad 1$$

$$1 \qquad 2 \qquad 1$$

$$1 \qquad 3 \qquad 3 \qquad 1$$

$$1 \qquad 4 \qquad 6 \qquad 4 \qquad 1$$

Figure 15

Note that any given binomial coefficient is the sum of the two binomial coefficients immediately above it in the preceding row. In other words,

$$\binom{n+1}{k} = \binom{n}{k-1} + \binom{n}{k}$$

This property, which was considered in Exercise 58 of Section 10.3, can be used to complete the fifth and sixth rows of Pascal's triangle, as shown in Figure 16. Note that the first and last entry of each row is 1 and that all other entries are obtained by adding together the entries immediately above the given entry, as suggested by the arrows.

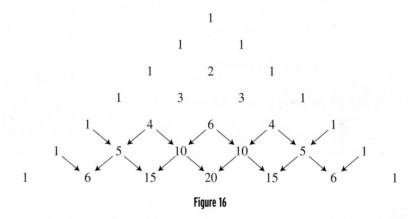

Figure 16

We refer to the top row of Pascal's triangle—the row consisting of a single 1—as row 0. The row consisting of two 1s is row 1, and so on. Since the nth row of Pascal's triangle consists of the binomial coefficients $\binom{n}{k}$, Pascal's triangle can be used to obtain binomial expansions, as illustrated in the following example.

EXAMPLE 6

Expanding with Pascal's Triangle

Expand using Pascal's triangle.

a. $(a + b)^6$ **b.** $(3x - 2y)^5$ **c.** $\left(1 + \sqrt{2}\right)^4$

Solution

a. The sixth row of Pascal's triangle in Figure 16 consists of the numbers 1, 6, 15, 20, 15, 6, and 1. Thus

$$(a + b)^6 = 1a^6 + 6a^5b + 15a^4b^2 + 20a^3b^3 + 15a^2b^4 + 6ab^5 + 1b^6$$

b. The fifth row of Pascal's triangle in Figure 16 consists of the numbers 1, 5, 10, 10, 5, and 1. Thus

$$(3x - 2y)^5 = 1(3x)^5 + 5(3x)^4(-2y) + 10(3x)^3(-2y)^2 + 10(3x)^2(-2y)^3 + 5(3x)(-2y)^4 + 1(-2y)^5$$
$$= 243x^5 - (5 \cdot 81 \cdot 2)x^4y + (10 \cdot 27 \cdot 4)x^3y^2 - (10 \cdot 9 \cdot 8)x^2y^3 + (5 \cdot 3 \cdot 16)xy^4 - 32y^5$$
$$= 243x^5 - 810x^4y + 1080x^3y^2 - 720x^2y^3 + 240xy^4 - 32y^5$$

c. From Figure 16, we see that the fourth row of Pascal's triangle consists of the numbers 1, 4, 6, 4, and 1. Thus, we have

$$\left(1 + \sqrt{2}\right)^4 = 1 \cdot 1^4 + 4 \cdot 1^3 \cdot \sqrt{2} + 6 \cdot 1^2 \cdot \left(\sqrt{2}\right)^2 + 4 \cdot 1 \cdot \left(\sqrt{2}\right)^3 + 1 \cdot \left(\sqrt{2}\right)^4$$
$$= 1 + 4\sqrt{2} + 6 \cdot 2 + 4 \cdot 2\sqrt{2} + 4$$
$$= 17 + 12\sqrt{2}$$

Understanding and Mastery Checklists

Concepts to Understand	Skills to Master
Binomial	Compute a binomial coefficient using combinations.
⬦	⬦
Binomial coefficient	Find the coefficient of a specified term of a binomial expansion.
⬦	⬦
Binomial Theorem	Use the Binomial Theorem to expand a binomial.
⬦	⬦
Symmetry property of binomial coefficients	Use Pascal's triangle to find binomial coefficients.
⬦	
Pascal's triangle	

Exercises 10.5

Exercises 1-12 *Find the coefficient of the indicated term in the expansion of the binomial.*

1. $(A + B)^4$; AB^3

2. $(C - D)^3$; CD^2

3. $(x + y)^5$; x^3y^2

4. $(x + y)^6$; x^2y^4

5. $(z + 1)^6$; z^5

6. $(x + 2)^3$; x^3

7. $(2x - 3y)^3$; xy^2

8. $(2a - b)^6$; a^3b^3

9. $\left(x + \dfrac{1}{x}\right)^4$; the constant term

10. $\left(z - \dfrac{1}{z}\right)^4$; z^2

11. $(1 + w^2)^5$; w^4

12. $(x + 2w^3)^4$; xw^9

Exercises 13-28 *Expand the given expression.*

13. $(x + y)^5$

14. $(a - b)^5$

15. $(2x - y)^4$

16. $(3a + b)^4$

17. $(3x + 5y)^3$

18. $(u - 2w)^4$

19. $(x + 1)^5$

20. $(2s + 1)^4$

21. $(10w - 1)^6$

22. $(z - 2)^6$

23. $\left(1 + \sqrt{2}\right)^3$ **24.** $\left(3 + \sqrt{3}\right)^4$

25. $(x^2 - 1)^3$ **26.** $(y^3 + 1)^3$

27. $\left(x + \dfrac{1}{x}\right)^5$ **28.** $\left(y - \dfrac{1}{y}\right)^4$

Exercises 29-34 *Simplify. (Recall that $i^2 = -1$.)*

29. $(1 + i)^3$ **30.** $(2 - 3i)^3$

31. $(2 + 2i)^4$ **32.** $(1 - 3i)^4$

33. $\left(\dfrac{\sqrt{2}}{2} + \dfrac{\sqrt{2}}{2}i\right)^2$ **34.** $\left(\dfrac{3}{5} + \dfrac{4}{5}i\right)^3$

Exercises 35-42 *Verify the given identity involving factorials or binomial coefficients.*

35. $n! \cdot n = (n + 1)! - n!$ **36.** $(n^2 - n)(n - 2)! = n!$

37. $\dfrac{(n + 1)!}{n!} = n + 1$ **38.** $\dfrac{(n - 1)!}{(n + 1)!} = \dfrac{1}{n^2 + n}$

39. $\dbinom{n - 1}{k} + \dbinom{n - 1}{k - 1} = \dbinom{n}{k}$ **40.** $\dbinom{n}{r} = \dfrac{n - r + 1}{r}\dbinom{n}{r - 1}$

41. $\dbinom{n}{r} = \dfrac{n}{n - r}\dbinom{n - 1}{r}$ **42.** $n\dbinom{n - 1}{r} = (r + 1)\dbinom{n}{r + 1}$

Concepts and Critical Thinking

Exercises 43-46 *Answer true or false.*

43. The symbols $\dbinom{n}{r}$ and $_nP_r$ have the same meaning.

44. $\dbinom{1,000,000}{999,998} = \dbinom{1,000,000}{2}$

45. All binomial coefficients can be found within Pascal's triangle.

46. The symbols $\dbinom{n}{r}$ and $_nC_r$ have the same meaning.

Exercises 47-50 *Give an example of each.*

47. Another context in which binomial coefficients arise (other than in expanding powers of binomials)

48. Two numbers x and y such that $(x + y)^2 = x^2 + y^2$.

49. A method for evaluating binomials that does not involve computation of factorials

50. A binomial coefficient $\dbinom{n}{r}$ that evaluates to 17

51. An extremely common error is for students to simplify $(x + y)^n$ as $x^n + y^n$. Why do you think this error is so common? Are there any values of x and y such that $(x + y)^n = x^n + y^n$ for all n? Graph the set of points (x, y) such that $(x + y)^2 = x^2 + y^2$.

52. Explain how
$$(x + y)^4 - 4(x + y)^3y + 6(x + y)^2y^2 - 4(x + y)y^3 + y^4$$
can be simplified within 10 seconds.

53. Write 11^7 as $(10 + 1)^7$, and explain how the Binomial Theorem can be used to compute 11^7 without using a calculator. Will this technique work for other similar computations? If so, give some examples. If not, explain why not.

Questions for Discussion or Essay

54. Under what circumstances is Pascal's triangle easier to use than the Binomial Theorem?

55. There was a time when nearly every high school student studying algebra would learn the Binomial Theorem. Recently, however, an increasing number of college students have never been exposed to the Binomial Theorem. In fact, many instructors using this text will not include this section in their course. How important do you feel this topic is? Why might it have been considered more important 40 years ago than it is today?

Projects for Enrichment

56. The Extended Binomial Theorem The Binomial Theorem tells us how to expand expressions of the form $(a + b)^n$, where n is a positive integer. In this project, we consider a generalization of the Binomial Theorem to the case where n is not an integer. We begin by extending our definition of binomial coefficients. For a fixed integer k with $0 \le k \le n$, the expression $\dbinom{n}{k}$ is a polynomial in the variable n. For example,

$$\dbinom{n}{2} = \dfrac{n(n - 1)}{2}$$

Clearly, this polynomial is defined whether n is an integer or not. Thus, for example, we could define

$$\dbinom{\frac{1}{2}}{2} = \dfrac{\frac{1}{2}\left(\frac{1}{2} - 1\right)}{2} = -\dfrac{1}{8}$$

Similarly, we define for any real number n

$$\binom{n}{3} = \frac{n(n-1)(n-2)}{3 \cdot 2 \cdot 1},$$

$$\binom{n}{4} = \frac{n(n-1)(n-2)(n-3)}{4 \cdot 3 \cdot 2 \cdot 1}, \cdots$$

a. Compute each of the following extended binomial coefficients:

i. $\binom{\frac{1}{3}}{2}$ **ii.** $\binom{\frac{3}{2}}{3}$

iii. $\binom{\frac{6}{5}}{4}$

Using this extended notion of a binomial coefficient, it can be shown that if $\left|\frac{b}{a}\right| < 1$, then

$$(a+b)^n = a^n + \binom{n}{1}a^{n-1}b + \binom{n}{2}a^{n-2}b^2$$

$$+ \binom{n}{3}a^{n-3}b^3 + \binom{n}{4}a^{n-4}b^4 + \cdots$$

Notice here that if n is not a natural number, the expansion has an *infinite* number of terms of the form

$$\binom{n}{k}a^{n-k}b^k$$

As an example, consider the binomial expansion for $(a+b)^{1/2}$.

$$(a+b)^{1/2} = a^{1/2} + \binom{\frac{1}{2}}{1}a^{-1/2}b + \binom{\frac{1}{2}}{2}a^{-3/2}b^2$$

$$+ \binom{\frac{1}{2}}{3}a^{-5/2}b^3 + \cdots$$

$$= a^{1/2} + \frac{1}{2}a^{-1/2}b - \frac{1}{8}a^{-3/2}b^2 + \frac{1}{16}a^{-5/2}b^3 + \cdots$$

b. Find the first four terms of the expansion for the given binomial expression.

i. $(a+b)^{1/3}$ **ii.** $(a+b)^{3/2}$

iii. $(a+b)^{6/5}$

One application of the extended Binomial Theorem is approximating roots, such as $\sqrt{10}$. We first write 10 as the sum of a perfect square and a "small" number—namely, $9 + 1$. Now, if we set $a = 9$ and $b = 1$ in the binomial expansion of $(9+1)^{1/2}$, we obtain

$$(9+1)^{1/2} = (9)^{1/2} + \frac{1}{2}(9)^{-1/2}(1) - \frac{1}{8}(9)^{-3/2}(1)^2$$

$$+ \frac{1}{16}(9)^{-5/2}(1)^3 + \cdots$$

$$= 3 + \frac{1}{2} \cdot \frac{1}{3} - \frac{1}{8} \cdot \frac{1}{27} + \frac{1}{16} \cdot \frac{1}{243} + \cdots$$

Taking only the first four terms, we thus have

$$\sqrt{10} \approx 3 + \frac{1}{6} - \frac{1}{216} + \frac{1}{3888} \approx 3.162$$

Note that a and b were chosen so that a was a perfect square. If we were to approximate $(a+b)^{1/3}$ using this procedure, we would want a to be a perfect cube. In addition, it is necessary for a and b to be such that $\left|\frac{b}{a}\right| < 1$. Finally, as we will see shortly, the accuracy of the approximation is much better if b is small compared to a.

c. Use the first four terms of an appropriate binomial expansion to approximate the following roots:

i. $\sqrt{26}$ **ii.** $\sqrt{102}$

iii. $\sqrt[3]{220}$

Any square root between 9 and 16 can be approximated by writing it in the form $(9+x)^{1/2}$ and then expanding, as we just did. For example, we can approximate $\sqrt{14}$ by expanding $(9+5)^{1/2}$. However, if we use only the first four terms of the expansion, the accuracy may not be very good. To see this, consider the expansion of $(9+x)^{1/2}$, as follows:

$$(9+x)^{1/2} = (9)^{1/2} + \frac{1}{2}(9)^{-1/2}(x) - \frac{1}{8}(9)^{-3/2}(x)^2$$

$$+ \frac{1}{16}(9)^{-5/2}(x)^3 + \cdots$$

$$= 3 + \frac{1}{6}x - \frac{1}{216}x^2 + \frac{1}{3888}x^3 + \cdots$$

The first four terms give us the polynomial

$$p(x) = 3 + \frac{1}{6}x - \frac{1}{216}x^2 + \frac{1}{3888}x^3$$

d. Use a graphing calculator to plot the graphs of

$$f(x) = \sqrt{9+x}$$

and

$$p(x) = 3 + \frac{1}{6}x - \frac{1}{216}x^2 + \frac{1}{3888}x^3$$

on the same coordinate axes. Then use the trace feature to determine the values of x for which $p(x)$ approximates $f(x)$ to at least one decimal place.

57. Finding the Number of Subsets of a Set   A set with 3 elements, such as $S = \{a, b, c\}$, has 2^3 subsets. The subsets of S are Ø (the empty set), $\{a\}, \{b\}, \{c\}, \{a, b\}, \{a, c\}, \{b, c\}$, and $\{a, b, c\}$. In this project, we demonstrate that a set with n elements has 2^n subsets.

a. Recall from Section 10.3 that $\binom{n}{k}$ represents the number of subsets of size k of a set with n elements. Show that the *total* number of subsets of a set with n elements is given by

$$\binom{n}{0} + \binom{n}{1} + \binom{n}{2} + \cdots + \binom{n}{n-1} + \binom{n}{n}$$

b. Now use the Binomial Theorem to express the sum in part a as a binomial to the nth power.

c. Finally, simplify the expression obtained in part b to show that the number of subsets of a set with n elements is 2^n.

d. How many subsets does a set with 50 elements have?

Section 10.6

Mathematical Induction

⚙ What does a cascading line of dominoes have in common with a mathematical technique for proving statements involving the natural numbers?

⚙ How can a statement such as $n^n \geq n!$ be shown to be true for *every* natural number n?

⚙ How can it be shown that any natural number debt of $8 or greater can be paid using only $3 and $5 bills?

⚙ How can the incorrect use of a mathematical technique be used to prove that all horses are the same color?

The Principle of Mathematical Induction

Many important mathematical formulas and theorems involving the natural numbers— that is, the numbers 1, 2, 3, . . .—can be proven using a technique known as *mathematical induction*. Before we give a formal statement of mathematical induction, let us consider the following nonmathematical examples of induction:

- A group of first graders are standing in a line. It is known that any first grader who gets pushed will push the first grader in front of him. It is also known that a second-grade bully pushes the first grader at the back of the line. We can thus conclude that every first grader in the line gets pushed.
- Dominoes are lined up in such a way that if a domino falls, it will knock over the domino in front of it. The first domino is pushed over. It then follows that all of the dominoes will fall.
- Sufferers of a certain genetic disorder will pass the disorder along to all of their children. It is known that Stan Edbury has this disorder. We can thus conclude that all of Stan Edbury's descendents will have the disorder.

Notice how in each of these examples we conclude that every member of a list (a line of first graders, a row of dominoes, the descendents of Stan Edbury) satisfies a certain property (being pushed, falling down, suffering from a genetic disorder) provided that two conditions are satisfied:

1. The property is possessed by the first member of the list.
2. If a member of the list has the property, then the next member of the list will have the property as well.

Clearly, if either of the conditions is not satisfied, then our conclusion is invalid. For example, if the dominoes are spaced sufficiently far apart that a falling domino *doesn't* knock over the next domino, then even if the first domino is toppled, it doesn't follow that all of the dominoes will fall. Likewise, even if our genetic disorder is inherited by 100% of one's offspring, if Stan Edbury himself doesn't have the disorder, then we cannot conclude that all of his descendents will have the disorder.

Mathematical induction is the application of this line of reasoning to statements made about the natural numbers. If we can show that a statement is true for the first natural number (the number 1) and that the truth of the statement for a natural number implies the truth of the statement for the next natural number, then we can conclude that the statement is, in fact, true for all natural numbers. More formally, we have the following.

Principle of Mathematical Induction

If a statement about the natural numbers is true for the number 1, and if the truth of the statement for the natural number k implies the truth of the statement for $k + 1$ (the next natural number), then the statement is true for all natural numbers.

Proofs by mathematical induction consist of two steps.

Step 1   Show that the statement is true for $n = 1$.

Step 2   Show that if the statement is true for a natural number k, then it is true for $k + 1$.

EXAMPLE 1

Establishing a Formula by Mathematical Induction

Use mathematical induction to prove that

$$1 + 2 + 3 + \cdots + n = \frac{n(n + 1)}{2}$$

for all natural numbers n.

Solution

Step 1   For $n = 1$, the statement reduces to

$$1 = \frac{1(1 + 1)}{2}$$

$$1 = 1$$

which is clearly true.

Step 2   Assume that the statement is true for the natural number k; that is, assume that

$$1 + 2 + 3 + \cdots + k = \frac{k(k + 1)}{2} \qquad \text{Replacing } n \text{ with } k \text{ in the original statement}$$

We must show that this implies the truth of the statement for $k + 1$; that is, we must show that

$$1 + 2 + \cdots + k + (k + 1) = \frac{(k + 1)[(k + 1) + 1]}{2} \qquad \begin{array}{l}\text{Replacing } n \text{ with } k + 1 \\ \text{in the original statement}\end{array}$$

We proceed as follows:

$$1 + 2 + \cdots + k + (k + 1)$$

$$= (1 + 2 + \cdots + k) + (k + 1)$$

$$= \frac{k(k + 1)}{2} + (k + 1) \qquad \begin{array}{l}\text{Using the assumption that} \\ 1 + 2 + \cdots + k = \frac{k(k + 1)}{2}\end{array}$$

$$= \frac{k(k + 1)}{2} + \frac{2(k + 1)}{2}$$

$$= \frac{(k + 1)(k + 2)}{2}$$

$$= \frac{(k + 1)[(k + 1) + 1]}{2}$$

We can thus conclude from the Principle of Mathematical Induction that

$$1 + 2 + \cdots + n = \frac{n(n + 1)}{2}$$

for all natural numbers n.

EXAMPLE 2

Establishing Divisibility by Mathematical Induction

Use mathematical induction to prove that $n^3 - n + 3$ is divisible by 3 for all natural numbers n.

Solution

Step 1 For $n = 1$, our statement reads "$1^3 - 1 + 3$ is divisible by 3," which is true since $1^3 - 1 + 3 = 3$.

Step 2 Assume that the statement is true for the natural number k. That is, assume $k^3 - k + 3$ is divisible by 3. We must show that $(k + 1)^3 - (k + 1) + 3$ is divisible by 3 also. We start by rewriting $(k + 1)^3 - (k + 1) + 3$.

$$(k + 1)^3 - (k + 1) + 3 = (k^3 + 3k^2 + 3k + 1) - k - 1 + 3$$
$$= (k^3 - k + 3) + (3k^2 + 3k)$$
$$= (k^3 - k + 3) + 3(k^2 + k)$$

Now $3(k^2 + k)$ is divisible by 3, and we are assuming that $k^3 - k + 3$ is divisible by 3. Thus, their sum, $(k^3 - k + 3) + 3(k^2 + k)$, is divisible by 3 also.

We can thus conclude from the Principle of Mathematical Induction that $n^3 - n + 3$ is divisible by 3 for all natural numbers n.

EXAMPLE 3

Establishing an Inequality by Mathematical Induction

Use mathematical induction to prove that $n^n \geq n!$ for all natural numbers n.

Solution

Step 1 For $n = 1$, we have $1^1 \geq 1!$ or $1 \geq 1$, which is true.

Step 2 Assume that the statement is true for the natural number k; that is, assume that $k^k \geq k!$. It must be shown that

$$(k + 1)^{k+1} \geq (k + 1)!$$

We proceed as follows:

$$(k + 1)^{k+1} = (k + 1)^k(k + 1)$$
$$> k^k(k + 1) \qquad \text{Using the fact that } (k + 1)^k > k^k$$
$$\geq k!(k + 1) \qquad \text{Using the assumption that } k^k \geq k!$$
$$= (k + 1)! \qquad \text{Since } (k + 1)! = (k + 1) \cdot k!$$

So $(k + 1)^{k+1} \geq (k + 1)!$.

Thus, we can conclude that $n^n \geq n!$ for all natural numbers n.

Generalized Principle of Mathematical Induction

Let us return to the scenario described earlier, in which dominoes are lined up so that falling is contagious: When one domino falls, the next one falls also. This time, however, instead of upsetting the *first* domino, we knock over the fifth. What happens? Well, we can't say anything about the first four dominoes, but since the fifth will knock over the sixth and the sixth will knock over the seventh, and so forth, we can conclude that all dominoes from the fifth on will fall. This line of reasoning suggests the following generalization of the Principle of Mathematical Induction.

Generalized Principle of Mathematical Induction

If a statement about the natural numbers is true for n_0 and if the truth of the statement for any natural number $k \geq n_0$ implies the truth of the statement for $k + 1$, then the statement is true for all natural numbers greater than or equal to n_0.

The Generalized Principle of Mathematical Induction is invaluable when proving statements that do not hold true for small natural numbers n, but rather are true from some point on. As with the ordinary Principle of Mathematical Induction, we prove a statement using the Generalized Principle of Mathematical Induction by proceeding in two steps. In step 1 we show that the statement is true for a particular natural number n_0, which is usually stated in the problem. In step 2, we assume that the statement is true for some natural number k greater than or equal to n_0, and we then show that this implies the statement is also true for $k + 1$.

EXAMPLE 4

Applying the Generalized Principle of Mathematical Induction

Show that $n! > 20n$ for $n \geq 5$.

Solution

Step 1 When $n = 5$, we have $n! = 5! = 120$, and $20n = 20 \cdot 5 = 100$. Since $120 > 100$, the statement is true for $n = 5$.

Step 2 Now assume that the statement is true for some natural number $k \geq 5$; that is, that $k! > 20k$ for some $k \geq 5$. We must show that $(k + 1)! > 20(k + 1)$.

$$
\begin{aligned}
(k + 1)! &= (k + 1) \cdot k! \\
&> (k + 1) \cdot 20k &&\text{Using the assumption that } k! > 20k \\
&= k \cdot 20(k + 1) \\
&> 20(k + 1) &&\text{Since } k > 1
\end{aligned}
$$

Thus, $(k + 1)! > 20(k + 1)$.

We can thus conclude from the Generalized Principle of Mathematical Induction that $n! > 20n$ for all $n \geq 5$.

Understanding and Mastery Checklists

Concepts to Understand	Skills to Master
Principle of Mathematical Induction	Use the Principle of Mathematical Induction to prove a statement about natural numbers.
❖	❖
Generalized Principle of Mathematical Induction	Use the Generalized Principle of Mathematical Induction to prove a statement about natural numbers.

Exercises 10.6

Exercises 1-24 *Use the Principle of Mathematical Induction to show that the statement is true for all natural numbers n.*

1. $1 + 3 + 5 + 7 + 9 + \cdots + (2n + 1) = (n + 1)^2$

2. $2 + 4 + 6 + 8 + \cdots + (2n) = n(n + 1)$

3. $1^2 + 2^2 + 3^2 + \cdots + n^2 = \dfrac{n(n + 1)(2n + 1)}{6}$

4. $1^3 + 2^3 + 3^3 + \cdots + n^3 = \dfrac{n^2(n + 1)^2}{4}$

5. $5 + 10 + 15 + \cdots + 5n = \dfrac{5n(n + 1)}{2}$

6. $6 + 10 + 14 + \cdots + (4n + 2) = 2n(n + 2)$

7. $1 + 2 + 2^2 + 2^3 + \cdots + 2^n = 2^{n+1} - 1$

8. $1 - 2 + 2^2 - 2^3 + \cdots - 2^{2n-1} = \dfrac{1 - 2^{2n}}{3}$

9. $1 - 2 + 2^2 - 2^3 + \cdots - 2^{2n-1} + 2^{2n} = \dfrac{1 + 2^{2n+1}}{3}$

10. $3 + 3^2 + 3^3 + \cdots + 3^n = \dfrac{3^{n+1} - 3}{2}$

11. $\dfrac{1}{1 \cdot 2} + \dfrac{1}{2 \cdot 3} + \dfrac{1}{3 \cdot 4} + \cdots + \dfrac{1}{n(n + 1)} = 1 - \dfrac{1}{n + 1}$

12. $\dfrac{1}{1 \cdot 3} + \dfrac{1}{3 \cdot 5} + \dfrac{1}{5 \cdot 7} + \cdots + \dfrac{1}{(2n - 1) \cdot (2n + 1)} = \dfrac{n}{2n + 1}$

13. $1 \cdot 2 + 3 \cdot 4 + 5 \cdot 6 + \cdots + (2n - 1) \cdot 2n = \dfrac{n(n + 1)(4n - 1)}{3}$

14. $1 \cdot 2 + 2 \cdot 3 + 3 \cdot 4 + \cdots + n \cdot (n + 1) = \dfrac{n(n + 1)(n + 2)}{3}$

15. $1 \cdot 2^1 + 2 \cdot 2^2 + 3 \cdot 2^3 + \cdots + n \cdot 2^n = 2^{n+1}(n - 1) + 2$

16. $\dfrac{1}{2} + \dfrac{3}{2^2} + \dfrac{5}{2^3} + \cdots + \dfrac{2n - 1}{2^n} = 3 - \dfrac{2n + 3}{2^n}$

17. $n^2 + n$ is even. **18.** $n^2 - n + 1$ is odd.

19. $n^3 - n$ is divisible by 3. **20.** $n^3 + 2n$ is divisible by 3.

21. $n \le 2^{n-1}$ **22.** $1 + 2n \le 3^n$

23. If $0 \le a < 1$, then $a^n < 1$. **24.** If $a > 1$, then $a^n > 1$.

Exercises 25-30 *Use the Generalized Principle of Mathematical Induction to show that the statement is true.*

25. $2^n < n!$ for $n \ge 4$ **26.** $4n < 2^n$ for $n \ge 5$

27. $n + 12 \le n^2$ for $n \ge 4$ **28.** $n^2 + 18 \le n^3$ for $n \ge 3$

29. $n^2 + 4 < (n + 1)^2$ for $n \ge 2$ **30.** $n^3 > (n + 1)^2$ for $n \ge 3$

Exercises 31-34 *These exercises deal with the Fibonacci sequence defined in Section 10.1 by $F_1 = 1$, $F_2 = 1$, and $F_n = F_{n-1} + F_{n-2}$ for $n > 2$. Use the Principle of Mathematical Induction (or the Generalized Principle) to prove the given property of the Fibonacci sequence for all natural numbers n.*

31. $F_1 + F_2 + \cdots + F_n = F_{n+2} - 1$

32. $F_2 + F_4 + \cdots + F_{2n} = F_{2n+1} - 1$

33. $F_1 + F_3 + \cdots + F_{2n-1} = F_{2n}$

34. $F_1^2 + F_2^2 + F_3^2 + \cdots + F_n^2 = F_n F_{n+1}$

Concepts and Critical Thinking

Exercises 35-38 *Answer true or false.*

35. When proving statements by mathematical induction, we actually assume that what we are trying to prove is true.

36. If we prove that a statement is true for $n = 7$ and we prove that if it is true for $n = k$ then it is also true for $n = k + 1$, then we have proven that the statement is true for all positive integers.

37. If we prove that a statement is true for $n = 7$ and we prove that if it is true for $n = k$ then it is also true for $n = k + 1$, then we have proven that the statement is true for all positive integers greater than or equal to 7.

38. If it is the case that whenever a certain statement is true for $n = k$ it is also true for $n = k + 1$, then it follows that the statement is true for all positive integers n.

Exercises 39-42   *Give an example of each.*

39. A statement about the natural numbers that is true for $n = 1$ and 2 but is false for $n = 3$

40. A statement about the integers that is true for integers $n \geq 3$ but is false for $n = 1$ and 2

41. An alternative method (other than the induction proof given in Exercise 17) for establishing that $n^2 + n$ is even for all natural numbers n

42. An alternative method (other than the induction proof given in Exercise 1) for establishing that
$$1 + 3 + 5 + 7 + 9 + \cdots + (2n + 1) = (n + 1)^2$$

Questions for Discussion or Essay

43. What is wrong with the following proof that all odd numbers are even?

> *Any odd number can be written in the form $2n + 1$. Thus, we must show that all numbers of the form $2n + 1$ are even. Suppose that the statement is true for n. Then $2n + 1$ is even and hence divisible by 2. There is, therefore, a natural number m such that $2n + 1 = 2 \cdot m$. Now we must show that $2 \cdot (n + 1) + 1 = 2n + 3$ is even as well. But $2n + 3 = (2n + 1) + 2 = 2 \cdot m + 2 = 2 \cdot (m + 1)$. So $2n + 3$ is divisible by 2 and is thus even. By induction, we have shown that all odd numbers are even.*

44. In this exercise we investigate a mysterious use of mathematical induction that seemingly proves that all horses are the same color! The "proof" of the statement *All horses are of the same color* is as follows:

> *Suppose there is 1 horse. Obviously, it is of one color. Now suppose it is true that any group of n horses have the same color. We will demonstrate that any group of $n + 1$ horses will also be of one color, as follows:*
>
> *Consider a group of $n + 1$ horses. Remove 1 horse (whom we will call Silver) from the group so that n horses are left. By our earlier assumption, all of these horses are the same color. We need only show that Silver is the same color as the rest of the horses.*
>
> *To do this, return Silver to the group and remove a different horse so that again n horses are left. By hypothesis, these n horses are all colored the same. Since Silver is one of the horses and the rest of the horses are part of the original group of n horses that had the same color, Silver is the same color as the other horses. Thus, by induction, all horses are the same color.*

Since the theorem is obviously false, there must be an error somewhere in the proof. Explain the error in detail.

45. It has been said that mathematical induction is not a method for discovering mathematical statements but a technique for rigorously proving a statement that has already been discovered. Do you agree? Why or why not?

46. Inductive reasoning is a method of reasoning whereby specific examples lead to a general conclusion. An example would be the argument, "Trevor studied hard and got an A on the test; LaShonda studied hard and got an A on the test; therefore, everyone who studies hard will get an A on the test." A mathematical example would be the argument "$5^2 < 2^5$, $6^2 < 2^6$, and $7^2 < 2^7$; therefore, $n^2 < 2^n$ for all natural numbers n." What do you see as a major problem with inductive reasoning? Under what circumstances is inductive reasoning useful? What is the connection between inductive reasoning and mathematical induction?

47. It is obviously not legitimate to prove a theorem by assuming that which is to be proven. For example, consider the following "proof" that all mathematicians are males:

> *Suppose that all mathematicians are men. Let Pat be a mathematician. Since all mathematicians are men, Pat is a man. The same will be true of all other mathematicians. Thus, all mathematicians are men.*

Of course, the reasoning is absurd, but isn't this precisely the same reasoning that is used when, in the course of a proof by mathematical induction, we assume that the statement is true for k? If not, how is it different?

Projects for Enrichment

48. Proving the Binomial Theorem   In this project, we use the technique of mathematical induction to establish the Binomial Theorem:

$$(a + b)^n = \binom{n}{0}a^n + \binom{n}{1}a^{n-1}b + \binom{n}{2}a^{n-2}b^2$$
$$+ \cdots + \binom{n}{n-1}ab^{n-1} + \binom{n}{n}b^n$$

a. Show that the Binomial Theorem is true for $n = 1$.

b. Assume that the Binomial Theorem is true for $n = k$. By writing $(a + b)^{k+1}$ as $(a + b)^k(a + b)$, show that

$$(a + b)^{k+1} = \binom{k}{0}a^{k+1} + \binom{k}{1}a^kb + \binom{k}{2}a^{k-1}b^2$$
$$+ \cdots + \binom{k}{k-1}a^2b^{k-1} + \binom{k}{k}ab^k$$
$$+ \binom{k}{0}a^kb + \binom{k}{1}a^{k-1}b^2 + \binom{k}{2}a^{k-2}b^3$$
$$+ \cdots + \binom{k}{k-1}ab^k + \binom{k}{k}b^{k+1}$$

c. Combine like terms in the expression obtained in part b. Then use this result to show that the coefficient of $a^{(k+1)-j}b^j$ in the expansion of $(a + b)^{k+1}$ is given by

$$\binom{k}{j} + \binom{k}{j-1}$$

d. Use the fact that

$$\binom{k}{j} + \binom{k}{j-1} = \binom{k+1}{j}$$

to complete the induction proof.

49. Paying Debts with $3 and $5 Bills In this project, we show that any $\$n$ debt, where n is a natural number greater than or equal to 8, can be paid using only $3 and $5 bills.

a. Prove by induction that any natural number debt of the form $3n + 5$, where $n \geq 1$, can be paid using only $3 and $5 bills.

b. Prove by induction that any natural number debt of the form $3n + 6$, where $n \geq 1$, can be paid using only $3 and $5 bills.

c. Prove by induction that any natural number debt of the form $3n + 7$, where $n \geq 1$, can be paid using only $3 and $5 bills.

d. Show that the results of parts a through c imply that any natural number debt greater than or equal to $8 can be paid using only $3 and $5 bills.

e. There is a slightly more general version of the induction principle that states that if a statement is true for a natural number k, and if it can be shown that the truth of the statement for all natural numbers between k and n implies the truth of the statement for $n + 1$, then the statement is true for all natural numbers $n \geq k$. Use this more general induction principle to prove that all $\$n$ debts can be paid using only $3 and $5 bills, provided that n is a natural number greater than or equal to 8.

Chapter 10 Review

Exercises 1–8 *List the first five terms of the sequence with the given nth term. If the sequence is arithmetic, indicate its common difference. If it is geometric, indicate its common ratio.*

1. $u_n = 1 - 5n$

2. $a_n = n + \dfrac{(-1)^n}{n}$

3. $c_n = n\,e^{-n}$

4. $b_n = \dfrac{n+3}{2}$

5. $x_n = \dfrac{1}{2} + (-1)^n$

6. $a_n = \dfrac{2^n}{3^n}$

7. $a_1 = 2$ and $a_n = 3a_{n-1}$ for $n \geq 2$

8. $u_1 = 256$ and $u_n = \sqrt{u_{n-1}}$ for $n \geq 2$

Exercises 9–16 *Find an expression for the nth term of a sequence whose first few terms are given. If the sequence is arithmetic, indicate its common difference. If it is geometric, indicate its common ratio.*

9. $2, 10, 50, 250, \ldots$

10. $e, 2e^2, 3e^3, 4e^4, \ldots$

11. $1, \dfrac{1}{2}, \dfrac{1}{3}, \dfrac{1}{4}, \ldots$

12. $-3, 2, 7, 12, \ldots$

13. $1, -\dfrac{3}{2}, -4, -\dfrac{13}{2}, \ldots$

14. $24, -6, \dfrac{3}{2}, -\dfrac{3}{8}, \ldots$

15. $1, -8, 27, -64, \ldots$

16. $-\dfrac{1}{2}, \dfrac{2}{3}, -\dfrac{3}{4}, \dfrac{4}{5}, \ldots$

Exercises 17–22 *Use the given information about the sequence to find the indicated term.*

17. $a_1 = 3$, common difference $d = 4$; $a_{12} =$ _____

18. $a_2 = 1$, common ratio $r = 2$; $a_{10} =$ _____

19. $a_1 = 1, a_2 = 4$, the sequence is geometric; $a_6 =$ _____

20. $a_1 = 1, a_2 = 4$, the sequence is arithmetic; $a_6 =$ _____

21. $a_1 = 1, a_{n+1} = (a_n + 1)^2$; $a_4 =$ _____

22. $a_1 = 2, a_2 = 3, a_{n+2} = a_n + a_{n+1}$; $a_7 =$ _____

Exercises 23–26 *Find the indicated sum.*

23. $\displaystyle\sum_{k=1}^{4} (2k + 5)$

24. $\displaystyle\sum_{n=1}^{5} 3(2^n)$

25. $\displaystyle\sum_{n=2}^{6} \dfrac{1}{n^2}$

26. $\displaystyle\sum_{k=0}^{4} \dfrac{k}{k+1}$

Exercises 27–30 *Express the series using summation notation.*

27. $[5(1) - 2] + [5(2) - 2] + [5(3) - 2] + \cdots + [5(11) - 2]$

28. $1^3 + 2^3 + 3^3 + \cdots + 16^3$

29. $4\left(\dfrac{2}{3}\right)^3 + 4\left(\dfrac{2}{3}\right)^4 + 4\left(\dfrac{2}{3}\right)^5 + 4\left(\dfrac{2}{3}\right)^6 + \cdots + 4\left(\dfrac{2}{3}\right)^{13}$

30. $\dfrac{1}{2} - \dfrac{1}{3} + \dfrac{1}{4} - \dfrac{1}{5} + \cdots - \dfrac{1}{15}$

Exercises 31–38 *Evaluate the given arithmetic or geometric series.*

31. The sum of the first 10 terms of the arithmetic sequence $2, 8, 14, 20, \ldots$

32. The sum of the first 11 terms of the geometric sequence 2, 8, 32, 128, . . .

33. $10 + \dfrac{10}{5} + \dfrac{10}{5^2} + \cdots + \dfrac{10}{5^6}$

34. $10 + 2 - 6 - 14 - \cdots - 62$

35. $\displaystyle\sum_{n=1}^{9} 5(0.3)^n$ **36.** $\displaystyle\sum_{k=1}^{12} (8 + 0.6k)$

37. $\displaystyle\sum_{k=3}^{10} (1000 - 9k)$ **38.** $\displaystyle\sum_{n=8}^{20} \dfrac{1}{3}(-3)^n$

Exercises 39–46 *Evaluate the given expression.*

39. $_9C_2$ **40.** $\dfrac{6!}{(6 - 2)!}$

41. $_8P_5$ **42.** $\dbinom{5}{4}$

43. $\dfrac{5!}{3!(5 - 3)!}$ **44.** $_{10}P_6$

45. $\dbinom{10}{3}$ **46.** $_7C_3$

Exercises 47–50 *A single card is drawn from a standard 52-card deck. Compute the probability of the given event.*

47. Drawing a diamond

48. Drawing a face card

49. Drawing a diamond or face card

50. Drawing a diamond face card

Exercises 51–54 *A pair of standard 6-sided dice are rolled. Compute the probability of the given event.*

51. Rolling an even number on both dice

52. Rolling the number 6 on exactly one die

53. Rolling a total of 5

54. Rolling a total that is either odd or less than 5

Exercises 55–56 *Find the probability that in a family of 3 children, the indicated outcome will occur. Assume that the probability of a male birth is 0.5.*

55. All of the children are boys.

56. At least 2 of the children are boys.

Exercises 57–60 *Use Figure 17 to determine the probability that a randomly selected member of the armed services will satisfy the given criterion.*

57. Is in the Army

58. Is in the Marines

59. Is in the Marines or Navy

60. Is not in the Air Force

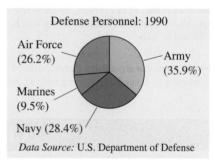

Defense Personnel: 1990

Air Force (26.2%)
Army (35.9%)
Marines (9.5%)
Navy (28.4%)

Data Source: U.S. Department of Defense

Figure 17

Exercises 61–64 *Find the coefficient of the indicated term in the expansion of the binomial.*

61. $(x + y)^8$; x^5y^3 **62.** $(a - b)^5$; a^3b^2

63. $(2a - 5b)^4$; ab^3 **64.** $(3 + v^2)^7$; v^6

Exercises 65–70 *Expand the given expression.*

65. $(x - y)^4$ **66.** $(2a + 3)^5$

67. $\left(r + \dfrac{1}{s}\right)^3$ **68.** $(x^2 - 3y)^4$

69. $\left(\sqrt{z} + 2\right)^5$ **70.** $\left(4 + \sqrt{2}\right)^3$

Exercises 71–76 *Use the Principle of Mathematical Induction to show that the statement is true for all natural numbers n.*

71. $1 + 4 + 7 + \cdots + (3n - 2) = \dfrac{n(3n - 1)}{2}$

72. $3 + 9 + 15 + \cdots + (6n - 3) = 3n^2$

73. $1^3 + 3^3 + 5^3 + \cdots + (2n - 1)^3 = n^2(2n^2 - 1)$

74. $\dfrac{1}{1 \cdot 4} + \dfrac{1}{4 \cdot 7} + \dfrac{1}{7 \cdot 10} + \cdots + \dfrac{1}{(3n - 2)(3n + 1)} = \dfrac{n}{3n + 1}$

75. $2^{2n} - 1$ is divisible by 3.

76. $5^n - 1$ is divisible by 4.

Exercises 77-80 *Express your answer using permutation or combination notation, if possible, and then evaluate.*

77. A club consisting of 20 members wishes to elect a president, vice president, secretary, and treasurer. In how many ways can this be done if no member holds more than one office?

78. A club consisting of 20 members wishes to form a 4-member committee. In how many ways can this be done?

79. Jermaine's CD wish list has 6 rock groups and 4 rap groups. Unfortunately, he has only enough money to buy 5 of the CDs on the list. Determine the number of different purchases he can make if

  a. there are no other restrictions

  b. he would like to buy 3 of the rock groups and 2 of the rap groups

80. Alia's CD player can hold 6 CDs at a time. Her CD collection consists of 10 classical and 8 jazz CDs. Determine the number of ways in which she can program her CD player to play 6 CDs if

  a. there are no other restrictions

  b. she would like to listen to 4 classical and 2 jazz CDs

81. Exam Variations A true–false portion of an exam consists of 10 questions. In how many different ways can this portion of the exam be completed if no questions are left blank?

82. Exam Variations A multiple-choice portion of an exam consists of 10 questions, each with 4 possible answers. In how many different ways can this portion of the exam be completed?

83. Population Projection The population of a certain city is increasing at a rate of 5% per year. Assuming a population of 200,000 in the year 2000 and that the growth rate remains the same, find the population for the next 3 years, and show that these values form a geometric sequence. What will the population be in 2012?

84. Car Price Inflation The price of a certain model of car has increased 3% per year. If the price was initially $15,000, find the price for the next 3 years, and show that these values form a geometric sequence. What will the price of the car be in the tenth year?

85. Income Projection Suppose you have been offered a job with a starting salary of $26,000 and guaranteed annual raises of 6%. Use a geometric series to determine the total income for the first 10 years.

86. Sales Projection At a year-end stockholders' meeting, a company reports that sales for 2001 totaled $1.5 million. Based on a market analysis, the company predicts annual increases in sales of 8% through the year 2010. Use a geometric series to determine the total sales for the 10-year period from 2001 to 2010.

87. Exam Possibilities A mathematics instructor decides to write a final exam by choosing 20 of the 80 problems that appeared on exams during the semester. Use combination notation to write an expression for the number of final exams that are possible if

  a. there are no additional restrictions

  b. there were four exams during the semester, each with 20 problems, and the final exam must have 5 problems from each exam

88. Basketball Recruiting A college basketball coach wishes to recruit 6 high school players to replace the current seniors on the team when they graduate. She has narrowed her list to a group of 15 prospective players, consisting of 4 guards, 6 forwards, and 5 centers. Determine the number of ways she can rank the top 6 players if

  a. there are no restrictions on positions

  b. she would like to recruit 2 guards, 3 forwards, and 1 center

89. Game Show Probability On a certain game show, you are given the 5 digits of the price of a car, but you must arrange them in the correct order to win the car. Find the probability that you will win the car if you order the digits at random. Assume no digits are repeated.

90. Lottery Probability In a certain state lottery, a player selects 4 numbers from 1 through 30. Find the probability that a single choice of 4 numbers will win the lottery, assuming the order of the numbers is not important.

Chapter 10 Test

Problems 1-8   *Answer true or false.*

1. No sequence is both arithmetic and geometric.

2. Some sequences are neither arithmetic nor geometric.

3. A permutation takes order into account, whereas a combination does not.

4. If $n \geq r$, then the value of $_nC_r$ is a natural number.

5. If E is an event in a sample space S, then $0 < P(E) < 1$.

6. If E and F are events in the same sample space S, then $P(E \cup F) = P(E) + P(F) - P(E \cap F)$.

7. Each term in the expansion of $(x + y)^n$ has degree $n + 1$.

8. Mathematical induction can be used to prove that a statement is true for all real numbers.

Problems 9-14   *Give an example of each.*

9. A formula for an arithmetic sequence with common difference -4

10. A formula for a geometric sequence with common ratio $\dfrac{1}{3}$

11. A series that is neither arithmetic nor geometric

12. An experiment involving coins that has a sample space S with $n(S) = 8$

13. An event that has probability $\dfrac{1}{3}$

14. A row from Pascal's triangle with at least 5 numbers

15. Write out the first five terms of the sequence $a_n = 3n^2 + 1$.

16. Write an expression for the nth term of a sequence whose first few terms are $\dfrac{2}{1}, \dfrac{3}{2}, \dfrac{4}{3}, \dfrac{5}{4}, \ldots$

17. Evaluate $\displaystyle\sum_{k=1}^{7}(5k - 2)$.

18. Write the geometric series $\dfrac{5}{2} + \dfrac{5}{4} + \dfrac{5}{8} + \cdots + \dfrac{5}{64}$ using summation notation and find the sum.

19. How many different ways are there to answer the first 8 problems of this exam? What is the probability that you will get them all correct if you randomly assign answers?

20. A Little League baseball team consists of 7 second graders and 8 first graders. The coach must select 9 children to start the next game. Assuming the coach is not concerned with batting order or playing position, determine the number of ways this can be done if

   a. there are no restrictions

   b. there must be 5 second graders and 4 first graders on the field at all times

   How would your solutions in part a change if the coach were arranging the 9 children in a batting order?

21. A pair of standard 6-sided dice are rolled. Find the probability that the sum is a multiple of 3.

22. A single card is drawn from a standard 52-card deck. Find the probability that the card is a black card or a face card.

23. Find the coefficient of z^4 in the expansion of $(z + 5)^9$.

24. Expand the expression $(2x - 3y)^4$.

25. Use the Principle of Mathematical Induction to show that $1 + 3 + 5 + \cdots + (2n - 1) = n^2$ for all natural numbers n.

Answers

Chapter 1

Section 1.1, page 12

1. Natural, whole, integer, rational, real, complex
3. Irrational, real, complex **5.** Complex
7. Rational, real, complex **9.** Complex
11. Integer, rational, real, complex **13.** Rational, real, complex
15. $>$ **17.** $<$ **19.** $-23, -22.9, 22.9, 23$
21. $3 + \dfrac{1}{10} + \dfrac{4}{100}, \pi, \dfrac{22}{7}, 3.15$
23. $(-2, 4]$

25. $\left(-7, -\dfrac{1}{2}\right)$

27. $(-\infty, -3)$

29. $[\sqrt{2}, \infty)$

31. $\{x \mid 3 < x < 5\}$ **33.** $\{x \mid x > 0\}$ **35.** $\{x \mid x > 100\}$
37. -7 **39.** -6.5 **41.** 16 **43.** 19 **45.** 6 **47.** 2
49. -1 **51.** $x - 1$ **53.** $-(y + 2)$ **55.** $|-3 - 7| = 10$
57. $|3x - 2|$ **59.** $\left|\left(x + \dfrac{2}{3}\right) - \left(-\dfrac{2}{3}\right)\right| = \left|x + \dfrac{4}{3}\right|$
61. Answers may vary. **63.** Answers may vary.
65. Answers may vary. **67.** Answers may vary.
69. a. $\$5.67$ **b.** $\$5.67$
c. $0.1(3.50 + 2.80) = 0.1(3.50) + 0.1(2.80)$
71. False **73.** False

Section 1.2, page 28

1. $\dfrac{8}{125}$ **3.** $-\dfrac{1}{1000}$ **5.** $\dfrac{1}{64}$ **7.** 1 **9.** $-\dfrac{1}{9}$ **11.** $\dfrac{1}{4096}$
13. a^9 **15.** $\dfrac{1}{r^2}$ **17.** $\dfrac{a}{b^6}$ **19.** $p^6 q^4$ **21.** $\dfrac{3x^6}{y^3}$ **23.** $\dfrac{z^3}{y^2}$
25. $-i$ **27.** 1 **29.** 0.00304 **31.** 0.27 **33.** 4×10^6
35. 9.43×10^{-2} **37.** 1×10^8 **39.** 1.3176×10^{25}
41. 3 **43.** $\dfrac{1}{2}$ **45.** $3\sqrt{2}$ **47.** 625 **49.** $xy^3\sqrt{xy}$
51. $5a^{15}b^{10}c^5\sqrt{2c}$ **53.** $x^3(y + z)^2\sqrt{x}$ **55.** $3s^2 t^2\sqrt[3]{2t}$
57. $\dfrac{3\sqrt{7}}{7}$ **59.** $\sqrt[6]{x}$ **61.** $5i$ **63.** $3\sqrt{5}$ **65.** $2xy\sqrt[3]{5y^2z^2}$
67. $\sqrt{5}$ **69.** $3\sqrt{6}$ **71.** $x^{1/3}$ **73.** $x^{-1/2}$ **75.** $2\sqrt{2}$
77. $\sqrt[3]{b^2}$ **79.** $\dfrac{1}{4}$ **81.** 4 **83.** x **85.** $3ab^3$ **87.** $z^{1/2}$
89. The second option, since the first pays $\$9.3$ million for the whole month, whereas the second pays $\$10.7$ million on the last day alone

91. a. Approximately 969 feet **b.** Approximately 646 feet
c. $1454\left(\dfrac{2}{3}\right)^n$ feet **d.** About 24 bounces
93. $\$58.38$ **95.** Approximately 1461 feet per second
97. a. $v_0 = 16t$
b. Approximately 16.6 feet per second; approximately 4.3 feet (51.5 inches)
c. Yes
99. 0.0085 second **101.** False **103.** True **105.** True

Section 1.3, page 42

1. Trinomial; degree 2 **3.** Monomial; degree 4
5. Not a polynomial **7.** Trinomial; degree 5
9. $4x^2 + 3x - 8$ **11.** $\dfrac{7}{2}x^2 + 4x$ **13.** $6a^2 - a - 2$
15. $k^2 - \dfrac{4}{9}$ **17.** $4x^4 - 1$ **19.** $9a^2 + 12ab + 4b^2$
21. $2x^3 - 11x^2 - 2x + 2$ **23.** $x^2 - y^2 + 4y - 4$
25. $a^3 - 3a^2 b + 3ab^2 - b^3$ **27.** $(w + 2)(w + 5)$
29. $2(t - 2)(t + 2)$ **31.** $(3n - 2)(2n - 5)$
33. $(z - 1)^2(z + 1)^2$ **35.** $(x + 7)(x - 6)$
37. $2(4a + 3)(2a + 1)$ **39.** $(t - 4)(t - 3)$ **41.** $(2x - 3y)^2$
43. $6(xy + 2)(x^2 y^2 - 2xy + 4)$ **45.** $t^2(2t^2 - 5)$
47. $x - 2$
49. $\dfrac{r + 5}{2(r - 3)}$ **51.** y **53.** $\dfrac{9}{16x}$ **55.** $\dfrac{2a}{(a + 1)(a - 1)}$
57. $\dfrac{1 - q}{q}$ **59.** $\dfrac{x - 4}{x^2 - 3x + 9}$ **61.** $\dfrac{2(b - a)(a + 2b)}{a - 2b}$
63. $\dfrac{4x}{x^2 + 6x + 8}$ **65.** $\dfrac{2 - t}{t^2}$ **67.** $\dfrac{1}{x^2 - 5x + 25}$
69. $x + y$
71. $-\dfrac{1}{xc}$ **73.** $\dfrac{3 + \sqrt{5}}{2}$ **75.** $3 - 2i$ **77.** $-6 + 17i$
79. 18 **81.** $2 - i$ **83.** $-4 + i$ **85.** 455 cubic inches
87. 324 board feet; approximately 10,050 board feet
89. 64 feet; 96 feet; $t = 5$ seconds **91.** 7680 calories
93. a. $99.7°F$ **b.** $158.1°F$
95. a. i. 10 minutes **ii.** 15 minutes
iii. $23\dfrac{1}{3}$ minutes **iv.** 40 minutes **v.** 90 minutes
As s and c get closer together, the waiting time increases without bound.
b. If c is less than s, the waiting time is negative. The waiting-time model is valid only for $c > s$.
97. True **99.** False

Section 1.4, page 57

1.

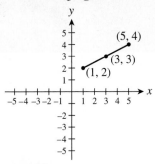

$d = \sqrt{20}$

Midpoint $(3, 3)$

3.

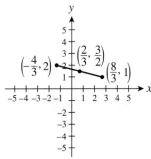

$d = \sqrt{17}$

Midpoint $\left(\dfrac{2}{3}, \dfrac{3}{2}\right)$

5.

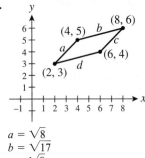

$a = \sqrt{8}$
$b = \sqrt{17}$
$c = \sqrt{8}$
$d = \sqrt{17}$
$a = c, b = d$

7.

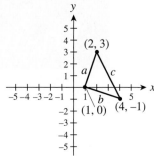

$a = \sqrt{10}$
$b = \sqrt{10}$
$c = \sqrt{20}$
$a^2 + b^2 = c^2$

9. $8^2 + 6^2 = 100$
$0^2 + (-10)^2 = 100$
$(5\sqrt{2})^2 + (5\sqrt{2})^2 = 100$

11. $\dfrac{-2}{(-2) + 1} = 2$

$\dfrac{\left(-\frac{4}{3}\right)}{\left(-\frac{4}{3}\right) + 1} = 4$

$\dfrac{\sqrt{2}}{\sqrt{2} + 1} = 2 - \sqrt{2}$

13.

x	$y = 1000(x^3 + 1)$
-0.1	999
0.0	1000
0.1	1001
0.2	1008
0.3	1027

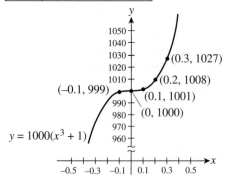

15.

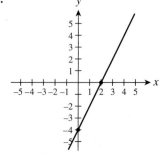

17.

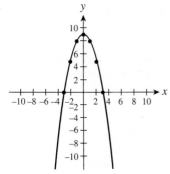

19.

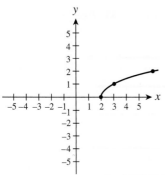

21. Center: (0, 0); radius: 5

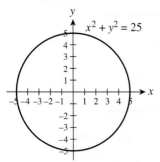

23. Center: (0, 0); radius: $2\sqrt{2}$

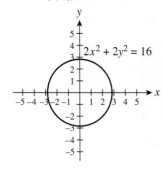

25. Center: $(-4, 1)$; radius: 2

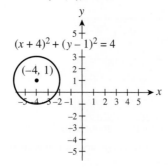

27. Center: $(2, -3)$; radius: 2

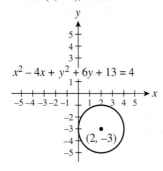

29. Center: $\left(-\dfrac{3}{2}, -1\right)$; radius: 2

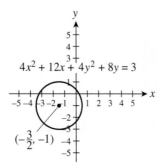

31. $(x + 1)^2 + (y - 3)^2 = 16$
33. $(x - 4)^2 + (y + 3)^2 = 25$
35. $(x + 1)^2 + (y - 5)^2 = 18$
37. ii   **39.** iii   **41.** vi
43. a. ii   **b.** i   **c.** iv   **d.** iii
45. a. i   **b.** iii   **c.** iv   **d.** ii
47. a. $x = -2$
    b. $(-\infty, -2)$
    c. $(-2, \infty)$
49. a. $x = -3, x = 2$
    b. $(-3, 2)$
    c. $(-\infty, -3) \cup (2, \infty)$
51. x-intercepts: $x \approx -10.41, x \approx 2.75, x \approx 10.05$
    y-intercept: $y = 12$
53. $(-4, -12)$
55. $x \approx 10.24, x \approx 5.76$

57.

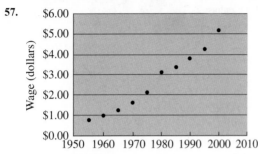

Approximately \$3.89 in 1991 (answers may vary), which is \$0.36 below the actual value of \$4.25

59. $\sqrt{117} + 2 \approx 12.82$ blocks traveling on Lloyd Avenue from point A to the point $(3, 4)$ and then right to point B; 17 blocks traveling from point A to the point $(5, -2)$ and then to point B.

61.

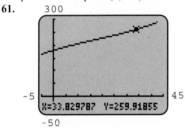

Approximately 1994

63.

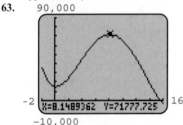

Approximately 1993

65. a. $x^2 + (y - 35)^2 = 30^2$ **b.** Approximately 63.3 feet
67. False   **69.** False

Section 1.5, page 78

1. $x = 2$ **3.** $m = -\dfrac{15}{2}$ **5.** $w = \dfrac{95}{13}$

7. $a = -3, a = 4, a = -6$ **9.** $x = -3, x = -2, x = 2$
11. $s = 0, s = 3$ **13.** $y = -1 \pm \sqrt{3}$ **15.** $u = \pm 2$
17. $y = 10$ **19.** $s = -6$ **21.** No solution

23. $x = -5 \pm 2\sqrt{7}$ **25.** $x = 9$ **27.** $x = \dfrac{7}{8}$ **29.** $t = -\dfrac{1}{4}$

31. $x = 4$ **33.** $x = -3$ **35.** $x = \dfrac{-2 \pm \sqrt{2}}{2}$

37. $w = 2 \pm i$

39. $m = \dfrac{E}{c^2}$ **41.** $r = \dfrac{A}{P} - 1$ **43.** $t = \dfrac{d_1 + d_2}{d_2 - d_1}$

45. $t = \dfrac{v_0 \pm \sqrt{v_0{}^2 - 64s + 64s_0}}{32}$ **47.** $R_1 = \dfrac{RR_2}{R_2 - R}$

49. $x \approx 1.5$ **51.** $x \approx -1.0, x \approx -2.0$ **53.** $x \approx 0.75$
55. $x \approx -11.99, x \approx 11.99$ **57.** $y = \pm \sqrt{4 - 3x}; x \approx 1.33$

59. $y = x^2 - 2x - \dfrac{15}{4}; x \approx 3.18, x \approx -1.18$

61. $y = -1 \pm \dfrac{1}{2}\sqrt{2x - 6}; x = 5$

63. 97   **65.** 2 hours 45 minutes after the second car leaves
67. Approximately 9.5 seconds   **69.** Approximately 19.8 miles
71. September 1998

73. a. $t = \dfrac{-1 + \sqrt{2}}{4} \approx 0.1$ second   **b.** $t = \dfrac{2 - \sqrt{3}}{4} \approx 0.07$ second

75. a. $s = \dfrac{dh}{l - h}$ **b.** $s = \dfrac{110}{23} \approx 4.78$ feet

77. a. 6 hours   **b.** $5\dfrac{1}{3}$ hours   **c.** $13\dfrac{1}{3}$ hours

79. $26\dfrac{2}{3}$ miles per hour   **81.** True   **83.** True

Section 1.6, page 92

1. $(-\infty, 13)$

3. $(-1, \infty)$

5. $\left[-\dfrac{1}{4}, \infty\right)$

7. $(-\infty, \infty)$

9. $\left(\dfrac{4}{3}, \infty\right)$

11. $(-2, 6)$

13. $\left[1, \dfrac{3}{2}\right]$

15. $\left[3, \dfrac{9}{2}\right]$

17. $\left[\dfrac{5}{2}, \infty\right)$

19. $x = \pm 6$ **21.** $u = 2, u = 3$ **23.** No solution
25. $(-3, 3)$ **27.** $(-\infty, -12) \cup (12, \infty)$ **29.** $[3, 7]$
31. $\left(-\dfrac{7}{2}, \dfrac{9}{2}\right)$ **33.** $\left[-1, \dfrac{7}{3}\right]$ **35.** $\left(-\dfrac{1}{6}, \dfrac{3}{2}\right)$

37. $|x - 1| < 5; (-4, 6)$ **39.** $|2x + 3| > 6; \left(-\infty, -\dfrac{9}{2}\right) \cup \left(\dfrac{3}{2}, \infty\right)$

41.

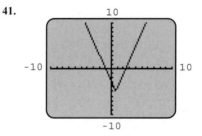

The x-intercepts; x-values where y is positive, x-values where y is negative
43. $x = 2.4, x = 6$ **45.** $[-3, 1]$ **47.** $[-5, 1]$
49. $(-2\sqrt{2}, 2\sqrt{2})$ **51.** Approximately $(-\infty, -1.22) \cup (0.72, \infty)$
53. $(-\infty, -1)$ **55.** $x = -\dfrac{3}{2}, x = \dfrac{2}{7}$ **57.** $[-2, 0] \cup [2, \infty)$

59. From 141.15 to 205.26 miles

61. Between 13.4 and 13.6 inches
63. From 98.6 to 102.2 degrees Fahrenheit
65. Between 2 and 3 seconds
67. From 49.3 seconds to 200.7 seconds after blast-off
69. False   **71.** True

Section 1.7, page 108

1. 4   **3.** $-\dfrac{11}{5}$ **5.** $\dfrac{9}{5}$ **7.** -1 **9.** $\dfrac{22}{13} \approx 1.69$

11. 1,000,000   **13.** 3   **15.** -4 **17.** -3 **19.** $-\dfrac{5}{3}$

21. Slope of $l_1 = -2$

Slope of $l_2 = \dfrac{2}{3}$

Slope of $l_3 = 3$
Slope of $l_4 = 0$

23. Slope of $l_1 = \dfrac{3}{7}$

Slope of $l_2 = -\dfrac{7}{3}$

Slope of $l_3 = \dfrac{3}{7}$

25. $y = x + 2$ **27.** $y = -x + 10$ **29.** $x = 3$

31. $y = 5x - \dfrac{11}{6}$ **33.** $y = 1$ **35.** $y = \dfrac{1}{3}x + \dfrac{11}{3}$

37. $y = -2x + 4$ **39.** $x = 3$ **41.** $y = 0.7353x + 2.5176$

43. $y = \dfrac{3}{20}x + \dfrac{17}{40}$ **45.** $x = \pi$

47. $m = \dfrac{2}{3}, b = -\dfrac{8}{3}$

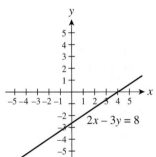

$2x - 3y = 8$

49. $m = \dfrac{1}{3}, b = -\dfrac{4}{3}$

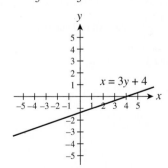

$x = 3y + 4$

51. $m = -2, b = -3$

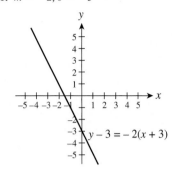

$y - 3 = -2(x + 3)$

53. $y = \dfrac{2}{3}x + 2$ **55.** $y = 4x + 5$ **57.** $y = -3x + 2$

59. $x = 3$ **61.** $3x - y = -4$ **63.** $2x - y = \dfrac{1}{3}$ or $6x - 3y = 1$

65. $\left(-\dfrac{13}{5}, -\dfrac{14}{5}\right)$ **67.** $(2, 5)$ **73.** $C = 5N + 1000$

75. a. $h = \dfrac{264}{67}t$ **b.** Approximately 3 inches

c. Approximately 50 inches
77. \$146,000
79. a. 271.4 million   **b.** 313 million
81. False   **83.** True   **85.** False

Chapter 1 Review, page 113

1. Rational, real, complex   **3.** Complex
5. $[-5, -1)$

7. $\left(-\infty, \dfrac{1}{2}\right]$

9. 81   **11.** $\dfrac{a^2}{b^4}$ **13.** 3.14×10^7 **15.** $a^2 b \sqrt[3]{b}$ **17.** $3\sqrt{3}$

19. $\dfrac{x + \sqrt{2}}{x^2 - 2}$ **21.** $2s^3 t\sqrt{3}$ **23.** $11\sqrt{7}$ **25.** $t^{1/2}$

27. $\sqrt[3]{x^2}$ **29.** $x^{1/12}$ **31.** $\dfrac{b^2}{a}$ **33.** $-t^3 - t^2 + 5t$

35. $3w^2 - 10w + 8$ **37.** $(3u - 2)(3u + 2)$ **39.** $(t - 4)(t + 8)$

41. $(3x + 1)^2$ **43.** $\dfrac{1}{x - 3}$ **45.** $\dfrac{t^2 + t + 1}{t(t + 1)}$ **47.** $\dfrac{2(y - 1)}{y(y + 2)}$

49. $-\dfrac{x}{(x + 1)(x + 3)}$ **51.** $\dfrac{1}{2t}$

53.

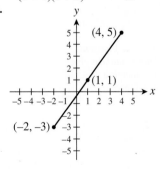

$d = 10$
Midpoint $(1, 1)$

55. Center: (0, 0); radius: 4

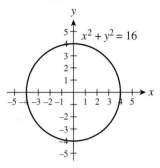

$x^2 + y^2 = 16$

57. Center: (−5, 4); radius: 3

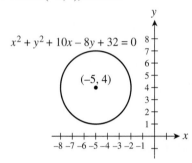

$x^2 + y^2 + 10x − 8y + 32 = 0$

(−5, 4)

59. $(x − 2)^2 + (y − 5)^2 = 9$ **61.** iv   **63.** ii
65. x-intercepts: $x \approx −3.79, x \approx 1.03, x \approx 10.26$; y-intercept: $y = 5$
67. (2, 3)   **69.** $x = \dfrac{5}{3}$ **71.** $x = \dfrac{7}{4}$

73. $x = −3, x = 0, x = \dfrac{1}{2}$ **75.** $x = 3, x = −4$

77. $w = −2 \pm \sqrt{2}$ **79.** $y = −\dfrac{1}{2}, y = \dfrac{2}{3}$ **81.** $x = 2$

83. $x = −\dfrac{1}{3}, x = 2$ **85.** $x = 2$ **87.** $a = 2$

89. $x \approx −1.14$ **91.** $x \approx −0.33, x \approx 1.54$

93. $y = 4 − \dfrac{2}{3}x; x = 6$ **95.** $y = \pm\sqrt[4]{x + 5}; x = −5$

97. $y = 2 \pm \sqrt{x − 1}; x = 5$
99. $(−4, \infty)$
101. $(−\infty, 10)$
103. $(−\infty, \infty)$
105. $\left(\dfrac{2}{3}, 2\right)$
107. $(−\infty, −5) \cup (−1, \infty)$

109. $(−0.47, 1.40)$ **111.** $\dfrac{1}{3}$ **113.** 4   **115.** $y = 3x + 2$
117. $y = −2x + 4$ **119.** $y = −2x + 2$ **121.** $y = −3x + 22$
123. $y = −\dfrac{3}{2}x − 4$
125. a. $4x^2 − 12xy + 9y^2$ **b.** $2x − 3y$
127. a. (0, 7)   **b.** (2, 11)
129. a. $(x − 3)(x + 2)$ **b.** $x = 3, x = −2$
131. a. $x^3 + 9x^2 − 9x − 81$ **b.** $Z = −9, Z = −3, Z = 3$
133. 84,375,000 cubic feet; approximately 1055   **135.** $11.50
137. Between 25 million and 75 million units
139. Approximately 1.3 hours for Roberto and 2.3 hours for Daniel

Chapter 1 Test, page 118

1. True   **2.** True   **3.** True   **4.** False   **5.** True
6. True   **7.** False   **8.** False   **9.** $\sqrt{2}$, for example
10. $x^2 + 2x + 1$, for example   **11.** $x^2 − 6x + 9 = 0$, for example
12. $|x + 1| < 0$, for example   **13.** $\dfrac{b^{10}}{a^{10}}$ **14.** $3x^2y^3\sqrt{3x}$

15. $\sqrt[3]{2}$ **16.** $6x^2 + 5x − 6$ **17.** $\dfrac{x − 1}{x(x − 2)}$

18. $(2x − 3)(2x + 3)$ **19.** $(u + 2v)(u^2 − 2uv + 4v^2)$
20. $(−\infty, 1) \cup (2, \infty)$ **21.** $x \approx −1.71, x \approx 1.35$
22. $x = 1 \pm \sqrt{5}$ **23.** [−1, 5]   **24.** $x = 4$ **25.** $x = −2$
26. $y = \dfrac{3}{4}x + \dfrac{19}{4}$
27. a. ii   **b.** i   **c.** iii
28. 24 feet; the ball hits the ground after 2.75 seconds

Chapter 2

Section 2.1, page 130

1. Function   **3.** Not a function   **5.** Function
7. y is a function of x.   **9.** y is not a function of x.
11. y is a function of x.   **13.** y is a function of x.
15. y is not a function of x.   **17.** y is a function of x.
19. a. 12   **b.** 2   **c.** 2
21. a. $\dfrac{14}{11}$ **b.** $\dfrac{17}{15}$ **c.** $\dfrac{−3t + 2}{−4t − 5}$
23. a. $3t^4$ **b.** $3t^2 − 6t + 3$ **c.** $\dfrac{75}{t^2}$
25. a. x^2 **b.** $x^4 + 2x^2 + 1$ **c.** $\dfrac{1}{a^2}$
27. a. $\dfrac{1}{8}$ **b.** $\dfrac{1}{t^2 − 1}$ **c.** $\dfrac{1}{x^2}$

29. $h(x) = |x|$ **31.** $f(x) = \dfrac{1}{x + 3}$ **33.** All real numbers
35. All real numbers except 4
37. All real numbers except −3 and 3
39. $[−1, \infty)$
41. All real numbers in $[−2, \infty)$ except 1 and 3
43. All real numbers except −3 and 3
45. $f(x) = 2x + 1$, for example
47. $f(x) = \dfrac{1}{x + 1}$, for example
49. $f(x) = \sqrt{7 − x}$, for example
51. $f(x) = \dfrac{1}{x(x − 3)}$, for example

53.

x	$f(x) = -2x + 5$
0	5
2	1
4	-3

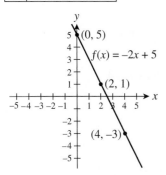

55.

x	$f(x) = (x + 1)^2 - 3$
-3	1
-2	-2
-1	-3
0	-2
1	1

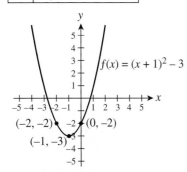

57. a.

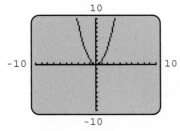

Domain: $(-\infty, \infty)$
Range: $[0, \infty)$

b.

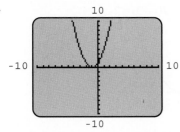

Domain: $(-\infty, \infty)$
Range: $[0, \infty)$

c.

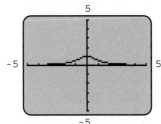

Domain: $(-\infty, \infty)$
Range: $[1, \infty)$

d.

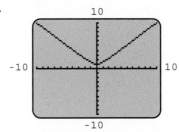

Domain: $(-\infty, \infty)$
Range: $(0, 1]$

59. a.

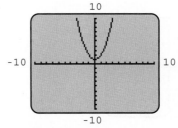

Domain: $(-\infty, \infty)$
Range: $[1, \infty)$

b.

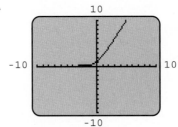

Domain: $(-\infty, \infty)$
Range: $(0, \infty)$

c.

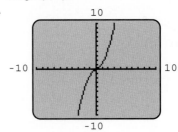

Domain: $(-\infty, \infty)$
Range: $(-\infty, \infty)$

d.

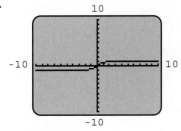

Domain: $(-\infty, \infty)$
Range: $(-1, 1)$

61. 4   **63.** 7

65. a. $x = -2 \pm \sqrt{y - 2}$ **b.** $y \geq 2$ **c.** $[2, \infty)$

67. $V = \dfrac{\pi}{6}d^3$ **69.** $A = 2r^2$ **71.** $A = \dfrac{a^2}{2(a - 2)}$

73. $V = x(16 - 2x)^2$; the domain is the set $(0, 8)$.

75. $V = (8.603 \times 10^{38})t^3$; 4.358×10^{43} cubic miles

77. $R(x) = \dfrac{2}{5}x - 220$ **79.** False   **81.** False

Section 2.2, page 152

1. $x = -\dfrac{1}{3}$ **3.** $x = -2, x = 4$ **5.** $x = -\dfrac{5}{2}$

7. $x = -2, x = 2$ **9.** $a = 3$

11. $x \approx -2.56, x \approx 0.10, x \approx 3.96$

13. $x \approx -2.34, x \approx -0.61, x \approx 0.14$

15. $x \approx -1.24, x \approx 3.24$

17. a. $g(x) = |x - 2|$ **b.** $g(x) = |x| + 1$
   c. $g(x) = |x + 2| - 2$

19. a. $g(x) = -\dfrac{1}{4}x^4 + \dfrac{1}{3}x^3 + x^2$

   b. $g(x) = \dfrac{1}{4}x^4 + \dfrac{1}{3}x^3 - x^2$

21. a.

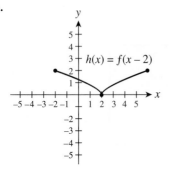

b.

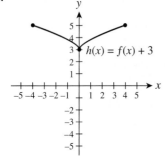

c.

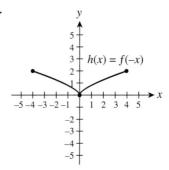

d.

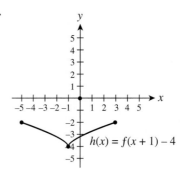

e.

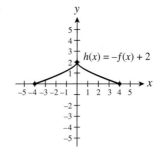

f.

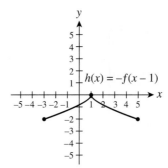

23. a.

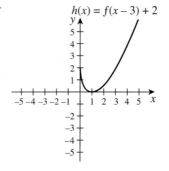

b.

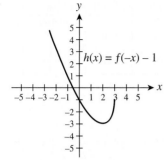

$$h(x) = f(-x) - 1$$

b.

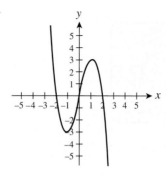

c.

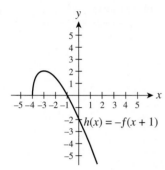

$$h(x) = -f(x + 1)$$

33. a.

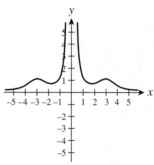

d.

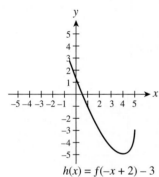

$$h(x) = f(-x + 2) - 3$$

b.

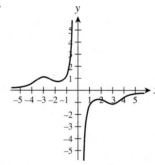

25. Even   **27.** Odd   **29.** Neither

31. a.

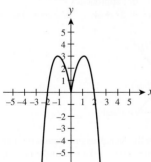

35. Increasing: $(2, \infty)$
Decreasing: $(-\infty, 2)$ Turning point: local minimum at $(2, 1)$

37. Increasing: $(-\infty, -0.62), (1.62, \infty)$
Decreasing: $(-0.62, 1.62)$
Turning points: local maximum at $(-0.62, 7.09)$, local minimum at $(1.62, -4.09)$

39. Increasing: $(-\infty, 2)$
Decreasing: $(2, \infty)$
Turning point: local maximum at $(2, 5)$

41. Increasing: $(-1.30, 0.17), (1.13, \infty)$
Decreasing: $(-\infty, -1.30), (0.17, 1.13)$
Turning points: local minimum at $(-1.3, -3.51)$, local maximum at $(0.17, 0.08)$, local minimum at $(1.13, -1.07)$

43. $Q(x) = (x + 5)^2 + 14(x + 5) + 100$

45. $g(t) = -16t^2 + 80t + 20$; the graph of g is the graph of f shifted 20 units upward

47. Approximately 30 or 270 systems per month to break even; 150 systems per month to maximize profit

49. $\left(2, \dfrac{3}{2}\right)$ **51.** 50 feet $\times$ 100 feet

53. a.

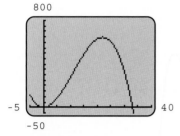

Approximately 645.33 feet

b. $g(t) = 33t$

c.

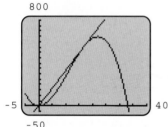

They meet at $t = 16.5$ seconds; the support car is furthest from the checkpoint at $t = 22$ seconds.

d. 33 feet per second

55. Turning points: $(0.26, 365.82)$, $(0.88, 369.89)$, $(1.45, 368.05)$, $(2.11, 372.70)$. The concentration is lowest in late summer/early fall and highest in late spring/early summer.

57. Increasing: $(19.55, 24)$
Decreasing: $(0, 19.55)$
Reserves were lowest at $x \approx 19.55$, which corresponds to midway through 1994.

59. True　　**61.** True

Section 2.3, page 168

1. a. $(f + g)(x) = x + 5$　　**b.** $(f - g)(x) = x - 5$

c. $(fg)(x) = 5x$　**d.** $(f/g)(x) = \dfrac{x}{5}$

3. a. $(f + g)(x) = x^2 + 3x + 1$
b. $(f - g)(x) = x^2 - 3x - 1$

c. $(fg)(x) = 3x^3 + x^2$　**d.** $(f/g)(x) = \dfrac{x^2}{3x + 1}, x \neq -\dfrac{1}{3}$

5. a. $(f + g)(x) = 2x$　**b.** $(f - g)(x) = 0$
c. $(fg)(x) = x^2$
d. $(f/g)(x) = 1, x \neq 0$

7. a. $(f + g)(x) = 2x - 2 + \dfrac{2}{x + 5}, x \neq -5$

b. $(f - g)(x) = 2x - 2 - \dfrac{2}{x + 5}, x \neq -5$

c. $(fg)(x) = \dfrac{4x - 4}{x + 5}, x \neq -5$

d. $(f/g)(x) = x^2 + 4x - 5, x \neq -5$

9. a. $(f + g)(x) = \sqrt{x - 2} + x - 4, x \geq 2$
b. $(f - g)(x) = \sqrt{x - 2} - x + 4, x \geq 2$
c. $(fg)(x) = (x - 4)\sqrt{x - 2}, x \geq 2$

d. $(f/g)(x) = \dfrac{\sqrt{x - 2}}{x - 4}, x \geq 2, x \neq 4$

11.

x	$(f + g)(x)$	$(f - g)(x)$	$(fg)(x)$	$(f/g)(x)$
0	7	7	0	Undefined
1	4	-4	0	0
2	6	-10	-16	$-\dfrac{1}{4}$

Domain $f + g$: $\{0, 1, 2\}$
Domain $f - g$: $\{0, 1, 2\}$
Domain fg: $\{0, 1, 2\}$
Domain f/g: $\{1, 2\}$

13. $(f \circ g)(x) = 8x - 7$　　**15.** $(f \circ g)(x) = 4x^2 + 28x + 49$
$(g \circ f)(x) = 8x + 7$　　$(g \circ f)(x) = 2x^2 + 7$
17. $(f \circ g)(x) = 9x, x \neq 0$　　**19.** $(f \circ g)(x) = x^3 - 3x + 1$
$(g \circ f)(x) = \dfrac{x}{9}, x \neq 0$　　$(g \circ f)(x) = x^3 + 3x^2 - 2$

21. $(f \circ g)(x) = 7$
$(g \circ f)(x) = 71$
23. $(f \circ g)(x) = x^2$
$(g \circ f)(x) = x^2, x \geq 0$
25. $(f \circ g)(x) = 3\sqrt{x - 2} + 5, x \geq 2$
$(g \circ f)(x) = \sqrt{3x + 3}, x \geq -1$
27. $(f \circ g)(x) = x, x \neq 0$
$(g \circ f)(x) = x, x \neq -2$

29.

x	$(f \circ g)(x)$	$(g \circ f)(x)$
1	2	2
2	4	8
3	8	Undefined
4	Undefined	Undefined

Domain $f \circ g$: $\{1, 2, 3\}$
Domain $g \circ f$: $\{1, 2\}$

31.

x	$f(x)$	$g(x)$	$(f \circ g)(x)$	$(g \circ f)(x)$
0	2	1	0	0
1	0	2	1	1
2	1	0	2	2

33. $f^3(x) = x + 15$　　**35.** $f^4(x) = 16x$
37. $f^{10}(0.98) \approx 1.03633 \times 10^{-9}$; $f^{10}(1.02) \approx 6.40583 \times 10^8$;
$f^n(0.98)$ approaches 0; $f^n(1.02)$ approaches ∞
39. a. 3　　**b.** 3　　**41. a.** $x^2 + 2x + 4$　　**b.** $3x^2 + 3hx + h^2$
43. a.

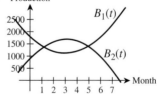

The first refinery gradually decreases production until about the third month, and then increases for the rest of the period. The second refinery does the opposite—gradually increasing production until about the third month and declining thereafter.

	Maximum	Minimum
Refinery 1	1800	1125
Refinery 2	1665	900

b. $B(t) = -10t^2 + 60t + 2700$

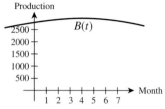

Total production is virtually constant.
Maximum: 2790
Minimum: 2700

45. a. $f(t) = \dfrac{25}{6}t, 0 \le t \le 12$

b. $g(n) = 113 + \dfrac{1}{4}n$ (assuming a hot dog and bun weigh one fourth of a pound)

c. $(f \circ g)(t) = \dfrac{25}{24}t + \dfrac{2825}{6}$; $(g \circ f)(t) = \dfrac{25}{24}t + 113$

d. $(g \circ f)(t)$ represents Kobayashi's weight after t minutes.

47. False   **49.** False

Section 2.4, page 178

1. Inverses   **3.** Not inverses
5. Inverses   **7.** Inverses
9. Inverses   **11.** 1–1
13. Not 1–1   **15.** 1–1
17. Not 1–1   **19.** Not 1–1
21. 1–1; $f^{-1}(x) = 3x + 3$
23. Not 1–1
25. 1–1; $f^{-1}(x) = \sqrt{x - 1}, x \ge 1$
27. 1–1; $f^{-1}(x) = \dfrac{1}{2}(x^2 - 5), x \ge 0$ **29.** Not 1–1
31. 1–1;

x	$f^{-1}(x)$
4	0
7	1
10	2
13	3
16	4

33. Not 1–1
35.

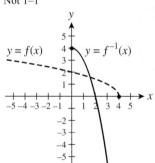

37. f^{-1} does not exist.

39.

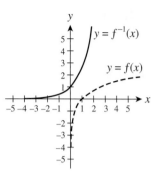

41. a. $M(t) = 98 + 0.12t$ **b.** 110°F   **c.** $M^{-1}(t) = 8.33t - 817$
d. $M^{-1}(t)$ gives the year (after 1935) in which the maximum temperature is t°F. For example, since $M^{-1}(120) \approx 183$, this model predicts that the maximum temperature will be 120°F in 2117.

43. a. 4 years
b. $f^{-1}(t) = \dfrac{72}{t}$ approximates the interest rate at which an amount takes t years to double.
c. 9%

45. a. Yes; the formula suggests a maximal weight of 195 pounds.
b. Assume $h \ge 0$; $f^{-1}(w) = 5.30548\sqrt{w}$ gives the maximum height required for a person of a given weight to not be overweight.
c. About $16\frac{1}{2}$ feet tall   **d.** Answers may vary.

47. False   **49.** True

Section 2.5, page 192

1. Quadratic   **3.** None   **5.** None   **7.** Linear
9. Linear   **11.** Piecewise
13.

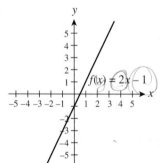

Slope 2; y-intercept -1

15.

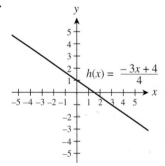

Slope $-\dfrac{3}{4}$; y-intercept 1

17. $h(x) = -\dfrac{1}{2}x + 5$ **19.** $g(x) = \dfrac{4}{3}x - 4$

21. $(0, -4)$

23. $\left(\dfrac{3}{2}, \dfrac{11}{2}\right)$ **25.** $(-8, -54)$ **27.** $(1, 0)$

29. $f(x) = 2(x - 1)^2 + 3$

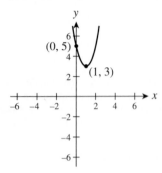

31. $g(x) = -(x - 3)^2 - 1$

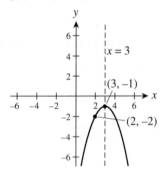

33. $f(-2) = -2; f(0) = -4; f(3) = 2$

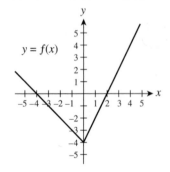

35. $h(0) = -1; h(2) = 2; h(4) = 3$

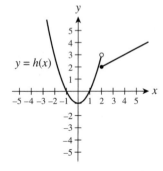

37. $f(-3) = -3; f(-1) = 1; f(4) = 6$

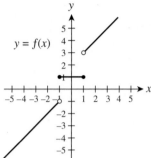

39.

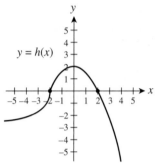

41.

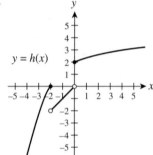

43.

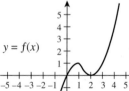

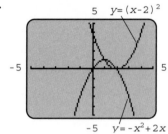

45.

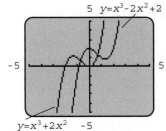

$$5 \quad y = x^3 - 2x^2 + 2$$

$$y = x^3 + 2x^2 \quad -5$$

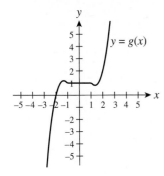

$y = g(x)$

47. $a = 2$ **49.** $a = 2$

51.

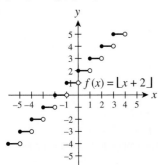

$f(x) = \lfloor x + 2 \rfloor$

53. a.

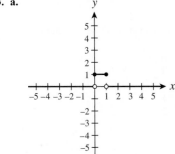

b.

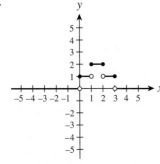

c.

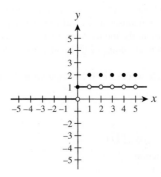

55. a. i. $g(2.8) = 0.8$ **ii.** $g(4.7) = 0.7$ **iii.** $g\left(\dfrac{23}{5}\right) = \dfrac{3}{5}$

b.

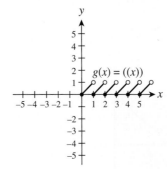

$g(x) = ((x))$

c. x

57. a. $C(x) = 80x + 5000$, where x is the number of units rented
  b. $p = -5x + 550$ **c.** $R(x) = -5x^2 + 550x$
  d. $P(x) = -5x^2 + 470x - 5000$ **e.** 47 units at \$315 per month

59. a. 81.13 feet   **b.** Yes

61. In 1960; approximately 1.14 billion acres

63. Cost

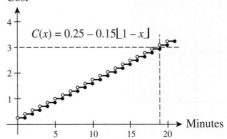

$$C(x) = 0.25 - 0.15 \lfloor 1 - x \rfloor$$

Minutes

19 minutes

65. \$34,211.75 for filing separately; \$34,210.50 for filing jointly

67. MATH ROCKS   **69.** False   **71.** True

Section 2.6, page 210

1. 4   **3.** 48   **5.** 10   **7.** 12   **9.** 60   **11.** $\dfrac{243}{64}$

13. $f(x) = \dfrac{18}{x^2}$ **15.** $f(x) = x^2 - 4x$ **17.** $h(x) = x^2 - 3x + 5$

19. $y = 0.5x + 2.33$ **21.** $y = 1.6x - 4.8$

23. $y = 2.5x + 90$; $y = 152.5$ when $x = 25$

25. $y = -31.55x + 77.51$; $y = -17.14$ when $x = 3$

27. 2592 board feet   **29.** 800,000 joules

31. 5.75×10^{-25} newton   **33.** 3000 pounds

35. $y = 2.0822x + 73.181$; 281.4 kilograms

37. $y = 9.5714x + 177.14$, where y is the CFC-11 concentration in parts per trillion and x is the number of years after 1980; approximately 387.7 parts per trillion; the model doesn't take into account the effects of global effort to cut usage of CFCs to preserve the ozone layer.

39. $y = 0.75x^2 + 34.8x + 320$, where $x = 0$ corresponds to 1980; 1,659,000 inmates

41. True

43. False

Chapter 2 Review, page 216

1. Function   **3.** Not a function

5. Defines y as a function of x

7. Does not define y as a function of x

9. a. 21   **b.** 28

11. a. $\dfrac{x^2}{x^2 + 1}$ **b.** $\dfrac{x^2 + 2x + 1}{x^2 + 2x + 2}$

13. a. 4   **b.** -1 **c.** $-a + 4$

15. All real numbers

17. $\left(-\dfrac{1}{2}, \infty\right)$

19.

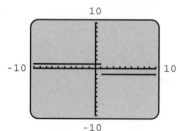

Domain: $(-\infty, -2] \cup [2, \infty)$
Range: $[0, \infty)$

21.

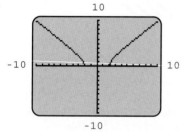

Domain: all real numbers except 1
Range: $\{-1, 1\}$

23. a. $g(x) = -2x^3 + 6x - 1$
    b. $g(x) = 2x^3 + 6x^2 - 3$

25. a.

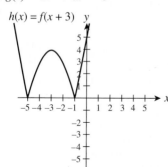

$h(x) = f(x + 3)$

b.

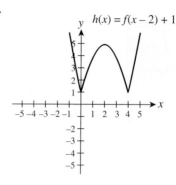

$h(x) = f(x - 2) + 1$

c.

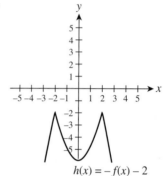

$h(x) = -f(x) - 2$

27. Odd

29. Neither

31. $x = \dfrac{9}{4}$

33. $x = \dfrac{8}{3}$

35. a. $x \approx -1.88, x \approx 0.35, x \approx 1.53$
    b. Turning points: local maximum at $(-1, 3)$, local minimum at $(1, -1)$
    c. Increasing: $(-\infty, -1) \cup (1, \infty)$
        Decreasing: $(-1, 1)$

37. a. $x \approx -1.73, x \approx 1.73$
    b. Turning points: local maximum at $(1, 1)$, local minimum at $(3, 3)$
    c. Increasing: $(-\infty, 1) \cup (3, \infty)$
        Decreasing: $(1, 2) \cup (2, 3)$

39. a. $(f + g)(x) = x^2 + 2x - 1$
    b. $(fg)(x) = 2x^3 - x^2$
    c. $(f/g)(x) = \dfrac{x^2}{2x - 1}; x \neq \dfrac{1}{2}$
    d. $(f \circ g)(x) = 4x^2 - 4x + 1$
    e. $(g \circ f)(x) = 2x^2 - 1$

41. a. $(f + g)(x) = \dfrac{1}{x - 4} + x^3; x \neq 4$
    b. $(fg)(x) = \dfrac{x^3}{x - 4}; x \neq 4$
    c. $(f/g)(x) = \dfrac{1}{x^4 - 4x^3}; x \neq 4, x \neq 0$
    d. $(f \circ g)(x) = \dfrac{1}{x^3 - 4}; x \neq \sqrt[3]{4}$
    e. $(g \circ f)(x) = \dfrac{1}{(x - 4)^3}; x \neq 4$

43. a. $(f + g)(x) = 2x^2 + \sqrt{2x + 4}; x \ge -2$
   b. $(fg)(x) = 2x^2\sqrt{2x + 4}; x \ge -2$
   c. $(f/g)(x) = \dfrac{2x^2}{\sqrt{2x + 4}}; x > -2$
   d. $(f \circ g)(x) = 4x + 8; x \ge -2$
   e. $(g \circ f)(x) = \sqrt{4x^2 + 4}$

45.

x	$(f + g)(x)$	$(fg)(x)$	$(f/g)(x)$	$(f \circ g)(x)$	$(g \circ f)(x)$
0	4	3	$\frac{1}{3}$	0	2
1	4	4	1	3	0
2	3	0	Undefined	1	1
3	1	0	0	2	3

Domain $f + g$: $\{0, 1, 2, 3\}$
Domain fg: $\{0, 1, 2, 3\}$
Domain f/g: $\{0, 1, 3\}$
Domain $f \circ g$: $\{0, 1, 2, 3\}$
Domain $g \circ f$: $\{0, 1, 2, 3\}$

47. Inverses   **49.** Not inverses   **51.** 1–1; $f^{-1}(x) = \dfrac{x + 2}{3}$

53. Not 1–1   **55.** 1–1; $f^{-1}(x) = x^2 + 4, x \ge 0$

57. 1–1;

x	$f^{-1}(x)$
2	1
5	2
10	3
17	4
26	5

59.

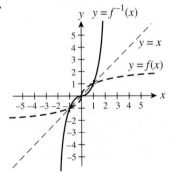

61. Quadratic

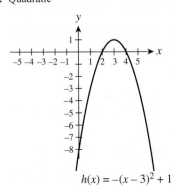

$h(x) = -(x - 3)^2 + 1$
x-intercepts $(2, 0)$ and $(4, 0)$; y-intercept $(0, -8)$

63. None

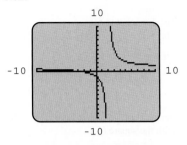

x-intercept $(-2, 0)$; y-intercept $(0, -1)$

65. Piecewise

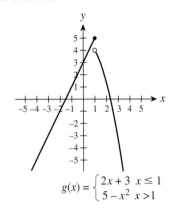

$g(x) = \begin{cases} 2x + 3 & x \le 1 \\ 5 - x^2 & x > 1 \end{cases}$

x-intercepts $\left(-\dfrac{3}{2}, 0\right)$ and $(\sqrt{5}, 0)$; y-intercept $(0, 3)$

67.

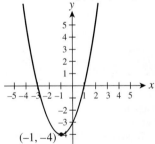

$f(x) = (x + 1)^2 - 4$

Vertex $(-1, -4)$; x-intercepts $(-3, 0)$ and $(1, 0)$; y-intercept $(0, -3)$

69.

$g(x) = -x^2 + 8x - 17$

Vertex $(4, -1)$; no x-intercepts; y-intercept $(0, -17)$

71. $f(x) = -\dfrac{1}{2}x + 3$ **73.** $h(x) = (x - 2)^2 - 1$

75. 15　　**77.** 36　　**79.** $f(x) = -x^3 + 3$
81. $y = 0.605x + 1.316$
83. $y = 2.37x + 42.6$; $y = 161.1$ when $x = 50$
85. a. $x = -3, x = -2$　　**b.** $-3, -2$
87. a. $x = \dfrac{1}{y} - 2$　　**b.** $f^{-1}(x) = \dfrac{1}{x} - 2$
89. a. $8x^3 - 12x^2 + 6x - 1$　　**b.** $f(x) = x^3, g(x) = 2x - 1$
91. a. $\dfrac{-x - 26}{(x + 5)(x - 2)}$
　　b. Domain: all real numbers except -5 and 2; Zero: $x = -26$
93. a. k^4x^2　　**b.** z varies with the square of x.
95. $A = \dfrac{1}{4\pi}C^2$　　**97.** $h = \sqrt{d^2 - 40{,}000}$; $d \geq 200$
99. a. 32 feet　　**b.** Yes; approximately 117 feet
101. $y = 4.98x - 41.34$; 48.3%
103. $y = 2.0952x^2 + 40.095x + 23$; 2335 million

Chapter 2 Test, page 222

1. False　　**2.** True　　**3.** True　　**4.** True　　**5.** True
6. False　　**7.** True　　**8.** False
9. $f(x) = 2x$; $f^{-1}(x) = \frac{1}{2}x$, for example
10. $f(x) = x^2$; $(0, 0)$, for example
11. $f(x) = \begin{cases} -x & x < -1 \\ 0 & -1 \leq x \leq 1, \text{ for example} \\ x & x > 1 \end{cases}$
12. $(0, 0)$, $(1, 1)$, $(2, 2)$, for example
13. $f(x) = x^3$, for example
14. $f(x) = x + 1$, for example
15. y defines a function of x; the set of all real numbers except 2
16. Inverses since $f(g(x)) = g(f(x)) = x$
17. a. 7　　**b.** 7　　**c.** 49

18. $[2, 3) \cup (3, \infty)$
19.

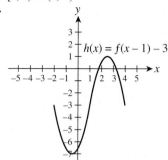

20. Zeros: $x = 0, x = 6$
Turning points: local maximum at $(0, 0)$, local minimum at $(4, -32)$
Increasing: $(-\infty, 0) \cup (4, \infty)$
Decreasing: $(0, 4)$

21.

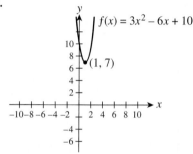

Vertex $(1, 7)$
22. $z = \dfrac{40}{9}$　　**23.** -3　　**24.** 1 minute 37.19 seconds

Chapter 3

Section 3.1, page 235

1. $f(x) \to \infty$ as $x \to \infty$ or $-\infty$; graph: ii
3. $p(x) \to -\infty$ as $x \to \infty$ or $-\infty$; graph: i
5. $f(x) \to \infty$ as $x \to \infty$ or $-\infty$　　**7.** $p(x) \to -\infty$ as $x \to \infty$ or $-\infty$
9. $h(x) \to \infty$ as $x \to -\infty$; $h(x) \to -\infty$ as $x \to \infty$
11. $g(x) \to -\infty$ as $x \to -\infty$; $g(x) \to \infty$ as $x \to \infty$
13. a. 1　　**b.** 0
15. a. 5　　**b.** 4
17. $f(x) = (x - 4)(x + 2)$; Zeros: $4, -2$
19. $f(x) = 3x(x - 1)(x + 1)$; Zeros: $0, 1, -1$
21. $f(x) = -x^2(x + 6)^2$; Zeros: $0, -6$
23. Zeros: $-1.05, 3.36, 3.69$; Turning points: local maximum at $(0.47, 14.13)$, local minimum at $(3.53, -0.13)$
25. Zero: -15.04; Turning points: local maximum at $(-10.14, 58.04)$, local minimum at $(-0.53, 2.48)$
27. Zeros: $-1.70, 0, 1.78, 11.91$; Turning points: local maximum at $(-1.00, 6.25)$, local minimum at $(1.00, -5.75)$, local maximum at $(9.00, 506.25)$
29. Zeros: $-4.16, -0.80, 11.96$; Turning points: local maximum at $(-2.58, 4.07)$, local minimum at $(7.24, -43.29)$
31. Zeros: $1.64, 10.00$; Turning point: local maximum at $(8.33, 668.79)$

33. Deaths peak in approximately 1994; deaths drop to 13,000 in approximately 2000; $f(x) \to \infty$ as $x \to \infty$, and so values of $f(x)$ would eventually get too large; $f(1.5)$ could be interpreted as the number of deaths between July 1, 1982, and June 30, 1983.
35. Turning point: $(7.5, 17.2)$; recycling percentage increased from 1968 to 2000
37. a. $f(t) \to -\infty$ as $t \to \infty$, whereas $g(t) \to \infty$ as $t \to \infty$
　　b. g; $f(t)$ eventually becomes negative, whereas $g(t)$ continues to increase
　　c. $t \approx 0.2$ (early 1990), $t \approx 2.6$ (mid-1992), $t \approx 4.6$ (mid-1994)
　　d. Approximately 5.3 in early 1991
39. True　　**41.** False　　**43.** True

Section 3.2, page 245

1. $q(x) = 2x + 3, r(x) = 0$　　**3.** $q(x) = 4x - 3, r(x) = 0$
5. $q(x) = x^2 - 3x - 3, r(x) = 18$　　**7.** $q(x) = x - 1, r(x) = 2$
9. $q(x) = 3x, r(x) = 5x + 2$
11. $q(x) = x^2 - 1, r(x) = -x^2 + x + 1$
13. $q(x) = 2x - 3, r(x) = 0$　　**15.** $q(x) = x^2 - 2, r(x) = 0$
17. $q(x) = 3x^3 + x^2 + 3x - 1, r(x) = 0$
19. $q(x) = x - 1, r(x) = -1$　　**21.** $q(x) = x^2 - 2x, r(x) = 2$

23. $q(x) = -4x^3 - \frac{1}{3}x^2 - \frac{1}{9}x - \frac{1}{27}$, $r(x) = \frac{80}{81}$

25. $f(x) = (x + 5)(x + 2)(x - 1)$; Zeros: $-5, -2, 1$

27. $p(x) = (2x + 1)(x - 2)(x - 3)$; Zeros: $-\frac{1}{2}, 2, 3$

29. $g(x) = (x + 4)^2(x + 2)(x - 1)$; Zeros: $-4, -2, 1$

31. $h(x) = (3x + 2)(2x + 1)(x - 1)(x - 8)$

Zeros: $-\frac{2}{3}, -\frac{1}{2}, 1, 8$

33. a. 39   **b.** -6 **c.** 4

35. a. 12   **b.** 1608   **c.** 88.2663

37. a. 14   **b.** 33

39. False   **41.** True

Section 3.3, page 256

1. $f(x) = (x + 3)(x + 2)(x - 2)$; Zeros: $-3, -2, 2$

3. $f(x) = (x + 1)(2x + 1)(x - 5)$; Zeros: $-1, -\frac{1}{2}, 5$

5. $f(x) = (x + 4)(3x + 1)(x - 1)^2$; Zeros: $-4, -\frac{1}{3}, 1$

7. $f(x) = (x + 2)(x - 1)(5x - 2)$; Zeros: $-2, 1, \frac{2}{5}$

9. $f(x) = (x - 2)(2x - 1)(3x + 1)$; Zeros: $2, \frac{1}{2}, -\frac{1}{3}$

11. $f(x) = (x + 1)(x - 2)(x^2 - 3)$; Zeros: $-1, 2, \pm\sqrt{3}$

13. $f(x) = x^2 - x - 12$, for example

15. $f(x) = \frac{1}{2}(x - 5)(x - 1)(x + 2)$

17. $f(x) = \frac{1}{3}x^2(x - 2)^2$

19. $f(x) = (x + 1)^3\left(x - \frac{1}{2}\right)^2$, for example

21. $f(x) = -x^2 + 2x + 3$ **23.** $f(x) = \frac{1}{4}(x + 4)(x - 2)^2$

25. $f(x) = -(x + 2)(x + 1)(x - 1)(x - 2)$

27. $f(x) = (x - 3)(x + 2i)(x - 2i)$; Zeros: $3, -2i, 2i$

29. $f(x) = (x - 1 - i)(x - 1 + i)(x - 4)(x + 1)$;
Zeros: $1 + i, 1 - i, 4, -1$

31. Zeros: $i, -i$; $f(x) = (x - i)(x + i)$

33. Zeros: $0, 1 + 2i, 1 - 2i$; $f(x) = x(x - 1 - 2i)(x - 1 + 2i)$

35. Zeros: $-1, 1, -2i, 2i$; $f(x) = (x + 1)(x - 1)(x + 2i)(x - 2i)$

37. Zeros: $0, 1, 2 - i, 2 + i$; $f(x) = x^2(x - 1)(x - 2 + i)(x - 2 - i)$

39. $f(x) = (x + 2)(x^2 + 4)$, for example

41. $f(x) = (x^2 + 1)(x^2 - 2x + 2)$, for example

43. $f(x) = x(x^2 - 2)(x^2 + 4x + 8)$, for example

45. Zeros: $0, 8, 10$; Domain: $(0, 8)$; Original dimensions: $20'' \times 16''$

47. Profit is zero at 100 or 300 frames per month and is positive between these two values.

49. $s(t) = -16t^2 + 80t - 64$; 36 feet   **51.** False   **53.** False

Section 3.4, page 267

1. Possible rational zeros: $\pm1, \pm3, \pm9, \pm27$; Actual rational zeros: $-3, 3$

3. Possible rational zeros: $\pm1, \pm2, \pm3, \pm4, \pm6, \pm12$; Actual rational zeros: $-2, 2, 3$

5. Possible rational zeros: $\pm1, \pm2, \pm\frac{1}{2}, \pm\frac{1}{4}$; Actual rational zeros: $-1, -\frac{1}{2}, \frac{1}{2}, 2$

7. Possible rational zeros: $\pm1, \pm3, \pm\frac{1}{2}, \pm\frac{3}{2}$

Actual rational zeros: $-\frac{3}{2}, -1, 1$

9. Possible rational zeros: $\pm1, \pm2, \pm3, \pm4, \pm6, \pm12, \pm\frac{1}{2}, \pm\frac{3}{2}$

Actual rational zeros: $-1, \frac{3}{2}, 4$

11. Number of positive zeros: 0; Number of negative zeros: 1

13. Number of positive zeros: 0; Number of negative zeros: 0

15. Possible number of positive zeros: 1 or 3;   Actual number of positive zeros: 1; Number of negative zeros: 0

17. Number of positive zeros: 1
Possible number of negative zeros: 0 or 2
Actual number of negative zeros: 0

19. Possible number of positive zeros: 0 or 2
Actual number of positive zeros: 2
Number of negative zeros: 1

27. $-6, -1, 1$ **29.** $1, 12$ **31.** $-3, \frac{1}{6}, \frac{1}{2}$

33. $-\frac{1}{8}, -\sqrt{2}, \sqrt{2}$ **35.** $-11, 11$

37. $x = -2, x = -1, x = 2$

39. $x = -3, x = 3$ **41.** $x = -1, x = -\frac{1}{2}, x = 1$ **43.** -0.32

45. $-1.49, 0.80$ **47.** 11.00

49. 1.5 inches $\times$ 1.5 inches $\times$ 4 inches

51. The square should have a side length of either 0.5 inch or approximately 1.7 inches.

53. 1968, 1984, and 1998

55. a. $1100v^1 + 1100v^2 + 1100v^3$
   b. Solutions of the equation $1100v^1 + 1100v^2 + 1100v^3 = 3000$ correspond to zeros of
   $f(v) = 1100v^1 + 1100v^2 + 1100v^3 - 3000$
   By Descartes' Rule of Signs, there is exactly one positive value for v.
   c. $v \approx 0.953$, $r \approx 4.9\%$

57. a. $p_1v^1 + p_2v^2 + p_3v^4$ **b.** $1000v^0 + dv^2$
   c. Either 2 or 4 interest rates
   d. $r \approx -15.9\%$, $r \approx 197.3\%$

59. True   **61.** False

Section 3.5, page 281

1. Domain: all real numbers except 2; Zeros: none; graph: vii

3. Domain: all real numbers except 0; Zero: -1; graph: iii

5. Domain: all real numbers except -2; Zeros: none; graph: iv

7. Domain: all real numbers except 1 and -1; Zeros: ±2; graph: ii

9. a.

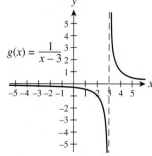

$g(x) = \dfrac{1}{x - 3}$

b.

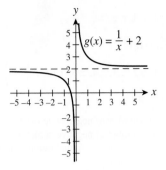

$g(x) = \dfrac{1}{x} + 2$

c.

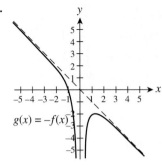

$g(x) = -f(x)$

c.

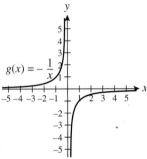

$g(x) = -\dfrac{1}{x}$

13.

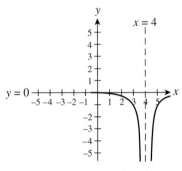

$x = 4$

$y = 0$

11. a.

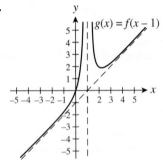

$g(x) = f(x - 1)$

15.

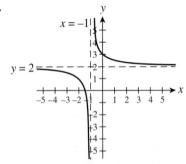

$x = -1$

$y = 2$

b.

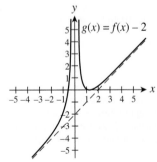

$g(x) = f(x) - 2$

17. Vertical asymptote: $x = -3$
Horizontal asymptote: $y = 0$

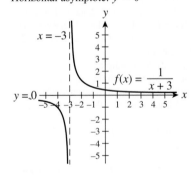

$x = -3$

$f(x) = \dfrac{1}{x + 3}$

$y = 0$

19. Vertical asymptote: $x = -2$
Horizontal asymptote: $y = 1$

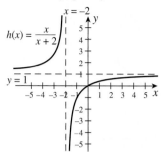

21. Vertical asymptote: $x = 2$
Horizontal asymptote: $y = 3$

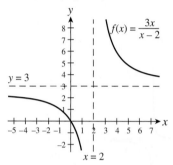

23. Vertical asymptotes: $x = 3, x = -3$
Horizontal asymptote: $y = 0$

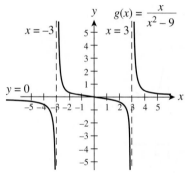

25. Vertical asymptote: none
Horizontal asymptote: $y = 5$

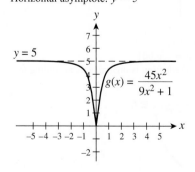

27. Vertical asymptote: $x = -1$
Horizontal asymptote: $y = 0$

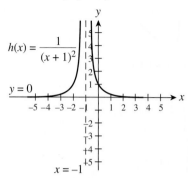

29. Vertical asymptote: $x = 2$
Horizontal asymptote: $y = 1$

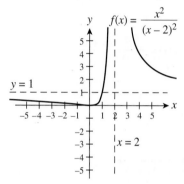

31. Vertical asymptote: $x = 0$
Inclined asymptote: $y = 2x$

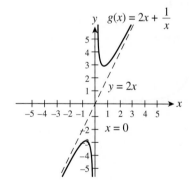

33. Vertical asymptote: $x = 1$
Inclined asymptote: $y = x + 1$

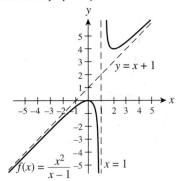

$$f(x) = \frac{x^2}{x - 1}$$

35. Vertical asymptote: $x = -2$
Inclined asymptote: $y = -3x$

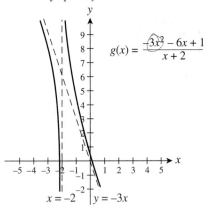

$$g(x) = \frac{-3x^2 - 6x + 1}{x + 2}$$

37. Horizontal asymptote: $y = 0$

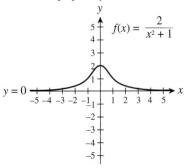

$$f(x) = \frac{2}{x^2 + 1}$$

39. Vertical asymptote: $x = 15$
Horizontal asymptote: $y = 12$

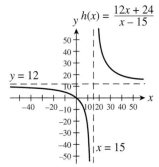

$$h(x) = \frac{12x + 24}{x - 15}$$

41. Vertical asymptote: $x = 3$
Inclined asymptote: $y = x - 1$

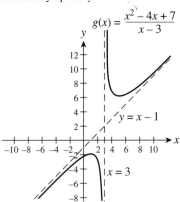

$$g(x) = \frac{x^2 - 4x + 7}{x - 3}$$

43. Vertical asymptotes: $x = 3, x = -3$
Horizontal asymptote: $y = 0$

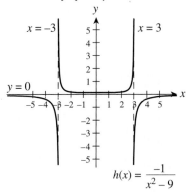

$$h(x) = \frac{-1}{x^2 - 9}$$

45. Vertical asymptote: $x = 3$

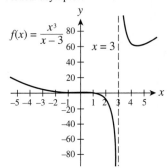

$$f(x) = \frac{x^3}{x - 3}$$

47. $y = 2$ **49.** $y = 3$

51. Vertical asymptote: $x = 100$
Horizontal asymptote: $y = -0.001$
A 75% reduction; 100% reduction is unattainable according to this model.

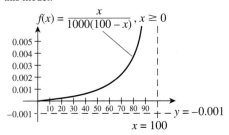

$$f(x) = \frac{x}{1000(100 - x)}, x \geq 0$$

53. Vertical asymptote: none
Horizontal asymptote: $y = 0$
Approximately 7 and 13 years; unemployment rate decreases as education increases; 0% unemployment is unattainable according to this model.

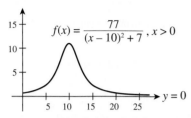

$$f(x) = \frac{77}{(x - 10)^2 + 7}, \, x > 0$$

55. Horizontal asymptote: $y = 100$
The average cost approaches $100 as the number of bicycles increases.

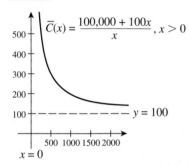

$$\overline{C}(x) = \frac{100{,}000 + 100x}{x}, \, x > 0$$

57. a. 58 balls should be ordered 10 times per year.

   b. Inclined asymptote: $y = \frac{3}{2}x + 600$; as the number of balls ordered gets very large, the inventory cost increases by roughly $1.50 for each additional ball ordered.

59. False   **61.** True

Chapter 3 Review, page 287

1. $g(x) \to -\infty$ as $x \to \infty$ or $-\infty$
3. $f(x) \to -\infty$ as $x \to -\infty$, $f(x) \to \infty$ as $x \to \infty$
5. $h(x) \to -\infty$ as $x \to -\infty$ or ∞
7. Zero: 9.11; Turning points: local maximum at $(0.00, -3.00)$, local minimum at $(6.00, -39.00)$
9. Zeros: -13.12, 3.12; Turning points: local minimum at $(-9.61, -417.63)$, local maximum at $(0.13, -3.94)$, local minimum at $(1.98, -8.40)$
11. $q(x) = 2x - 1$, $r(x) = 0$
13. $q(x) = x^2 - 4x + 11$, $r(x) = -44x + 2$
15. $q(x) = 3x^2 - 5x + 2$, $r(x) = 0$
17. $q(x) = -2x^3 + 6x^2 - 14x + 43$, $r(x) = -132$
19. a. $\frac{255}{8}$ **b.** 10
21. a. -610 **b.** 2
23. Zeros: $-3, \frac{1}{2}$ **25.** Zeros: $\pm 3, \pm\sqrt{2}$ **27.** Zeros: $-1, -\frac{1}{2}, \frac{1}{3}$
29. $f(x) = x^3 - 6x^2 - x + 30$, for example
31. $f(x) = 4(x - 1)^2(x + 1)^2$
33. $f(x) = x^3 - 3x^2 - x + 3$
35. $f(x) = x^3 + 3x^2 + 16x + 48$, for example
37. $f(x) = (x - 2 - i)(x - 2 + i)$; Zeros: $2 - i, 2 + i$
39. $h(x) = (x - 3)(x + 3)(x - 4i)(x + 4i)$; Zeros: $3, -3, -4i, 4i$
41. Possible rational zeros: $\pm 1, \pm 2, \pm 3, \pm 6$; actual rational zeros: $-2, 1, 3$

43. Possible rational zeros: $\pm 1, \pm 3, \pm\frac{1}{2}, \pm\frac{1}{4}, \pm\frac{3}{2}, \pm\frac{3}{4}$

Actual rational zeros: $-1, \frac{1}{2}, 1, \frac{3}{2}$

45. Number of positive zeros: 0; Number of negative zeros: 1
47. Number of positive zeros: 1; Possible number of negative zeros: 2 or 0; Actual number of negative zeros: 2

51. $x = -1, x = 2, x = 6$ **53.** $x = -\frac{2}{3}, x = -\frac{1}{4}, x = 4$

55. $-2.13, -0.20, 2.33$ **57.** $-21.00, 1.52$

59. Domain: all real numbers except 3; Zeros: none; graph: iv
61. Domain: all real numbers except 3 and -3; Zero: -2; graph: i
63. Vertical asymptote: $x = -2$
Horizontal asymptote: $y = 0$

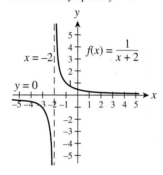

$$f(x) = \frac{1}{x + 2}$$

65. Vertical asymptote: $x = \frac{1}{3}$

Horizontal asymptote: $y = \frac{2}{3}$

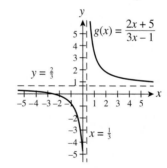

$$g(x) = \frac{2x + 5}{3x - 1}$$

67. Vertical asymptote: $x = 0$
Horizontal asymptote: $y = 0$

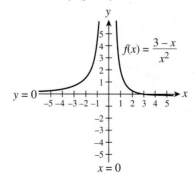

$$f(x) = \frac{3 - x}{x^2}$$

69. Vertical asymptote: none
Horizontal asymptote: $y = 2$

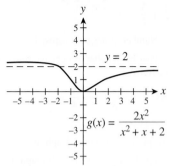

71. Vertical asymptote: $x = 1$
Horizontal asymptote: $y = 0$

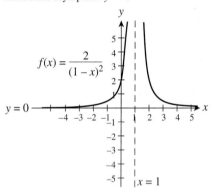

73. Vertical asymptote: $x = 0$
Inclined asymptote: $y = -2x$

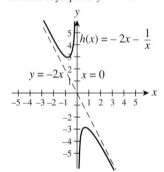

75. Vertical asymptote: $x = -3$
Inclined asymptote: $y = 3x - 1$

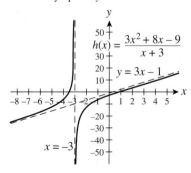

77. a. $\dfrac{x^2 - 4x + 5}{x - 1}$ **b.** $y = x - 3$

79. a. $x = \dfrac{-3 \pm \sqrt{13}}{2}$ **b.** $x = \dfrac{-3 \pm \sqrt{13}}{2}, x = 0$

81. a. $x^2 + 4x + 4$ **b.** $(x + 5) + \dfrac{9}{x - 1}$

83. a. Vertical asymptote: $x = 3$; Horizontal asymptote: $y = -2$
b. $h(x) = \dfrac{1}{x - 3} - 2$

85. a. $(2 - 3i) + (2 + 3i) = 4,\ (2 - 3i)(2 + 3i) = 13$
b. $f(x) = x^2 - 4x + 13$

87. $f(x) = -\dfrac{3}{1000}x^2 + 18x - 15{,}000$
The largest profit of $12,000 occurs when $x = 3000$.

89. Zeros: 2, 10.3; The months when the average temperature is zero

91.

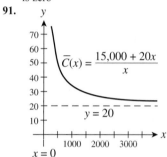

The average cost approaches $20.

Chapter 3 Test, page 292

1. True   **2.** True   **3.** False   **4.** False
5. True   **6.** True   **7.** False   **8.** True
9. $f(x) = -x$, for example
10. $f(x) = x^3 - 6x^2 + 5x + 12$, for example
11. $f(x) = x^3 - 2x^2 + 9x - 18$, for example
12. $f(x) = x^4 + 1$, for example
13. $f(x) = \dfrac{x}{x + 1}$, for example
14. $f(x) = x - 3 + \dfrac{1}{x - 2}$, for example
15. $f(x) = x^2(x - 4)(2x + 1)$; Zeros: $0, 4, -\dfrac{1}{2}$
16. $f(x) = (x + 2)(3x - 1)(2x - 3)$
Zeros: $-2, \dfrac{1}{3}, \dfrac{3}{2}$
17. $q(x) = 3x^3 - 6x^2 + 5, r(x) = 4$
18. $f(x) = x(5x - 2)(x + 3i)(x - 3i)$
Zeros: $0, \dfrac{2}{5}, -3i, 3i$
19. Possible rational zeros: $\pm 1, \pm 3, \pm 9, \pm\dfrac{1}{2}, \pm\dfrac{3}{2}, \pm\dfrac{9}{2}$

Actual rational zeros: $-3, 1, \dfrac{3}{2}$

20. Possible number of positive zeros: 3 or 1
Actual number of positive zeros: 1
Number of negative zeros: 1
21. Approximately 1.52

22. Vertical asymptote: $x = \dfrac{3}{2}$

Horizontal asymptote: $y = \dfrac{1}{2}$

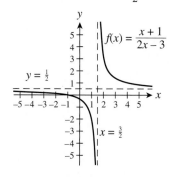

23. a. Approximately 1988 and 1994
b. The rate peaked at approximately 9.6 per 100,000 in 1991.
c. The crime rate dropped for 1991 through 2000.

Chapter 4

Section 4.1, page 304

1.

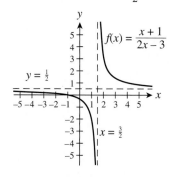

3. $a = \dfrac{1}{3}$ **5.** $a = 10$

7.

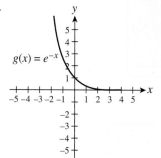

a.

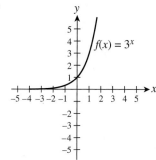

b.

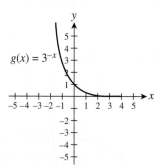

9.

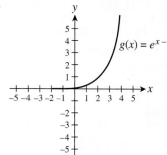

11. $t \approx 11.79$ **13.** $t \approx 49.70$ **15.** $x = 4$ **17.** $x = 3$
19. $x = -3$ **21.** $x = -5$ **23.** $x = -2, x = 1$
25. a. \$1790.85   **b.** \$1814.02   **c.** \$1819.40
d. \$1822.03   **e.** \$1822.12
27. 20 gerbils initially present;

t	$f(t)$
0	20
1	30
2	45
3	68
4	101

approximately 228 gerbils after 6 months;
approximately 735 billion gerbils after 5 years

c.

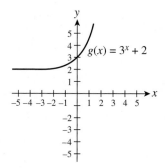

d.

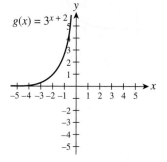

29. $20,805.70; $21,103.48

31. Approximately 25.96 quadrillion Btu in 1940; approximately 101.59 quadrillion Btu in 2000; approximately 291.3% increase; approximately 127.53 quadrillion Btu in 2010

33. Approximately 4.64 hours

35.

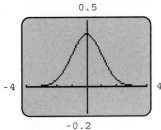

Maximum at approximately (0, 0.40); $f(x)$ approaches 0 as x approaches $\pm\infty$

37. a. Approximately 130 students after 2 days; approximately 190 students after 5 days

b.

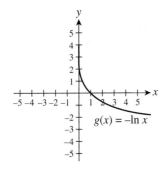

c. Approximately 24 days

d. $N(t)$ approaches 1000 as t approaches ∞.

39. True   **41.** True

Section 4.2, page 318

1. 2.72130   **3.** -0.17609 **5.** 0.53959   **7.** -0.70877

9. 4   **11.** Undefined   **13.** 2   **15.** -3 **17.** -2

19. -3 **21.** 100   **23.** 7   **25.** 2.24   **27.** 5.64

29. -6.75 **31.** $2^{-2} = \dfrac{1}{4}$ **33.** $10^3 = 1000$

35. $e^2 = x + 1$

37. $x^3 = 10$ **39.** $\log_3 9 = 2$ **41.** $\log_{1/2} 8 = -3$

43. $\ln 5 = x$ **45.** $\ln\dfrac{1}{2} = -0.013t$

47. Domain: $(0, \infty)$

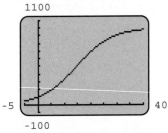

49. Domain: $(3, \infty)$

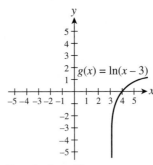

51. Domain: $(-\infty, 0)$

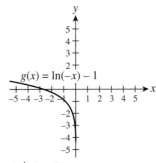

53. $f^{-1}(x) = 3^x$

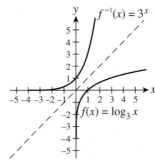

55. $f^{-1}(x) = \ln(x + 5)$

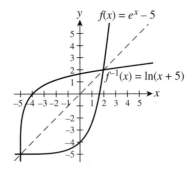

57. $f^{-1}(x) = \dfrac{10^x}{1000} = 10^{x-3}$

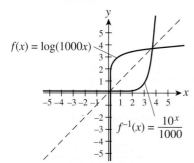

$f(x) = \log(1000x)$

$f^{-1}(x) = \dfrac{10^x}{1000}$

59.

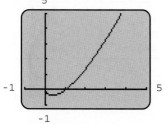

Domain: $(0, \infty)$

61.

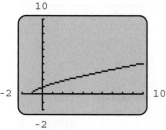

Domain: $(-1, \infty)$

63. Approximately 49.7 minutes   **65.** Approximately 13.4 hours
67. Approximately 20.4 decibels   **69.** 100 decibels
71. Approximately 3.55   **73.** Approximately 4.5
75. 25 times
77. a. 60 wpm in approximately 4.6 weeks; 80 wpm in approximately
8 weeks
b. The time gets larger without bound; no
c.

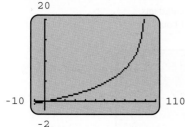

Domain: $[0, 100)$

79. True   **81.** False   **83.** True

Section 4.3, page 330

1. $2\ln x + \ln y$ **3.** $3\log_5 x + 4\log_5 y + 3\log_5 z$

5. $1 - \ln 7 - 3\ln x$ **7.** $\dfrac{1}{4}\log x + \dfrac{1}{2}\log y - \log z$ **9.** $\log\dfrac{x}{y^2}$

11. $\ln\dfrac{x^3 y^4}{z}$ **13.** $\ln\sqrt[3]{\dfrac{x-z}{(x+z)^2}}$ **15.** $\log\dfrac{64}{3}$ **17.** $\ln\dfrac{x+3}{x+2}$

19. $x = 27$ **21.** $x = 2$ **23.** $x = \sqrt[3]{25}$ **25.** $x = -\dfrac{15}{32}$

27. $x = -\dfrac{1}{3}$ **29.** $x = 2$ **31.** $x = \sqrt{6} - 1$ **33.** $x = 95$

35. $x = 1$ **37.** No solution   **39.** $x = \pm 999$ **41.** All $x > 0$

43. $x = e^e$ **45.** $\dfrac{\ln 16}{\ln 3} \approx 2.52372$ **47.** $\dfrac{\ln 0.341}{\ln 12} \approx -0.43296$

49. $\dfrac{\ln 10}{\ln \pi} \approx 2.01147$ **51.** 0.16, 3.15   **53.** 0.04, 1.50

55. $-11.42, -2.97, 0.59$ **57.** 10,000 times more intense
59. 30 decibels   **61.** Approximately 586.8 tons
63. Approximately 177.8 times more energy   **65.** False
67. True   **69.** True   **71.** True

Section 4.4, page 342

1. $x = 4$ **3.** $x = 1$ **5.** $x = \dfrac{\log 10}{\log 7} \approx 1.18329$

7. $x = -\dfrac{\log 11}{\log 3} \approx -2.18266$ **9.** $x = \dfrac{\log 36}{\log \dfrac{18}{5}} \approx 2.79758$

11. $x = \dfrac{\ln 2}{1 - \ln 2} \approx 2.25889$ **13.** $x = \dfrac{1}{2(1 - \ln \pi)} \approx -3.45471$

15. $x = 1$ **17.** $t = \dfrac{3}{\log 1.06} \approx 118.54959$

19. $x \approx 0.69, x \approx 1.10$ **21.** $x \approx 0.57$
23. $x \approx -4.48, x \approx -2.11, x \approx 0.35$
25. Approximately 10.2 years   **27.** $r \approx 5.78\%$
29. February 2008   **31.** February 2061
33. Approximately 237.63; the year 2017
35. At approximately 6:33 A.M. on July 7, 2003
37. Approximately 122,269   **39.** Approximately 11,520 years
41. Approximately 5.2 years   **43.** True   **45.** False

Section 4.5, page 355

1. a.

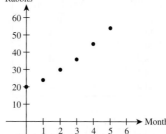

b.

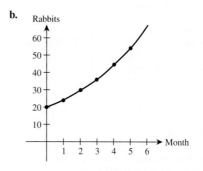

c.   Rabbits

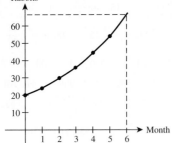

Approximately 66 rabbits (Answers may vary.)

3. $P(t) = 20e^{0.1823t}$

t	Actual data	$P(t) = 20e^{0.1823t}$	Absolute error
0	20	20.0	0.0
1	24	24.0	0.0
2	30	28.8	1.2
3	36	34.6	1.4
4	45	41.5	3.5
5	54	49.8	4.2

Largest absolute error: 4.2
Sum of the absolute errors: 10.3

5. $P(t) = 20e^{0.1987t}$

t	Actual data	$P(t) = 20e^{0.1987t}$	Absolute error
0	20	20	0
1	24	24.4	0.4
2	30	29.8	0.2
3	36	36.3	0.3
4	45	44.3	0.7
5	54	54	0

Largest absolute error: 0.7
Sum of the absolute errors: 1.6
This model is more accurate.

7. $P(t) = 19.9e^{0.201t}$

t	Actual data	$P(t) = 19.9e^{0.201t}$	Absolute error
0	20	19.9	0.1
1	24	24.3	0.3
2	30	29.7	0.3
3	36	36.4	0.4
4	45	44.5	0.5
5	54	54.4	0.4

Largest absolute error: 0.5
Sum of the absolute errors: 2.0
The largest error is smaller, but the sum of the errors is larger.
Both models are extremely accurate.

9. $P(t) = 232,506e^{0.0264t}$

t	Actual data	$P(t) = 232,506e^{0.0264t}$	Absolute error
0	226,000	232,506	6506
10	303,000	302,753	247
20	402,000	394,223	7777
30	550,000	513,329	36,671
40	627,000	668,420	41,420

Largest absolute error: 41,420
Sum of the absolute errors: 92,621
The population in 2010 will be approximately 870,368.
The population will first exceed 1,500,000 in 2031.

11. $S(t) = 78.7e^{-0.0809t}$

t	Actual data	$S(t) = 78.7e^{-0.0809t}$	Absolute error
0	80	78.7	1.3
1	72	72.6	0.6
2	66	66.9	0.9
3	61	61.7	0.7
4	58	56.9	1.1

Largest absolute error: 1.3
Sum of the absolute errors: 4.6
The sales for month 6 will be approximately $48,400.
The sales will reach $20,000 in month 17.

13. $p(t) = 9.2e^{0.5142t}$

t	Actual data	$p(t) = 9.2e^{0.5142t}$	Absolute error
0	8	9.2	1.2
1	19	15.4	3.6
2	26	25.7	0.3
3	40	43.0	3.0

Largest absolute error: 3.6
Sum of the absolute errors: 8.1
90% will have encountered the virus approximately halfway through the 4th quarter.
About 5.74×10^{12}%

15. $N(t) = 2.7e^{0.3226t}$

t	Actual data	$N(t) = 2.7e^{0.3226t}$	Absolute error
0	2	2.7	0.7
3	6	7.1	1.1
7	29	25.8	3.2
11	134	93.9	40.1
14	275	247.1	27.9
18	1200	897.9	302.1
22	3100	3,263.2	163.2
26	7500	11,859.1	4359.1
28	24,000	22,607.8	1392.2
30	42,000	43,098.9	1098.9

Largest absolute error: 4359.1
Sum of the absolute errors: 7388.5
There will be approximately 156,632 thousand (156,632,000) transistors in 2005.
The number of transistors will exceed 100 billion in 2046.

17. a. 4 students
b. Approximately 77 students
c. Approximately 17 hours
d.

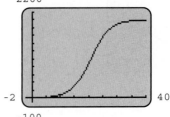

Approximately 2000 students
e. The 21st hour

19. a. 1945
b. 700 miles per hour
c. No; the model predicts 674.37 miles per hour for 1997 and has an upper bound of 700 miles per hour.

21. $v = 0.81 \log P + 0.3$

P	Actual data	$v = 0.81 \log P + 0.3$	Absolute error
5,500	3.3	3.33	0.03
14,000	3.7	3.66	0.04
71,000	4.3	4.23	0.07
138,000	4.4	4.46	0.06
342,000	4.8	4.78	0.02

Largest absolute error: 0.07
Sum of the absolute errors: 0.22
Little Rock: 4.6 feet/second
New York: 5.9 feet/second
Mexico City: 6.2 feet/second
23. False   **25.** True

Chapter 4 Review, page 360

1.

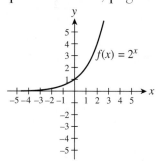

a.

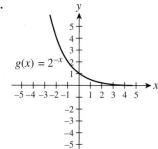

b.

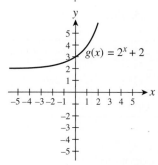

c.

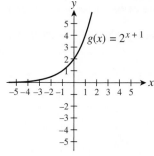

3. $x = 3$ **5.** $x = -\dfrac{1}{2}$ **7.** -2 **9.** -2 **11.** $\dfrac{1}{2}$

13. $2^{-4} = \dfrac{1}{16}$ **15.** $10^4 = 2x + 1$ **17.** $\ln x = 5$

19. $\log_2 5 = x + 3$

21.

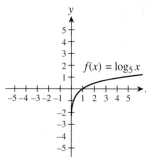

Domain: $(0, \infty)$

23.

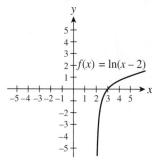

Domain: $(2, \infty)$

25. $\log_2 x + 2\log_2 y$ **27.** $\dfrac{2}{3}\log x + \dfrac{1}{3}\log y$ **29.** $\log(x^2 y^3)$

31. $\ln\left(\dfrac{\sqrt[3]{xy^2}}{z}\right)$ **33.** $\dfrac{\ln 10}{\ln 4} \approx 1.66096$ **35.** $x = \dfrac{1}{2}$

37. $x = 3$ **39.** $y = \dfrac{12}{5}$ **41.** $x = 2$

43. $x = \dfrac{\ln 10}{\ln 5} \approx 1.43068$ **45.** $t = \dfrac{\ln 10}{\ln 1.05} \approx 47.19363$

47. $x = 10 \ln 5 \approx 16.09438$ **49.** $x = \dfrac{2\ln 3}{2\ln 2 - \ln 3} \approx 7.63768$

51. $x \approx -2.82, x \approx 1.66$ **53.** $x \approx 0.17, x \approx 5.05, x \approx 45.48$

55. a. $\left(\dfrac{2}{3}\right)^x$ **b.** $x = \dfrac{\ln 2}{\ln(2/3)} \approx -1.70951$

57. a. $x = -2, x = 1$ **b.** $x = -2, x = 1$

59. a.

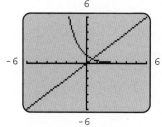

b.   One solution

61. a. $x \le \dfrac{3}{2}$ **b.** $\left(\dfrac{3}{2}, \infty\right)$

63. a. $u = 1$ **b.** $x = 0$

65. a.

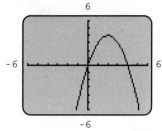

b. $(0, 4)$

67. $56.34   **69.** Approximately 23.1 years

71. Approximately 12.6 hours

73. a. Approximately 149 astronauts
b. Approximately 28.8 days
c.

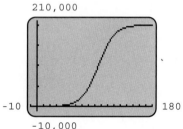

Approximately 200,000 astronauts
d. The 96th day

75. $V(t) = 27{,}957e^{-0.1495t}$

t	Actual data	$V(t) = 27{,}957e^{-0.1495t}$	Absolute error
0	28,000	27,957	43
1	24,000	24,075	75
2	20,800	20,732	68
3	17,800	17,853	53
4	15,400	15,374	26

Largest absolute error: 75
Sum of the absolute errors: 265
The SUV will be worth $4000 in 13 years.

Chapter 4 Test, page 363

1. False   **2.** True   **3.** False

4. False   **5.** True   **6.** True

7. False   **8.** True   **9.** $f(t) = e^{-t}$, for example

10. 1, for example   **11.** 0.1, for example

12. Limited food, for example

13. $x = -2, x = 1$

14. $x = 10^{12} - 1 = 999{,}999{,}999{,}999$

15. $x = \dfrac{\ln 3}{\ln 3 - \ln 2} \approx 2.70951$

16. $x = \dfrac{\ln 2}{0.3} \approx 2.31049$

17. 5   **18.** -3

19.

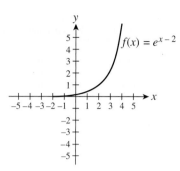

20.

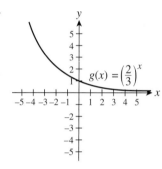

21.

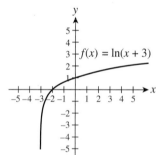

22. $4 \log x + 2 \log y + 2 \log z$ **23.** $\ln\left(\dfrac{x^3 z^{10}}{y^4}\right)$

24. Approximately 8.8 years   **25.** 2029

26. $N(t) = 2243e^{0.6115t}$

t	Actual data	$N(t) = 2243e^{0.6115t}$	Absolute error
0	2,100	2,243	143
1	4,400	4,134	266
2	8,200	7,620	580
3	13,100	14,046	946

Largest absolute error: 946
Sum of the absolute errors: 1935
Approximately 25,900 new cases in 1987
The approximation overestimates the number of new cases by almost 5000.
The model predicts approximately 250 million in 2002, which is an enormous overestimate.

Chapter 5

Section 5.1, page 376

1. v   **3.** viii   **5.** x   **7.** iii   **9.** iv

11. 390° and −330°, for example

13. 225° and −495°, for example

15. $\dfrac{7\pi}{3}$ and $-\dfrac{5\pi}{3}$, for example   **17.** $\dfrac{\pi}{2}$ and $-\dfrac{7\pi}{2}$, for example

19. Complement: 70°; Supplement: 160°

21. Complement: $\dfrac{\pi}{3}$; Supplement: $\dfrac{5\pi}{6}$

23. 25.27°   **25.** 173.34°   **27.** 132°24′   **29.** 15°37′30″

31. $\dfrac{\pi}{6}$ **33.** $\dfrac{5\pi}{4}$ **35.** $-\dfrac{5\pi}{6}$ **37.** 0.262   **39.** −2.049

41. 270°   **43.** −240°   **45.** 165°   **47.** −51.429°

49. 114.592°

51. a. $\dfrac{100\pi}{3}$ centimeters   **b.** $\dfrac{1000\pi}{3}$ square centimeters

53. a. $\dfrac{5\pi}{3}$ inches   **b.** $\dfrac{10\pi}{3}$ square inches

55. a. Approximately 251.3 inches   **b.** Approximately 773.5°

57. Approximately 25,133 feet per minute

59. Approximately 0.0131 mile per hour

61. Approximately 736.7 revolutions per minute

63. a. Approximately 150.8 square inches
   b. Approximately 49.13 inches

65. Approximately 8.590×10^{-6} radians or 4.922×10^{-4} degrees

67. Approximately 501 miles   **69.** Approximately 2827 miles

71. Approximately 181.79° or 3.17 radians

73. False   **75.** True

Section 5.2, page 391

1. $\sin \theta = \dfrac{3}{5}$ $\cos \theta = \dfrac{4}{5}$

   $\tan \theta = \dfrac{3}{4}$ $\cot \theta = \dfrac{4}{3}$

   $\sec \theta = \dfrac{5}{4}$ $\csc \theta = \dfrac{5}{3}$

3. $\sin \theta = \dfrac{2\sqrt{2}}{3}$ $\cos \theta = \dfrac{1}{3}$

   $\tan \theta = 2\sqrt{2}$ $\cot \theta = \dfrac{\sqrt{2}}{4}$

   $\sec \theta = 3$ $\csc \theta = \dfrac{3\sqrt{2}}{4}$

5. $\sin \theta = \dfrac{6\sqrt{61}}{61}$ $\cos \theta = \dfrac{5\sqrt{61}}{61}$

   $\tan \theta = \dfrac{6}{5}$ $\cot \theta = \dfrac{5}{6}$

   $\sec \theta = \dfrac{\sqrt{61}}{5}$ $\csc \theta = \dfrac{\sqrt{61}}{6}$

7. $\cos \theta = \dfrac{12}{13}$

   $\tan \theta = \dfrac{5}{12}$ $\cot \theta = \dfrac{12}{5}$

   $\sec \theta = \dfrac{13}{12}$ $\csc \theta = \dfrac{13}{5}$

9. $\sin \theta = \dfrac{\sqrt{15}}{4}$ $\cos \theta = \dfrac{1}{4}$

   $\tan \theta = \sqrt{15}$ $\cot \theta = \dfrac{\sqrt{15}}{15}$

   $\csc \theta = \dfrac{4\sqrt{15}}{15}$

11. $\sin \theta = \dfrac{2\sqrt{13}}{13}$ $\cos \theta = \dfrac{3\sqrt{13}}{13}$

   $\cot \theta = \dfrac{3}{2}$

   $\sec \theta = \dfrac{\sqrt{13}}{3}$ $\csc \theta = \dfrac{\sqrt{13}}{2}$

13. $\dfrac{1}{2}$ **15.** $\dfrac{2\sqrt{2}}{3}$ **17.** $\dfrac{24}{7}$ **19.** 8   **21.** $\dfrac{1}{2}$ **23.** $\sqrt{3}$

25. $\dfrac{2\sqrt{3}}{3}$ **27.** $\sqrt{3}$ **29.** $\dfrac{1}{4}(\sqrt{6} - \sqrt{2})$ **31.** $\sqrt{2} + 1$

33. 0.3420   **35.** 0.0279   **37.** 1.0690   **39.** 0.9239

41. 1.1884   **43.** 78.4630°   **45.** 77.0822°   **47.** 10.2866°

49.   **51.**

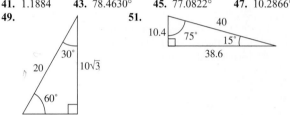

53.   **55.**

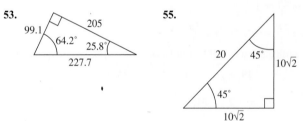

57.

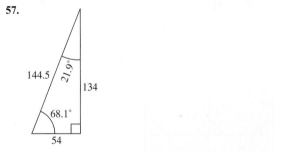

59. Approximately 368 feet   **61.** Approximately 2352 feet

63. Approximately 16,271 square feet

65. Approximately 95 feet or 32 yards

67. Approximately 12,108 cubic feet

69. Approximately 1.36 miles

71. Approximately 7:05 P.M.

73. The video shows the angle of elevation to be approximately 34.6°.

75. False   **77.** False

Section 5.3, page 405

1. $\theta = \pi$

$\sin \pi = 0$ $\cos \pi = -1$

$\tan \pi = 0$ $\cot \pi$ is undefined.

$\sec \pi = -1$ $\csc \pi$ is undefined.

3. $\theta = \dfrac{5\pi}{3}$

$\sin \dfrac{5\pi}{3} = -\dfrac{\sqrt{3}}{2}$ $\cos \dfrac{5\pi}{3} = \dfrac{1}{2}$

$\tan \dfrac{5\pi}{3} = -\sqrt{3}$ $\cot \dfrac{5\pi}{3} = -\dfrac{\sqrt{3}}{3}$

$\sec \dfrac{5\pi}{3} = 2$ $\csc \dfrac{5\pi}{3} = -\dfrac{2\sqrt{3}}{3}$

5. $\theta = \dfrac{3\pi}{4}$

$\sin \dfrac{3\pi}{4} = \dfrac{\sqrt{2}}{2}$ $\cos \dfrac{3\pi}{4} = -\dfrac{\sqrt{2}}{2}$

$\tan \dfrac{3\pi}{4} = -1$ $\cot \dfrac{3\pi}{4} = -1$

$\sec \dfrac{3\pi}{4} = -\sqrt{2}$ $\csc \dfrac{3\pi}{4} = \sqrt{2}$

7. $(0, -1)$

$\sin(-90°) = -1$ $\cos(-90°) = 0$

$\tan(-90°)$ is undefined. $\cot(-90°) = 0$

$\sec(-90°)$ is undefined. $\csc(-90°) = -1$

9. $\left(-\dfrac{\sqrt{2}}{2}, -\dfrac{\sqrt{2}}{2}\right)$

$\sin \dfrac{5\pi}{4} = -\dfrac{\sqrt{2}}{2}$ $\cos \dfrac{5\pi}{4} = -\dfrac{\sqrt{2}}{2}$

$\tan \dfrac{5\pi}{4} = 1$ $\cot \dfrac{5\pi}{4} = 1$

$\sec \dfrac{5\pi}{4} = -\sqrt{2}$ $\csc \dfrac{5\pi}{4} = -\sqrt{2}$

11. Fourth quadrant   **13.** Third quadrant   **15.** Second quadrant

17. $-\dfrac{2\sqrt{2}}{3}$ **19.** $\dfrac{\sqrt{2}}{4}$ **21.** $-\dfrac{\sqrt{5}}{5}$ **23.** $60°$ **25.** $50°$

27. $\dfrac{\pi}{4}$ **29.** $\pi - 2$ **31.** $-\dfrac{1}{2}$ **33.** $\dfrac{\sqrt{3}}{2}$ **35.** $-\dfrac{\sqrt{3}}{3}$

37. 1 **39.** $\sqrt{2}$ **41.** 0.9397 **43.** -1.3331 **45.** 0.8935

47. a. 1 **b.** $\dfrac{\sqrt{2}}{2}$ **c.** -0.4161

49. a. $-\dfrac{2\sqrt{3}}{3}$ **b.** Undefined   **c.** -1.2015

51.

Domain: All real numbers
Range: $[-1, 1]$

53.

Domain: All real numbers except multiples of π
Range: $(-\infty, -1] \cup [1, \infty)$

55. a. Approximately 61.2, 280.7, and 0 feet, respectively

b. Approximately 0.7760 or 44.45°; no

57. a. 60; 60; the average temperatures for March 2003 and March 2004 are the same

b. Highest: Approximately 87° in August
Lowest: Approximately 59° in February

59. True   **61.** False

Section 5.4, page 422

1. Period: 2
Amplitude: 3

3. Period: $\dfrac{2\pi}{3}$
Amplitude: 5

5. Period: 2π
Amplitude: approximately 7.2

7. Period: 12π
Amplitude: approximately 4.5

9. Period: $\dfrac{\pi}{2}$
Amplitude: 5

11. Period: 2π
Amplitude: approximately 2.78

13. Period: 2π
Amplitude: approximately 1.87

15.

Amplitude: 1
Period: π
Phase shift: 0

17.

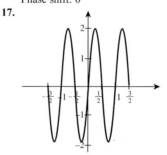

Amplitude: 2
Period: 1
Phase shift: 0

19.

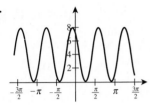

Amplitude: 3
Period: π

Phase shift: $-\dfrac{1}{2}$

21.

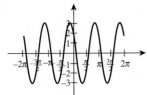

Amplitude: 4

Period: $\dfrac{2\pi}{3}$

Phase shift: $\dfrac{\pi}{12}$

23. Amplitude: 3
Period: 2
Phase shift: 0
Maximum: 3
Minimum: -3

25. Amplitude: 4
Period: π

Phase shift: $-\dfrac{1}{2}$

Maximum: 4
Minimum: -4

27. Amplitude: 6
Period: π

Phase shift: $-\dfrac{\pi}{4}$

Maximum: 6
Minimum: -6

29. Amplitude: 2

Period: $\dfrac{2\pi}{3}$

Phase shift: $\dfrac{2}{3}$

Maximum: 7
Minimum: 3

31. Decreasing   **33.** Changes direction at $(0, -2)$
35. Increasing   **37.** $a = 1, b = \pi, c = 0$

39. $a = 2, b = 1, c = -\dfrac{\pi}{4}$ **41.** $a = 5, b = 2, c = -\pi$

43. Not an identity   **45.** Not an identity   **47.** Identity

49. $f(x) = 3\sin\left(2x - \dfrac{2\pi}{3}\right)$ **51.** $f(x) = \sqrt{2}\sin(8x - \pi)$

53. 0.89, 1.59   **55.** 0.045, 0.27

57. Amplitude: 0.002   Period: $\dfrac{1}{264}$

59. a. Period: 0.5 second   **b.** 5 centimeters

61. a. Amplitude: $110\sqrt{2}$

Period: $\dfrac{1}{60}$

b. $\dfrac{1}{480}$ second

63. a. 5 feet
b. Maximum: 105 feet at $t = 15$ seconds
Minimum: 5 feet at $t = 0$ seconds
c. 30 seconds

65. 24-hour period; minimum of 97.6° at 6:23 A.M., maximum of 99.2° at 6:23 P.M.

67. True   **69.** True

Section 5.5, page 437

1. vii   **3.** ii   **5.** vi   **7.** v

9.

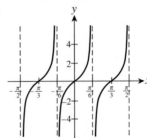

Period: $\dfrac{\pi}{3}$

Asymptotes: $\dfrac{\pi}{6} + \dfrac{\pi k}{3}$, k an integer

11.

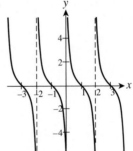

Period: 2
Asymptotes: $2k$, k an integer

13.

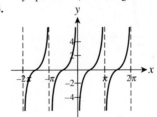

Period: π
Asymptotes: πk, k an integer

15.

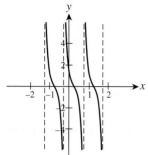

Period: 1

Asymptotes: $-\dfrac{1}{4} + k$, k an integer

17.

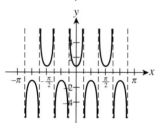

Period: $\dfrac{\pi}{2}$

Asymptotes: $\dfrac{\pi}{8} + \dfrac{\pi k}{4}$, k an integer

19.

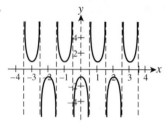

Period: 2

Asymptotes: $\dfrac{1}{2} + k$, k an integer

21.

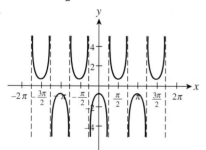

Period: π

Asymptotes: $\dfrac{\pi}{4} + \dfrac{\pi k}{2}$, k an integer

23. 1.05   **25.** No zeros

27. a. $h = 5\tan\theta$ **b.** $0 \le \theta < \dfrac{\pi}{2}$

c.

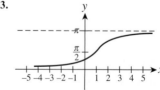

d. h approaches infinity.

29. True   **31.** True

Section 5.6, page 450

1. $\dfrac{\pi}{2}$ **3.** $\dfrac{2\pi}{3}$ **5.** $\dfrac{\pi}{4}$ **7.** $\dfrac{\pi}{3}$ **9.** $-\dfrac{\pi}{4}$ **11.** π **13.** 0

15. Undefined   **17.** $x \approx 0.3398$, $x \approx 2.8018$

19. $x \approx 1.6208$, $x \approx 4.7623$ **21.** $x \approx 0.7954$, $x \approx 5.4878$

23. No solution   **25.** $\dfrac{1}{3}$ **27.** $-\dfrac{2\pi}{5}$ **29.** -7 **31.** $\dfrac{3\pi}{4}$

33. $-\dfrac{\pi}{2}$ **35.** $\dfrac{4}{5}$ **37.** -8 **39.** $\dfrac{1}{2}$ **41.** $\dfrac{\sqrt{7}}{7}$

43. $\dfrac{\sqrt{1-x^2}}{x}$, $0 < x \le 1$ **45.** $\dfrac{x}{\sqrt{x^2-4}}$, $x > 2$ **47.** $\dfrac{2}{x}$, $x > 2$

49.

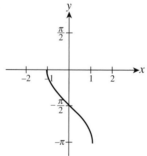

51.

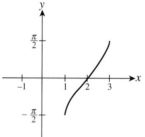

53.

55. $x \approx 1.00$ **57.** $x \approx -1.40$, $x \approx 0.65$

59. Not an identity   **61.** Identity

63. a. 48 feet   **b.** Approximately 25.64°

c. $\theta = \tan^{-1}\left(\dfrac{6t}{100}\right)$, t in seconds

65. a. Approximately 59.04°   **b.** $\theta = \tan^{-1}\left(\dfrac{2000}{x}\right)$, x in feet

c. Approximately 1154.7 feet

67. a.

x (feet)	20	50	100
θ (radians)	1.6307	1.9401	1.7985

The angle increases and then decreases.

b. Approximately 61.8 feet

69. True   **71.** False

Chapter 5 Review, page 454

1. iv   **3.** ii   **5.** 420° and −300°, for example

7. $\dfrac{\pi}{2}$ and $-\dfrac{7\pi}{2}$, for example   **9.** $\dfrac{\pi}{3}$ **11.** $-\dfrac{7\pi}{6}$ **13.** −150°

15. 22.5°   **17.** 4π

19. $\sin\theta = \dfrac{4}{5}$ $\cos\theta = \dfrac{3}{5}$

$\tan\theta = \dfrac{4}{3}$ $\cot\theta = \dfrac{3}{4}$

$\sec\theta = \dfrac{5}{3}$ $\csc\theta = \dfrac{5}{4}$

21. $\sin\theta = \dfrac{2}{3}$ $\cos\theta = \dfrac{\sqrt{5}}{3}$

$\tan\theta = \dfrac{2\sqrt{5}}{5}$ $\cot\theta = \dfrac{\sqrt{5}}{2}$

$\sec\theta = \dfrac{3\sqrt{5}}{5}$ $\csc\theta = \dfrac{3}{2}$

23. $\sin\theta = \dfrac{8}{17}$

$\tan\theta = \dfrac{8}{15}$ $\cot\theta = \dfrac{15}{8}$

$\sec\theta = \dfrac{17}{15}$ $\csc\theta = \dfrac{17}{8}$

25. $\sin\theta = \dfrac{1}{2}$ $\cos\theta = \dfrac{\sqrt{3}}{2}$

$\tan\theta = \dfrac{\sqrt{3}}{3}$ $\cot\theta = \sqrt{3}$

$\sec\theta = \dfrac{2\sqrt{3}}{3}$

27. 74.4759°   **29.** 18.6629°

31.

A right triangle with an angle of 33° adjacent to side 25, angle 57°, legs labeled 20.97 and 13.62.

33.

A right triangle with right angle at top, sides 3.25 and 7.3, angles 66° and 24°, and base 7.99.

35. 30°   **37.** $\dfrac{\pi}{4}$ **39.** $\dfrac{\sqrt{3}}{2}$ **41.** $-\dfrac{2\sqrt{3}}{3}$ **43.** 1

45. 0.3420   **47.** −1.7013   **49.** −0.7047   **51.** $\dfrac{2\sqrt{5}}{5}$

53. $-\dfrac{\sqrt{5}}{2}$ **55.** vi   **57.** vii   **59.** i   **61.** viii

63. Period: 4π **65.** Period: 2
Amplitude: 3   Amplitude: 2.25

67. Period: $\dfrac{2\pi}{3}$; Amplitude: 2   **69.** Period: 2π; Amplitude: 2.25

71.

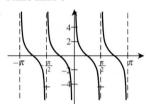

Amplitude: 1
Period: 4
Phase shift: 0

73.

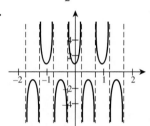

Period: $\dfrac{\pi}{2}$

Phase shift: $\dfrac{\pi k}{2}$, k an integer

75.

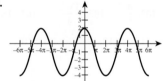

Period: 1

Asymptotes: $\dfrac{1}{4} + \dfrac{k}{2}$, k an integer

77.

A sinusoidal graph.

Amplitude: 3
Period: 4π
Phase shift: 0

79.

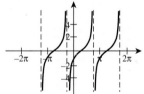

Period: π

Asymptotes: $\dfrac{3\pi}{4} + \pi k$, k an integer

81.

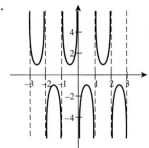

Period: 2
Asymptotes: k, k an integer

83.

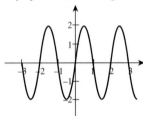

Amplitude: 2
Period: 2
Phase shift: $-\dfrac{1}{2}$

85. $a = 1, b = 2, c = 0$ **87.** $a = 3, b = 1, c = \dfrac{\pi}{4}$

89. $f(x) = 4\sin\left[4\left(x - \dfrac{\pi}{4}\right)\right]$ **91.** 0.45 **93.** 0 **95.** $\dfrac{\pi}{3}$

97. $-\dfrac{\pi}{6}$ **99.** $x \approx 1.8235, x \approx 4.4597$

101. $x \approx 0.6155, x \approx 2.5261$ **103.** $\dfrac{4}{5}$ **105.** $\dfrac{3\sqrt{2}}{4}$ **107.** 16

109. 0 **111.** $\dfrac{1}{\sqrt{1 - x^2}}, 0 < x \leq 1$ **113.** $\dfrac{3}{\sqrt{x^2 + 9}}, x > 0$

115.

117.

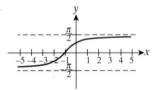

119. a. $2\sqrt{x}$ **b.** $\dfrac{2\sqrt{x}}{x + 1}$

121. a. x **b.** 2001

123. 120 radians or approximately 19.1 revolutions
125. Approximately 0.049 mile per hour
127. Approximately 171.38 feet
129. a. 12 months; the dieter's weight depends only on the date, not on the year
b. 10 pounds **c.** 210 pounds; January 1
d. 190 pounds; July 1

Chapter 5 Test, page 459

1. False **2.** True **3.** False **4.** False **5.** True
6. False **7.** True **8.** False **9.** True **10.** True

11. $f(x) = 3\sin 4x$, for example **12.** $\dfrac{5\pi}{4}$, for example

13. $\cot x$ or $\csc x$

14. 1 and 0.8391, for example (the ratio of the smaller side to the larger side should be 0.8391)

15. $\dfrac{25\pi}{36} \approx 2.1817$

16. $\sin\theta = \dfrac{2\sqrt{5}}{5}$ $\cos\theta = \dfrac{\sqrt{5}}{5}$

$\tan\theta = 2$ $\cot\theta = \dfrac{1}{2}$

$\sec\theta = \sqrt{5}$ $\csc\theta = \dfrac{\sqrt{5}}{2}$

17. Approximately 186.6 feet **18.** $-\dfrac{\sqrt{3}}{2}$

19.

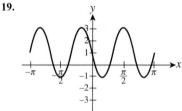

Amplitude: 2
Period: $\dfrac{2\pi}{3}$
Phase shift: $\dfrac{\pi}{3}$

20. Amplitude: $5\dfrac{1}{3}$; Period: 4

21.

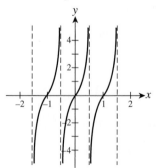

Period: 1
Asymptotes: $\dfrac{1}{2} + k$, k an integer

22. $\dfrac{2\pi}{5}$ **23.** $x \approx 0.3398$ or $x \approx 2.8018$

Chapter 6

Section 6.1, page 468

1. Conditional equation   **3.** Contradiction   **5.** Identity
7. Contradiction   **9.** Identity   **11.** Conditional equation
13. Identity   **15.** Contradiction   **17.** Conditional equation; 2.65
19. Identity   **21.** Conditional equation; 3.14   **59.** sec x

61. csc x **63.** sin x **65.** $\tan x = \dfrac{1}{\cot x}$

67. $\cos x = \pm\sqrt{1 - \sin^2 x}$ **69.** $\cot x = \pm\dfrac{\cos x}{\sqrt{1 - \cos^2 x}}$

71. $\sec x = \pm\dfrac{1}{\sqrt{1 - \sin^2 x}}$

73. **a.** $v = \sqrt{32r\tan\theta}$
   b. Approximately 82 feet per second, or 56 miles per hour
75. True   **77.** True

Section 6.2, page 479

1. **a.** $-\dfrac{16}{65}$ **b.** $\dfrac{56}{65}$ **c.** $-\dfrac{63}{65}$ **d.** $\dfrac{33}{65}$

3. **a.** $\dfrac{63}{65}$ **b.** $-\dfrac{33}{65}$ **c.** $-\dfrac{16}{65}$ **d.** $\dfrac{56}{65}$

5. **a.** $-\dfrac{24}{25}$ **b.** 0 **c.** $-\dfrac{7}{25}$ **d.** -1

7. **a.** $-\dfrac{56}{65}$ **b.** $\dfrac{16}{65}$ **c.** $\dfrac{33}{65}$ **d.** $-\dfrac{63}{65}$

9. $\sin 30° = \dfrac{1}{2}$ **11.** $\cos\dfrac{3\pi}{4} = -\dfrac{\sqrt{2}}{2}$ **13.** $\sin\dfrac{4\pi}{15}$

15. $\cos 3y$ **17.** $\tan 2a$ **19.** $\dfrac{\sqrt{2} - \sqrt{6}}{4}$ **21.** $\sqrt{3} - 2$

23. $-\cos x$ **25.** $\cos x$ **27.** $\dfrac{1}{2}\cos x + \dfrac{\sqrt{3}}{2}\sin x$

29. $-\dfrac{1}{2}\cos x - \dfrac{\sqrt{3}}{2}\sin x$ **31.** $\dfrac{\tan x - 1}{1 + \tan x}$ **33.** 1 **35.** $-\dfrac{1}{49}$

51. True   **53.** False

Section 6.3, page 489

1. **a.** $-\dfrac{24}{25}$ **b.** $\dfrac{7}{25}$ **c.** $-\dfrac{24}{7}$

3. **a.** $\dfrac{24}{25}$ **b.** $\dfrac{7}{25}$ **c.** $\dfrac{24}{7}$

5. **a.** $\dfrac{240}{289}$ **b.** $\dfrac{161}{289}$ **c.** $\dfrac{240}{161}$

7. **a.** $-\dfrac{120}{169}$ **b.** $\dfrac{119}{169}$ **c.** $-\dfrac{120}{119}$

9. **a.** $\dfrac{3}{5}$ **b.** $\dfrac{4}{5}$ **c.** $\dfrac{3}{4}$

11. **a.** $\dfrac{\sqrt{5}}{3}$ **b.** $\dfrac{2}{3}$ **c.** $\dfrac{\sqrt{5}}{2}$

13. **a.** $\dfrac{1}{3}$ **b.** $-\dfrac{2\sqrt{2}}{3}$ **c.** $-\dfrac{\sqrt{2}}{4}$

15. **a.** $\dfrac{1}{4}$ **b.** $-\dfrac{\sqrt{15}}{4}$ **c.** $-\dfrac{\sqrt{15}}{15}$

17. $\dfrac{1}{2}\sqrt{2 - \sqrt{3}}$ **19.** $\dfrac{2}{\sqrt{2 + \sqrt{2}}}$ **31.** $\sin^2 2x$ **33.** $2\cot 2x$

43. **a.** $A = \dfrac{d^2}{2}\sin 2\theta$ **b.** Approximately 359 square inches

47. True   **49.** False

Section 6.4, page 501

1. Solution   **3.** Not a solution   **5.** $\dfrac{\pi}{2}, \dfrac{3\pi}{2}$ **7.** 1.11, 4.25

9. $\dfrac{4\pi}{3}, \dfrac{5\pi}{3}$ **11.** $\dfrac{\pi}{6}, \dfrac{\pi}{3}, \dfrac{5\pi}{6}, \dfrac{5\pi}{3}$ **13.** $0, \dfrac{\pi}{2}, \pi, \dfrac{3\pi}{2}$ **15.** $0, \pi$

17. 0.90, 2.81, 4.04, 5.95   **19.** $\dfrac{3\pi}{2}$ **21.** πk, k an integer

23. $\dfrac{\pi}{3} + 2\pi k, \dfrac{5\pi}{3} + 2\pi k$, k an integer

25. $\dfrac{\pi}{4} + \pi k, \dfrac{3\pi}{2} + 2\pi k$, k an integer

27. $\dfrac{\pi}{2} + \pi k, \dfrac{\pi}{6} + 2\pi k, \dfrac{5\pi}{6} + 2\pi k$, k an integer

29. πk, k an integer   **31.** $\pi k, \dfrac{\pi}{4} + \dfrac{\pi}{2}k$, k an integer

33. $\dfrac{\pi}{3} + 2\pi k, \dfrac{5\pi}{3} + 2\pi k, \pi + 2\pi k$, k an integer   **35.** $\dfrac{5\pi}{4}, \dfrac{7\pi}{4}$

37. $\dfrac{\pi}{4}, \dfrac{5\pi}{4}$ **39.** $0, \pi, 0.34, 2.80$ **41.** No solution

43. $\dfrac{\pi}{6}, \dfrac{\pi}{2}, \dfrac{5\pi}{6}, \dfrac{7\pi}{6}, \dfrac{3\pi}{2}, \dfrac{11\pi}{6}$ **45.** 0.81, 2.33, 4.11, 5.31

47. 0.84, 1.91, 4.37, 5.44   **49.** $\dfrac{3\pi}{4}, \dfrac{7\pi}{4}, 1.11, 4.25$

51. No solution   **53.** $\dfrac{\pi}{12}, \dfrac{5\pi}{12}, \dfrac{13\pi}{12}, \dfrac{17\pi}{12}$ **55.** 0.35, 2.79

57. 0.74   **59.** 0.60, 3.74

61. **a.** Period: 1.00; zero: 0.31   **b.** $0.31 + k$, k an integer
63. **a.** Period: 12.00; zeros: 3.69, 8.31
   b. $3.69 + 12k, 8.31 + 12k$, k an integer
65. **a.** 10 and 20 seconds   **b.** 30 seconds
67. Approximately 0.58 and 1 second   **69.** Approximately 32.08°
71. 0° or approximately 75.96°   **73.** False   **75.** True

Chapter 6 Review, page 504

1. Conditional equation; approximately 0.6435
3. Contradiction   **5.** Identity   **17.** $\cos x = \pm\sqrt{1 - \sin^2 x}$

19. **a.** $\dfrac{77}{85}$ **b.** $-\dfrac{13}{85}$ **c.** $-\dfrac{36}{85}$ **d.** $\dfrac{84}{85}$

21. $\cos 60° = \dfrac{1}{2}$ **23.** $\sin 3x$ **25.** $\cos x$

27. $\dfrac{\sqrt{2}}{2}(\cos x + \sin x)$

31. **a.** $\dfrac{24}{25}$ **b.** $\dfrac{7}{25}$ **c.** $\dfrac{24}{7}$

33. **a.** $\dfrac{\sqrt{10}}{10}$ **b.** $-\dfrac{3\sqrt{10}}{10}$ **c.** $-\dfrac{1}{3}$

35. $\dfrac{1}{2}\sqrt{2 + \sqrt{3}}$ **37.** $\sqrt{3} + 2$ **45.** π **47.** $\dfrac{\pi}{6}, \dfrac{5\pi}{6}, \dfrac{\pi}{3}, \dfrac{4\pi}{3}$

49. $0, \pi, \dfrac{\pi}{3}, \dfrac{5\pi}{3}$ **51.** $0, \dfrac{\pi}{3}, \dfrac{5\pi}{3}$ **53.** $0, \dfrac{2\pi}{3}, \pi, \dfrac{4\pi}{3}$

55. 0.65, 2.49, 3.42, 6.01   **57.** $\dfrac{\pi}{3}, \pi, \dfrac{5\pi}{3}$

59. $\dfrac{\pi}{3} + 2\pi k, \dfrac{5\pi}{3} + 2\pi k$, k an integer

61. $\dfrac{\pi}{3} + 2\pi k, \dfrac{2\pi}{3} + 2\pi k, \dfrac{\pi}{2} + \pi k$, k an integer

63. $\frac{\pi}{2} + \pi k$, k an integer   **65.** $\frac{\pi}{4} + \pi k$, k an integer

67. 0, 0.45, 3.70, 4.50

69. a. Period: 2; zeros: 0, 0.58, 1.00, 1.42
 b. k, 0.58 + 2k, 1.42 + 2k, k an integer

71. Approximately 31.37° or 58.63°

73. $\frac{\pi}{90} + \frac{\pi k}{30}, \frac{\pi k}{30}$, k an integer

Chapter 6 Test, page 507

1. False   **2.** False   **3.** True   **4.** False   **5.** True
6. False   **7.** False   **8.** False

9. $a = 2$, $b = 1$, or $a = 0$, $b = 2$
10. $a = 1$, $b = 1$, for example
11. $a = 2$, $b = 2$, for example
15. $\frac{\tan x - 1}{1 + \tan x}$ **16.** $\frac{1}{2}\sqrt{2 - \sqrt{3}}$
17. Approximately 1.25, 4.39
18. $\frac{\pi}{6} + 2\pi k, \frac{5\pi}{6} + 2\pi k$, k an integer   **19.** Contradiction
20. $\frac{2}{3}$ second

Chapter 7

Section 7.1, page 517

1. $a \approx 11.79$, $b \approx 9.35$, $B = 50°$
3. $a \approx 3.21$, $A \approx 16.15°$, $C \approx 43.85°$
5. $b \approx 9.25$, $c \approx 5.49$, $A = 70°$
7. $a \approx 8.20$, $b \approx 10.46$, $C = 120°$
9. $a \approx 5.28$, $b \approx 17.58$, $B = 77°$
11. $a \approx 17.32$, $A = 60°$, $B = 90°$
13. $b \approx 6.04$, $c \approx 38.49$, $A = 122°$
15. $b \approx 10.09$, $B \approx 21.85°$, $C \approx 36.15°$ **17.** No triangle possible
19. $b \approx 18.66$, $A \approx 48.58°$, $B \approx 91.42°$;
 $b \approx 2.79$, $A \approx 131.42°$, $B \approx 8.58°$
21. No triangle possible   **23.** $a \approx 68.20$, $c \approx 61.68$, $B = 65°$
25. $b \approx 391.94$, $c \approx 150.45$, $C = 20°$
27. $a \approx 0.97$, $A \approx 121.46°$, $B \approx 6.54°$
29. 142.65   **31.** 3,528,057   **33.** 5142.13
35. Approximately 200.2 feet   **37.** Approximately 990 feet
39. Approximately 8.4 miles from the southern station and 6.9 miles from the northern station; approximately 5.7 miles from shore
41. Approximately 242,340 square feet; approximately $11,130
43. False   **45.** True

Section 7.2, page 527

1. $c \approx 10.67$, $A \approx 74.67°$, $B \approx 46.33°$
3. $A \approx 17.61°$, $B \approx 28.96°$, $C \approx 133.43°$
5. $a \approx 22.84$, $B \approx 139.24°$, $C \approx 5.77°$ **7.** No triangle possible
9. $a \approx 269.63$, $B \approx 24.68°$, $C \approx 35.32°$
11. $A \approx 36.18°$, $B \approx 43.53°$, $C \approx 100.29°$
13. $b \approx 22.05$, $c \approx 16.90$, $A = 71°$
15. $c \approx 16.78$, $B \approx 29.21°$, $C \approx 35.79°$
17. $c \approx 1198.32$, $A \approx 22.26°$, $B \approx 130.74°$
19. $A \approx 53.11°$, $B \approx 35.68°$, $C \approx 91.21°$
21. $a \approx 21.29$, $A \approx 134.43°$, $C \approx 17.57°$ **23.** No triangle possible
25. No triangle possible   **27.** $a \approx 1.12$, $c \approx 1.17$, $A = 37°$
29. $x \approx 53.2°$ **31.** $x \approx 1181.2$ **33.** $x \approx 39.6°$
35. $x \approx 47.0°$ **37.** $x \approx 360.5$ **39.** $x \approx 33.8°$
41. 82.65   **43.** 105.5   **45.** 0.5402
47. Approximately 50.6 miles
49. A horizontal distance of approximately 3647 feet (toward Elizabeth) and a vertical distance of approximately 4036 feet
51. Approximately 13.9 square inches; $\angle A \approx 72°$, $\angle B \approx 48°$, $\angle C \approx 60°$
53. a. Approximately 42.4 miles
 b. $p(t) = 80t, f(t) = 100 - 90t$
 c. $d \approx \sqrt{23,756.1t^2 - 28,284.6t + 10,000}$
 d. Approximately 0.6 hour
55. False   **57.** False

Section 7.3, page 538

1. 5   **3.** $\sqrt{10}$ **5.** 2   **7.** 15   **9.** 1

11.

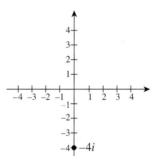

$4\left(\cos \frac{3\pi}{2} + i \sin \frac{3\pi}{2}\right)$

13.

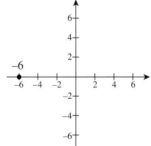

$6(\cos \pi + i \sin \pi)$

15.

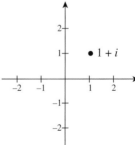

$\sqrt{2}\left(\cos \frac{\pi}{4} + i \sin \frac{\pi}{4}\right)$

17.

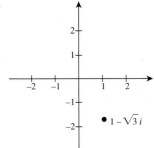

$$2\left(\cos\frac{5\pi}{3} + i\sin\frac{5\pi}{3}\right)$$

19.

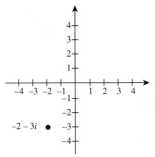

Approximately $\sqrt{13}(\cos 4.1244 + i\sin 4.1244)$

21.

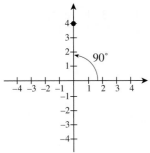

$4i$

23.

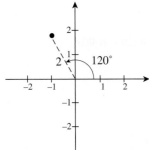

$-1 + \sqrt{3}i$

25.

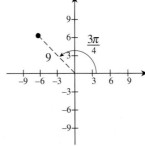

$$-\frac{9\sqrt{2}}{2} + \frac{9\sqrt{2}}{2}i$$

27.

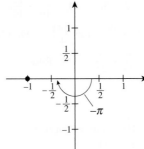

-1

29.

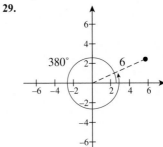

Approximately $5.6382 + 2.0521i$

31.

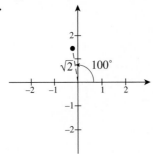

Approximately $-0.2456 + 1.3927i$

33.

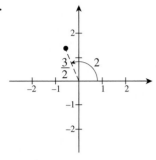

Approximately $-0.6242 + 1.3639i$

35. $6(\cos 70° + i \sin 70°)$ **37.** $1 - 3i$

39. $\dfrac{1}{2}\left(\cos\dfrac{23\pi}{12} + i \sin\dfrac{23\pi}{12}\right)$ **41.** $-2 + 8i$ **43.** $-\dfrac{7}{10} + \dfrac{11}{10}i$

45. $15\left(\cos\dfrac{5\pi}{4} + i \sin\dfrac{5\pi}{4}\right)$ **47.** $7\left(\cos\dfrac{3\pi}{4} + i \sin\dfrac{3\pi}{4}\right)$ **49.** $12i$

51. -4 **53.** $4\sqrt{3} - 4i$ **55.** $-\dfrac{6}{25} + \dfrac{8}{25}i$ **57.** False

59. False

Section 7.4, page 545

1. $32(\cos 150° + i \sin 150°)$ **3.** $16(\cos 80° + i \sin 80°)$

5. 256 **7.** $64\sqrt{3} - 64i$ **9.** $117 + 44i$

11. $\dfrac{\sqrt{3}}{10} + \dfrac{1}{10}i,\ -\dfrac{\sqrt{3}}{10} - \dfrac{1}{10}i$

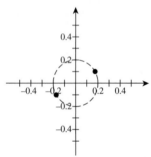

13. $10i,\ -5\sqrt{3} - 5i,\ 5\sqrt{3} - 5i$

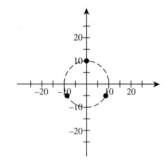

15. $3\sqrt{2} + 3\sqrt{2}i,\ -3\sqrt{2} - 3\sqrt{2}i$

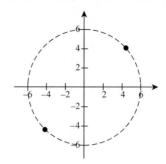

17. $3,\ -\dfrac{3}{2} + \dfrac{3\sqrt{3}}{2}i,\ -\dfrac{3}{2} - \dfrac{3\sqrt{3}}{2}i$

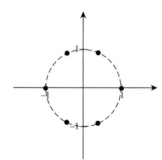

19. $1,\ \dfrac{1}{2} + \dfrac{\sqrt{3}}{2}i,\ -\dfrac{1}{2} + \dfrac{\sqrt{3}}{2}i,\ -1,\ -\dfrac{1}{2} - \dfrac{\sqrt{3}}{2}i,\ \dfrac{1}{2} - \dfrac{\sqrt{3}}{2}i$

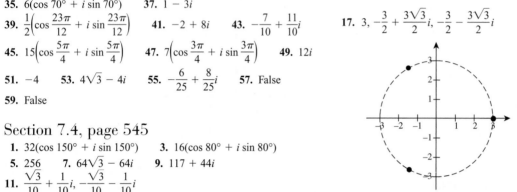

21. $\dfrac{\sqrt{2}}{4} + \dfrac{\sqrt{2}}{4}i,\ -0.483 + 0.129i,\ 0.129 - 0.483i$

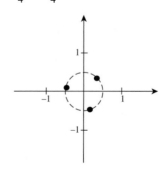

23. $1.755 - 0.285i, -1.755 + 0.285i$
25. $0.924 + 0.383i, -0.924 - 0.383i,$
$0.383 - 0.924i, -0.383 + 0.924i$
27. $1.617 + 0.677i, -1.395 + 1.061i, -0.222 - 1.739i$
29. $x = \dfrac{\sqrt{2}}{2} + \dfrac{\sqrt{2}}{2}i, x = -\dfrac{\sqrt{2}}{2} + \dfrac{\sqrt{2}}{2}i,$

$x = -\dfrac{\sqrt{2}}{2} - \dfrac{\sqrt{2}}{2}i, x = \dfrac{\sqrt{2}}{2} - \dfrac{\sqrt{2}}{2}i$

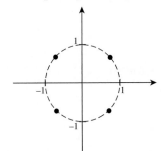

31. $x = i, x = -\dfrac{\sqrt{3}}{2} - \dfrac{1}{2}i, x = \dfrac{\sqrt{3}}{2} - \dfrac{1}{2}i$

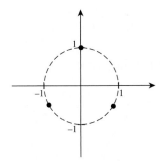

33. $x = \sqrt{3} + i, x = -\sqrt{3} - i$

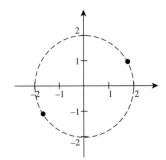

35. $\left(x + \sqrt{3} + i\right)\left(x + \sqrt{3} - i\right)\left(x - \sqrt{3} + i\right)$
$\left(x - \sqrt{3} - i\right)(x + 2i)(x - 2i)$
39. False   **41.** True

Section 7.5, page 558

1.

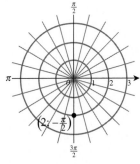

$(0, -2)$

3.

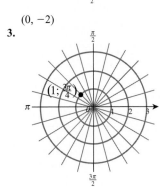

$\left(-\dfrac{\sqrt{2}}{2}, \dfrac{\sqrt{2}}{2}\right)$

5.

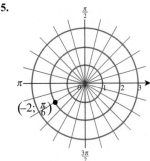

$\left(-\sqrt{3}, -1\right)$

7.

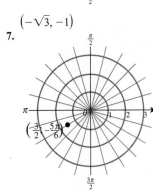

$\left(-\dfrac{3\sqrt{3}}{4}, -\dfrac{3}{4}\right)$

9.

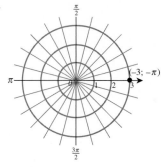

$(3, 0)$

11.

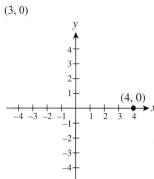

$(4; 0), (-4; \pi)$

13.

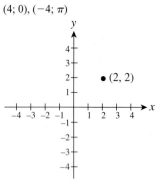

$\left(2\sqrt{2}; \dfrac{\pi}{4}\right), \left(-2\sqrt{2}; \dfrac{5\pi}{4}\right)$

15.

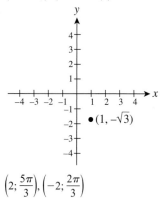

$\left(2; \dfrac{5\pi}{3}\right), \left(-2; \dfrac{2\pi}{3}\right)$

17.

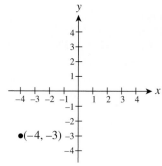

Approximately $(5; 3.7851), (-5; 0.6435)$

19.

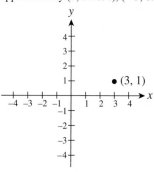

Approximately $\left(\sqrt{10}; 0.3218\right), \left(-\sqrt{10}; 3.4633\right)$

21.

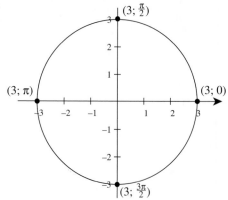

23.

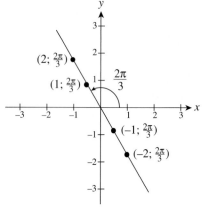

25.

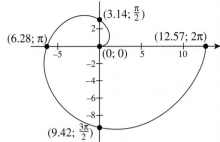

27.

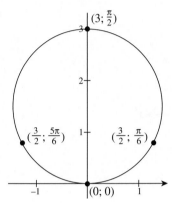

29.

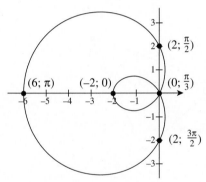

31.

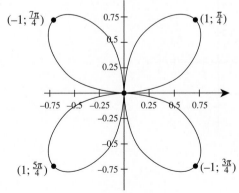

33.

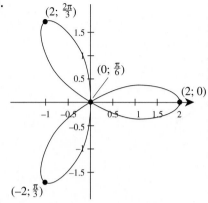

35. $r \sin \theta = 3$ or $r = 3 \csc \theta$

37. $r(\sin \theta + \cos \theta) = 0$ or $\theta = \dfrac{3\pi}{4}$

39. $r = 3$ **41.** $\tan \theta = e$ **43.** $x^2 + y^2 = 49$

45. $y = -\sqrt{3}x$ **47.** $x = 1$ **49.** $x^2 + y^2 = 2y$

51. $(x^2 + y^2)^3 = (x^2 - y^2)^2$

53. Approximately $(2; 0.52)$ and $(2; 2.62)$ (Answers may vary due to the nonuniqueness of polar coordinates.)

55. Approximately $(0; 0)$ and $(1.6; -0.64)$ (Answers may vary due to the nonuniqueness of polar coordinates.)

57. Approximately $(0; 0)$, $(0.87; 1.05)$, and $(-0.87; -1.05)$ (Answers may vary due to the nonuniqueness of polar coordinates.)

59. True   **61.** False

Section 7.6, page 571

1. a.

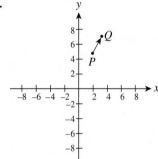

b. $\langle 1, 2 \rangle$

c.

3. a.

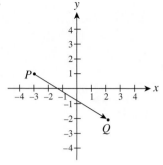

b. $\langle 5, -3 \rangle$

c.

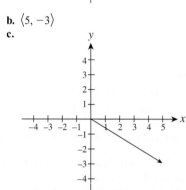

5. a.

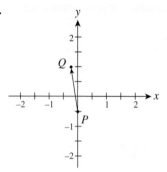

b. $\left\langle -\dfrac{1}{4}, \dfrac{3}{2} \right\rangle$

c.

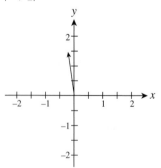

7. 4   **9.** 13   **11.** $\sqrt{34}$ **13.** $2\sqrt{3}$

15.

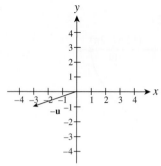

17.

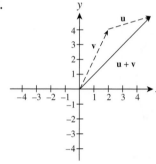

19.

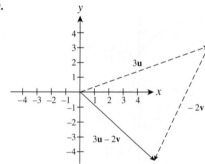

21.

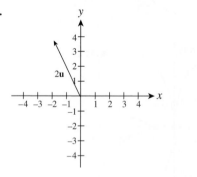

23.

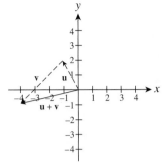

25.

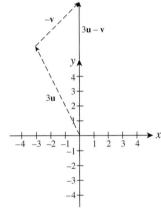

27. a. $\langle -1, -3 \rangle$ **b.** $\langle 3, 8 \rangle$ **c.** $\langle 0, -1 \rangle$
29. a. $3\sqrt{2}\mathbf{i} - \sqrt{2}\mathbf{j}$ **b.** $-7\mathbf{i} - \mathbf{j}$ **c.** $31\mathbf{i} - 2\mathbf{j}$
31. a. $\mathbf{i} - \mathbf{j}$ **b.** $2\mathbf{i} - 3\mathbf{j}$ **c.** $\dfrac{1}{5}\mathbf{i}$
33. $\mathbf{i}$ **35.** $\left\langle -\dfrac{3}{5}, \dfrac{4}{5} \right\rangle$ **37.** $-\dfrac{2\sqrt{13}}{13}\mathbf{i} - \dfrac{3\sqrt{13}}{13}\mathbf{j}$
39. $\left\langle \dfrac{1}{2}, \dfrac{\sqrt{3}}{2} \right\rangle$ **41.** $\left\langle -\dfrac{1}{4}, -\dfrac{\sqrt{3}}{4} \right\rangle$ **43.** $\langle 0, -5 \rangle$
45. Approximately $\langle -0.6840, 1.8794 \rangle$
47. $7[(\cos 0)\mathbf{i} + (\sin 0)\mathbf{j}]$
49. $4\left[\left(\cos \dfrac{11\pi}{6}\right)\mathbf{i} + \left(\sin \dfrac{11\pi}{6}\right)\mathbf{j}\right]$
51. $\dfrac{\sqrt{2}}{8}\left[\left(\cos \dfrac{5\pi}{4}\right)\mathbf{i} + \left(\sin \dfrac{5\pi}{4}\right)\mathbf{j}\right]$
61. Approximately 217.5 miles per hour in the direction 350.65° (clockwise from north)
63. Approximately 15.9 miles per hour on a course of 33.7° (clockwise from north)
65. Approximately 217 pounds and 10.9° off course
67. Approximately 500 pounds   **69.** True   **71.** True

Chapter 7 Review, page 576
1. $a \approx 11.14, b \approx 15.04, B = 81°$
3. $c \approx 1.49, A \approx 53.41°, C \approx 84.59°;$
$c \approx 0.296, A \approx 126.59°, C \approx 11.41°$
5. $a \approx 4.97, B \approx 138.36°, C \approx 31.64°$
7. $A \approx 33.56°, B \approx 62.18°, C \approx 84.26°$
9. $A \approx 28.41°, B \approx 21.67°, C \approx 129.91°$
11. No triangle possible

13.

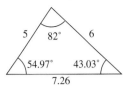

15.

17. Approximately 67.62   **19.** Approximately 2993.33
21. 5   **23.** $\sqrt{82}$
25.

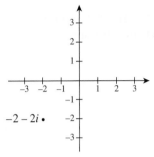

$2\sqrt{2}\left(\cos \dfrac{5\pi}{4} + i \sin \dfrac{5\pi}{4}\right)$

27.

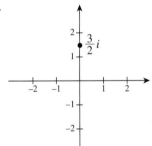

$\dfrac{3}{2}\left(\cos \dfrac{\pi}{2} + i \sin \dfrac{\pi}{2}\right)$

29.

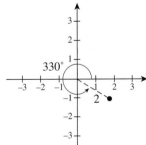

$\sqrt{3} - i$

31.

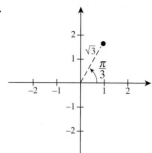

$$\frac{\sqrt{3}}{2} + \frac{3}{2}i$$

33. $10\left(\cos\dfrac{7\pi}{12} + i\sin\dfrac{7\pi}{12}\right)$

35. $3(\cos 200° + i\sin 200°)$ **37.** $-21 - 21i$ **39.** $5i$

41. $625(\cos 40° + i\sin 40°)$ **43.** $16 + 16\sqrt{3}i$

45. $1 + \sqrt{3}i, -1 - \sqrt{3}i$

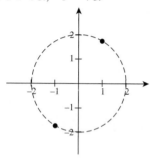

47. $3, -3, 3i, -3i$

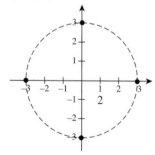

49. $1 + 2i, -1 - 2i$

51. $-1, \dfrac{1}{2} + \dfrac{\sqrt{3}}{2}i, \dfrac{1}{2} - \dfrac{\sqrt{3}}{2}i$

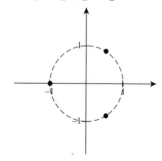

53.

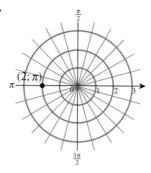

$(-2, 0)$

55.

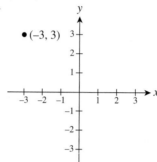

$\left(3\sqrt{2}; \dfrac{3\pi}{4}\right), \left(-3\sqrt{2}; \dfrac{7\pi}{4}\right)$

57.

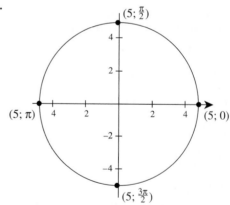

59.

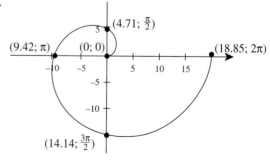

61.

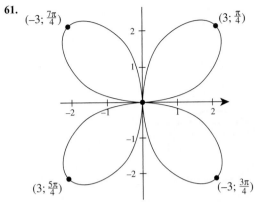

$\left(-3; \frac{7\pi}{4}\right)$ $\left(3; \frac{\pi}{4}\right)$

$\left(3; \frac{5\pi}{4}\right)$ $\left(-3; \frac{3\pi}{4}\right)$

63. $\tan \theta = 2$ **65.** $r = 1$ **67.** $x^2 + y^2 = 1$
69. $x^2 + y^2 = 3y$
71. Approximately $(3; 0.72)$ and $(3; -0.72)$
73. a.

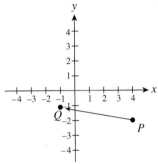

b. $\langle -5, 1 \rangle$
c.

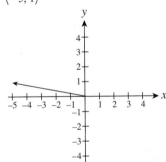

75. 10
77.

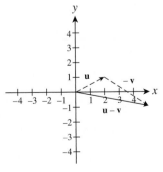

79. a. $\langle 8, -4 \rangle$ **b.** $\langle 0, -4 \rangle$ **c.** $\left\langle -3, \frac{5}{2} \right\rangle$

81. $\frac{12}{13}\mathbf{i} + \frac{5}{13}\mathbf{j}$ **83.** $\langle \sqrt{3}, 1 \rangle$ **85.** $4\left[\left(\cos\frac{\pi}{2}\right)\mathbf{i} + \left(\sin\frac{\pi}{2}\right)\mathbf{j}\right]$

87. Approximately 15 minutes and 49 seconds
89. Approximately 8.37 miles per hour

Chapter 7 Test, page 578

1. False   **2.** True   **3.** False   **4.** True   **5.** False
6. False   **7.** False   **8.** True   **9.** False   **10.** False
11. Answers will vary.   **12.** $-i$, for example
13. $\mathbf{u} = \langle 2, 0 \rangle$, $\mathbf{v} = \langle -2, 0 \rangle$, for example
14. $-1 + i$, for example   **15.** $a \approx 9.89$, $B \approx 83.52°$, $C \approx 41.48°$
16. $b \approx 4.41$, $A \approx 41.47°$, $B \approx 76.53°$
17. a.

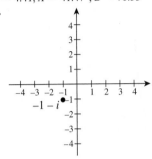

$-1 - i$

$$\sqrt{2}\left(\cos\frac{5\pi}{4} + i\sin\frac{5\pi}{4}\right)$$

b. $-8i$ **c.** $-0.455 + 1.099i$, $0.455 - 1.099i$
18. a. $\mathbf{v} = \mathbf{i} + \mathbf{j}$ **b.** $\langle 1, 5 \rangle$
19.

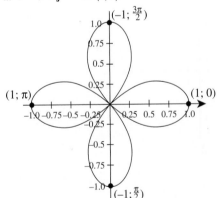

$\left(-1; \frac{3\pi}{2}\right)$

$(1; \pi)$ $(1; 0)$

$\left(-1; \frac{\pi}{2}\right)$

20. Approximately 29.2 miles

Chapter 8

Section 8.1, page 592

1. a.

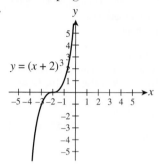

$y = (x + 2)^3$

b.

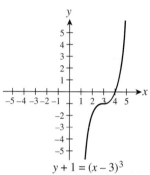

$y + 1 = (x - 3)^3$

c.

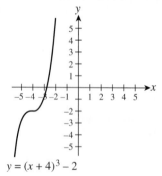

$y = (x + 4)^3 - 2$

3. a.

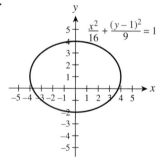

$\dfrac{x^2}{16} + \dfrac{(y-1)^2}{9} = 1$

b.

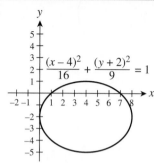

$\dfrac{(x-4)^2}{16} + \dfrac{(y+2)^2}{9} = 1$

5. $y - 3 = \sqrt{x + 2}$ **7.** $(x + 1)(y + 2) = 4$

9. a.

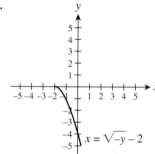

$-x = \sqrt{y} - 2$

b.

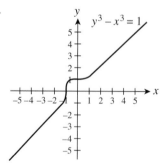

$x = \sqrt{-y} - 2$

11. a.

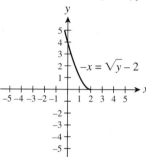

$y^3 - x^3 = 1$

b.

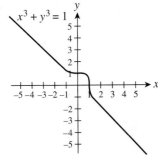

$x^3 + y^3 = 1$

13. $\sqrt{x} + \sqrt{y} = 2$

15 a.

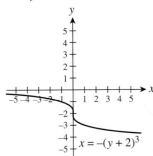

$x = -(y + 2)^3$

b.

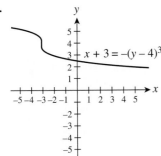

$x + 3 = -(y - 4)^3$

17. Symmetric with respect to the y-axis
19. Symmetric with respect to the x-axis
21. Symmetric with respect to the x-axis
23. Symmetric with respect to the origin
25. Not symmetric with respect to the x-axis, y-axis, or origin
27. Symmetric with respect to the origin
29. Symmetric with respect to the y-axis

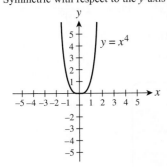

$y = x^4$

31. Symmetric with respect to the origin

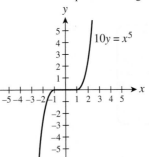

$10y = x^5$

33. Symmetric with respect to the x-axis, the y-axis, and the origin

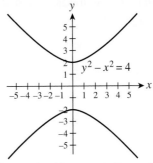

$y^2 - x^2 = 4$

35. Symmetric with respect to the y-axis

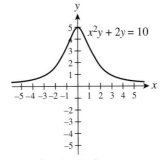

$x^2y + 2y = 10$

37. a. Translate 2 units down.
   b. Translate 3 units to the right.
   c. Answers may vary. One possibility is to reflect about the x-axis and then translate 2 units down.

39. a. Translate 3 units up.
   b. Answers may vary. One possibility is to reflect about the x-axis and then translate 1 unit to the left.
   c. Translate 2 units to the left and 2 units up.

41. $y = \pm\dfrac{4}{5}\sqrt{25 - x^2}$

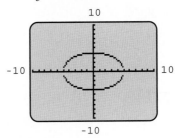

Highest point: $(0, 4)$
Lowest point: $(0, -4)$

43. $y = \pm\sqrt{5 - x - x^3}$

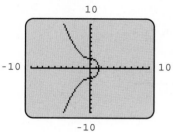

x-intercept: $x \approx 1.52$
y-intercepts: $y \approx -2.24$, $y \approx 2.24$

45. $\dfrac{1}{2}$ **47. a.** 0.1587   **b.** 0.1587   **c.** 0.6826   **d.** 0.8413

49. True   **51.** False   **53.** False

Section 8.2, page 614

1. Vertex: $(0, 0)$; focus: $\left(0, \dfrac{1}{8}\right)$; directrix: $y = -\dfrac{1}{8}$

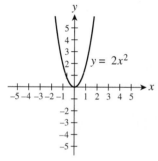

3. Vertex: $(0, 0)$; focus: $\left(\dfrac{3}{4}, 0\right)$; directrix: $x = -\dfrac{3}{4}$

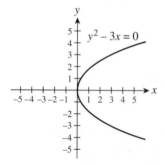

5. Vertex: $(0, 0)$; focus: $\left(-\dfrac{7}{16}, 0\right)$; directrix: $x = \dfrac{7}{16}$

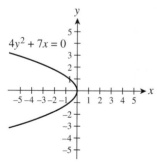

7. Vertex: $(-5, 3)$; focus: $\left(-5, \dfrac{37}{12}\right)$; directrix: $y = \dfrac{35}{12}$

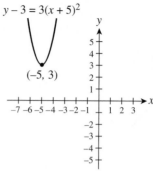

9. Vertex: $(2, 1)$; focus: $(2, 2)$; directrix: $y = 0$

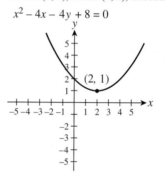

11. Vertex: $\left(-\dfrac{5}{2}, -\dfrac{3}{2}\right)$; focus: $\left(-\dfrac{19}{8}, -\dfrac{3}{2}\right)$; directrix: $x = -\dfrac{21}{8}$

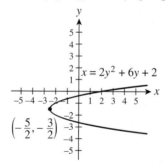

13. $x = \dfrac{1}{8}y^2$ **15.** $y = -\dfrac{1}{16}x^2$ **17.** $x + 2 = \dfrac{1}{4}y^2$

19. $y - 1 = -\dfrac{1}{8}(x - 4)^2$ **21.** $y + 2 = 2(x - 3)^2$

23. $y = x^2 - 4x$; vertex: $(2, -4)$

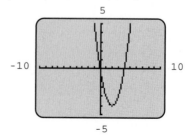

25. $y = \frac{1}{2}x^2 + 4x + 11$; vertex: $(-4, 3)$

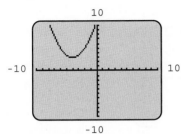

27. $y = \pm\sqrt{4x + 8}$; vertex: $(-2, 0)$

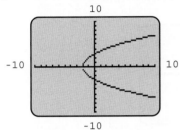

29. $y = 1 \pm \sqrt{-x - 1}$; vertex: $(-1, 1)$

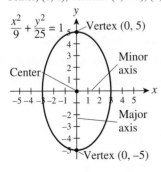

31. Maximum height: 50 feet   Horizontal distance traveled: 200 feet
33. a. $y - 115 = -0.128x^2$ **b.** 0.68 foot
35. Approximately 374 feet above the vertex
37. True   **39.** False

Section 8.3, page 627

1. Center; $(0, 0)$; vertices: $(0, -5), (0, 5)$; foci: $(0, -4), (0, 4)$

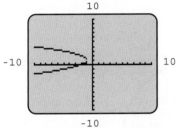

3. Center: $(0, 0)$; vertices: $(-3, 0), (3, 0)$; foci: $\left(-\sqrt{5}, 0\right), \left(\sqrt{5}, 0\right)$

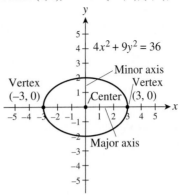

5. Center: $(0, 0)$; vertices: $\left(0, -\frac{1}{2}\right), \left(0, \frac{1}{2}\right)$; foci: $\left(0, -\sqrt{5}/6\right), \left(0, \sqrt{5}/6\right)$

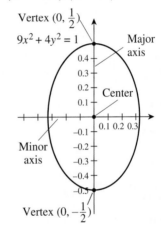

7. Center: $(0, 0)$; vertices: $\left(-2\sqrt{2}, 0\right), \left(2\sqrt{2}, 0\right)$; foci: $\left(-\sqrt{3}, 0\right), \left(\sqrt{3}, 0\right)$

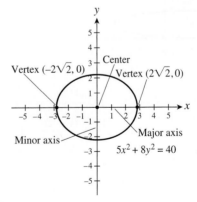

9. Center: $(-3, 1)$; vertices: $(-3, -3)$, $(-3, 5)$; foci:
$\left(-3, 1 - 2\sqrt{3}\right), \left(-3, 1 + 2\sqrt{3}\right)$

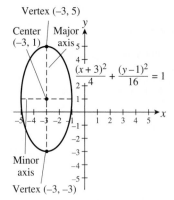

11. Center: $(1, 3)$; vertices: $(1, 0)$, $(1, 6)$; foci: $\left(1, 3 - \sqrt{5}\right), \left(1, 3 + \sqrt{5}\right)$

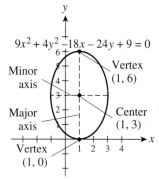

13. Center: $(-2, 1)$; vertices: $(-8, 1)$, $(4, 1)$; foci:
$\left(-2 - \sqrt{35}, 1\right), \left(-2 + \sqrt{35}, 1\right)$

$x^2 + 36y^2 + 4x - 72y + 4 = 0$

15. $\dfrac{x^2}{16} + \dfrac{y^2}{9} = 1$ **17.** $\dfrac{x^2}{4} + \dfrac{y^2}{6} = 1$ **19.** $\dfrac{x^2}{36} + \dfrac{y^2}{36/5} = 1$

21. $\dfrac{(x-2)^2}{4} + \dfrac{(y+1)^2}{9} = 1$ **23.** $\dfrac{(x-1)^2}{25} + \dfrac{(y+2)^2}{16} = 1$

25. $\dfrac{(x+3)^2}{1} + \dfrac{(y-1)^2}{9} = 1$

27. $y = \pm\sqrt{-\dfrac{1}{4}x^2 - x}$; vertices: $(-4, 0)$, $(0, 0)$

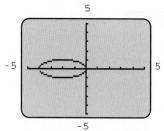

29. $y = 1 \pm \sqrt{4 - 4x^2}$; vertices: $(0, -1)$, $(0, 3)$

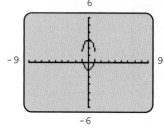

31. $\dfrac{x^2}{93^2} + \dfrac{y^2}{92.9^2} = 1$

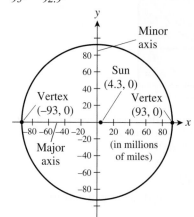

33. a. $\dfrac{x^2}{25^2} + \dfrac{y^2}{20^2} = 1$ **b.** 12 feet   **c.** Yes

35. Perihelion: approximately 0.3 A.U.; Aphelion: approximately 4.1 A.U.

37. a. $\dfrac{x^2}{84.15^2} + \dfrac{y^2}{15.1^2} = 1$ **b.** Approximately 14.1 centimeters

39. True   **41.** True

Section 8.4, page 642

1. Center: $(0, 0)$; vertices: $(-3, 0)$, $(3, 0)$; foci: $(-5, 0)$, $(5, 0)$;

asymptotes: $y = \pm\frac{4}{3}x$

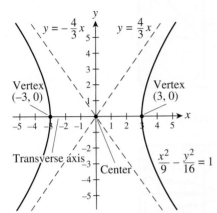

3. Center: $(0, 0)$; vertices: $(0, -3)$, $(0, 3)$; foci: $\left(0, -\sqrt{13}\right)$, $\left(0, \sqrt{13}\right)$;

asymptotes: $y = \pm\frac{3}{2}x$

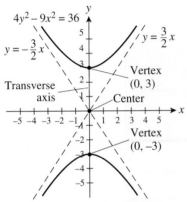

5. Center: $(0, 0)$; vertices: $\left(0, -2\sqrt{2}\right)$, $\left(0, 2\sqrt{2}\right)$; foci:

$\left(0, -\sqrt{10}\right)$, $\left(0, \sqrt{10}\right)$; asymptotes: $y = \pm 2x$

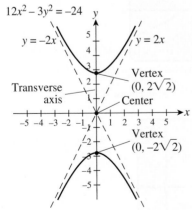

7. Center: $(1, 4)$; vertices: $(-1, 4)$, $(3, 4)$; foci:

$\left(1 - \sqrt{13}, 4\right)$, $\left(1 + \sqrt{13}, 4\right)$; asymptotes:

$y = -\frac{3}{2}x + \frac{11}{2}$, $y = \frac{3}{2}x + \frac{5}{2}$

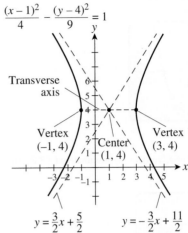

9. Center: $(2, 3)$; vertices: $(-1, 3)$, $(5, 3)$; foci:

$\left(2 - \sqrt{13}, 3\right)$, $\left(2 + \sqrt{13}, 3\right)$; asymptotes:

$y = -\frac{2}{3}x + \frac{13}{3}$, $y = \frac{2}{3}x + \frac{5}{3}$

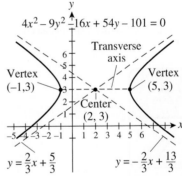

11. Center: $(4, 0)$; vertices: $(4, -2)$, $(4, 2)$; foci: $\left(4, -\sqrt{29}\right)$, $\left(4, \sqrt{29}\right)$;

asymptotes: $y = -\frac{2}{5}x + \frac{8}{5}$, $y = \frac{2}{5}x - \frac{8}{5}$

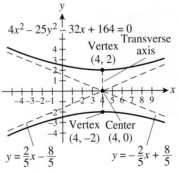

13. $\dfrac{y^2}{16} - \dfrac{x^2}{9} = 1$ **15.** $\dfrac{y^2}{4} - \dfrac{x^2}{16} = 1$

17. $\dfrac{(x-2)^2}{9} - \dfrac{(y-3)^2}{16} = 1$ **19.** Parabola

21. Circle **23.** Ellipse **25.** Hyperbola

27. $y = x^2 - 5x$; parabola; vertex: $(2.5, -6.25)$

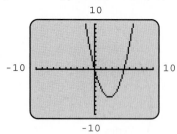

29. $y = \pm\sqrt{5 - \dfrac{5}{8}x^2}$; ellipse; vertices: approximately $(\pm2.83, 0)$

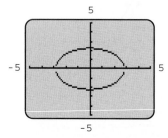

31. $y = \pm\sqrt{3x^2 - 4x - 9}$; hyperbola; vertices: approximately $(-1.19, 0)$ and $(2.52, 0)$

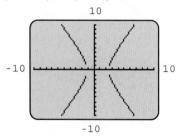

33. $y = \pm\sqrt{8 + 5x - 2x^2}$; ellipse; vertices: approximately $(1.25, \pm3.34)$

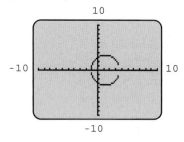

35. $y = -4 \pm \sqrt{3x^2 + 8x + 9}$; hyperbola; vertices: approximately $(-1.33, -2.09)$, $(-1.33, -5.91)$

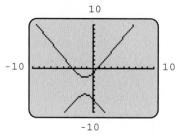

37. $y^2 - x - 3 = 0$; parabola

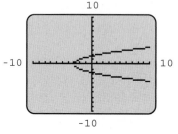

39. $x^2 + y^2 + 6x - 4y - 12 = 0$; circle

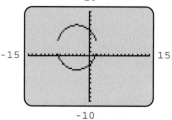

41. $x^2 - 3y^2 - 6 = 0$; hyperbola

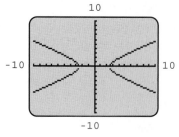

43. $2x^2 + y^2 + 8x - 2y - 1 = 0$; ellipse

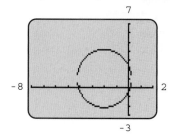

45. a. $\dfrac{y^2}{74.24^2} - \dfrac{x^2}{66.99^2} = 1$

   b. Approximately 60.5 miles east of transmitting station A

47. False   **49.** False

Section 8.5, page 654

1. $\left(\dfrac{3\sqrt{2}}{2}, -\dfrac{3\sqrt{2}}{2}\right)$ **3.** $(1, 1)$ **5.** $(-3, -5)$

7. $\left(-\sqrt{2}, \sqrt{2}\right)$ **9.** $(4, 4)$ **11.** $(2, 0)$ **13.** $v = e^u$

15. $2u + 3v = \sqrt{2}$

17. $u^2 - 2\sqrt{3}\,uv - v^2 - 2 = 0$

19. $\theta = \dfrac{\pi}{4}; \dfrac{u^2}{4} + \dfrac{v^2}{9} = 1$

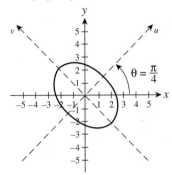

21. $\theta = \dfrac{\pi}{3}; \dfrac{u^2}{4} - v^2 = 1$

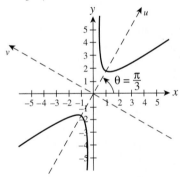

23. $\theta = \dfrac{\pi}{6}; u = 2v^2$

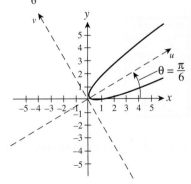

25. $\theta = \dfrac{\pi}{3}; \dfrac{u^2}{9} + v^2 = 1$

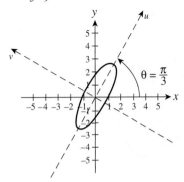

27.

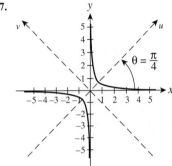

Vertices: $\left(\dfrac{\sqrt{2}}{2}, \dfrac{\sqrt{2}}{2}\right), \left(-\dfrac{\sqrt{2}}{2}, -\dfrac{\sqrt{2}}{2}\right)$; foci: $(1, 1), (-1, -1)$;

asymptotes: $x = 0, y = 0$

29.

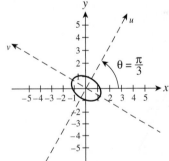

Vertices: $\left(-\dfrac{\sqrt{6}}{2}, \dfrac{\sqrt{2}}{2}\right), \left(\dfrac{\sqrt{6}}{2}, -\dfrac{\sqrt{2}}{2}\right)$; foci: $\left(-\dfrac{\sqrt{3}}{2}, \dfrac{1}{2}\right), \left(\dfrac{\sqrt{3}}{2}, -\dfrac{1}{2}\right)$

31. Hyperbola   **33.** Ellipse   **35.** Parabola

37. $\theta = \dfrac{\pi}{3}; \dfrac{u^2}{4} + v^2 = 1$

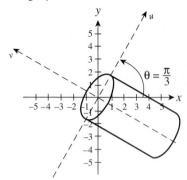

39. True   **41.** False

Section 8.6, page 665

1.

t	x	y
-1	-4	-1
0	-3	1
1	-2	3
2	-1	5

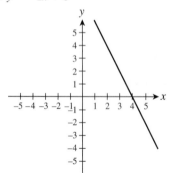

3. v   **5.** vi   **7.** i

9. $y = -2x + 8$

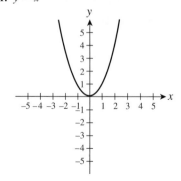

11. $y = x^2$

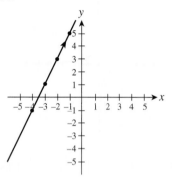

13. $x^2 + y^2 = 16$

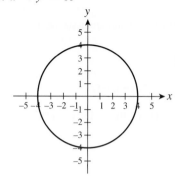

15. $\dfrac{(x-1)^2}{4} + \dfrac{(y-2)^2}{9} = 1$

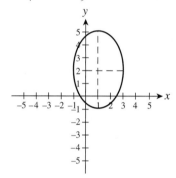

17. a. $y = x + 2$

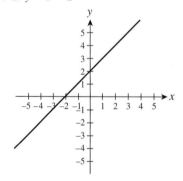

b.

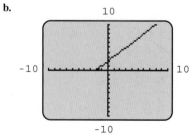

c. Since $y = t^2$ and $t^2 \geq 0$, it follows that $y \geq 0$ for every point (x, y) on the graph.

19. a. $y = x^2$

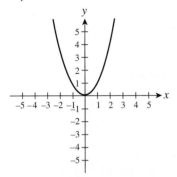

b.

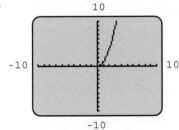

c. Since $x = e^t$ and $e^t > 0$, it follows that $x > 0$ for every point (x, y) on the graph.

21. a. $y = 4x$

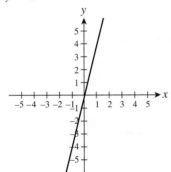

b.

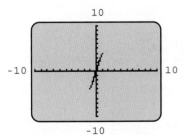

c. Since $x = \cos t$ and $-1 \le \cos t \le 1$, it follows that $-1 \le x \le 1$ for every point (x, y) on the graph.

23. $(2, 3)$

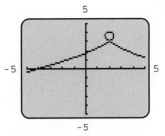

25. x-intercepts: $(-2, 0)$, $(2, 0)$
y-intercepts: $(0, -4)$, $(0, 4)$

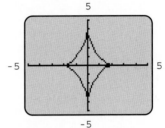

27. Maximum height: approximately 11.32 inches
Minimum height: approximately 4.48 inches

29. a. $(0, 5)$ **b.** 20 seconds
c. Approximately 59.7 feet at $t \approx 7.8$ seconds and again at $t \approx 12.2$ seconds

31. False   **33.** True

Chapter 8 Review, page 668

1 a.

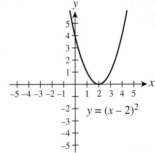

$y = (x - 2)^2$

b.

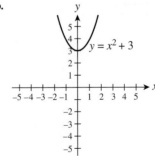

$y = x^2 + 3$

c.

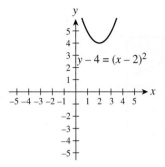

$$y - 4 = (x - 2)^2$$

d.

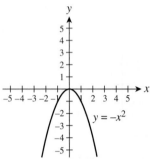

$$y = -x^2$$

3. a.

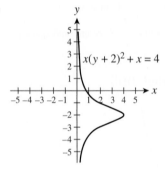

$$x(y + 2)^2 + x = 4$$

b.

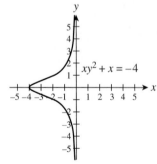

$$xy^2 + x = -4$$

c.

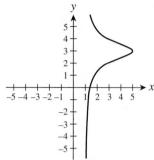

$$(x - 1)(y - 3)^2 + x - 1 = 4$$

5. $(x + 1)^2 + (y - 3)^4 = 16$ **7.** $-y = x^4 - 4x^2 + 2$
9. Symmetric with respect to the x-axis   **11.** No symmetry
13. Symmetric with respect to the x-axis

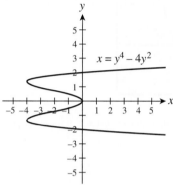

$$x = y^4 - 4y^2$$

15. Symmetric with respect to the origin

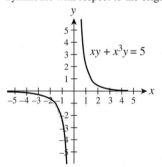

$$xy + x^3y = 5$$

17. Vertex: $(0, 0)$; focus: $\left(\dfrac{1}{8}, 0\right)$; directrix: $x = -\dfrac{1}{8}$

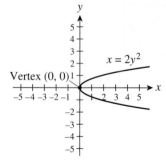

Vertex $(0, 0)$

$$x = 2y^2$$

19. Vertex: $(3, -4)$; focus: $\left(3, -\dfrac{33}{8}\right)$; directrix: $y = -\dfrac{31}{8}$

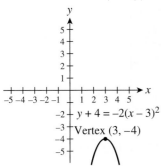

21. Vertex: $(4, -1)$; focus: $\left(\dfrac{49}{12}, -1\right)$; directrix: $x = \dfrac{47}{12}$

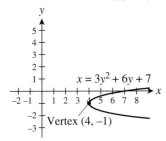

23. $y = x^2$ **25.** $y - 2 = -\dfrac{1}{16}(x + 1)^2$

27. Center: $(0, 0)$; vertices: $(0, -8)$, $(0, 8)$; foci: $\left(0, \sqrt{15}\right)$, $\left(0, -\sqrt{15}\right)$

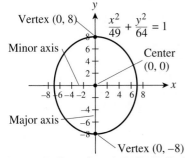

29. Center: $(3, 2)$; vertices: $(-2, 2)$, $(8, 2)$; foci: $(7, 2)$, $(-1, 2)$

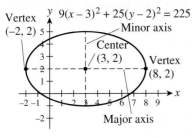

31. Center: $(1, -3)$; vertices: $(-3, -3)$, $(5, -3)$; foci: $\left(1 - 2\sqrt{3}, -3\right)$, $\left(1 + 2\sqrt{3}, -3\right)$

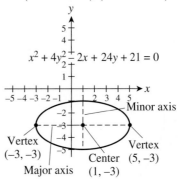

33. $\dfrac{x^2}{9} + \dfrac{y^2}{4} = 1$ **35.** $\dfrac{x^2}{9} + \dfrac{y^2}{25} = 1$

37. Center: $(0, 0)$; vertices: $(-5, 0)$, $(5, 0)$; foci: $\left(\sqrt{34}, 0\right)$, $\left(-\sqrt{34}, 0\right)$; asymptotes: $y = \pm\dfrac{3}{5}x$

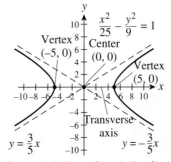

39. Center: $(-5, 1)$; vertices: $(-5, -2)$, $(-5, 4)$; foci: $\left(-5, 1 + \sqrt{13}\right)$, $\left(-5, 1 - \sqrt{13}\right)$; asymptotes: $y = -\dfrac{3}{2}x - \dfrac{13}{2}$, $y = \dfrac{3}{2}x + \dfrac{17}{2}$

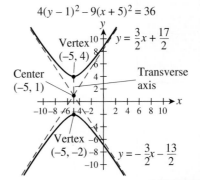

41. Center: $(3, 2)$; vertices: $(-1, 2)$, $(7, 2)$;
foci: $\left(3 - \sqrt{41}, 2\right)$, $\left(3 + \sqrt{41}, 2\right)$;
asymptotes: $y = -\dfrac{5}{4}x + \dfrac{23}{4}$, $y = \dfrac{5}{4}x - \dfrac{7}{4}$

$$25x^2 - 16y^2 - 150x + 64y - 239 = 0$$

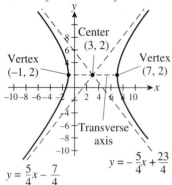

43. $\dfrac{x^2}{25} - \dfrac{y^2}{9} = 1$

45. $y = \pm\sqrt{6 - 4x}$; parabola; vertex: $(1.5, 0)$

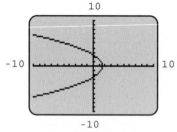

47. $y = \pm\sqrt{6 + 7x - 2x^2}$; ellipse; vertices: $(1.75, -3.48)$, $(1.75, 3.48)$

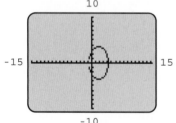

49. Circle: $x^2 + y^2 - 16 = 0$

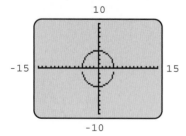

51. Parabola; $y^2 - x - 4y + 8 = 0$

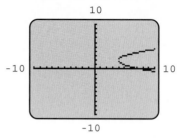

53. $\left(2\sqrt{2}, 0\right)$ **55.** $(1, 2)$ **57.** $\dfrac{\sqrt{2}}{2}(u + v) = (u - v)^2$

59. $\theta = \dfrac{\pi}{4}$; $v = 2u^2$

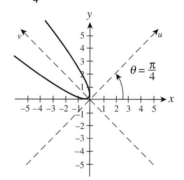

61. $\theta = \dfrac{\pi}{3}$; $u^2 - v^2 = 1$

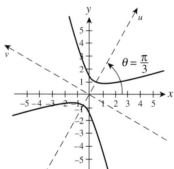

63. Ellipse   **65.** Parabola   **67.** Hyperbola   **69.** Circle

71. $y = -\dfrac{3}{2}x - \dfrac{11}{2}$

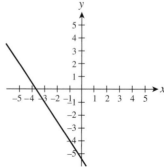

73. $y = 2x$

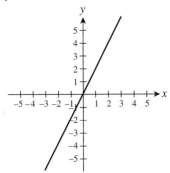

75. $x^2 + y^2 = 4$

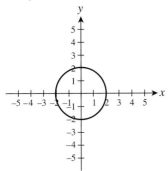

77. (2, 3)

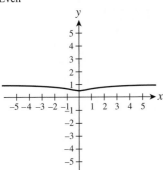

79. a. Even
b.

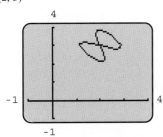

81. a. (1.5, 3.25)   **b.** (3.25, 1.5)

83. b. $x^2 - y^2 = 1$

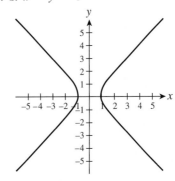

85. a. Midpoint: (1, 2); slope: 1
   b. (0, 3)

87.

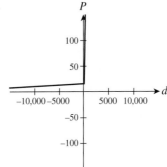

89. 2 seconds

Chapter 8 Test, page 673

1. True   **2.** True   **3.** True   **4.** False   **5.** False
6. True   **7.** True   **8.** False   **9.** $x = y^2$, for example
10. $x = t, y = 2t$ and $x = -t, y = -2t$, for example
11. $x = -y^2$, for example
12. $\dfrac{x^2}{16} + \dfrac{y^2}{4} = 1$, for example
13. A parabolic satellite dish, for example
14. a.

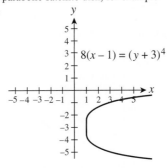

$8(x - 1) = (y + 3)^4$

b.

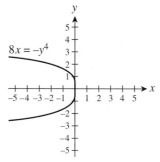

$8x = -y^4$

15. Symmetric with respect to the *x*-axis

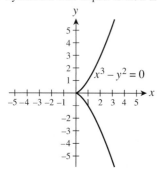

$x^3 - y^2 = 0$

16. Parabola; vertex: $(-4, -2)$; focus: $\left(-\dfrac{15}{4}, -2\right)$

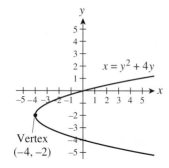

$x = y^2 + 4y$

Vertex
$(-4, -2)$

17. Hyperbola; center: $(0, 0)$; vertices: $(-3, 0), (3, 0)$ foci: $\left(-\sqrt{13}, 0\right),$
$\left(\sqrt{13}, 0\right)$; asymptotes: $y = \pm\dfrac{2}{3}x$

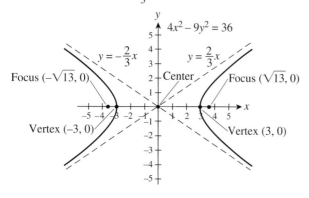

$4x^2 - 9y^2 = 36$

$y = -\dfrac{2}{3}x$ $y = \dfrac{2}{3}x$

Focus $(-\sqrt{13}, 0)$ Center   Focus $(\sqrt{13}, 0)$

Vertex $(-3, 0)$ Vertex $(3, 0)$

18. Ellipse; center: $(2, -3)$; vertices: $(2, -7), (2, 1)$;
foci: $\left(2, -3 - 2\sqrt{3}\right), \left(2, -3 + 2\sqrt{3}\right)$

$$4x^2 + y^2 - 16x + 6y + 9 = 0$$

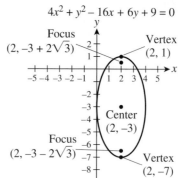

Focus
$(2, -3 + 2\sqrt{3})$ Vertex
$(2, 1)$

Center
$(2, -3)$

Focus
$(2, -3 - 2\sqrt{3})$ Vertex
$(2, -7)$

19. $\theta = \dfrac{\pi}{6}; v = \dfrac{1}{4}u^2$

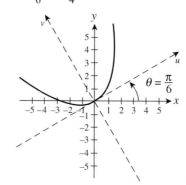

$\theta = \dfrac{\pi}{6}$

20. $y = 3x - 4$

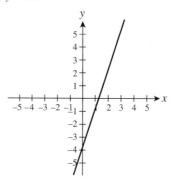

21. Approximately 3.3 A.U

Chapter 9

Section 9.1, page 686

1. $x = 3, y = 4$ **3.** $x = -2, y = 1$ **5.** Inconsistent

7. $m = 3, n = 7$ **9.** $x = 2, y = -\dfrac{3}{5}$ **11.** Inconsistent

13. $a = \dfrac{12}{5}, b = -7$ **15.** $x = 1.3, y = 2.4$

17. $x = 3.4, y = 1.7$ **19.** $x = 1.05 \times 10^7, y = 1.2 \times 10^7$

21. $x = 3, y = 5$
$x = 5, y = 3$

23. $p = 3, q = -2$
$p = -\dfrac{9}{5}, q = \dfrac{14}{5}$

25. $a = 4, b = -6$ **27.** $x = 0, y = 3$
$x = -\dfrac{15}{13}, y = -\dfrac{36}{13}$

29. $x = 3, y = 2$
$x = -3, y = 2$
$x = 3, y = -2$
$x = -3, y = -2$

31. $x = \dfrac{9}{2}, y = \dfrac{1}{2}, z = -\dfrac{3}{2}$ **33.** $x = -2, y = 1, z = 3$

35. $u = 2, v = -1, w = 0$ **37.** $(2, 3)$ **39.** $(-2, 0), (1, 3)$

41. $(1, 1), (1, -1)$ **43.** $(-1.25, 1.31), (0.45, 1.64), (1.80, 6.05)$

45. No points of intersection

47. All points on the line segment from $(2, 0)$ to $(4, 6)$

49. 8 nickels and 4 dimes

51. $\dfrac{50}{9}$ ounces of yogurt and $\dfrac{40}{9}$ ounces of dessert topping

53. Mark is 33 and Benjamin is 3.

55. 350 \$200 tickets and 50 \$500 tickets

57. Hans's time would be approximately 25.8 hours, and Franz's time would be approximately 35.8 hours.

59. The older brother's time is 10 seconds, and the younger brother's time is 12.5 seconds.

61. Approximately 15.2 miles east and 13 miles north of point A

63. $a = 2, b = -3, c = 4$

65. True   **67.** False

Section 9.2, page 703

1. 2×3 **3.** 4×1 **5.** $\begin{bmatrix} 2 & 0 \\ 6 & 10 \end{bmatrix}$ **7.** $\begin{bmatrix} -2 & 11 \\ 6 & 6 \end{bmatrix}$

9. Not defined   **11.** $\begin{bmatrix} 5 \\ -1 \\ 6 \end{bmatrix}$ **13.** $\begin{bmatrix} 4a & 3b \\ 2c & 6d \end{bmatrix}$

15. $\begin{bmatrix} 1 & 7 & 4 \\ 7 & 9 & 18 \\ 7 & 2 & -3 \end{bmatrix}$ **17.** $\begin{bmatrix} 2-a & 0 & 1-a \\ -a & 1-a & 2 \end{bmatrix}$

19. $\begin{bmatrix} 3 & 3 \\ 7 & 14 \end{bmatrix}$ **21.** $\begin{bmatrix} -8 & -1 \\ -14 & -35 \end{bmatrix}$ **23.** $X = \begin{bmatrix} 2 & 3 \\ 4 & 1 \end{bmatrix}$

25. $X = \begin{bmatrix} -\frac{1}{2} & -1 & -\frac{3}{2} \\ -3 & -\frac{5}{2} & -2 \end{bmatrix}$ **27.** Not defined   **29.** Defined; 2×3

31. Not defined   **33.** $\begin{bmatrix} -6 \\ 6 \end{bmatrix}$ **35.** Not defined   **37.** $\begin{bmatrix} 10 & 15 \\ 4 & 6 \end{bmatrix}$

39. $\begin{bmatrix} 1 & 0 \\ 0 & 1 \end{bmatrix}$ **41.** $\begin{bmatrix} 6 & 10 \\ 7 & 33 \\ 7 & 17 \end{bmatrix}$ **43.** $\begin{bmatrix} 39.12 \\ 33.94 \\ 16.35 \end{bmatrix}$

45. $\begin{bmatrix} a & b \\ ax+c & bx+d \end{bmatrix}$ **47.** Not defined   **49.** $\begin{bmatrix} 1 & 0 \\ 0 & 1 \end{bmatrix}$

51. $\begin{bmatrix} p & q \\ r & s \\ t & u \end{bmatrix}$ **53.** $\begin{bmatrix} a_{31} & a_{32} & a_{33} \\ a_{21} & a_{22} & a_{23} \\ a_{11} & a_{12} & a_{13} \end{bmatrix}$ **55.** $\begin{bmatrix} 2 & -3 \\ -1 & 1 \end{bmatrix}\begin{bmatrix} x \\ y \end{bmatrix} = \begin{bmatrix} 5 \\ 6 \end{bmatrix}$

57. $\begin{bmatrix} 1 & 1 & 0 \\ -1 & 0 & 2 \\ 1 & 1 & 1 \end{bmatrix}\begin{bmatrix} r \\ s \\ t \end{bmatrix} = \begin{bmatrix} 1 \\ 3 \\ 4 \end{bmatrix}$ **59.** $\begin{bmatrix} 1 & 1 & 1 & 1 \\ 0 & 1 & 1 & 1 \\ 0 & 0 & 1 & 1 \end{bmatrix}\begin{bmatrix} x_1 \\ x_2 \\ x_3 \\ x_4 \end{bmatrix} = \begin{bmatrix} 4 \\ 3 \\ 2 \end{bmatrix}$

61. $\begin{bmatrix} 1 & 4 \\ 2 & 5 \\ 3 & 6 \end{bmatrix}$ **63.** $\begin{bmatrix} 1 & 2 & 3 \\ 2 & 5 & 6 \\ 3 & 6 & 4 \end{bmatrix}$ **65.** $\begin{bmatrix} 1 & 4 \\ 0 & 1 \end{bmatrix}$ **67.** $\begin{bmatrix} 1 & 0 \\ 0 & 0 \end{bmatrix}$

69. $\begin{bmatrix} 1 & 4 & 14 \\ 0 & 1 & 8 \\ 0 & 0 & 1 \end{bmatrix}$ **71.** $\begin{bmatrix} 1 & 64 & 4064 \\ 0 & 1 & 128 \\ 0 & 0 & 1 \end{bmatrix}$

77. Hourly: $\begin{bmatrix} 230,000 & 320,000 \\ 280,000 & 300,000 \end{bmatrix}$

Salaried: $\begin{bmatrix} 125,000 & 250,000 \\ 400,000 & 500,000 \end{bmatrix}$

Sum: $\begin{bmatrix} 355,000 & 570,000 \\ 680,000 & 800,000 \end{bmatrix}$

Total major medical claims were \$355,000 for employees and \$570,000 for dependents, whereas total comprehensive claims were \$680,000 for employees and \$800,000 for dependents.

79. $S = \begin{bmatrix} 5 & 40 & 18 \\ 4 & 37 & 26 \end{bmatrix}; T = \begin{bmatrix} 3 \\ 2 \\ 1 \end{bmatrix}; ST = \begin{bmatrix} 113 \\ 112 \end{bmatrix}$

So the final score is Mavericks 113, Magic 112.

81. False   **83.** True   **85.** True   **87.** False   **89.** True

Section 9.3, page 714

1. Linear   **3.** Not linear   **5.** Not linear

7. $\begin{bmatrix} 3 & 4 & | & 10 \\ 2 & -7 & | & 4 \end{bmatrix}$ **9.** $\begin{bmatrix} 1 & 0 & | & 3 \\ 0 & 1 & | & 4 \end{bmatrix}$ **11.** $\begin{bmatrix} 1 & -1 & 2 & | & 1 \\ 0 & 2 & 6 & | & 3 \end{bmatrix}$

13. 0 solutions   **15.** Infinitely many solutions

17. 1 solution   **19.** $x = 2, y = 1$ **21.** $x = 3, y = 7, z = 1$

23. $q = -1, r = 3, s = 4, t = -2$ **25.** $x = 2, y = 3$

27. $x = -6, y = 10, z = -4$ **29.** $x = 1, y = -2, z = -1$

31. Salary: \$40,000
Income from municipal bonds: \$10,000
Income from U.S. treasury bonds: \$8000

33. Approximately 2 bananas, 1.5 servings of milk, and 3.75 slices of wheat toast

35. False   **37.** False

Section 9.4, page 726

1. Yes   **3.** No; $a_{31} \neq 0$ **5.** No; $a_{32} \neq 0, a_{33} \neq 1$

7. $\begin{bmatrix} 1 & 2 \\ 3 & 0 \end{bmatrix}$ **9.** $\begin{bmatrix} 1 & 4 & 6 \\ 0 & -9 & -9 \end{bmatrix}$ **11.** $\begin{bmatrix} 1 & 2 & | & 3 \\ 0 & 1 & | & 4 \end{bmatrix}$

13. $\begin{bmatrix} 1 & 0 & 3 & | & 4 \\ 0 & 1 & -2 & | & 4 \end{bmatrix}$ **15.** $\begin{bmatrix} 1 & 3 & 4 & | & -1 \\ 0 & 1 & -2 & | & -\frac{7}{2} \\ 0 & 0 & 1 & | & 3 \end{bmatrix}$

17. $\begin{bmatrix} 1 & -3 & -1 & | & 2 \\ 0 & 1 & \frac{4}{5} & | & \frac{1}{5} \\ 0 & 0 & 1 & | & \frac{7}{2} \end{bmatrix}$ **19.** $\begin{bmatrix} 1 & 3 & 0 & | & 4 \\ 0 & 1 & 2 & | & -2 \end{bmatrix}$

21. $\begin{bmatrix} 1 & 0 & 1 & 2 & | & 1 \\ 0 & 1 & 0 & 1 & | & 0 \\ 0 & 0 & 1 & 1 & | & 0 \\ 0 & 0 & 0 & 1 & | & 2 \end{bmatrix}$ **23.** $x = 2, y = 1$

25. $p = \dfrac{1}{2}, q = \dfrac{2}{3}$ **27.** $x = 1, y = 0, z = 3$

29. $p = 0, q = 1, r = 2$ **31.** $w = \dfrac{3}{2}, x = -\dfrac{1}{2}, y = 0, z = 0$

33. $\begin{bmatrix} 1 & 0 & | & -1 \\ 0 & 1 & | & 2 \end{bmatrix}$ **35.** $\begin{bmatrix} 1 & 0 & | & 2 \\ 0 & 1 & | & -3 \end{bmatrix}$ **37.** $\begin{bmatrix} 1 & 0 & 0 & | & 2 \\ 0 & 1 & 0 & | & 1 \\ 0 & 0 & 1 & | & -1 \end{bmatrix}$

39. $x = 3, y = -2$ **41.** $x = \dfrac{1}{2}, y = -\dfrac{1}{3}$

43. $x = 4, y = -1, z = 3$ **45.** $x = 0, y = 3, z = -4$

47. $\begin{bmatrix} 2 & 1 & | & 5 \\ 4 & 2 & | & 10 \end{bmatrix}; \left(\dfrac{5}{2} - \dfrac{1}{2}y, y\right)$ **49.** $\begin{bmatrix} -1 & 3 & | & -7 \\ 3 & -2 & | & 7 \end{bmatrix}; x = 1, y = -2$

51. $\begin{bmatrix} -3 & -1 & 1 & | & -1 \\ 1 & 4 & -1 & | & 3 \\ -5 & 2 & 1 & | & 2 \end{bmatrix}$; inconsistent

53. $\begin{bmatrix} 2 & 4 & 2 & | & 0 \\ 3 & -1 & 1 & | & 1 \\ 1 & -5 & -1 & | & 1 \end{bmatrix}; \left(\dfrac{2}{7} - \dfrac{3}{7}z, -\dfrac{1}{7} - \dfrac{2}{7}z, z\right)$

55. $y = 2x^2 - 4x + 7$ **57.** 3 nickels, 7 dimes, and 5 quarters

59. The bar weighs 15 pounds, the small disk weighs 7.5 pounds, and the large disk weighs 10 pounds. A barbell with 4 large disks and 2 small disks on each weighs 125 pounds.

61. Hinduism was founded in 1500 B.C., Buddhism in 525 B.C., and Islam in A.D. 622.

63. 4 schools of the first type, 3 schools of the second type, and 1 school of the third type

65. False   **67.** False

Section 9.5, page 737

1. Yes   **3.** No   **5.** $\begin{bmatrix} 1 & 0 \\ -2 & 1 \end{bmatrix}$ **7.** $\begin{bmatrix} -5 & 4 \\ 5 & -2 \end{bmatrix}$

9. No inverse   **11.** $\begin{bmatrix} 0 & \dfrac{1}{b} \\ \dfrac{1}{a} & 0 \end{bmatrix}$ **13.** $\begin{bmatrix} \dfrac{1-x}{2x} & \dfrac{x+1}{2x} \\ \dfrac{1}{2} & -\dfrac{1}{2} \end{bmatrix}$

15. No inverse   **17.** $\begin{bmatrix} 1 & -2 & 0 \\ \dfrac{3}{4} & -1 & \dfrac{1}{4} \\ \dfrac{3}{2} & -3 & \dfrac{1}{2} \end{bmatrix}$ **19.** $\begin{bmatrix} 1 & 2 & 3 \\ 3 & 2 & 1 \\ 2 & 3 & 1 \end{bmatrix}$

21. $\begin{bmatrix} 1 & -x & x^2 \\ 0 & 1 & -x \\ 0 & 0 & 1 \end{bmatrix}$ **23.** No inverse   **25.** $\begin{bmatrix} \dfrac{1}{a} & 0 & 0 & 0 & 0 \\ \dfrac{1}{a} & \dfrac{1}{a} & 0 & 0 & 0 \\ \dfrac{1}{a} & \dfrac{1}{a} & \dfrac{1}{a} & 0 & 0 \\ \dfrac{1}{a} & \dfrac{1}{a} & \dfrac{1}{a} & \dfrac{1}{a} & 0 \\ \dfrac{1}{a} & \dfrac{1}{a} & \dfrac{1}{a} & \dfrac{1}{a} & \dfrac{1}{a} \end{bmatrix}$

27. a. $x = \dfrac{10}{3}, y = \dfrac{13}{3}$ **b.** $x = -3, y = 1$ **c.** $x = \dfrac{1}{2}, y = -\dfrac{1}{2}$

29. a. $x = 5, y = 5$ **b.** $x = 4, y = 1$ **c.** $x = -2, y = 3$

31. a. $x = 2, y = 3, z = 4$ **b.** $x = -2, y = 0, z = 1$
c. $x = 5, y = 5, z = 5$

33. a. Approximately 35.1 minutes walking, 40.2 minutes jogging, and 44.7 minutes running
b. Approximately 38.6 minutes walking, 39.7 minutes jogging, and 11.7 minutes running

35. $f(x) = 2x^3 - 3x^2 + 5x + 1$ **37.** True   **39.** False

Section 9.6, page 750

1. -2 **3.** 0   **5.** -25 **7.** x **9.** -10 **11.** -1
13. -51.406 **15.** a
17. $(a_2b_3 - a_3b_2)i - (a_1b_3 - a_3b_1)j + (a_1b_2 - a_2b_1)k$
19. -561 **21.** a^5 **23.** Invertible   **25.** Not invertible
27. Not invertible   **29.** Invertible   **31.** $-\dfrac{1}{2}$ **33.** $3, -3$
35. Invertible for all real x **37.** $x = 9, y = 5$
39. $x = 1, y = -3, z = 4$ **41.** $x = \dfrac{87}{23}, y = -\dfrac{25}{23}, z = \dfrac{106}{23}$
43. $x = 1, y = 0, z = 0, w = 0$ **45.** $x = \dfrac{ce - bf}{ae - bd}, y = \dfrac{af - cd}{ae - bd}$
47. Approximately \$9722.22
49. Approximately 185.6 grams of milk
51. False   **53.** True

Section 9.7, page 764

1. $(0, 0)$: no; $(4, 1)$: no; $(-1, 3)$: yes
3. $(1, 4)$: yes; $(3, 0)$: no; $(-2, -1)$: no
5. $(0, 2)$: yes; $(-1, 3)$: yes; $(5, -1)$: no
7. iv   **9.** v   **11.** vi
13.

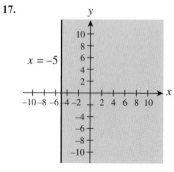

15.

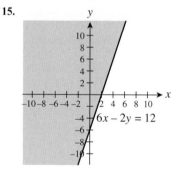

17.

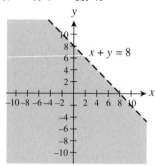

19.

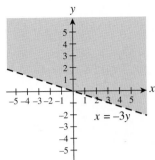

21.

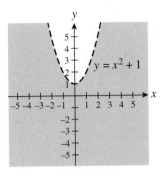

23.

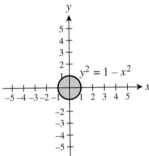

25.

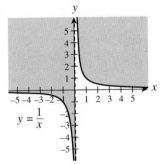

27.

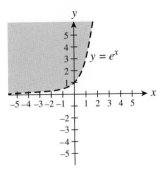

29. ii   **31.** i

33.

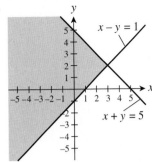

35.

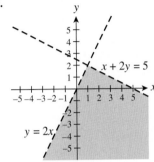

37.

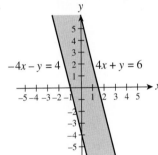

39.

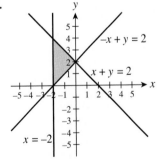

41.

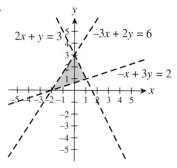

$2x + y = 3$ $-3x + 2y = 6$ $-x + 3y = 2$

43.

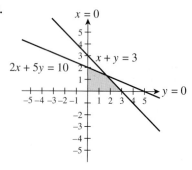

$x = 0$ $x + y = 3$ $2x + 5y = 10$ $y = 0$

45.

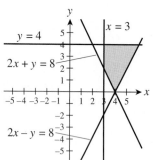

$y = 4$ $x = 3$ $2x + y = 8$ $2x - y = 8$

47.

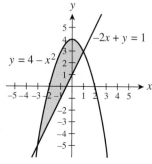

$-2x + y = 1$ $y = 4 - x^2$

49.

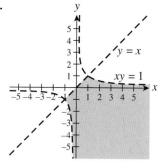

$y = x$ $xy = 1$

51.

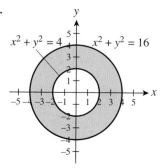

$x^2 + y^2 = 4$ $x^2 + y^2 = 16$

53.

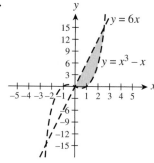

$y = 6x$ $y = x^3 - x$

55. $y \leq \dfrac{3}{4}x - 3$

57. $y \geq x + 2$

$y \leq -\dfrac{2}{5}x + 2$

59. $x \geq 0$

$y \leq -\dfrac{2}{3}x + 4$

$y \geq \dfrac{2}{3}x$

61. $y \geq \dfrac{2}{3}x - \dfrac{4}{3}$

$y \leq -\dfrac{3}{2}x + \dfrac{19}{2}$

$y \geq -\dfrac{3}{2}x + 3$

$y \leq \dfrac{2}{3}x + 3$

63. $x^2 + y^2 \leq 9$
$\qquad\qquad x \geq 0$

65. x = Amount in Redbook Growth
y = Amount in Redbook Small Cap
$\qquad x \geq 6000$
$\qquad y \geq 6000$
$\qquad y \leq \dfrac{x}{2}$
$\qquad x + y \leq 24{,}000$

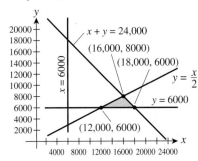

67. x = Number of acres of corn
y = Number of acres of soybeans
$\qquad x + y \leq 1000$
$\qquad 100x + 80y \leq 88{,}000$
$\qquad 2x + y \leq 1600$
$\qquad x \geq 200$
$\qquad y \geq 0$

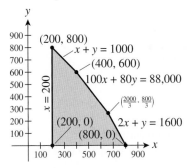

69. False   **71.** True

Section 9.8, page 778

1. Maximum of 76 at (7, 2); minimum of 44 at (3, 2)
3. Maximum of 50 at (5, 6); minimum of 0 at (0, 0)
5. Maximum of 28 at (7, 0); minimum of -32 at (0, 8)
7. Maximum of 85 at (4, 3)   **9.** Minimum of 150 at (10, 0)
11. Maximum of 144 at (12, 2)   **13.** Minimum of 104 at (4, 1)
15. Maximum of 88 at (4, 8)
17. The company should make 22 of style A and 3 of style B.
19. Refinery 1 should be operated for 4 days and refinery 2 for 3 days.
21. Steve should travel 100 miles by car and 200 miles by truck. This will cost $16.00.
23. The show should allot 18 minutes for the senator and 12 minutes for the governor.
25. The food service should serve 4 ounces of Swiss steak and $5\frac{1}{2}$ ounces of peas.
27. The company should have 6 employees from the Bozeman office and 4 from New Haven fly to Orlando; the remaining employees from each office should fly to Columbus.
29. True   **31.** False

Chapter 9 Review, page 784

1. $x = 3, y = 2$ **3.** Inconsistent   **5.** $x = 1, y = 3, z = 4$
7. $\left(\dfrac{9}{7}, \dfrac{8}{7}\right)$ **9.** $(-1, 6), (3.5, 10.5)$ **11.** $\begin{bmatrix} 6 & -3 \\ -15 & 9 \end{bmatrix}$
13. $\begin{bmatrix} 3 & -1 & 2 \\ 3 & 5 & 5 \end{bmatrix}$ **15.** $\begin{bmatrix} 5 & -2 \\ 6 & 9 \end{bmatrix}$
17. Not defined; the number of columns of C is not equal to the number of rows of D.
19. $\begin{bmatrix} 9 & 1 \\ -8 & 8 \\ 10 & 0 \end{bmatrix}$
21. Not defined; the number of rows of C is not equal to the number of columns of C (C isn't square).
23. $\begin{bmatrix} -9 \\ -1 \end{bmatrix}$ **25.** The product is undefined.   **27.** [2]
29. $\begin{bmatrix} -12 & 4 \\ -8 & -1 \\ 6 & 9 \end{bmatrix}$ **31.** $\begin{bmatrix} b - c + 2a \\ 3b - a \\ 4b + 2c \end{bmatrix}$ **33.** $\begin{bmatrix} 5 & 1 & | & 3 \\ 1 & -3 & | & 4 \end{bmatrix}$
35. $\quad x - 4y + z = 3$
$\qquad\quad y - 2z = 5$
$\qquad\qquad\quad z = -2$
$\quad x = 9, y = 1, z = -2$
37. $\begin{bmatrix} 1 & -3 & | & 2 \\ 0 & 1 & | & -\frac{7}{3} \end{bmatrix}$ **39.** $\begin{bmatrix} 1 & 2 & -\frac{3}{2} & | & \frac{1}{2} \\ 0 & 1 & -\frac{3}{2} & | & \frac{1}{2} \end{bmatrix}$ **41.** $x = 3, y = 5$
43. $\left(\dfrac{3}{2}y - 3, y\right)$ **45.** $x = -5, y = -4, z = -3$
47. $x = 0, y = -4, z = -5$ **49.** $\begin{bmatrix} 1 & 0 & | & -2 \\ 0 & 1 & | & 3 \end{bmatrix}$
51. $x = 5, y = -4$ **53.** $x = 2, y = 5, z = 4$
55. $\begin{bmatrix} -1 & -1 \\ 3 & 2 \end{bmatrix}$ **57.** $\dfrac{1}{a^2 - b^2} \cdot \begin{bmatrix} a & -b \\ -b & a \end{bmatrix}$ **59.** $\begin{bmatrix} 1 & 2 & 1 \\ 0 & 1 & 2 \\ 0 & 0 & 1 \end{bmatrix}$
61. No inverse
63. a. $x = -7, y = 10$ **b.** $x = -2, y = 4$
65. a. $x = \dfrac{39}{4}, y = -\dfrac{5}{4}$ **b.** $x = \dfrac{9}{2}, y = -\dfrac{3}{2}$
67. -10 **69.** 1   **71.** $x^3 - x$ **73.** 35   **75.** $x = 5$
77. $x = 2, x = 3$ **79.** $x = \dfrac{1}{4}, y = -\dfrac{3}{4}$
81. $x = 1, y = -1, z = 5$
83.

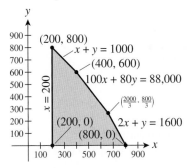

85.

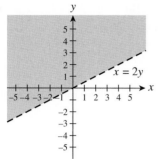

87.

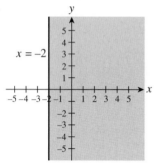

89.

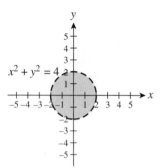

91.

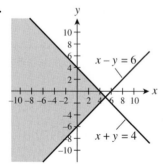

93.

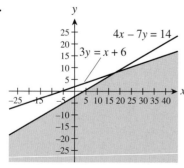

95.

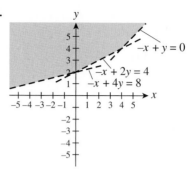

97.

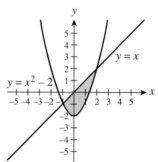

99.

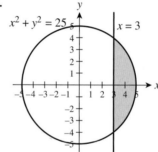

101. Minimum of 46 at $(6, 2)$; Maximum of 70 at $(6, 5)$

103. Maximum of 68 at $(4, 2)$ **105.** Minimum of 60 at $(5, 2)$

107. Side lengths 24 and 45 inches

109. $40,000 in money market, $35,000 in mutual fund, $25,000 in savings account

111. $\dfrac{13}{2}$

113. To maximize either revenue or profit, plant 327.3 acres of corn and 90.9 acres of hay.

115. $6666.67 in municipal bonds, $3333.33 in mutual funds

Chapter 9 Test, page 788

1. True   **2.** False   **3.** False   **4.** False   **5.** True

6. False   **7.** True   **8.** False   **9.** $\begin{bmatrix} 1 & 2 & 3 \\ 4 & 5 & 6 \end{bmatrix}$, for example

10. $\begin{bmatrix} 0 & 0 \\ 0 & 0 \end{bmatrix}$, for example   **11.** $\begin{bmatrix} 1 & 1 & | & 2 \\ 0 & 1 & | & 3 \end{bmatrix}$, for example

12. $xy = 1$
      $x - y = 2$, for example

13. Switching two rows, for example   **14.** $x = 3, y = 5$

15. a. $\begin{bmatrix} -6 & -7 & 9 \\ 16 & 6 & 21 \\ -14 & -9 & -4 \end{bmatrix}$ **b.** $\begin{bmatrix} 4 & -4 & 5 \\ 12 & 11 & -3 \\ 4 & -6 & 1 \end{bmatrix}$

c. $\begin{bmatrix} 2 & 1 & -5 \\ -3 & 1 & -3 \\ -4 & -3 & 2 \end{bmatrix}$

16. $\begin{bmatrix} 3 & -5 & 1 \\ 2 & -8 & 0 \\ 3 & 0 & -1 \end{bmatrix} \begin{bmatrix} x \\ y \\ z \end{bmatrix} = \begin{bmatrix} 2 \\ -2 \\ 0 \end{bmatrix}$; $x = \dfrac{13}{19}, y = \dfrac{8}{19}, z = \dfrac{39}{19}$

17. a. $2x - 2y + 2z = 0$
$y - 3z = -2$
$4z = 8$
$x = 2, y = 4, z = 2$

b. $\begin{bmatrix} 1 & 0 & 0 & | & 2 \\ 0 & 1 & 0 & | & 4 \\ 0 & 0 & 1 & | & 2 \end{bmatrix}$

18. $x = 2, y = 3, z = -5$ **19.** $b = 11$

20. Approximately 71.0 million tons of waste paper, 33.2 million tons of yard waste, and 75.4 million tons of miscellaneous waste

21. Maximum of 0 at $(0, 0)$

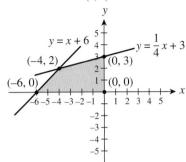

22. $y > \dfrac{3}{2}x$
$y < -x + 5$
$y > -6x$

Chapter 10

Section 10.1, page 800

1. 4, 7, 10, 13   **3.** $-4, 16, -64, 256$ **5.** $1, \dfrac{1}{4}, \dfrac{1}{9}, \dfrac{1}{16}$

7. 0, 2, 0, 2   **9.** $1, -\dfrac{1}{2}, \dfrac{1}{3}, -\dfrac{1}{4}$ **11.** $\sqrt{2}, \sqrt{6}, 2\sqrt{3}, 2\sqrt{5}$

13. $2, \dfrac{9}{4}, \dfrac{64}{27}, \dfrac{625}{256}$ **15.** 1, 3, 6, 10   **17.** $a_n = 2n$

19. $a_n = \left(-\dfrac{1}{2}\right)^n$ **21.** $a_n = \dfrac{n}{n+1}$ **23.** $a_n = 1 - \dfrac{1}{n}$

25. Arithmetic; $d = 3$ **27.** Neither   **29.** Arithmetic; $d = -4$

31. Neither   **33.** Geometric; $r = \dfrac{1}{10}$ **35.** $a_n = 6n + 1$

37. $a_n = 115 - 15n$ **39.** $a_n = \dfrac{5n - 13}{4}$ **41.** $a_{10} = 75$

43. $a_{15} = -60$ **45.** $a_{31} = 176$ **47.** $a_n = 3^{n-1}$

49. $a_n = 32\left(-\dfrac{1}{2}\right)^{n-1}$ **51.** $a_n = 4\left(\dfrac{5}{4}\right)^{n-1}$ **53.** $a_8 = 4374$

55. $a_7 = \dfrac{1}{81}$ **57.** $2, -3, 7, -13, 27$ **59.** 1, 2, 6, 24, 120

61. $2, \sqrt{2}, \sqrt[4]{2}, \sqrt[8]{2}, \sqrt[16]{2}$ **63.** 1, 2, 2, 4, 8

65. \$29,355; \$30,591; \$31,827; $d = 1236$; \$36,771   **67.** 74

69. $a_n = 1000(1.055)^{n-1}$; \$1619.09; 5.5%

71.

Month	Pairs at the beginning of the month	Pairs born during the month	Pairs at the end of the month
1	1	0	1
2	1	1	2
3	2	1	3
4	3	2	5
5	5	3	8
6	8	5	13
⋮	⋮	⋮	⋮
12	144	89	233

73. True   **75.** True

Section 10.2, page 810

1. 75   **3.** $\dfrac{71}{20}$ **5.** 51   **7.** -15 **9.** 60   **11.** $\dfrac{19}{20}$

13. $\dfrac{155}{16}$ **15.** $\sum\limits_{k=1}^{6} \dfrac{1}{k^3}$ **17.** $\sum\limits_{k=1}^{12} 5k + 2$ **19.** $\sum\limits_{k=1}^{7} 3^k$

21. $\sum\limits_{k=1}^{16} \dfrac{(-1)^{k-1}}{k^2}$ **23.** $\sum\limits_{k=1}^{6} \dfrac{2^k - 1}{2^k}$ **25.** 1640   **27.** 159.5

29. -240 **31.** 3725   **33.** 3210   **35.** 724   **37.** 147,620

39. Approximately 4.5   **41.** Approximately 3.6364   **43.** 262,140

45. Approximately 1.7143   **47.** Approximately -0.0012

49. 500,500   **51.** 10,000   **53.** 294   **55.** 900

57. Approximately \$276.85 billion   **59.** \$5323.58

61. Approximately \$18 billion   **63.** \$150

65. Approximately \$50.8 million   **67.** False   **69.** True

Section 10.3, page 823

1. 720   **3.** 720   **5.** 30   **7.** 120   **9.** 15   **11.** 8
13. 1   **15.** 28

17.

College Algebra
— Elementary Physics
 — General Biology
 — Zoology
— University Physics I
 — General Biology
 — Zoology

Calculus I
— Elementary Physics
 — General Biology
 — Zoology
— University Physics I
 — General Biology
 — Zoology

} 8 possibilities

19.

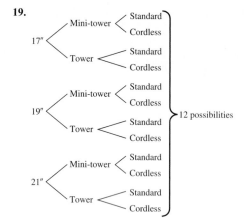

17″ — Mini-tower — Standard / Cordless; Tower — Standard / Cordless
19″ — Mini-tower — Standard / Cordless; Tower — Standard / Cordless
21″ — Mini-tower — Standard / Cordless; Tower — Standard / Cordless
} 12 possibilities

21. a. 125   **b.** 60

23. 70,304,000   **25.** $_8P_3 = 336$

27. a. $_{13}P_9 = 259,459,200$ **b.** $_2P_2 \cdot _{11}P_7 = 3,326,400$

29. $_8C_3 = 56$ **31. a.** $_{12}C_6 = 924$ **b.** $_5C_2 \cdot _7C_4 = 350$

33. $_{10}P_4 = 5040$ **35.** $_{40}C_5 = 658,008$

37. a. $_{40}P_3 = 59,280$ **b.** $40^3 = 64,000$

39. $_6C_4 = 15$; $_6C_4 \cdot _{15}C_{10} = 45,045$

41. $2 \cdot _{30}C_2 = 870$ seconds; $2 \cdot _{60}C_2 = 3540$ seconds

43. $4 \cdot _{13}C_5 = 5148$ **45. a.** $_{5000}P_{1127}$ **b.** 5000^{1127}

47. False   **49.** False

Section 10.4, page 836

1. $S = \{$red, blue, green, white$\}$; $n(S) = 4$

3. $S = \{HHH, HHT, HTH, HTT, THH, THT, TTH, TTT\}$; $n(S) = 8$

5. $\dfrac{1}{6}$ **7.** 1   **9.** $\dfrac{1}{4}$ **11.** $\dfrac{7}{12}$ **13.** $\dfrac{1}{52}$ **15.** $\dfrac{1}{13}$ **17.** $\dfrac{3}{26}$

19. $\dfrac{1}{4}$ **21.** $\dfrac{3}{8}$ **23.** $\dfrac{5}{18}$ **25.** $\dfrac{5}{9}$ **27.** $\dfrac{1}{6}$ **29.** $\dfrac{1}{12}$

31. $\dfrac{1}{54,145} \approx 1.85 \times 10^{-5}$ **33.** $\dfrac{33}{16,660} \approx 0.002$ **35.** $\dfrac{1}{16}$

37. $\dfrac{1}{8}$ **39.** $\dfrac{5}{16}$

41. $P(E) = \frac{1}{2}$; $P(F) = \frac{3}{13}$; $E \cap F$ is the event that a red face card is drawn, $P(E \cap F) = \frac{3}{26}$; $E \cup F$ is the event that a red card or a face card is drawn, $P(E \cup F) = \frac{8}{13}$; E' is the event that a black card is drawn, $P(E') = \frac{1}{2}$; F' is the event that a non–face card is drawn, $P(F') = \frac{10}{13}$

43. $P(E) = \frac{1}{2}$; $P(F) = \frac{7}{12}$; $E \cap F$ is the event that the sum is even and less than 8, $P(E \cap F) = \frac{1}{4}$; $E \cup F$ is the event that the sum is even or less than 8, $P(E \cup F) = \frac{5}{6}$; E' is the event that the sum is odd, $P(E') = \frac{1}{2}$; F' is the event that the sum is 8 or greater, $P(F') = \frac{5}{12}$

45. 0.35

47. $\dfrac{1}{658,008} \approx 1.52 \times 10^{-6}$; $\dfrac{5}{164,502} \approx 3.04 \times 10^{-5}$; 329,004

49. a. 0.066   **b.** 0.571

51. $\dfrac{9}{28} \approx 0.32$

53. a. 0.206   **b.** 0.039   **c.** 0.88   **d.** 0.081   **e.** 0.086

55. False   **57.** True

Section 10.5, page 846

1. 4   **3.** 10   **5.** 6   **7.** 54   **9.** 6   **11.** 10

13. $x^5 + 5x^4y + 10x^3y^2 + 10x^2y^3 + 5xy^4 + y^5$

15. $16x^4 - 32x^3y + 24x^2y^2 - 8xy^3 + y^4$

17. $27x^3 + 135x^2y + 225xy^2 + 125y^3$

19. $x^5 + 5x^4 + 10x^3 + 10x^2 + 5x + 1$

21. $1,000,000w^6 - 600,000w^5 + 150,000w^4 - 20,000w^3 + 1500w^2 - 60w + 1$

23. $7 + 5\sqrt{2}$ **25.** $x^6 - 3x^4 + 3x^2 - 1$

27. $x^5 + 5x^3 + 10x + \dfrac{10}{x} + \dfrac{5}{x^3} + \dfrac{1}{x^5}$ **29.** $-2 + 2i$

31. -64 **33.** i **43.** False   **45.** True

Section 10.6, page 853

35. False   **37.** True

Chapter 10 Review, page 855

1. $-4, -9, -14, -19, -24$; arithmetic; $d = -5$

3. $e^{-1}, 2e^{-2}, 3e^{-3}, 4e^{-4}, 5e^{-5}$ **5.** $-\dfrac{1}{2}, \dfrac{3}{2}, -\dfrac{1}{2}, \dfrac{3}{2}, -\dfrac{1}{2}$

7. $2, 6, 18, 54, 162$; geometric; $r = 3$

9. $a_n = 2(5)^{n-1}$; geometric; $r = 5$

11. $a_n = \dfrac{1}{n}$ **13.** $a_n = \dfrac{7}{2} - \dfrac{5}{2}n$; arithmetic; $d = -\dfrac{5}{2}$

15. $a_n = (-1)^{n-1}n^3$ **17.** 47   **19.** 1024   **21.** 676

23. 40   **25.** Approximately 0.4914   **27.** $\displaystyle\sum_{k=1}^{11} 5k - 2$

29. $\displaystyle\sum_{k=3}^{13} 4\left(\dfrac{2}{3}\right)^k$ **31.** 290   **33.** Approximately 12.5

35. Approximately 2.1428   **37.** 7532   **39.** 36   **41.** 6720

43. 10   **45.** 120

47. $\dfrac{1}{4}$ **49.** $\dfrac{11}{26}$ **51.** $\dfrac{1}{4}$ **53.** $\dfrac{1}{9}$ **55.** $\dfrac{1}{8}$ **57.** 0.359

59. 0.379   **61.** 56   **63.** -1000

65. $x^4 - 4x^3y + 6x^2y^2 - 4xy^3 + y^4$ **67.** $r^3 + \dfrac{3r^2}{s} + \dfrac{3r}{s^2} + \dfrac{1}{s^3}$

69. $z^2\sqrt{z} + 10z^2 + 40z\sqrt{z} + 80z + 80\sqrt{z} + 32$

77. $_{20}P_4 = 116,280$

79. a. $_{10}C_5 = 252$ **b.** $_6C_3 \cdot _4C_2 = 120$

81. 1024

83. The populations for 2001–2003 are 210,000, 220,500, and 231,525, respectively; the common ratio is 1.05; the population in 2012 will be approximately 359,171.

85. Approximately \$342,700.67

87. a. $_{80}C_{20}$ **b.** $\left(_{20}C_5\right)^4$ **89.** $\dfrac{1}{120}$

Chapter 10 Test, page 858

1. False   **2.** True   **3.** True   **4.** True   **5.** False

6. True   **7.** False   **8.** False   **9.** $a_n = -4n$, for example

10. $a_n = \left(\dfrac{1}{3}\right)^n$, for example   **11.** $\displaystyle\sum_{k=1}^{5} \dfrac{1}{k}$, for example

12. Tossing three coins, for example

13. A red ball is drawn from a bag containing one red ball, one white ball, and one blue ball, for example

14. 1, 4, 6, 4, 1, for example   **15.** 4, 13, 28, 49, 76

16. $a_n = \dfrac{n+1}{n}$ **17.** 126   **18.** $\displaystyle\sum_{k=1}^{6} 5\left(\dfrac{1}{2}\right)^k \approx 4.9219$

19. $256; \dfrac{1}{256}$

20. a. $_{15}C_9 = 5005$

b. $_7C_5 \cdot _8C_4 = 1470$; the combinations would be replaced with permutations

21. $\dfrac{1}{3}$ **22.** $\dfrac{8}{13}$ **23.** 393,750

24. $16x^4 - 96x^3y + 216x^2y^2 - 216xy^3 + 81y^4$

Index of Applications

Index

Credits

This page constitutes an extension of the copyright page. We have made every effort to trace the ownership of all copyrighted material and to secure permission from copyright holders. In the event of any question arising as to the use of any material, we will be pleased to make the necessary corrections in future printings. Thanks are due to the following authors, publishers, and agents for permission to use the material indicated.

Chapter 1. 1: Media Fusion, Inc./NASA **4:** Photo Chamoix Initiative/Liaison Agency, Inc. **13:** NASA/SPL/ Photo Researchers, Inc. **44:** Greg Vaughn/Tom Stack & Associates **60:** Oscar White/Corbis **65:** Yann Arthus-Bertrand/Corbis **68:** Professor Peter Goddard/SPL/Photo Researchers, Inc. **80: left,** AP/Wide World Photos **80: right,** USC Sports Information **83: left,** Wolfgang Kaehler Photography **83: right,** Viviane Moos/The Stock Market **87:** AP/ Wide World Photos **111:** NASA/Phototake, Inc. **117:** Frank Siteman/Rainbow

Chapter 2. 120: Sygma/Corbis **129:** Peter Marlow/Sygma/Corbis **135:** ©1996 Jonathan Bowen. Reprinted with permission. Reprinted from: http://www.cs.reading.ac.uk/archive/hypercubes **136: right,** Altan/Sygma/ Corbis **136: left,** Reuters/Corbis **167:** Monique Salaber/Liaison/Getty **181: right,** Globe Photos, Inc. **181: left,** NASA Jet Propulsion Laboratory/NGS Image **187:** Tom Hauk/Getty Images **202:** Betmann/ Corbis **212: left,** J. P. Varin/Jacana/Photo Researchers, Inc. **212: right,** NASA **222:** AP/Wide World Photos

Chapter 3. 224: IBM Research/Peter Arnold **232:** AP/Wide World Photos **240:** Gregory Sams/SPL/Photo Researchers, Inc. **252:** Kuroda Lee/SuperStock, Inc. **285:** Getty Images

Chapter 4. 293: National Snow and Ice Data Center/SPL/Photo Researchers, Inc. **307: left,** James D. Wilson/ Woodfin Camp & Associates **307: right,** Robert Frerck/Woodfin Camp & Associates **317:** Flip Nicklin/ Minden Pictures **321:** Luiz C. Marigo/Peter Arnold, Inc. **327:** Photori **337:** Tom Ebenhoh/Image Quest **340:** Douglas Burrows/Liaison/Getty Images **341:** Patrick Messner/Liaison/Getty Images **344:** Hinterleitner/ Liaison/Getty Images **347:** Marta Serra-Jovenich/Bruce Coleman **352:** R&S Michaud/Woodfin Camp & Associates **353:** Hurricane Entertainments Limited **356:** Central Point Software, Inc. **357:** Peter Brock/ Liaison/Getty Images

Chapter 5. 364: Philip Crossley/Hurricane Entertainments Limited **375:** Courtesy of Steve Grohe/Thinking Machines, Corp. **379: top,** Peter Brock/Liaison/Getty Images **379: bottom,** Royal Astronomical Society Library **381: left,** Corbis **381: right,** Courtesy of Trek Bicycles **393:** P. Touraire/Sygma/Corbis **406:** AFP/ Corbis **410: top,** BSIP Agency/Index Stock **410: bottom,** Kevin Schafer/Corbis **413:** Julian Baum/SPL/Photo Researchers, Inc. **418:** Alexander Tsiara/SPL/Photo Researchers, Inc. **438:** NASA/SPL/Photo Researchers, Inc.

Chapter 6. 460: Dana Berry/STSci (Hubble Space Telescope Science Institute) **471: top,** Gary D. McMichael/ Photo Researchers, Inc. **471: bottom,** Rick McClain/Liaison/Getty Images **478: left,** Hank Morgan/Science Source/Photo Researchers, Inc. **478: right,** NASA/SPL/Photo Researchers, Inc. **491:** Michael P. Gadowinski/ Science Source/Photo Researchers, Inc. **495:** M. Hans/Vandystadt/Photo Researchers, Inc.

Chapter 7. 508: Irwin Karnick/Photo Researchers, Inc. **528:** Chuck Solomon/Sports Illustrated/Time Life **568:** Stone/Getty Images **573:** Ross Harrison Koty/Stone/Getty Images

Chapter 8. 580: Stephanie Maze/Woodfin Camp & Associates **599:** Photori **600: top right** and **bottom left,** Norbert Wu **600: top left,** Tim Davis/Photo Researchers, Inc. **600: bottom right,** Gary Braasch/Woodfin Camp Associates **607:** Sovfoto/Eastfoto **612:** Doug Johnson/SPL/Photo Researchers, Inc. **615:** P. Delacroix/ Liaison/Getty Images **626:** Architect of the Capitol **629:** Photori **656:** NASA **661:** Richard Megna/ Fundamental Photographs **666:** Donald Dietz/Stock Boston

Chapter 9. 675: Allen Tannenbaum **690:** Crown Copyright/Health &Safety Laboratory/SPL/Photo Researchers, Inc. **729:** Jed Jacobson/Getty Images **777:** Courtesy of AT&T Archives

Chapter 10. 789: Armen Kachaturian/Liaison/Getty Images **791:** National Science Foundation **798: left,** Ned Haines/Photo Researchers, Inc. **798: center left,** Rod Plank/Photo Researchers, Inc. **798: right,** Dave Nagel/ Liaison/Getty Images **798: center right,** Snowdon/Hoyer Focus/Woodfin Camp & Associates **825:** Robert Silvers **840:** Corbis

Geometric Formulas

Right triangle

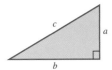

Pythagorean Theorem: $a^2 + b^2 = c^2$

Triangle

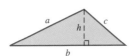

$\text{Area} = \dfrac{1}{2} bh$ $\text{Perimeter} = a + b + c$

Equilateral triangle

$h = \dfrac{\sqrt{3}}{2} s$ $\text{Area} = \dfrac{\sqrt{3}}{4} s^2$

Rectangle

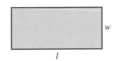

$\text{Area} = lw$ $\text{Perimeter} = 2l + 2w$

Trapezoid

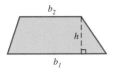

$\text{Area} = \dfrac{1}{2}(b_1 + b_2)h$

Circle

$\text{Area} = \pi r^2$ $\text{Circumference} = 2\pi r$

Rectangular box

$\text{Volume} = lwh$ $\text{Surface area} = 2lw + 2wh + 2lh$

Right circular cylinder

$\text{Volume} = \pi r^2 h$ $\text{Lateral surface area} = 2\pi rh$

Sphere

$\text{Volume} = \dfrac{4}{3} \pi r^3$ $\text{Surface area} = 4\pi r^2$

Right square pyramid

$\text{Volume} = \dfrac{1}{3} b^2 h$ $\text{Lateral surface area} = b\sqrt{4h^2 + b^2}$

Right circular cone

$\text{Volume} = \dfrac{1}{3} \pi r^2 h$ $\text{Lateral surface area} = \pi r \sqrt{r^2 + h^2}$

Trigonometry

Right Triangle Definitions

$$\sin \theta = \frac{\text{opp}}{\text{hyp}} \quad \cos \theta = \frac{\text{adj}}{\text{hyp}} \quad \tan \theta = \frac{\text{opp}}{\text{adj}}$$

$$\csc \theta = \frac{\text{hyp}}{\text{opp}} \quad \sec \theta = \frac{\text{hyp}}{\text{adj}} \quad \cot \theta = \frac{\text{adj}}{\text{opp}}$$

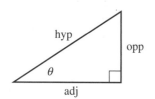

Unit Circle Definitions

$$\sin \theta = y \qquad \cos \theta = x \qquad \tan \theta = \frac{y}{x}, x \neq 0$$

$$\csc \theta = \frac{1}{y}, y \neq 0 \quad \sec \theta = \frac{1}{x}, x \neq 0 \quad \cot \theta = \frac{x}{y}, y \neq 0$$

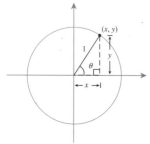

Special Angles

θ	θ (radians)	$\sin \theta$	$\cos \theta$	$\tan \theta$
0°	0	0	1	0
30°	$\frac{\pi}{6}$	$\frac{1}{2}$	$\frac{\sqrt{3}}{2}$	$\frac{\sqrt{3}}{3}$
45°	$\frac{\pi}{4}$	$\frac{\sqrt{2}}{2}$	$\frac{\sqrt{2}}{2}$	1
60°	$\frac{\pi}{3}$	$\frac{\sqrt{3}}{2}$	$\frac{1}{2}$	$\sqrt{3}$
90°	$\frac{\pi}{2}$	1	0	undefined
180°	π	0	-1	0
270°	$\frac{3\pi}{2}$	-1	0	undefined
360°	2π	0	1	0

Ratio Identities

$$\sin \theta = \frac{1}{\csc \theta} \quad \cos \theta = \frac{1}{\sec \theta} \quad \tan \theta = \frac{1}{\cot \theta}$$

$$\csc \theta = \frac{1}{\sin \theta} \quad \sec \theta = \frac{1}{\cos \theta} \quad \cot \theta = \frac{1}{\tan \theta}$$

$$\tan \theta = \frac{\sin \theta}{\cos \theta} \quad \cot \theta = \frac{\cos \theta}{\sin \theta}$$